Handbook for Chemic Research and Development, Second Edition

This fully updated second edition reflects the significant changes in process chemistry since the first edition and includes more common process issues such as safety, cost, robustness, and environmental impact. Some areas have made notable progress such as process safety, stereochemistry, new reagents, and reagent surrogates. Forty years ago there were few process research and development activities in the pharmaceutical industry, partly due to the simplicity of drug molecules. With increasing structural complexity especially the introduction of chiral centers into drug molecules and stricter regulations, process research and development (R&D) has become one of the most critical departments for pharmaceutical companies.

Features:

- This unique volume, now in its second edition, is designed to provide readers with an unprecedented strategy and approach which will help chemists and graduate students develop chemical processes in an efficient manner.
- Promotes an industrial mindset concerning safety and commercial viability when developing methods.
- The author discusses development strategies with case studies and experimental procedures.
- Focuses on mechanism-guided process development which provides readers with practical strategies and approaches.
- Addresses more common process issues such as safety, cost, robustness, and environmental impact.

Handbook for Chemical Process Research and Development, Second Edition

Authored By
Wenyi Zhao
Senior Research Scientist
Jacobus Pharmaceutical Company, Princeton

CRC Press
Taylor & Francis Group
Boca Raton London New York

CRC Press is an imprint of the
Taylor & Francis Group, an **informa** business

Second edition published 2023
by CRC Press
6000 Broken Sound Parkway NW, Suite 300, Boca Raton, FL 33487-2742

and by CRC Press
4 Park Square, Milton Park, Abingdon, Oxon, OX14 4RN

©2023 Wenyi Zhao

CRC Press is an imprint of Taylor & Francis Group, LLC

ISBN: 978-1-032-25927-7 (hbk)
ISBN: 978-1-032-26461-5 (pbk)
ISBN: 978-1-003-28841-1 (ebk)

DOI: 10.1201/9781003288411

Typeset in Times
by Deanta Global Publishing Services, Chennai, India

Contents

Preface

According to the 12 principles of green chemistry, a good chemical process shall satisfy the following key elements: low cost, use available raw materials and reagents, simple workup, robustness, high throughput (fast reaction with high concentration), good product purity, and minimal environmental impact. The scale-up of a chemical process from laboratory to large production scales is by no means a simple linear process. Compared with laboratory-scale reactions, large-scale operations are expected to have an expanded time scale, insufficient mixing, and less-efficient heat transfer, which may potentially lead to runaway reactions. The key responsibility of process chemists is to develop chemical processes that are feasible for manufacturing pharmaceutical intermediates and final drug substances (active pharmaceutical ingredients–APIs) for support of clinical studies and, eventually, for commercial production.

Based on the first edition, this new edition continues to address the process issues such as safety, solvent selection, various reagent surrogates, product purification, stereochemistry, solid form, environmental impacts, etc. In addition, this book introduces (a) new reactions, for example, photoredox catalysis for C–C and C–N bond formations, catalytic C–H activations, multicomponent reactions, Grignard regent formation, and its related reactions, (b) new purification techniques to address the challenges in the control of mutagenic impurities, crystallization of low melting solids, and polymorph transformations, (c) design of efficient chemical processes, (d) newly developed reagent surrogates, such as ammonia surrogates, carbon dioxide surrogates, sulfur dioxide surrogates, and cyanide surrogates.

P.1 PROCESS EVALUATION

P.1.1 PROCESS SAFETY

Process safety refers to thermal/reactive hazards and health hazards. Thermal or reactive hazards are associated with reactions that are exothermic and/or generating gases or reactions that are involved shock and/or heat sensitive, pyrophoric, flammable, or corrosive materials. Health hazards refer to exposure to toxic chemicals that can cause acute or chronic negative health effects. It is most important to have a hazard assessment for a given process, particularly when using materials or intermediates without available Material Safety Data Sheet (MSDS).

Although the advent of flow chemistry has brought much attention recently, most chemical processes in the pharmaceutical industry are developed based on batchwise operations. There are several motivations for developing semibatch processes, such as avoidance of the accumulation of reactive reactants and control of the heat production rate (exothermic reactions). Thus, most exothermic reactions are conducted in a semibatch fashion in order to mitigate the exothermic event and prevent runaway reactions from occurring.

P.1.2 PROCESS COST

Process costs depend largely on the following aspects: materials, labor, equipment, and waste disposal. An economic process will use less expensive, commercially available materials as much as possible. Chromatographic purification is not an ideal process on large scale due to the burden of intensive labor. As per the reduction of process cost, a one-pot process is frequently employed to minimize process wastes, time-consuming isolations, and handling losses. In addition, cryogenic reactions or reactions that require high temperature or pressure should be avoided as much as possible. These reaction conditions usually need special equipment and large amounts of energy, which, in turn, will increase process costs.

P.1.3 Environmental Impact

Green chemistry addresses environmentally benign chemical synthesis, encouraging the design of chemical processes that minimize the use and generation of hazardous substances. An ideal reaction would incorporate all of the atoms of the reactants into the product with limited wastes, which, in turn, effectively reduces environmental pollution and improves efficiency. For instance, the process (shown in Equation P.1) developed by Pfizer uses the Baylis–Hillman reaction in the synthesis of the allyl alcohol, an intermediate for sampatrilat (an inhibitor of the zinc metalloprotease). The inherently environmentally friendly, atom-efficient Baylis–Hillman reaction not only incorporates all the atoms of the two starting materials into the product, but also adds environmental benefit since it allows simple reuse of the 3-quinuclidinol and generated much less waste stream.

P.2 DESIGN OF NEW SYNTHETIC ROUTES

At the end of the process evaluation, a decision has to be made on whether the existing process needs to be redesigned. When designing new synthetic routes, a rule of thumb should be followed:

- Use commercially available and less expensive materials.
- Use catalytic systems.
- Limit protecting group manipulations.
- Use convergent routes over linear ones.
- Use addition reactions.
- Use multicomponent reactions (MCRs).
- Use tandem or cascade processes, etc.

A catalytic reaction can be performed at relatively mild conditions, which is desired for large-scale production in terms of process safety and costs. Catalytic reactions are considered environmentally friendly due to the reduced amount of waste generated, as opposed to stoichiometric reactions. Classical olefinations, such as the Wittig reaction and Julia olefination, employ ketones or aldehydes as starting materials which are typically prepared by oxidation of the corresponding alcohols. A direct catalytic olefination of alcohols was realized using the thermal stable Ru-pincer catalyst (Equation P.2). This approach represents a step-economical synthesis, which avoided the alcohol oxidation step.

The one-pot process is an economically favorable method by performing a series of bond-formation steps in a single reaction vessel without the isolation of intermediates. The use of one-pot synthesis can greatly improve the process efficiency by minimizing isolation and purification steps. The development of one-pot synthesis is summarized in a review article which highlights various telescoping techniques such as multicomponent reaction (MCR), cascade (or domino) reaction, and tandem reaction.

P.3 PROCESS OPTIMIZATION

Prior to scaling up, a number of process parameters need to be identified so that reactions can be carried out under optimal conditions. These parameters include the mode of addition of starting materials/reagents/solvents, temperature, solvent/concentration, pressure (for some cases), agitation rate, etc.

P.3.1 Reaction Temperature

A reaction temperature is established based primarily on the reaction rate and impurity profile. Ideally, the reaction temperature shall be within –20 to 100 °C range, too low or too high will require additional energy and time at scale, and sometimes, special equipment is needed. Generally, high reaction temperature will lead to poor selectivity thereby the formation of impurities. Large jumps in temperature shall be avoided.

P.3.2 SOLVENT AND CONCENTRATION

Several solvent evaluation tools are developed as solvent selection guides. Generally, solvents shall be selected and assessed based on three aspects: (a) toxicity (including carcinogencity, mutagenicity, reprotoxicity, and skin absorption/sensitization), (b) process safety (including flammability, emission, electrostatic charge, and potential for peroxide formation), and (c) environmental and regulatory considerations (including ecotoxicity, groundwater contamination, and ozone depletion potential). Class 3 solvents, as proposed in the International Conference on Harmonization (ICH) guidelines, are preferred, especially at the end of the synthesis because of their low toxic potential (see Chapter 1).

In general, high-concentration reactions are desired because not only do the reactions at high concentrations afford high throughput, but they also produce less downstream waste.

Anhydrous reaction conditions can be reached by using anhydrous reagents and solvents. In addition, azeotropic distillation is the most commonly used technique to remove moisture from a reaction system. In the case of the presence of temperature-sensitive species, a moisture scavenger, such as acetic anhydride, is employed.

P.3.3 ISOLATION AND PURIFICATION

Direct isolation and extractive workup are two commonly used isolation approaches. Direct isolation is preferred over extractive workup in terms of process wastes, processing times, and costs.

An isolated reaction product usually needs to be purified in order to meet the predetermined purity criteria. The purification methods include distillation, crystallization/precipitation, and column chromatography. Owing to the intensive labor requirement, column chromatography is generally not recommended at large scales.

Obviously, the product yield and quality, including chemical/chiral purity and solid form (for solid materials), are two important parameters in determining the efficiency of a given process. Generally, reaction product yields of around 100% are considered quantitative, yields between 90 and 100% are considered excellent, yields between 80 and 90% are considered very good, yields between 60 and 80% are considered good, yields between 40 and 50% are considered moderate, and yields below 40% are considered poor. A product failed to meet the predetermined purity criteria may contain impurities such as residual processing solvents, undesired products, or metals (See Chapters 18 to 20 for various isolation/purification strategies).

Author Biography

Dr. Wenyi Zhao, Senior Research Scientist (Process Chemistry), Member of the American Chemical Society. 1980–1990: Ph.D., Organic/Physical Organic Chemistry, Nanjing University, China. 1990–1992: Postdoctoral fellow at the Institute of Chemistry, Chinese Academy of Sciences, Beijing. 1992–1995: Associate Professor at the Institute of Photographic Chemistry, Chinese Academy of Sciences, Beijing (now the Institute of Chemistry, Chinese Academy of Sciences, Beijing). 1995–2001: Senior Research Associate in the Department of Chemistry and Biochemistry, Texas Tech University, Lubbock, Texas. Early in 2001, Dr. Zhao moved to the private sector to become a process chemist in the pharmaceutical industry in the United States. He is the author of numerous journal articles and holds several patents.

List of Abbreviations

ACE-Cl:	1-Chloroethyl chloroformate
API:	Active pharmaceutical ingredient
ARC:	Accelerating rate calorimeter
9-BBN:	9-Borabicyclo[3.3.1]nonane
BCS:	Biopharmaceutics classification system
BDMAEE:	Bis[2-(*N,N*-dimethylamino)ethyl] ether
BHT:	Butylated hydroxy toluene (2,6-bis(1,1-dimethylethyl)-4-methylphenol)
BINAP:	2,2′-Bis(diphenylphosphino)-1,1′-binaphthyl
Boc:	*tert*-Butyloxycarbonyl
BOP:	(Benzotriazol-1-yloxy)tris-(dimethylamino)phosphonium hexafluorophosphate
BQ	1,4-Benzoquinone
BSA:	Bis(trimethylsily)acetamide (or TMCS)
***m*CBA:**	*m*-Chlorobenzoic acid
***m*CBPO:**	*m*-Chlorobenzoyl peroxide
Cbz:	Benzyloxycarbonyl
CDI:	1,1′-Carbonyldiimidazole
CDMT:	2-Chloro-4,6-dimethoxy-1,3,5-triazine
CIDR:	Crystallization-induced dynamic resolution
CLD:	Chord length distribution
CMCS:	Chloromethyl chlorosulfate
COBC:	Continuous oscillatory baffled crystallizer
***m*CPBA:**	*m*-Chloroperbenzoic acid
CPME:	Cyclopentyl methyl ether
CSA:	Camphorsulfonic acid
CSD:	Crystal size distribution
CSI:	*N*-Chlorosulfonylisocyanate
CTP:	4-Chlorothiophenol
DABCO:	1,4-Diazabicyclo[2.2.2]octane
DABSO:	1,4-Diazabicyclo[2.2.2]octane (DABCO)–sulfur dioxide charge-transfer complex
1,2-DAP:	1,2-Diaminopropane
DAS:	Dipolar aprotic solvent
DAST:	Diethylaminosulfur trifluoride
DBDMH:	Dibromodimethylhydantoin
DBH	1,3-Dibromo-5,5-dimethylhydantoin
DBN:	1,5-Diazabicyclo[4.3.0]non-5-ene
DBUL	1,8-Diazabicyclo[5.4.0]undec-7-ene
DCC:	Dicyclohexyl carbodiimide
DCE:	Dichloroethane
DCH:	1,3-Dichloro-5,5-dimethylhydantoin
DCM:	Dichloromethane
DEA:	Diethanolamine
DEAD:	Diethyl azodicarboxylate
DEAN:	*N,N*-Diethylaniline
DEG:	Diethylene glycol
DEM:	Diethoxymethane
DEMS:	Diethoxy(methyl)silane

Deoxo-Fluor:	Bis(2-methoxyethyl)aminosulfur trifluoride
DFI:	2,2-Difluoro-1,3-dimethylimidazolidine
DHP:	3,4-Dihydro-2H-pyran
DIAD:	Diisopropyl azodicarboxylate
DIBAL-H:	Diisobutylaluminum hydride
DIC:	1,3-Diisopropylcarbodiimide
DIPEA:	Diisopropylethylamine
DIPT:	Diisopropyl tartrate
DKR:	Dynamic kinetic resolution
DMAc:	Dimethylacetamide
DMAP:	4-Dimethylaminopyridine
DMC:	Dimethyl carbonate
DMCC:	Dimethylcarbamoyl chloride
DME:	Dimethoxyethane
DMEDA:	N,N-Dimethylethylenediamine
DMF:	Dimethyl formamide
DMI:	Dimethylimidazolidinone
DMP:	Dess–Martin periodinane
2,2-DMP:	2,2-Dimethoxypropane
DMPU:	1,3-Dimethyl tetrahydropyrimidin-2(1H)-one
DMS:	Dimethyl sulfide
DMSO:	Dimethyl sulfoxide
DPEphos:	Bis[(2-diphenylphosphino)phenyl]ether
DPPA:	Diphenylphosphoryl azide
DPPB:	1,4-Bis(diphenylphosphino)butane
DPPE:	1,2-Bis(diphenylphosphino)ethane
DPPF:	Bis(diphenylphosphino)ferrocene
DPPH:	O-(Diphenylphosphinyl)hydroxylamine
DPPP:	1,3-Bis(diphenylphosphino)propane
DSC:	Differential scanning calorimetry
DTA:	Differential thermal analysis
DTAD:	Di-*tert*-butyl azodicarboxylate
DTBP:	2,6-Di-*tert*-butyl pyridine
DTTA:	Di-*p*-toluoyl-tartaric acid
DVS:	Dynamic vapor sorption
EDC:	1-Ethyl-3-(3-dimethylaminopropyl)carbodiimide
EDCI:	1-Ethyl-3-(3-dimethylaminopropyl)-carbodiimide hydrochloride
EDTA:	Ethylenediamine tetraacetic acid
EEDQ:	2-Ethoxy-1-ethoxycarbonyl-1,2-dihydroquinoline
EH&S:	Environment, health and safety
EMA:	European Medicines Agency
EPA:	Environmental Protection Agency
FBRM:	Focused beam reflectance measurement
FDA:	Food and Drug Administration
FTIR:	Fourier transform infrared spectroscopy
GSK:	GlaxoSmithKline Pharmaceuticals
HATU:	1-[Bis(dimethylamino)methylene]-1H-1,2,3-triazolo[4,5-*b*]pyridinium 3-oxide hexafluorophosphate
HBTU:	O-(Benzotriazol-1-yl)-N,N,N',N'-tetramethyluronium hexafluorophosphate
HDMT:	2-Hydroxy-4,6-dimethoxy-1,3,5-triazine
HFIP:	Hexafluoro-2-propanol

HKR:	Hydrolytic kinetic resolution
HMDS:	Hexamethyldisilazane
HMPA:	Hexamethylphosphoramide
HMTA:	Hexamethylenetetramine
HNB:	2-Hydroxy-5-nitrobenzaldehyde
HOAt:	1-Hydroxy-7-azabenzotriazole
HOBt:	1-Hydroxybenzotriazole
HOSu:	*N*-Hydroxysuccinimide
HPLC:	High-performance liquid chromatography
HAS:	Hydroxylamine-*O*-sulfonic acid
HWE:	Horner–Wadsworth–Emmons olefination
IBCF:	Isobutyl chloroformate
ICH:	International Conference on Harmonization
IMS:	Industrial methylated spirit
INCB:	International Narcotics Control Board
IPA:	2-Propanol
IPAc:	Isopropyl acetate
IPE:	Diisopropyl ether
LAH:	Lithium aluminum hydride
LDA:	Lithium diisopropylamide
LiHMDS:	Lithium bis(trimethylsilyl)amide
LiTMP:	Lithium 2,2,6,6-tetramethylpiperidin-1-ide
MCR:	Multicomponent reaction
MEK:	Methyl ethyl ketone
MEMCl:	2-Methoxyethoxymethyl chloride
MeTHF:	2-Methyl tetrahydrofuran
MIBK:	Methyl isobutyl ketone
MIDA:	*N*-Methyl iminodiacetic acid
MP:	Melting point
MPD:	Mechanism-guided process development
MP-TMT:	Macroporous polystyrene-2,4,6-trimercaptotriazine
MPMN:	2-Methyl-2-phenyl malononitrile
MPV:	Meerwein–Ponndorf–Verley reduction
MSA:	Methanesulfonic acid
MSDS:	Material safety data sheet
MTBE:	Methyl *tert*-butyl ether
MW:	Molecular weight
NBS:	*N*-Bromosuccinimide
NCP:	*N*-Chlorophthalimide
NCS:	*N*-Chlorosuccinimide
NFSI:	*N*-Fluorobenzenesulfonimide
NIS:	*N*-Iodosuccinimide
NMM:	*N*-Methylmorpholine
NMO:	*N*-Methylmorpholine *N*-oxide
NMP:	*N*-Methyl-2-pyrrolidone
NPTC:	*p*-Nitrophenol tempo carbonate
NSFI:	*N*-Fluorobenzenesulfonimide
P-BiAA:	Polymer-supported bis(2-aminoethyl)-amine
PBPB:	Pyridinium bromide perbromide
PBS:	Phosphate-buffered saline
PCC:	Pyridinium chlorochromate

PCP:	Purity control point
P-EDA:	Polymer-supported ethylenediamine
L-PGA:	L-Pyroglutamic acid
PGIs:	Potential genotoxic impurities
PGME:	Propylene glycol monomethyl ether
PICB:	2-Picoline borane
PKA:	Porcine kidney
PLE:	Pig liver esterase
PMB:	*p*-Methoxybenzyl
PMP	*p*-Methoxyphenyl
PNB:	*p*-Nitrobenzyl
PPTS:	Pyridinium *p*-toluenesulfonate
PSD:	Particle size distribution
PTAB:	Phenyltrimethyl ammonium tribromide
PTC:	Phase-transfer catalyst
P-TriAA:	Polymer-supported tris(2-aminoethyl)-amine
PVE:	Propyl vinyl ether
PVM:	Particle vision measurement
PYBOP:	(Benzotriazol-1-yloxy)-trispyrrolidinophosphonium hexafluorophosphate
RCM:	Ring-closing metathesis
R&D:	Research and development
RSST:	Reactive system screening tool
SAS:	Sodium anthraquinone-2-sulfonate
SEM:	Scanning electron microscope
S$_N$Ar:	Aromatic nucleophilic substitution
S$_N$1:	Nucleophilic unimolecular substitution
S$_N$2:	Nucleophilic two-molecular substitution
STAB:	Sodium triacetoxyborohydride
T3P:	*n*-Propanephosphonic acid cyclic anhydride
TATP:	Triacetone triperoxide
TBAB:	Tetrabutylammonium bromide
TBACl:	Tetrabutylammonium chloride
TBAF:	Tetrabutylammonium fluoride
TBHP:	*tert*-Butyl hydroperoxide
TBS:	*tert*-Butyldimethylsilyl
TCAN:	Trichloroacetonitrile
TCCA:	Trichloroisocyanuric acid
TDA:	Tris(3,6-dioxaheptyl)amine
TEA:	Triethylamine
TEAB:	Tetraethylammonium bromide
TEAHC:	Tetraethylammonium hydrogen carbonate
TEBA:	Triethylbenzylammonium chloride
TEBAC:	Triethylbenzylammonium chloride
TEMP:	2,2,6,6-Tetramethylpiperidine
TEMPO:	2,2,6,6-Tetramethyl-1-piperidine-*N*-oxide
TFA:	Trifluoroacetic acid
TFB:	(Trifluoromethyl)benzene
TFBen:	Benzene-1,3,5-triyl triformate
TFE:	Trifluoroethanol
TFFH:	*N,N,N′,N′*-Tetramethylfluoroformamidinium hexafluorophosphate
TGA:	Thermogravimetric analysis

THF:	Tetrahydrofuran
THP:	Tetrahydropyran
TIPS:	Triisopropylsilyl
TMAF:	Tetramethylammonium fluoride
TMCS:	*N,O*-Bis(trimethylsilyl)acetamide (or BSA)
TMDS:	Tetramethyl disiloxane
TMEDA:	Tetramethylethylenediamine
TMG:	Tetramethyl guanidine
TMM:	Trimethylenemethane
TMP:	2,2,6,6-Tetramethylpiperidine
TMS:	Trimethylsilyl
TMSCl:	Chlorotrimethylsilane
TMSCN:	Trimethylsilyl cyanide
TMSI:	Iodotrimethylsilane
TMSOK:	Potassium trimethylsilanolate
TMT:	Trimercaptotriazine (or trithiocyanuric acid)
TMU:	Tetramethyl urea
TPPMS:	Sodium 3-(diphenylphosphino)benzenesulfonate
TPPTS:	3,3′,3″-Phosphanetriyltris(benzenesulfonic acid) trisodium salt
TRIS:	Tris(hydroxymethyl)aminomethane
***p*-TSA:**	*para*-Toluenesulfonic acid
UHP:	Urea–hydrogen peroxide
VDKR:	Vinylogous dynamic kinetic resolution
VOC:	Volatile organic compound
WFE:	Wiped film evaporation
XRPD:	X-ray powder diffraction
XtalFluor-E:	Diethylaminodifluorosulfinium tetrafluoroborate
XtalFluor-M:	Difluoro(morpholino)sulfonium tetrafluoroborate

1 Reaction Solvent Selection

Process chemists face numerous challenges, such as dealing with issues including process safety, waste streams, chemical toxicity, recycling of solvents/catalysts, process economics, and a multitude of engineering/technology considerations. In a chemical process, solvents are employed for multiple purposes: to achieve the desired reaction rate and selectivity and to improve the process safety by efficient mass transfer, taking up heat generated by the reaction, or offering a safety barrier via refluxing. Solvents are also required during product isolations such as extractions and crystallizations. Therefore, the solvent plays a critical role in the synthetic process; the appropriate selection of solvent for processing a drug substance may enhance the yield, allow isolation of a preferred crystal form, or improve product purity.

The International Council for Harmonization of Technical Requirements for Pharmaceuticals for Human Use guidance for industry (ICH Q3C) makes recommendations as to what amounts of residual solvents are considered safe in pharmaceuticals and commonly used solvents have been grouped by toxicity under the ICH Q3C guidance. The most toxic solvents (Class 1, Table 1.1) should be avoided in the production of drug substances or excipients unless their use can be strongly justified in a risk-benefit assessment. Solvents with less severe toxicity (Class 2) should be limited in the drug production process in order to protect patients from potential adverse effects. Ideally, less toxic solvents (Class 3) should be used. In general, process solvents cannot be completely removed by practical manufacturing techniques; therefore, besides the reactivity and compatibility, the solvent selection should be also based on the assessment of health hazards, such as carcinogenicity, mutagenicity, skin sensitization, etc.,[1] and all residual solvents should be removed from the drug product to the extent possible to meet product specifications or other quality-based requirements.

Generally, in addition to ICH Q3C the selection of an appropriate solvent for a given reaction should meet the following criteria:

- be inert to the reaction conditions.
- be able to dissolve the reactants and reagents.
- have an appropriate boiling point.
- be easily removed at the end of the reaction.

In order to help process chemists to choose sustainable solvents, some pharmaceutical companies have elaborated solvent selection guides.[2] Figure 1.1 shows the top 10 most frequently used solvents by GlaxoSmithKline Pharmaceuticals (GSK) in 2005 and during the time period of 1990–2000.

An indication, from these data, is that there is a trend toward decreasing tetrahydrofuran (THF), toluene, and dichloromethane use. In general, in order to achieve a high material throughput, a solvent with a solubility of 100 mg/mL or greater is preferred.

Solvent use is responsible for 60% of the overall energy used in a pharmaceutical process and accounts for 50% of the post-treatment green house gas emissions.[3] In addition, 80% of waste generated during the manufacture of a typical active pharmaceutical ingredient (API) is related to solvent use. Given these figures, vigilant solvent selection to maximize efficiency and potential recovery can have a huge impact on the process costs and the environment. In general, a single solvent in a process is preferred because of the simplicity of recycling.

DOI: 10.1201/9781003288411-1

TABLE 1.1
Some Class 1, Class 2, and Class 3 Solvents[a]

Class	Solvent	Concentration Limit (in ppm)	Concern
1	Benzene	2	Carcinogen
	Carbon tetrachloride	4	Toxic and environmental hazard
	1,2-dichloroethane	5	Toxic
	1,1-dichloroethene	8	Toxic
	1,1,1-Trichloroethane	1500	Environmental hazard
2	Acetonitrile	410	–
	Chlorobenzene	360	–
	Chloroform	60	–
	Cyclohexane	3880	–
	1,2-Dichloroethene	1870	–
	Dichloromethane	600	–
	Dimethoxyethane	100	–
	N,N-Dimethylacetamide	1090	–
	N,N-Dimethylformamide	880	–
	1,4-Dioxane	380	–
	2-Ethoxyethanol	160	–
	Ethylene glycol	620	–
	Hexane	290	–
	Methanol	3000	–
	2-Methoxyethanol	50	–
	N-Methyl pyrrolidone	530	–
	Nitromethane	50	–
	Pyridine	200	–
	Sulfolane	160	–
	Tetrahydrofuran	720	–
	Toluene	890	–
	Xylene	2170	–
3	Acetic acid	5000	–
	Acetone	5000	–
	Anisole	5000	–
	1-Butanol	5000	–
	2-Butanol	5000	–
	Butyl acetate	5000	–
	tert-Butyl methyl ether	5000	–
	Dimethyl sulfoxide	5000	–
	Ethanol	5000	–
	Ethyl acetate	5000	–
	Formic acid	5000	–
	Heptane	5000	–
	Isobutyl acetate	5000	–
	1-Propanol	5000	–
	2-Propanol	5000	–
	Propyl acetate	5000	–
	Triethylamine	5000	–

[a] Cited from: https://www.fda.gov/media/71737/download.

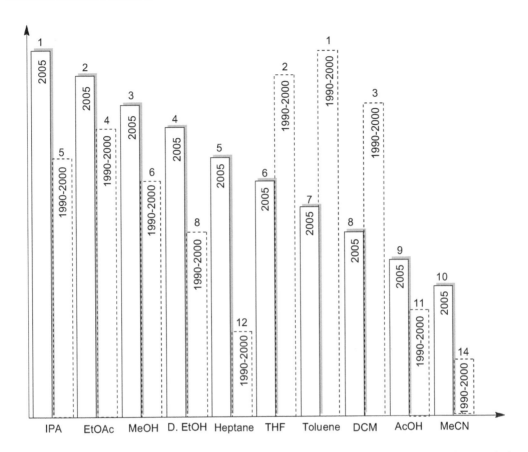

FIGURE 1.1 Comparison of solvent use in GSK in 2005 and from 1990–2000 (reprinted with permission from Constable, D.J.C. et al., *Org. Process Research & Development*, 11, 2007, 133. Copyright 2007 American Chemical Society).

1.1 ETHEREAL SOLVENTS

Prudent selection of solvents is important for developing a safe and robust chemical process. It is well-known that solvents with low boiling points and low flash points,[4] such as diethyl ether, are not suitable for scaling up. Tetrahydrofuran (THF), 2-methyl tetrahydrofuran (MeTHF), and methyl *tert*-butyl ether (MTBE) are frequently used as substituents of diethyl ether. Some physical properties of commonly used ether solvents can be found in the literature.[5]

Notably, the peroxide content in ether solvents is also a process safety concern, as most ethereal solvents have a propensity of forming explosive organic peroxides after storing for a period of time. Therefore, commercially available ether solvents are usually blended with anti-oxidant, 2,6-di-*tert*-butyl-4-methylphenol (BHT).

MeTHF (CAS 96-47-9) and cyclopentyl methyl ether (CPME: CAS 5614-37-9) are being increasingly used[6] as alternatives to their analogs such as THF and MTBE. These two solvents offer superior physical (higher flash point, water azeotrope, phase cuts, and lower volatility) and chemical properties (greater acid/base stability). A toxicological assessment of MeTHF and CPME was made[7] in support of their use in pharmaceutical chemical process development.

1.1.1 CYCLOPENTYL METHYL ETHER

CPME has become available in commercial quantities since late 2005. Compared with other ethereal solvents, CPME has a high boiling point (106 °C), low peroxide level, and relative stability under acidic and basic conditions. Having a relatively low water content (the water solubility is 0.3 g per 100 g of CPME) among the ethereal solvents, CPME can be used directly as the solvent for reactions including Grignard reactions, enolate formations, Claisen condensations, reductions, and Pd-mediated transformations. Other characteristics of CPME, such as a low level of peroxides coupled with a narrow explosion range, render CPME a good alternative to other ethereal solvents such as THF, MeTHF, dioxane (carcinogenic), and dimethoxyethane (DME). In addition, relatively low solubility in water and low vaporization energy make the recovery of CPME more efficient. However, the use of solvents such as CPME and MeTHF should be with caution, especially in the last processing steps, as these solvents are not classified for use by ICH guidelines.[8]

1.1.1.1 Brook Rearrangement

CPME shows a unique solvent effect in stabilizing a reactive enolate **2** in Brook rearrangement-mediated [4+3] cycloaddition for the stereoselective construction of 2-oxabicyclo[3.3.2]decenone derivatives **4** (Scheme 1.1).[9] As the enolate **2** of cycloheptenone is unstable at −80 °C, replacement of THF with CPME was able to perform this cycloaddition at −98 °C. As a result, the yields of cycloadducts **4** were improved.

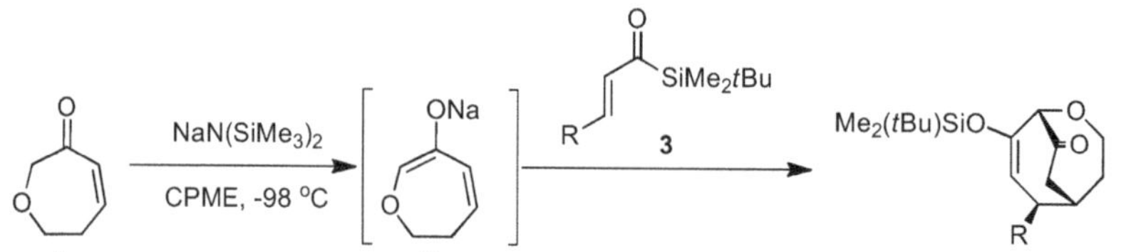

R	Yield, %
SiMe$_3$	83
SiMe$_2$Ph	73
SiMe$_2$tBu	77
iPr	79
tBu	77

SCHEME 1.1

REMARKS

(a) It was found that enantioselective α-methylation of chiral amides in CPME gave the highest enantiomer excess among other ethereal solvents.[10] The asymmetric induction in CPME was assumed to proceed via a mixed aggregate during the chirality transfer from an undeprotonated chiral amide into an achiral enolate.

(b) CPME as an additive for the carbenoid chemistry led to a high yield of enynes (Equation 1.1).[11]

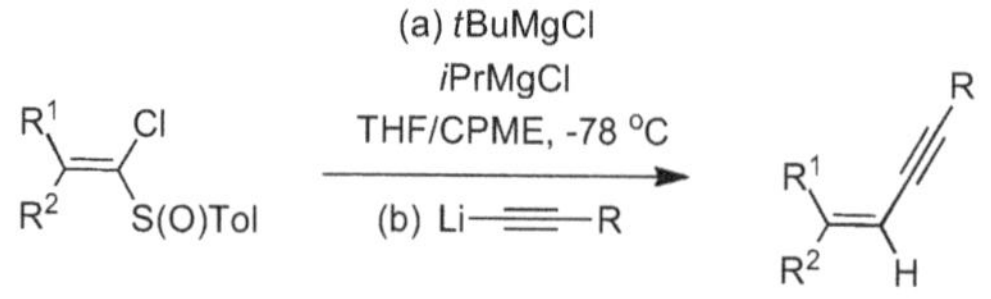

(1.1)

1.1.1.2 N-Alkylation Reaction

Having a low solubility in water, CPME was chosen as the reaction solvent for the alkylation reaction (Equation 1.2).[12] In this CPME/water two-phase reaction system, CPME served as the reaction solvent and the solvent for the extraction during the workup. This would avoid the need for solvent removal as in the case of THF or acetonitrile before extracting the free base into solvents such as DCM or ethyl acetate.

(1.2)

Procedure

Stage (a)

- Charge 404.55 g (2.93 mol, 4.2 equiv) of K_2CO_3 into a solution of **5** (300 g, 0.7 mol) in a biphasic mixture of water (900 mL) and CPME (1.2 L) at 25–30 °C.
- Stir for 30 min.
- Charge 185.17 g (0.82 mol, 1.2 equiv) of **6** at 25–30 °C.
- Heat the batch to 70–75 °C and stir at that temperature for 12 h.
- Cool to 25–30 °C.
- Add CPME (1.2 L) and water (900 mL) and stir for 30 min.
- Filter through a bed of hyflow.
- Separate layers and extract the aqueous layer with CPME (600 mL).
- Wash the combined CPME layers with water.
- Distill under reduced pressure.
- Add toluene and continue to distill.

Stage (b)

- Add acetone (1.5 L).
- Cool to 0–5 °C.
- Add 117.3 mL of 48% (in water) HBr.
- Stir at 0–5 °C for 2 h and at room temperature for 5 h.
- Distill under reduced pressure.
- Add acetone (2.4 L) and stir for 4 h.
- Filter to give a wet solid.
- Dissolve the wet solid in n-butanol (2.1 L) at reflux.
- Cool to room temperature and stir for 3 h.
- Filter and wash with acetone.
- Dry at 50–60 °C under vacuum.
- Yield: 265 g (75%).

1.1.2 Tetrahydrofuran

1.1.2.1 Grignard Reagent Formation

Ethereal solvents, such as THF and diether ether, are frequently chosen as solvents for the preparation of Grignard reagents. One of the most common problems in Grignard preparation is the Wurtz-type coupling forming a dimer side product. Attempts to prepare the Grignard reagent from a reaction of benzyl chloride with magnesium metal (Equation 1.3)[13] in THF led to the almost exclusive formation of the ethane dimer. Except for diethyl ether, other ethereal solvents failed to give the desired Grignard reagent.

(1.3)

R=H, SMe

In order to avoid dimer formation, a protocol of adding the chloride **8** THF solution to Mg (1.5 equiv) in toluene at <35 °C provided the desired Grignard **9**.

1.1.2.2 Bromination of Ketone

THF is susceptible to acid-promoted degradation, which would lead to unexpected results for a reaction. Bromination of methylandrostanolone **10** by using phenyltrimethylammonium tribromide (PTAB) in THF furnished the desired α-bromoketone **11** in good yield (Equation 1.4).[14] However, upon drying at 35–45 °C, the bromide **11** turned from white to dark green in color along with the formation of a dehydrated impurity **12**.

$$(1.4)$$

The impurity **12** was generated from a Wagner–Meerwein rearrangement in the presence of HBr in the wet cake. The source of the HBr was 4-bromo-1-butanol, which is produced from the ring opening of THF by HBr during the bromination reaction (Equation 1.5).

$$(1.5)$$

Ultimately, a new bromination condition (PyHBr$_3$, EtOH/H$_2$O) was developed using aqueous ethanol as the solvent.[15]

1.1.3 2-METHYL TETRAHYDROFURAN

MeTHF[15] is a commercially available solvent and is produced from furfural, which is derived from renewable resources (corn cobs, sugar cane bagasse, oat hulls) (Equation 1.6).

$$(1.6)$$

MeTHF is a versatile solvent that can directly substitute for volatile dichloromethane or dichloroethane in many of the production processes. Compared with THF, MeTHF has limited solubility in water (14%) and, in the event, MeTHF is a good substitute for THF in aqueous isolation without the need of adding another solvent. Besides, the organic–water phase separations have little tendency to form emulsions or rag layers. MeTHF can be used to azeotropically dry reaction products. For example, products isolated from MeTHF can be conveniently dried because of the favorable MeTHF–water azeotrope (boiling point of 71 °C with a composition of 89.4 wt% of MeTHF and 10.6 wt% of water).

Because of the low melting points (–136 °C) and low viscosity (1.85 cp at –70 °C), MeTHF is increasingly used in organometallic chemical processes, such as in Grignard reactions, low-temperature lithiations, and lithium aluminum hydride reductions.

1.1.3.1 Control of Impurity Formation

Utilization of the solubility difference of water between THF and MeTHF to control the impurity formation was demonstrated nicely in the preparation of sulfonamide **14** (Equation 1.7).[16] Initially, the product **14** was prepared by carrying out the reaction of sulfonyl chloride **13** with aqueous ammonia in THF, resulting in a dimer impurity **13** in 2–4% yield depending on the THF concentration. Under diluted conditions (6 volumes of THF), the level **15** was reduced to 2% compared with 4% in 2 volumes of THF.

$$(1.7)$$

In contrast, the reaction in MeTHF afforded the desired sulfonamide **14** with the impurity **15** below 0.5%. As water has limited solubility in MeTHF (4.4%), the concentration of ammonia in MeTHF is significantly higher than in THF (see Equation 1.8).

$$NH_3 + H_2O \rightleftharpoons NH_4OH \qquad (1.8)$$

Thereby the abundant ammonia in MeTHF suppressed the competitive side reaction.[17]

REMARK

Due to the low solubility in water, MeTHF was utilized as an extraction solvent to separate organic product from the aqueous mixture[18] and the resulting MeTHF solution was directly used in the subsequent step.[19]

1.1.3.2 Enhancing Reaction Rate

Reactions would proceed faster in MeTHF than in THF due to the relatively high boiling point of MeTHF (80 °C). For example, the reaction of 1-(4-methoxy-2-methylphenyl)pyrrolidin-2-imine hydrobromide **16** with **17** and the subsequent cycloaddition completed in MeTHF under reflux (80 °C) within 17 h, while it required about 28 h in THF (Equation 1.9).[20]

$$(1.9)$$

Procedure

- Charge 17.5 L of 10% NaOH aqueous solution to a solution of **16** (10 kg) in 70 L of MeTHF.
- Wash the organic layer with 15% NaCl aqueous solution.
- Distill the organic layer azeotropically.
- Charge 4.6 L (1.1 equiv) of **17**.
- Heat the content under reflux for 8 h.
- Cool the batch to 20 °C.
- Charge a pre-prepared solution of KOtBu (10 kg, 2.5 equiv) in 90 L of MeTHF.

- Heat the batch under reflux for 9 h.
- Concentrate the batch to 120 L.
- Cool the batch to 20 °C.
- Wash with 40 L of 20% NH$_4$Cl aqueous solution and 20 L of 15% NaCl aqueous solution.
- Concentrate the batch to 40 L.
- Cool the batch to 20 °C.
- Seed the batch.
- Filter and wash with 20 L of MeTHF.
- Dry at 50 °C under vacuum for 18 h.
- Yield: 5.2 kg (55%).

1.1.3.3 Improving Layer Separation

Owing to its low water solubility, the use of MeTHF will result in an easy aqueous workup with limited loss of product in the aqueous phase. This beneficial effect was observed during the aqueous workup of the Wadsworth–Emmons reaction. As the inorganic impurities in **21** had a detrimental effect on the efficiency and reproducibility of the subsequent enantioselective hydrogenation, the aqueous workup was required to remove inorganic impurities. Thus, the use of MeTHF in the Wadsworth–Emmons reaction made the workup operation easy with good phase separations (Scheme 1.2).[21]

Procedure

Stage (a)
- Charge a solution of **19** (177 g, 324 mmol) in MeTHF (338 mL) and water (18 mL) to a solution of **20** (173 g, 279 mmol) in MeTHF (850 mL).
- Cool the resulting suspension to between -5 °C and 0 °C.
- Charge 98 g (818 mmol) of tetramethylguanidine over 1.5 h.
- Warm the batch to 20 °C and stir for 18 h.

Stage (b)
- Add water (700 mL).
- Add concentrated H$_2$SO$_4$ to pH 2.0.
- Separate layers.
- Wash the organic layer with water (2×700 mL).
- Filter through a bed of zeolite.

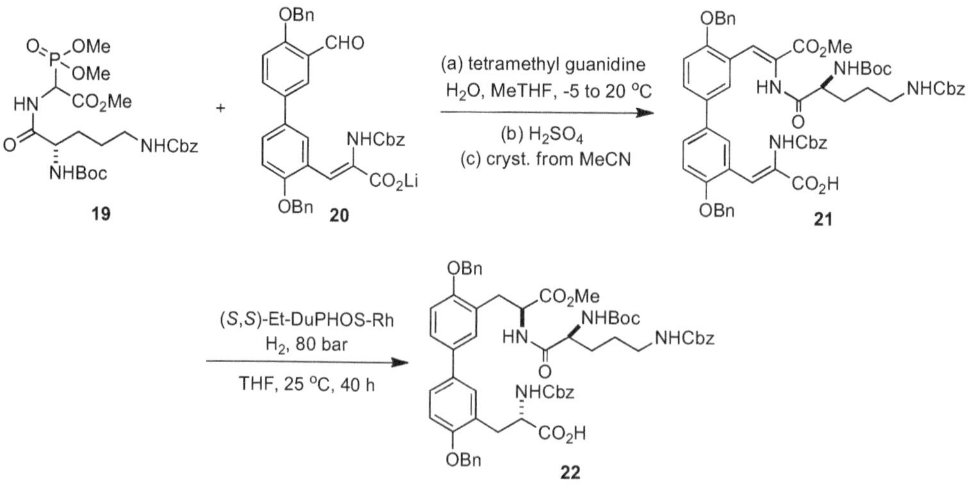

SCHEME 1.2

Stage (c)
- Distill off ca 900 mL of solvent under reduced pressure.
- Add MeCN (3.0 L).
- Heat to 50 °C over 3 h.
- Cool to 20 °C over 18 h then to 0 °C.
- Filter, then rinse with cold MeCN (3×400 mL) and water (1.0 L).
- Dry at 45 °C under vacuum.
- Yield of **21**: 196 g (68%).

REMARKS

(a) The lack of reproducibility of enantioselective hydrogenation of **21** could be due to halide impurities introduced during the workup with hydrochloric acid. Accordingly, the hydrochloric acid was replaced by sulfuric acid.

(b) This detrimental effect of chloride was also observed in the asymmetric hydrogenation of 2-methylenesuccinamic acid (Scheme 1.3).[22] Thus, a chloride-free 2-methylenesuccinamic acid material was required for the asymmetric reduction and the hydrochloric acid was replaced by sulfuric acid during the workup.

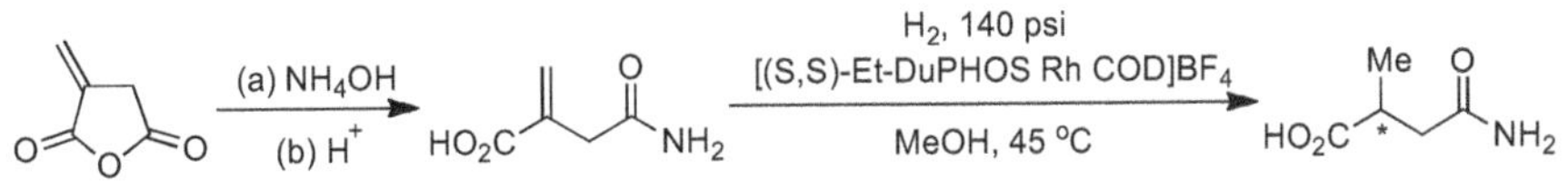

SCHEME 1.3

1.1.4 METHYL *TERT*-BUTYL ETHER

1.1.4.1 Chlorination Reaction

Initially, a chlorination reaction of **23** was conducted with 1.6 equiv of thionyl chloride in MTBE at ambient temperature, affording the corresponding chloride **24** in excellent yield and purity (Equation 1.10).[23] However, the reaction became very sluggish at the 2.0-mole scale, requiring several hours to complete.

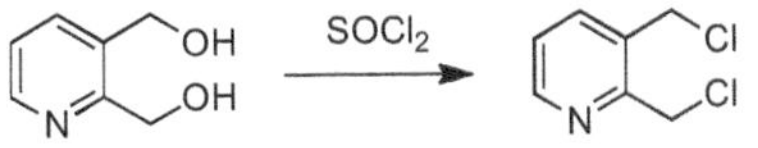

(1.10)

Concerned about prolonged exposure of MTBE to acidic reaction conditions, hazardous evaluation was initiated using reaction calorimetry. It was demonstrated that under the reaction conditions MTBE decomposed in the presence of HCl (generated in situ) to isobutylene as large amounts off-gas. In the event, toluene (in the presence of 1% v/v dimethylformamide (DMF) as a catalyst) was identified as the reaction solvent to replace MTBE, and the chlorination produced >3.0 kg of **24** in 89% yield in a safe and environmentally acceptable manner.

1.1.4.2 Darzens Reaction

The methyl glycidate intermediate **27** was obtained via Darzens reaction by treating benzophenone **25** with methyl chloroacetate **26** in the presence of sodium methoxide in MTBE (Equation 1.11).[24] Although the calorimetric evaluation showed no safety issues and the reaction was carried out successfully on laboratory scales, the first pilot scale synthesis was accompanied by an unwelcome surprise. The reaction was initiated very violently, after half of **26** had been added, resulting in most of the MTBE evaporating within seconds.

$$
\textbf{25} + \textbf{26} \xrightarrow[\text{solvent}]{\text{NaOMe}} \textbf{27} \tag{1.11}
$$

Investigations found that the poor solubility of NaOMe in MTBE was the cause of the problem, leading to the accumulation of **26**. As the reaction produced methanol as the byproduct which helped to dissolve NaOMe, a self-accelerated reaction was thus observed. Therefore, the replacement of MTBE with THF could solve the problem.

1.1.5 DIETHOXYMETHANE AND DIMETHOXYETHANE

Diethoxymethane (DEM) is a useful organic solvent and can be used under basic conditions for a variety of applications,[25] including sodium hydride reactions, organolithium chemistry, copper-catalyzed conjugate additions, and phase-transfer reactions. It has unique properties: a moderate boiling point (88 °C), azeotropes with water, and a very low affinity for water.

Like DEM, dimethoxyethane (DME) is also used as a reaction solvent. For instance, the S_N2 reaction of enolate of 4-fluoroacetopheone **28** gave two products, ketone **30** and aryl vinyl ether **31** (Equation 1.12).[26] A significantly solvent effect was observed when conducting the reaction in DME (or THF) compared to in DMSO, indicating that high solvating solvents, such as DMSO, promoted reaction at oxygen, while DME enhanced reaction at carbon.

$$
\textbf{28} + \textbf{29} \xrightarrow{\text{NaH}} \textbf{30} + \textbf{31} \tag{1.12}
$$

X=F, Cl

1.2 PROTIC SOLVENTS

Because of their low toxicity, commercial availability, and good solubility for most organic compounds, methanol, ethanol, and 2-propanol are three commonly used solvents in various applications in the pharmaceutical industry.

1.2.1 METHANOL AS A SOLVENT

Due to their water miscibility and good solubility with most organic compounds, alcoholic solvents are commonly used in catalytic hydrogenations.

1.2.1.1 Leak of Palladium Catalyst

When using alcohol solvents, the isolated product may be contaminated with black solids because of the leaching of palladium catalyst from the carbon support or partial digestion of the carbon support. For example, the palladium-catalyzed hydrogenation of nitro acetal **32** in alcohols led to the product amino acetal **33** being contaminated with black solids (Equation 1.13).[27]

$$
\textbf{32} \xrightarrow[\text{toluene}]{5\%\ \text{Pd/C, H}_2} \textbf{33} \tag{1.13}
$$

To address this issue, toluene was chosen as the reaction solvent to replace alcoholic solvents, providing 93.6% of the product **33**.

1.2.1.2 Side Product Formation

The selection of reaction solvents can impact the product stability, and thus the isolated product yield and purity. For example, using methanol as the solvent for the hydrogenation of nitroindazole **34** resulted in the contamination of the desired product **35** with 30% of deprotected amino indazole **36** (Equation 1.14).[28]

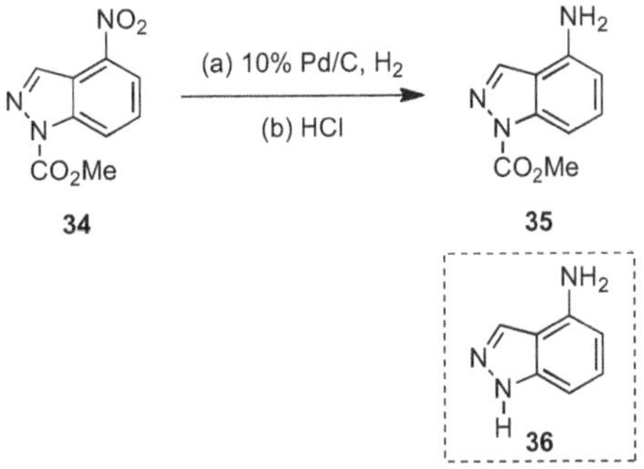

$$(1.14)$$

Ultimately, the use of THF or ethyl acetate as the reaction solvent suppressed the formation of **36**.

1.2.1.3 Palladium-Catalyzed Methylation Reaction

In addition to being a reaction solvent, methanol was utilized as a surrogate for the methylation reagent in a highly efficient, palladium-catalyzed methylation of nitroarenes (Scheme 1.4).[29]

Generally, *N*-methylation can be prepared by either methylation of amines with toxic reagents, such as methyl iodide, dimethyl sulfate, and dimethyl carbonate or reductive methylation with formaldehyde. The use of methanol as an inexpensive reductive methylation reagent requires a challenging, controlled oxidation. A phosphine-based ligand containing a pyridyl moiety activates the O–H bond in methanol, allowing the oxidative addition of Pd(0) to the O–H bond leading to a palladium hydride complex. The subsequent β-hydride elimination would generate formaldehyde and hydrogen in situ. Thus, methanol in this reaction acts as a solvent, hydrogen source, and methylation reagent.

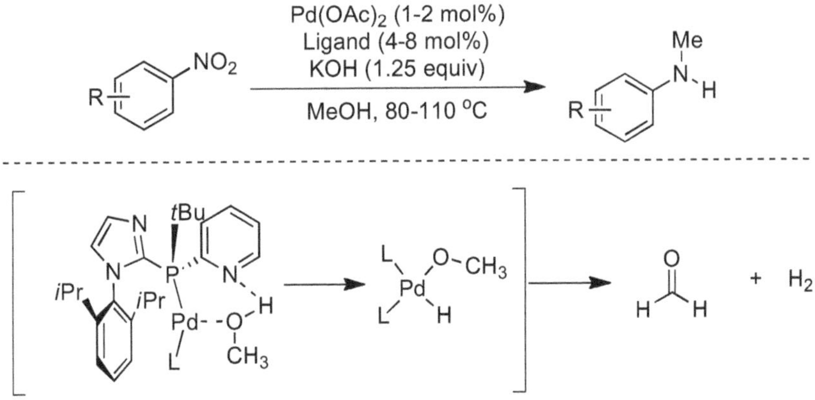

SCHEME 1.4

1.2.2 ETHANOL AS A SOLVENT

1.2.2.1 Catalytic Reduction of Diaryl Methanol

Solvent selection sometimes has a significant effect on reaction outcomes, for example, the selectivity of the catalytic reduction of diaryl methanol **37** varied drastically when switching alcoholic solvent to a non-protic solvent (Equation 1.15).[30]

(1.15)

Solvent	ZnBr$_2$ (mol%)	Conversion (%)	Selectivity (**38:39**)
EtOH	none	80	3.3:1
EtOH	100	22	40:1
EtOAc	1	99	>200:1

(I) Reaction problems

Although a good conversion of diarylmethanol **37** was achieved in acidic ethanol, this reaction suffered poor selectivity, leading to a mixture of the product **38** and a dechlorinated side product **39** in a ratio of 3.3:1 (**38:39**).

(II) Solutions

The use of less polar solvents or halide additives can reduce the rate of reductive dehalogenation.[31] Thus, using ethyl acetate as the reaction solvent with ZnBr$_2$ as the additive gave a 99% conversion in excellent selectivity (**38:39**, >200:1).

1.2.2.2 S$_N$2 Reaction

Ethanol could replace DMF in the S$_N$2 nucleophilic substitution reaction to synthesize alkyl aryl ether **53** (Equation 1.16).[32]

(1.16)

Compared with the 84% yield in DMF, the reaction in EtOH gave an excellent yield (95%) of **53** reproducibly on a 10 kg scale.

1.2.3 2-PROPANOL AS A SOLVENT

1.2.3.1 Reaction of Acyl Hydrazine with Trimethylsilyl Isocyanate

An intriguing solvent-related effect during the installation of the urea moiety on the basic nitrogen of acyl hydrazine **43** was observed (Equation 1.17).[33] In isopropyl alcohol (IPA) the reaction of

43 with trimethylsilyl isocyanate (TMS-NCO) gave the desired product **44** in 81% yield, while in dichloromethane no product was formed.

$$(1.17)$$

Solvent	Yield, %
DCM	0
IPA	81

REMARK

Although IPA is frequently used in laboratories and manufacturing plants, the likelihood of per-oxide (or triacetone triperoxide) formation in IPA and its associated hazards is overlooked. There have been several reports of explosions occurring during the distillation of IPA that have resulted in injury to researchers. Thus, a research paper[34] suggests a few safety practices that should be imple-mented in the laboratory environment:

1. Store IPA bottles in a flammable cabinet and away from light.
2. Dispose of old bottles of IPA.
3. Take extra caution when distilling IPA. Always check for peroxides before distillation, and do not distill when peroxides are detected. Never distill the solvent to dryness.

1.2.3.2   Classical Resolution of Racemic Acid

Solvent selection for classical resolution affects both the yield and optical purity. It was observed that the resolution of racemic hexynoic acid (±)-**45** with (1S,2R)-1-amino-2-indanol **46** by forming a diastereomeric salt was solvent dependent, and different yield and optical purity were obtained when using three different solvents (Equation 1.18).[35]

$$(1.18)$$

Solvent	Equiv of 46	Optical Purity	Yield
2-propanol	1.0	~93% ee (free acid)	28%
1-propanol	0.55	>98% de	28 34%
acetonitrile	0.55	98% de	39 41%

Resolution in 2-propanol gave the chiral hexynoic acid **45** (after the salt break) in 28% yield with ~93% ee, while in 1-propanol or acetonitrile provided the diastereomeric salt **47** in 28–34% yield (>98% de) and 39–41% yield (98% de), respectively.

REMARK

Glycerol is immiscible with hydrocarbons and is considered a biodegradable, inexpensive, and non-toxic solvent. It can form strong hydrogen bonds and dissolve inorganic compounds. Glycerol was used as a solvent in the metal-catalyzed cycloisomerization of (Z)-2-en-4-yn-1-ols into furans.[36]

1.2.3.3 Nickel-Catalyzed Addition Reaction

A nickel-catalyzed addition reaction was developed to access chiral β-hydroxynitriles (Scheme 1.5).[37]

The reaction proceeded in 2-propanol via an enantioselective addition of acetonitrile to aldehydes in the presence of nickel pincer complex catalyst and potassium *tert*-butoxide (2 mol%). Although with a high pKa of 31.3 (in DMSO), acetonitrile, being coordinated with the Ni(II) pincer complex, is deprotonated with potassium *tert*-butoxide. The resulting coordinated cyanocarbanion adds to the aldehyde affording a zwitterionic intermediate. This process features a high atom economy and uses inexpensive acetonitrile as a nitrogen and two-carbon source.

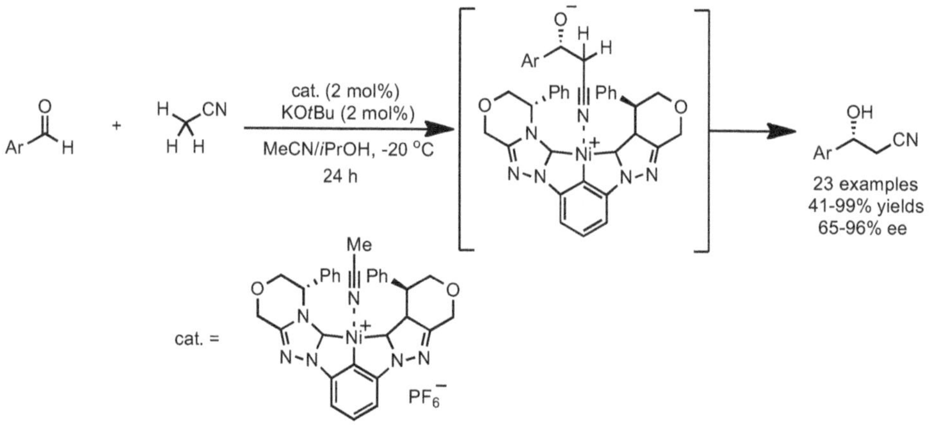

SCHEME 1.5

1.2.4 1-PENTANOL

The debenzylation of dibenzylated amine hydrochloride **48** was carried out under catalytic transfer hydrogenation conditions (Equation 1.19)[38] with ammonium formate as the hydrogen source.

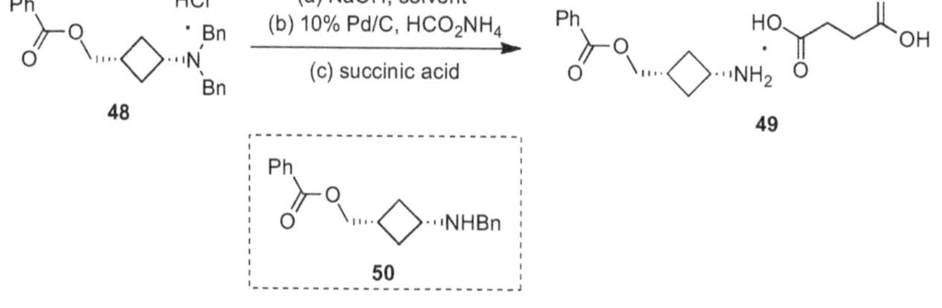

(1.19)

(I) Reaction problems

It was observed that the catalytic debenzylation in ethanol at 40–45 °C was very slow even with 5 equiv of ammonium formate, giving a considerable amount (73%) of mono-benzyl side product **50**.

(II) Solutions

A significant improvement was achieved when switching to 1-pentanol by conducting the reaction in a two-phase system. After 12 h, 100% conversion was achieved. The amount of the mono-benzyl intermediate **50** was reduced to less than 1%.

Procedure

Stage (a) free base formation
- Charge 121.0 g (0.286 mol) of **48** and 405.5 g (500 mL) of 1-pentanol.
- Charge 312.0 g of 1 N NaOH.
- Heat to 45–50 °C and stir for 5 min.
- Separate layers and cool the organic layer to 20–25 °C.

Stages (b) hydrogenolysis
- Charge 24.2 g of 10% Pd/C to another reactor.
- Charge the 1-pentanol solution of the free base of **48**.
- Prepare, in a separate container, an aqueous solution by dissolving 90.2 g (1.43 mol) of ammonium formate in 150 mL of water.
- Charge the ammonium formate solution over 30–40 min to the reactor at 45–50 °C.
- Stir at 45–50 °C for 16 h.
- Cool to 20–25 °C prior to filtration.
- Wash the cake with 100 mL of 1-pentanol and 50 mL of water.
- Separate the lower aqueous layer from the filtrate.
- Wash the organic layer with 2×100 mL of water at 45–50 °C.
- Distill at atmospheric pressure at 102 °C until ~100 mL of distillate is collected.
- Cool to 30 °C.

Stage (c) formation of succinic acid salt
- Charge 33.8 g (0.29 mol) of succinic acid.
- Heat to 95–100 °C.
- Cool to 20–25 °C over 2 h and stir at that temperature for ≥90 min.
- Add 250 mL of heptane and stir for 1.5 h.
- Filter and wash with 150 mL of 1:1 mixture of heptane and 1-pentanol.
- Dry at 55–60 °C under vacuum.
- Yield: 77.5 g (80%).

1.2.5 ETHYLENE GLYCOL

Ethylene glycol is a good solvent for the addition/elimination reaction of ester **51** with 4-chlorobenzylamine **52** (Equation 1.20).[39]

(1.20)

In ethylene glycol, a homogeneous solution was obtained by heating the slurry of **51** and **52**, and the reaction was performed at 132 °C for 8 h, furnishing the desired amide **53**.

Procedure

- Charge 800 g (2.1 mol) of **51** into a 22-L, three-necked flask under nitrogen, followed by adding ethylene glycol (4.8 L) and 755 mL (6.2 mol, 3 equiv) of **52**.
- Heat the resulting suspension at 132 °C for 8 h.
- Cool to 102 °C.
- Add toluene (2.4 L).
- Cool to 75 °C.
- Add MeCN (1.2 L), followed by adding 215 mL (2.1 mol, 1.0 equiv) of pivaldehyde.
- Cool to room temperature
- Filter, rinse with EtOH (2 L) and MeCN (2×1 L), and dry at 55 °C under vacuum.
- Yield: 1318 g (132%) of crude **53**.

REMARK

During the isolation, 4-chlorobenzylamine readily forms an insoluble carbonate salt. Therefore, the addition of pivaldehyde was used to convert the excess amine to the soluble imine and removed via filtration.

1.3   WATER AS REACTION SOLVENT

When designing organic syntheses, the selection of reaction solvents is considered one of the most important aspects, especially at scale. Regarding the negative environmental impact of using volatile organic compounds (VOCs) as solvents, water becomes a highly recommended reaction solvent. Due to its high heat capacity, water is an ideal solvent for exothermic reactions. In addition, the hydrogen-bonding network can influence the reactivity of substrates.[40] When conducting a reaction in water, other interesting features of water can be applied, such as adding additives or co-solvents and adjusting pH.

1.3.1   IODINATION REACTION

Approaches to the synthesis of iodinated molecules include diazotization–iodination reaction and electrophilic iodination using iodine or iodide in combination with an oxidant. Using molecular iodine and hydrogen peroxide in an aqueous medium is one of the most useful protocols,[41] which addresses the growing concern over environmental pollution and sustainable development.

Water as a reaction solvent was employed in the iodination of N-methylpyrazole **54** with iodine in the presence of hydrogen peroxide (0.6 equiv) (Equation 1.21).[42] The 4-iodo-N-methylpyrazole product **55** was conveniently isolated directly by filtration from the reaction mixture in 91% yield (16.2 kg).

$$
\underset{\mathbf{54}}{\underset{\text{Me}}{\text{N-N}}} \quad \xrightarrow[\text{H}_2\text{O}]{\begin{array}{c}\text{I}_2\ (0.5\ \text{equiv})\\ \text{H}_2\text{O}_2\ (0.6\ \text{equiv})\end{array}} \quad \underset{\mathbf{55}}{\underset{\text{Me}}{\text{N-N}}}\text{-I} \tag{1.21}
$$

Procedure

- Charge, into a 100-L reactor, 7.0 kg (85.3 mol) of **54** and water (30 L) followed by adding 11.0 kg (43.5 mol) of iodine.
- Stir for 10 min.
- Charge 15 g of **55** (as seeds).
- Charge 5.2 L of 30% H_2O_2 (51.2 mol) over 30 min (the temperature increased from 18 to 32 °C).

- Stir for 42 h.
- Cool to 10 °C.
- Add 20 L of 10% of NaHSO$_3$ at <20 °C.
- Stir the resulting slurry at room temperature for 3 h.
- Filter and wash with water (5×20 L).
- Dry at 30–35 °C under vacuum.
- Yield: 16.2 kg (91%).

1.3.2 Synthesis of Quinazoline-2,4-dione

A pilot plant scale synthesis of 5-substituted-6,7-dimethoxy-1-H-quinazoline-2,4-diones **58** was carried out in water (Scheme 1.6).[43] The process is unique in that all stages of transformations were affected by the manipulation of the pH of the reaction mixture. The products were isolated in near-quantitative yields with good purity. The process was demonstrated on scales up to 5 kg.

Procedure (for the synthesis of 58a)

Stage (a)
- Charge 2.35 kg (9.1 mol) of **56a** into a reactor followed by adding 245 g of Pd/C under N$_2$.
- Charge 23.5 L of water and 870 g of 50% NaOH (1.2 equiv).
- Inert the vessel by evacuating and flushing with nitrogen three times.
- Hydrogenate under 5 psig of hydrogen at ambient temperature.
- Filter over solka-floc.

Stage (b)
- Charge the filtrate into a clean reactor followed by adding 964 g (1.3 equiv) of potassium cyanate in water (2 L).
- Charge AcOH (560 g, 1.2 equiv) slowly to maintain pH at 6.8–8.0 at <10 °C.

Stage (c)
- Charge a solution of 50% NaOH (2.1 kg, 2.9 equiv) to the batch.
- Heat the resulting mixture to 50 °C.
- Charge 565 g additional 50% NaOH (0.8 equiv) to pH 13.
- Stir the batch at 50 °C for 18 h.
- Cool to 20 °C.
- Add 2.9 kg of AcOH to pH 8.0.
- Cool the resulting slurry to 5 °C.
- Filter, wash with water, and dry at 70 °C overnight.
- Yield of **58a**: 2.06 kg (90%) in 99.9% HPLC purity.

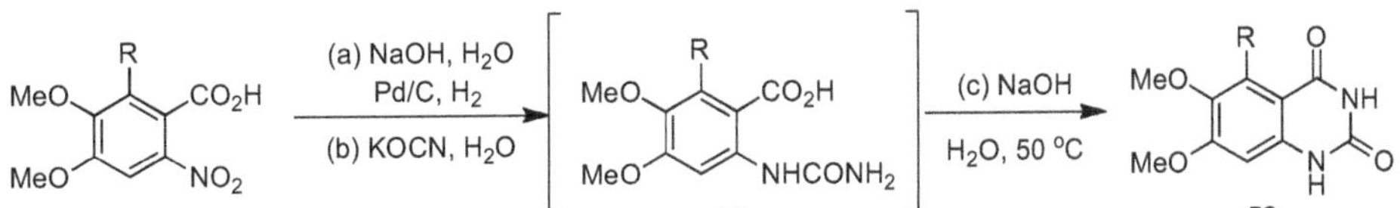

R = OMe, **58a**, 90%
Me, **58b**, 98%

SCHEME 1.6

1.3.3 Synthesis of Pyrrolocyclohexanone

An aqueous system was utilized to prepare pyrrolocyclohexanone **61** in a two-step, one-pot process, starting with cyclohexane-1,3-dione **59** (Scheme 1.7).[44] The protocol involves alkylation of **59** with chloroacetone in the presence of potassium hydroxide, followed by Paal–Knorr pyrrole synthesis. On a scale of 130 kg, the alkylation with chloroacetone (1.17 equiv) finished within 20 h at 30–35 °C; the subsequent pyrrole formation took place at 50–55 °C by adding MTBE, sodium acetate, and glycine ethyl ester.

Procedure

Stage (a)
- Charge 143 L of water into a 1600-L glass-lined reactor, followed by adding 130 kg (1.16 kmol) of **59**.
- Cool the mixture to 12 °C.
- Charge a solution of 90 wt% KOH (76.6 kg, 1.23 kmol, 1.06 equiv) in water (117 L) at 11 °C.
- Stir at 11 °C for 30 min.
- Charge 126.2 kg (1.36 kmol, 1.17 equiv) of chloroacetone at 11.5 °C over 4 h.
- Warm the batch to 32 °C and stir at this temperature for 20 h.

Stage (b)
- Charge 180.7 kg (1.29 kmol, 1.11 equiv) of glycine ethyl ester hydrochloride and MTBE (390 L).
- Stir the resulting mixture at 30±3 °C for 10 min.
- Charge 104.7 kg (1.28 kmol, 1.10 equiv) of anhydrous NaOAc and water (260 L).
- Warm the batch to 52±3 °C and maintain at this temperature for 6 h.
- Separate layers.
- Extract the aqueous layer with MTBE (260 L) at 52±3 °C.
- Wash the combined organic layers with 5% aqueous $NaHCO_3$ (520 L).
- Extract the aqueous layer with MTBE (130 L) at 53 °C.
- Cool the combined organic layers to 0–5 °C and maintain at the temperature for 12 h.
- Filter and wash with water (260 L) followed by cold MTBE (260 L).
- Dry at 40–45 °C under vacuum.
- Yield: 183.7 kg (67 %).

REMARKS

(a) It has been demonstrated that Suzuki–Miyaura coupling of aryl tosylates and mesylates could be carried out in water as the reaction solvent in combination with water-soluble catalysts.[45]

(b) The use of water as a solvent was also used in Diels–Alder reactions.[46]

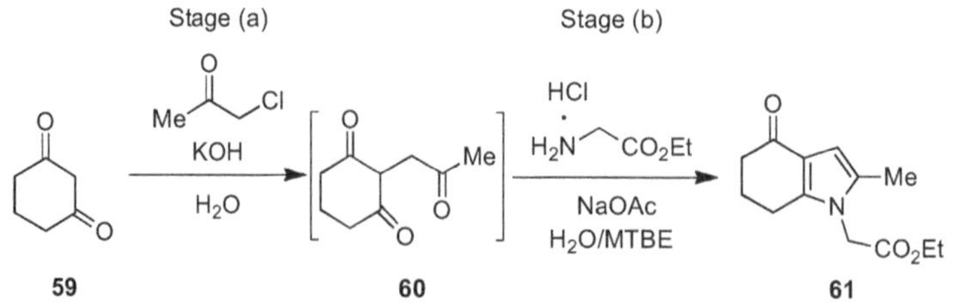

SCHEME 1.7

1.3.4 SYNTHESIS OF THIOUREA

Water as a solvent sometimes can cause process problems. An illustrative example is shown in Equation 1.22[47] wherein the preparation of thiourea **63** was achieved by a reaction of *o*-anisidine **62** with potassium isothiocyanate under aqueous acidic conditions in the presence of ammonium chloride.

$$(1.22)$$

(I) Reaction problems

- This transformation was plagued by incomplete conversion even when an excess (6 equiv) of potassium isothiocyanate was used.
- The dimeric thiourea side product **64** had to be removed by recrystallization.
- Product **63** was in a moderate yield of 65%.

(II) Solutions

- It was surmised that water as a solvent under the reaction conditions triggered significant decomposition of potassium isothiocyanate.
- Isopropyl acetate (or 2-butyl acetate) was proved the solvent of choice.
- The reaction proceeded smoothly in the presence of potassium isothiocyanate (1.5 equiv) and trifluoroacetic acid (2.5 equiv) and completed at 80 °C within 6–7 h.
- The isolated product yield was 74% yield (>99% HPLC purity) along with <5% of the dimeric thiourea impurity which was rejected in the mother liquor.

1.3.5 SYNTHESIS OF AMIDE

Amide bond formation is one of the most investigated reactions in organic synthesis. Polar aprotic solvents, such as DMF, DMAc, DMSO, and NMP, are frequently used for the C–N amide bond formation, and among them, 47% of amidation reactions use DMF as the reaction medium.[48] However, the extensive consumption of polar aprotic solvents poses cost and environmental issues. A report[49] described a selective amidation of unprotected amino alcohols (Equation 1.23).

$$(1.23)$$

TPGS-750-M

The reaction was conducted using EDC (1.5 equiv), HOBt (1.2 equiv), N-methylmorpholine (NMM) (3.0 equiv) as a base in the presence of 2 wt% of surfactant (TPGS-750-M) in water at 40 °C. The solid products were isolated directly from the reaction mixture by simple filtration.

1.3.6 SYNTHESIS OF 1,3/1,4-DIKETONES

Generally, multi-component reactions (MCRs) are step-economic, allowing the construction of complex molecules in an efficient and economic manner. To achieve MCRs, each individual reaction must be chemoselective and occur sequentially. A copper-catalyzed three-component reaction was developed to access 1,3/1,4-diketone products (Equation 1.24).[49] The reaction was carried out simply by mixing α-ketoaldehydes and 1,3-dicarbonyl compounds with aryl boronic acids in the presence of nano CuO catalyst under reflux in water. The initial aldol condensation reaction occurs between α-ketoaldehyde and 1,3-dicarbonyl compound generating α,β-unsaturated intermediate. Subsequently, the intermediate reacts with a key nucleophilic intermediate (ArCuOB(OH)$_2$), formed in situ via a migratory insertion of CuO into a C–B bond, furnishing the Michael addition product.

$$(1.24)$$

1.4 NON-POLAR SOLVENTS

A safety issue should be addressed when using non-polar solvents (generally with dielectric constants of less than 5) such as hydrocarbons. An electrostatic charge can build up within the liquids, especially those with low conductivity such as hydrocarbons, especially where there is a flow of liquid or a settling process is occurring. A two-phase system can also generate a charge on stirring. High flow (or agitation) rates with low conductivity can generate a great electric charge. This charge can ultimately lead to a discharge or spark that could result in a fire if a flammable atmosphere is present. Therefore, special attention is required when handling such low-conductivity hydrocarbon solvents.

1.4.1 CONDENSATION OF KETONE WITH *TERT*-BUTYL HYDRAZINE CARBOXYLATE

Compared to hexane (dielectric constant of 1.9), heptane has several advantages as a reaction solvent, such as the higher boiling point (98 °C vs 69 °C of hexane) and higher conductivity (3×10^{-2} vs 1×10^{-5} pS/m of hexane), which make heptane a preferred processing solvent over hexane. However, exposure of a reaction mixture to heptane under reflux conditions led to a concern about the stability of the product (and starting materials/reagents). For example, the yield of product hydrazone **67** dropped from 85% to 54% under reflux in heptane when the reaction time was increased from 4 to 16 h, when the reaction in hexanes occurred during 3 to 16 h, the yield only decreased to 90% from 95% (Equation 1.25).[50] Therefore, hexanes were chosen as the process solvent.

(1.25)

The conductivity issue was addressed by adding 2-propyl alcohol (10% w/w, relative to hexanes) to the slurry prior to filtration. The presence of 2-propyl alcohol did not adversely affect product recovery.

Procedure

- Charge 800 g (6.05 mol) of **66**, 713 g (1.2 equiv) of **65**, and 2.0 kg of hexanes into a 10-L reactor.
- Heat the reaction mixture under reflux for 3.5 h with a Dean–Stark trap to remove byproduct water.
- Cool the contents to 0–5 °C.
- Add 2-propanol (0.2 kg).
- Filter to afford the crude product **67** (1.26 kg) that was used directly in the subsequent step.

NOTE
The boiling point of the azeotropic mixture of *n*-hexane with water is 62 °C with 6% water.[51]

1.4.2 ACID-CATALYZED ESTERIFICATION

Acid-catalyzed esterification of acids with alcohols is a reversible process and the removal of the water byproduct is the driving force for the completion of the reaction. Under reflux (110 °C) in toluene, for example, the esterification (Equation 1.26)[52] of α-enol acid **68** with D-borneol **69**, catalyzed by TsOH (10 mol%), occurred to furnish the desired ester product **70** in 76% yield after 12 h.

(1.26)

Interestingly, heating the reaction mixture in a mixture of THF/hexane (5:1, v/v) under reflux (80 °C) for 12 h produced 80% product yield, and a 95% yield of product was obtained following 27 h of reflux. Such rate enhancement was presumably attributed to the effective water removal from the reaction system.

1.5 POLAR APROTIC SOLVENTS

Polar aprotic solvents are commonly used in reactions that involve charge-separation transition states such as substitution reactions ($S_N Ar$, $S_N 1$, $S_N 2$). Normally, reaction rates can be improved significantly in polar aprotic solvents compared to other solvents. However, the use of polar aprotic solvents, such as dimethylformamide (DMF), dimethyl sulfoxide (DMSO), *N*-methyl-2-pyrrolidone (NMP), and dimethyl acetamide (DMAc) has significant issues. For example, DMF, DMAc, and NMP have caused effects on embryo-fetal development in animals; DMSO has some process safety issues.[53] Despite their exceptional effects on the reaction reactivity, polar aprotic solvents are difficult to be removed by distillation from reaction mixtures due to their high boiling points. Extractive workup is usually used to reject polar aprotic solvents into aqueous layers.[54]

1.5.1   ACETONE AS A SOLVENT

1.5.1.1   Michael Addition Reaction with Acetone Cyanohydrin

Solvent selection can impact waste management and the selection of appropriate solvent allows facile waste disposal especially when the waste stream contains environmentally hazardous chemicals. For example, the conduction of chalcone hydrocyanation reaction in acetone avoided mixed organic solvents in the waste stream since acetone was released in the reaction of acetone cyanohydrin (Equation 1.27).[55]

$$ (1.27) $$

More importantly, acetone could be separated from the aqueous phase easily by atmospheric distillation, because it does not form an azeotrope with water. This allowed the destruction of the residual cyanide by treating the aqueous phase with sodium hypochlorite.

1.5.1.2 S_N2 Alkylation Reaction

Considering the toxicity of DMF,[53] acetone was used to replace DMF in the S_N2 alkylation of amine **72** with mesylate **71** (Equation 1.28).[53] An additional benefit of using acetone as the reaction solvent is the enantiomeric purity of the product **73** was improved from 7.6% ee in DMF to 97.8% ee.

$$ (1.28) $$

1.5.1.3   Multi-Component Reactions

A three-component reaction was developed for the synthesis of substituted oxazolidines under photoredox catalysis conditions (Equation 1.29).[56] In this three-component reaction, acetone was used as both the reaction solvent and the oxygen source in the oxazolidine rings. Mechanistically, the reaction proceeds through a radical addition followed by [3+2] annulations in a one-pot manner.

$$ (1.29) $$

R = H, Ar = Ph: 88% yield
R = Me, Ar = 2-Cl-Ph: 88% yield

1.5.1.4   Amidation Reaction

Conversion of amino ester to the corresponding amide in neat pyrrolidine suffered significant decomposition via retro-aldol reaction and a variable amount of C2 epimerization (5–10%) (Scheme 1.8).[57]

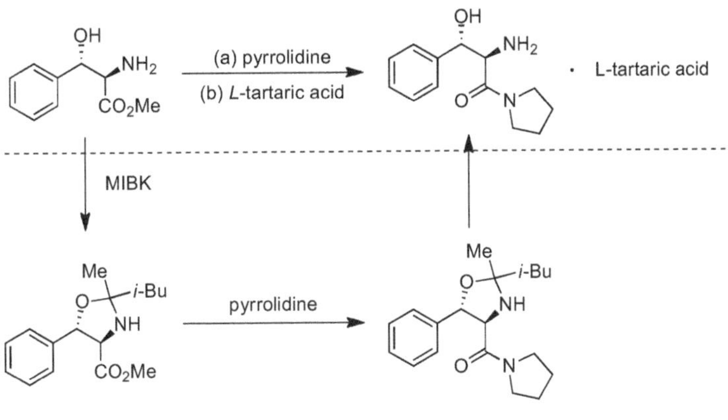

SCHEME 1.8

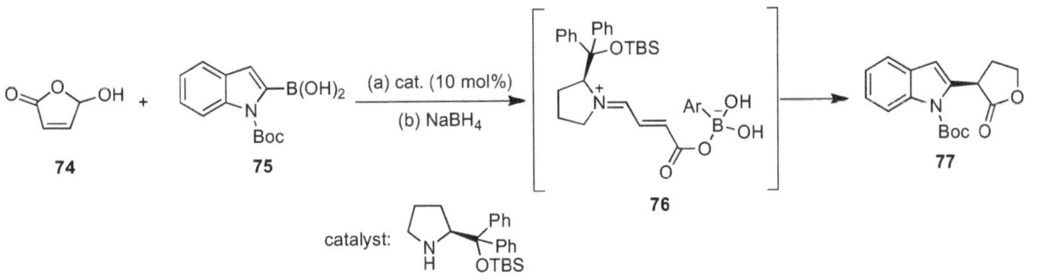

SCHEME 1.9

Therefore, acetone or methyl isobutyl ketone (MIBK) could be employed as the reaction solvent, affording the desired product with minimal decomposition or epimerization. The reaction was monitored by NMR and showed that mixing the ester with pyrrolidine in MIKB generated a readily oxazolidine intermediate.

1.5.2 ACETONITRILE AS A SOLVENT

Acetonitrile is regarded as a key solvent in the pharmaceutical industry. Acetonitrile has excellent solvation ability with respect to a wide range of polar and non-polar solutes and favorable properties such as low freezing/boiling points, low viscosity, and relatively low toxicity.

1.5.2.1 Intramolecular Michael Addition Reaction

Acetonitrile was shown to be a good solvent for the reaction of 5-hydroxyfuran-2(5H)-one **74** with indole boronic acid **75** (Scheme 1.9).[58] The reaction was proposed to occur via an iminium intermediate **76** which undergoes an intramolecular Michael addition.

Table 1.2 demonstrates clearly that acetonitrile is the ideal solvent for this aryl migration process.

In addition to acetonitrile, the reaction worked well in other solvents such as ethyl acetate and dichloromethane, but proceeded slower in MTBE and EtOH and was not complete after prolonged reaction times in DMF and DMSO.

1.5.2.2 Synthesis of Imidazolines

Acetonitrile (and other nitriles, e.g., cyclopropanecarbonitrile) was utilized as a reaction solvent in a three-component reaction to synthesize imidazoline under mild photoredox catalysis conditions.[56] In this reaction (Scheme 1.10), acetonitrile (and cyclopropanecarbonitrile) also provided one of the nitrogen sources in the imidazoline rings.

TABLE 1.2

Solvent Effects

Solvent	Reaction time	Yield
MeCN	2 h	98%
EtOAc	6 h	76%
CH$_2$Cl$_2$	2 h	67%
MTBE	48 h	60%
EtOH	24 h	62%
DMF	240 h	42%
DMSO	240 h	20%

Source: Reprinted with permission from Bos, M. and Riguel, E., *J. Org. Chem.*, **2014**, *79*, 10881. Copyright 2014 American Chemical Society.

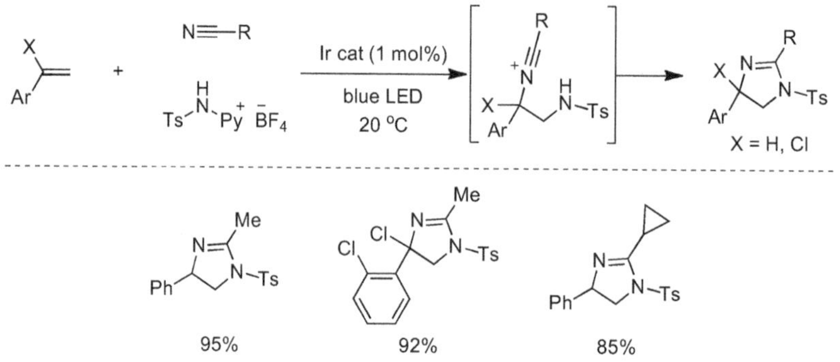

SCHEME 1.10

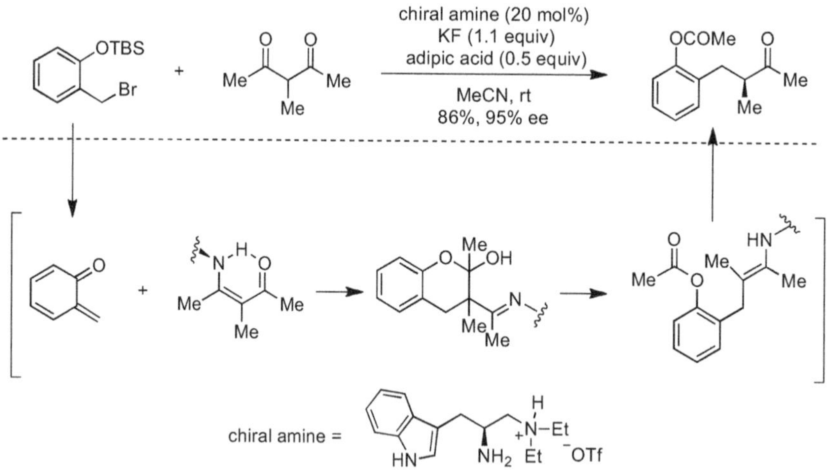

SCHEME 1.11

1.5.2.3 Synthesis of α-Alkylated Ketones

A significant solvent effect was observed for the synthesis of α-methyl ketone employing an enamine-mediated retro-Claisen reaction (Scheme 1.11).[59] Using acetonitrile as the reaction solvent provided 86% yield of the chiral ketone product with 95% ee, while no reaction occurred in methylene chloride otherwise under the same conditions. Mechanistically, the reaction proceeds through a sequential process of stereoselective C–C bond formation, C–C bond cleavage, and a highly stereospecific enamine protonation.

REMARK

Mixing acetonitrile with strong acids (Equation 1.30)[60] or strong bases (Equation 1.31),[61] such as NaOH, KOH, and LiOH may pose significant safety concerns due to runaway hydrolysis reactions of acetonitrile.

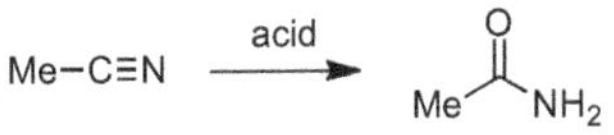

$$ \text{Me}-\text{C}\equiv\text{N} \xrightarrow{\text{acid}} \underset{\text{Me}}{\overset{\text{O}}{\|}}\text{NH}_2 \qquad (1.30) $$

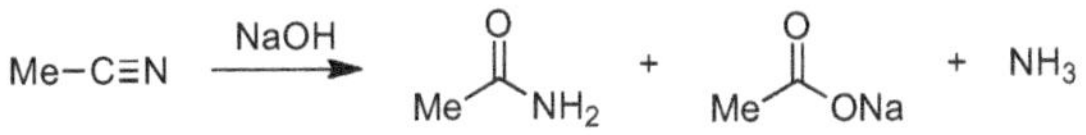

$$ \text{Me}-\text{C}\equiv\text{N} \xrightarrow{\text{NaOH}} \underset{\text{Me}}{\overset{\text{O}}{\|}}\text{NH}_2 + \underset{\text{Me}}{\overset{\text{O}}{\|}}\text{ONa} + \text{NH}_3 \qquad (1.31) $$

1.5.2.4   Chlorosulfonylation Reaction

Installation of sulfonamide onto a benzene ring required three distinct chemical steps using sulfolane as the solvent (Scheme 1.12).[62] The use of highly toxic and hazardous chlorosulfonic acid and POCl$_3$ renders this process not sustainable. In addition, sulfolane, a Class 2 solvent, is not desired on a manufacturing scale.

It was found that conducting chlorosulfonylation in acetonitrile allowed for the elimination of POCl$_3$ (Scheme 1.13). These changes provide a greener and operationally simpler process.

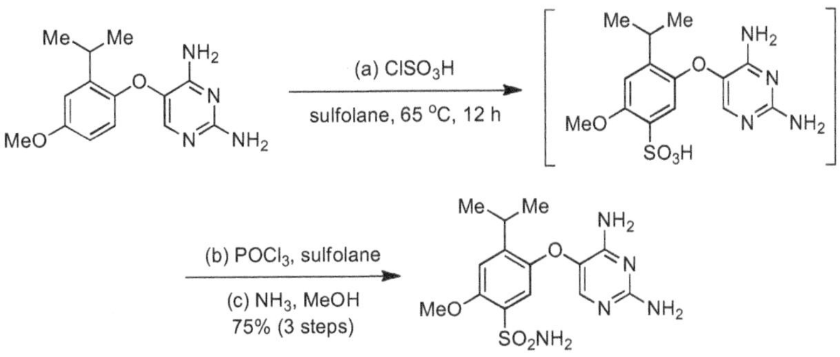

SCHEME 1.12

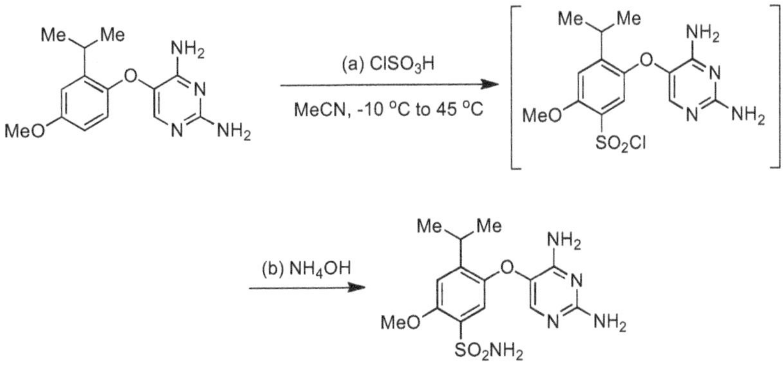

SCHEME 1.13

It was hypothesized that acid-promoted hydrolysis of acetonitrile occurs during the chlorosulfo-nylation process, which may be important to drive the reaction to the desired sulfonyl chloride by consuming the water byproduct.

1.5.3   N,N-Dimethylformamide as a Solvent

1.5.3.1   Preparation of Alkyl Aryl Ether

Etherification of pyrazolopyrazine **78** with 2-pyridinemethanol **79** under S_NAr reaction conditions gave alkyl aryl ether **80** (Equation 1.32).[63]

(1.32)

(I) Reaction problems

The reaction in DMF with 2 equiv of 50% NaOH at 45 °C proceeded rapidly (>99% conversion after 10 min), but the product **80** was hydrolyzed to **81** (16% after 25 min) under the reaction conditions.

(II) Solutions

To address the instability of **80**, DMF was replaced with THF. The reaction was carried out in THF, producing the product **80** with negligible decomposition impurity **81** (<1% after 15 h), albeit requiring longer reaction times (27 h).

1.5.3.2   Preparation of Bisaryl Ether

Initially, the S_NAr reaction of aryl fluoride **82** with phenol **83** was conducted in DMF at 120 °C to give the desired product **84** in 63% isolated yield (Equation 1.33).[64]

(1.33)

After solvent and base screening, it was found that the reaction could be carried out at 78 °C in MeTHF in the presence of potassium *t*-butoxide as the base to furnish **87** in 62% isolated yield, albeit with a long reaction time (5 days).

1.6   HALOGENATED SOLVENTS

1.6.1   Dichloromethane

1.6.1.1   Reaction with Pyridine

When selecting a reaction solvent, solvent compatibility under the reaction conditions has to be considered. The preparation of pyridine derivatives is commonly carried out in dichloromethane as a solvent. However, dichloromethane slowly reacts with pyridine and its derivatives to form methy-lenebispyridinium dichloride compounds under ambient conditions (Scheme 1.14).[65]

Because of the higher reaction rate of the second substitution, the monosubstitution product **86** could not be detected during the course of the reaction and only the methylenebispyridinium dichloride **87** was isolated.

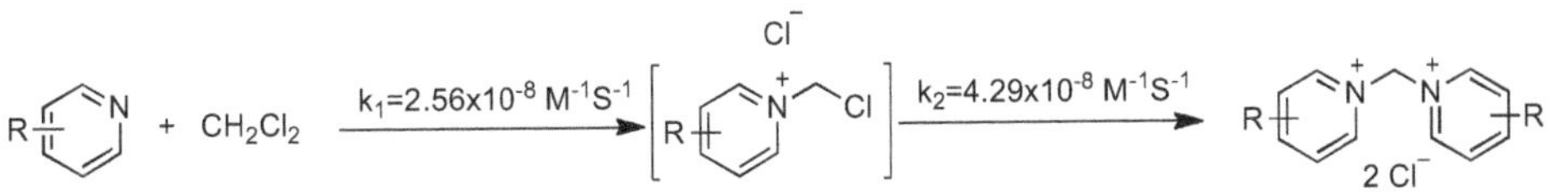

SCHEME 1.14

REMARKS

(a) It appears that the steric effect can suppress S_N2 substitution reactions, for example, no reaction was observed between 2- (or 2,6-) substituted pyridines and dichloromethane.
(b) Emulsion is usually encountered during extractive workup with dichloromethane. Warming the mixture to above room temperature sometimes helps to break the emulsion.

1.6.1.2 Synthesis of Benzo[d]isothiazolone

(I) Reaction problems

The conversion of disulfide **88** to benzo[*d*]isothiazolone **89** in dichloromethane was plagued by widely variable yields (25–90%) (Equation 1.34).[66]

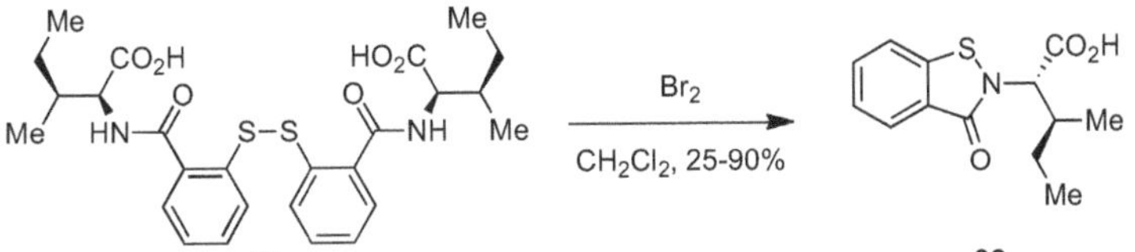

(1.34)

(II) Chemistry diagnosis

This transformation starts with oxidative cleavage of the disulfide bond with bromine via bromo disulfanium intermediate **90** (Scheme 1.15). Obviously, dichloromethane as the reaction solvent will not be able to stabilize the ionic species **90**.

(III) Solutions

Accordingly, the replacement of dichloromethane with acetic acid allowed the production of **89** in consistent yields. In addition, the isolation of hydrobromide salt **92** as a yellow solid provided a very effective purification procedure for the production of the final pharmaceutical compound **89** (Scheme 1.16). An aqueous workup was necessary to remove traces of residual acetic acid.

Procedure

Stage (a)
- Charge 50.0 kg (94 mol) of **88** into a reactor, followed by adding 332 kg of AcOH.
- Charge 16.5 kg (103 mol, 1.1 equiv) of bromine at 22–30 °C.
- Stir the resulting slurry at 22–30 °C for 6 h.
- Filter and rinse with AcOH (24 kg) and heptanes.
- Yield of **92**: 60.4 kg (8.1% residual solvent).

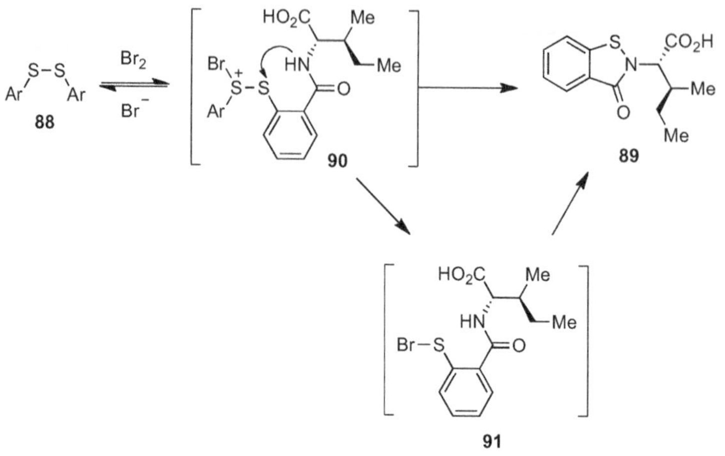

SCHEME 1.15

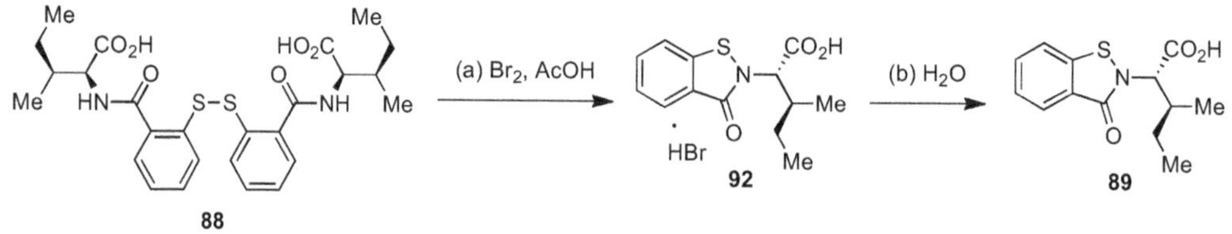

SCHEME 1.16

Stage (b)
- Charge 60.4 kg of **92** into a reactor followed by adding MTBE (198 kg) and water (140 L).
- Stir for 15 min.
- Separate layers.
- Extract the aqueous layer with MTBE (29 kg).
- Wash the combined MTBE layer with aqueous NaCl (32 kg in 106 L of water).
- Dry the organic layer with $MgSO_4$ (5.6 kg).
- Filter and distill until ca. 170 L.
- Cool the batch to 45–50 °C.
- Add heptane (80 kg).
- Cool to 10–15 °C.
- Filter and rinse with heptane (42 kg).
- Dry at 45–50 °C under vacuum.
- Yield of **89**: 36.9 kg (74.1%).

REMARK

The salt break was effected simply by treating the hydrobromide salt **92** with water.

1.6.2 1,2-DICHLOROETHANE

Despite its toxicity 1,2-dichloroethane (Class 1 solvent) was employed as the solvent for a decarboxylative Blaise reaction for the synthesis of 2,6-dichloro-5-fluoronicotinoylacetate **96** (Scheme 1.17).[67] The synthesis was accomplished in a single spot with 3-cyano-2,6-dichloro-5-fluoropyridine **94** and potassium ethyl malonate **93** in the presence of zinc chloride.

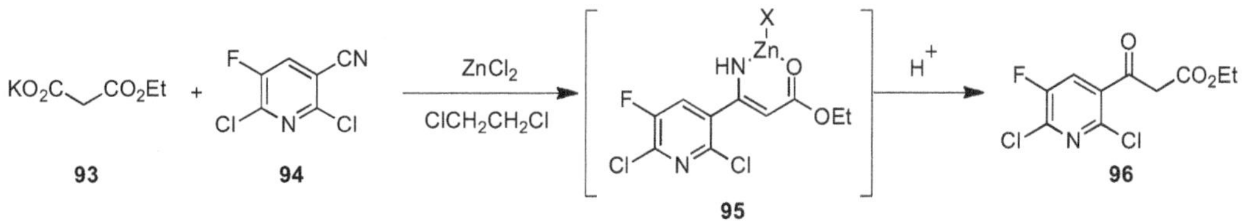

SCHEME 1.17

Procedure

- Add 20.0 kg (105 mol) of **94** into a 250-L reactor containing 1,2-dichloroethane (100 L), followed by adding sequentially zinc chloride (7.14 kg, 52.4 mol) and **93** (21.4 kg, 126 mol).
- Heat the mixture at reflux for 3 h.
- Add slowly 6 N HCl (150 L).
- Continue to reflux for 30 min.
- Cool the batch to ambient temperature.
- Separate layers.
- Concentrate the organic layer.
- Add EtOH/H$_2$O (5/1, 100 L) to the residue.
- Heat the mixture to 70 °C to dissolve the solid.
- Cool the solution to 40 °C, then to 0 °C.
- Stir for 1 h.
- Filter and wash with EtOH/H$_2$O (5/1, 100 L).
- Yield of **96**: 22.0 kg (75.0%) as a white solid.

REMARKS

(a) The use of toxic 1,2-dichloroethane as the reaction solvent streamlined the workup process; the 1,2-dichloroethane could be conveniently recycled by distillation.
(b) The use of DMF, 1,3-dimethyl tetrahydropyrimidin-2(1H)-one (DMPU), or DMSO is not recommended in the presence of alkali metal hydrides. A mixture of DMF and NaH can undergo uncontrollable exothermic decomposition even at 26 °C.[68]
(c) The poor thermal stability of sodium borohydride (NaBH$_4$) in DMF can cause a violent runaway reaction.[69]

1.6.3 TRIFLUOROACETIC ACID

The cleavage of the N-toluenesulfonyl group generally requires harsh reaction conditions, for instance, the deprotection of the Ts-group in N-toluenesulfonamide **97** was achieved under conditions of either 4-hydroxybenzoic acid (3 equiv) in 10 volumes of HBr (33 wt% solution in acetic acid) or concentrated sulfuric acid at 75 °C (Equation 1.35).[47]

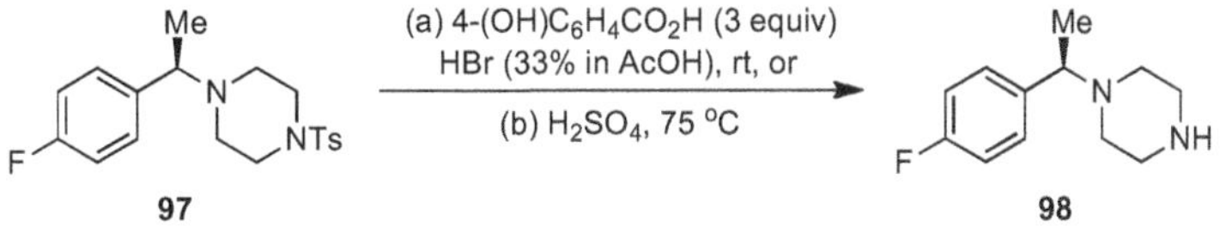

$$(1.35)$$

(I) Problems

The reaction with 4-hydroxybenzoic acid (3 equiv) in 10 volumes of HBr (33 wt% solution in acetic acid) was sluggish upon scaling up, requiring 3–4 days with 3–5% of unreacted **97**. Under conditions (b), the large excess of sulfuric acid had to be neutralized with sodium hydroxide, leading to a strongly exothermic reaction and large workup volumes.

(II) Solutions

- Trifluoroacetic acid (TFA) was chosen as an acidic co-solvent to reduce the equivalents of sulfuric acid.
- Using H_2SO_4 (6.7 equiv) in TFA led to complete conversion within 5–6 h. giving the product a 70% yield and 97–98% purity.

Procedure

- Charge 8.4 L of TFA into a glass-lined reactor.
- Cool to 5 °C.
- Charge 4.18 kg (11.5 mol) of **97** portionwise at <20 °C.
- Add slowly 4.18 L (78.5 mol, 6.7 equiv) of H_2SO_4.
- Heat to 75 °C and maintain at 75 °C for 6 h.
- Cool to 5 °C.
- Add deionized water (33.0 L) at <20 °C.
- Add 10 N NaOH to pH 8.0±0.5.
- Extract with isopropyl acetate (4×10.5 L).
- Wash the combined organic layers with 1 N NaOH (2×10.5 L).
- Distill the organic layer under reduced pressure until 12–13 L remains.
- Add heptane (12.6 L).
- Continue to distill until 12–13 L remains.
- Add heptane (12.6 L).
- Continue to distill until 7 L remains.
- Cool to –10 °C and stir at this temperature for 2 h.
- Filter and rinse with cold heptane (2 L).
- Dry at 45 °C under vacuum.
- Yield: 1.68 kg (70%).

1.6.4 (Trifluoromethyl)benzene

(Trifluoromethyl)benzene (TFB) (or trifluorotoluene) is a colorless fluorocarbon (bp 102 °C) with relatively low toxicity and is used as a specialty solvent in organic synthesis to replace dichloromethane.[70] TFB could also be used as a DMF surrogate in the synthesis of sulfonamide **102** (Scheme 1.18).[71] This

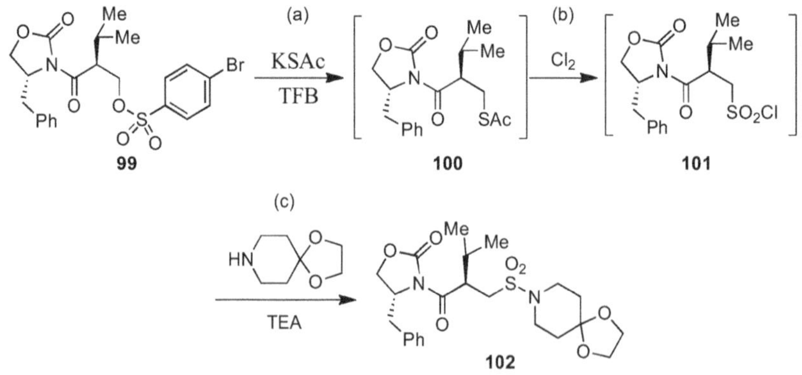

SCHEME 1.18

one-pot synthesis involves S_N2 substitution, oxidative chlorination, and coupling reaction between sulfonyl chloride **101** and 1,4-dioxa-8-azaspiro[4.5]decane. The product **102** was crystallized directly from the reaction mixture.

TFB was also used as the solvent in the palladium-catalyzed amination of triflate **103** with benzylamine (Equation 1.36).[72]

(1.36)

REMARK

Fluorobenzene is a relatively inert compound because the C–F bond is very strong. Thereby fluorobenzene is a useful solvent for reactions that involve highly reactive species.[73] Fluorobenzene is a colorless liquid with a boiling point of 84–85 °C and a melting point of –44 °C.

1.6.5 HEXAFLUOROISOPROPANOL

Fluorinated alcohols present unique properties, including high hydrogen bonding donor ability, low nucleophilicity, high ionizing power, and the ability to solvate water. Hexafluoroisopropanol (HFIP) is a highly polar solvent, commercially available in multi-ton quantities from several suppliers. HFIP appears as a colorless and volatile liquid (bp 58 °C) and can be used as a solvent in a variety of reactions.[74]

1.6.5.1 Selective Oxidation of Sulfide

Because of the fact that HFIP can activate hydrogen peroxide by facilitating the O–O bond cleavage and is resistant to oxidation, HFIP was selected as the solvent for selective oxidation of sulfide **104** with urea–hydrogen peroxide (UHP) (Equation 1.37).[75]

(1.37)

However, HFIP is expensive and the development of a suitable recycling process is very important for a reduction in manufacturing costs. As HFIP is immiscible with heptanes, upon completion of the reaction in neat HFIP, heptanes were added. The resulting biphasic solution could then be distilled under reduced pressure. The distillate was collected as a biphasic mixture, and the lower HFIP layer could be easily separated and reused without further purification.

REMARKS

(a) HFIP is volatile and corrosive. Its vapors can cause severe respiratory problems and immediate, irreversible eye damage. Care should be taken to avoid contact by wearing appropriate gloves, safety glasses, and other protective equipment as needed. HFIP may also contain a very low level of hexafluoroacetone, which is a potential reproductive hazard.

(b) Trifluoroethanol (TFE) is a colorless and water-miscible liquid (bp 74 °C) and can be used as a solvent in a variety of reactions.

1.6.5.2 Cycloaddition Reaction

It was demonstrated that HFIP could promote cycloaddition of 1,2,3-triazine **108** with enamine **107**, giving the cycloadduct product **110** in 95% yield (Scheme 1.19).[76]

However, the reaction in other solvents, such as chloroform, ethanol, or isopropyl alcohol provided poor yield (8%) or no product under otherwise the same conditions. This profound effect of HFIP is attributed to the ability of hydrogen bonding, mild acidic character to promote the aromatization of the intermediate **109** via loss of pyrrolidine, and inability to serve as a nucleophile.

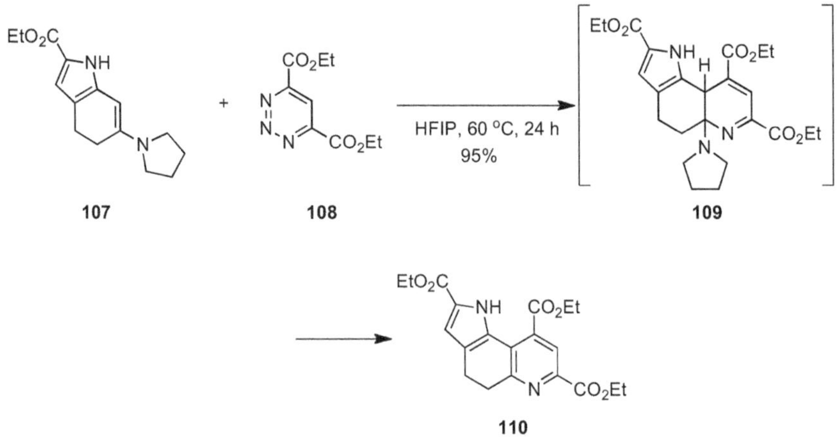

SCHEME 1.19

1.7   CARCINOGEN SOLVENT

Despite its toxicity, benzene (bp 80 °C) is occasionally used as a reaction solvent in laboratory research. For example, aldehyde **111** protection with ethylene glycol proceeds smoothly in benzene, leading to 1,3-dioxolane acetal **112** in 95% yield (Equation 1.38).[77]

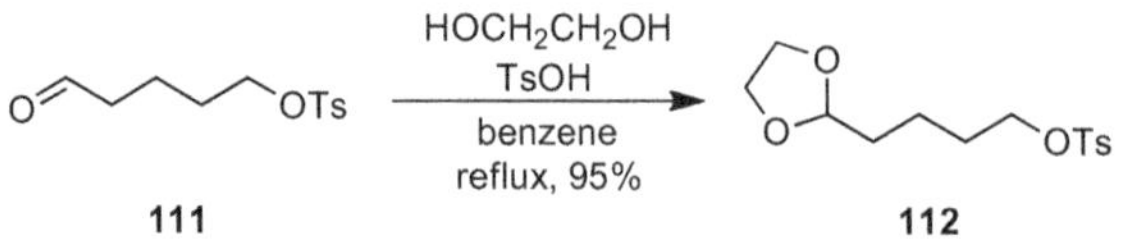

$$(1.38)$$

Toluene (bp 111 °C) is often considered for use as a substitute solvent for benzene. Considering the toxicity of benzene and the stability of starting material **111** and product **112**, MTBE (bp 55 °C) was chosen as the desired solvent and the protection step was performed on a scale with 97% product yield.

1.8   OTHER SOLVENTS

1.8.1   DW-THERM

DW-therm, a commercial heat-exchange fluid, comprises a mixture of triethoxyalkylsilanes. It is a non-viscous, non-hazardous, and environmentally friendly liquid with a boiling point of 240 °C and a flash point of 101 °C. With excellent stability at high temperatures and insoluble in water,

DW-therm is an ideal reaction solvent for high-temperature reactions. The synthesis of bicyclic lactam intermediate **114** requires cyclization of aziridine **113** under high temperatures (Equation 1.39).[78]

$$(1.39)$$

From a safety and temperature control perspective, DW-therm was chosen as the cyclization solvent and the reaction was conducted at 240 °C. After the reaction, the DW-therm was separated from the products by vacuum distillation.

1.8.2 DOWTHERM A

Dowtherm A, a heat transfer fluid, is a eutectic mixture of two very stable compounds, biphenyl and diphenyl oxide with a boiling point of 257 °C.

1.8.2.1 Synthesis of 6-Chlorochromene

It was found that the Claisen rearrangement of **115** and subsequent cyclization occurred at 220 °C in Dowtherm A to give 6-chlorochromene **116** (Equation 1.40).[79]

$$(1.40)$$

Procedure

- Charge 9.61 kg (35.5 mol) of **115** to a hot Dowtherm A (36.8 kg of diphenyl ether and 15.8 kg of biphenyl) at 195–220 °C.
- Heat the batch at 220 °C for 50 min.
- Cool to 40 °C.
- Add heptane (96 L).
- Cool the resulting suspension to 20 °C and stir overnight.
- Filter, rinse with heptane (2×6.8 L) and MIBK (2×6.8 L), and dry under a nitrogen stream.
- Yield: 5.44 kg.

1.8.2.2 Conrad–Limpach Synthesis of Hydroxyl Naphthyridine

Dowtherm A was also employed as Conrad–Limpach reaction solvent for the synthesis of hydroxyl naphthyridine **118** (Equation 1.41).[80]

$$(1.41)$$

For large-scale preparations, the use of Dowtherm A as the solvent offers an opportunity for recycling the solvent, as the product was readily isolated by filtration after cooling the reaction mixture.

Procedure

- Charge 1.3 L of Dowtherm A into a 4-L glass reactor and heat to 255 °C under N_2.
- Charge a hot (80 °C) solution of **117** (161 g) in 1,3-dimethylimidazolidinone (DMI) (0.5 L) at 235–255 °C over 35 min (the acetone byproduct was distilled into a receiver).
- Upon completion of the reaction, cool the reaction mixture to 20 °C and filter.
- Slurry the crude product in EtOH (0.8 L) at 80 °C.
- Cool to 20 °C.
- Filter, rinse with EtOH (0.15 L), and dry at 50 °C at 5 mbar.
- Yield: 69.6 g (68%).

1.8.2.3 Conrad–Limpach Synthesis of Quinolone

The synthesis of quinolones such as **120**, which has recently been accepted as a preclinical candidate by the Medicines for Malaria Venture for potential use in the prevention and treatment of malaria, was also realized via the Conrad–Limpach reaction in Dowtherm A (Equation 1.42).[81]

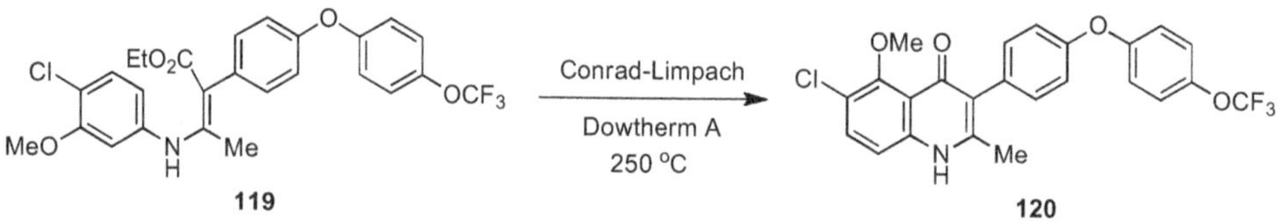

$$(1.42)$$

1.8.3 POLYETHYLENE GLYCOL

Sevoflurane, 1,1,1,3,3,3-hexafluoro-2-(fluoromethoxy)propane, is one of the most widely used inhalation anesthetics. Based on the current commercial manufacturing process, Abbott Laboratories developed a safe and efficient process for the synthesis of sevoflurane (shown in Scheme 1.20).[82] Solvent screening for the fluoride exchange revealed that using NMP, DMF, or DMSO as the reaction solvent was problematic, making the product isolation difficult.

Ultimately, poly(ethylene glycol)-400 (PEG-400) was identified as the solvent of choice. Consequently, the fluoride exchange could be affected under much milder conditions, producing the sevoflurane in good yields. In addition, this reaction system could be tolerant of up to 3.2% water, which allowed telescoping the two steps into a one-pot operation by using the crude intermediate **122** without drying.

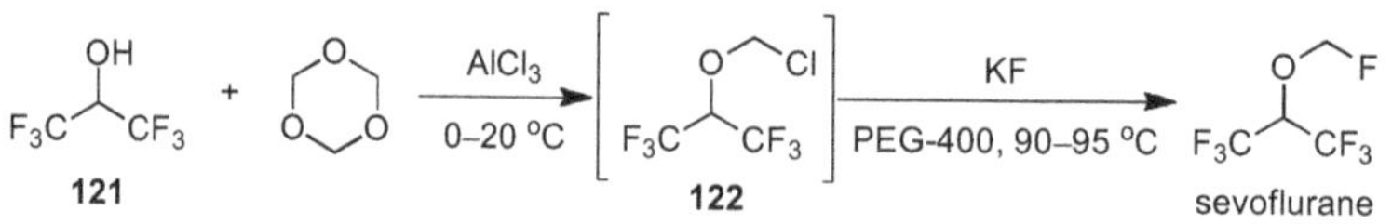

SCHEME 1.20

1.8.4 PROPYLENE GLYCOL MONOMETHYL ETHER

Propylene glycol monomethyl ether (PGME) (bp 118 °C) is used as the solvent for cellulose, acrylics, dyes, inks, and stains. The use of PGME is anticipated to increase due to its low systemic toxicity. One application of PGME was found in the literature, in which PGME was identified as the crystallization solvent for the purification of pharmaceutical intermediate **125** (Scheme 1.21).[83] Gastrazole (JB95008) is a potent and highly selective cholecystokinin-2 (CCK$_2$) receptor antagonist that was developed for the treatment of pancreatic cancer.

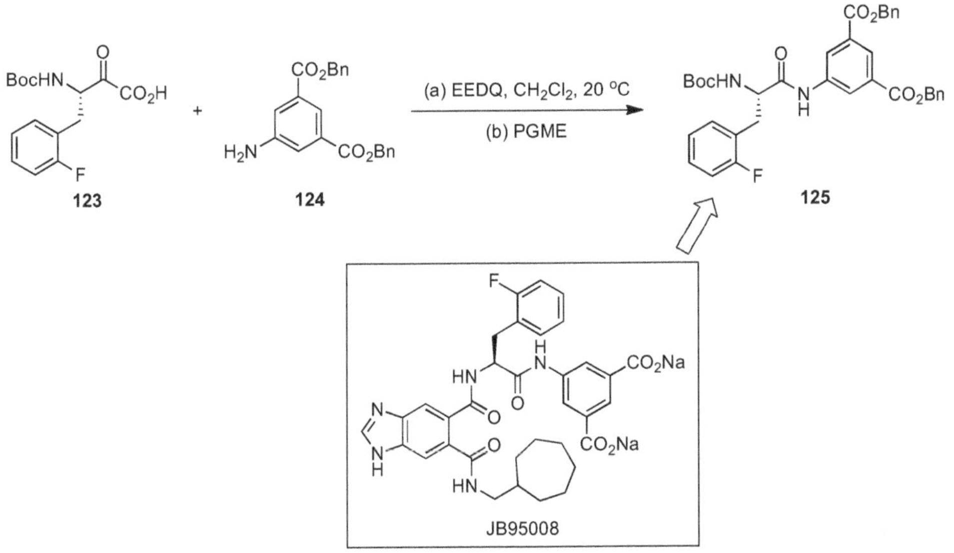

SCHEME 1.21

Procedure

Stage (a)
- Charge 212.2 L of dichloromethane into a reaction vessel under nitrogen.
- Charge 10.74 kg (37.9 mol) of **123**, 13.71 kg (37.9 mol) of **124**, and 9.4 kg (37.9 mol) of 2-ethoxy-1-ethoxycarbonyl-1,2-dihydroquinoline (EEDQ).
- Stir the batch at 20 °C for 22 h.
- Distill CH_2Cl_2 (136 L) under atmospheric pressure.

Stage (b)
- Add 190 L of PGME.
- Distill until the internal temperature reaches 115 °C.
- Cool to 22 °C over 15 h.
- Filter, rinse with PGME, and dry at 50 °C under vacuum.
- Yield: 21.87 kg (92%) as a white solid.

1.8.5 SULFOLANE

Sulfolane (tetrahydrothiophene sulfone or tetrahydrothiophene-1,1-dioxide) is a clear, colorless liquid, and miscible with water and polar organic solvents. Sulfolane is not miscible with alkanes and methyl *tert*-butyl ether (MTBE). It is very stable at elevated temperatures in air; for example, at 200 °C the rate of decomposition is only 0.009%/h. Like other dipolar aprotic solvents, sulfolane is capable of strong solvation of cations by the oxygen atoms on the sulfone group. A review article[84] summarizes the utilization of sulfolane as a reaction solvent in reactions such as Friedel–Crafts acylation/alkylation, annulation,[85] nitration, halogen-exchange reaction, etc.

1.8.5.1 Bromination/Esterification

Preparation of methyl 2,6-dibromoisonicotinate **126** was accomplished in two stages involving bromination of citrazinic acid with phosphorus oxybromide in sulfolane at 125 °C followed by esterification with methanol (Equation 1.43).[86]

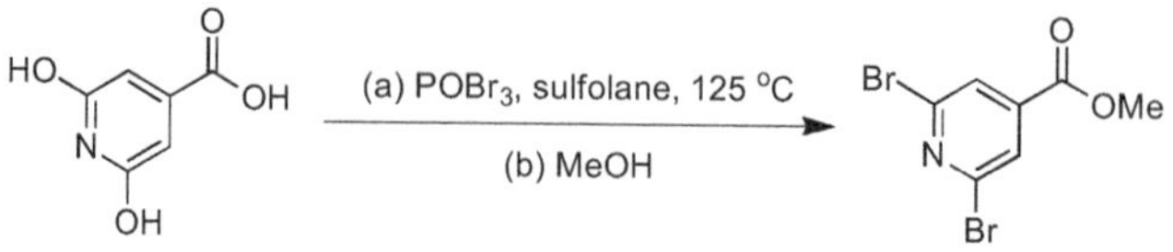

citrazinic acid   126

(1.43)

REMARK

Filtration of product slurry in sulfolane (bp 285 °C, mp 27.5 °C, and viscosity 0.01007 Pa·s at 25 °C) shall be conducted above its melting point to avoid a slow filtration rate.[87]

1.8.5.2 Fluorine-Exchange Reaction

A fluorine-exchange reaction of tetrachlorophthalimide **128** with spray-dried KF was conducted in sulfolane at 200 °C (shown in Scheme 1.22)[88] giving tetrafluorophthalimide **129** in 85% yields.

Several other fluorine-exchange reactions were realized in sulfolane to afford the fluorinated products.[89] Most of those transformations were carried out at high temperatures (140–260 °C) using potassium fluoride as the fluorine source. Decarboxylation of heterocyclic carboxylic acids could occur at 300 °C in sulfolane as well.[90]

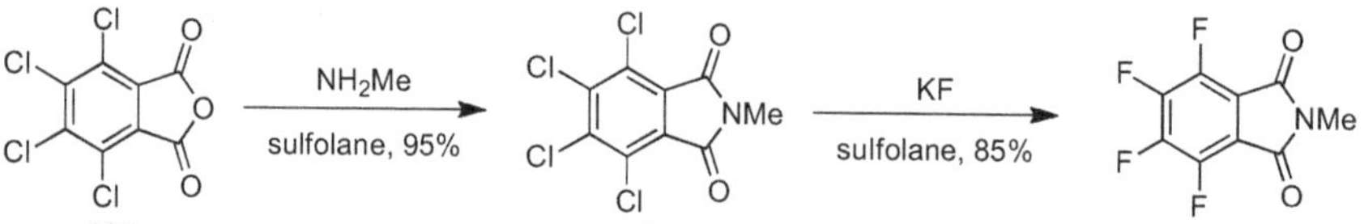

127   128   129

SCHEME 1.22

1.8.6 Ionic Liquids

Ionic liquids are low melting salts in which the ions are poorly coordinated, which results in these solvents being liquid below 100 °C, or even at room temperature. Compared to conventional organic solvents, ionic liquids are very attractive as alternative reaction media because of their properties, such as their low vapor pressure, high thermal stability, and recyclability. Butyl-3-methylimidazolium tetrafluoroborate ([bmim][BF$_4$]), was employed as the reaction solvent for the S_N2 transformation of mesylate **130** with CsF, affording cleanly the fluoride **131** (Equation 1.44).[91]

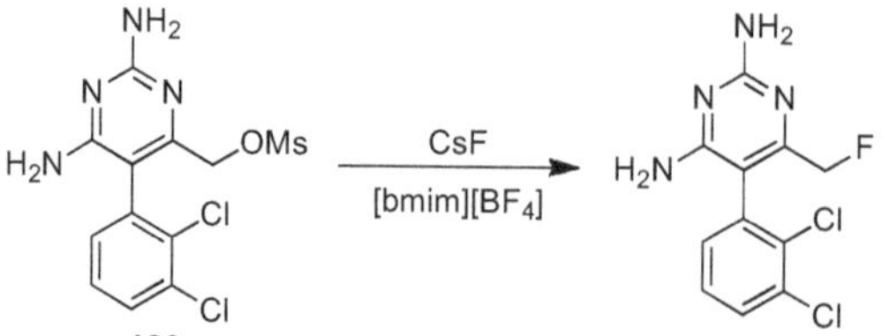

130   131

(1.44)

In contrast, the reaction in other solvents failed to provide the desired product **131**.

REMARK

It was also noticed that the reaction could be conducted in MeCN/H$_2$O with [bmim][BF$_4$] as a co-solvent.

1.9 MIXTURE OF SOLVENTS

1.9.1 ALDOL CONDENSATION REACTION

Attention should be paid when developing a process which involves reagents with opposite solubility behavior. For instance, there is a remarkable solubility difference between the sodium salt of 3,5-dimethyl-4-nitroisoxazole **132** and 3-methylbutanal **133**: the former is very soluble in water and not soluble in organic solvents, while the latter is very soluble in organic solvent and not in water. In order to accommodate the solubility difference between **132** and **133**, a ternary solvent system was identified and the aldol condensation was conducted in a mixture of water, methanol, and THF in a ratio of 3:7:4 (Equation 1.45).[92]

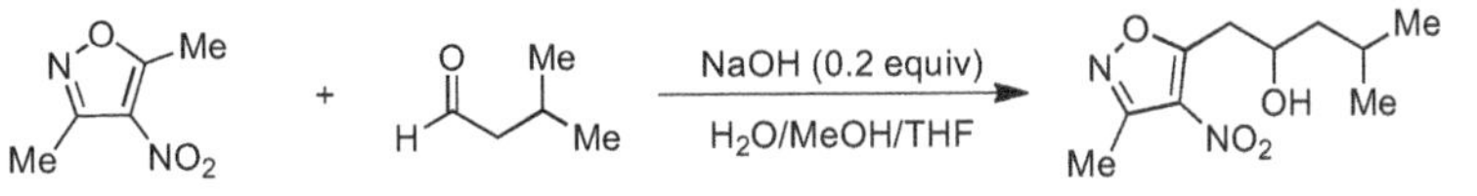

$$(1.45)$$

1.9.2 VISIBLE-LIGHT MEDIATED REDOX NEUTRAL REACTION

A visible-light mediated redox neutral reaction was developed to access γ-functionalized amines (Scheme 1.23).[93]

The reaction was carried out in a ternary solvent system of $PhCF_3/PhCl/H_2O$ (1:1:1) under mild conditions. In the presence of iridium photocatalyst and base (K_2HPO_4), visible-light irradiation of N-protected γ,γ-bis(aryl)-substituted amines generated the corresponding sulfamidyl radicals, which underwent intramolecular aryl migration to give corresponding carbon-centered radicals. A subsequent radical addition to alkenes followed by a SET reaction to furnish the desired amine products.

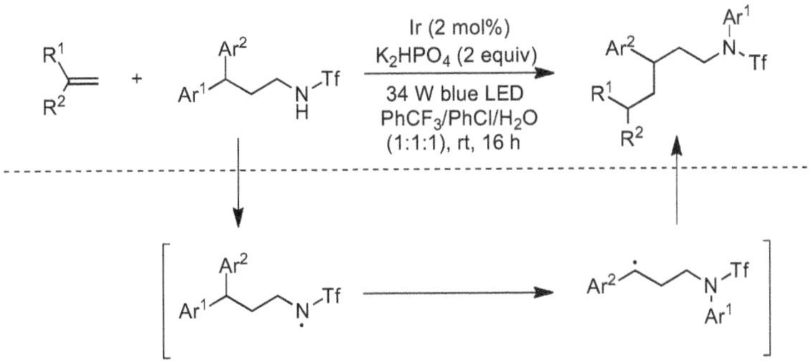

SCHEME 1.23

1.10 SOLVENT-FREE REACTION

The demethylation of 4-methoxyphenylbutyric acid **135** was achieved under solvent-free conditions. At 185–195 °C the reaction with molten pyridinium hydrochloride afforded the product 4-hydroxyphenylbutyric acid **136** in high yield and purity (Equation 1.46).[94]

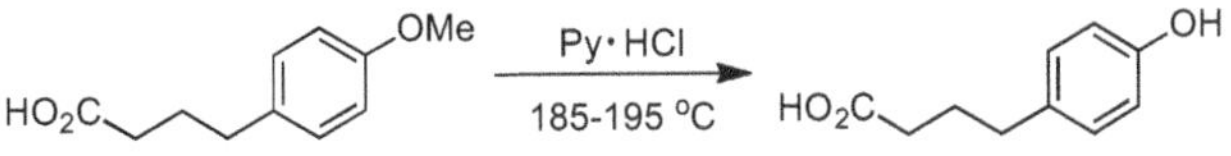

$$(1.46)$$

The workup involved the addition of aqueous HCl to the reaction mixture followed by MTBE extraction.

Procedure

- Charge 2.25 kg (11.58 mol) of **135** into a 22-L reactor.
- Charge 5.36 kg (46.34 mol, 4 equiv) of pyridinium hydrochloride.
- Heat the mixture to 185–195 °C (start to stir at 50 °C).
- Stir at 185–195 °C for 2 h.
- Cool to 90 °C.
- Add 5 N HCl (2.9 L, 14.5 mol, 1.25 equiv) followed by adding water (2.7 L).
- Extract with MTBE (3×4 L).
- Wash with combined organic layers with 5 N HCl (750 mL).
- Dry (over Na$_2$SO$_4$) and filter.
- Distill under reduced pressure.
- Yield: 2.02 kg (96.8%) of **136** as a white solid.

NOTES

1. For solvent selection guides developed by Pfizer and GSK, see: (a) Alfonsi, K.; Colberg, J.; Dunn, P. J.; Fevig, T.; Jennings, S.; Johnson, T. A.; Kleine, H. P.; Knight, C.; Nagy, M. A.; Perry, D. A.; Stefaniak, M. *Green Chem.* **2008**, *10*, 31. (b) Henderson, R. K.; Jiménez-González, C.; Constable, D. J. C.; Alston, S. R.; Inglis, G. G. A.; Fisher, G.; Sherwood, J.; Binks, S. P.; Curzons, A. D. *Green Chem.* **2011**, *13*, 854.
2. Prat, D.; Pardigon, O.; Flemming, H.-W.; Letestu, S.; Ducandas, V.; Isnard, P.; Guntrum, E.; Senac, T.; Ruisseau, S.; Cruciani, P.; Hosek, P. *Org. Process Res. Dev.* **2013**, *17*, 1517.
3. Constable, D. J. C.; Curzons, A. D.; Cunningham, V. L.; Mortimer, D. N. *Green Chem.* **2001**, *3*, 1.
4. Kong, D. *Chem. Eng.* **2004**, *12*, 50.
5. Watanabe, K.; Yamagiwa, N.; Torisawa, Y. *Org. Process Res. Dev.* **2007**, *11*, 251.
6. (a) Giubellina, N.; Stabile, P.; Laval, G.; Perboni, A. D.; Cimarosti, Z.; Westerduin, P.; Cooke, J. W. B. *Org. Process Res. Dev.* **2010**, *14*, 859. (b) Houpis, I. N.; Shilds, D.; Nettekoven, U.; Schnyder, A.; Bappert, E.; Weerts, K.; Canters, M.; Vermuelen, W. *Org. Process Res. Dev.* **2009**, *13*, 598. (c) Kuethe, J. T.; Childers, K. G.; Humphrey, G. R.; Journet, M.; Peng, Z. *Org. Process Res. Dev.* **2008**, *12*, 1201. (d) Datta, K. G.; Ellman, J. A. *J. Org. Chem.* **2010**, *75*, 6283. (e) Wakayama, M.; Ellman, J. A. *J. Org. Chem.* **2009**, *74*, 2646. (f) Fei, Z.; Wu, Q.; Zhang, F.; Cao, Y.; Liu, C.; Shieh, W.-C.; Xue, S.; McKenna, J.; Prasad, K.; Prashad, M.; Baeschlin, D.; Namoto, K. *J. Org. Chem.* **2008**,*73*, 9016. (g) Busacca, C. A.; Raju, R.; Grinberg, N.; Haddad, N.; James-Jones, P.; Lee, H.; Lorenz, J. C.; Saha, A.; Senanayake, C. H. *J. Org. Chem.* **2008**, *73*, 1524.
7. Antonucci, V.; Coleman, J.; Ferry, J. B.; Johnson, N.; Mathe, M.; Scott, J. P.; Xu, J. *Org. Process Res. Dev.* **2011**, *15*, 939.
8. Ritter, S. *Chemical & Engineering News*, **2014**, *March 24*, p 30.
9. Sawada, Y.; Sasaki, M.; Takeda, K. *Org. Lett.* **2004**, *6*, 2277.
10. Kawabata, T.; Ozturk, O.; Chen, J.; Fuji, K. *Chem. Commun.* **2003**, 162.
11. Watanabe, M.; Nakamura, M.; Satoh, T. *Tetrahedron* **2005**, *61*, 4409.
12. Pramanik, C.; Bapat, K.; Chaudhari, A.; Tripathy, N. K.; Gurjar, M. K. *Org. Process Res. Dev.* **2012**, *16*, 1591.
13. Davies, I. W.; Marcoux, J.-F.; Corley, E. G.; Journet, M.; Cai, D.-W.; Palucki, M.; Wu, J.; Larsen, R. D.; Rossen, K.; Pye, P. J.; DiMichele, L.; Dormer, P.; Reider, P. J. *J. Org. Chem.* **2000**, *65*, 8415.
14. Cabaj, J. E.; Kairys, D.; Benson, T. R. *Org. Process Res. Dev.* **2007**, *11*, 378.
15. For applications of MeTHF, see: Aycock, D. F. *Org. Process Res. Dev.* **2007**, *11*, 156.
16. Yates, M. H.; Kallman, N. J.; Ley, C. P.; Wei, J. N. *Org. Process Res. Dev.* **2009**, *13*, 255.
17. The authors hypothesized that "As the reaction progressed, the sulfonamide quickly saturated the organic layer but could be easily deprotonated and taken into the aqueous phase, where it competed with the ammonia."
18. de Koning, P. D.; Castro, N.; Gladwell, I. R.; Morrison, N. A.; Moses, I. B.; Panesar, M. S.; Pettman, A. J.; Thomson, N. M. *Org. Process Res. Dev.* **2011**, *15*, 1256.

19. Chen, K.; Eastgate, M. D.; Zheng, B.; Li, J. *Org. Process Res. Dev.* **2011**, *15*, 886.
20. Ribecai, A.; Baccki, S.; Delpogetto, M.; Guelfi, S.; Manzo, A. M.; Perboni, A.; Stabile, P.; Westerduin, P.; Hourdin, M.; Rossi, S.; Provera, S.; Turco, L. *Org. Process Res. Dev.* **2010**, *14*, 895.
21. Berwe, M.; Jöntgen, W.; Krüger, J.; Cancho-Grande, Y.; Lampe, T.; Michels, M.; Paulsen, H.; Raddatz, S.; Weigand, S. *Org. Process Res. Dev.* **2011**, *15*, 1348.
22. Cobley, C. J.; Lennon, I. C.; Praquin, C.; Zanotti-Gerosa, A.; Appell, R. B.; Goralski, C. T.; Sutterer, A. C. *Org. Process Res. Dev.* **2003**, *7*, 407.
23. Grimm, J. S.; Maryanoff, C. A.; Patel, M.; Palmer, D. C.; Sorgi, K. L.; Stefanick, S.; Webster, R. R. H.; Zhang, X. *Org. Process Res. Dev.* **2002**, *6*, 938.
24. Jansen, R.; Knopp, M.; Amberg, W.; Bernard, H.; Koser, S.; Müller, S.; Münster, I.; Pfeiffer, T.; Riechers, H. *Org. Process Res. Dev.* **2001**, *5*, 16.
25. Boaz, N. W.; Venepalli, B. *Org. Process Res. Dev.* **2001**, *5*, 127.
26. Renga, J. M.; McLaren, K. L.; Ricks, M. J. *Org. Process Res. Dev.* **2003**, *7*, 267.
27. Boini, S.; Moder, K. P.; Vaid, R. K.; Kopach, M.; Kobierski, M. *Org. Process Res. Dev.* **2006**, *10*, 1205.
28. Lukin, K.; Hsu, M. C.; Chambournier, G.; Kotecki, B.; Venkatramani, C. J.; Leanna, M. R. *Org. Process Res. Dev.* **2007**, *11*, 578.
29. Wang, L.; Neumann, H.; Beller, M. *Angew. Chem. Int. Ed.* **2019**, *58*, 5417.
30. Whiting, M.; Harwood, K.; Hossner, F.; Turner, P. G.; Wilkinson, M. C. *Org. Process Res. Dev.* **2010**, *14*, 820.
31. Wu, G.; Huang, M.; Richards, M.; Poirier, M.; Wen, X.; Draper, R. W. *Synthesis* **2003**, *11*, 1657.
32. Cook, D. C.; Jones, R. H.; Kabir, H.; Lythgoe, D. J.; McFarlane, I. M.; Pemberton, C.; Thatcher, A. A.; Thompson, D. M.; Walton, J. B. *Org. Process Res. Dev.* **1998**, *2*, 157.
33. Braden, T. M.; Coffey, D. S.; Doecke, C. W.; LeTourneau, M. E.; Martinelli, M. J.; Meyer, C. L.; Miller, R. D.; Pawlak, J. M.; Pedersen, S. W.; Schmid, C. R.; Shaw, B. W.; Staszak, M. A.; Vicenzi, J. T. *Org. Process Res. Dev.* **2007**, *11*, 431.
34. Cismesia, M. A.; Céspedes, S. V. *Org. Process Res. Dev.* **2022**, *26*, 1558.
35. Walker, S. D.; Borths, C. J.; DiVirgilio, E.; Huang, L.; Liu, P.; Morrison, H.; Sugi, K.; Tanaka, M.; Woo, J. C. S.; Faul, M. M. *Org. Process Res. Dev.* **2011**, *15*, 570.
36. Francos, J.; Cadierno, V. *Green Chem.* **2010**, *12*, 1552.
37. Saito, A.; Adachi, S.; Kumagai, N.; Shibasaki, M. *Angew. Chem. Int. Ed.* **2021**, *60*, 8739.
38. Slade, J.; Bajwa, J.; Liu, H.; Parker, D.; Vivelo, J.; Chen, G.-P.; Calienni, J.; Villhauer, E.; Prasad, K.; Repič, O.; Blacklock, T. J. *Org. Process Res. Dev.* **2007**, *11*, 825.
39. Dorow, R. L.; Herrinton, P. M.; Hohler, R. A.; Maloney, M. T.; Mauragis, M. A.; McGhee, W. E.; Moeslein, J. A.; Strohbach, J. W.; Veley, M. F. *Org. Process Res. Dev.* **2006**, *10*, 493.
40. Engberts, J. B. F. N.; Blandamer, M. J. *Chem. Commun.* **2001**, 1701.
41. Gallo, R. D. C.; Ferreira, I. M.; Casagrande, G. A.; Pizzuti, L.; Oliveira-Silva, D.; Raminelli, C. *Tetrahedron Lett.* **2012**, *53*, 5372.
42. Ruck, R. T.; Huffman, M. A.; Stewart, G. W.; Cleator, E.; Kandur, W. V.; Kim, M. M.; Zhao, D. *Org. Process Res. Dev.* **2012**, *16*, 1329.
43. Connolly, T. J.; McGarry, P.; Sukhtankar, S. *Green Chem.* **2005**, *7*, 586.
44. Sulur, M.; Sharma, P.; Ramakrishnan, R.; Naidu, R.; Merifield, E.; Gill, D. M.; Clarke, A. M.; Thomson, C.; Butters, M.; Bachu, S.; Benison, C. H.; Dokka, N.; Fong, E. R.; Hose, D. R. J.; Howell, G. P.; Mobberley, S. E.; Morton, S. C.; Mullen, A. K.; Rapai, J.; Tejas, B. *Org. Process Res. Dev.* **2012**, *16*, 1746.
45. Pschierer, J.; Plenio, H. *Eur. J. Org. Chem.* **2010**, 2934.
46. (a) Li, C.-J.; Chan, T.-K. *Organic Reactions in Aqueous Media*; Wiley, New York, 1997. (b) *Organic Synthesis in Water*; Grieco, P. A., Ed.; Blackie Acad. Professional Publication, London, 1998. (c) Li, C.-J. *Chem. Rev.* **1993**, *93*, 2023. (d) Li, C.-J.; Chen, L. *Chem. Soc. Rev.* **2006**, *35*, 68. (e) Otto, S.; Engberts, J. B. F. N. *Pure Appl. Chem.* **2000**, *72*, 1365.
47. Thiel, O. R.; Bernard, C.; King, T.; Dilmeghani-Seran, M.; Bostick, T.; Larsen, R. D.; Faul, M. M. *J. Org. Chem.* **2008**, *73*, 3508.
48. Parmentier, M.; Wagner, M. K.; Magra, K.; Gallou, F. *Org. Process Res. Dev.* **2016**, *20*, 1104.
49. Xia, Q.; Li, X.; Fu, X.; Zhou, Y.; Peng, Y.; Wang, J.; Song, G. *J. Org. Chem.* **2021**, *86*, 9923.
50. Connolly, T. J.; Crittall, A. J.; Ebrahim, A. S.; Ji, G. *Org. Process Res. Dev.* **2000**, *4*, 526.
51. McConville, F. X. "The Pilot Plant Real Book: A Unique Handbook for the Chemical Process Industry", FXM Engineering and Design, Worcester, MA, 2002, Chapter 6, pp 34–35.
52. Bai, Y.; Zhang, Q.; Jia, P.; Yang, L.; Sun, Y.; Nan, Y.; Wang, S.; Meng, X.; Wu, Y.; Qin, F.; Sun, Z.; Gao, X.; Liu, P.; Luo, K.; Zhang, Y.; Zhao, X.; Xiao, C.; Liao, S.; Liu, J.; Wang, C.; Fang, J.; Wang, X.; Wang, J.; Gao, R.; An, X.; Zhang, X.; Zheng, X. *Org. Process Res. Dev.* **2014**, *18*, 1667.

53. Ashcroft, C. P.; Dunn, P. J.; Hayler, J. D.; Wells, A. S. *Org. Process Res. Dev.* **2015**, *19*, 740.
54. For survey of removal of polar aprotic solvents by extractive workup, see: Delhaye, L.; Ceccato, A.; Jacobs, P.; Köttgen, C.; Merschaert, A. *Org. Process Res. Dev.* **2007**, *11*, 160.
55. Ellis, J. E.; Davis, E. M.; Brower, P. L. *Org. Process Res. Dev.* **1997**, *1*, 250.
56. Chen, J.-Q.; Yu, W.-L.; Wei, Y.-L.; Li, T.-H.; Xu, P.-F. *J. Org. Chem.* **2017**, *82*, 243.
57. Schmidt, M. A.; Reiff, E. A.; Qian, X.; Hang, C.; Truc, V. C.; Natalie, K. J.; Wang, C.; Albrecht, J.; Lee, A. G.; Lo, E. T.; Guo, Z.; Goswami, A.; Goldberg, S.; Pesti, J.; Rossano, L. T. *Org. Process Res. Dev.* **2015**, *19*, 1317.
58. Bos, M.; Riguet, E. *J. Org. Chem.* **2014**, *79*, 10881.
59. Zhu, Y.; Zhang, L.; Luo, S. *J. Am. Chem. Soc.* **2016**, *138*, 3978.
60. Bretherick, L. *Handbook of Reactive Chemical Hazards*, 6th ed.; Butterworth-Heinemann Ltd., Boston, MA, 1999; p 281.
61. Wang, Z.; Richter, S. M.; Rozema, M. J.; Schellinger, A.; Smith, K.; Napolitano, J. G. *Org. Process Res. Dev.* **2017**, *21*, 1501.
62. Di Maso, M. J.; Ren, H.; Zhang, S.-W.; Liu, W.; Desmond, R.; Alwedi, E.; Narsimhan, K.; Kalinin, A.; Larpent, P.; Lee, A. Y.; Ren, S.; Maloney, K. M. *Org. Process Res. Dev.* **2020**, *24*, 2491.
63. Gontcharov, A.; Dunetz, J. R. *Org. Process Res. Dev.* **2014**, *18*, 1145.
64. Tao, Y.; Widlicka, D. W.; Hill, P. D.; Couturier, M.; Young, G. R. *Org. Process Res. Dev.* **2012**, *16*, 1805.
65. Rudine, A. B.; Walter, M. G.; Wamser, C. C. *J. Org. Chem.* **2010**, *75*, 4292.
66. Fiore, P. J.; Puls, T. P.; Walker, J. C. *Org. Process Res. Dev.* **1998**, *2*, 151.
67. Lee, J. H.; Choi, B. S.; Chang, J. H.; Kim, S. S.; Shin, H. *Org. Process Res. Dev.* **2007**, *11*, 1062.
68. (a) Bretherick, L. *Handbook of Reactive Chemical Hazards*, 4th ed.; Butterworth–Heinemann, Oxford, **1990**, p 1181. (b) For an example on the risks of thermal runaway reactions using the NaH/DMF combination see: DeWall, G. *Chem. Eng. News* **1982**, *60*, 5.
69. (a) Young, I. S.; Ortiz, A.; Sawyer, J. R.; Conlon, D. A.; Buono, F. G.; Leung, S. W.; Burt, J. L.; Sortore, E. W. *Org. Process Res. Dev.* **2012**, *16*, 1558. (b) Shimizu, S.; Osato, H.; Imamura, Y.; Satou, Y. *Org. Process Res. Dev.* **2010**, *14*, 1518.
70. Ogawa, A.; Curran, D. P. *J. Org. Chem.* **1997**, *62*, 450.
71. Frampton, G. A.; Hannah, D. R.; Henderson, N.; Katz, R. B.; Smith, I. H.; Tremayne, N.; Watson, R. J.; Woollam, I. *Org. Process Res. Dev.* **2004**, *8*, 415.
72. Anderson, J. T.; Ting, A. E.; Boozer, S.; Brunden, K. R.; Crumrine, C.; Danzig, J.; Dent, T.; Faga, L.; Harrington, J. J.; Hodnick, W. F.; Murphy, S. M.; Pawlowski, G.; Perry, R.; Raber, A.; Rundlett, S. E.; Stricker-Krongrad, A.; Wang, J.; Bennani, Y. L. *J. Med. Chem.* **2005**, *48*, 7096.
73. Liwosz, T. W.; Chemler, S. R. *J. Am. Chem. Soc.* **2012**, *134*, 2020.
74. Bégué, J.-P.; Bonnet-Delpon, D.; Crousse, B. *Synlett* **2004**, *1*, 18.
75. (a) Brenek, S. J.; Caron, S.; Chisowa, E.; Delude, M. P.; Drexler, M. T.; Ewing, M. D.; Handfield, R. E.; Ide, N. D.; Nadkarni, D. V.; Nelson, J. D.; Olivier, M.; Perfect, H. H.; Phillips, J. E.; Teixeira, J. J.; Weekly, R. M.; Zelina, J. P. *Org. Process Res. Dev.* **2012**, *16*, 1348. For more examples of sulfide oxidation in HFIP, see: (b) Ravikumar, K. S.; Zhang, Y. M.; Bégué, J.-P.; Bonnet-Delpon, D. *Eur. J. Org. Chem.* **1998**, *12*, 2937. (c) Ravikumar, K. S.; Bégué, J.-P.; Bonnet-Delpon, D. *Tetrahedron Lett.* **1998**, *39*, 3141.
76. Glinkerman, C. M.; Boger, D. L. *J. Am. Chem. Soc.* **2016**, *138*, 12408.
77. Wennekes, T.; Lang, B.; Leeman, M.; van der Marel, G. A.; Smits, E.; Weber, M.; van Wiltenburg, J.; Wolberg, M.; Aerts, J. M. F. G.; Overkleeft, H. S. *Org. Process Res. Dev.* **2008**, *12*, 414.
78. Shieh, W.-C.; Chen, G.-P.; Xue, S.; McKenna, J.; Jiang, X.; Prasad, K.; Repič, O.; Straub, C.; Sharma, S. K. *Org. Process Res. Dev.* **2007**, *11*, 711.
79. Walker, A. J.; Adolph, S.; Connell, R. B.; Laue, K.; Roeder, M.; Rueggeberg, C. J.; Hahn, D. U.; Voegtli, K.; Watson, J. *Org. Process Res. Dev.* **2010**, *14*, 85.
80. Abele, S.; Schmidt, G.; Fleming, M. J.; Steiner, H. *Org. Process Res. Dev.* **2014**, *18*, 993.
81. Pou, S.; Dodean, R. A.; Frueh, L.; Liebman, K. M.; Gallagher, R. T.; Jin, H.; Jacobs, R. T.; Nilsen, A.; Stuart, D. R.; Doggett, J. S.; Riscoe, M. K.; Winter, R. W. *Org. Process Res. Dev.* **2021**, *25*, 1841.
82. Ramakrishna, K.; Behme, C.; Schure, R. M.; Bieniarz, C. *Org. Process Res. Dev.* **2000**, *4*, 581.
83. Ormerod, D.; Willemsens, B.; Mermans, R.; Langens, J.; Winderickx, G.; Kalindjian, S. B.; Buck, I. M.; McDonald, I. M. *Org. Process Res. Dev.* **2005**, *9*, 499.
84. Tilstam, U. *Org. Process Res. Dev.* **2012**, *16*, 1273.
85. Withbroe, G. J.; Seadeek, C.; Girard, K. P.; Guinness, S. M.; Vanderplas, B. C.; Vaidyanathan, R. *Org. Process Res. Dev.* **2013**, *17*, 500.

86. Daver, S.; Rodeville, N.; Pineau, F.; Arlabosse, J.-M.; Moureou, C.; Muller, F.; Pierre, R.; Bouquet, K.; Dumais, L.; Boiteau, J.-G.; Cardinaud, I. *Org. Process Res. Dev.* **2017**, *21*, 231.
87. Hose, D. R. J.; Hopes, P.; Steven, A.; Herber, C. *Org. Process Res. Dev.* **2018**, *22*, 241.
88. Fertel, L. B. *Org. Process Res. Dev.* **1998**, *2*, 111.
89. (a) Moore, R. M., Jr. *Org. Process Res. Dev.* **2003**, *7*, 921. (b) Cantrell, G. L. U.S. Pat. 4,849,552, 1989. (c) O'Reilly, N. J.; Derwin, W. S.; Lin, H. C. Int. Pat. Appl. WO/91/16308, 1991. (d) Ritzer, E.; Meier, J.-D.; Käss, V. Ger. Pat. Appl. DE4445543A1, 1994.
90. Tilstam, U. *Org. Process Res. Dev.* **2012**, *16*, 1449.
91. Walker, M. D.; Albinson, F. D.; Clark, H. F.; Clark, S.; Henley, N. P.; Horan, R. A. J.; Jones, C. W.; Wade, C. E.; Ward, R. A. *Org. Process Res. Dev.* **2014**, *18*, 82.
92. Moccia, M.; Cortigiani, M.; Monasterolo, C.; Torri, F.; Del Fiandra, C.; Fuller, G.; Kelly, B.; Adamo, M. F. A. *Org. Process Res. Dev.* **2015**, *19*, 1274.
93. Shu, W.; Genoux, A.; Li, Z.; Nevado, C. *Angew. Chem. Int. Ed.* **2017**, *56*, 10521.
94. Schmid, C. R.; Beck, C. A.; Cronin, J. S.; Staszak, M. A. *Org. Process Res. Dev.* **2004**, *8*, 670.

2 Reagent Selection

2.1 INORGANIC BASE

Careful selection of bases is important in chemical process development. There are a number of inorganic bases available for applications toward chemical transformations.

2.1.1 SODIUM BICARBONATE

Sodium bicarbonate is a weak inorganic base and is frequently used to quench acid by-products. A common issue associated with using $NaHCO_3$ is foaming, due to CO_2 off-gassing. For instance, during the preparation of sulfone a significant amount of CO_2 off-gassing was generated when treating sulfonyl chloride with Na_2SO_3 and sodium bicarbonate in water (Scheme 2.1).[1]

Replacement of $NaHCO_3$ with Na_2HPO_4 avoided the CO_2 gas evolution, thereby increasing the process safety.

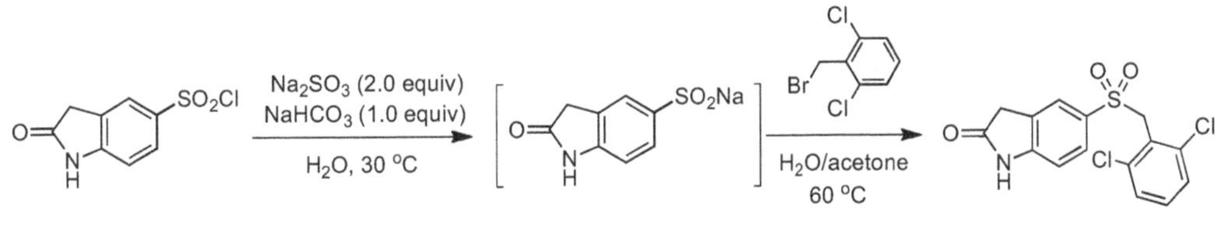

SCHEME 2.1

2.1.2 POTASSIUM CARBONATE

2.1.2.1 Boc Protection of Amino Group

Selective protection of sterically hindered amino groups in the presence of a hydroxyl group proved to be rather challenging due to the nucleophilicity of the phenol. Boc protection of the amino group with Boc_2O using potassium carbonate in THF and water led to reasonable success, but it was not possible to drive the reaction past 90% conversion without forming significant levels of hydroxyl-protected phenol (Equation 2.1).[2]

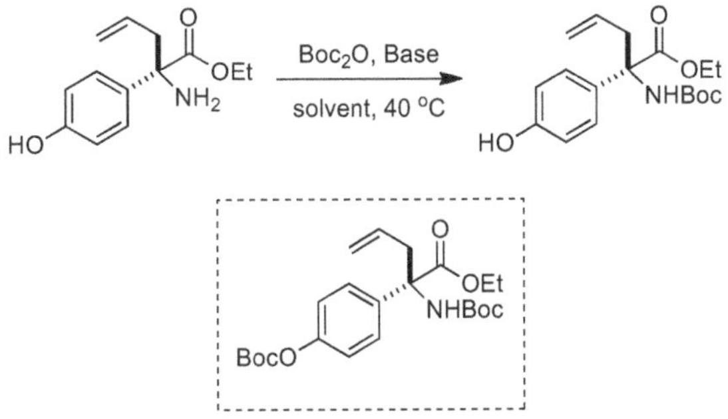

(2.1)

Ultimately, the use of lithium carbonate in place of potassium carbonate was able to drive the reaction to a conversion of >95% while limiting the impurity level to <5%. Changing the reaction solvent from THF to ethyl acetate enabled an efficient phase separation after the reaction and an effective crystallization from EtOAc/heptane afforded an 80% yield of the desired Boc-protected amine product.

DOI: 10.1201/9781003288411-2

2.1.2.2   Ring-Opening Iodination

During the optimization of the cyclopropane ring-opening iodination,[3] there were two concerns: the moderate product yield (60%) because of the side reaction (to generate β-elimination side product) and the use of relatively expensive iodotrimethylsilane (1.5 equiv) (Condition a, Scheme 2.2).

To mitigate the elimination side reaction, a weaker sodium hydrogen carbonate was used to replace potassium carbonate. As shown in Scheme 2.2, TMSI was applied to activate the cyclopropane ring via chelating the oxygen atom of the dioxolane. Thereby, a less expensive chlorotrimethylsilane combined with NaI, as a TMSI surrogate, was used (Condition b, Scheme 2.2). Water (0.1 equiv) was added to the reaction mixture to increase the solubility of NaI. Ultimately, the optimized process improved the product yield from 60% to 88%.

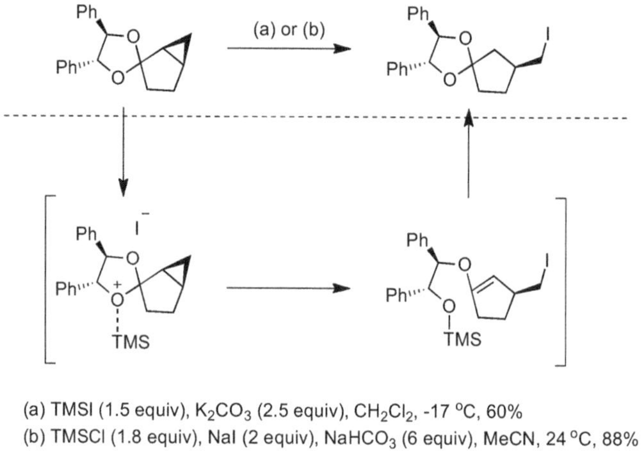

(a) TMSI (1.5 equiv), K_2CO_3 (2.5 equiv), CH_2Cl_2, -17 °C, 60%
(b) TMSCl (1.8 equiv), NaI (2 equiv), $NaHCO_3$ (6 equiv), MeCN, 24 °C, 88%

SCHEME 2.2

2.1.3   Sodium Hydride

2.1.3.1 S_N2 Reaction

Sodium hydride (NaH) is not compatible with DMF and the mixture of NaH and DMF can undergo uncontrollable exothermic decomposition even at 26 °C.[4] Using NaH in DMF (or NMP) as shown in Equation 2.2[5] to prepare N-alkylated iodopyrazole is deemed unsafe on large scale.

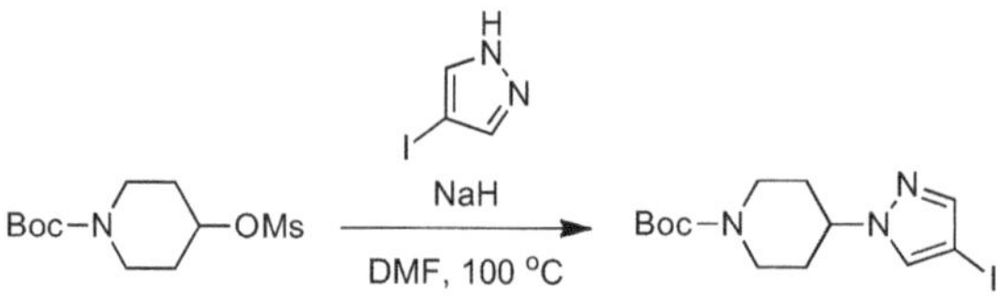

(2.2)

In order to address the safety issue, cesium carbonate was identified as an alternative base and the reaction was carried out by heating a mixture of the mesylate, 4-iodopyrazole, and Cs_2CO_3 in NMP at 80 °C. Under these conditions, N-alkylated iodopyrazole was obtained in 85–95% yields.

REMARK

NaOtBu can also serve as NaH surrogate to avoid using the hazardous NaH/DMF reaction system.[6]

2.1.3.2 Addition/Elimination Reaction

NaH could be utilized on a large scale in THF for the preparation of β-ketoester (Equation 2.3).[7]

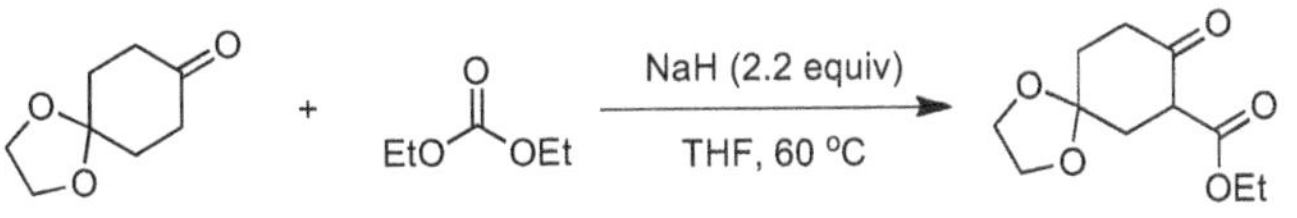

(2.3)

A catalytic amount (5 mol%) of ethanol was necessary with an addition-controlled hydrogen release and a mild heat evolution ($\Delta H = -160$ kJ/mol). The in situ generated, more soluble sodium ethoxide may be the base for the ketone enolization.

Procedure

- Add 2090 kg of anhydrous THF (KF <150 ppm) into a 6 m³ reactor at 20 °C, followed by adding 131.4 kg (3.28 kmol, 2.2 equiv) of 60 wt% NaH dispersion in mineral oil, 351.9 kg (2.98 kmol, 2.0 equiv) of diethyl carbonate, and 3.4 kg (74 mol, 0.05 equiv) of EtOH,
- Heat the above mixture to 60 °C.
- Add a solution of 232.6 kg (1.49 kmol) of 1,4-cyclohexanedione monoethylene ketal in THF (700 kg) over 2 h.
- Stir the batch at 60 °C for 3 h.
- Cool to 0 °C.
- Transfer the batch into a mixture of AcOH (273.1 kg, 4.55 kmol, 3.0 equiv) and water (1721 kg) at <10 °C.
- Separate layers.
- Add heptane (510 kg) to the organic layer, followed by adding 7.5 wt% aqueous solution of NaHCO₃ (1162 kg).
- Separate layers.
- Distill the organic layer under reduced pressure (~200 mmHg) at <40 °C to ~950 L.
- Add MeOH (922 kg).
- Distill under reduced pressure (~150 mmHg) at <67 °C until 1180 kg of distillate is collected.
- Cool the resulting methanolic solution (529 kg) to rt.
- Yield of the β-ketoester: 316 kg (~93% solution yield) (used directly in the subsequent step).

2.1.3.3 Nucleophilic Addition Reaction

As a strong Brønsted base, sodium hydride (NaH) is commonly used to abstract protons for the generation of carbon anions. Besides, NaH can be activated through the addition of a soluble iodide source, thus a composite of NaH–NaI can serve as a nucleophilic hydride to add hydride (H⁻) to unsaturated electrophiles such as nitriles, imines, and carbonyl compounds. Treatment of the tertiary amides with NaH–NaI composite gives anionic hemiaminal intermediates which are trapped by trimethylsilyl chloride (TMSCl) followed by reaction with nucleophiles such as Grignard reagents or cyanide to afford α-branched amines (Scheme 2.3).[8]

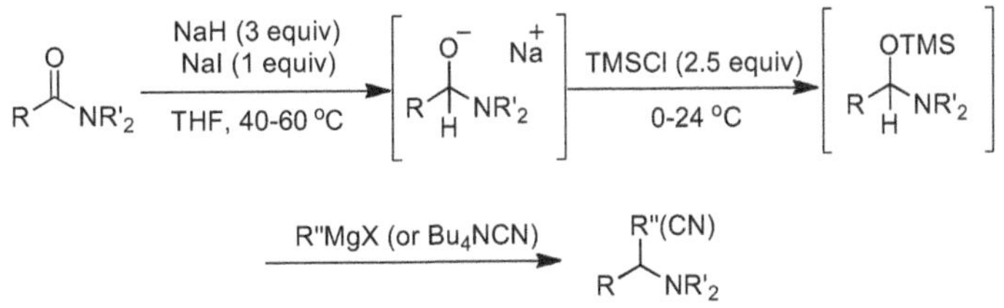

SCHEME 2.3

2.1.4   LiOH/H₂O₂ Combination

Ester hydrolysis can occur under either acidic or basic conditions. However, these conditions are not compatible with molecules bearing stereogenic centers due to the racemization in the presence of acids or bases. In order to avoid erosion of chirality, a combination of lithium hydroxide with hydrogen peroxide (LiOH/H₂O₂)[9] has been developed and proven to be the reagent of choice for the hydrolysis of chiral molecules such as esters and amides.[10]

2.1.4.1   Hydrolysis of Chiral Pentanoate

The success of the conversion of ester **1** into the desired (*R*)-acid **2** is dependent on reaction conditions. For instance, the hydrolysis of **1** to the corresponding (*R*)-acid **2** under acidic conditions (AcOH/HCl) was not successful, resulting in either a sluggish reaction or a 10% reduction in ee; alkaline conditions (LiOH/MeOH or THF/water) led to significant racemization. Ultimately, the use of literature conditions (LiOH/H₂O₂/MeOH/Na₂SO₃)[11] afforded **2** in 93% yield with minimal racemization(Equation 2.4),[12] which is presumably due to the lower basicity and enhanced nucleophilicity of HOOLi.

$$(2.4)$$

Procedure

- Dissolve 1.92 kg (6.72 mol) of **1** in 22 L of MeOH.
- Cool to 0 °C.
- Charge 1.9 L (18.4 mol) of 30% H₂O₂.
- Charge 1.85 L (7.4 mol) of 4 N LiOH over 45 min at <5 °C.
- Stir for 2 h.
- Add 6 N HCl (1.4 L) to pH 1.
- Add 8.74 L (17.5 mol) of 2 M aqueous Na₂SO₃ solution over 45 min at <20 °C.
- Add 6 N HCl (4 L) to pH 6.
- Add toluene (30 L) and stir.
- Separate layers and wash the organic layer with H₂O (2×5 L then 2×3 L).
- Yield (solution yield) of **2**: 1.7 kg (93% yield, 83% ee).

REMARKS

(a) It was recommended[13] that aqueous H₂O₂ and acetic anhydride should never be combined due to the possible formation of diacetyl peroxide.

(b) A mixture of acetone and hydrogen peroxide in the presence of a catalytic amount of H₂SO₄ was reported to form triacetone triperoxide (TATP, Figure 2.1).[14] TATP is an extremely explosive material as a white crystalline solid.

2.1.4.2   Hydrolysis of Chiral Propanoate

The use of the LiOH/H₂O₂ reagent for the hydrolysis of chiral ester **3** avoided the erosion of the stereocenter (Equation 2.5).[15] An additional benefit of using LiOH/H₂O₂ reagent is both the hydrolysis

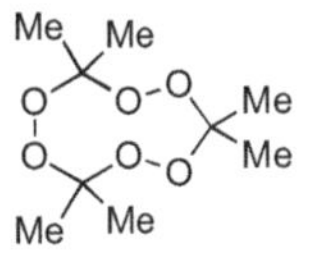

FIGURE 2.1 The structure of triacetone triperoxide (TATP).

and sulfide oxidation could be finished in one-pot operation. Thus, exposure of **3** to LiOH (4 equiv) and H_2O_2 (8 equiv) in a mixture of THF/water enabled complete hydrolysis with concomitant sulfide oxidation, yielding sulfone acid **4** without erosion of the stereocenter.

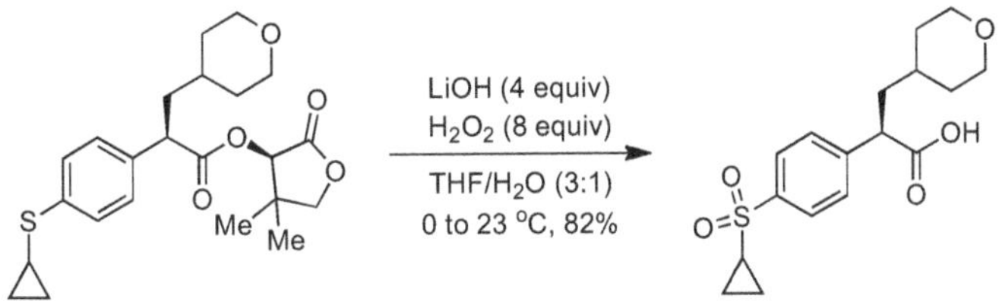

$$(2.5)$$

2.1.4.3 Hydrolysis of Chiral Amide

The LiOH/H_2O_2 reagent was also utilized to hydrolyze chiral amides, such as amides **5** and **6**, to remove the chiral auxiliary (*R*)-4-phenyloxazolidin-2-one (Scheme 2.4).[16a] Thus, treatment of **5** or **6** with 30% aqueous H_2O_2 and LiOH afforded the desired (−)-**7** and (+)-**7** in 72% yield, respectively.

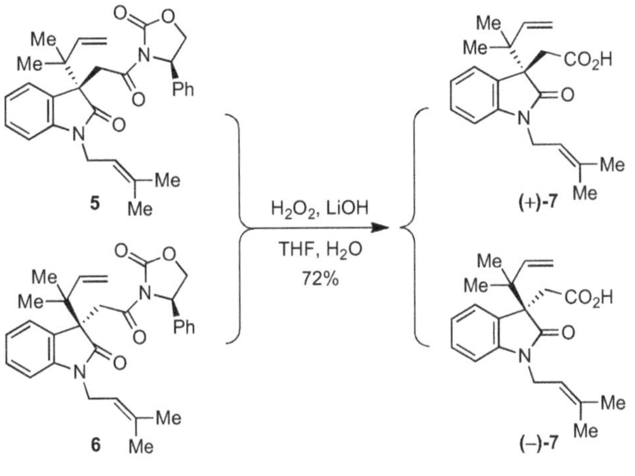

SCHEME 2.4

REMARKS[11]

(a) It was found that LiOOH prefers an exocyclic cleavage for all classes of oxazolidone-derived carboxamides. In an aqueous solution, HOO⁻ should be an effectively smaller nucleophile than HO⁻ as a result of smaller solvation energy. Accordingly, HOO⁻ should be comparatively insensitive to steric effects.

(b) However, the handling of hydrogen peroxide in ethereal solvents such as THF is not recommended due to the possible formation of thermally unstable organo-peroxide derivatives.[17]

(c) Studies[18] show that the cleavage of an Evans auxiliary using the LiOH/H$_2$O$_2$ reagent releases a stoichiometric amount of oxygen (Scheme 2.5), thereby care must be taken when using LiOH/H$_2$O$_2$ system.

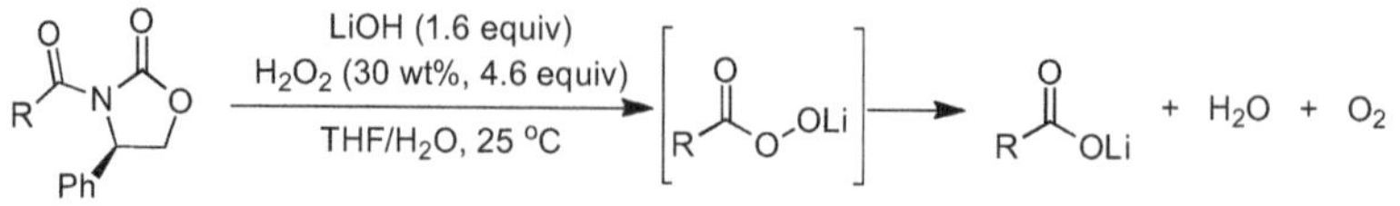

SCHEME 2.5

Accordingly, a method for a safe scale-up of the O$_2$-releasing reaction was developed by a combination of N$_2$ sweeping and control of the LiOH addition rate.[19]

2.2 ORGANIC BASE

Acid byproducts often need to be neutralized/or removed from the reaction mixture otherwise it may have detrimental effects. Organic amines are frequently utilized as acid scavengers.

2.2.1 TRIALKYLAMINE

Tertiary amines such as triethylamine, diisopropylethylamine (DIPEA), and N-methylmorpholine are normally selected due to their lack of nucleophilicity. Another application of tertiary amines is salt-exchange to release the base as an in situ nucleophile.

2.2.1.1 Diisopropylethylamine

2.2.1.1.1 Trap Hydrogen Chloride

Cbz-protected sulfamide **10** was prepared from the coupling reaction of amine·HCl salt **8** with sulfonic chloride **9** in dichloromethane in the presence of an organic base (Equation 2.6).[20] Diisopropylethylamine (DIPEA or Hünig's base) was used to release the amine from **8** and to scavenge the HCl byproduct as well. Other organic bases, such as Et$_3$N or Cy$_2$NMe, will form insoluble HCl salts that would require an additional filtration step.

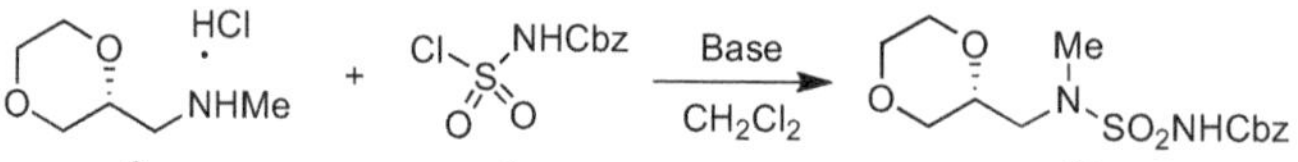

(2.6)

The use of Hünig's base for the amination of 4,6-dichloro-5-methoxypyrimidine with piperazine allowed the desired N-arylated product to crystallize out from the reaction mixture directly in good yield (85%) and purity (100%) (Equation 2.7).[21] The by-product DIPEA·HCl remained in the mother liquor after filtration. However, when using Et$_3$N for this N-arylation, the product was contaminated with Et$_3$N·HCl due to the poor solubility of Et$_3$N·HCl in acetonitrile.

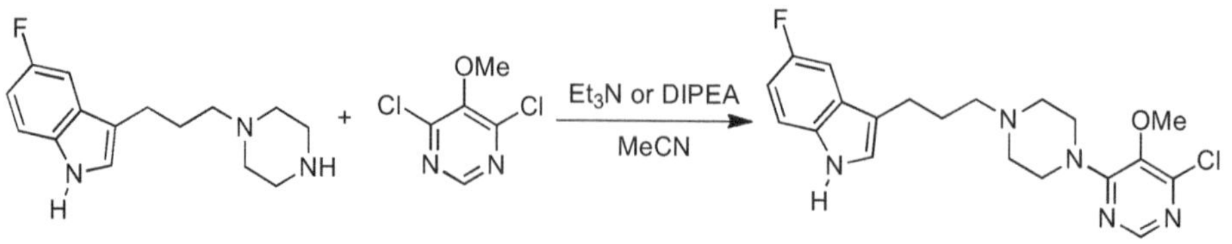

(2.7)

2.2.1.1.2 Trap Triflic Acid

In contrast, it would be desirable that the DIPEA-related byproduct precipitates out from the reaction mixture when the product is soluble under the reaction conditions. Utilization of DIPEA in the silylation of oxindole thiosulfonate **11** with TBS triflate provided the TBS-protected product **12** along with the triflate byproduct as a solid (Equation 2.8),[22] which allowed efficient removal of the byproduct by filtration.

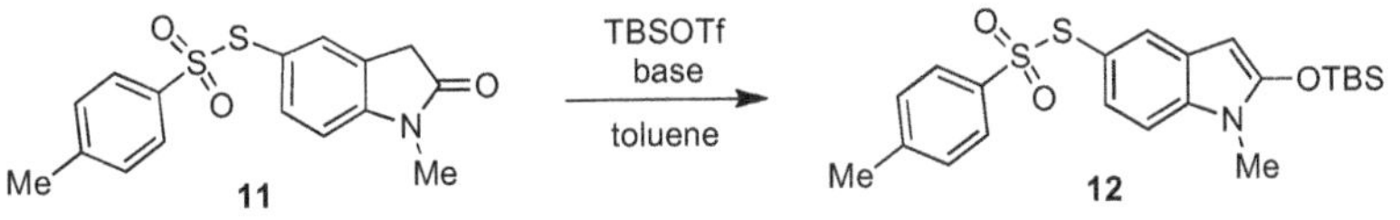

$$(2.8)$$

Procedure

- Charge 0.95 kg (7.35 mol, 1.13 equiv) of DIPEA to a suspension of **11** (2.16 kg, 6.48 mol) in toluene (13.7 L) at 0±2 °C over 13 min.
- Charge 1.85 kg (7.00 mol, 1.08 equiv) of TBS triflate at –5 to –3 °C over 1 h.
- Stir the resulting mixture at –3 to –1 °C for 30 min.
- Warm to 20–25 °C over 30 min.
- Filter and rinse with toluene (2.1 L).
- The combined filtrates were used directly in the subsequent step.

REMARK

The solid DIPEA·OTf salt would have facilitated the recovery and recycling of the expensive triflic acid.

2.2.1.1.3 Trap Hydrogen Bromide

The use of the base to suppress impurity formation was demonstrated from the reaction of 2-bromoisobutyric acid **13** in the presence of sodium ethoxide or Hünig's base for the preparation of 2-ethoxyisobutyric acid **15** (Scheme 2.6).[23] When using sodium ethoxide, a significant amount (15–20%) of acrylate **16** was generated.

Presumably, the formation of **15** occurred in a stepwise fashion via an intermediate **14** followed by ethanolysis. The acrylate byproduct **16** may be generated in an E2 pathway (a) or in the pathway (b) via the intermediate **14**. Obviously, both pathways (a) and (b) need a relatively strong base. Switching sodium ethoxide to Hünig's base reduced the level of byproduct to <5%.

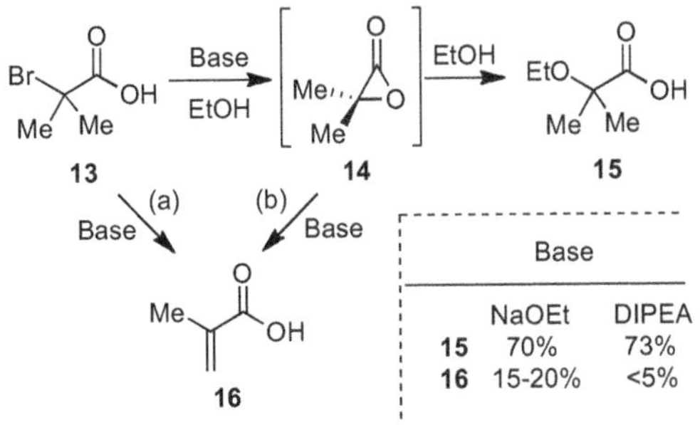

Base		
	NaOEt	DIPEA
15	70%	73%
16	15-20%	<5%

SCHEME 2.6

2.2.1.2 Triethylamine

2.2.1.2.1 pH Adjustment

Soluble amine salts are preferred during the isolation of solid products. For instance, the use of triethylamine to adjust the pH of a cyclization reaction mixture in ethanol resulted in soluble triethylamine salts (Equation 2.9).[24] Thereby product **19** was directly isolated by cooling crystallization in 86% yield.

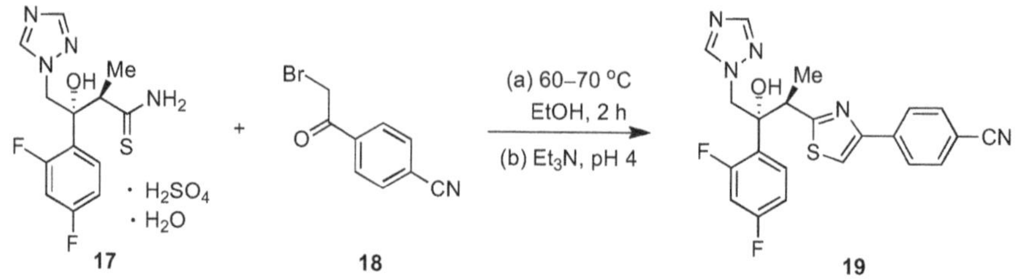

$$(2.9)$$

Procedure

- Heat a mixture of **17** (25.8 kg, 60.2 mol) and **18** (13.8 kg, 61.9 mol) in 95% EtOH (104 kg) at 60–70 °C for 2 h.
- Add water (122 kg) and 95% EtOH (104 kg) at 55–70 °C.
- Add 15.6 kg (154.2 mol) of TEA to pH 3.5–4.0.
- Stir the resulting solution at 55–65 °C for 20 min.
- Cool to 20 °C over 2 h.
- Stir at 20–25 °C for 10 h.
- Filter, rinse with 50 v/v% EtOH/water (98 kg, 123 kg), and dry at 50–55 °C for 8 h.
- Yield of **19**: 22.7 kg (85.8%) as a white crystalline solid.

REMARK

The use of hydroxide and bicarbonate bases for the removal of acids produced insoluble hydrogen sulfate salts which contaminated the isolated product.

2.2.1.2.2 Decarboxylative Alkylation

A reagent ($MgCl_2/Et_3N$),[25] formed by a combination of anhydrous magnesium chloride and triethylamine, was successfully employed in the preparation of ketoester **22** (Equation 2.10).[26]

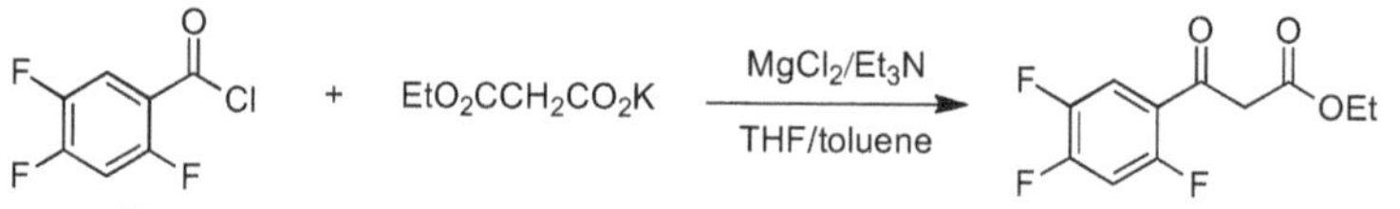

$$(2.10)$$

2.2.1.2.3 Selective Ester Hydrolysis

A combination of triethylamine with lithium bromide ($Et_3N/LiBr$) was demonstrated to be an effective system for the chemoselective hydrolysis of esters with limited epimerization. The preparation of chiral piperidine-4-carboxylic acid (2R,4S)-**24** via hydrolysis of the methyl ester **23** may possess two issues: chemoselectivity and epimerization. Chemoselective hydrolysis using NaOH in MeOH or LiOH in THF resulted in epimerization to give 5% (2R,4R)-**24** (Equation 2.11).[27]

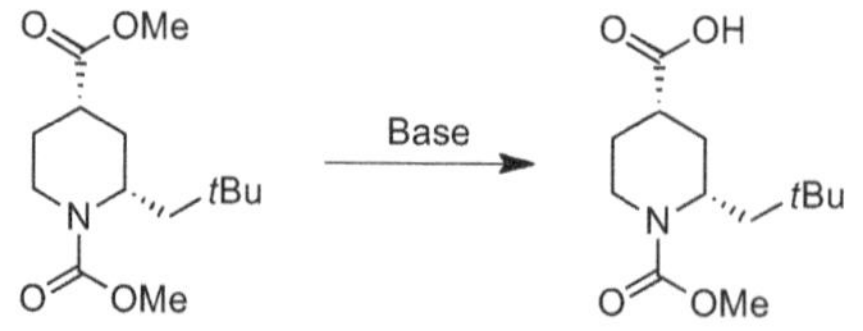

$$(2.11)$$

To address these issues, mild Et$_3$N/LiBr hydrolysis conditions were utilized which reduced the epimerization to 1%.

The Et$_3$N/LiBr hydrolysis system was also applied in the hydrolysis of methyl ester **25** to suppress the hydrolysis of the amide bond, which occurred when using LiOH (Equation 2.12).[28] Analogously, selective hydrolysis of ethyl ester was achieved as well using the Et$_3$N/LiBr hydrolysis system (Equation 2.13).[29]

(2.12)

(2.13)

REMARK

The Et$_3$N/LiBr was utilized to selectively hydrolyze methyl ester in the presence of an isopropyl ester group.[30]

2.2.2 IMIDAZOLE

Direct coupling of pyridinium oxalate salt **27** with phenol **28** required a high reaction temperature (180 °C), leading to degradation of the pyridine-free base of **27** (Equation 2.14).[31] To address this issue, the removal of the oxalate byproduct appeared necessary prior to the coupling reaction.

(2.14)

Imidazole was identified as the base of choice for the salt exchange. Treatment of a suspension of the oxalate salt **27** in NMP with two equivalents of imidazole at ambient temperature resulted in insoluble bis-imidazole salt of oxalic acid leaving the free base of **27** in NMP solution after filtration.

Procedure

- Charge a solution of imidazole (6.48 kg, 95.2 mol) in NMP (27.9 kg) to a suspension of **27** (13.50 kg, 47.7 mol) in NMP (62.5 kg) at 13 °C.
- Stir the resulting mixture at 22–25 °C for 2 h.
- Filter and rinse with NMP (18 kg).
- Charge a solution of **28** (14.9 kg, 47.8 mol) in NMP (44.2 kg) to the combined filtrates.
- Charge 15.5 kg (47.6 mol) of Cs$_2$CO$_3$.
- Heat the resulting mixture at 100 °C for 12 h.
- Cool to 15 °C.
- Add EtOAc (126 kg) and water (324 kg) and stir for 75 min.
- Separate layers.

- Extract the aqueous layer with EtOAc (106 kg).
- Combine the EtOAc layers to give **29** as the EtOAc solution.

2.2.3   2,6-DIMETHYLPIPERIDINE

A three-step synthesis of D-phenylalanine derivative **33** involves mesylation/amination of benzyl alcohol **30**, followed by Boc protection and removal of the acetyl group (Scheme 2.7).[32] The mesylation/amination step utilized fourfold *cis*-2,6-dimethylpiperidine **31** and 1.6 equiv of methanesulfonic anhydride in dichloromethane, in which **31**serves as both a base and a reagent. Using this protocol, the isolated product **33** was contaminated with 0.52% of an impurity **34** that was formed from a reaction with 2-methylpiperidine, a known impurity in the *cis*-2,6-dimethylpiperidine **31**. However, the level of the 2-methylpiperidine impurity in **31** presents at only 0.25%.

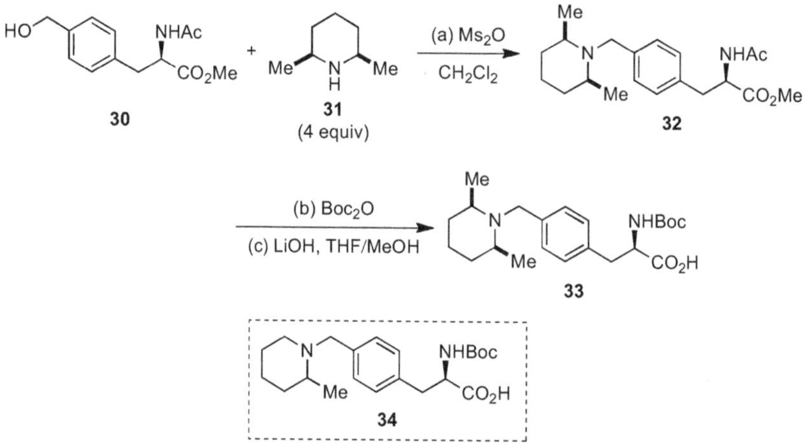

SCHEME 2.7

The much higher level of this impurity **34** in **33** is presumably attributed to the relatively higher reactivity of the less hindered 2-methylpiperidine. To mitigate the level of impurity **34**, it was suggested, by the published report,[32] that an external base is needed to reduce the amount of **31** in the mesylation/amination step.

2.2.4   2-(N,N-DIMETHYLAMINO)PYRIDINE

During the synthesis of amide **38** (Scheme 2.8)[28] using in situ generated acid chloride from a reaction of **35** with Ghosez's reagent, followed by coupling with amine **36** in the presence of DIPEA, three impurities (**39–41**) were observed in a combined level of 30%.

It was proposed that these impurities were formed via elimination and Michael addition pathways from an intermediate **42**. Such elimination to **39** may be promoted by DIPEA (pKa 10.98). A weaker base, 2-(N,N-dimethylamino)pyridine (pKa 7.04), was identified as the base of choice and used in this coupling reaction. Using this protocol, compound **38** obtained an excellent yield (98.4%).

NOTE

The base has to be stronger than **36** (its calculated pKa 6.89). Table 2.1 lists several frequently used organic amine bases with their pKa data.

TABLE 2.1

List of pKa of Several Common Organic Amine Bases.

Amine	pKa	Amine	pKa
Me₂N tBu / Me₂N =N (Barton's base)	14	N N (DABCO)	8.8
(DBU)	12	iPr₂EtN (DIPEA)	10.92
N NMe₂ [2-(N,N-dimethylamino)pyridine]	7.04	N N Me Me [4-(N,N-dimethylamino)pyridine] (DMAP)	9.6
HN N (Imidazole)	14.5	O N–Me (N-methylmorpholine)	7.58
N–Me (N-methylpyrrolidine)	10.7	N (pyridine)	5.23
Et₃N (triethylamine)	10.7	iPr₃N (triisopropylamine)	11.4

2.2.5 METAL ALKOXIDE BASE

2.2.5.1 Potassium *tert*-Pentylate

For starting materials with base-sensitive groups, such as esters, a bulky base is usually the base of choice. The alkylation of Mannich base **43** using ethyl 3-(2-methylphenyl)-3-oxopropano-ate **44** under biphasic reaction conditions afforded the desired ketone **45** in yields of 40–50% (Equation 2.15).[33]

$$(2.15)$$

(a) NaOH (2.0 equiv), toluene-H₂O, reflux 8 h.
(b) citric acid, n-BuOH, H₂O, 53%.
(c) aq NH₃ (25% wt%), CH₂Cl₂, H₂O, 82%.

This transformation is a multiple-step tandem process, involving a base-catalyzed retro-Michael addition, Michael addition of **46** (to afford lactone **48**), and subsequent base-catalyzed lactone ring-opening and decarboxylation (Scheme 2.9). Under the biphasic conditions (Equation 2.15), the ethyl

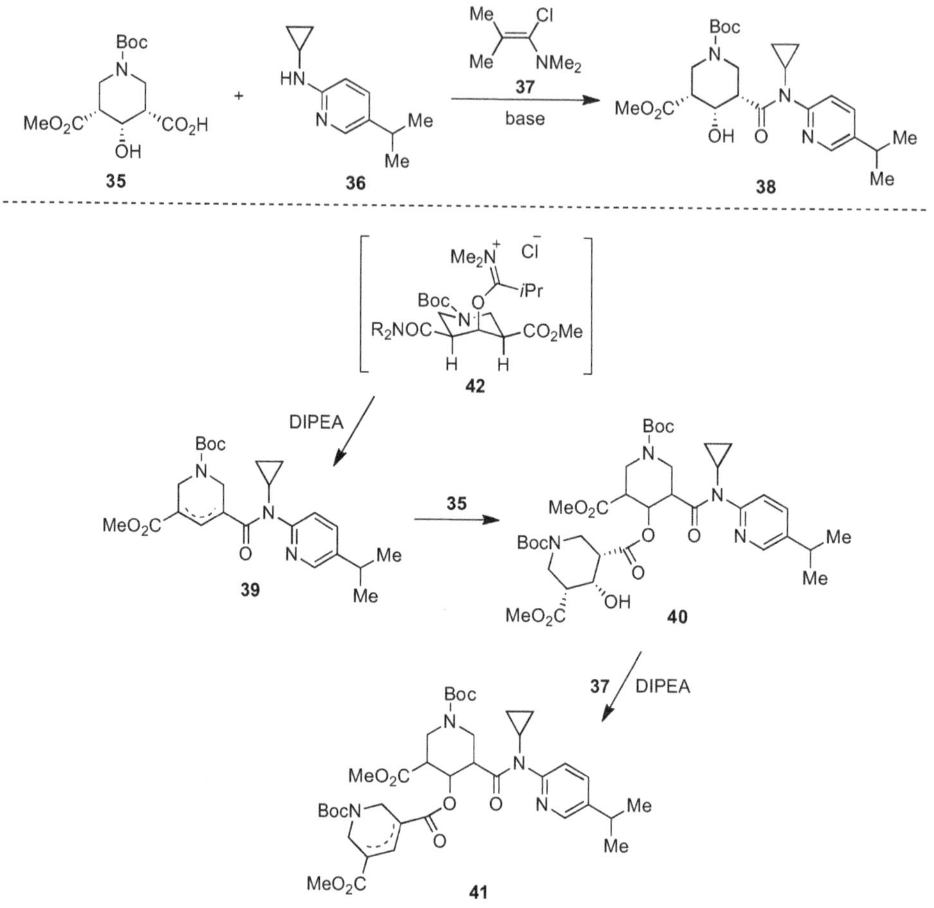

SCHEME 2.8

ester **44** would, besides the desired enolation to **47**, hydrolyze to form acid **49** or ketone **50**, resulting in the observed poor product yields.

The use of hindered potassium *tert*-pentylate base under anhydrous conditions prevented the hydrolysis of the ester group in **44**. Consequently, the product was obtained with a good yield (90%) and purity.

Procedure

- Charge 5.0 kg (15.3 mol) of **43** into an inerted reaction vessel, followed by adding 3.8 kg (18.4 mol) of **44** and 35 L of toluene.
- Heat the resulting suspension to 55–62 °C.
- Charge a solution of potassium *tert*-pentylate (25% in toluene, 17.4 kg, 34.5 mol) in DMF (10 L) over 2–3 h.
- Heat to 75–85 °C and stir for 0.5–1 h.
- Add water (40 L).
- Continue to heat at reflux for 2–3 h.
- Distill to remove 54 L of solvents.
- Cool to 15–25 °C and stir for 2–5 h.

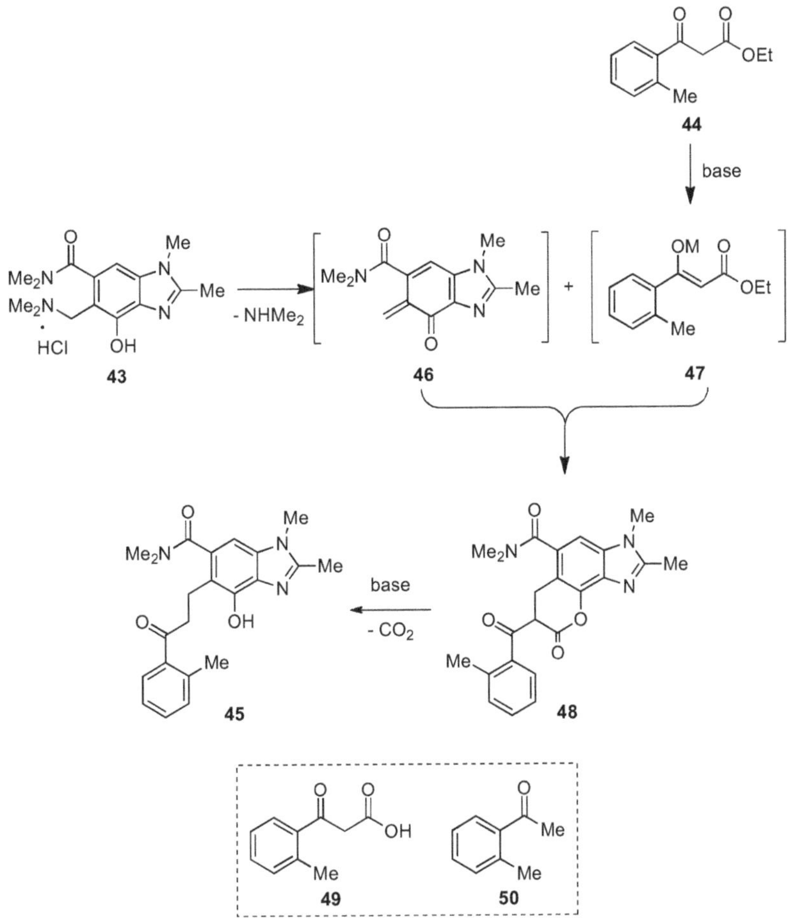

SCHEME 2.9

- Cool to 7–15 °C and stir for 1–2 h.
- Filter and dry at 50 °C under vacuum.
- Yield: 5.2 kg (90%).

2.2.5.2 Lithium *tert*-Butoxide

The base selection can improve the chemoselectivity and eliminate protection/deprotection operations. The allylation of the imine with allyl bromide (Equation 2.16)[2] presents a competitive side reaction with the unprotected hydroxyl group.

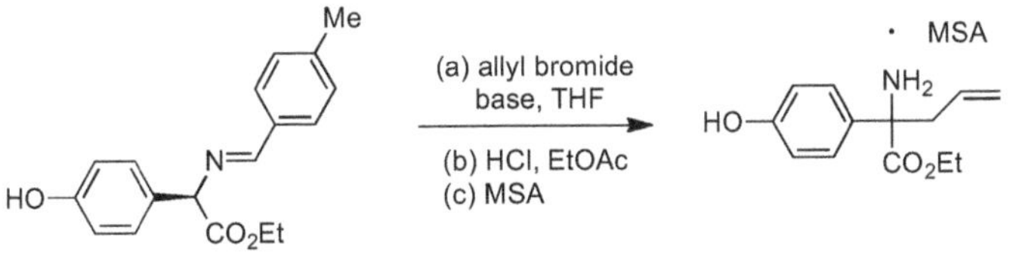

$$(2.16)$$

The use of lithium *t*-butoxide minimized allyl ether formation, allowing an effective *C*-allylation without the protection of the phenol group.

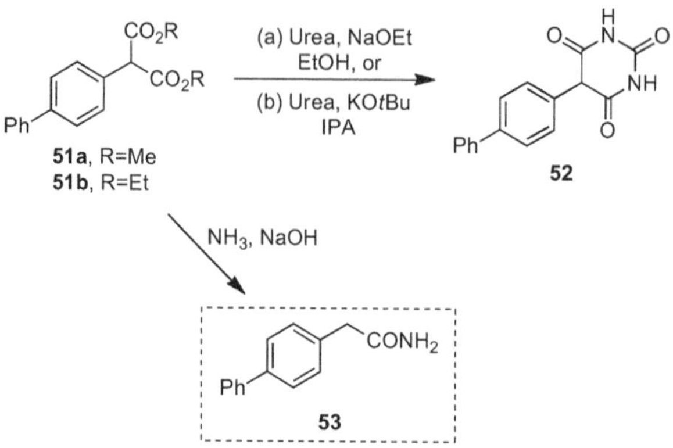

SCHEME 2.10

2.2.5.3 Potassium *tert*-Butoxide

2.2.5.3.1 Improve Product Yields

To prepare barbituric acid **52**, initially, a standard condensation protocol was adopted via refluxing a mixture of diethyl malonate **51b** and urea (1.5 equiv) with sodium ethoxide (2.0 equiv) in ethanol (Scheme 2.10, Condition (a)).[34]

(I) Reaction problems (for Condition (a))

- A rather low yield (36%) of **52** was obtained on a 100-gram scale.
- Heavy precipitation of the sodium salt of **52** necessitated a large volume of ethanol (27 volumes).
- A major side product obtained from this reaction was 4-biphenylacetamide **53** which was difficult to remove from the product.

(II) Solutions

A system, using potassium *tert*-butoxide in THF as the base and 2-propanol (bp 82 °C vs 78 °C of ethanol) as the solvent, was found to be the best in terms of minimizing the impurity formation and reaction time. Under the optimized conditions, the reaction provided an 81% yield of **52**. Moreover, the potassium salt of the product did not precipitate as heavily as the sodium salt, thus enabling a reduction in the amount of solvent (IPA) to six volumes.

REMARKS

(a) The reflux condition reduces the level of ammonia, generated from the decomposition of urea,[35] in the reaction mixture, which, in turn, reduces the formation of the impurity **53**.

(b) The metal alkoxides, such as potassium *tert*-butoxide and sodium *tert*-butoxide, are moisture-sensitive, the moisture in the air is detrimental to the quality of moisture-sensitive bases, forming NaOH. One way to remove the NaOH in NaOtBu is to dissolve the base in THF, followed by filtration as NaOH has very low solubility in THF.

2.2.5.3.2 Improve Optical Purity

The base selection can impact the product's enantiopurity. Since its discovery in 1893, Nef reaction has been widely used in organic synthesis to convert nitro-containing compounds into aldehydes and ketones. Various modifications of the Nef reaction are made to meet the synthetic needs. Driven by green chemistry, a new version of the Nef reaction was developed,[36] in which the nitro-to-ketone group transformation used DBU as the base at room temperature with molecular oxygen as the oxidant. Because of mild conditions, this Nef reaction could tolerate a wide range of functional groups, which lent itself readily accessible to versatile ketones. However, with an optically pure substrate such as **54** in the presence of DBU the reaction gave the product **55** in poor enantiopurity (Scheme 2.11).

It is envisioned that the deprotonation with the weak organic base (DBU) would be reversible and small amounts of DBU may catalyze the enolization of the ketone product **55**. In light of this consideration, potassium *tert*-butoxide was selected to replace DBU. In the event, by employing KO*t*Bu base the reaction afforded **55** in good yield and enantiopurity.

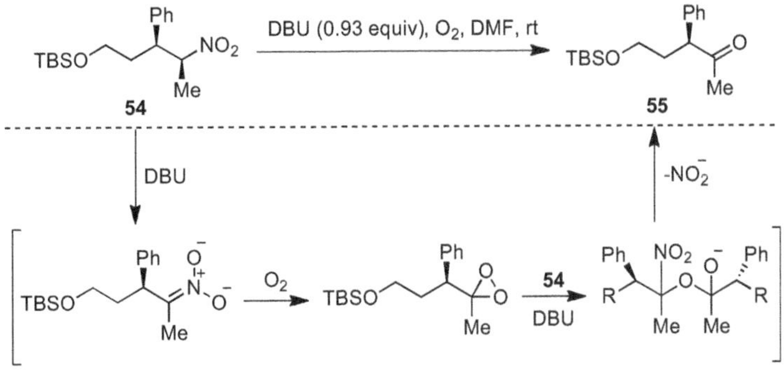

SCHEME 2.11

In contrast, the use of bulky and strong potassium *tert*-butoxide as a base in the synthesis of pyrrolocyclohexanone via alkylation of cyclohexane-1,3-dione and subsequent Paal–Knorr reaction proved to be interior to potassium hydroxide in terms of impurity profile. The use of KO*t*Bu in ethanol led to the formation of unacceptably high levels of furan impurity (Scheme 2.12).[37]

(I) Chemistry diagnosis

Three possible pathways for the formation of furan side product are outlined in Scheme 2.13. Under conditions (a), the alkylated intermediate trione may be in equilibrium with enolate **56** and **57**, and unlikely to form the furan from the latter with the relatively weak base KOH under aqueous conditions. However, the conditions such as (b) would allow the formation of the furan side product via the intermediacy of **56/57** or **58/59**.

(II) Solutions

To suppress such side reactions, KOH was chosen as the base for the alkylation reaction using water as the solvent.

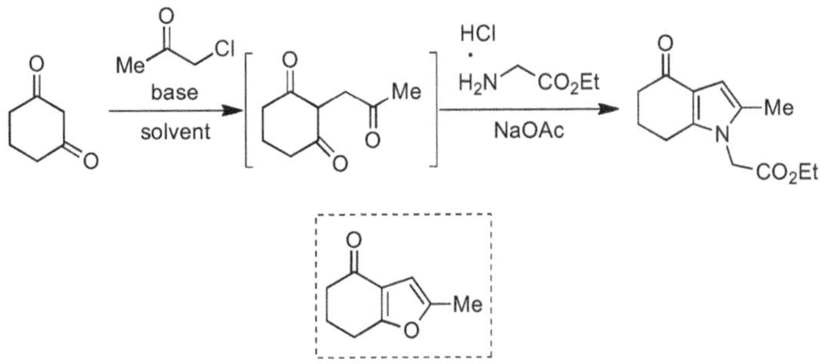

SCHEME 2.12

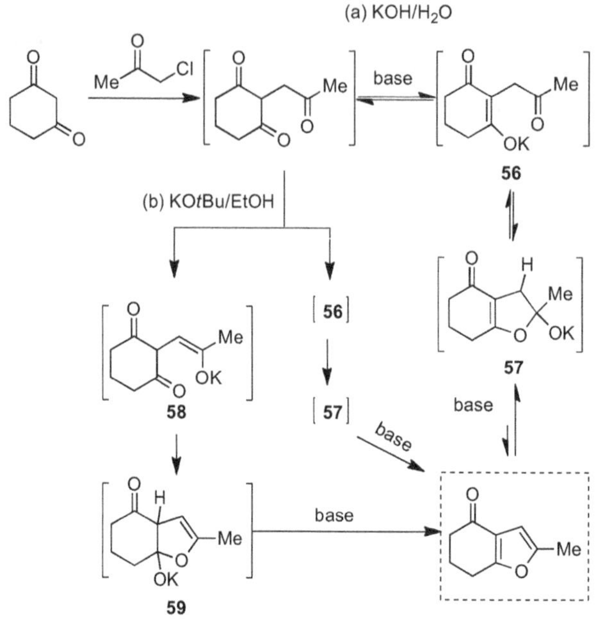

SCHEME 2.13

2.2.5.4 Potassium Trimethylsilanoate

Occasionally, ester hydrolysis is carried out under anhydrous conditions where other base-sensitive functional groups are present in the molecules. Potassium trimethylsilanolate (TMSOK) is a selective and mild base and enables hydrolyzing of ester groups selectively.

2.2.5.4.1 Hydrolysis of Ethyl Ester

BMS 986020 is a small molecule drug candidate, developed by Bristol-Myers Squibb Company for possible treatment of pulmonary fibrosis. The final hydrolysis of the ethyl ester with sodium hydroxide (NaOH) in ethanol produced significant amounts of solvolysis side products (Scheme 2.14).[38] Therefore, potassium trimethylsilanolate (TMSOK) was used as a mild alternative to NaOH for the hydrolysis of ethyl ester to form the desired carboxylic acid.

2.2.5.4.2 Hydrolysis of Methyl Ester

Potassium trimethylsilanolate (TMSOK) was chosen for the hydrolysis of methyl ester in acetonitrile to avoid the potential hydrolysis of nitrile group under aqueous strong basic conditions.[39] Thus, using TMSOK as a base gave the acid product a high chemoselectivity and the nitrile group remained intact (Equation 2.17). Furthermore, the acid product directly precipitated out of the reaction mixture as a crystalline potassium salt.

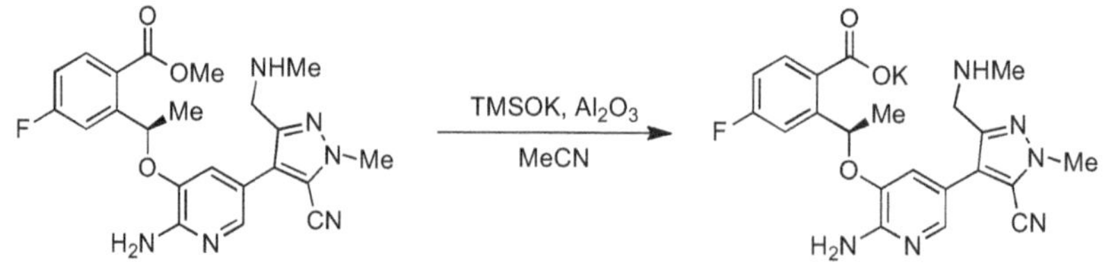

$$(2.17)$$

Because of the presence of KOH in commercial TMSOK, neutral aluminum oxide (alumina) was used to neutralize KOH.

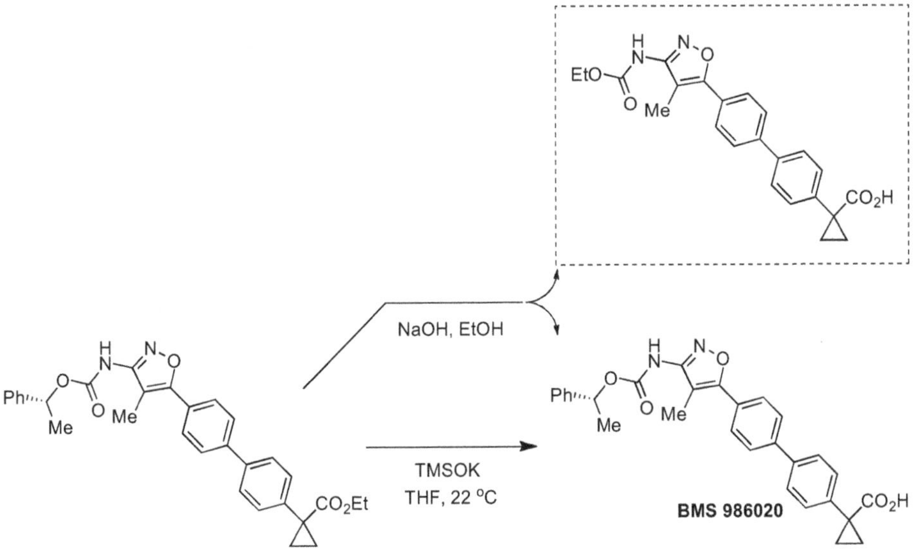

SCHEME 2.14

Procedure

- Add 3.3 kg (25.4 mol) of TMSOK (technical grade) into a reactor followed by adding MeCN (16 kg) and aluminum oxide (6.6 kg).
- Stir the contents at 20 °C under N_2 for 3 h, then filter.
- Transfer the filtrate into the ester (6.88 kg, 15.7 mol) MeCN (10.5 kg) solution in a 200-L reactor.
- Stir the batch at 20 °C for 2 h.
- Filter, wash with heptane (13 kg), and dry at 50 °C/60 mmHg.
- Yield: 4.80 kg.

2.2.5.5 Combination of Potassium *tert*-Butoxide with *tert*-Butyllithium

Regioselective deprotonation of compounds such as picolylpyrrolidinone **60** at the methylene group α to the pyridine ring, rather than α to the carbonyl proved challenging. For instance, attempts to make pyrroloimidazole **62** via deprotonation of **60** with one equivalent of butyllithium followed by the addition of nitrile **61** provided rather poor yields (10–20%) (Equation 2.18).[40] Using two equivalents of butyllithium; however, gave a 75% yield of enamine **63**.

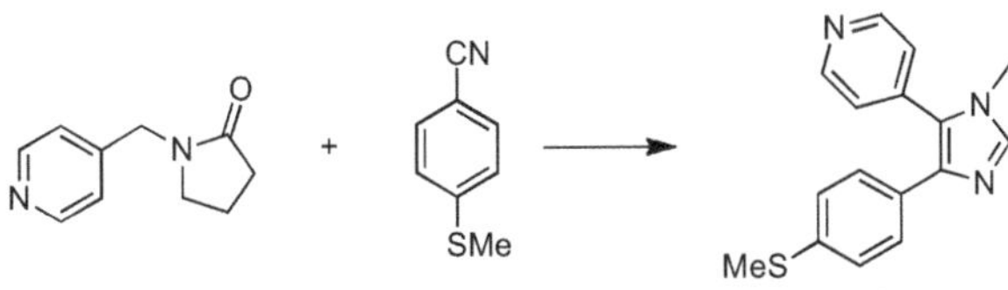

(2.18)

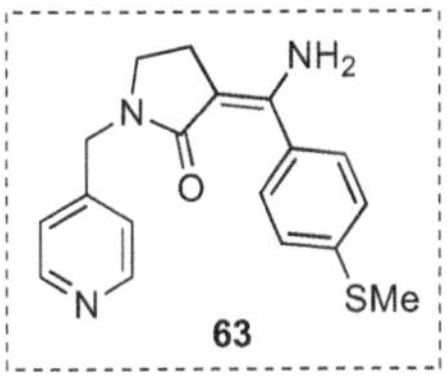

A "superbase", prepared from potassium *tert*-butoxide and *tert*-butyllithium, has been used for selective deprotonation of weakly acidic compounds. Thus, the addition of one equivalent of potassium *tert*-butoxide to the preformed monolithium salt of **60** increased the product yield to 85%.

2.2.5.6 Sodium Methoxide

Using sodium *tert*-butoxide or sodium *tert*-pentylate for amidation of (S)-ethyl lactate with N-benzylpiperazine led to erosion of the enantiomeric purity of the product **66** (Equation 2.19)[41] due to presumably the deprotonation of the α-hydrogen of the amide **66** by the bulky strong bases.

$$\text{(2.19)}$$

    64   **65**   **66**

In order to address this issue, sodium methoxide (25 wt% in MeOH) was identified as the base of choice.

2.3 REAGENTS FOR AMIDE C(O)–N BOND FORMATION

N-Acylation of nitrogen for the synthesis of amides is ubiquitous in both academia and the pharmaceutical industry. Generally, in order to form the amide C(O)–N bond[42] the acids needs to be activated prior to coupling with amines. A recent review paper[43] by Magano from Pfizer includes applications of 36 coupling reagents in the amide preparations. Among various activation methods, the conversion of the acids into the corresponding acid chlorides is by far the most common approach. The use of mixed anhydrides provides an inexpensive and readily scalable process for the preparation of amides, which is particularly valuable for compounds prone to epimerization.

There have been increasing safety concerns related to the use of triazole-based coupling agents, such as 1-[bis(dimethylamino)methylene]-1H-1,2,3-triazolo[4,5-b]pyridinium 3-oxid hexafluorophosphate (HATU) and O-(benzotriazol-1-yl)-N,N,N',N'-tetramethyluronium hexafluorophosphate (HBTU), on large scale due to the explosive nature of the reagents (>1500 J/g; maximum operating temperature approximately 50 °C).

2.3.1 CDI-MEDIATED AMIDE FORMATION

Amide bond formation is one of the most valuable tools used to construct drug molecules. N,N'-carbonyldiimidazole (CDI) as a coupling agent for amide bond formation is increasingly applied to the large-scale synthesis of a number of pharmaceutical products, for example, sildenafil,[44] darifenacin,[45] and sunitinib.[46] The evolution of carbon dioxide byproducts provides a driving force for the reaction. In addition, the imidazole byproduct can be removed by acidic wash. A disadvantage of using CDI is that the resulting imidazolide intermediate is less reactive than the corresponding acid chloride. Consequently, CDI-mediated amide bond formation often requires a catalyst to promote the reaction.

CDI[47] is a white crystalline solid and convenient to handle at scale. Compared with acid chlorides, CDI is less reactive and often used in the coupling of amino acids for peptide synthesis. CDI-mediated amide formation is usually required with no external base as the imidazole byproduct can act as the base.

2.3.1.1 Preparation of Nicotinic Acid Amide

Attention should be paid during the synthesis of amide **69** due to the presence of a hydrolytic labile ethoxy group at the 2-position on the pyridine ring of nicotinic acid **67**. Medicinal chemists utilized

1-(3-dimethylaminopropyl)-3-ethyl-carbodiimide hydrochloride (EDCI)-mediated coupling reaction between nicotinic acid **1** and aminopyrazole **68** in the presence of hydroxybenzotriazole (HOBt) to provide **69** in moderate yield (70%). However, EDCI, a suspect sensitizer, is expensive, and anhydrous HOBt is explosive. In order to address process safety and cost issues, CDI was chosen as the coupling reagent (Equation 2.20).[48]

In addition, this CDI-mediated approach allowed the product amide **69** to be crystallized and filtered directly from the reaction mixture in 91% yield, rejecting the imidazole byproduct in the mother liquor.

Procedure

- Dissolve 0.875 kg (2.55 mol) of **67** in EtOAc (7 L).
- Distill off 1.75 L of EtOAc at atmospheric pressure to dry the reaction system.
- Cool the batch to room temperature.
- Charge 0.43 kg (2.65 mol) of CDI in one portion.
- Heat the resulting slurry at 35 °C for 30 min, 45–50 °C for 30 min, and at reflux for 1 h.
- Cool to 45–50 °C.
- Charge 0.59 kg (2.42 mol) of **68** in one portion.
- Heat at reflux and distill off 875 mL of EtOAc at atmospheric pressure.
- Heat at reflux for 16 h.
- Cool to 10–15 °C and granulate for 1 h.
- Filter, rinse with EtOAc, and dry at 50 °C in vacuo.
- Yield: 1.252 kg (90.7%).

2.3.1.2 Preparation of Ureas

CDI-promoted unsymmetrical urea preparation was accomplished in a two-step process, involving a reaction of amine hydrochloride **70** with CDI to give carbonyl imidazole intermediate **71** whose reaction with dehydrocoronamate hydrochloride **72** gave the desired urea **73** (Scheme 2.15).[49] It should be noted that the isolation of **71** was necessary in order to remove imidazole byproduct and excess CDI since the excess CDI would react with **72** to generate hydantoin impurity **74**.

REMARKS

(a) The addition of a base, such as triethylamine, yielded no signs of product, presumably due to the fact that protonation of the imidazole nitrogen was required to promote the coupling reaction with **72**.

(b) Activation of imidazolyl carboxamides can also be achieved by *N*-methylation of imidazole nitrogen with MeI (Scheme 2.16):[48]

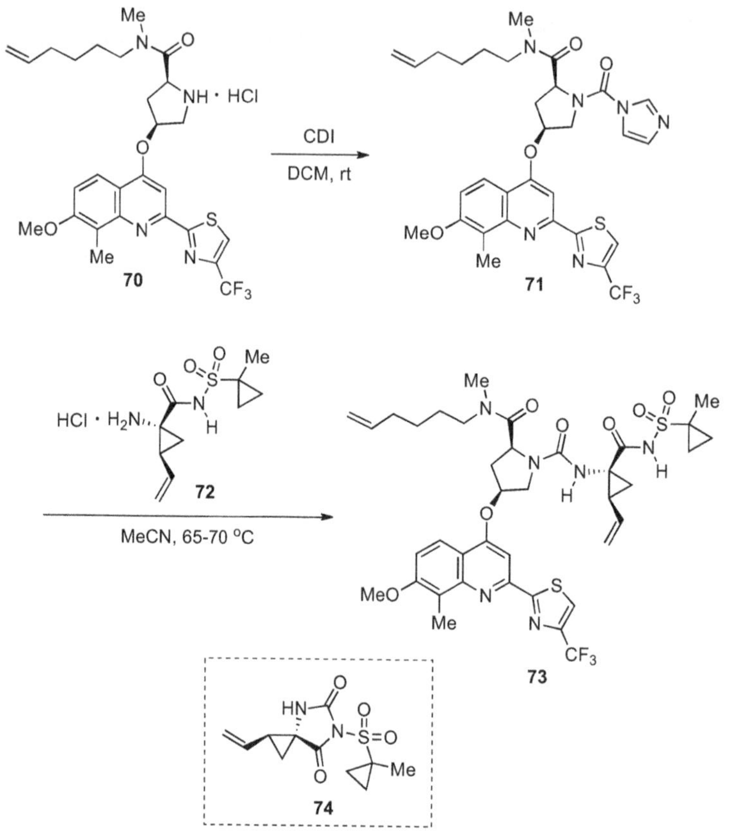

SCHEME 2.15

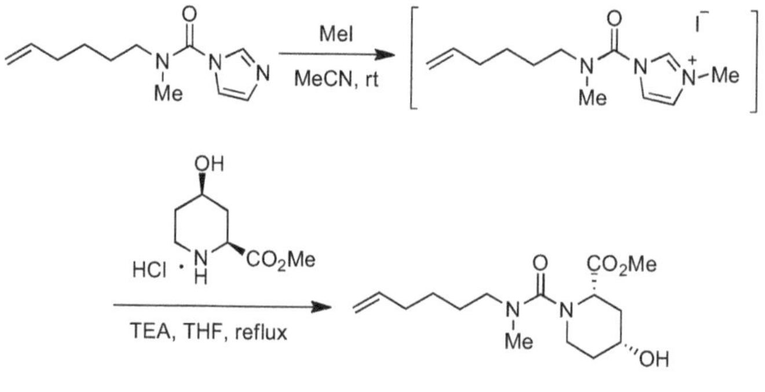

SCHEME 2.16

2.3.2 Thionyl Chloride-Mediated Amide Formation

Amide synthesis can be mediated with thionyl chloride through an acid chloride intermediate (Scheme 2.17).

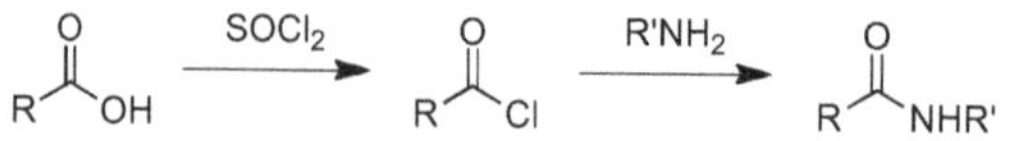

SCHEME 2.17

The reactions of carboxylic acids with thionyl chloride are usually catalyzed by DMF, which can be conducted in neat using an excess of thionyl chloride as the reaction solvent or in an inert solvent. However, the use of DMF as the catalyst has been questioned recently on safety concerns, owing to the formation of highly toxic byproducts, dimethylcarbamoyl chloride (DMCC) (Scheme 2.18).[50] DMCC[51] is a known animal carcinogen and a potential human carcinogen, necessitating control of exposure to very stringent standards, measured in terms of only a few parts per billion.

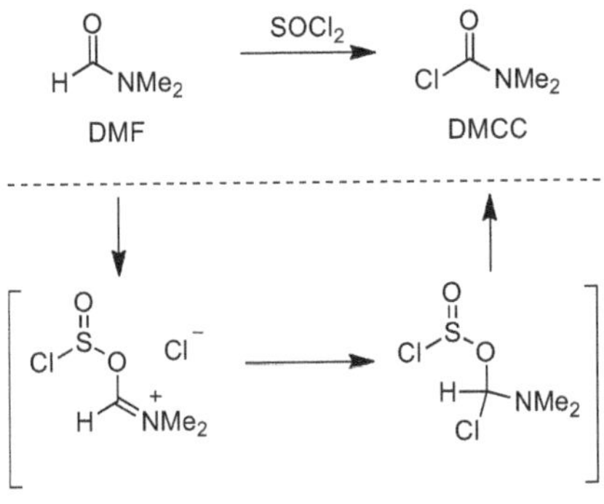

SCHEME 2.18

2.3.2.1 Tetramethylurea-Catalyzed Acid Chloride Formation

5-Amino-2,4,6-triiodoisophthalic acid dichloride **76** is a common intermediate for the synthesis of water-soluble, non-ionic iodinated X-ray contrast agents, such as iopamidol,[52] iohexol,[53] and ioversol,[54] for imaging soft tissues (Scheme 2.19).

Because of the potential formation of DMCC and unacceptably high levels of other byproducts, a protocol for the preparation of acid dichloride **76** was developed. Tetramethylurea (TMU) was identified as the catalyst to substitute for DMF and the reaction was carried out by mixing the acid **75** with thionyl chloride at 50–80 °C (Scheme 2.20).[55]

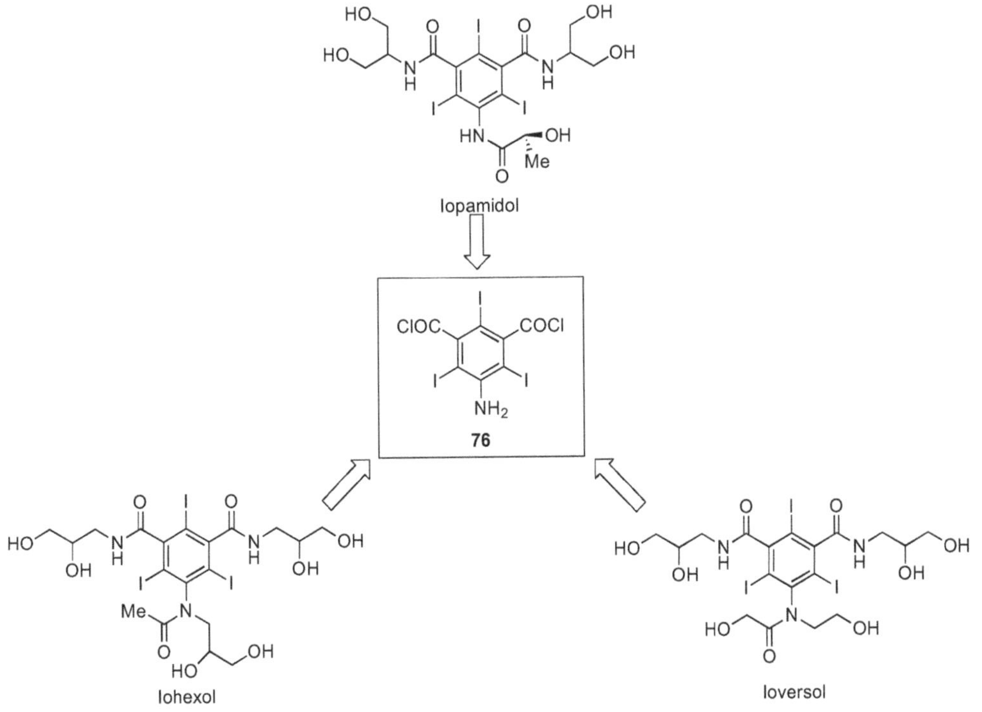

SCHEME 2.19

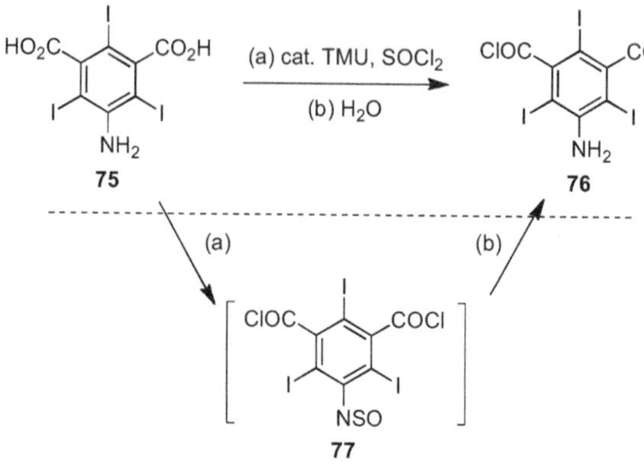

SCHEME 2.20

Hydrolysis of the *N*-sulfinyl intermediate **77** and simultaneous crystallization of **76** was realized with the addition of acetone and water.

Procedure

Stage (a)
- Charge 1.43 kg (1.2 mol) of thionyl chloride into a reactor that contains 600 g (1.07 mol) of **75**.
- Charge 2.9 g (25 mmol) of TMU.
- Stir the resulting slurry at 50–60 °C for 6 h followed by heating at 80 °C for 0.5 h.
- Distill off the excess $SOCl_2$ under reduced pressure.
- Dissolve the residual solid in 500 mL of toluene at 50–60 °C
- Distill under reduced pressure to remove residual $SOCl_2$.

Stage (b)
- Add acetone (738 mL) to the batch and warm to 45 °C to form a solution.
- Add dropwise a mixture of acetone/water (9:1, w/w, 550 mL) at 45 °C over 1 h.
- Stir for 1 h and cool to 25 °C.
- Add dropwise water (487 mL) over 2 h.
- Stir for 24 h.
- Filter and wash with a mixture of acetone/water (1:1, w/w, 400 mL) followed by water (400 mL).
- Dry at 40 °C under vacuum.
- Yield: 547 g (85.5%).

2.3.2.2 *N*-Sulfinylaniline-Involved Amide Preparation

Preparation of amides of α-hydroxyl acids proves to be challenging and often requires converting the hydroxyl group in the α-hydroxyl acids into their corresponding TMS-protected derivatives (Equation 2.21). Such protection/coupling/deprotection sequence renders an inefficient amide preparation process.

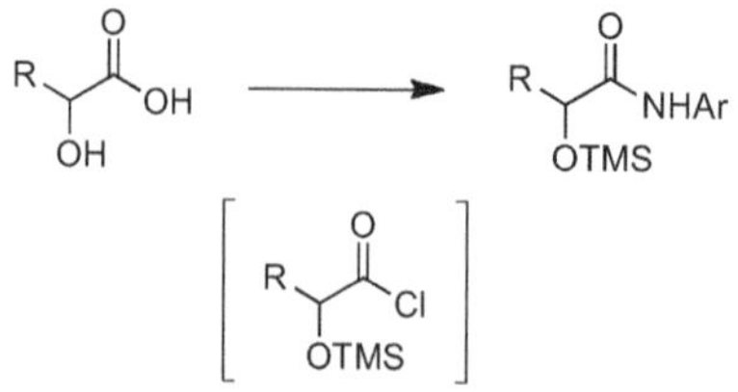

(2.21)

Therefore, an approach was developed[56] in which amides were prepared via reactions of the α-hydroxyl acids with "activated" anilines, that is, N-sulfinylanilines, obtained from a reaction of anilines with thionyl chloride. Two intermediates, **78** and **79**, are presumably formed and the latter would be attacked by anilines to give amides with the release of SO_2 (Scheme 2.21).

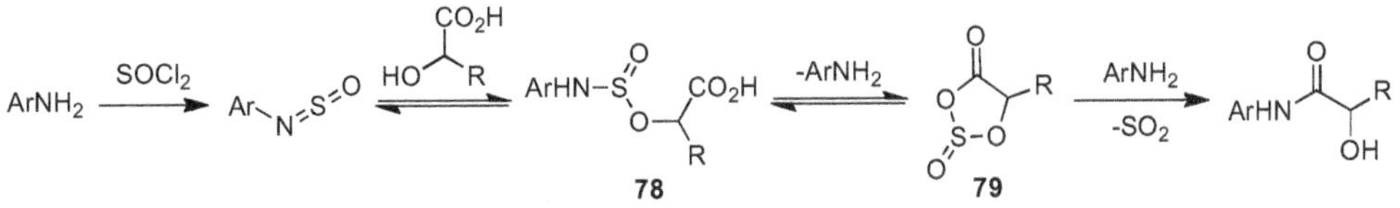

SCHEME 2.21

This approach using N-sulfinyl amine-mediated reaction lent itself readily to the preparation of chiral amide **83**, which avoided the protection of the chiral acid **82** (Scheme 2.22).[57]

Analogously, this coupling protocol was adopted by scientists of Bristol-Myers Squibb to prepare enantiomerically pure α-hydroxyl amide **88** (Scheme 2.23).[58] Activation of the electron-deficient anilines **85** with thionyl chloride followed by a reaction of the resulting N-sulfinylaniline derivative **86** with acid **84** furnished via intermediate **87** the desired chiral amide **88** in good yields.

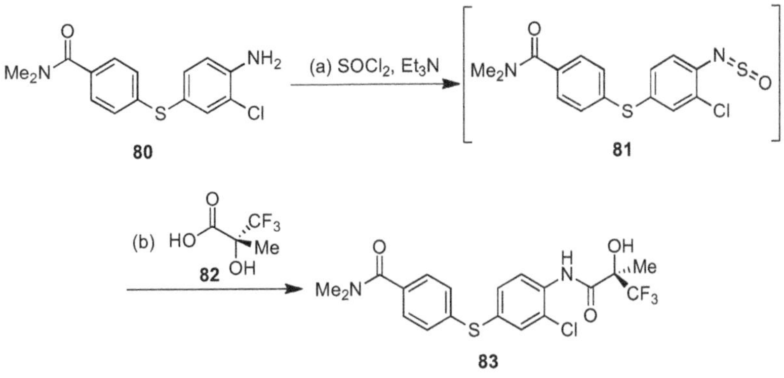

SCHEME 2.22

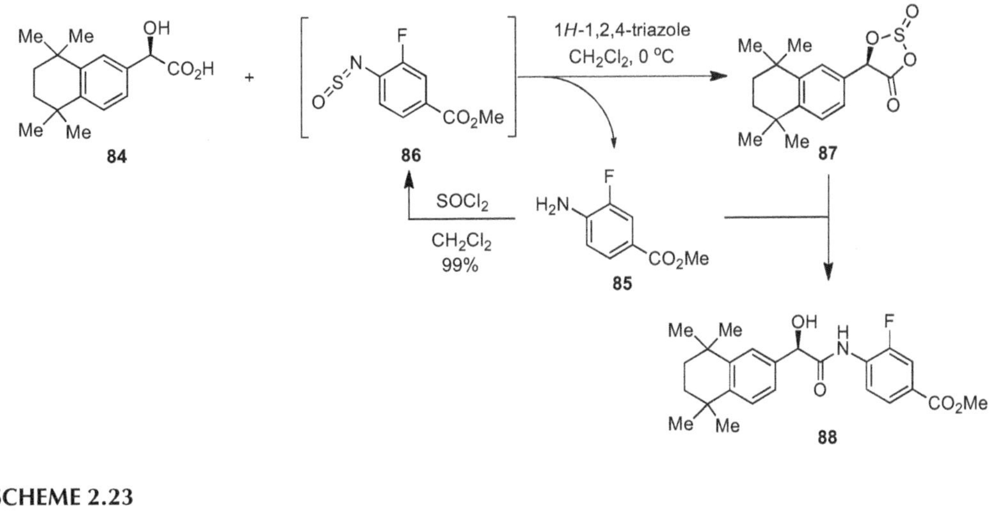

SCHEME 2.23

2.3.3 BOC₂O-MEDIATED AMIDE FORMATION

Conversion of azaindole-7-carboxylic acid **89** to its corresponding amide **90** was realized by activation of the acid **89** with di-*tert*-butyl dicarbonate (Boc₂O) in the presence of pyridine followed by reaction with ammonium bicarbonate (Equation 2.22).[59]

(2.22)

DISCUSSIONS

Attempts to convert acid **89** to amide **90** by activation of **89** under a variety of standard conditions (acid chloride, CDI, EDC, etc.) failed due to either cleavage of the Boc protecting group (under acid chloride conditions) or low reactivity of the hindered acid.

2.3.4 SCHOTTEN–BAUMANN REACTION

Poor solubility of reagents often results in a slow reaction or incomplete conversion. For the preparation of amide **93** (Equation 2.23),[60] the reaction of acid chloride **91** with 2-bromoethylamine hydrobromide **92** in the presence of pyridine or triethylamine in organic solvents such as THF or MeCN gave a poor conversion.

(2.23)

A Schotten–Baumann reaction is a reaction between amines and acid chlorides that is carried out under two-phase conditions. Utilization of the Schotten–Baumann reaction conditions (EtOAc/water) in the presence of potassium carbonate allowed a complete conversion within 1 h. In comparison with other solvents such as toluene, MeCN, MTBE, and MIBK, EtOAc gave the best reaction results.

Procedure

- Charge 27.3 kg (133.2 mol, 1.0 equiv) of **92** to a solution of K₂CO₃ (36.8 kg, 266.3 mol, 2.0 equiv) in water (110 L) at < –5 °C.
- Add 100 L of EtOAc followed by adding 21.1 kg (133.2 mol) of **91** at ≤5 °C.
- Stir the reaction mixture for 1 h.
- HPLC showed a purity of 85.2% and the product solution was used directly in the subsequent step.

Table 2.2 lists some frequently used acid activation reagents for acid-amine coupling reactions.

TABLE 2.2
Acid Activation Reagents for Amide Bond Formation

Entry	Reagent	Structure	Molecular formula	Molecular weight	Mp (or Bp)	Storage	Comments
1	1-[Bis(dimethylamino)-methylene]-1H-1,2,3-triazolo[4,5-b]pyridinium 3-oxide hexafluorophosphate (HATU)		$C_{10}H_{15}F_6N_6OP$ (white crystal powder)	380.23	Mp 183–185 °C	Store in a cool and dry place (with 24 months shelf life)	HATU is incompatible with strong oxidizing agents.
2	O-Benzotriazole-N,N,N',N'-tetramethyluronium hexafluorophosphate (HBTU)		$C_{11}H_{17}F_6N_5OP$ (white to light yellow powder)	380.25	Mp 200 °C (dec)	Store in cool and dry place (with 24 months shelf life)	Incompatible with strong oxidizing agents.
3	Vilsmeier Reagent[a]		$C_3H_7Cl_2N$	128.00	Mp132°C(dec)–	–	Vilsmeier reagent is generated in situ by a reaction of DMF with POCl₃.
4	1,1'-Carbonyl diimidazole (CDI)[b]		$C_7H_6N_4O$ (white crystal powder)	162.15	Mp 117–122 °C	2–8 °C	CDI is stable, but moisture-sensitive. It is incompatible with acids, strong oxidizing agents, and water.
5	1-Ethyl-3-(3-dimethyl-aminopropyl)carbodiimide (EDC)[c]		$C_7H_{17}N_3$	155.24	Liquid (d=0.877 g/mL at 20 °C)	–20 °C	EDC is soluble in water and typically employed in the 4.0–6.0 pH range. It's a suspected sensitizer.[47]
6	Dicyclohexylcarbodiimide (DCC)		$C_{13}H_{22}N_2$	206.33	Mp 34 °C (Bp 122 °C/6 mmHg)	Keep the container tightly closed in a dry and well-ventilated place	DCC is used for coupling reactions, dehydration, etc. DCC is a potent allergen and a sensitizer, often causing skin rashes.
7	Diisopropylcarbodiimide (DIC)		$C_7H_{14}N_2$ (liquid)	126.2	(Bp 145–148 °C)	Store in cool place and under inert gas. Moisture sensitive	DIC is easier to handle than the commonly used DCC.

(Continued)

TABLE 2.2 (CONTINUED)
Acid Activation Reagents for Amide Bond Formation

Entry	Reagent	Structure	Molecular formula	Molecular weight	Mp (or Bp)	Storage	Comments
8	(Benzotriazol-1-yloxy)tris-(dimethylamino) phosphonium hexafluorophosphate (BOP)		$C_{12}H_{22}F_6N_6OP_2$	422.28	Mp >130 °C (dec)	Recommended storage temperature: 2–8 °C. Keep in a dry place	—
9	(Benzotriazol-1-yloxy)-tripyrrolidinophosphonium hexafluorophosphate (PYBOP)		$C_{18}H_{28}F_6N_6OP_2$	520.39	Mp 154–156 °C (dec)	2–8 °C	—
10	Cyanuric chloride[d]		$C_3Cl_3N_3$ (White, odorless solid)	184.41	Mp 145–147 °C (Bp 190 °C)	Recommended storage temperature: 2–8 °C. Air and moisture sensitive. Keep container tightly closed in a dry and well-ventilated place	The cyanuric acid byproduct is classified as "essentially non-toxic".
11	2-Chloro-4,6-dimethoxy-1,3,5-triazine (CDMT)		$C_5H_6ClN_3O_2$	175.57	Mp 71–74 °C	Keep container tightly closed in a dry and well-ventilated place	N-(4,6-Dimethoxy-1,3,5-triazin-2-yl)-N-methylmorpholinium chloride, prepared in situ from CDMT and N-methylmorpholine, is the species used to activate the acid for the C–N bond formation.[e]
12	2-Chloro-1-methylpyridinium iodide[f]		C_6H_7ClIN	255.48	Mp 200 °C (dec)	Keep container tightly closed in a dry and well-ventilated place	—
13	N-Hydroxybenzotriazole hydrate (HOBt)		$C_6H_5N_3O \cdot xH_2O$ (White crystal powder)	–	Mp 155–158 °C	Keep container tightly closed in a dry and well-ventilated place	HOBt could suppress racemization in peptide synthesis.[g] Commercial HOBt is the HOBt monohydrate because anhydrous HOBt is explosive. After the reaction, HOBt can be rejected with citrate wash.

(Continued)

TABLE 2.2 (CONTINUED)
Acid Activation Reagents for Amide Bond Formation

Entry	Reagent	Structure	Molecular formula	Molecular weight	Mp (or Bp)	Storage	Comments
14	1-Hydroxy-7-azabenzotriazole (HOAt)		$C_5H_4N_4O$	136.11	Mp 213–216 °C (dec)	–	–
15	Ethyl (hydroxyimino)-cyanoacetate (Oxyma)		$C_5H_6N_2O_3$	142.11	Mp 130–132 °C	–	Oxyma is an efficient additive for peptide synthesis to replace the HOBt and HOAt with a lower risk of explosion.[h]
16	Isobutyl chloroformate (IBCF)[i]		$C_5H_9ClO_2$	136.58	(Bp 128.8 °C)	2–8 °C	IBCF is a well-known coupling agent in peptide synthesis.
17	Pivaloyl chloride		C_5H_9ClO	120.58	(Bp 105–106 °C)	Keep the container tightly closed in a dry and well-ventilated place	–
18	Di-tert-butyl dicarbonate (Boc₂O)		$C_{10}H_{18}O_5$	218.25	Mp 23 °C (Bp 56–57 °C/0.5 mmHg)	Recommended storage temperature: 2–8 °C with the container tightly closed. Avoid heating above 40 °C	Boc₂O hydrolyzes to t-butanol and CO₂ if exposed to moisture.
19	Dioxolanedione[j]		$C_{17}H_{20}O_4$	288.34	–	–	–
20	Acetonide[k]		$C_{19}H_{26}O_3$	302.41	–	–	–

(Continued)

TABLE 2.2 (CONTINUED)
Acid Activation Reagents for Amide Bond Formation

Entry	Reagent	Structure	Molecular formula	Molecular weight	Mp (or Bp)	Storage	Comments
21	2-Ethoxy-1-ethoxycarbonyl-1,2-dihydroquinoline (EEDQ)[l]		$C_{14}H_{17}NO_3$	247.29	Mp 62–67 °C	2–8 °C	Using EEDQ as activating agent could avoid racemization.[m]
22	p-Toluenesulfonyl chloride (TsCl)[n]		$C_7H_7ClO_2S$	190.65	Mp 65–69 °C (134 °C/10 mmHg)	Store under inert gas and moisture-free conditions	—
23	n-Propanephosphonic acid cyclic anhydride (T3P)[o]		$C_9H_{21}O_6P_3$	318.18	(65 °C)	Store under nitrogen and moisture-free conditions	T3P is commercially available as ≥50 wt% EtOAc solution. The postreaction workup required simple water washes to remove T3P-related byproducts.

[a] (a) Bosshard, H. H.; Zollinger, H. C. H. Angew. Chem. 1959, 71, 375. (b) Zhang, B.; Crispino, G. A.; Springer, D. M.; Wichtowski, J. A.; Zhang, Y.; Goodrich, J.; Ueda, Y.; Luh, B. Y.; Burke, B. D.; Brown, M.; Dutka, A. P.; Zheng, B.; Hsieh, D.-M.; Humora, M. J.; North, J. T.; Pullockaran, A. J.; Livshits, J.; Swaminathan, S.; Gao, Z.; Schierling, P.; Ermann, P.; Perrone, R. K.; Lai, M. C.; Gougoutas, J. Z.; DiMarco, J. D.; Bronson, J. J.; Heikes, J. E.; Grosso, J. A.; Kronenthal, D. R.; Denzel, T. W.; Mueller, R. H. Org. Process Res. Dev. 2000, 4, 488.

[b] Ashcroft, C. P.; Dessi, Y.; Entwistle, D. A.; Hesmondhalgh, L. C.; Longstaff, A.; Smith, J. D. Org. Process Res. Dev. 2012, 16, 470.

[c] (a) Ito, T.; Ikemoto, T.; Isogami, Y.; Wada, H.; Sera, M.; Mizuno, Y.; Wakimasu, M. Org. Process Res. Dev. 2002, 6, 238. (b) Dillon, B. R.; Roberts, D. F.; Entwistle, D. A.; Glossop, P. A.; Knight, C. J.; Laity, D. A.; James, K.; Praquin, C. F.; Stang, R. S.; Watson, C. A. L. Org. Process Res. Dev. 2012, 16, 195. (c) Lafrance, D.; Caron, S. Org. Process Res. Dev. 2012, 16, 409.

[d] Rayle, H. L.; Fellmeth, L. Org. Process Res. Dev. 1999, 3, 172.

[e] (a) Barnett, C.; Wilson, T. M.; Kobierski, M. E. Org. Process Res. Dev. 1999, 3, 184. (b) Graham, M. A.; Raw, S. A.; Andrews, D. M.; Good, C. J.; Matusiak, Z. S.; Maybury, M.; Stokes, E. S. E.; Turner, A. T. Org. Process Res. Dev. 2012, 16, 1283.

[f] Karlsson, S.; Sörensen, J. H. Org. Process Res. Dev. 2012, 16, 586.

[g] Lee, K. W.; Hwang, S. Y.; Kim, C. R.; Nam, D. H.; Chang, J. H.; Choi, S. C.; Choi, B. S.; Lee, K. K.; So, B.; Cho, S. W.; Shin, H. Org. Process Res. Dev. 2003, 7, 839.

[h] Subirós-Funosas, R.; Prohens, R.; Barbas, R.; El-Faham, A.; Albericio, F. Chem. Eur. J. 2009, 15, 9394.

[i] Chaudhary, A.; Girgis, M.; Prashad, M.; Hu, B.; Har, D.; Repič, O.; Blacklock, T. J. Tetrahedron Lett. 2003, 44, 5543.

[j] Toyooka, K.; Takeuchi, Y.; Kubota, S. Heterocycles 1989, 29, 975.

[k] Khalij, A.; Nahid, E. Synthesis 1985, 12, 1153.

[l] Belleau, D.; Malek, G. J. Am. Chem. Soc. 1968, 90, 1651.

[m] Ormerod, D.; Willemsens, B.; Mermans, R.; Langens, J.; Winderickx, G.; Kalindjian, S. B.; Buck, I. M.; McDonald, I. M. Org. Process Res. Dev. 2005, 9, 499.

[n] Campeau, L.-C.; Dolman, S. J.; Gauvreau, D.; Corley, E.; Liu, J.; Guidry, E. N.; Ouellet, S. G.; Steinhuebel, D.; Weisel, M.; O'Shea, P. D. Org. Process Res. Dev. 2011, 15, 1138.

[o] (a) Patterson, D. E.; Powers, J. D.; LeBlanc, M.; Sharkey, T.; Boehler, E.; Irdam, E.; Osterhout, M. H. Org. Process Res. Dev. 2009, 13, 900. (b) Cimarosti, Z.; Giubellina, N.; Stabile, P.; Laval, G.; Tinazzi, F.; Maton, W.; Pachera, R.; Russo, P.; Morett, R.; Russo, S.; Cooke, J. W. B.; Westerduin, P. Org. Process Res. Dev. 2011, 15, 1287.

2.3.5 OTHER AMIDE FORMATION METHODS

2.3.5.1 Copper (II)-Catalyzed Transamidation

A copper-catalyzed transamidation protocol was developed for the synthesis of functionalized amides (and ureas) (Scheme 2.24).[61]

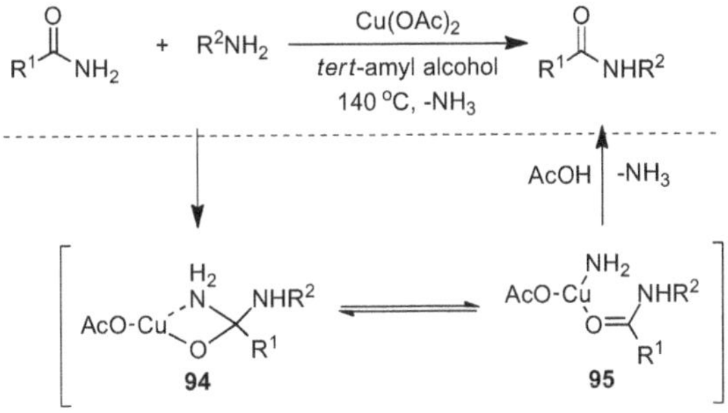

SCHEME 2.24

The transformation was performed by exposure of primary carboxamides (or ureas) to amines in the presence of $Cu(OAc)_2$ (10 mol%) as a catalyst in *tert*-amyl alcohol at 140 °C. Mechanistically, the reaction proceeds through several intermediates such as **94** and **95**, and the release of ammonia is the driving force.

2.3.5.2 Cross-Coupling between Acyltrifluoroborates and Hydroxylamines

A mild and environmentally friendly protocol for amide synthesis was developed (Scheme 2.25).[62] The reaction was performed simply by mixing acyltrifluoroborates with *O*-benzoyl hydroxylamines in a mixture of water and *tert*-butanol (1:1) at room temperature for 30 min, affording the amides in good to excellent yields.

The reaction is compatible with various functional groups in the acyltrifluoroborates and *O*-benzoyl hydroxylamine substrates.

The acyltrifluoroborates could be prepared through either method A or method B described in the literature (Scheme 2.26).[63]

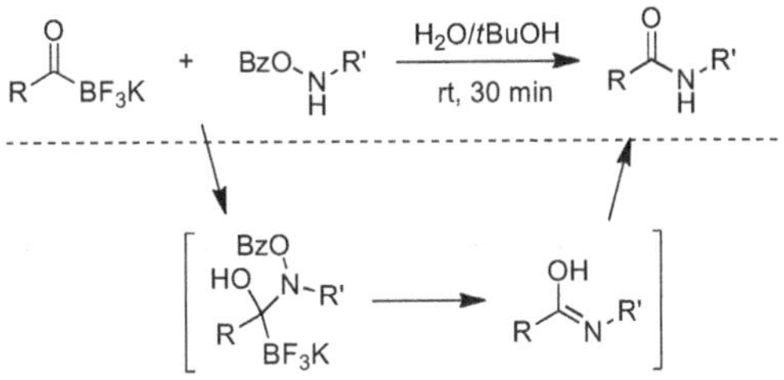

SCHEME 2.25

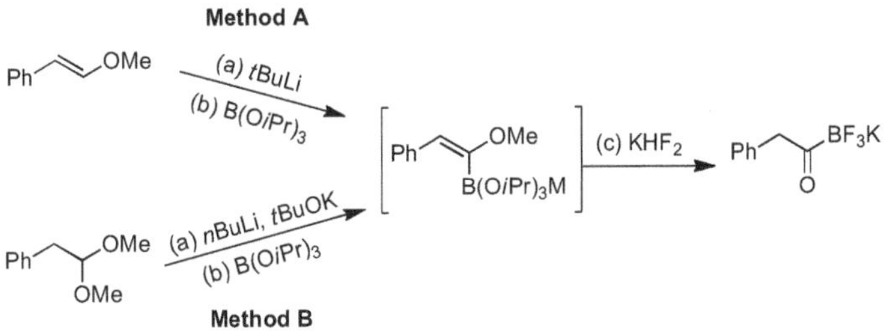

SCHEME 2.26

2.3.5.3   Catalytic Aminolysis of Ester

An amide formation using a direct coupling of amine **96** and carboxylic acid **97a** (R=H) was not possible because this acid **97a** was prone to undergo decarboxylation. The C(O)–N bond formation between amine **96** and carboxylic acid sodium salt **97b** (R=Na) was challenging due to its low solubility in most organic solvents. Ultimately, scientists from AstraZeneca developed a catalytic aminolysis using carboxylic ester **97c** (R=Et) (Equation 2.24).[64] A condition of a slight excess of ethyl ester **97c**, a catalytic amount of 3-nitrophenol (21 mol%), and K_2CO_3 (21 mol%) as the base in CH_3CN at 70 °C rendered a full and clean conversion of **96** to the amide product **98**.

(2.24)

97a (R = H)
97b (R = Na)
97c (R = Et)

96

98

NOTES

1. Chen, J. J.; Nugent, T. C.; Lu, C. V.; Kondapally, S.; Giannousis, P.; Wang, Y.; Wilmot, J. T. *Org. Process Res. Dev.* **2003**, *7*, 313.
2. Savage, S. A.; Waltermire, R. E.; Campagna, S.; Bordawekar, S.; Toma, J. D. R. *Org. Process Res. Dev.* **2009**, *13*, 510.
3. Yamashita, Y.; Morinaga, Y.; Kasai, M.; Hashimoto, T.; Takahama, Y.; Ohigashi, A.; Yonishi, S.; Akazome, M. *Org. Process Res. Dev.* **2017**, *21*, 346.
4. (a) Bretherick, L. *Handbook of Reactive Chemical Hazards*, 4th ed.; Butterworth–Heinemann, Oxford, **1990**, p 1181. (b) For an example on the risks of thermal runaway reactions using the NaH/DMF combination see: DeWall, G. *Chem. Eng. News* **1982**, *60*, 5.
5. de Koning, P. D.; McAndrew, D.; Moore, R.; Moses, I. B.; Boyles, D. C.; Kissick, K.; Stanchina, C. L.; Cuthberton, T.; Kamatani, A.; Rahman, L.; Rodriguez, R.; Urbina, A.; Sandoval, A.; Rose, P. R. *Org. Process Res. Dev.* **2011**, *15*, 1018.
6. Yu, R. H.; Schultze, L. M.; Rohloff, J. C.; Dudzinski, P. W.; Kelly, D. E.; *Org. Process Res. Dev.* **1999**, *3*, 53.
7. Deerberg, J.; Prasad, S. J.; Sfouggatakis, C.; Eastgate, M. D.; Fan, Y.; Chidambaram, R.; Sharma, P.; Li, L.; Schild, R.; Müslehiddinoğlu, J.; Chung, H.-J.; Leung, S.; Rosso, V., *Org. Process Res. Dev.* **2016**, *20*, 1949.
8. Ong, D. Y.; Fan, D.; Dixon, D. J.; Chiba, S. *Angew. Chem. Int. Ed.* **2020**, *59*, 11903.
9. Leighty, M. W.; Shen, B.; Johnston, J. N. *J. Am. Chem. Soc.* **2012**, *134*, 15233.
10. Ghosh, S. K.; Chaubey, N. R. *Tetrahedron: Asymmetry* **2012**, *23*, 1206.
11. Evans, D. A.; Britton, T. C.; Ellman, J. A. *Tetrahedron Lett.* **1987**, *28*, 6141.
12. Chen, C.; Dagneau, P.; Grabowski, E. J. J.; Oballa, R.; O'Shea, P.; Prasit, P.; Robichaud, J.; Tillyer, R.; Wang, X. *J. Org. Chem.* **2003**, *68*, 2633.
13. Hupp, J. T.; Evanston, S. T. N., III, *Chem. Eng. News* **2011**, *89*(2), 2.
14. Dubnikova, F.; Kosloff, R.; Almog, J.; Zeiri, Y.; Boese, R.; Itzhaky, H.; Alt, A.; Keinan, E. *J. Am. Chem. Soc.* **2005**, *127*, 1146.
15. (a) DeBaillie, A. C.; Magnus, N. A.; Laurila, M. E.; Wepsiec, J. P.; Ruble, J. C.; Petkus, J. J.; Vaid, R. K.; Niemeier, J. K.; Mick, J. F.; Gunter, T. Z. *Org. Process Res. Dev.* **2012**, *16*, 1538. (b) Yamagami, T.; Moriyama, N.; Kyuhara, M.; Moroda, A.; Uemura, T.; Matsumae, H.; Moritani, Y.; Inoue, I. *Org. Process Res. Dev.* **2014**, *18*, 437.
16. (a) Kawasaki, T.; Shinada, M.; Ohzono, M.; Ogawa, A.; Terashima, R.; Sakamoto, M. *J. Org. Chem.* **2008**, *73*, 5959. (b) Ouyang, J.; Yan, R.; Mi, X.; Hong, R. *Angew. Chem. Int. Ed.* **2015**, *54*, 10940.
17. Butters, M.; Catterick, D.; Craig, A.; Curzons, A.; Dale, D.; Gillmore, A.; Green, S. P.; Mariziano, I.; Sherlock, J.-P.; White, W. *Chem. Rev.* **2006**, *106*, 3002.
18. Beutner, G. L.; Cohen, B. M.; DelMonte, A. J.; Dixon, D. D.; Fraunhoffer, K. J.; Glace, A. W.; Lo, E.; Stevens, J. M.; Vanyo, D.; Wilbert, C. *Org. Process Res. Dev.* **2019**, *23*, 1378.
19. Glace, A. W.; Cohen, B. M.; Dixon, D. D.; Beutner, G. L.; Vanyo, D.; Akpinar, F.; Rosso, V.; Fraunhoffer, K. J.; Delmonte, A. J.; Santana, E.; Wilbert, C.; Gallo, F.; Bartels, W. *Org. Process Res. Dev.* **2020**, *24*, 172.

20. Stewart, G. W.; Brands, K. M. J.; Brewer, S. E.; Cowden, C. J.; Davies, A. J.; Edwards, J. S.; Gibson, A. W.; Hamilton, S. E.; Katz, J. D.; Keen, S. P.; Mullens, P. R.; Scott, J. P.; Wallace, D. J.; Wise, C. S. *Org. Process Res. Dev.* **2010**, *14*, 849.

21. Anderson, N. G.; Ary, T. D.; Berg, J. L.; Bernot, P. J.; Chan, Y. Y.; Chen, C.-K.; Davies, M. L.; DiMarco, J. D.; Dennis, R. D.; Deshpande, R. P.; Do, H. D.; Droghini, R.; Early, W. A.; Gougoutas, J. Z.; Grosso, J. A.; Harris, J. C.; Haas, O. W.; Jass, P. A.; Kim, D. H.; Kodersha, G. A.; Kotnis, A. S.; LaJeunesse, J.; Lust, D. A.; Madding, G. D.; Modi, S. P.; Moniot, J. L.; Nguyen, A.; Palaniswamy, V.; Phillipson, D. W.; Simpson, J. H.; Thoraval, D.; Thurston, D. A.; Tse, K.; Polomski, R. E.; Wedding, D. L.; Winter, W. J. *Org. Process Res. Dev.* **1997**, *1*, 300.

22. Alcaraz, M.-L.; Atkinson, S.; Cornwall, P.; Foster, A. C.; Gill, D. M.; Humphries, L. A.; Keegan, P. S.; Kemp, R.; Merifield, E.; Nixon, R. A.; Noble, A. J.; O'Beirne, D.; Patel, Z. M.; Perkins, J.; Rowan, P.; Sadler, P.; Singleton, J. T.; Tornos, J.; Watts, A. J.; Woodland, I. A. *Org. Process Res. Dev.* **2005**, *9*, 555.

23. Ragan, J. A.; Ide, N. D.; Cai, W.; Cawley, J. J.; Colon-Cruz, R.; Kumar, R.; Peng, Z.; Vanderplas, B. C. *Org. Process Res. Dev.* **2010**, *14*, 1402.

24. Pasti, J.; Chen, C.-K.; Spangler, L.; DelMonte, A. J.; Benoit, S.; Berglund, D.; Bien, J.; Brodfuehrer, P.; Chan, Y.; Corbett, E.; Costello, C.; DeMena, P.; Discordia, R. P.; Doubleday, W.; Gao, Z.; Gingras, S.; Grosso, J.; Haas, O.; Kacsur, D.; Lai, C.; Leung, S.; Miller, M.; Muslehiddinoglu, J.; Nguyen, N.; Qiu, J.; Olzog, M.; Reiff, E.; Thoraval, D.; Totleben, M.; Vanyo, D.; Vemishetti, P.; Wasylak, J.; Wei, C.; *Org. Process Res. Dev.* **2009**, *13*, 716.

25. (a) Clay, R. J.; Collom, T. A.; Karrick, G. L.; Wemple, J. *Synthesis* **1993**, *3*, 290. (b) Rathke, M. W.; Cowan, P. J. *J. Org. Chem.* **1985**, *50*, 2622.

26. Barnes, D. M.; Christesen, A. C.; Engstrom, K. M.; Haight, A. R.; Hsu, M. C.; Lee, E. C.; Peterson, M. J.; Plata, D. J.; Raje, P. S.; Stoner, E. J.; Tedrow, J. S.; Wagaw, S. *Org. Process Res. Dev.* **2006**, *10*, 803.

27. Andersen, S. M.; Bollmark, M.; Berg, R.; Fredriksson, C.; Karlsson, S.; Liljeholm, C.; Sörensen, H. *Org. Process Res. Dev.* **2014**, *18*, 952.

28. Shieh, W.-C.; Du, Z.; Kim, H.; Liu, Y.; Prashad, *Org. Process Res. Dev.* **2014**, *18*, 1339.

29. Karlsson, S.; Brånalt, J.; Halvarsson, M. Ö.; Bergman, J. *Org. Process Res. Dev.* **2014**, *18*, 969.

30. Karlsson, S.; Gardelli, C.; Lindhagen, M.; Nikitidis, G.; Svensson, T. *Org. Process Res. Dev.* **2018**, *22*, 1174.

31. Allsop, G. L.; Cole, A. J.; Giles, M. E.; Merifield, E.; Noble, A. J.; Pritchett, M. A.; Purdie, L. A.; Singleton, J. T. *Org. Process Res. Dev.* **2009**, *13*, 751.

32. Fox, M. E.; Jackson, M.; Meek, G.; Willets, M. *Org. Process Res. Dev.* **2011**, *15*, 1163.

33. Palmer, A. M.; Webel, M.; Scheufler, C.; Haag, D.; Müller, *Org. Process Res. Dev.* **2008**, *12*, 1170.

34. Daniewski, A. R.; Liu, W.; Okabe, M. *Org. Process Res. Dev.* **2004**, *8*, 411.

35. Chen, J. P.; Isa, K. *J. Mass Spectrom. Soc. Jpn.* **1998**, *46*, 299.

36. Umemiya, S.; Nishino, K.; Sato, I.; Hayashi, Y. *Chem. Eur. J.* **2014**, *20*, 15753.

37. Sulur, M.; Sharma, P.; Ramakrishnan, R.; Naidu, R.; Merifield, E.; Gill, D. M.; Clarke, A. M.; Thomson, C.; Butters, M.; Bachu, S.; Benison, C. H.; Dokka, N.; Fong, E. R.; Hose, D. R. J.; Howell, G. P.; Mobberley, S. E.; Morton, S. C.; Mullen, A. K.; Rapai, J.; Tejas, B. *Org. Process Res. Dev.* **2012**, *16*, 1746.

38. Smith, M. J.; Lawler, M. J.; Kopp, N.; Mcleod, D. D.; Davulcu, A. H.; Lin, D.; Katipally, K.; Sfouggatakis, C. *Org. Process Res. Dev.* **2017**, *21*, 1859.

39. Li, B.; Barnhart, R. W.; Hoffman, J. E.; Nematalla, A.; Raggon, J.; Richardson, P.; Sach, N.; Weaver, J. *Org. Process Res. Dev.* **2018**, *22*, 1289.

40. (a) Rathman, T. L.; Bailey, W. F. *Org. Process Res. Dev.* **2009**, *13*, 144. (b) Hayes, J. F.; Mitchell, M. B.; Proctet, G. *Tetrahedron Lett.* **1994**, *35*, 273.

41. Tian, Q.; Hoffmann, U.; Humphries, T.; Cheng, Z.; Hidber, P.; Yajima, H.; Guillemot-Plass, M.; Li, J.; Bromberger, U.; Babu, S.; Askin, D.; Gosselin, F. *Org. Process Res. Dev.* **2015**, *19*, 416.

42. For review paper on peptide coupling reagets, see: El-Faham, A.; Albericio, F. *Chem. Rev.* **2011**, *111*, 6557.

43. Magano, J. *Org. Process Res. Dev.* **2022**, *26*, 1562.

44. Dale, D. J.; Dunn, P. J.; Golightly, C.; Hughes, M. L.; Levett, P. C.; Pearce, A. K.; Searle, P. M.; Ward, G.; Wood, A. S. *Org. Process Res. Dev.* **2000**, *4*, 17.

45. Dunn, P. J.; Newbury, T. N.; Matthews, J. M.; O'Connor, G. World Patent WO 03/080599.

46. Vaidyanathan, R.; Kalthod, V. G.; Ngo, D. P.; Manley, J. M.; Lapekas, S. P. *J. Org. Chem.* **2004**, *69*, 2565.

47. Staab, H. A.; Wendel, K. *Org. Syn. Coll. Vol.* **1973**, *5*, 201.

48. Dale, D. J.; Draper, J.; Dunn, P. J.; Hughes, M. L.; Hussain, F.; Levett, P. C.; Ward, G. B.; Wood, A. S. *Org. Process Res. Dev.* **2002**, *6*, 767.

49. Arumugasamy, J.; Arunachalam, K.; Bauer, D.; Becker, A.; Caillet, C. A.; Glynn, R.; Latham, G. M.; Lim, J.; Liu, J.; Mayes, B. A.; Moussa, A.; Rosinovsky, E.; Salanson, A. E.; Soret, A. F.; Stewart, A.; Wang, J.; Wu, X. *Org. Process Res. Dev.* **2013**, *17*, 811.

50. Levin, D. *Org. Process Res. Dev.* **1997**, *1*, 182.

51. (a) Stare, M.; Laniewski, K.; Westermark, A.; Sjögren, M.; Tian, W. *Org. Process Res. Dev.* **2009**, *13*, 857. (b) Robinson, D. I. *Org. Process Res. Dev.* **2010**, *14*, 946. (c) Bollyn, M. *Org. Process Res. Dev.* **2005**, *9*, 982. (d) Miyake, A.; Suzuki, M.; Sumino, M.; Iizuka, Y.; Ogawa, T. *Org. Process Res. Dev.* **2002**, *6*, 922. (e) Dyer, U. C.; Henderson, D. A.; Mitchell, M. B.; Tiffin, P. D. *Org. Process Res. Dev.* **2002**, *6*, 311.
52. (a) Felder, E. *InVest. Radiol.* **1984**, *19*, S164. (b) Felder, E.; Grandi, M.; Pitre, D.; Vittadini, G. *Anal. Profiles Drug Subst.* **1988**, *17*, 115.
53. (a) Haria, M.; Brogden, R. N. *CNS Drugs* **1997**, *7*, 229. (b) Haavaldsen, J.; Nordal, V.; Kelly, M. *Acta Pharm. Suec.* **1983**, *20*, 219.
54. Thesen, R. *Pharm. Ztg.* **1992**, *137*, 36.
55. Gijsen, H. J. M.; van Bakel, H. C. C. K.; Zwaan, W.; Hulshof, L. A. *Org. Process Res. Dev.* **1999**, *3*, 38.
56. (a) Shin, J. M.; Kim, Y. H. *Tetrahedron Lett.* **1986**, *27*, 1921. (b) Kim, Y. H.; Shin, J. M. *Tetrahedron Lett.* **1985**, *26*, 3821.
57. Patel, B.; Firkin, C. R.; Snape, E. W.; Jenkin, S. L.; Brown, D.; Chaffey, J. G. K.; Hopes, P. A.; Reens, C. D.; Butters, M.; Moseley, J. D. *Org. Process Res. Dev.* **2012**, *16*, 447.
58. (a) Chidambaram, R.; Zhu, J.; Penmetsa, K.; Kronenthal, D.; Kant, J. *Tetrahedron Lett.* **2000**, *41*, 6017. (b) Chidambaram, R.; Kant, J.; Zhu, J.; Lajeunesse, J.; Sirard, P.; Ermann, P.; Schierling, P.; Lee, P.; Kronenthal, D. *Org. Process Res. Dev.* **2002**, *6*, 632.
59. Wallace, D. J.; Baxter, C. A.; Brands, K. J. M.; Bremeyer, N.; Brewer, S. E.; Desmond, R.; Emerson, K. M.; Foley, J.; Fernandez, P.; Hu, W.; Keen, S. P.; Mullens, P.; Muzzio, D.; Sajonz, P.; Tan, L.; Wilson, R. D.; Zhou, G.; Zhou, G. *Org. Process Res. Dev.* **2011**, *15*, 831.
60. Yoshida, S.; Marumo, K.; Takeguchi, K.; Takahashi, T.; Mase, T. *Org. Process Res. Dev.* **2014**, *18*, 1721.
61. Zhang, M.; Imm, S.; Bähn, S.; Neubert, L.; Neumann, H.; Beller, M. *Angew. Chem. Int. Ed.* **2012**, *51*, 3905.
62. Dumas, A. M.; Molander, G. A.; Bode, J. W. *Angew. Chem. Int. Ed.* **2012**, *51*, 5683.
63. Molander, G. A.; Raushel, J.; Ellis, N. M. *J. Org. Chem.* **2010**, *75*, 4304.
64. Karlsson, S.; Gardelli, C.; Lindhagen, M.; Nikitidis, G.; Svensson, T. *Org. Process Res. Dev.* **2018**, *22*, 1174.

3 Various Reagent Surrogates

3.1 AMMONIA SURROGATES

The amino group is one of the most important functional groups and is ubiquitous in natural products, active pharmaceutical ingredients, and agrochemicals. The amino group is usually introduced via the reduction of the nitro group, which is installed by nitration. The traditional nitration/reduction method used in the preparation of aromatic amines suffers several limitations including safety issues with scale, poor functional group compatibility, and large amounts of waste. To address these issues, a transition metal-catalyzed nitration was recently developed via C–H bond activation.[1] Activated with an appropriate chelating group, the palladium-catalyzed *ortho* aromatic C–H nitration with silver nitrate gave the corresponding nitro product in good yields (Equations 3.1 and 3.2). It was demonstrated that quinoxaline and *O*-methyl oxime were two chelating groups for achieving the *ortho* C–H activation, with the latter as the removable one. The method tolerates various functional groups including bromide and iodide. The only limitation of this nitration approach is the use of expensive silver nitrate as the nitro source.

$$\text{(3.1)}$$

$$\text{(3.2)}$$

Despite recent advances in nitration strategies and the development of flow chemistry for some hazardous reactions, the development of the nitration process remains challenging due to its hazardous nature.

Apparently, ammonia would be the best reagent in amination reactions. However, ammonia (bp −33.3 °C) is a toxic gas, which is not convenient to handle. In order to address these issues, a number of ammonia surrogates are developed.

3.1.1 AMMONIUM HYDROXIDE

Reductive amination reactions can be carried out using various ammonia sources such as NH_4OAc, HCO_2NH_4, $(NH_4)_2SO_4$, $(NH_4)_2CO_3$, hydroxylamine hydrochloride ($NH_2OH \cdot HCl$), and NH_4OH, and reducing reagents including $NaBH(OAc)_3$, $NaB(CN)H_3$, $NaBH_4$, or Pd/C under H_2.

The amino group in piperidine **2** was installed via reductive amination of ketone **1** with concentrated ammonium hydroxide, followed by a catalytic imine reduction (Equation 3.3).[2]

$$\text{(3.3)}$$

DOI: 10.1201/9781003288411-3

Procedure

- Charge 35.0 g of 10% Pd/C in water (0.4 L) and 28% NH_4OH (2.8 L) to a solution of **1** (311 g, 97.2%, 1.15 mol) in 1,4-dioxane (10.5 L).
- Hydrogenate at 21–28 °C under balloon pressure of H_2 for 12 h.
- Filter and wash with a mixture of 1,4-dioxane/MeOH (1:1) (1.0 L).
- Distill the combined filtrate and washings to ca. 2.8 L.
- Add water (2.8 L).
- Distill again to ca. 3.5 L.
- Add 520 mL of 2 N HCl to pH 6.9–7.1.
- Wash the aqueous layer with nBuOAc (3×1.8 L).
- Add NaCl (899 g) and 980 mL of 2 M K_2CO_3 to pH 9.8.
- Extract the resulting aqueous layer with EtOAc (2×2.6 L).
- Distill the combined organic layers under reduced pressure.
- Dry at 50 °C for 16 h.
- Yield: 276 g (90.4%).

REMARK

Ammonium hydroxide was also able to participate in S_NAr reaction with 4-chloropyridine deriva-tive to form 4-amino-substituted pyridine product (Equation 3.4):[3]

$$(3.4)$$

3.1.2 AMMONIUM ACETATE

Compared with ammonia, weak acid ammonium salts, such as ammonium formate and ammonium acetate, are easy-to-handle and suitable ammonia surrogates.

3.1.2.1 Condensation with β-Keto Amide

Reactions of ammonium acetate with ketones proceed through an addition/elimination sequence to furnish imines or enamine after tautomerization.

Sitagliptin, a selective, potent DPP-4 inhibitor developed by Merck, is the active ingredient in JANUVIA and JANUMET (a fixed-dose combination with the anti-diabetic agent metformin), which received approval for the treatment of type 2 diabetes from the FDA in 2006. A telescoped process was developed for the synthesis of enamine intermediate **7**. The preparation of **7** was real-ized in an easily operated one-pot process (Scheme 3.1).[4] The last step in the one-pot synthesis is the amination of β-keto amide **6** with ammonium acetate as the ammonia equivalent. Ultimately, this chemistry afforded **7** in 82% overall isolated yield with 99.6 wt% purity through a simple filtration.

This highly efficient manufacturing process received the Presidential Green Chemistry Challenge Award (2006).

Procedure

Stage (a)
- Charge 2.5 kg (13.15 mol) of **3** into a 50-L three-neck flask, followed by adding 2.09 kg (14.46 mol, 1.1 equiv) of **4**, DMAP (128.5 g, 1.052 mol, 8 mol%), and MeCN (7.5 L).
- Charge 4.92 L (28.27 mol, 2.15 equiv) of DIPEA at <50 °C portionwise.

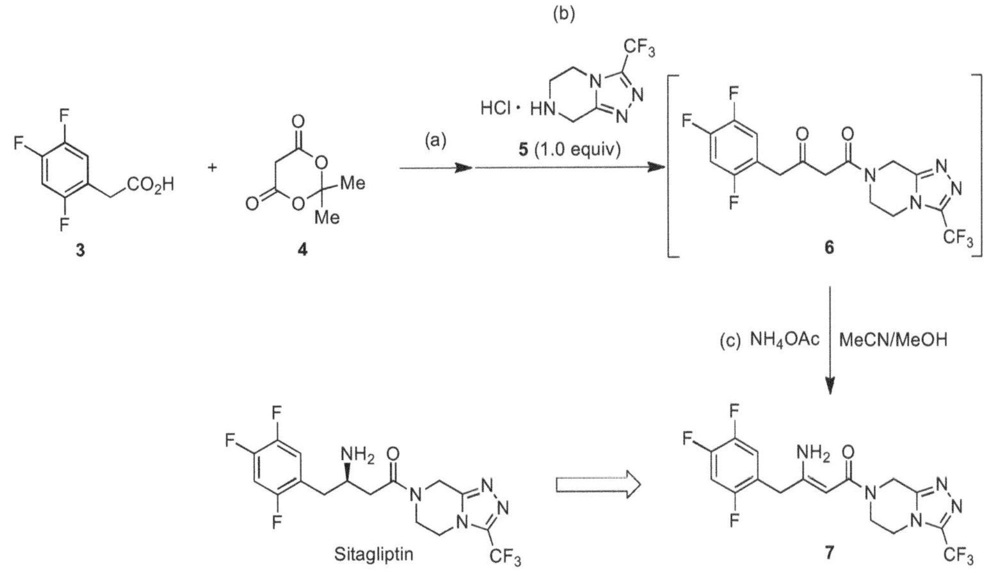

SCHEME 3.1

- Charge 1.78 L (14.46 mol, 1.1 equiv) of pivaloyl chloride at <55 °C over 1–2 h.
- Age the resulting mixture at 45–50 °C for 2–3 h.

Stage (b)
- Charge 3.01 kg (13.2 mol, 1.0 equiv) of **5** at 40–50 °C in one portion.
- Charge 303 mL (3.95 mol, 0.3 equiv) of TFA dropwise.
- Age the batch at 50–55 °C for 6 h.
- Yield (solution assay) of **6**: 4.81 kg (90%).

Stage (c)
- Charge 10% of the ketoamide **6** solution to a solution of NH$_4$OAc (0.91 kg, 11.81 mol) in MeOH (27 L) at 45 °C dropwise over 30 min.
- Stir the resulting mixture at 45 °C for 1.5 h.
- Add 140 g of seeds.
- After 30 min at 45 °C, charge the remaining **6** solution dropwise over 3–6 h.
- Age the batch for 3 h.
- Add MeOH (12 L) at 40–45 °C over 2 h.
- Cool the slurry to 0–5 °C over 3–4 h and age for 1 h.
- Filter and rinse with cold (0 °C) MeOH (29 L).
- Dry at ambient temperature under vacuum.
- Yield: 4.37 kg (82%).

3.1.2.2 Condensation with Cyclohexanones

A simple, streamlined process for the synthesis of anilines and indoles was developed as shown in Scheme 3.2.[5]

The reaction was carried out by mixing cyclohexanone and ammonium acetate in the presence of Pd/C catalyst, ethylene, and potassium carbonate in acetonitrile. The initial condensation of cyclohexanones with in situ formed ammonia generates imine intermediates, which undergo double palladium-catalyzed hydrogen transfer reactions to furnish the aniline products. The reaction

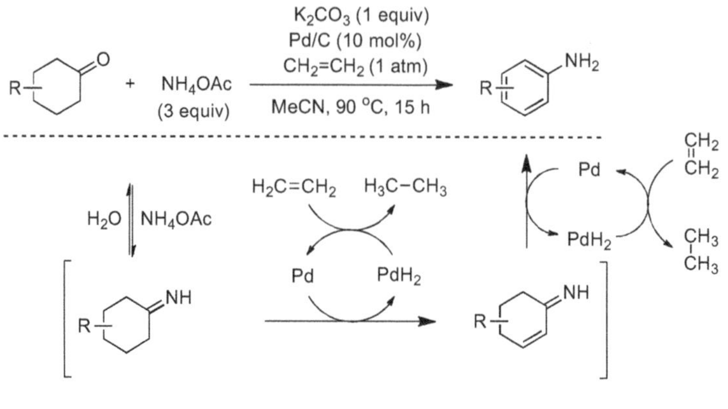

SCHEME 3.2

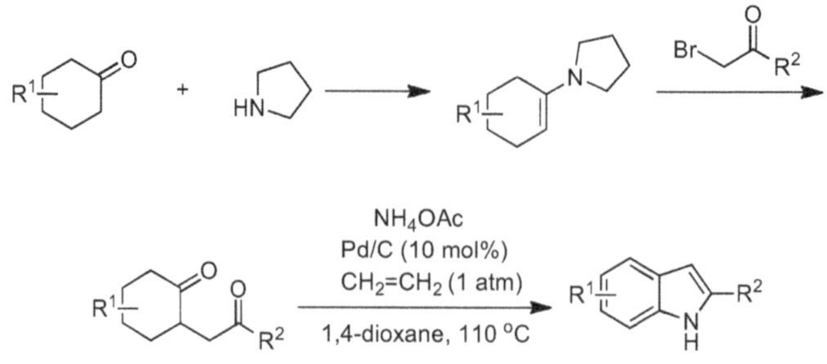

SCHEME 3.3

tolerates a wide range of functional groups in the starting cyclohexanones with various substitution patterns including not only *ortho-*, *para-*substituted substrates but also *meta-* and 2,6-disubstituted cyclohexanones.

Reaction of 5-methyl-1,3-cyclohexanedione in xylene under similar conditions provided the corresponding 3-amino-5-methylphenol in 65% yield (Equation 3.5). Scheme 3.3 describes the application of this protocol toward the synthesis of indole derivatives.

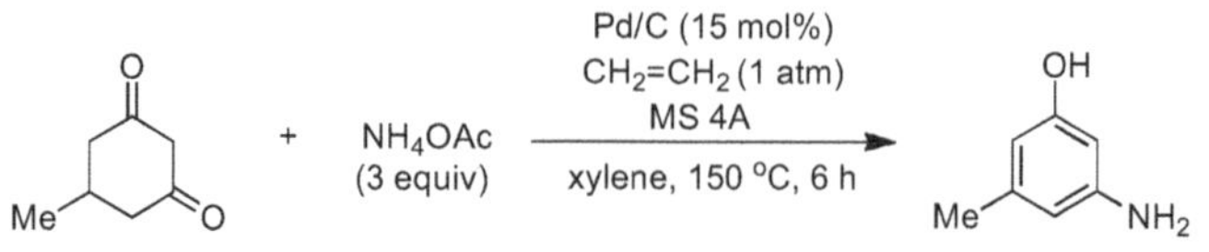

$$(3.5)$$

Additionally, the palladium catalyst could be recycled and reused in the hydrogen transfer process. Safety should be addressed especially on large-scale preparation due to the presence of two flammable gases (ethylene and ethane) in the reaction system.

3.1.2.3 Consecutive Reductive Amination Reactions

The synthesis of benzyl-protected 1-deoxynojirimycin **9** was realized using ammonium acetate as the ammonia source via an intermolecular reductive amination followed by an intramolecular reductive amination (Equation 3.6).[6]

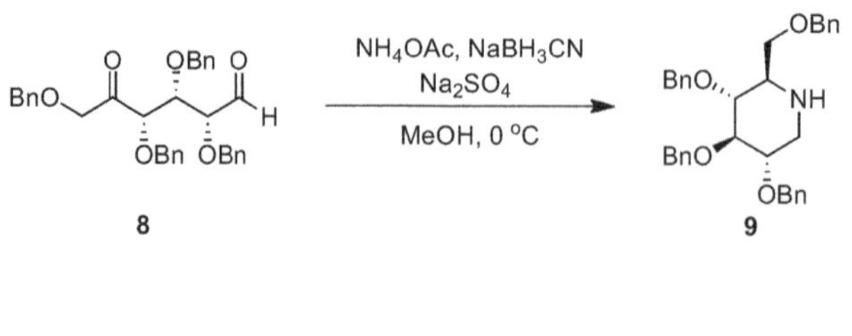

$$(3.6)$$

NOTE

The excess of ammonium acetate was able to reduce to 10 equiv from 20 equiv when using ammonium formate.

3.1.3 AMMONIUM CHLORIDE

Ammonium chloride was utilized as the ammonia source for the synthesis of 4-phenylquinazoline (Equation 3.7).[7] This multi-component reaction was catalyzed by iodine in the presence of tetramethylethylenediamine (TMEDA) under an oxygen atmosphere in dimethyl sulfoxide (DMSO) at 120 °C, furnishing the desired product in excellent yield. TMEDA serves as a base to release ammonia from ammonium chloride and quench the HI byproduct, and also as one carbon source to form the 4-phenylquinozoline product (Scheme 3.4). Mechanistically, this process involves multiple intermediates: an iminium salt, formimidamides, dihydroquinazolines, etc.

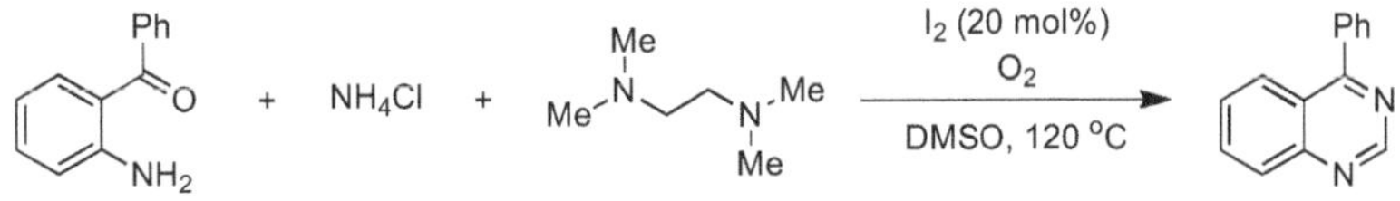

(3.7)

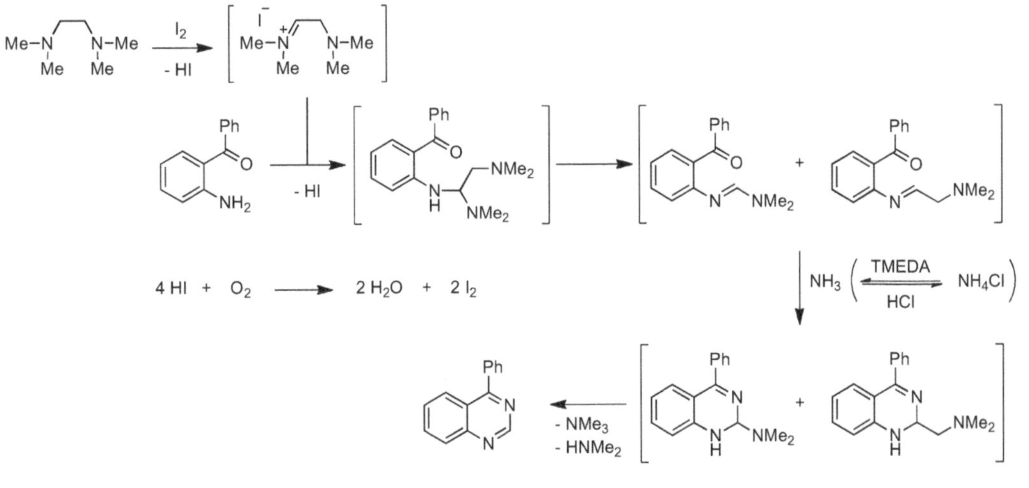

SCHEME 3.4

3.1.4 HYDROXYLAMINE HYDROCHLORIDE

Condensation of hydroxylamine hydrochloride with aldehydes or ketones generates the corresponding oxime intermediates. Two approaches were developed for the conversion of oxime intermediates to amines: reduction and aromatization.

3.1.4.1 Reductive Amination

Scheme 3.5[8] displays an application of hydroxylamine hydrochloride as the ammonia source toward the preparation of cyclohex-3-en-1-ylmethanamine. This reductive amination approach occurred through condensation of the aldehyde with hydroxylamine to give oxime intermediate, followed by LiAlH$_4$ reduction.

3.1.4.2 Aromatization

An efficient transformation of hydroxylamine to the amino group was demonstrated during the conversion of dihydrothiophen-3(2H)-one **10** to aminothiophene **11**, in which the oxime

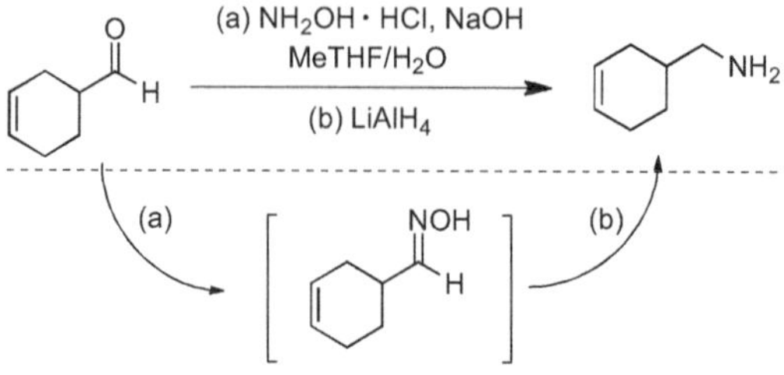

SCHEME 3.5

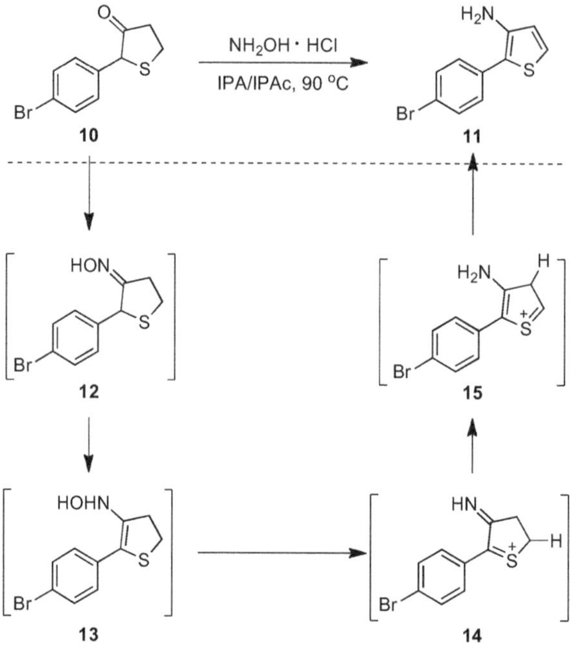

SCHEME 3.6

intermediate **12** was converted to the corresponding amino group via aromatization process (Scheme 3.6).[9]

3.1.5 O-BENZYLHYDROXYLAMINE

In the first generation for the synthesis of sitagliptin, the O-benzylhydroxylamine was employed as an ammonia surrogate and the amino group was released in the penultimate step (Scheme 3.7).[4b]

3.1.6 HYDROXYLAMINE-O-SULFONIC ACID

Hydroxylamine-O-sulfonic acid (NH_2OSO_3H, HSA) is a commercially available and inexpensive nitrogen source. It is a white, water-soluble, and hygroscopic solid (MP 210 °C). HSA can be used for various chemical transformations, for example, for conversions of aldehydes to nitriles[10] and alicyclic ketones to lactams.[11]

3.1.6.1 S_N2 Reaction with Sulfinate

HSA was employed as the ammonia surrogate to access sulfonamides **18**. The synthesis involves an S_N2 reaction of sodium sulfinate **17** with hydroxylamine sulfonate in water, affording the desired

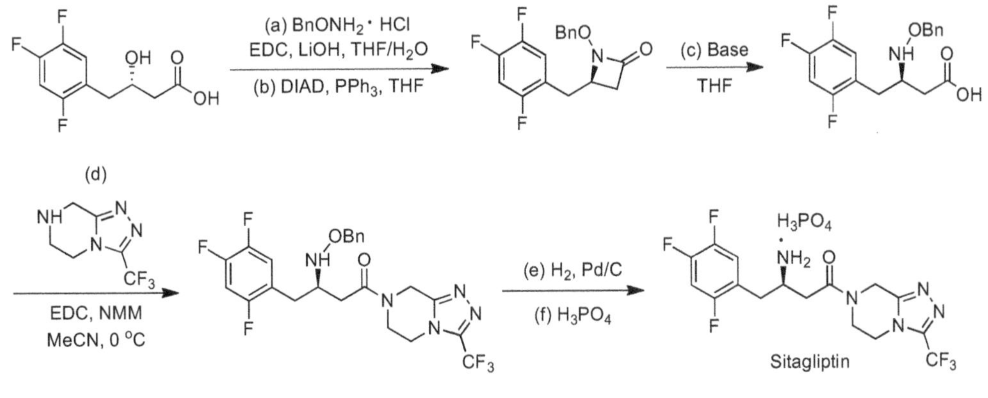

SCHEME 3.7

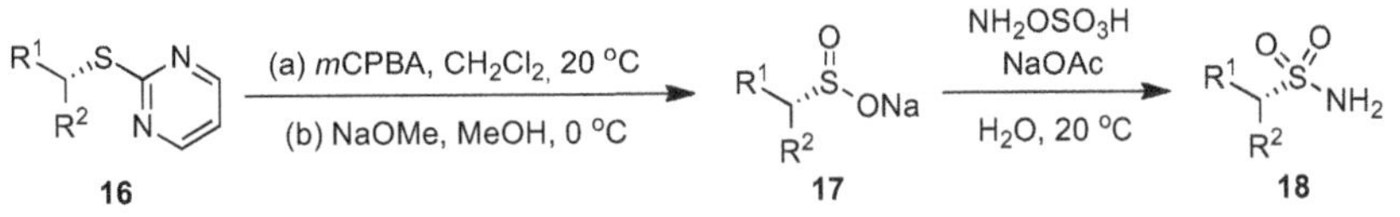

SCHEME 3.8

sulfonamides **18** (Scheme 3.8),[12] which crystallized directly from the aqueous reaction mixture. **17** was obtained from pyrimidinyl sulfides **16** via oxidation and deprotection.

Using this protocol, chiral pyrimidinyl sulfides could be converted into optically active sulfonamides without loss of enantiopurity.

3.1.6.2 Reaction with Boronic Acid

Analogously, HSA was utilized as an ammonia equivalent to react with arylboronic acids, furnishing the corresponding anilines in good yields (Equation 3.8).[13]

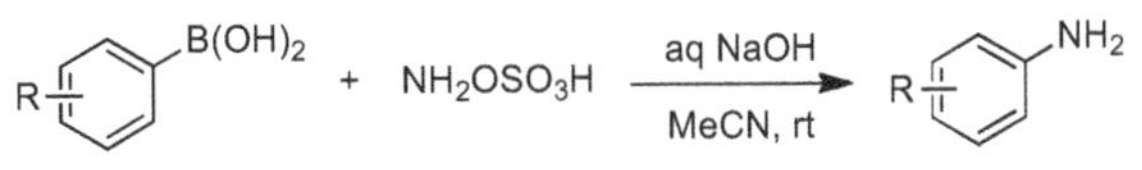

$$(3.8)$$

With a base-mediated activation of HSA, the reaction was carried out under mild, metal-free conditions.

3.1.7 HEXAMETHYLENE TETRAMINE

As an inexpensive nitrogen source, hexamethylene tetramine (HMTA) was utilized to convert allylic bromide **19** to the corresponding ammonium salt **20**. Upon acid hydrolysis, the desired amine **21** was isolated as the HCl salt (Scheme 3.9).[14]

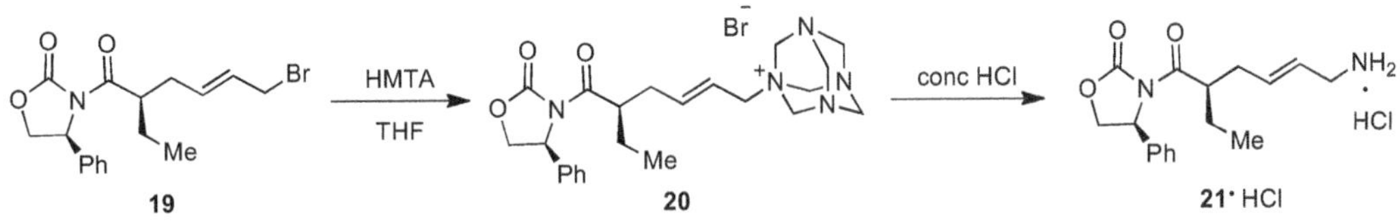

SCHEME 3.9

3.1.8 ACETONITRILE

The amino group can also be installed using acetonitrile as the ammonia surrogate via a Ritter reaction under acidic conditions. *Cis*-Aminoindanol **24**, as shown in Scheme 3.10, is the key intermediate for the synthesis of HIV protease inhibitor indinavir (Crixivan®)[15] and the development of a practical synthetic route to **24** was the focus of intensive research efforts.[16] The use of a modified Ritter reaction to convert indene oxide **22** to the corresponding *cis*-amino alcohol constitutes the most direct and economical route.

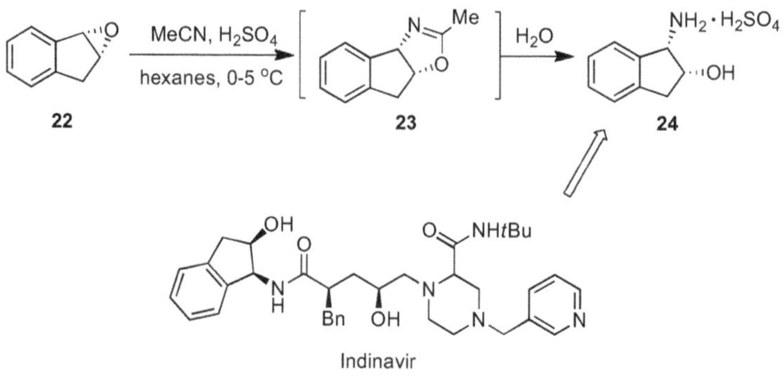

SCHEME 3.10

Procedure

- Charge 100 mL of dry MeCN into a 1-L three-necked flask.
- Cool to –5 °C.
- Charge slowly 10% of 20 mL (0.4 mol, 2 equiv) of fuming sulfuric acid, followed by dropwise addition of a solution of **22** (26.0 g, 0.197 mol) in dry hexanes (200 mL) and the rest of the acid simultaneously at 0–5 °C.
- Warm the batch to room temperature after complete addition.
- Stir for 1 h.
- Add water (100 mL).
- Separate layers.
- Add 100 mL of water to the aqueous layer.
- Distill at atmospheric pressure until a head temperature of 100 °C.
- Heat at reflux for 3 h.
- Cool to rt (the resulting crude aqueous solution of **24** is used in the subsequent step).

3.1.9 CHLOROACETONITRILE

The use of chloroacetonitrile, as an ammonia surrogate in a Ritter reaction, allowed the conversion of tertiary alcohol **25** to chloroacetamide **26** (Scheme 3.11).[17] The subsequent treatment of **26** with thiourea furnished the desired amine **27**.

3.1.10 *TERT*-BUTYL CARBAMATE

The synthesis of organic amines is largely limited to nucleophilic substitution and reductive amination. Most of those methods suffer from limited substrate scope and poor functional group tolerance. For instance, the Ullmann coupling reaction usually requires harsh reaction conditions that limit its application at large scales. In 1995, Buchwald and Hartwig concurrently published[18] results

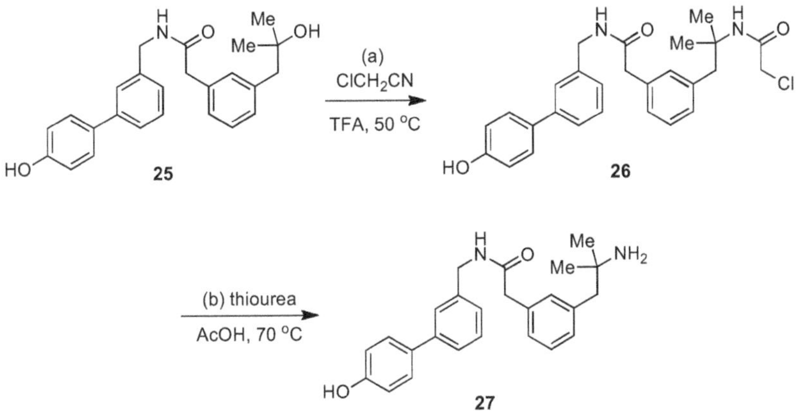

SCHEME 3.11

on aminations of aryl halides. The Buchwald–Hartwig amination reactions provide a means to construct aryl amines from aryl halides and amines in the presence of a palladium catalyst and a base. Despite progress made thus far, cross-coupling with ammonia directly remains challenging due to its tight binding with transition metal catalysts. The use of ammonia surrogates can address these issues.

tert-Butyl carbamate, as the ammonia surrogate, participated in palladium-catalyzed amination with 4-iodo-2-(trifluoromethyl)nicotinic acid, followed by deprotection of the Boc group, giving 4-amino-2-(trifluoromethyl)nicotinic acid as its hydrochloride salt (Equation 3.9).[19]

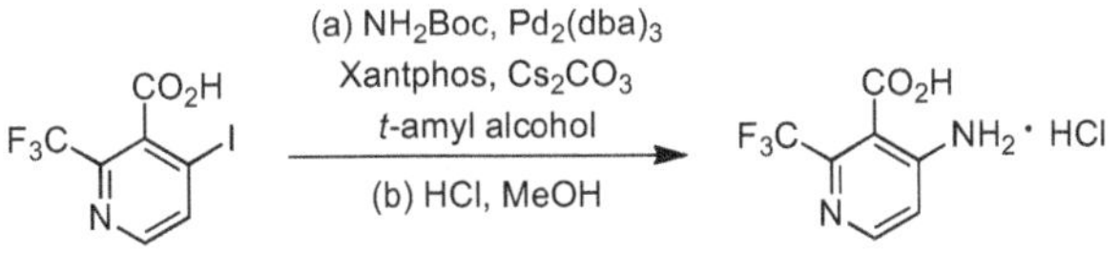

$$(3.9)$$

3.1.11 DIPHENYLMETHANIMINE

Commercially available diphenylmethanimine can serve as a convenient ammonia equivalent in palladium-catalyzed amination (Scheme 3.12).[20] This Pd-catalyzed coupling reaction of aromatic substrates with benzophenone imines proved to be efficient. The diphenyl ketimine products showed enhanced crystallinity which allowed for facile purification by crystallization.

Three methods were developed for the cleavage of diphenyl ketimines to the anilines.

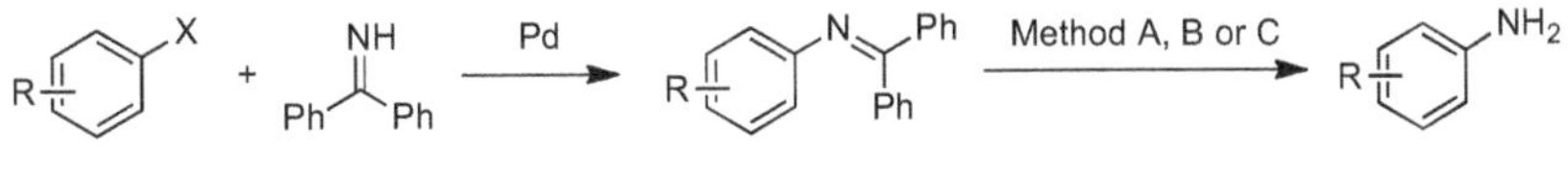

SCHEME 3.12

Procedure

Method A (acidic hydrolysis)
- Charge 2.0 M HCl to a solution of the imine adduct in THF.
- Stir the resulting mixture until hydrolysis is complete (5–20 min).
- Partition the reaction mixture between 0.5 M HCl and hexane/EtOAc (2:1).
- Basify the aqueous layer.

- Extract with CH_2Cl_2 and dry over Na_2SO_4.
- Concentrate under reduced pressure to give the product.

Method B (hydrogenolysis)
- Heat the mixture of the imine adduct, HCO_2NH_4 (15 equiv), and 5% Pd/C (10 mol%) in MeOH at 60 °C for 2 h.
- Cool to room temperature.
- Add CH_2Cl_2 (5 volumes).
- Filter, wash the filtrate with 0.1 M NaOH, and dry over Na_2SO_4.
- Concentrate under reduced pressure.
- Purify by chromatography.

Method C (transamination with hydroxylamine)
- Charge NaOAc (2.4 equiv) and hydroxylamine hydrochloride (1.8 equiv) to a solution of the imine adduct in MeOH.
- Stir for 15–30 min until the oxime formation complete.
- Partition the solution between 0.1 M NaOH and CH_2Cl_2.
- Separate layers.
- Dry the organic layer over Na_2SO_4.
- Concentrate under reduced pressure.
- Purify by chromatography.

Benzophenone imine was employed as the ammonia equivalent for the palladium-catalyzed amination reaction of triflate **28** to furnish aniline **29** after acidic hydrolysis (Equation 3.10)[21] in 66% overall yield.

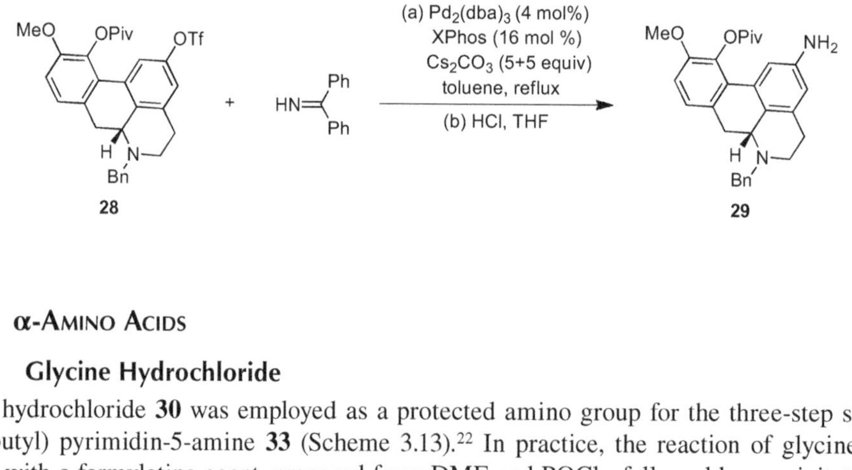

$$(3.10)$$

3.1.12   α-AMINO ACIDS

3.1.12.1   Glycine Hydrochloride

Glycine hydrochloride **30** was employed as a protected amino group for the three-step synthesis of 2-(t-butyl) pyrimidin-5-amine **33** (Scheme 3.13).[22] In practice, the reaction of glycine hydrochloride with a formylating agent, prepared from DMF and $POCl_3$, followed by precipitation with HPF_6, afforded vinamidinium hexafluorophosphate salt **31** in high yield after treatment of the crude

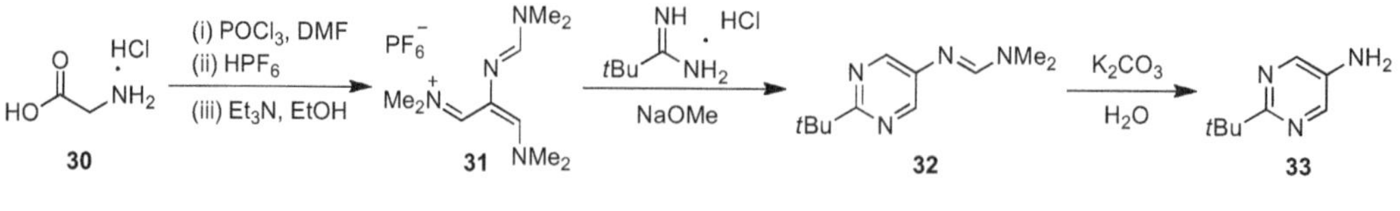

SCHEME 3.13

product with Et_3N in ethanol. Pyrimidine **32** was obtained from a base-catalyzed condensation of **31** with *tert*-butylcarbamidine hydrochloride. Finally, the hydrolysis of **32** afforded the desired aminopyrimidine **33** in good yield.

Procedure

Stage (a)_(i)
- Charge 6.18 kg (40.3 mol) of $POCl_3$ to DMF (8.1 L) at 0–5 °C followed by adding 1.5 kg (13.5 mol) of **30** at the same temperature in three portions within 10 min.
- Heat to 125 °C within 6 h and hold for 2.5 h.
- Cool to 30 °C and stir overnight.
- Add water (6.75 L) slowly (exotherm).
- Transfer the mixture to water (8.5 L) and rinse with DMF (2 L).
- Cool to 0–5 °C.

Stage (a)_(ii)
- Charge 10.25 kg (64% in water) of hexafluorophosphoric acid within 1 h.
- Stir the resulting suspension at 0 °C for 1.75 h.
- Filter and rinse with EtOH (1×7 L; 5×2 L).
- Dry at 40 °C under reduced pressure.
- Yield: 8.48 kg (of a mixture of mono- and dihexafluorophosphates).

Stage (a)_(iii)
- Charge 2.12 L of Et_3N to a suspension of the mixture obtained in Stage (a)_(ii) (6 kg) in EtOH (31.8 L).
- Heat to 80 °C within 45 min and hold at 80 °C for 4 h.
- Add EtOH (5 L).
- Cool to 0 °C and stir overnight.
- Filter and rinse with EtOH (2×2 L).
- Dry at 45 °C under reduced pressure to give 4.49 kg of crude **31**.
- Dissolve 4.40 kg of crude **31** in EtOH (66.2 L) under reflux.
- Filter and rinse with warm EtOH (13 L).
- Cool the filtrate to 0 °C within 1.5 h.
- Stir the resulting suspension at 0 °C overnight.
- Filter and rinse with EtOH (4 L).
- Dry at 40 °C under reduced pressure.
- Yield of **31**: 2.99 kg (93%).

Stage (b)
- Charge 690 g (5.05 mol) of *t*-butylcarbamidine hydrochloride to a suspension of **31** (2.26 kg, 6.6 mol) in MeOH (25 L) at room temperature within 30 min, followed by adding a solution of NaOMe (2.28 kg, 30% w/w, 12.7 mol).
- Heat the batch at reflux for 3 h.
- Cool 45 °C.
- Add water (185 mL).
- Distill at 45 °C under reduced pressure to remove MeOH.
- Add 15% NaCl solution (15 kg) to the residue.
- Extract with MTBE (3×10 kg).
- Wash the combined MTBE layers with 15% NaCl (5 kg).
- Dry the organic layer over $MgSO_4$.
- Filter and distill to dryness at 45 °C under reduced pressure.
- Yield of **32**: 584 g (56%).

Stage (c)

- Heat a mixture of **32** (574 g, 2.78 mol) and 5% aqueous solution of K_2CO_3 (5.8 kg) at 91 °C for 24 h.
- Cool to room temperature.
- Extract with isopropyl acetate (1×10 L, 2×5 L).
- Wash the combined organic layers with brine (5 L) and dry over $MgSO_4$.
- Distill to dryness.
- Yield of **33**: 377 g (89.7%).

REMARKS

(a) HPF_6 is highly corrosive and the use of such a reagent should be with prudence.

(b) Due to the limited availability of *t*-butylcarbamidine hydrochloride and the use of highly corrosive HPF_6, the Buchwald amination approach was employed for the synthesis of **33** using the commercially available 2-bromo-5-*tert*-butylpyrimidine as the starting material and benzophenone imine as the ammonia surrogate (Scheme 3.14).[22] The resulting cross-coupling intermediate was not isolated and hydrolyzed with aqueous HCl readily to the desired product **33**.

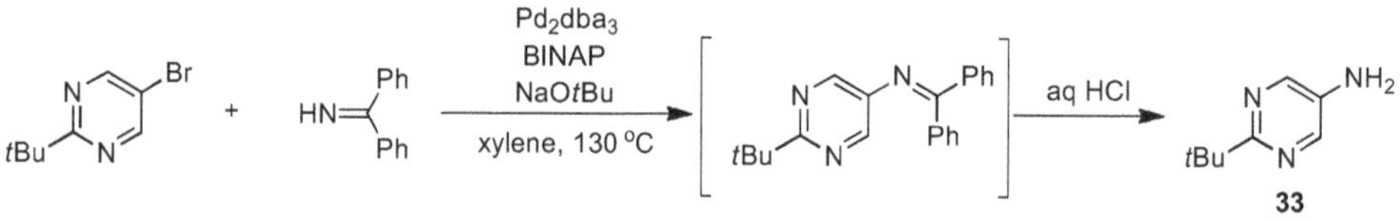

SCHEME 3.14

3.1.12.2   2,2-Diphenylglycine

2,2-Diphenylglycine was utilized as the ammonia surrogate in an asymmetric transamination with α-keto acids in the presence of a chiral pyridoxamine catalyst, giving various α-amino acids (Equation 3.11).[23]

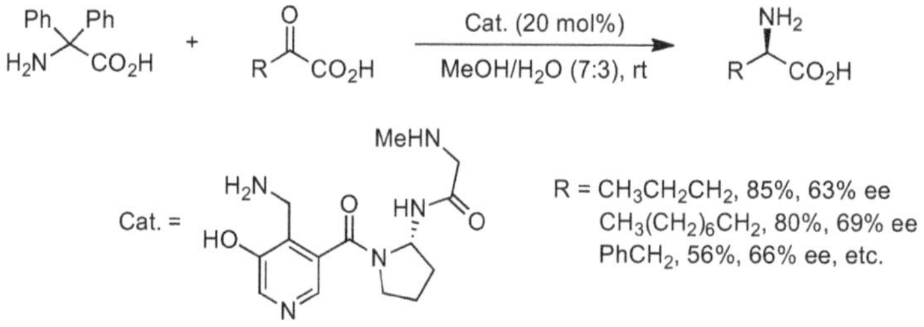

$$(3.11)$$

R = $CH_3CH_2CH_2$, 85%, 63% ee
$CH_3(CH_2)_6CH_2$, 80%, 69% ee
$PhCH_2$, 56%, 66% ee, etc.

In this process (Scheme 3.15), the chiral pyridoxamine condenses with α-keto acids to form corresponding Schiff bases which undergo an asymmetric 1,3-proton transfer followed by hydrolysis to give the amino acid products along with chiral pyridoxal. The chiral pyridoxal would react with 2,2-diphenylglycine, then water to regenerate the pyridoxamine catalyst along with CO_2 and benzophenone byproducts.

3.1.13   Silylated Amines

Lithium bis(trimethylsilyl)amide (LiHMDS) is commercially available as solutions in either hydrocarbon or ether solvents or as solids. Besides being primarily used as a strong non-nucleophilic base; LiHMDS is also utilized as an ammonia surrogate for palladium-catalyzed amination reactions to convert aryl halides to anilines (Equation 3.12).[24] After the complete reaction, a simple

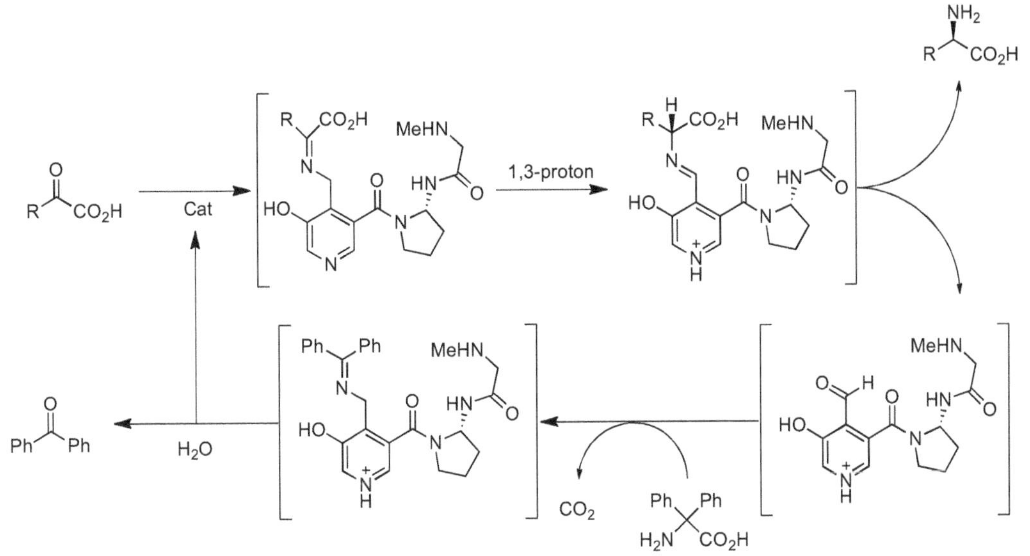

SCHEME 3.15

aqueous workup under acidic conditions or with tetrabutylammonium fluoride (TBAF) removes the protecting group to afford the corresponding aniline products.

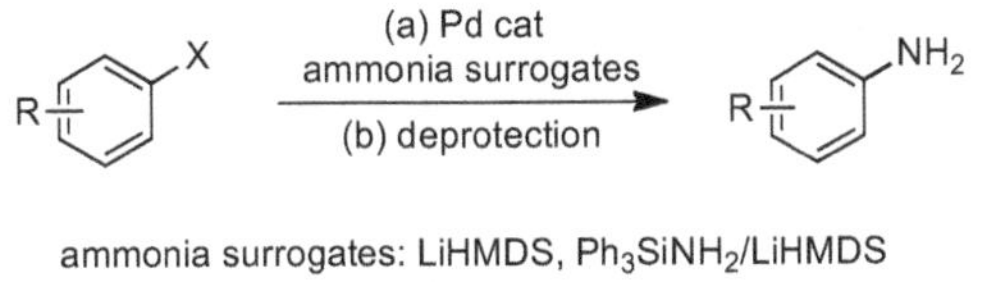

$$\text{(3.12)}$$

ammonia surrogates: LiHMDS, Ph_3SiNH_2/LiHMDS

Palladium-catalyzed amination reactions of aryl bromide **34** (Equation 3.13) and heteroaryl chloride **36** (Equation 3.14) with LiHMDS as the ammonia source furnished, upon deprotection, the corresponding amine products **35** and **37** in yields of 80% and 86%, respectively.[25]

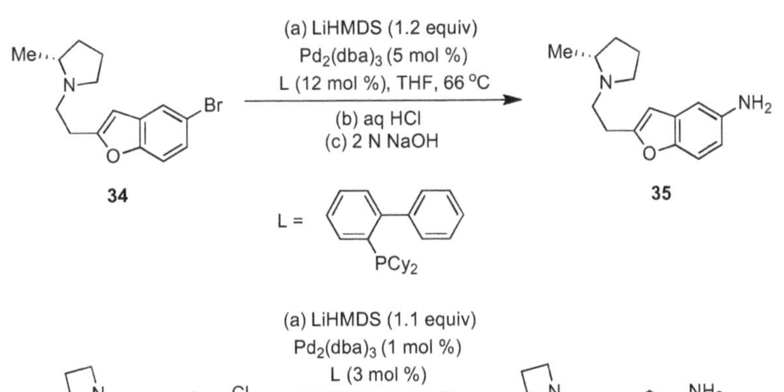

$$\text{(3.13)}$$

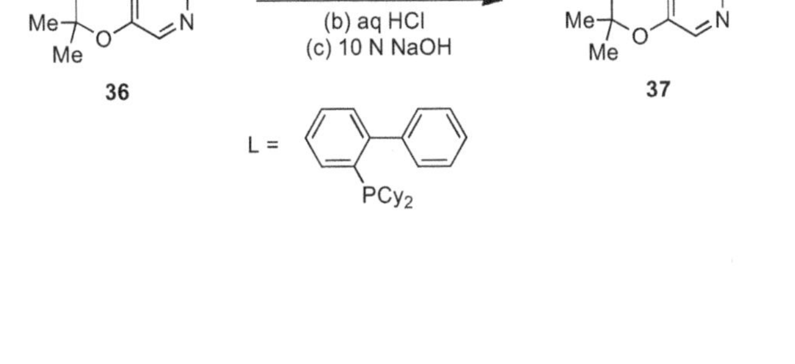

$$\text{(3.14)}$$

Procedure (for the preparation of **37**)[25b]

Stage (a)
- Charge 21.4 g (23.3 mmol, 1 mol%) of Pd$_2$(dba)$_3$ to a degassed solution of **36** (607 g, 2.33 mol) in toluene (2.43 L), followed by adding 24.5 g (70.0 mmol, 3 mol%) of 2-(dicyclohexyphosphino)biphenyl.
- Heat the resulting mixture to 45 °C.
- Charge 0.65 L of 1 M solution (0.65 mol) of LiHMDS over 5 min.
- Heat the batch to 62 °C.
- Charge 1.92 L of 1 M solution (1.92 mol) of LiHMDS over 90 min.
- Heat the batch at 63 °C overnight.
- Cool to 30 °C.
- Add ice/water (1.4 L) and separate layers.
- Filter the organic layer through Celite and wash with THF.
- Concentrate the combined filtrates.
- Dilute the resulting dark oil with MTBE (1.8 L).

Stage (b)
- Add slowly 3 N HCl (1.6 L) at <34 °C.
- Separate layers.
- Extract the organic layer with 1 N HCl (2.2 L).
- Add MTBE (3.6 L) to the combined aqueous layers.

Stage (c)
- Add 0.82 L of 10 N NaOH to the bi-layer mixture to pH 11.5 with cooling.
- Add NaCl (1.5 kg), MTBE (1.0 L), and MeTHF (1.0 L).
- Separate layers.
- Extract the aqueous layer with MeTHF (3.0 L).
- Distill the combined organic layers under reduced pressure to half volume.
- Co-evaporate with MeTHF (2×1.0 L) to ca. 4 L.
- Filter the resulting solution through Celite and wash with MeTHF (1.0 L).
- Distill under reduced pressure.
- Dilute the residue with cyclohexane (2.0 L) followed by adding seed crystals.
- Distill the resulting slurry until near dryness.
- Add cyclohexane (0.6 L) and heat to 45 °C.
- Add heptane (2.4 L).
- Cool to room temperature.
- Filter and rinse with 50% cyclohexane/heptane (1.0 L) and heptane (2×0.5 L).
- Dry at 40 °C under vacuum.
- Yield of **37**: 443 g (85.8%) as a dark-brown solid.

3.1.14 ALLYLAMINES

Both allylamine and diallylamine were employed as the ammonia source in the palladium-catalyzed amination reactions with aryl halides (Equation 3.15).[26]

$$Ar-X \;+\; HN \xrightarrow[\substack{(b)\ 10\ \%\ Pd/C\ (w/w)\\ MeSO_3H\ (1.0\ equiv)\\ EtOH,\ reflux}]{(a)\ Pd} Ar-NH_2 \qquad (3.15)$$

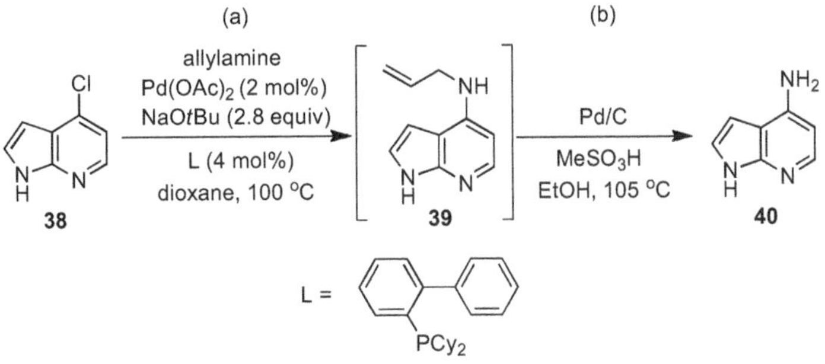

SCHEME 3.16

The deallylation of the amination intermediates was typically effected via noble metal-catalyzed isomerization of the allyl groups to the hydrolytically labile enamines followed by acid-promoted hydrolysis of the enamines. The two-step deallylation was accomplished in one-pot in good to excellent yields with catalytic amounts of 10% palladium on carbon in boiling ethanol containing an equivalent of methanesulfonic acid.

Using allylamine as the ammonia source in Pd-catalyzed amination of 4-chloro-1H-pyrrolo[2,3-b]pyridine **38** gave allylamine **39** (Scheme 3.16).[27] Subsequent deallylation using palladium on carbon in an acidic alcohol solution provided amine **40** in 75% yield over two steps.

Procedure

Stage (a)
- Charge 20 g (131 mmol) of **38** into a 350-mL flask followed by adding 35.2 g (367 mmol) of NaO*t*Bu, 589 mg of Pd(OAc)$_2$ (2.62 mmol), and 1.83 g (5.24 mmol) of (*o*-biphenyl)PCy$_2$.
- Inert the flask by evacuating and backfilling with Argon.
- Charge 1,4-dioxane (250 mL) and 29 mL (393 mmol) of allylamine.
- Inert the contents by bubbling Argon through the mixture for 20 min.
- Heat at 100 °C for 16 h.
- Cool the batch to room temperature.
- Add Et$_2$O (500 mL).
- Filter and distill the filtrate under reduced pressure.
- Dissolve the resulting oil in CH$_2$Cl$_2$ (250 mL).
- Wash with water and dry over Na$_2$SO$_4$.
- Distill under reduced pressure to give **39** as a brown gum.

Stage (b)
- Charge 22.69 g (131 mmol) of **39** into a 500-mL flask followed by adding EtOH (262 mL), 15 g of 10% Pd/C, and 8.5 mL (131 mmol) of MsOH.
- Heat the batch at 105 °C for 72 h.
- Cool to room temperature.
- Filter and distill under reduced pressure.
- Purify with column chromatography.
- Yield of **40**: 13.15 g (75% over two steps).

3.1.15 ISOAMYL NITRITE

Electrophilic isoamyl nitrite can serve as a nitrogen source in the amination process via condensation/reduction sequence. As demonstrated in Scheme 3.17, the installation of an amino group into the

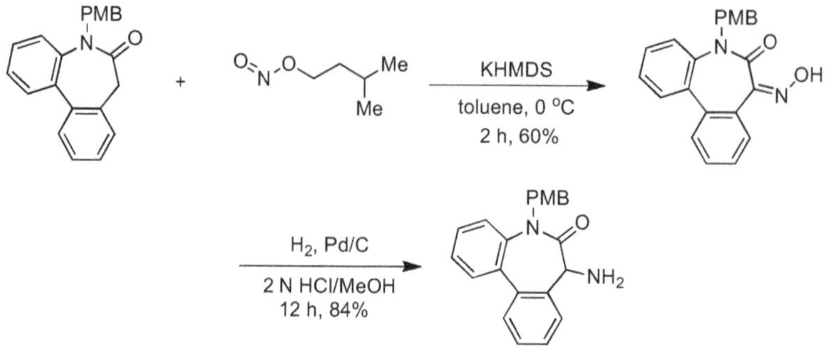

SCHEME 3.17

α-position in a lactam was achieved by a reaction of the lactam with an isoamyl nitrite under basic conditions. This process involves the formation of *N*-hydroxyl imine followed by hydrogenation.[28]

3.1.16 1,2-Benzisoxazole

The Buchwald laboratory discovered that 1,2-benzisoxazole **42** can be used as an electrophilic ammonia equivalent in CuH-catalyzed hydroamination of alkenes.[29] Under mild reaction conditions, a wide range of chiral α-branched primary amines were obtained from aryl alkenes or alkynes. For instance, hydroamination of 2-(methylthio)-5-vinylpyrimidine **41** with **42** as ammonia surrogate furnished the desired (*R*)-1-(2-methylthiopyrimidin-5-yl)ethanamine **45** through a chiral Schiff base intermediate **44** in a two-step process. The chiral Schiff base **44** was formed via a substitution reaction of **42** with the Cu-alkyl intermediate **43** (Scheme 3.18).

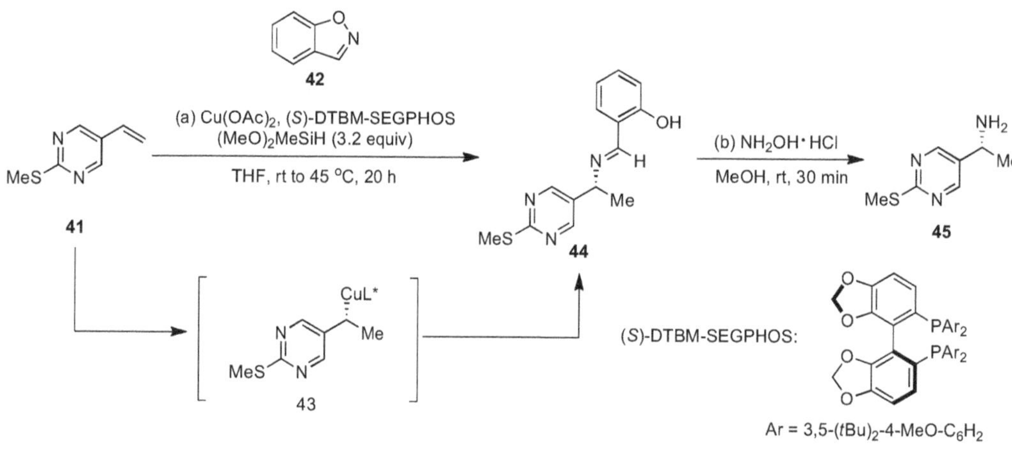

SCHEME 3.18

Hydroamination of alkyl-substituted terminal alkenes or alkynes with **42** afforded linear primary amines in good to excellent yields with high levels of enantioselectivity. Equation 3.16 shows the conversion of alkene **46** to its corresponding liner primary amine **47** in 79% overall yield.[29]

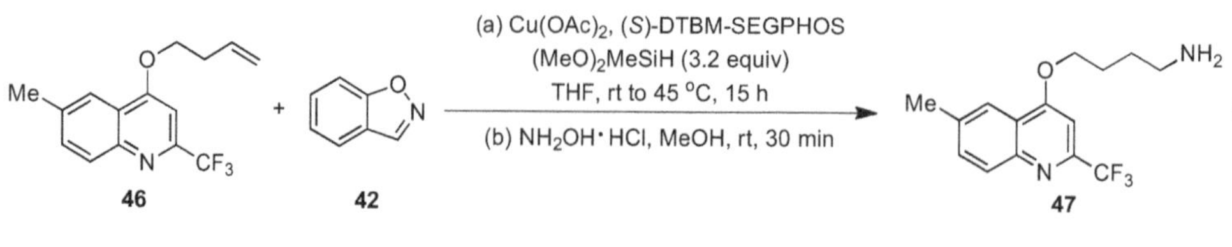

(3.16)

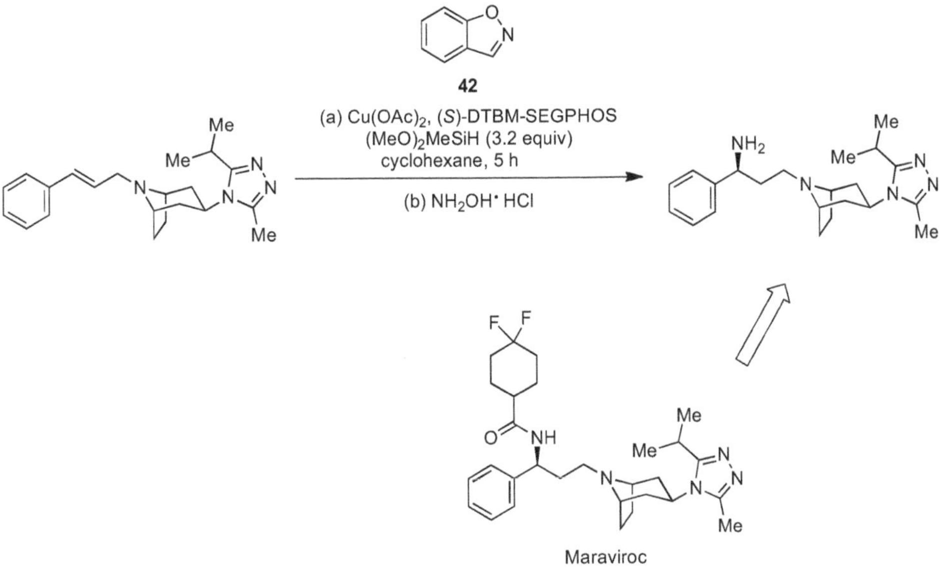

SCHEME 3.19

An important advantage of this mild enantioselective hydroamination process using 1,2-benzisoxazole as an ammonia surrogate lies in its ability to install a chiral primary amine unit onto highly functionalized late-stage synthetic intermediates. An application of this approach toward the synthesis of the anti-retroviral drug maraviroc, which is used for the treatment of HIV infections, is illustrated in Scheme 3.19.[29]

3.2 CARBON MONOXIDE SURROGATES

Owing to its toxicity and gaseous nature, carbon monoxide is not a desired reagent for organic transformations. Consequently, CO-free carbonylation has attracted great attention and some progress has been made over the last three decades. Various compounds, such as formic acid derivatives and metal carbonyl compounds have already been developed as alternatives to toxic CO gas.

3.2.1 N-FORMYLSACCHARIN

Aromatic aldehydes can participate in various organic transformations including oxidation, reduction, condensation, and the Wittig reaction. Three general methods have been used to prepare aldehydes: oxidation of alcohols, reduction of carboxylic acid derivatives, and direct formylation. Reduction of acids usually requires activation of the acids prior to reduction, while oxidation of alcohols is often plagued by over-oxidation problems. The direct formylation (halogen–metal exchange/quenching with DMF) suffers poor functional group tolerance. To overcome these limitations, a report[30] described a palladium-catalyzed reductive carbonylation of aryl halides using N-formylsaccharin as the CO source (Equation 3.17).

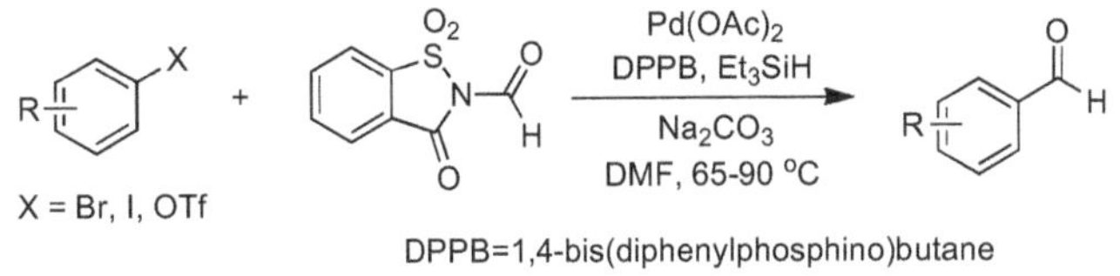

$$ (3.17) $$

X = Br, I, OTf

DPPB=1,4-bis(diphenylphosphino)butane

It was confirmed that *N*-formylsaccharin decarbonylates rapidly to generate CO and saccharin at room temperature (Equation 3.18).[30]

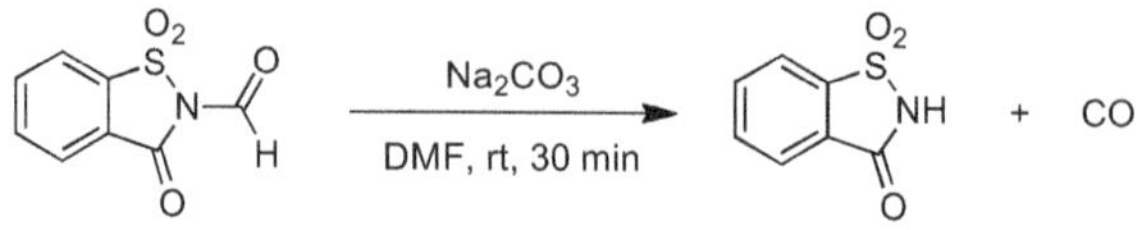

$$\qquad (3.18)$$

Compared to other methods, this reaction features mild reaction conditions and tolerance of a wide range of functional groups.

3.2.2 PARAFORMALDEHYDE

Paraformaldehyde is a readily available and inexpensive formylating reagent. A recent study found that paraformaldehyde can generate CO under palladium-catalyzed conditions (Equation 3.19).[31]

$$(CH_2O)_n \xrightarrow[\text{DMF, 100 °C}]{\substack{Pd(CH_3CN)_2Cl_2 \\ (4.0\ mol\%) \\ DPPB\ (8.0\ mol\%) \\ Na_2CO_3\ (2.0\ equiv)}} CO \qquad (3.19)$$

Thus, paraformaldehyde was utilized as the CO source in palladium-catalyzed carbonylation of aryl bromides to access aryl aldehydes (Scheme 3.20).[31]

Compared to conventional carbonylations using syngas or CO, this approach offers a convenient alternative to access aryl aldehydes. Using this method, various aryl bromides with different substitution patterns were transformed into their corresponding aldehydes in moderate to good yields. This method tolerates electron-neutral, electron-rich, or electron-poor functional groups. Reactions of heterocycles containing sulfur and nitrogen atoms led to moderate yields of the corresponding products.

In addition, using paraformaldehyde as the CO surrogate a palladium-catalyzed reductive alkoxycarbonylation of aryl bromides was also achieved (Scheme 3.21).[31]

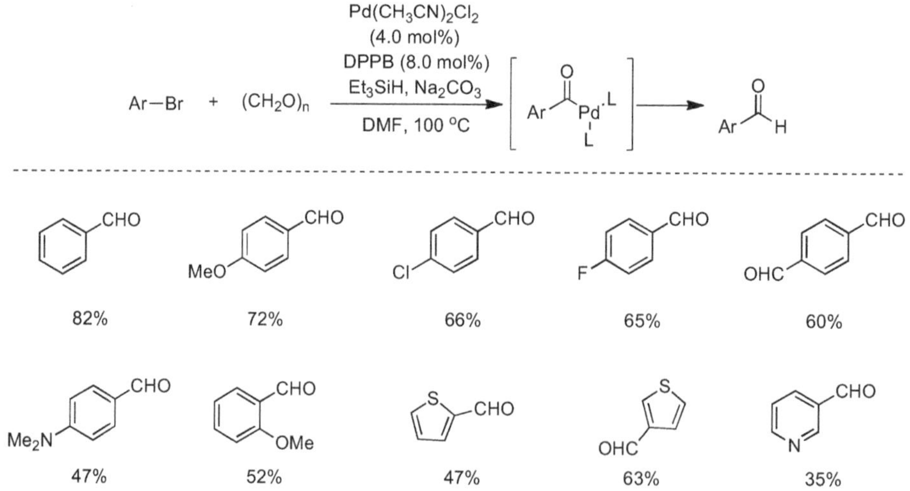

SCHEME 3.20

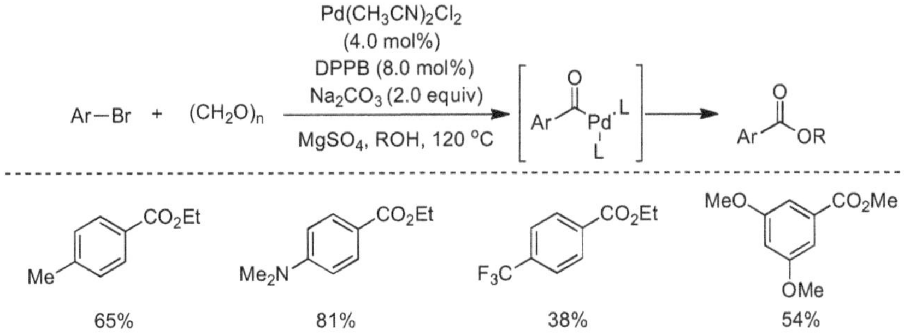

SCHEME 3.21

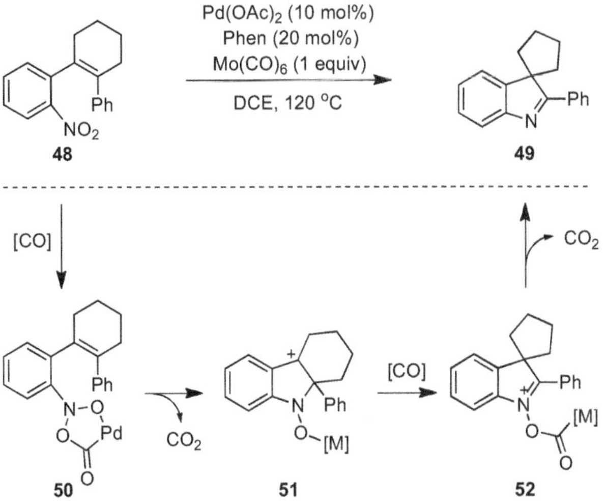

SCHEME 3.22

3.2.3 MOLYBDENUM CARBONYL

Carbon monoxide can act as a ligand to form a stable complex with transition metals. Recently, molybdenum carbonyl ($Mo(CO)_6$) was employed as a CO surrogate in the reductive transformation of nitrostyrene **48** into a spiro indole **49**. In this transformation, carbon monoxide, released from $Mo(CO)_6$ on heating, functions as a reductant to participate in palladium-catalyzed indole synthesis. According to the proposed mechanism (Scheme 3.22),[32] the reduction of the nitro group with CO was accomplished via three intermediates, **50–52**.

$Mo(CO)_6$-involved nitro reduction was also demonstrated in the synthesis of 3-phenylquinolin-2(1*H*)-ones. As shown in Scheme 3.23,[33] 2-nitrobenzaldehye is reduced to 2-aminobenzaldehyde with $Mo(CO)_6$ in the presence of a palladium catalyst. Besides serving as a reductant, the in situ formed CO also reacts with benzylpalladium **53** to give acylpalladium complex **54**.

3.2.4 PHENYL FORMATE

The decomposition of aryl formate generates carbon monoxide under mild reaction conditions (Equation 3.20).[34]

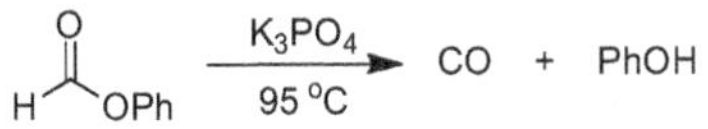

$$\underset{H}{\overset{O}{\|}}\text{C}-\text{OPh} \xrightarrow[95\,°C]{K_3PO_4} CO + PhOH \qquad (3.20)$$

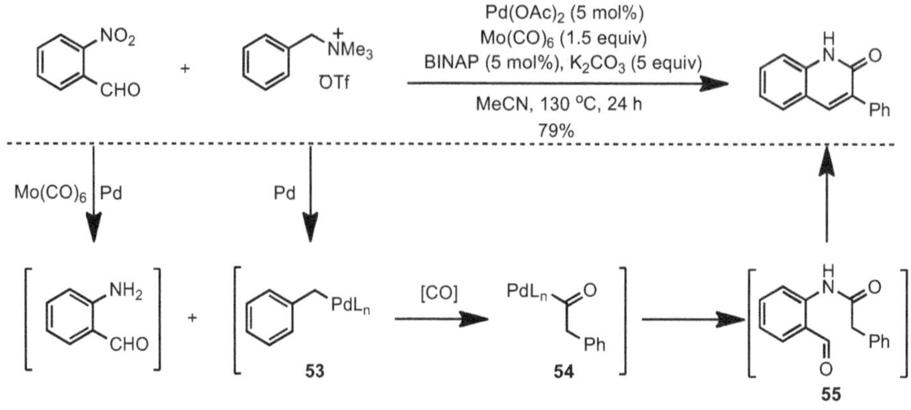

SCHEME 3.23

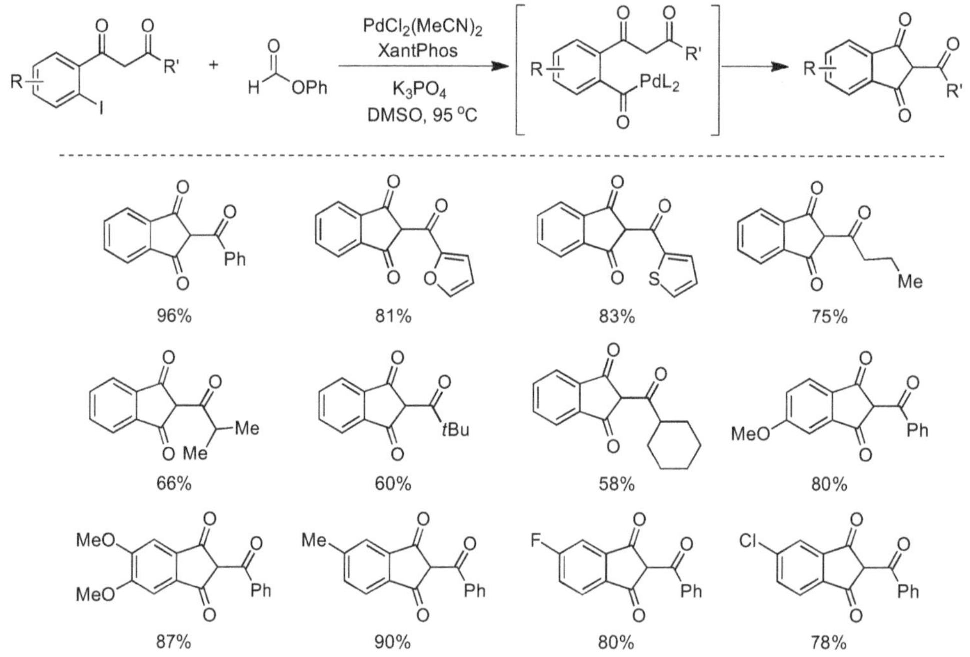

SCHEME 3.24

Therefore, phenyl formate was used as a CO source to synthesize 2-substituted indene-1,3(2H)-diones via a palladium-catalyzed carbonylative annulation (Scheme 3.24).

3.2.5   BENZENE-1,3,5-TRIYL TRIFORMATE (TFBEN)

A report[35] described a Pd-catalyzed C–H carbonylation of benzylamines with benzene-1,3,5-triyl triformate (TFBen) as the CO surrogate to produce isoindolinones (Equation 3.21).

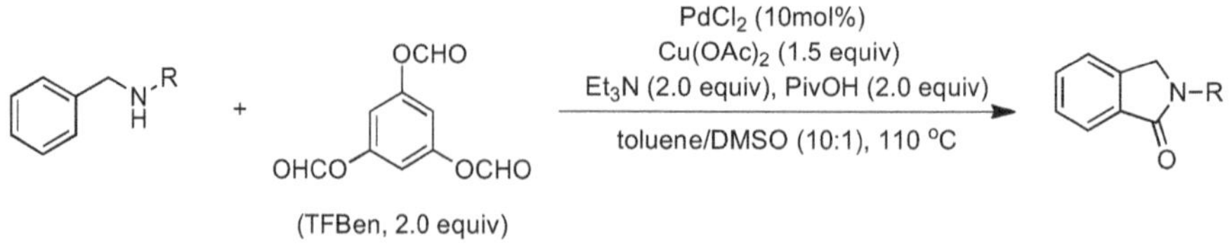

TFBen is a white solid (mp 53.2–55.6 °C) and can be conveniently prepared from trihydroxybenzene (Equation 3.22):

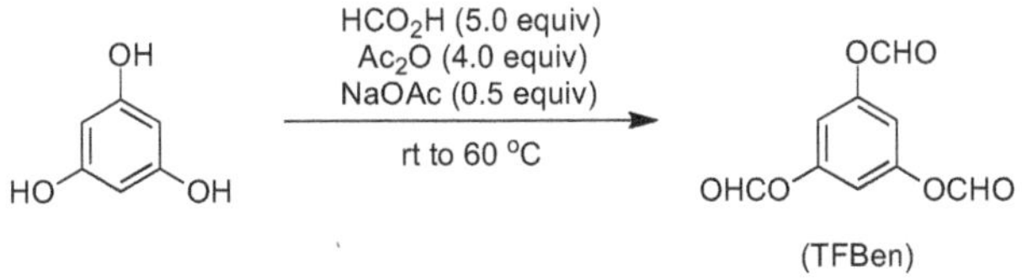

$$(3.22)$$

Procedure

- Add formic acid (8.4 mL, 222.8 mmol, 5.0 equiv) to acetic anhydride (16.8 mL, 178.2 mmol, 4.0 equiv) at room temperature.
- Stir the resulting mixture at 60 °C for 1 h.
- Cool to room temperature.
- Pour the reaction mixture into a mixture of 1,3,5-trihydroxybenzene (5.62 g, 44.6 mmol, 1.0 equiv) and sodium acetate (1.83 g, 22.3 mmol, 0.5 equiv).
- Stir the batch for 4 h in a water bath.
- Add toluene (100 mL) and separate layers.
- Wash the toluene layer with water (2×50 mL).
- Cool the toluene layer (2–8 °C).
- Filter and dry under vacuum.
- Yield of TFBen: 5.1 g (55%) as a white solid.

3.2.6 FORMIC ACID

In general, the decomposition of formic acid produces either hydrogen and CO_2 or water and CO depending on reaction conditions (Scheme 3.25). The former decomposition is frequently utilized in transfer hydrogenations.

A palladium-catalyzed decomposition of formic acid enabled to selectively generate CO which was utilized for alkoxycarbonylation of alkenes (Scheme 3.26).[36]

In the presence of the in situ formed CO, a fast alkene isomerization and regioselective alkoxycarbonylation could occur to produce the carboxylates in high yields.

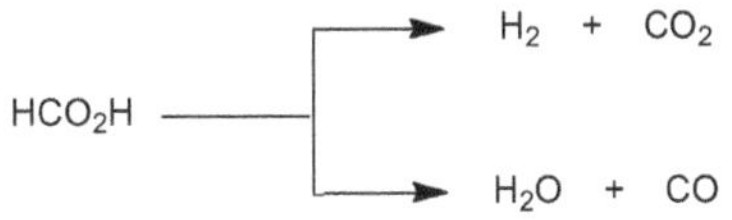

SCHEME 3.25

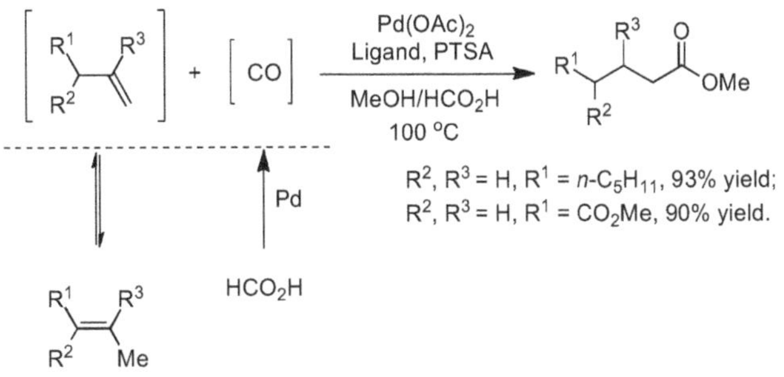

SCHEME 3.26

3.3   CARBON DIOXIDE SURROGATES

Although CO_2 represents an inexpensive and readily accessible carbon source, the inert nature of CO_2 limits its applicability in organic synthesis. A protocol[37] was developed for the activation of CO_2 by a reaction of CO_2 with tetraethylammonium hydroxide which can be prepared by treatment of tetraethylammonium chloride with KOH in methanol (Scheme 3.27).

The resulting tetraethylammonium hydrogen carbonate (TEAHC) could be utilized as a CO_2 surrogate for the carboxylation of diamine **56** to give the corresponding dibenzyl hexane-1,6-di-yldicarbamate **58** in a two-step process (Scheme 3.28). Notably, the tetraethylammonium chloride byproduct can be recovered and used to regenerate the carboxylation agent TEAHC.

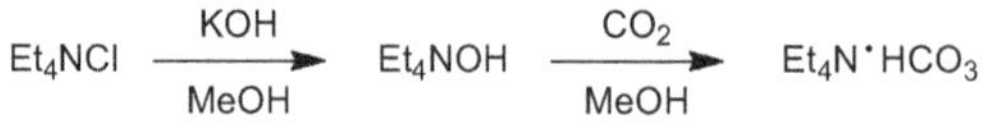

$$Et_4NCl \xrightarrow[\text{MeOH}]{\text{KOH}} Et_4NOH \xrightarrow[\text{MeOH}]{CO_2} Et_4N \cdot HCO_3$$

SCHEME 3.27

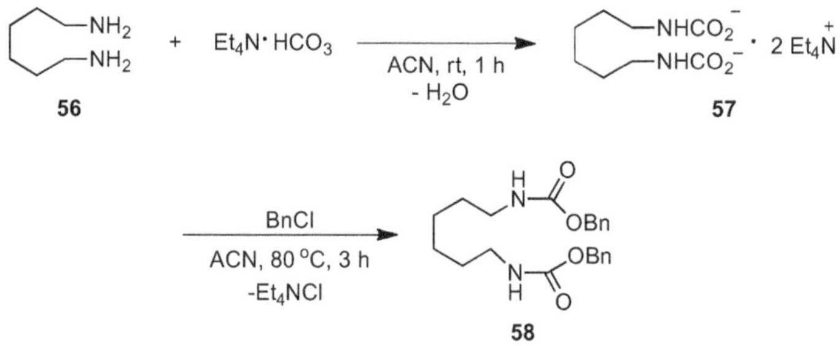

SCHEME 3.28

3.4   α-HYDROXYSULFONATES AS ALDEHYDE SURROGATES

Aldehydes are useful synthetic intermediates and have many applications. Aldehydes are commonly prepared by the oxidation of alcohols. Acetaldehyde is produced industrially via the Wacker process using copper and palladium catalysts. Aldehydes, however, are highly reactive and prone to oxidation to the corresponding acids. To address these issues, a strategy was developed that involved converting the aldehyde intermediates to the corresponding α-hydroxysulfonates (or α-hydroxysulfonic acids) by reactions of aldehydes with sodium bisulfite ($NaHSO_3$) (or sulfurous acid H_2SO_3). Most of the bisulfite adducts are crystalline solids possessing good inherent physical properties, such as stability and non-hygroscopicity, and are easy to handle on scales.

3.4.1   OXIDATION OF ALDEHYDE TO ACID

The transformation of aldehydes such as **59** to the corresponding acids or acid derivatives suffered from low product yields under common oxidation conditions due to side reactions associated with C–C bond cleavage (Equation 3.23).[38] In all cases, side products, ketones **61**, were observed.

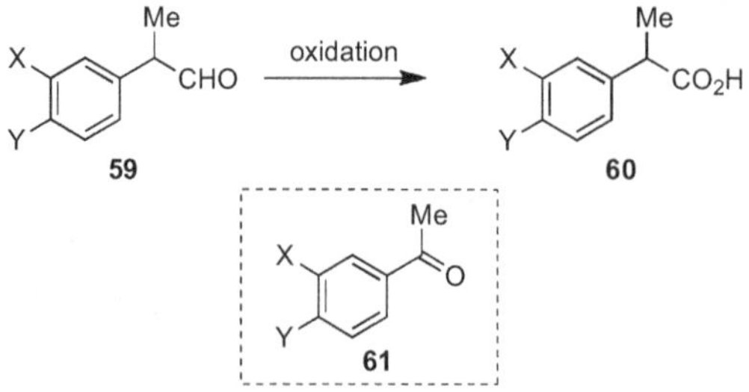

(3.23)

Solutions

Instead of using aldehydes **59**, bisulfite adducts **62**, readily obtained from the reaction of **59** with sodium bisulfite, were used as aldehyde surrogates in the oxidation reaction. Employing DMSO/Ac$_2$O the adducts **62** were conveniently oxidized to **60** (Scheme 3.29).

Depending on the methods of quenching, acids **60a**, methyl esters **60b**, or amides **60c** could be obtained in good yields, by reactions of ketosulfonates **63** with an aqueous base, NaOMe/MeOH, or ammonia.

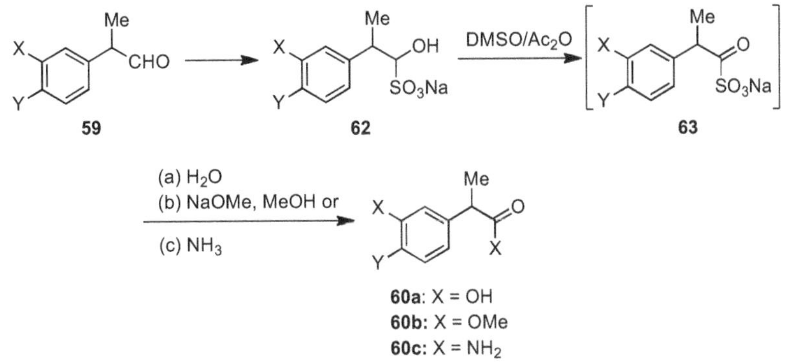

SCHEME 3.29

3.4.2 REDUCTIVE AMINATION

The synthesis of amino alcohol **66** from alkene **64** was envisioned to proceed through an aldehyde intermediate **65**. Due to the instability and non-crystallinity of **65**, isolation of **65** might be problematic.

Bisulfite adduct **68** was viewed as an attractive intermediate in the synthesis of **66**. Thus, treatment of the primary ozonide with methanol generates the methoxy–hydroperoxide **67**, whose subsequent reaction with sodium bisulfite (NaHSO$_3$) to effect the peroxide reduction and simultaneous formation of bisulfite adduct **68**. Treatment of **68** with an aqueous base generates, in situ, the aldehyde **65**, which could be used directly in the reductive amination to generate the product **66** (Scheme 3.30).[39] Although the efficiency of isolating bisulfite adduct **68** is less than going directly to **66**, it avoids the isolation of the unstable and non-crystalline aldehyde **65**, and is thus viewed as the preferred method for preparing kilogram quantities of **66**.

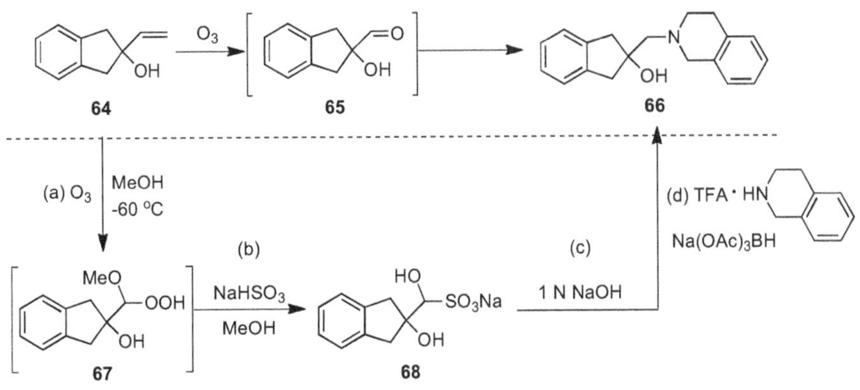

SCHEME 3.30

DISCUSSIONS

(a) Due to the instability of aldehydes, it is convenient to store aldehydes as their corresponding bisulfite adducts.[40]

(b) Although bisulfite adducts confer chemical stability to the parent aldehydes and exhibit desirable physical properties such as crystallinity, thus facilitating the isolation and purification, subsequent use of the bisulfite adducts often requires treatment under aqueous conditions to reveal the aldehyde functionality for further utilization. Direct use of bisulfite adducts in synthetic transformations is also possible and demonstrated in the reaction (Equation 3.24).[41] Aldehyde bisulfite adduct **70**, derived from its unstable parent aldehyde, condensed directly with amine hydrochloride salt **69** in the presence of NaOAc (3.0 equiv) followed by in situ reduction with 2-picoline-borane (PICB)[42] (0.6 equiv).

(3.24)

(c) Many aldehydes or other functional groups do not tolerate aqueous acid or base conditions. In order to avoid such aqueous conditions, an anhydrous release method was developed by using an organic base to liberate the aldehyde in situ (Equation 3.25).[43]

(3.25)

3.4.3 DIELS–ALDER REACTION

In addition to the reductive amination, the aldehyde bisulfite adducts have been successfully employed in chiral *N*-heterocyclic carbene (NHC)-catalyzed hetero-Diels–Alder reactions (Equation 3.26).[44]

(3.26)

$R1 = H$, alkyl
$R^2 = Me$, *c*-Hex, Ph, *p*-BrC$_6$H$_4$, *p*-OMeC$_6$H$_4$
$R^3 = CO_2Et$

REMARK

α-Chloroaldehyde bisulfite adducts as substrates, however, would necessitate the use of aqueous conditions to decompose the bisulfite adducts to the corresponding aldehydes in the presence of the NHC catalyst. In the event, a biphasic system was employed for this Diels–Alder reaction.

3.4.4 STRECKER REACTION

The preparation of α-amino nitrile **73** via Stecker reaction of aldehyde **72** with benzylamine and TMSCN (Conditions (a), Scheme 3.31)[45] suffers two major drawbacks:

- Problems in the purification of **72** due to its instability.
- Using expensive TMSCN.

Thereby conversion of the aldehyde **72** to water-soluble bisulfite adduct **74** with aqueous sodium bisulfite (1.1 equiv) (Conditions (b), Scheme 3.31) allowed a simple purification of **72** by extraction and the use of less expensive NaCN under biphasic reaction conditions.

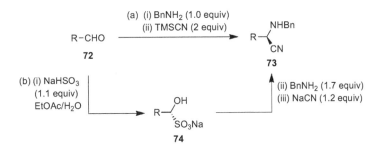

SCHEME 3.31

3.4.5 TRANSAMINASE DKR OF ALDEHYDE

The limited stability of aldehyde **75** poses some concerns over isolation and storage. To address the stability issue, the aldehyde **75** was converted to the corresponding bisulfite adduct **77**, which could be used directly in the transaminase DKR process (Scheme 3.32).[46]

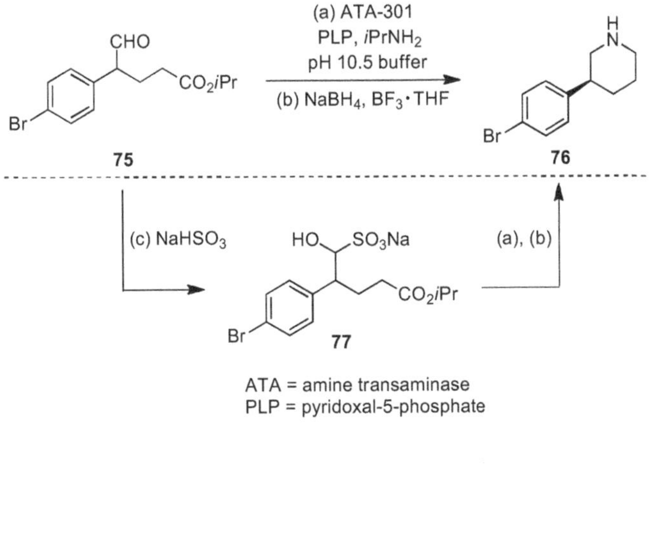

ATA = amine transaminase
PLP = pyridoxal-5-phosphate

SCHEME 3.32

REMARK

In addition to sodium α-hydroxysulfonates, lactol **79** was employed as an alternative aldehyde surrogate and proved to be an excellent substrate for the transaminase DKR (Scheme 3.33).[46]

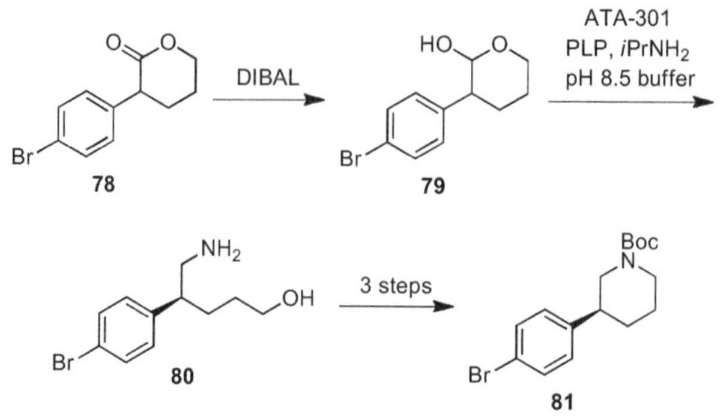

SCHEME 3.33

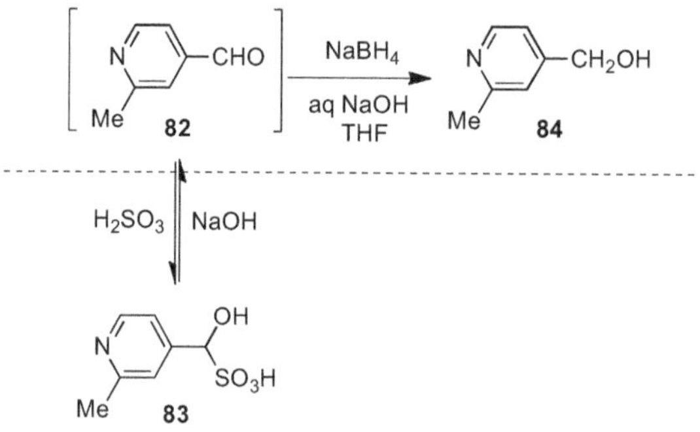

SCHEME 3.34

3.4.6   REDUCTION OF ALDEHYDE TO ALCOHOL

α-Hydroxysulfonic acid **83**, obtained by a reaction of 2-methylisonicotinaldehyde **82** with sulfurous acid (H_2SO_3), was used as the aldehyde surrogate for the synthesis of 4-hydroxymethyl-2-methyl-pyridine **84** (Scheme 3.34).[47]

   The bisulfite adduct **83** is a well-behaved crystalline, free-flowing solid; in the event, the use of **83** allowed the development of a convenient and chromatography-free process for the synthesis of **84**.

Procedure

- Charge a solution of **82** (78 mol) in CH_2Cl_2 (24 L) into a 200-L glass-lined reactor containing 94 L of THF.
- Charge 84 L (79 mol) 6% H_2SO_3 (SO_2 saturated water) slowly at below 35 °C.
- Filter the resulting precipitate and rinse with isopropyl ether (1 L).
- Dry under vacuum.
- Yield of **83**: 10.10 kg (49.7 mol, 63.7%).

REMARK

The sodium bisulfite adduct was proved to be inferior to the "free acid" bisulfite adduct.

3.5   SULFUR DIOXIDE SURROGATE

Sulfony-derived functional groups can be utilized in many chemical transformations. Introducing such groups by using sulfur dioxide (SO_2) gas directly presents some limitations because of the inconvenience of handling the corrosive and toxic SO_2 gas. The 1,4-diazabicyclo[2.2.2]octane

(DABCO)–sulfur dioxide charge-transfer complex (DABSO) is a convenient and easy-to-handle solid and selected as the SO_2 surrogate.

DABSO could be readily prepared in multigram scales via a reaction of DABCO with ex situ generated SO_2 (Equation 3.27).[48]

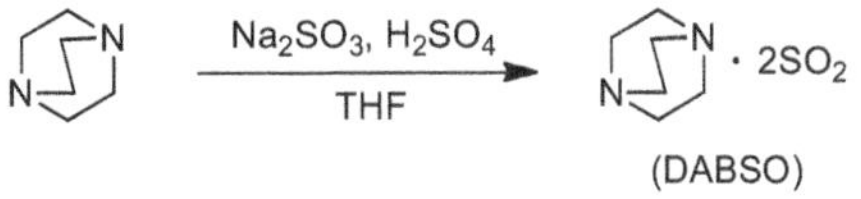

$$(3.27)$$

The reaction was conducted in a two-chamber reactor in which the SO_2 is generated in one chamber and consumed in the other.

3.5.1 SYNTHESIS OF SULFONES

As the SO_2 surrogate, DABSO was employed to synthesize sulfones. This two-step process involves a palladium-catalyzed reaction of aryl iodides (or boronic acid) with DABSO to generate ammonium sulfonate intermediates **86** (Scheme 3.35).[49]

The subsequent S_N2 reaction of the in situ formed ammonium sulfinates **86** with bromoacetate furnished aryl alkyl sulfones **87**. Using this protocol, a variety of sulfonyl derivatives were prepared including sulfones, sulfonamides, and sulfonyl chloride.

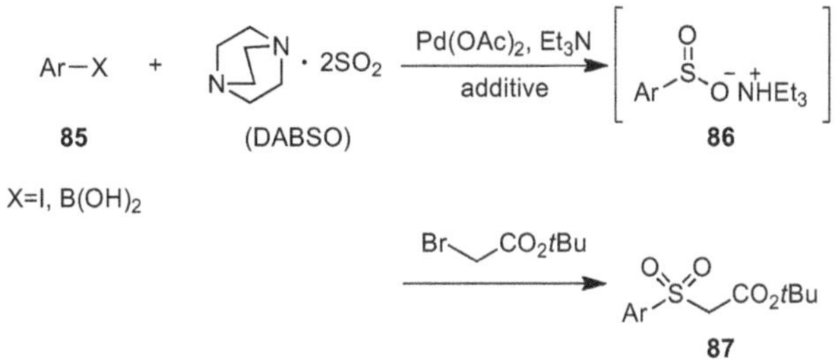

SCHEME 3.35

REMARKS

(a) When using aryl iodides, the additive was not needed and isopropanol was used as both solvent and reductant, which makes the approach attractive for large-scale preparations.

(b) When using aryl boronic acids, tetra-*n*-butylammonium bromide (TBAB, 0.3 equiv) was employed.

3.5.2 SYNTHESIS OF SULFOXIDES

DABSO was also utilized in the synthesis of unsymmetrical sulfoxides using organometallic nucleophiles (Scheme 3.36).[50]

This two-step process was conducted in one-pot wherein the sulfinate silyl ester intermediates were generated by trapping the initially formed sulfinates with TMSCl. Both organolithium or Grignard reagents could be employed, delivering sulfoxides in good to excellent yields.

3.5.3 SYNTHESIS OF SULFONAMIDES

Analogously, the DABSO complex was utilized in the synthesis of sulfonamides (Scheme 3.37).[51] This practical protocol allowed to prepare of various alkyl, alkenyl, and (hetero)aryl sulfonamides via a reaction of DABSO with Grignard reagents followed by quenching the resulting metal sulfinate intermediates with in situ generated *N*-chloramines.

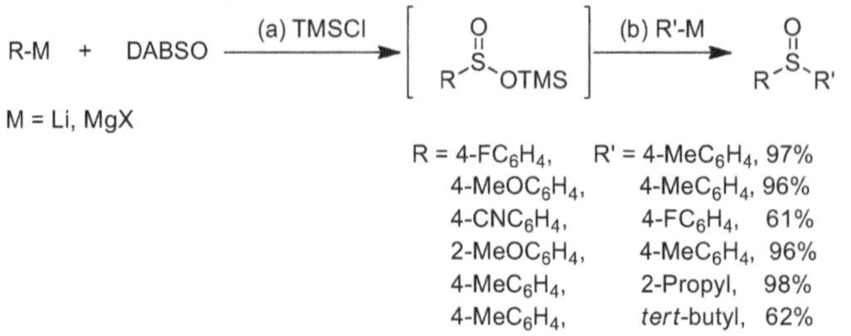

R-M + DABSO $\xrightarrow{\text{(a) TMSCl}}$ $\left[\underset{\text{R}}{\overset{\text{O}}{\|}}\overset{\|}{\underset{\text{OTMS}}{\text{S}}} \right]$ $\xrightarrow{\text{(b) R'-M}}$ $\underset{\text{R}}{\overset{\text{O}}{\|}}\overset{\|}{\underset{\text{R'}}{\text{S}}}$

M = Li, MgX

R = 4-FC$_6$H$_4$,	R' = 4-MeC$_6$H$_4$,	97%
4-MeOC$_6$H$_4$,	4-MeC$_6$H$_4$,	96%
4-CNC$_6$H$_4$,	4-FC$_6$H$_4$,	61%
2-MeOC$_6$H$_4$,	4-MeC$_6$H$_4$,	96%
4-MeC$_6$H$_4$,	2-Propyl,	98%
4-MeC$_6$H$_4$,	tert-butyl,	62%

SCHEME 3.36

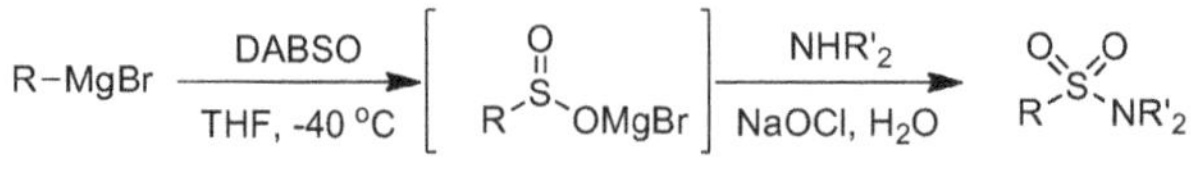

R–MgBr $\xrightarrow[\text{THF, -40 °C}]{\text{DABSO}}$ $\left[\underset{\text{R}}{\overset{\text{O}}{\|}}\overset{\|}{\underset{\text{OMgBr}}{\text{S}}} \right]$ $\xrightarrow[\text{NaOCl, H}_2\text{O}]{\text{NHR'}_2}$ $\underset{\text{R}}{\overset{\text{O}\,\,\text{O}}{\|\|}}\overset{}{\underset{\text{NR'}_2}{\text{S}}}$

SCHEME 3.37

$\underset{\text{(or HetAr-Bpin)}}{\text{[Bpin arene]}}$ $\xrightarrow[\text{DMI, 50 °C}]{\substack{\text{CuI (10 mol%), LiI (20 mol%)}\\ \text{PO(OMe)}_3 \text{ (1.1 equiv)}\\ \text{LiO}t\text{Bu (1.1 equiv)}}}$ $\underset{\text{(or HetAr-Me)}}{\text{[Me arene]}}$

DMI = 1,3-dimethyl-2-imidazolidinone

ArBpin + LiOtBu $\rightleftharpoons$ Li[ArBpin(OtBu)]

PO(OMe)$_3$ $\xrightarrow[\text{rate-determining}\\\text{step}]{\text{Li}^+,\,\text{I}^-}$ Li[OP(O)(OMe)$_2$] + MeI

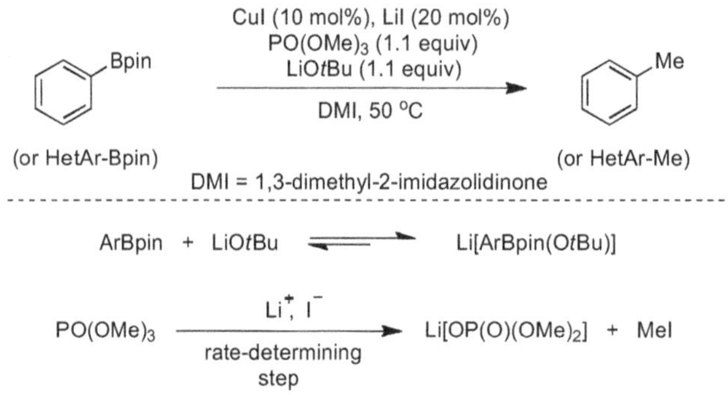

SCHEME 3.38

3.6 MISCELLANEOUS SURROGATES

3.6.1 METHYL IODIDE SURROGATE

Methyl halides or methyl pseudo halides are frequently employed as methylating agents for cross-coupling in the preparation of methylarenes. The use of these agents, however, has a serious concern in the pharmaceutical industry associated with their genotoxicity, especially at the late-stage synthesis. Hence, a synthetic strategy for the selective installation of methyl groups without using toxic methyl halides is highly desirable.

In the event, trimethylphosphate was utilized as the methylating agent[52] in the copper-mediated cross-coupling reaction (Scheme 3.38). This catalytic methylation process features a slow-release mechanism in which the reaction of trimethylphosphate with iodide is the rate-determining step. The in situ formed MeI would subsequently react rapidly with arylcopper species. Using the trimethylphosphate as the inexpensive and non-toxic methylating agent, this catalytic process is amenable to large-scale and late-stage installation of the methyl group to pharmaceutical intermediates.

3.6.2 CYANIDE SURROGATES

Aromatic nitriles are important intermediates in synthetic chemistry and are prevalent in active pharmaceutical ingredients. A common synthetic approach for the preparation of aromatic nitriles employs toxic nucleophilic cyanides, such as NaCN, KCN, CuCN, TMSCN, etc. However, utilization of the metal cyanides on large production scales is a serious concern due to their high toxicity.

3.6.2.1 2-Methyl-2-Phenyl Malononitrile (MPMN)

A nickel-catalyzed reductive cyanation of aryl halides and phenol derivatives using 2-methyl-2-phenyl malononitrile (MPMN) as the cyanide surrogate allows to access various aryl nitriles (Scheme 3.39).[53] MPMN is a bench-stable, carbon-bound electrophilic CN reagent that does not release cyanide under reaction conditions.

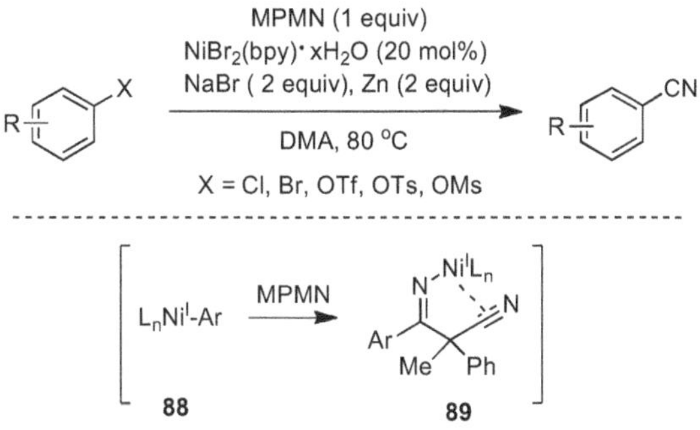

SCHEME 3.39

The reaction proceeds through a unique mechanistic pathway involving a reversible oxidative addition of Ni(0) catalyst to aryl bromide to generate a Ni(II) intermediate. A subsequent Zn reduction generates (aryl)Ni(I) intermediate **88** which is trapped by MPMN to give Ni(I)-imine **89** via a key 1,2-migratory insertion of the nitrile into the Ni(I)-aryl bond in **88**. The following reductive β–C elimination of **89** would afford the desired benzonitrile products. Additionally, NaBr can be used as a halide additive to enable the functionalization of otherwise inert aryl–X bonds including aryl chlorides, mesylates, tosylates, and aryl triflates. Arenes with various electronic and steric properties and heteroarenes are good substrates for this reaction.

3.6.2.2 2-Cyanoisothiazolidine 1,1-Dioxide

Aromatic nitriles could be obtained via a nickel-catalyzed C–H cyanation using 2-cyanoisothiazolidine 1,1-dioxide as the electrophilic cyanation reagent (Scheme 3.40).[54]

The C–H cyanation requires substrates bearing a nitrogen-containing heterocyclic functionality including pyridine, (iso)quinoline, and benzimidazole as the directing groups.

3.6.3 Ethylene Surrogates

Ethylene is an inexpensive gas (annual production >100 million tons),[55] but its use in organic synthesis is rare and typically requires specialized equipment (high-pressure reactors or UV-photoreactors). A vinyl thianthrenium tetrafluoroborate[56] was developed as a practical and versatile vinylating reagent. This thianthrenium salt enables its participation in various synthetic transformations, such as N-vinylation of heterocycles and Suzuki-type cross-coupling reactions (Scheme 3.41).

The vinyl thianthrenium tetrafluoroborate salt can be prepared on a multigram scale (50 mmol) from a reaction of thianthrene-S-oxide with ethylene (in a balloon at 1 atm) in the presence of Tf_2O as an activator (Equation 3.28). The isolation can be carried out by simple precipitation to furnish the product with an 86% yield. The vinyl thianthrenium salt is a non-hygroscopic crystalline solid that can be stored in the presence of air and moisture. Differential scanning calorimetry (DSC)–thermogravimetric analysis (TGA) reveals that the salt has a desirable safety profile with no decomposition below 280 °C.

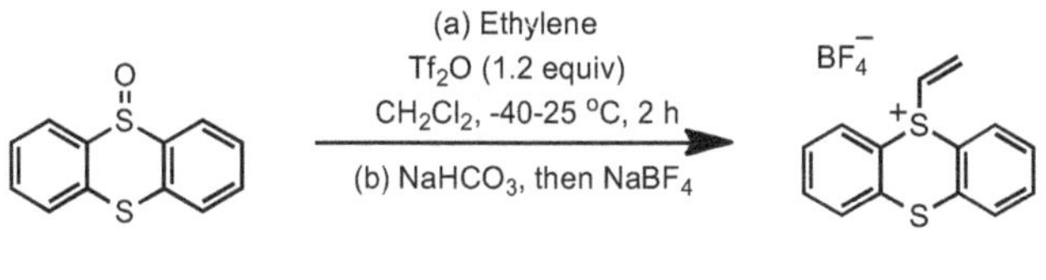

(3.28)

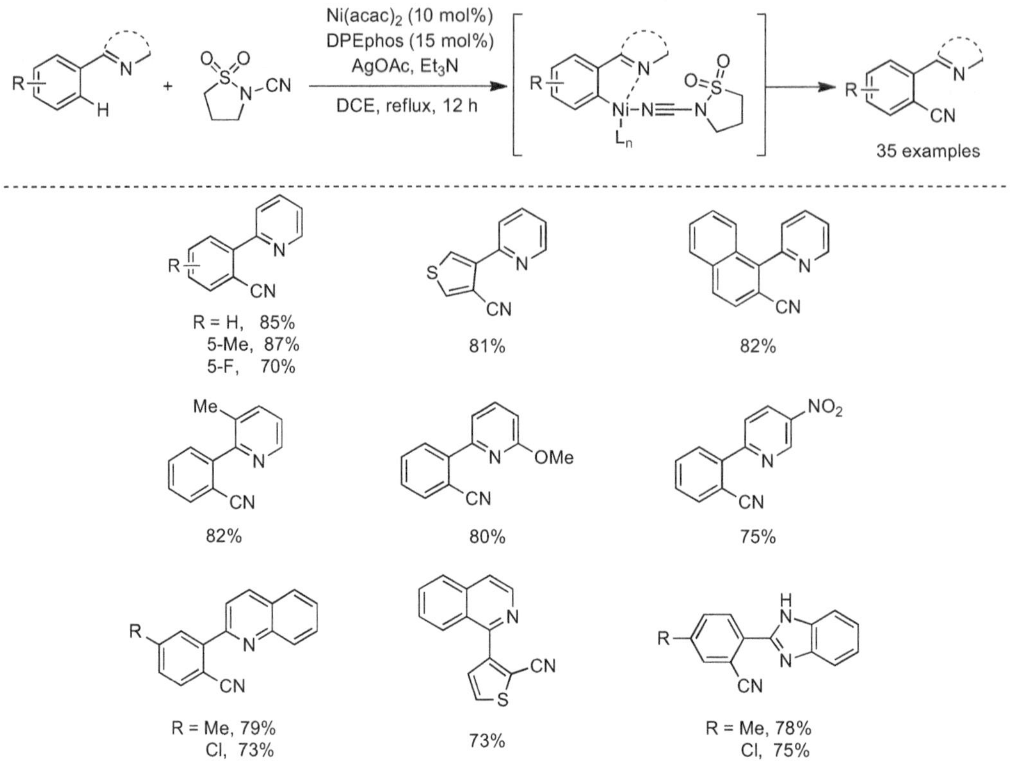

SCHEME 3.40

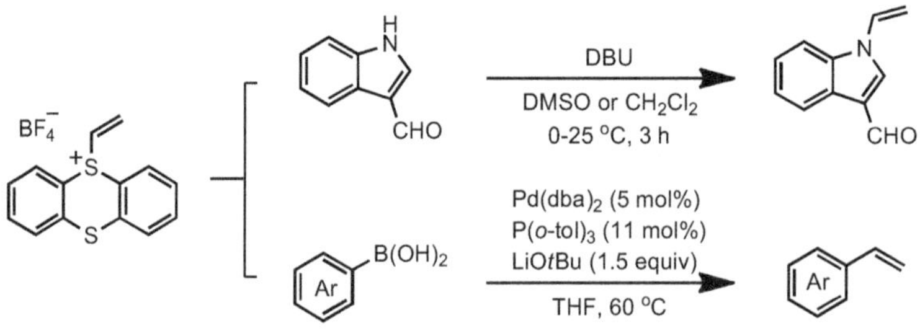

SCHEME 3.41

NOTES

1. Zhang, W.; Lou, S.; Liu, Y.; Xu, Z. *J. Org. Chem.* **2013**, *78*, 5932.
2. Ishikawa, M.; Tsushima, M.; Kubota, D.; Yanagisawa, Y.; Hiraiwa, Y.; Kojima, Y.; Ajito, K.; Anzai, N. *Org. Process Res. Dev.* **2008**, *12*, 596.
3. The author's unpublished process development work.
4. (a) Hansen, K. B.; Hsiao, Y.; Xu, F.; Rivera, N.; Clausen, A.; Kubryk, M.; Krska, S.; Rosner, T.; Simmons, B.; Balsells, J.; Ikemoto, N.; Sun, Y.; Spindler, F.; Malan, C.; Grabowski, E. J. J.; Armstrong III, J. D. *J. Am. Chem. Soc.* **2009**, *131*, 8798. (b) Hansen, K. B.; Balsells, J.; Dreher, S.; Hsiao, Y.; Kubryk, M.; Palucki, M.; Rivera, N.; Steinhuebel, D.; Armstrong III, J. D.; Askin, D.; Grabowski, E. J. J. *Org. Process Res. Dev.* **2005**, *9*, 634.
5. Maeda, K.; Matsubara, R.; Hayashi, M. *Org. Lett.* **2021**, *23*, 1530.
6. Wennekes, T.; Lang, B.; Leeman, M.; van der Marel, G. A.; Smits, E.; Weber, M.; van Wiltenburg, J.; Wolberg, M.; Aerts, J. M. F. G.; Overkleeft, H. S. *Org. Process Res. Dev.* **2008**, *12*, 414.

7. Yan, Y.; Xu, Y.; Niu, B.; Xie, H.; Liu, Y. *J. Org. Chem.* **2015**, *80*, 5581.
8. Breining, S. R.; Genus, J. F.; Mitchener, J. P.; Cuthbertson, T. J.; Heemstra, R.; Melvin, M. S.; Dull, G. M.; Yohannes, D. *Org. Process Res. Dev.* **2013**, *17*, 413.
9. Mitchell, D.; Coppert, D. M.; Moynihan, H. A.; Lorenz, K. T.; Kissane, M.; McNamara, O. A.; Maguire, A. R. *Org. Process Res. Dev.* **2011**, *15*, 981.
10. Fizet, C.; Streith, J. *Tetrahedron Lett.* **1974**, *15*, 3187.
11. Olah, G. A.; Fung, A. P. *Org. Synth.* **1985**, *63*, 188.
12. Johnson, M. G.; Gribble, M. W., Jr.; Houze, J. B.; Paras, N. A. *Org. Lett.* **2014**, *16*, 6248.
13. Voth, S.; Hollett, J. W.; McCubbin, J. A. *J. Org. Chem.* **2015**, *80*, 2545.
14. Michida, M.; Takayanagi, Y.; Imai, M.; Furuya, Y.; Kimura, K.; Kitawaki, T.; Tomori, H.; Kajino, H. *Org. Process Res. Dev.* **2013**, *17*, 1430.
15. Thompson, W. J.; Fitzgerald, P. M. D.; Holloway, M. K.; Emini, E. A.; Darke, P. L.; McKeever, B. M.; Schlief, W. A.; Quintero, J. C.; Zugay, J. A.; Tucker, T. J.; Schwering, J. E.; Homnick, C. F.; Nunberg, J.; Springer, J. P.; Huff, J. R. *J. Med. Chem.* **1992**, *35*, 1685.
16. Larrow, J. F.; Roberts, E.; Verhoeven, T. R.; Ryan, K. M.; Senanayake, C. H.; Reider, P. J.; Jacobsen, E. N. *Org. Synth.* **2004**, *10*, 29.
17. Content, S.; Dupont, T.; Fédou, N. M.; Smith, J. D.; Twiddle, S. J. R. *Org. Process Res. Dev.* **2013**, *17*, 193.
18. (a) Guram, A. S.; Buchwald, S. L. *J. Am. Chem. Soc.* **1994**, *116*, 7901. (b) Guram, A. S.; Rennels, R. R.; Buchwald, S. L. *Angew. Chem. Int. Ed.* **1995**, *34*, 1348. (c) Paul, F.; Patt, J.; Hartwig, J. F. *J. Am. Chem. Soc.* **1994**, *116*, 5969. (d) Louie, J.; Hartwig, J. F. *Tetrahedron Lett.* **1995**, *36*, 3609.
19. Li, B.; Samp, L.; Sagal, J.; Hayward, C. M.; Yang, C.; Zhang, Z. *J. Org. Chem.* **2013**, *78*, 1273.
20. Wolfe, J. P.; Åhman, J.; Sadighi, J. P.; Singer, R. A.; Buchwald, S. L. *Tetrahedron Lett.* **1997**, *38*, 6367.
21. Søndergaard, K.; Kristensen, J. L.; Gillings, N.; Begtrup, M. *Eur. J. Org. Chem.* **2005**, *20*, 4428.
22. Denni-Dischert, D.; Marterer, W.; Bänziger, M.; Yusuff, N.; Batt, D.; Ramsey, T.; Geng, P.; Michael, W.; Wang, R.-M. B.; Taplin, F., Jr.; Versace, R.; Cesarz, D.; Perez, L. B. *Org. Process Res. Dev.* **2006**, *10*, 70.
23. (a) Lan, X.; Tao, C.; Liu, X.; Zhang, A.; Zhao, B. *Org. Lett.* **2016**, *18*, 3658. (b) Shi, L.; Tao, C.; Yang, Q.; Liu, Y. E.; Chen, J.; Chen, J.; Tian, J.; Liu, F.; Li, B.; Du, Y.; Zhao, B. *Org. Lett.* **2015**, *17*, 5784.
24. (a) Lee, S.; Jørgensen, M.; Hartwig, J. F. *Org. Lett.* **2001**, *3*, 2792. (b) Huang, X.; Buchwald, S. L. *Org. Lett.* **2001**, *3*, 3417.
25. (a) Sun, M.; Zhao, C.; Gfesser, G. A.; Thiffault, C.; Miller, T. R.; Marsh, K.; Wetter, J.; Curtis, M.; Faghih, R.; Esbenshade, T. A.; Hancock, A. A.; Cowart, M. *J. Med. Chem.* **2005**, *48*, 6482. (b) Alabanza, L. M.; Dong, Y.; Wang, P.; Wright, J. A.; Zhang, Y.; Briggs, A. J. *Org. Process Res. Dev.* **2013**, *17*, 876. (c) See also: Thiemann, F.; Piehler, T.; Haase, D.; Saak, W.; Lützen, A. *Eur. J. Org. Chem.* **2005**, *10*, 1991.
26. Jaime-Figueroa, S.; Liu, Y.; Muchowski, J. M.; Putman, D. G. *Tetrahedron Lett.* **1998**, *39*, 1313.
27. Thibault, C.; L'Heureux, A.; Bhide, R. S.; Ruel, R. *Org. Lett.* **2003**, *5*, 5023.
28. Karmakar, S.; Byri, V.; Gavai, A. V.; Rampulla, R.; Mathur, A.; Gupta, A. *Org. Process Res. Dev.* **2016**, *20*, 1717.
29. Guo, S.; Yang, J. C.; Buchwald, S. L. *J. Am. Chem. Soc.* **2018**, *140*, 15976.
30. Ueda, T.; Konishi, H.; Manabe, K. *Angew. Chem. Int. Ed.* **2013**, *52*, 8611.
31. Natte, K.; Dumrath, A.; Neumann, H.; Beller, M. *Angew. Chem. Int. Ed.* **2014**, *53*, 10090.
32. Jana, N.; Zhou, F.; Driver, T. G. *J. Am. Chem. Soc.* **2015**, *137*, 6738.
33. Liu, Y.; Qi, X.; Wu, X.-F. *J. Org. Chem.* **2021**, *86*, 13824.
34. Zhang, Y.; Chen, J. L.; Chen, Z.-B.; Zhu, Y.-M.; Ji, S.-J. *J. Org. Chem.* **2015**, *80*, 10643.
35. Fu, L.-Y.; Ying, Y.; Qi, X.; Peng, J. B.; Wu, X.-F. *J. Org. Chem.* **2019**, *84*, 1421.
36. Sang, S.; Kucmierczyk, P.; Dong, K.; Franke, R.; Neumann, H.; Jackstell, R.; Beller, M. *J. Am. Chem. Soc.* **2018**, *140*, 5217.
37. Forte, G.; Chiarotto, I.; Richter, F.; Trieu, V.; Feroci, M. *Org. Process Res. Dev.* **2018**, *22*, 1323.
38. Wuts, P. G. M.; Bergh, C. L. *Tetrahedron Lett.* **1986**, *27*, 3995.
39. Ragan, J. A.; Ende, D. J.; Brenek, S. J.; Eisenbeis, S. A.; Singer, R. A.; Tickner, D. L.; Teixeira, J. J.; Vanderplas, B. C.; Weston, N. *Org. Process Res. Dev.* **2003**, *7*, 155.
40. Monte, W. T.; Lindbeck, A. C. *Org. Process Res. Dev.* **2001**, *5*, 267.
41. Faul, M.; Larsen, R.; Levinson, A.; Tedrow, J.; Vounatsos, F. *J. Org. Chem.* **2013**, *78*, 1655.
42. Compared to NaBH(OAc)$_3$, the use of PICB produced reproducible results.
43. Pandit, C. R.; Mani, N. S. *Synthesis* **2009**, *23*, 4032.
44. He, M.; Beahm, B. J.; Bode, J. W. *Org. Lett.* **2008**, *10*, 3817.
45. (a) Seki, M.; Hatsuda, M.; Mori, Y.; Yoshida, S.; Yamada, S.; Shimizu, T. *Chem. Eur. J.* **2004**, *10*, 6102. (b) Seki, M.; Hatsuda, M.; Yoshida, S. *Tetrahedron Lett.* **2004**, *45*, 6579.

46. Chung, C. K.; Bulger, P. G.; Kosjek, B.; Belyk, K. M.; Rivera, N.; Scott, M. E.; Humphrey, G. R.; Limanto, J.; Bachert, D. C.; Emerson, K. M. *Org. Process Res. Dev.* **2014**, *18*, 215.
47. Ragan, J. A.; Jones, B. P.; Meltz, C. N.; Teixeira, J. J., Jr. *Synthesis* **2002**, *4*, 483.
48. Mileghem, S. V.; De Borggraeve, V. M. *Org. Process Res. Dev.* **2017**, *21*, 785.
49. (a) Emmett, E. J.; Hayter, B. R.; Willis, M. C. *Angew. Chem. Int. Ed.* **2014**, *53*, 10204. (b) Deeming, A. S.; Russell, C. J.; Willis, M. C. *Angew. Chem. Int. Ed.* **2016**, *55*, 747.
50. Lenstra, D. C.; Vedovato, V.; Flegeau, E. F.; Maydom, J.; Willis, M. C. *Org. Lett.* **2016**, *18*, 2086.
51. Deeming, A. S.; Russell, C. J.; Willis, M. C. *Angew. Chem. Int. Ed.* **2015**, *54*, 1168.
52. He, Z.-T.; Li, H.; Haydl, A. M.; Whiteker, G. T.; Hartwig, J. F. *J. Am. Chem. Soc.* **2018**, *140*, 17197.
53. Mills, L. R.; Graham, J. M.; Patel, P.; Rousseaux, S. A. L. *J. Am. Chem. Soc.* **2019**, *141*, 19257.
54. Ma, J.; Liu, H.; Chen, Z.; Liu, Y.; Hou, C.; Sun, Z.; Chu, W. *Org. Lett.* **2021**, *23*, 2868.
55. Zimmermann, H.; Walzl, R.; *Ethylene. Ullmann's Encyclopedia of Industrial Chemistry* **2009**, https://doi.org/10.1002/14356007.a10_045.pub3
56. Juliá, F.; Yan, J.; Paulus, F.; Ritter, T. *J. Am. Chem. Soc.* **2021**, *143*, 1992.

4 Modes of Reagent Addition
Control of Impurity Formation

There are two types of impurities: process impurity and degradation (of the product) impurity. The former is generated from the manufacturing process and can be carried through the process to the final API if the isolation/purification is not effective. The latter refers to the decomposition of the product during processing or storage. In general, two strategies are used to address impurity issues, that is, minimizing the impurity formation and removing the impurity from the product.

One of the challenges in the process development is to suppress side reactions. Side products can be generated at three different reaction stages as shown in Figure 4.1.

During the early reaction stage, side products are usually produced via competitive side reactions. For instance, a regioisomer may be generated due to poor regioselectivity. At this stage, the mechanism for the formation of side products appears to be relatively simple, as there are no intermediates or products involved. Given that a reaction involves an intermediate, impurities may arise from side reactions of the intermediate. Reducing the level and the number of impurities at the intermediate stage is a formidably challenging task due to the fact that the intermediate, being formed in situ, can participate in a number of competitive side reactions in addition to the desired reaction pathway. Further reactions of the desired product would occur at the final stage such as hydrolysis of an ester group in the product during the aqueous workup. Nonetheless, undesired side products can be minimized through the optimization of reaction conditions, such as temperature, pressure, and solvent.

In addition to the control of the reaction temperature and pressure, modes of reagent additions are valuable tools that are frequently used by process chemists to control impurity formation. Not only can an addition sequence significantly improve the impurity profile of a given process, especially for impurities generated during the early and intermediate stages, but it would also mitigate exothermic runaway reactions. This chapter reveals various opportunities for the optimization of addition sequences including direct addition (DA), reverse addition (RA), sequential addition, portionwise addition, alternate addition, and concurrent addition (CA). Many case studies include the Grignard reaction, the amide formation, the nitration and cyclization, and the Sonogashira reaction.

Charging starting materials, reagents, and solvents into a reaction vessel in proper order has substantial effects on the process safety and impurity profile. In addition, an appropriate addition order will result in reducing the batch volume and improving the reaction selectivity. The "all-in" addition is to charge the starting material, reagent, and solvent one after another without control of the addition order and rate. This "all-in" addition mode shall be avoided as an exothermic reaction can create a thermal runaway situation at large scales where heat removal rates are much slower compared to laboratory reactions. In order to manipulate reactions in a safe and controllable manner, in many cases, reactions have to be run in a semibatch manner, that is, charging the starting material, followed by introducing the reagent (and vice versa).

The semibatch operation can be performed by a DA[1] or a RA. DA is a way of conducting a reaction by adding a reagent to a mixture of starting material and solvent in a dose-controlled manner; in contrast, RA is to add the starting material to the mixture of the reagent and solvent. The CA mode of addition is to introduce the starting material and reagent simultaneously.

Each mode of addition can be further differentiated into three categories: solid addition, liquid addition, and gas addition. The addition of solids to a reactor containing flammable solvents

DOI: 10.1201/9781003288411-4

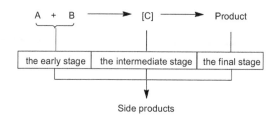

FIGURE 4.1   The formation of side products at three reaction stages.

should be avoided as the static electric charge of the solid dust may ignite the flammable solvents. Therefore, solids should be charged as a solution or slurry if the solids are not fully soluble or are charged into the inert reactor first. Liquid addition is the most commonly used method to introduce the starting material and reagent into the reactor. In the liquid addition, the reagent (or the starting material) (either in solid, liquid, or gas state) is usually dissolved in an appropriate solvent and, if the reagent is liquid, addition as neat may be used. The addition of gas directly is less common except in catalytic hydrogenation.

4.1   DIRECT ADDITION

As a development approach, the DA is commonly used to improve process safety by controlling the heat release rate. In addition, an impurity profile can also be improved by controlling the concentration of reagents.

4.1.1   SONOGASHIRA REACTION

The Sonogashira reaction is a valuable tool for functionalizing aromatic and heteroaromatic halides with a versatile acetylene functional group.[2] One of the common difficulties in the Sonogashira reaction is the polymerization of acetylene, particularly when the cross-coupling reaction is conducted at an elevated temperature, as in the case of aromatic bromides or chlorides when used as starting materials.

(I) Problematic "all-in" conditions
- Mixing **1** with **2** (1.1–1.3 equiv) in the presence of $BuNH_2$ (1.2 equiv) and CuI (2 mol%)/PPh_3 (4 mol%)/$Pd_2(dba)_3$ (0.5 mol%) in 1,3-dimethyl tetrahydropyrimidin-2(1H)-one (DMPU), then heating at 70 °C (Equation 4.1).[3]
- Scale: 100 g.
- Observations: Abrupt temperature rise to 145 °C along with a formation of dark polymeric materials.

$$\text{(Equation 4.1)}$$

(4.1)

(II) Solutions: Semibatch conditions (DA)
- Addition of **2** in NMP over 1 h to a mixture of **1**, $BuNH_2$ (1.2 equiv), CuI (4 mol%), and $Pd(PPh_3)_4$ (2 mol%) in NMP at 50 °C.
- The product **3** was isolated in a yield of 74% with little polymerization.

DISCUSSIONS

(a) The polymerization of the accumulated acetylene is exothermic, which is a big safety concern on a large scale. The DA sequence and temperature (50 °C) are crucial in controlling the concentration of acetylene at a relatively low level in the reaction system.

(b) In order to remove the residual palladium and copper from the product, the workup included washing the product solution with diluted aqueous ammonium hydroxide, followed by treating the solution with Silica*Bond*-Thiol and activated carbon. As a result, the residual metals in the product **3** are 120–150 ppm of Pd and <20 ppm of Cu.

4.1.2 MICHAEL REACTION

Control of impurity formation at the intermediate stage is challenging, because, at this stage, reaction intermediates may react with the starting material/reagent, product, or byproduct. It is well-known that a copper-mediated addition of Grignard reagent onto a conjugated ketone favors 1,4- or 1,6-Michael addition over 1,2-addition. The addition order was found to be critical for the chemoselectivity of the Michael reaction (Equation 4.2).[4]

(I) Problematic reaction conditions (RA Mode)

- Adding **4** to a mixture of **5**, CuCl, Me$_2$S (2.0 equiv) in THF at –30 °C (Equation 4.2).
- Producing only 15% of the desired product **6** along with significant quantities of 1,2-addition side product.

(4.2)

(II) Chemistry diagnosis

The results indicate that under the reaction conditions, the turnover rate of the catalytic cycle leading to cuprate intermediate was insufficient to compete effectively with the direct un-conjugated addition reaction (1,2-addition).

(III) Solutions

In this event, an alternative addition profile was adopted, in which the preformed Grignard reagent **5** was added to a mixture of dienone **4**, copper(I) chloride, and dimethylsulfide. This methodology proved to be very effective to lower the concentration of the Grignard reagent, leading to the desired bis-alkylated enone **6** in 75% isolated yield.

DISCUSSIONS

Dimethyl sulfide has an important effect on stabilizing the active organo-dicuprate species formed by a reaction of Cu(I)Cl with di(bromomagnesio)nonane. In the absence of dimethly sulfide, 1,2-addition of the Grignard and loss of acetate from the sterol are the dominant processes.

4.1.3 FISHER INDOLE SYNTHESIS

Generally, the Fisher indole synthesis by conversion of aryl hydrazones to indoles requires elevated temperatures and acid catalysts. For instance, the synthesis of indole **8** was realized via the reaction of (4-fluorophenyl)hydrazine hydrochloride **7** with dihydropyran (DHP) at 100 °C (Equation 4.3).[5]

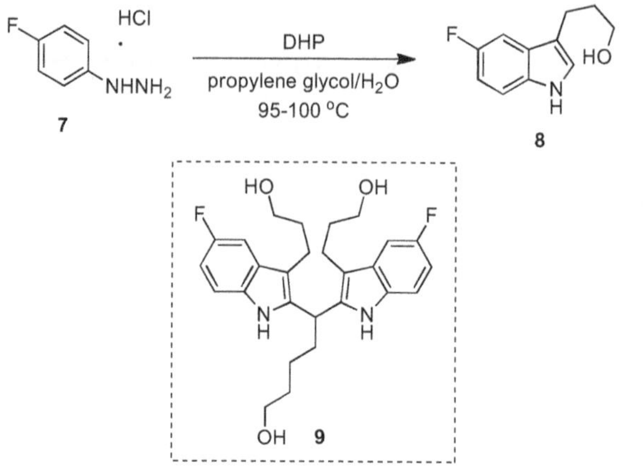

(4.3)

(I) Reaction problems

Initially, the reaction adopted the "all-in" procedure from the literature[6], resulting in several side products; among them, triol **9** was identified as the major impurity and formed in the final stage via a reaction of product **8** with DHP.

(II) Solutions

In order to preclude the formation of **9**, a DA protocol was developed and the reaction was conducted by adding DHP (0.98 equiv) to a solution of **7** in propylene glycol and water at 95–100 °C.

REMARK

Due to its simplicity, the "all-in" protocol is preferred if safety and impurity are not concerns. For instance, the diaryl ether **12** was prepared by mixing phenol **10** and 4,6-dichloropyrimidine **11** with potassium carbonate in dimethyl acetamide (DMAc) prior to heating to 55 °C (Scheme 4.1).[7]

Procedure

- Charge DMAc (8.8 L), 1.377 kg (9.24 mol, 1.05 equiv) of **11**, 1.832 kg (8.8 mol) of **10**, and 1.338 kg (9.68 mol, 1.1 equiv) of K_2CO_3 into a glass-lined reactor.
- Heat the batch to 55 °C and stir for 2 h (<0.5% of **10** by HPLC).

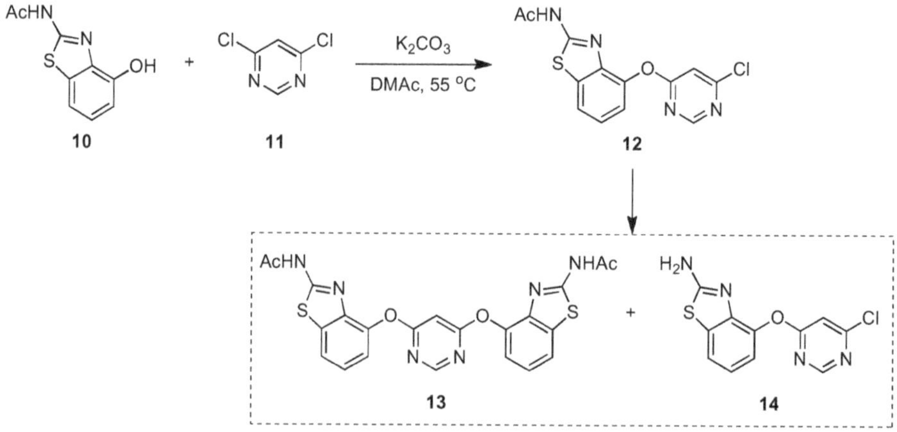

SCHEME 4.1

- Cool to 10 °C.
- Add water (8.8 L) at <20 °C.
- Cool the resulting mixture to 10 °C and age for 2 h.
- Filter, wash with water (1×4.4 L, 4×2.2 L), and dry at 60 °C under vacuum.
- Yield of **12**: 2.54 kg (90%).

4.1.4 AMIDE FORMATION

Amides can be prepared by treatment of carboxylic acids with amines in the presence of activation agents under mild conditions. A number of acid activation agents[8] (see Chapter 2), such as dicyclohexyl carbodiimide (DCC), POCl$_3$, (COCl)$_2$,[9] TiCl$_4$, 1,1-carbonyldiimidazole (CDI), and molecular sieves, are available and applicable for the use of amide preparations on a large scale. It is noteworthy that the addition order of such reagents can have a significant impact on the outcome of amide formations.

4.1.4.1 EEDQ-Promoted Amide Formation

(I) Problematic addition order

When a coupling reaction is performed by mixing the activation agent, 2-ethoxy-1-ethoxycarbonyl-1,2-dihydroquinoline (EEDQ), with acid **15**, followed by adding aniline **16**, an ethyl ester side product **18** could be formed as much as a yield of 20% (Scheme 4.2).[10]

(II) Chemistry diagnosis

Activation of the acid **15** with EEDQ produces, in addition to the mixed anhydride[11] **20**, two by-products: quinoline and ethanol. The formation of **18** (at the intermediate stage) is due to a side reaction of the intermediate **20** with ethanol (Scheme 4.3). Therefore, in order to control the intermediate **20** at a minimum level, the activation of **15** should be performed in the presence of aniline **16**.

(III) Solutions

The use of the DA approach proved to be the method of choice in suppressing the formation of **18**. Thus, the reaction was performed by dissolving the amino acid **15** and aniline **16** in dichloromethane, followed by adding the coupling agent. This protocol resulted in a robust reaction that always gave reproducible product yields of good quality (99.6%). Upon completion of the reaction,

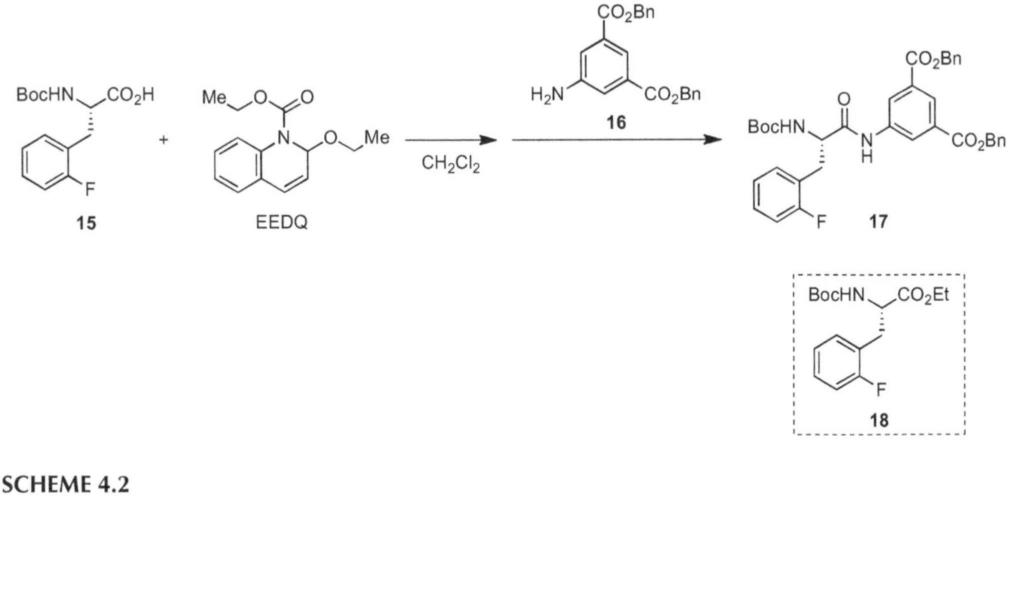

SCHEME 4.2

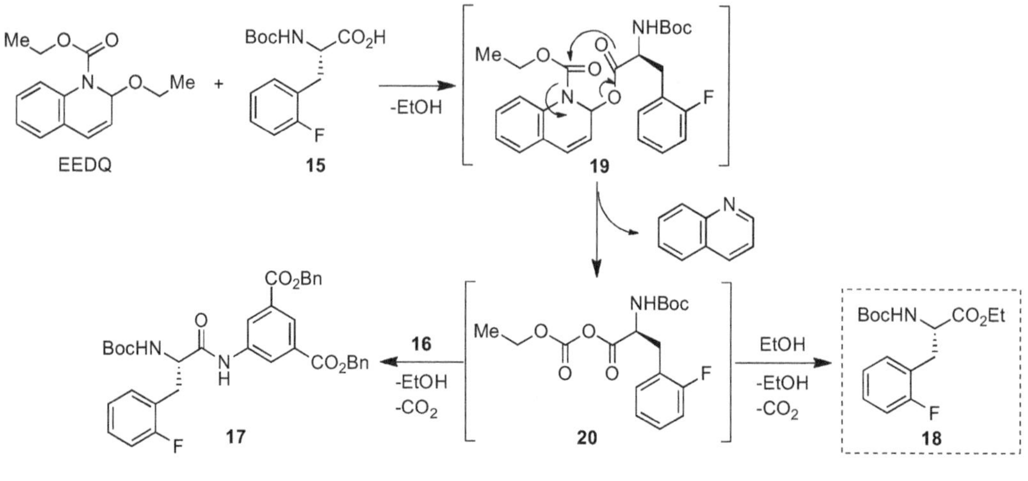

SCHEME 4.3

the solvent was changed from dichloromethane to propylene glycol monomethyl ether (PGME), a solvent from which the reaction product crystallized.

Procedure

- Charge CH_2Cl_2 (212.2 L) into a reaction vessel, followed by adding **15** (10.74 kg, 37.9 mol) and **16** (13.71 kg, 37.9 mol).
- Charge 9.4 kg (37.9 mol) of EEDQ.
- Stir the resulting solution at 20 °C for 22 h.
- Distill off CH_2Cl_2 (136 L) under 1 atm.
- Add PGME (190 L).
- Continue to distill until the internal temperature reached 115 °C.
- Cool the batch to 22 °C over 15 h.
- Filter and wash with PGME.
- Dry at 50 °C under vacuum.
- Yield: 21.87 kg (92%) as a white solid.

4.1.4.2 CDI-Promoted Amide Formation

(I) Problematic addition order

Amide **24** is an intermediate for the synthesis of $\alpha_v\beta_3$ integrin antagonist **26** (Scheme 4.4).[12] Initially, a coupling reaction between amine salt **21** and activated glycine **22** was carried out by adding the solid amine salt **21** to a solution of **22**, providing the desired intermediate **24** in 93% crude yield. A diglycine impurity **25**, derived from *O*-acylation of **24** at the final state, was also observed in 3%.

(II) Solutions

To suppress the formation of impurity **25**, an alternative addition strategy was adopted by mixing the acid with an activation agent prior to the coupling with the amine.

- Controlled-addition of the solution of **22** to the solution of **21** to limit the concentration of **22** in the reaction system.
- 85% yield of **24** with 98.9% purity.

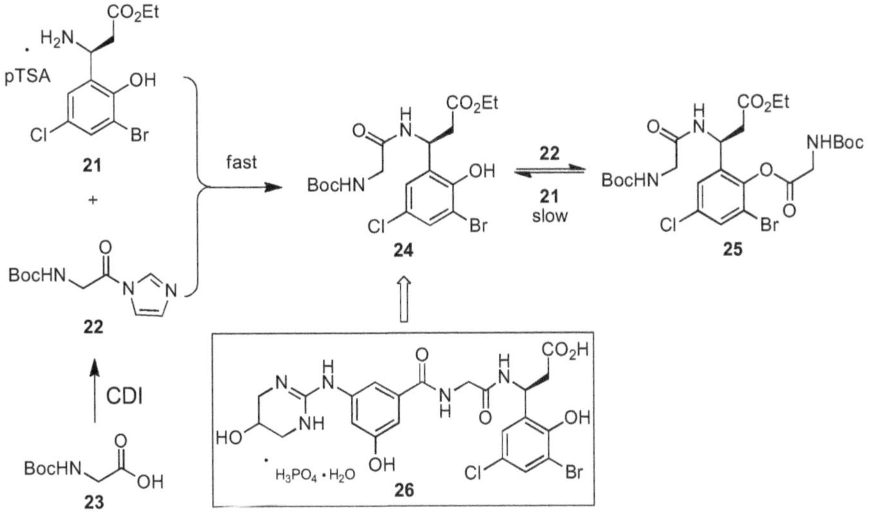

SCHEME 4.4

4.1.5 THIOAMIDE FORMATION

The DA approach is also useful in controlling the levels of toxic gas present in the reaction system. Examination of the transformation of nitrile **27** to thioamide **28** showed that reactions with sodium sulfide or $P_{10}S_4$ suffered low yields of product. Ultimately, diethyl dithiophosphate was chosen for the conversion of **27** to **28** (Equation 4.4),[13] though diethyl dithiophosphate is unpleasant and difficult to handle.

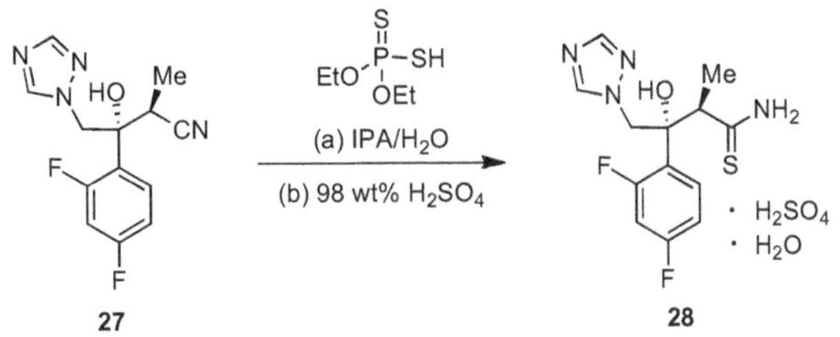

$$(4.4)$$

(I) Problems

The use of diethyl dithiophosphate also produced H_2S, which was flammable, toxic, and malodorous. H_2S was generated from a reaction of by-product, $(EtO)_2P(O)SH$, with water. However, water proved to be essential for this reaction as the reaction in the absence of water became much slower and the formation of side products interfered with the subsequent isolation.

(II) Solutions

To address the H_2S-associated safety issues, diethyl dithiophosphate was added to a preheated aqueous isopropanol slurry of **27** at 80 °C. Thereby, the formation of H_2S was controlled by the rate of the addition of the diethyl dithiophosphate.

Procedure

- Heat a mixture of **27** (43.0 kg, 154.5 mol), water (17.2 L), and IPA (54.0 kg) to 78 °C to give a solution.

- Charge 86.2 kg (462.9 mol, 3.0 equiv) of diethyl dithiophosphate at 75–80 °C over 1.5 h.
- Rinse the addition line with IPA (2 kg).
- Heat the resulting mixture at 75–80 °C for 7.5 h.
- Cool the batch to 20 °C over 1 h.
- Add EtOAc (194 kg) and water (215 kg) over 50 min.
- Cool the resulting mixture to 2 °C over 80 min.
- Add 3N NaOH (284 kg) at 10–20 °C over 75 min (with final pH 8.1 at 15 °C).
- Separate layers.
- Treat the organic layer with 20 wt% NaOH (259 kg) and stir at ambient temperature for 30 min.
- Separate layers.
- Treat the organic layer with 98 wt% H_2SO_4 (15.3 kg, 152.9 mol) at 40 °C for 3 min.
- Concentrate the resulting mixture at 50–60 °C under reduced pressure to 87 L over 6 h.
- Cool the resulting slurry to 26 °C over 70 min.
- Add MTBE (128 g) over 9 min.
- Cool to 5 °C over 1.5 h and age at 0.2–5 °C for 1 h.
- Filter, wash with MTBE (136 kg), and dry at 60 °C under vacuum.
- Yield of **28**: 58.2 kg (91.2%).

4.1.6   C–O BOND FORMATION

4.1.6.1 $S_{RN}2$ Reaction

An *O*-arylation of azetidinol **30** with aryl iodide **29** furnished the corresponding ether **31** in the presence of potassium *tert*-butoxide (Equation 4.5).[14]

(4.5)

(I) Problematic addition order

- Addition of the aryl iodide **29** to the solution of the azetidine alkoxides generated in situ by treating **30** with KO*t*Bu.
- The des-iodo purine **32** was observed as the major product.

(II) Solutions

Conducting the etherification step by slow addition of the alkoxide of **30** to the solution of aryl iodide **29** gave the desired aryl ether **31** in a quantitative yield.

DISCUSSIONS

(a) The reaction may proceed through an $S_{RN}2$ reaction (radical nucleophilic bimolecular substitution) pathway[15] as described in Scheme 4.5.

In the case of the RA (adding **29** to alkoxide of **30**), an alkoxyaryl anion radical would have a chance to dissociate into alkoxide and aryl radical which reacts with *tert*-butanol (generated in situ from the reaction of KO*t*Bu with **30**) giving the des-iodo side product **32**. In contrast, adding the solution of alkoxide of **30** to the solution of **29** would suppress the radical proton abstraction due to the relatively low concentration of *tert*-butanol. Furthermore, the excess of aryl iodide **29** would react with the alkoxyaryl anion radical rapidly via the SET process to furnish the desired product **31** along with the regeneration of anion radical of **29**.

(b) It would be difficult to know when and how the impurity **32** was generated without the knowledge of the reaction mechanism. Therefore the mechanism-guided process development methodology is critical in resolving this reaction problem (see Chapter 12 for more information).

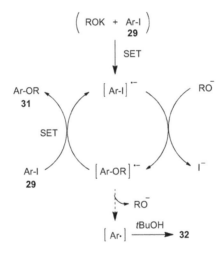

SCHEME 4.5

4.1.6.2 Mitsunobu Reaction

Mitsunobu reaction is an organic reaction that converts an alcohol into a variety of functional groups using triphenylphosphine and an azodicarboxylate, such as diethyl azodicarboxylate (DEAD), diisopropyl azodicarboxylate (DIAD), or di-*tert*-butyl azodicarboxylate (DTAD). The order of reagent addition in Mitsunobu reaction can be important. Typically, the DEAD solution is slowly added to a mixture of alcohol, phenol (for preparation of ether), and triphenylphosphine in a suitable solvent. Alternatively, the addition of the alcohol to a preformed betaine intermediate may give better results. For instance, the preparation of alkyl aryl ether **35** (a potent SRC kinase inhibitor) via the Mitsunobu reaction was performed by adding the alcohol **34** to a mixture of **33**, DTAD, and triphenylphosphine in THF (Scheme 4.6).[16]

Procedure

- Charge a solution of triphenylphosphine (3.057 kg, 11.66 mol, 2.47 equiv) in THF (8 L) over 15 min into a 100-L vessel containing a slurry of **33** (2.139 kg, 4.73 mol) and DTAD (2.771 kg, 12.03 mol, 2.54 equiv) in THF (31 L) at ambient temperature.

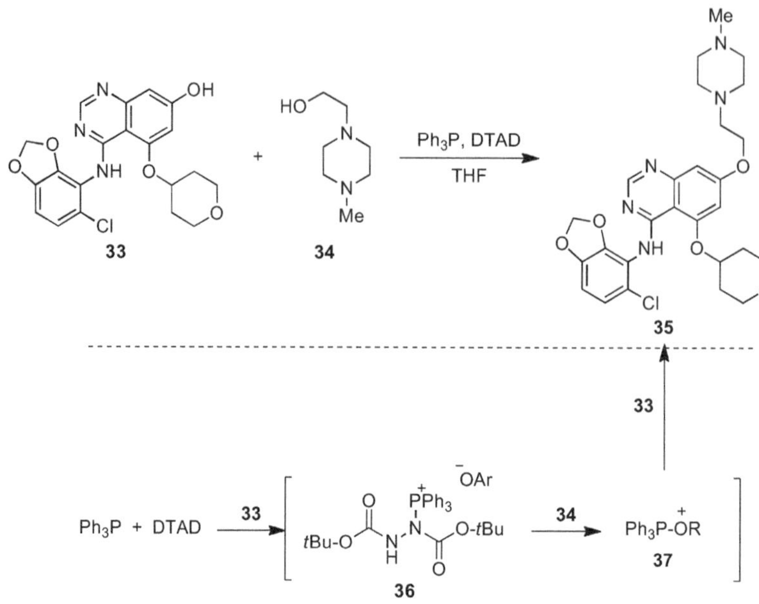

SCHEME 4.6

- Rinse the line with THF (2 L) and stir for 10 min.
- Cool to 15 °C.
- Charge a solution of **34** (1.050 L, 1.049 kg, 7.27 mol, 1.54 equiv) in THF (3 L) at below 20 °C over 12 min.
- Rinse the line with THF (1 L).
- Stir the resulting mixture at ambient temperature for 50 min.
- Charge a solution of **34** (348.5 mL, 348.2 g, 2.41 mol, 0.51 equiv) in THF (1 L).
- Stir the batch for 1.5 h.
- Workup.

DISCUSSIONS

The formation of the ion pair **36** is fast, while the formation of the oxyphosphonium intermediate **37** is a slow and rate-determining step.[17]

4.2   REVERSE ADDITION

Reverse addition (RA) is a mode of addition by adding the starting material to a mixture of reagents. This mode of addition allows control of the concentration of the starting material in the reaction system at the desired level. Enolation is a side reaction that occurs frequently in Grignard reaction of ketones bearing α-protons. Maintaining a low concentration of ketone starting materials by the RA could suppress this side reaction. Such an RA strategy was also used to inhibit a side reaction during the copper-catalyzed epoxide-ring opening. Nitration of electron-rich arenes often produces multiple products due to poor regioselectivity. Using the RA approach could improve the regioselectivity of the nitration reaction. Not surprisingly, a preference for low concentration for an intra-molecular reaction is based on the fact that many undesired intermolecular reactions can be prohibited under such low-concentration conditions. However, those diluted reaction conditions render poor volume efficiency and should be avoided on production scales. The RA approach proved to be the method of choice to meet the concentration criterion of intramolecular reactions.

In addition to the desired coupling with amine, the acylimidazolide, an intermediate from a CDI-mediated amide synthesis, can participate in various side reactions such as hydrolysis back to the acid. Similar to the DA approach, the RA, by adding the acylimidazolide intermediate to a solution of amine, prevented the hydrolysis of acylimidazolide. Choosing the right addition mode sometimes is not obvious and requires understanding the reaction mechanism. The example in Section 4.2.6 illustrates the application of the RA to inhibit the side reaction during an unusual ketone reduction to hydrocarbon with sodium borohydride.

The RA can also create a situation wherein the reagent was maintained in excess to ensure that the labile, starting material-derived intermediate was consumed as fast as possible to prevent side reactions such as dimerization (Section 4.2.7.1). In addition to the desired reaction, a reactive starting material can also react with a by-product, as shown in the case of [3+2] cycloaddition (Section 4.2.7.2), to form an undesired side product. Therefore, the reverse mode of addition was used to control the concentration of the starting material.

4.2.1   Grignard Reaction

The addition of Grignard reagents to aldehydes or ketones to form, upon hydrolysis, alcohols is known as the Grignard reaction. Several potential side reactions[18] occur mostly with hindered ketones and with bulk Grignard reagents. Enolization and reduction[19] (as shown in Equations 4.6 and 4.7) are the side reactions that take place frequently.

$$R^1 \overset{O}{\underset{R^3}{\overset{R^2}{\diagup}}}\!\!\!\!H \quad \xrightarrow{RMgX} \quad R^1 \overset{OMgX}{\underset{R^3}{\diagup}} R^2 \quad + \quad RH \tag{4.6}$$

38   **39**

$$R^1 \overset{O}{\diagup} R^2 \; + \; R^4 \overset{R^3}{\underset{H}{\diagup}}\overset{R^5}{\underset{MgX}{\diagup}} R^6 \quad \longrightarrow \quad R^1 \overset{H}{\underset{R^2}{\diagup}} OMgX \; + \; R^4 \overset{R^3}{\diagup}\overset{R^5}{\diagup} R^6 \tag{4.7}$$

40   **41**   **42**   **43**

Additionally, enolates **39**, generated from deprotonation of an α-hydrogen of the aldehydes or ketones, can participate in a condensation reaction with aldehydes or ketones **38** (aldol reaction) to give condensation side products **44** (Equation 4.8).

$$R^1 \overset{O}{\underset{R^3}{\diagup}} R^2 \; + \; R^1 \overset{OMgX}{\underset{R^3}{\diagup}} R^2 \quad \longrightarrow \quad R^1 \overset{O \; R^2}{\underset{HO \; R^1 \; R^3}{\diagup}} \overset{R^3}{\underset{R^2}{\diagup}} \tag{4.8}$$

38   **39**   **44**

In order to restrain this enolization side reaction, several approaches[20] have been developed including the use of lower temperature or $CeCl_3$. The subsequent condensation of the enolate can be suppressed by utilizing controlled RA of the aldehyde or ketone to minimize the concentration of **38** in the reaction mixture.

4.2.1.1   Reaction with Alkyl Aryl Ketone

The enolates formed in the Grignard reaction of alkyl aryl ketones may lead to not only low product yields, but also problems in the isolation and purification. For instance, the Grignard reaction of ketone **45** with vinyl Grignard **46** was carried out in THF at 5 °C to produce a tertiary alcohol **47** along with a considerable amount of a self-aldol adduct **48** (Equation 4.9).[21]

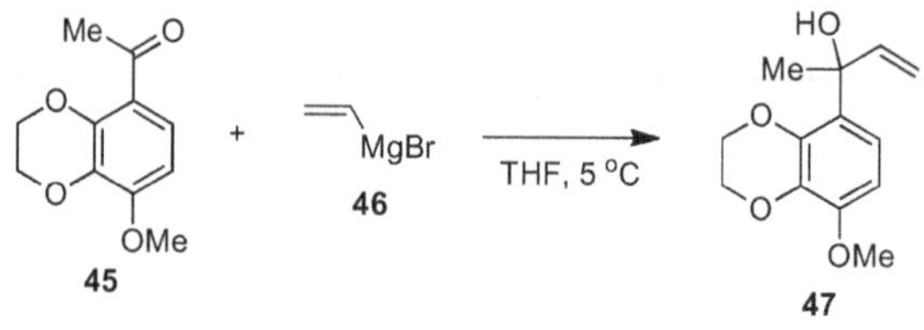

(4.9)

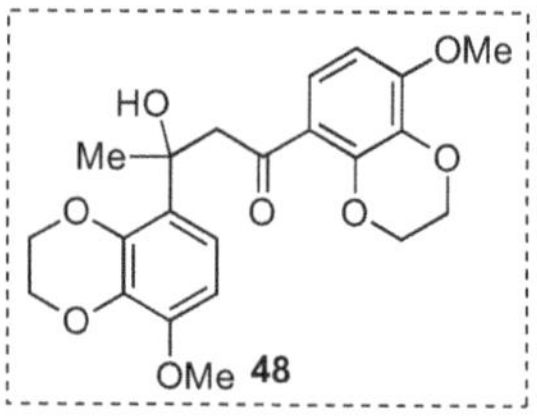

The formation of **48** was caused by the DA of **46** to the solution of ketone **45**. To suppress this impurity formation, the reaction was carried out by the RA giving the desired product **47** in an assay yield of 97%.

4.2.1.2   Grignard Reaction with Aldehydes

(I) Problematic addition order
- Addition of MeMgCl solution (3M in THF) to a solution of **49** (or **51**).
- The reduction product **50b** (or **52b**) in yields of 6–8% (Equations 4.10 and 4.11).[22]

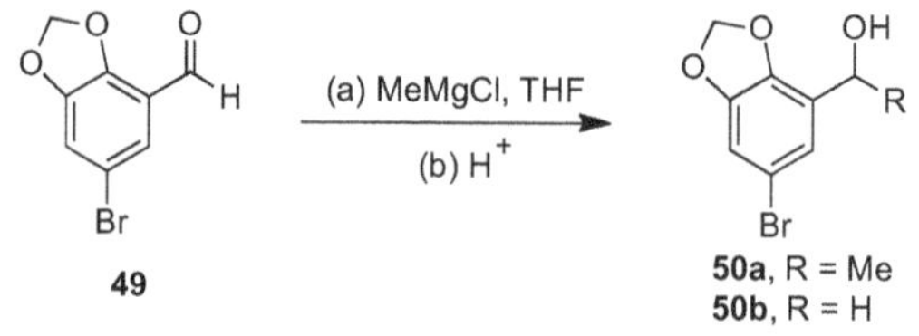

(4.10)

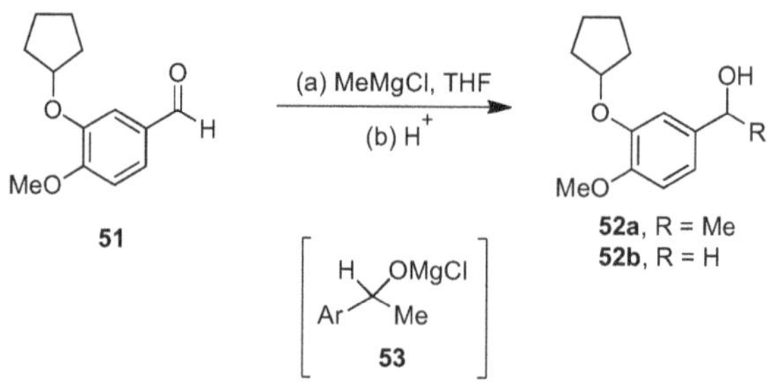

(4.11)

(II) Solutions

Unlike the reduction as described in the reaction (Equation 4.7), the reductive side product **50b** (or **52b**) is formed via a Cannizzaro-type reaction between intermediate **53** and aldehyde **49** (or **51**). To address the formation of these side products, an RA approach was adopted by charging the solution of aldehyde **49** (or **51**) in THF to a cold solution of MeMgCl.

4.2.1.3   Reaction of Grignard Reagent with Ester

The transformation of ester **54** (or **56**) to drug substance **55** (or tertiary alcohol **57**) was realized via the addition of methyl Grignard (Equations 4.12 and 4.13).[23]

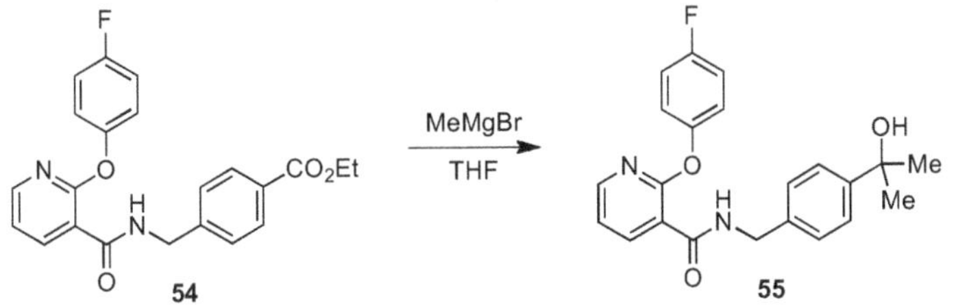

(4.12)

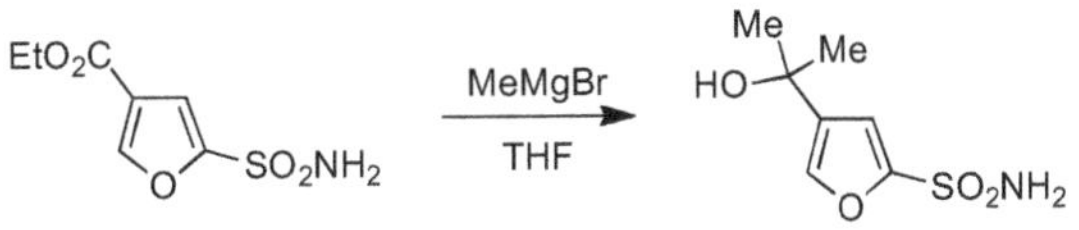

(4.13)

(I) Reaction problems

These reaction protocols resulted in a generation of >20% of diol compound **58** (or **59**) (Scheme 4.7).

(II) Chemistry diagnosis

These dimeric diols are the results of enolization of ketone intermediates, followed by aldol-type condensation and the Grignard addition (Scheme 4.8).

(II) Solutions

In order to improve the reaction selectivity and suppress the undesired enolization, the reaction was conducted using RA (adding the substrate to an excess of the Grignard) at room temperature, resulting in a good yield of **55** with only 2% of **58**.

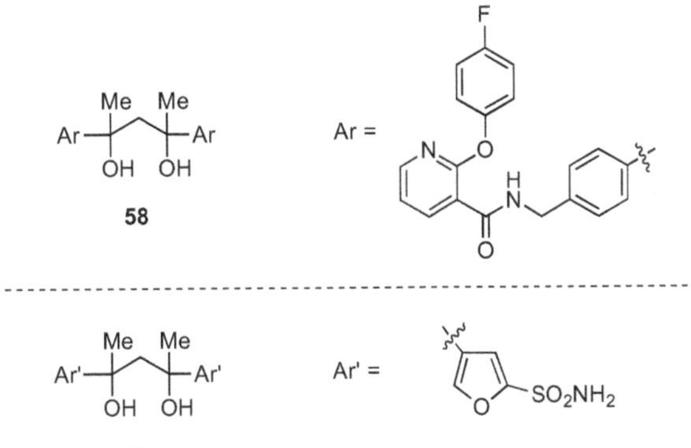

SCHEME 4.7

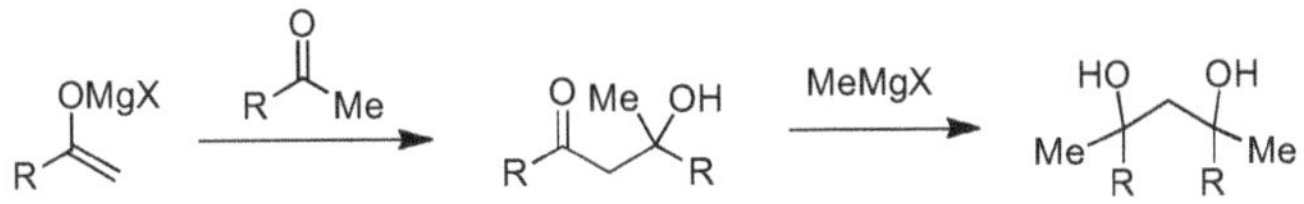

SCHEME 4.8

4.2.2 COPPER-CATALYZED EPOXIDE RING OPENING

Epoxide ring-opening with Grignard reagent mediated by CuCl produced the desired ring-opening product **62**. In addition to the desired **62**, this reaction may generate chlorohydrin side product **63** (Equation 4.14).[24]

$$(4.14)$$

Solutions

- RA of the epoxide **60** to the mixture of vinylmagnesium chloride **61** and copper catalyst.
- The level of chlorohydrin impurity **63** was controlled below 1%.

Procedure

- Charge a solution of **60** (20.7 kg, 126.1 mol) in THF (105 L) to a mixture of **61** (117 kg, 1.6M, 1.5 equiv) and CuCl (0.63 kg, 6.4 mol) in THF over 1 h while maintaining the temperature at –10 °C to –5 °C.
- Warm the batch to 0 °C.
- Add MeOH (10.2 kg) followed by 2N HCl (186 L) at 0–10 °C.
- Age the batch for 1 h.
- Add MTBE (105 L).
- Separate layers.
- Wash the organic layer with 2N HCl (80 L), water (40 L), 10% sodium thiosulfate (80 L), and water (40 L).
- Yield: 22.2 kg (94% solution yield).

4.2.3 NITRATION REACTION

Nitration of aromatics requires strong acidic conditions in order to generate nitration species (NO_2^+) in situ, and concentrated sulfuric acid is the most commonly used strong acid in conjunction with nitric acid. A challenging issue related to the use of nitric acid is to control the nitric acid in strict one equivalent or the un-reacted starting material/or over-nitration side product will be produced. In order to maintain precisely the 1:1 ratio of the starting material to the nitric acid, the 4-(4'-methoxyphenyl)morpholine nitric acid salt **64** was chosen as the nitration starting material. However, it was noticed that the additional order in the nitration of **64** was critical in controlling the formation of regioisomer **66** (Equation 4.15).[25]

$$(4.15)$$

(I) Problematic addition order

- The addition of H_2SO_4 to the solution of **64** in CH_2Cl_2 resulted in a significant amount of the regioisomer **66**.

(II) Chemistry diagnosis

The addition of sulfuric acid to the solution of **64** in CH_2Cl_2 would lead to nitric acid via a salt-exchange process with a rather low level of nitration species (NO_2^+). As a consequence, the nitric acid would attack the C2′ position of the aromatic ring to form nitro iminium intermediate **69** whose deprotonation would afford **66** (Scheme 4.9).[26]

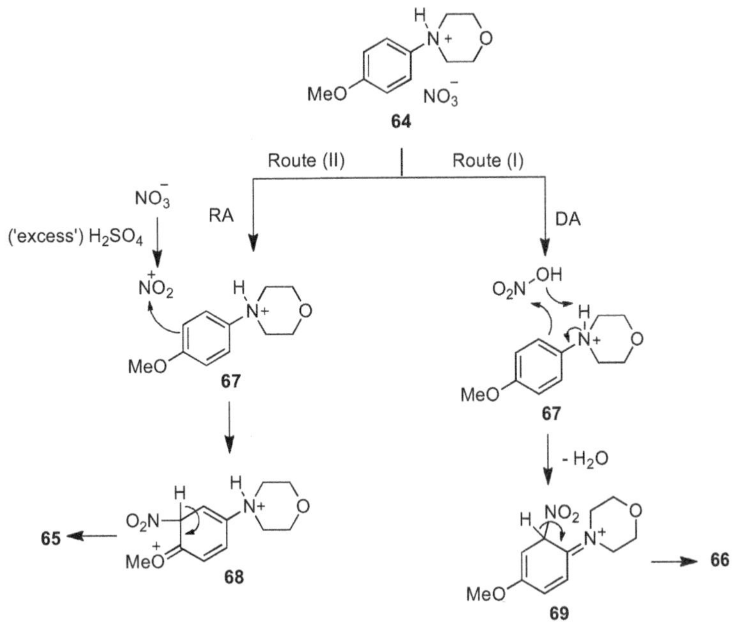

SCHEME 4.9

(III) Solutions

In contrast, the addition of **64** to sulfuric acid would allow the initially formed nitric acid to react with an excess of sulfuric acid to generate NO_2^+. Thus, the nitration occurs at the desired C3′ position to generate intermediate **68** which, upon deprotonation, furnishes the desired product **65**.

Procedure

- Charge 80 g (0.815 mol) of 95% H_2SO_4 into a reactor and cool to 0 °C.
- Charge a solution of **64** in CH_2Cl_2 (125 mL) to the acid while maintaining the batch temperature at 0±5 °C.
- Stir for 30 min.
- Separate the bottom acid layer.
- Add the acid layer to water (200 mL) at <10 °C.
- Add 190 mL of 28% NH_4OH to the resulting acid solution at <10 °C until pH>10.
- Stir the batch at 5±5 °C for 1 h.
- Filter and wash with water (50 mL).
- Dry at 45 °C under vacuum.
- Yield: 17.5 g (94%).

4.2.4 CYCLIZATION REACTION

Generally, low-concentration reaction conditions are preferred for intramolecular reactions in order to avoid undesired intermolecular reactions. These reaction conditions, however, render the process non-economical, which precludes it from large-scale production. Therefore, a strategy for controlling the concentration of the starting material is to adopt the RA protocol.

An intramolecular epoxide ring-opening reaction of **70** was carried out by a slow addition of the solution of **70** in dimethyl acetamide (DMAc) to a slurry of cesium carbonate (0.5 equiv) in DMAc at 65 °C (Equation 4.16).[27]

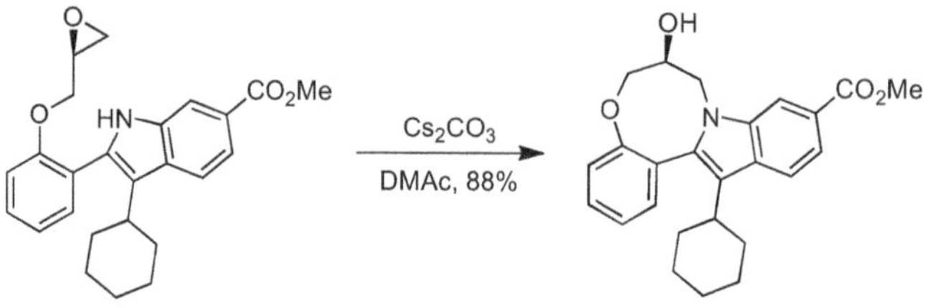

$$(4.16)$$

Procedure

- Charge a solution of **70** (8.6 kg) in DMAc (24.4 kg) to a slurry of Cs$_2$CO$_3$ (3.48 kg) in DMAc (56.7 kg) at 65 °C over 2 h.
- After the addition, stir the resulting slurry at 65 °C for 30 min.
- Cool the batch to 45 °C.
- Add isopropyl acetate (7.5 kg) followed by a slow addition of water (86.5 L) over 30 min while maintaining the temperature at 45 °C.
- Cool the batch to 21 °C and age for 30 min.
- Filter and wash with DMAc/water (1:3, 12 kg) followed by water (12 kg)
- Dry at 60 °C under vacuum.
- Yield: 8.3 kg (88%, 92.7% ee).

DISCUSSIONS

This RA approach has its limitations, for instance, an RA of the pyrrole anion of **72** to a solution of **73** could only reduce the level of **76** to 1–2% from 2–5% generated from DA protocol (Scheme 4.10).[28]

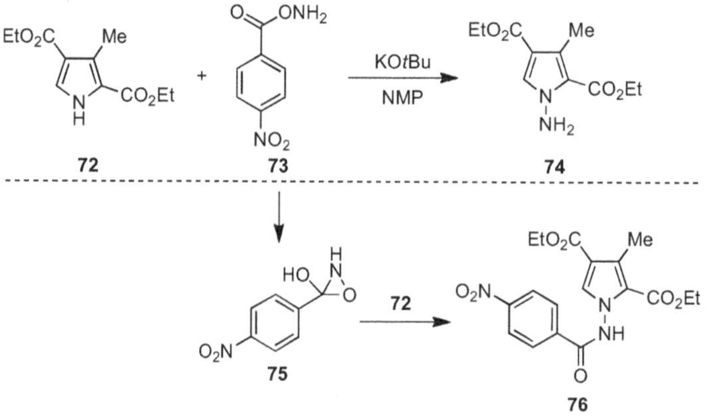

SCHEME 4.10

4.2.5 AMIDE FORMATION

4.2.5.1 CDI-Promoted Amide Formation

An issue associated with amide formation via CDI-promoted acid activation is the hydrolysis of the acylimidazolide intermediate back to the acid.

(I) Problematic addition order

The addition of hygroscopic ethylamine hydrochloride to a solution of **78** (obtained in situ) resulted in significant hydrolysis of **78** (back to **77**) (Scheme 4.11).[29]

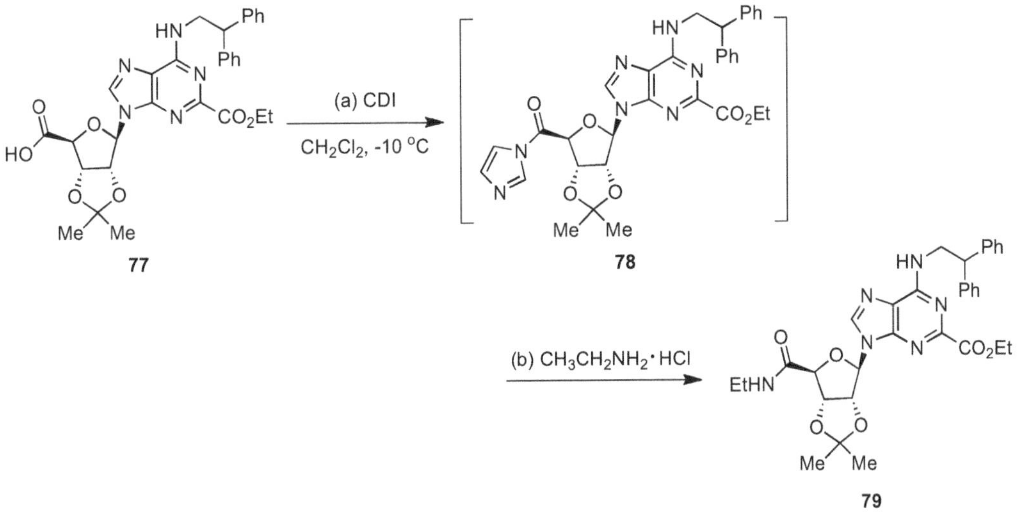

SCHEME 4.11

(II) Solutions

Utilization of a protocol by adding the solution of **78** to the slurry of ethylamine hydrochloride in CH$_2$Cl$_2$ successfully suppressed the hydrolysis of **78**.

Procedure

Stage (a)
- Charge 16.28 kg (100.4 mol, 1.2 equiv) of CDI into a solution of **77** (~48.0 kg, 83.6 mol, 1.0 equiv) in CH$_2$Cl$_2$ (480 L) at room temperature.
- Stir for 3 h to form a solution of **78** and cool the mixture to 5 °C.

Stage (b)
- Prepare, in a separate vessel, a slurry of ethylamine hydrochloride (8.2 kg, 100 mol, 1.2 equiv) in CH$_2$Cl$_2$ (144 L) and cool to −15 °C.
- Charge the solution of **78** into the slurry of ethylamine hydrochloride at <0 °C.
- Rinse the line with CH$_2$Cl$_2$ (30 L).
- Stir the batch at 0 °C for 16 h.
- Warm the batch to 20 °C.
- Replace CH$_2$Cl$_2$ with methanol and use the methanol solution directly in the next step.

4.2.5.2 Phenyl Chloroformate-Promoted Urea Formation

The order of addition was important for coupling of the carbamate **81** with the spirolactone **82** as **81** was unstable to base (diisopropylethylamine [DIPEA]) in the absence of the amine coupling partner. Thus, the reaction was carried out by mixing the HCl salt of spirolactone **82** and DIPEA for 0.5 h followed by the addition of the carbamate **81** (Scheme 4.12).[30]

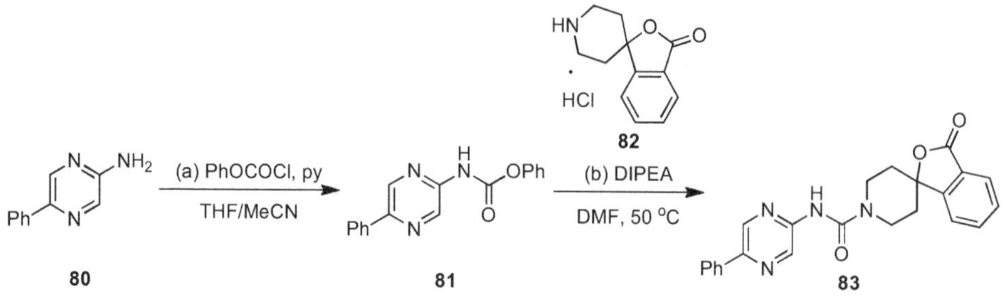

SCHEME 4.12

Procedure

Stage (b)

- Stir a mixture of **82** (1.68 kg, 7.5 mol), DIPEA (990 mL, 5.67 mol) in DMF (13.9 L) for 0.5 h.
- Charge 1.98 kg (6.8 mol) of **81**.
- Heat the resulting mixture at 40–45 °C for 2 h.
- Cool to ambient temperature.
- Add 119 mL (2.08 mol) of AcOH followed by adding water (4.2 L).
- Stir the batch for 1 h to initiate the crystallization.
- Add water (5.1 L) over 2–4 h and age the batch for 2 h.
- Filter and wash with DMF/H$_2$O (1:1) and water.
- Yield: 2.76 kg (97%) as monohydrate.
- Convert the monohydrate to the anhydrous form by suspending it in MeCN (7.9 L) overnight to give 2.5 kg of **83**.

4.2.6 REDUCTION OF KETONE TO HYDROCARBON

Sodium borohydride (NaBH$_4$) is a mild and selective reducing agent for the selective reduction of aldehydes and ketones to alcohols in the presence of functional groups such as NO$_2$, Cl, CO$_2$R, and CN and can be used in water or alcoholic solvents.

A synthetic strategy was designed to access 5-(benzyloxy)-2-ethylphenol **85** via the reduction of ketone **84** with NaBH$_4$ in aqueous THF. This reductive protocol involves two stages: (a) acylation of the hydroxyl group with ethyl chloroformate and (b) reduction with NaBH$_4$ by DA protocol (Equation 4.17).[31]

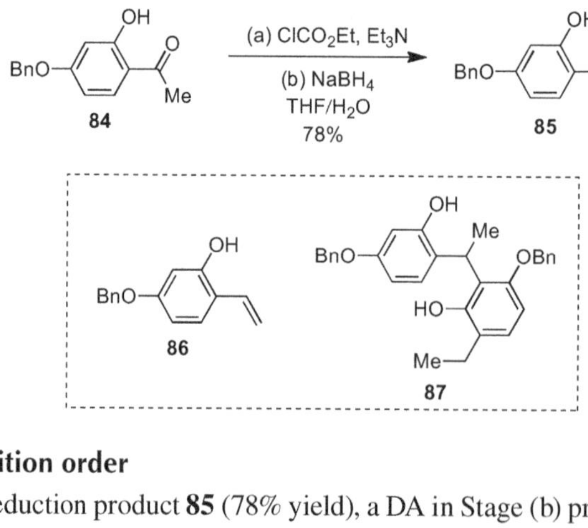

$$(4.17)$$

(I) Problematic addition order

Besides the desired reduction product **85** (78% yield), a DA in Stage (b) produced two side products, **86** and **87**.

(II) Chemistry diagnosis

A detailed reaction mechanism revealed that quinone methide **91** is the key intermediate for the formation of the product **85**, and also responsible for the formation of **86** and **87** (Scheme 4.13) at the intermediate/final stages.

As described in Scheme 4.13, the intermediate **91** would undergo, besides the pathway to the desired reduction product **85**, tautomerization or 1,4-addition with the product **85** to afford the 5-(benzyloxy)-2-vinylphenol **86** or dimer **87**, respectively. Therefore, the presence of an excess of $NaBH_4$ in the reaction system to maintain a low concentration of **91** is essential in minimizing the formation of impurities **86** and **87**.

(II) Solutions

An addition protocol was adopted in Stage (b) by the addition of the solution of the intermediate **88** to $NaBH_4$.

Procedure

Stage (a)
- Charge 53.3 kg (491 mol) of ethyl chloroformate to a solution of **84** (91.0 kg, 375.6 mol) and Et_3N (50 kg, 500 mol) in THF (91 kg) at 20–25 °C.
- Stir at 20–25 °C for 1 h before quenching with water (220 L).
- Separate layers.

Stage (b)
- Charge the resulting organic phase (containing **88**) to a solution of $NaBH_4$ (30 kg, 793 mol).
- Charge two additional portions of $NaBH_4$ to the reactor: $NaBH_4$ (20 kg)/water (70 kg) and $NaBH_4$ (10 kg)/water (60 kg).
- Separate layers (the resulting organic solution of **85** was directly used in the subsequent transformation).

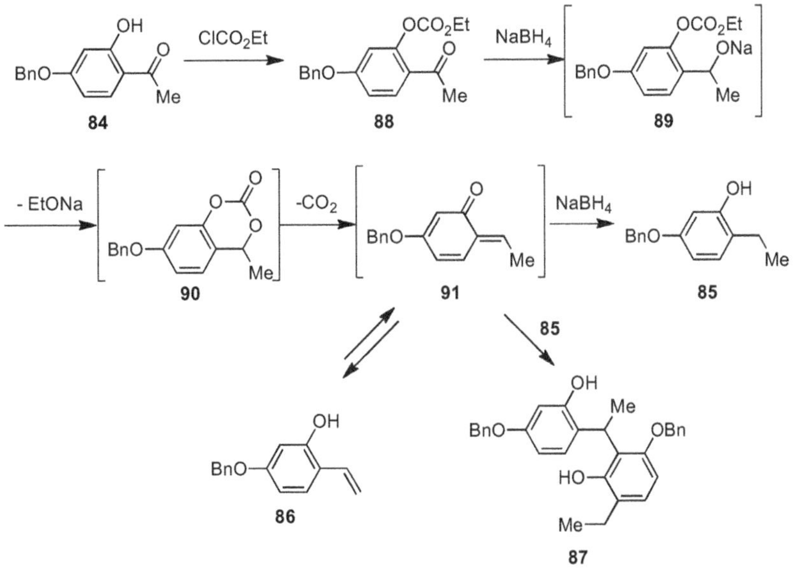

SCHEME 4.13

REMARK

This acylation/reduction method was also used for the reduction of ketone **92** (Equation 4.18).[32]

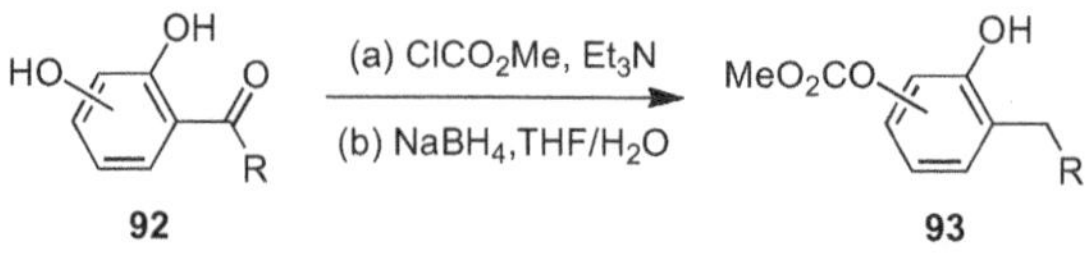

$$(4.18)$$

4.2.7   1,3-Dipole-Involved Reactions

4.2.7.1   Addition–Elimination/Cyclization

The mode of addition is also critical for some base-sensitive starting materials. A two-step, one-pot process for the synthesis of indazole **96** involves a coupling of a base-sensitive chloroimidate **94** with piperazine **95**, followed by an intramolecular cyclization (Scheme 4.14).[33]

(I) Reaction problems

Tetrazene **99** was observed as the major product when adding **95** to **94**.

(II) Solutions

The reaction between **94** and **95** would generate dipole intermediate **97** along with piperazine hydrochloride (**95**·HCl) salt. Thus, the addition of **95** to **94** would make the concentration of the free base **95** relatively low, allowing **97** to dimerize to the dimer **99**. To suppress the formation of **99** and maximize the yield of **96**, the reaction was performed by adding solid chloroimidate **94** to a solution of piperazine **95** (2.37 equiv).

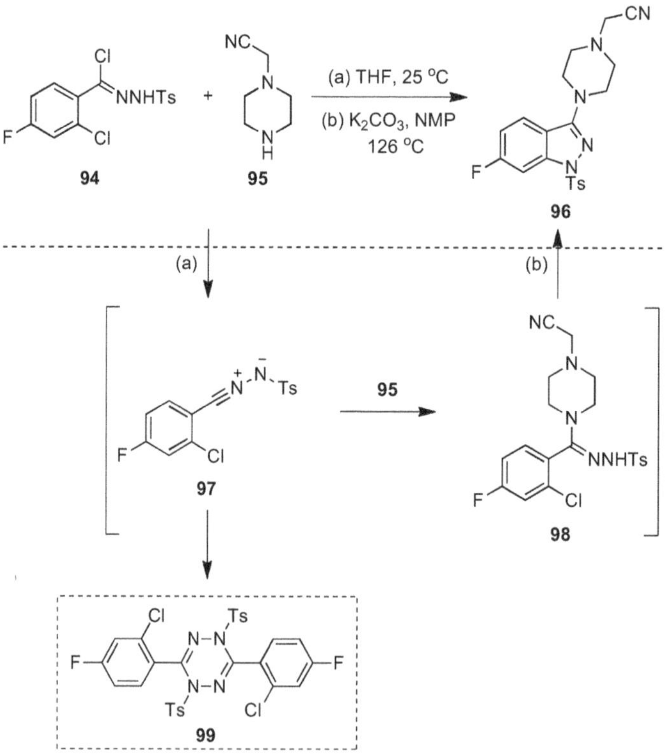

SCHEME 4.14

Procedure

Stage (a)
- Charge 3.5 kg (9.7 mol) of **94**, as solid portionwise, to a mixture of **95** (2.9 kg, 23 mol, 2.37 equiv) in THF (32 L) at 21–29 °C over 2 h.
- Stir at 25 °C for 30 min.
- Distill at 35 °C under 50 Torr to remove THF.

Stage (b)
- Add NMP (14 L) followed by finely milled K_2CO_3 (5.5 kg, 39.8 mol, 4.1 equiv).
- Heat the mixture at 126 °C for 1 h.
- Cool to 40 °C.
- Add MeOH (14 L) followed by water (56 L) at 25 °C over 45 min.
- Age the resulting suspension for 18 h.
- Filter and wash with water (60 L).
- Dry at 70 °C/50 Torr for 18 h.
- Yield: 2.9 kg (73%).

4.2.7.2   [3+2] Cycloaddition

[3+2]-Cycloadditions are important routes toward the synthesis of five-membered heterocycles such as triazoles, furans, isoxazoles, and pyrazoles. Scheme 4.15[34] illustrates a [3+2]-cycloaddition between bromide **100** and enamine **101** to afford pyrazole **102**.

(I) Reaction problems

Apart from the desired pyrazole product **102**, a side product **105** was observed which was stable under the reaction conditions and its elimination to the dipole intermediate **103** did not occur.

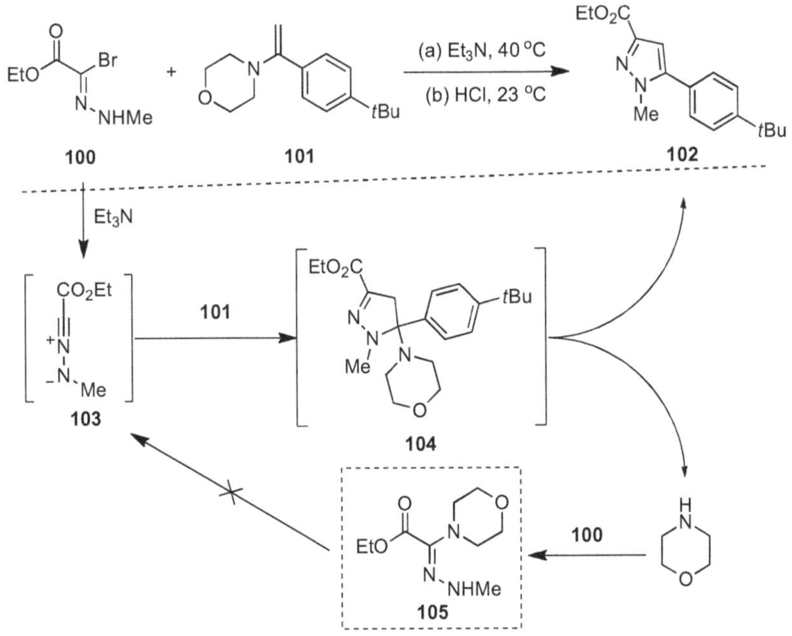

SCHEME 4.15

(II) Solutions

The formation of **105** is attributed to a reaction of **100** with morpholine by-product. Thereby, in order to suppress the formation of **105** during this 1,3-dipolar cycloaddition, an additional protocol was adopted by adding the bromide **100** to a mixture of **101** and triethylamine.

Procedure

Stage (a)
- Charge a solution of **100** (31.2 kg, 149 mol, 1.5 equiv) in EtOAc (116.5 L)/CH$_2$Cl$_2$ (77.5 L) to a mixture of enamine **101** (24.4 kg, 99 mol, 1.0 equiv) and triethylamine (58.3 kg, 576 mol, 5.8 equiv) in EtOAc (73.5 L) while maintaining the temperature at 40–50 °C over 30–45 min,
- Stir the resulting mixture at 40–43 °C for 3.5 h.
- Cool the batch to 0–5 °C.

Stage (b)
- Add 4 N HCl (213.2 kg, 8.0 equiv) at <30 °C.
- Stir at 20–23 °C for 1 h.
- Separate layers.
- Wash with an organic layer with deionized water (104.9 L) and a solution of LiOH (8.4 kg, 200 mol) in water (83.2 L).
- Distill under reduced pressure to remove ~373 L of EtOAc.
- Use the resulting mixture in the subsequent step.

REMARK

In addition to the RA, temperature control was also critical in this process. For example, the morpholine elimination rate from **104** was faster at 70 °C, generating as much as 10–20% of **105**. On the other hand, when the reaction was conducted at 40 °C, the formation of morpholine adduct **105** was minimized (3–5%), while an acceptable rate of the cycloaddition reaction was still maintained.

4.3   OTHER ADDITION MODES

In addition to the direct and reverse modes of addition, scientists have developed various other modes of addition in order to resolve complex impurity issues. Sequential addition (Section 4.3.1) was developed to improve the chemoselectivity of 2-chloro-4,6-dimethoxy-1,3,5-triazine (CDMT)-mediated amide formation. Portionwise addition is another approach to control side reactions (Section 4.3.2). It was demonstrated that portionwise addition can manipulate the temperature fluctuation and off-gassing (due to the fluctuation of concentration) to avoid the accumulation of the reagent/starting material. The side product formation could be limited by a slow-releasing mode addition in which the concentration of the starting material/reagent (in the form of salts) was controlled by a controlled-addition of acid or base (Section 4.3.3). An alternate addition approach was employed (Section 4.3.4) to deal with low conversion and decomposition of the base-sensitive starting material/reagent. Concurrent addition (CA) (Section 4.3.5) was used to limit the formation of side products and control off-gassing.

4.3.1   SEQUENTIAL ADDITION

The addition sequence during the synthesis of amide **108** is crucial in controlling side product formation. For instance, the level of side product **110** could be reduced from 10% to 1.5% by alternating the order of addition sequence (Scheme 4.16).[35]

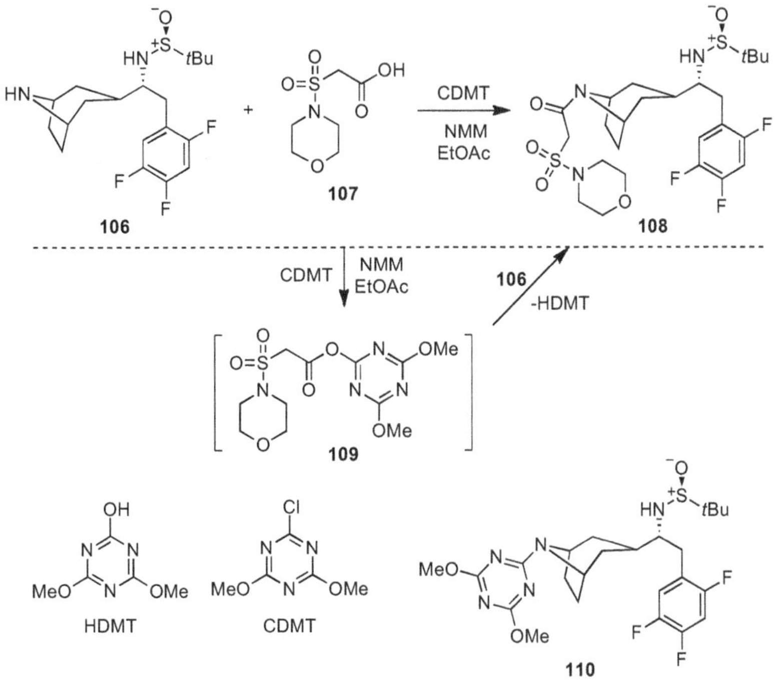

SCHEME 4.16

(I) Problematic addition sequence

- Addition order: **106**, **107**, CDMT, *N*-methylmorpholine (NMM).
- A high level (10%) of **110** was observed.

(II) Solutions (to control the concentration of CDMT)

- Addition order: **106**, **107**, NMM, CDMT.
- The level of **110** was reduced to 1.5%.

Procedure

- Charge, into a 2-L flask, in the following sequence: 30.0 g (77.0 mmol) of **106**, 17.1 g (81.0 mmol) of **107**, 1.4 L of ethyl acetate, and 15.0 g (148 mmol) of NMM.
- Stir at room temperature for 15 min to obtain a hazy solution.
- Charge 20.0 g (112 mmol) of CDMT.
- Stir at room temperature for 3 h.
- Wash with 5% $NaHCO_3$ (400 mL) and saturated NaCl (400 mL).
- Distill the organic layer under reduced pressure to dryness.
- Yield: 44.2 g (99%) in 96% purity.

REMARK

The major side product, 2-hydroxy-4,6-dimethoxy-1,3,5-triazine (HDMT), was easily removed by aqueous $NaHCO_3$ wash.

4.3.2 PORTIONWISE ADDITION

4.3.2.1 Cyclization

The synthesis of imidazothione **112** was realized via a reaction of **111** with phenyl isothiocyanate through thiourea intermediate **114** (Scheme 4.17).[36]

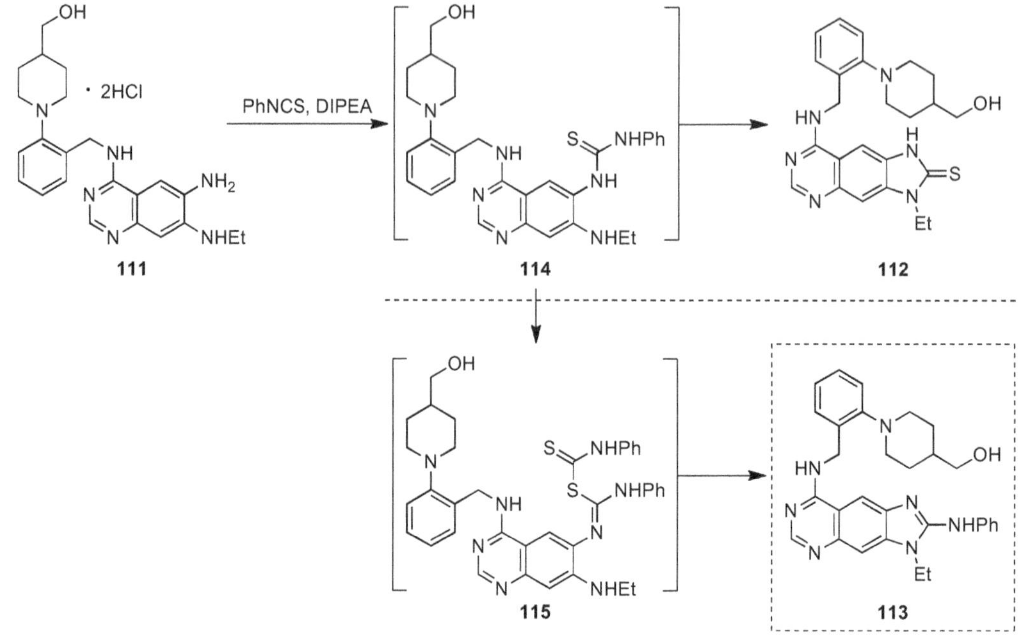

SCHEME 4.17

(I) Reaction problems

In addition to the desired product **112**, imidazole **113** was observed as a major side product.

(II) Solutions

To prevent the formation of **113**, the reaction was conducted in a semibatch mode by adding phenyl isothiocyanate portionwise.

Procedure

- Under reflux, charge phenyl isothiocyanate portionwise (5×16.2 mL, total 81 mL, 0.70 mol, 2.5 equiv) at 1 h intervals to a suspension of **111** (145 g, 0.28 mol) and DIPEA (163 mL, 0.98 mol, 3.5 equiv) in nPrOH (2.2 L).
- Stir the resulting mixture under reflux for 3 h.
- Cool and stir in an ice bath for 5 h.
- Filter and wash with cold nPrOH (220 mL).
- Dry under vacuum.
- Yield of the crude product: 120 g.

4.3.2.2 Deoxychlorination

Deoxychlorination of hydroxyl heteroaromatic derivatives is a well-established approach to access chlorinated arenes. Conversion of **116** to anilinoquinazoline **119** was carried out using phosphorus oxychloride and DIPEA in toluene to generate chloride **117** followed by chloride replacement with aniline **118** (Scheme 4.18).[16] In the deoxychlorination stage, using uncontrolled addition or addition of POCl$_3$ to **116** resulted in a significant amount of dimeric side product **120**.

In order to inhibit the formation of **120**, a controlled, portionwise addition approach was employed by adding **116** to a solution of POCl$_3$ and DIPEA in anhydrous toluene.

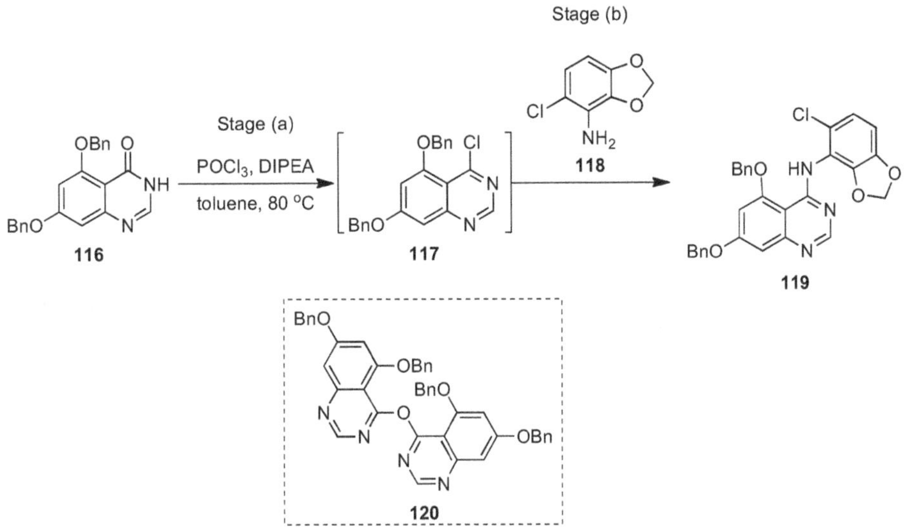

SCHEME 4.18

Procedure

Stage (a)
- Charge 4.5 kg (95%, 11.96 mol) of **116** in ten equal portions to a mixture of POCl$_3$ (1.437 L, 15.42 mol, 1.3 equiv) and DIPEA (3.195 L, 18.34 mol, 1.5 equiv) in anhydrous toluene (63.5 L) at 80 °C over 1 h.
- Stir the batch at 80 °C for 3 h.
- Cool to ambient temperature and hold overnight.

Stage (b)
- Reheat the batch to 80 °C.
- Charge a solution of **118** (2.364 kg, 92%, 12.68 mol, 1.06 equiv) in toluene (8.5 L) over 30 min.
- Stir the batch at 80 °C for 4 h.
- Cool to ambient temperature and hold overnight.
- Add IPA (10 L) and stir for 10 min.
- Filter and wash with IPA.
- Dry at 50 °C under reduced pressure.
- Yield: 4.994 kg (69%).

4.3.3 SLOW RELEASE OF STARTING MATERIAL/REAGENT

Owing to the instability or undesired physical property, some materials are stored and transported as their corresponding salts. At large-scale synthesis, it is preferred to generate the free base (or free acid) of such material in situ using an appropriate acid (or base). The slow-release mode of addition refers primarily to a controlled addition of the acid (or base) to the salt of the material in order to maintain the material at the desired concentration during the course of the reaction.

4.3.3.1 Synthesis of Urea

The synthesis of urea **123** was accomplished by coupling an amine hydrochloride salt **121** with Boc-protected ethylenediamine **122** in the presence of DIPEA using CDI as the coupling reagent.

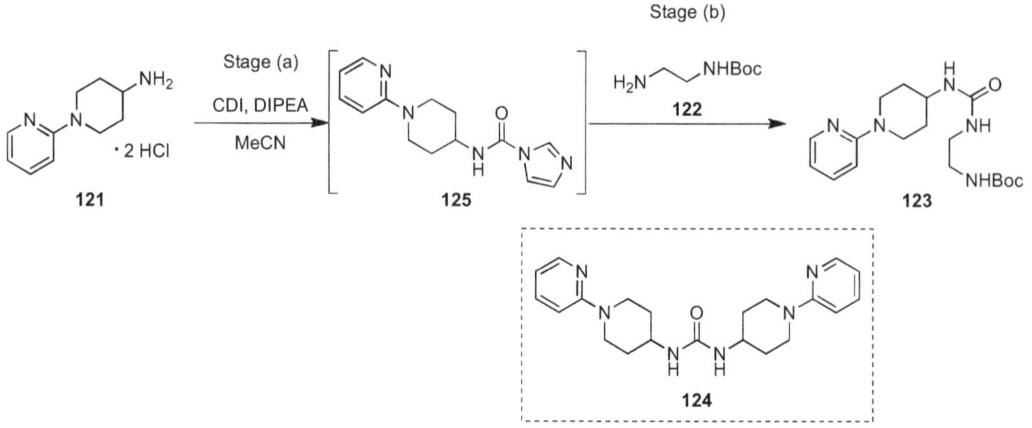

SCHEME 4.19

However, the presence of the base during the preparation of **123** presented some problems due to the formation of the symmetrical urea derivative **124** (Scheme 4.19).[29]

Apparently, the formation of **124** is the result of a competing reaction of a carbamoyl imidazolide intermediate **125** with the free base of **121**. To suppress the formation of **124**, the reaction had to be run stepwise: the preparation of the intermediate **125**, followed by coupling of **125** with **122**. Furthermore, a slow addition of DIPEA to a cooled suspension of the dihydrochloride salt **121** and CDI in acetonitrile was necessary, which resulted in the slow release of the free base of **121** to generate the required intermediate **125**. Under the optimized conditions, the undesired symmetrical urea side product **124** was reduced to a minimum level. Subsequent addition of the Boc-protected ethylenediamine **122** furnished the desired urea **123**.

Procedure

Stage (a)
- Suspend 23.0 kg (91.9 mol) of **121** and 16.4 kg (101 mol) of CDI in MeCN (118 kg).
- Cool to 5 °C.
- Charge 24.4 kg (189 mol) of DIPEA over 6 h with maximum stirring at <10 °C.
- Warm the mixture to room temperature upon full consumption of **121**.

Stage (b)
- Charge a solution of **122** (15.5 kg, 96.7 mol) in MeCN (9 kg) to the above reaction mixture.
- Heat at reflux for 2 h.
- Distill to remove MeCN (110 kg).
- Add CH_2Cl_2 (230 kg).
- Wash the reaction mixture with water (2×92 L).
- Exchange CH_2Cl_2 to EtOAc.
- Cool to 5 °C.
- Filter and wash with chilled EtOAc.
- Dry at 61 °C under a stream of N_2.
- Yield: 30.2 kg (90%).

REMARK

In order to maintain a homogeneous reaction mixture, DIPEA is the base of choice as the by-product (DIPEA·HCl) is soluble in MeCN.

4.3.3.2 Preparation of Alkylamine

The same addition strategy was employed in the conversion of aldehyde **126** to primary amine **127** (Equation 4.19).[37] The reaction proceeded through an oxime intermediate **128**, followed by reduction with LiAlH$_4$. A hydroxylamine free base was generated in situ by a controlled addition of sodium hydroxide to a binary solution of **126** in 2-methyl tetrahydrofuran (MeTHF) and hydroxylamine hydrochloride in water. Extraction of the resulting oxime **128** from the aqueous mixture with MeTHF gave a solution suitable for use directly in the subsequent reduction step.

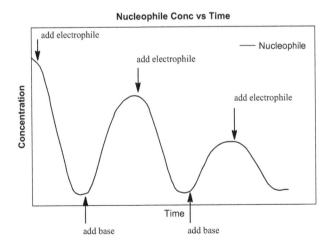

(4.19)

4.3.4 ALTERNATE ADDITION

The alternate addition is employed to manipulate concentrations of both the starting material and reagent. Figure 4.2 depicts the concentration oscillation of the starting material caused by an alternate addition of base (to deprotonate the starting material to generate a nucleophile in situ) and electrophile. This mode of addition is useful when the electrophile is base-sensitive.

The coupling of amine **129** with aryl bromide **130** by treating a solution of **129** in THF with one equivalent of a strong base, such as sodium *tert*-pentoxide, sodium *tert*-butoxide, or potassium *tert*-butoxide, followed by a slow addition of **130** resulted in only 50% conversion (Equation 4.20).[38]

(4.20)

FIGURE 4.2 The oscillation of nucleophile concentration versus time.

(I) Chemistry diagnosis

The incomplete reaction is attributed to the consumption of base by deprotonation of product **131** as it bears a more acidic N–H proton. Therefore, a second equivalent of sodium *tert*-pentoxide was introduced leading to complete conversion, but the levels of side products were also increased significantly. The formation of side products is due to the decomposition of aryl bromide **130** as it is not stable in the presence of sodium *tert*-pentoxide.

(II) Solutions

In order to avoid the direct contact of **130** with sodium *tert*-pentoxide, an alternate addition was employed as follows:

- Treat compound **129** in THF with 1.0 equiv of sodium *tert*-pentoxide in THF at room temperature
- Add 0.5 equiv of **130** in THF dropwise at ≤35 °C
- Add a further 0.5 equiv of sodium *tert*-pentoxide (after **130** was consumed, HPLC analysis indicated a formation of essentially a 1:1 mixture of **129** and **131**) followed by adding 0.25 equiv of **130** dropwise
- Repeat this alternate addition sequence two more times, each with 0.25 equiv of sodium *tert*-pentoxide and 0.125 equiv of **130**, to give a complete reaction according to HPLC analysis.

The reaction was then quenched with acetic acid and water, and the product **131** was isolated by filtration in 87% yield as a white solid.

DISCUSSIONS

Quenching with AcOH would avoid potential filtration issues. The use of weak acetic acid would allow small particles to dissolve due to their relatively higher solubility[39] (See Chapter 17).

4.3.5   Concurrent Addition

4.3.5.1   Bromination Reaction

Controlling impurity formation during the monobromination of aminopyrazine remains one of the most challenging tasks in process development. The bromination of aminopyrazine **132** with *N*-bromosuccinimide (NBS) yielded only 30% of the desired monobrominated product **133** with large amounts of a dark gummy solid. In addition, the reaction was exothermic with an induction period of several minutes (Scheme 4.20).[30] Three side products were identified as dibrominated compound **134**, dimer **135**, and azo-type compound **136**. Albeit having a similar reaction profile (compared with NBS), 1,3-dibromo-5,5-dimethylhydantoin (DBH) was chosen as the bromination agent due to its more robust solution stability. In addition, the process was atom-economical as both bromine atoms were utilized in the bromination.

(I) Chemistry diagnosis

The dibromo side product **134** is formed from bromination of the mono-bromopyrazine **133** while the dimer **135** is presumably derived from *N*-bromo intermediate **137** or **138**. The azo-type side product **136** was generated by *N*-bromination of **135** followed by elimination of HBr. Protonation of the nitrogen atom in **132** by HBr would occur with slow mode addition of DBH, leading to diminished reactivity of **132**. Accordingly, the further reaction of **133** with DBH would become significant to produce the dibrominated side product **134**.

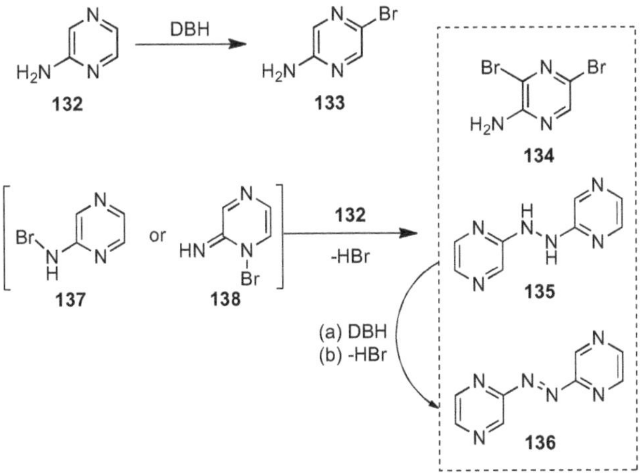

SCHEME 4.20

(II) Solutions

In the event, the bromination was performed by adding the solutions of DBH and aminopyrazine **132** in CH_3CN/DMF simultaneously into the reaction flask at 0–5 °C.

Procedure

- Charge a solution **132** (3.5 kg, 36.8 mol) in DMF/MeCN (7 L/22.8 L) and a solution of DBH (5.5 kg, 19.3 mol, 0.52 equiv) in DMF/MeCN (3.5 L/33.9 L) simultaneously to a reactor containing 3.5 L of MeCN at 0 °C over 1 h.
- Stir the batch for 0.5 h.
- Quench with aq $Na_2S_2O_3\cdot5H_2O$ (10%, 7.0 L).
- Concentrate under reduced pressure.
- Filter the batch through Solka-Floc (1.1 kg).
- Add 33.0 L of 2.0 wt% of aq Na_2CO_3 to the filtrate and stir for 1 h prior to filtration.
- Add NaCl (~10.0 kg) to the filtrate.
- Extract with a mixture of EtOAc/heptane (3:2, 35.0 L).
- Wash with brine and treat the organic layer with Darco-KB overnight.
- Switch the solvent to heptane.
- Filter to give 4.48 kg of **133** (65%) as a yellow solid.

DISCUSSIONS

An alternative approach is to convert the side product **134** to the desired monobromide **133** via a catalytic regioselective debromination (Equation 4.21).[30]

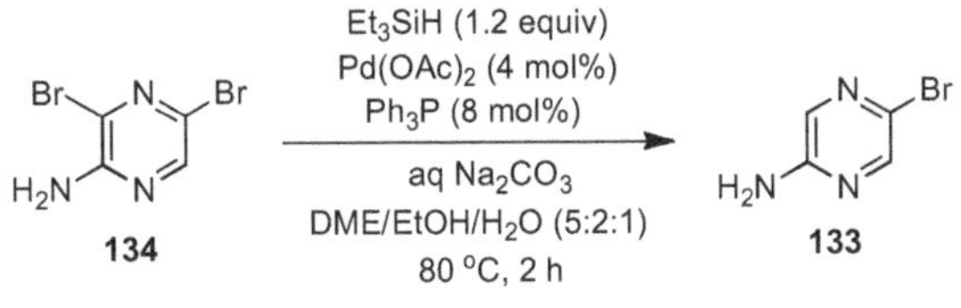

$$(4.21)$$

4.3.5.2 Difluoromethylation

Difluoromethylation of phenol **139** was achieved with sodium chlorodifluoroacetate **140** in the presence of potassium carbonate in DMF at 95 °C (Equation 4.22).[40]

(4.22)

(I) Problematic reactions

- Two side products, dimer **142** and trimer **143** were observed with the former as the major one (~10%).
- A significant amount of CO_2 off-gassing was noticed.

(II) Chemistry diagnosis

The product **141** is formed through a reaction of the starting material **139** with difluorocarbene intermediate, generated from the decomposition of **140** along with CO_2. Apparently, keeping the concentration of **140** higher than the concentration of **139** would minimize the formation of the dimer **142** and trimer **143**. Therefore, a controlled CA of **139** and **140** is critical in minimizing the side reactions and controlling CO_2 off-gassing.

(III) Solutions

- Addition of a solution of **139** (1.0 equiv) and **140** (2.5 equiv) into a hot (95 °C) suspension of K_2CO_3 over 4 h
- The desired product **141** in a near quantitative yield and 99% HPLC purity

Procedure

- Charge a solution of **139** (7.0 kg, 25.2 mol, 1.0 equiv) and **140** (9.6 kg, 63.0 mol, 2.5 equiv) in DMF (22.1 kg) to a hot (95 °C) suspension of K_2CO_3 (5.2 kg, 37.6 mol, 1.5 equiv) in DMF (13.2 kg) over 4 h at 93–98 °C.
- Stir for 15 min.
- Cool to 30 °C.
- Add water (17.5 kg).
- Transfer the batch to a 180-L reactor.
- Add water (52.5 kg).
- Cool to 10–15 °C and stir for 1 h.
- Filter and wash with water (3×22.4 kg)
- Yield: 8.24 kg (99.7% yield and 99.6% purity).

4.3.5.3   Diels–Alder Reaction

The Diels–Alder reaction between bromocoumalate **144** and benzyne **148**, derived from an unstable benzenediazonium-2-carboxylate **147**, involves a strained bridged-lactone **149** intermediate. A controlled CA of anthralinic acid **145** and isoamyl nitrite into the solution of the starting material **144** under reflux allowed to maintain concentrations of unstable reaction intermediates **147**, **148**, and **149** at safe levels (Scheme 4.21).[41]

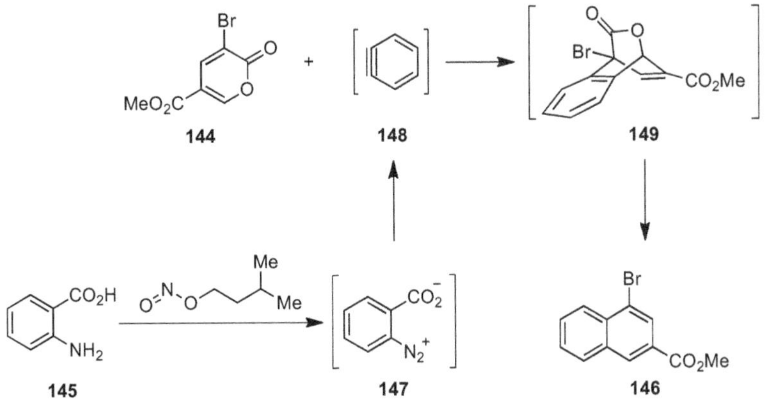

SCHEME 4.21

An additional benefit of such an addition mode is to minimize the formation of undesired side products such as **150** and **151** (Equations 4.23 and 4.24).

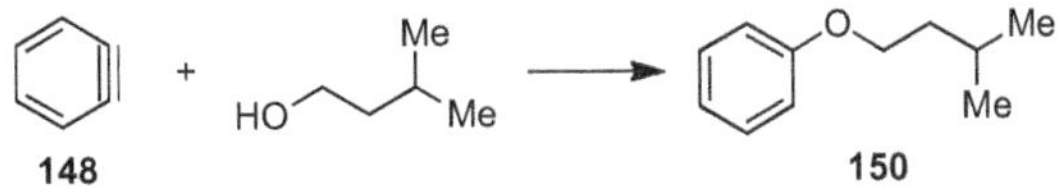

$$(4.23)$$

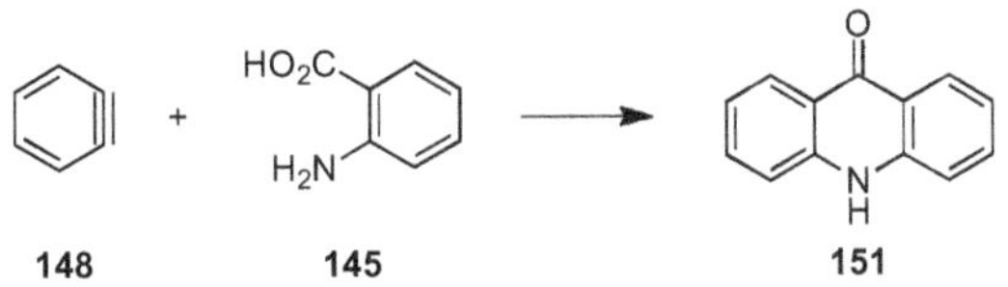

$$(4.24)$$

NOTES

1. Direct addition is an addition of a reagent to a starting material. In contrast, Soffer and Katz defined, in a reduction of nitrile with lithium aluminum hydride (Soffer, L. M.; Katz, M. *J. Am. Chem. Soc.* **1956**, *78*, 1705), the direct addition as the addition of the starting material to the reagent.
2. Sonogashira, K. *Metal-Catalyzed Reactions*; Wiley-VCH, New York, 1998; pp 203–229.
3. Houpis, I. N.; Shilds, D.; Nettekoven, U.; Schnyder, A.; Bappert, E.; Weerts, K.; Canters, M.; Vermuelen, W. *Org. Process Res. Dev.* **2009**, *13*, 598.
4. Powell, L.; Mahmood, A.; Robinson, G. E. *Org. Process Res. Dev.* **2011**, *15*, 49.
5. Anderson, N. G.; Ary, T. D.; Berg, J. L.; Bernot, P. J.; Chan, Y. Y.; Chen, C.-K.; Davies, M. L.; DiMarco, J. D.; Dennis, R. D.; Deshpande, R. P.; Do, H. D.; Droghini, R.; Early, W. A.; Gougoutas, J. Z.; Grosso, J. A.; Harris, J. C.; Haas, O. W.; Jass, P. A.; Kim, D. H.; Kodersha, G. A.; Kotnis, A. S.; LaJeunesse, J.; Lust, D. A.; Madding, G. D.; Modi, S. P.; Moniot, J. L.; Nguyen, A.; Palaniswamy, V.; Phillipson, D. W.; Simpson, J. H.; Thoraval, D.; Thurston, D. A.; Tse, K.; Polomski, R. E.; Wedding, D. L.; Winter, W. J. *Org. Process Res. Dev.* **1997**, *1*, 300.
6. Demerson, C. A.; Humber, L. G.; Philipp, A. *J. Med. Chem.* **1976**, *19*, 391.
7. Thiel, O. R.; Bernard, C.; King, T.; Dilmeghani-Seran, M.; Bostick, T.; Larsen, R. D.; Faul, M. M. *J. Org. Chem.* **2008**, *73*, 3508.
8. Smith, M. B.; March, J. *Advanced Organic Chemistry–Reactions, Mechanisms, and Structure*, 5th ed., John Wiley & Sons, Inc., New York, 2001, p 509.
9. Beylin, V.; Boyles, D. C.; Curran, T. T.; Macikenas, D.; Parlett, R. V. IV; Vrieze, D. *Org. Process Res. Dev.* **2007**, *11*, 441.

10. Ormerod, D.; Willemsens, B.; Mermans, R.; Langens, J.; Winderickx, G.; Kalindjian, S. B.; Buck, I. M.; McDonald, I. M. *Org. Process Res. Dev.* **2005**, *9*, 499.

11. Belleau, B.; Malek, G. *J. Am. Chem. Soc.* **1968**, *90*, 1651.

12. (a) Weisenburger, G. A.; Anderson, D. K.; Clark, J. D.; Edney, A. D.; Karbin, P. S.; Gallagher, D. J.; Knable, C. M.; Pietz, M. A. *Org. Process Res. Dev.* **2009**, 13, 60. see related example, (b) Jiang, X.; Prasad, K.; Repič, O. *Syn. Comm.* **2009**, *39*, 2640.

13. Pasti, J.; Chen, C.-K.; Spangler, L.; DelMonte, A. J.; Benoit, S.; Berglund, D.; Bien, J.; Brodfuehrer, P.; Chan, Y.; Corbett, E.; Costello, C.; DeMena, P.; Discordia, R. P.; Doubleday, W.; Gao, Z.; Gingras, S.; Grosso, J.; Haas, O.; Kacsur, D.; Lai, C.; Leung, S.; Miller, M.; Muslehiddinoglu, J.; Nguyen, N.; Qiu, J.; Olzog, M.; Reiff, E.; Thoraval, D.; Totleben, M.; Vanyo, D.; Vemishetti, P.; Wasylak, J.; Wei, C.; *Org. Process Res. Dev.* **2009**, *13*, 716.

14. Carrera, D. E.; Sheng, P.; Safina, B. S.; Li, J.; Angelaud, R. *Org. Process Res. Dev.* **2013**, *17*, 138.

15. Rossi, R. A.; Palacios, S. M. *Tetrahedron* **1993**, *49*, 4485.

16. Ford, J. G.; Pointon, S. M.; Powell, L.; Siedlecki, P. S. *Org. Process Res. Dev.* **2010**, *14*, 1078.

17. Hughes, D. L.; Reamer, R. A.; Bergan, J. J.; Grabowski, E. J. J. *J. Am. Chem. Soc.* **1988**, *110*, 6487.

18. Lajis, N. H.; Khan, M. N.; Hassan, H. A. *Tetrahedron* **1993**, *49*, 3405.

19. (a) Morrison, J. D.; Tomaszewski, J. E.; Mosher, H. S.; Dale, J.; Miller, D.; Elsenbaumer, R. L. *J. Am. Chem. Soc.* **1977**, *99*, 3167. (b) Okuhara, K. *J. Am. Chem. Soc.* **1980**, *102*, 244.

20. Smith, M. B.; March, J. *Advanced Organic Chemistry, Reactions, Mechanisms, and Structure*, 5th ed. John Wiley & Sons, New York, 2001, p 1208.

21. Yanagisawa, A.; Nishimura, K.; Ando, K.; Nezu, T.; Maki, A.; Kato, S.; Tamaki, W.; Imai, E.; Mohri, S. *Org. Process Res. Dev.* **2010**, *14*, 1182.

22. (a) Jacks, T. E.; Belmont, D. T.; Briggs, C. A.; Horne, N. M.; Kanter, G. D.; Karrick, G. L.; Krikke, J. J.; McCabe, R. J.; Mustakis, J. G.; Nanninga, T. N.; Risedorph, G. S.; Seamans, R. E.; Skeean, R.; Winkle, D. D.; Zennie, T. M. *Org. Process Res. Dev.* **2004**, *8*, 201. (b) Wilk, B. K.; Helom, J. L.; Coughlin, C. W. *Org. Process Res. Dev.* **1998**, *2*, 407.

23. Tucker, J. L.; Couturier, M.; Leeman, K. R.; Hinderaker, M. P.; Andresen, B. M. *Org. Process Res Dev.* **2003**, *7*, 929.

24. Alam, M.; Wise, C.; Baxter, C. A.; Cleator, E.; Walkinshaw, A. *Org. Process Res. Dev.* **2012**, *16*, 435.

25. Zhang, P.; Shankar, A.; Cleary, T. P.; Cedilote, M.; Locklear, D.; Pierce, M. E. *Org. Process Res. Dev.* **2007**, *11*, 861.

26. (a) It was considered by the authors that the formation of regioisomer **66** is attributed to the nitration of the free base of **64**. (b) Cu(O-NO$_2$)X was suggested as the nitration species for the regioselective nitration of arenes: Sadhu, P.; Alla, S. K.; Punniyamurthy, T. *J. Org. Chem.* **2014**, *79*, 8541.

27. Scott, J. P.; Alam, M.; Bremeyer, N.; Goodyear, A.; Lam, T.; Wilson, R. D.; Zhou, G. *Org. Process Res. Dev.* **2011**, *15*, 1116.

28. Shi, Z.; Kiau, S.; Lobben, P.; Hynes, J., Jr.; Wu, H.; Parlanti, L.; Discordia, R.; Doubleday, W. W.; Leftheris, K.; Dyckman, A. J.; Wrobleski, S. T.; Dambalas, K.; Tummala, S.; Leung, S.; Lo, E. *Org. Process Res. Dev.* **2012**, *16*, 1618.

29. Ashcroft, C. P.; Dessi, Y.; Entwistle, D. A.; Hesmondhalgh, L. C.; Longstaff, A.; Smith, J. D. *Org. Process Res. Dev.* **2012**, *16*, 470.

30. Itoh, T.; Kato, S.; Nonoyama, N.; Wada, T.; Maeda, K.; Mase, T.; Zhao, M. M.; Song, J. Z.; Tschaen, D. M.; McNamara, J. M. *Org. Process Res. Dev.* **2006**, *10*, 822.

31. Yates, M. H.; Koenig, T. M.; Kallman, N. J.; Ley, C. P.; Mitchell, D. *Org. Process Res. Dev.* **2009**, *13*, 268.

32. Mitchell, D.; Doecke, C. W.; Hay, L. A.; Koenig, T. M.; Wirth, D. D. *Tetrahedron Lett.* **1995**, *36*, 5335.

33. Watson, T. J.; Ayers, T. A.; Shah, N.; Wenstrup, D.; Webster, M.; Freund, D.; Horgan, S.; Carey, J. P. *Org. Process Res. Dev.* **2003**, *7*, 521.

34. Oh, L. M.; Wang, H.; Shilcrat, S. C.; Herrmann, R. E.; Patience, D. B.; Spoors, P. G.; Sisko, J. *Org. Process Res. Dev.* **2007**, *11*, 1032.

35. Fei, Z.; Wu, Q.; Zhang, F.; Cao, Y.; Liu, C.; Shieh, W.-C.; Xue, S.; McKenna, J.; Prasad, K.; Prasad, M.; Baeschlin, D.; Namoto, K. *J. Org. Chem.* **2008**, *73*, 9016.

36. Fujino, K.; Takami, H.; Atsumi, T.; Ogasa, T.; Mohri, S.; Kasai, M. *Org. Process Res. Dev.* **2001**, *5*, 426.

37. Breining, S. R.; Genus, J. F.; Mitchener, J. P.; Cuthbertson, T. J.; Heemstra, R.; Melvin, M. S.; Dull, G. M.; Yohannes, D. *Org. Process Res. Dev.* **2013**, *17*, 413.

38. Shu, L.; Wang, P.; Gu, C.; Garofalo, L.; Alabanza, L. M.; Dong, Y. *Org. Process Res. Dev.* **2012**, *16*, 1870.

39. For Ostwald ripening, see: Noorduin, W. L.; Vlieg, E.; Kellog, R. M.; Kaptein, B. *Angew. Chem. Int. Ed.* **2009**, *48*, 9600.

40. (a) Sperry, J. B.; Farr, R. M.; Levent, M.; Ghosh, M.; Hoagland, S. M.; Varsolona, R. J.; Sutherland, K. *Org. Process Res. Dev.* **2012**, *16*, 1854. (b) Sperry, J. B.; Sutherland, K. *Org. Process Res. Dev.* **2011**, *15*, 721.

41. Ashworth, I. W.; Bowden, M. C.; Dembofsky, B.; Levin, D.; Moss, W.; Robinson, E.; Szczur, N.; Virica, J. *Org. Process Res. Dev.* **2003**, *7*, 74.

5 Process Optimization with Additives

Process optimization focuses mainly on improving process safety, reaction profile, and product yield. This chapter addresses various opportunities for the optimization of problematic reactions by adding additives. Among a number of methods/technologies that have been developed to suppress side reactions, the addition of additives is one of the valuable tools that are frequently used by process chemists. Additives can be categorized into three classes: acid additives, base additives, and other miscellaneous additives that include various inorganic and organic salts, scavengers, and other organic compounds.

In general, the addition of additives can solve a number of process problems, such as unstable reaction intermediate, slow reaction, poor solubility, and impurity formation. Additives can change reaction kinetics or completely shut down undesired side reactions. A reaction with highly reactive species may allow a reaction to complete in a short time frame, while it may take several days for a slow reaction to complete. Acid (or base) additives are able to tune the reactivity by protonation (or deprotonation) of a reagent. Impurities are often generated due to poor selectivity, residual acid (or base), etc. In most cases, the residual acid (or base) is responsible for impurity formation. Thus, a base (or acid) additive is frequently used to neutralize the residual acid (or base). Hinging on each individual reaction, additives can be used in either catalytic or stoichiometric amounts.

Residual water in wet solvents can hydrolyze ester under acidic or basic conditions. Accordingly, esters are often selected as sacrificial agents to protect ester functionality. Apparently, the selected ester additive has to be more reactive or have less steric hindrance compared to the ester substrate. Water, on the other hand, can be utilized as an additive to improve reaction profiles.

5.1 ACID ADDITIVES

Generally, most side reactions are attributed to poor selectivity. To improve the reaction selectivity, a number of parameters can be optimized, such as temperature, pressure, solvent, and concentration. Acid additives have been employed to maximize the desired reactions and, at the same time, suppress undesired side reactions.

5.1.1 HYDROCHLORIC ACID

Hydrochloric acid is a strong inorganic acid (pKa −6.3) and has many applications. In chemical laboratories, hydrochloric acid is frequently used to form hydrochloride salts with basic organic compounds such as amines and to neutralize basic reaction mixtures.

5.1.1.1 S_NAr Reaction

An S_NAr amination of dichloropyrimidine **1** with 4-chloroaniline **2** produced arylaniline **3** along with various amounts of side product **4** (Equation 5.1).[1] For instance, **4** was generated in 2–3% in the presence of 0.4 equiv of 3N HCl.

DOI: 10.1201/9781003288411-5

(5.1)

In general, a semibatch would suppress the formation of **4** by adding **2** to **1**. Alternatively, an increase in the amount of HCl from 0.4 to 2.0 equiv minimized the levels of **4** from 2–3% to 0.2–0.3%. With 2.0 equiv of HCl, the reaction was conveniently carried out in an "all in" mode.

Procedure

- Charge, into a 12-L reactor, 1.0 kg (6.1 mol) of **1**, followed by 930 g (7.3 mol, 1.2 equiv) of **2** and 2B ethanol (anhydrous EtOH is denatured with toluene) (5.8 L).
- Charge 4.0 L (12 mol, 2.0 equiv) of 3 N HCl.
- Heat the resulting slurry at reflux for 36 h.
- Cool, filter, and wash with EtOAc (3.75 L).
- Dry at 40 °C under vacuum.
- Yield: 1.5 kg (84%).

DISCUSSIONS

(a) The concentration of aniline **2** was controlled by the acid–base equilibrium, which avoided the semibatch operation.
(b) Washing with EtOAc (instead of heptane) led to a more efficient displacement of residual water from the filter cake, resulting in a shorter drying time.

5.1.1.2 Deoxychlorination

Deoxychlorination is a method that is frequently used to convert hydroxyl heteroaromatic compounds to their corresponding heteroaromatic chlorides. Phosphorus oxychloride ($POCl_3$) is often selected as the deoxychlorination agent. For instance, deoxychlorination of 4-chloro-2-hydroxy-3-nitropyridine **5** to the corresponding 2,4-dichloro-3-nitropyridine **6** was accomplished with $POCl_3$ (Equation 5.2).[2]

(5.2)

It was found that the addition of HCl/LiCl enhanced the reaction rate, and the presence of these additives was presumably able to maintain the level of chloride ion in the reaction mixture, enabling the reaction to go to completion in a favorable time frame.

Procedure

- Charge 29.4 kg (191.9 mol) of POCl$_3$ into a vessel, followed by charging 6.72 kg (38.4 mol) of **5** and 1.7 kg of LiCl.
- Heat the mixture to 105 °C and stir for 2.5 h.
- Cool to 89 °C.
- Charge 70 g of 37% HCl.
- Reheat the batch to 105 °C and stir for 1 h.
- Cool to 88 °C.
- Charge 80 g of 37% HCl.
- Reheat the batch to 105 °C and stir for 1 h.
- Charge another 2 portions of 37% HCl until <5% of **1**.
- Cool the reaction mixture to 44 °C.
- Add MeCN (43.2 L).
- Stir the batch at ambient temperature overnight.
- Cool to 0 °C.
- Add water (4×0.25 L, 18×1 L) at 22–28 °C and stir for 5.5 h.
- Distill at 28 °C under reduced pressure to remove MeCN.
- Filter, wash with water (16.7 L), and dry at 55 °C under vacuum.
- Yield: 7.36 kg (88%).

5.1.2 PHOSPHORIC ACID

The addition of phosphoric acid (pKa 2.14, 7.20, 12.37) was reported to lower the deoxychlorination temperature. For example, a deoxychlorination of 3-hydroxybenzo[d]isoxazole **7** to 3-chlorobenzo[d] isoxazole **8** (Equation 5.3)[3] was initially conducted at 100–120 °C. Under such conditions, chlorination of the benzisoxazole ring was observed, and these side products were found to be difficult to purge in the downstream process.

$$(5.3)$$

The addition of phosphoric acid[4] (0.2 equiv) to the reaction mixture enabled the reaction to proceed at a relatively lower temperature (80–85 °C).

Procedure

Stage (a)
- Charge 22.9 kg (2.0 equiv) of POCl$_3$ to **7** (10.0 kg) followed by adding 1.02 L (0.2 equiv) of 85% H$_3$PO$_4$.
- Charge 6.10 L (1.0 equiv) of pyridine at 20–50 °C over 1 h.
- Heat the resulting mixture at 80–85 °C for 12 h.
- Cool to 35–50 °C.

Stage (b)
- Add the reaction mixture to a mixture of MTBE (35 L), heptanes (45 L), and water (50 L) at 35–50 °C.

- Stir the batch at 35 °C for 1 h.
- Cool to 20–25 °C.
- Separate layers.
- Wash with an organic layer with brine (5 kg of NaCl in 40 L of water).
- Yield: 11.4 kg (in MTBE/heptanes solution).

5.1.3 SULFURIC ACID

As a strong inorganic acid, sulfuric acid (pKa –2.8, 1.99) was also utilized as an additive in the chemical process development.

5.1.3.1 Iodination Reaction

4'-Hydroxy-3'-iodo-biphenyl-4-carbonitrile **10** was prepared via iodination of **9** with N-iodosuccinimide (NIS) (Equation 5.4).[5]

$$(5.4)$$

(a) NIS, AcOH
(b) NIS, H_2SO_4 (0.5 equiv), AcOH

(I) Problematic reaction

This iodination in acetic acid under condition (a) afforded the desired product **10** in 65% yield along with 25% of di-iodinated product.

(II) Solutions

- With the addition of 0.5 equiv of concentrated sulfuric acid, the yield of **10** was increased to 93%.
- Only 2% of the di-iodinated undesired product was formed.

Procedure

- Charge 2.15 kg (10.90 mol) of **9** followed by adding 17.89 kg of AcOH and 533 g (5.4 mol, 0.5 equiv) of conc. H_2SO_4.
- Charge 2.40 kg (10.36 mol, 0.95 equiv) of NIS (97%) portionwise at ca. 20 °C.
- Stir the resulting suspension for 20 h.
- Add water (34.4 kg) and stir at 20 °C for 1 h.
- Filter and wash with water (32.2 kg) and heptane (15.0 kg).
- Dry at 55 °C under vacuum.
- Yield: 3.27 kg (93%).

DISCUSSIONS

Strong acids such as sulfuric acid and TFA were hypothesized to facilitate the formation of proton-solvated iodinating species.[6]

5.1.3.2 Chlorination Reaction with N-Chlorosuccinimide

Analogously, chlorination of the deactivated aromatic ring with N-chlorosuccinimide (NCS) requires an acid additive to activate NCS.[7] The synthesis of delafloxacin, 6-fluoroquinolone antibiotic, involves selective chlorination in the presence of 3.5 mol% of H_2SO_4 on the 8-position of the functionalized quinolone **11** as shown in Scheme 5.1.[8]

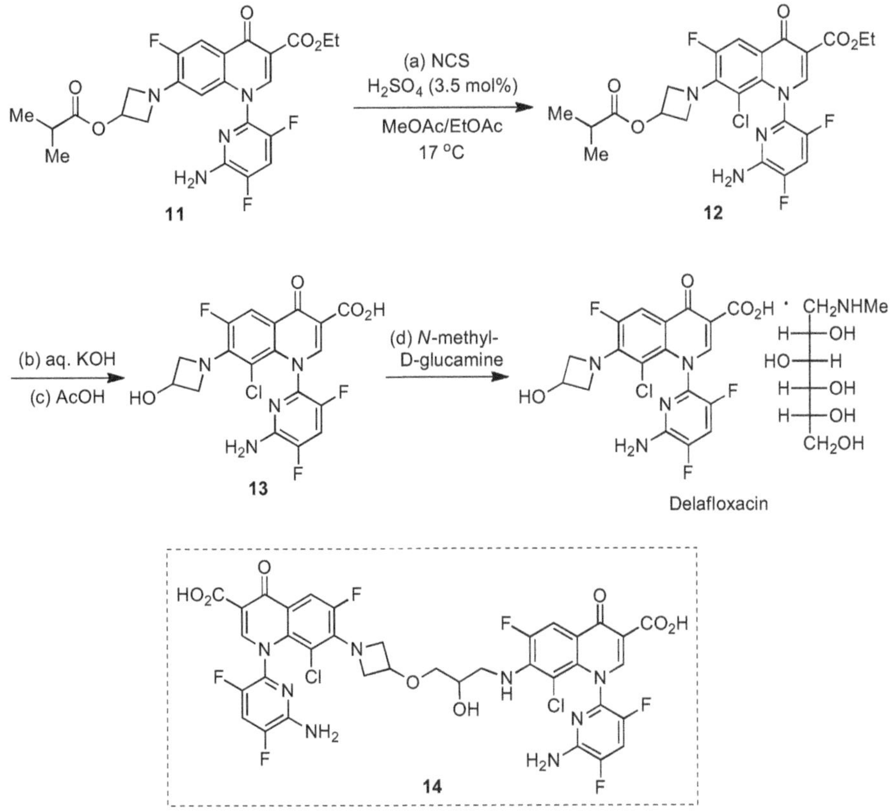

SCHEME 5.1

(I) Problems

The isolated chloroquinolone **13** was contaminated with 0.43 area% of dimeric adduct impurity **14** in the scale-up batch. Additionally, this impurity turned out to be difficult to purge during the final salt formation step.

(II) Chemistry diagnosis

It was found that the impurity levels increase upon holding the chlorination mixture (Stage (a)) prior to saponification (Stage (b)). Scheme 5.2 outlines a hypothetical reaction pathway for the formation of **14** that involves iminium intermediate **16**.

The iminium **16** would undergo deprotonation/aromatization to give the product **12**. **12** would be protonated in the presence of acid to form intermediates **17** which would rearrange to give **18**. The intermediate **18** reacts with **13** to form **19**. Treatment of **19** with KOH ultimately to form the dimer **14**.

(III) Solutions

In order to inhibit and control the dimer formation, the chlorination conditions were optimized by reducing H_2SO_4 from 3.5 mol% to 1.0 mol%. Under the optimized conditions, the level of **14** in the isolated **13** was well controlled below 0.07%.

5.1.3.3 Chlorination Reaction with Phosphorus Trichloride

Significant variations in reaction time were observed for the chlorination of Cbz-methoxyglycin ester **20** with phosphorus trichloride (Scheme 5.3).[9] With the addition of a catalytic amount of sulfuric acid (0.3%), the chlorination was completed within 16 h with high reproducibility independent of the batch history of **20**.

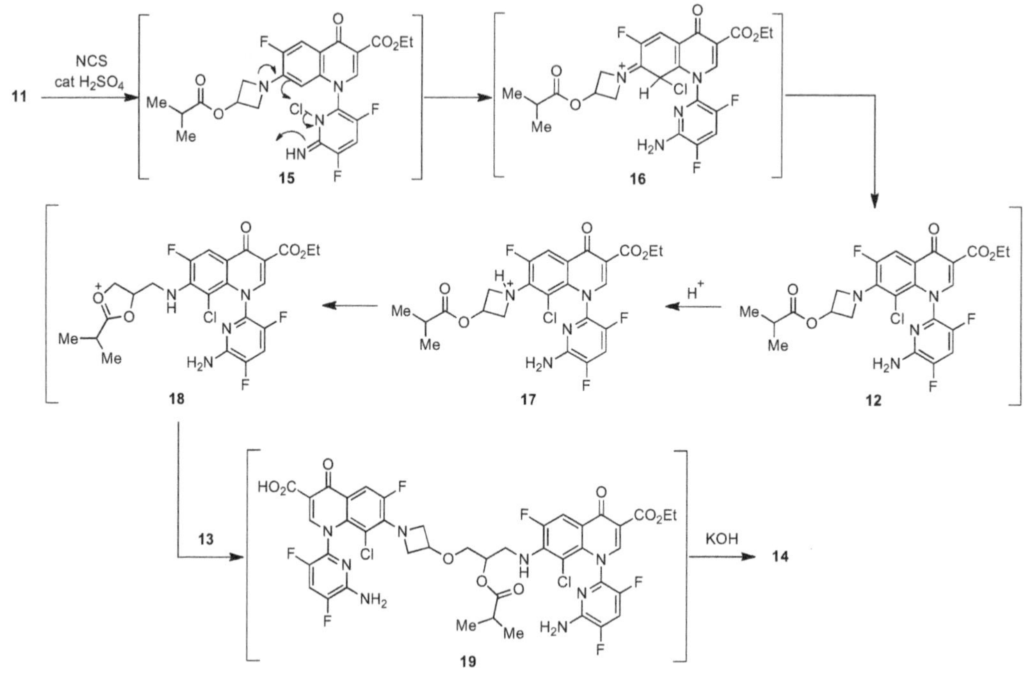

SCHEME 5.2

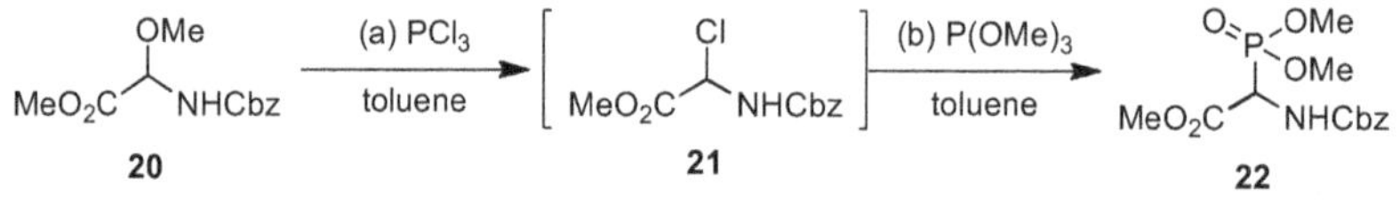

SCHEME 5.3

Procedure

Stage (a)

- Charge 24.0 kg (174.8 mol) of PCl_3 over a period of 1 h into a solution of 0.05 kg (0.51 mol) of H_2SO_4 and 37.5 kg (148.1 mol) of **20** in 210 kg of toluene.
- Stir the resulting mixture at 75 °C for 16 h.
- Distill at 70 °C under 250 mbar to remove excessive PCl_3 and part of the toluene.

Stage (b)

- Charge 20.7 kg (166.8 mol) of $P(OMe)_3$ at 75 °C within 1 h.
- Stir at 90 °C for 1.5 h.
- Cool to 20 °C and add aqueous ammonia to pH 6.6.
- Separate layers and extract the aqueous layer with toluene.
- Wash the combined organic layers with H_2O (30 kg).
- Distill at 60 °C under 200 mbar to remove part of the solvent.
- Add diisopropyl ether and cool to 5 °C.
- Filter and wash with diisopropyl ether.
- Dry at 50 °C under 30 mbar.
- Yield: 152.6 kg (79%).

5.1.3.4 Hydrogenation Reaction

It was noted that deactivation of an amine nucleophile with a weak acid such as acetic acid was not sufficient to prevent undesired side reactions, because such acid/base interaction would be a reversible process. For instance, the hydrogenation of the unsaturated nitrile **23** using acetic acid as the additive at 50 °C generated up to 50% of side product **26** acid (Scheme 5.4).[10].

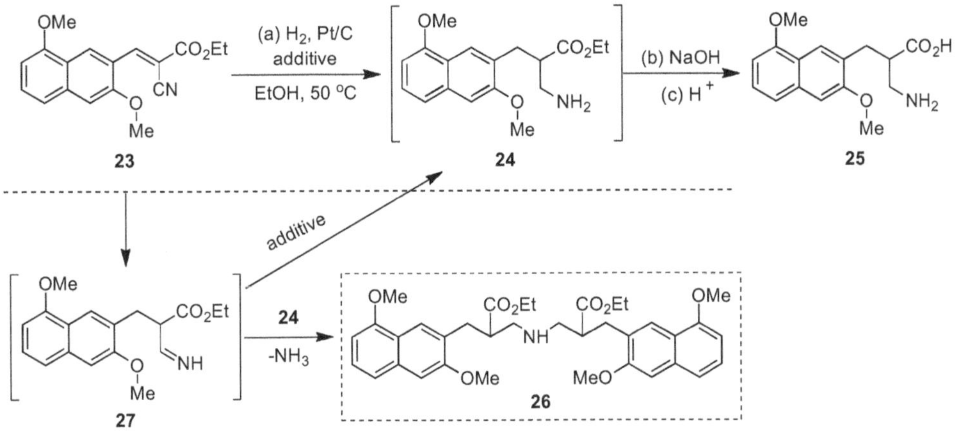

SCHEME 5.4

Solutions

- The formation of **26** is the result of condensation of imine intermediate **27** with the hydrogenation product **24**.
- A strong acid additive such as sulfuric acid is the acid of choice.
- With H_2SO_4 (1.6 equiv) **26** was reduced to a range of 5–10% (relative to the desired product **24**) which was removed after hydrolysis.
- Another benefit of the H_2SO_4 additive is the prevention of the amino group in **24** from binding to the active catalyst.

5.1.4 METHANESULFONIC ACID

Analogously, methanesulfonic acid (MSA) was selected as an additive in the optimization of the α-chlorination of alkyl aryl ketone **28**. Thus, the addition of catalytic amounts (10 mol%) of MSA to the mixture of **28** and NCS in methanol significantly improved the reaction rate and yield of the product **29** (Equation 5.5).[11]

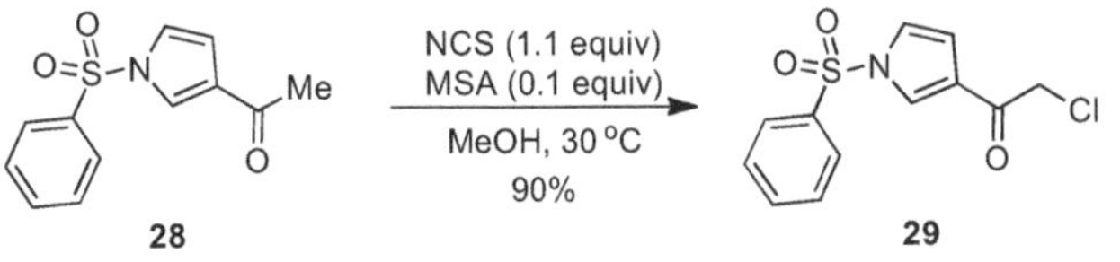

$$(5.5)$$

5.1.5 ACETIC ACID

Weak acids, as additives, are frequently employed to improve the reaction selectivity by controlling the concentration of nucleophiles (usually amines) in the reaction mixture through the acid–base equilibrium (Equation 5.6):

$$(5.6)$$

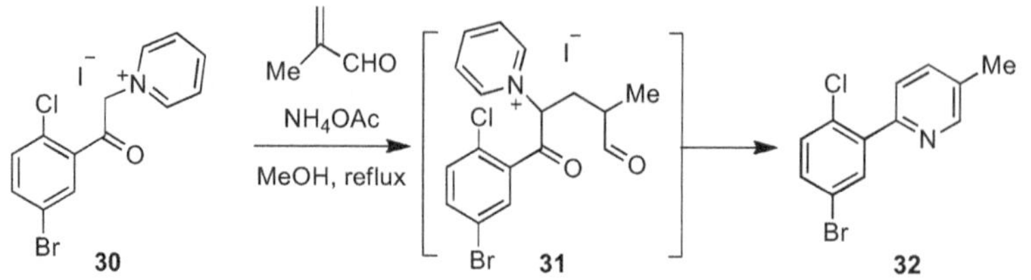

SCHEME 5.5

5.1.5.1 Condensation Reaction

Compared with ammonia (bp –33 °C and mp –78 °C) ammonium acetate (NH_4OAc) is a solid and easy to handle and is frequently used as the ammonia surrogate in organic transformations. The preparation of pyridine derivative **32** was realized by refluxing a mixture of pyridinium salt **30** and methacrolein in the presence of ammonium acetate in MeOH (Scheme 5.5).[12]

(I) Problems

Disappointingly, the only trace of the desired pyridine derivative **32** was obtained and most of the starting material **30** was recovered.

(II) Solutions

The poor yield of **32** is presumably due to the loss of volatile NH_3 generated from the decomposition of NH_4OAc under refluxing conditions (Equation 5.7):

$$NH_4OAc \rightleftharpoons NH_3 + HOAc \tag{5.7}$$

In this context, the addition of acetic acid would shift the equilibrium reaction (Equation 5.7) to the "left" suppressing the release of NH_3. Thus, the addition of AcOH (2.5 equiv) improved the yield of **32** to 93%.

5.1.5.2 S$_N$2 Reaction

Some base-sensitive reactions cannot tolerate residual bases, carried over from materials obtained from the previous step. As a weak organic acid, acetic acid (pKa 4.76) is able to scavenge the residual base that otherwise will have a detrimental effect on the reaction. In addition to trapping the residual base, the introduction of acetic acid as an additive will allow certain acid-labile functional groups to present as well, while maintaining the reaction system in slightly acidic conditions:

- Reaction conditions (Scheme 5.6):[13] **33** and **34**, toluene, 20 °C for 5–6 h.
- Scale: 590 g (2.15 mol).

(I) Problems

- Two side products, *O*-alkylated **36** and bisalkylated **37**, were observed in 5.1% and 8.5%, respectively.
- The yield of **35** was not reproducible, ranging from 35 to 60%.

(II) Solutions

The residual base NaOH in reagent **34** is the cause for the formation of **37**. With the addition of AcOH (0.06 equiv) a consistent and good solution yield (81%) of **35** was obtained; **37** was reduced to 3.7%.

Procedure

- Charge 7.4 mL (0.13 mol, 0.06 equiv) of AcOH to a slurry of **34** (516 g, 2.33 mol, 1.08 equiv) in toluene (8.0 L) at 20 °C.

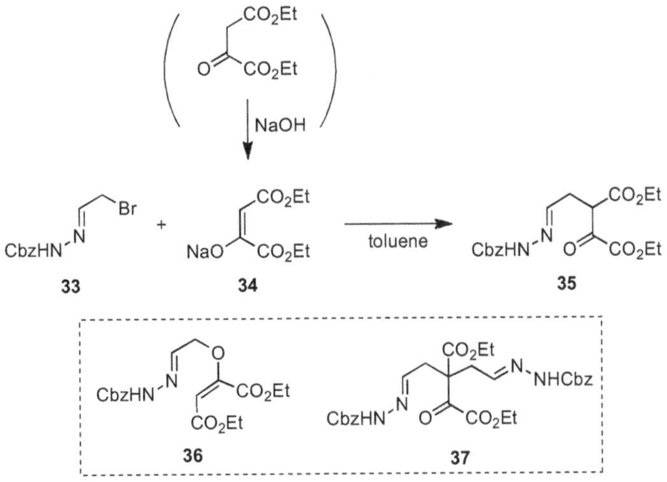

SCHEME 5.6

- Stir for 30 min.
- Charge 590 g (2.15 mol) of **33** in four equal portions over 1 h followed by adding 850 mL of toluene.
- Stir for 5–6 h until complete consumption of **33** (≤0.5% by HPLC).
- The product **35** in toluene was directly used in the subsequent step without isolation.

5.1.5.3 Mitsunobu Reaction

If reaction vessels and transfer lines are not rinsed thoroughly after cleaning with detergents, the residual basic soap may have a detrimental effect on the subsequent operations. For example, during the solvent switch from THF to methanol prior to the crystallization of oxoazetidine **39**, a complete methanolysis of **39** was observed giving the corresponding methyl ester **40** (Scheme 5.7).[14]

A solution to this methanolysis issue was to add acetic acid (0.05 equiv) to maintain the reaction mixture in acidic prior to the solvent switch to methanol.

Procedure

Stage (a)
- Charge 4.26 kg (22.2 mol, 1.3 equiv) of EDC·HCl to a slurry of **41** (4.0 kg, 17.1 mol, 1.0 equiv), benzyloxyamine hydrochloride (3.0 kg, 18.8 mol, 1.1 equiv), and LiOH (0.72 kg, 17.1 mol, 1.0 equiv) in a mixture of THF (10 L) and water (30 L) at 20 °C.
- Stir the batch for 1 h.
- Add MTBE (32 L) and separate layers.
- Distill the organic layer under reduced pressure.
- Add THF and distill until all MTBE is removed and water content is KF<2000 ppm.
- Add THF to form a THF solution of **38** (in 17 L).

Stage (b)
- Charge 3.70 L (18.8 mol, 1.1 equiv) of DIAD into a separate vessel which contains a solution of PPh$_3$ (4.93 kg, 18.8 mol, 1.1 equiv) at 0–10 °C.
- Charge the solution of **38** slowly at <10 °C.
- Upon completion of the addition, warm to 20 °C and age for 18 h.

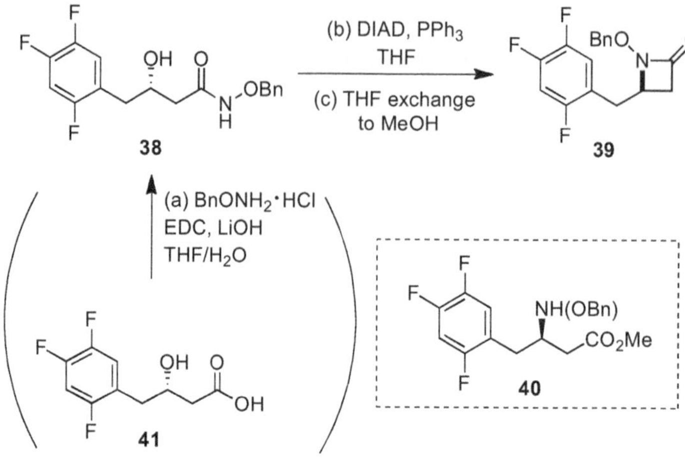

SCHEME 5.7

Stage (c)
- Charge 51.6 g (0.86 mol, 0.05 equiv) of AcOH.
- Solvent switch to MeOH.
- Add MeOH to adjust the volume to 43 L.
- Add water (2.8 L).
- Heat the mixture to 35 °C to redissolve the solids.
- Cool to −20 °C and age for 1 h.
- Filter and rinse with cold (−20 °C) 10% H_2O/MeOH (v/v, 2×4.5 L).
- Yield: 4.7 kg (82% over two steps).

5.1.6 BENZOIC ACID

Besides scavenging residual bases, acid additives can also be used to modify the reactivity of basic nucleophiles to prevent undesired reactions from occurring.

(I) Problematic reactions

Amine **43** intends to dimerize during the cycloaddition with ketone **42** in heptane at 82 °C (Equation 5.8).[15]

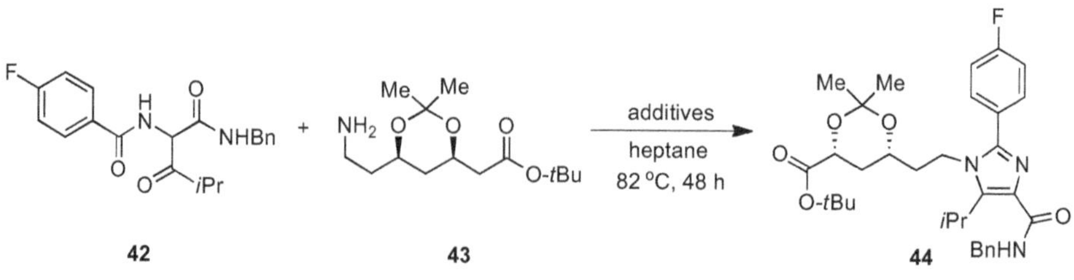

(II) Solutions

In searching for additives in an attempt to prevent amine **43** from dimerization, benzoic acid was identified as a mild acid additive to stabilize amine **43**.

Procedure

- Charge, into a 5-L round-bottomed flask, 356.7 g (1.0 mol) of **42**, followed by adding 410.0 g (1.5 mol) of **43**, 183.0 g (1.5 mol) of benzoic acid, and heptane (3 L).
- Heat at reflux for 48 h (water by-product was removed via Dean–Stark apparatus).
- Cool to 40 °C.
- Add MTBE (1 L).
- Wash the resulting solution with sat. Na_2CO_3 (2×750 mL) followed by 0.5 N HCl (1 L).
- The crude **44** as an MTBE solution was carried forward without further purification.

REMARKS

(a) The use of acetic acid resulted in poor conversion to **44**, because acetic acid was removed from the reaction system via its azeotrope with heptane and was trapped in the aqueous layer in the Dean–Stark apparatus.

(b) Benzoic acid was easily washed away during the workup.

5.1.7 TRIFLUOROACETIC ACID

Trifluoroacetic acid is a strong organic acid (pKa −0.3) that can be utilized as an additive to improve a reaction profile. Enamine **49** is an intermediate for the synthesis of sitagliptin, developed by Merck scientists as a potent and selective DPP-4 inhibitor for the treatment of type 2 diabetes mellitus. A highly efficient one-pot process for the synthesis of **49** was developed (Scheme 5.8).[16]

The reaction sequence shown in Scheme 5.8 involves three notable reaction intermediates: isolable salt **50** and free acid **51**, and **52**.

Kinetic studies[17] showed that the formation of the enamine **49** occurred via the degradation of the free acid **51** to the oxo-ketene intermediate **52**. Owing to the thermal instability of the free acid

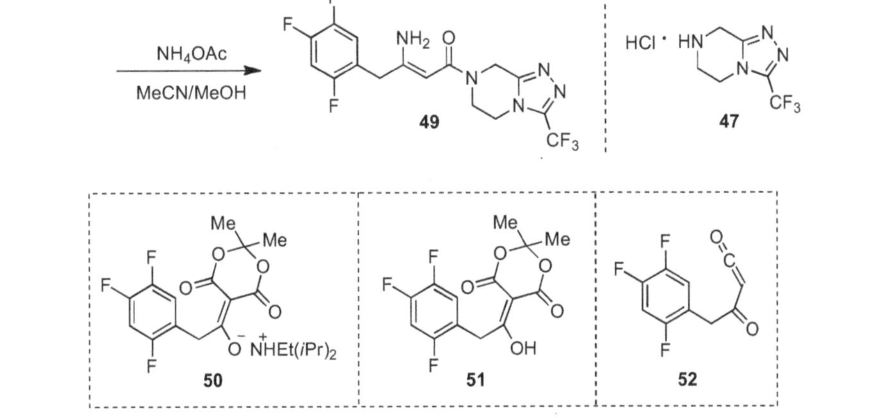

SCHEME 5.8

51, 2.1 equiv of the base is employed to maintain **51** as its stable anionic form **50** to minimize its decomposition. Following the full consumption of acid **45**, a catalytic amount of CF_3CO_2H (0.3 equiv) was used to facilitate the liberation of free acid **51**. The oxo-ketene **52** was captured by triazole·HCl salt **47**, furnishing the amide intermediate **48**.

5.1.8   TOLUENESULFONIC ACID

Strong toluenesulfonic acid was also utilized during reaction optimizations to facilitate the desired reaction, thereby suppressing side reactions.

(I) Problematic reactions

When a Grignard reaction mixture was quenched into methanol and then hydrogenated in the presence of a Pd/C catalyst, a mixture of **54** and **55** (1.7:1) was obtained (Scheme 5.9).[18]

(II) Chemistry diagnosis

The Grignard adduct intermediate **56** may undergo two reaction pathways: dehydration leading to an iminium intermediate **57** or equilibration with ring-opened ketone **58**. The subsequent hydrogenation of **57** and **58** would give **54** and **55**, respectively. Apparently, acid additives would facilitate the dehydration of **56**.

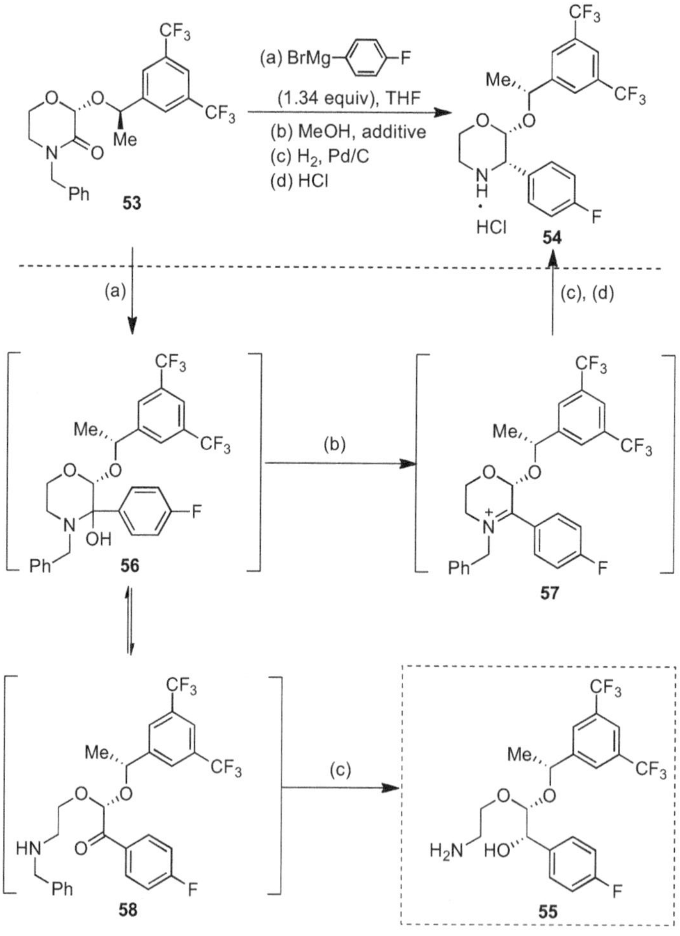

SCHEME 5.9

(III) Solutions

- Quenching the Grignard adduct mixture into methanol followed by adding toluenesulfonic acid (2.0 equiv).
- Hydrogenating immediately the resulting mixture containing the iminium intermediate **57**.

Procedure

Stage (a)
- Charge 150 mL of 4-fluorophenylmagnesium bromide THF solution (0.99 M, 148.2 mmol, 1.34 equiv) to a solution of **53** (49.5 g, 110.6 mmol) in THF (50 mL) at 20–25 °C.
- Stir for 30 min.

Stages (b)–(d)
- Transfer the resulting reaction mixture into ice-cold MeOH (100 mL) at 0–10 °C.
- Add a solution of TsOH (42.1 g, 221.3 mmol, 2.0 equiv) in MeOH (50 mL) followed by 5% Pd/C (45.5 wt%, 5.4 g).
- Hydrogenate at 20–25 °C under 20 psi of hydrogen for 3 h.
- Filter and wash with MeOH (150 mL).
- Concentrate the combined filtrates to dryness.
- Partition the solid between 4-methyl-2-pentanone (350 mL) and aqueous solution of $NaHCO_3$ (35.1 g) and sodium citrate dihydrate (42.0 g) in water (500 mL).
- Separate layers.
- Add 37% HCl (13.0 g) to the organic layer.
- Distill at atmospheric pressure to a volume of 200 mL.
- Cool to room temperature over 3 h.
- Filter and wash with 4-methyl-2-pentanone (2×50 mL).
- Dry at 75 °C under vacuum.
- Yield: 47.6 g (91% overall yield).

NOTES

Reduction of the nitro group in 3-bromo-5-nitroaniline **59** using catalytic amounts of platinum doped with vanadium provided the corresponding amino group (Equation 5.9).[19]

$$(5.9)$$

Notably the addition of TsOH (1.0 equiv) was important to minimize the levels (<1%) of debrominated side product.

5.2 BASE ADDITIVES

Not surprisingly, bases, including organic and inorganic bases, are extensively utilized in chemical process development. Depending on each individual reaction condition, base additives can play several key roles in process optimization. For instance, the addition of a base may alter the reaction kinetics, thereby suppressing side reactions.

5.2.1 POTASSIUM CARBONATE

In addition to the common regioselectivity issues during the halogenation of heteroaryl compounds, an acid-promoted dimerization[20] was observed during the bromination of imidazole **61** (Equation

5.10).[21] The use of potassium carbonate suppressed the dimerization, leading to a much-improved product yield (83%).

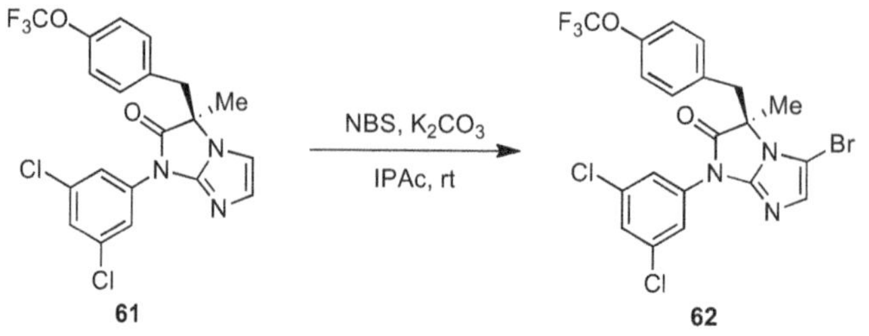

(5.10)

Procedure

- Charge 15 g (0.11 mol, 4.9 mol%) of K_2CO_3 to a solution of **61** (2.22 mol) in IPAc.
- Charge a suspension of NBS (395 g, 2.22 mol, 1.0 equiv) in IPAc (2 L) at 10–20 °C.
- Stir the mixture at room temperature for 1 h.
- Filter and concentrate the filtrate to 2 L.
- Add IPA (4 L) and concentrate to 2 L.
- Add IPA (4 L) and H_2O (0.6 L).
- Stir for 1 h.
- Filter, wash with IPA (0.5 L), and dry at 40 °C/vacuum.
- Yield: 990 g (83%).

REMARKS

(a) Triethylamine was also effective albeit requiring 1.4 equiv of NBS due to its reaction with NBS.

(b) A rate enhancement was observed when using K_2CO_3 additive in a S_NAr reaction (Equation 5.11).[22] The reaction of 4-[chloro(2-pyridyl)]-*N*-methylcarboxamide **63** with 4-aminophenol **64** in the presence of potassium *tert*-butoxide (1.04 equiv) and potassium carbonate (0.5 equiv) completed within 6 h with 87% yield, while the reaction required 16 h without K_2CO_3.

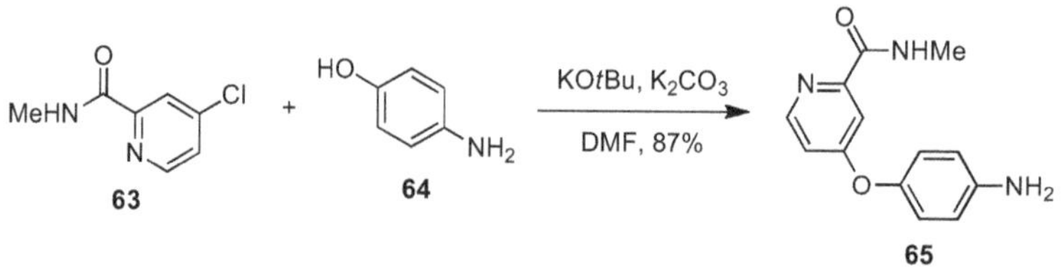

(5.11)

This rate enhancement was presumably caused by a palladium-catalyzed cross-coupling reaction as some commercial metal carbonates such as K_2CO_3 and Na_2CO_3 may be contaminated with low levels of Pd.[23]

5.2.2 SODIUM HYDROGEN CARBONATE

A base is frequently used to capture in situ generated acid by-product that otherwise may have a detrimental effect on the reaction. In order to suppress TBS-deprotection promoted by HBr, sodium hydrogen carbonate (3 equiv) was used to scavenge HBr (Equation 5.12).[24]

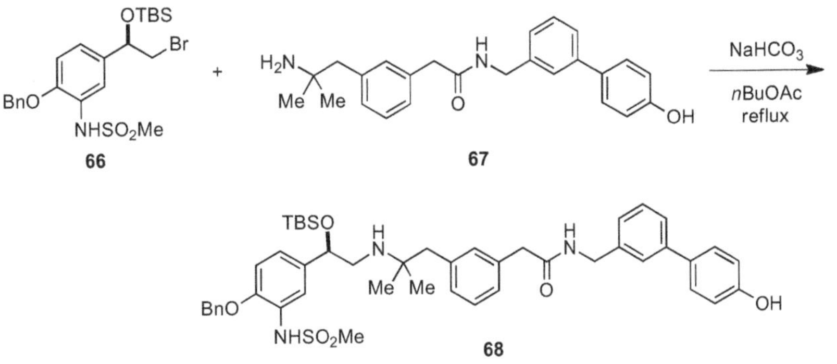

(5.12)

Procedure

- Mix 22.9 kg (44.6 mol) of **66** and 16.5 kg (42.5 mol) of **67** with 7.1 kg (85 mol) of $NaHCO_3$ in *n*BuOAc (165 L).
- Heat the resulting slurry at reflux for 15 min.
- Distill to ca. 80 L (120 L was distilled out over 2 h).
- Heat the resulting slurry at reflux for further 37 h.
- Cool to 50 °C, dilute with EtOAc (165 L), and cool to 20 °C.
- Wash with water and proceed to the subsequent step.

5.2.3 DIISOPROPYLETHYLAMINE

5.2.3.1 Neutralizing AcOH/Replacing Ammonia

A reaction of methyl 2-amino-4,6-difluorobenzoate **69** with formamidine acetate **70** was carried out in 2-methoxyethanol under Dean–Stark conditions to remove water and methanol by-products for the preparation of difluoroquinazolinone **71**. Under these reaction conditions, a significant level of impurity **72** was observed over a period of 10 hours (Scheme 5.10).[25]

(I) Chemistry diagnosis

Refluxing under the Dean–Stark conditions could disproportionate ammonium acetate by-product resulting in a reduction in pH of the reaction mixture via loss of ammonia. The reduction in pH

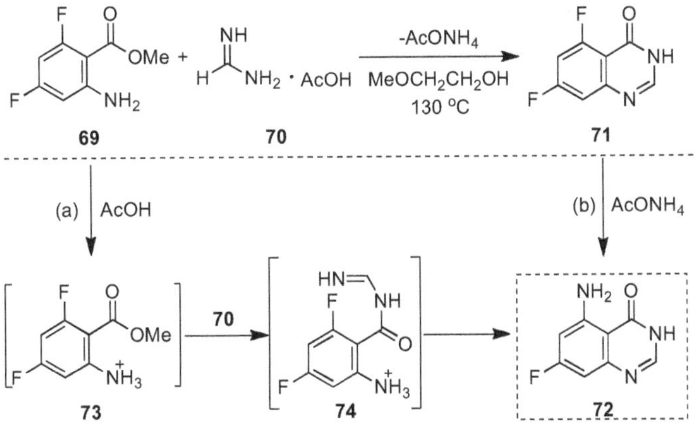

SCHEME 5.10

could lead to the formation of ammonium salt **73**, shutting down the desired reaction pathway. In turn, **73** could react with formamidine to generate **74** whose intramolecular S_NAr displacement of the *ortho* fluorine would form aniline **72** (Pathway A).[26] Alternatively, aniline **72** could be formed via a direct S_NAr reaction of **71** with ammonium acetate (Pathway B).

(II) Solutions

Because of the two potential side reactions, basic diisopropylethylamine (DIPEA) was used with the intention to increase the pH of the reaction system. In addition to increasing the pH, DIPEA is also expected to replace ammonia. Therefore, both pathways A and B could be shut down by the addition of DIPEA.

Procedure

- Mix **69** (2.65 kg, 14 mol in 1 L of toluene) with 25.6 kg of 2-methoxyethanol to form a slurry.
- Charge 1.5 kg (14 mol) of formamidine acetate and 1.84 kg (14 mol) of DIPEA.
- Heat the content to reflux (123 °C).
- Charge 3.32 kg (32 mol) of formamidine acetate and 4.20 kg (32 mol) of DIPEA in four equal portions at 2 h intervals at 40 °C.
- Reheat to reflux between additions.
- Cool and distill (at 60–70 °C) under reduced pressure (24 kg distillate collected).
- Charge 34 kg of water and 3.5 kg of IPA at 50 °C.
- Heat to 96 °C.
- Cool to ambient temperature over 4 h.
- Filter and wash with a mixture of water (11.5 kg)/IPA (1 kg), then with 6 kg of water.
- Dry at 50 °C under vacuum.
- Yield: 2.4 kg (89%).

5.2.3.2   Stabilizing Intermediate

Some reaction problems are caused by unstable reaction intermediates, leading to side products, thereby low product yields. In order to suppress such undesired reactions, a base can be employed to stabilize labile reaction intermediates.

The ability of readily available acyl Meldrum's acid adducts to react with various nucleophiles allows quick access to a variety of functionalized compounds. The preparation of β -keto esters and amides, versatile intermediates in organic synthesis, is often achieved by using acyl Meldrum's acid adducts.[26] This method involves the reaction of Meldrum's acid with activated carboxylic acids, followed by decarboxylation in the presence of nucleophiles such as alcohols or amines.

A practical, one-pot process for a three-component reaction furnished a β -keto amide **79** (Scheme 5.11).[18] The synthetic process consists of three key steps: carboxylic acid activation, the reaction of the activated acid with Meldrum's acid, followed by reaction with amine and decarboxylation.

As **77** (pKa 3.1) was unstable in acidic media at ambient temperature, the addition of DIPEA (2.1 equiv) into the reaction mixture mitigated the degradation of **77** via quenching the HCl by-product and neutralizing the hydrochloride salt **78** used in the subsequent reaction.

DISCUSSIONS

It was found, however, that the base (DIPEA) had the opposite effects on the process. Although more base favors the formation of **77**, the decarboxylation was slow and could not reach full conversion at 70 °C with >2.2 equiv of DIPEA. Now that the overall reaction process is quite sensitive to the DIPEA, the amount of DIPEA was considered one of the critical process parameters and must be carefully controlled (2.0–2.1 equiv) to achieve maximum conversion and yield. Similar to the decarboxylation, the amination reaction of **77** with **78** required slightly acidic conditions.[27] In the

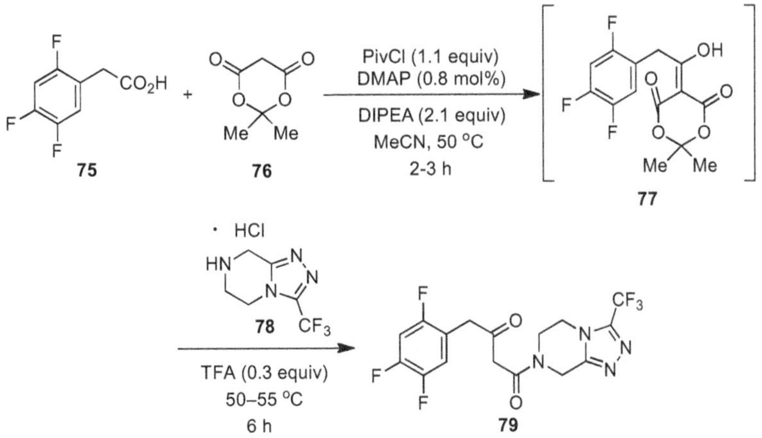

SCHEME 5.11

event, the reaction was further optimized by adding CF_3CO_2H (TFA) as TFA has an appropriate acidity.

5.2.4 1,4-DIAZABICYCLO[2.2.2]OCTANE

Phosphorous oxychloride ($POCl_3$) is extensively used to chlorinate hetero aromatics. Common reaction conditions include $POCl_3$/base (e.g., DIPEA) /solvent (e.g., toluene, or neat) under relatively high temperatures. Usually, excess $POCl_3$ was employed in order to drive the reaction to completion in a desired time frame. Reactions (5.13)[28] and (5.14)[13] represent two deoxychlorination examples.

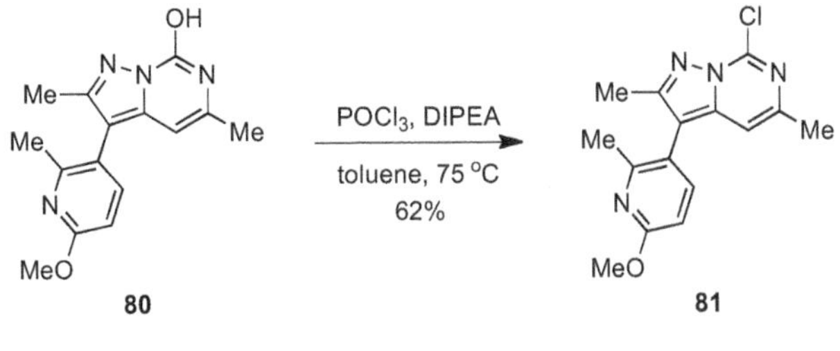

(5.13)

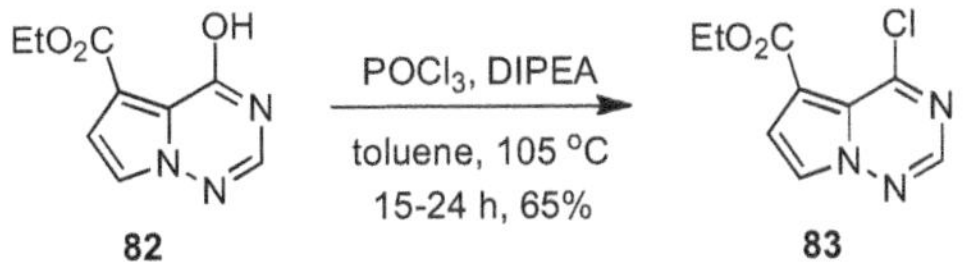

(5.14)

(I) Reaction problems

Because of the relatively high temperature, these deoxychlorinations could generate impurities such as dimers **84** (>1%) and **85** (up to 10%) as shown in Figure 5.1.

(II) Solutions

1,4-diazabicyclo[2.2.2]octane (DABCO) could be used as a catalyst to accelerate deoxychlorination reactions under relatively mild reaction conditions.

- Reaction conditions.
  - For the preparation of **81**: **80**/$POCl_3$ (1.3 equiv)/DIPEA (5.0 equiv)/DABCO (1 mol%)/ MeCN, 20 °C for 1 h.

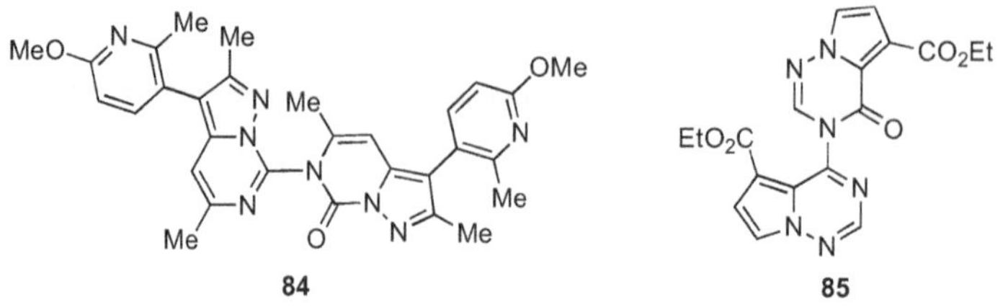

FIGURE 5.1   The dimeric side products.

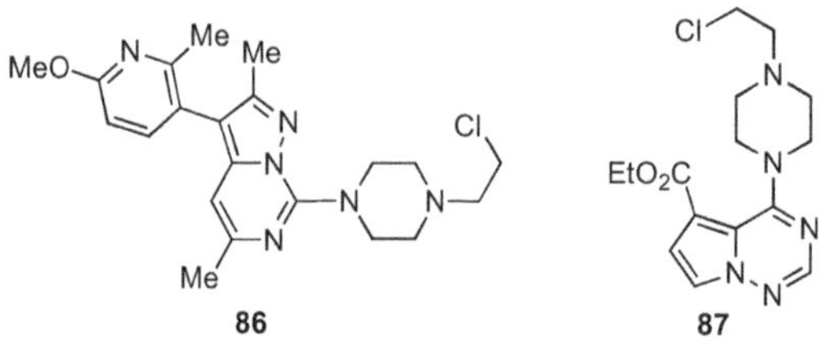

FIGURE 5.2   Structures of **86** and **87**.

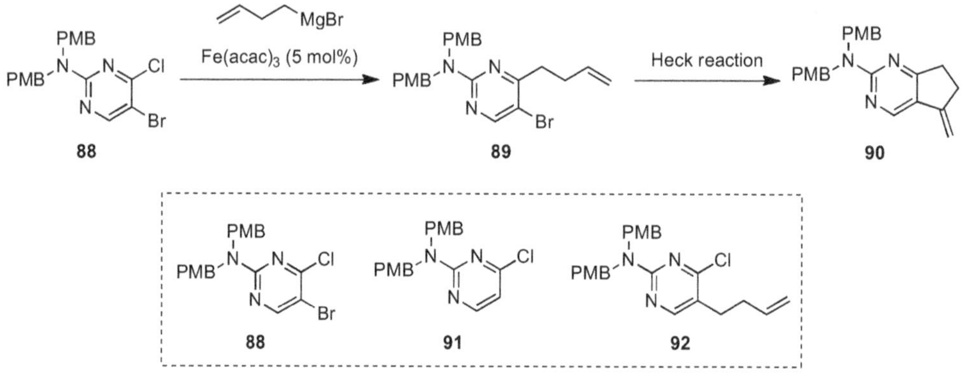

SCHEME 5.12

- For preparation of **83**: **82**/POCl$_3$ (1.3 equiv)/DIPEA (1.3 equiv)/DABCO (10 mol%)/toluene, 45–75 °C for 1.5 h, then 95 °C for 2–3 h.
- The yields of the chlorinated products were improved to 78% from 62% for **81** and to 90% from 65% for **83**, in which side products **84** and **85** were reduced to 0.2% and <3%, respectively.

DISCUSSIONS

Two side products, **86** and **87** (Figure 5.2), could be detected due to reactions with DABCO. Therefore, *N*-methylmorpholine was considered as an alternative base additive.

5.2.5   *N,N′*-DIMETHYLETHYLENEDIAMINE

Catalytic reactions sometimes have strict requirements for the levels of impurities present in the starting materials due to the possibility of catalyst poisoning by impurities. It was observed that the intramolecular Heck reaction of pyrimidyl bromide **89** tended to stall if a crude **89** was used (Scheme 5.12).[29]

The crude intermediate **89** was contaminated with two side products, **91** and **92**, plus unreacted starting material **88**. It was found that these chloro impurities (**88**, **91**, and **92**) in the crude **89** were responsible for the incomplete Heck reaction. To address these impurity issues, *N,N'*-dimethylethylene diamine (DMEDA) was identified as an additive to scavenge the chloro impurities via S_NAr displacements. The resulting S_NAr replacement products were easily removed by washing with aqueous acid.

5.2.6   POTASSIUM *TERT*-BUTOXIDE

Rofecoxib (Vioxx) was developed by Merck & Co. and used in the treatment of pain, which had been removed from the market in 2004 due to the increased risk of heart attack. Compound **95** is a prodrug for rofecoxib and the synthesis of **95** involves a key intermediate **94**. Scheme 5.13[30] highlights a three-step strategy for the preparation of **94**.

(I) Reaction problems

Using excess acetic anhydride afforded the desired acid **94** in only ca. 50% yield along with ca. 25% of butenolide **96**.

(II) Chemistry diagnosis

A possible reaction pathway (as shown in Scheme 5.14) was hypothesized to explain the formation of **96**. The alkoxide anion of **98** would react with excess CO_2 in the reaction system to form a carbonate species **99** that in essence protects the allylic alkoxide from the subsequent acetylation. Apparently, carbonate species **99** is responsible for the formation of **96**.

(III) Solutions

In the event, an approach to scavenge excess CO_2 was adopted and KO*t*Bu was proved to be the additive of choice. The addition of KO*t*Bu (0.5 equiv) to the reaction system suppressed the formation of the side product **96** (Equation 5.15).

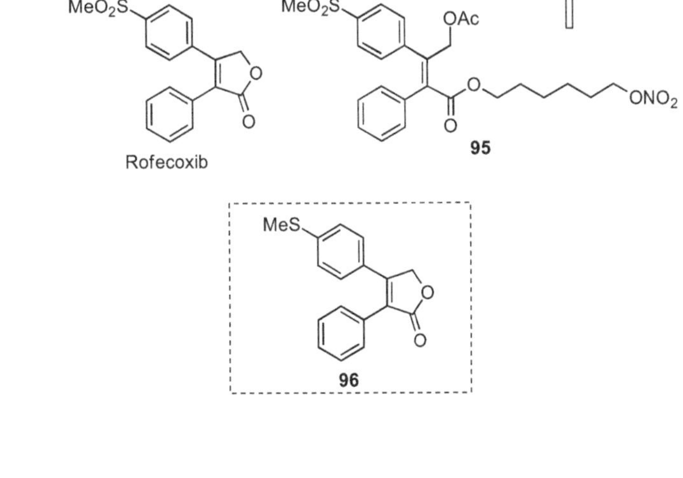

SCHEME 5.13

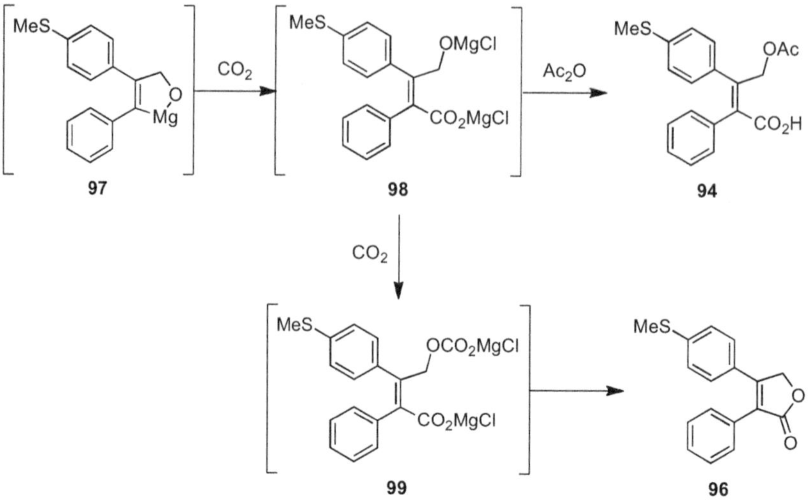

SCHEME 5.14

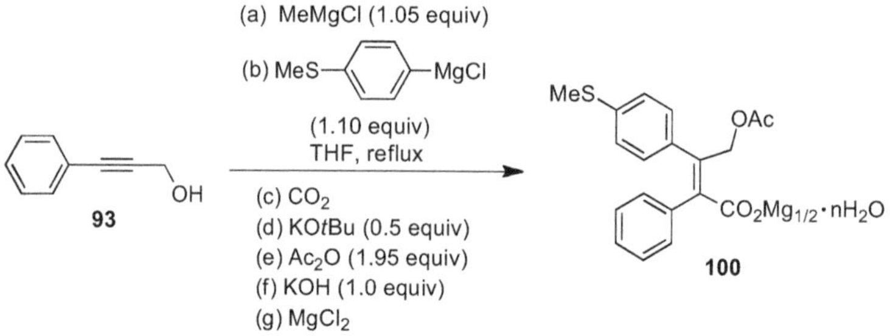

$$(5.15)$$

Procedure

Stage (a)
- Charge 300 mL of THF into a reactor and flush the reactor with N_2.
- Charge 66.1 g (0.50 mol) of **93**.
- Cool the batch to 5 °C.
- Charge 175 mL of 3 M of MeMgCl (0.525 mol, 1.05 equiv) at 25–30 °C over 30 min.

Stage (b)
- Charge 268 mL of 2.05 M of 4-(methylthio)phenylmagnesium chloride (0.549 mol, 1.10 equiv) in THF.
- Heat the batch at reflux (65–70 °C) for 3 h.
- Cool the batch to 18 °C.

Stage (c)
- Charge CO_2 in subsurface and age at 30–35 °C for 2 h.
- Purge the vessel to remove residual CO_2 in the headspace.

Stage (d)
- Charge 250 mL of 1 M of KOtBu (0.25 mol, 0.50 equiv) in THF.
- Stir the batch at 32 °C for 1 h.
- Adjust the batch temperature to 23 °C.

Stage (e)
- Add 94.5 mL of Ac$_2$O (1.00 mol, 2.00 equiv) at 30–45 °C.
- Stir the batch for 1.5 h.
- Cool to 10 °C.

Stage (f)
- Add 93.5 mL of 30 wt% KOH (1.0 equiv) followed by water (300 mL).
- Age the batch at 40 °C for 7 h.

Stage (g)
- Add aqueous solution of MgCl$_2$ (2 M).
- Stir the batch at 40 °C for 15 min.
- Separate layers.
- Add water (100 mL) to the organic layer followed by isopropyl acetate at 20 °C over 2 h.
- Cool the batch to 0 °C.
- Filter and wash with water (500 mL) and cold (0 °C) isopropyl acetate (500 mL).
- Yield: 160.3 g (73.8%).

REMARKS

(a) The amount of KOtBu added to the carbometalation reaction was optimized on the basis of the volume of THF in the reaction and the solubility of CO$_2$ in THF at ambient temperature (ca. 0.25M).
(b) MeMgCl was used as a "sacrificial base" to deprotonate propargyl alcohol **93**.
(c) The magnesium salt **100** was isolated as a crystalline solid without resorting to chromatographic purification.

5.2.7   SODIUM METHOXIDE

As a strong base (pKa 16), sodium methoxide can participate in a number of reactions including acid/base reaction, deprotonation, and so on.

5.2.7.1   Increasing Reaction Rate

Although preparation of amide **101** could be realized at kilogram scale in ca. 90% yield, the reaction required excess morpholine (3 equiv) at 85 °C for 30–60 h and product **101** was contaminated with morpholine (Equation 5.16).[31] The addition of 15 mol% of sodium methoxide permitted the morpholine charge to be reduced to 1.03 equiv. Under these modified conditions, the reaction proceeded to completion at ambient temperature in only 4–6 h instead of 30–60 h.

$$(5.16)$$

101

5.2.7.2   Improving Product Yield

Reduction of pyridinium **102** in methanol required 74 equiv of NaBH$_4$, giving a product **103** in a low yield (24%) due to the incomplete reaction and the formation of a secondary amine **104**.[32]

Scheme 5.15 illustrates the possible reaction pathway for the formation of **104**. A Michael-type addition of **103** onto the pyridinium **102** followed by a reduction of the resulting adduct **105** is responsible for the formation of **104**. The addition of sodium methoxide is presumably to convert **102** to the corresponding pyridiniumyl methanide **106** that is anticipated to be less reactive toward nucleophilic addition. In the event, the reduction of **102** was carried out by the addition of NaOMe (1.0 equiv) to the reaction mixture of **102** and NaBH$_4$ (2.0 equiv) in 2-propanol and methanol followed by treatment with HCl, giving **103** in 80% isolated yield (98.6% of purity) (Scheme 5.16).

Procedure

- Charge MeOH (5.2 L) dropwise to a solution of **102** (2 kg, 6.82 mol), NaBH$_4$ (520 g, 13.75 mol, 2.0 equiv), and NaOMe (368 g, 6.81 mol, 1.0 equiv) in IPA (32 kg) under refluxing.
- Stir the resulting mixture under refluxing for 2 h.
- Cool to 20–30 °C.
- Add H$_2$O (26 L) and 6 N HCl (8 L) and stir for 1.5 h.
- Add 30% NaOH (6.8 L).
- Distill and extract the residue with EtOAc (3×20 L).
- Wash the combined organic layers with H$_2$O (2×10 L) and dry over Na$_2$SO$_4$ (2.5 kg).
- Distill and dissolve the residue in EtOH (4 L).
- Add 3.96 N HCl in EtOH (3 L) dropwise at 0–10 °C and stir for 0.5 h.
- Warm to 20–30 °C.

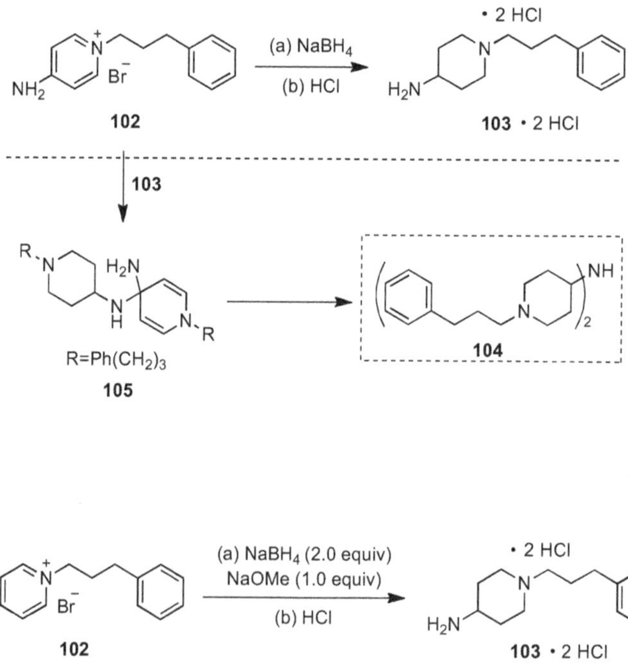

SCHEME 5.15

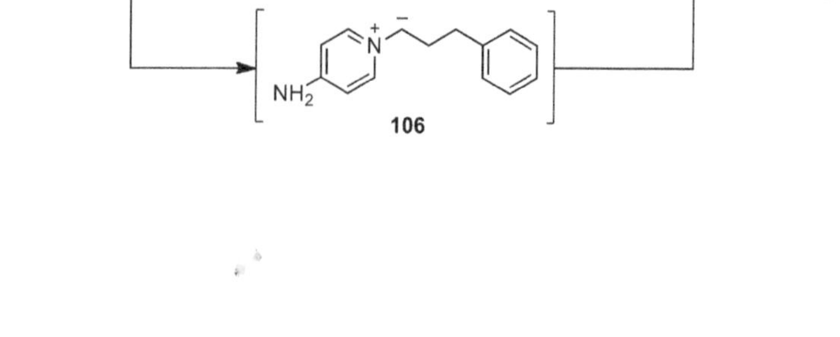

SCHEME 5.16

- Add isopropyl ether (10.6 L) and stir for 2 h.
- Cool to 0–10 °C and stir for 1 h.
- Filter and wash with isopropyl ether (2×0.5 L).
- Dry at 40 °C under reduced pressure.
- Yield: 1.58 kg (80%).

DISCUSSIONS

The effect of alkoxide, generated in situ, could also be observed during the Rh(I)-catalyzed hydrogenation. For Wilkinson's catalyst, the hydroxyl group is not strong enough to associatively displace the chloride such that the trisubstituted olefin **107** does not undergo hydrogenation. In the presence of KH, an alkoxide is generated that can displace chloride anion, leading to effective hydrogenation exclusively from the top of the olefin (Scheme 5.17).[33]

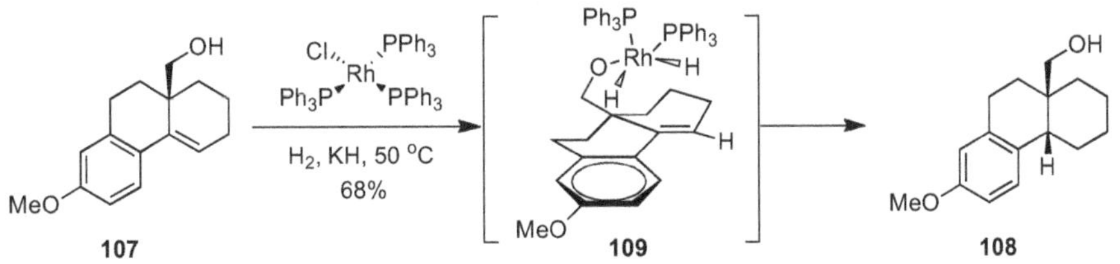

SCHEME 5.17

5.2.8 SODIUM ACETATE

5.2.8.1 Trapping Hydrogen Iodide

In general, solid salts are relatively more stable with higher melting points than their corresponding free bases. In solution, however, such salts may show poor reactivity and stability. For instance, a decomposition of 3-aminoisoxazoline **110** was observed due to the instability of the in situ formed **110·HI** salt under refluxing conditions in chlorobenzene (Scheme 5.18).[34]

In order to circumvent the decomposition of **110**, the addition of NaOAc was proved to be effective in trapping HI, affording the desired isoxazole **111** in 86% assay yield.

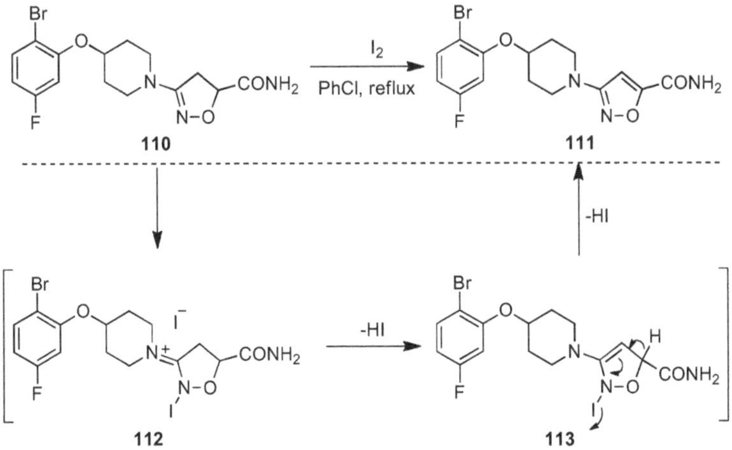

SCHEME 5.18

Procedure

- Charge 3.68 kg (14.5 mol) of iodine to a suspension of **110** (4.31 kg, 11.2 mol) and NaOAc (2.38 kg, 29.0 mol) in chlorobenzne (21.6 L).
- Heat the mixture at reflux for 3 h 45 min.
- Cool to 73 °C over 1 h.
- Add MeTHF (43 L) and cool to 45 °C.
- Add $Na_2S_2O_3$ (21.6 L, 10 wt%).
- Stir for 5 min and filter through Solka-Floc.
- Wash the cake with MeTHF (21.6 L).
- Separate layers.
- Wash the organic layer with brine (21.6 L).
- Stir the organic layer with Darco-KB (2.47 kg) for 3 h.
- Filter through Solka-Floc and wash with THF (20 L).
- Switch solvent to toluene/THF and treat with Darco-KB again.
- Filter, concentrate, and use in the next step.

DISCUSSIONS
Other bases, such as Et_3N and $KHCO_3$, were unsuitable since these bases inhibited the reaction.

5.2.8.2 Trapping Hydrogen Chloride
3-Amino-4-fluorophenol **116** was prepared by catalytic hydrogenation of nitrophenol **114**, followed by dechlorination as shown in Scheme 5.19.[35] The nitro group reduction underwent readily to give the amino phenol intermediate **115** within 1 h. The subsequent catalytic dechlorination of **115**, however, proved to be sluggish. The reaction was plagued with a moderate product yield (64%) and the unreacted intermediate **115** (4%), over-reduction **117**, and other unidentified side products.

Solutions
The slow dechlorination was due to the hydrogen chloride by-product, generated in situ from the dechlorination. Accordingly, 1.1 equiv of NaOAc was added as an additive to neutralize the HCl, resulting in an 80% product yield with 95 area% purity after crystallization.

Procedure

- Charge MeOH (108 L) into a 300 L glass-lined vessel, followed by adding 5.40 kg (28.2 mol) of **114** and 2.54 kg (30.6 mol) of NaOAc.
- Inert the vessel with nitrogen by evacuation and backfilling with nitrogen.

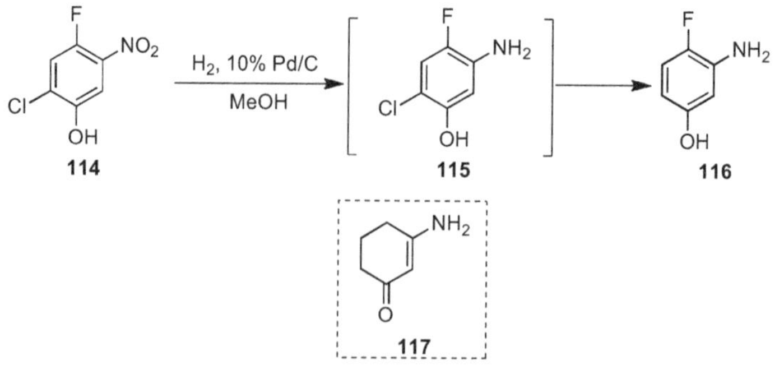

SCHEME 5.19

- Charge 1.32 kg (0.578 mol of Pd) of 10% Pd/C (52.7% water).
- Inert the vessel again.
- Stir the batch at 15–25 °C for 4 h under 0.1 MPa of hydrogen.
- Vent the H_2 and inert the vessel with N_2.
- Filter and rinse with MeOH (32 L).
- Distill under reduced pressure.
- Add EtOAc (22 L).
- Distill under reduced pressure.
- Add EtOAc (54 L) and water (27 L) and stir.
- Separate layers.
- Wash the organic layer with 5% NaCl (27 L).
- Distill the organic layer under reduced pressure to 10.8 L.
- Add EtOAc (43 L).
- Distill under reduced pressure to 10.8 L.
- Add n-heptane (43 L).
- Stir the resulting slurry at 15–25 °C for 30 min.
- Cool to 0–10 °C and stir for 2 h.
- Filter and wash with EtOAc/n-heptane (1:4, 10.8 L).
- Dry at 50 °C under vacuum.
- Yield of **116**: 2.85 kg (80%).

5.2.8.3 Improving Conversion

The performance of some reactions depends largely on the pH of reaction mixtures. For instance, the conversion of amine hydrobromide salt **118** to tetracyclic hydrochloride salt **120** with ferric chloride was pH-dependent: with the reaction progressing, the pH of the reaction mixture drops rapidly, which would retard the ionization required for the ring-closing step (via an intermediate **119**), resulting in poor conversions (40–50%) (Scheme 5.20).[36]

To overcome this problem, buffering the reaction system with sodium acetate (1.05 equiv) provided the product **120** in consistent yields of 76–82%.

Procedure

Stage (a)
- Mix 1.0 kg (2.62 mol) of **118** with 225 g (2.74 mol) of NaOAc in 2.5 L of water.
- Heat to 85 °C to form a solution.
- Cool to 80 °C and add 2.5 L of EtOH.
- Cool to 0–5 °C.
- Charge a pre-prepared solution of $FeCl_3$ (1.225 kg, 7.55 mol) in water (1.5 L)/EtOH (1.5 L).
- Stir the batch at ambient temperature for 15–18 h.

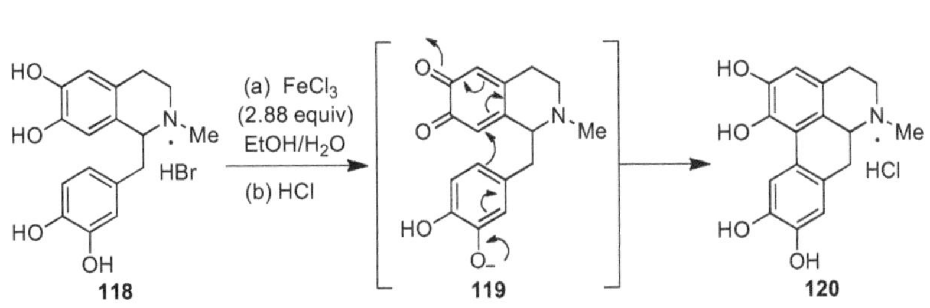

SCHEME 5.20

Stage (b)

- Charge 1.0 L of 37% HCl and stir for 30 min.
- Filter, wash with acetone, and dry at 50–60 °C under vacuum.
- Yield: 668 g (76%).

5.2.9   SODIUM ACRYLATE

Like sodium acetate, sodium acrylate is a weak base and also a Micheal addition acceptor. The synthesis of 1-*tert*-butylpiperidin-4-one **122** was realized via transamination of 1,1-dimethyl-4-oxo-piperidinium iodide **121** with *tert*-butylamine (Scheme 5.21).[37]

However, this reaction was plagued by a low product yield owing to side reactions caused by dimethylamine by-product.

Removal of dimethylamine from the reaction was the key to the success of the transformation. The use of sodium acrylate (5 equiv) to trap dimethylamine proved to be an efficient and practical method. This protocol allowed the synthesis to be performed in water, and the subsequent extractive workup rejected successfully the adduct **123** into the aqueous layer, producing the product **122** in high yield and purity.

Procedure

- Charge 230 mL (2.3 mol) of 10 N NaOH to a solution of acrylic acid (176 g, 2.5 mol) in water (500 mL).
- Charge 128 g (0.5 mol) of **121** and *t*-butylamine (1000 mL).
- Heat the resulting mixture at reflux (65 °C) for 90 min.
- Distill at 25 °C/40 Torr to remove excess *t*-butylamine.
- Extract with EtOAc (500 mL, 2×250 mL).
- Wash the combined organic layers with aq NaCl (200 mL).
- Distill the organic layer at 20 °C under reduced pressure.
- Yield: 70 g (87% assay yield) (99.7% by GC).

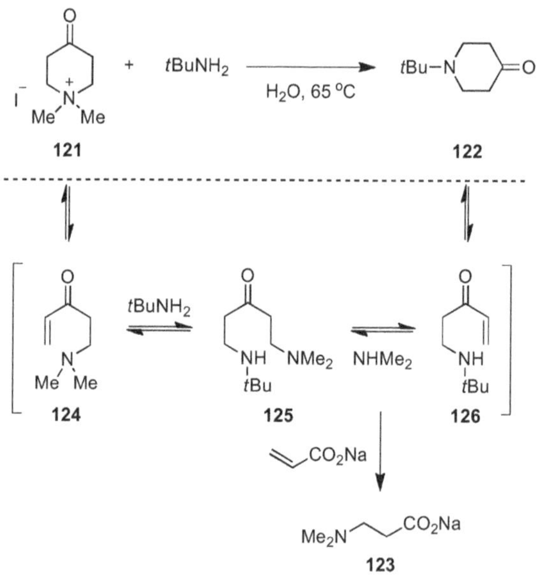

SCHEME 5.21

REMARK

Removal of dimethylamine by distillation and displacement with *tert*-butylamine was inefficient, requiring large amounts of *tert*-butylamine.

5.3   INORGANIC SALTS

5.3.1   Lithium Chloride

5.3.1.1   Increasing Reaction Rate

It is well-known that lithium chloride as an additive can improve the reactivity of Grignard reagents (RMgCl) through a mechanism which involves the breakup of polymeric RMgCl by LiCl and the formation of the magnesium-lithium ate complex (Equation 5.17).

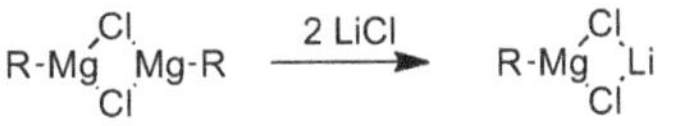

$$\underset{\text{Cl}}{\overset{\text{Cl}}{R\text{-Mg}\diagdown\text{Mg-R}}} \xrightarrow{\text{2 LiCl}} \underset{\text{Cl}}{\overset{\text{Cl}}{R\text{-Mg}\diagdown\text{Li}}} \qquad (5.17)$$

In addition to the reactivity improvement for Grignard reagents, the addition of lithium chloride could increase the reaction rate of ester hydrolysis (Equation 5.18).[38] For instance, treating methyl ester **127** with wet triethylamine (5 equiv) resulted in no reaction, while the same reaction was carried out by adding lithium chloride (5 equiv) under otherwise the same conditions led to 74% conversion.

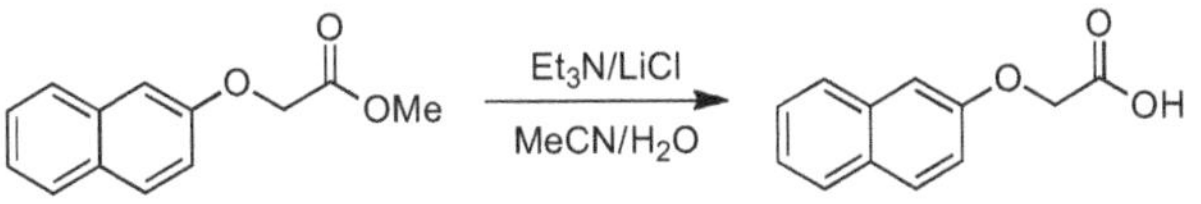

$$\text{127} \xrightarrow[\text{MeCN/H}_2\text{O}]{\text{Et}_3\text{N/LiCl}} \text{128} \qquad (5.18)$$

The reaction protocol has been utilized to selectively hydrolyze carboxylic esters (see Chapter 2).

5.3.1.2   Improving Stereoselectivity

The stereochemical outcome of palladium-catalyzed diacetoxylation of 1,3-dienes is related to the internal or external attack of the nucleophile on the (π-allyl)palladium complex (Scheme 5.22).[39]

Without LiCl additive, the reaction of the cyclohexa-1,3-diene gave a *trans* diacetate isomer via an internal attack by acetate on the allyl group, while with catalytic amounts of LiCl, an external attack occurred affording a *cis* isomer.

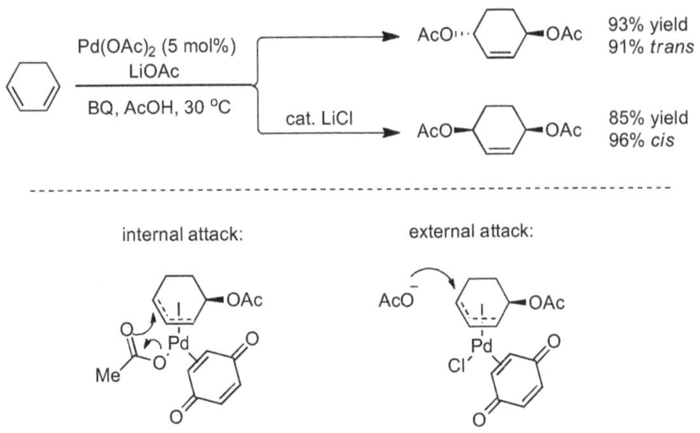

SCHEME 5.22

5.3.1.3   Improving Conversion

Organozinc compounds are frequently employed in the Negishi cross-coupling reactions with aryl bromides. Organozinc reagents are often prepared from a redox reaction of alkyl (or aryl) halides with zinc metal. The LiCl additive used in the preparation of organozinc reagent **130** had a significant impact on the subsequent Negishi cross-coupling reaction with aryl bromide **131**. For example, the Negishi coupling using **130**, prepared from zinc dust in the presence of TMSCl as a zinc activator without LiCl, gave a poor conversion (20%), while using LiCl (0.5 equiv) a full conversion was achieved (Scheme 5.23).[40]

Procedure

Preparation of alkylzinc
- Add TMSCl (50.1 g, 58.6 mL, 0.462 mol, 0.05 equiv) to a mixture of anhydrous LiCl (196 g, 4.61 mol, 0.5 equiv), zinc dust (726 g, 11.1 mol, 1.2 equiv), and **129** (1.80 kg, 1.32 L, 9.23 mol) in THF (2.5 L) under nitrogen.
- Heat the mixture to 55 °C over 0.3 h and age for 2 h (exothermic).
- Stir at 60–65 °C for 19 h.
- Cool to 25 °C (the supernatant (~4 L) was used directly for the Negishi coupling).

Negishi coupling
- Add 0.4–0.5 L of the solution of alkylzinc **130** to a solution of **131** (6.20 mol) in THF (3.7 L).
- Stir the resulting mixture at 35–40 °C for 0.5 h under nitrogen.
- Prepare a solution of the catalyst by dissolving Pd(OAc)$_2$ (1.39 g, 6.10 mmol, 0.10 mol%) and SPhos (10.2 g, 24.7 mmol, 0.40 mol%) in THF (62 mL) at 20 °C under nitrogen.
- Add the catalyst solution into the batch (exothermic).
- After the initial exotherm subsided, add the remaining **130** at 45–60 °C over 1–2 h.
- Age the batch until the reaction completion.
- Cool to 20 °C.
- Add EtOH (359 mL, 1.0 equiv) and age for 0.5 h.
- Concentrate the batch to 3–4 L.
- Add toluene (5 L), water (3.7 L), and 50% aqueous citric acid (150 mL).
- Separate layers, wash the organic layer with water (2×2.3 L).
- Concentrate to ~3 L, flush with toluene (3×1.3 L).
- Yield: 2.20 kg (crude product).

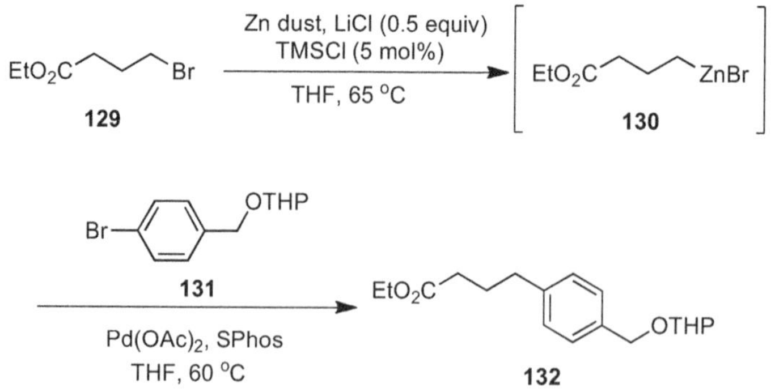

SCHEME 5.23

5.3.2   Lithium Bromide

The application of organolithium reagents is frequently plagued by insolubility and aggregation of lithium species. The use of lithium bromide in tetrahydrofuran is able to modify lithium aggregates, thereby improving the organolithium reagent solubility. An effort in the preparation of ketone **134** from a reaction of **133** with isobutyllithium led to a sparingly soluble salt. The use of LiBr (2.6 equiv) in this reaction allowed to produce **134** in 87% yield. The function of LiBr is presumably to improve the solubility of the lithium salt intermediate (Equation 5.19).[41]

$$(5.19)$$

5.3.3   Sodium Bromide

5.3.3.1   Lowering Reaction Temperature

In addition to high energy costs, high reaction temperature usually leads to undesired side reactions. The preparation of ketone **137** was realized by heating a mixture of **135** and **136** at high temperatures (120 °C) (Equation 5.20, condition (a)).[42] The product **137** produced under such conditions was highly colored.

$$(5.20)$$

The addition of NaBr (1.0 equiv) permitted the reaction to be carried out under mild conditions (35 °C), affording an 84% isolated yield of **137** (Equation 5.20, condition (b)). Furthermore, the isolation was operationally simple involving the addition of water to the reaction mixture and filtration. Consequently, the color of the isolated product **137** was significantly improved.

5.3.3.2   Improving Product Yield

An improved product yield was realized when using sodium bromide additive in the preparation of imidazopyridazine **140**. The reaction of 6-chloropyridazin-3-amine **138** with ethyl 2-chloroaceto-acetate **139** was plagued with low yields (Equation 5.21).[43]

$$(5.21)$$

An improvement in the product yield was achieved by adding substoichiometric NaBr (0.7 equiv) and H_2O (4 equiv), affording 91% conversion and 61% yield of **140** with 99.3 % purity.

Procedure

- Heat a slurry of **138** (1.20 kg, 9.26 mol), NaBr (672 g, 6.53 mol, 0.7 equiv), water (672 mL, 37.3 mol, 4.0 equiv), DMF (7.2 L), and **139** (762 mL, 5.6 mol, 0.6 equiv) at 100 °C for 4 h.

- Charge 762 mL (5.6 mol, 0.6 equiv, total 1.2 equiv) of **139** over 1 h
- Stir the resulting mixture at 100 °C for additional 10 h.
- Cool to 20 °C.
- Add water (25 L) over 2 h and stir at 20 °C for 1 h.
- Filter to collect a wet cake.
- Heat the wet cake in EtOAc (21.6 L) at 60 °C to form a solution.
- Wash the organic solution with water (2×6 L).
- Distill to 3.6 L.
- Add heptane (6 L).
- Filter and dry.
- Yield: 1.38 kg (61 %).

5.3.4 MAGNESIUM CHLORIDE

Tetrahydroquinoline **143a** was synthesized via a protocol outlined in Scheme 5.24,[44] including reduction of *N*-acyl carbamate **141a** to aminal **142** followed by acid-catalyzed cyclization.

(I) Reaction problems
- Reduction with $NaBH_4$ in MeOH required large amounts of borohydride (15 equiv)
- Multiple charges are needed
- Over-reduction of product **144** was observed
- Using ethanol resulted in only small amounts of desired **142** along with a large amount of **144**.

(II) Solutions

N-acyl carbamate (**141b**) was reduced with $NaBH_4$ in the presence of $MgCl_2$ (Scheme 5.25). The magnesium salt served dual purposes of activating the imide carbonyl for reduction (no reaction was observed with $NaBH_4$ alone) and preventing over-reduction by stabilizing the hemiaminal salt **145** via chelating of the carbonyl oxygen with Mg.

Procedure

Stage (a)
- Charge 7.474 kg of **141b** (containing 7% solvent, 6.951 kg of **141b**, 21.84 mol) into a 100-L glass reactor.
- Charge EtOH (46 L) and water (2.35 L).
- Charge 620 g (16.4 mol) of $NaBH_4$ (as 11-mm pellets).

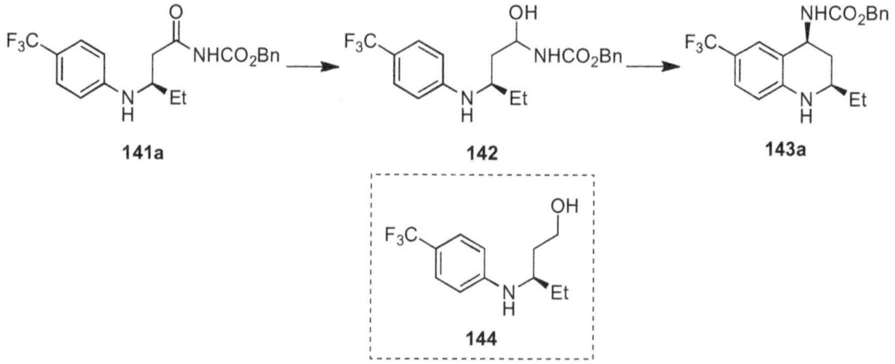

SCHEME 5.24

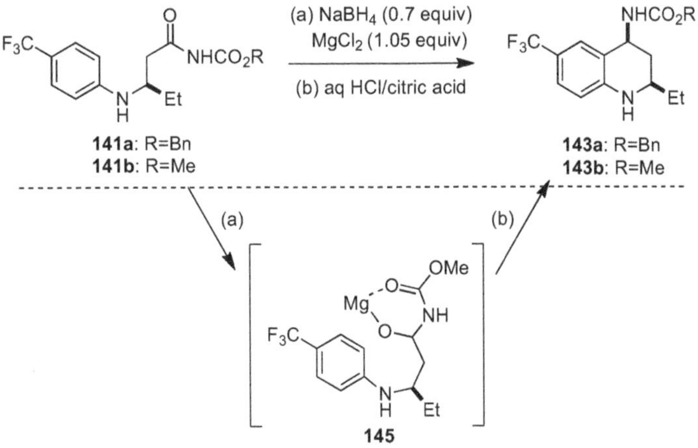

SCHEME 5.25

- Stir the mixture at room temperature for 20 min.
- Cool to –10 °C.
- Charge an aqueous solution of MgCl$_2$ (4.68 kg of MgCl$_2$·6H$_2$O, 23.0 mol in 7 L of water) while maintaining the batch temperature at <–5 °C.
- Stir the batch at 0 °C for 45 min.

Stage (b)
- Transfer the batch into a 200-L reactor containing a mixture of CH$_2$Cl$_2$ (70 L), concentrated HCl (5.8 L), citric acid (10.5 kg), and water (64 L).
- Stir at room temperature for 2 h.
- Remove the aqueous layer.
- Wash the organic layer with an aqueous citric acid solution (6.3 kg of citric acid in 34 L of water).
- Add Darco-activated carbon (G-60 grade, 700 g) to the organic layer.
- Stir for 30 min.
- Filter through Celite and rinse with CH$_2$Cl$_2$.
- Exchange solvent to hexanes via distillation.
- Cool to room temperature.
- Filter and wash with hexanes (14 L).
- Dry at 40 °C under vacuum.
- Yield: 5.291 kg (80%).

DISCUSSIONS

(a) Acid activation was also evaluated, but the addition of acid led to a significant evolution of hydrogen.
(b) Although the addition of CaCl$_2$ improved the reaction rate, over-reduction was a concern at scale up.

5.3.5 MAGNESIUM BROMIDE

Usually, less attention is paid to inorganic impurities in isolated intermediates as they can be rejected easily by aqueous wash. However, the inorganic impurity may turn out to be a valuable additive to catalyze the desired reaction.

It was found that the residual bromide in an isolated thiophthalimide **147** had a significant impact on the reaction profiles: the presence of 1000 ppm of bromide gave 80–86% yields of the desired product **148**, while with 40 ppm of bromide, the yield of **148** diminished to 50% along with 45% of side product **149** (Scheme 5.26).[45]

(I) Chemistry diagnosis

A plausible mechanism for the formation of **148** and **149** is illustrated in Scheme 5.27. The bromide would react with **147** to generate in situ a reactive sulfenyl bromide whose reaction with **146** affords the desired product **148**. In the absence of bromide, **146** would equilibrate with **150**, which may undergo reactions with less reactive **147** to lead to **151**. The subsequent 1,3-arylthiolate migration of **151** generates the side product **149**.

(II) Solutions

The addition of $MgBr_2$ as a bromide source would convert **147** to the reactive sulfenyl bromide which is expected to react with **146** readily to furnish **148**. Thereby, the undesired side reaction was suppressed.

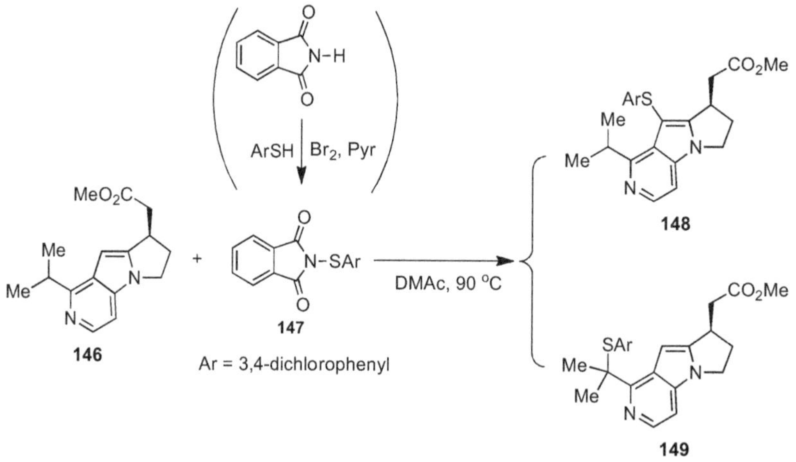

SCHEME 5.26

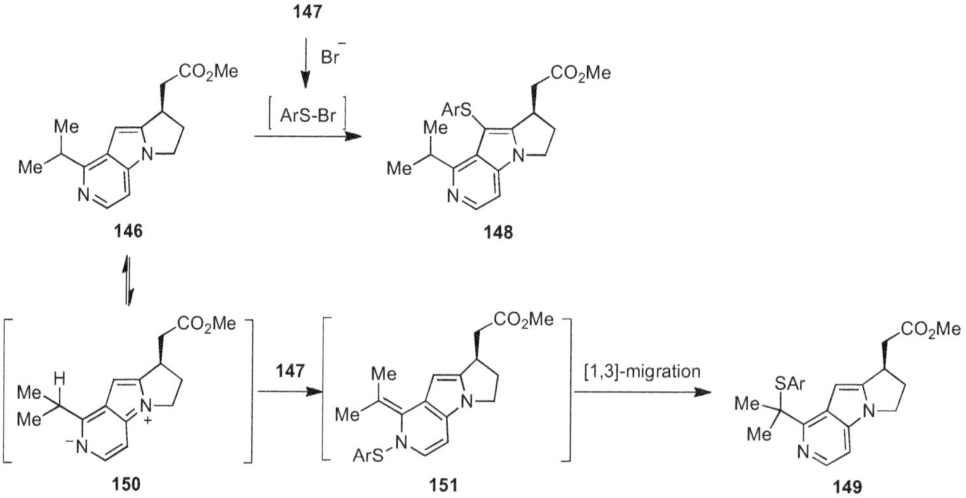

SCHEME 5.27

Procedure

- Charge 2.79 kg (8.6 mol, 1.07 equiv) of **147** to a solution of **146** (2.2 kg, 8.02 mol) in DMAc (12.6 kg) followed by $MgBr_2$ (7.5 g, 0.04 mol).
- Heat the batch at 90 °C for 4 h.
- Cool to room temperature.
- Add 85% H_3PO_4 (610 mL, 8.9 mol) and stir for 10 min.
- Add water (8.6 L) and phthalimide (30 g).
- Add additional water (20 L) over 50 min and Solka-Floc (1 kg).
- Cool the resulting slurry to 10 °C.
- Filter and wash with 17 wt% H_3PO_4 (2×2.8 L).
- Add TEA (1.2 L) to the combined filtrates.
- Add seeds (20 g) followed by 3.5 L of additional TEA over 45 min until pH 8.
- Cool to 10 °C.
- Filter and wash with $MeOH/H_2O$ (2:1, 18 L).
- Yield: 3.66 kg (96%).

5.3.6 CALCIUM CHLORIDE

5.3.6.1 Reaction with $NaBH_4$

$Ca(BH_4)_2$-mediated reduction of optical active ester **152** for the synthesis of fumarate salt **154**, a Ca^{2+} antagonist with potent cardioprotective activity, was developed to avoid racemization of the product **153** due to the basicity of reductant $NaBH_4$ (Scheme 5.28).[46] Treatment of methyl ester **152** with $Ca(BH_4)_2$, generated in situ from a reaction of $NaBH_4$ and $CaCl_2$, in EtOH at 0 °C, afforded **153** in excellent yield without any epimerization.

REMARK

Reduction of ester **155** to alcohol **156** could also be achieved with $CaCl_2/NaBH_4$ (Equation 5.22):[47]

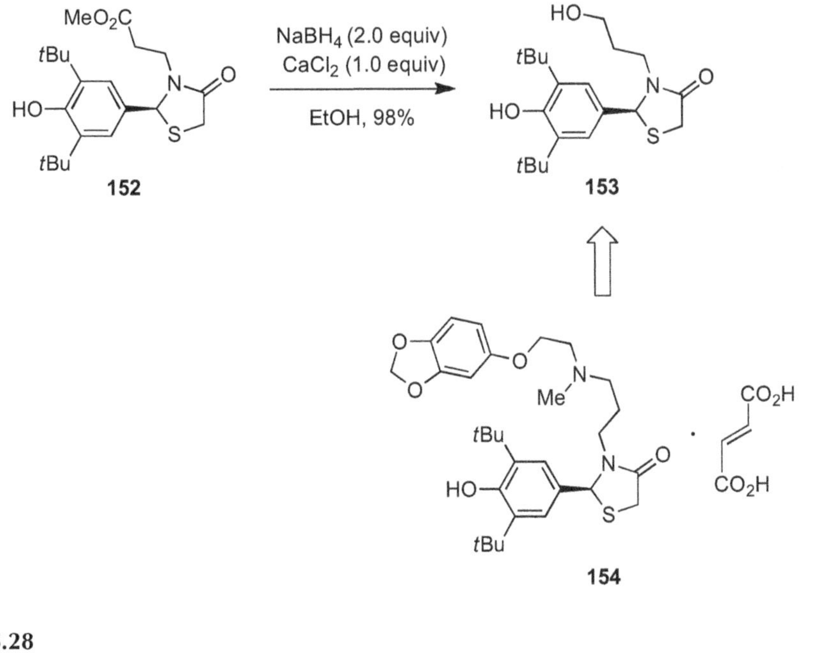

SCHEME 5.28

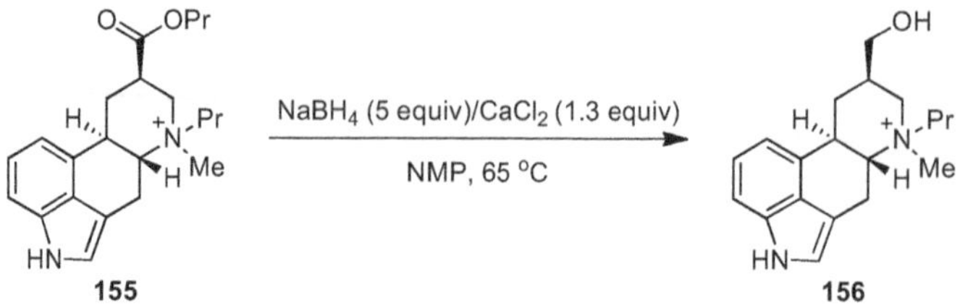

$$(5.22)$$

5.3.6.2 Trapping Fluoride

In order to protect the glass-lined reactor from etching by fluoride, calcium chloride was used to trap fluoride, generated from an S_NAr reaction of 4-Bromo-1-fluoro-2-nitrobenzene with alkylamine (Equation 5.23).[48]

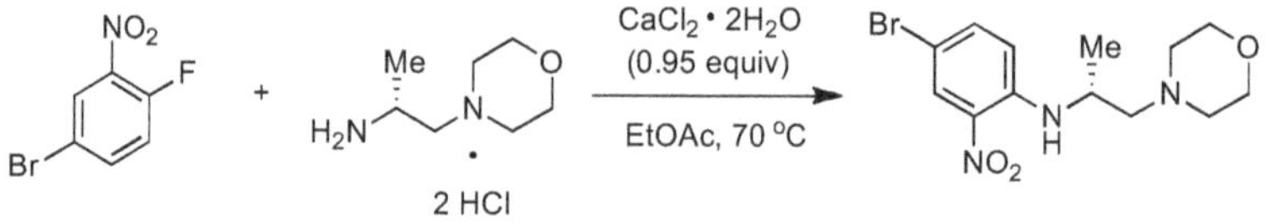

$$(5.23)$$

Note

Besides calcium chloride, calcium propionate $(Ca(EtCO_2)_2)$ was also employed as a trapping agent to trap fluoride generated in S_NAr reaction.[49]

5.3.7 ZINC CHLORIDE

Bromination of aromatic compounds often encounters selectivity issues. For example, a typical *ortho*-lithiation/bromination of bromoarenes **157** with lithium 2,2,6,6-tetramethylpiperidin-1-ide (LiTMP) resulted in a desired dibromoarenes **159** in low to moderate yields (Scheme 5.29).[50] Such unsatisfied yields were rationalized by secondary deprotonation of the product **159** with the reactive lithiated intermediate **158**.

To suppress this undesired lithium exchange, a transmetalation strategy was designed to convert the reactive **158** to a less reactive intermediate. Therefore, a transmetalation of aryllithium **158** with $ZnCl_2$ generated a less reactive arylzinc species **162** (Scheme 5.30).[51] Accordingly, the yields of dibromoarenes **159** were improved dramatically.

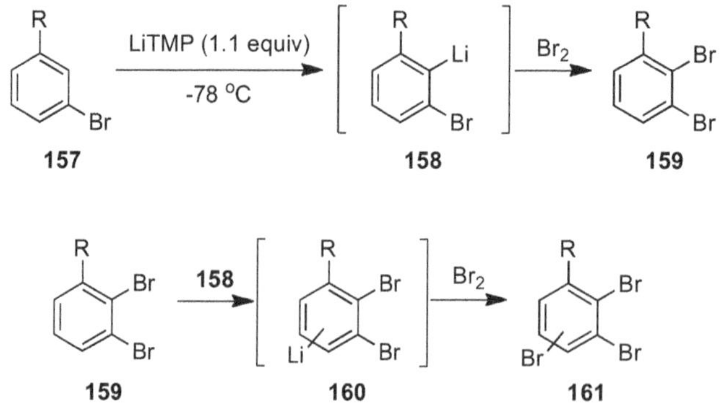

SCHEME 5.29

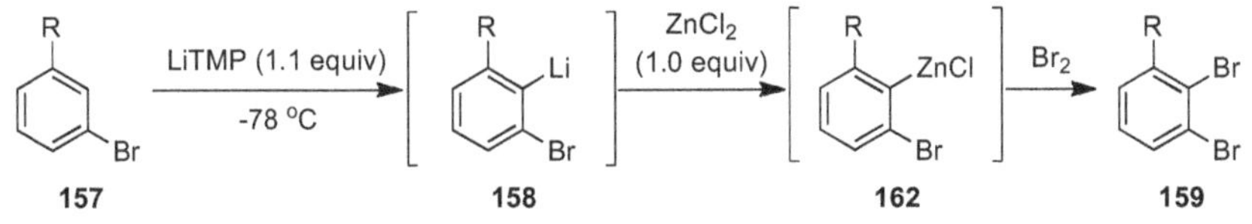

SCHEME 5.30

5.3.8 Zinc Acetate

Aminolysis of 2*H*-imidazole-4-thiol **163** using ammonia gave the corresponding aminoimidazole **164** along with hydrogen sulfide (H$_2$S) (Equation 5.24).[51] In order to limit the release of toxic H$_2$S into the atmosphere, zinc acetate was used to quench the H$_2$S by-product. Subsequently, the solid ZnS precipitate was conveniently removed by filtration.

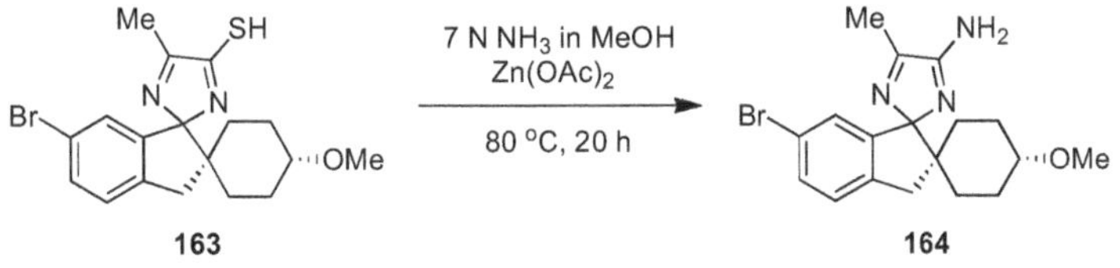

$$(5.24)$$

5.4 ASSORTMENT OF SCAVENGERS

Residual water, acids, or other reactive species in the reaction mixture may have a detrimental effect on the reaction performance, leading to impurities, poor quality of isolated products, or low product yields. To address these issues, various scavengers have been employed to remove or trap the residual water, acids, or reactive species.

5.4.1 Catechol as Methyl Cation Scavenger

(I) Problematic reactions

Deprotection of the methyl group in aryl methyl ether **165** was accomplished by using hydrobromic acid. Although the reaction on a 50-kg scale was relatively clean, the formation of the ring methylated impurity **167** (~2.5%) was problematic on a 200-kg scale (Equation 5.25).[52] This side product probably arises through Friedel–Crafts alkylation with either the bromomethane by-product or in situ formed methanol, which could serve as precursors to methyl cation under the strongly acidic reaction conditions.

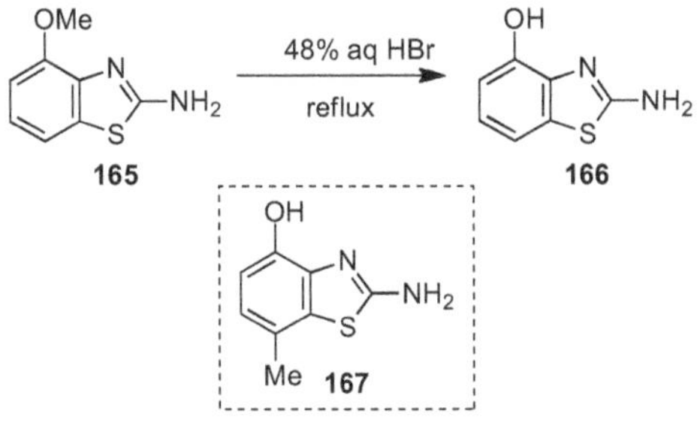

$$(5.25)$$

(II) Solutions

The addition of catechol (0.3 equiv) to the reaction mixture led to a cleaner reaction profile, reducing the impurity **167** to <0.3%. The catechol was completely rejected during the isolation, and the product thus obtained was in excellent yield (93%).

5.4.2 ANISOLE AS QUINONE METHIDE SCAVENGER

The final step in the synthesis of **169**, a JAK2 inhibitor for the treatment of several myeloprolif-erative disorders, was initially involved in TFA-mediated deprotection of *p*-methoxybenzyl (PMB) group (Scheme 5.31).[43] Under such conditions, an alkylated impurity **172** was observed, which can be rationalized by an attack of the electron-rich pyrazole nucleus on *p*-methoxybenzyl cation **170** or quinone methide-type species **171**, the deprotection by-product.

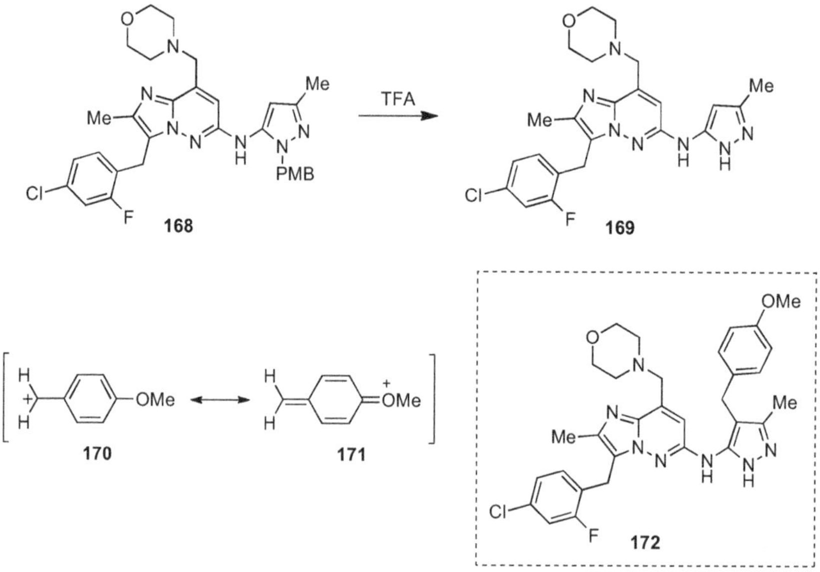

SCHEME 5.31

Using anisole as an additive addressed the concern associated with the reactive species **170**/and or **171**, which eliminated the formation of impurities such as **172**. As illustrated in Reaction (5.26), the reactive species, **170** and/or **171**, could be successfully trapped by anisole, forming **173** and **174**.

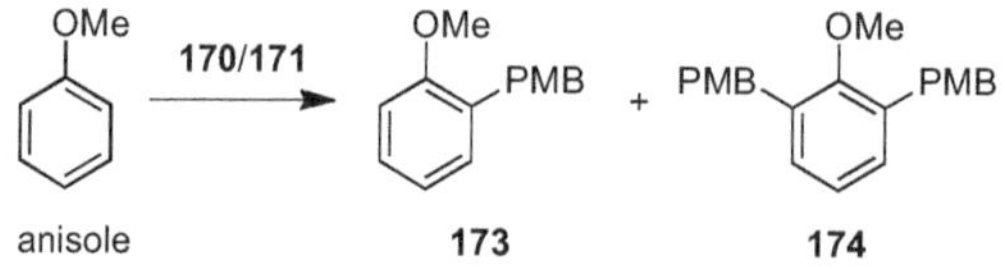

$$(5.26)$$

The **173** and **174** could be readily removed along with anisole by washing with heptane.

REMARKS

(a) Carbon cations such as *p*-methoxybenzyl cation **170** and quinone methides such as **171** are very unstable species and react very rapidly with nucleophiles, or in self-polymerization. Therefore, the removal of PMB protecting group could form a sticky gumball because of polymerizations of **170**/and or **171**. 1,3-Dimethoxybenzene was used to prevent such unde-sired polymerizations.[53]

(b) Benzyl and benzyl-derived protecting groups are widely used to protect alcohols, amines, and other hetero-atom functionalities. Benzyl halides, lacking an electron-withdrawing group on the aromatic ring, are prone to undergo autocatalytic Friedel–Craft polymeriza-tions (Equation 5.27).[54]

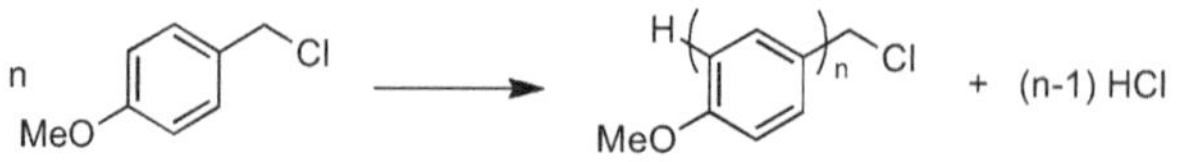

$$(5.27)$$

This polymerization is accelerated under acidic conditions and can also be catalyzed by iron. The runaway polymerization with concomitant release of HCl gas and overpressurization can result in the rupture of storage containers.[55] Therefore, with the aim of preventing conditions in the storage container from becoming acidic, 4-methoxybenzyl chloride from commercial suppliers is stabilized with solid potassium or calcium carbonate, which neutralizes any HCl produced, thereby, preventing the potential uncontrolled polymerization.

5.4.3 Ethyl Acetate as Hydroxide Scavenger

In cases where base-sensitive functional groups are present under anhydrous basic conditions, additives such as carboxylic esters are usually utilized to scavenge hydroxide formed from reactions of the base with residual water.

A commercial solution of NaOEt in EtOH should be used carefully, because residual water in the reagent may cause impurity problems. For instance, a diacid side product **177** was observed when the preparation of monoacid **176** using NaOEt in EtOH (Equation 5.28).[56]

$$(5.28)$$

This hydrolysis issue was addressed by the addition of a sacrificial ester. Compared with bulky ester group in the product **176**, ethyl acetate, acting as the sacrificial ester, would react readily with NaOH to produce EtOH and NaOAc. In this event, the addition of the less expensive ethyl acetate (1.0 equiv) successfully suppressed the ester hydrolysis, affording **176** in yields of 85% with 95–98% potency.

Procedure

- Charge EtOAc (450 mL, 4.58 mol, 1.0 equiv) into a 12-L flask followed by 3318 mL (8.87 mol, 2.0 equiv) of 21 wt% NaOEt solution in EtOH.
- Heat the mixture at reflux for 30 min (to destroy NaOH).
- Cool the content to slightly below reflux.
- Charge 800 g (4.43 mol) of **175**.
- Heat to reflux again and hold for 30 min.
- Charge 1954 mL (13.32 mol, 3.0 equiv) of 2-bromoisobutyrate.
- Reflux the resulting mixture for 1 h.
- Charge 1660 mL (4.43 mol, 1.0 equiv) of 21 wt% NaOEt solution in EtOH dropwise over 1 h.
- Continue to reflux for 30 min.
- Cool the batch to 5–10 °C.
- Transfer the batch to a 22-L flask.
- Add 6.0 L of 0.5 M H_3PO_4 to quench the reaction.
- Distill under reduced pressure to remove EtOH.
- Hold the resulting slurry at 5–10 °C overnight.

- Filter and wash with water (2.0 L).
- Re-slurry the wet cake in water (2.0 L) at ambient temperature for 30 min.
- Filter and wash with heptane (2×2.5 L).
- Dry at 45 °C under vacuum.
- Yield: 1.247 kg (95.5 %).

5.4.4 ETHYL TRIFLUOROACETATE AS HYDROXIDE SCAVENGER

Ethyl trifluoroacetate can also be utilized as a sacrificial ester in condensation reactions to mitigate hydrolyses of ester products.

Triazine **178** was chosen as a development drug candidate with a potential therapy for rheumatoid arthritis and other inflammatory diseases.[57] A substituted pyrrole **180** was identified as a key intermediate for access to **178** and the synthesis of **180** is outlined in Scheme 5.32.[58]

(I) Problematic reactions
- Reaction conditions: Potassium *tert*-amylate in toluene at 60 °C.
- Three side products **181–183** (due to the hydrolysis of the ester groups caused by water by-product).

(II) Solutions
Due to the electron-withdrawing ability of trifluoromethyl group, ethyl trifluoroacetate was used as a sacrificial ester to scavenge potassium hydroxide generated in situ from the reaction of potassium *tert*-amylate with water.

Procedure
- Charge a solution of potassium *tert*-amylate (1.7M in toluene, 61.3 L, 53.9 kg, 104 mol) to a solution of **179** (12.0 kg, 49.3 mol) and ethyl trifluoroacetate (6.8 kg, 47.9 mol) in toluene (13.0 kg) containing ethanol (25.0 kg) over 30 min at <27 °C.

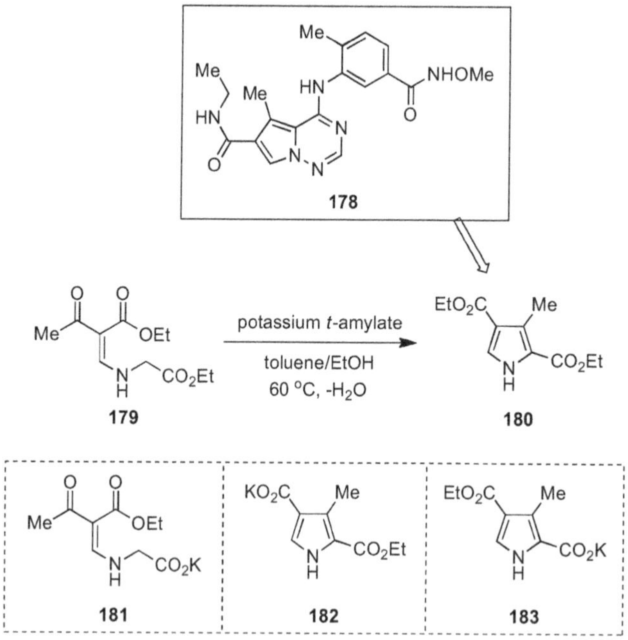

SCHEME 5.32

- Heat the batch to 60–63 °C over 1 h and stir at that temperature for 9 h.
- Cool to 15–20 °C over 1 h.
- Add AcOH (4.2 kg) at <25 °C.
- Add 64 L of water.
- Separate layers.
- Wash the organic layer with aqueous K_2HPO_4 (6%, 80 kg).
- Filter through a Zeta Pad and rinse with toluene (10 kg).
- Distill the combined filtrates at 35–45 °C under reduced pressure to ca. 20 L.
- Cool to 25 °C.
- Add heptane (68 kg) over 20 min.
- Stir the resulting slurry at 20 °C for 2 h.
- Filter and wash with heptane (21 kg).
- Dry at 20–30 °C under vacuum.
- Yield: 7.5 kg (67%).

5.4.5 ETHYL TRIFLUOROACETATE AS BENZYLAMINE SCAVENGER

Compound **184** is a potent and selective antagonist for the human NK-1 tachykinin receptor discovered by Novartis. Amide **186** is an intermediate in the synthesis of **184** (Scheme 5.33).[59] Studies revealed that **184** was contaminated with a desmethyl impurity **187**, which was attributed to the benzylamine impurity in the *N*-benzylmethylamine.

To suppress this impurity formation, ethyl trifluoroacetate (3%) was used as an additive to block its coupling reaction with acid **185**.

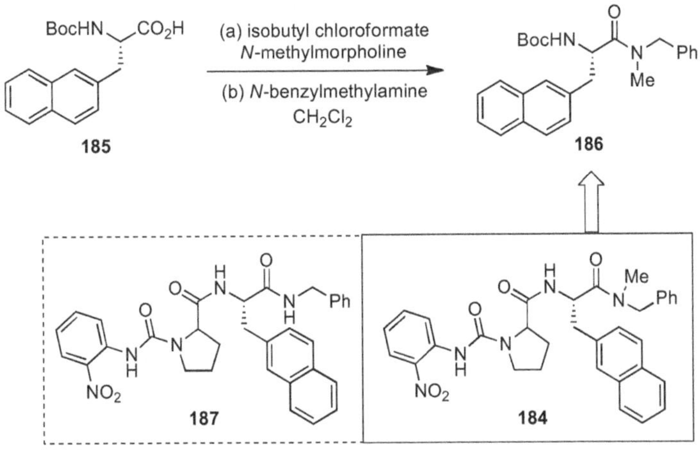

SCHEME 5.33

5.4.6 ETHYL PIVALATE AS HYDROXIDE SCAVENGER

Under basic anhydrous conditions, water by-products may destroy moisture-sensitive functional groups in the starting material/and or product, leading to impurities. For example, an intramolecular condensation of amide **188** produced a desired pyrimidone **189** along with a mixture of 2-pyridone impurities **190** and **191** in a combined level of 35% (Scheme 5.34).[60]

To suppress the formation of **190** and **191**, ethyl pivalate was selected as an additive to trap the hydroxide generated in situ from water. In the event, the addition of ethyl pivalate reduced the level of **190** and **191** to ca. 6%.

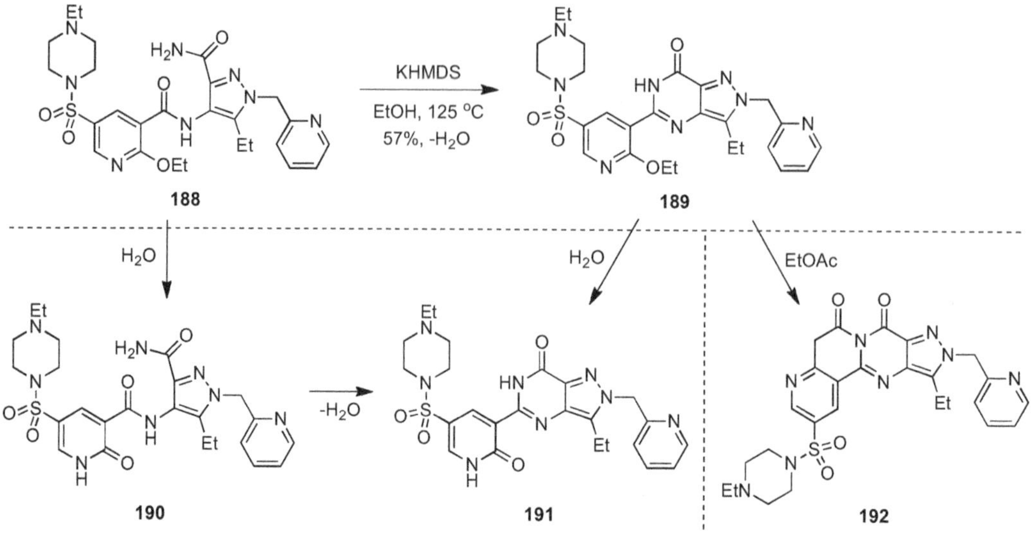

SCHEME 5.34

The use of less bulky ethyl acetate as an additive to scavenge the water by-product, however, generated another impurity **192** (Scheme 5.34) whose formation presumably was the result of reaction of **189** with ethyl acetate.

5.4.7 Trimethyl Orthoformate as Water Scavenger

2-Oxopropanethioamide **194** was prepared via deprotection of ketal **193** under acidic conditions. The optimal conditions were 10 mol% of aqueous 4 M HCl (0.08 mol equiv of HCl and 1.41 mol equiv of water) and 10 vol of isopropanol (relative to **193**) at 30 °C for 2 h (Equation 5.29).[61]

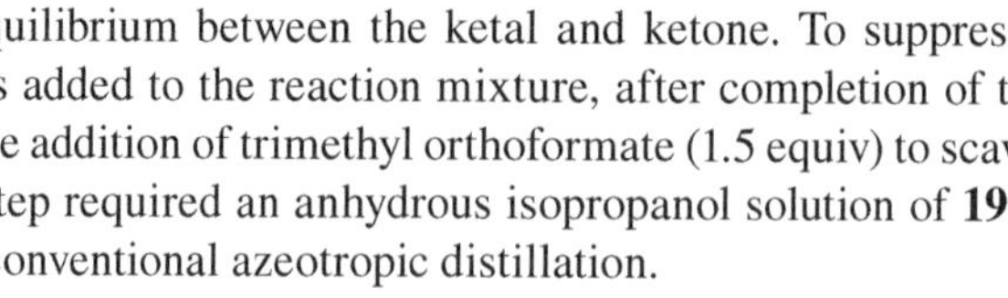

$$(5.29)$$

Under these conditions, the reaction reached >98% conversion (87% solution yield). Reduction in the water amounts resulted in an incomplete reaction and/or formation of isopropanol ketal by-products due to the equilibrium between the ketal and ketone. To suppress the reverse reaction, DIPEA (0.2 equiv) was added to the reaction mixture, after completion of the reaction, to quench the HCl, followed by the addition of trimethyl orthoformate (1.5 equiv) to scavenge the excess water. Since the subsequent step required an anhydrous isopropanol solution of **194**, the use of trimethyl orthoformate avoided conventional azeotropic distillation.

Note

2-Oxopropanethioamide **194** is inherently unstable and prone to polymerization. Safety assessment revealed that **194** has a low on-set temperature of 30 °C, therefore, it should be stored under cold and low-concentration conditions

5.4.8 Thionyl Chloride as Water Scavenger

Often, water, even in small quantities, is detrimental to a chemical process. Hence, careful control of water contents in starting materials, reagents, and solvents is critical to the success of a chemical reaction. Albeit not common, residual water can be generated from the reaction of by-products, which makes the identification of the residual water source challenging.

(I) Problems

A reaction of chloroiodomethane **195** with chlorosulfonic acid, as illustrated in Reaction (5.30),[62] gave the desired chloromethyl chlorosulfate **196** in a moderate yield (~60%) along with 9% of bischloromethylsulfate **197**.

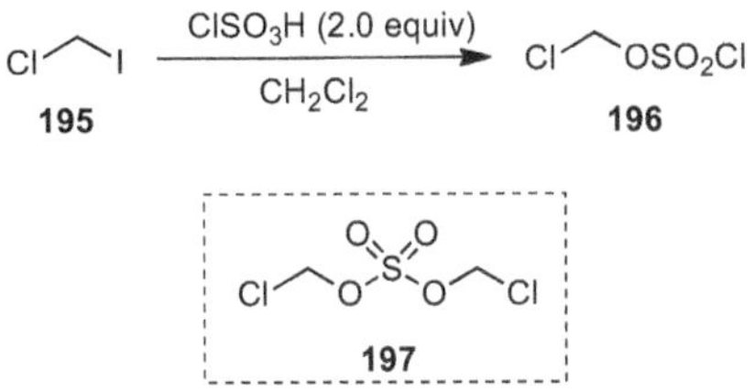

$$Cl\diagup\diagdown I \xrightarrow[\text{CH}_2\text{Cl}_2]{\text{ClSO}_3\text{H (2.0 equiv)}} Cl\diagup\diagdown OSO_2Cl \tag{5.30}$$

195   **196**

$$\begin{array}{c} \underset{O}{\overset{O}{\underset{\parallel}{\overset{\parallel}{S}}}} \\ Cl\diagup\diagdown O\diagdown\diagup O\diagdown\diagup Cl \\ \textbf{197} \end{array}$$

(II) Chemistry diagnosis

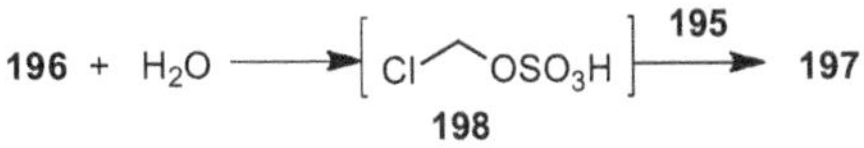

$$\textbf{196} + \text{H}_2\text{O} \longrightarrow \left[Cl\diagup\diagdown OSO_3H \right] \xrightarrow{\textbf{195}} \textbf{197}$$

198

SCHEME 5.35

The side product **197** is presumably formed through a reaction of chloroiodomethane **195** with sulfonic acid **198** that was formed in situ from hydrolysis of **196** with water (Scheme 5.35). Oxidation of by-product HI by chlorosulfonic acid would produce iodine, SO_2, and HCl along with water (Equation 5.31).

$$2\,\text{HI} + \text{ClSO}_3\text{H} \longrightarrow \text{I}_2 + \text{SO}_2 + \text{HCl} + \text{H}_2\text{O} \tag{5.31}$$

(III) Solutions

Thionyl chloride was identified as an optimal water scavenger.

Procedure

- Charge, into a 5-L reactor, CH_2Cl_2 (1.44 L), chlorosulfonic acid (170.9 g, 1.47 mol, 1.67 equiv), and $SOCl_2$ (84.6 g, 0.711 mol, 0.8 equiv) at 20 °C.
- Charge 160 g (0.889 mol, 1.0 equiv) of **195**.
- Stir at 20–25 °C for 1 h.
- Heat to 35 °C over 30 min and stir until reaction completion.
- Workup.

DISCUSSIONS

Thionyl chloride may be able to convert any hydrolyzed sulfuric acid or chloromethylsulfonic acid back to the active sulfonyl chloride or **196**. Furthermore, no additional treatment is required to remove the excess thionyl chloride since it readily decomposes during the aqueous bleach quench.

5.4.9 1-Hexene

5.4.9.1 As HCl Scavenger

Cefaclor **203** is an important wide-spectrum antibiotic used in the treatment of a broad range of bacterial infections. A key step in the production[63] of **203** is the ring expansion of penicillin V sulfoxide p-nitrobenzyl ester **199** to 3-exomethylene cephalosporin V sulfoxide p-nitrobenzyl ester **200**, involving sulfinyl chloride intermediate **202** (Scheme 5.36).[64] The formation of sulfinyl chloride **202** occurred via a reaction of penicillin sulfoxide **199** with N-chlorophthalimide (NCP) in refluxing toluene.

(I) Problematic reactions

- A capricious side reaction often intervened in which the reaction mixture rapidly turned dark with the concomitant evolution of HCl gas. In the event, compounds **204** and **205** were identified as side products which compose the majority of reaction products.
- In addition to side products **204** and **205** generated during the preparation of sulfinyl chloride **202**, the formation of chlorine-substituted side product **201** was also observed during the ring-closure of **202**, which is attributed chlorination of **200** with Cl_2 (generated in situ from a reaction of HCl with NCP).

(II) Solutions

The by-product HCl was identified as the major cause of these side reactions. Thus poly(vinylpyridine-divinylbenzene) was used to minimize the concentration of HCl during the preparation of **202** and 1-hexene was employed to scavenge HCl in the ring-closure step (Scheme 5.37).

REMARK

The formation of **204** and **205** was hypothesized[65] through rearrangement as outlined in Scheme 5.38. Penicillin sulfoxide **199** undergoes NCP-mediated O-chlorination to **206** whose subsequent cascade involves loss of proton and CO followed by S–C bond cleavage to give intermediate **207**. A subsequent ring enclosure gives episulfonium ion **208**. Loss of HCl from **208** gives compound **209**. The HCl salt of **209** reacts with Cl_2 (generated in situ by the reaction of HCl and NCP) to give compound **204** that, on heating, cyclizes with loss of HCl to give **205**.

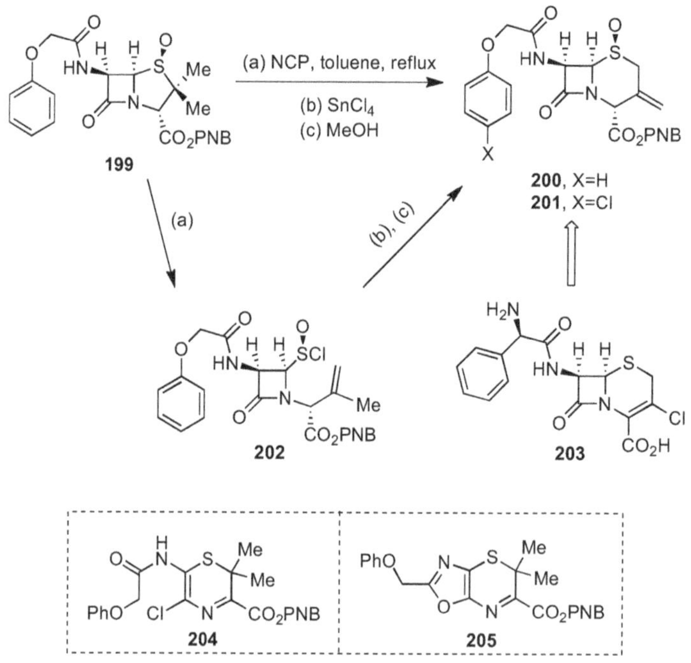

SCHEME 5.36

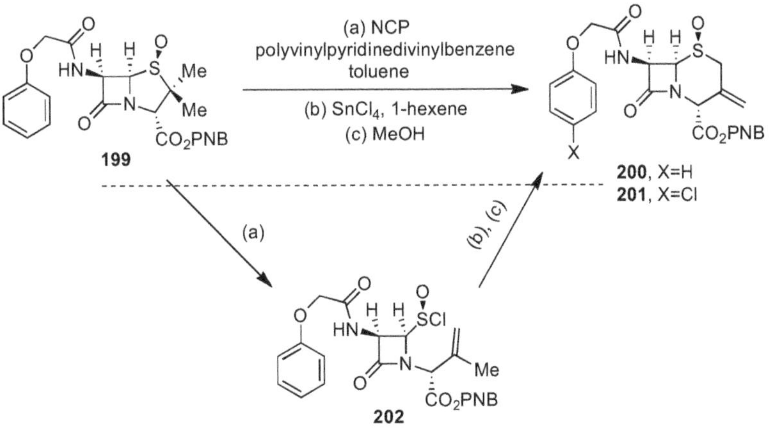

SCHEME 5.37

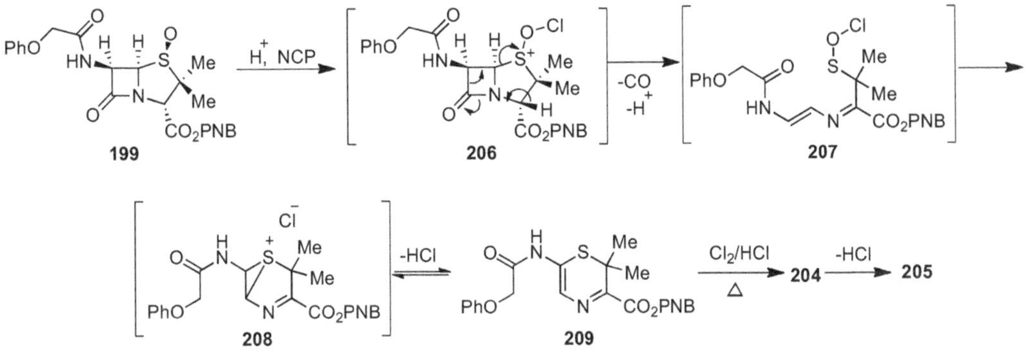

SCHEME 5.38

5.4.9.2 As Diimide (NH=NH) Scavenger

The preparation of allyl amine **211** via hydrazinolysis of *N*-substituted phthalimide **210** was plagued by the formation of an alkyl amine impurity **212** in levels of 1.5–2.0% (Equation 5.32).[65] The alkyl amine **212** was generated from presumably hydrogenation of the double bond of **210** by diimide, formed in situ from oxidation by air. To suppress such an undesired reaction, 1-hexene was used as a diimide scavenger.

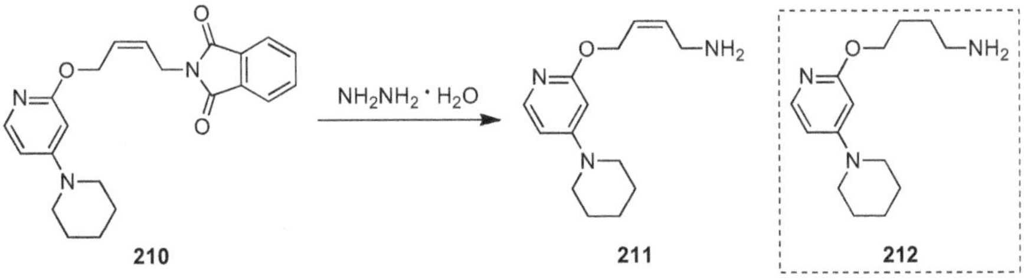

(5.32)

Procedure

- Add MeOH (120.0 kg) into a 250 L reactor followed by adding **210** (17.0 kg, 43.4 mol) and 1-hexene (5.5 kg, 65.1 mol).

- Add 5.5 kg (86.8 mol) of 85% hydrazine hydrate.
- Heat under reflux for 4 h.
- Cool to room temperature.
- Filter to remove insoluble solids.
- Concentrate the filtrate to 1/5 of the original volume.
- Filter, and dry at 45 °C under reduced pressure.
- Yield: 10.5 kg (92.5%).

5.4.10 Epoxyhexene as HBr Scavenger

Radical cyclization of amide **213a** under reflux in toluene afforded **214** in 92% yield, while the reaction of **213b** under the same conditions led to only 38% yield of product **215** (Scheme 5.39).[66] It appeared that the HBr by-product from the reaction of **213b** had a detrimental effect on the radical reaction. Thus, the addition of epoxyhexene (10 equiv) to the reaction mixture of **213b** as an acid trap significantly improved the yield of product **215** to 90%.

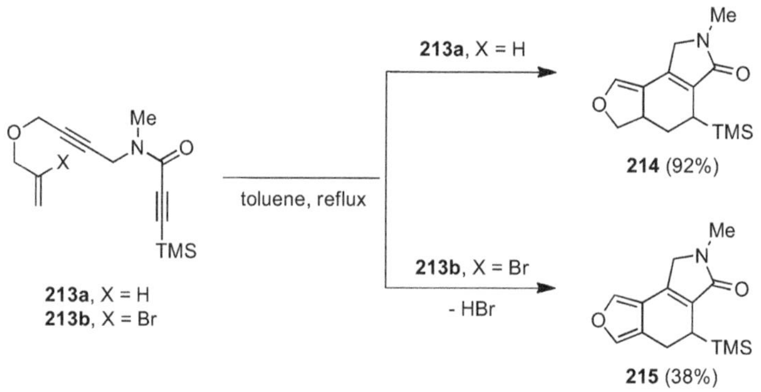

SCHEME 5.39

DISCUSSION

1,2-Epoxybutane had been used[67] to scavenge HBr generated during the Wittig reaction, though it is highly flammable (bp 63 °C).

5.4.11 Acetic Anhydride as Aniline Scavenger

A Rh(III)-catalyzed amination/cyclization/aromatization cascade process provided **218** in 57% yield (Equation 5.33).[68] This modest yield might be due to the interaction of the catalyst with aniline by-product released upon cyclization and aromatization.

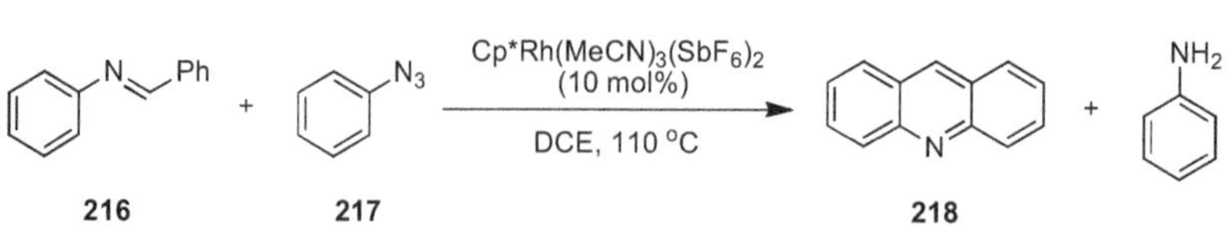

To address this issue, acetic anhydride (or $CF_3CO_2CH_2CF_3$) was used as the aniline scavenger, which improved the yield to 77% (or 71%).

DISCUSSIONS

The use of imine **216** was necessary to direct the Rh(III)-catalyzed C–H amination. A revised cascade involved in situ generation of the imine **216** via reaction of benzaldehyde with benzylamine (40 mol%) eliminating the need for acetic anhydride additive (Equation 5.34):

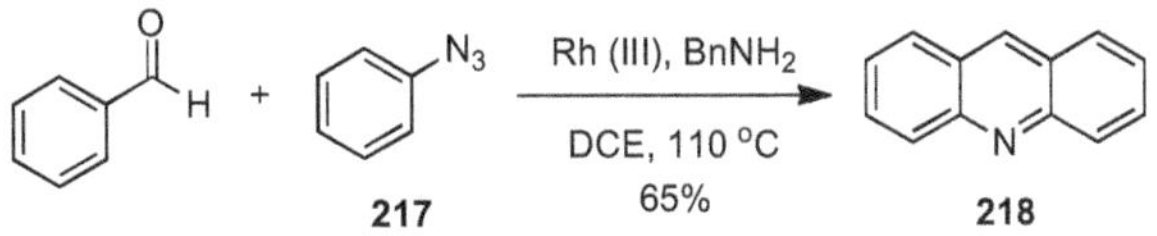

$$(5.34)$$

5.4.12 AMBERLITE CG50 AS AMMONIA SCAVENGER

An asymmetric Fischer indolization of 4-substituted cyclohexanone-derived phenylhydrazone **219** gave product **220** in only 9% of yield due to the poison of the catalyst by ammonia by-product (Equation 5.35).[69] Amberlite CG50 is a weakly acidic cation exchange resin with a high concentration of carboxylic groups. In the event, Amberlite CG50 was used to scavenge the ammonia by-product.

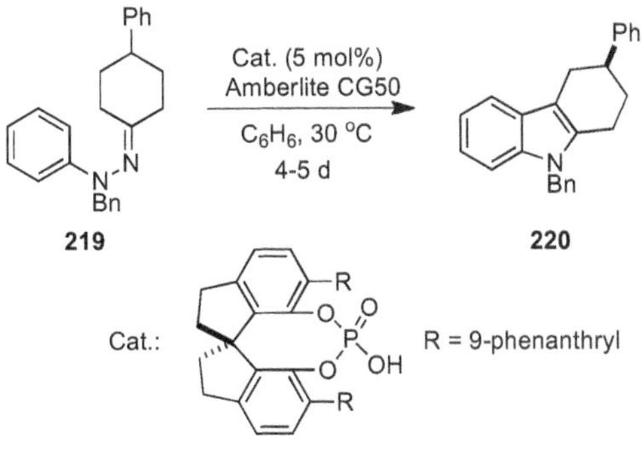

$$(5.35)$$

5.4.13 3-PENTANONE AS HCN SCAVENGER

5-Iodo-7-azaindazole **223** could be conveniently prepared via tandem reaction sequence involving an inverse-electron-demand intramolecular Diels–Alder reaction of 2-substituted pyrimidine **221** followed by retro-Diels–Alder reaction of **222** releasing hydrogen cyanide (HCN) by-product (Scheme 5.40).[70] HCN is a highly volatile, flammable, and extremely poisonous liquid (boiling point

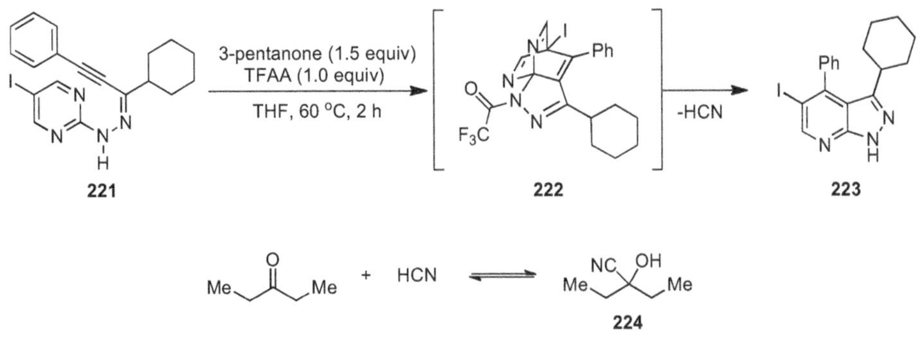

SCHEME 5.40

26 °C, freezing point –14 °C). To address the hazardous HCN, 5-pentanone, an efficient HCN scavenger,[71] was used to trap the toxic HCN by-product likely via the formation of cyanohydrin **224**.

5.5   OTHER ADDITIVES

Besides acids, bases, salts, and various scavengers, other additives also play indispensable roles in the chemical process development. These additives include, but are not limited to, imidazole, triethylamine hydrochloride, methyl triocylammonium chloride, chlorotrimethylsilane (TMSCl), chlorotriethylsilane (TESCl), bis(trimethylsilyl)acetamide (BSA), water, hydroquinone, trimethyl borate (B(OMe)$_3$), isobutanoic anhydride, 1,1-dimethyl-2-phenylethyl acetate, alcohols 1,4-dioxane, benzotriazole, 1-hydroxybenzotriazole, 1,4-dibromobutane, diethanolamine, trimethyl phosphite. diethyl phosphite, and 4-trifluoromethyl benzaldehyde.

5.5.1   IMIDAZOLE

It has been observed that amidation of aromatic amines with *N,N'*-carbonyldiimidazole (CDI) proceeds at relatively slow reaction rates, which limits its application scope. For instance, the CDI-mediated amidation of acid **225** with cyclopropylamine resulted in an incomplete reaction with 2–4% of **225** remaining (by HPLC), and the addition of excess CDI or heating of the reaction mixture at elevated temperatures (50 °C) did not increase the conversion. The introduction of imidazole (1.0 equiv) into the reaction mixture facilitated the reaction (Scheme 5.41).[72]

This rate enhancement was due to the increased solubility of **225** by imidazole as the coupling reaction was dissolution-controlled. Furthermore, the addition of imidazole has two other beneficial effects:

- Suppressing the formation of the anhydride impurity **227.**
- Reducing the level of residual Pd, carried over from the previous Suzuki coupling reaction (Pd was reduced from 100–150 ppm to <10 ppm).

DISCUSSIONS

(a) CDI is widely used for amide preparations by coupling carboxylic acids with amines. Compared with acid chlorides, imidazolide intermediates are easy to handle and can be isolated if necessary. The use of CDI as an acid activation reagent offers several benefits. It is relatively less expensive and readily available in kilogram quantities. CDI-mediated amidation generates two by-products, carbon dioxide and imidazole, that can be easily removed from the reaction mixture by acidic wash. In addition, imidazole is relatively

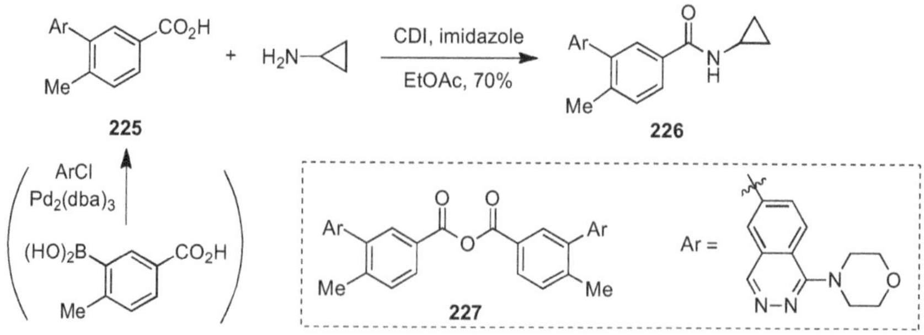

SCHEME 5.41

benign to the environment, which is unlikely to cause problems on scale-up. However, excess of CDI reagent may result in the formation of symmetrical urea.[73]

(b) It was reported that the presence of imidazole·HCl[74] or CO_2[75] salts enhanced the amidation rate.

5.5.2 Triethylamine Hydrochloride

The mechanism-guided process development strategy is a great tool in solving tough reaction problems. For instance, the conversion of **228** to **229** was effected by using sulfuryl chloride in dichloromethane at −20 °C (Scheme 5.42). This transformation, however, had the following problems.[76]

(I) Problems

- The isolation provided a mixture of the desired chloromethyl ester **229** and chlorinated side product **230** (2:1 ratio) in a combined yield of 67%.
- Removal of **230** had to use distillation.

(II) Chemistry diagnosis

As illustrated in Scheme 5.42, the formation of **230** involves Chloro–Pummerer rearrangement in which the elimination of HCl is the rate-determining step, followed by a rapid addition of chloride to quench the reactive sulfur carbenium ion **232**. In contrast, the desired reaction is expected to proceed via an S_N2 displacement of methylsulfenyl chloride (ClSMe) with chloride. Thus, the chloride would be involved in the rate-determining step for the desired pathway (A) but not for the undesired pathway (B). Ultimately, a practical approach is to accelerate the rate of the desired step (A) by introducing an external chloride source to the reaction system.

(III) Solutions

The addition of triethylamine hydrochloride as the chloride source completely suppressed the competing chlorination pathway (B), affording the desired product **229** in good yield and purity.

Procedure

- Charge triethylamine hydrochloride (27.5 g, 0.2 mol, 1.25 equiv) to a solution of **228** (37.0 g, 0.16 mol) in 148 mL of CH_2Cl_2.
- Cool the mixture to −30 °C.
- Charge 16.2 mL (0.2 mol, 1.25 equiv) of sulfuryl chloride.
- Stir at −20 °C for 1 h and at 0 °C for 2 h.
- Add 33.8 g (0.192 mol, 1.2 equiv) of L-ascorbic acid, followed by water (185 mL).

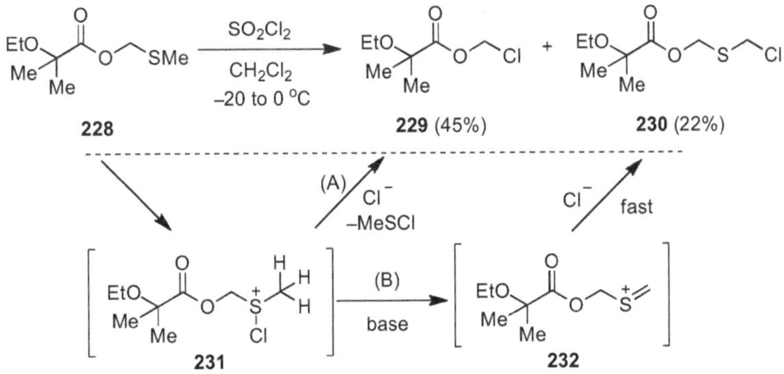

SCHEME 5.42

- Wash the organic layer with 10% aq NaHCO$_3$ (400 mL) and water (185 mL).
- Concentrate the organic layer to give 30 g of yellow oil.
- Distill at 66–68 °C/15 mbar.
- Yield: 20.5 g (71%) as a colorless oil.

REMARK

Ascorbic acid was added during the aqueous workup in order to scavenge the methylsulfenyl chloride (ClSMe) generated during the reaction.

5.5.3 METHYL TRIOCTYLAMMONIUM CHLORIDE

Frequently, S$_N$2 reaction is complicated by E2 side reactions. Depending on reaction conditions, these E2 side reactions can become major reaction pathways. For instance, a nucleophilic substitution reaction of alkyl bromide **233** with sodium azide (3 equiv) at an elevated temperature (80 °C) in DMF gave only 35% of the desired product **234** along with a mixture of elimination side products **235** and **236** in 63% combined yield (Equation 5.36).[77]

(5.36)

To suppress the side reactions, methyl trioctylammonium chloride (0.1 equiv) was employed. As a result, the yield of the desired product **234** was improved to 95% without any elimination of side products.

5.5.4 CHLOROTRIMETHYLSILANE

5.5.4.1 Julia Olefination

Julia olefination of aldehyde **238** with sulfone **237** was selected for the preparation of olefin **239** (Scheme 5.43).[78]

(I) Problems

The major hurdle in this reaction was the instability of both the aldehyde **238** and the product **239** under olefination conditions, leading to a low product yield.

(II) Solutions

The use of chlorotrimethylsilane (TMSCl) (or BF$_3$ etherate) as an additive showed great benefit for the olefination with significant improvement in the product yield. It is postulated that the alkoxide adduct **242** was trapped by TMSCl to form **240**. Thus, avoiding the exposure of the starting

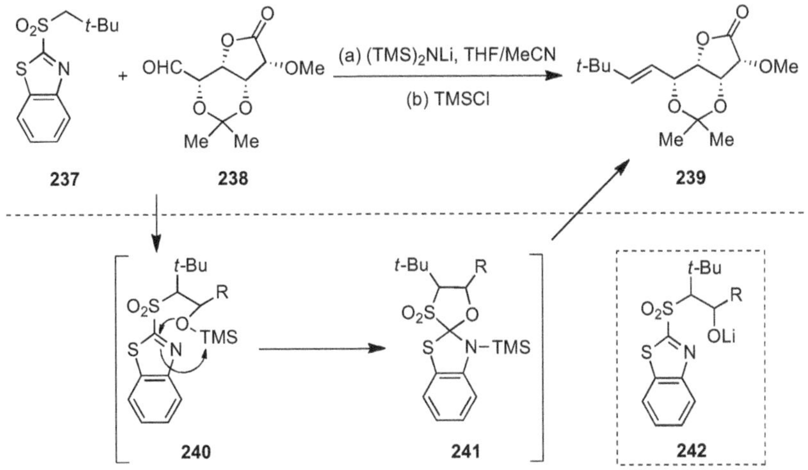

SCHEME 5.43

aldehyde and the reaction product to the carbon anion of sulfone or the alkoxide adduct **242** and the decomposition of these compounds through the elimination pathway was minimized.

Procedure

- Charge 677.6 mL (0.68 mol) of 1 M (TMS)$_2$NLi in THF to a solution of **237** (168.5 g, 0.63 mol) in 2 L of anhydrous THF.
- Cool the resulting solution to –78 °C and stir for 1.5 h.
- Charge 86 mL (0.68 mol) of TMSCl.
- Charge the above solution to a cold (0–15 °C) mixture of **238** (120 g, 0.52 mol) and NaHCO$_3$ (1.96 g, 0.029 mol) in anhydrous MeCN (1.96 L).
- Warm the resulting mixture to 0 °C and stir for 1 h.
- Warm to 22 °C and stir for 1 h.
- Warm to 30 °C and stir for 1 h.
- Cool to 10 °C.
- Add H$_2$O (1.5 L) and concentrate under reduced pressure.
- Add MTBE (4.0 L) and separate layers.
- Wash the organic layer with H$_2$O (0.8 L).
- Concentrate to 640 mL and cool to 22 °C.
- Cool to 0 °C and filter.
- Wash with MTBE (200 mL).
- Yield: 65.9 g (44.5%).

5.5.4.2 Michael Addition

TMSCl was also used as an additive to mask the anion intermediate **245** obtained from Cu-mediated Michael addition (Scheme 5.44).[79]

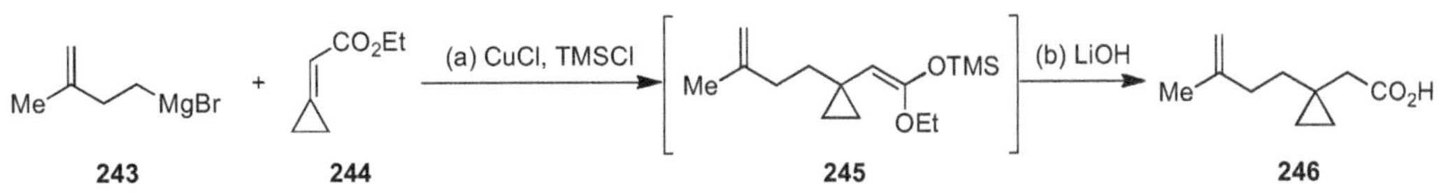

SCHEME 5.44

Thus, treatment of the freshly prepared Grignard reagent **243** with ester **244** in the presence of CuCl and TMSCl, followed by the addition of LiOH, afforded the acid product **246** in 60% yield.

REMARK
TMSCl was also used for the activation of zinc dust in the preparation of organozinc reagent.[41]

5.5.5 CHLOROTRIETHYLSILANE

Chlorotriethylsilane (TESCl) can react with manganese oxide to form hexaethyl disiloxane and $MnCl_2$ that is soluble in polar organic solvents such as DMF (Equation 5.37).

$$2 \text{ (TESCl)} \quad + \quad \text{MnO} \quad \longrightarrow \quad \text{(TES)}_2\text{O} \quad + \quad \text{MnCl}_2 \qquad (5.37)$$

Accordingly, TESCl was employed to activate manganese dust reductant in the nickel-catalyzed cross-coupling reaction of substituted 2-chloropyridine **247** with 3-chloropropionic acid ethyl ester **248** (Equation 5.38).[80]

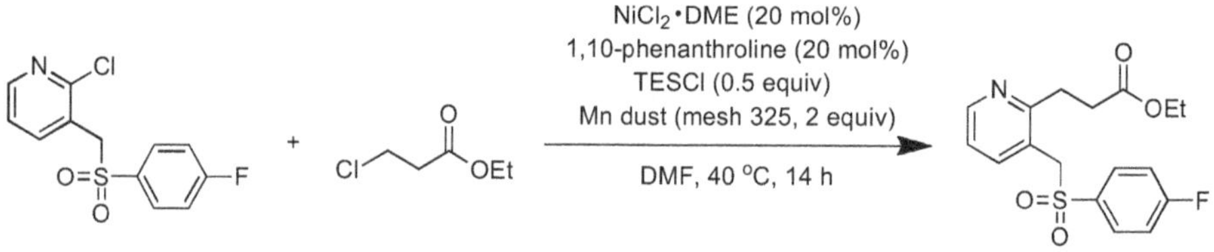

$$(5.38)$$

247   **248**   **249**

5.5.6 BIS(TRIMETHYLSILYL)ACETAMIDE

Palladium-catalyzed hydrogenolysis (or hydrogenation) is frequently used to remove protective groups, such as benzyl or *N*-benzyloxy carbonyl (Cbz) groups (or to reduce nitro and azide into amino groups). The spent palladium catalysts are usually removed by filtration leaving the products in the filtrates. Apparently, these operations are not suitable for products that are less soluble in the reaction media, for example, hydroxyurea **251** is poorly soluble in isopropyl acetate (2.1 mg/mL). Thus, an in situ protection approach was developed in the hydrogenolysis of benzyl-protected hydroxyurea **250** (Scheme 5.45).[81]

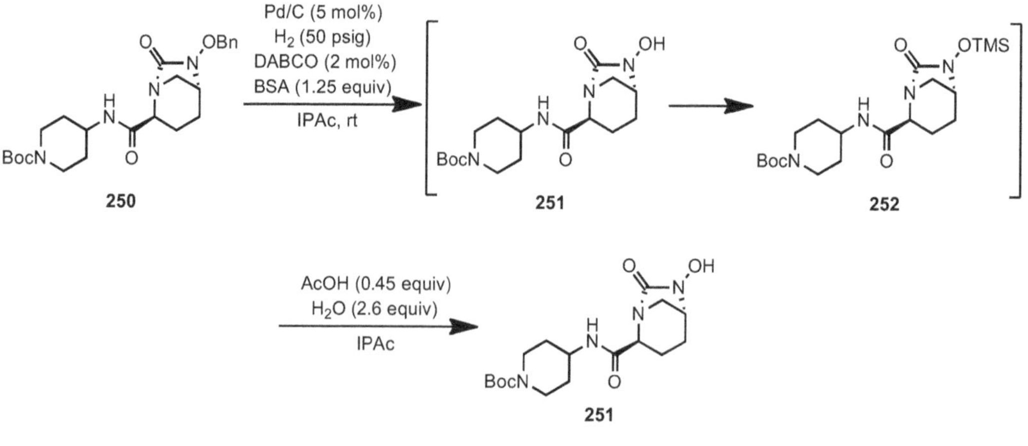

250   **251**   **252**

251

SCHEME 5.45

Under hydrogenolysis conditions, the initially formed hydroxyurea **251** was converted in situ to the corresponding, more soluble TMS-hydroxyurea **252** with bis(trimethylsilyl)acetamide (BSA), avoiding the precipitation of the less soluble **251** in the heterogeneous hydrogenolysis mixture. Upon completion of the reaction and removal of the catalyst by filtration, the TMS-protective group in **252** was removed by treatment with AcOH/water. As such, hydroxyurea **251** was isolated through a reactive crystallization.

Procedure

- Add 34.4 g of 5 wt% Pd/C into a 10-gallon Hastelloy vessel followed by adding 687 g (1.50 mol) of **250**, 385 g (1.87 mol, 1.25 equiv) of BSA, 3.4 g (30 mmol, 2 mol%) of DABCO.
- Add 8.6 L of IPAc.
- Stir at 50 psig of H_2 at room temperature for 4 h (HPLC showed <1.0% of **250**).
- Filter through Celite under N_2 and rinse with IPAc (2.7 L).
- Concentrate the combined filtrate to 9.6 L under reduced pressure.
- Add 47.3 wt% aq AcOH (53 mL) to a portion (2.7 L) of the batch at 15 °C.
- Add 6.8 g of hydroxyurea **251** as seeds and age for 1 h.
- Add 2.3 L of the batch to the resulting suspension over 1 h followed by adding 47.3 wt% aqueous AcOH (40 mL).
- Add 2.3 L of the batch to the resulting suspension over 1 h followed by adding 47.3 wt% aqueous AcOH (40 mL).
- Add the remaining batch to the resulting suspension over 1 h and age at 15 °C for 2 h.
- Add MTBE (2.1 L) over 1 h and stir for 1 h.
- Filter and wash with a mixture of IPAc and MTBE (4:1, 2 L) twice.
- Dry 40 °C in a vacuum oven.
- Yield of **251**: 486.4 g (87% yield, 98.7 wt% purity) as a white crystalline solid.

5.5.7   WATER

In contrast with the negative effects that were discussed in this chapter, water as an additive plays an important role in optimization of reactions.

5.5.7.1   Wadsworth–Emmons Reaction

The preparation of **255** was realized using tetramethyl guanidine-mediated Wadsworth–Emmons protocol (Equation 5.39).[10]

(5.39)

(I) Problems

In addition to the desired product **255**, diester **256** was formed as the side product.

(II) Solutions

Methyl phosphonate species is hypothesized to be the only reasonable cause for the methyl ester formation. Accordingly, the addition of water to the reaction mixture allowed in situ quenching of the alkylating phosphonate species, suppressing the formation of the diester.

REMARK

An additional benefit of the water additive was the considerable improvement in the solubility of **254** and **255**, which resulted in an increase in volume capacity as well as the reaction rate.

5.5.7.2   O-Alkyation

Water content in the reaction medium was a crucial factor for suppressing side reactions during the O-alkylation of phenol **257**. Depending on the water content, side product **260** was observed in various levels from 0.2% to 12% (Equation 5.40).[82] For example, the addition of water (2%, v/v) reduced the level of **260** to 0.46% from 12% (in the absence of water). In contrast, the presence of water could increase the level of diol **261**. Thus, MeCN containing 1.0% (v/v) of water was selected as a suitable solvent system for the O-alkylation, giving the alcohol **259** with >95% HPLC purity suitable for the next step without further purification.

(5.40)

5.5.7.3   Methylation Reaction

Alkylation is usually accomplished via deprotonation with base in a polar aprotic solvent followed by S_N2 reaction with an electrophile. The reaction is typically carried out under anhydrous conditions.

(I) Reaction problems

A side product **264** was generated from a base-mediated alkylation of sulfone **262** with MeI under anhydrous conditions (Equation 5.41).[83]

(5.41)

(II) Solutions

The formation of **264** is due to the poor chemoselectivity when using a strong base (NaO*t*Bu) for deprotonation. Development efforts to inhibit the formation of **264** led to optimized reaction conditions in which the reaction was carried out in a wet solvent. The addition of water (1.3 equiv) suppressed the formation of **264**. It was rationalized that the in situ generated sodium hydroxide was the base to selectively deprotonate benzylic protons.

Procedure

- Dissolve 1.648 kg (5.49 mol) of **262** in 24 L of DMF (KF=5280 μg/mL, 7.04 mol (1.3 equiv) of water relative to NaO*t*Bu).
- Degas by vacuum and nitrogen backfilling.
- Cool the resulting solution to –10 to –5 °C.
- Charge 528 g (5.49 mol) of solid NaO*t*Bu at –10 to –5 °C.
- Charge 779 g (5.49 mol) of MeI at –10 to –5 °C.
- Charge 528 g (5.49 mol) of solid NaO*t*Bu at –10 °C.
- Charge 781 g (5.5 mol) of MeI at –10 to –3 °C.
- Charge 79 g (0.83 mol) of solid NaO*t*Bu followed by 120 g (0.85 mol) of MeI.
- Stir the batch at –10 to –5 °C for 100 min.
- Add aqueous AcOH (2 vol%, 25 L) slowly.
- Stir the resulting slurry at –5 to 0 °C for 6 h.
- Filter and wash with a cold (10–14 °C) mixture of water and DMF (3:1, v/v, 2×5 L)
- Dry at 35 °C under vacuum with nitrogen sweep.
- Yield: 1.647 kg (88%).

5.5.7.4 Enolization Reaction

The enolization of trifluoromethyl ketone **265** was affected with NaH in THF on the laboratory scale (Equation 5.42).[84]

$$(5.42)$$

(I) Problematic reaction

On scaling to a 4-kg reaction, however, the enolization was found to be much slower. After 18 h at room temperature, only 50% enolization occurred.

(II) Solutions

It was found that the water level in THF plays a critical role in the enolization. The THF used in the laboratory contained 100–500 ppm water, while the THF from 200 L drum was extremely dry (ca. 10 ppm water). Presumably, NaH reacts with the residual water in THF to form NaOH, which acts as a base to react with **265**. Therefore, the water level in THF was controlled between 300–500 ppm, giving reproducible results.

5.5.7.5 Pyrazole Synthesis

Pyrazole synthesis was effected via an oxidation of propargylamine **267** followed by treatment with hydrazine (Equation 5.43).[85]

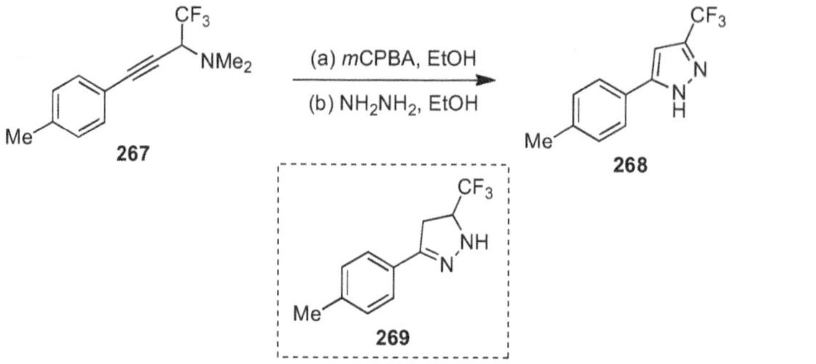

(5.43)

In addition to the desired pyrazole product **268**, a partially reduced side product **269** was also observed.

Solutions

As described in Scheme 5.46, the formation of **268** and **269** occurs via the same hypothesized intermediate **271**. The pyrazole product **268** is formed through the pathway (i) involving deprotonation of the relatively acidic proton at 3-position of 2,2-dimethyl-2-isoxazolium **271** to lead to **272** whose subsequent cycloaddition with hydrazine delivers the desired **268**. The pathway (ii) suggests that the side product **269** is formed via enone intermediate **273** whose formation is presumably caused by deprotonation of the less acidic proton at N-methyl group of **271**.

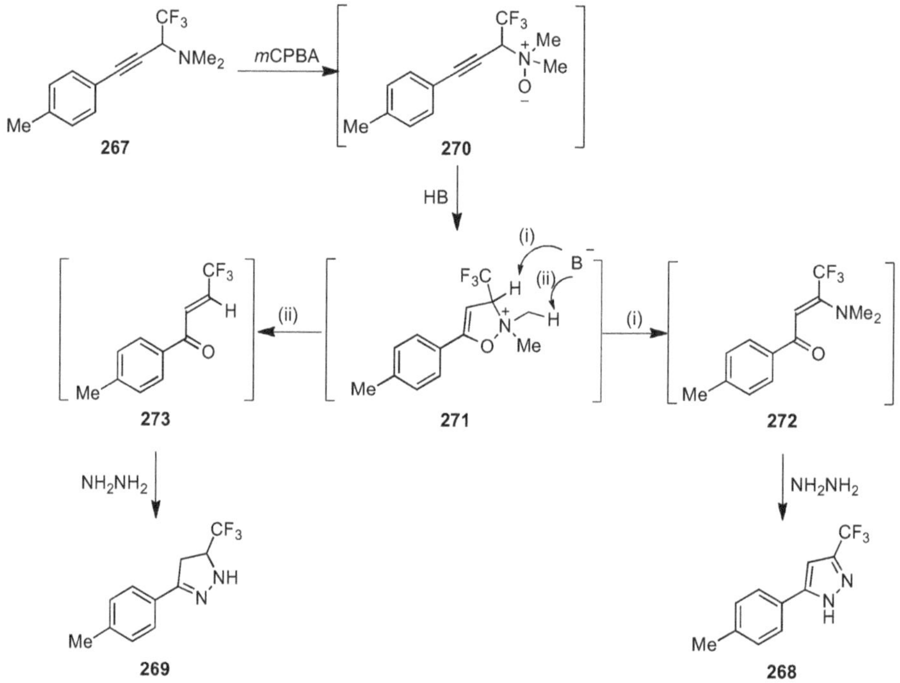

SCHEME 5.46

To suppress this side reaction, water (10% volume) was introduced as an additive to adjust the pH of the reaction system.

5.5.7.6 Copper-Mediated Intramolecular Cyclization

Water as an additive showed a positive effect on the copper-mediated intramolecular cyclization of 5-iodopyrimidine derivative **274** (Equation 5.44).[30] An incomplete conversion was observed if the reaction was carried out under anhydrous conditions due to presumably the poor solubility of Cs_2CO_3 in the reaction mixture. Thereby, addition of water (1.0 equiv) allowed a complete conversion within 3 h at 60 °C.

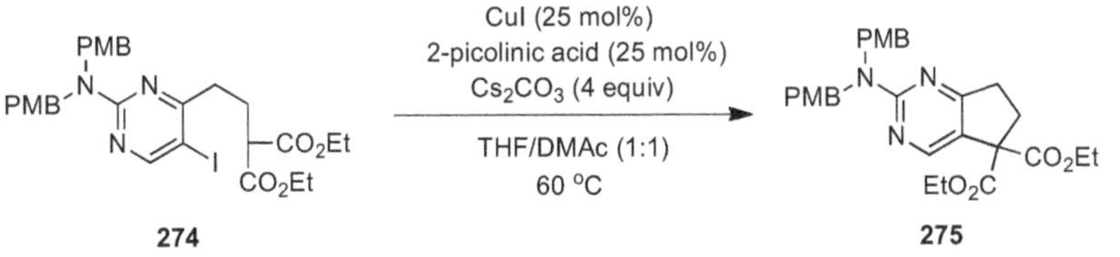

$$(5.44)$$

5.5.7.7 Crystallization-Induced Dynamic Resolution of Amine

Crystallization-induced dynamic resolution of amine *rac*-**276** employed Boc-D-phenylalanine **277** as the chiral resolving agent and 3,5-dichlorosalicylaldehyde (2 mol%) as the racemization catalyst (Scheme 5.47).[86] The resolution was carried out in toluene under reflux in the presence of catalytic amounts of water (1.5–2.0 mol%) producing the resolved salt **278** in 82% yield and 99.4% ee. In the absence of water, however, no racemization took place.

Water presumably promotes the regeneration of the aldehyde catalyst from the imine intermediate **280**.

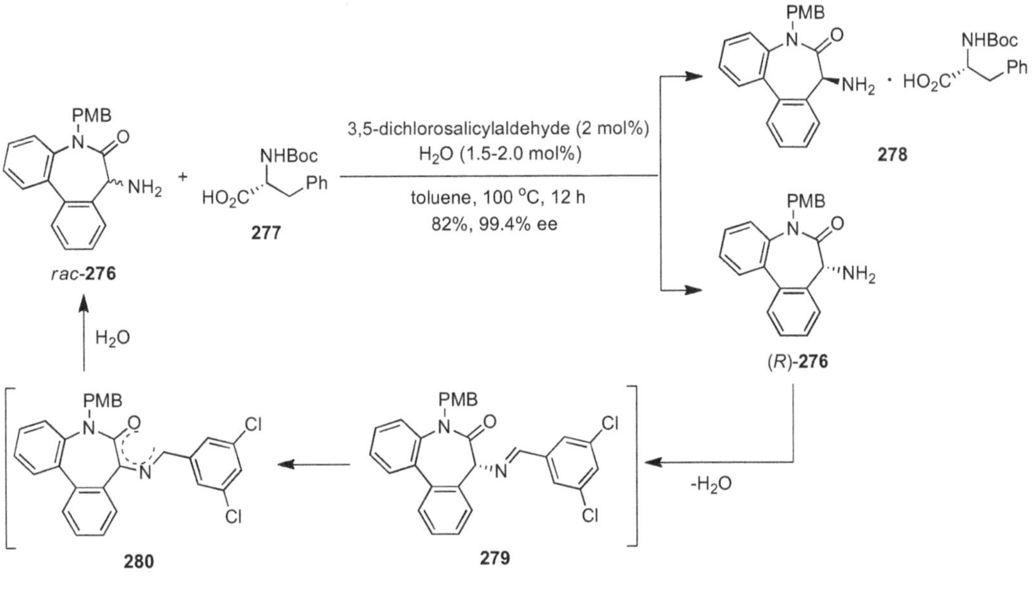

SCHEME 5.47

5.5.8 HYDROQUINONE

5.5.8.1 Preclusion of Oxidation

Thiolactone **282** is an intermediate for the synthesis of biotin (Vitamin H) and was prepared by thiolactonization of lactone **281** with potassium thioacetate (KSAc) in the presence of hydroquinone (0.01 equiv) in *N,N*-dimethylacetamide (DMAc) at 150 °C (Equation 5.45).[87]

$$(5.45)$$

Hydroquinone was used as an anti-oxidant to limit the oxidative degradation of KSAc during the reaction.

5.5.8.2 Preclusion of Polymerization

Baylis–Hillman reaction is inherently an environmentally friendly, atom-efficient reaction in which all the atoms of both starting materials get incorporated into the product. One of the concerns during the scaling up of the Baylis–Hillman reaction is the safety hazard due to the potential exothermic polymerizations and/or decompositions of the starting materials and the product. For instance, a differential scanning calorimetry (DSC) study of the Baylis–Hillman product mixture showed an exotherm event starting at 110 °C, while the starting *tert*-butyl acrylate **283** showed no evidence of exothermic decomposition until 200 °C (Equation 5.46).[88]

$$(5.46)$$

It was found that the presence of phenolic stabilizers such as dihydroquinone and methylhydroquinone in the reaction product was critical or polymerization of the product was observed. Therefore, 50–100 ppm of methylhydroquinone was added prior to stripping the toluene solution (DSC showed, in the presence of the stabilizer, an exotherm at 177 °C).

Procedure

Stage (a)
- Charge 459 kg of H_2O and 12.65 kg (99.4 mol) of 3-quinuclidinol.
- Charge 19.1 kg (636.7 mol) of paraformaldehyde.
- Heat the mixture to 40 °C.
- Charge 201.5 kg of MeCN.
- Heat to reflux (77 °C).
- Charge cautiously 51 kg (397.9 mol) of **283**.
- Heat the mixture for 1.5 h.
- Add toluene (255 L) cautiously.
- Cool to 30 °C.
- Separate layers and extract the aq layer with toluene (76.5 L).
- Wash the combined toluene layers with brine (2×150 L).
- Wash with ctiric acid aqueous solution (pH 4, 100 L), followed by brine (2×100 L).

Stage (b)
- Charge 150 g of methylhydroquinone to the toluene solution.
- Distill at 50 °C under reduced pressure to give a final volume of 120 L.
- Yield (solution yield): 39.4 kg (64.1%).

REMARK
The stabilizer should be added at the beginning of stage (b) because the stabilizer present in the reaction mixture could be removed during the aqueous work-up.

5.5.9 TRIMETHYL BORATE

Reduction of 3-[4'-(trifluoromethyl)phenyl]benzoic acid **285** with borane (generated in situ) gives the desired benzyl alcohol **286** (Equation 5.47).[89]

$$ (5.47) $$

During the reaction, the reaction mixture became a thick gel, resulting in difficult agitation. It was postulated that the in situ borane reduction may produce a polymeric material such as **287** that is responsible for the formation of the gels.

The addition of trimethyl borate (B(OMe)$_3$) resolved the thick gels-related problem. The use of B(OMe)$_3$ may prevent such polymer formation or cleave the polymeric chain into smaller fragments by disproportionation as shown in Reaction (5.48).

$$ (5.48) $$

Procedure

- Charge a solution **285** (19.7 kg, 74.0 mol) in THF (56.0 kg) to a mixture of NaBH$_4$ (3.15 kg, 83.3 mol) in THF (35.0 kg) at <25 °C over 2 h.
- Charge 2.31 kg (22.2 mol) of B(OMe)$_3$.
- Heat to 44 °C.
- Charge slowly 13.2 kg (87.6 mol) of 45% BF$_3$·THF at <50 °C.
- Age at 41–43 °C for 2 h.
- Add 8.59 kg of acetone followed by water (47.2 kg).
- Distill under atmospheric pressure to remove THF.
- Add toluene (51.1 kg) and water (9.9 kg).
- Separate layers.
- Wash the organic layer with a solution of NaHCO$_3$ (1.42 kg) in water (23.6 kg).
- Distill the organic layer under reduced pressure to remove half of the toluene (26.9 kg distillate collected).
- Add toluene (21.3 kg) (55 ppm of water content reached).
- Filter and wash with toluene (8.5 kg).
- Distill the combined filtrates to afford a clear solution of **286**.
- Yield of **286**: 18.5 kg (assay) (99% yield).

5.5.10   Isobutanoic Anhydride

Initially, 2-phenyl-4-phenanthrol **293** was synthesized using the process as described in Scheme 5.48,[90] and the product **293** was obtained in only 27% overall yield. The process involved [2+2] cycloaddition, electrocyclic ring-opening, electrocyclic ring-closure, and tautomerization cascade followed by saponification.

Solutions

As shown in Scheme 5.48, the process contains several intermediates such as acid chloride **291** and ketene **292**. Due to the high activity of **291** and **292**, the product **293** could further react with **291/292** to form ester **294**. After consumption of all **291/292**, the product **293** could be released via KOH hydrolysis along with the starting material **290**. Apparently, the low yield was on account of the fact that half of the starting 2-naphthaleneacetic acid **290** was wasted in the formation of **294**. Therefore, a protecting group strategy was considered and isobutanoic anhydride was identified as the reagent of choice. As a result, the use of isobutanoic anhydride (2.0 equiv) in the optimized process improved the yield of **293** to 74%.

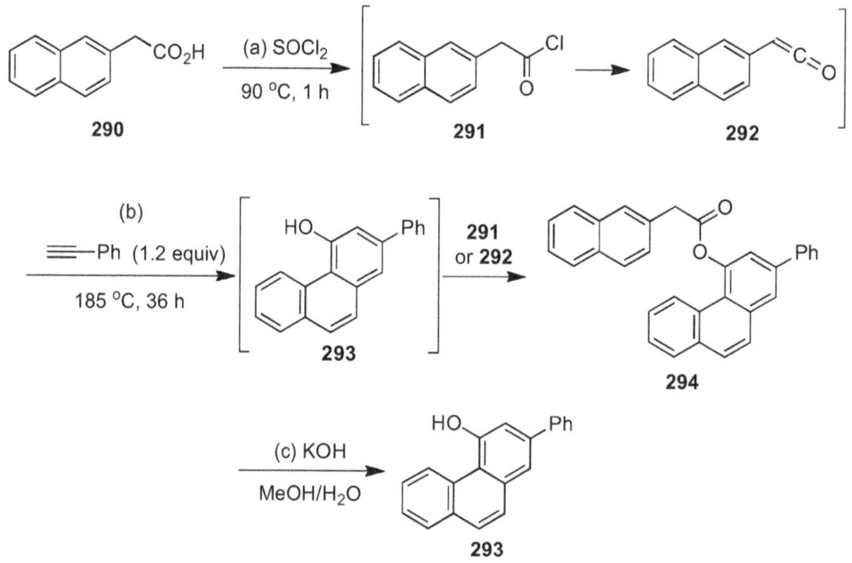

SCHEME 5.48

5.5.11   1,1-Dimethyl-2-Phenylethyl Acetate

Olefination of ester **295** with excess dimethyltitanocene **296** at 80 °C produced olefin **297** (as shown in Scheme 5.49)[91] typically in 98% assay yield within 5 h. However, the yield would drop by as much as 10% if heating continued for a further 30 min. Therefore, the time required to sample, assay, and cool on a large-scale reaction became critical.

Mechanistically, the olefination proceeds through an in situ–generated titanium carbene intermediate **298** whose reaction with ester **295** provides the desired product **297**. Once the ester starting material is consumed, the unreacted carbene **298** would react with vinyl ether product **297** to form side products. To address this issue, 1,1-dimethyl-2-phenylethyl acetate was introduced as a sacrificial ester that could protect **297** for as much as 24 h of additional heating once the completion of the reaction.

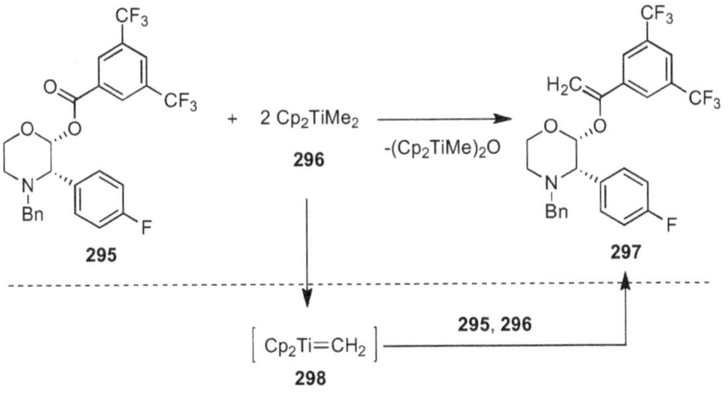

SCHEME 5.49

5.5.12 ALCOHOLS

5.5.12.1 Ethanol

Gas evolution reaction can be a hurdle for process development because it can not only create unsafe high pressure, but also sometimes generate undesired foams. For example, foaming was noticed in the decarboxylation step during the Hofmann rearrangement of isonicotinamide **299** (Equation 5.49).[92]

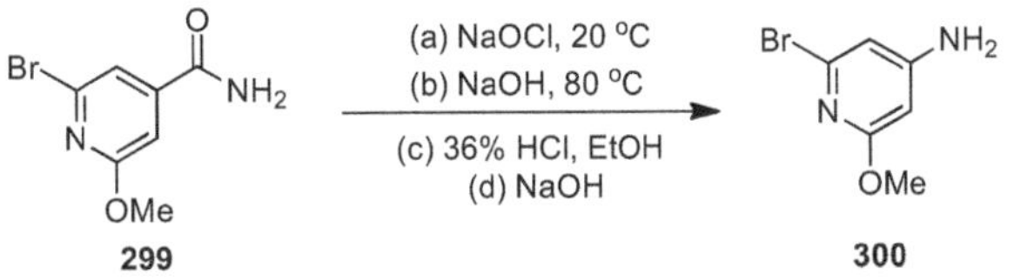

$$(5.49)$$

The foam formation may be caused by the high surface tension of the aqueous solution, thereby, the addition of ethanol (one volume) reduced dramatically the foam formation.

5.5.12.2 2-Propanol

5.5.12.2.1 Improving Agitation

The synthesis of oxindole **302** via palladium-catalyzed C–H functionalization of chloroacetanilide **301** in MeTHF generated sticky solids that prevented good agitation.

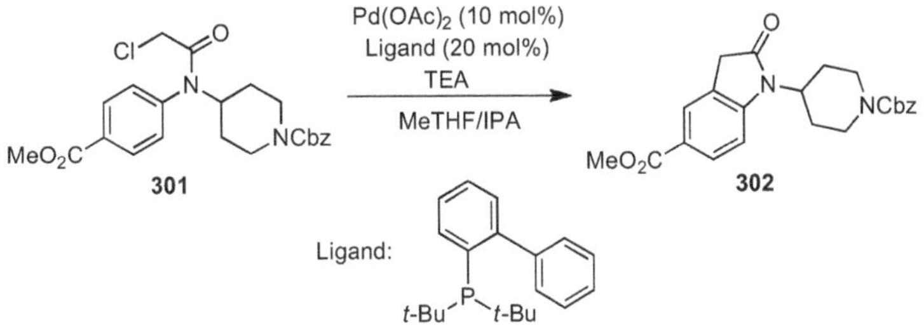

$$(5.50)$$

The addition of 2-propanol to the reaction mixture resulted in a more homogeneous and stirrable mixture (Equation 5.50).[93]

Procedure

• Sparge the subsurface of a solution of **301** (10.95 kg, 24.6 mol), Pd(OAc)$_2$ (0.553 kg, 2.46 mol), and 2-(di-*tert*-butylphosphino)biphenyl (1.47 kg, 4.93 mol) in MeTHF (71.2 L) and IPA (17.7 L) with nitrogen for 30 min.

- Charge 3.74 kg (37.0 mol) of triethylamine.
- Heat the reaction mixture at 71 °C for 2.5 h.
- Filter through Celite and rinse with hot (70 °C) MeTHF (11 L).
- Distill at 51 °C under reduced pressure to 20 L.
- Add IPA (153 L) and heat to reflux to dissolve solids.
- Cool to 25 °C over 3 h and stir for 5 h.
- Filter and wash with IPA (32.9 L).
- Dry at 35 °C under vacuum.
- Yield: 7.70 kg (76%).

5.5.12.2.2 Improving Product Yield

Reduction/deprotection of **303** in alcohol solvent such as 2-propanol (IPA) led to an isolated product **304** in only 36% yield (Equation 5.51).[94] In addition to the disappointing yield, this reaction required 10 equiv of $NaBH_4$.

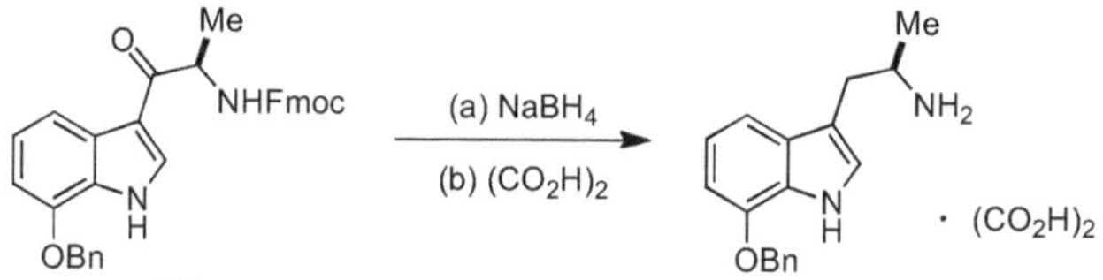

$$(5.51)$$

Solutions

Since $NaBH_4$ tends to decompose in protic solvents, acetonitrile was identified as a non-protic solvent for this reaction. 2-Propanol was chosen as an additive and exposure of **303** to $NaBH_4$ (3 equiv) in MeCN (5 v/w) in the presence of IPA (4 equiv) at reflux for 5 h giving **304** in 60% isolated yield with 98% ee.

5.5.12.3 *tert*-Butanol

The Birch reduction of compound **305** with lithium (2.4 equiv) was performed in liquid ammonia and THF (as cosolvent) to afford iminium hydrochloride **307** with **308** as an over-reduction side product (Scheme 5.50).[11]

The use of *tert*-butyl alcohol (2.0 equiv) as the proton source in addition to the proton of the carboxylic acid was able to control the level of **308** at <3%.

Procedure

Stage (a)
- Charge 40 g (138.2 mmol) of **305** into a mixture of THF (400 mL) and *t*-BuOH (20.48 g, 276.3 mmol, 2.0 equiv).

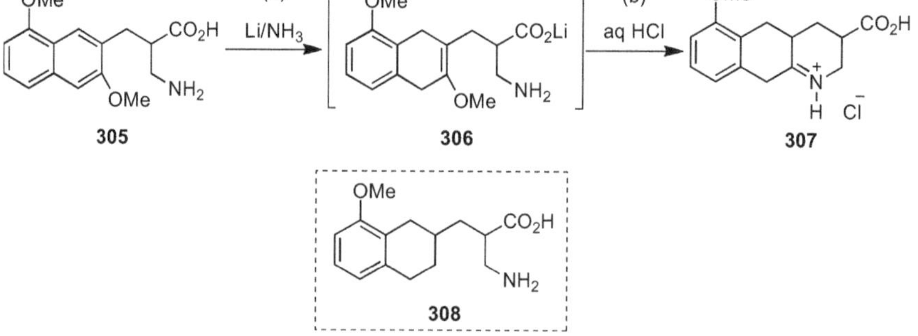

SCHEME 5.50

- Cool the resulting suspension to –70 °C.
- Condense 150 g of ammonia into the mixture.
- Charge portionwise 2.3 g (331.4 mmol, 2.4 equiv) of lithium metal.
- Stir at that temperature for 1.5 h.
- Evaporate ammonia.
- Neutralize the reaction mixture with 35% H_2SO_4.
- Add water (270 mL).
- Distill to remove THF and t-BuOH at 50 °C under reduced pressure to afford **306**.

Stage (b)
- Pour the aq solution of **306** into concentrated HCl (160 g) at <10 °C.
- Stir the resulting slurry in an ice bath for 4 h.
- Filter and wash with 2 M HCl.
- Dry at 40 °C under vacuum.
- Yield: 39.7 g (97%).

REMARK

In the solid form, **307** is stable for at least 3 months; in a solution **307** decomposes rapidly, probably due to aromatization.

5.5.12.4   Ethylene Glycol

5.5.12.4.1   Stabilization of Tetrahydroxydiboron

Tetrahydroxydiboron was used in a palladium-catalyzed borylation of 5-Bromo-benzooxazol-2-ylamine **309** for the preparation of the corresponding arylboronic acid **310** (Equation 5.52).[95]

(5.52)

However, tetrahydroxydiboron was not stable under the reaction conditions, which required excess of tetrahydroxydiboron (2.0 equiv) in order to achieve complete conversion. The addition of ethylene glycol (3.0 equiv) allowed the formation of the more stable bidentate boronate ester and, as a result, the amount of tetrahydroxydiboron could be reduced to 1.25 equiv.

5.5.12.4.2   Removal of 3-Pentanone

The preparation of lactone **313** was accomplished through a cascade involving deprotection of diethyl-dioxolane **311** to release diol **312** followed by cyclization in the presence of BF_3 etherate (Scheme 5.51).[96] Because the 3-pentanone by-product may interfere with the reaction, ethylene glycol was added to the reaction mixture to trap the 3-pentanone by-product.

Thus, the use of ethylene glycol led to a complete reaction.

5.5.12.5   1,2-Propanediol

A potential small molecule drug candidate **315** was developed by Bristol Myers Squibb Company for the treatment of a variety of autoimmune and inflammatory diseases such as rheumatoid arthritis and psoriasis.

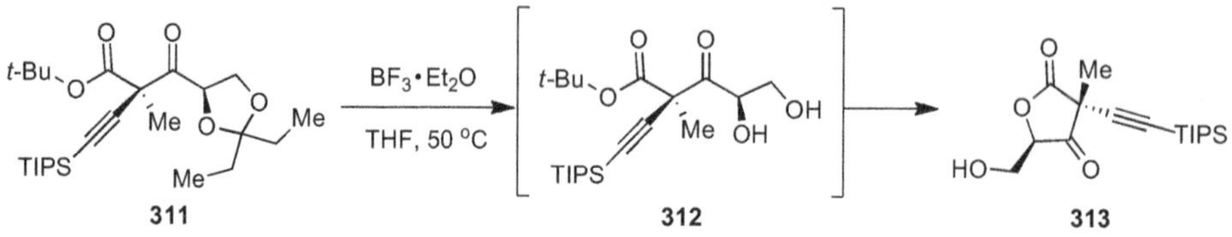

SCHEME 5.51

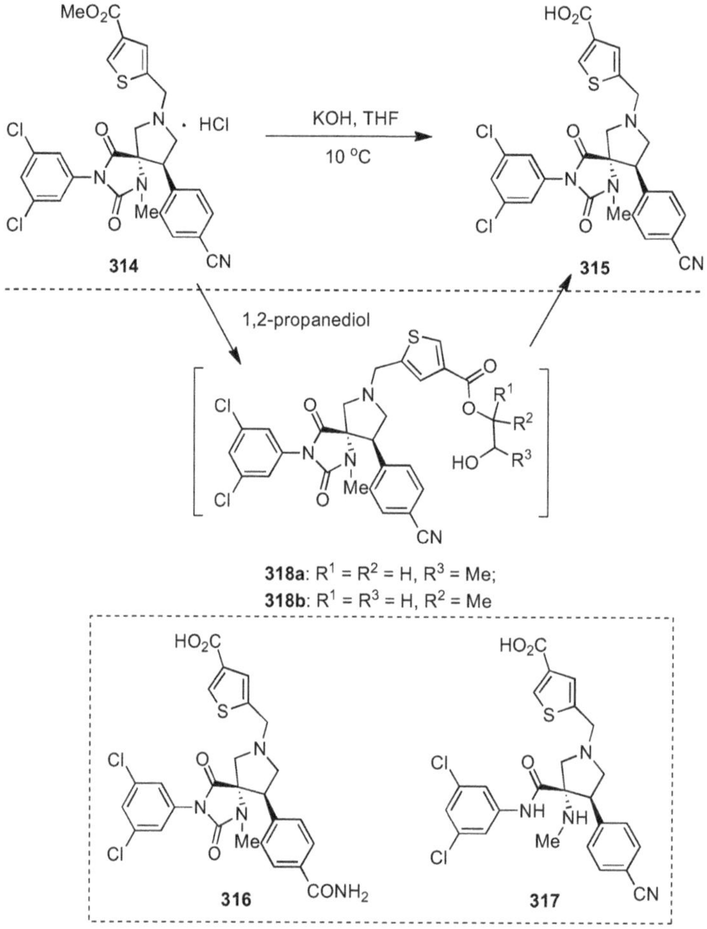

318a: $R^1 = R^2 = H$, $R^3 = Me$;
318b: $R^1 = R^3 = H$, $R^2 = Me$

SCHEME 5.52

(I) Reaction problems

The final hydrolysis step was initially carried out in THF with KOH, resulting in the desired acid **315** along with two major impurities, CN-group hydrolysis product **316** and spiro ring-opening product **317** (Scheme 5.52).[97] Although the hydrolysis at low temperatures (–10 °C) effectively controlled the level of impurity formation, the rate of reaction was extremely slow.

(II) Solutions

It was found that the presence of 1,2-propanediol (2 equiv) was able to accelerate the hydrolysis presumably proceeding via the transesterification intermediates **318**, which provide intramolecular anchimeric assistance in the ester hydrolysis.

Procedure

- Charge a cold (0–10 °C) solution of 1 N KOH (244.5 kg) to a solution of **314** (46.9 kg, 77.6 mol) and 1,2-propanediol (11.8 kg) in THF (41.7 kg) and water (266.8 kg) at 8–12 °C.
- Stir the mixture at 8–12 °C for 18–24 h.
- Wash the reaction mixture with heptane (385.7 kg).
- Add 1.5 M of citric acid (22.9 kg) to pH 7.5.
- Add isopropyl acetate (81.7 kg) and 1.5 M of citric acid (22.9 kg) o pH 6.5.
- Agitate for 15 min and hold for 30 min.
- Separate layers.
- Wash the organic layer with water (470 kg).
- Polish, filter, and rinse with isopropyl acetate (52.2 kg).
- Distill under reduced pressure (240 Torr) at <45 °C to 718 L.
- Add 500 g of seeds.
- Continue to distill until 207 L is attained.
- Add heptane (117.8 kg).
- Cool to 20 °C over 1.5 h.
- Mill at wet until d_{90} <60 μm.
- Hold the batch for >2 h.
- Filter and wash with isopropyl acetate/heptane (1:1, 109.7 kg).
- Dry at 35–40 °C under vacuum.
- Yield: 39.6 kg (91.5%).

5.5.12.6 Neopentyl Glycol

The Rh-catalyzed Hayashi asymmetric reaction of cyclohex-2-enone with (isopropenyl)pinacolboronate **319** allowed to access (S)-3-isopropenyl-cyclohexan-1-one **320** in one step (Scheme 5.53).[98] During the process optimization, it was found that the rate of conversion of the enone starting material slowed dramatically as the reaction progressed.

(I) Chemistry diagnosis

As shown in Scheme 5.53, (isopropenyl)pinacolboronate **319** would undergo hydrolytic and methanolic cleavages to form isopropenylboronic acid **321a** and methyl isopropenylboronate **321b**, respectively. Due to the reduced steric congestion around the boron atom as compared to **319**, both **321a** and **321b** would proceed with a more rapid transmetalation step to generate isopropenylrhodium

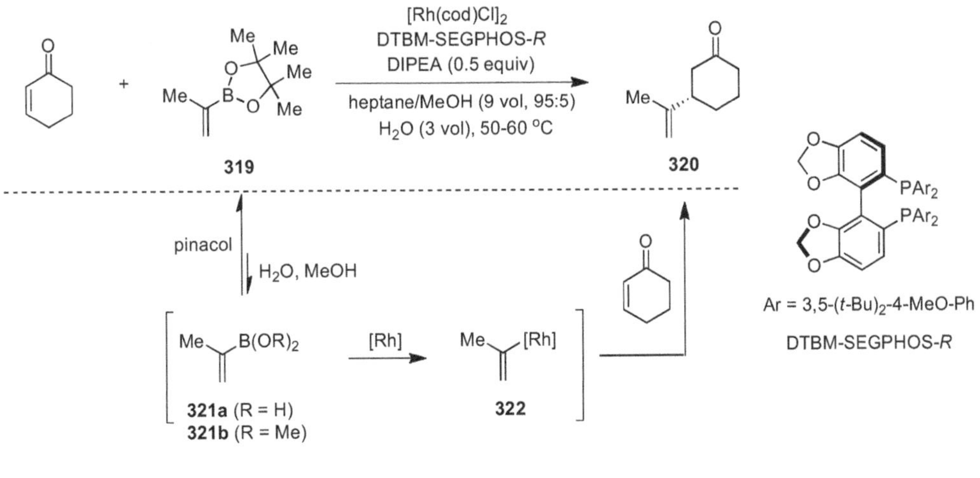

SCHEME 5.53

322. However, as the reaction progresses, the buildup of pinacol by-products in the reaction mixture would reduce the concentration of **321a/321b** because of the disfavored equilibrium. Therefore, pinacol by-product is responsible for the observed decrease in reaction rate.

(II) Solutions

Scientists at Bristol Myers Squibb Company identified neopentyl glycol as an additive to promote the reaction (Scheme 5.54).[99] The use of neopentyl glycol (1.1 equiv) enabled a more favorable diol exchange with pinacol, leading to a higher concentration of the corresponding (isopropenyl) boronate **323**. Furthermore, because of its high solubility in heptane and less steric hindrance, the transmetalation of **323** with Rh catalyst is faster. Ultimately, the optimized process was successfully scaled up to a 100-kg scale to produce >600 kg of **320** with 82% average yield and 99.4–99.5% ee.

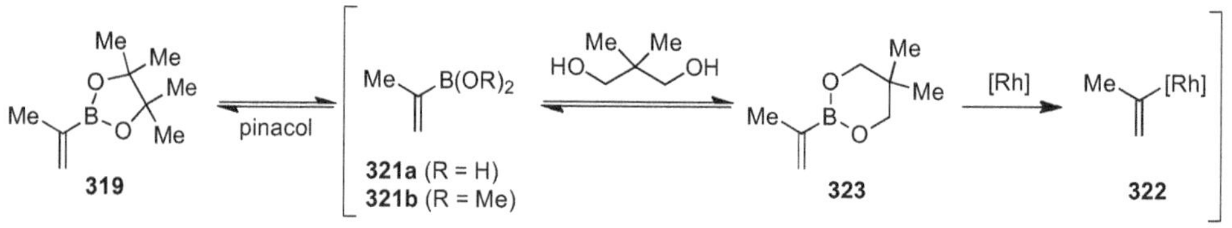

SCHEME 5.54

5.5.13   1,4-Dioxane

The synthesis of amide **325** was realized via a one-pot process involving the enolation of aryl bromide **324** with NaH, bromine/metal exchange with Turbo Grignard, and trapping the resulting aryl Grignard with isocyanate (Scheme 5.55).[85]

The bromine/metal exchange of enolate **326** using either isopropylmagnesium chloride or isopropylmagenesium·lithium chloride complex (Turbo Grignard) resulted in low conversion.

(I) Chemistry diagnosis

The poor conversion of the bromine/metal exchange appears to be the low activity of i-PrMgCl or i-PrMgCl·LiCl. Knochel reported that the addition of 1,4-dioxane to the Turbo Grignard can generate

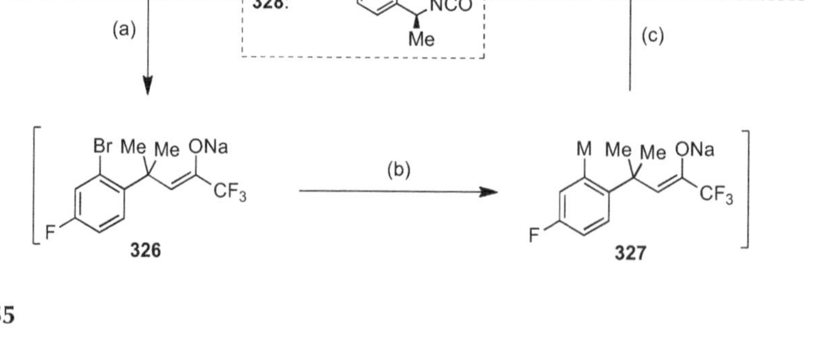

SCHEME 5.55

the highly active exchange reagent "$i\text{-Pr}_2\text{Mg·LiCl}$". The precipitation of the MgCl_2–dioxane complex was the driving force for the Schlenk equilibrium (Equation 5.53) toward the dialkylmagnesium.

$$2\ i\text{-PrMgCl} \rightleftharpoons i\text{-Pr}_2\text{Mg} + \text{MgCl}_2 \tag{5.53}$$

(II) Solutions

By adding 1,4-dioxane to the reaction mixture, a dramatic acceleration in the exchange rate was observed, and a complete consumption of the bromide **324** was achieved after 5 h at room temperature.

Procedure

Stage (a)
- Charge 0.690 kg (60% wt% in mineral oil, 17.26 mol, 1.2 equiv) of NaH and THF (18.5 L, 500 ppm water).
- Cool the resulting slurry to ca. 0 °C.
- Charge 5.25 kg (89.6 wt%, 14.38 mol) of a solution of **324** in THF (2.6 L) at ≤10 °C.
- After the addition, warm the batch to 25 °C and stir for 18 h.

Stage (b)
- Cool the batch to 0 °C.
- Charge 11.66 kg of 1.33 M in THF (15.82 mol, 1.1 equiv) of $i\text{-PrMgCl·LiCl}$ at ≤20 °C.
- Charge 4.0 L (46.94 mol, 3.3 equiv) of 1,4-dioxane.
- Stir the batch at 25 °C for 2–4 h.

Stage (c)
- Cool the batch to 0 °C.
- Charge a solution of **328** (2.80 kg, 15.82 mol, 1.1 equiv) in THF (2.6 L) at ≤15 °C.
- Stir the batch at 5–15 °C for 30 min.
- Work-up.

Note

1,4-Dioxane is a Class 2 solvent, with permissible daily exposure of 3.8 mg/day and a maximum level of 380 ppm in the drug substance.

5.5.14 BENZOTRIAZOLE

The initial synthesis of tetrahydroquinoline **332** employed a two-step, one-pot process involving an aza Diels–Alder reaction between an in situ formed *N*-arylimine **330** and a vinyl carbamate **331** (Scheme 5.56).[99] This aza Diels-Alder reaction was considered an ideal approach since it delivered a high degree of *cis* selectivity with no detectable amount of the *trans* isomer (by NMR). The *N*-arylimine **330** was generated in CH_2CH_2 from *p*-trifluoromethylaniline **329** and *n*-propanal using TiCl_4 as a water scavenger.

(I) Reaction problems

The process gave unsatisfied yields (40–60%) with modest purity (ca. 70%) due to the formation of significant amounts of aminal **333**.

(II) Solutions

To suppress the formation of **333**, benzotriazole was used as the additive to trap the in situ formed *N*-arylimine **330** giving a crystalline adduct **334** (Scheme 5.57). Next, treating **334** and **331** with toluenesulfonic acid (1 mol%) resulted in a 78% yield of **332**. Again, no *trans* isomer was detectable

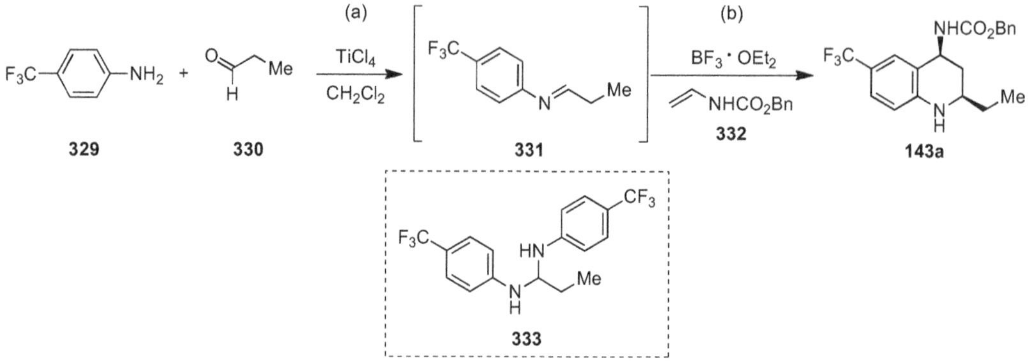

SCHEME 5.56

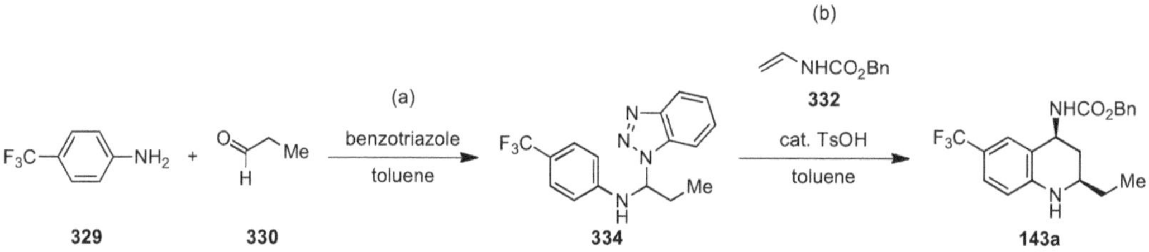

SCHEME 5.57

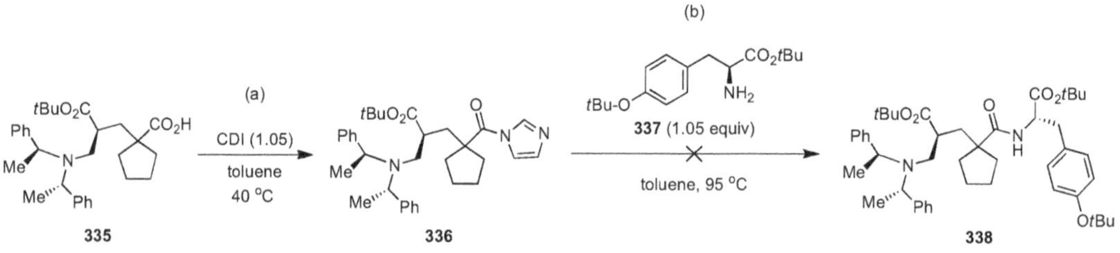

SCHEME 5.58

REMARK

This method also has the advantage of not requiring the use of the moisture-sensitive Lewis acids, TiCl$_4$ and BF$_3$ etherate.

5.5.15 1-HYDROXYBENZOTRIAZOLE

Amides are commonly prepared from carboxylic acids and amines. Usually, these reactions involve the activation of carboxylic acid. *N,N′*-Carbonyldiimidazole (CDI) is frequently used as the activation agent. The activation of acid **335**, for example, was realized through a reaction with CDI to give imidazolide **336** (Stage (a), Scheme 5.58).[89]

Although the reaction of **335** with CDI cleanly gave the imidazolide **336**, this imidazolide would not react with di-*tert*-butyl tyrosine **337** even at elevated temperatures (Stage (b), Scheme 5.58).

The unsuccessful amide coupling is presumably due to the steric hindrance. The addition of a catalytic quantity of 1-hydroxybenzotriazole (0.1 equiv) solved this problem, and the imidazolide **336** reacted cleanly with **337** to give the desired amide **338**.

5.5.16   1,4-Dibromobutane

Many cyclopropane-containing pharmaceutical and natural products exhibit unique biological activities. In addition to existing methods for the synthesis of cyclopropane derivatives, Scheme 5.59[100] illustrates a practical and convenient approach to access cyclopropane derivatives.

In an alkaline environment, ketones **339** were converted into cyclopropane products **340** in good yields. As shown in Scheme 5.59, this transformation employed 1,4-dibromobutane to trap the thiolate anion in **341** to give the intermediate **342** in which the sulfonium moiety serves as a good leaving group to promote the cyclopropanation process.

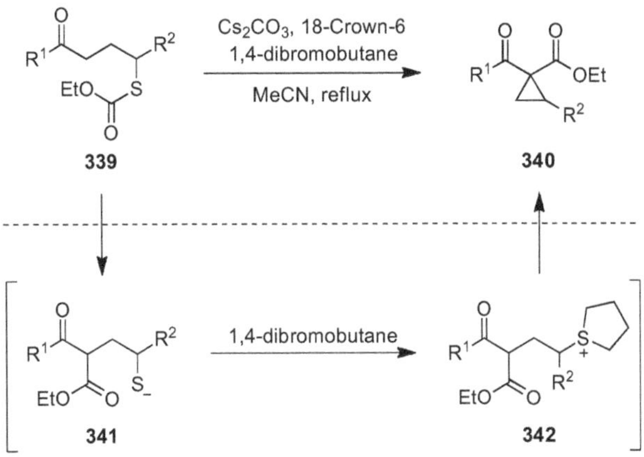

SCHEME 5.59

5.5.17   Diethanolamine

5.5.17.1   Improving Selectivity

Functional group protection is frequently used in order to achieve high selectivity in organic synthesis. However, using such an approach in large production scales is not an ideal option. Therefore, developing a protection-free synthesis has become a relentless effort for process chemists. Equation 5.54[101] demonstrates a protection-free O-alkylation reaction to synthesize diol **344**, γ-secretase modulator (GSM) as a potential treatment for Alzheimer's disease.

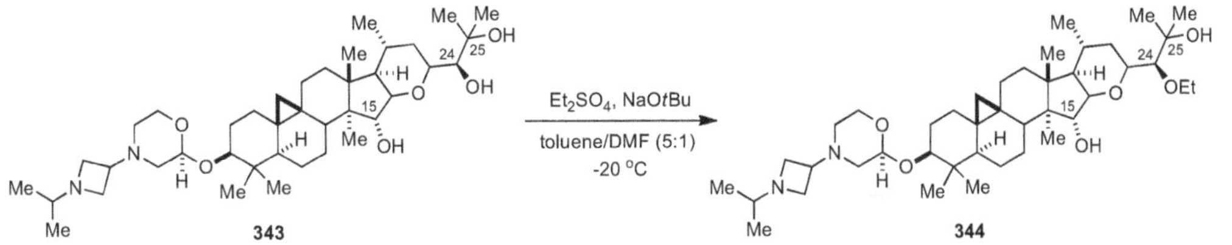

(5.54)

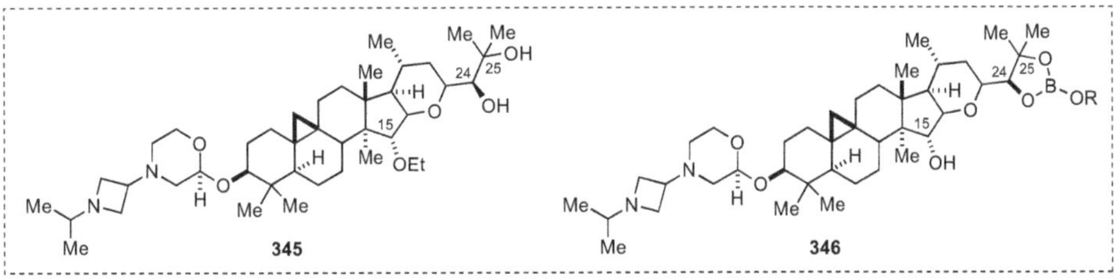

(I) Reaction problems

It was observed that using **343**, obtained by triturating the crude triol **343** from NaBH$_4$-reduction of C15 ketone, led to a sluggish reaction affording a C15 O-alkylated side product **345**.

(II) Solutions

The formation of this side product **345** was presumably caused by an O-alkylation of a cyclic boronate **346** carried over from the ketone reduction step with NaBH$_4$. To circumvent this issue associated with the cyclic boronate ester, the C15 ketone reduction mixture was treated with dietha-nolamine (DEA) to decomplex the cyclic boronate ester **346**.

5.5.17.2   Boranate Ester Exchange

DEA can be used for the preparation of boronic acids via a two-step procedure: deprotection of pinacolyl boronate esters via transesterification with diethanolamine followed by hydrolysis.[102] This procedure has the advantages of tolerance to various functional groups, short reaction time, and ease of product isolation. This two-step procedure was successfully utilized in the preparation of an aryl boronic acid **349** from the corresponding pinacolyl boronate **347** that contains an acid-sensitive tertiary hydroxyl group (Scheme 5.60).[103]

The intermediate diethanolamine-boronate adduct **348** was isolated by precipitation from iso-propanol-hexane. The subsequent hydrolysis with hydrochloric acid and crystallization from the reaction mixture afforded the boronic acid **349** in 56% yield over two steps on a >5-kg scale.

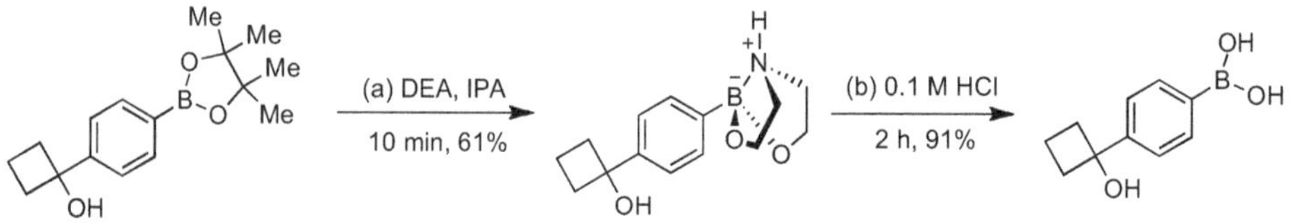

SCHEME 5.60

Procedure

Stage (a)
- Add 1.38 kg (13.1 mol) of DEA and 5 L of IPA into a 22 L 4-necked round bottom flask.
- Add 1.80 kg (6.57 mol) of crude **347** and rinse with IPA (1 L).
- Stir the batch for 10 min.
- Add hexane (3 L) and stir for 18 h.
- Filter, wash with hexane (3×2 L), and dry to give 1.05 kg (61%) DEA adduct.

Stage (b)
- Transfer 2.0 kg (7.66 mol, combined two batches) into a 12-L reactor.
- Add 3.0 M HCl (6.0 L, 18 mol) and stir the resulting slurry for 2 h.
- Filter, wash with water (2×2 L), hexane (3×2 L), and dry.
- Crude yield: 1.34 kg (91%).

5.5.18   TRIMETHYL PHOSPHITE

The synthesis of ethyl 3-(trifluoromethyl)pyrazine-2-carboxylate **351** was realized via a cyclization of 2-hydroxyimino-4,4,4-trifluoro-3-oxobutanoate **350** with ethylenediamine in the presence of two additives: benzoic acid (3.5 equiv) and trimethyl phosphite (P(OMe)$_3$) (1.4 equiv) (Scheme 5.61).[104]

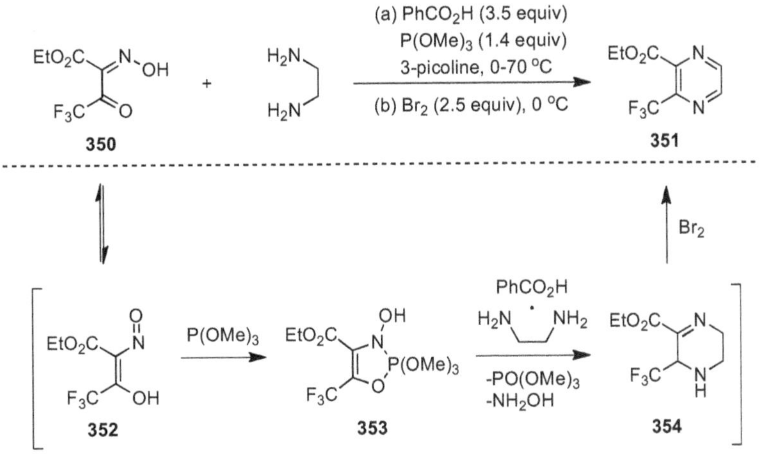

SCHEME 5.61

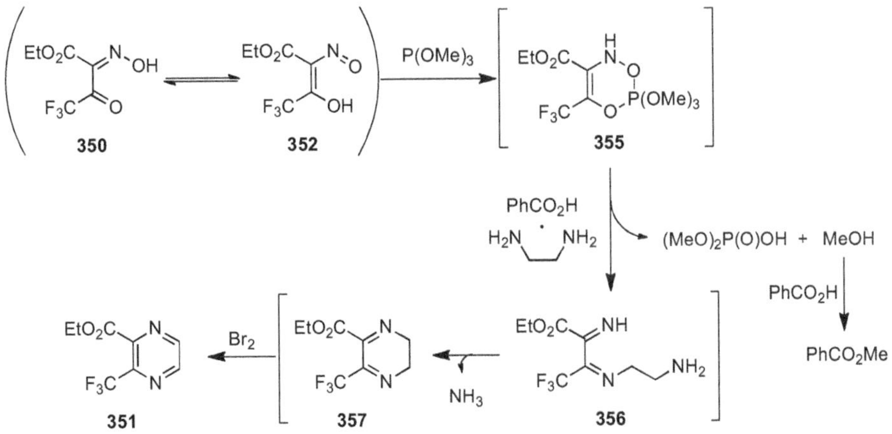

SCHEME 5.62

Benzoic acid (3.5 equiv) would form a salt in situ with ethylenediamine to deactivate and suppress its side reactions. The function of P(OMe)₃ is presumably to mediate the cyclization involving a highly electrophilic oxazaphosphole intermediate **353**.

DISCUSSIONS

Besides the hypothesized reaction pathway,[105] an alternative reaction mechanism may also exist (Scheme 5.62). The formation of methanol is convinced by the fact that a small amount of methyl benzoate was formed during the reaction.

5.5.19 DIETHYL PHOSPHITE

A radical bromination of 2-fluoro-3-nitrotoluene **358** produced a mixture of the desired 3-(bromomethyl)-2-fluoro-nitrobenzene **359** (65%) and the *gem*-dibromide side product **360** (35%). To address the issue associated with impurity **360**, a selective debromination approach was developed to reduce **360** to **359**. Thus, treatment of the mixture of **359** and **360** (in a ratio of 13:7) with diethyl phosphite (HP(O)(OEt)₂) (0.46 equiv) (Scheme 5.63).[105]

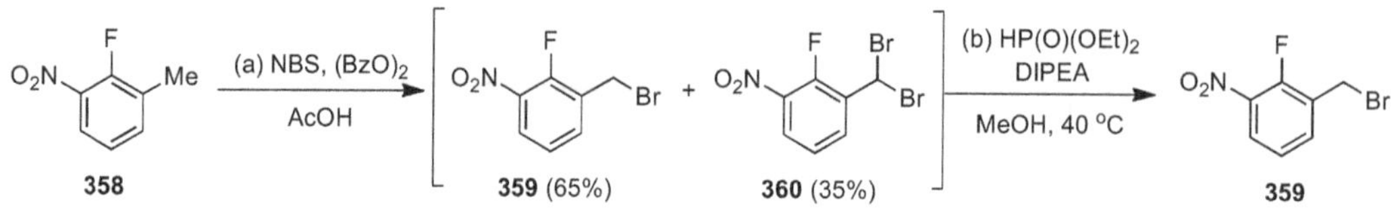

SCHEME 5.63

Procedure

Stage (a)
- Heat a mixture of **358** (225.0 kg, 1450 mol, 1.0 equiv), NBS (25.8 kg, 145 mol, 0.1 equiv), benzoyl peroxide (14.1 kg, 44.0 mol, 0.03 equiv, containing 25 wt% water), and AcOH (450 L, 2.0 vol) in a 6000-L glass-lined reactor at 83 °C under N_2 for 1.5 h.
- Add a suspension of NBS (408 kg, 2292 mol, 1.58 equiv) in AcOH (792 L, 3.52 vol) in three portions (0.25, 0.25, and 0.5 of the total mass) over 5 h at 83 °C.
- Rinse with AcOH (225 L, 1.0 vol).
- Stir the batch as 83 °C for 12 h.
- Add a solution of H_3PO_3 (17.0 kg, 145 mol, 0.1 equiv, containing 30 wt% water) and AcOH (22.5 L, 0.1 vol) and stir for 2.5 h.
- Cool the batch to 25 °C.
- Add deionized water (1240 L, 5.5 vol) and toluene (1800 L, 8 vol).
- Stir for 30 min and separate layers.
- Add aq 1.6 N NaOH (1600 L, 7.1 vol) into the organic layer at <25 °C (pH 12.5).
- Filter and separate layers.

Stage (b)
- Add DIPEA (99.35 kg, 769 mol, 0.53 equiv) and MeOH (20 L, 494 mol, 0.34 equiv) to the organic layer.
- Heat the mixture to 40 °C.
- Add a solution of HP(O)(OEt)$_2$ (91.5 kg, 667 mol, 0.46 equiv) in MeOH (82 L, 2024 mol, 1.4 equiv) at 40 °C ± 5 °C over 2.5 h.
- Stir the batch at 40 °C for 5 h (to complete the conversion of **360** to **359**).
- **359** was used directly in the following step without purification.

5.5.20   4-TRIFLUOROMETHYL BENZALDEHYDE

The synthesis of aminoketone **362** proceeded in a two-step process involving a decyanation reaction followed by reduction of the imine intermediate **365**, without impacting the ketone (Scheme 5.64).[106]

(I) Reaction problems

However, under the reaction conditions, the desired aminoketone product **362** was obtained in only <20% yield and most of the starting material **361** was converted instead to an aromatized side product **363**.

(II) Solutions

As shown in Scheme 5.64, the imine intermediate **365** would react with the in situ generated hydrogen cyanide (HCN) to form the corresponding cyanohydrin **366** whose aromatization leads to the side product **363** along with 2-hydroxypropionitrile **367**. In this context, a HCN scavenger was needed to trap the in situ–formed HCN. Ultimately, 4-trifluoromethyl benzaldehyde was selected

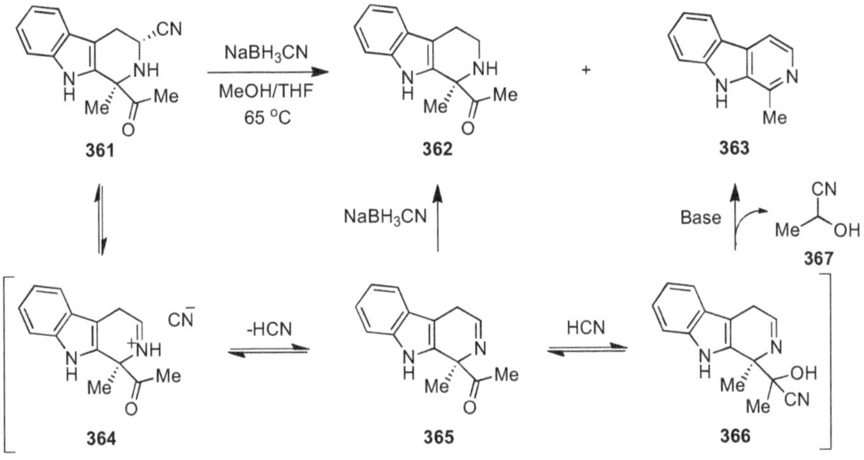

SCHEME 5.64

as the HCN scavenger and the addition of 4-trifluoromethyl benzaldehyde (3.0 equiv) improved the product yield to 71%.

NOTES

1. Ragan, J. A.; Bourassa, D. E.; Blunt, J.; Breen, D.; Busch, F. R.; Cordi, E. M.; Damon, D. B.; Do, N.; Engtrakul, A.; Lynch, D.; McDermott, R. E.; Monggillo, J. A.; O'Sullivan, M. M.; Rose, P. R. *Org. Process Res. Dev.* **2009**, *13*, 186.

2. Cleator, E.; Scott, J. P.; Avalle, P.; Bio, M. M.; Brewer, S. E.; Davies, A. J.; Gibb, A. D.; Sheen, F. J.; Stewart, G. W.; Wallace, D. J.; Wilson, R. D. *Org. Process Res. Dev.* **2013**, *17*, 1561.

3. Sieser, J. E.; Singer, R. A.; McKinley, J. D.; Bourassa, D. E.; Teixeira, J. J.; Long, J. J. *Org. Process Res. Dev.* **2011**, *15*, 1328.

4. The use of phosphoric acid to improve the dehydrochlorination has been reported previously: Andersen, K.; Begtrup, M. *Acta Chem. Scand.* **1992**, *46*, 1130.

5. Pu, Y.-M.; Grieme, T.; Gupta, A.; Plata, D.; Bhatia, A. V.; Cowart, M.; Ku, Y.-Y. *Org. Process Res. Dev.* **2005**, *9*, 45.

6. Castanet, A.-S.; Colobert, F.; Broutin, P.-E. *Tetrahedron Lett.* **2002**, *43*, 5047.

7. Prakash, G. K. S.; Mathew, T.; Hoole, D.; Esteves, P. M.; Wang, Q.; Rasul, G.; Olah, G. A. *J. Am. Chem. Soc.* **2004**, *126*, 15770.

8. Hanselmann, R.; Johnson, G.; Reeve, M. M.; Huang, S.-T. *Org. Process Res. Dev.* **2009**, *13*, 54.

9. Berwe, M.; Jöntgen, W.; Krüger, J.; Cancho-Grande, Y.; Lampe, T.; Michels, M.; Paulsen, H.; Raddatz, S.; Weigand, S. *Org. Process Res. Dev.* **2011**, *15*, 1348.

10. Bänziger, M.; Cercus, J.; Stampfer, W.; Sunay, U. *Org. Process Res. Dev.* **2000**, *4*, 460.

11. Beutner, G. L.; Albrecht, J.; Fan, J.; Fanfair, D.; Lawler, M. J.; Bultman, M.; Chen, K.; Ivy, S.; Schild, R. L.; Tripp, J. C.; Murugesan, S.; Dambalas, K.; McLeod, D. D.; Sweeney, J. T.; Eastgate, M. D.; Conlon, D. A. *Org. Process Res. Dev.* **2017**, *21*, 1122.

12. Betti, M.; Castagnoli, G.; Panico, A.; Coccone, S. S.; Wiedenau, P. *Org. Process Res. Dev.* **2012**, *16*, 1739.

13. Zheng, B.; Conlon, D. A.; Corbett, R. M.; Chau, M.; Hsieh, D.-M.; Yeboah, A.; Hsieh, D.; Müslehiddinoğlu, J.; Gallagher, W. P.; Simon, J. N.; Burt, J. *Org. Process Res. Dev.* **2012**, *16*, 1846.

14. Hansen, K. B.; Balsells, J.; Dreher, S.; Hsiao, Y.; Kubryk, M.; Palucki, M.; Rivera, N.; Steinhuebel, D.; Armstrong III, J. D.; Askin, D.; Grabowski, E. J. J. *Org. Process Res. Dev.* **2005**, *9*, 634.

15. Bowles, D. M.; Bolton, G. L.; Boyles, D. C.; Curran, T. T.; Hutchings, R. H.; Larsen, S. D.; Miller, J. M.; Park, W. K. C.; Ritsema, K. G.; Schineman, D. C. *Org. Process Res. Dev.* **2008**, *12*, 1183.

16. Hansen, K. B.; Hsiao, Y.; Xu, F.; Rivera, N.; Clausen, A.; Kubryk, M.; Krska, S.; Rosner, T.; Simmons, B.; Balsells, J.; Ikemoto, N.; Sun, Y.; Spindler, F.; Malan, C.; Grabowski, E. J. J.; Armstrong, J. D., III. *J. Am. Chem. Soc.* **2009**, *131*, 8798.

17. Xu, F.; Armstrong, J. D., III.; Zhou, G. X.; Simmons, B.; Hughes, D.; Ge, Z.; Grabowski, E. J. J. *J. Am. Chem. Soc.* **2004**, *126*, 13002.
18. Brands, K. M. J.; Payack, J. F.; Rosen, J. D.; Nelson, T. D.; Candelario, A.; Huffman, M. A.; Zhao, M. M.; Li, J.; Craig, B.; Song, Z. J.; Tschaen, D. M.; Hansen, K.; Devine, P. N.; Pye, P. J.; Rossen, K.; Dormer, P. G.; Reamer, R. A.; Welch, C. J.; Mathre, D. J.; Tsou, N. N.; McNamara, J. M.; Reider, P. J. *J. Am. Chem. Soc.* **2003**, *125*, 2129.
19. Maddess, M. L.; Scott, J. P.; Alorati, A.; Baxter, C.; Bremeyer, N.; Brewer, S.; Campos, K.; Cleator, E.; Dieguez-Vazquez, A.; Gibb, A.; Gibson, A.; Howard, M.; Keen, S.; Klapars, A.; Lee, J.; Li, J.; Lynch, J.; Mullens, P.; Wallace, D.; Wilson, R. *Org. Process Res. Dev.* **2014**, *18*, 528.
20. (a) Monguchi, D.; Yamamura, A.; Fujiwara, T.; Somete, T.; Mori, A. *Tetrahedron Lett.* **2010**, *51*, 850. (b) Stachel, S. J.; Habeeb, R. L.; Van Vranken, D. L. *J. Am. Chem. Soc.* **1996**, *118*, 1225.
21. Wang, X.-J.; Frutos, R. P.; Zhang, L.; Sun, X.; Xu, Y.; Wirth, T.; Nicola, T.; Nummy, L. J.; Krishnamurthy, D.; Busacca, C. A.; Yee, N.; Senanayake, C. H. *Org. Process Res. Dev.* **2011**, *15*, 1185.
22. Bankston, D.; Dumas, J.; Natero, R.; Riedl, B.; Monahan, M.-K.; Sibley, R. *Org. Process Res. Dev.* **2002**, *6*, 777.
23. (a) Gonda, Z.; Tolnai, G. L.; Novák, Z. *Chem. Eur. J.* **2010**, *16*, 11822. (b) Arvela, R. K.; Leadbeater, N. E.; Sangi, M. S.; Williams, V. A.; Granados, P.; Singer, R. D. *J. Org. Chem.* **2005**, *70*, 161. (c) Lauterbach, T.; Livendahl, M.; Rosellón, A.; Espinet, P.; Echavarren, A. M. *Org. Lett.* **2010**, *12*, 3006.
24. de Koning, P. D.; Castro, N.; Gladwell, I. R.; Morrison, N. A.; Moses, I. B.; Panesar, M. S.; Pettman, A. J.; Thomson, N. M. *Org. Process Res. Dev.* **2011**, *15*, 1256.
25. Ford, J. G.; Pointon, S. M.; Powell, L.; Siedlecki, P. S. *Org. Process Res. Dev.* **2010**, *14*, 1078.
26. For reviews, see: (a) Far, A. D. *Angew. Chem., Int. Ed.* **2003**, *42*, 2340. (b) Gaber, A. E. M.; McNab, H. *Synthesis* **2001**, *14*, 2059. (c) Chen, B. C. *Heterocycles* **1991**, *32*, 529. (d) Huang, X. *Youji Huaxue* **1986**, *6*, 329.
27. For examples, see: (a) Pak, C. S.; Yang, H. C.; Choi, E. B. *Synthesis* **1992**, *12*, 1213. (b) Svetlik, J.; Goljer, I.; Turecek, F. *J. Chem. Soc., Perkin Trans. 1* **1990**, *10*, 1315. (c) Sato, M.; Yoneda, N.; Katagiri, N.; Watanabe, H.; Kaneko, C. *Synthesis* **1986**, *8*, 672. (d) Oikawa, Y.; Sugano, K.; Yonemitsu, O. *J. Org. Chem.* **1978**, *43*, 2087.
28. Broxer, S.; Fitzgerald, M. A.; Sfouggatakis, C.; Defreese, J. L.; Barlow, E.; Powers, G. L.; Peddicord, M.; Su, B.-N.; Yue, T.-Y.; Pathirana, C.; Sherbine, J. P. *Org. Process Res. Dev.* **2011**, *15*, 343.
29. Mullens, P.; Cleator, E.; McLaughlin, M.; Bishop, B.; Edwards, J.; Goodyear, A.; Andreani, T.; Jin, Y.; Kong, J.; Li, H.; Williams, M.; Zacuto, M. *Org. Process Res. Dev.* **2016**, *20*, 1075.
30. Engelhardt, F. C.; Shi, Y.-J.; Cowden, C. J.; Conlon, D. A.; Pipik, B.; Zhou, G.; McNamara, J. M.; Dolling, U.-H. *J. Org. Chem.* **2006**, *71*, 480.
31. Pasti, J.; Chen, C.-K.; Spangler, L.; DelMonte, A. J.; Benoit, S.; Berglund, D.; Bien, J.; Brodfuehrer, P.; Chan, Y.; Corbett, E.; Costello, C.; DeMena, P.; Discordia, R. P.; Doubleday, W.; Gao, Z.; Gingras, S.; Grosso, J.; Haas, O.; Kacsur, D.; Lai, C.; Leung, S.; Miller, M.; Muslehiddinoglu, J.; Nguyen, N.; Qiu, J.; Olzog, M.; Reiff, E.; Thoraval, D.; Totleben, M.; Vanyo, D.; Vemishetti, P.; Wasylak, J.; Wei, C. *Org. Process Res. Dev.* **2009**, *13*, 716.
32. Ito, T.; Ikemoto, T.; Isogami, Y.; Wada, H.; Sera, M.; Mizuno, Y.; Wakimasu, M. *Org. Process Res. Dev.* **2002**, *6*, 238.
33. Thompson, H. W.; McPherson, E. *J. Am. Chem. Soc.* **1974**, *96*, 6232.
34. Girardin, M.; Dolman, S. J.; Lauzon, S.; Ouellet, S. G.; Hughes, G.; Fernandez, P.; Zhou, G.; O'Shea, P. D. *Org. Process Res. Dev.* **2011**, *15*, 1073.
35. Ishimoto, K.; Fukuda, N.; Nagata, T.; Sawai, Y.; Ikemoto, T. *Org. Process Res. Dev.* **2014**, *18*, 122.
36. Goralski, C. T.; Hasha, D. L.; Henton, D. R.; Krauss, R. C.; Pfeiffer, C. D.; Williams, B. M. *Org. Process Res. Dev.* **1997**, *1*, 273.
37. Amato, J. S.; Chung, J. Y. L.; Cvetovich, R. J.; Reamer, R. A.; Zhao, D.; Zhou, G.; Gong, X. *Org. Process Res. Dev.* **2004**, *8*, 939.
38. Mattsson, S.; Dahlström, M.; Karlsson, S. *Tetrahedron Lett.* **2007**, *48*, 2497.
39. (a) Bäckvall, J.-E.; Byström, S. E.; Nordberg, *J. Org. Chem.* **1984**, *49*, 4619. Related examples can also be found in (b) Itami, K.; Palmgren, A.; Thorarensen, A.; Bäckvall, J.-E. *J. Org. Chem.* **1998**, *63*, 6466 and (c) Bäckvall, J.-E.; Andersson, P. G. *J. Am. Chem. Soc.* **1992**, *114*, 6374.
40. Sirois, L. E.; Zhao, M. M.; Lim, N.-K.; Bednarz, M. S.; Harrison, B. A.; Wu, W. *Org. Process Res. Dev.* **2019**, *23*, 45.
41. Schwindt, M. A.; Belmont, D. T.; Carlson, M.; Franklin, L. C.; Hendrickson, V. S.; Karrick, G. L.; Poe, R. W.; Sobieray, D. M.; Vusse, J. V. D. *J. Org. Chem.* **1996**, *61*, 9564.

42. Mitchell, D.; Cole, K. P.; Pollock, P. M.; Coppert, D. M.; Burkholder, T. P.; Clayton, J. R. *Org. Process Res. Dev.* **2012**, *16*, 70.
43. Campbell, A. N.; Cole, K. P.; Martinelli, J. R.; May, S. A.; Mitchell, D.; Pollock, P. M.; Sullivan, K. A. *Org. Process Res. Dev.* **2013**, *17*, 273.
44. Damon, D. B.; Dugger, R. W.; Hubbs, S. E.; Scott, J. M.; Scott, R. W. *Org. Process Res. Dev.* **2006**, *10*, 472.
45. (a) Tudge, M.; Savarin, C. G.; DiFelice, K.; Maligres, P.; Humphrey, G.; Reamer, B.; Tellers, D. M.; Hughes, D. *Org. Process Res. Dev.* **2010**, *14*, 787. (b) Tudge, M.; Tamiya, M.; Savarin, C.; Humphrey, G. R. *Org. Lett.* **2006**, *8*, 565.
46. Kato, T.; Ozaki, T.; Tsuzuki, K.; Ohi, N. *Org. Process Res. Dev.* **2001**, *5*, 122.
47. Cabri, W.; Roletto, J.; Olmo, S.; Fonte, P.; Ghetti, P.; Songia, S.; Mapelli, E. Alpegiani, M.; Paissoni, P. *Org. Process Res. Dev.* **2006**, *10*, 198.
48. Hansen, M. M.; Kallman, N. J.; Koenig, T. M.; Linder, R. J.; Richey, R. N.; Rizzo, J. R.; Ward, J. A.; Yu, H.; Zhang, T. Y.; Mitchell, D. *Org. Process Res. Dev.* **2017**, *21*, 208.
49. Blacker, A. J.; Moran-Malagon, G.; Powell, L.; Reynolds, W.; Stones, R.; Chapman, M. R. *Org. Process Res. Dev.* **2018**, *22*, 1086.
50. Menzel, K.; Fisher, E. L.; DiMichele, L.; Frantz, D. E.; Nelson, T. D.; Kress, M. H. *J. Org. Chem.* **2006**, *71*, 2188.
51. Znidar, D.; Cantillo, D.; Inglesby, P.; Boyd, A.; Kappe, C. O. *Org. Process Res. Dev.* **2018**, *22*, 633.
52. Thiel, O. R.; Bernard, C.; King, T.; Dilmeghani-Seran, M.; Bostick, T.; Larsen, R. D.; Faul, M. M. *J. Org. Chem.* **2008**, *73*, 3508.
53. Thaisrivongs, D. A.; Morris, W. J.; Tan, L.; Song, Z. J.; Lyons, T. W.; Waldman, J. H.; Naber, J. R.; Chen, W.; Chen, L.; Zhang, B.; Yang, J. *Org. Lett.* **2018**, *20*, 1568.
54. Brewer, S. E.; Vickery, T. P.; Backert, D. C.; Brands, K. M. J.; Emerson, K. M.; Goodyear, A.; Kumke, K. J.; Lam, T.; Scott, J. P. *Org. Process Res. Dev.* **2005**, *9*, 1009.
55. Harger, M. J. P.; Sreedharan-Menon, R. *J. Chem. Soc., Perkin Trans. 1,* **1997**, *4*, 527.
56. Braden, T. M.; Coffey, D. S.; Doecke, C. W.; LeTourneau, M. E.; Martinelli, M. J.; Meyer, C. L.; Miller, R. D.; Pawlak, J. M.; Pedersen, S. W.; Schmid, C. R.; Shaw, B. W.; Staszak, M. A.; Vicenzi, J. T. *Org. Process Res. Dev.* **2007**, *11*, 431.
57. Hynes, J., Jr.; Dyckman, A. J.; Lin, S.; Wrobleski, S. T.; Wu, H.; Gillooly, K. M.; Kanner, S. B.; Lonial, H.; Loo, D.; McIntyre, K. W.; Pitt, S.; Shen, D. R.; Shuster, D. J.; Yang, X.; Zhang, R.; Behnia, K.; Zhang, H.; Marathe, P. H.; Doweyko, A. M.; Tokarski, J. S.; Sack, J. S.; Pokross, M.; Kiefer, S. E.; Newitt, J. A.; Barrish, J. C.; Dodd, J.; Schieven, G. L.; Leftheris, K. *J. Med. Chem.* **2008**, *51*, 4.
58. Shi, Z.; Kiau, S.; Lobben, P.; Hynes, J., Jr.; Wu, H.; Parlanti, L.; Discordia, R.; Doubleday, W. W.; Leftheris, K.; Dyckman, A. J.; Wrobleski, S. T.; Dambalas, K.; Tummala, S.; Leung, S.; Lo, E. *Org. Process Res. Dev.* **2012**, *16*, 1618.
59. Prashad, M.; Prasad, K.; Repic, O.; Blacklock, T. J.; Prikoszovich, W. *Org. Process Res. Dev.* **1999**, *3*, 409.
60. Dale, D. J.; Draper, J.; Dunn, P. J.; Hughes, M. L.; Hussain, F.; Levett, P. C.; Ward, G. B.; Wood, A. S. *Org. Process Res. Dev.* **2002**, *6*, 767.
61. Taylor, B.; Barrat, C. F.; Woodward, R. L.; Inglesby, P. A. *Org. Process Res. Dev.* **2017**, *21*, 1404.
62. Zheng, B.; Sugiyama, M.; Eastgate, M. D.; Fritz, A.; Murugesan, S.; Conlon, D. A. *Org. Process Res. Dev.* **2012**, *16*, 1827.
63. Chou, T. S.; Spitzer, W. A.; Dorman, D. E.; Kukolja, S.; Wright, I. G.; Jones, N. S.; Chaney, M. O. *J. Org. Chem.* **1978**, *43*, 3835.
64. Copp, J. D.; Tharp, G. A. *Org. Process Res. Dev.* **1997**, *1*, 92.
65. Wu, C.; Li, Z.; Wang, C.; Zhou, Y.; Sun, T. *Org. Process Res. Dev.* **2018**, *22*, 1081.
66. Parsons, P. J.; Waters, A. J.; Walter, D. S.; Board, J. *J. Org. Chem.* **2007**, *72*, 1395.
67. Shu, L.; Wang, P.; Radinov, R.; Dominique, R.; Wright, J.; Alabanza, L. M.; Dong, Y. *Org. Process Res. Dev.* **2013**, *17*, 114.
68. Lian, Y.; Hummel, J. R.; Bergman, R. G.; Ellman, J. A. *J. Am. Chem. Soc.* **2013**, *135*, 12548.
69. Müller, S.; Webber, M. J.; List, B. *J. Am. Chem. Soc.* **2011**, *133*, 18534.
70. Brach, N.; Fouler, V. L.; Bizet, V.; Lanz, M.; Gallou, F.; Bailly, C.; Hoehn, P.; Parmentier, M.; Blanchard, N. *Org. Process Res. Dev.* **2020**, *24*, 776.
71. Fouler, V. L.; Chen, Y.; Gandon, V.; Bizet, V.; Salomé, C.; Fessard, T.; Liu, F.; Houk, K. N.; Blanchard, N. *J. Am. Chem. Soc.* **2019**, *141*, 15901.
72. Achmatowicz, M.; Thiel, O. R.; Wheeler, P.; Bernard, C.; Huang, J.; Larsen, R. D.; Faul, M. M. *J. Org. Chem.* **2009**, *74*, 795.

73. Bradley, P. A.; Carroll, R. J.; Lecouturier, Y. C.; Moore, R.; Noeureuil, P.; Patel, B.; Snow, J.; Wheeler, S. *Org. Process Res. Dev.* **2010**, *14*, 1326.

74. Woodman, E. K.; Chaffey, J. G. K.; Hopes, P. A.; Hose, D. R. J.; Gilday, J. P. *Org. Process Res. Dev.* **2009**, *13*, 106.

75. Vaidyanathan, R.; Kalthod, V. G.; Ngo, D. P.; Manley, J. M.; Lapekas, S. P. *J. Org. Chem.* **2004**, *69*, 2565.

76. Ragan, J. A.; Ide, N. D.; Cai, W.; Cawley, J. J.; Colon-Cruz, R.; Kumar, R.; Peng, Z.; Vanderplas, B. C. *Org. Process Res. Dev.* **2010**, *14*, 1402.

77. Pan, X.; Xu, S.; Huang, R.; Yu, W.; Liu, F. *Org. Process Res. Dev.* **2015**, *19*, 611.

78. Xu, D. D.; Waykole, L.; Calienni, J. V.; Ciszewski, L.; Lee, G. T.; Liu, W.; Szewczyk, J.; Vargas, K.; Prasad, K.; Repič, O.; Blacklock, T. J. *Org. Process Res. Dev.* **2003**, *7*, 856.

79. Hou, S.-H.; Tu, Y.-Q.; Wang, S.-H.; Xi, C.-C.; Zhang, F.-M.; Wang, S.-H.; Li, Y.-T.; Liu, L. *Angew. Chem. Int. Ed.* **2016**, *55*, 4456.

80. Nimmagadda, S. K.; Korapati, S.; Dasgupta, D.; Malik, N. A.; Vinodini, A.; Gangu, A. S.; Kalidindi, S.; Maity, P.; Bondigela, S. S.; Venu, A.; Gallagher, W. P.; Aytar, S.; González-Bobes, F.; Vaidyanathan, R. *Org. Process Res. Dev.* **2020**, *24*, 1141.

81. Kim, J.; Itoh, T.; Xu, F.; Dance, Z. F. X.; Waldman, J. H.; Wallace, D. J.; Wu, F.; Kats-Kagan, R.; Ekkati, A. R.; Brunskill, A. P. J.; Peng, F.; Fier, P. S.; Obligacion, J. V.; Sherry, B. D.; Liu, Z.; Emerson, K. M.; Fine, A. J.; Jenks, A. V.; Armenante, M. E. *Org. Process Res. Dev.* **2021**, *25*, 2249.

82. Nishida, T.; Yoshinaga, H.; Toyoda, T.; Toshima, M. *Org. Process Res. Dev.* **2012**, *16*, 625.

83. Conlon, D. A.; Drahus-Paone, A.; Ho, G.-J.; Pipik, B.; Helmy, R.; McNamara, J. M.; Shi, Y.-J.; Williams, J. M.; Macdonald, D.; Deschênes, D.; Gallant, M.; Mastracchio, A.; Roy, B.; Scheigetz, J. *Org. Process Res. Dev.* **2006**, *10*, 36.

84. Reeves, J. T.; Fandrick, D. R.; Tan, Z.; Song, J. J.; Rodriguez, S.; Qu, B.; Kim, S.; Niemeier, O.; Li, Z.; Byrne, D.; Campbell, S.; Chitroda, A.; DeCroos, P.; Fachinger, T.; Fuchs, V.; Gonnella, N. C.; Grinberg, N.; Haddad, N.; Jäger, B.; Lee, H.; Lorenz, J. C.; Ma, S.; Narayanan, B. A.; Nummy, L. J.; Premasiri, A.; Roschangar, F.; Sarvestani, M.; Shen, S.; Spinelli, E.; Sun, X.; Varsolona, R. J.; Yee, N.; Brenner, M.; Senanayake, C. H. *J. Org. Chem.* **2013**, *78*, 3616.

85. Chen, J.; Properzi, R.; Uccello, D. P.; Young, J. A.; Dushin, R. G.; Starr, J. T. *Org. Lett.* **2014**, *16*, 4146.

86. Karmakar, S.; Byri, V.; Gavai, A. V.; Rampulla, R.; Mathur, A.; Gupta, A. *Org. Process Res. Dev.* **2016**, *20*, 1717.

87. Warm, A.; Naughton, A. B.; Saikali, E. A. *Org. Process Res. Dev.* **2003**, *7*, 272.

88. Dunn, P. J.; Hughes, M. L.; Searle, P. M.; Wood, A. S. *Org. Process Res. Dev.* **2003**, *7*, 244.

89. Walker, S. D.; Borths, C. J.; DiVirgilio, E.; Huang, L.; Liu, P.; Morrison, H.; Sugi, K.; Tanaka, M.; Woo, J. C. S.; Faul, M. M. *Org. Process Res. Dev.* **2011**, *15*, 570.

90. Ding, Z.; Osminski, W. E. G.; Ren, H.; Wulff, W. D. *Org. Process Res. Dev.* **2011**, *15*, 1089.

91. Payack, J. F.; Huffman, M. A.; Cai, D.; Hughes, D. L.; Collins, P. C.; Johnson, B. K.; Cottrell, I. F.; Tuma, L. D. *Org. Process Res. Dev.* **2004**, *8*, 256.

92. Daver, S.; Rodeville, N.; Pineau, F.; Arlabosse, J.-M.; Moureou, C.; Muller, F.; Pierre, R.; Bouquet, K.; Dumais, L.; Boiteau, J.-G.; Cardinaud, I. *Org. Process Res. Dev.* **2017**, *21*, 231.

93. Kiser, E. J.; Magano, J.; Shine, R. J.; Chen, M. H. *Org. Process Res. Dev.* **2012**, *16*, 255.

94. Ikunaka, M.; Kato, S.; Sugimori, D.; Yamada, Y. *Org. Process Res. Dev.* **2007**, *11*, 73.

95. Gurung, S. R.; Mitchell, C.; Huang, J.; Jonas, M.; Strawser, J. D.; Daia, E.; Hardy, A.; O'Brien, E.; Hicks, F.; Papageorgiou, C. D. *Org. Process Res. Dev.* **2017**, *21*, 65.

96. Zhong, Y.-L.; Cleator, E.; Liu, Z.; Yin, J.; Morris, W. J.; Alam, M.; Bishop, B.; Dumas, A. M.; Edwards, J.; Goodyear, A.; Mullens, P.; Song, Z. J.; Shevlin, M.; Thaisrivongs, D. A.; Li, H.; Sherer, E. C.; Cohen, R. D.; Yin, J.; Tan, L.; Yasuda, N.; Limanto, J.; Davies, A.; Campos, K. R. *J. Org. Chem.* **2019**, *84*, 4780.

97. Delmonte, A. J.; Waltermire, R. E.; Fan, Y.; McLeod, D. D.; Gao, Z.; Gesenberg, K. D.; Girard, K. P.; Rosingana, M.; Wang, X.; Kuehne-Willmore, J.; Braem, A. D.; Castoro, J. A. *Org. Process Res. Dev.* **2010**, *14*, 553.

98. Simmons, E. M.; Mudryk, B.; Lee, A. G.; Qiu, Y.; Razler, T. M.; Hsiao, Y. *Org. Process Res. Dev.* **2017**, *21*, 1659.

99. Damon, D. B.; Dugger, R. W.; Magnus-Aryitey, G.; Ruggeri, R. B.; Wester, R. T.; Tu, M.; Abramov, Y. *Org. Process Res. Dev.* **2006**, *10*, 464.

100. Li, S.-G.; Zard, S. Z. *Org. Lett.* **2014**, *16*, 6180.

101. Fuller, N. O.; Hubbs, J. L.; Austin, W. F.; Shen, R.; Ives, J.; Osswald, G.; Bronk, B. S. *Org. Process Res. Dev.* **2014**, *18*, 683.

102. Sun, J.; Perfetti, M. T.; Santos, W. L. *J. Org. Chem.* **2011**, *76*, 3571.

103. Smith, A. C.; Kung, D. W.; Shavnya, A.; Brandt, T. A.; Dent, P. D.; Genung, N. E.; Cabral, S.; Panteleev, J.; Herr, M.; Yip, K. N.; Aspnes, G. E.; Conn, E. L.; Dowling, M. S.; Edmonds, D. J.; Edmonds, I. D.; Fernando, D. P.; Herrinton, P. M.; Keene, N. F.; Lavergne, S. Y.; Li, Q.; Polivkova, J.; Rose, C. R.; Thuma, B. A.; Vetelino, M. G.; Wang, G.; Weaver, III, J. D.; Widlicka, D. W.; Wiglesworth, K. E. P.; Xiao, J.; Zahn, T.; Zhang, Y. *Org. Process Res. Dev.* **2018**, *22*, 681.
104. Zaragoza, F.; Gantenbein, A. *Org. Process Res. Dev.* **2017**, *21*, 448.
105. Caille, S.; Allgeier, A. M.; Bernard, C.; Correll, T. L.; Cosbie, A.; Crockett, R. D.; Cui, S.; Faul, M. M.; Hansen, K. B.; Huggins, S.; Langille, N.; Mennen, S. M.; Morgan, B. P.; Morrison, H.; Muci, A.; Nagapudi, K.; Quasdorf, K.; Ranganathan, K.; Roosen, P.; Shi, X.; Thiel, O. R.; Wang, F.; Tvetan, J. T.; Woo, J. C. S.; Wu, S.; Walker, S. D. *Org. Process Res. Des.* **2019**, *23*, 1558.
106. Zhou, Z.; Gao, A. X.; Snyder, S. A. *J. Am. Chem. Soc.* **2019**, *141*, 7715.

6 Process Optimization of Catalytic Reactions

A catalytic reaction proceeds through a transition state with lower activation energy, resulting in a higher reaction rate than an uncatalyzed reaction under otherwise the same reaction conditions. Thereby a catalytic reaction can be performed at relatively mild conditions, which is desired in large-scale production in terms of process costs and safety. Catalytic reactions are considered environmentally friendly due to the reduced amount of waste generated, as opposed to stoichiometric reactions. Catalysis is ubiquitous in chemical and pharmaceutical industries with estimates of 90% of commercially produced chemicals and most pharmaceutical products involving catalysts at some stage in the process of their manufacture.

Depending on the reaction medium, a catalytic reaction is classified as either homogeneous or heterogeneous.

6.1 SUZUKI–MIYAURA REACTION

Transition metal-catalyzed cross-coupling reactions are an indispensable tool for the construction of carbon–carbon and carbon–heteroatom bonds, consisting of 25% of all reactions performed in the pharmaceutical industry. And among these, Suzuki–Miyaura reaction has become one of the most efficient methods for the construction of diaryl compounds and has been used extensively in the synthesis of natural products, nucleoside analogs, and pharmaceuticals.

Suzuki cross-coupling reaction offers several advantages including high tolerance to most functional groups and relative stability of boronic acids/esters to heat, oxygen, and moisture. In addition, the ease of handling and separation of boron-containing by-products, and the commercial availability of various organic boronic acids and their less toxic nature make Suzuki crossing-coupling one of the most attractive approaches on a production scale. Suzuki reaction is generally carried out in a mixture of organic solvent and water in the presence of a palladium catalyst and base.

Despite compatibility with a variety of substrates containing various functional groups, the use of organoboron derivatives has shown some limitations: (1) reactions using boronic acids are often carried out in the presence of water because boronic acids dimerize and trimerize to form anhydrides and boroxines under anhydrous conditions;[1] (2) the diols used to generate stable boronate esters such as catechol, pinacol, and diethanolamine (DEA) add considerable expense to the overall process and must be separated from the final product; (3) boronate esters, as Lewis acids and electrophiles, interact extensively with Lewis bases and nucleophiles (consequently, elaboration of functionalized boronate esters via nucleophilic reactions may prove challenging); (4) most palladium catalysts used in Suzuki reaction are air-sensitive and may be poisoned easily. This chapter provides effective approaches to deal with various challenges.

6.1.1 CATALYST POISONING

Catalyst poisoning refers to the effect that the ability to enhance the reaction rate is lost due to the formation of an inert catalyst-contained species. Common poisons for these palladium and platinum catalysts are sulfur and nitrogen heterocycles such as pyridine and quinoline.

6.1.1.1 Catalyst Poisoning by Sulfhydryl Group

Generally, palladium-catalyzed cross-couplings for the C–C bond formation in the presence of sulfhydryl group present formidable challenges primarily due to the propensity of poisoning the Pd catalyst by the formation of $L_2Pd(II)(SAr)_2$ species that are either too stable to be broken down or decompose to inactive complex.[2] Therefore, compounds bearing sulfhydryl functional groups are not suitable for the Suzuki–Miyaura reaction.

As expected, a cross-coupling of arylboronic acid **1** with 4-bromobenzenethiol **2** resulted in a poor yield of a cross-coupling product **3** (Equation 6.1) along with a disulfide side product.

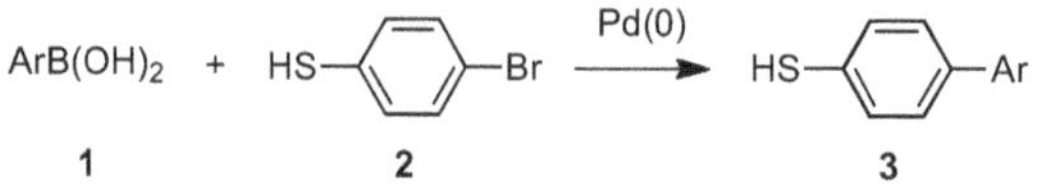

$$\text{ArB(OH)}_2 \ + \ \text{HS}\!\!-\!\!\bigcirc\!\!-\!\!\text{Br} \ \xrightarrow{\text{Pd(0)}} \ \text{HS}\!\!-\!\!\bigcirc\!\!-\!\!\text{Ar} \qquad (6.1)$$

$$\quad \textbf{1} \qquad\qquad\qquad \textbf{2} \qquad\qquad\qquad\qquad \textbf{3}$$

In order to address this issue, a protection strategy was developed by conducting the Suzuki–Miyaura reaction of arylboronic acids with protected arylthiols. It was demonstrated that the Suzuki–Miyaura reaction of the protected arylthiols **4** afforded the cross-coupling products **5** in high yields (Scheme 6.1).[3] The resulting thioethers **5** could be deprotected via β-elimination to generate the corresponding thiols **3** in quantitative yields.

DISCUSSIONS

(a) It has been postulated that the mercapto dimer **6**, formed via oxidation by oxygen, is the culprit of the palladium poisoning since the dimer is likely to undergo oxidative addition with Pd(0) to afford a stable complex **7** (Equation 6.2):[4]

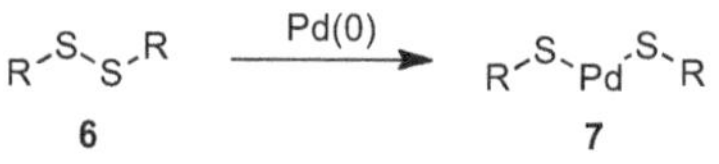

$$\text{R}^{\diagup\text{S}}\!\diagdown_{\text{S}}\diagdown^{\text{R}} \ \xrightarrow{\text{Pd(0)}} \ \text{R}^{\diagup\text{S}}\diagdown_{\text{Pd}}\diagup^{\text{S}}\diagdown_{\text{R}} \qquad (6.2)$$

$$\qquad\quad \textbf{6} \qquad\qquad\qquad\qquad \textbf{7}$$

(b) The use of bidentate phosphine ligands would prevent the formation of such a palladium-dithiolate complex.

(c) The palladium catalyst may be poisoned by a high concentration of cyanide ions when conducting cross-coupling with potassium cyanide. Therefore, zinc cyanide (with lower solubility) is usually chosen to replace KCN.[5]

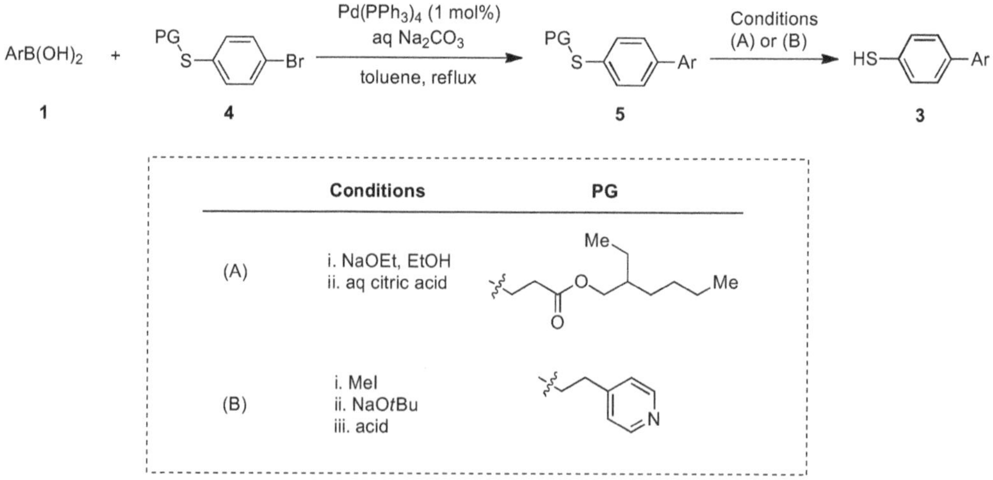

	Conditions	PG
(A)	i. NaOEt, EtOH ii. aq citric acid	(structure)
(B)	i. MeI ii. NaOtBu iii. acid	(structure)

SCHEME 6.1

(d) It was observed[6] that the catalyst loading in a two-phase system such as toluene/water is lower than in other solvent systems such as DME/water. It was demonstrated that the Suzuki cross-coupling reaction of **8** and **9** in toluene/water in conjunction with a phase-transfer catalyst (Bu$_4$NBr) enabled a significant reduction in the catalyst loading (to 0.8 mol% from 3.8 mol%) (Equation 6.3).[7]

(6.3)

(A) PdCl$_2$(dppf)$_2$·CH$_2$Cl$_2$ (3.8 mol%) Cs$_2$CO$_3$, DME/H$_2$O	(B) PdCl$_2$(dppf)$_2$·CH$_2$Cl$_2$ (0.8 mol%) Bu$_4$NBr, Cs$_2$CO$_3$, toluene/H$_2$O

(e) It was observed[8] that the order of addition was critical to the success of the following Suzuki reaction (Equation 6.4):

(6.4)

Procedure

- Degas a mixture of diethoxymethane (80 mL) and THF (20 mL).
- Charge 262 mg (1.0 mmol) of Ph$_3$P.
- Charge 56 mg (0.25 mmol) of Pd(Oac)$_2$.
- Stir the mixture at room temperature for 30 min.
- Charge 14.2 g (26.3 mmol) of **12** and stir for 30 min.
- Add 1.1 mL (61 mmol) of H$_2$O and stir at room temperature for 30 min.
- Charge 8.6 g (62.2 mmol) of powdered K$_2$CO$_3$ followed by adding 8.97 g (25.0 mmol) of **11**.
- Degas the mixture three times.
- Heat at reflux for 3–6 h.
- Cool to <76 °C.
- Add THF (25 mL) and H$_2$O (30 mL).
- Stir the mixture at 55–60 °C.
- Wash the isolated organic layer with H$_2$O (15 mL) at 55–60 °C.
- Yield: 15.5 g (93% after removal of Pd with Bu$_3$P).

(f) Palladium on carbon (Pd/C) was successfully employed as a catalyst in the Suzuki coupling reaction in the absence of organophosphorus ligands. Pd/C is inexpensive and readily available and can be readily removed by Celite filtration at the end of the reaction.

In addition, the Pd/C-catalyzed ligand-free Suzuki reaction proceeds smoothly without degassing the solvents. These robust reaction conditions lent itself readily to the preparation of some cross-coupling products (Equations 6.5 and 6.6).[9]

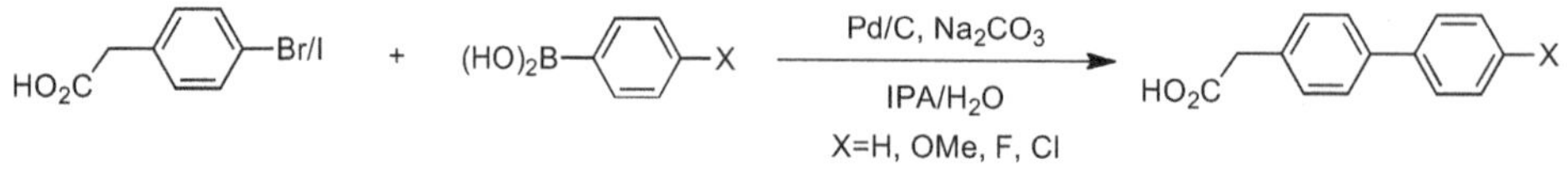

(6.5)

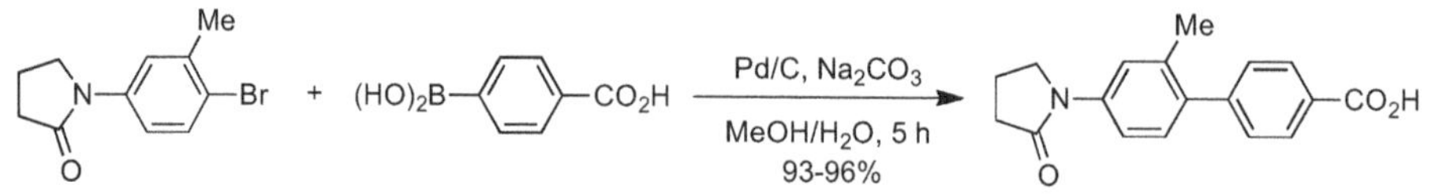

(6.6)

(g) A report[10] described a solvent effect on such heterogeneous catalysis. The Suzuki cross-coupling of aryl bromide 14 with boronic acid 15 worked well in alcohols with more than two carbons, while the reaction in DMF failed to give the desired product 16 (Equation 6.7).

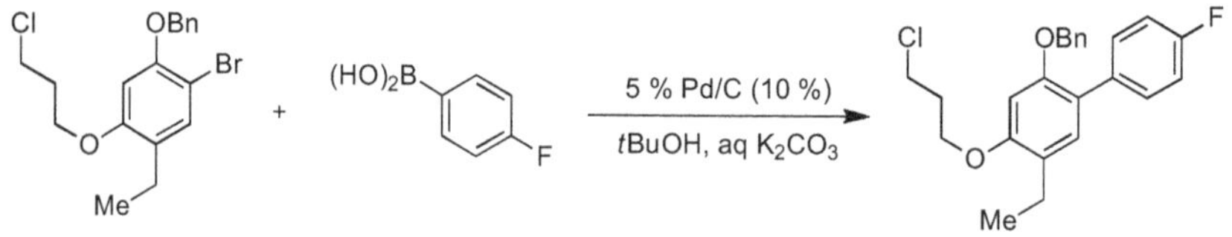

(6.7)

6.1.1.2 Catalyst Poisoning by Unknown Impurities

2-Aryl substituted indole 19 was accessed by yjr Suzuki–Miyaura reaction of 2-bromo indole 17 with boronic acid 18 (Equation 6.8).[11]

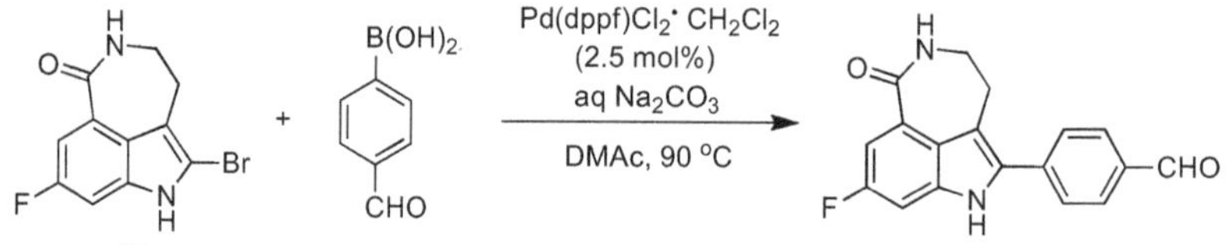

(6.8)

However, the cross-coupling with 2.5 mol% of palladium catalyst at 90 °C overnight led to only 28% conversion.

Solutions

It was suggested that a low-level unknown impurity in the boronic acid 18 deactivated the catalyst. Therefore, dosing of 18 should limit the amount of poison that was present in the initial stage of the reaction, allowing the boronic acid to react as it was added, thereby, ensuring the completion of the reaction prior to catalyst deactivation.

Procedure

- Charge 0.539 kg (0.66 mol, 2.5 mol%) of Pd(dppf)Cl$_2$·CH$_2$Cl$_2$ to a mixture of 7.41 kg (26.2 mol) of 17 in degassed DMAc (119 L).
- Stir at 20 °C for 1 h.

- Heat the mixture to 95 °C and stir for 1 h.
- Charge, in a separate vessel, 4.71 kg (31.4 mol, 1.2 equiv) of **18** followed by adding DMAc (30 L) and stirring to form a solution prior to adding an aqueous solution of Na_2CO_3 (5.6 kg, 52.8 mol, 2.0 equiv), and stir at 20 °C for 3 h.
- Charge the solution of **18**/Na_2CO_3 into the solution of **17**/Pd(dppf)Cl$_2$·CH$_2$Cl$_2$ at >90 °C.
- Stir the batch at 90 °C for 2 h.
- Cool the batch to room temperature and stir for 4 h.
- Add water (74 L) to give crude **19**.
- Charge the crude **19** into MeOH (59 L).
- Stir the mixture at 60 °C for 1 h.
- Cool to 20 °C and stir for 1 h.
- Filter and wash with MeOH (15 L).
- Dry under vacuum.
- Yield: 7.44 kg (92%).

6.1.2 CATALYST PRECIPITATION

The performance of Suzuki cross-coupling is also dependent on the stability of the catalytic system and an unstable catalytic system may require high catalyst loading in order to achieve good conversion or the reaction will stall with a poor conversion. For instance, the Suzuki coupling reaction between iodide **20** and 4-chlorophenylboronic acid catalyzed with Pd(PPh$_3$)$_4$ stalled after ca. 30 min (Equation 6.9).[12]

$$(6.9)$$

Oxidation of triphenylphosphine generated a ligand deficient Pd(0) species. Initially, the reaction, catalyzed by the active ligand deficient Pd(0), proceeded fast until no ligand is available to stabilize the palladium, leading to the observed initial fast reaction and subsequent stalling.

Solutions

Reaction optimization led to replacement of Pd(PPh$_3$)$_4$/NaOH with Pd(dppf)Cl$_2$·CH$_2$Cl$_2$/K$_2$CO$_3$ (dppf = bis(diphenylphosphino)ferrocene) in 20% aqueous dimethoxyethane (DME). As a result of this optimization, the catalyst load could be reduced to 0.5 mol% (as compared to 5 mol%).

REMARK

One of the common problems in palladium-catalyzed cross-coupling reactions is the instability of the Pd catalyst, leading to the precipitation of significant amounts of insoluble Pd black. PEPPSI is an air- and moisture-stable precatalysts and is useful for industrially intermolecular cross-coupling reactions, aminations, and Heck transformations (Figure 6.1).[13]

6.1.3 INSTABILITY OF ARYLBORONIC ACIDS

Many arylboronic acids are prone to protodeboronation[14] during the coupling process. The protodeboronation becomes a concern particularly for heteroarylboronic acids bearing the boron atom *ortho* to the heteroatom. Electron-deficient aromatic boronic acids **22** and 2-heteroary boronic acids

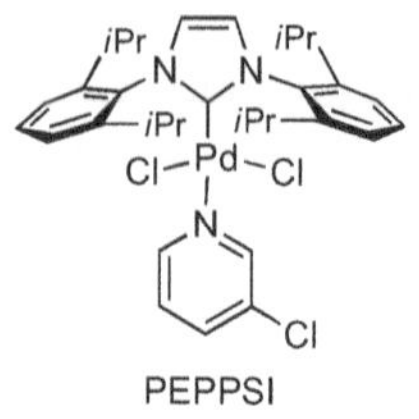

FIGURE 6.1   The structure of PEPPSI.

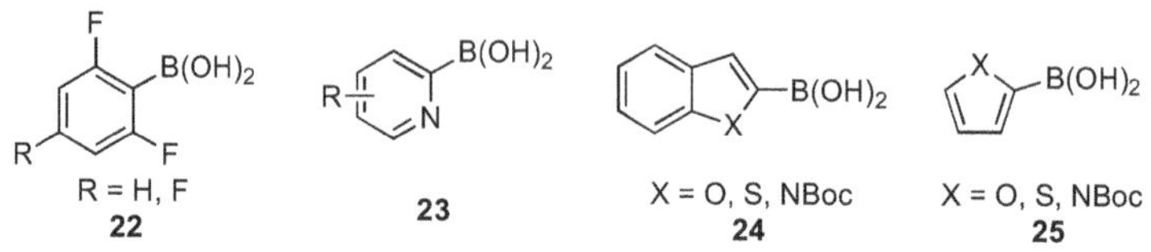

FIGURE 6.2   Unstable arylboronic acids.

23–25 (Figure 6.2) are unstable under basic conditions due to the sensitivity of the C–B bond to protodeboronation. Thereby, Suzuki–Miyaura coupling reactions of these boronic acids often suffer from low product yields.

One viable solution to this problem is using N-methyliminodiacetic acid (MIDA) boronates, boronic esters, cyclic triolborates,[15] or trifluoroborate salts[16] as the cross-coupling partners. Under the reaction conditions, these reagents are hydrolyzed[16a] to the corresponding free boronic acids. However, the apparent disadvantage of using masked boronic acids is the lack of atom- and step-economy.

6.1.3.1   Buchwald's Precatalyst

Buchwald and coworkers at the Massachusetts Institute of Technology (MIT) developed a mild cross-coupling system (Equation 6.10)[17] allowing direct use of such unstable arylboronic acids as the cross-coupling partners.

$$ArB(OH)_2 \; + \; Ar'X \quad \xrightarrow[\text{THF. } K_3PO_4,\ rt]{\text{Cat. (2 mol\%)}} \quad Ar\text{-}Ar' \tag{6.10}$$

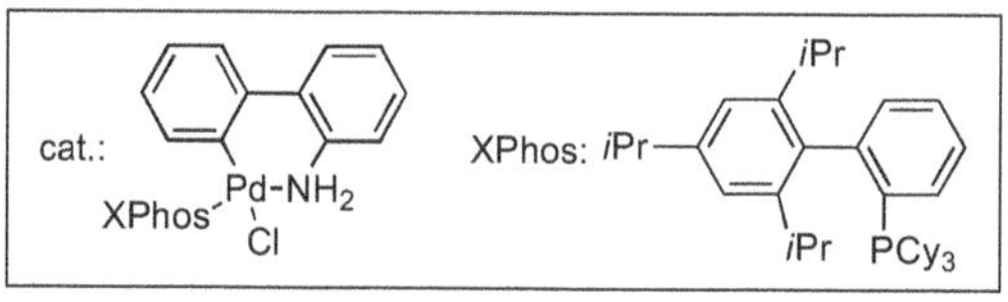

Using the 2-aminobiphenyl/XPhos-based palladium precatalyst, the catalytically active XPhosPd(0) species is formed rapidly at room temperature, thereby, allowing the successful coupling of these unstable boronic acids.

6.1.3.2   Tridentate Ligand

6.1.3.2.1   N-Phenyldiethanolamine

The preparation and application of 2-pyridyl boranes is challenging due mainly to the sensitivity of the 2-pyridyl–boron bond to protodeborylation.[15a] It was found that the stability of organoboron compounds can be improved by introducing tridentate ligands to form tetra-coordinated ate complex, such as N-phenyldiethanolamine-2-pyridylboronate **28**. Compound **28** could be conveniently prepared via a one-pot process in which 2-lithiopyridine is formed by a reaction of 2-bromopyridine with *n*-butyllithium at −70 °C, followed by trapping in situ with triisopropyl borate. The intermediate boronate **27** reacted with *N*-phenyldiethanolamine to give the product boronate **28** in approximately 75% yield on a 4 molar scale (Scheme 6.2).[18]

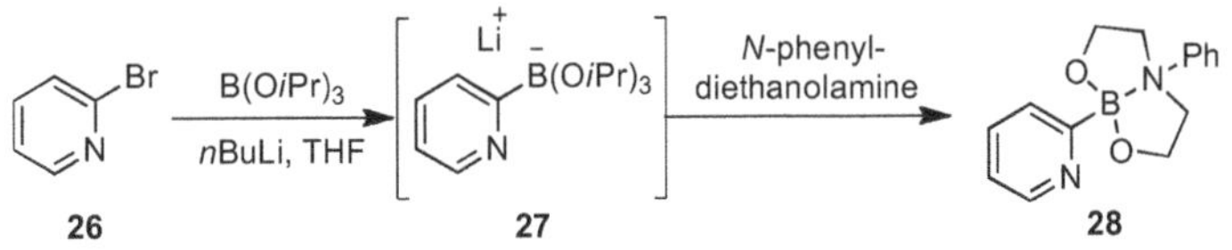

SCHEME 6.2

Procedure

- Charge 4.0 L (6.40 mol, 1.2 equiv) of 1.6 M solution *n*-butyllithium in hexanes into a stirred solution of 2-bromopyridine **26** (843 g, 5.33 mol) and triisopropylborate (1.2 kg, 6.40 mol, 1.2 equiv) in THF (6.74 L) at −67 °C to −75 °C
- Upon addition, warm the batch to room temperature and stir for 16 h.
- Charge a solution of *N*-phenyldiethanolamine (966 g, 5.33 mol, 1.0 equiv) in THF (966 mL).
- Heat to reflux and stir at reflux for 4 h.
- Distill to replace the solvent with IPA until the head temperature reached 76 °C (distilled out 11.3 L of solvent and 8.4 L of IPA added).
- Cool to room temperature and stir for 12 h.
- Filter and wash with IPA (1.7 L).
- Dry at 40 °C under vacuum.
- Yield: 1.605 kg (75%).

6.1.3.2.2 Diethanolamine

Delanzomib (CEP-18770, **30**) is a peptide boronic acid for the treatment of solid tumors. Delanzomib **30** is an amorphous solid that is both unstable, requiring storage at −20 °C, and very potent with an occupational exposure limit of 0.3 µg/m3. The synthesis of **30** involved transesterification of the pinane diol ester **29** with isobutyl boronic acid under acid catalysis conditions using a two-phase MeOH/heptane mixture to provide **30** (Scheme 6.3).[19] The crude product **30** was thus obtained with purities as low as 85% and 75% de required column chromatographic purification.

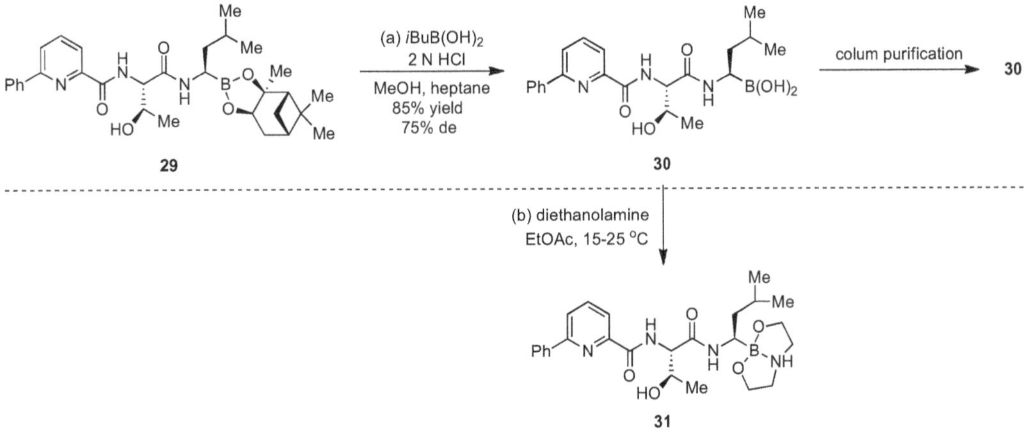

SCHEME 6.3

To address these stability and purification issues, diethanolamine (DEA) was used to convert the unstable boronic acid **30** to **31**, which avoided chromatography purification giving **31** in high yield and purity. The stable, crystalline diethanolamine adduct **31** can potentially be used as a prodrug.

Procedure

Stage (a)
- Mix 490 g (88.2% of purity, 0.79 mol) of **29** with MeOH (2.35 L), heptanes (4.35 L), (2-methylpropyl)boronic acid (283 g, 2.78 mol), a solution of 37% HCl (65 mL), and water (300 mL).
- Stir the resulting mixture at 20 °C for 20 h.
- Separate layers and wash the lower MeOH phase with heptanes (4.35 L).
- Distill at 38 °C/54 Torr.
- Add EtOAc (4.35 L) to the resulting aqueous slurry followed by 8% aqueous NaHCO$_3$ (3.5 L).
- Separate layers and wash the EtOAc layer with 8% aqueous NaHCO$_3$ (3.5 L) and brine (2.5 L).
- Yield of **30**: 306.6 g (0.744 mol, solution yield of 94.2%).

Stage (b)
- Transfer the organic layer into a clean reactor.
- Charge 78.7 g (0.75 mol, 1.0 equiv) of DEA into the reactor.
- Stir the resulting slurry at 20 °C for 12–18 h.
- Filter in a sealed filter and wash with EtOAc (1.3 L).
- Dry at 50 °C/50 Torr.
- Crude yield of **31**: 289 g (88% yield in 97.1 A% purity).
- Recrystallize the crude product from absolute EtOH (7 vol).
- Yield of **31**: 243 g (84.2% yield in 99.8 A% purity and >99.8% de).

6.1.3.2.3 N-Methyliminodiacetic Acid

In addition to the improved stability, the formation of the tetra-coodinated ate complexes enhanced the nucleophilicity of the arylboronic acids as well. For instance, 2-pyridyl *N*-methyliminodiacetic acid (MIDA) boronate and its derivatives are air-stable and can be isolated in chemically pure forms.[20] Applications of the air-stable 2-pyridyl MIDA boronates toward cross-coupling reactions were described (Equation 6.11).[21]

$$\text{(6.11)}$$

Generally, the reaction was carried out by mixing 1.0 equiv aryl halide with 1.5 equiv MIDA boronate in the presence of XphosPd cycle (5 mol%), Cu(OAc)$_2$ (50 mol%), DEA (1.0 equiv), and K$_3$PO$_4$ (5.0 equiv) in DMF at 100 °C.

REMARKS

(a) The MIDA group is sensitive to strong bases such as LiHMDS and hydrolysis of the MIDA group was observed under such strong basic conditions.[22]

(b) Lithium triisopropyl 2-pyridylborates, prepared readily from the corresponding 2-bromo- or 2-iodopyridine derivatives, could react with aryl halide under palladium-catalyzed Suzuki–Miyaura reaction conditions with dialkyl- or diarylphosphine oxide ligand to afford the diaryl products in excellent yields.[23]

6.1.3.3 Alternative Negishi Coupling Reaction

The sensitivity of the aryl–boron bond to the protodeboronation poses problems for other hetero-aromic boronic acids. For instance, the cross-coupling between 5-(2-chloro)pyrimidyl boronic acid **34** and the 2-bromopyridine **35** under standard Suzuki conditions did not afford the expected product **36** (Equation 6.12).[24] The protodeborylated compound **37** was the major undesired product, besides other degradation products.

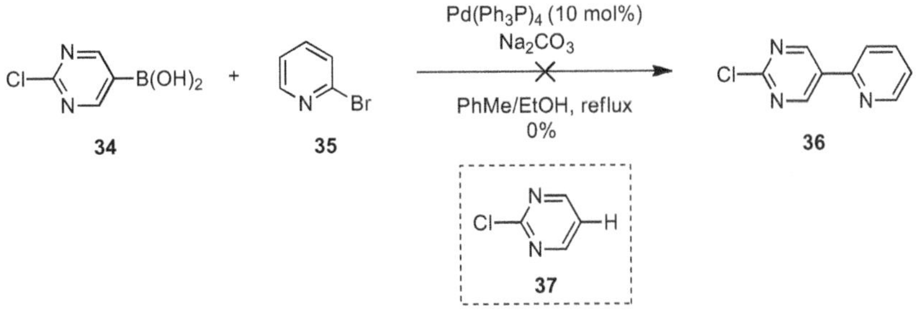

$$(6.12)$$

An alternative Negishi cross-coupling between 5-iodo-2-chloropyrimidine **38** and the in situ prepared 2-pyridylzinc chloride **39** was developed to overcome this issue (Equation 6.13).[24]

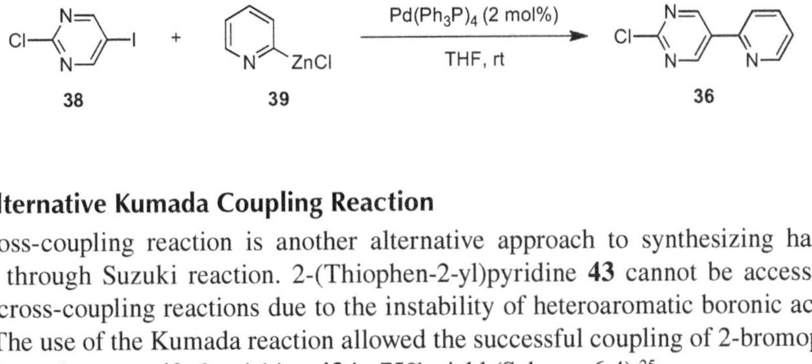

$$(6.13)$$

6.1.3.4 Alternative Kumada Coupling Reaction

Kumada cross-coupling reaction is another alternative approach to synthesizing hard-to-make compounds through Suzuki reaction. 2-(Thiophen-2-yl)pyridine **43** cannot be accessed through the Suzuki cross-coupling reactions due to the instability of heteroaromatic boronic acids such as **40** and **41**. The use of the Kumada reaction allowed the successful coupling of 2-bromopyridine **35** with the Grignard reagent **42**, furnishing **43** in 75% yield (Scheme 6.4).[25]

Kumada reaction is usually carried out by adding a small aliquot of Grignard solution in order to form active nickel catalytic species, followed by the addition of the remainder Grignard reagent.

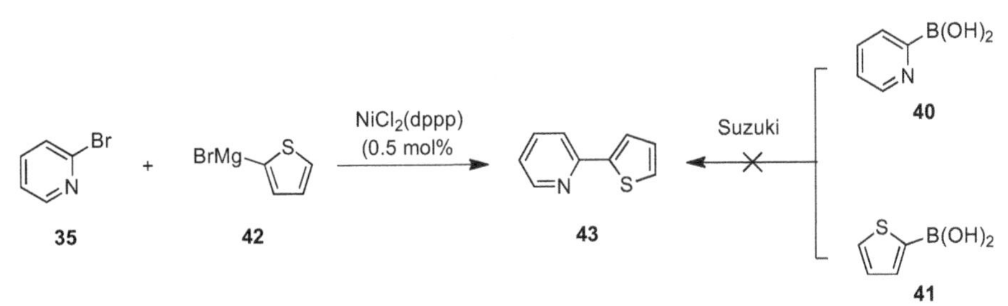

SCHEME 6.4

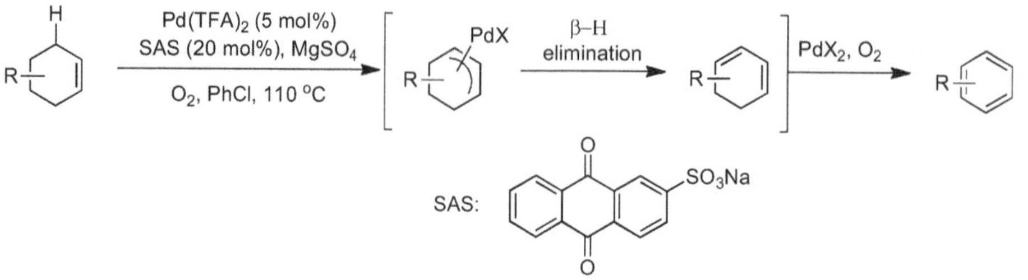

SCHEME 6.5

REMARKS

(a) β-Hydride elimination may occur in the case of nickel or palladium-catalyzed couplings of aryl halides with alkyl Grignard reagents. Strategies to suppress this side reaction typically involve the use of bulky alkyl or bidentate ligands to produce a metal–alkyl intermediate that cannot obtain the requisite geometry for the β-hydride elimination.

(b) Interestingly, the β-hydride elimination was utilized to synthesize substituted arene derivatives (Scheme 6.5).[26a]

This novel approach was catalyzed by Pd(II) using O_2 as an oxidant and sodium anthraquinone-2-sulfonate (SAS) as a cocatalyst. The use of SAS avoided the disproportionation of cyclohexene into cyclohexane and benzene.

6.1.3.5 Using Protecting Group

Similarly, pyrazole boronic acids or esters, having short shelf-lives, are not ideal Suzuki coupling partners due to rapid competing protodeboronations. Therefore, using a THP-protected pyrazole boronic ester **45** solved the protodeboronation issues allowing the Suzuki coupling to proceed smoothly (Scheme 6.6).[27]

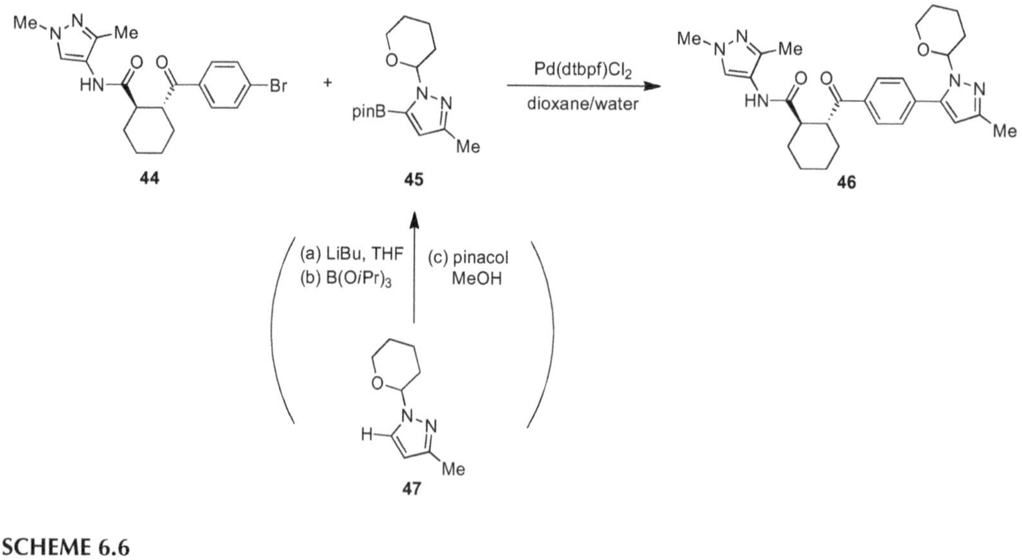

SCHEME 6.6

REMARK

Because of the directing effect of the THP group, the THP-protected pyrazole boronate **45** could be prepared conveniently by direct lithiation with butyllithium followed by reactions with B(OiPr)$_3$ and pinacol in a one-pot fashion.

6.1.3.6 Using Trifluoroborate Salt

To alleviate the protodeboronation, a trifluoroborate salt of **49**, formed in situ, was employed in the Suzuki reaction with aryl chloride **48** (Equation 6.14).[28]

$$(6.14)$$

The reaction, however, has to be conducted in Hastelloy vessels because KHF_2 etches glass-lined reactors severely.

Procedure

- Add a solution of $Pd(OAc)_2$ (58.1 g, 259 mmol) and PCy_3 (143 g, 510 mmol) in degassed IPA (32 L) to a degassed mixture of **48** (6.40 kg, 12.8 mol), **49** (3.42 kg, 14.1 mol), K_3PO_4 (3.26 kg, 15.4 mol), and KHF_2 (2.20 kg, 28.2 mol) in IPA (32 L)/water (32 L) at 80 °C.
- Stir the batch at 80 °C for 17 h.
- Cool to 20 °C and age for 1 h.
- Filter and wash with a mixture of IPA/water (1:1, 2×12 L) and then water (2×13 L).
- Dry at 60 °C/vacuo.
- Yield of **50**: 6.12 kg (82% yield) as a tan solid.

6.1.4 PROBLEMS ASSOCIATED WITH BASE

It has been noticed that the base plays an important role in Suzuki cross-coupling reactions, presumably facilitating the otherwise slow transmetalation of the boronic acids. In general, the inorganic base works better than the organic base, and the reactivity is in the following order: $Cs_2CO_3 > CsF > K_2CO_3 > KOAc > Na_2CO_3$.[29]

6.1.4.1 Protodeboronation

Deboronation followed by protonation occurs frequently in Suzuki coupling reactions, resulting in protodeboronation side products. This side reaction is both acid- and base-catalyzed; it may also be catalyzed by metals. Thus, in laboratory-scale reactions, excess boronic acid is often used in order to drive the reaction to completion. However, it can be an important issue in production scale as it will not only increase production costs, but also lead to downstream purification problems. Therefore, it is necessary to suppress this side reaction prior to scale-up.

The Suzuki cross-coupling of aryl boronic acid **51** with 6-chlorophthalazin-1-ol **52** gave 50% of *p*-toluic acid **54** as a protodeboronation side product when using inorganic base Cs_2CO_3 (Scheme 6.7).[30]

It was found that the use of an organic base, such as dicyclohexylamine, suppressed considerably the protodeboronation reaction with much lower levels of *p*-toluic acid **54** (0.5–10%).

The protodeboronation was also observed for the Suzuki coupling reaction of *O*-alkylated chlorophthalazine **55** with the arylboronic acid **51** under $Pd_2(dba)_3$/S-Phos/K_3PO_4 conditions. Similarly, the reaction using dicyclohexylamine as the base was devoid of the protodeboronation issue. Under these reaction conditions (Equation 6.15),[31] a complete conversion could be obtained within 4 h at 80 °C with minimal protodeboronation (<1%) and palladium black formation.

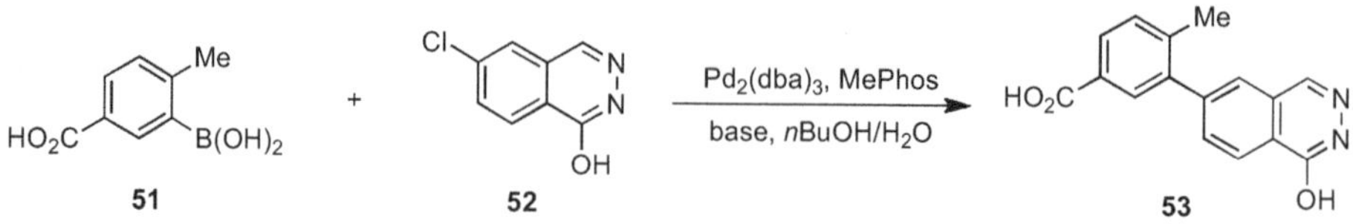

SCHEME 6.7

Base	**54**, mol %
Cs$_2$CO$_3$	50
dicyclohexylamine	0.5–10

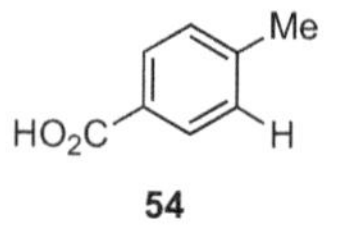

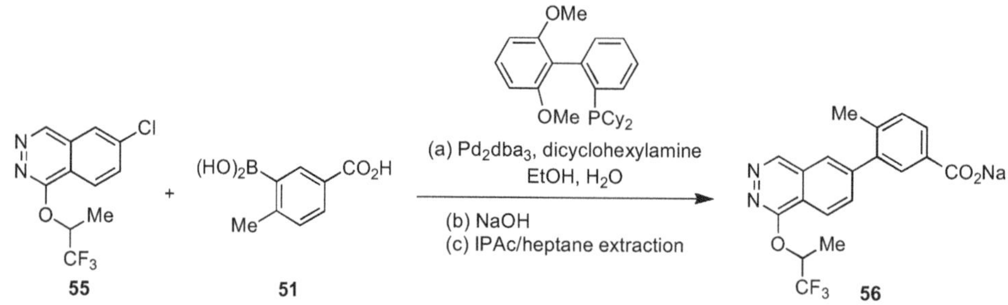

$$(6.15)$$

Product isolation was accomplished by the formation of a solution of sodium carboxylate using aqueous NaOH and washing with IPAc/heptane to remove dicyclohexylamine.

Notes

(a) Inorganic bases[30a] such as Na$_2$CO$_3$, K$_2$CO$_3$, KHCO$_3$, K$_3$PO$_4$, and NaOH led to substantial protodeboronation and deposition of palladium black on the reactor walls prior to achieving complete conversion.

(b) Using non-polar solvents such as toluene and a high ligand-to-palladium ratio can mitigate protodeboronation.[32]

6.1.4.2 Formation of Carbamate

Besides protodeboronation side products, the use of an inorganic base such as potassium carbonate can cause the formation of carbamate impurities. For instance, a carbamate side product **60** was detected at a level of up to 3% (based on HPLC area%) when using K$_2$CO$_3$ as the base for the Suzuki coupling of aryl iodide **58** (Equation 6.16).[33]

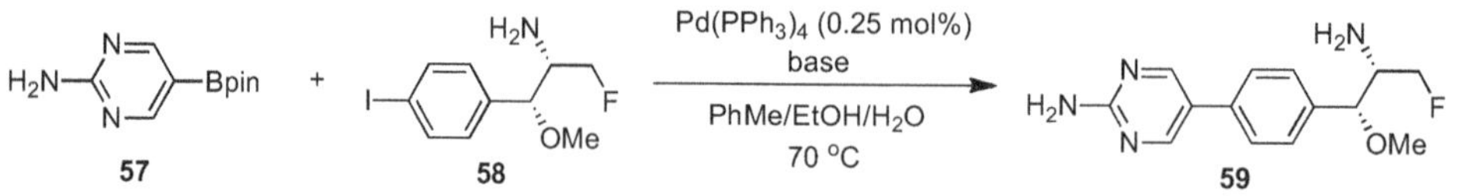

$$(6.16)$$

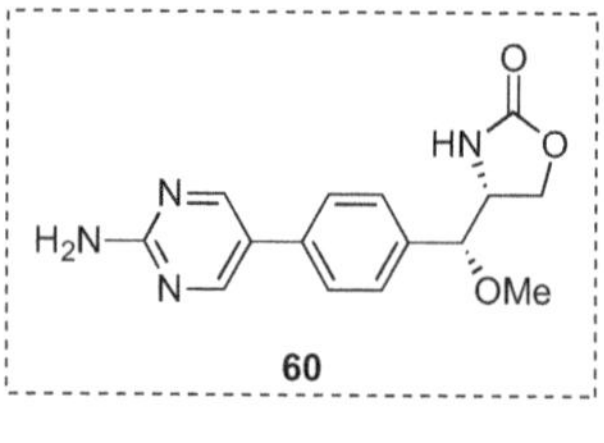

Solutions

This impurity **60** is presumably produced from the reaction of the carbonate (or liberated CO_2) with **59**, followed by a ring closure. Switching the base to K_3PO_4 eliminated this side reaction entirely.

6.1.5 DIMER IMPURITY

Suzuki cross-coupling reaction may accompany the homocoupling of arylboronic acids leading to symmetrical biaryls as side products. It was proposed that the formation of the homocoupling dimer in the Suzuki cross-coupling reaction is catalyzed by palladium(II) species (Equation 6.17).[34]

$$2\ ArB(OH)_2 \xrightarrow{\text{Pd(II), base}} Ar-Ar \tag{6.17}$$

Thus, two approaches have been developed to reduce levels of homocoupling dimeric impurities: to reduce the concentration of arylboronic acid in the reaction mixture and to minimize the concentration of free Pd(II).

6.1.5.1 Reducing Arylboronic Acid Concentration

The Suzuki coupling of boronic acid **61** with **62** was conducted by mixing the reagents (1:1 **61/62**, 2.5 mol% $Pd(OAc)_2$, 7.5 mol% PPh_3, Hünig's base) in toluene/water at room temperature and then heating the mixture to 70 °C (Equation 6.18).[35]

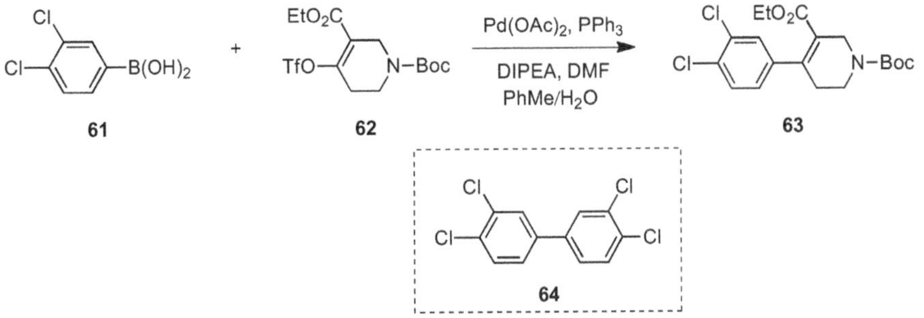

Under reaction conditions, a dimer impurity **64** was observed in the isolated product **63** at levels of 1–5%.

Solutions

In order to suppress the Pd(II)-catalyzed homocoupling, the process was optimized as follows:

*   The active Pd(0) catalyst was generated in situ by mixing 1 equiv of vinyl triflate **62** with 2.5 mol% of $Pd(OAc)_2$, 7.5 mol% of Ph_3P, Hünig's base, and water in toluene at room temperature; then heating to 70 °C to generate Pd(0).
*   The boronic acid **61** (0.95 equiv) as DMF–toluene solution was then added over a period of 1 h.
*   Under these conditions the dimer **64** was reduced to levels of <30 ppm.

REMARK

Alternatively, the addition of a mixture of starting materials into a catalytic system under reflux was employed to suppress the dimer formation.[36]

6.1.5.2 Reducing Free Pd(II) Concentration

It was observed that minimizing boronic acid concentration will not efficiently control levels of dimeric impurity **68** during the synthesis of LY 451395 (**67**) (Equation 6.19).[37] Compound **67** is one of a series of AMPA potentiators of interest in the treatment of cognitive deficits associated with Alzheimer's disease. Although the standard Suzuki coupling conditions produced the desired product **67** in a good yield, the dimer **68** was observed at an unacceptable level (2.6%).

(6.19)

Consequently, a protocol, Pd black (1.0 mol%)/HCO$_2$K (0.46 equiv)/K$_2$CO$_3$, was developed to suppress the formation of the homocoupling of aryl boronic acid **66**.

- For reduction in levels of the dimer **68**, potassium formate was used to minimize the concentration of free Pd(II).
- The exclusion of dissolved oxygen from the reaction mixture could be easily and efficiently achieved by subsurface sparging nitrogen.

Procedure

- Dissolve potassium formate (52.6 g, 0.62 mol), potassium carbonate (303 g, 2.19 mol), aryl iodide **65** (501 g, 1.36 mol), and boronic acid **66** (325 g, 1.34 mol) in a mixture of 2.8 L of deionized water and 3.0 L of 1-propanol.
- Stir at 24 °C to form a clear solution.
- Sparge nitrogen in subsurface for 30 min.
- Charge a suspension of palladium black (1.45 g, 0.0136 mol) in water.
- Continue sparging nitrogen for an additional 15 min.
- Heat the batch to reflux at 90 °C for 8 h.
- Cool to 80 °C and filter.
- Dilute the filtrate with 0.65 L of deionized water.
- Cool to 4 °C.
- Filter the product.
- Wash with cold 1-propanol (1.3 L) and then deionized water (1.3 L).
- Dry at 45 °C.
- Yield: 542 g (92.2%).

DISCUSSIONS

(a) It was found that refluxing the solvent under nitrogen or distilling 1-2 volumes of solvent are superior to degassing by sparging with nitrogen gas.[38]
(b) The use of palladium black as a catalyst in Suzuki cross-coupling reactions showed additional benefits such as low levels of residual Pd in the isolated product and relatively low costs as no ligand was required (Equation 6.20).[39]

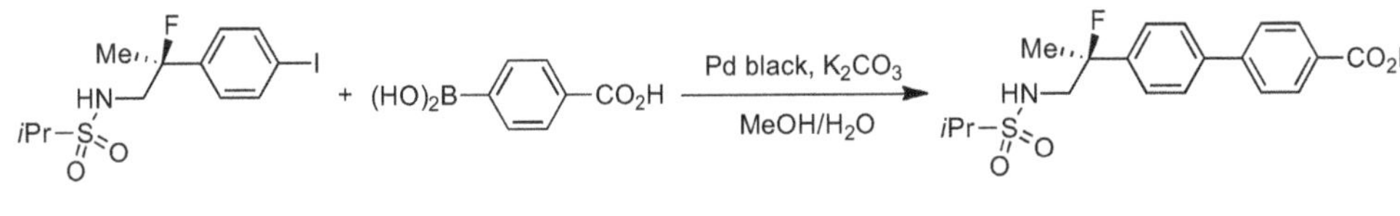

(6.20)

6.2 NEGISHI REACTION

6.2.1 POOR PRODUCT YIELD

The Negishi reaction is well described in the literature for cross-coupling of aromatic heterocycles, including substituted pyridines. For instance, the Negishi reaction was used in the final synthetic step for benzoxadiazole **72**,[40] a potent, selective, and orally active inhibitor for phosphodiesterase. Initially, a sequential addition protocol was employed by adding bromide **71** to a solution of organozinc intermediate **70** followed by Pd(Ph$_3$P)$_4$ catalyst (Scheme 6.8).

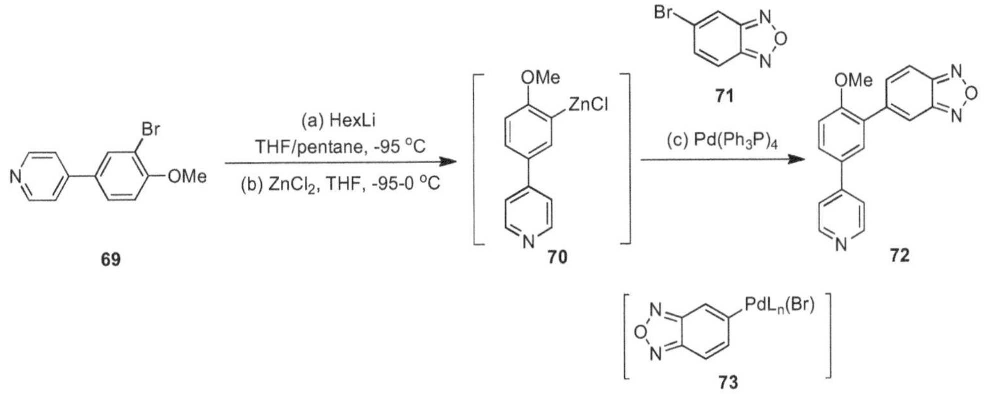

SCHEME 6.8

(I) Problems

- The yield of the product **72** was poor (38%).
- Due to the strong complexation of **72** with Pd(0), the isolated product **72** was contaminated with high levels (300–800 ppm) of Pd.

(II) Solutions

In addition to high levels of Pd, the complexation of **72** with Pd(0) may account for the slow reaction rate and low product yields. To obviate these issues, the reaction was carried out by adding a solution of complex **73**, performed from the bromide **71** and Pd(Ph$_3$P)$_4$, to a solution of arylzinc **70**. The optimized protocol allowed the reduction of Pd(Ph$_3$P)$_4$ from 8% to 0.8%, giving **72** in 79% yield. The residual palladium in **72** was reduced by the formation of hemi-maleate salt.

6.2.2 THICK REACTION MIXTURE

Transmetalation of aromatic Grignard reagent (ArMgX) with ZnCl$_2$ is frequently employed to prepare the corresponding organic zinc compound (ArZnX) for Negishi cross-coupling reactions. However, the arylzinc halide reagent, prepared from the transmetalation, is usually in a thick suspension, which is difficult to stir or transfer. To overcome these issues, a practical approach was developed by employing ZnCl$_2$ in catalytic quantities (Scheme 6.9).[41] Accordingly, the Negishi

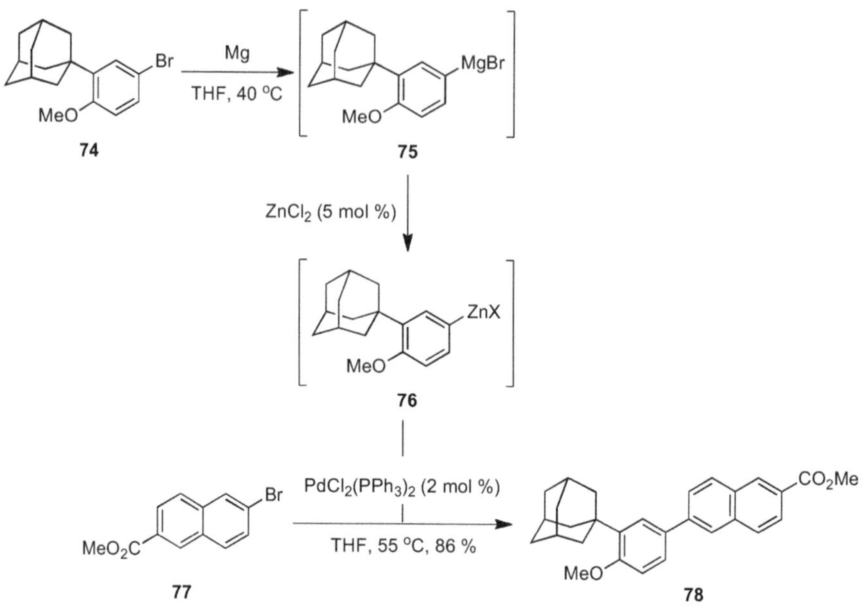

SCHEME 6.9

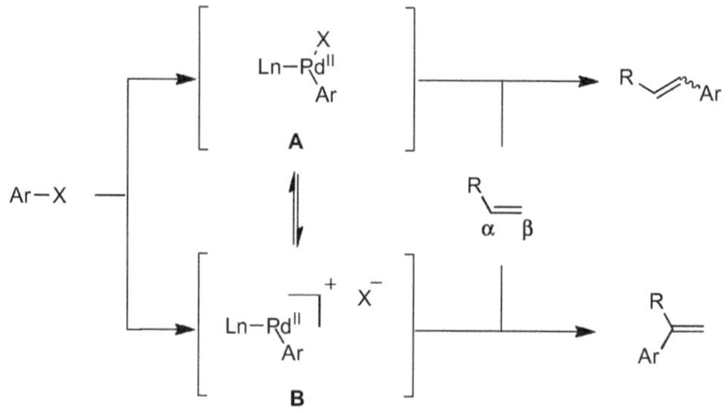

SCHEME 6.10

cross-coupling reaction between aryl bromide **77** and in situ generated aryl zinc **76** was carried out by adding the Grignard reagent **75** to a mixture solution of **77**, PdCl$_2$(PPh$_3$)$_2$ (2 mol%), and catalytic amounts of ZnCl$_2$ (5 mol%) in THF at 55 °C.

6.3   HECK REACTION

The palladium-catalyzed Heck–Mizoroki reaction has been well-known as one of the most versatile methods for the formation of C–C bonds in synthetic chemistry. In general, the catalytic cycle involves oxidative addition, migratory insertion, and β-hydride elimination to give olefin products along with HPd(II)X complex. The final reductive elimination of HPd(II)X regenerates the Pd(0) catalyst. Scheme 6.10[42] describes two reaction pathways involving ArPd(II)(L$_n$)X complex (**A**) and [ArPd(II)(L$_n$)]X complex (**B**) from the oxidative addition step. Complexes **A** and **B** may exist as an equilibrium mixture in solution and the presence of halide scavengers would facilitate the dissociation of anion halide X. Migratory insertions of the equilibrium mixture of **A** and **B** with alkene would lead to the formation of a mixture of linear and branched olefins, in which **A** is presumably responsible for the formation of the linear olefin (β-arylation) while the formation of branched olefin is the result of α-arylation of an alkene with ionic complex **B**.

6.3.1 ENHANCING PALLADIUM-CATALYST STABILITY

Heck cross-coupling reaction between aryl bromide **79** and allyl alcohol **80** occurred through intermediate **A** (Scheme 6.10) providing ketone **81** following enol tautomerization (Equation 6.21).[43]

$$(6.21)$$

Although the reaction proceeded with simple Pd(OAc)$_2$/base conditions, the catalyst was much more stable in the presence of added chloride ligands. Ultimately, the use of LiCl (2.0 equiv) in combination with Pd(OAc)$_2$ and LiOAc (0.5 equiv) made the reaction more robust.

6.3.2 IMPROVING SELECTIVITY

Heck cross-coupling reaction of *t*-butyl vinyl ether **83** with 2-bromonaphthalene **82** was plagued by poor chemoselectivity giving a mixture of branched and linear olefins (**84** and **85**) (Equation 6.22).[42]

$$(6.22)$$

Solvent screening found that the reaction in DMSO instead of other dipolar aprotic solvents such as DMAc and DMF produced exclusively the branched olefin **84** in high yield. DMSO has high polarity and good coordination capability to Pd(II). It was hypothesized that the use of DMSO as the reaction solvent favored the ionic intermediate **B** (Scheme 6.10), thus, improving the α-arylation selectivity.

6.4 SONOGASHIRA REACTION

Sonogashira reaction is a powerful tool for the installation of internal alkyne functional groups into a molecule. This coupling of terminal alkynes with aryl or vinyl halides is performed with a palladium catalyst, a copper(I) cocatalyst, and an amine base. Similar to the Suzuki–Miyaura reaction, the Sonogashira reaction consists of oxidative addition, transmetallation, and reductive elimination.

6.4.1 REDUCING PALLADIUM-CATALYST LOADING

Sonogashira reaction typically needs palladium-catalyst loading ranging from 0.1 mol% to 5 mol%. High loading of expensive transition-metal catalysts will impact directly on production costs and be a burden with the removal of residual palladium from the product as well. The Sonogashira coupling of 4-bromobenzonitrile **86** with phenylacetylene could achieve a high product yield (94%) with low catalyst loading (0.001 mol%) in the presence of hydrazone ligand (Equation 6.23).[44]

$$(6.23)$$

6.4.2 IMPROVING REACTIVITY

Although useful in forging C(sp)–C(sp^2) bonds, Sonogashira reaction is not compatible with electron-poor terminal alkynes or sterically hindered electron-rich aryl chlorides. In an effort to address these challenges, a deacetonative coupling approach was developed using aryl propargyl alcohols as the masked terminal alkynes. For instance, Sonogashira reaction of deactivated terminal alkyne **88a** with aryl chloride **89** gave less than 5% product **90**, while the masked terminal alkyne **88b** reacted with **89** smoothly providing **90** in 75% yield (Equation 6.24).[45]

$$(6.24)$$

In addition, a masked terminal alkyne **91** could react with bulky 2,6-dimethylphenyl chloride **92** readily under the palladacycle/Xphos copper-free catalysis conditions producing diaryl acetylene **93** in excellent yield (99%) (Equation 6.25).

$$(6.25)$$

Table 6.1 summarizes various transition metal-catalyzed cross-coupling reactions.

6.5 CATALYTIC DEPROTECTION

Benzyl and benzyl-derived protecting groups are widely utilized to protect amino and hydroxyl groups in organic synthesis.[46] Removal of such protecting groups is frequently achieved under catalytic hydrogenolysis conditions because of good selectivity and clean reaction profiles.

6.5.1 DEBENZYLATION

6.5.1.1 Catalyst Poisoning

Catalyst poisoning may occur during the catalytic deprotection of amino groups due to the strong chelating effect of the amine product.[47] Debenzylation of 3,3-dimethyl-1-(phenylmethyl)piperidine

TABLE 6.1
Comparison of Various Transition Metal-Catalyzed Cross-Coupling Reactions

Reaction	Starting materials	Reaction conditions	Comments
Heck[a]	Unsaturated halides/or triflates and alkenes are the starting materials.	Heck reaction needs Pd catalyst, ligand, a base. Ligand-free conditions are also possible.	This coupling reaction is stereoselective, producing branched or linear olefins with a propensity of forming *trans* isomer.
Suzuki[b]	aryl/or vinyl boronic acids or esters and aryl/or vinyl halide are the starting materials.	The reaction requires transition metal catalyst such as palladium and nickel, ligand, and a base. Usually, a mixture of organic solvent and water is used as the reaction media.	Aryl halides (ArX) and ArOTf have the following reactivity order: ArI>ArOTf>ArBr>>ArCl.
Negishi	Aryl halides and organozinc salts are the starting materials.	Negishi reaction needs transition metal catalyst and ligand. Base is not necessary.	Negishi reaction is inherently sensitive to moisture and air, and not tolerant with most of functional groups. Organic zinc reagent is usually prepared freshly under relatively low temperature.
Sonogashira	Terminal alkynes and aryl or vinyl halide are the starting materials.	The reaction typically needs two catalysts, Pd and Cu, ligand, and amine as base/solvent.	As alkynes are less stable and prone to dimerization at elevated reaction temperature, Sonogashira reaction is often carried out under mild conditions with more reactive aryl iodides as the coupling partners.
Stille	Aryl halides and organostannanes are the starting materials.	Transition metal catalyst and ligand are required. Base is not necessary.	Compared with trimethylstannyl compound, tributylstannyl compound is preferred because of the higher toxicity of the former. Stille reaction is rarely used on large scale due to the toxicity of organotin reagent.

[a] Reviews: (a) Heck, R. F. In *Comprehensive Organic Synthesis*; Trost, B. M., Ed.; Pergamon, New York, 1991; Vol. 4, Chapter 4.3. (b) Bräse, S.; deMeijere, A. In *Metal-Catalyzed Cross-Coupling Reactions*; Deiderich, F., Stang, P. J., Eds.; Wiley, New York, 1998; Chapter 3. (c) Cabri, W.; Candiani, I. *Acc. Chem. Res.* **1995**, *28*, 2. (d) deMeijere, A.; Meyer, F. E. *Angew. Chem. Int. Ed.* **1994**, *33*, 2379.

[b] (a) Suzuki, A. In *Metal-Catalyzed Cross-Coupling Reactions*; Deiderich, F., Stang, P. J., Eds.; Wiley, New York, 1998; Chapter 3. (b) Miyamura, N.; Suzuki, A. *Chem. Rev.* **1995**, *95*, 2457. (c) Suzuki reactions of aryl chlorides: (d) Littke, A. F.; Fu, G. C. *Angew. Chem. Int. Ed.* **1998**, *37*, 3387. (e) Wolfe, J. P.; Buchwald, S. L. *Angew. Chem. Int. Ed.* **1999**, *38*, 2413. (f) Littke, A. F.; Dai, C.; Fu, G. C. *J. Am. Chem.Soc.* **2000**, *122*, 4020.

94 had previously been accomplished using 1-chloroethyl chloroformate (ACE-Cl)[48] in 60% yield. This chemistry was considered undesirable for a pilot-plant campaign due to the moderate yield and environmental issues associated with handling the ACE-Cl reagent and its by-products, benzyl chloride and acetaldehyde (Equation 6.26).[49]

$$(6.26)$$

Catalytic debenzylation of **94** was, however, stalled due to the catalyst (Pd/C) poisoning by the amine product **95**. Thus, maintaining the reaction medium in acidic is a highly effective way of preventing catalyst poisoning. Practically, this issue was readily overcome by the addition of acetic acid (1.1 equiv) to the reaction mixture to protonate the product **95**.

REMARK

Palladium-catalyzed debenzylation of **96** suffered dehalogenation side reactions and the resulting dehalogenated side products were hard to remove (Condition (a), Equation 6.27).[50] Alternatively, ACE-Cl was employed in the debenzylation of **96** (Condition (b), Equation 6.27).

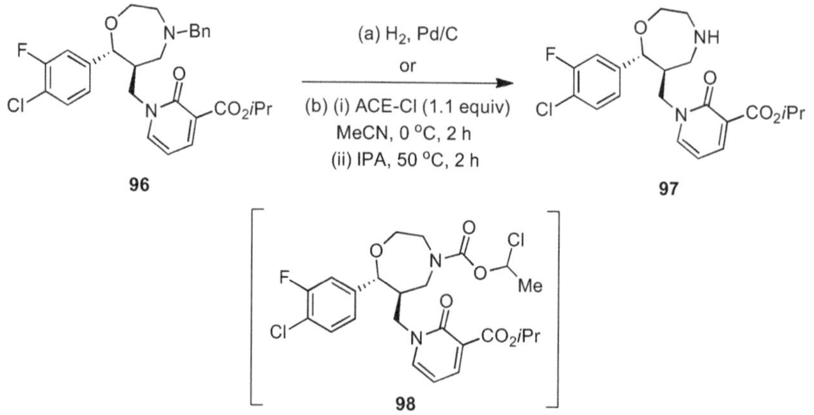

(6.27)

Following the reaction with ACE-Cl in MeCN at 0 °C, the resulting intermediate carbamate **98** was decomposed by the addition of 2-propanol with heating at 50 °C.

6.5.1.2 Erosion of Chiral Purity

The preparation of chiral amine (*R*)-**100** required a selective cleavage of the C–N bond to remove the auxiliary 1-phenylethyl group in **99** (Scheme 6.11).[51] The use of Pd(OH)$_2$/C catalyst in the hydrogenolysis of **99** (97% ee) provided the corresponding debenzylation product (*R*)-**100** with only ca. 80% ee.

Solutions

Plausible hydrogenolysis pathways are illustrated in Scheme 6.11 involving a key oxidative addition palladium complex **101**. Further reaction of **101** with H$_2$ (route A) would lead to the desired product

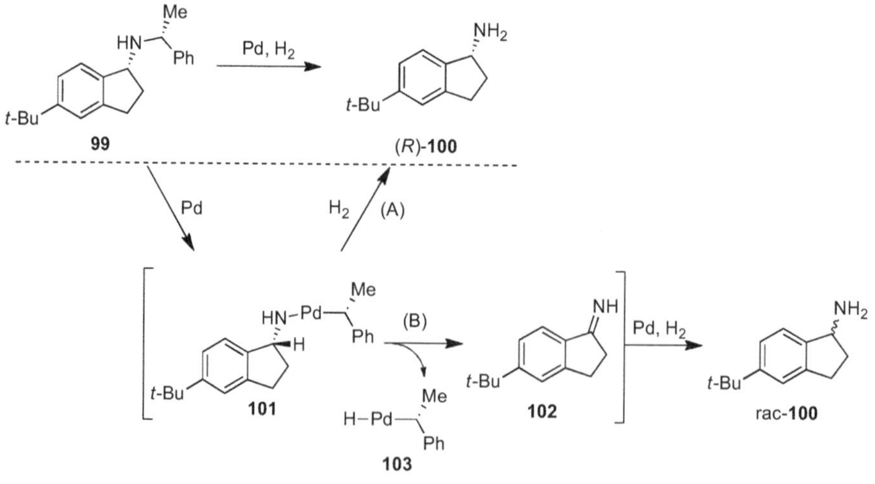

SCHEME 6.11

(R)-**100**. The intermediate **101** may undergo β-hydride elimination as well to form imine **102** (route B) whose reduction would provide racemic rac-**100**. Therefore, the partial racemization through pathway B deteriorates the amine chiral purity. Ultimately, the utilization of less active catalyst Pd/C and the interruption of the reaction at ca. 96% conversion helped to maintain the chiral purity of (R)-**100** in >90% ee.

6.5.2 CATALYTIC REMOVAL OF CBZ GROUP

The benzyloxycarbonyl (Cbz) group is one of the most common amine-protecting groups in many synthetic transformations. The Cbz group is quite stable under relatively mild acidic as well as basic conditions and is orthogonal to many other common amine-protecting groups. Frequently, Cbz deprotections are carried out under catalytic hydrogenolytic conditions using metal catalysts.

6.5.2.1 Impurity Formation

Traditionally, HPLC, GC, and NMR spectrometers are employed as analytic tools to monitor the progress of a reaction. Albeit effective, these off-line methods show some limitations for monitoring fast reactions or reactions with labile functional groups. Brivanib alaninate (Equation 6.28)[52] is a prodrug with a broad spectrum anti-tumor activity developed by the Bristol Myers Squibb Company. The prodrug molecule has ester and nucleophilic primary amino functional groups plus a labile heteroaryl–O bond. Removal of the Cbz group from amine **104** via hydrogenolysis to prepare the prodrug was challenging, resulting in two impurities: **105**, and **107**, plus the parent drug **106**.

(6.28)

Solutions

A transamidation would generate dialanine impurity **105** and parent drug **106**. Over-reduction may produce impurity **107**. In addition, a slow reaction routinely led to >5% hydrolysis of the prodrug to form **106** as well as the formation of the dialanine **105**. The hydrolysis of the prodrug is a

CO_2-mediated process; however, full conversion was required due to poor purging of the intermediate **104** in the final crystallization.

Therefore, an in-line analytic method is critical in controlling impurity formation. Ultimately, FT-IR was selected to monitor the real-time concentration of both the starting material **104** and the CO_2 by-product, allowing to conduct Cbz-deprotection successfully on a multi-kilogram scale.

REMARKS

(a) THF was chosen as the reaction solvent as the labile heteroaryl–O bond is not compatible with alcohol solvents.

(b) A 45% headspace had to be maintained in order to mitigate the CO_2 concentration in the reaction mixture.

6.5.2.2   Pd(OAc)₂/Charcoal System

An active Pd/C catalyst from Pd(OAc)$_2$ and charcoal was prepared in situ in methanol for hydrogenation and hydrogenolysis. Using this in situ prepared Pd/C catalyst, Cbz protecting groups were readily cleaved furnishing the corresponding free amines in excellent yields (Equation 6.29).[53]

(6.29)

This protocol features mild reaction conditions (25 °C), low catalyst loading (0.05 mol%), and the absence of contamination of the product by palladium residues (<4 ppb).

6.6   CATALYTIC HYDROGENATION

More than 80% of commercial drug substances contain nitrogen atoms, in which amino groups are the major contributors of the nitrogen source. Among various methods, catalytic hydrogenation of nitro groups is one of the most frequently used methods to introduce amino groups. Generally, this reaction gives a relatively clean reaction profile, providing the desired amine products in good yields. Palladium or platinum on carbon and Raney nickel are frequently employed in the hydrogenation process.[54] The hydrogenation of aromatic nitro groups can be performed easily under mild conditions, while hydrogenation of aliphatic nitro groups is usually slower, requiring slightly elevated temperatures and high pressures. In general, for a given catalyst, the hydrogenation selectivity is influenced by various factors such as the presence of additives, solvents, reaction temperature, and pressure.

Besides nitro group reduction, catalytic hydrogenation is utilized to reduce pyridine ring, imine, nitrile, etc. A common problem associated with catalytic hydrogenation is catalyst poisoning, leading to sluggish or incomplete reactions.

6.6.1 REDUCTION OF NITRO GROUP

6.6.1.1 Palladium-Catalyzed Hydrogenation

A catalytic reduction of the nitro group of nitropyridine **108** in aqueous HCl/MeOH with Pd/C catalyst at atmospheric pressure required 9 days to complete. Such a slow reaction posed a safety concern, as a significant amount of hydroxylamine intermediate was accumulated during the course of the reaction. For example, after 2 days, the reaction mixture consisted of 3% of the desired amine **109** and 97% of the corresponding hydroxylamine intermediate.

Solutions

It was discovered that by adding a small amount of iron powder to the reaction mixture, hydrogen uptake was enhanced to the point where little or no hydroxylamine intermediate was observed.[55] Acceptable hydrogenation rates were achieved with a Pd/C catalyst and catalytic quantities of iron powder (Pd:Fe = 40:1) in EtOAc at 65 °C (Equation 6.30).[56]

(6.30)

Procedure

Stage (a)
- Charge a mixture of 242 g of 10% Pd/C (50% wet Degussa E101 NE/W) and 6 g of iron powder to a solution of **108** (ca. 2.28 mol) in EtOAc (ca. 7.1 L).
- Evacuate and purge with H_2.
- Stir the reaction mixture at 65 °C under a hydrogen atmosphere for 8 h.
- Evacuate the reactor and flush with nitrogen.
- Filter the reaction mixture through Celite.

Stage (b)
- Add a solution of 10% HCl (5.3 L).
- Distill at 50 °C/150–200 mmHg to remove most of the EtOAc.
- Add toluene (1.5 L).
- Cool to 2 °C and stir for 1 h.
- Filter and rinse with EtOAc.
- Yield: 523 g (77%).

DISCUSSIONS

(a) Reduction of the nitro group via catalytic hydrogenation proceeds through unstable intermediates such as nitroso and hydroxylamine derivatives with precious metal catalysts (Pd and Pt) azo and azoxy derivatives with Raney nickel.[57] When the hydrogenation reaction is fast, the probability of the accumulation of unstable intermediates is low. If the hydrogenation reaction exhibits an unexpected slow rate, the unstable intermediate accumulation hazard is high and one should immediately investigate the cause of the abnormal situation.

(b) The spent palladium on carbon has the risk of self-ignition, especially when using alcoholic solvents. Thus, alcoholic solvents should be avoided in the palladium-catalyzed hydrogenation reactions if possible and use other solvents such as THF, MeTHF, ethyl acetate, isopropylacetate, toluene, DMF, DMAc, or acetone since the ignitions were not observed with these solvents.[58] It was also suggested to rinse the filter cake with water or with a thiophene solution to prevent Pd self-ignition.

6.6.1.2 Nickel-Catalyzed Hydrogenation

Although palladium or platinum is frequently used as a catalyst in the reduction of the nitro group, dehalogenation is a common side reaction during the catalytic hydrogenation of halogen-containing nitroaromatic compounds. With Pd/C catalyst, the ease of dehalogenation depends on the nature of the halogen (I > Br > Cl > F) and its position (*ortho* > *para* > *meta*) relative to the nitro group.

Nickel is an alternative hydrogenation catalyst that is less reactive toward dehalogenation than the corresponding palladium or platinum catalysts. Given the possibility of the dehalogenation during hydrogenation of **110**, sponge-nickel was selected as the catalyst (Equation 6.31).[7]

(6.31)

REMARK

The use of iron/HCl to reduce nitropyridine **110** to aminopyridine **111** would be challenging to operate at scale, and cleaning the iron salts from the reactor was laborious and time-consuming.

6.6.1.3 Platinum-Catalyzed Hydrogenation

Initial reduction of the nitro group in **112** using Raney nickel in THF at 90 °C was plagued by the formation of a significant amount of des-chloro side product **114** (Equation 6.32).[59]

(6.32)

In general, dehalogenation can be suppressed by the use of additives. Triphenyl phosphite (P(Oph)$_3$) was the most effective additive and the des-chloro **114** was not observed under these conditions (Pt/Al$_2$O$_3$/P(Oph)$_3$, Pt/Cu/C/P(Oph)$_3$, or Pt/Fe/C/P(Oph)$_3$). In addition, a study showed also that the use of Ir/C generally resulted in high yields of **113** with low levels of de-chloro **114**.

6.6.1.4 Catalytic Transfer Hydrogenation

Unlike catalytic hydrogenation, a transfer hydrogenation reaction employs a hydrogen source to generate hydrogen in situ. These hydrogen sources include cyclohexene, cyclohexadiene, dihydronaphthalene, dihydroanthracene, isopropanol, diimide (generated from hydrazine), formic acid, ammonium formate, and sodium borohydride.

6.6.1.4.1   Iron-Catalyzed Transfer Hydrogenation

An iron-based catalyst system for the transfer hydrogenation of nitroarenes **115** to anilines **116** was developed (Scheme 6.12).[60] The reduction was conducted at 40 °C in the presence of $Fe(BF_4)_2$ (4 mol%) and tetraphos ligand (4 mol%) with 4.5 equiv of formic acid as the hydrogen source in EtOH.

More importantly, for halogenated nitrobenzenes, full conversions were achieved without dehalogenation by-products. Similar to other methods, the Fe-catalyzed nitro reduction underwent through two intermediates, nitrosoarenes **117** and hydroxylamines **118**. Hydroxylamines are known carcinogens and potentially explosive at higher concentrations due to their thermal instability. Using the reaction protocol, no hydroxylamines were observed in the reaction mixture, indicating that the reductions of **118**→ **116** proceed promptly.

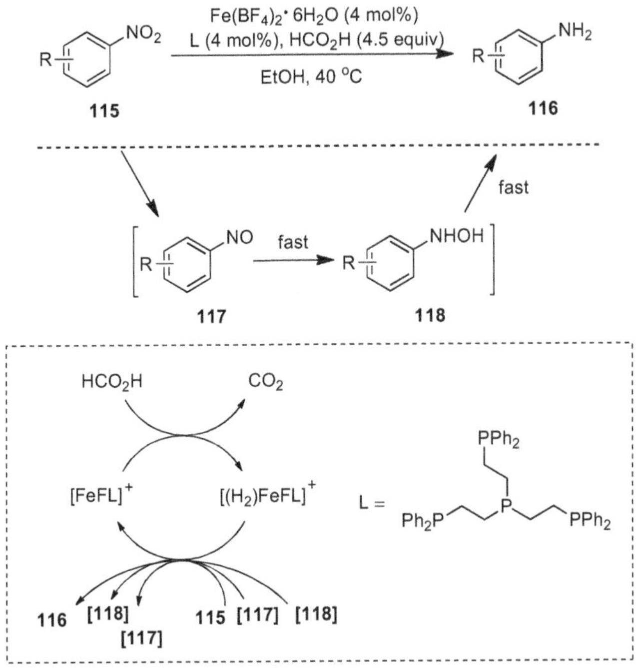

SCHEME 6.12

6.6.1.4.2   Palladium-Catalyzed Transfer Hydrogenation

Occasionally, halogen substituents need to be removed at a certain stage; thereby, dehalogenation is a desired reaction. A simultaneous reduction of the nitro group and ptimizationn can be achieved using catalytic transfer hydrogenation. As shown in the reaction (Equation 6.33),[61] 6-amino-7-ethylamino-quinazoline dihydrochloride salt **120** was obtained via a palladium-catalyzed transfer hydrogenation of 2-chloro-6-nitroquinazoline **119** with sodium formate in aqueous DMF at 80 °C in 79% yield after treatment with HCl in aqueous EtOH.

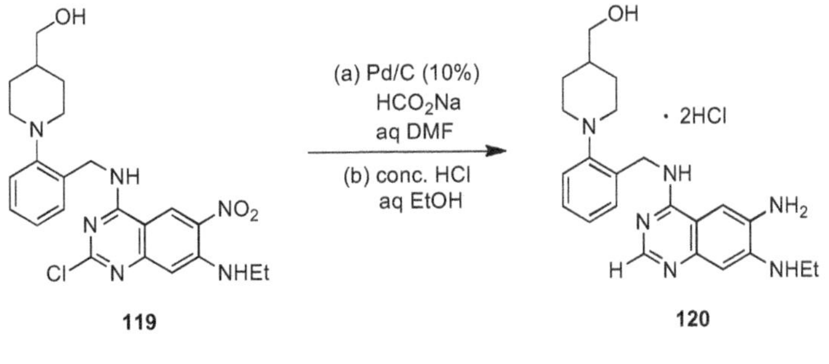

(6.33)

REMARK

Palladium-catalyzed transfer hydrogenation using sodium dihydrogenphosphite (NaH_2PO_2) (or NH_2NH_2) as the hydrogen source could be applied in a reductive removal of the benzylic chlorine substituent in **121**, providing oxindole **122** in 95% yield (Equation 6.34).[62]

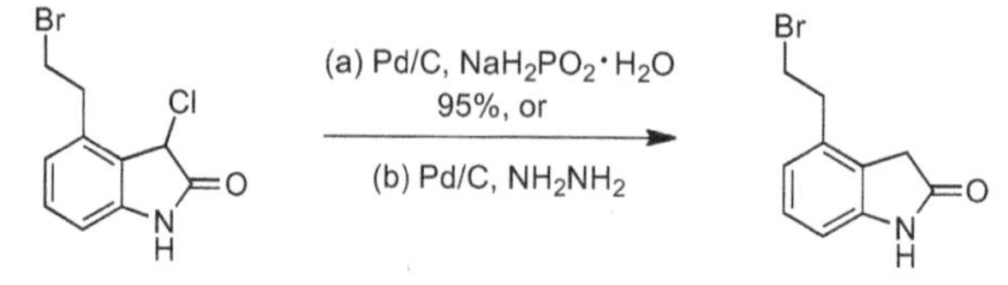

(6.34)

121   **122**

6.6.2 REDUCTION OF PYRIDINE RING

Compound **125**, developed by GlaxoSmithKline (GSK), is a potent fibrinogen receptor antagonist that prevents platelet aggregation. The manufacturing route for **125** involved an aminocarbonylation reaction as a key step to assemble the penultimate intermediate **124** (Scheme 6.13).[63]

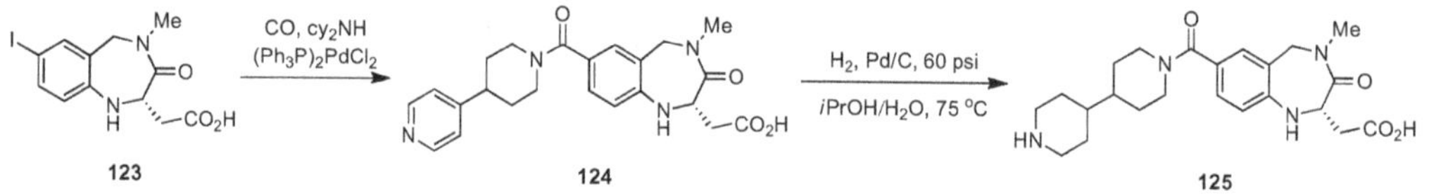

123   **124**   **125**

SCHEME 6.13

Reaction Problems

The aminocarbonylation reaction of aryl iodide **123** with 4-(piperidin-4-yl)pyridine generated 1 equivalent of HI, which had a poisoning effect on the subsequent catalytic hydrogenation of the pyridine ring. Although the use of dicyclohexylamine (cy_2NH, 2.5 equiv) could remove ca. 95% of HI as an insoluble salt with cy_2NH, the remaining iodide was present at a level of ca. 2000 ppm, which was much higher than 100 ppm (a level of iodide that was considered acceptable for the subsequent hydrogenation).

(II) Solutions

The most efficient way to remove the residual iodide is to oxidize iodide to iodine followed by extraction. Thus, treating the reaction mixture with excess hydrogen peroxide at pH 5 followed by extraction could easily reduce the residual iodide to 10 ppm or less.

REMARKS

Removal of excess H_2O_2 was initially done by adding inorganic reducing agents such as sodium metabisulfite or sodium sulfite. Albeit effective, sulfur residues made the resulting solutions very difficult to hydrogenate.[64] An alternative approach for the destruction of H_2O_2 was to use Pd/C as a catalyst to decompose the excess peroxide to oxygen and water. However, 2-propanol had to be used as a cosolvent to avoid the crystallization of **124**. Apparently, it was not ideal having an organic solvent mixture heated in the presence of a Pd catalyst, hydrogen, and oxygen. Ultimately, a biotransformation process was developed[65b] that utilizes a polymer-supported catalase enzyme to remove peroxide from reaction mixtures. The biocatalytic approach provides an economic and environmentally friendly solution to peroxide removal when compared to the chemical process.

6.6.3   REDUCTION OF A,β-UNSATURATED COMPOUNDS

Organocatalytic transfer hydrogenation was developed by the List group[65] in 2004 using Hantzsch ester **126** as the hydrogen source. This metal-free catalytic transfer hydrogenation of α,β-unsaturated aldehydes occurred in a highly efficient and remarkably chemoselective manner (Scheme 6.14).

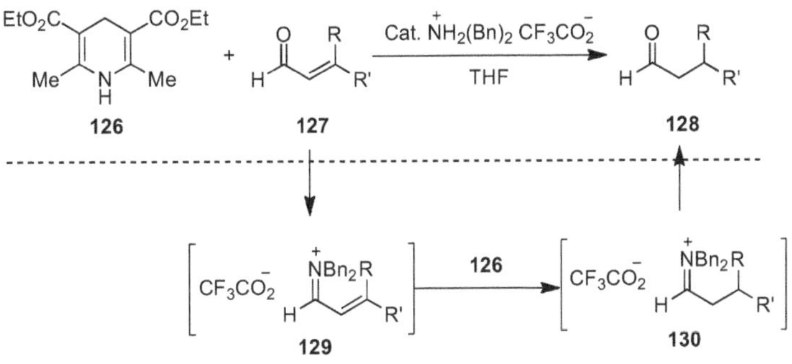

SCHEME 6.14

This Hantzsch ester-involved transfer hydrogenation lent itself readily to the kilogram-scale GMP preparation of lobeglitazone **132**, a drug candidate developed by Chong Kun Dang Pharmaceutical Cooperation in Korea for the treatment of type-2 diabetes (Equation 6.35).[66] This reduction requires anhydrous conditions as the water molecules may deteriorate the role of silica gel as an acid catalyst possibly by binding with the silica gel. Therefore, a Dean–Stark water-trapping apparatus was used to remove moisture.

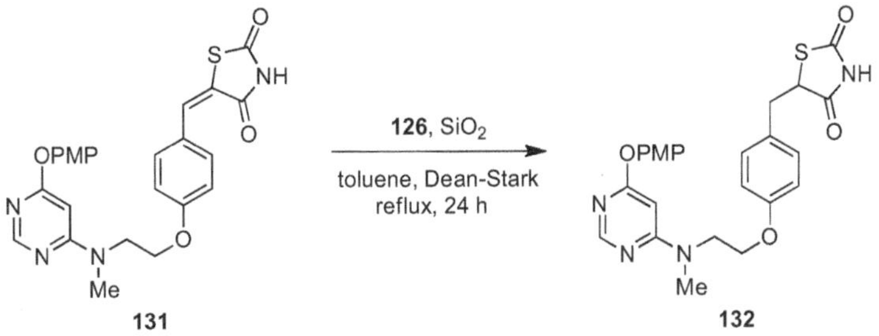

(6.35)

Procedure

- Charge 2.1 kg (4.3 mol) of **131** into a dry, 70-L reactor followed by adding 50 L of toluene.
- Charge 1.4 kg (5.6 mol, 1.3 equiv) of **126** and 6 kg of silica gel.
- Heat at reflux for 12 h.
- Charge 6 kg of silica gel.
- Continue to reflux for 1h.
- Filter to remove silica gel and wash with EtOAc (2×5.0 L).
- Distill under reduced pressure.
- Purify the crude **132** by recrystallization.

6.6.4   ENANTIOSELECTIVE REDUCTION OF QUINOLINES

Initiated by a catalytic protonation of quinolines **133**, a cascade reaction sequence that constitutes a 1,4-hydride addition, isomerization, and 1,2-hydride addition (Equation 6.36).[67] In general, good yields of tetrahydroquinolines **134** were obtained in high enantioselectivities. The R group varies from aromatic and heteroaromatic to aliphatic substituents. Notably, this metal-free hydrogenation protocol is compatible with halogen substituents, such as 3-bromophenyl or chloromethyl groups.

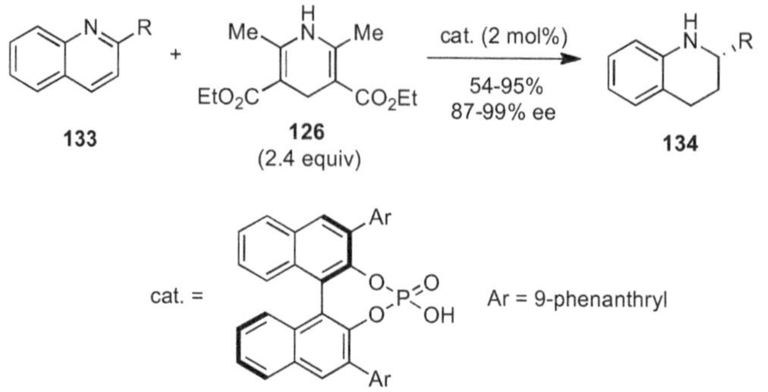

(6.36)

A protonation of quinoline **133** by Brønsted acid catalyst ((ArO)$_2$PO(OH)) would generate the iminium salt **135** (Scheme 6.15).[69] Reduction of the resulting iminium salt **135** via hydride transfer from dihydropyridine **126** affords the corresponding enamine **136** along with the regenerated Brønsted acid. Further reduction of **137** with **126** in the presence of Brønsted acid furnishes the desired product **134**. The mild reaction conditions, operational simplicity, as well as low catalyst loading render this transformation an attractive and practical approach for access to optically active tetrahydroquinolines.

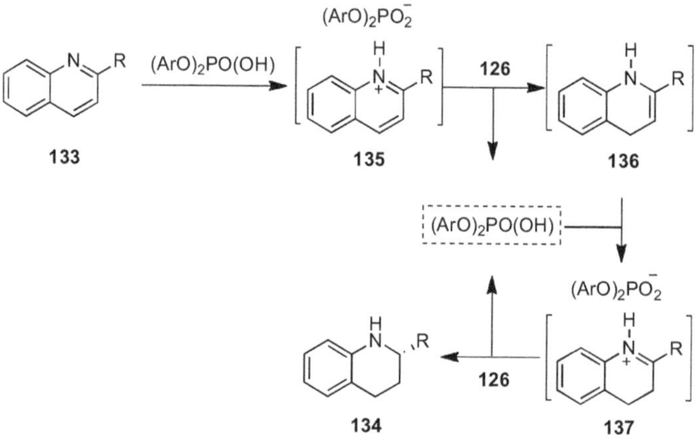

SCHEME 6.15

6.6.5 REDUCTION OF NITRILE

Catalytic hydrogenation of benzonitrile with mesoporous Al$_2$O$_3$ supported Ni catalyst to prepare benzylamine produced significant amounts of dibenzylamine side product (Scheme 6.16).[68]

Solutions

The reaction may proceed through two imine intermediates (**138** and **139**) that are responsible for the formation of benzylamine and dibenzylamine, respectively. To mitigate the N-benzylidenebenzylamine **139** formation, aqueous ammonia was used as an additive. In the event, the addition of aqueous ammonia gave a quantitative yield of the desired benzylamine product.

Ammonia was also used in the catalytic hydrogenation of nitrile **140** with Raney nickel (Equation 6.37).[69]

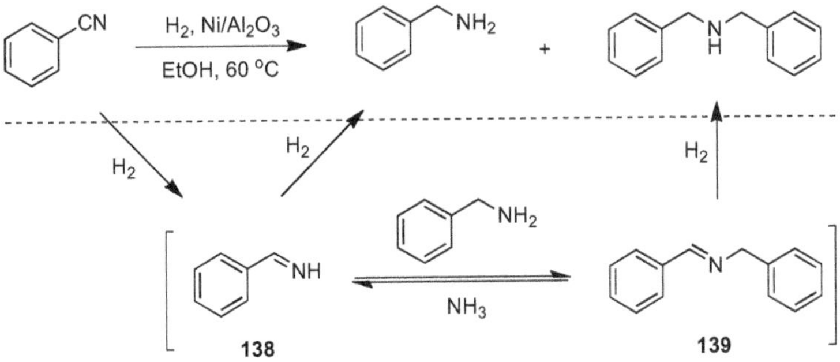

138 139

SCHEME 6.16

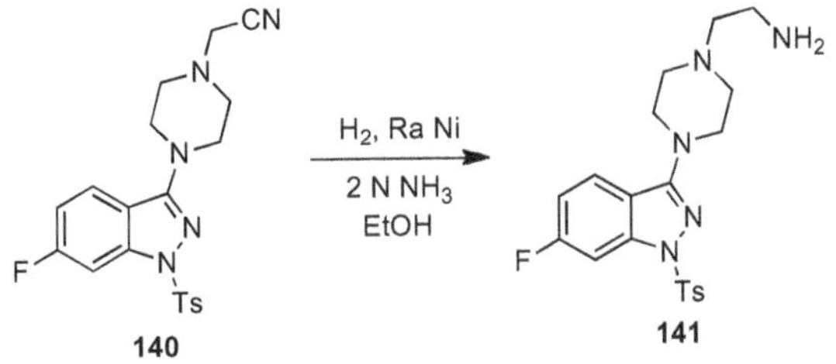

140 141

(6.37)

Procedure

- Heat a mixture of 1.32 kg (3.2 mol) of **140**, 840 g of Ra Ni, and 2 N NH_3 in EtOH (12 L) under 200 psi of H_2 at 80 °C for 4 h.
- Cool, filter through Celite, and wash with EtOH (8 L).
- Concentrate at 35 °C under reduced pressure.
- Dissolve the residue in CH_2Cl_2 (20 L) and dry over $MgSO_4$ (1 kg).
- Concentrate to give the product **141**.

6.6.6 Reduction of Azide

During catalytic hydrogenation of azide **142** to amine **143** using Pt/C catalyst, a significant amount of dimer **144** was formed (Equation 6.38).[70]

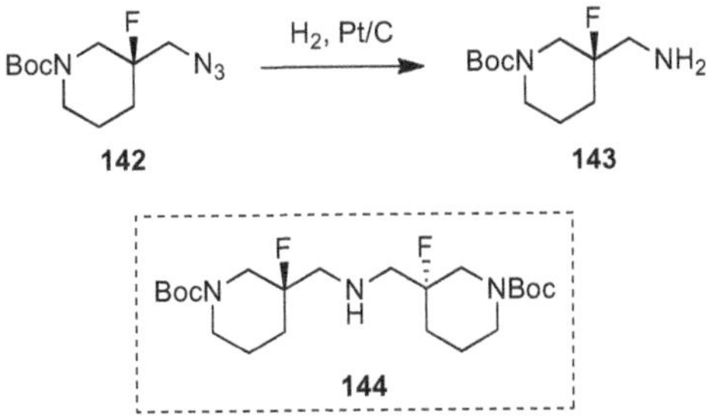

142 143

144

(6.38)

In order to suppress the formation of **144**, aqueous ammonia was used as an additive in the hydrogenation. As a result, the dimer **144** was successfully controlled below 0.3%.

6.7 OTHER CATALYTIC REACTIONS

A low product yield can be attributed to an incomplete conversion due to the lack of reactivity from the reagent. To overcome this reaction problem, several reaction parameters, such as temperature, concentration, solvent, and catalyst, are normally considered by process chemists. Besides these commonly used approaches, other methods are also applied in the process development.

6.7.1 Cu(I)-Catalyzed Reaction

Scheme 6.17[71] outlines a conversion of tertiary amine **145** to tertiary amine **148**. This three-step process involves the amino group migration via aziridinium intermediate **146**, furnishing the desired product **148** in overall yields of 60–65% along with several side products.

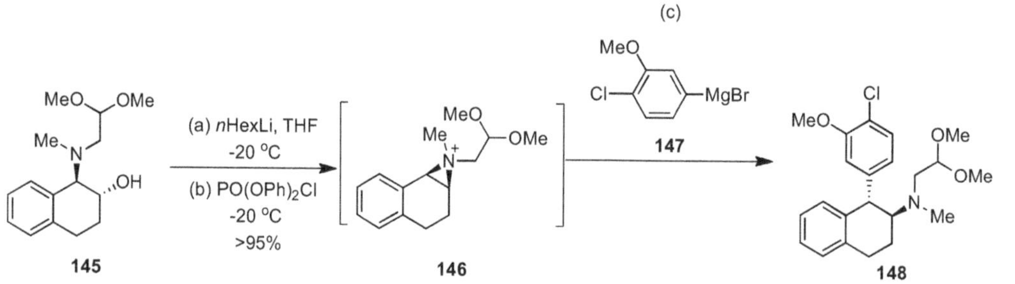

SCHEME 6.17

Solutions

The moderate product yields were presumably caused by the low nucleophilic reactivity of the Grignard reagent **147**. To improve the nucleophilicity of **147**, a catalytic amount of CuCl·2LiCl complex was employed to convert **147** into the corresponding organocopper **149** (Equation 6.39). Ultimately, the reaction was optimized by adding 5 mol% of CuCl·2LiCl complex to the reaction mixture, giving an overall isolated product yield of 70%.

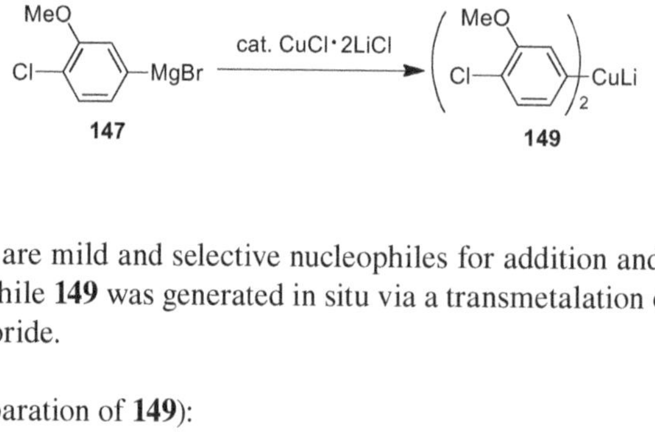

(6.39)

DISCUSSIONS

Organocopper reagents are mild and selective nucleophiles for addition and substitution reactions. The arylcopper nucleophile **149** was generated in situ via a transmetalation of the Grignard reagent **147** with copper (I) chloride.

Procedure (for the preparation of **149**):

- Add anhydrous THF (50 mL) to a mixture of copper(I) chloride (1.9 g, 19 mmol) and LiCl (1.6 g, 38 mmol).
- Stir the mixture for 15 min at rt to give a golden-yellow solution.
- Transfer the solution via a cannula to the Grignard solution of **147** at room temperature.
- Stir for 5 min prior to mixing with **146**.

6.7.2 Decarboxylative Bromination

The preparation of pyridone bromide **151** was realized through decarboxylative bromination (Equation 6.40).[72]

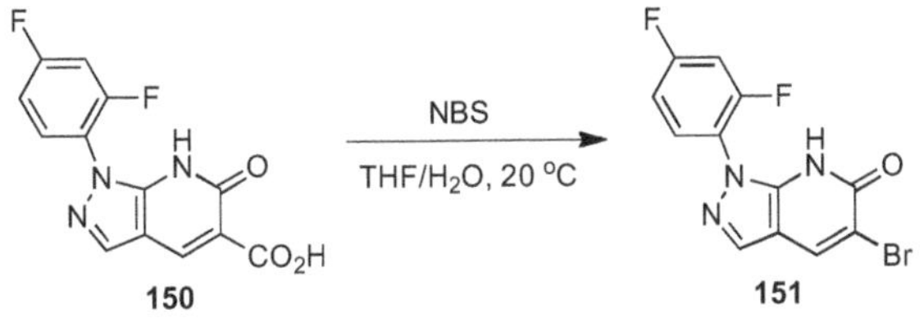

(6.40)

Reaction problems

This decarboxylative bromination had the following issues:

- Excess of NBS (2–3 equiv) was required to achieve a full conversion in solvents such as MeCN, DMF, N-methyl-2-pyrrolidone (NMP), and AcOH due to the background reaction.
- Using THF as the solvent resulted in a slow reaction and low conversion.

(II) Solutions

The addition of a catalytic amount of lithium acetate[73] allowed the reaction to complete with 1.1 equiv of NBS in aqueous THF. The lithium acetate-mediated decarboxylative bromination may proceed through β-lactone intermediate **153** (Scheme 6.18).

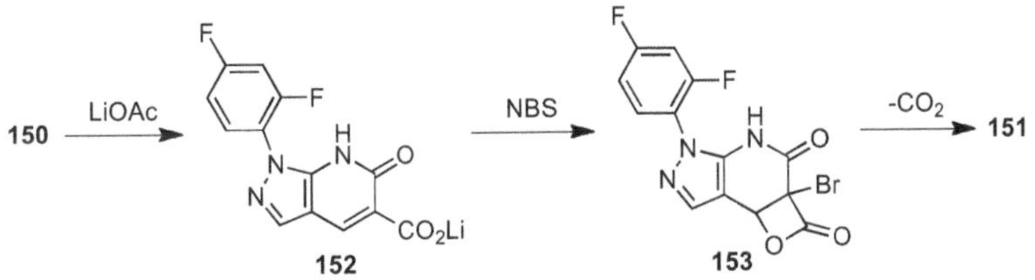

SCHEME 6.18

Procedure

- Charge 3.32 kg (11.39 mol) of **150** into a dry 60-L jacketed reactor, followed by adding 75.1 g (1.14 mol, 10 mol%) of LiOAc, THF (20 L), and water (1.66 L).
- Charge 2.23 kg (12.53 mol, 1.1 equiv) of NBS at 20 °C in four portions over 3 h.
- Stir at 20 °C for 2 h.
- Heat to 50 °C and stir for 1 h.
- Cool to 20 °C.
- Add water (33.16 L) over 70 min.
- Stir the resulting mixture at 20 °C for 2 h.
- Filter, wash the reactor, and the filter cake with water (3×6.63 L), and dry.
- Yield: 3.30 kg (89%).

6.7.3 FORMATION OF ACID CHLORIDE

Acid chlorides are normally prepared by a reaction of acids with oxalyl chloride in the presence of catalytic amounts of DMF. For example, DMF was used as the catalyst for the preparation of acid chloride **155** on a kilogram scale.[74] Although effective at this scale of preparation, this experimental protocol raised concerns due to the potential generation of carcinogenic dimethyl carbamyl chloride **156** (Equation 6.41). Thus, alternative reaction conditions were sought for this transformation by evaluating NMP and 4-dimethylaminopyridine (DMAP) as alternative catalysts for this reaction. Ultimately, NMP was chosen as the catalyst to replace DMF.

$$(6.41)$$

R: H, Et, OMe

cat.: DMF, NMP, or DMAP

The optimized reaction conditions were to employ 1.2 mol equiv of oxalyl chloride and 10 mol% of NMP in methylene chloride.

6.7.4 CATALYTIC DECHLORINATION

A sluggish reaction was encountered during the catalytic dechlorination of **157** with palladium on carbon (Equation 6.42).[75] In the absence of a base the reaction was stalled, as the product hydrochloride salt **158** crystallized on the surface of the catalyst.

$$(6.42)$$

In the presence of calcium hydroxide as a base, the reaction proceeded readily and completed in 4 h at 55–65 °C under 20 psi of hydrogen.

6.7.5 TWO-PHASE REACTIONS

The application of phase transfer catalyst (PTC) in organic synthesis is well-documented in the literature. A phase-transfer catalyst is used to facilitate the migration of a reactant from one phase into another phase where the reaction occurs. The use of PTC allows the achieving of faster reactions in higher product yields with fewer side products. PTC-catalyzed reactions are especially desired in the green chemistry perspective as the phase-transfer catalysis process permits the use of water to reduce the organic solvent quantity, thereby minimizing the process waste. Quaternary ammonium and phosphonium salts are commonly utilized as PTC for nucleophilic aliphatic substitution. Typical catalysts include tetrabutylammonium bromide (TBAB), benzyltrimethylammonium chloride, and hexadecyltributylphosphonium bromide.

6.7.5.1 Enhancement of Reaction Rate

The major advantage of using PTC is that it increases the reaction rate, which leads to a more efficient process due to a shorter processing cycle time on a larger scale. For instance, the Gabriel reaction (Equation 6.43)[76] of bromide **159** with potassium phthalimide **160** required 18 h to complete without PTC, while in the presence of TBAB it is completed within 4 h.

$$(6.43)$$

S_NAr sulfonylation of halogen-substituted pyridine derivatives **162** with sodium solfinate **163** suffered from a slow reaction rate and low conversion due to the poor solubility of **163** in the reaction media (Equation 6.44).[77]

$$(6.44)$$

Ultimately, the addition of tetrabutylammonium chloride (TBACl) phase transfer catalyst accelerated the reaction rate significantly through the in situ formed n-Bu$_4$NSO$_2$Tol.

REMARK
Ammonium chloride was found to be able to catalyze the cyclization of arylhydrazides with orthoesters during the synthesis of 1,3,4-oxadiazoles (Equation 6.45).[78]

$$(6.45)$$

Compared with uncatalyzed cyclization, NH$_4$Cl-catalyzed cyclization was completed in a much shorter time.

6.7.5.2 Suppressing Side Reactions
The use of a phase-transfer catalyst can suppress the impurity formation. For instance, the preparation of alkyl aryl ether **167** was achieved by an S_N2 reaction of phenol **165** with alkyl chloride **166** (Equation 6.46).[79] Under homogeneous reaction media,[80] the formation of undesired quaternary ammonium salt **168** was unavoidable. Employing triethylbenzylammonium chloride (TEBAC) (5%) in the presence of 2.2 equiv of 30% sodium hydroxide, 4-bromophenol **165** was converted selectively (>95%) into **167** within 18 h at 30–35 °C.

$$(6.46)$$

A water-miscible organic solvent can be utilized as a non-charged PTC. For instance, acetonitrile was utilized as a PTC to mitigate the formation of side product **172** during the preparation of potassium phosphate **170** (Scheme 6.19).[81]

Initially, the oxidation of di-*tert*-butylphosphite **169** with iodine, generated in situ via oxidation of KI with H$_2$O$_2$, in toluene produced a low yield (39%) of the corresponding potassium phosphate

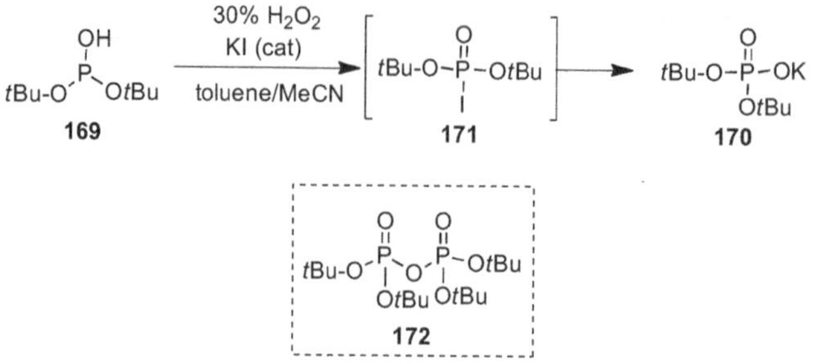

SCHEME 6.19

170 upon treatment with potassium *tert*-butoxide. Besides product **170**, this reaction generated a significant amount of pyrophosphate **172**.

It was hypothesized that the low water content (KF<0.1%) in the toluene phase led to a slow hydrolysis of the phosphoryl iodide intermediate **171**. Consequently, the relatively high concentration of **171** in the organic phase resulted in high levels of **172**.

To address this issue, acetonitrile was used and this oxidation reaction was carried out with a catalytic amount of KI (0.15 equiv) in the presence of 30% H_2O_2 in a mixture of toluene and acetonitrile (3:1), in which acetonitrile transferred water from the aqueous phase into the organic phase where the reaction occurred. Consequently, the product yield was improved from 39% to 86%.

REMARKS

(a) Ethereal solvents, such as MeTHF, are less attractive at scale because of the potential peroxide formation.

(b) Utilization of the toluene/water biphasic system in the presence of acetonitrile allowed to control water content in the organic phase in order to suppress the disproportionation of iodine (Equation 6.47):

$$3\,I_2 + 3\,H_2O \longrightarrow 5\,I^- + IO_3^- + 6\,H^+ \tag{6.47}$$

6.7.5.3 Reducing the Amount of Toxic Sodium Cyanide

In addition to the rate enhancement, the use of PTC can reduce the amount of excess reagent. For instance, the amount of sodium cyanide in the S_N2 displacement reaction of benzyl chloride **173** (Equation 6.48)[82] could be reduced from 1.7 equiv to 1.05 equiv under phase-transfer conditions.

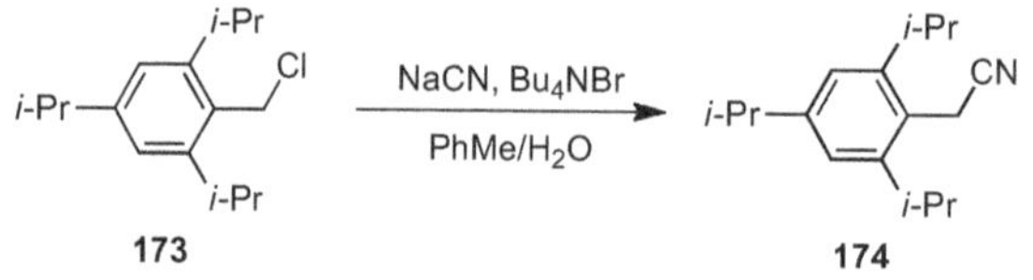

$$\tag{6.48}$$

6.7.5.4 Replacing DMSO Solvent in S_NAr Reaction

Polar aprotic solvents are preferred reaction solvents for nucleophilic substitution reactions due to their excellent dissolution power and ability to accelerate reaction rates. However, the use of polar aprotic solvents, such as dimethylformamide (DMF), dimethyl sulfoxide (DMSO), *N*-methyl-2-pyrrolidone (NMP), and dimethyl acetamide (DMAc) has some concerns. For example, DMF, DMAc, and NMP have effects on embryo-fetal development in animals; DMSO has some process safety issues.[83]

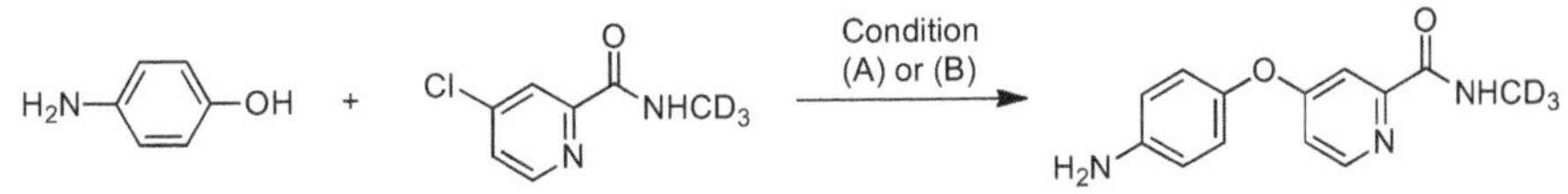

$$(6.49)$$

(A) KO*t*Bu, DMSO, 80 °C, 78%.
(B) NaOH, Bu$_4$NHSO$_4$, THF/H$_2$O, 67 °C, 82%.

Therefore, a phase-transfer catalysis system has been developed to replace DMSO in the S$_N$Ar reaction between aminophenol **175** and chloropyridine **176** (Equation 6.49).[85] The S$_N$Ar reaction was carried out successfully by using tetrabutylammonium bisulfate as the PTC in aqueous THF, affording the desired bisaryl ether **177** in good yields.

6.7.5.5 Two-Phase Reactions without PTC

6.7.5.5.1 Formation of Ylide

Two-phase reactions can occur rapidly without using PTC. For instance, the formation of ylide **179** proceeded smoothly in a bisphasic system consisting of aqueous sodium carbonate and xylenes without PTC, presumably the phosphonium salt **178** acts as PTC (Scheme 6.20).[84]

Procedure

- Charge water (1185 L) into a reactor, followed by adding 120 kg (1700 mol) of Na$_2$CO$_3$, 370 kg (813 mol) of **178**, and xylenes (1350 kg).
- Stir the mixture for 30 min.
- Remove the lower aqueous layer.
- Wash the organic layer with Na$_2$CO$_3$ (60 kg in 1185 L of water).
- Charge, to the organic layer, 158 kg (507 mol) of **180**.
- Heat the batch at 136 °C for 24 h (initially under distillation conditions to remove residual water and then under reflux).
- Cool to 80 °C and distill under reduced pressure to remove xylenes.
- Charge IPA (2284 L).
- Heat the resulting slurry to reflux and then cool to 20 °C.
- Filter, wash with IPA (600 L) and MeOH (300 L), and dry at 60 °C under vacuum.
- Yield of **181**: 170 kg (82%).

6.7.5.5.2 Formation of Pyrazole

Interestingly, PTC, such as benzyl triethylammoniun chloride, has no effect on the rates or product yields of the cycloaddition between alkyne **182** and 1-aminopyridin-1-ium **183** (Equation 6.50).[85]

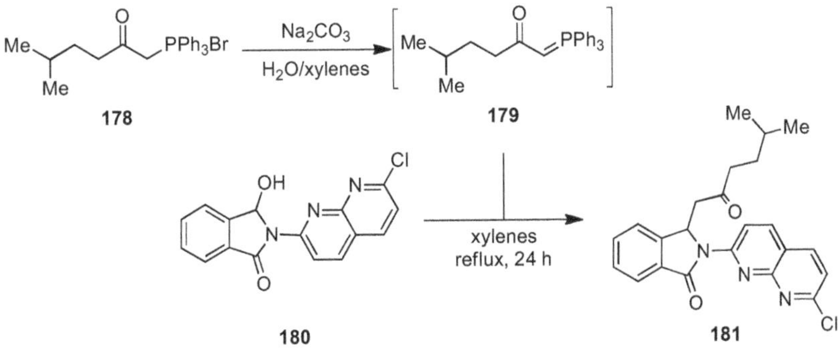

SCHEME 6.20

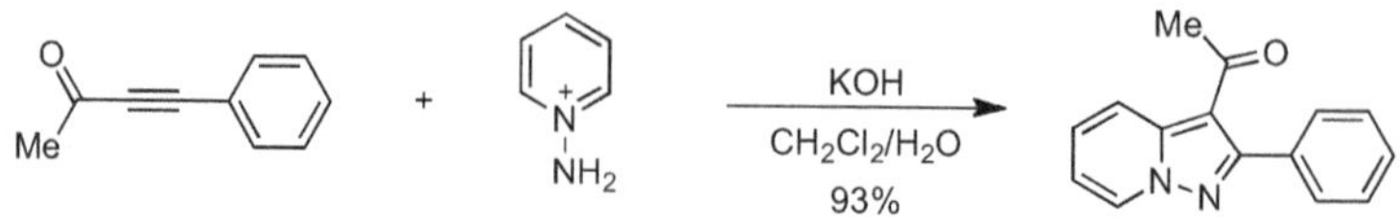

As shown in Scheme 6.21, the equilibrium between of **183** and the dimer **185** allows the transfer of **183** from the aqueous phase into the organic phase wherein the cycloaddition occurs.

Simply by adoption of a two-phase system (CH$_2$Cl$_2$–H$_2$O) without PTC, the reaction with 2.0 equiv of 1-aminopyridin-1-ium **183** proceeded cleanly with complete consumption of the alkyne **182**. This protocol allowed an easy layer separation followed by the exchange of methylene chloride in the system for 2-propanol, rejecting the by-product **187** into the aqueous stream.

An alternative reaction pathway involves an intermediate **188**, formed via a reaction between **182** and **183**, whose oxidation by **183** would give the product **184** and by-product **187** (Scheme 6.22).

6.7.6 DEOXYBROMINATION

The deoxybromination of 5-bromo-2-hydroxy-3-nitropyridine **189** with phosphorus oxybromide (POBr$_3$) was carried out under refluxing in toluene giving the corresponding 2,5-dibromo-3-nitropyridine **190** in 80% yield. In contrast, the addition of DMF (10 mol%) as a catalyst was able to run the reaction at a lower temperature (90 °C) and the product yield was increased to 92% (Equation 6.51).[86]

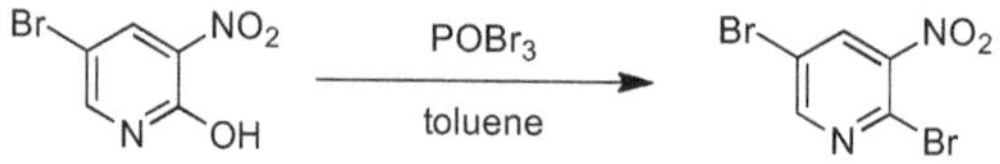

DMF, mol%	Temperature, °C	Yield, %
0	110	80
10	90	92

(6.51)

Procedure

- Suspend 4.50 kg (20.55 mol) of **189** in toluene (11.25 L).
- Charge 158.8 mL (2.05 mol, 10 mol%) of DMF.
- Heat to 90 °C with protection from light by aluminum foil.

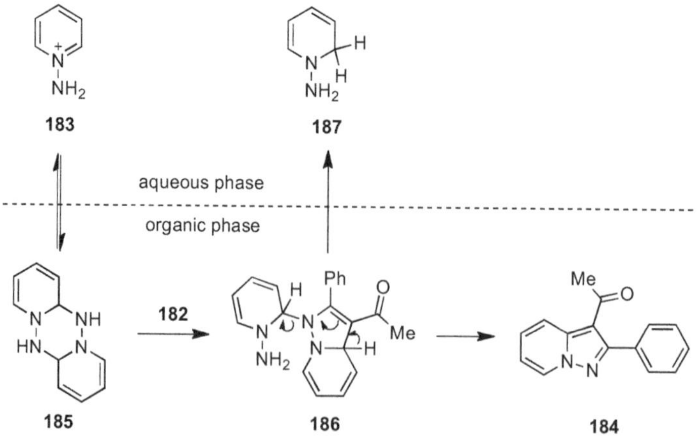

SCHEME 6.21

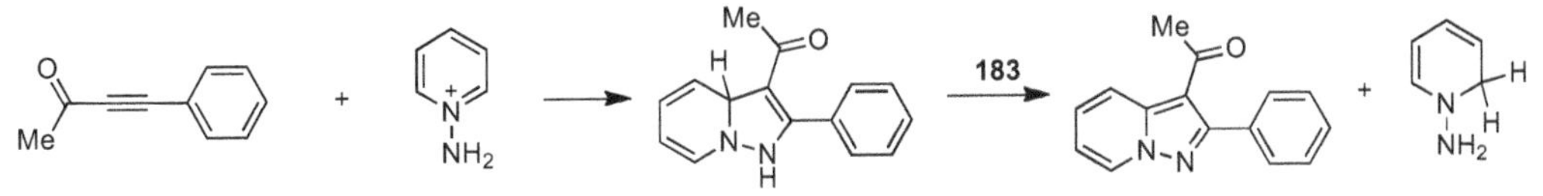

SCHEME 6.22

- Charge a solution of phosphorus oxybromide (7.07 kg, 24.66 mol) in toluene (11.25 L) over 1.5 h.
- Heat the resulting mixture at 90 °C for 15 h.
- Cool to room temperature.
- Add toluene (22.5 L) and water (45 L).
- Separate layers.
- Wash the organic layer with 1 N NaOH (2×45 L) and water (22.5 L).
- Distill at 45 °C under reduced pressure.
- Yield: 5.31 kg (92%).

6.7.7 REGIOSELECTIVE CHLORINATION

Normally, chlorination may suffer two drawbacks when using reactive chlorination reagents such as 1,3-dichloro-5,5-dimethylhydantoin (DCH): poor regioselectivity and over chlorination. Less reactive reagents usually end up with either longer reaction times or incomplete conversion. It was found that the chlorination of **191** with DCH gave variable isolated yields (<85%) of **192** due to the over chlorination, while with N-chlorosuccinimide (NCS) resulted in no reaction. However, the chlorination of **191** with NCS in the presence of a catalytic amount of sulfuric acid (3 mol%) provided the product with 91% yield and 99% purity (Equation 6.52).[87] Following a solvent switch to 2-propanol and subsequent saponification, the product **192** was isolated in 78% yield over two steps.

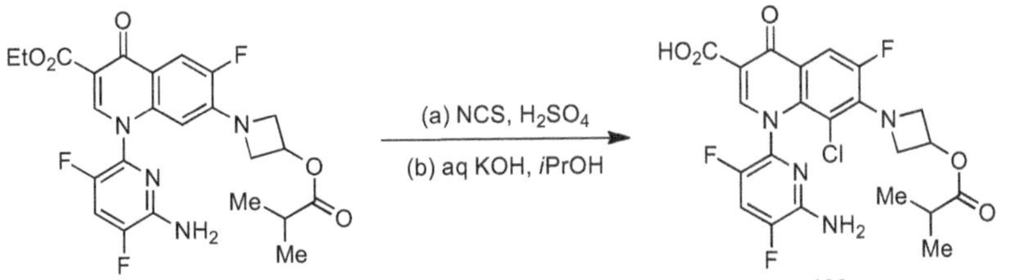

$$(6.52)$$

Procedure

Stage (a)
- Charge 0.56 kg (5.7 mol, 0.03 equiv) of H_2SO_4 into a solution of NCS (25.3 kg, 190 mol, 1.03 equiv) in MeOAc (419 kg) at 17 °C.
- Transfer the resulting solution into a slurry of **191** (92.7 kg, 184 mol) in EtOAc (244 kg) at 17 °C.
- Quench the reaction with 370 kg of 1.5% aqueous $NaHCO_3$.
- Separate layers.
- Wash the organic layer with 10% Na_2SO_3 (200 kg).
- Concentrate the organic layer.

Stage (b)
- Add IPA, followed by adding 750 kg of 4% aqueous KOH.
- Stir at 50 °C until completing the hydrolysis.
- Filter and add 410 kg 12% aqueous AcOH.
- Filter, wash with water, and dry at 50 °C.
- Yield: 73 kg (77.7%).

6.7.8 Regioselective Magnesiation

Deprotonation of aromatics by lithiation is a commonly used approach to generate Csp^2 nucleophiles. However, this approach usually requires cryogenic temperature and suffers poor selectivity.

Scheme 6.23[88] describes a catalytic protocol for the regioselective metalation of 3-methylthiophene **193**.

Utilizing a catalytic amount of 2,2,6,6-tetramethylpiperidine (TMP), a regioselective magnesiation of **193** was realized with a commercial Grignard reagent via TMP magnesium chloride ((TMP)MgCl). The in situ formed, steric demanding (TMP)MgCl intermediate enabled to metalate selectively at the 5-position on the thiophene ring in **193**. Under reflux in THF, which avoided the undesired cryogenic temperature.

Procedure

- Charge 25.2 mL (0.15 mol) of TMP into a solution of **193** (147 g, 1.5 mol) in 576 mL of THF.
- Charge a solution of *i*PrMgCl (2.0 M THF solution, 633 mL, 1.27 mol) at <30 °C.
- Heat the resulting solution to reflux at 66 °C.
- Yield: 98% conversion (after 23 h).

6.7.9 Amide Preparation

6.7.9.1 NaOMe as Catalyst

The preparation of amide **196a** from methyl lactate **195a** required excess morpholine (3.0 equiv) at 85 °C for 30–60 h. The isolated amide **196a** contained up to 17 wt% morpholine. Reduction

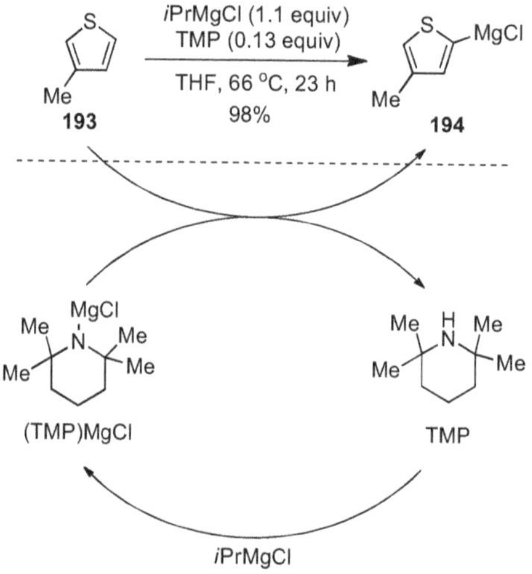

SCHEME 6.23

in morpholine charge was the focus for process ptimization. Accordingly, sodium methoxide (15 mol%) was employed as a catalyst. This modification permitted the morpholine charge to be reduced to 1.04 equiv. Under the modified conditions, the reaction was completed at ambient temperature in only 4–6 h (Equation 6.53).[89]

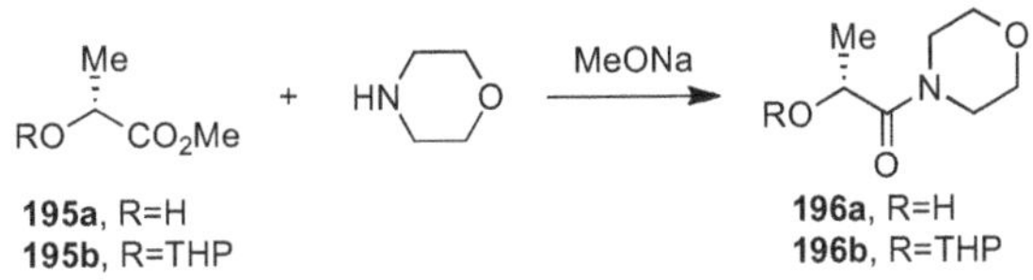

Me
RO
CO2Me
+
HN
O
MeONa
Me
O
N
RO
O
195a, R=H
195b, R=THP
196a, R=H
196b, R=THP

(6.53)

Procedure

- Charge 86.2 kg (1000 mol) of morpholine to a cold (10 °C) **195a** (100 kg, 961 mol).
- Cool the resulting mixture to 5 °C.
- Charge 31.2 kg (149 mol) of 25 wt% NaOMe in MeOH at <5 °C.
- Warm the batch to 20 °C and stir for 16 h.
- Add HCl in MeOH to pH 6.8.
- Filter and wash with THF (50 kg).
- Heat the filtrate to 60–65 °C to exchange MeOH with THF (total of 880 kg THF added).
- The THF solution of **196a** was used in the subsequent step directly.

DISCUSSIONS

With catalytic amounts of NaOMe, the amide **196a** formation could proceed through a three-membered lactone intermediate **197** (Scheme 6.24).

In comparison, when submitting the THP ether **195b** to the reaction conditions, the amidation rate was slow and epimerization was observed.

6.7.9.2 HOBt as Catalyst

Process development toward the coupling of acid **198** with amine **199** to prepare amide **200** began by surveying a number of coupling reagents.[90] Among the coupling reagents, 1,3-diisopropyl carbodiimide (DIC) appeared to be the ideal coupling reagent due to availability, cost, and safety.[91] The coupling reaction required 23 h at 20 °C with 1.3 equiv of DIC (Equation 6.54).[92] It was found that the use of an auxiliary nucleophile (1-hydroxybenzotriazole (HOBt), 0.2 equiv) allowed the reaction to complete within 4 h.

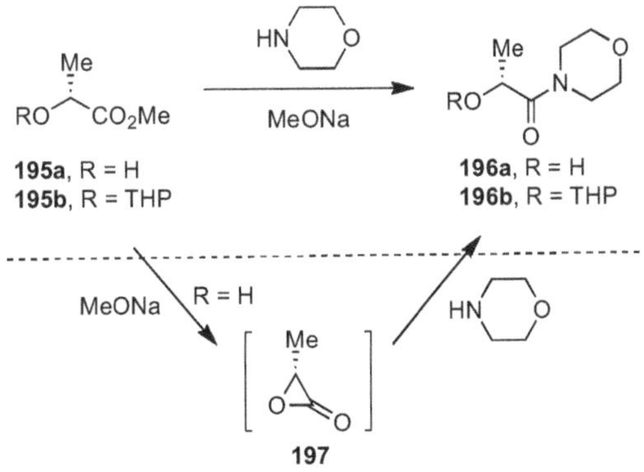

HN
O
Me
RO
CO2Me
MeONa
Me
O
N
RO
O
195a, R = H
195b, R = THP
196a, R = H
196b, R = THP
MeONa
R = H
Me
O
O
197
HN
O

SCHEME 6.24

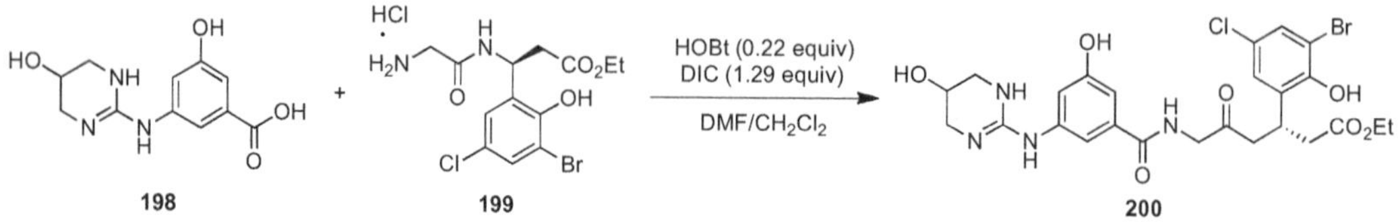

(6.54)

Procedure

- Charge 14.8 kg (58.9 mol) of **198** into a reaction vessel, followed by adding 24.5 kg (58.9 mol) of **199**, DMF (89 kg), a 12 wt% solution of HOBt (13 mol, 0.2 equiv) in DMF (15.0 kg), and CH_2Cl_2 (162 kg).
- Stir the resulting mixture at 20 °C.
- Charge 9.6 kg (76 mol) of DIC.
- Stir the batch at 20 °C for 6 h.
- Add 9.4 wt% NaOH (188.3 kg) and stir for 2.5 h.
- Add CH_2Cl_2 (325 kg).
- Workup.

6.7.10 SYNTHESIS OF INDOLE

5-Methoxy-1*H*-indole-2-carboxylic acid **201** (Figure 6.3)[93] was a key intermediate for the preparation of anti-viral agents such as U-87201E (atevirdine mesylate), a non-nucleoside, reverse transcriptase inhibitor of HIV.

The synthesis of **201** was commenced with 4-methoxyaniline **202** as the starting material involving a two-step conversion of diazo intermediate **203** to α-imino ester **205**, a precursor for the Fisher cyclization (Scheme 6.25).[94] The coupling of **203** with diethyl 2-methylmalonate provided (*E*)-diazene **204** followed by decarboxylation affording **205**. It was found that NaOEt as a base was able to affect decarboxylation. Eventually, using NaOEt (0.2 equiv)/ethanol catalytic system produced selectively the desired **205** in 15 min at room temperature. An additional benefit of using catalytic amounts of NaOEt was that this decarboxylation could be telescoped into the subsequent Fisher indole cyclization without isolation of **205**.

Procedure

Stage (c)
- Charge 65.0 g (0.2 mol) of 21% NaOEt solution to a solution of crude **204** (325 g, 1.05 mol) in EtOH (400 mL) dropwise at 25–30 °C.
- Stir the resulting mixture at rt for 15 min.

Stage (d)
- Heat the reaction mixture of **205** to reflux.
- Charge 120 g (3.29 mol) of gaseous HCl over 2 h.

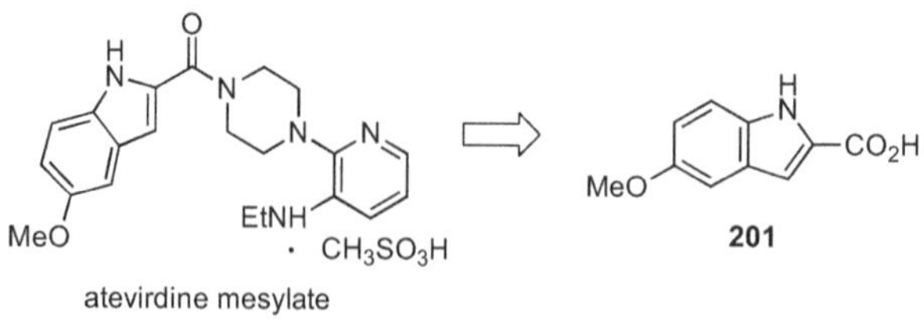

atevirdine mesylate

FIGURE 6.3 The structure of atevirdine mesylate

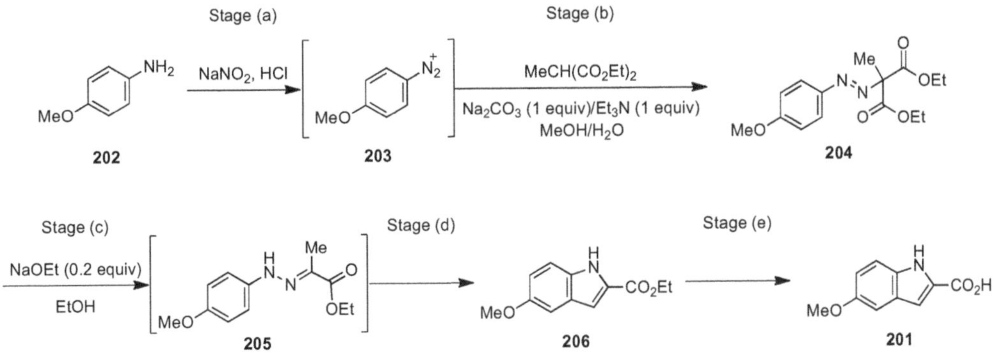

SCHEME 6.25

- Maintain reflux for 15 min after the addition.
- Cool to room temperature.
- Add water (100 mL).
- Cool the resulting mixture to 0 °C over 2.5 h.
- Filter and wash with precooled (0 °C) EtOH (4×100 mL) and then with water (2×250 mL).
- Yield of **206**: ca. 146 g.

6.7.11 N-METHYLATION REACTION

N-Methylation of amino acid derivative **207** (Equation 6.55)[95] was carried out in THF with dimethyl sulfate in the presence of sodium hydride. It was found that a catalytic amount of water was required and no reaction was observed without water. The reaction of water with sodium hydride generated highly reactive dry sodium hydroxide, which led to much faster reaction rates than powdered sodium hydroxide itself. After the reaction, the excess dimethyl sulfate could be destroyed with ammonium hydroxide.

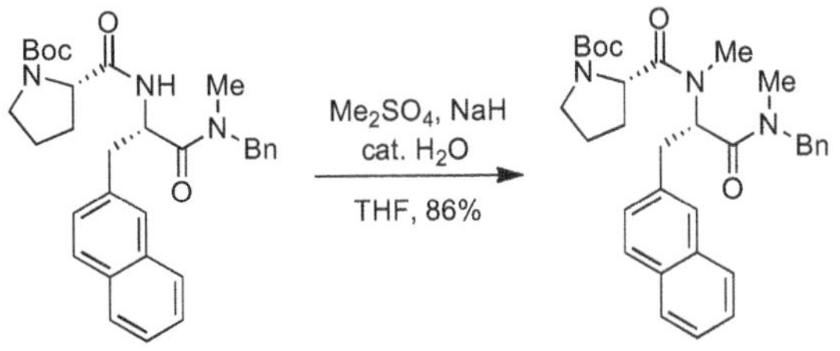

$$(6.55)$$

Procedure

- Charge 6.188 g (154.7 mmol, 60% dispersion in mineral oil) of NaH followed by adding THF (295 mL).
- Cool the suspension to 10 °C.
- Charge a solution of 46.93 g (91.0 mmol) of **207** in toluene, THF (90.9 mL), 17.71 g (140.4 mmol) of dimethyl sulfate, and 0.3276 g (18.2 mmol) of H_2O over 1 h.
- Stir at 10 °C for 3 h.
- Add 49.14 g of 28% NH_4OH over 30 min at 10 °C and stir for 1 h.
- Add water (72.6 mL) and toluene (182.0 mL).
- Stir for 10 min.
- Separate layers.

- Wash the organic layer with water (91 mL).
- Distill the organic layer at 28–45 °C under reduced pressure to collect ca. 374 mL of solvents.
- Heat the resulting residue to 57 °C.
- Add 35.3 mg of **208**.
- Cool the batch to 21 °C over 1 h and stir for 10 h.
- Filter, wash with toluene/heptane (1:5), and dry at 45–50 °C under vacuum.
- Yield: 41.4 g.

6.7.12 BAYLIS–HILLMAN REACTION

The Baylis–Hillman reaction of aldehyde with α,β-unsaturated compound bearing electron-with-drawing group is catalyzed by organic bases such as 1,4-diazabicyclo[2.2.2]octane (DABCO), 4-dimethylaminopyridine (DMAP), 1,8-diazabicycloundec-7-ene (DBU), or phosphines to give allyl alcohol product. Scientists at Pfizer utilized the Baylis–Hillman reaction to access intermediate **210** in the synthesis of sampatrilat, an inhibitor of the zinc metalloprotease. The reaction was carried out in aqueous acetonitrile in the presence of 3-quinuclidinol (0.25 equiv) to produce allyl alcohol **210** in 80–85% (Equation 6.56).[96]

$$
\begin{array}{ccc}
\text{CO}_2t\text{Bu} & \xrightarrow[\substack{\text{H}_2\text{O/MeCN}}]{\substack{\text{paraformaldehyde} \\ \text{3-quinuclidinol} \\ (0.25\ \text{equiv})}} & \overset{\text{OH}}{\diagup}\text{CO}_2t\text{Bu} \\
\mathbf{209} & & \mathbf{210}
\end{array}
\tag{6.56}
$$

Procedure

- Charge water (459 kg) and 12.65 kg (99.4 mol) of 3-quinuclidinol into a reactor and stir to form a clear solution.
- Charge 19.1 kg (636.7 mol) of paraformaldehyde.
- Heat the mixture to 40 °C.
- Add MeCN (201.5 kg).
- Heat to reflux (77 °C).
- Charge cautiously 51 kg (397.9 mol) of **209**.
- Rinse with MeCN (50 L).
- Heat at reflux for 1.5 h.
- Add toluene (255 L) and cool to 30 °C.
- Separate layers.
- Extract the aqueous layer with toluene (76.5 L).
- Wash the combined organic layers with sat. NaCl (2×150 L).
- Add water (100 L).
- Add 2 M citric acid solution to pH 4.
- Separate layers.
- Wash the toluene layer with sat. NaCl (2×100 L).
- Add 150 g of monomethylhydroquinone (as stabilizer).
- Distill at 50 °C under reduced pressure until the final volume of 120 L.
- Yield of **210**: 64.1% (by GC assay).

6.7.13 CATALYTIC WITTIG REACTION

Since the discovery in 1953 by Georg Wittig, Wittig reaction has been used extensively in organic synthesis. Wittig reaction, however, suffers from the removal of the phosphine oxide by-product, which usually requires chromatography. In order to address this issue, the first Wittig reaction

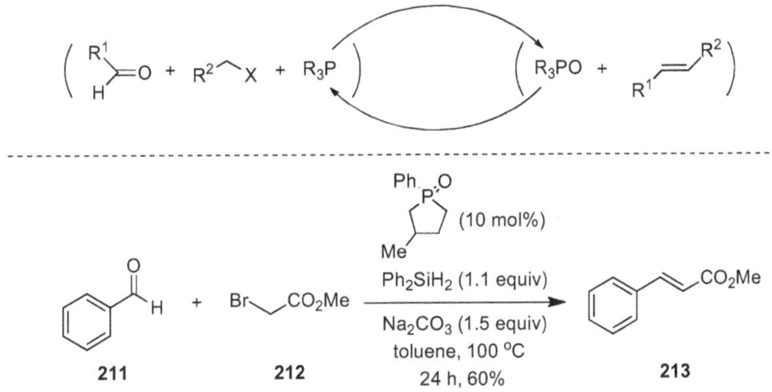

SCHEME 6.26

catalytic in phosphine was developed. Employing 3-methyl-1-phenylphospholane-1-oxide (10 mol%) as a catalyst, the Wittig reaction of benzaldehyde **211** with methyl bromoacetate **212** afforded alkene **213** in 60% yield with 95% of *E*-isomer selectivity (Scheme 6.26).[97]

6.7.14 PALLADIUM-CATALYZED REARRANGEMENT

The Pd-catalyzed rearrangement of epoxides was first introduced by Noyori et al[98] to prepare β-diketones. The rearrangement was carried out with tetrakis(triphenylphosphine)palladium(0) catalyst in the presence of 1,2-bis(diphenylphosphino)ethane ligand. Considering the air-sensitive Pd(PPh$_3$)$_4$, a modified protocol was to combine Pd(OAc)$_2$ with triphenylphosphine to generate Pd(PPh$_3$)$_4$ in situ in the presence of triethylamine as a reducing agent.[99] Pd(PPh$_3$)$_4$ served as the pre-catalyst which would generate the active catalytic species upon reacting with rac-2,2'-bis(di phenylphosphino)-1,1'-binaphthyl (BINAP). The catalyst which was formed by this protocol proved highly active, so the catalyst loading was able to be reduced from 3 mol % to 1 mol% without seeing detrimental effects on the reaction rate. Using the optimized conditions, the conversion of epoxy-hexanone **214** into β-diketone **215** proceeded smoothly at kilogram scale (Equation 6.57).[100]

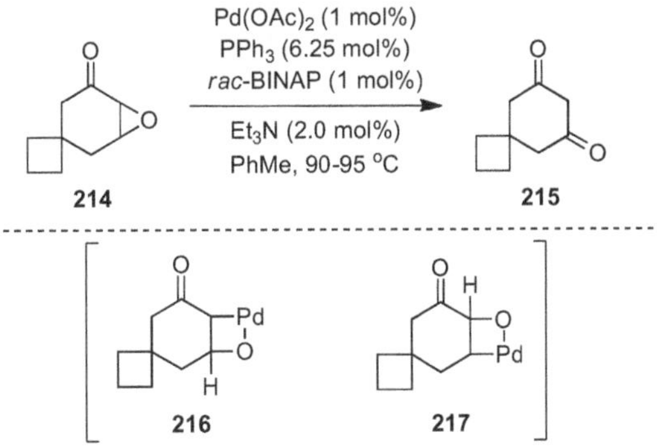

(6.57)

Mechanistically, the reaction may proceed through two oxidative addition intermediates **216** and **217**. Subsequently, **216/217** would undergo β-hydride elimination, reductive elimination, and tautomerization to give the desired product **215**.

Procedure

Stage (a) Catalyst activation
- Charge 8.0 kg solution of **214** in toluene (5% of total **214**) into a 250-L Alloy 59 reaction vessel, followed by adding 103 kg of toluene.

- Charge 3.0 kg (12 mol) of Ph_3P, followed by adding 1.2 kg (1.9 mol) of rac-BINAP, 0.43 kg (1.9 mol) of $Pd(OAc)_2$, and 0.39 kg (3.9 mol) of Et_3N in toluene (5.0 kg) at 20 °C.
- Heat the resulting mixture at 90–95 °C for 3 h to activate the catalytic system.

Stage (b) Pd-catalyzed rearrangement
- Charge the remaining solution of **214** in toluene at 90–92 °C within 50 min.
- Stir the resulting mixture at 91–94 °C for 4.5 h.

Stage (c) Workup
- Cool to 23 °C.
- Add Celite (11.7 kg) and stir for 1 h.
- Filter and wash with toluene (35 kg).
- Add H_2O (175 kg) to the filtrate.
- Add 13.8 kg of 22.5% aqueous NaOH to pH 8.6.
- Separate the layers and wash the aqueous layer with toluene (47.0 kg).
- Add ethyl acetate (34.0 kg) to the aqueous layer and adjust pH to 7.0 with 1.4 kg of 20.5% HCl.
- Separate layers.
- Cool the aqueous layer to 4 ± 2 °C.
- Add 21.0 kg of 12% aqueous HCl to pH 2.0.
- Isolate the solid product with a centrifuge and wash with H_2O (200 kg).
- Dry at 45 °C under 50 mbar for 12 h.
- Yield: 9.9 kg.

REMARK

The catalyst activation was considered to be necessary in order to achieve a consistent reaction rate.

NOTES

1. Onak, T. *Organoborane Chemistry*, Academic Press, New York, 1975.
2. Farmer, J. L.; Pompeo, M.; Lough, A. J.; Organ, M. G. *Chem. Eur. J.* **2014**, *20*, 15790.
3. Itoh, T.; Mase, T. *J. Org. Chem.* **2006**, *71*, 2203.
4. Chekal, B. P.; Guinness, S. M.; Lillie, B. M.; McLaughlin, R. W.; Palmer, C. W.; Post, R. J.; Sieser, J. E.; Singer, R. A.; Sluggett, G. W.; Vaidyanathan, R.; Withbroe, G. J. *Org. Process Res. Dev.* **2014**, *18*, 266.
5. (a) Kim, B. C.; Hwang, S. Y.; Lee, T. H.; Chang, J. H.; Choi, H.; Lee, K. W.; Choi, B. S.; Kim, Y. K.; Lee, J. H.; Kim, W. S.; Oh, Y. S.; Lee, H. B.; Kim, K. Y.; Shin, H. *Org. Process Res. Dev.* **2006**, *10*, 881. (b) Shevlin, M. *Tetrahedron Lett.* **2010**, *51*, 4833.
6. Lipton, M. F.; Mauragis, M. A.; Maloney, M. T.; Veley, M. F.; VanderBor, D. W.; Newby, J. J.; Appell, R. B.; Daugs, E. D. *Org. Process Res. Dev.* **2003**, *7*, 385.
7. de Koning, P. D.; McAndrew, D.; Moore, R.; Moses, I. B.; Boyles, D. C.; Kissick, K.; Stanchina, C. L.; Cuthberton, T.; Kamatani, A.; Rahman, L.; Rodriguez, R.; Urbina, A.; Sandoval, A.; Rose, P. R. *Org. Process Res. Dev.* **2011**, *15*, 1018.
8. Larsen, R. D.; King, A. O.; Chen, C. Y.; Corley, E. G.; Foster, B. S.; Roberts, F. E.; Yang, C.; Lieberman, D. R.; Reamer, R. A.; Tschaen, D. M.; Verhoeven, T. R.; Reider, P. J.; Lo, Y. S.; Rossano, L. T.; Brookes, S.; Meloni, D.; Moore, J. R.; Arnett, J. F. *J. Org. Chem.* **1994**, *59*, 6391.
9. (a) Gala, D.; Stamford, A.; Jenkins, J.; Kugelman, M. *Org. Process Res. Dev.* **1997**, *1*, 163. (b) Ennis, D. S.; McManus, J.; Wood-Kaczmar, W.; Richardson, J.; Smith, G. E.; Carstairs, A. *Org. Process Res. Dev.* **1999**, *3*, 248.
10. Yates, M. H.; Koenig, T. M.; Kallman, N. J.; Ley, C. P.; Mitchell, D. *Org. Process Res. Dev.* **2009**, *13*, 268.
11. Gillmore, A. T.; Badland, M.; Crook, C. L.; Castro, N. M.; Critcher, D. J.; Fussell, S. J.; Jones, K. J.; Jones, M. C.; Kougoulos, E.; Mathew, J. S.; McMillan, L.; Pearce, J. E.; Rawlinson, F. L.; Sherlock, A. E.; Walton, R. *Org. Process Res. Dev.* **2012**, *16*, 1897.
12. Brandt, T. A.; Caron, S.; Damon, D. B.; DiBrino, J.; Ghosh, A.; Griffith, D. A.; Kedia, S.; Ragan, J. A.; Rose, P. R.; Vanderplas, B. C.; Wei, L. *Tetrahedron* **2009**, *65*, 3292.

13. Houpis, I. N.; Shilds, D.; Nettekoven, U.; Schnyder, A.; Bappert, E.; Weerts, K.; Canters, M.; Vermuelen, W. *Org. Process Res. Dev.* **2009**, *13*, 598.
14. Protodeboronation was recently shown to proceed via thermal degradation process: Lee, C.-Y.; Ahn, S.-J.; Cheon, C.-H. *J. Org. Chem.* **2013**, *78*, 12154.
15. (a) Yamamoto, Y.; Takizawa, M.; Yu, X.-Q.; Miyaura, N. *Angew. Chem. Int. Ed.* **2008**, *47*, 928. (b) Gravel, M.; Thompson, K. A.; Zak, M.; Bérubé, C.; Hall, D. G. *J. Org. Chem.* **2002**, *67*, 3.
16. (a) Molander, G. A.; Biolatto, B. *J. Org. Chem.* **2003**, *68*, 4302. (b) Molander, G. A.; Canturk, B.; Kennedy, L. E. *J. Org. Chem.* **2009**, *74*, 973. (c) Butters, M.; Harvey, J. N.; Jover, J.; Lennox, A. J. J.; Lloyd-Jones, G. C.; Murray, P. M. *Angew. Chem. Int. Ed.* **2010**, *49*, 5156.
17. Kinzel, T.; Zhang, Y.; Buchwald, S. L. *J. Am. Chem. Soc.* **2010**, *132*, 14073.
18. Hodgson, P. B.; Salingue, F. H. *Tetrahedron Lett.* **2004**, *45*, 685.
19. Roemmele, R. C.; Christie, M. A. *Org. Process Res. Dev.* **2013**, *17*, 422.
20. Knapp, D. M.; Gillis, E. P.; Burke, M. D. *J. Am. Chem. Soc.* **2009**, *131*, 6961.
21. Dick, G. R.; Woerly, E. M.; Burke, M. D. *Angew. Chem. Int. Ed.* **2012**, *51*, 2667.
22. Grob, J. E.; Dechantsreiter, M. A.; Tichkule, R. B.; Connolly, M. K.; Honda, A.; Tomlinson, R. C.; Hamann, L. C. *Org. Lett.* **2012**, *14*, 5578.
23. Billingsley, K. L.; Buchwald, S. L. *Angew. Chem. Int. Ed.* **2008**, *47*, 4695.
24. Pérez-Balado, C.; Willemsens, A.; Ormerod, D.; Aelterman, W.; Mertens, N. *Org. Process Res. Dev.* **2007**, *11*, 237.
25. Tamao, K.; Kodama, S.; Nakajima, I.; Kumada, M.; Minato, A.; Suzuki, K. *Tetrahedron*, **1982**, *38*, 3354. For other examples, see: Martin, R.; Buchwald, S. L. *J. Am. Chem. Soc.* **2007**, *129*, 3844 and IIa, H.; Baron, O.; Wagner, A. J.; Knochel, P. *Chem. Commun.* **2006**, 583.
26. (a) Iosub, A. V.; Stahl, S. S. *J. Am. Chem. Soc.* **2015**, *137*, 3454. (b) Mfuh, A. M.; Zhang, Y.; Stephens, D. E.; Vo, A. X. T.; Arman, H. D.; Larionov, O. V. *J. Am. Chem. Soc.* **2015**, *137*, 8050.
27. Karlsson, S.; Bergman, R.; Broddefalk, J.; Löfberg, C.; Moore, P. R.; Stark, A.; Emtenäs, H. *Org. Process Res. Dev.* **2018**, *22*, 618.
28. Edney, D.; Hulcoop, D. G.; Leahy, J. H.; Vernon, L. E.; Wipperman, M. D.; Bream, R. N.; Webb, M. R. *Org. Process Res. Dev.* **2018**, *22*, 368.
29. (a) Zhang, C.; Huang, J.; Trudell, M. L.; Nolan, S. P. *J. Org. Chem.* **1999**, *64*, 3804. (b) Bei, X.; Crevier, T.; Guram, A. S.; Jandeleit, B.; Powers, T. S.; Turner, H. W.; Uno, T.; Weinberg, W. H. *Tetrahedron Lett.* **1999**, *40*, 3855.
30. For a related example, see: Payack, J. F.; Vazquez, E.; Matty, L.; Kress, M. H.; McNamara, J. *J. Org. Chem.* **2005**, *75*, 175.
31. Thiel, O. R.; Achmatowicz, M.; Bernard, C.; Wheeler, P.; Savarin, C.; Correll, T. L.; Kasparian, A.; Allgeier, A.; Bartberger, M. D.; Tan, H.; Larsen, R. D. *Org. Process Res. Dev.* **2009**, *13*, 230.
32. Busacca, C. A.; Cerreta, M.; Dong, Y.; Eriksson, M. C.; Farina, V.; Feng, X.; Kim, J.-Y.; Lorenz, J. C.; Sarvestani, M.; Simpson, R.; Varsolona, R.; Vitous, J.; Campbell, S. J.; Davis, M. S.; Jones, P.-J.; Norwood, D.; Qiu, F.; Beaulieu, P. L.; Duceppe, J.-S.; Haché, B.; Brong, J.; Chiu, F.-T.; Curtis, T.; Kelley, J.; Lo, Y. S.; Powner, T. H. *Org. Process Res. Dev.* **2008**, *12*, 603.
33. Hanselmann, R.; Job, G. E.; Johnson, G.; Lou, R.; Martynow, J. G.; Reeve, M. M. *Org. Process Res. Dev.* **2010**, *14*, 152.
34. (a) Adamo, C.; Amatore, C.; Ciofini, I.; Jutand, A.; Lakmini, H. *J. Am. Chem. Soc.* **2006**, *128*, 6829. (b) Yamamoto, Y.; Suzuki, R.; Hattori, K.; Nishiyama, H. *Synlett* **2006**, *7*, 1027. (c) Moreno-Manas, M.; Perez, M.; Pleixats, R. *J. Org. Chem.* **1996**, *61*, 2346.
35. Elitzin, V. I.; Harvey, K. A.; Kim, H.; Salmons, M.; Sharp, M. J.; Tabet, E. A.; Toczko, M. A. *Org. Process Res. Dev.* **2010**, *14*, 912.
36. Li, B.; Barnhart, R. W.; Hoffman, J. E.; Nematalla, A.; Raggon, J.; Richardson, P.; Sach, N.; Weaver, J. *Org. Process Res. Dev.* **2018**, *22*, 1289.
37. Miller, W. D.; Fray, A. H.; Quatroche, J. T.; Sturgill, C. D. *Org. Process Res. Dev.* **2007**, *11*, 359.
38. Rassias, G.; Hermitage, S. A.; Sanganee, M. J.; Kincey, P. M.; Smith, N. M.; Andrews, I. P.; Borrett, G. T.; Slater, G. R. *Org. Process Res. Dev.* **2009**, *13*, 774.
39. Magnus, N. A.; Aikins, J. A.; Cronin, J. S.; Diseroad, W. D.; Hargis, A. D.; LeTourneau, M. E.; Parker, B. E.; Reutzel-Edens, S. M.; Schafer, J. P.; Staszak, M. A.; Stephenson, G. A. Tameze, S. L.; Zollars, L. M. H. *Org. Process Res. Dev.* **2005**, *9*, 621.
40. Manley, P. W.; Acemoglu, M.; Marterer, W.; Pachinger, W. *Org. Process Res. Dev.* **2003**, *7*, 436.
41. (a) Liu, Z.; Xiang, J. *Org. Process Res. Dev.* **2006**, *10*, 285. (b) Miller, J. A.; Farrell, R. P. *Tetrahedron Lett.* **1998**, *39*, 7275.
42. Liu, S.; Berry, N.; Thomson, N.; Pettman, A.; Hyder, Z.; Mo, J.; Xiao, J. *J. Org. Chem.* **2006**, *71*, 7467.

43. Camp, D.; Matthews, C. F.; Neville, S. T.; Rouns, M.; Scott, R. W.; Truong, Y. *Org. Process Res. Dev.* **2006**, *10*, 814.

44. Mino, T.; Suzuki, S.; Hirai, K.; Sakamoto, M.; Fujita, T. *Synlett* **2011**, *9*, 1277.

45. Hu, H.; Yang, F.; Wu, Y. *J. Org. Chem.* **2013**, *78*, 10506.

46. Greene, T. W.; Wuts, P. G. M. *Protective Groups in Organic Synthesis*, 3rd ed.; John Wiley & Sons, New York, 1999.

47. Baumeister, P.; Studer, M.; Roessler, F. *Handbook of Heterogeneous Catalysis*; VCH, New York, 1997.

48. Olofson, R. A.; Martz, J. T.; Senet, J.-P.; Piteau, M.; Malfroot, T. *J. Org. Chem.* **1984**, *49*, 2081.

49. Bret, G.; Harling, S. J.; Herbal, K.; Langlade, N.; Loft, M.; Negus, A.; Sanganee, M.; Shanahan, S.; Strachan, J. B.; Turner, P. G.; Whiting, M. P. *Org. Process Res. Dev.* **2011**, *15*, 112.

50. Ishimoto, K.; Yamaguchi, K.; Nishimoto, A.; Murabayashi, M.; Ikemoto, T. *Org. Process Res. Dev.* **2017**, *21*, 2001.

51. Lukin, K.; Hsu, M. C.; Chambournier, G.; Kotecki, B.; Venkatramani, C. J.; Leanna, M. R. *Org. Process Res. Dev.* **2007**, *11*, 578.

52. Lobben, P. C.; Barlow, E.; Bergum, J. S.; Braem, A.; Chang, S.-Y.; Gibson, F.; Kopp, N.; Lai, C.; LaPorte, T. L.; Leahy, D. K.; Müslehiddinoğlu, J.; Quiroz, F.; Skliar, D.; Spangler, L.; Srivastava, S.; Wasser, D.; Wasylyk, J.; Wethman, R.; Xu, Z. *Org. Process Res. Dev.* **2015**, *19*, 900.

53. Felpin, F.-X.; Fouquet, E. *Chem. Eur. J.* **2010**, *16*, 12440.

54. (a) Augustine, R. L. *Heterogeneous Catalysis for the Synthetic Chemist*; Marcel Dekker, New York, 1996. (b) Auer, E.; Berweiler, M.; Gross, M.; Pietsch, J.; Ostgard, D.; Panster, P. In *Catalysis of Organic Reactions*; Ford, M. E., Ed.; Marcel Dekker, New York, 2000; Vol. 82, p 293. (c) Nishimura, S. *Handbook of Heterogeneous Catalytic Hydrogenation for Organic Synthesis*; Wiley, New York, 2001.

55. (a) For elimination of hydroxylamine by metal salts (including iron), see: Baumeister, P.; Blaser, H. U.; Studer, M. *Catal. Lett.* **1997**, *49*, 219. (b) For hydrogenation of polynitroaromatic compounds by noble metal catalysts in the presence of catalytic amounts of iron; see: Theodoridis, G.; Manfredi, M. C.; Krebs, J. D. *Tetrahedron Lett.* **1990**, *31*, 6141.

56. Boros, E. E.; Burova, S.; Erickson, G. A.; Johns, B. A.; Koble, C. S.; Kurose, N.; Sharp, M. J.; Tabet, E. A.; Thompson, J. B.; Toczko, M. A. *Org. Process Res. Dev.* **2007**, *11*, 899.

57. Gustin, J.-L. *Org. Process Res. Dev.* **1998**, *2*, 27.

58. Fannes, C.; Verbruggen, S.; Janssen, B.; Egle, B. *Org. Process Res. Dev.* **2021**, *25*, 2438.

59. Hoogenraad, M.; van der Linden, J. B.; Smith, A. A.; Hughes, B.; Derrick, A. M.; Harris, L. J.; Higginson, P. D.; Pettman, A. J. *Org. Process Res. Dev.* **2004**, *8*, 469.

60. Wienhöfer, G.; Sorribes, I.; Boddien, A.; Westerhaus, F.; Junge, K.; Junge, H.; Llusar, R.; Beller, M. *J. Am. Chem. Soc.* **2011**, *133*, 12875.

61. Fujino, K.; Takami, H.; Atsumi, T.; Ogasa, T.; Mohri, S.; Kasai, M. *Org. Process Res. Dev.* **2001**, *5*, 426.

62. Hayler, J. D.; Howie, S. L. B.; Giles, R. G.; Negus, A.; Oxley, P. W.; Walsgrove, T. C.; Whiter, M. *Org. Process Res. Dev.* **1998**, *2*, 3.

63. (a) Atkins, R. J.; Banks, A.; Bellingham, R. K.; Breen, G. F.; Carey, J. S.; Etridge, S. K.; Hayes, J. F.; Hussain, N.; Morgan, D. O.; Oxley, P.; Passey, S. C.; Walsgrove, T. C.; Wells, A. S. *Org. Process Res. Dev.* **2003**, *7*, 663. (b) Alston, M.; Willetts, A.; Wells, A. *Org. Process Res. Dev.* **2002**, *6*, 505.

64. Residual Na$_2$S could be detrimental to the catalytic debenzylation: Lee, G. M.; Eckert, J.; Gala, D.; Schwartz, M.; Renton, P.; Pergamen, E.; Whittington, M.; Schumacher, D.; Heimark, L.; Shipkova, P. *Org. Process Res. Dev.* **2001**, *5*, 622.

65. Yang, J. W.; Fonseca, M. T. H.; List, B. *Angew. Chem. Int. Ed.* **2004**, *43*, 6660.

66. Lee, H. W.; Ahn, J. B.; Kang, S. K.; Ahn, S. K.; Ha, D.-C. *Org. Process Res. Dev.* **2007**, *11*, 190.

67. Rueping, M.; Antonchick, A. P.; Theissmann, T. *Angew. Chem. Int. Ed.* **2006**, *45*, 3683.

68. Wang, J.; Tang, Q.; Jin, S.; Wang, Y.; Yuan, Z.; Chi, Q.; Zhang, Z. *New J. Chem.* **2020**, *44*, 549.

69. Watson, T. J.; Ayers, T. A.; Shah, N.; Wenstrup, D.; Webster, M.; Freund, D.; Horgan, S.; Carey, J. P. *Org. Process Res. Dev.* **2003**, *7*, 521.

70. Curtis, N. R.; Davies, S. H.; Gray, M.; Leach, S. G.; McKie, R. A.; Vernon, L. E.; Walkington, A. J. *Org. Process Res. Dev.* **2015**, *19*, 865.

71. Gala, D.; Dahanukar, V. H.; Eckert, J. M.; Lucas, B. S.; Schumacher, D. P.; Zavialov, I. A. *Org. Process Res. Dev.* **2004**, *8*, 754.

72. Milburn, R. R.; Thiel, O. R.; Achmatowicz, M.; Wang, X.; Zigterman, J.; Bernard, C.; Colyer, J. T.; DiVirgilio, E.; Crockettt, R.; Correll, T. L.; Nagapudi, K.; Ranganathan, K.; Hedley, S. J.; Allgeier, A.; Larsen, R. D. *Org. Process Res. Dev.* **2011**, *15*, 31.

73. (a) Chowhury, S.; Roy, S. *J. Org. Chem.* **1997**, *62*, 199. (b) Kuang, C.; Yang, Q.; Senboku, H.; Tokuda, M. *Synthesis* **2005**, 1319.

74. Parker, J. S.; Bowden, S. A.; Firkin, C. R.; Moseley, J. D.; Murray, P. M.; Welham, M. J.; Wisedale, R.; Young, M. J.; Moss, W. O. *Org. Process Res. Dev.* **2003**, *7*, 67.

75. Anderson, N. G.; Ary, T. D.; Berg, J. L.; Bernot, P. J.; Chan, Y. Y.; Chen, C.-K.; Davies, M. L.; DiMarco, J. D.; Dennis, R. D.; Deshpande, R. P.; Do, H. D.; Droghini, R.; Early, W. A.; Gougoutas, J. Z.; Grosso, J. A.; Harris, J. C.; Haas, O. W.; Jass, P. A.; Kim, D. H.; Kodersha, G. A.; Kotnis, A. S.; LaJeunesse, J.; Lust, D. A.; Madding, G. D.; Modi, S. P.; Moniot, J. L.; Nguyen, A.; Palaniswamy, V.; Phillipson, D. W.; Simpson, J. H.; Thoraval, D.; Thurston, D. A.; Tse, K.; Polomski, R. E.; Wedding, D. L.; Winter, W. J. *Org. Process Res. Dev.* **1997**, *1*, 300.

76. Magnus, N. A.; Aikins, J. A.; Cronin, J. S.; Diseroad, W. D.; Hargis, A. D.; LeTourneau, M. E.; Parker, B. E.; Reutzel-Edens, S. M.; Schafer, J. P.; Staszak, M. A.; Stephenson, G. A. Tameze, S. L.; Zollars, L. M. H. *Org. Process Res. Dev.* **2005**, *9*, 621.

77. Maloney, K. M.; Kuethe, J. T.; Linn, K. *Org. Lett.* **2011**, *13*, 102.

78. Gnanasekaran, K. K.; Nammalwar, B.; Murie, M.; Bunce, R. A. *Tetrahedron Lett.* **2014**, *55*, 6776.

79. Larkin, J. P.; Webrey, C.; Boffelli, P.; Lagraulet, H.; Lemaitre, G.; Nedelec, A.; Prat, D. *Org. Process Res. Dev.* **2002**, *6*, 20.

80. (a) Lednicer, D.; Babcock, J. C.; Marlatt, P. E.; Lyster, S. C.; Duncan, G. W. *J. Med. Chem.* **1965**, *8*, 52. (b) Short, J. H.; Biermacher, U.; Dunnigan, D. A.; Lambert, G. F.; Martin, D. L.; Nordeen, C. W.; Wright, H. B. *J. Med. Chem.* **1965**, *8*, 223. (c) Robertson, D. W.; Katzenellenbogen, J. A. *J. Org. Chem.* **1982**, *47*, 2387.

81. Zheng, B.; Fox, R. J.; Sugiyama, M.; Fritz, A.; Eastgate, M. D. *Org. Process Res. Dev.* **2014**, *18*, 636.

82. Dozeman, G. J.; Fiore, P. J.; Puls, T. P.; Walker, J. C. *Org. Process Res. Dev.* **1997**, *1*, 137.

83. Ashcroft, C. P.; Dunn, P. J.; Hayler, J. D.; Wells, A. S. *Org. Process Res. Dev.* **2015**, *19*, 740.

84. Stuk, T. L.; Assink, B. K.; Bates, Jr., R. C.; Erdman, D. T.; Fedij, V.; Jennings, S. M.; Lassig, J. A.; Smith, R. J.; Smith, T. L. *Org. Process Res. Dev.* **2003**, *7*, 851.

85. (a) Zanka, A.; Hashimoto, N.; Uematsu, R.; Okamoto, T. *Org. Process Res. Dev.* **1998**, *2*, 320. (b) Zanka, A.; Uematsu, R.; Morinaga, Y.; Yasuda, H.; Yamazaki, H. *Org. Process Res. Dev.* **1999**, *3*, 389.

86. O'Shea, P. D.; Gauvreau, D.; Gosselin, F.; Hughes, G.; Nedeau, C.; Roy, A.; Shultz, S. *J. Org. Chem.* **2009**, *74*, 4547.

87. Barnes, D. M.; Christesen, A. C.; Engstrom, K. M.; Haight, A. R.; Hsu, M. C.; Lee, E. C.; Peterson, M. J.; Plata, D. J.; Raje, P. S.; Stoner, E. J.; Tedrow, J. S.; Wagaw, S. *Org. Process Res. Dev.* **2006**, *10*, 803.

88. Asselin, S. M.; Bio, M. M.; Langille, N. F.; Ngai, K. Y. *Org. Process Res. Dev.* **2010**, *14*, 1427.

89. Pesti, J.; Chen, C-K.; Spangler, L.; DelMonte, A. J.; Benoit, S.; Berglund, D.; Bien, J.; Brodfuehrer, P.; Chan, Y.; Corbett, E.; Costello, C.; DeMena, P.; Discordia, R. P.; Doubleday, W.; Gao, Z.; Gingras, S.; Grosso, J.; Haas, O.; Kacsur, D.; Lai, C.; Leung, S.; Miller, M.; Muslehiddinoglu, J.; Nguyen, N.; Qiu, J.; Olzog, M.; Reiff, E.; Thoraval, D.; Totleben, M.; Vanyo, D.; Vemishetti, P.; Wasylak, J.; Wei, C. *Org. Process Res. Dev.* **2009**, *13*, 716.

90. Bodanszky, M. *Peptide Chemistry*, 2nd ed.; Springer-Verlag: New York, 1993.

91. Hayes, B. B.; Gerber, P. C.; Griffey, S. S.; Meade, B. J. *Drug Chem. Toxicol.* **1998**, *24* (2), 195.

92. Clark, J. D.; Anderson, D. K.; Banaszak, D. V.; Brown, D. B.; Czyzewski, A. M.; Edeny, A. D.; Forouzi, P. S.; Gallagher, D. J.; Iskos, V. H.; Kleine, H. P.; Knable, C. M.; Lantz, M. K.; Lapack, M. A.; Moore, C. M. V.; Muellner, F. W.; Murphy, J. B.; Orihuela, C. A.; Pietz, M. A.; Rogers, T. E.; Ruminski, P. G.; Santhanam, H. K.; Schilke, T. C.; Shah, A. S.; Sheikh, A. Y.; Weisenburger, G. A.; Wise, B. E. *Org. Process Res. Dev.* **2009**, *13*, 1088.

93. (a) Léon, P.; Garbay-Jaureguiberry, C.; Barsi, M. C.; Le Pecq, J. B.; Roques, B. P. *J. Med. Chem.* **1987**, *30*, 2074. (b) Caubere, P.; Jamart-Gregoire, B.; Caubere, C.; Bizot-Espiard, J. G.; Renard, P.; Adam, G. Eur. Pat. Appl. EP 624,575 (*Chem. Abstr.* **1995**, *122*, 81177f). (c) Morales-Rios, M. S.; Joseph-Nathan, P. *Rev. Soc. Quim. Méx.* **1989**, *33*, 331.

94. Bessard, Y. *Org. Process Res. Dev.* **1998**, *2*, 214.

95. (a) Prashad, M.; Har, D.; Hu, B.; Kim, H. Y.; Repič, O.; Blacklock, T. J. *Org. Lett.* **2003**, *5*, 125. (b) Prashad, M.; Har, D.; Hu, B.; Kim, H.-Y.; Girgis, M. J.; Chaudhary, A.; Repič, O.; Blacklock, T. J. *Org. Process Res. Dev.* **2004**, *8*, 330.

96. Dunn, P. J.; Hughes, M. L.; Searle, P. M.; Wood, A. S. *Org. Process Res. Dev.* **2003**, *7*, 244.

97. O'Brien, C. J.; Tellez, J. L.; Nixon, Z. S.; Kang, L. J.; Carter, A. L.; Kunkel, S. R.; Przeworski, K. C.; Chass, C. A. *Angew. Chem. Int. Ed.* **2009**, *48*, 6836.

98. Suzuki, M.; Watanabe, A.; Noyori, R. *J. Am. Chem. Soc.* **1980**, *102*, 2095.

99. Amatore, C.; Jutand, A.; M'Barki, M. A. *Organometallics* **1992**, *11*, 3009.

100. Lehmann, T. E.; Kuhn, O.; Krüger, J. *Org. Process Res. Dev.* **2003**, *7*, 913.

7 Process Optimization of Problematic Reactions

Temperature is an important factor in the optimization of chemical processes. In general, a reaction rate doubles for every 10 °C rise in reaction temperature. Although high temperature makes a reaction faster, in many cases, an elevated temperature can have negative impacts on the reaction outcomes, such as inferior selectivity and degradation of the product. High reaction temperatures will also lead to high energy costs and potential runaway reactions. Therefore, extreme reaction conditions (–20 °C> T >150 °C) shall be avoided at large production scales. The pressure effect is normally observed in gas-involved heterogeneous reactions such as catalytic hydrogenation and Kolbe–Schmit reactions. In general, high gas pressure has a beneficial effect on the reaction rate. However, high pressure may also generate over-reduction side products during catalytic hydrogenation.

Most chemical reactions proceed through intermediates. Unstable reaction intermediates may, in addition to the desired reaction, participate in undesired reaction pathways to lead to side products. Organic transformation frequently needs functional group protection in order to mitigate side reactions and achieve the desired reaction pathway. The selection of protecting group is an important step in organic synthesis, especially when a molecule has two or more functional groups.

This chapter will discuss various opportunities associated with the reaction temperature, pressure, unstable reaction intermediates, and functional group protections during the chemical process development.

7.1 TEMPERATURE EFFECT

7.1.1 METAL–HALOGEN/HYDROGEN EXCHANGE

Functionalization of the C–H (X) bond normally employs a two-step protocol including metalation followed by trapping the resulting organometallic intermediate with electrophiles. Lithium–hydrogen/halogen exchange with alkyllithium reagents becomes one of the most useful methods in synthetic chemistry to generate aryllithium species. Strongly polarized by the electron-positive character of lithium, organolithium compounds are highly reactive nucleophiles and react with almost all types of electrophiles. Some simple organolithium compounds that are commercially available include LDA, n-BuLi, sec-BuLi, t-BuLi, MeLi, HexLi, and PhLi. For safety reasons, a semibatch process is usually used on scale-up by adding the reactive organolithium reagent to the pre-cooled reaction mixture. On occasion, organolithium intermediates are insoluble in the reaction media and lithium halide salts are often used as additives to improve the solubility, reactivity, and stability of organolithium reagents. The lithium–halogen exchange is a reversible process with the rates decreasing in the order of I > Br > Cl. Due to strong C–F bonds, aryl fluorides do not undergo lithium–fluorine exchange. Both the alkyllithium and the resulting aryllithium are highly reactive and unstable under normal reaction conditions, thereby cryogenic temperature is frequently required for such lithium–halogen exchange process. Deprotonation can be a side reaction with enolizable carbonyl compounds, especially with hindered organolithium reagents such as $tert$-butyllithium.

REMARK

Due to their high reactivity, organolithium compounds are incompatible with water, oxygen, and carbon dioxide, and must be handled under a protective atmosphere, such as nitrogen or argon. Organolithium compounds are often corrosive and flammable (t-BuLi is pyrophoric). Alkyllithium reagents can also undergo thermal decomposition to form the corresponding alkyl species and lithium hydride. Organolithium reagents are typically stored below 2–8 °C.

7.1.1.1 Magnesium–Bromine Exchange

Generally, Grignard reagents are prepared by refluxing a heterogenous mixture of aryl bromides and magnesium metal in THF. However, these reactions frequently suffer from incomplete conversion and hard-to-control exotherm events that may lead to runaway reactions. Magnesium–halogen exchange has emerged as a viable synthetic tool used to generate arylmagnesium in situ. For instance, magnesium–bromine exchange was employed to prepare aryl Grignard reagent **2** from 2-Bromo-4-fluoroanisole **1** (Equation 7.1).[1] The protocol used 2.0 equiv of isopropyl magnesium chloride in THF under reflux and the exchange was completed in 1 h.

$$(7.1)$$

(I) Problems

Besides the desired Mg–Br exchange Grignard product **2**, a significant amount of side product **3** was also observed under reflux conditions. The formation of **3** may be attributed to the reaction of **2** with THF followed by reacting with isopropyl bromide.

(II) Solutions

When the Mg–Br exchange was performed at 30 °C, levels of **3** were dramatically reduced from 20.8% to 0.4%.

7.1.1.2 Lithium–Bromine Exchange

Due to the high reactivity of organolithium compounds, lithium–halogen exchange is carried out at cryogenic temperature. However, cryogenic reaction conditions are undesired on a large-scale preparation because such conditions will not only generate a viscous reaction mixture that may create agitation problems but also significantly increase the process costs.

The synthesis of spironolactone **6** was carried out by treatment of 2-bromobenzoic acid **4** with 2.2 equiv of *n*-BuLi at −78 °C, followed by the addition of *N*-benzylpiperidone **5** and acidification (Equation 7.2).[2]

$$(7.2)$$

Besides the requirement of costly cryogenic reaction conditions, two side products, benzoic acid **7** and tertiary alcohol **8**, were generated.

Solutions

An approach was developed to overcome the aforementioned issues. The combination of the Grignard reagent with *n*-BuLi was envisioned to convert the less stable aryllithium **9** to the thermally more stable arylmagnesium **10**, and thus could potentially obviate the costly cryogenic conditions (Scheme 7.1).

Therefore, the sequential treatment of 2-bromobenzoic acid **4** with *n*-BuMgCl (0.9 equiv) and *n*-BuLi (1.3 equiv) at –20 to 0 °C, followed by adding *N*-benzyl piperidone **5** afforded the spironolactone **6** in 74% yield.

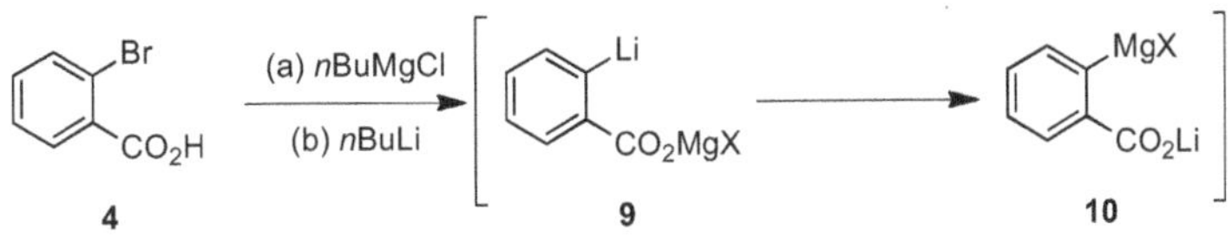

SCHEME 7.1

Procedure

- Charge 4.5 L of *n*-BuMgCl (2.0 M in THF, 9.0 mol) slowly to a solution of **4** (2.0 kg, 9.95 mol) in 20 L of THF, followed by adding 8.3 L of *n*-BuLi (1.56 M in hexane, 12.9 mol) at –15 to –5 °C.
- Stir for 0.5 h.
- Charge a solution of **5** (2.02 L, 10.9 mol) in 6.0 L of heptane over 1 h.
- Stir the mixture for 1 h.
- Add 10.0 L of MTBE and 3.14 L (54.8 mol) of AcOH in 20 L of water.
- Heat to 35 to 40 °C and stir for 3 h.
- Workup.

7.1.1.3 Lithium–Hydrogen Exchange

An isolated, apolar C–H bond in a molecule has a very low reactivity owing to the large kinetic barrier associated with the C–H bond cleavage. Therefore, metal–hydrogen exchange typically occurs at the C–H bond adjacent to a heteroatom such as oxygen and nitrogen or at the C–H bond where the hydrogen is relatively acidic. Lithium organic compounds are frequently used in the metalation reaction under cryogenic conditions especially when a reaction involves unstable intermediates. The synthesis of thienopyrimidine carbaldehyde **12** was achieved via a two-step, one-pot process involving lithiation of thienopyrimidine **11** with *n*-butyllithium at –70 to –50 °C followed by quenching the resulting lithium intermediate **13** with DMF (Equation 7.3).[3] Apparently, such cryogenic conditions and instability of the organolithium species **13** should be precluded on large scale.

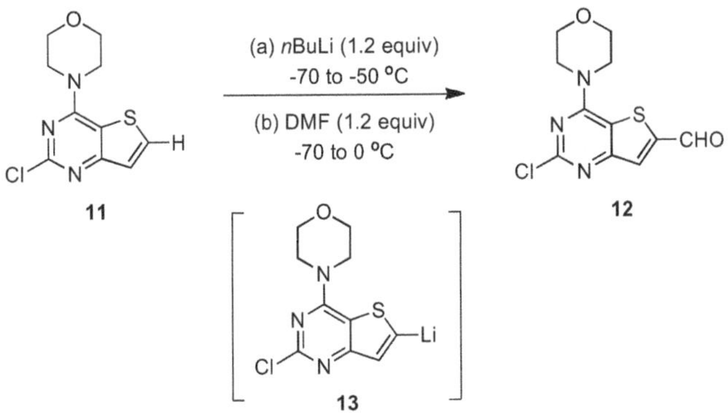

(7.3)

Solutions

In an effort to address the cryogenic condition and instability of **13**, lithium triarylmagnesiate **14** (Figure 7.1) was considered to replace less stable **13**.

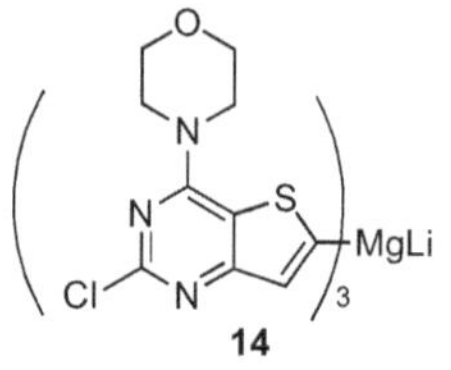

FIGURE 7.1   The structure of lithium triarylmagnesiate **14**.

Compared with **13**, **14** was more stable and could be conveniently prepared with LiMg(nBu)$_2$(i-Pr) at −10 °C. LiMg(nBu)$_2$(i-Pr) reagent could be generated in situ by a reaction of isopropylmagnesium chloride with n-butyllithium. Ultimately, an operationally simple process was developed and demonstrated on a 20-kg scale, producing the product **12** with 87% yield.

Procedure

- Charge 20.1 kg (39.1 mol, 0.51 equiv) of 20% solution of i-PrMgCl in THF to a cold (−10 °C) mixture of **11** (19.8 kg, 77.4 mol) in THF (197 L) over 1.5 h.
- Charge 32.6 kg (76.3 mol, 0.99 equiv) of 15% solution of n-BuLi in hexanes at ≤−10 °C over 1.5 h.
- Stir the resulting mixture at −10 °C for 1 h.
- Charge slowly 8.8 kg (120 mol, 1.6 equiv) of anhydrous DMF at −10±5 °C.
- Stir the batch for 4 h.
- Transfer the batch to a cold mixture of AcOH (58.7 kg), 35% aqueous HCl (21.3 kg), and water (159 kg) over 1.5 h.
- Stir the resulting mixture for 1 h.
- Heat to 55 °C over 4 h and stir for 3 h.
- Cool to 20–30 °C over 1 h and age for 1 h.
- Filter and wash with water (4×25 kg).
- Dry at 50 °C under vacuum.
- Yield: 19.2 kg (87%).

7.1.2   RING EXPANSION

The beneficial effect of higher reaction temperature was observed during the ring expansion of hydroxyisoinolinone **15** in the presence of a stoichiometric amount of hydrazine, affording 6-chlorophthalazin-1-ol **16** (Equation 7.4).[4] It was found that the reaction at a lower temperature (<90 °C) generated impurity **17**.

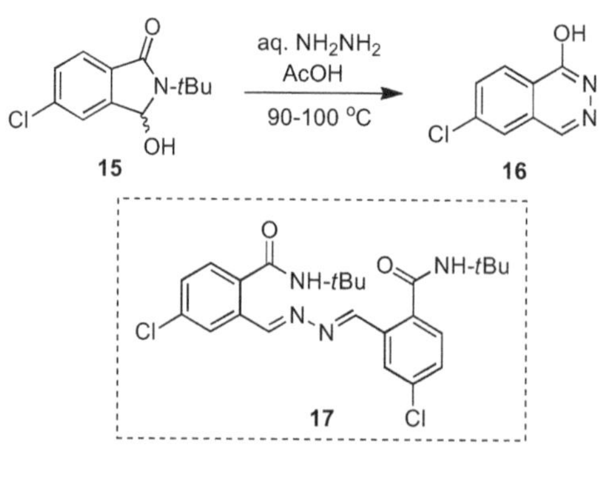

$$(7.4)$$

Scheme 7.2 indicates that the formation of the desired product **16** and impurity **17** occurs via the same intermediate **19**. Higher reaction temperature (>90 °C) and diluted reaction mixture would suppress the formation of impurity **17**.

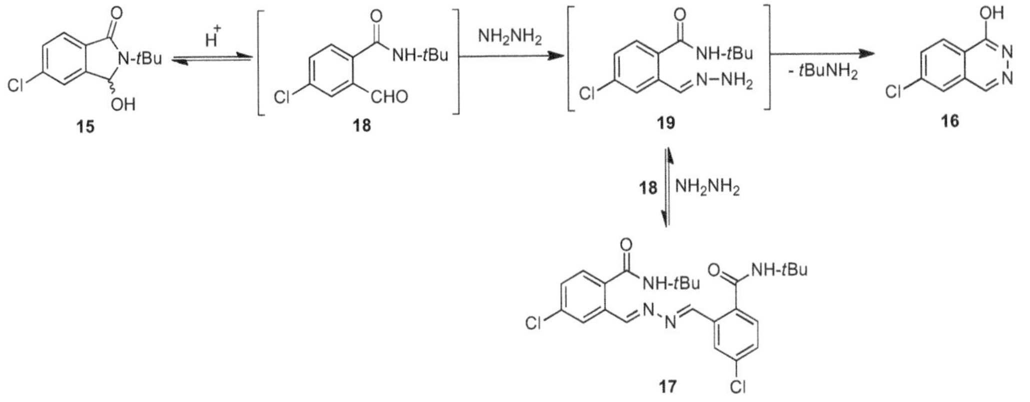

SCHEME 7.2

Procedure

- Charge 2.06 kg (8.59 mol) of **15** into a 30-L jacketed reactor under nitrogen, followed by adding glacial AcOH (6.4 L).
- Heat the resulting thick slurry to 90 °C.
- Charge, dropwise, 0.53 kg (9.0 mol, 1.05 equiv) of 54 wt% hydrazine hydrate (exotherm) while keeping the internal temperature at 90–93 °C over 3 h.
- Continue to stir at 90 °C for 1.5 h (until **15** is <1%).
- Charge preheated water (80 °C, 12.8 L) at 80–90 °C.
- Cool to 20 °C over 4 h.
- Filter, wash with water (2×3.0 L), and dry to give a crude **16**.
- Charge 3.59 kg of crude **16** (from two batches) into a 30-L reactor.
- Add CH_2Cl_2 (18.0 L) and stir at 20 °C for 1 h.
- Filter, wash with CH_2Cl_2 (4.0 L), and dry at 50 °C.
- Yield: 2.76 kg (89.0%).

7.1.3 SYNTHESIS OF PYRAZOLE

The synthesis of pyrazole **22** comprises two stages: 1,3-dipolar cycloaddition between the in situ generated nitrile imine dipole **23** and enamine **20** to form dihydropyrazole **24** and morpholine elimination (Scheme 7.3).[5]

(I) Reaction problems
Reactions at stage (a) at 70 °C generated as much as 10–20% of a side product **25** resulting from a reaction of **21** (or **23**) with the released morpholine.

(II) Solutions
In order to minimize the elimination of morpholine at stage (a), reactions of stage (a) were conducted at a lower temperature (40 °C), reducing levels of **25** to 3–5%.

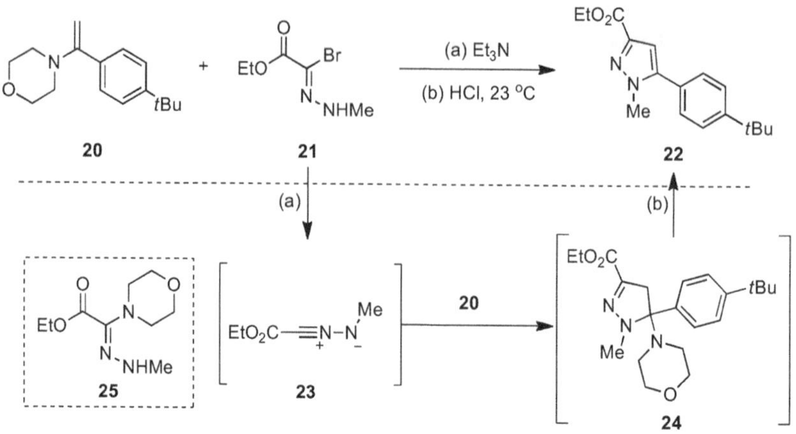

SCHEME 7.3

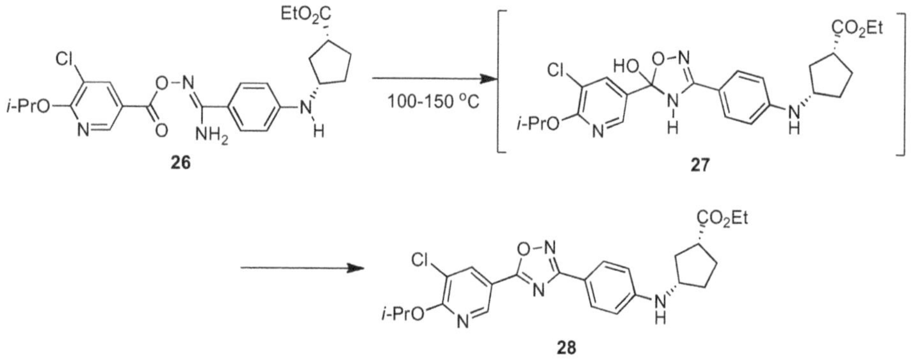

SCHEME 7.4

7.1.4 SYNTHESIS OF OXADIAZOLE

A cyclization of acylamidoxime **26** under thermal conditions (100–150 °C) led to a low yield of oxadiazole **28** with various amounts of degradation products (Scheme 7.4).[6]

Initial evaluation of amine bases such as triethylamine or diisopropylethylamine showed that both bases failed to convert **26** to **28** at 60 °C in THF. However, it was found that using a stronger base, that is, 1,8-diazabicyclo[5.4.0]undec-7-ene (DBU) or tetramethylguanidine (TMG) provided a remarkable acceleration of the reaction rate, allowing the cyclization to occur at 60 °C to give the oxadiazole **28**. DBU or TMG presumably promotes the dehydration of the intermediate **27**. Thereby, this relatively low reaction temperature with DBU as the base suppressed the formation of degradation-related side products.

7.1.5 CROSS-COUPLING REACTION

The temperature effect was also observed in palladium-catalyzed cross-coupling reactions. In general, the oxidative insertion of palladium catalyst into the aryl–halogen bond occurs at an elevated temperature. The elevated reaction temperature may have detrimental effects on the reaction selectivity especially when the substrate bears other halogen substituents.

Early process development found that the cross-coupling between ketone **29** and aryl bromide **30** at 80 °C produced two side products: *tert*-butylbenzene **32** (from debromination of **30**) and ketone **33** (as a result of further cross-coupling between product **31** and **29**) (Equation 7.5).[7]

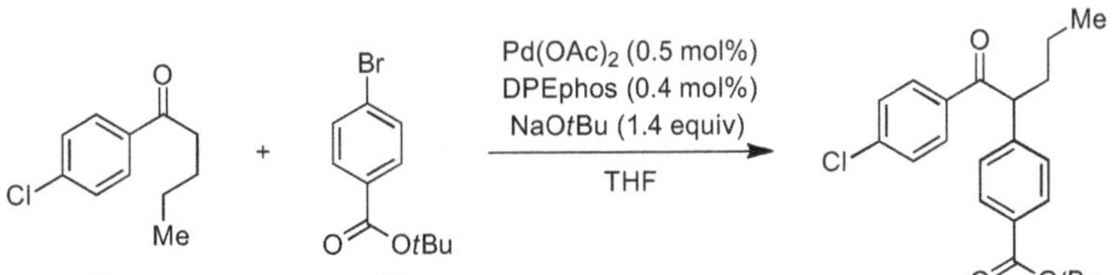

(7.5)

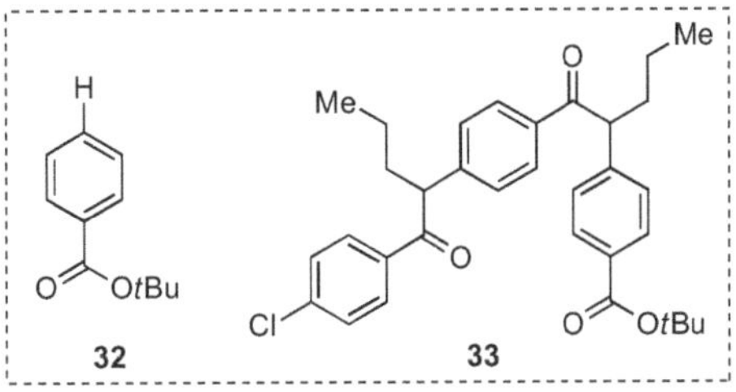

These side reactions were suppressed successfully by lowering the reaction temperature to 58–64 °C.

Procedure

- Degas a solution of NaOtBu (53 kg, 552 mol, 1.4 equiv) in THF (460 L) via three vacuum/ N₂ purge cycles.
- Charge 454 g (2 mol, 0.5 mol%) of Pd(OAc)₂, followed by adding (oxidi-2,1-phenlene)bis(d iphenylphosphine) (DPEphos) (1.08 kg, 1.59 mol, 0.4 mol%).
- Degas the batch via three vacuum/N₂ purge cycles.
- Stir for 30 min.
- Charge 79 kg (98.4%, 395 mol, 1.0 equiv) of **29** and 160.7 kg (71.3 wt% THF solution, 446 mol, 1.13 equiv) of **30**.
- Degas the batch three vacuum/N₂ purge cycles.
- Heat to 58–64 °C and stir at the temperature for 8 h.
- Cool to 15–25 °C.
- Work up (involving reverse quenching by transferring the batch into a mixture of heptane and an aqueous solution of NaHCO₃).

7.1.6 VILSMEIER REACTION

Phosphoryl trichloride (POCl₃) is a powerful deoxychlorination agent and is commonly used to prepare chlorinated heteroaromatics or generate in situ Vilsmeier–Haack reagent from reaction with DMF. For instance, the formylation of indole **34** via the Vilsmeier reaction with POCl₃ (3 equiv) and DMF (6 equiv) at 90 °C furnished the desired aldehyde **35** (Equation 7.6).[8]

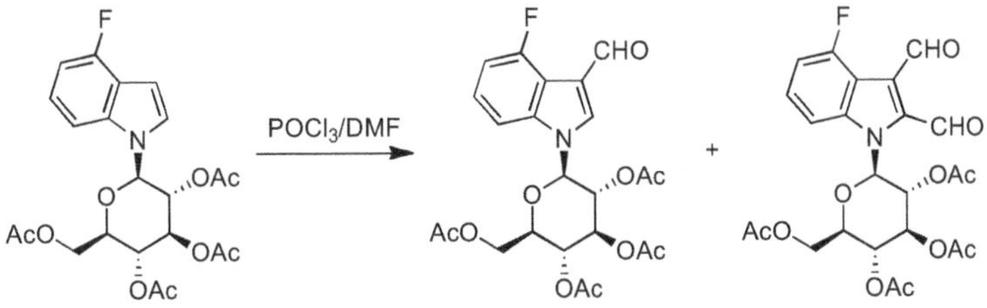

(7.6)

In addition to the desired product **35**, the reaction produced also diformylated **36** as the major product.

It was found that the diformylation rate was highly dependent on the reaction temperature. At 40 °C, **36** was not detectable even after extended aging. Ultimately, the formylation reaction was carried out at 40 °C with POCl$_3$ (1.4 equiv) and DMF (2.5 equiv), which mitigated **36** to undetectable levels.

7.1.7 OXIDATIVE HYDROLYSIS

Following the literature procedure, the conversion of isatin **37** to anthranilic acid **38** was carried out using 10 equiv of NaOH (18%, w/w) and 3 equiv of hydrogen peroxide at 60–65 °C (Equation 7.7).[9] At scale, due to the prolonged heating and cooling operations, a significant amount of 3-hydroxy-anthranilic acid **39** was formed.

Evaluation of the reaction found that conducting the reaction at a lower temperature (25 °C) with 1.4 equiv of hydrogen peroxide mitigated the formation of **39**.

(7.7)

Procedure

Stage (a)
- Charge 4.6 kg (26.6 mol) of **37** portionwise to a solution of NaOH (11.5 kg, 287.5 mol, 10.8 equiv) in water (59 L) at ambient temperature over 40 min.
- Charge 4.3 L (38 mol, 1.4 equiv) of 30% H$_2$O$_2$ solution at 20–25 °C over 2.5 h.
- Stir for 4 h.

Stage (b)
- Add water (23 L).
- Add ca. 10 kg of 36% HCl over 1 h at <15 °C to pH 2.
- Stir for 30 min.
- Filter and wash with water (2×10 L).
- Dry at 50–60 °C under reduced pressure.
- Yield: 3.5 kg (70%).

Note: See the reference[10] for plausible reaction pathways for the formation of the product **38** and side product **39**.

7.2 PRESSURE EFFECT

7.2.1 NITRILE REDUCTION

The pressure effect was observed in the hydrogenation of nitrile **40** during the synthesis of aldehyde **42**. Scheme 7.5[11] illustrates the two-step, one-pot process involving hydrogenation of nitrile **40** to the corresponding imine **41**, followed by hydrolysis.

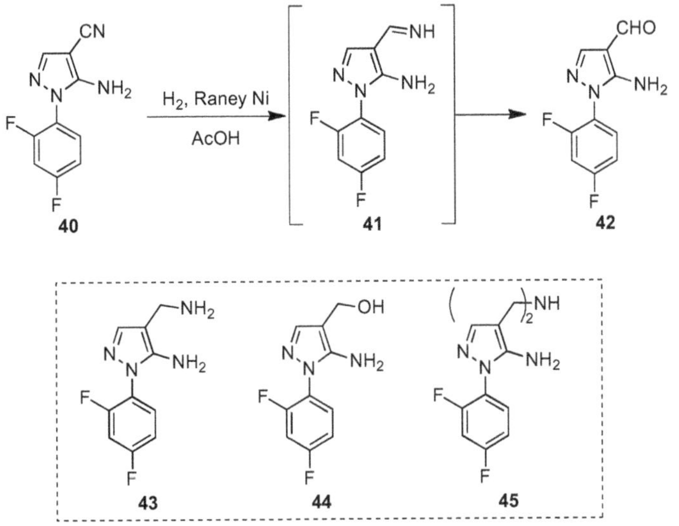

SCHEME 7.5

(I) Reaction problems

Initially, the hydrogenation reaction was carried out in an acetic acid/water mixture with a Raney nickel catalyst under 100 psig of hydrogen. Under these conditions, three potential side products may be generated: amine **43** from the reduction of imine **41**, alcohol **44** from the reduction of product **42**, and amine **45** from the condensation of **43** with **41** and/or **42** followed by reduction.

(II) Solutions

As over-reduction is the cause for the formation of the three side products, maintaining a low steady-state concentration of **41** and a low hydrogen pressure would mitigate the over-reduction. Low hydrogen pressure would decrease the rate of the hydrogenation reaction relative to the rate of hydrolysis. In this event, using 15 psig of hydrogen provided a significantly cleaner reaction profile, generating only 4% of **43** compared to 25% at 100 psig H_2.

Procedure

- Charge 1.02 kg (4.63 mol) of **40** into a 10-L autoclave.
- Charge 4.5 L of AcOH and a slurry of Raney nickel 2400 (45 g) in water (0.25 L).
- Rinse with water (1.25 L).
- Purge the autoclave with hydrogen (15 psig) then evacuate.
- Repeat five times.
- Pressurize the autoclave with 15 psig hydrogen.
- Stir (800 rpm) for 9 h.
- Purge the reactor with argon.
- Charge AcOH (1.5 L) to the reactor.
- Filter and rinse with AcOH (0.5 L).
- Transfer the filtrate (from two batches) into a 60-L jacketed reactor and rinse with AcOH (1.0 L).
- Adjust the temperature to 6 °C.
- Add 5 N aqueous NaOH (6 L) and stir the resulting mixture for 1 h.
- Add 5 N aqueous NaOH (15 L) over 1.5 h and stir at 20 °C for 75 min.
- Filter and wash with water (2×4 L).
- Dry under a stream of nitrogen.
- Yield: 1.56 kg (70.8%).

7.2.2   [3+2]-CYCLOADDITION

A pressure effect on regioselectivity was reported for palladium-catalyzed [3+2]-cycloadditions (Scheme 7.6).[12] Under 1 bar of pressure, cycloadduct **51** was observed as the major product along with **50** as a regioisomer. The preferable formation of **51** occurred through a thermodynamically favored trimethylenemethane (TMM) complex **48**. However, only **50** was formed under 10 bar in 71% yield. This increase in regioselectivity is presumably due to the rate acceleration in the bimolecular cycloaddition of **47** and **49** under high pressure compared to the isomerization between TMM complexes **47** and **48**.

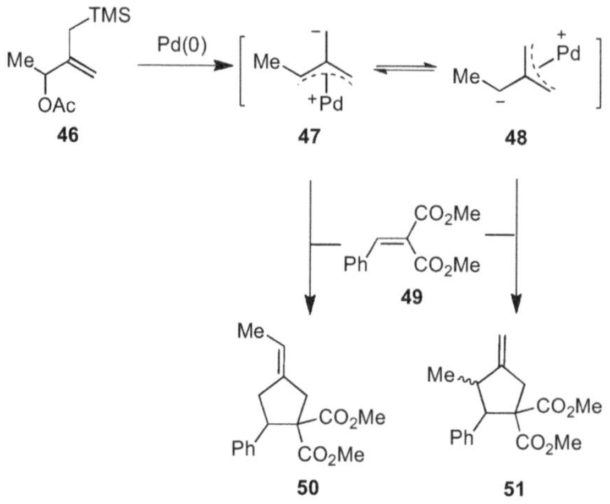

SCHEME 7.6

7.3   LOW PRODUCT YIELDS

Low product yields can be caused by a number of factors, including incomplete reactions, side reactions, loss of products during isolation and purification, etc. To address these issues and improve product yields, various strategies have been developed.

7.3.1   INCOMPLETE REACTIONS

Besides temperature and pressure, incomplete reactions may result from conditions such as poor mass transfer, poor solubility of starting materials, and high flow rate of inert nitrogen.

7.3.1.1   Poor Mass Transfer

Non-proper reactor configuration can cause incomplete reactions due to the less efficient mass transfer, particularly under heterogeneous conditions.

7.3.1.1.1   Schotten–Baumann Reaction

Acetylation of amine **52** under Schotten–Baumann reaction conditions in the laboratory gave a complete conversion, while under the same conditions the reaction failed to complete in the pilot plant (Equation 7.8)[13]. The failure of this heterogeneous reaction was caused by the poor mass transfer in the pilot plant reactor.

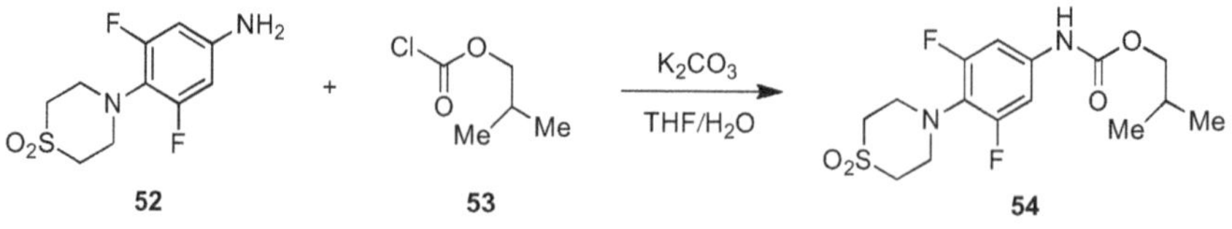

(7.8)

REMARK

Attention should be paid to the difference in reactor configurations between the laboratory and the plant. Ideally, process development parameters (such as temperature, pressure, heating/cooling profiles, agitation rate, etc) shall be generated from the plant-mimicked vessel.

7.3.1.1.2 Hydrogenation

Efficient agitation can be crucial for reaction rate, conversion, impurity profile, etc. The impact of agitation on the outcome of a heterogeneous reaction is apparent. It was observed that the conversion of 5-(hydroxymethyl)furan-2-carbaldehyde **55** to diol **57** was sluggish in a 2-L Parr pressure reactor (Scheme 7.7).[14] At a lower agitation rate (<200 rpm), a large portion of Raney nickel remained on the bottom of the reactor, which led to a slow and incomplete conversion. With a higher agitation rate (600–700 rpm), the reduction was completed within 4 h.

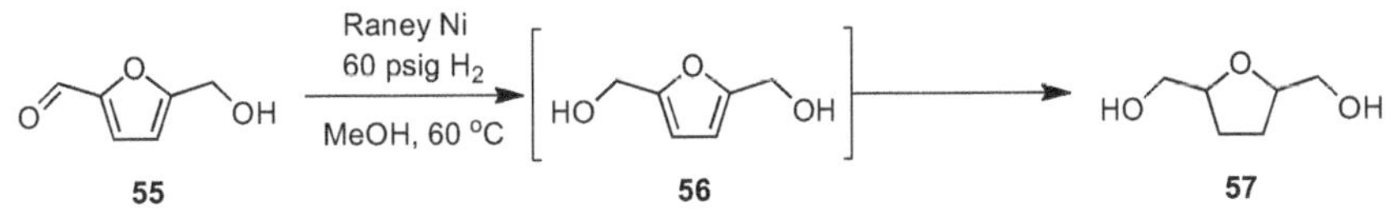

SCHEME 7.7

Notes

(a) The use of an axial flow impeller is advisable when a heterogeneous reaction involves an insoluble high-density solid or liquid.

(b) For some reactions, mixing of starting materials with reagents may result in an unstirrable mass which would damage the agitator. Therefore, the agitator has to be turned off temporarily. For instance, large-scale conversion of acid **58** to its corresponding phosphate salt **59** occurred by mixing dried **58** with water and acetonitrile, followed by the addition of H_3PO_4 without agitation (because mixing the slurry at ambient temperature resulted in a cement-like material that literally stopped mechanical agitation) (Equation 7.9).[15]

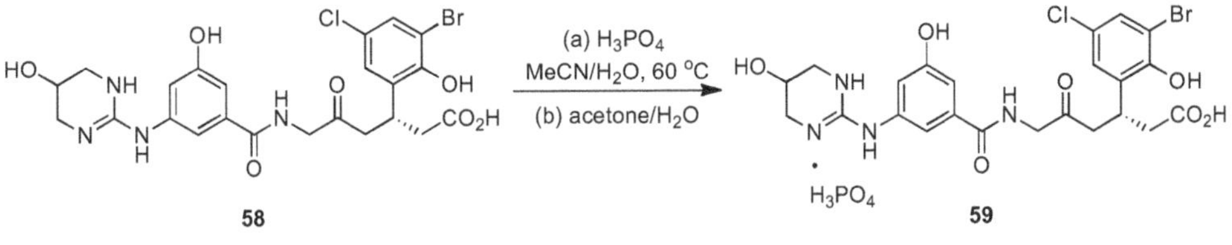

Procedure

Stage (a)

- Charge 86.0 kg of water into a crystallization vessel followed by adding 21.5 kg of **58**.
- Without agitation, charge a solution of phosphoric acid (4.25 kg) in MeCN (67.6 kg).
- Heat the resulting mixture to 65 °C to dissolve all the solids.
- Start the agitation.
- Add an additional 135 kg of MeCN at 60 °C.
- Stir the batch at 60 °C for 16 h.
- Cool to 5 °C over 5 h.
- Filter, wash with MeCN (33.8 kg), and dry under vacuum.
- Yield of crude **59**: 21.5 kg.

Stage (b)

- Charge 21.5 kg of the crude **59** into a reactor followed by adding acetone (153 kg) and water (21.5 kg).
- Stir the mixture at 40 °C for 3 h.
- Cool to 20 °C.
- Filter and dry.
- Yield: 21.0 kg (84%).

7.3.1.2   Undesired Physical Properties

7.3.1.2.1   Undesired Particle Size

The physical property of reagents can influence reaction performance. For instance, the reduction of amide **60** with in situ generated diborane in THF resulted in variable conversion: using the granular $NaBH_4$ gave only 60% of conversion after 24 h, while the powder $NaBH_4$ gave a conversion higher than 90% in a few hours (Equation 7.10).[16]

(7.10)

REMARKS

(a) Tetrabutylammonium bromide (TBAB) was employed to help the dissolution of the mandelate salt **60** in THF at room temperature by increasing the ionic character of the solvent.

(b) An engineering assessment was carried out with the aid of computer modeling techniques, and it suggested that adopting a definite reactor configuration would ensure homogenization of $NaBH_4$ particles by mechanical shearing, resulting in a reproducible conversion. The prescribed measures were: the use of a dish base vessel, the use of a high-shear impeller such as a Rushton turbine, the use of four baffles, the need to work at relatively low vessel occupancy (liquid height at max equal to tank diameter), and the use of a high stirring speed.

7.3.1.2.2   Undesired Solid Form

Poor solubility of starting material/reagent may give rise to a slow reaction rate, thereby incomplete conversion. The solubility of a solid organic compound in a given solvent is dependent not only on the temperature, but on the solid state as well. For instance, the Suzuki–Miyaura cross-coupling between arylboronic acid **62** and acetyl chloride **63** proceeded smoothly with the crude amorphous boronic acid **62**, while with a crystalline **63** the reaction progressed sluggishly (Equation 7.11).[17]

(7.11)

7.3.1.2.3   Pool Solubility

Due to the poor solubility of paraformaldehyde in dichloromethane, the Pictet–Spengler reaction of protected ketopyrrole **65** at scale suffered lower isolated product yield and increased levels of impurities during the synthesis of ketone product **67** (Scheme 7.8).[18]

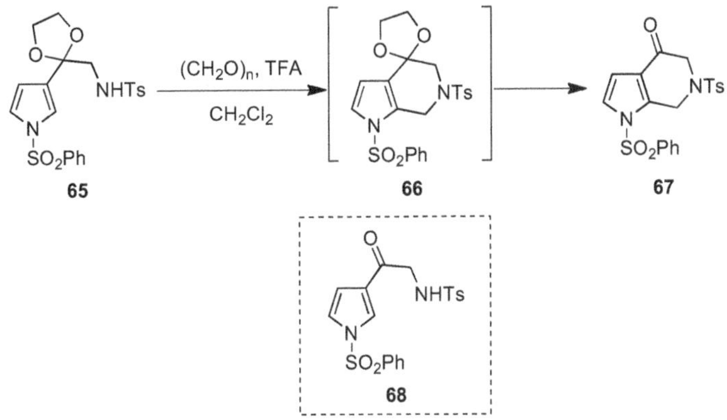

SCHEME 7.8

Besides the desired Pictet–Spengler cyclization, an undesired hydrolysis of **65** would form unprotected ketopyrrole **68**. It was demonstrated that **68** did not undergo productive cyclization and is the source of the majority of the impurities.

Solutions
In order to suppress the hydrolysis of **65** and to increase the desired cyclization reaction rate, enhancement of the solubility of paraformaldehyde in the reaction mixture would be a good choice. In the event, this solubility issue was resolved by premixing the paraformaldehyde with TFA in dichloromethane to form a clear solution.

Procedure

- Add 1113 kg of CH_2Cl_2 into a 8000-L reactor followed by the addition of 88.5 kg (3.1 equiv) of paraformaldehyde and 379 kg (3.5 equiv) of TFA.
- Rinse the charge line with CH_2Cl_2 (57 kg) and stir at 20 °C for 2–7 h to form a clear solution.
- Add 1170 kg of CH_2Cl_2 followed by the addition of 440 kg of **65**.
- Stir at 20 °C for 1 h until complete reaction.
- Quench with aqueous 10% K_2HPO_4 (pH of the resulting aqueous phase is >3) and separate layers.
- Add 1390 L of IPA to the organic layer and age for 2 h.
- Add 2789 L of IPA over 2 h.
- Cool the slurry to 0 °C and age for 2 h.
- Filter, wash with 1390 L of IPA, and dry at 50 °C/vacuum.
- Yield: 336 kg (81.3%) as a white solid.

REMARK
The Csp^3 hybridized substrate at the C4 position, obtained by a reaction of **68** with ethylene glycol, is required for the Pictet–Spengler cyclization (likely due to both electronic and conformational advantages).

7.3.1.3 High Flow Rate of Nitrogen
Most of the organic transformations are carried out in an inert atmosphere by filling the reactor with an inert gas such as nitrogen (or argon) or maintaining a constant flow of nitrogen gas. However, a high flow rate of nitrogen can create unexpected scale-up problems. For example, deoxychlorination of

hydroxypyridone **69** with phenylphosphonic dichloride was met with failure upon scale-up into a 1-L reactor (Scheme 7.9).[19] Evaluation of the process found that the low concentration of hydrogen chloride in the headspace under the high flow rate of nitrogen was the cause of the incomplete conversion.

Accordingly, the reaction could proceed to completion upon reduction of nitrogen flow rates.

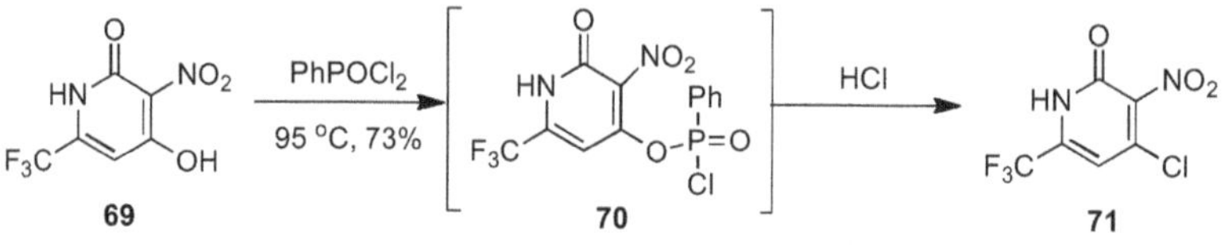

SCHEME 7.9

7.3.2 Loss of Product During Isolation

Low product yields could be caused by loss of product during the workup/isolation. Amino ester hydrochloride **73** could be obtained from catalytic hydrogenolysis of Cbz-protected amino ester **72** with palladium on carbon (Equation 7.12).[20]

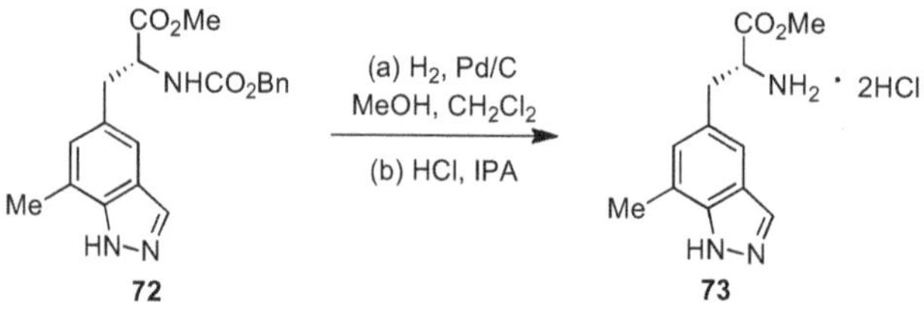

$$(7.12)$$

This deprotection of the Cbz group, however, suffered inconsistent isolated product yields ranging from 53% to 89%.

(I) Chemistry diagnosis

Investigations found that carbon dioxide by-product was the culprit for the inconsistency of the isolated product yields. An amine carbonate salt precipitate, formed from a reaction of the amine-free base of **73** with CO_2, was inadvertently removed during the filtration of the heterogeneous catalyst.

(II) Solutions

Purging the reaction mixture at the end of the hydrogenolysis with nitrogen removed the carbon dioxide, and the carbonate salt disproportionated to regenerate amine free base of **73**.

Procedure

Stage (a)
- Charge 0.2 kg (0.1 mol) of Pd/C followed by adding 2.6 kg (7.1 mol) of **72**, MeOH (13 L), and EtOAc (7 L) into a hydrogenation reactor.
- Purge with N_2.
- Stir at 25 °C for 2 h under 45 psig of hydrogen.
- Add EtOAc (13 L).
- Purge with N_2 for 30 min.
- Filter and rinse with a mixture of EtOAc (10 L) and MeOH (11 L).

Stage (b)
- Charge 3.3 kg (15.6 mol) of HCl solution in IPA (4.5 M) over 1 h.
- Stir the resulting slurry at 20 °C for 3 h.
- Filter and wash with EtOAc (4 L).

- Dry at 35 °C/25 mmHg.
- 1.85 kg (85.6%).

7.3.3   Side Reactions of Starting Materials

Low product yields can be caused by the instability of starting materials under reaction conditions such as high reaction temperature, leading to low product yields. To overcome these instability issues, a few approaches were developed.

7.3.3.1   Decomposition of *N*-Oxide

Transformation of *N*-oxide **74** to 5-bromobenzofurazan **75** was carried out via deoxygenation with triphenylphosphine in xylene at 130 °C (Condition A, Equation 7.13).[21]

$$ \text{(7.13)} $$

74

(A) Ph$_3$P, xylene, 130 °C, 55%
(B) P(OEt)$_3$, EtOH, 70-75 °C, 77%

75

(I) Problems

- The high reaction temperature (130 °C) led to a moderate yield (55%) of the product **75** due to the decomposition of the starting material **74**.
- Purification of **75** required chromatography or steam distillation to remove triphenylphosphine oxide by-product.

(II) Solutions

To obviate these problems, triethyl phosphite was chosen as the deoxygenation agent, which allowed the reaction to be conducted in ethanol at 70–75 °C (Condition B, Equation 7.13). The product **75** crystallized directly from the reaction mixture upon the addition of water.

7.3.3.2   Decomposition of Hydrazone

As described in Equation 7.14,[22] pyrazole **78** was synthesized by treating a mixture of hydrazone **76** and nitroolefin **77** in the presence of air. This preparation required several equiv of **76** in order to drive the reaction to completion.

$$ \text{(7.14)} $$

76

77

78

(I) Chemistry diagnosis

Mechanistically, this reaction involves Michael-type addition to form nitro intermediate **79** whose slow air oxidation leads to nitroolefin **80**, followed by a fast elimination of HNO$_2$ to furnish pyrazole **78** (Scheme 7.10). Presumably, the in situ formed HNO$_2$ caused decomposition of **76**.

(II) Solutions

The addition of a base to capture the nitrous acid could be helpful in preventing this decomposition. Accordingly, using triethylamine as the solvent eliminated this undesirable decomposition and the reaction went to completion with 1.2 equiv of **76**.

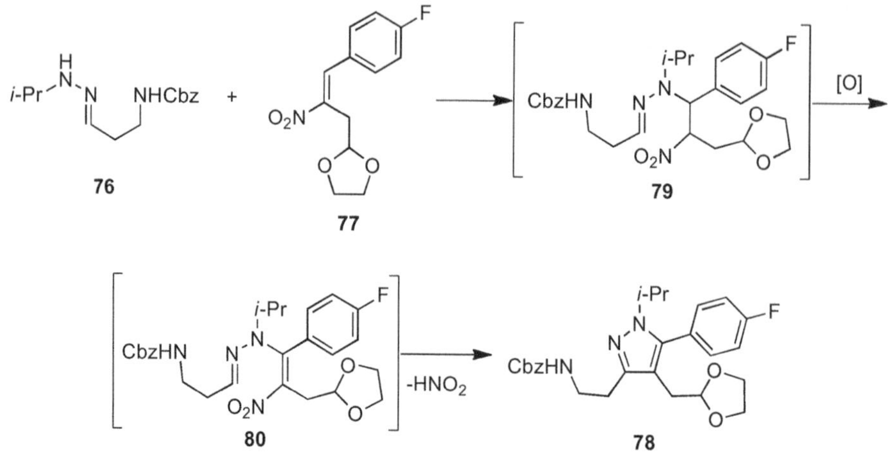

SCHEME 7.10

7.3.3.3 Hydrolysis of Chlorotriazine

Compound CE-178,253 is a CB1 antagonist discovered by Pfizer. The synthesis of CE-178,253 in a Kilo-Lab campaign (2 kg scale) was achieved by coupling chlorotriazine **81** with azetidine bis HCl salt **82** using i-Pr$_2$NEt in 3:1 dimethoxyethane (DME)/water (Equation 7.15).[23]

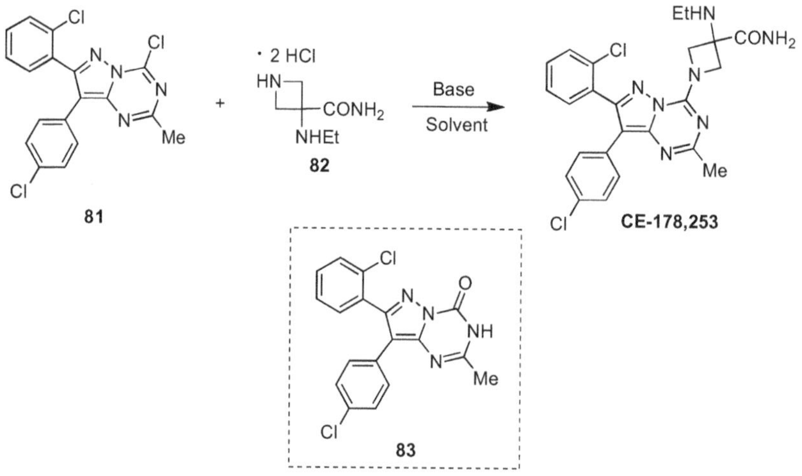

(7.15)

Reaction problems

- Under these conditions, significant levels of hydrolysis side product **83** and other unidentified impurities were observed.
- A rework of the crude product by recrystallization from dichloromethane/MTBE was required.
- As a result, the isolated yield was low (42%).

(II) Solutions

In order to avoid the hydrolysis of chlorotriazine **81**, a two-phase system was developed. The coupling reaction was conducted in a two-phase solvent mixture of 2-methyltetrahydrofuran and aqueous NaHCO$_3$ cleanly providing CE-178,253 in 79% isolated yield with undetectable **83**.

7.3.4 SIDE REACTIONS OF INTERMEDIATES

7.3.4.1 Sandmeyer Reaction

Sandmeyer reaction is frequently plagued by side reactions arising from the intermediate. Attempts to convert 2-amino substituted pyrazine **84** to the corresponding 2-chloropyrazine **86a** via the Sandmeyer reaction (CuCl, aq HCl, NaNO$_2$) failed to give the desired chloride **86a**. Under such conditions, 2-hydroxypyrazine **87** was formed. Although a condition in which the diazonium salt **85** was prepared under anhydrous conditions using *t*-BuONO as the oxidant in the presence of cupric chloride (CuCl$_2$, *t*-BuONO, MeCN) gave chloride **86a** as the major product, side products such as **87** and acetamide **88** (resulting from Ritter reaction of MeCN) were also observed (Scheme 7.11).[24]

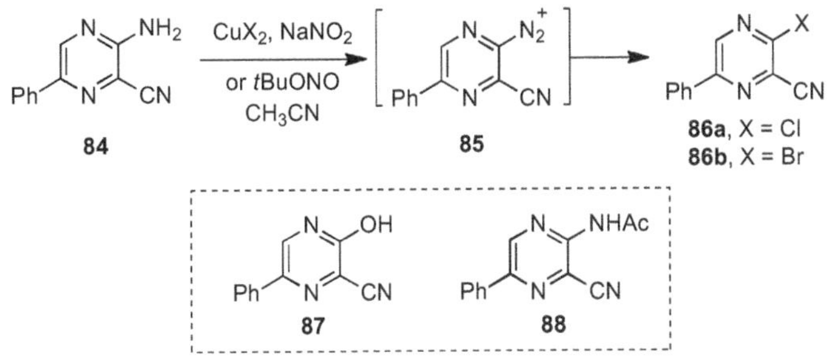

SCHEME 7.11

The formation of **87** and **88** suggests that chloride was not reactive enough, which allowed side reactions to occur. Therefore, a strategy was adopted to rapidly consume the reactive diazonium salt **85** with more reactive CuBr$_2$. Accordingly, switching cupric chloride to cupric bromide as the halide source and MeCN to DMF as the solvent was able to afford the product **86b** in a 73% isolated yield with 98% purity.

7.3.4.2 Hofmann Rearrangement

Hofmann rearrangement of primary amides under basic conditions in the presence of bromine is utilized to prepare primary amines. Such reactions proceed through *N*-bromoamide intermediates that rearrange to isocyanates followed by hydrolysis to afford amine products, giving off carbon dioxide at the same time.

Besides the desired aniline **91**, the Hofmann rearrangement of **89** gave a brominated aniline **92** as a major impurity (Scheme 7.12).[25]

N-Bromoamides are known brominating agents. Presumably, the formation of **92** was caused by a reaction of the aniline product **91** with the *N*-bromoamides intermediate **90**. In order to mitigate the formation of **92**, an excess of the base was necessary to reduce the concentration of the *N*-Bromo

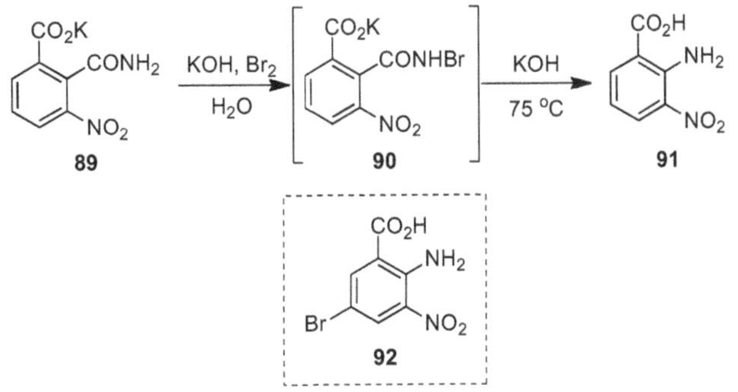

SCHEME 7.12

amide intermediate **90** during the reaction. Therefore, a good product yield (90%) was achieved with high purity (>99%) by using an excess of KOH (5 equiv).

Procedure

- Charge a solution of KOH (13.0 kg, 232 mol, 5 equiv) in 110 kg of H_2O into a 190 L glass-lined reactor.
- Cool to 15 ± 5 °C.
- Charge 11.0 kg (46.5 mol) of **89**.
- Cool the resulting solution to 0 ± 5 °C.
- Charge slowly 7.4 kg (46.4 mol, 1.0 equiv) of bromine at <5 °C.
- Stir at 0 ± 5 °C for not less than 30 min.
- Heat to 75 ± 5 °C at a rate of 18 °C/h.
- Stir at 75 ± 5 °C for not less than 1 h.
- Cool to 50 °C.
- Add slowly a concentrated HCl (21.5 kg, 218 mol, 4.7 equiv) to pH<2.5.
- Cool the resulting yellow slurry to 20 °C.
- Filter and wash with H_2O.
- Dry at ≤70 °C under vacuum.
- Yield: 7.6 kg (90%).

7.3.4.3 Tosylation/Amination Reactions

The conversion of diol **93** to a drug candidate **94** was performed via tosylation and subsequent amination (Equation 7.16).[26] This transformation resulted in the formation of a significant amount of amidine **95**, in addition to the amino alcohol **94**.

$$(7.16)$$

(I) Chemistry diagnosis

Since the intermolecular S_N2 reaction of monotosylate intermediate **96** with methylamine (aqueous or ethanolic solution) is a relatively slow process, a competitive intramolecular S_N2 displacement of **96** would occur to give an intermediate **97** whose subsequent reaction with methylamine generates the side product **95** (Scheme 7.13).

(II) Solutions

Conversion of the tosylate **96** into a more stable epoxide intermediate **98** is a strategy that was used to inhibit the formation of **95** (Scheme 7.14).

Therefore, once the completion of the tosylation, **96** was converted readily into epoxide **98** by adding a concentrated aqueous NaOH solution. The epoxide **98** exhibited enhanced stability toward

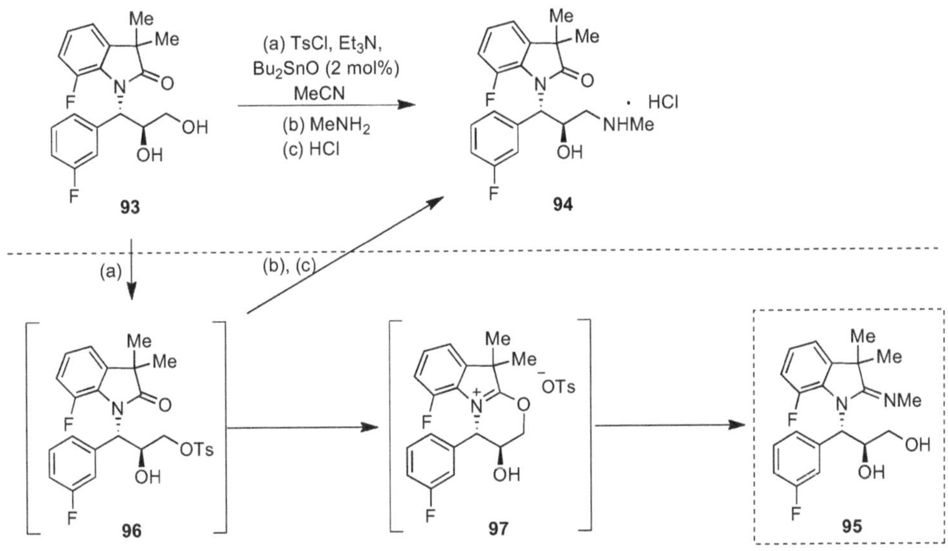

SCHEME 7.13

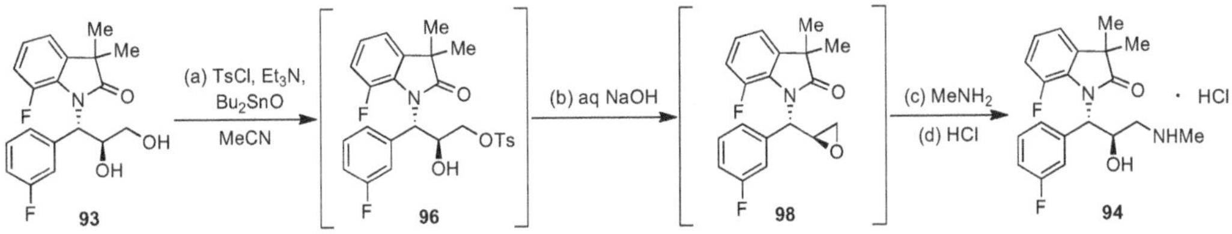

SCHEME 7.14

degradation versus the monotosylate. Therefore, the optimized protocol involved monotosylation of **93** with TsCl in acetonitrile in the presence of triethylamine and catalytic dibutyltin oxide at 0 °C for 1 h, followed by treatment with aqueous NaOH at 0 °C for 1 h. After a solvent switch to toluene, the resulting terminal epoxide **98** was treated with a solution of methylamine in EtOH, leading to a slow epoxide-opening at the terminal carbon to furnish the desired amino alcohol **94**. The amino alcohol **94** was crystallized from a mixture of MTBE/EtOH as a hydrochloride salt in 54–58% yield with >99% purity and 99.1% ee.

7.3.4.4 Synthesis of Cyclic Sulfimidate

Using literature conditions for the synthesis of cyclic sulfimidate **101** suffered from a side reaction, yielding α-chloroacetophenone **102** which is a severe lachrymator (Scheme 7.15).[27] α-Chloroacetophenone **102** was generated from a reaction of the *O*-sulfamoyl intermediate **100** with chloride.

Solutions

It was found that the use of pyridine as a base led to the formation of pyridinium hydrochloride byproduct that has a relatively high solubility in the reaction medium. To limit the concentration of chloride ions in the reaction medium, triethylamine was found to be the base of choice and the resulting triethylamine hydrochloride salt precipitated out from the reaction medium (Scheme 7.16). The new process was conducted in one-pot involving two steps: a reaction of in situ formed *N*-Boc-sulfamoyl chloride (from a reaction of chlorosulfonylisocyanate with *tert*-butanol) with α-hydroxyketone **99** to generate *N*-Boc-sulfamoyl intermediate **103** and Boc-deprotection/cycloaddition.

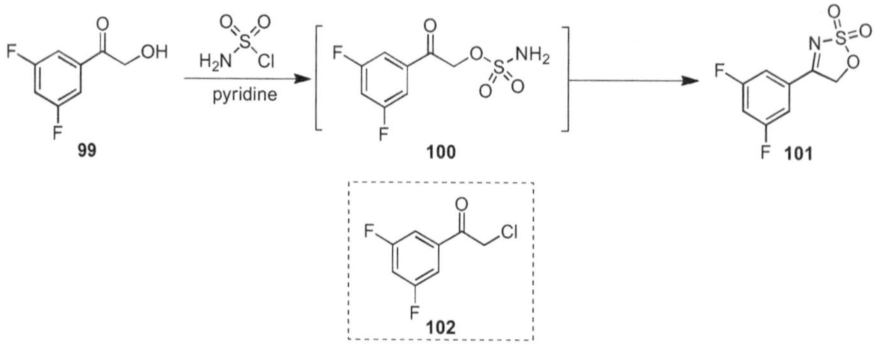

SCHEME 7.15

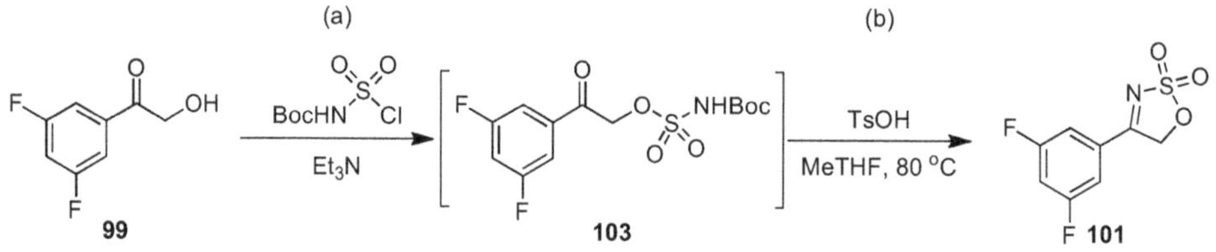

SCHEME 7.16

REMARKS

(a) The unstable sulfamoyl chloride was prepared in situ from a reaction of chlorosulfonyliso-cyanate with formic acid (95% aqueous). This reaction is considered relatively hazardous: highly exothermic and releasing 2 moles of gases (carbon monoxide and carbon dioxide) (Equation 7.17):

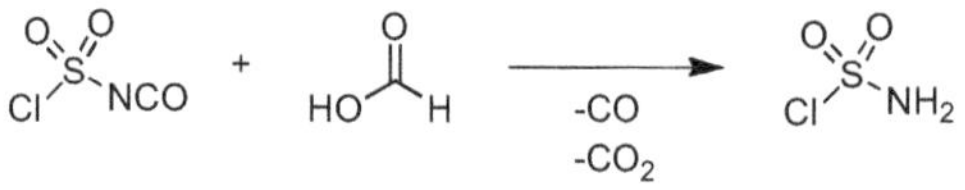

$$(7.17)$$

(b) Compared to sulfamoyl chloride, N-Boc-sulfamoyl chloride has good stability. Its preparation is outlined in Reaction (Equation 7.18):

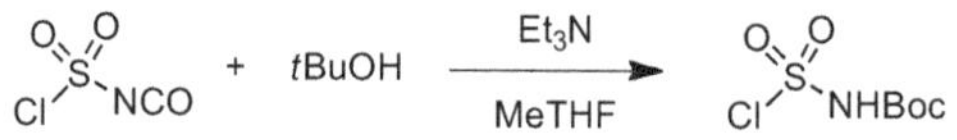

$$(7.18)$$

7.3.4.5 Cyclization/Ring Expansion

As outlined in Scheme 7.17,[28] the synthesis of lactam lactol **107** involves two stages: a condensation of aqueous glyoxylic acid **104** with N-benzyl ethanol amine **105** at ambient temperature provided an intermediate **106**, followed by a ring rearrangement to the product **107** under reflux in a mixture of water and THF.

Under such conditions, the reaction suffered a low product yield presumably due to the decomposition of **106** to **108** via decarboxylation.

(I) Chemistry diagnosis

At reflux, the initially formed intermediate **106** may undergo ring expansion through ring-opening intermediates **109** and **110** to **111** whose tautormerization furnishes the product **107** (Scheme 7.18). The intermediate **110** may also decouple into **104** and **105**, and the former would dimerize to form **112** whose further reaction with **105** produces the product **107**.

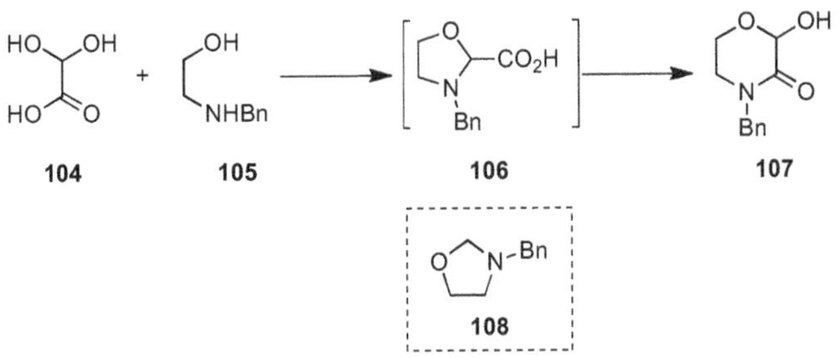

SCHEME 7.17

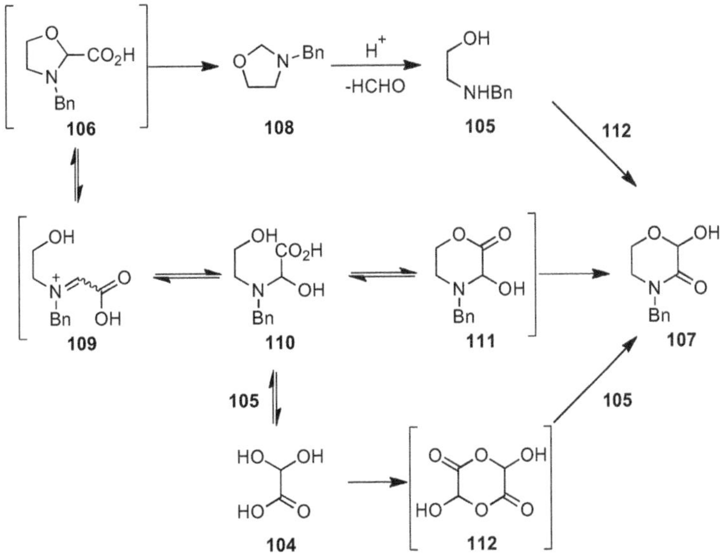

SCHEME 7.18

Besides the desired reaction pathways, **106** deccarboxylates to form the side product **108**. It was found that heating **108** provided formaldehyde and **105** and the latter in the presence of aqueous glyoxylic acid **104** could be converted to the product **107** via a reaction with **112**. Overall, the decomposition of **106** consumed **104** with regeneration of **105**.

(II) Solutions

Based on the aforementioned chemistry diagnosis, the reaction conditions were optimized by using an excess of **104**. As a result, compound **107** could be obtained in 76% yield via heating the solution of **104** (2.3 equiv) with **105** in aqueous THF at reflux.

Procedure

- Charge 7.8 kg (96 wt%, 49.5 mol) of **105** to a solution of **104** (50%, 16.9 kg, 114.1 mol) in THF (27.0 L) under reflux over 45 min.
- Heat the resulting mixture at reflux for 21 h.
- Distill out THF under atmospheric pressure while maintaining a constant volume by simultaneous addition of water (27.0 L) (residual THF <8 vol%).
- Cool to 80 °C and add 250 g of **107** as seeds.
- Cool to room temperature.

- Filter, wash with water, and dry at 60 °C under vacuum.
- Yield of **107**: 8.05 kg (76%).

7.3.4.6 Michael Addition

A preparation of aminopyrazole **116**, outlined in Scheme 7.19, involves two stages: a rapid condensation reaction of arylhydrazine **113** with 2-ethyoxymethylene malononitirle **114** to give an intermediate **115**, followed by a slow cycloaddition.[11]

(I) Reaction problems

Initially, the reaction was carried out by the addition of triethylamine to a suspension of **113** and **114** in ethanol. A study revealed that the reaction was extremely sensitive to the rate of triethylamine addition and a rapid addition (<1 min) gave a dark-brown slurry from which no desired product **116** could be isolated. A side product **117**, however, could be observed by LC/MS. Apparently, the high sensitivity to the addition rate renders this process not robust.

(II) Solutions

The formation of **117** could be attributed to the Michael addition of the free base of **113** with the intermediate **115**. Although this problem can be mitigated by the slow addition of triethylamine to control the release rate of the free base of **113**, a better solution was to perform the reaction in acetic acid in the presence of triethylamine and sodium acetate.

Procedure

Stage (a)
- Charge 2.0 kg (10.85 mol) of **113** into a dry 60-L jacketed reactor, followed by adding 445 g (5.43 mol) of NaOAc.
- Charge AcOH (10.0 L) and adjust the batch temperature to 15 °C.
- Charge 1.65 kg (16.28 mol) of triethylamine at <30 °C.
- Charge a solution of **114** (1.4 kg, 11.4 mol) in AcOH (4.0 L).
- Stir the resulting mixture at 20 °C for 1.5 h (<1 A% of **113** by HPLC).

Stage (b)
- Heat the batch to 60 °C and stir at that temperature for 2 h (<1 A% of **115** by HPLC)
- Cool to 20 °C.
- Add a solution of 5 N aqueous NaOH (10.0 L) at <27 °C over 30 min.
- Age the batch at 20 °C for 8 h.

SCHEME 7.19

- Filter, wash with water (2×4 L), and dry.
- Yield of crude **116**: 2.2 kg (89.8%).

7.3.5 SIDE REACTIONS OF PRODUCTS

The instability of products can lead to side reactions or decomposition, resulting in low isolation yields and impurities. To address these issues, several approaches were developed.

7.3.5.1 Decomposition of Amide

Initial debenzylation of benzyl ester **118** was carried out under catalytic conditions with 10% Pd/C in wet IPA to furnish the desired acid **119** (Equation 7.19).[29]

(7.19)

(I) Problems

An unexpected sensitivity of the amide bond within **119** to hydrolysis and solvolysis during the hydrogenolysis in wet IPA (ca. 50% water-wet Pd/C) led to the degradation of the product. Two acid impurities, **122** (0.1%) and **123** (0.5%), were observed in the isolated product **119**. In addition, acid **120** (5%) and isopropyl ester **121** (5%) were found in the mother liquor.

(II) Solutions

To address these degradation issues, the initial conditions were replaced with transfer hydrogenolysis in ethyl acetate using formic acid (13 equiv) as a hydrogen source in the presence of 10% Pd/C (20 wt%) at 55 °C for 6 h. Under these conditions, the isolated product was in 81% yield without racemization and no amide hydrolysis was observed.

Procedure

- Charge a slurry of 10% Pd/C (8.26 kg, 20 wt% relative to **118**) in EtOAc (145 L) to a stirred solution of **118** (41.28 kg, 84.85 mol) in EtOAc (230 L).
- Charge 41 L (13 equiv) of formic acid.
- Heat the resulting mixture at 55 °C for 6 h.
- Cool to 20 °C.
- Add Arbocel (6 kg).
- Filter and rinse with EtOAc (290 L).
- Wash the filtrate with water (3×620 L).
- Filter the organic layer through a speck-free filter.
- Distill the filtrate at 20 °C under 0.1 bar to 130 L.
- Dilute with EtOAc (210 L).

- Distill at 20 °C under 0.1 bar to 130 L.
- Repeat the dilution and distillation again until <0.3% water.
- Add heptane (210 L) at 20 °C over 2 h.
- Stir at 20 °C for 20 h.
- Filter and wash with heptane (66 L).
- Dry at 50 °C with N_2 sweep.
- Yield: 27.22 kg (81%) as a white solid.

7.3.5.2   Side Reactions of 1,4-Isochromandione

The synthesis of 1,4-isochromandione **126** was achieved by a two-step protocol involving bromination of 2-acetylbenzoic acid **124** to **125a/125b** followed by ring expansion to deliver the desired product **126** (Scheme 7.20).[30]

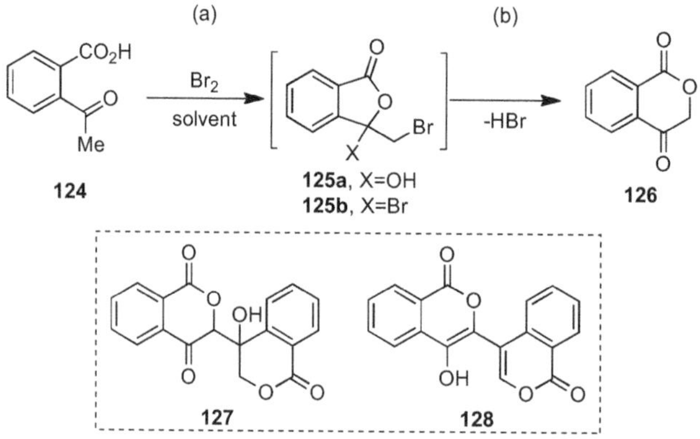

SCHEME 7.20

(I) Reaction problems

The thermal ring expansion of **125a/125b** was proved to be problematic due to the acid-catalyzed enolization of the product **126**:

- Attempts to convert the crude brominated intermediate **125a** to 1,4-isochromandione **126** using sodium acetate in methanol led to an immediate red coloration and the formation of **128**.
- Treatment of crude bromomethyl phthalides **125a** and **125b** in water under reflux delivered the target isochromandione **126** in poor overall yield (12%) with a dimeric compound **127** being identified as the major side product.

(II) Solutions

The formation of compounds **127** and **128** was attributed to the acid (HBr)-catalyzed enolization of **126**, which condensed with another molecule of **126** to give aldol product **127**. Subsequent acid-mediated dehydration of **127** generated **128**. In order to mitigate the acid-catalyzed side reactions, the thermal ring expansion was conducted under biphasic conditions in which the hydrobromic acid would be effectively removed from the organic phase. Utilization of biphasic conditions (PhCl/H_2O) suppressed the problematic acid-catalyzed dimerization.

Procedure

Stage (a)
- Charge 5 L of 5.5 M HBr in AcOH and 25.2 L (78.6 kg, 492 mol) of Br_2 into a stirred mixture of **124** (82 kg, 500 mol) in chlorobenzene (905 kg).
- Warm the resulting mixture to 30 °C and stir for 3 h.

Stage (b)

- Charge water (820 L).
- Heat to reflux and hold at reflux for 3 h.
- Cool to 60 °C.
- Separate layers.
- Extract the aqueous layer with PhCl (328 L).
- Distill the combined organic layers under reduced pressure to 260 L.
- Add IPA (410 L) and stir at 0 °C.
- Filter and wash with IPA.
- Dry at 50 °C under vacuum.
- Yield: 59 kg (73%).

7.3.5.3 Side Reactions of Oxirane

Synthesis of 2-((2-methoxyphenoxy)methyl)oxirane **131** via a reaction of 2-methoxyphenol **129** with epichlorohydrin is anticipated to generate two side products: diol **132** and dimer **133** (Scheme 7.21).[31]

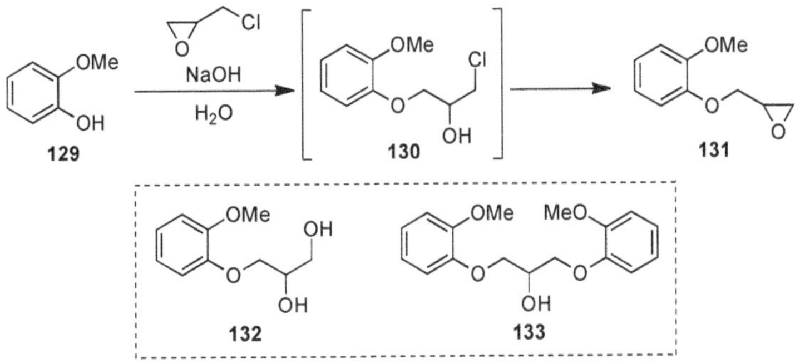

SCHEME 7.21

(I) Chemistry diagnosis

The side products **132** and **133** would be generated through reactions of the product **131** (or the intermediate **130**) with hydroxide and 2-methylphenoxide, respectively.

(II) Solutions

- In order to mitigate the formation of **132** and **133**, a biphasic reaction system (excess of epichlorohydrin/aqueous) was employed to reduce the concentrations of the two anions (especially hydroxide) in the organic phase.
- To further minimize the formation of **132** and **133**, sodium hydroxide was added in two portions. The first portion (0.5 equiv) of NaOH employed led to intermediate **130**. The aqueous layer containing the unreacted 2-methoxyphenoxide was removed from the reaction mixture by layer separation to avoid the formation of dimer impurity **133**. The rest (1.0 equiv) of NaOH was added to the organic phase containing the intermediate **130** to effect the cyclization reaction.

Procedure

- Charge 1.61 kg (40.25 mol, 0.5 equiv) of NaOH to a stirred solution of **129** (10 kg, 80.55 mol) in water (40 L) followed by adding water (10 L) at 30 °C.
- Stir for 30–45 min.

- Charge 22.35 kg (241.62 mol, 3.0 equiv) of epichlorohydrin.
- Stir at 25–35 °C for 10–12 h.
- Separate layers.
- Add water (40 L) to the organic layer followed by NaOH (3.22 kg, 80.5 mol, 1.0 equiv) and water (10 L).
- Stir at 27 °C for 5–6 h.
- Separate layers and wash the organic layer with a solution of NaOH (3.0 kg) in water (30 L).
- Distill out excess epichlorohydrin at <90 °C under reduced pressure (650–700 mmHg).
- Yield: 13.65 kg (94%).

7.4   REACTION PROBLEMS ASSOCIATED WITH IMPURITIES

Impurities affect not only the quality of products, but also the subsequent reactions. These impurities can be residual solvents including MTBE and water, residual metals such as zinc, and so on.

7.4.1   RESIDUAL MTBE

A reaction can be influenced by the presence of residual solvents carried over from the previous process. It was observed in the laboratory that the amidomethylation was sluggish due to the presence of residual MTBE in **134** (Scheme 7.22).[5]

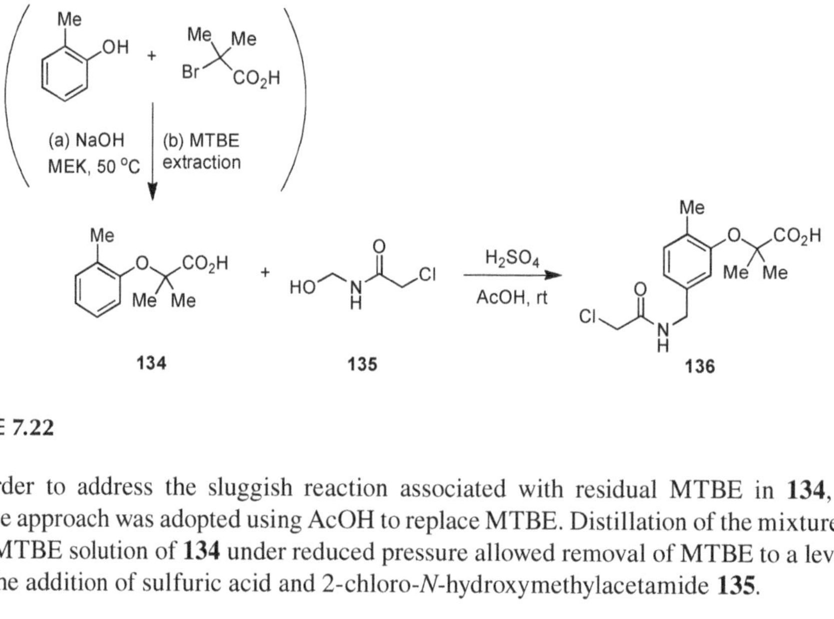

SCHEME 7.22

In order to address the sluggish reaction associated with residual MTBE in **134**, a solvent exchange approach was adopted using AcOH to replace MTBE. Distillation of the mixture of AcOH and an MTBE solution of **134** under reduced pressure allowed removal of MTBE to a level of 0.5% before the addition of sulfuric acid and 2-chloro-*N*-hydroxymethylacetamide **135**.

Procedure

- Charge AcOH (49.4 L) to a concentrated MTBE solution of **134** (19.7 kg, 101 mol).
- Distill under reduced pressure until the MTBE level is 0.5%.
- Add AcOH (18.2 kg).
- Charge a mixture of AcOH (49.4 L) and H_2SO_4 (9.9 L) to achieve a 10:1 ratio of AcOH and H_2SO_4.
- Charge 13.8 kg (112 mol, 1.1 equiv) of **135**.
- Stir the resulting mixture at room temperature overnight.
- Quench the reaction with aqueous NaOAc (18.4 kg, 224 mol) in water (197.5 L).
- Extract with MTBE (2×99 L) and wash the combined organic layers with water (99 L).

- Distill to a minimum stir volume.
- Add toluene (198 L) and distill to minimum stir volume.
- Add toluene (198 L) and repeat the distillation to a minimum stir volume.
- Add toluene (238 L) and heat the mixture to 90 °C to achieve complete dissolution.
- Cool to 50 °C over 30 min and hold at 50 °C for 1 h.
- Cool to room temperature over 20 min and hold at rt for 1 h.
- Filter, wash with cyclohexane (2×40 L), and dry at 45 °C under vacuum.
- Yield: 24.1 kg (79%).

7.4.2 RESIDUAL WATER

Water, as a common process impurity, can present in a reaction system as a result of moisture in the starting materials/reagents or solvents, or poor layer cut during the reaction work-up. Water can also present as a reaction by-product (or side product). Azeotropic distillation is commonly used on large scale to remove residual water from a reaction system.[32]

7.4.2.1 Bromination Reaction

The presence of water can have a detrimental effect on a reaction profile. During the pilot-plant production of bromoimidazole 139, an occasion when bromination of crude 137 in isopropyl acetate (IPAc) only 40% yield of the product 139 was obtained on a 50-kg scale. The poor product yield is due to the formation of a ring-cleavage side product 140 (Scheme 7.23).[33]

(I) Chemistry diagnosis

Scheme 7.24 illustrates that water could be the possible cause for the formation of 140. The reaction of the bromination intermediate 138 reacts with water would form epoxide 142 via the elimination of HBr from 141. The subsequent epoxide-ring opening with water would give 4.5-dihydroxy-4,5-dihydroimidazole 143 whose ring cleavage and loss of oxalaldehyde may produce the side product 140.

(II) Solutions

In order to improve the process robustness, the water content (a specification of <1000 ppm) was controlled by azeotropic distillation.

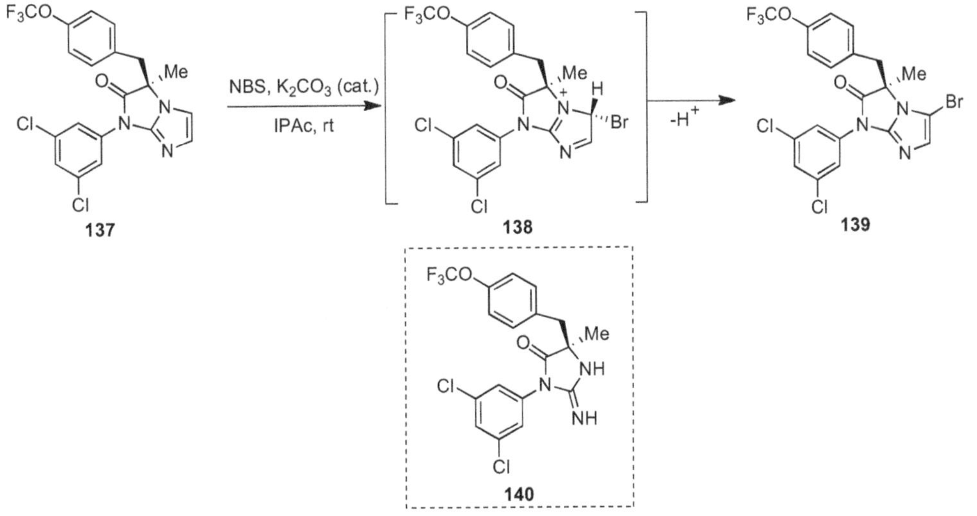

SCHEME 7.23

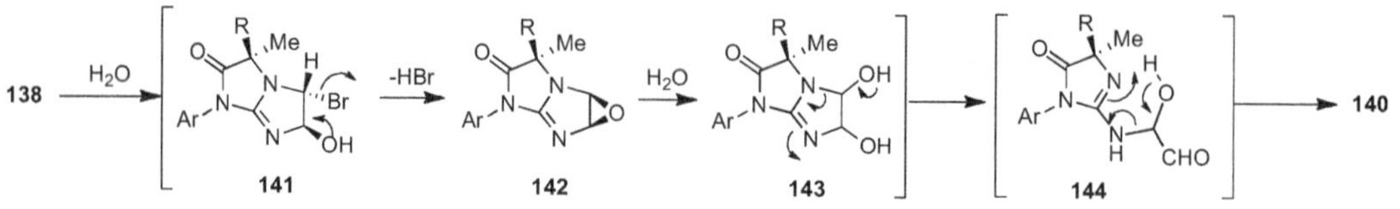

SCHEME 7.24

7.4.2.2 S_NAr Fluorination Reaction

Generally, hydrate water is difficult to be removed as it bonds to the molecule via hydrogen bonds. S_NAr fluorination of 4-chlorothiazole **145** requires anhydrous fluoride, but reaction with CsF resulted in low conversion and product decomposition. Using tetramethylammonium fluoride tetrahydrate (TMAF·4H_2O), as an alternative fluorinating agent, presented a challenging issue when an attempt to obtain anhydrous TMAF. Drying the commercially available TMAF·4H_2O via azeotropic distillation with toluene failed to give satisfied Karl Fischer (KF) results (0.50 wt% water in DMF solution) (Equation 7.20).

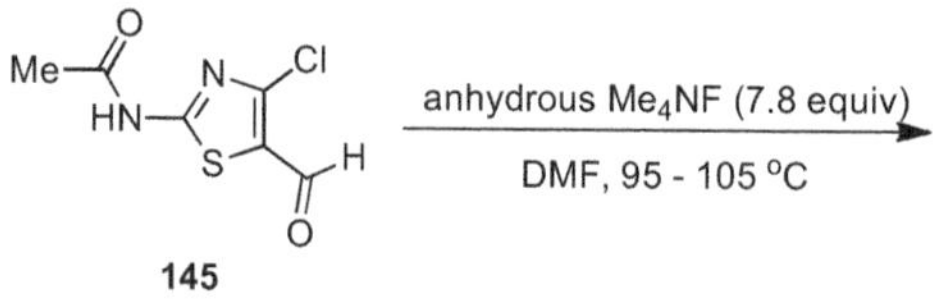

$$(7.20)$$

Ultimately, distillation with isopropyl alcohol and then with DMF at elevated temperature provided anhydrous TMAF containing <0.2 wt% water and <60 ppm isopropanol. Using the dried TMAF in the S_NAr fluorination on a 45.1-kg scale afforded the product **146** in 89% yield.[34]

7.4.2.3 Copper-Catalyzed C–N Bond Formation

It was noticed that the moisture levels in the headspace impacted the reaction rate during the C–N bond formation via coupling 2-fluoro-4-iodophenylamine **147** with pyrazole (Equation 7.21).[35]

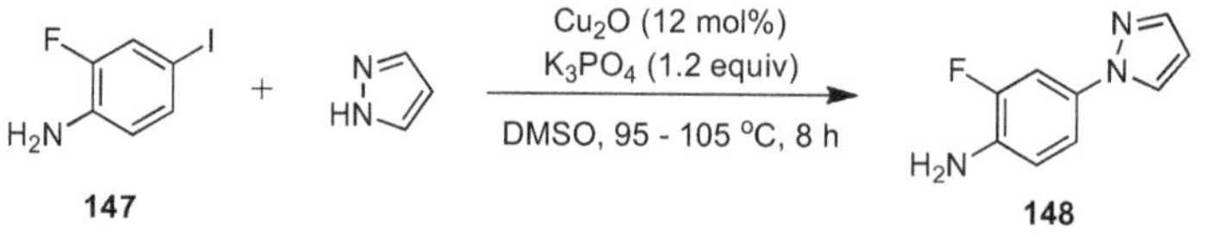

$$(7.21)$$

The reaction time was longer for larger volumes of headspace, though the same amounts of starting material and reagents were used under the same rate of N_2 flow. On the other hand, the reaction rate decreased remarkably when an N_2 balloon was used. Ultimately, the headspace moisture levels were reduced by passing an inert argon at a flow rate of 13.3 L/min.

DISCUSSIONS

On the other hand, a small amount of water could have a positive impact on a reaction rate. It was observed that oxidation of alcohol **149** by Dess–Martin reagent in wet methylene chloride was completed within 3 h, while the reaction was sluggish in dry methylene chloride (Equation 7.22).[36]

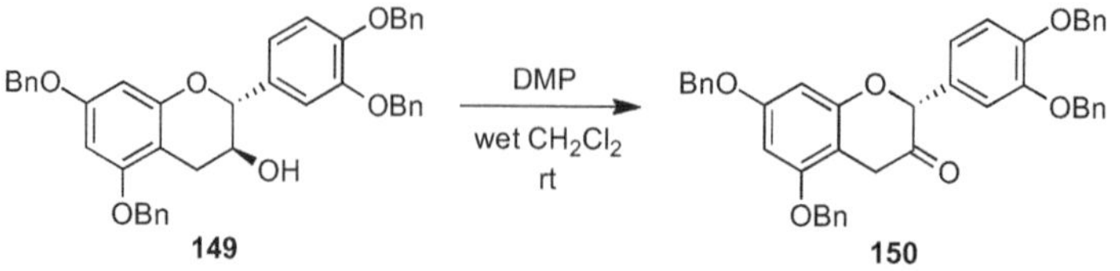

$$(7.22)$$

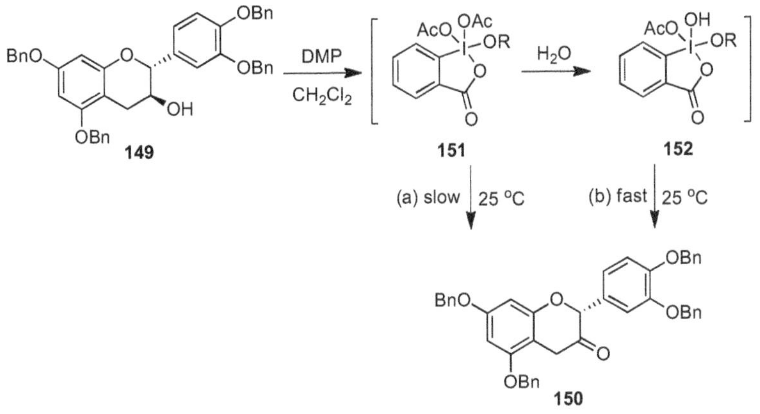

SCHEME 7.25

The rationale for the rate enhancement by water was proposed to involve two intermediates **151** and **152** (Scheme 7.25).[37] The decomposition of **152** to **150** is faster than **151** due to the fact that the increased electron-donating ability of a hydroxyl substituent in place of acetyl may enhance the rate of dissociation of the remaining acetate ligand.

Procedure

- Charge DMP (315.4 g, 0.74 mol, 1.1 equiv) to a solution of **149** (440 g, 0.675 mol) in 3.2 L of CH_2Cl_2.
- Add 242 mL of CH_2Cl_2 saturated with water dropwise over 1.5 h at 16 to 24 °C.
- Add 4 L of saturated $NaHCO_3$, followed by the addition of a 10% aqueous solution of sodium thiosulfate pentahydrate (161.5 g/1.6 L of water).
- Filter and separate layers.
- Extract the aqueous layer with 650 mL of CH_2Cl_2.
- Dry the organic layer with $MgSO_4$.
- Filter to remove $MgSO_4$ and the reduced DMP precipitate.
- Distill the filtrate under reduced pressure.
- Stir the distillation residue with 1.5 L of MeOH for 1 h.
- Filter and dissolve the crude product in 800 mL of boiling CH_2Cl_2.
- Add 1.5 L of MeOH.
- Cool in an ice bath.
- Filter and wash with MeOH (4×100 mL).
- Dry under vacuum.
- Yield: 336 g (76.6%).

7.4.3 RESIDUAL OXYGEN

Residual oxygen dissolved in reaction solvents can oxidize metal catalysts or cause undesired oxidation reactions. In order to reduce oxygen levels, several approaches are available including purging the reaction system with an inert gas such as nitrogen or argon, vacuuming the reaction system followed by backfilling with nitrogen, and boiling the reaction solvent.

7.4.3.1 Oxidative Dimerization

The solubility of a gas in solvents decreases with increases in temperature. Thus, boiling a reaction solvent is able to purge most of the air dissolved in the solvent. For instance, degassing the reaction

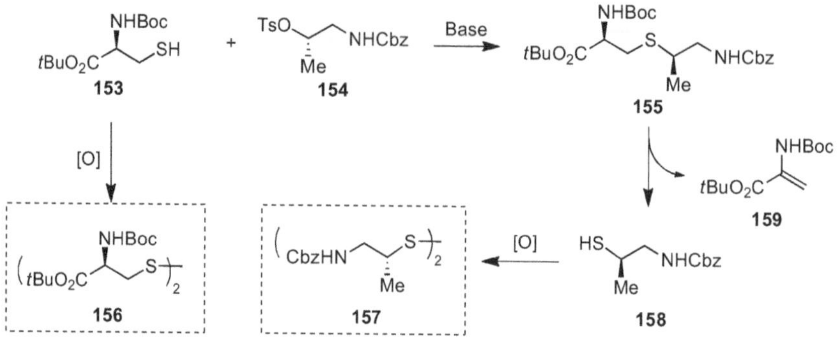

SCHEME 7.26

solvent by boiling MTBE prior to adding thiol **153** during the preparation of thioether **155** allowed controlling the oxidative dimerization side products **156** and **157** (Scheme 7.26).[38]

Procedure

- Degas MTBE (1500 L) by distillation to remove 150 L of MTBE.
- Charge 75 kg (270.4 mol, 1.31 equiv) of **153** and stir under N_2 to form a solution.
- Cool the resulting solution to 10±3 °C.
- Charge 120 kg (219.2 mol, 1.06 equiv) KO*t*Bu in THF in less than 10 min.
- Stir for ≤5 min.
- Immediately after, charge a solution of **154** (75 kg, 206.4 mol, 1.0 equiv) in THF (187.5 L) as fast as possible.
- Warm the batch to 22±3 °C and stir for 12 h.
- Add 600 L of 15% w/v aqueous NH_4Cl solution and stir for 10 min.
- Separate layers.
- Wash the organic layer with 8% w/v aq $NaHCO_3$ solution (450 L) followed by 30% w/v aqueous NaCl.solution (450 L).
- Dry the organic layer via circulating through a bed of Na_2SO_4 (75 kg) for ≥30 min.
- Distill at <50 °C under reduced pressure.
- Add MTBE (450 L) and continue to distill to ~187.5 L.
- Repeat this dilution/distillation cycle one more time.
- Dilute with MTBE to a total of 300 L.
- Add isooctane (450 L) at 20±3 °C over 1 h and age for 1.5 h.
- Filter and wash with a 1:5 mixture of MTBE/isooctane (225 L).
- Dry at 30±3 °C.
- Yield: 82.2 kg (85%).

7.4.3.2 Oxidation of Phenylenediamine

For air-sensitive materials such as phenylenediamine **160**, a combination of two methods was proved to be effective to prevent air oxidation. In addition to degassing, the addition of a reductive sacrificial agent, sodium sulfite, provided further protection of **160** from oxidation during the cyclization reaction (Equation 7.23).[39]

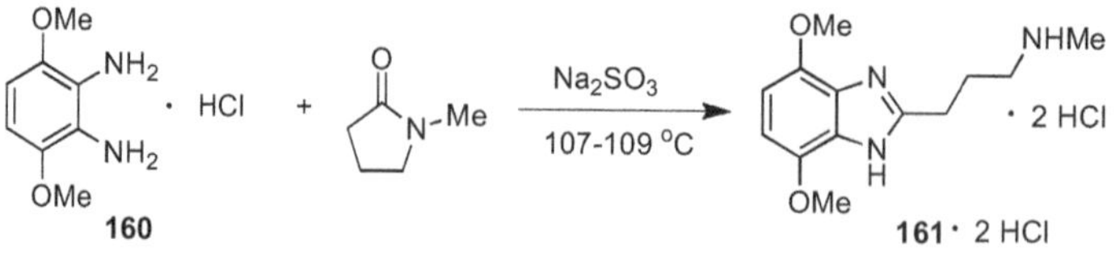

(7.23)

Procedure

- Charge 41.4 kg (4.1 equiv) of 21.5% HCl into a 64-L glass-lined reactor followed by adding 14.9 kg (2.54 equiv) of NMP.
- Degas the mixture with N_2 at 20 °C for 15 min.
- Charge 350 g (2.78 mol, 0.047 equiv) of sodium sulfite, followed by adding 12.1 kg (59.18 mol, 82.2 wt% of **160** and 17.8 wt% of regioisomer) of **160**.
- Heat the resulting mixture at 107–109 °C for 19 h.
- Transfer the batch into a 400-L stainless-steel reactor containing a degassed mixture of n-butanol (109 kg), 30% NaOH (52.4 kg), and water (20 kg) at 20–30 °C (pH >12).
- Separate layers.
- Wash the organic layer with water (2×33 kg).
- Transfer the organic layer into a 200-L glass-lined reactor.
- Add 10.4 kg (1.78 equiv) of 37% HCl at 20–30 °C to pH <3.
- Add 10.0 kg activated charcoal.
- Stir the mixture at 60–80 °C for 2 h.
- Cool to 20 °C.
- Add Celite (7.0 kg) and stir at 20–25 °C for 30 min.
- Filter and pump the filtrate into the 64-L reactor via a 5 μm PAL filter and wash with warm (50 °C) n-butanol (57 kg).
- Distill at 70–80 °C/300 mbar until ca. 40 L (total of 140–160 L solvents removed) (KF 2.5–3.5%).
- Cool to 0 °C over 3 h.
- Filter, wash with EtOAc (2×38 kg), and dry at 70 °C/<100 mbar for 15 h.
- Yield of **161**·2HCl·H_2O: 12.1 kg (59%) as an off-white powder.

7.4.4 RESIDUAL ZINC

Generally, metals used in a process (either as reagents or catalysts) need to be removed after the reaction as their presence may potentially interfere with subsequent reactions or contaminate the final product as impurities. For example, the residual Zn (II) from the reduction of the nitro group (Zn/TMSCl in EtOH) in **162** suppressed the subsequent K_2CO_3-promoted retro Michael reaction (Scheme 7.27).[40]

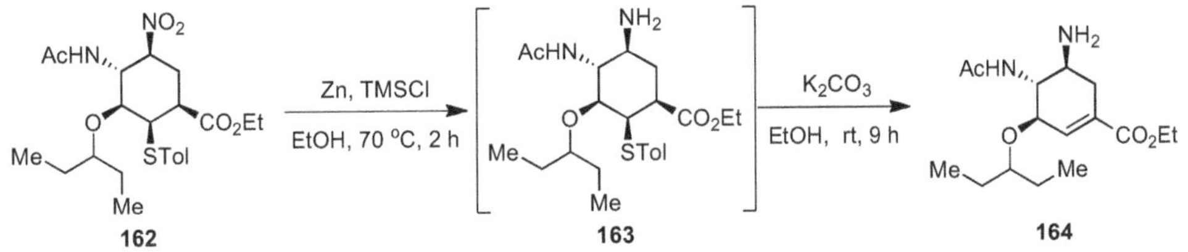

SCHEME 7.27

The presence of Zn(II) species may presumably chelate the amine moieties, which prevented the retro Michael reaction of **163**. To circumvent the problem, Zn(II) species have to be removed or captured by stronger chelating agents. Therefore, bubbling NH_3 gas through the reaction mixture prior to treatment with K_2CO_3 afforded a clean retro Michael reaction product **164** in a yield of 81% over two steps.

7.5 REACTIONS WITH POOR SELECTIVITY

Good selectivity is important in suppressing undesired side reactions; thus several methods are developed to improve reaction selectivity. Besides optimizing common reaction parameters (such as temperature, pressure, concentration, etc.), several other approaches have been developed in order to improve the reaction selectivity, including the use of crystallization-induced dynamic resolution (CIDR), sacrificial reagents, protecting groups, less reactive substrates, and enamine intermediate exchange. Also, a carry-over strategy was designed for a process in which the presence of a side product could be tolerated in the subsequent regiospecific reaction.

7.5.1   CIDR TO IMPROVE CIS/TRANS SELECTIVITY

Nitrile **166** was prepared by hydrocyanation of aryl olefin **165** using 2.1 equiv of triflic acid (TfOH) and 2.0 equiv of trimethylsilyl cyanide (TMSCN) in trifluorotoluene at –20 °C (Equation 7.24).[41]

(7.24)

(I) Reaction problems

Although an 89% assay yield of the product nitrile **166** was obtained, the reaction was plagued by poor *cis/trans* selectivity (62:38). Using conditions including varying temperature, solvent, and CN source[42] failed to improve the *cis/trans* selectivity.

(II) Solutions

As the desired *cis*-**166** is less soluble in ethanol than the *trans*-**166**, an approach of CIDR was designed to improve the *cis/trans* ratio. Using a catalytic amount of potassium *tert*-butoxide (KO*t*Bu) the CIDR process successfully increased the ratio of *cis/trans* to 99:1 with a 90% isolated yield.

Procedure

- Prepare a slurry of *cis/trans* mixture of **166** (*cis/trans*=61/39) by mixing 28.5 g (82.5 mmol) of **166** in 15 mL of anhydrous ethanol and 60 mL of hexane.
- Charge 0.93 g (8.26 mmol) of potassium *tert*-butoxide.
- Heat under reflux for 3 h.
- Charge 75 mL of hexane portionwise and reflux for 1 h.
- Cool the mixture to 20 °C.
- Filter to afford crude **166** (27.5 g, *cis/trans*=99.6/0.4).
- Dissolve the crude **166** in 82.6 mL of ethanol at 65 °C.
- Cool to 5 °C and filter.
- Yield: 25.8 g (90.4%, *cis/trans*=99.97/0.03).

7.5.2   TWO-STEP PROCESS TO MITIGATE RACEMIZATION

A complete conversion of ketone **167** into product **168** was effected via a catalytic reduction using Pd/C in a mixture of dichloromethane (3 volumes) and methanesulfonic acid (1 volume) (Scheme 7.28).[43]

However, it was found that the starting material **167** was not stable under reaction conditions, resulting in a significant racemization of **168**.

Solutions

To overcome the racemization problem, the one-step reduction was replaced with a two-step process, comprising the conversion of **167** into an intermediate alcohol **169** with sodium borohydride (NaBH₄) in methanol followed by catalytic reduction.

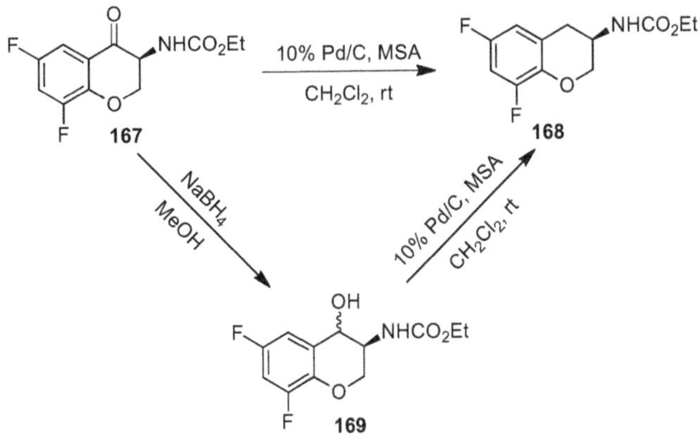

SCHEME 7.28

7.5.3 REDUCTION OF CARBOXYLIC ACID

NaBH$_4$ is a widely used reagent for the reduction of aldehydes and ketones to the corresponding alcohols. However, carboxylic acid derivatives are usually quite resistant to borohydrides. The reduction of carboxylic acid derivatives to the corresponding alcohols is most commonly effected using lithium aluminum hydride (LAH), which, in comparison to NaBH$_4$, is a much more powerful reducing agent. Due to its high reactivity, LAH is less discriminating and in addition to acid derivatives, it readily attacks other functionalities.

The conversion of the acid **170** to the corresponding alcohol **171** was effected using LAH in the discovery chemistry route (Scheme 7.29).[44]

In order to avoid using LAH, a two-step, one-pot protocol was developed to replace the LAH-involved one-step transformation. The two-step process[45] consisted of a conversion of the acid **170** to the corresponding acid imidazolides **172** with 1,1-carbonyldiimidazole (CDI), followed by a selective reduction with NaBH$_4$.

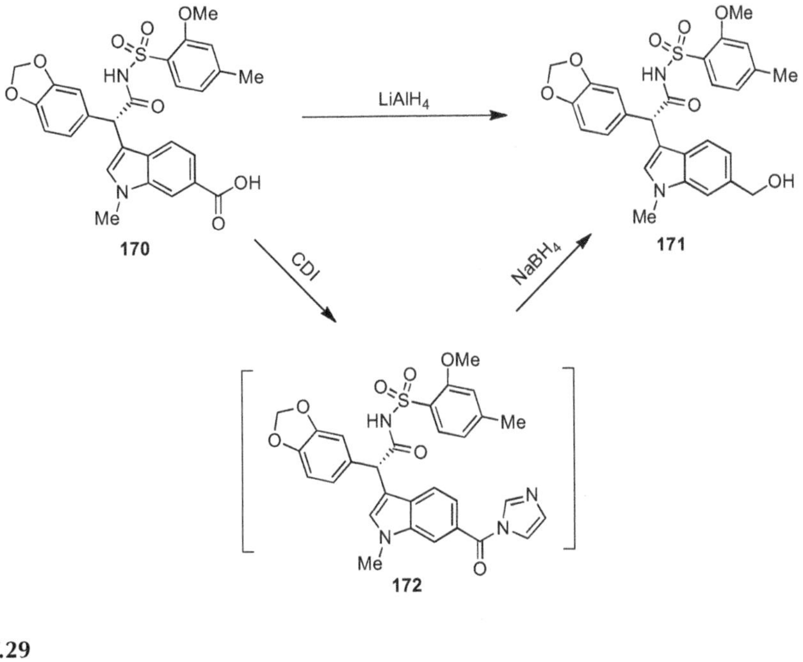

SCHEME 7.29

7.5.4 SACRIFICIAL REAGENT IN REGIOSELECTIVE ACETYLATION

The effect of the neighboring group was utilized to enhance acetylation selectivity. The synthesis of 2'-*O*-acetyl derivative **174** was initially conducted using Ac_2O (1.1 equiv) in the presence of $NaHCO_3$ (4.5 equiv) in methylene chloride at room temperature (Equation 7.25).[46]

(7.25)

Although this reaction condition provided the desired product **174**, a diacetyl side product **175** was observed in a level of 5%.

(I) Chemistry diagnosis

The regioselectivity of this acetylation is controlled by the reaction kinetics. Compared with the rate at 4''-OH, the desired 2'-OH acetylation rate is expected to be enhanced due to the catalytic effect of the vicinal dimethylamino group on the C-3' position.[47] Thus, it was surmised that the introduction of a secondary alcohol as a competitor to the less reactive hydroxyl group (4''-OH) would reduce the level of the side product **175**.

(II) Solutions

Switching methylene chloride to 2-propanol as the reaction solvent suppressed successfully the formation of **175** under otherwise the same reaction conditions, and only less than 0.1% of diacetyl by-product **175** was observed.

Procedure

- Charge 2.1 L of IPA, 700.0 g (0.875 mol) of azithromycin **173**, and 112.0 g (1.33 mol, 1.5 equiv) of $NaHCO_3$.
- Cool the mixture to 0–5 °C.
- Charge 99.4 mL (1.05 mol) of Ac_2O.
- Warm to room temperature and stir for 2 h.
- Add 1.4 L of water and 300 mL of 10% aqueous NaOH to pH 9.3–9.5.
- Add water (700 mL).
- Cool to 5–10 °C.
- Filter after stirring for 2 h.
- Wash with 500 mL of water/IPA (1:1), followed by water (2×350 mL).
- Dry at 45–50 °C, then 75–80 °C under vacuum.
- Yield: 609.9 g (88.3%) in 98.0% purity.

7.5.5 CONVERTING SIDE PRODUCT TO PRODUCT

Iodination of electron-rich phenol derivatives usually leads to over-iodination side products. A strategy of converting the over-iodiination side product to the product was developed by selective deiodination. As shown in Equation 7.26,[48] *ortho*-selective deiodination of 2,4-diiodinated phenol derivatives **176** was achieved using sodium bis(2-methyoxyethoxy)aluminium hydride (Red-Al) in CH_2Cl_2. The reaction conditions are mild (0 °C), yielding the desired product in good yields.

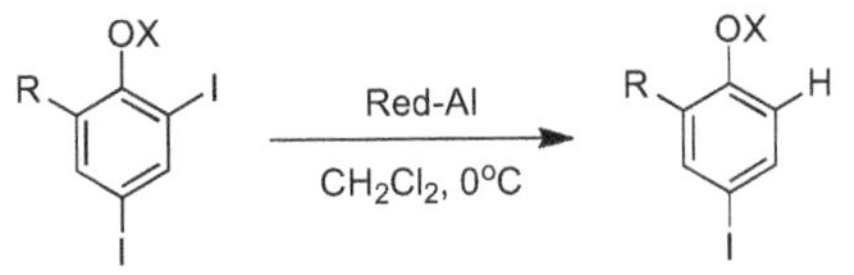

$$(7.26)$$

R = H, Me, *i*-Pr, *t*-Bu; X = H, Me, MOM, TBS

7.5.6 ENAMINE EXCHANGE

The reaction of diaryl-β-diketone **178** with hydroxylamine hydrochloride (5 equiv) in *N*-methylpyrrolidone at 60 °C afforded the desired isoxazole **179** in 63% isolated yield (Equation 7.27).[49] The reaction also produced a regioisomer **180** in 11%, which proved to be hard to remove.

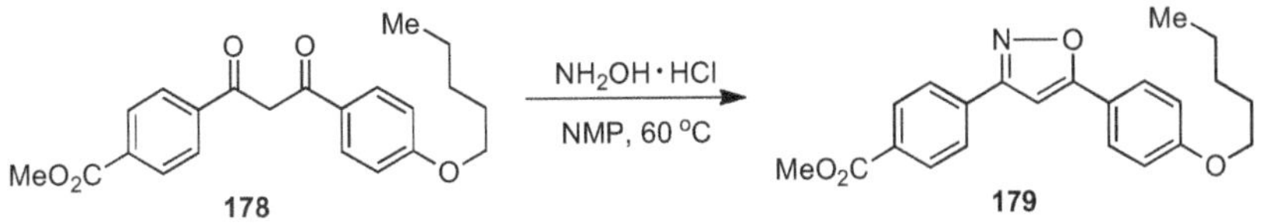

$$(7.27)$$

An enamine-exchange approach was developed by converting **178** to enamine **181** followed by treatment of **181** with hydroxylamine affording **179** in good yield and purity (Scheme 7.30). The side product **182** could be easily rejected by recrystallization from EtOAc/heptane (1:5) at below 0.2% level.

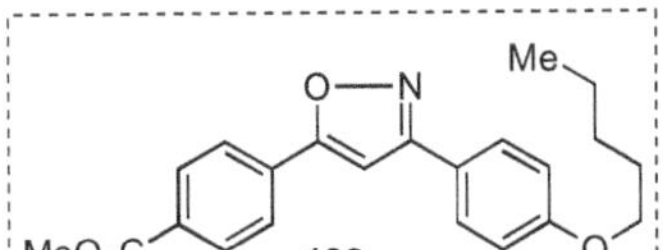

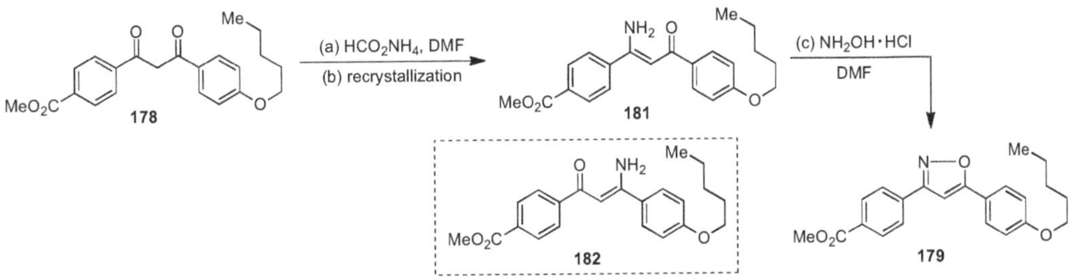

SCHEME 7.30

Procedure

Stage (a)
- Charge 83.3 kg of **178**, DMF (417 L), and 71.3 kg of HCO_2NH_4.
- Stir the mixture at 100–105 °C for 4 h.

- Cool to 20–30 °C and add EtOAc (2083 L) and water (2083 L).
- Separate layers and wash the organic layer with aqueous NaCl.
- Distill the organic layer under reduced pressure to ca. 417 L.
- Add heptane (2083 L) at 50–55 °C.
- Cool to 20–25 °C and stir for 1 h.
- Filter and wash with EtOAc/heptane (1:5).
- Dry at 40–60 °C under vacuum to give 58.2 kg of **181** (70%).

Stage (b)
- Dissolve 58.2 kg of **181** in 291 L of EtOAc at 50–70 °C.
- Add heptane (1455 L).
- Cool to 20–30 °C.
- Filter, wash, and dry to give 50.1 kg of **181**.

Stage (c)
- Charge 50.1 kg of **181**, DMF (401 L), and 18.9 kg of NH$_2$OH·HCl.
- Stir at 60–65 °C for 4 h.
- Cool to 20–40 °C.
- Add MeCN (501 L) followed by water (501 L).
- Cool to 20–25 °C.
- Filter, wash, and dry to give 47.8 kg of **179** (96%).

7.5.7 CARRY-OVER APPROACH

A product with poor solubility in the reaction system offers an opportunity to develop a direct crystallization/precipitation approach for product isolation. A process for the preparation of *p*-menth-2-ene-1,8-diol **187** involves olefin double bond migration/epoxidation leading to a mixture of epoxides **185** and **186** (Scheme 7.31).[50] The subsequent acid-catalyzed hydrolytic ring opening of the epoxide is highly selective, which tolerated epoxide **186** containing certain levels of regioisomer **185**. Furthermore, the desired diol product **187** was poorly soluble in the reaction mixture. Thus, a carry-over approach was developed and demonstrated well in the synthesis of diol **187**.

Ultimately, a mixture of epoxides **185** and **186** was directly subjected to the acid-catalyzed hydrolytic ring-opening reaction, providing the desired diol product **187** as an insoluble solid that was isolated simply by filtration.

7.5.8 FRIEDEL–CRAFTS REACTION

It was observed that Friedel–Crafts acylation of anisole **188** by treating **188** with acetyl chloride in the presence of 2 equiv of AlCl$_3$ at −40 °C gave, in addition to the desired product **189**, regioisomer **190** and demethylated regioisomer **191** (Equation 7.28).[51]

$$(7.28)$$

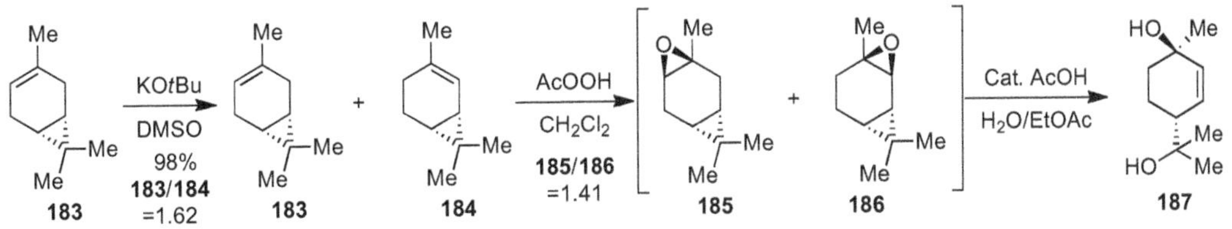

SCHEME 7.31

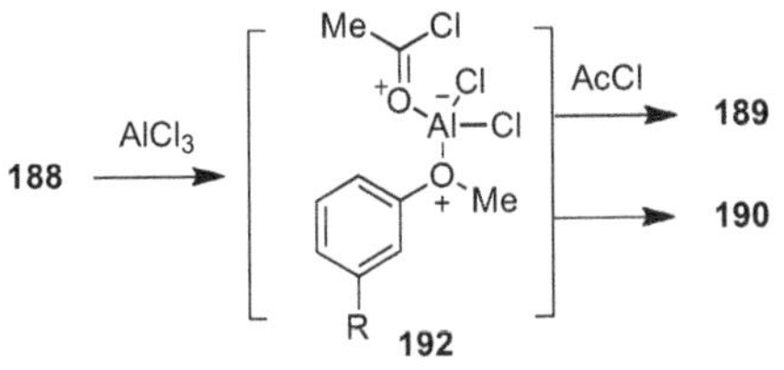

SCHEME 7.32

The Lewis acid-mediated acylation was speculated to proceed through a complex **192** as illustrated in Scheme 7.32. Apparently, the higher temperature might help the dissociation of the acylium ion from the Lewis acid. To achieve high regioselectivity, it is beneficial to have a bulky Lewis acid, such as AlBr₃, coordinate with the aryl ether. To further suppress the intramolecular formation of the regioisomer **190**, excess Lewis acid is desired.

Accordingly, the reaction was conducted using AlBr₃ (2.3 equiv) by adding **188** to a cold solution (0 °C) of AlBr₃ and AcCl, assuring a large excess of the Lewis acid.

7.5.9   Reduction of Carbon–Carbon Double Bond

Reduction of carbon–carbon double bond in the presence of other functional groups may lead to side products in addition to the desired one. For instance, hydrogenation of the carbon–carbon double bond in **195** with Pd/C in EtOAc or THF provided a mixture of the desired product **196** and amino acid **197** in a 3:1 ratio (Scheme 7.33).[52]

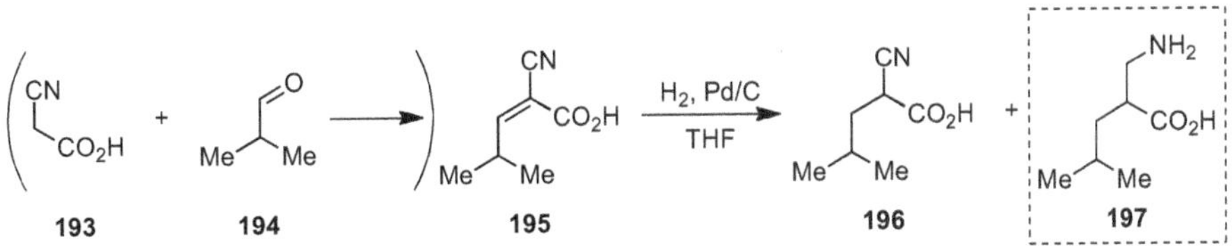

SCHEME 7.33

It was proposed that the hydrogen-bonding of the cyano group with carboxylic acid facilitated the nitrile reduction. The interruption of the hydrogen-bonding interaction between the acid and the cyano group may prevent the reduction of the cyano group. Therefore, using the sodium salt of **195** in the hydrogenation reaction provided the saturated cyano acid **196** as the sole product. Eventually, a one-pot Knoevenagel condensation/transfer hydrogenation was developed. The reaction was carried out in water by mixing **193** and **194** with aqueous NaOH in the presence of Pd/C and sodium hypophosphite (Equation 7.29).

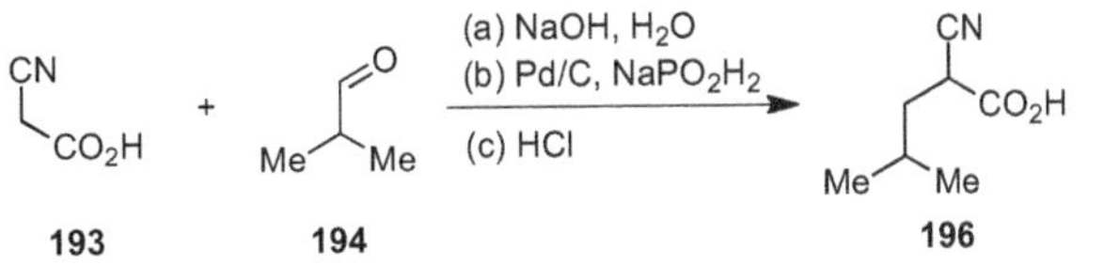

$$(7.29)$$

Procedure

Stage (a)
- Charge 25 g (0.29 mol) of **193** to an aqueous solution of NaOH (13 g, 0.33 mol, 1.1 equiv, in 250 mL of water) at 0 °C.
- Charge 22.2 g (0.31 mol, 1.1 equiv) of **194** at 0 °C over 15 min.
- Stir the resulting mixture at 0 °C for 30 min.

Stage (b)
- Charge 5% Pd/C (Escat-147) (50% wet, 2.5% loading based on intermediate **195**), followed by adding NaPO$_2$H$_2$ (78 g, 0.87 mol) in water (50 mL) at 0 °C over 5 min.
- Heat the batch to 60–65 °C.

Stage (c)
- Add conc HCl to pH 1 (80 mL, 0.96 mol).
- Extract with EtOAc (2×400 mL).
- Distill to dryness.
- Yield of **196**: 39.0 g (94%).

REMARK

Transfer hydrogenation eliminated the need for specialized hydrogenation equipment.

7.5.10 REDUCTION OF NITRILE

A general problem for nitrile hydrogenation to the corresponding primary amine is the inevitable side reaction towards the secondary amine. To address this issue, various catalysts have been used including ruthenium, rhodium, and iridium.[53]

Raney nickel is frequently used to reduce nitriles to primary amines. This reduction proceeds through an imine intermediate **198** and the product distribution in the hydrogenation of nitriles is determined by the fate of the intermediate imine **198**. The imine **198** may be hydrogenated further to produce the primary amine or transamination with the primary amine product to form the imine **199** whose further hydrogenation generates a secondary amine (Scheme 7.34). To mitigate such side reactions during the hydrogenation of nitriles with Raney nickel, ammonia is often used as an additive.

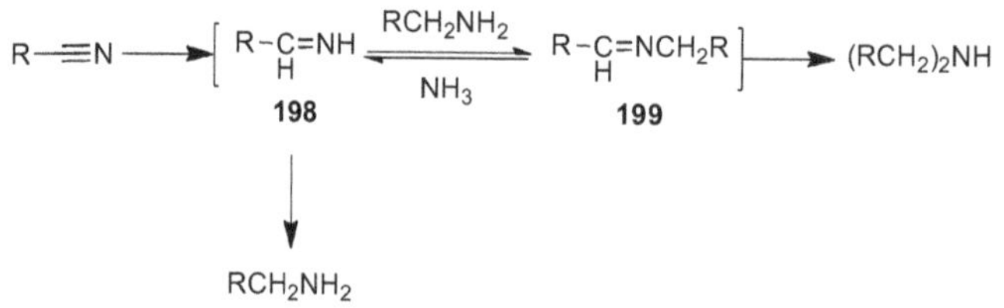

SCHEME 7.34

In addition to the aforementioned approach by adding additives, steric bulky groups can also suppress the undesired imine side reaction. For example, the hydrogenation of nitrile **200** using Raney nickel occurred in the absence of the ammonia additive without the formation of the secondary amine (Equation 7.30),[54] presumably due to the adjacent bulky isobutyl and phenyl groups.

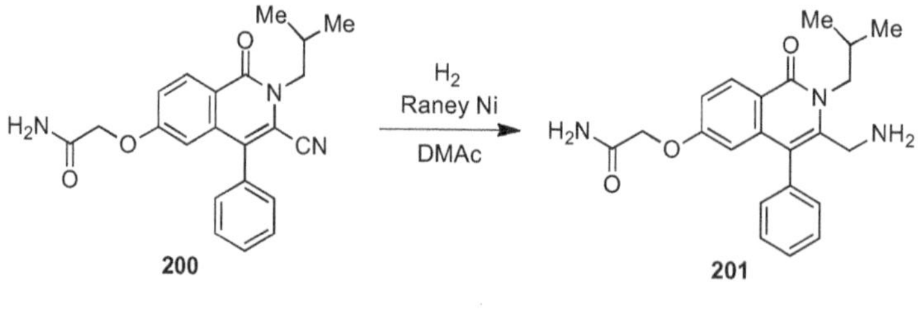

(7.30)

REMARKS

(a) Hydrogenation may be stopped at the intermediate imine stage under carefully controlled conditions. Rapid hydrolysis of the imine and further hydrogenation may be used to produce the primary alcohol.

(b) Raney nickel is not poisoned by sulfur.

(c) Raney nickel is ferromagnetic, which precludes the use of magnetic stirring.

7.6 PROTECTING GROUP

Organic transformation frequently needs functional group protection and deprotection. The selection of protecting group is an important step in organic synthesis. When a molecule has two or more functional groups, protecting groups are often employed to improve selectivity by preventing undesired reactions from occurring. Besides, the protecting group is also utilized to optimize reactions.

7.6.1 ACETYL GROUP

The hydroxyl group in phenol activates the aromatic ring to lead to undesired reactions. The protecting group is used to deactivate the aromatic ring in order to avoid a phenol product from further reactions under reaction conditions. For instance, protection of phenol with an acetyl group (such as **203**) would greatly reduce its reactivity toward electrophilic bromination, and the formation of brominated side products **205** could effectively be suppressed during the aromatization (Scheme 7.35).[55]

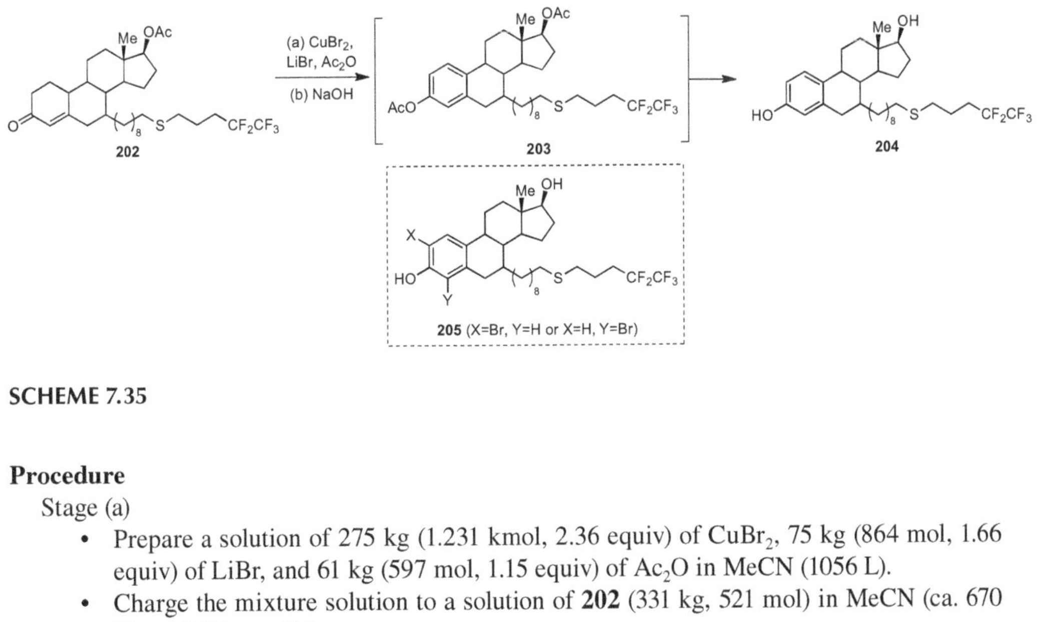

SCHEME 7.35

Procedure

Stage (a)

- Prepare a solution of 275 kg (1.231 kmol, 2.36 equiv) of $CuBr_2$, 75 kg (864 mol, 1.66 equiv) of LiBr, and 61 kg (597 mol, 1.15 equiv) of Ac_2O in MeCN (1056 L).
- Charge the mixture solution to a solution of **202** (331 kg, 521 mol) in MeCN (ca. 670 L) at 20 °C over 3 h.
- Charge 45 kg (441 mol, 0.85 equiv) of Ac_2O.
- Stir the batch at 20 °C for 4 h.
- Quench and go to stage (b).

7.6.2 TRIMETHYLSILYL GROUP

Trimethylsilyl (TMS) groups are frequently used as temporary protecting groups in chemical synthesis to functionalize alcohols, phenols, and terminal alkynes. The TMS group can be readily cleavaged with HF-based reagents such as tetrabutylammonium fluoride and HCl.

7.6.2.1 Protection of Terminal Alkyne

As discussed in Chapter 4, impurities can be formed as a result of side reactions of an intermediate and/or product. The formation of **209** from the condensation of amine **206** and aldehyde **207a** via either direct reductive amination with NaBH(OAc)$_3$ or step-wise reductive amination with NaBH$_4$ was plagued with a significant quantity of side product **210** (Scheme 7.36).[56] The formation of **210** is presumably via a reaction of **208a** with the product **209** through diamine **211** and iminium **212**.

Protection of the terminal alkyne **207a** with the TMS group may alter the electronic property, leading to the rate increase in the imine (**208b**) reduction step. Ultimately, the desired product **209** was generated exclusively upon using the TMS-protected **207b**. Compound **209** was isolated in 90% yield with good purity.

Procedure

- Dissolve 10 g (49 mmol, 1.0 equiv) of **207b** in MeOH (200 mL).
- Charge 7.4 g (52 mmol, 1.05 equiv) of **206**.
- Stir at room temperature overnight.
- Charge slowly 1.9 g (49 mmol, 1.0 equiv) of NaBH$_4$.
- Stir the resulting mixture at rt for 1 h.
- Add 1 N aqueous HCl slowly.
- Evaporate MeOH.
- Partition the residue in CH$_2$Cl$_2$/sat NaHCO$_3$.
- Separate layers.
- Dry the organic layer over MgSO$_4$.
- Concentrate and purify the crude product by chromatography.
- Yield: 11 g (90%).

REMARK

The removal of the TMS protecting group was achieved in situ upon the addition of NaBH$_4$, which made the sequence highly efficient.

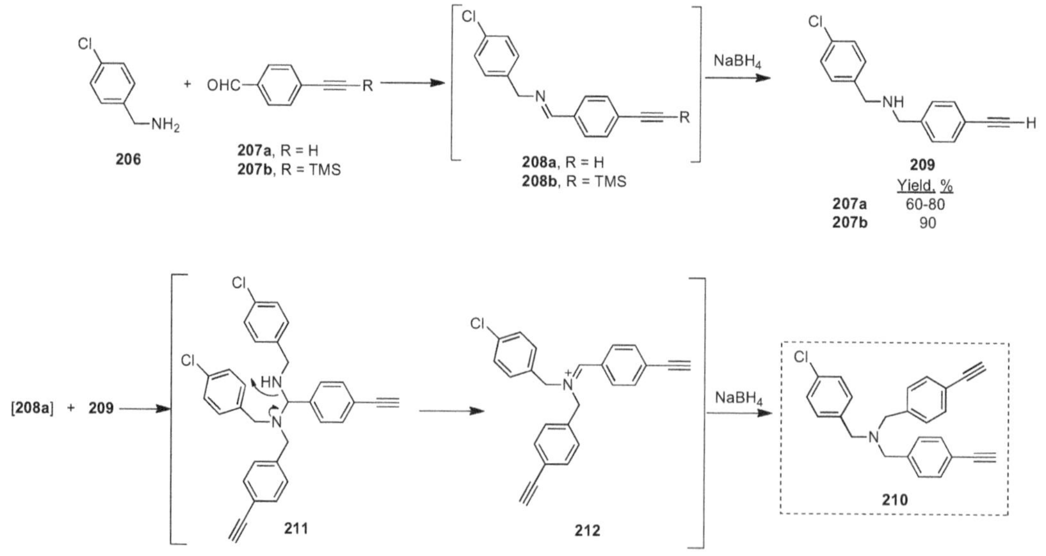

SCHEME 7.36

7.6.2.2 Protection of Enolate

Epimerization at C-8 of dihydrolysergic acid methyl ester **213** was performed via a base-induced isomerization, followed by protonation to afford the desired epimer **215** (Scheme 7.37).[57]

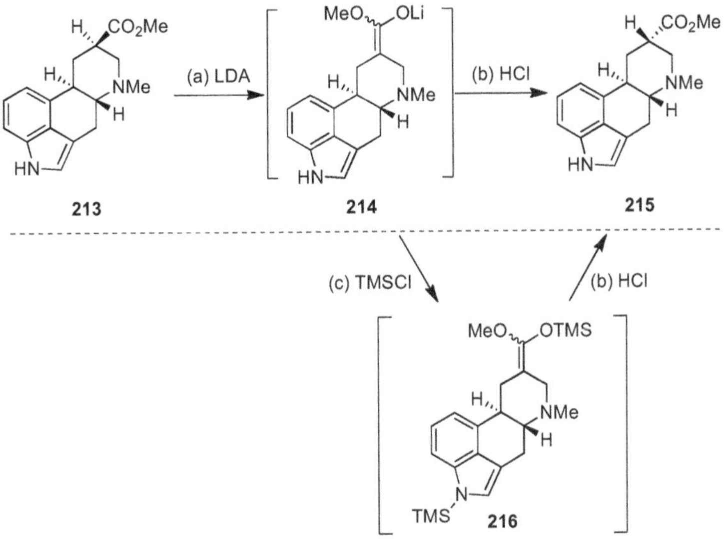

SCHEME 7.37

(I) Reaction problems

- Under conditions of (a)–(b), a mixture of the corresponding product **215** and the starting material **213** was obtained in a ratio of 4:1, which is controlled by kinetic protonation.
- Attempts to improve the ratio of **215** to **213** by variation of base or temperature proved to be unsuccessful.
- In the event, the isolated yields of pure **215** after crystallization were moderate (60–65%).

(II) Solutions

To improve the epimerization ratio, the enolate **214** was converted to silyl ketene acetal **216** in situ by treatment of **215** with trimethylsilyl chloride followed by reacting with HCl [(a)–(c)–(b)]. As a result, an improved epimeric ratio of 96:4 of **215** to **213** was obtained and **215** was isolated in yields of 91–93%.

7.6.2.3 Protection of Hydroxyl Group

Preparation of sulfonamide **218** with 2-thienylsulfonyl chloride ($ThSO_2Cl$) in the presence of hydroxyl group proved to be unsuccessful, leading to aziridine **219** as the major product (90%) (Equation 7.31).[58]

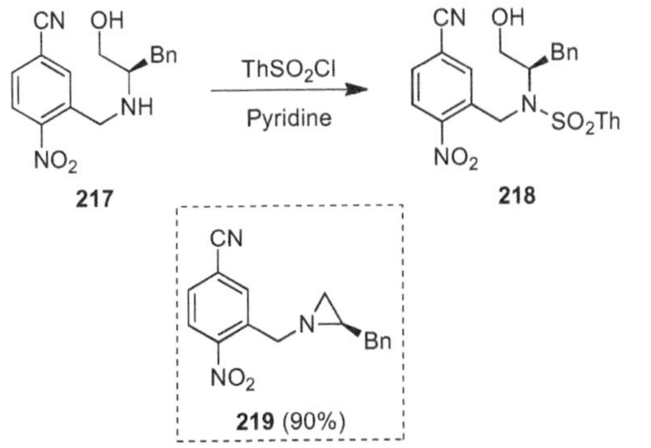

(7.31)

In situ protection approach with TMSCl, however, resulted in contamination of the product with hydroxyl sulfonamide impurity **222** (25%) (Scheme 7.38). The formation of **222** was hypothesized from a nucleophilic attack of chloride at the benzylic position of the TMS ether intermediate **220**.

In order to overcome this side reaction, the substitution of TMSCl with chloride-free *N,O*-bis(trimethylsilyl)acetamide (TMCS) was able to reduce the impurity from 25% to ~5%.

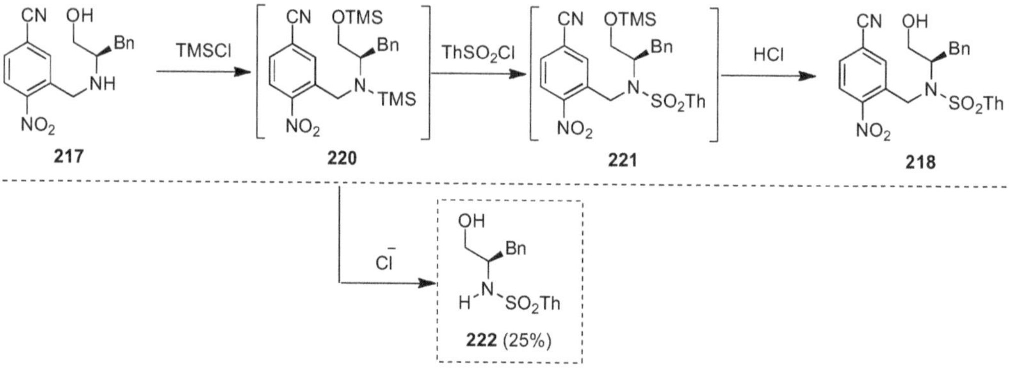

SCHEME 7.38

7.6.3 CYANOETHYL GROUP

The protecting group strategy was demonstrated during the process development for the synthesis of OPC-15161, a drug to prevent tissue damage caused by superoxide anion. The synthesis of OPC-15161 involved a crucial cyclization of oxime acid **223** into **224a** via activation by *N*-hydroxysuccimide (HOSu)/DCC.

The subsequent methylation, however, was plagued by the formation of undesired *O*-methylated isomer **225** (**225**/OPC-15161=2/1) (Scheme 7.39).[59]

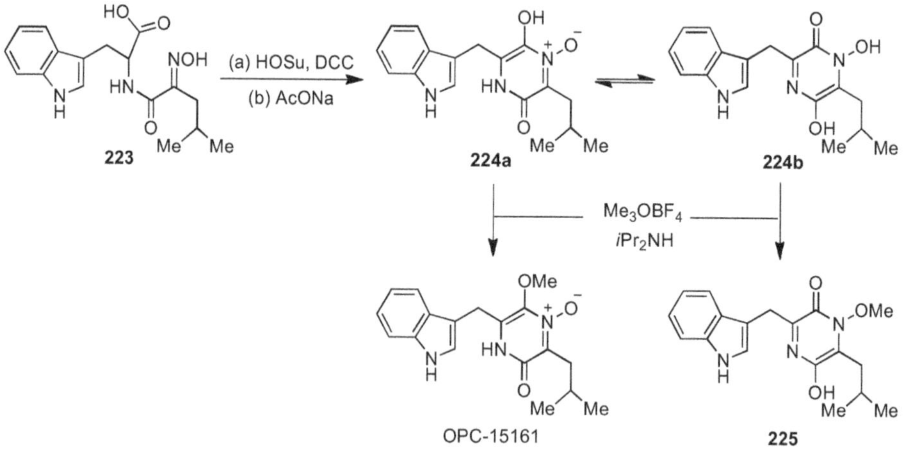

SCHEME 7.39

Solutions

The intermediate **224b**, an isomer of the cyclization product **224a**, is responsible for the formation of **225**. Ultimately, protection of the amide nitrogen with 2-cyanoethyl group suppressed the isomerization affording the desired methylated product **228** whose subsequent removal of 2-cyanoethyl protecting group furnished OPC-15161 (Scheme 7.40).

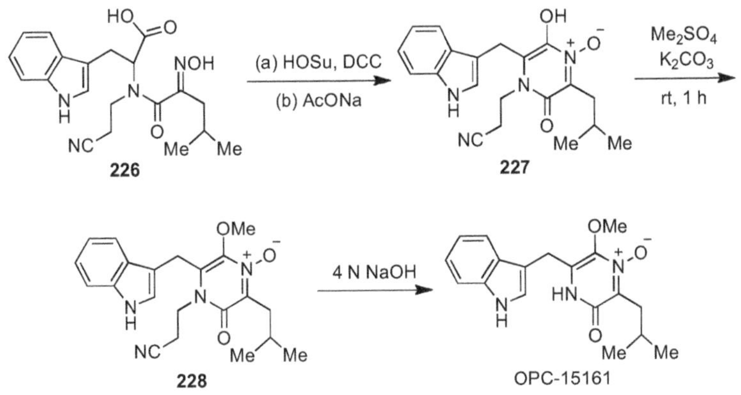

SCHEME 7.40

7.6.4 BENZHDRYL GROUP

The protecting group strategy was also utilized in a cyclization reaction to improve the regioselectivity. Synthesis of aminopyrazole **231** was initially accomplished via cyclic condensation of bromoketone **229** with thiosemicarbazide **230** in ethanol followed by sulfur extrusion (Equation 7.32).[23] Although the aminopyrazole **231** was the major product, this reaction also generated 25% of side product **232**.

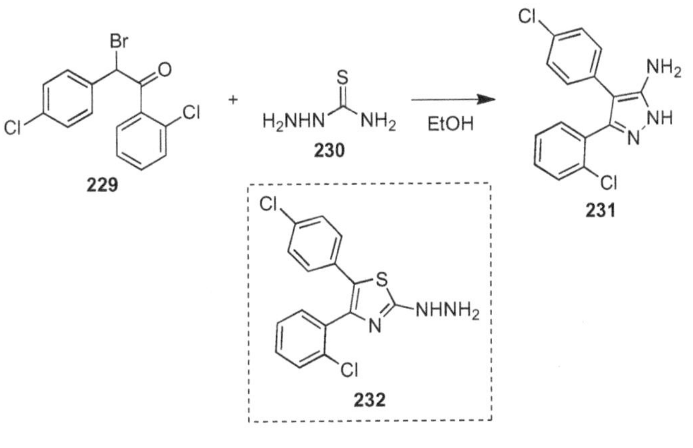

(7.32)

To overcome the regioselectivity issue, an approach of using protecting group was pursued. As anticipated, utilization of benzhdryl-protected substrate **233** afforded the desired product **231** in an overall yield of 70% (Equation 7.33).

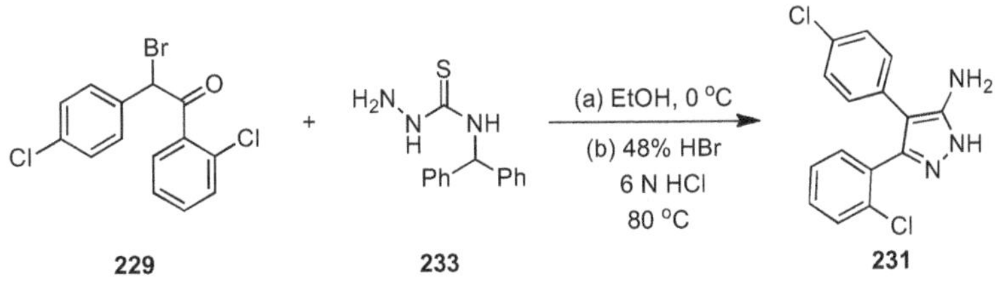

(7.33)

Procedure

- Charge 3.41 g (9.95 mmol) of **229** to EtOH (120 mL) at 0 °C followed by adding 2.58 g (10.0 mmol) of **233**.
- Warm the resulting slurry to room temperature and stir overnight.

- Charge 2 mL of 48% aqueous HBr.
- Heat to 80 °C and stir at that temperature for 4 h.
- Charge 10 mL of 6 N HCl.
- Continue to heat at 80 °C for 16 h.
- Cool to room temperature.
- Distill and dissolve the resulting residue in a mixture of EtOAc (100 mL) and 2N NaOH (50 mL).
- Filter and separate layers.
- Distill the organic layer to give a crude product.
- Purify the crude product by chromatography.
- Yield: 2.10 g (70%).

REMARKS

(a) The use of the benzhydryl protecting group allowed the deprotection step to be telescoped into a one-pot operation.
(b) Extrusion of sulfur solid from the reaction media was the driving force (Equation 7.34) to realize the desired ring contraction:

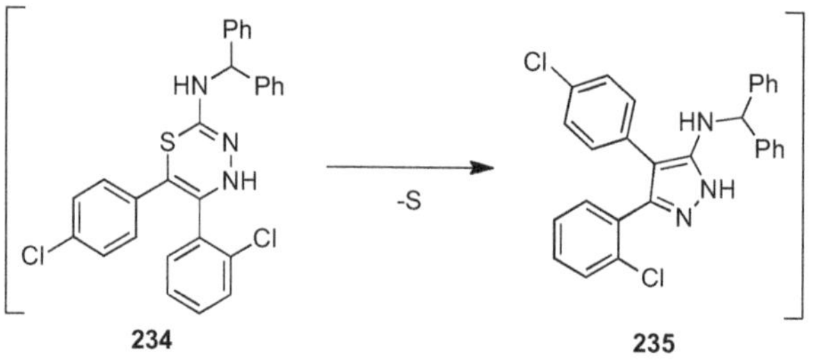

$$(7.34)$$

(c) Desulfonylation was also used as a reaction driving force to access 2-oxoindolines **238** (Scheme 7.41):[60]

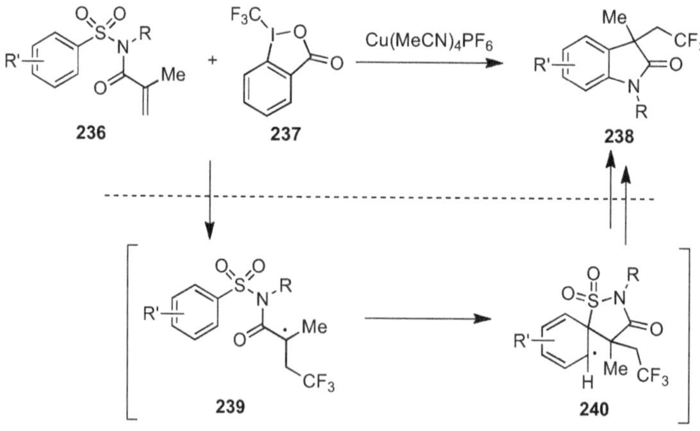

SCHEME 7.41

7.7 POLYMERIZATION ISSUES

Polymeric reaction side products can aggregate into a sticky mass which may lead to agitation failure, isolation problems, and loss of products.

7.7.1　Polymerization of Chloroacetyl Chloride

The base is often used as an acid scavenger in preparation of amide from the reaction of the acid chloride with an amine. Sodium carbonate was selected to trap the HCl by-product from N-acetylation of **241** with chloroacetyl chloride (Equation 7.35).[61] Unfortunately, the presence of sodium carbonate generated chloroketene side product that polymerized producing colored impurities during the N-acetylation.

$$(7.35)$$

Solutions

The colored impurity issue was resolved by employing Schotten–Baumann conditions (aqueous Na_2CO_3/toluene). The acylation under the Schotten–Baumann conditions gave a clean reaction profile and eliminated the formation of highly colored materials, providing a colorless product **242** in a quantitative yield.

Procedure

- Charge a solution of Na_2CO_3 (14.7 g, 0.139 mol) in 74 mL of water into a slurry of **241** (20.0 g, 0.069 mol) in 200 mL of toluene.
- Cool the mixture to 5 °C.
- Charge a solution of chloroacetyl chloride (13.36 g, 0.118 mol) in 66 mL of toluene over 1 h.
- Stir the reaction mixture until the reaction is complete.
- Add 200 mL of water.
- Heat to 50 °C.
- Wash the organic phase with water (200 mL) and brine (200 mL).
- Distill the organic layer under reduced pressure until KF <0.20% (the resulting solution was used directly in the subsequent step).

7.7.2　Polymerization of Acid Chloride

The peptide coupling of amine **243** with acid chloride **244** was utilized in the final step for the synthesis of ABT-378, a drug candidate as an HIV protease inhibitor (Equation 7.36).[62] It was observed that polymerization of the acyl chloride **244** to **245** was the main reaction pathway when less than 1 equiv of imidazole was used. When an excess of imidazole (3.3 equiv) was used, this undesired polymerization was suppressed completely. Obviously, imidazole plays dual roles in this peptide coupling: deactivation of the acid chloride by the formation of acyl imidazole and HCl scavenger.

$$(7.36)$$

7.7.3 Polymerization of Chloroacrylonitrile

Diels–Alder reaction is a useful organic transformation for access to 6-membered ring systems with good controls over regio- and stereochemical selectivities. The [4+2] cycloaddition between 1,3-butadiene derivatives and dienophiles presents an atom-economic process without any by-products. Despite these favorable features, Diels–Alder reaction often requires high reaction temperature that may lead to side reactions, such as polymerization.

Diels–Alder reaction of chloroacrylonitrile **246** with **247** led to extensive polymerization even with stabilizers such as BHT and hydroquinone(Scheme 7.42).[63]

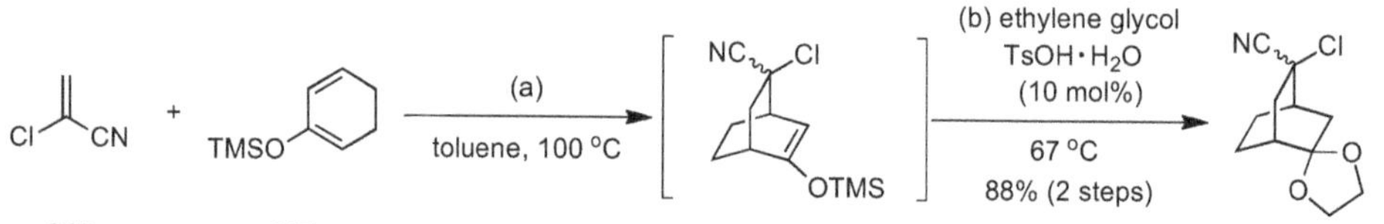

246   **247**   **248**   **249**

SCHEME 7.42

It was found that a catalytic amount of 2,2,6,6-tetramethylpiperidine-1-oxyl (TEMPO) could effectively prevent polymerization during the synthesis of ketal **249**.

Hence, the mixture of **246** (1.24 equiv) and **247** was heated at 100 °C in the presence of catalytic amounts of TEMPO (1.3 mol%) and sodium bicarbonate (0.31 equiv) for 19 h and 88% of **249** was obtained following deprotection and protection.

Note

The addition of solid $NaHCO_3$ was used to neutralize HCl contamination in **246**.

7.7.4 Polymerization of Enone

In general, dilute reaction conditions are preferred for an intramolecular reaction in order to suppress undesired intermolecular reactions. For instance, Nazarov cyclization had to be conducted at a fairly high dilution (0.2M) using concentrated sulfuric acid as the solvent to limit the polymerization of enone **251** (Scheme 7.43).[64]

Consequently, the workup with water created a very large exotherm and resulted in unacceptably high process volumes.

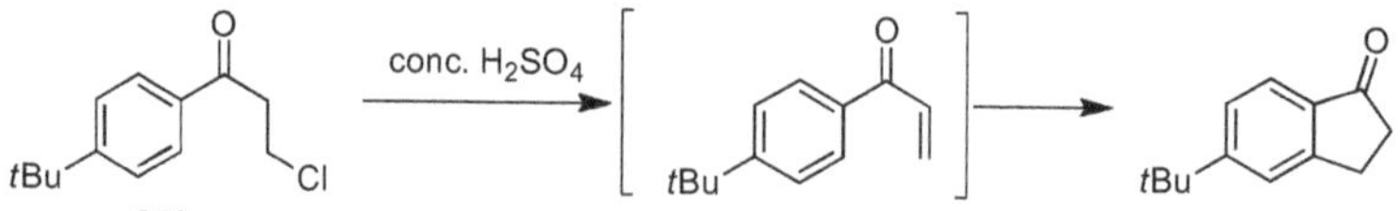

250   **251**   **252**

SCHEME 7.43

Solutions

The polymerization of **251** was minimized by adding chloroketone **250** to a pre-heated sulfuric acid at 90 °C at a rate comparable to the cyclization rate of enone **251**. With such modified reaction conditions, the reaction was conducted at higher concentrations (0.7M), and the polymerization of **251** was minimized.

Procedure

- Charge a solution of **250** (>21.5 mol, obtained from Friedel–Crafts acylation) in methylene chloride and heptanes to a hot (90 °C) conc. H_2SO_4 (59 kg) over 1 h.

- Stir the resulting mixture at 90 °C for 1 h.
- Cool the batch to 15–20 °C.
- Transfer slowly the batch to a mixture of water/MTBE/heptanes (30:5.5:5, 40.5 kg) at <10 °C.
- Separate layers and wash the organic layer with 5% aqueous K_2CO_3.
- Distill the organic layer under reduced pressure.
- Yield of **252**: 3.7 kg (83%).

7.7.5 POLYMERIZATION OF PENTASULFIDE

Phosphorus pentasulfide (exists as dimer P_4S_{10}) was used to prepare thioamide **254** from amide **253** (Equation 7.37).[65] Due to the self-polymerization of P_4S_{10}, the reaction mixture intends to aggregate to form a sticky mass. This issue was resolved by the addition of diatomaceous earth into the reaction mixture.

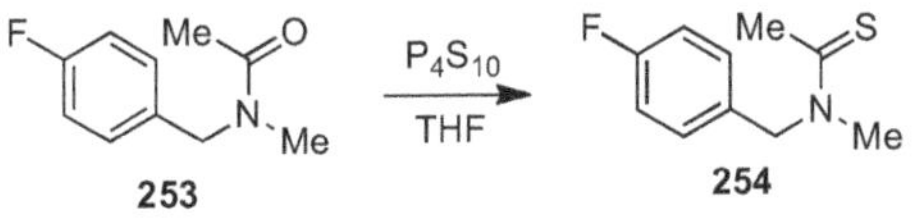

$$(7.37)$$

7.8 ACTIVATION OF FUNCTIONAL GROUPS

Although a direct organic transformation is desired from the perspective of process efficiency, it is, sometimes, necessary to carry out additional protection, and activation/deactivation steps in order to achieve the desired reactivity and selectivity.

7.8.1 ACTIVATION OF ANILINE NITROGEN

The nitro group is a valuable functional group and has a variety of applications in organic synthesis. Frequently, the nitro group, serving as an ammonia surrogate, is installed onto aromatic rings via nitration followed by reduction. Furthermore, the nitro group could be used as a unique functional group to activate an amino group for N-alkylation of aniline **255** (Scheme 7.44).[66] It was found that the N-alkylation of **255** with benzyl bromide **257** required activation of the aniline nitrogen. Attempts to activate the aniline nitrogen by acetylation with ethyl chloroformate, acetic anhydride, or acetyl chloride under conventional conditions resulted in the recovery of the starting compound.

In the event, the aniline activation was successfully realized by the installation of the nitro group onto the aniline nitrogen. The subsequent hydrogenation removed the nitro group providing the benzylated aniline product **259** with a yield of 81%.

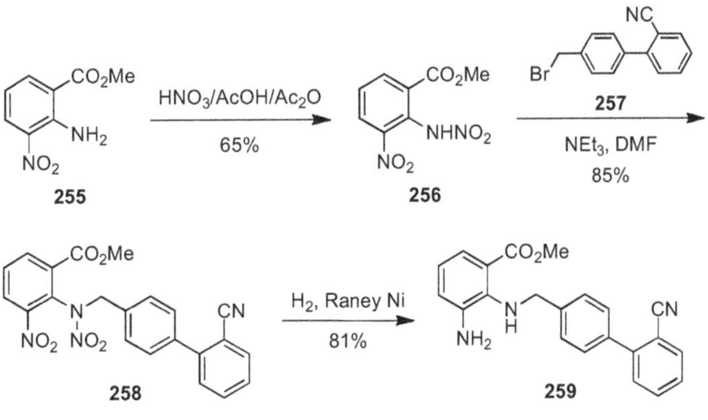

SCHEME 7.44

REMARK

DSC studies of both dinitro derivatives **256** and **258** indicate that they are stable.

7.8.2 Activation of Amide

The carbonyl group in amide is relatively less reactive compared with the one in ketone or ester. The synthesis of aminoindanol-derived M1 Agonist **262** was initially achieved under Vilsmeier conditions[67] in poor yields. The amidine **262** formation was optimized by coupling aniline **261** with activated thioamide **260** (Scheme 7.45).[65]

The activated thioamide **260** (as a mixture of *E* and *Z* isomers) was prepared in a two-step process involving the formation of thioamide **254** and a subsequent reaction with MeI (1.5 equiv).

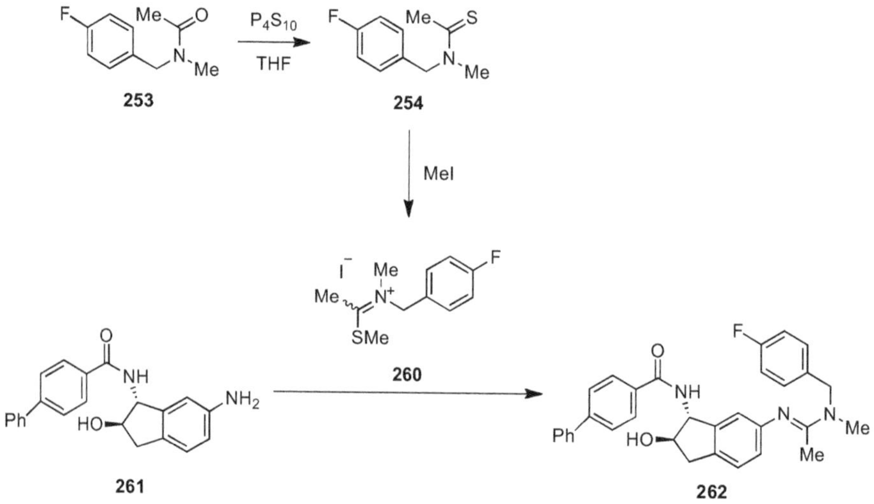

SCHEME 7.45

7.8.3 Activation of Lactol

The synthesis of *N*-benzyl lactam acetal **265a/265b** via a coupling reaction between *N*-benzyl lactol **107** and alcohol **264** requires the activation of **107** (Scheme 7.46).[28]

(I) Reaction problems

Initially, the coupling reaction of **264** with **263a**, generated via a reaction of **107** with trichloroacetonitrile in the presence of DBU, provided a 55:45 mixture of **265a** and **265b** in 90% yield. Although

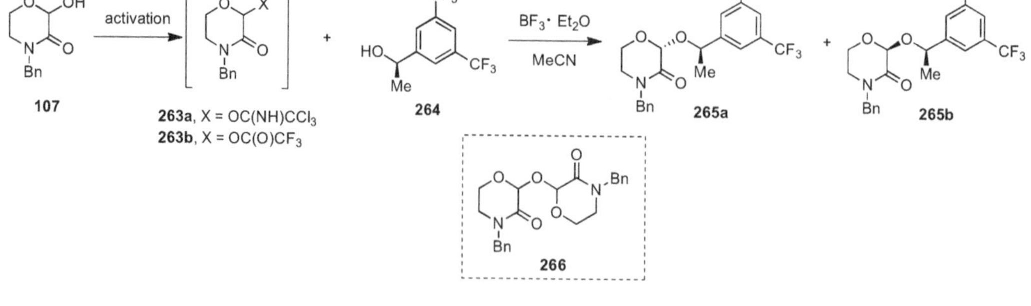

SCHEME 7.46

this protocol worked quite well on small scale, it proved difficult to scale. Invariably, significant amounts of "dimeric" **266** were also produced. Additionally, the handling of the rather unstable and moisture-sensitive **263a** was quite problematic.

(II) Solutions

In order to address these issues, several activation groups (X: Cl, Br, OC(O)CH$_3$, OC(O)tBu, OC(O)iPr, OC(O)Ph, OC(O)CF$_3$) were screened. As a result, trifluoroacetate ester (X: OC(O)CF$_3$) provided excellent activation, affording the desired product **265a/265b** in 95–98% overall yields (with 55:45 diastereomeric ratio). This process proved practical and scalable.

Procedure

- Charge 2.056 kg (9.8 mol) of trifluoroacetic anhydride to a slurry (KF<140 μg/mL) of **107** (2.03 kg, 9.8 mol) in MeCN (4.8 L) at 5–34 °C over 10 min.
- Stir the resulting solution at ambient temperature.
- Charge a solution of **264** (4.85 kg of the solution containing 2.40 kg of **264**, 9.3 mol) followed by adding 0.62 L (4.9 mol) of BF$_3$·Et$_2$O.
- Stir the resulting mixture at ambient temperature for 2 h.
- Add a 5 M aqueous solution of NaOH (8.1 L) at below 27 °C followed by 2.0 L (10.5 mol) of 3,7-dimethyloctan-3-ol.
- Distill at atmospheric pressure until reached 92 °C (8.1 L of distillate collected).
- Add water (4.1 L) and heptane (12.2 L).
- Warm the resulting mixture to 45 °C and separate layers.
- Wash the organic layer with water (12.2 L) at 45–50 °C.
- Distill until KF <130 μg/mL reached (7.8 L of distillate collected).
- Cool to rt and add **265a** (50 mg).
- Cool to –11 °C and add 1.09 kg of 48.7 wt% solution of potassium 3,7-dimethyloct-3-oxide in heptane (2.7 mol) over 10 min.
- Age the mixture at –12 to –7 °C for 5 h.
- Add 0.28 L of AcOH followed by a 5 wt% solution of NaHCO$_3$ in water (4.1 L).
- Warm the biphasic mixture to 45–50 °C prior to layer separation.
- Wash the organic layer with water (4.1 L) at 45–50 °C.
- Distill at atmospheric pressure to a total volume of 16 L.
- Add seeds at 45 °C.
- Cool to 5 °C and age for 1.5 h.
- Filter and wash with cold heptane (3.0 L).
- Dry under vacuum.
- Yield: 3.51 kg of **265a** (84%).

REMARK

The crystallization-induced dynamic resolution in 3,7-dimethyl-3-octanol (tetrahydrolinalool) using potassium 3,7-dimethyloct-3-oxide as a base allowed to convert the undesired diastereomer **265b** into **265a**.

7.8.4 C–H Bond Activation

Transition metal-catalyzed C–H activation represents an efficient approach to introducing functional groups in organic transformations. Room temperature Suzuki–Miyaura coupling reactions were realized via C–H bond activation under cationic palladium (II) catalysis conditions (Scheme 7.47).[68]

In this approach, the use of aromatic ureas **267** as Suzuki coupling partners to replace common aryl halides allows C–C bond formation under mild conditions in the presence of 1,4-benzoquinone (BQ). It was notable that this method involves nitrogen- or oxygen-based directing groups (DG) for *ortho*

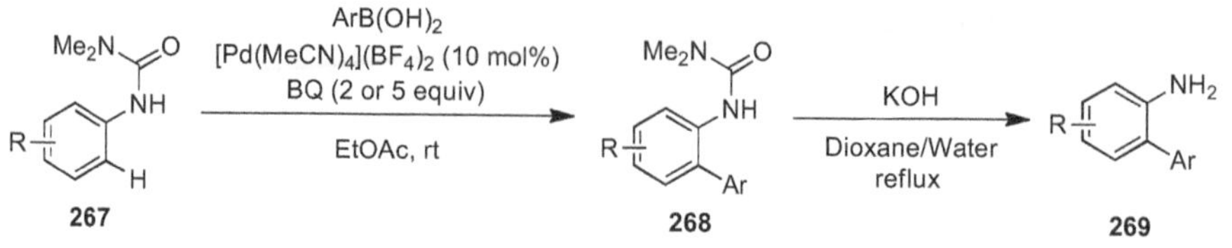

SCHEME 7.47

C–H activation. Various arylboronic acids having electron-donating or -withdrawing groups reacted smoothly with **267** in high yields. The *N,N*-dimethylcarbamoyl moiety in **268** was easily removed under general hydrolysis conditions to produce the corresponding amines **269** quantitatively.

7.9 DEACTIVATION OF FUNCTIONAL GROUPS

7.9.1 DEACTIVATION OF AMINO GROUP

Due to the basicity and nucleophility, the amino group can participate in various undesired reactions in organic transformations. For instance, a coupling reaction between pyridyl amine **270** and *N*-carboxylate anhydride **271** would form a polymeric side product **273** instead of the desired product **272**, because the amino group of **270** is a much weaker nucleophile than the amino group of glutamic acid (Scheme 7.48).[69]

Instead of using a protection approach, the amino group can be conveniently deactivated by forming their corresponding salts. This was achieved by the utilization of catalytic amounts of free base **270** (0.1 equiv) in combination with its hydrochloride salt (0.9 equiv).

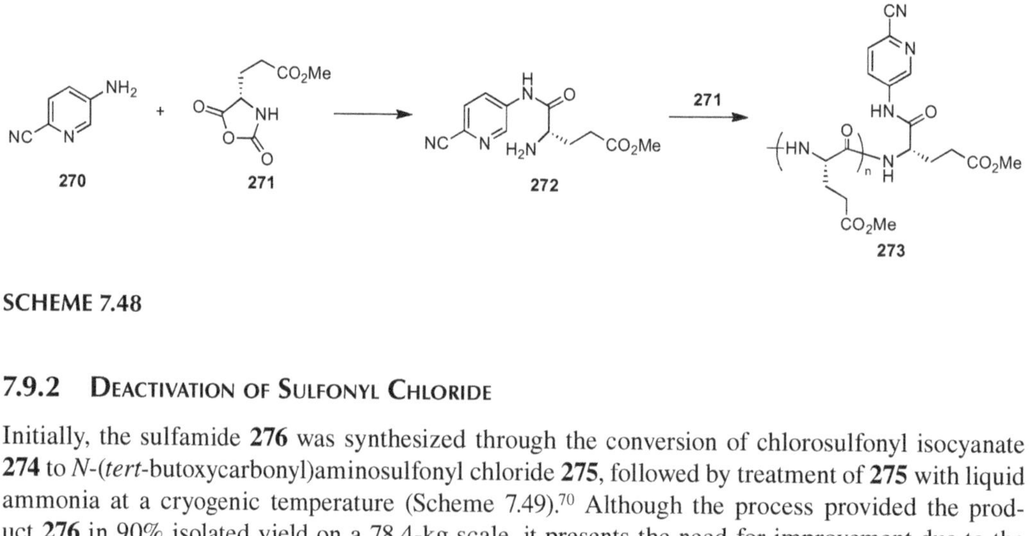

SCHEME 7.48

7.9.2 DEACTIVATION OF SULFONYL CHLORIDE

Initially, the sulfamide **276** was synthesized through the conversion of chlorosulfonyl isocyanate **274** to *N*-(*tert*-butoxycarbonyl)aminosulfonyl chloride **275**, followed by treatment of **275** with liquid ammonia at a cryogenic temperature (Scheme 7.49).[70] Although the process provided the product **276** in 90% isolated yield on a 78.4-kg scale, it presents the need for improvement due to the

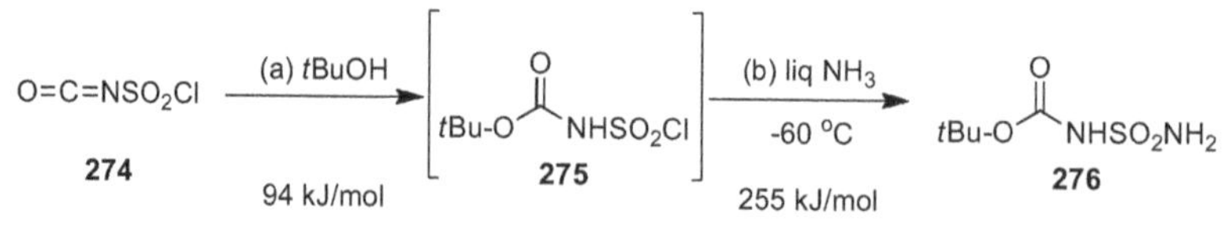

SCHEME 7.49

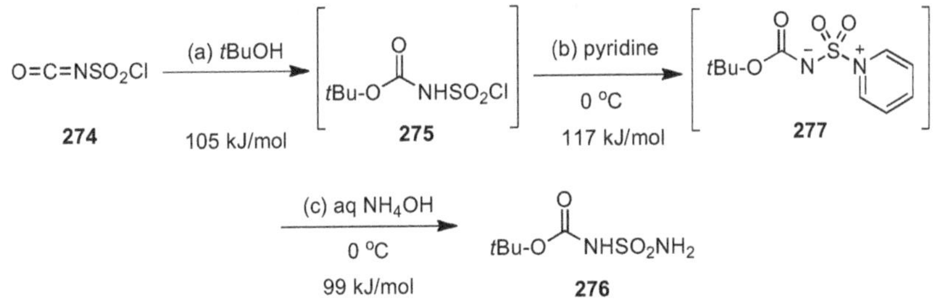

SCHEME 7.50

cryogenic conditions and instability of **275** against moisture. Furthermore, the ammonium chloride by-product appeared as a sticky, adhesive precipitate in the anhydrous reaction system, which renders the process unsuitable for further scaling up.

The instability of **275** is attributed to the high reactivity of the sulfonyl chloride group. Therefore, a strategy was adopted to deactivate the sulfonyl chloride by converting **275** to less water-sensitive *N*-(*tert*-butoxycarbonyl)aminosulfonylpyridinium salt **277** via a reaction of **275** with pyridine (Scheme 7.50).

Sulfonyl pyridinium salt **277** showed better stability toward moisture than the corresponding sulfonyl chloride **275**. Without isolation, treatment of **277** with aqueous ammonia furnished the desired **276** in 90–96% yields.

Procedure

Stage (a)
- Charge 21.0 g of *t*-BuOH to a solution of **274** (40.0 g, 283 mmol) in toluene (387 mL) at <0 °C.
- Stir the resulting mixture at 0 °C for 20 min.

Stage (b)
- Charge 49.2 g (2.2 equiv) of pyridine at <0 °C.
- Stir the resulting mixture at 0 °C for 40 min.

Stage (c)
- Charge aqueous ammonia (25%, 114 g, 5.9 equiv) at <0 °C.
- Add water (32 mL) at 0 °C and stir at 0 °C for 2.5 h.
- Separate layers and wash the aqueous layer with toluene (150 mL).
- Back-extract the organic layer with water (128 mL).
- Add 271 g of 22% sulfuric acid to the combined aqueous layers to pH 2.
- Filter the resulting slurry, wash with water, and dry.
- Yield: 53.4 g (96%).

REMARK

Measurement of reaction heats revealed that the optimized process is safer as the overall heat released was leveled out in each of the three individual steps.

7.10 OPTIMIZATION OF TELESCOPED PROCESS

As a telescoped process contains two or more individual steps, it may exist a number of problems, thereby optimization of such telescoped process usually needs a combination of a number of approaches.

A process shown in Scheme 7.51[71] was developed for the preparation of a tetrazole intermediate **282** for the synthesis of BMS-317180, a potent, orally active agonist of the human growth hormone secretagogue (GHS) receptor.

(I) Problems

- Stage (a)

   A by-product isobutanol had a negative effect on the mesylation step requiring excess MsCl (2.4 equiv); this byproduct, however, proved hard to remove.
- Stage (b)

   The use of triethylamine resulted in downstream issues leading to two impurities, **283** and **284** (Figure 7.2), from a base-catalyzed dimerization of amidrazone **281**. These dimeric impurities led to the coloration of the isolated product.
- Stage (c)

   A long processing time led to a higher level of racemization of oxazoline **280** due to the acid sensitivity of **280**.
- Stage (d)

   The diazotization reaction exhibited a substantial exotherm after an induction period, which posed a potential safety issue. In addition, a solvent exchange from methanol to toluene led to the cleavage of the Boc-protecting group, causing a loss of the product yield.
- Although a 100 g-scale run in the laboratory provided tetrazole **282** in 75% isolated yield, a 2.5-kg scale production in the pilot plant gave only less than 20% yield.

(II) Solutions

To address these scale-up issues, a modified synthetic route (Scheme 7.52)[71] involved: (i) the preparation of a derived amide **285**, avoiding the mesylation step, (ii) the use of solid sodium bicarbonate

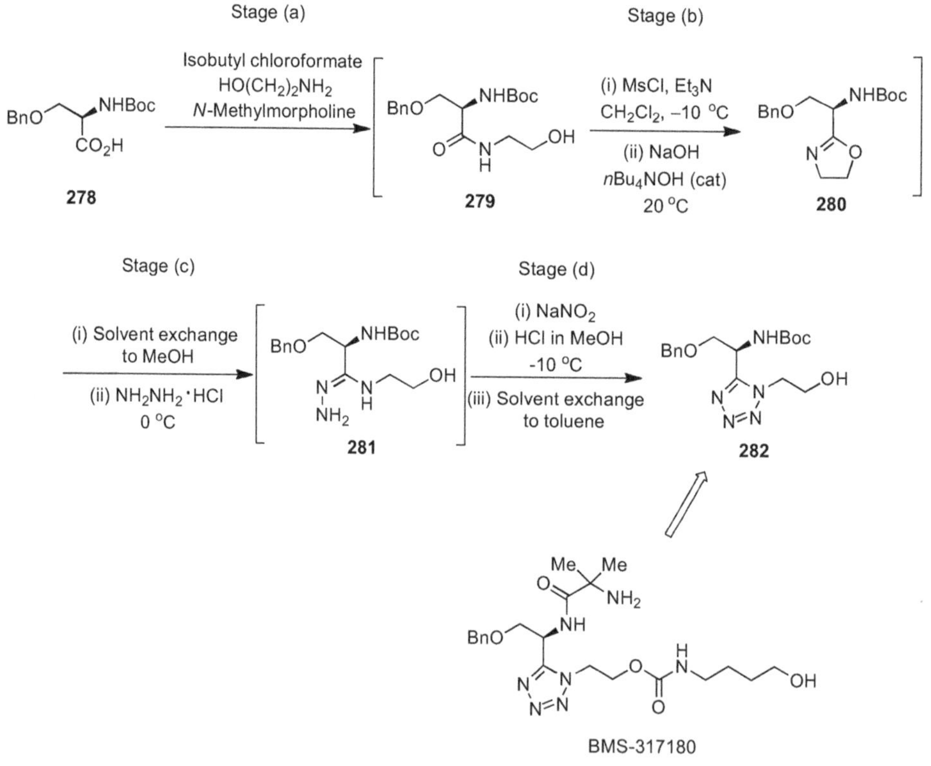

SCHEME 7.51

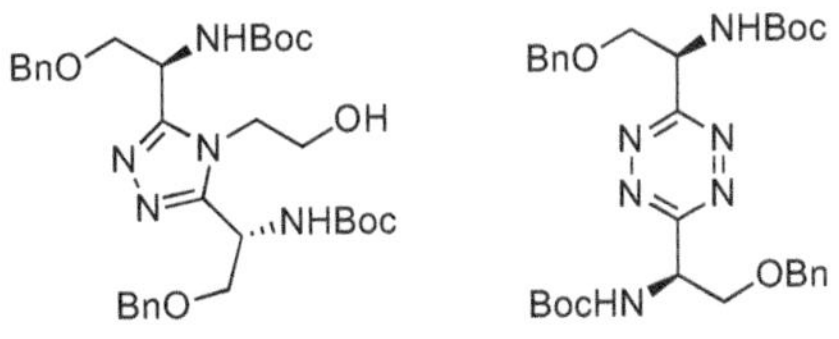

FIGURE 7.2 Impurities of **283** and **284**.

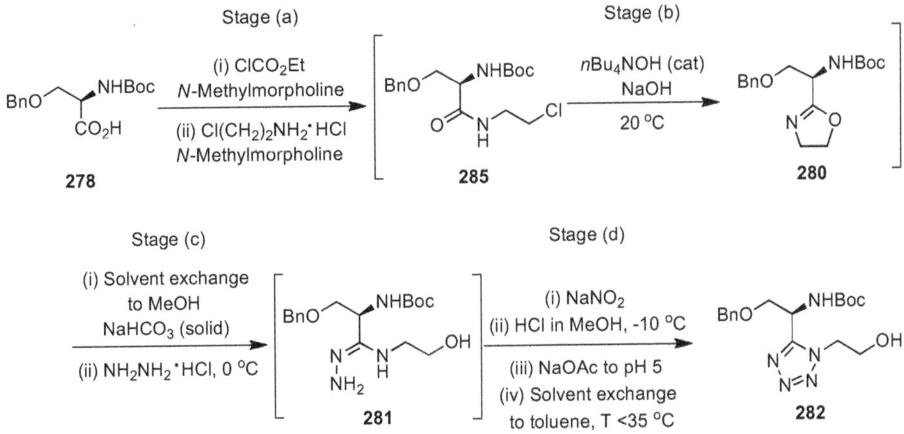

SCHEME 7.52

in the required solvent exchange to methanol allowed to control the racemization (≤2%), and (iii) in order to avoid product loss associated with cleavage of the Boc group in **282**, a weak base (sodium acetate) was selected to quench HCl prior to the solvent exchange to toluene.

Procedure

Stage (a) (**278→ 285**)

- Charge 40.0 kg (135 mol) of **278** to an 800-L Hastelloy reactor followed by adding CH$_2$Cl$_2$ (200 L).
- Cool the resulting solution to –35 °C.
- Charge 15.0 kg (148 mol, 1.1 equiv) of N-methylmorpholine at <–30 °C.
- Rinse with CH$_2$Cl$_2$ (10 L).
- Charge 16.3 kg (150 mol, 1.1 equiv) of ethyl chloroformate at <–30 °C.
- Rinse with CH$_2$Cl$_2$ (10 L).
- Age the resulting white slurry at –30±3 °C for 50 min.
- Charge 26.0 kg (257 mol, 1.9 equiv) of N-methylmorpholine at <–30 °C.
- Cool to –40 °C.
- Charge 17.3 kg (149 mol, 1.1 equiv) of milled chloroethylamine hydrochloride in one portion.
- Agitate the batch at –30 °C for 45 min.
- Warm the batch to 0 °C over 40 min and stir at that temperature until completion (≤2.0% of **278**).
- Add water (140 L) at 0 °C.
- Warm the batch to 20 °C over 30 min.
- Separate layers.
- Stir the organic layer with 0.5 N HCl solution (140 L) for 30 min.

- Separate layers.
- Wash the organic layer with water (140 L).

Stage (b) (**285→ 280**)
- Mix the CH_2Cl_2 solution of **285** with water (97 L), 105 L of 15% NaOH solution, and 4.4 kg (17 mol) of 40% tetrabutylammonium hydroxide solution.
- Stir the resulting biphasic mixture at 20 °C for 10.5 h.
- Separate layers.
- Wash with an organic layer with 5% $NaHCO_3$ (224 L).

Stage (c) (**280→ 281**)
- Add $NaHCO_3$ (7.9 kg) and distill at ≤35 °C/(200–500) mbar until reached the concentration of 1.2–1.5 L/kg of **280**.
- Add MeOH (100 L) and distill at ≤35 °C/(50–500 mbar) until ≤1.0% CH_2Cl_2.
- Add MeOH (80 L) and filter.
- Charge 12.1 kg (176.6 mol, 1.3 equiv) of hydrazine monohydrochloride and 520 L of MeOH to an 800-L Hastelloy reactor.
- Cool to 0 °C.
- Charge the solution of **280** at ≤5 °C and age at 0 °C for 1.5 h.

Stage (d) (**281→ 282**)
- Charge MeOH (400 L) to a 2000-L Hastelloy reactor at 0 °C.
- Charge 21.3 kg (271 mol, 2.0 equiv) of acetyl chloride at 0–5 °C.
- Age the resulting mixture at 0 °C for 30 min prior to cooling to –50 °C.
- Charge 16.4 kg (238 mol, 1.75 equiv) of $NaNO_2$ to the solution of **281** at 0 °C.
- Stir the mixture at 0 °C for 15 min prior to cooling to –30 °C.
- Transfer the cold (–30 °C) mixture into the 2000-L Hastelloy reactor containing the methanolic HCl at ≤–10 °C.
- Stir the batch at –10 °C for 20 min.
- Charge 26.0 kg (316 mol, 2.3 equiv) of anhydrous NaOAc.
- Warm to 20 °C (pH of 5–6).
- Add toluene (600 L) and distill at ≤40 °C under reduced pressure to ca. 530 L.
- Add toluene (600 L) and repeat the distillation to ca. 530 L (≤3.0% MeOH).
- Adjust the temperature to 20 °C.
- Add water (320 L) and stir for 15 min.
- Separate layers and wash the organic layer with 5% $NaHCO_3$ (320 L) and then 5% NaCl (320 L).
- Distill at ≤35 °C under reduced pressure to ca 266 L (KF≤500 ppm).
- Filter the resulting solution into an 800-L Hastelloy reactor.
- Heat to 50 °C.
- Add heptane (180 L) at 45–50 °C.
- Cool to 20 °C over 3 h.
- Cool the resulting slurry to 0 °C over 1 h and age for 1 h.
- Filter and wash with heptane (2×60 L).
- Dry at 35 °C/25 mmHg.
- Yield: 36.9 kg (75%) of **282** as a white crystalline solid.

NOTES

1. Slattery, C. N.; Deasy, R. E.; Maguire, A. R.; Kopach, M. E.; Singh, U. K.; Argentine, M. D.; Trankle, W. G.; Scherer, R. B.; Moynihan, H. *J. Org. Chem.* **2013**, *78*, 5955.

2. Itoh, T.; Kato, S.; Nonoyama, N.; Wada, T.; Maeda, K.; Mase, T. *Org. Process Res. Dev.* **2006**, *10*, 822.

3. (a) Tian, Q.; Cheng, Z.; Yajima, H. M.; Savage, S. J.; Green, K. L.; Humphries, T.; Reynolds, M. E.; Babu, S.; Gosselin, F.; Askin, D.; Kurimoto, I.; Hirata, N.; Iwasaki, M.; Shimasaki, Y.; Miki, T. *Org. Process Res. Dev.* **2013**, *17*, 97. (b) Tian, Q.; Hoffmann, U.; Humphries, T.; Cheng, Z.; Hidber, P.; Yajima, H.; Guillemot-Plass, M.; Li, J.; Bromberger, U.; Babu, S.; Askin, D.; Gosselin, F. *Org. Process Res. Dev.* **2015**, *19*, 416.

4. Achmatowicz, M.; Thiel, O. R.; Wheeler, P.; Bernard, C.; Huang, J.; Larsen, R. D.; Faul, M. M. *J. Org. Chem.* **2009**, *74*, 795.

5. Oh, L. M.; Wang, H.; Shilcrat, S. C.; Herrmann, R. E.; Patience, D. B.; Spoors, P. G.; Sisko, J. *Org. Process Res. Dev.* **2007**, *11*, 1032.

6. Lukin, K.; Kishore, V.; Gordon, T. *Org. Process Res. Dev.* **2013**, *17*, 666.

7. Chung, J. Y. L.; Steinhuebel, D.; Krska, S. W.; Hartner, F. W.; Cai, C.; Rosen, J.; Mancheno, D. E.; Pei, T.; DiMichele, L.; Ball, R. G.; Chen, C.; Tan, L.; Alorati, A. D.; Brewer, S. E.; Scott, J. P. *Org. Process Res. Dev.* **2012**, *16*, 1832.

8. Li, X.; Wells, K. M.; Branum, S.; Damon, S.; Youells, S.; Beauchamp, D. A.; Palmer, D.; Stefanick, S.; Russell, R. K.; Murray, W. *Org. Process Res. Dev.* **2012**, *16*, 1727.

9. Ford, J. G.; Pointon, S. M.; Powell, L.; Siedlecki, P. S. *Org. Process Res. Dev.* **2010**, *14*, 1078.

10. The formation of **38** and **39** may proceed through the following reaction pathways:

11. Milburn, R. R.; Thiel, O. R.; Achmatowicz, M.; Wang, X.; Zigterman, J.; Bernard, C.; Colyer, J. T.; DiVirgilio, E.; Crockettt, R.; Correll, T. L.; Nagapudi, K.; Ranganathan, K.; Hedley, S. J.; Allgeier, A.; Larsen, R. D. *Org. Process Res. Dev.* **2011**, *15*, 31.

12. (a) von Eldick, R.; Klärner, F.-G. in *High Pressure Chemistry, Synthetic, Mechanistic, and Supercritical Applications*, Wiley–VCH, Weinheim, 2002. (b) Trost, B. M.; Parquette, J. R.; Marquart, A. L. *J. Am. Chem. Soc.* **1995**, *117*, 3284.

13. Lu, C. V.; Chen, J. J.; Perrault, W. R.; Conway, B. G.; Maloney, M. T.; Wang, Y.; *Org. Process Res. Dev.* **2006**, *10*, 272.

14. Connolly, T. J.; Considine, J. L.; Ding, Z.; Forsatz, B.; Jennings, M. N.; MacEwan, M. F.; McCoy, K. M.; Place, D. W.; Sharma, A.; Sutherland, K. *Org. Process Res. Dev.* **2010**, *14*, 459.

15. Clark, J. D.; Anderson, D. K.; Banaszak, D. V.; Brown, D. B.; Czyzewski, A. M.; Edeny, A. D.; Forouzi, P. S.; Gallagher, D. J.; Iskos, V. H.; Kleine, H. P.; Knable, C. M.; Lantz, M. K.; Lapack, M. A.; Moore, C. M. V.; Muellner, F. W.; Murphy, J. B.; Orihuela, C. A.; Pietz, M. A.; Rogers, T. E.; Ruminski, P. G.; Santhanam, H. K.; Schilke, T. C.; Shah, A. S.; Sheikh, A. Y.; Weisenburger, G. A.; Wise, B. E. *Org. Process Res. Dev.* **2009**, *13*, 1088.

16. Guercio, G.; Manzo, A. M.; Goodyear, M.; Bacchi, S.; Curti, S.; Provera, S. *Org. Process Res. Dev.* **2009**, *13*, 489.

17. Wang, Y.; Przyuski, K.; Roemmele, R. C.; Hudkins, R. L.; Bakale, R. P. *Org. Process Res. Dev.* **2013**, *17*, 846.

18. Bultman, M. S.; Fan, J.; Fanfair, D.; Soltani, M.; Simpson, J.; Murugesan, S.; Soumeillant, M.; Chen, K.; Riatti, C.; La Cruz, T. E.; Buono, F. G.; Hung, V.; Schild, R. L.; Ivy, S.; Sweeney, J. T.; Conlon, D. A.; Eastgate, M. D. *Org. Process Res. Dev.* **2017**, *21*, 1131.

19. Adam, F. M.; Bish, G.; Calo, F.; Carr, C. L.; Castro, N.; Hay, D.; Hodgson, P. B.; Jones, P.; Knight, C. J.; Paradowski, M.; Parsons, G. C.; Proctor, K. J. W.; Pryde, D. C.; Rota, F.; Smith, M. C.; Smith, N.; Tran, T.-D.; Hitchin, J.; Dixon, R. *Org. Process Res. Dev.* **2011**, *15*, 788.

20. Cann, R. O.; Chen, C-P. H.; Gao, Q.; Hanson, R. L.; Hsieh, D.; Li, J.; Lin, D.; Parsons, R. L.; Pendri, Y.; Nielsen, R. B.; Nugent, W. A.; Parker, W. L.; Quinlan, S.; Reising, N. P.; Remy, B.; Sausker, J.; Wang, X. *Org. Process Res. Dev.* **2012**, *16*, 1953.

21. Manley, P. W.; Acemoglu, M.; Marterer, W.; Pachinger, W. *Org. Process Res. Dev.* **2003**, *7*, 436.

22. Liang, J. T.; Deng, X.; Mani, N. S. *Org. Process Res. Dev.* **2011**, *15*, 876.

23. Brandt, T. A.; Caron, S.; Damon, D. B.; DiBrino, J.; Ghosh, A.; Griffith, D. A.; Kedia, S.; Ragan, J. A.; Rose, P. R.; Vanderplas, B. C.; Wei, L. *Tetrahedron* **2009**, *65*, 3292.

24. Haight, A. R.; Bailey, A. E.; Baker, W. S.; Cain, M. H.; Copp, R. P.; DeMattei, J. A.; Ford, K. L.; Henry, R. F.; Hsu, M. C.; Keyes, R. F.; King, S. A.; McLaughlin, M. A.; Melcher, L. M.; Nadler, W. R.; Oliver, P. A.; Parekh, S. I.; Patel, H. H.; Seif, L. S.; Staeger, M. A.; Wayne, G. S.; Wittenberger, S. J.; Zhang, W. *Org. Process Res. Dev.* **2004**, *8*, 897.

25. Barkalow, J. H.; Breting, J.; Gaede, B. J.; Haight, A. R.; Henry, R.; Kotecki, B.; Mei, J.; Pearl, K. B.; Tedrow, J. S.; Viswanath, S. K. *Org. Process Res. Dev.* **2007**, *11*, 693.

26. Alimardanov, A.; Gontcharov, A.; Nikitenko, A.; Chan, A. W.; Ding, Z.; Ghosh, M.; Levent, M.; Raveendranath, P.; Ren, J.; Zhou, M.; Mahaney, P. E.; McComas, C. C.; Ashcroft, J.; Potoski, J. R. *Org. Process Res. Dev.* **2009**, *13*, 880.

27. McLaughlin, M.; Belyk, K.; Chen, C.; Xin, L.; Pan, J.; Qian, G.; Reamer, R. A.; Xu, Y. *Org. Process Res. Dev.* **2013**, *17*, 1052.

28. Brands, K. M. J.; Payack, J. F.; Rosen, J. D.; Nelson, T. D.; Candelario, A.; Huffman, M. A.; Zhao, M. M.; Li, J.; Craig, B.; Song, Z. J.; Tschaen, D. M.; Hansen, K.; Devine, P. N.; Pye, P. J.; Rossen, K.; Dormer, P. G.; Reamer, R. A.; Welch, C. J.; Mathre, D. J.; Tsou, N. N.; McNamara, J. M.; Reider, P. J. *J. Am. Chem. Soc.* **2003**, *125*, 2129.

29. Dunetz, J. R.; Berliner, M. A.; Xiang, Y.; Houck, T. L.; Salingue, F. H.; Chao, W.; Chen, Y.; Wang, S.; Huang, Y.; Farrand, D.; Boucher, S. J.; Damon, D. B.; Makowski, T. W.; Barrila, M. T.; Chen, R.; Martínez, I. *Org. Process Res. Dev.* **2012**, *16*, 1635.

30. Britton, H.; Catterick, D.; Dwyer, A. N.; Gordon, A. H.; Leach, S. G.; McCormick, C.; Mountain, C. E.; Simpson, A.; Stevens, D. R.; Urquhart, M. W. J.; Wade, C. E.; Warren, J.; Wooster, N. F.; Zilliox, A. *Org. Process Res. Dev.* **2012**, *16*, 1607.

31. Madivada, L. R.; Anumala, R. R.; Gilla, G.; Kagga, M.; Bandichhor, R. *Org. Process Res. Dev.* **2012**, *16*, 1660.

32. (a) McConville, F. X. *The Pilot Plant Real Book, a Unique Handbook for the Chemical Process Industry*, FXM Engineering and Design, Worcester, MA, 2002. (b) Chang, S.-J.; Fernando, D.; Fickes, M.; Gupta, A. K.; Hill, D. R.; McDermott, T.; Parekh, S.; Tian, Z.; Wittenberger, S. J. *Org. Process Res. Dev.* **2002**, *6*, 329.

33. Wang, X.-J.; Frutos, R. P.; Zhang, L.; Sun, X.; Xu, Y.; Wirth, T.; Nicola, T.; Nummy, L. J.; Krishnamurthy, D.; Busacca, C. A.; Yee, N.; Senanayake, C. H. *Org. Process Res. Dev.* **2011**, *15*, 1185.

34. Hawk, M. K.; Ryan, S. J.; Zhang, X.; Huang, P.; Chen, J.; Liu, C.; Chen, J.; Lindsay-Scott, P. J.; Burnett, J.; White, C.; Lu, Y.; Rizzo, J. R. *Org. Process Res. Dev.* **2021**, *25*, 1167.

35. Suzuki, A.; Fukuda, N.; Kajiwara, T.; Ikemoto, T. *Org. Process Res. Dev.* **2019**, *23*, 484.

36. Sharma, P. K.; Kolchinski, A.; Shea, H. A.; Nair, J. J.; Gou, Y.; Romanczyk, L. J., Jr.; Schmitz, H. H. *Org. Process Res. Dev.* **2007**, *11*, 422.

37. Meyer, S. D.; Schreiber, S. L. *J. Org. Chem.* **1994**, *59*, 7549.

38. Rassias, G.; Hermitage, S. A.; Sanganee, M. J.; Kincey, P. M.; Smith, N. M.; Andrews, I. P.; Borrett, G. T.; Slater, G. R. *Org. Process Res. Dev.* **2009**, *13*, 774.

39. Funel, J.-A.; Brodbeck, S.; Guggisberg, Y.; Litjens, R.; Seidel, T.; Struijk, M.; Abele, S. *Org. Process Res. Dev.* **2014**, *18*, 1674.

40. Ishikawa, H.; Suzuki, T.; Orita, H.; Uchimaru, T.; Hayashi, Y. *Chem. Eur. J.* **2010**, *16*, 12616.

41. Yanagisawa, A.; Nishimura, K.; Ando, K.; Nezu, T.; Maki, A.; Kato, S.; Tamaki, W.; Imai, E.; Mohri, S. *Org. Process Res. Dev.* **2010**, *14*, 1182.

42. Yanagisawa, A.; Nezu, T.; Mohri, S. *Org. Lett.* **2009**, *11*, 5286.

43. Beliaev, A.; Wahnon, J.; Russo, D. *Org. Process Res. Dev.* **2012**, *16*, 704.

44. (a) Challenger, S.; Dack, K. N.; Derrick, A. M.; Dickinson, R. P.; Ellis, D.; Hajikarimian, Y.; James, K.; Rawson, D. J. WO 99/20623, 1999.
    (b) Ashcroft, C. P.; Challenger, S.; Clifford, D.; Derrick, A. M.; Hajikarimian, Y.; Slucock, K.; Silk, T. V.; Thomson, N. M.; Williams, J. R. *Org. Process Res. Dev.* **2005**, *9*, 663.

45. Sharma, R.; Voynov, G. H.; Ovaska, T. V.; Marquez, V. E. *Synlett* **1995**, 839.

46. Štimac, V.; Škugor, M. M.; Jakopović, I. P.; Vinter, A.; Ilijaš, M.; Alihodžić, S.; Mutak, S. *Org. Process Res. Dev.* **2010**, *14*, 1393.

47. (a) Wessjohann, L. A.; Zhu, M. *Adv. Synth. Catal.* **2008**, *350*, 107. (b) Hoffmann, H. M. R.; Schrake, O. *Tetrahedron: Asymmetry* **1998**, *9*, 1051.
48. Ikoma, A.; Ogawa, N.; Kondo, D.; Kawada, H.; Kobayashi, Y. *Org. Lett.* **2016**, *18*, 2074.
49. Ohigashi, A.; Kanda, A.; Tsuboi, H.; Hashimoto, N. *Org. Process Res. Dev.* **2005**, *9*, 179.
50. Cabaj, J. E.; Lukesh, J. M.; Pariza, R. J.; Zizelman, P. M. *Org. Process Res. Dev.* **2009**, *13*, 358.
51. Caron, S.; Do, N. M.; Sieser, J. E.; Arpin, P.; Vazquez, E. *Org. Process Res. Dev.* **2007**, *11*, 1015.
52. Debarge, S.; McDaid, P.; O'Neill, P.; Frahill, J.; Wong, J. W.; Carr, D.; Burrell, A.; Davies, S.; Karmilowicz, M.; Steflik, J. *Org. Process Res. Dev.* **2014**, *18*, 109.
53. Werkmeister, S.; Junge, K.; Beller, M. *Org. Process Res. Dev.* **2014**, *18*, 289.
54. Sera, M.; Yamashita, M.; Ono, Y.; Tabata, T.; Muto, E.; Ouchi, T.; Tawada, H. *Org. Process Res. Dev.* **2014**, *18*, 446.
55. Brazier, E. J.; Hogan, P. J.; Leung, C. W.; O'Kearney-McMullan, A.; Norton, A. K.; Powell, L.; Robinson, G.; Williams, E. G. *Org. Process Res. Dev.* **2010**, *14*, 544.
56. Deng, X.; Liang, J. T.; Peterson, M.; Rynberg, R.; Cheung, E.; Mani, N. S. *J. Org. Chem.* **2010**, *75*, 1940.
57. Baenziger, M.; Mak, C.-P.; Muehle, H.; Nobs, F.; Prikoszovich, W.; Reber, J.-L.; Sunay, U. *Org. Process Res. Dev.* **1997**, *1*, 395.
58. Shi, Z.; Fan, J.; Kronenthal, D. R.; Mudryk, B. M. *Org. Process Res. Dev.* **2018**, *22*, 1534.
59. Tone, H.; Matoba, K.; Goto, F.; Torisawa, Y.; Nishi, T.; Minamikawa, J. *Org. Process Res. Dev.* **2000**, *4*, 312.
60. Kong, W.; Casimiro, M.; Merino, E.; Nevado, C. *J. Am. Chem. Soc.* **2013**, *135*, 14480.
61. Henegar, K. E.; Cebula, M. *Org. Process Res. Dev.* **2007**, *11*, 354.
62. Stoner, E. J.; Stengel, P. J.; Cooper, A. J. *Org. Process Res. Dev.* **1999**, *3*, 145.
63. Funel, J.-A.; Schmidt, G.; Abele, S. *Org. Process Res. Dev.* **2011**, *15*, 1420.
64. Lukin, K.; Hsu, M. C.; Chambournier, G.; Kotecki, B.; Venkatramani, C. J.; Leanna, M. R. *Org. Process Res. Dev.* **2007**, *11*, 578.
65. Hansen, M. M.; Borders, S. S. K.; Clayton, M. T.; Heath, P. C.; Kolis, S. P.; Larsen, S. D.; Linder, R. J.; Reutzel-Edens, S. M.; Smith, J. C.; Tameze, S. L.; Ward, J. A.; Weigel, L. O. *Org. Process Res. Dev.* **2009**, *13*, 198.
66. Porcs-Makkay, M.; Mezei, T.; Simig, G. *Org. Process Res. Dev.* **2007**, *11*, 490.
67. Jones, G.; Stanforth, S. P. *Org. React.* **1997**, *49*, 1.
68. Nishikata, T.; Abela, A. R.; Huang, S.; Lipshutz, B. H. *J. Am. Chem. Soc.* **2010**, *132*, 4978.
69. Ager, D. J.; Babler, S.; Erickson, R. A.; Froen, D. E.; Kittleson, J.; Pantaleone, D. P.; Prakash, I.; Zhi, B. *Org. Process Res. Dev.* **2004**, *8*, 72.
70. Masui, T.; Kabaki, M.; Watanabe, H.; Kobayashi, T.; Masui, Y. *Org. Process Res. Dev.* **2004**, *8*, 408.
71. Davulcu, A. H.; McLeod, D. D.; Li, J.; Katipally, K.; Little, A.; Doubleday, W.; Xu, Z.; McConlogue, C. W.; Lai, C. J.; Gleeson, M.; Schwinden, M.; Parsons, R. L., Jr. *J. Org. Chem.* **2009**, *74*, 4068.

8 Hazards of Oxidation and Reduction Reactions

Many challenges exist with using hazardous materials or conducting a chemical reaction under hazardous conditions (high temperature or/and pressure). A rapid generation of heat and gas during the decomposition of hazardous materials may lead to an explosive event. Organic substances with a decomposition energy of 800 J/g or more are classified as explosive materials.[1] Besides methods to evaluate the instability of materials, such as oxygen balance calculation,[2] differential scanning calorimetry (DSC), and thermogravimetric analysis (TGA), the rule of six is another qualitative approach for examining the potential of material explosive properties. The rule of six is as follows:[3] If a molecule presents at least six atoms of carbon (or other atoms of approximately the same size or greater) per energetic functionality, this should render the molecule relatively safe to handle. Table 8.1 lists the known explosive functional groups.[4]

8.1 OXIDATION REACTIONS

Compared to reduction reactions, oxidative transformations during drug substance synthesis are limited due to process safety, toxicity, and waste disposal concerns related to the use of oxidants. So, the raw materials for the manufacture of pharmaceuticals are often bought from external vendors at the required (or higher) oxidation state in order to avoid carrying out unnecessary hazardous oxidative reactions.

One of the most important goals for process chemists is to build safety into chemical processes during the course of process research and development. To achieve this goal, one must stick to the following guidelines:

1. to use less hazardous starting materials, reagents, and intermediates.
2. to design less energetic reactions.
3. to utilize catalysts that allow for less severe operating conditions or use less reactive starting materials.
4. to use water as reaction solvent whenever possible.
5. to develop fast reactions to prevent the high energy materials from accumulation.

8.1.1 Oxidation of Olefins

8.1.1.1 Oxidation with *m*CPBA

m-Chloroperoxybenzoic acid (*m*CPBA) is a widely used oxidant both in laboratory and pharmaceutical manufacturing. *m*CPBA is mainly used in the conversion of ketones to esters, epoxidation of alkenes, oxidation of sulfides to sulfoxides and sulfones, and oxidation of amines to amine oxides.

Compared with other peroxy acids, *m*CPBA is preferred because of its relative ease of handling. As a pure substance, *m*CPBA can be detonated by shock or by sparks. A report[5] describes that ethanol can stabilize *m*CPBA during the epoxidation reaction in methylene chloride (Equation 8.1). The epoxidation of olefin **1** with *m*CPBA (1.4–1.7 equiv) in the presence of 4-methylmorpholine-*N*-oxide

DOI: 10.1201/9781003288411-8

TABLE 8.1
List of Explosive Functional Groups

Structural Feature	Examples
C–C bond unsaturation	Acetylene, acetylides, 1,2-dienes (allenes)
C–metal, N–metal	Grignard reagents, organolithium species
Contiguous nitrogen atoms	Azides, aliphatic azo compounds, diazonium salts, hydrazines, sulfonyl hydrazides
Contiguous oxygen atoms	Peroxides, ozonides
N–O bonds	Nitro, nitroso, nitrates, hydroxyamines, N-oxides, 1,2-oxazoles
N-halogen, O-halogen	Chloramines, fluoroamines, chlorates, perchlorates, iodosyl compounds

(NMO) (2.0–2.3 equiv) and manganese catalyst (4 mol%) in methylene chloride at –65 °C was accomplished, affording epoxide **2** in greater than 80% yields with 70–74% ee.

$$(8.1)$$

Procedure

- Charge 1.57 kg (13.42 mol, 2.3 equiv) of NMO, 170 g (0.23 mol, 4 mol%) of catalyst, 850 g (5.82 mol) of **1**, and CH_2Cl_2 (7.7 L).
- Cool to –70 °C.
- Dissolve, in a separate reactor, 2.295 kg (9.31 mol, 1.6 equiv) of mCPBA (70%) in ethanol (3 L).
- Charge the solution of mCPBA while maintaining the temperature at –60 to –70 °C.
- Transfer the content, after the addition, to 1 M NaOH (14 L).
- Separate layers and extract the aqueous phase with CH_2Cl_2 (4 L).
- Wash the combined organic phase with H_2O (2×9 L).
- Distill the organic phase under reduced pressure to obtain **2** as a black oil.
- Yield: 1.019 kg (90%, 83.5% of purity).

REMARKS

(a) Although stable under room temperature with a melting point of 92 °C (decomposition), mCPBA is reported to undergo hazardous thermal decomposition near 60 °C.[6] Commercial mCPBA is in 70% purity (mp 69–71 °C) containing 10% of m-chlorobenzoic acid (mCBA) and 20% water and shall be stored at 2–8 °C in a plastic container. The presence of mCBA reduces the hazardous nature of mCPBA to shock and friction.

(b) An incident occurred[7] in a pilot plant when using mCPBA in DMF to oxidize sulfide to sulfoxide. It was revealed that mixing mCPBA with DMF could form a large amount of m-chlorobenzoyl peroxide (mCBPO), leading to instability of the system. Therefore, it was suggested that mCPBA should be used in methylene chloride[8] during scale-up.

(c) A report[9] described a method to remove the excess *m*CPBA and *m*CBA by-products from the reaction mixture by adding anhydrous KF.

8.1.1.2   Oxidation with Sodium Perborate

The epoxidation of olefins can also use sodium perborate ($NaBO_3$) as the oxidant. Compared with H_2O_2, $NaBO_3$ is a relatively safe and inexpensive source of peroxide and was reported to react smoothly with α,β-unsaturated ketones without the need for an additional base.[10] For example, epoxidation of enone **3** with $NaBO_3$ (1.6 equiv) as an aqueous emulsion at 40 °C gave epoxide **4** in a quantitative yield (Equation 8.2).[11]

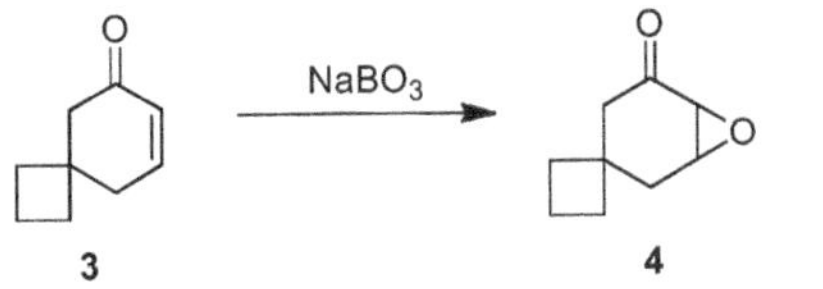

(8.2)

       3   **4**

8.1.1.3   Oxidation with Ozone

During process development for chemical production, two key environmental issues must be addressed: (1) environmental impact (human, plant, and animal) involving toxicity, ozone depletion, and climate change, and (2) sustainability which includes the reduction in depletion of natural resources, the mass of materials used and waste generated, and energy-inefficient processes.

Ozone is a reagent that is widely used in many organic transformations in both laboratory and industry conditions. Ozonolysis is a green oxidation process with high atom efficiency producing oxygen as the only by-product. However, the application of ozonolysis is limited due to the safety of handling the potentially explosive ozonide intermediates, the use of flammable solvents in the presence of oxygen, and the high exothermicity of the subsequent quenching step.

Ozone is the reagent of choice to cleave a C=C double bond to form alcohols, aldehydes or ketones, or carboxylic acids, depending on workup conditions (Scheme 8.1).

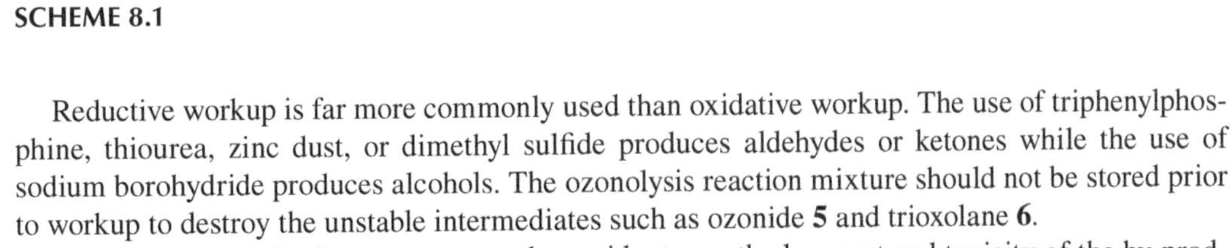

SCHEME 8.1

Reductive workup is far more commonly used than oxidative workup. The use of triphenylphosphine, thiourea, zinc dust, or dimethyl sulfide produces aldehydes or ketones while the use of sodium borohydride produces alcohols. The ozonolysis reaction mixture should not be stored prior to workup to destroy the unstable intermediates such as ozonide **5** and trioxolane **6**.

The advantages of using ozone over other oxidants are the low cost and toxicity of the by-product, that is, DMSO from DMS and Ph_3PO from PPh_3. Some of the drawbacks of using ozone at scale are high exotherms associated with the initial reaction. To address those process safety issues, capillary reactors are developed,[12] which can alleviate both the exotherm problem and the build-up of potentially explosive materials.

8.1.1.3.1   Oxidation to Diol

The ozonolysis of olefin **7** was employed to prepare diol **8**, an intermediate for the synthesis of a protein kinase C inhibitor (LY333531) developed by Elli Lilly (Scheme 8.2).[13]

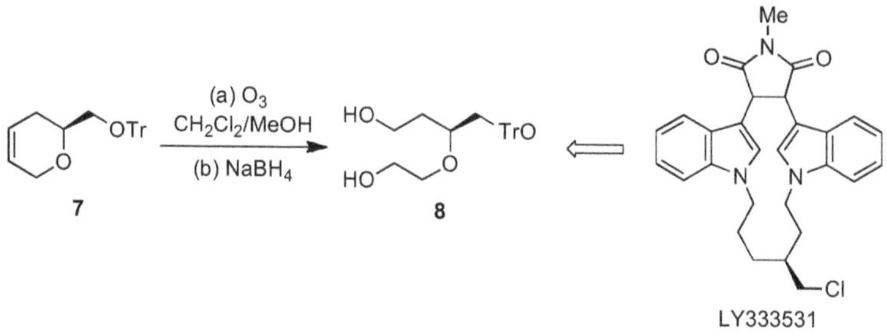

SCHEME 8.2

Procedure

Stage (a)
- Charge 71.5 g (0.2 mol) of **7**, 350 mL of CH_2Cl_2, and 250 mL of MeOH into a jacketed reactor.
- Cool the resulting solution to −40 °C.
- Bubble O_3 at −40 to −25 °C until a light blue solution was obtained.
- Bubble N_2 through the mixture at −45 to −50 °C until the blue color disappeared.

Stage (b)
- Charge the cold reaction mixture into a solution of $NaBH_4$ (prepared by mixing 17.5 g of $NaBH_4$ and 500 mL of 0.02 N NaOH).
- Warm slowly, after the addition, to room temperature and stir overnight.
- Separate layers.
- Distill the organic layer to dryness.
- Yield: 81.9 g (100% yield).

REMARK

Ozone is a pale blue gas, slightly soluble in water and much more soluble in inert non-polar solvents such as carbon tetrachloride or fluorocarbons, where it forms a blue solution. Ozone cannot be stored and transported like other industrial gases (because it quickly decays into diatomic oxygen) and must therefore be produced on-site.

8.1.1.3.2   Oxidation to Alcohol

Analogously, ozonolysis of olefin **9**, affording the desired alcohol **10** (Equation 8.3).[14] In this case, however, the addition of sodium acetate (1 equiv) and acetic acid (1 equiv) to the reaction mixture was required in order to achieve consistent reaction yields.

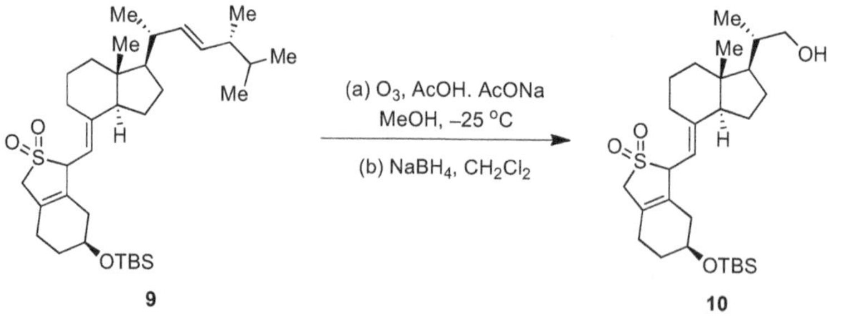

(8.3)

8.1.1.3.3 Oxidation to Aldehyde

The oxidation of α,β-unsaturated ketone **11** with ozone (O$_3$/MeOH, −30 to −40 °C) to prepare aldehyde **12** eliminated the use of OsO$_4$ or Pb(OAc)$_4$, which not only are highly toxic but also present
disposal problems (Equation 8.4).[15]

$$(8.4)$$

Procedure

Stage (a)
- Dissolve 250 g (0.828 mol) of **11** in 2.5 L of MeOH.
- Cool the resulting solution to −30 to −40 °C.
- Bubble O$_3$ into the solution for 6 h.

Stage (b)
- Warm the reaction mixture to −10 °C.
- Charge aqueous NaOH solution (109 mL of concentrated NaOH in 3.13 L of H$_2$O)
  dropwise over 2 h.
- Warm the mixture to room temperature.
- Distill MeOH under reduced pressure (2.8 L removed.
- Wash the aqueous solution with MTBE (2×1.0 L).

Stage (c)
- Add 6.0 N HCl to pH 4.
- Filter and wash with H$_2$O (2×0.3 L) and heptane (0.2 L).
- Dry at 40–60 °C under vacuum.
- Yield: 211 g (79%) as an off-white solid.

NOTE

The excess ozone was, after the reaction, destroyed by treating the effluent stream with an aqueous
solution of sodium metabisulfite (<1 ppm ozone detected after the quenching).

8.1.1.4 Oxidation with KMnO$_4$

Conventional methods for using permanganate as an oxidant are severely limited by the lack of solvents that can dissolve permanganate without co-oxidation of the solvent. A water/dichloromethane
solvent mixture in the presence of a catalytic amount of a quaternary ammonium salt proved to be a
safe and effective solvent system for the oxidation of α-pinene with KMnO$_4$ (Equation 8.5).[16]

$$(8.5)$$

8.1.1.5 Oxidation with 9-BBN/H$_2$O$_2$-NaOH

Conversion of olefin **13** to primary alcohol **14** could be realized via selective hydroboration with
9-borabicyclo[3.3.1]nonane (9-BBN) followed by oxidation with H$_2$O$_2$ (Equation 8.6).[17]

(8.6)

Procedure

Stage (a)
- Charge 1.10 kg (91% w/w, 3.36 mol) of **13** and THF (2 L) into a 25-L jacketed vessel.
- Inert the vessel with s strong flush of N_2.
- Add slowly a solution of 9-BBN (0.5 M in THF, 16.8 L, 8.40 mol) at 15–20 °C over 2.5 h.
- Stir to batch at 20 °C for 3 h.

Stage (b)
- Cool to 1 °C.
- Add water (1 L) at <4 °C over 15 min.
- Add a solution of 35% H_2O_2 (2.87 L, 3.36 mol) dropwise at <15 °C (–20 °C in the jacket) over 2.3 h.
- Add 3.8 M NaOH solution (2.05 L, 7.77 mol) at <18 °C (0 °C in the jacket) over 40 min until pH 9 is reached.
- Stir the mixture at 20 °C for 2.5 h.
- Set the jacket to 10 °C and add brine (2 L).
- Separate layers.
- Add a solution of 8.6% $Na_2S_2O_3$ (6 L) at 12 °C over 2.5 h to the organic layer to destroy excess H_2O_2.
- Wash the organic layer with brine (2.5 L and 1.5 L).
- Concentrate the organic layer under reduced pressure.
- Dissolve the residue in MTBE (10 L).
- Wash with 10% Na_2CO_3 (3 L) and water (4 L).
- Concentrate the organic layer under reduced pressure.
- Purify the oil residue with SFC.
- Yield: 953 g (90% w/w, 2.71 mol, 81% yield) as a yellow oil.

8.1.2 OXIDATION OF ALCOHOLS

Carbonyl compounds are of great importance in organic synthesis. Oxidation of alcohols is frequently used to prepare carbonyl compounds. As carbonyl compounds can be further oxidized to the corresponding acids, the oxidation conditions are critical in controlling the oxidation state. A viable oxidation reaction should meet several criteria: (1) safe and mild reaction conditions, (2) high conversion to avoid purification issues associated with unreacted alcohols, and (3) high volume efficiency with minimum wastes and environmental impacts.

Swern oxidation is one of the most popular methods for converting primary or secondary alcohols into aldehydes or ketones. The reaction is well-known for its mild nature and tolerance of various functional groups. Swern oxidation requires, however, cryogenic reaction temperature and generates two very toxic volatile by-products, dimethyl sulfide and carbon monoxide. Another issue of Swern oxidation is the use of oxalyl chloride which is moisture sensitive and dangerously toxic, and its vapor is a powerful irritant, particularly to the respiratory system and eyes. These drawbacks limit the application of Swern oxidation at a larger production scale. Consequently, attempts have been made to optimize the classical Swern oxidation conditions. Alternatives to Swern oxidation include NaClO/TEMP,[18] P_2O_5/DMSO/TEA,[19] and SO_3·Py/DMSO.

Another alternative is the use of cyanuric chloride to replace oxalyl chloride in the activation of DMSO (Equation 8.7).[20] This method is operationally simple and can be used under mild conditions. However, this procedure will generate by-products dimethyl sulfide (DMS) and 4,6-dichloro-[1,3,5] triazin-2-ol which renders this method less practical on a large scale.

$$
\underset{\substack{\text{R} \quad \text{R'} \\ }}{\overset{\text{OH}}{\bigwedge}} \quad
\begin{array}{c}
\text{(a) DMSO, cyanuric chloride (1.2 equiv)} \\
\text{THF, -30 °C, 30 min} \\
\hline
\text{(b) NEt}_3\text{, THF, -30 °C, 30 min}
\end{array}
\quad \underset{\text{R} \quad \text{R'}}{\overset{\text{O}}{\bigwedge}} \tag{8.7}
$$

Recently, a Swern-type oxidation system was reported[21] using 1,1-dichlorocycloheptatriene to activate DMSO. Such a DMSO/1,1-dichlorocycloheptatriene system allowed the oxidation of alcohols at a relatively high temperature (–30 °C) without the generation of CO_2 and toxic CO gases. Like the Swern oxidation, this protocol will generate a toxic dimethyl sulfide by-product as well plus a tropone by-product which requires column chromatography to remove it.

REMARKS

(a) DMS is a water-insoluble flammable liquid that boils at 37 °C. The DMS off-gas is usually scrubbed with 5–15% NaClO.
(b) Alternatively, dodecyl methyl sulfoxide was elected as a substitute for DMSO to avoid the emission of toxic dimethyl sulfide.[22]

8.1.2.1 Py·SO₃/DMSO System

Pyridine·sulfur trioxide complex is a widely used reagent in Swern-type (Parikh–Doering) oxidation on large scales. Oxidation of N,N-dibenzylphenylalaninol **15** with Py·SO₃/DMSO at 15 °C gave the corresponding aldehyde **16** in quantitative yield without racemization (Equation 8.8).[23] The dimethyl sulfide by-product was easily removed by nitrogen sparging and treatment of the effluent gas stream with bleach solution.

$$
\underset{\substack{\text{Bn} \\ \textbf{15}}}{\overset{\text{Bn}_2\text{N}}{\bigwedge}}\text{OH} \quad
\xrightarrow[\text{TEA, DMSO}]{\text{SO}_3\cdot\text{Py}} \quad
\underset{\substack{\text{Bn} \\ \textbf{16}}}{\overset{\text{Bn}_2\text{N}}{\bigwedge}}\text{O} \tag{8.8}
$$

Procedure

- Dissolve 190 kg (573 mol) of **15** in 322 L (2292 mol) of TEA.
- Cool to 12 °C.
- Charge a solution of Py·SO₃ complex (184 kg, 1156 mol) in 635 L of DMSO while maintaining the temperature at 8–18 °C.
- Stir the batch at ambient temperature for 45 min.
- Cool the batch to 12 °C.
- Add cold water (10–15 °C) (224 L).
- Sparge the mixture with N_2 for 6 h while maintaining the temperature below 20 °C.
- Add water (688 L).
- Extract with ethyl acetate and wash the organic layer with 5% citric acid (1777 L) followed by saturated NaCl solution (1376 L).
- Distill at 40 °C under reduced pressure (100 mmHg).
- Dilute the residue with THF (287 L).
- Distill at 40 °C under reduced pressure (100 mmHg).
- Dissolve the residue in THF (143 L) and **16** be used directly in the subsequent step.

REMARK

Compared with Swern oxidation, Parikh–Doering oxidation (DMSO/Py·SO$_3$) can be carried out at a much higher temperature (0 °C to room temperature) and is widely applied in organic synthesis.[24]

The commercial Py·SO$_3$ contains a pyridine·sulfuric acid salt (1:1) impurity **21**. The presence of **21** sometimes is problematic. For example, oxidation of alcohol **17** led to the formation of dimer side product **20** (Scheme 8.3).[25]

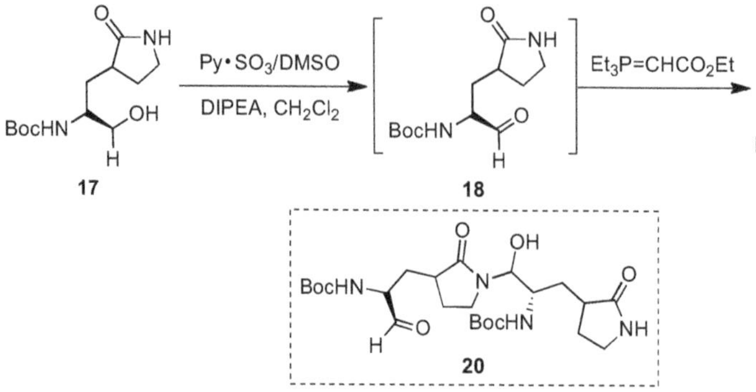

SCHEME 8.3

To suppress the formation of **20**, a simple but effective approach was developed by adding pyridine to convert the pyridine sulfuric acid salt (1:1) **21** to an inactive salt (2:1) **22** (Equation 8.9):

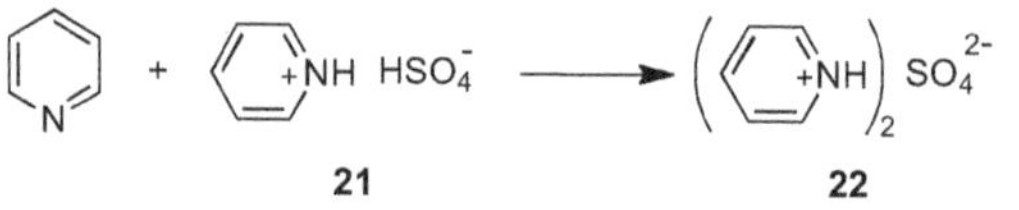

$$(8.9)$$

Procedure

- Charge 4.73 L (27.1 mol) of DIPEA to a solution of **17** (2.0 kg, 7.74 mol) in DMSO (2.75 L, 38.7 mol) and CH$_2$Cl$_2$ (10 L) at –5 °C.
- Charge a suspension of Py·SO$_3$, pre-prepared by mixing Py·SO$_3$ complex (2.46 kg, 15.5 mol) with 1.25 L (15.5 mol) of pyridine in DMSO (2.75 L, 38.7 mol) at –5 °C.
- Stir the reaction mixture at –5 °C for 1 h.
- Use the reaction mixture directly in the subsequent Wittig reaction.

Studies show that the Py·SO$_3$ adduct degrades over time in DMSO. Therefore, when using Py·SO$_3$/ TEA/DMSO system for the oxidation of alcohols, the following addition protocol should be adopted: the addition of the Py·SO$_3$ complex to a cold solution of alcohol **23**, TEA, and DMSO in methylene chloride in a dropwise manner (Equation 8.10).[26]

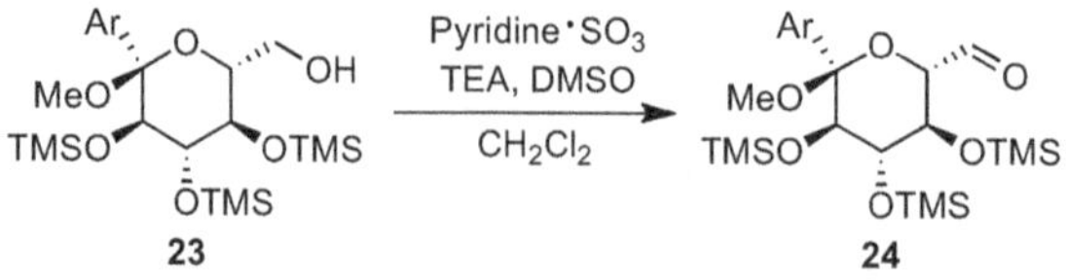

$$(8.10)$$

Procedure

- Charge 138.4 kg (1771.3 mol) of DMSO to a pre-cooled (10 °C) solution of **23** (32.25 kg, 49.2 mol) in CH$_2$Cl$_2$ (162 L), followed by adding 18.0 kg (177.9 mol) of TEA.
- Charge a slurry of sulfur trioxide-pyridine complex (22.0 kg, 138.2 mol) in CH$_2$Cl$_2$ (324 kg) at <12 °C over 45 min.

- Stir at 10 °C for 3 h.
- Add water (149 L) and warm the batch to room temperature.
- Separate layers.
- Wash the organic layer with saturated aqueous NH_4Cl (149 L).
- Concentrate the organic layer under reduced pressure to 57 L.
- Add EtOH (127 L).
- Concentrate under reduced pressure to 38 L.
- Yield: 27.33 kg (85%, HPLC assay).

NOTE

An in-line aqueous ozone scrubber was set up to trap dimethylsulfide liberated during processing.

8.1.2.2 Ac₂O/DMSO System

Anhydrides, such as trifluoroacetic anhydride and acetic anhydride,[27] can also be used to activate DMSO. Albright–Goldman procedure (Ac₂O/DMSO) utilizes inexpensive reagents at room temperature without an added base. Under these conditions, the oxidation reaction is typically complete within 12 and 24 h. For example, the oxidation of D-glucopyranose derivative **25** using the Albright–Goldman procedure provided a complete conversion within 18–24 h at 20–25 °C (Equation 8.11).[28]

$$(8.11)$$

Procedure

- Charge 75 mL (793.4 mmol, 17.2 equiv) of acetic anhydride to a solution of **25** (25.0 g, 46.2 mmol) in 100 mL of DMSO at 15–25 °C over 15 min.
- Stir the resulting mixture at 20–25 °C for 18–24 h.
- Add toluene (225 mL) and 3.3 N aqueous HCl (225 mL).
- Stir at 20–25 °C for 20 min.
- Separate layers.
- Wash the organic layer with 2 M aqueous phosphate buffer (pH 7, 225 mL).
- Dry over Na_2SO_4.
- Yield: 26.9 g (92%, solution yield).

8.1.2.3 TFAA/DMSO/TEA System

n alternative reagent combination of TFAA/DMSO/TEA[29] allowed the oxidation of alcohol **27** to ketonucleoside **28** under mild conditions (Equation 8.12).

$$(8.12)$$

8.1.2.4 TEMPO/NaOCl System

From economical and environmental standpoints, catalytic reaction systems are preferred owing to relatively mild reaction conditions, thereby fewer amounts of by-products and waste. Oxidation with

NaOCl, catalyzed by 2,2,6,6-tetramethyl-1-piperidine-*N*-oxide (TEMPO), is a well-recognized pro-cedure used to generate carbonyl compounds from corresponding alcohols. This oxidation protocol is attractive since it generates only NaCl and water as by-products. Generally, oxidations of primary alcohols give better results compared to secondary alcohols, due to the steric effect of TEMPO. Commonly used solvents include dichloromethane and ester solvents such as MeOAc, EtOAc, and *i*PrOAc. Standard conditions are: NaOCl (1.1–1.5 equiv) and TEMPO (1.0–10.0 mol%) under pH of 8.5–9.5 (adjusted with solid NaHCO$_3$) as NaOCl is unstable at lower pH. Additives such as KBr are required in catalytic amounts (0.1 equiv) to generate the reactive oxidant KOBr in situ. The TEMPO-mediated alcohol oxidation lent itself readily to preparations of aldehydes[30] and ketones.[31]

8.1.2.4.1   Oxidation of Primary Alcohol to Aldehyde

A synthetic route for the preparation of glucosylceramide synthase inhibitor *N*-[5-(adamantan-1-yl-methoxy)-pentyl]-1-deoxynojirimycin methanesulfonic acid salt (AMP-DNM) involves the oxida-tion of a primary alcohol **29** into the aldehyde **30** (Equation 8.13).[32] In order to avoid the formation of toxic dimethyl sulfide Swern oxidation, the oxidation was performed at 5 °C using TEMPO/ NaOCl, affording the desired aldehyde **30** in 91.5% yield.

$$(8.13)$$

Procedure

- Charge a solution of 0.43 kg (3.61 mol) of KBr in water (1.74 L) into a solution of **29** (7.93 kg, 30.69 mol) and TEMPO (52.4 g, 0.33 mol) in CH$_2$Cl$_2$ (43.65 L).
- Cool the mixture to 5 °C.
- Charge 30.55 L of a solution (pH 8.5–9.5) of NaOCl/NaHCO$_3$, pre-prepared by diluting NaOCl (ca 20.07 L, 12–15% in water, 30.69 mol) with an aqueous solution of NaHCO$_3$ (9%, ca 20.07 L) to reach pH 8.5–9.5, portionwise at 5 °C over 2 h.
- Stir the resulting mixture at 5 °C for 0.5 h.
- Check the progress of the reaction with HPLC by sampling the organic layer.
- Charge, portionwise, an additional (1–2 L) of the NaOCl/NaHCO$_3$ solution until <1% of **29** remaining.
- Warm to 20 °C.
- Separate layers.
- Add aqueous 2.87 N HCl (34.92 L) containing KI (87 g) to the organic layer.
- Wash the organic layer successively with aqueous Na$_2$S$_2$O$_3$ (34.92 L), saturated aqueous NaHCO$_3$ (34.9 L), and water (34.9 L).
- Distill at <40 °C.
- Yield: 7.93 kg (91.5%).

REMARK
Portionwise addition of NaOCl/NaHCO$_3$ is critical to avoid over-oxidations.

8.1.2.4.2   Oxidation of Secondary Alcohol to Ketone

Oxidation of a secondary alcohol **33** to synthesize (2-chlorophenyl)[2-(phenylsulfonyl)pyridin-3-yl] methanone **34** (Scheme 8.4)[33] required a large excess of NaOCl (3.5 equiv).

Procedure

- Charge 1.3 L of 10% aqueous KBr solution to a solution of **33** [obtained from a reaction of **32** (272.3 g, 1.94 mol) with **31** (400.5 g, 1.70 mol)] in EtOAc (ca 4 L).

- Charge 20.8 g (0.13 mol) of TEMPO.
- Cool the mixture to 3 °C.
- Add a mixture of 3.6 L of 12% aqueous NaOCl and 192.2 g of $NaHCO_3$ in 2.2 L of water over 1 h while keeping the batch temperature below 22 °C.
- Warm the content to 20–25 °C and stir at that temperature for 1 h (HPLC showed then <0.73% of **33** remaining).
- Cool the batch to 0 °C and stir for 3 h.
- Filter and wash with cold EtOAc (2 L) followed by water (2 L).
- Dry at 45 °C under vacuum.
- Yield: 571.8 g (87%) in 97.8% potency.

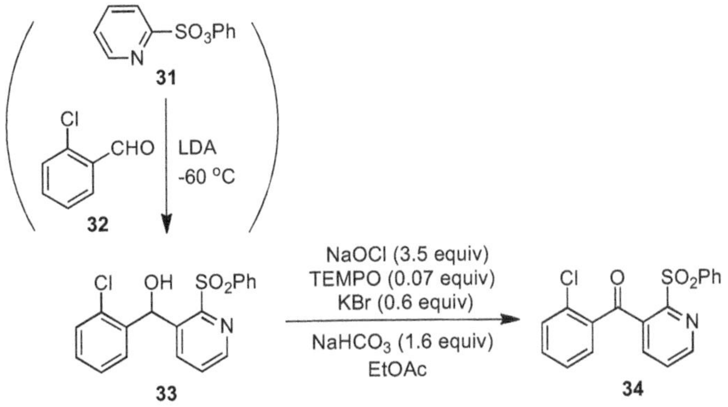

SCHEME 8.4

REMARK

Compared with CH_2Cl_2, the product **34** has a lower solubility in EtOAc. The use of EtOAc as the reaction solvent to replace CH_2Cl_2 allowed simple isolation via filtration.

Steric hindrance is a known limitation of TEMPO-mediated reactions and higher loading of TEMPO (0.35 equiv) was used to convert secondary alcohol **35** to its corresponding ketone **36** (Equation 8.14).[34]

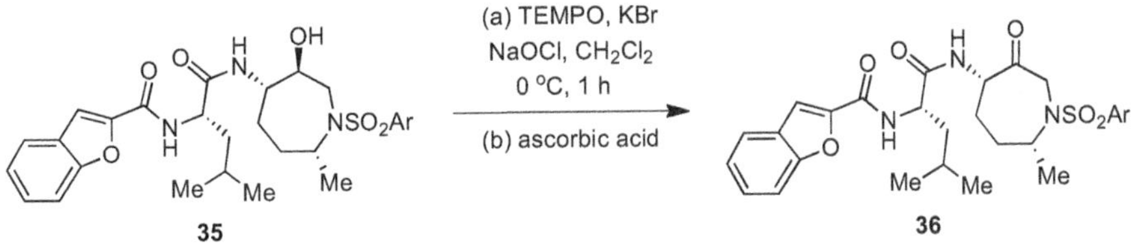

REMARKS

(a) Solid ascorbic acid was used to destroy the excess of NaOCl and remove some of the dark color.
(b) It was observed[35] that buffered bleach (pH 8.5 to 9.5) is less stable than non-buffered bleach and the potency decreased by ca 50% over a period of 12 h, while the potency of the concentrated NaOCl decreased by 20% per month during storage.

8.1.2.5 RuCl₃/NaOCl System

The RuCl₃/NaOCl represents another catalytic oxidation system that is amendable for large-scale production of ketones. Sodium hypochlorite is used as an inexpensive oxidant to convert $RuCl_3$ to

RuO$_4$ (the active oxidation species). Upon completion of the reaction, the reduced RuO$_2$ is normally rejected in the aqueous phase.

On a large scale, RuCl$_3$/NaOCl was used to oxidize secondary alcohol **37** to ketone **38** in high yield (Equation 8.15).[36]

$$(8.15)$$

Procedure

- Charge 42 g (0.2 mol, 1.1 mol%) of RuCl$_3$ to a solution of **37** (17.8 mol) in IPAc (36 L).
- Cool the resulting mixture to 5 °C.
- Charge 26 kg (23.6 mol) of 10–13 w/v% of NaOCl below the surface at 2–8 °C over 1.5 h.
- Stir the resulting biphasic mixture at 2–5 °C for 3 h.
- Add an aqueous solution of sodium bisulfite (1.0 kg) in water (5 L) over 5 min.
- Stir for 10 min.
- Add 6.6 L 3 N HCl.
- Separate layers.
- Extract the aqueous layer with IPAc (2×30 L).
- Wash the combined organic layers with brine (5 L) and dry over MgSO$_4$.
- Distill to 20 L.
- Solvent exchange to heptane (6×18 L).
- Filter, wash with heptane (2×5 L), and dry.
- Yield: 3.07 kg (90%).

8.1.2.6 Sulfinimidoyl Chloride

Given the high reactivity of aldehydes, partial oxidation of alcohols at the aldehyde stage is challenging. The oxidation of diol **39** to phthalaldehyde derivative **41** was performed with strict requirements for the oxidation system or lactone **42** was formed (Equation 8.16).[37]

$$(8.16)$$

Using Swern oxidation showed limited success for this transformation. Ultimately, using sulfinimidoyl chloride **40** proved to be the oxidant of choice, producing the phthaldehyde **41** in yields between 65% and 75%.

8.1.2.7   2-Amadamantane *N*-Oxide/CuCl

Japanese scientists identified a less hindered 2-amadamantane *N*-oxide (AZADO) radical/CuCl catalytic system for the oxidative conversion of alcohols to aldehydes or ketones.

$$(8.17)$$

This oxidation system is especially effective for both primary and secondary alcohols containing electron-rich, oxidation-labile functional groups, such as dithianes and sulfides. Equation 8.17[38] describes the oxidation of (4-(methylthio)phenyl)methanol **43** with AZADO/CuCl at room temperature in the presence of air furnishing the corresponding aldehyde **44** in 92% yield.

8.1.3   OXIDATION OF ALDEHYDES TO ACIDS

Many oxidation reactions can be hazardous on large scale not only due to their inherent exothermic nature, but also because of the subsequent workup of the reaction mixture, as oxidation reactions may accompany the formation of energetic by-products. For example, an oxidative conversion of 2-chloro-6-methylbenzaldehyde **45** with sodium chlorite to the corresponding benzoic acid **46** generated hazardous hypochlorite that had to be destroyed in situ as an accumulation of the hypochlorite in the reaction system is extremely dangerous (Equation 8.18).[39]

$$(8.18)$$

Accordingly, two reaction protocols were developed. In protocol (a), hydrogen peroxide was added to the reaction to scavenge hypochlorite; however, this led to the loss of an inert atmosphere in the reactor due to the generation of oxygen during the consumption of hypochlorite (Equation 8.19).

$$H_2O_2 + HOCl \longrightarrow HCl + O_2 + H_2O \qquad (8.19)$$

DMSO proved to be an effective HOCl scavenger (Protocol (b), Equation 8.18) and the use of DMSO was allowed to destroy the HOCl in situ (Equation 8.20).

$$(8.20)$$

Procedure

- Charge 40.0 g (257 mmol) of **45**, followed by CH_3CN (100 mL), DMSO (24.28 g, 311 mmol), and water (33 mL).
- Stir at 10 °C.
- Charge concentrated sulfuric acid (14 g).
- Charge a solution of sodium chlorite (34.85 g) in water (170 mL) at <10 °C.
- Workup once the reaction is complete.

REMARK

Mixing DMSO with aqueous hydrogen peroxide may lead to a violent exothermic reactions.[40]

8.1.4 Oxidation of Sulfides to Sulfoxides

Urea–hydrogen peroxide (UHP) is a relatively safe alternative to anhydrous hydrogen peroxide. UHP is a white crystalline solid and is commercially available. UHP is formed when urea is crystallized from aqueous hydrogen peroxide. Although UHP is relatively stable as judged by negative impact and pressure-time tests, it is not recommended to use UHP in combination with carboxylic anhydrides.[41]

UHP was selected as the oxidant for the selective oxidation of sulfide **47** to sulfoxide **48** in hexafluoro-2-propanol (HFIP) (Equation 8.21).[6] HFIP is an expensive solvent; thereby HFIP was recycled by distilling the biphasic solution of HFIP and heptanes.

(8.21)

REMARKS

(a) HFIP is volatile and corrosive. Its vapors can cause severe respiratory problems and immediate, irreversible eye damage. Care should be taken to avoid contact by wearing appropriate gloves, safety glasses, and other protective equipment as needed. HFIP may also contain a very low level of hexafluoroacetone, which is a potential reproductive hazard.

(b) Aqueous 30% H_2O_2 in HFIP is described as a facile, selective, and efficient oxidant for the conversion of sulfides to sulfoxides under neutral conditions. This method can be applied to the oxidation of sulfides without sulfone formation.[42]

8.1.5 Oxidation of Sulfides to Sulfones

8.1.5.1 Oxidation with Oxone

The oxidation of sulfide to sulfone can be carried out by using different oxidants, such as HIO_4 (cat. Cr_2O_3), UHP,[41] H_2O_2,[43] oxone,[44] etc. The use of HIO_4 with catalytic Cr_2O_3 has environmental and scale-up issues.[45]

Oxone is a versatile and inexpensive oxidant with the formula $2KHSO_5 \cdot KHSO_4 \cdot K_2SO_4$. Oxone can be used to oxidize aldehydes, alkenes, thioethers, tertiary amines, and phosphines. The oxidation of diaryl sulfide **49** to its corresponding sulfone **50** was accomplished by using oxone in a high yield (Equation 8.22).[45]

(8.22)

Similarly, alkyl aryl sulfide **51** was easily oxidized to sulfone **52** using oxone in acetone and THF (Equation 8.23).[46]

$$(8.23)$$

Procedure

- Dissolve **51** (≤134.5 mol) in acetone (53.75 kg) and THF (151.10 kg).
- Charge a 23.4% aqueous solution of oxone (531.00 kg, 404.2 mol) at 15–25 °C over 3.5 h.
- Stir for 2 h.
- Add EtOAc (245.00 kg).
- Filter the resulting suspension and wash the cake with EtOAc (61.80 kg).
- Add 9.5% aqueous $Na_2S_2O_3$ (143.30 kg) at 15–30 °C over 45 min to the combined filtrates.
- Separate layers.
- Extract the aqueous layer with EtOAc (122.10 kg).
- Wash the combined organic layers with 10% aqueous Na_2CO_3 (135.50 kg) and 10% aqueous NaCl (135.50 kg).
- Distill at 35–45 °C under reduced pressure.
- Dissolve the resulting residue in toluene (47.10 kg).
- Distill at 45–55 °C under reduced pressure.
- Dilute the residue with EtOAc (76.55 kg) and heat to 50–70 °C for dissolution.
- Cool the resulting solution to 40–45 °C over 1 h.
- Age the resulting slurry at 40–50 °C for 1.5 h.
- Cool to 5–15 °C over 3.5 h and stir for 3 h.
- Add *n*-heptane (116.05 kg) over 1 h and age for 1 h.
- Filter, rinse with a cold mixture of EtOAc (21.60 kg) and *n*-heptane (32.55 kg), and dry at 45–55 °C under vacuum.
- Yield: 28.90 kg (80%).

REMARK

When using oxone, it should be cognizant of potential product contamination with inorganics.[47]

8.1.5.2 Oxidation with Sodium Perborate

Sodium perborate can be generated in situ from a reaction of boric acid with hydrogen peroxide. Employment of $NaBO_3$ (generated in situ) as a catalyst, alkyl aryl sulfide **53** was converted into sulfone **54** (Equation 8.24).[48,45] The same transformation, however, would require 19 equiv of oxone to drive the reaction to completion.

$$(8.24)$$

REMARK

Quenching the excess hydrogen peroxide after the sulfide oxidation could be conveniently achieved by using a more reactive sulfoxide. For instance, the excess of H_2O_2 from the oxidation of mixed sulfide **55** to sulfone **56** was destroyed using DMSO (Equation 8.25).[49]

$$(8.25)$$

8.1.5.3   Oxidation with Sodium Periodate

Sodium periodate combined with catalytic amounts of ruthenium trichloride is proven to be an efficient oxidation system for the transformation of thioanisole **57** to sulfone **58** at a kilogram scale (Equation 8.26).[50] The oxidation involves two stages: sulfide **57** to sulfoxide intermediate and the ruthenium-catalyzed oxidation of sulfoxide to sulfone product **58**.

$$
\text{57} \xrightarrow[\text{MeCN/H}_2\text{O/CH}_2\text{Cl}_2]{\substack{\text{RuCl}_3\ (0.05\ \text{mol\%}) \\ \text{NaIO}_4\ (2.0\ \text{equiv})}} \text{58}
\tag{8.26}
$$

Procedure

- Charge CH_2Cl_2 (10 L), acetonitrile (10 L), and water (20 L) into a reactor.
- Charge 3.0 kg (18.9 mol) of **57** followed by adding 8.49 kg (39.7 mol, 2.1 equiv) of $NaIO_4$.
- Charge 2.0 g (19 mmol, 0.05 mol%) of $RuCl_3 \cdot H_2O$ portionwise (4×0.5 g).
- Stir at 25–50 °C for 5 h.
- Add water (60 L) and separate layers.
- Extract the aqueous layer with CH_2Cl_2 (2×10 L).
- Wash the combined organic layers with water (20 L) followed by brine (4 L).
- Distill under reduced pressure to produce a dark-green oil.
- Triturate with hexanes (4 L).
- Filter and wash with hexanes (1 L).
- Yield: 3.30 kg (91%).

REMARK

The portionwise addition of the ruthenium catalyst was necessary to control the mild but immediate exothermic reaction.

8.1.5.4   Oxidation with NaOCl

The Chapman–Stevens oxidation protocol was adopted for the oxidation of biaryl sulfides **59** to the corresponding sulfones **60** (Equation 8.27).[51]

$$
\text{59} \xrightarrow[\text{(b) NaOH, H}_2\text{O}]{\substack{\text{(a) NaOCl} \\ \text{AcOH, DMF}}} \text{60}
\tag{8.27}
$$

X = H; Y = 2-Cl, 3-Br, 4-Br, 2,3-DiCl, 2,3-DiBr, 2,6-DiCl, 2-Cl-4-Me
2-Cl-3-NO_2, 2-Cl-5-NO_2, 2-Cl-N-oxide, 2-Br-N-oxide
3-Cl-5-CN, 2-Cl-quinoxaline, 2-Cl-quinoline
Y = H; X = H, 3-Cl, 3-F, 3,5-bis(CF_3), 2-Br, 4-NO_2, 4-Me, 4-OMe, 2-CF_3

Procedure

- Charge 57 mL (0.995 mol, 1.26 equiv) of AcOH into a crude sulfide **59** (~0.79 mol, obtained from the previous step) solution in 645 mL of DMF.
- Heat to 40 °C.
- Charge 13 wt% NaOCl solution (850 mL, 1.7 mol, 2.15 equiv) over 2 h.
- Add 530 mL of water and stop heating.

- Add 250 mL of 20 w/v NaOH solution.
- Cool to <5 °C and hold for 1.5 h.
- Filter and wash with water.
- Dry at 55 °C.
- Yield: 149 g (86%) with >99% purity.

REMARKS

(a) For safety, it is important to add the NaOCl solution to a warm (40 °C) DMF solution of **59** to avoid the accumulation of oxidant in the reaction system.

(b) In the absence of AcOH, the amount of NaOCl required for complete conversion of sulfide to sulfone was inconsistent from run to run and generally required 3–4 equiv.

(c) An induction period was observed during the oxidation of dihydropyrazole **61** with sodium hypochlorite under biphasic solution in DCM and K_3PO_4 (Equation 8.28).[52]

$$(8.28)$$

The addition of KBr (0.5 equiv) into the reaction mixture eliminated the induction period. The in-situ-formed HOBr was hypothesized as the oxidative species.

8.1.5.5   Oxidation with H_2O_2/Na_2WO_4

Another oxidation protocol was developed to convert tetrahydrothiophene **63** to the corresponding sulfone **64** by using hydrogen peroxide in conjunction with sodium tungstate (Equation 8.29).[24d] Owing to the good solubility of the sodium salt of **63**, the oxidation could be carried out in an aqueous environment at 70 °C in presence of 0.5 mol% sodium tungstate dihydrate. This cost-efficient and environmentally benign process provides good conversion and high throughput, avoiding using a halogenated solvent.

$$(8.29)$$

Procedure

- Charge, to a mixture of H_2O (64.2 L) and 13.4 kg of 50% NaOH, 0.32 kg (0.97 mol) of $Na_2WO_4 \cdot 2H_2O$ at 20–30 °C followed by adding 43 kg (189 mol) of **63**.
- Heat the mixture to 65–75 °C.
- Charge 55.7 kg (573 mol) of 35% H_2O_2 over 4 h at 65–75 °C.
- Stir at 70 °C for 4 h.
- Add 32% HCl to a pH of 0–1.
- Cool the resulting slurry to 0 °C and stir for 1 h.
- Filter and wash with IPA (2×110 L).
- Dry at 50 °C under vacuum.
- Yield: 44.9 kg (91%).

REMARK

The tungstate-catalyzed H_2O_2 oxidation system was used to convert 2-(methylthio)pyrimidine **65** to its corresponding sulfone **66** (Equation 8.30).[53] During the workup, the excess H_2O_2 was destroyed by adding an aqueous sodium ascorbate solution at 80 °C.

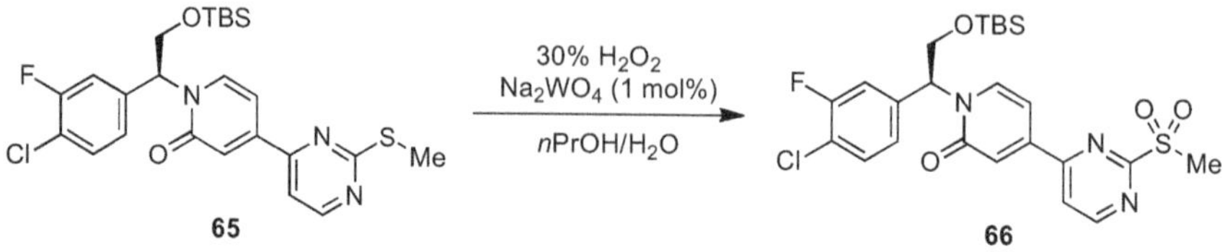

$$(8.30)$$

8.1.5.6   Oxidation with TMSCl/KNO₃

A mixture of nitrate salt and chlorotrimethylsilane is found to be a mild and efficient reagent for the direct oxidative conversion of sulfides and sulfoxides into the corresponding sulfones (Equation 8.31).[54]

$$(8.31)$$

Thiols and disulfides are also found to undergo oxidation under similar reaction conditions to afford sulfonyl chlorides (Equations 8.32 and 8.33).

$$Ar-SH \xrightarrow{\text{TMSCl/KNO}_3} Ar-SO_2Cl \qquad (8.32)$$

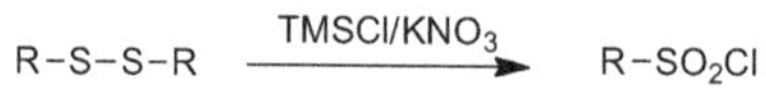

$$R-S-S-R \xrightarrow{\text{TMSCl/KNO}_3} R-SO_2Cl \qquad (8.33)$$

8.1.6   OTHER OXIDATIVE REACTIONS

8.1.6.1   Dakin Oxidation

Dakin oxidation of pyridyl aldehyde **67** using an excess of urea-hydrogen peroxide complex as a safe alternative to H_2O_2 gave the corresponding hydroxyl pyridine **68** in 48% yield (Equation 8.34)[55] on a 89-g scale.

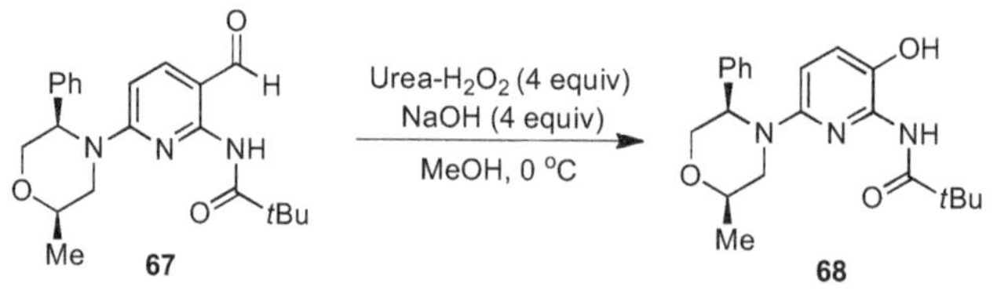

$$(8.34)$$

Procedure

- Charge 87.66 g (931.9 mmol, 4.0 equiv) of UHP to a cold (0 °C) solution of **67** (88.8 g, 232.9 mmol) in MeOH (300 mL) followed by adding NaOH in MeOH (1M, 930 mL).
- Stir the resulting mixture at 0 °C for 42 h.
- Add Na_2SO_3 (100 g) and stir at 0 °C for 30 min.
- Concentrate the reaction mixture.
- Add water (1 L) and EtOAc (500 mL).
- Separate layers.

- Extract the aqueous layer with EtOAc (2×300 mL).
- Wash the combined organic layers with brine (300 mL).
- Dry over MgSO$_4$, filter, and distill.
- Purify the crude via a silica column.
- Yield: 42.7 g (49%).

8.1.6.2   Hydroxylation

Hydroxylation of 5-bromopyrimidine **69** using a literature procedure[56] with peracetic acid and sulfuric acid in acetone provided pyrimidone **72** (Scheme 8.5).[57] However, during the isolation of the solid product from two-mole scale preparation (a 20-fold increase over the literature preparation), an explosion occurred.[58] The peroxide dimer of acetone was the most likely cause of the explosion. This material is reported to be both shock and friction sensitive and known to sublime at room temperature.

To address the safety concerns, a safer aqueous hydroxylation system was developed[57] which provided comparable yields to that reported in the literature.

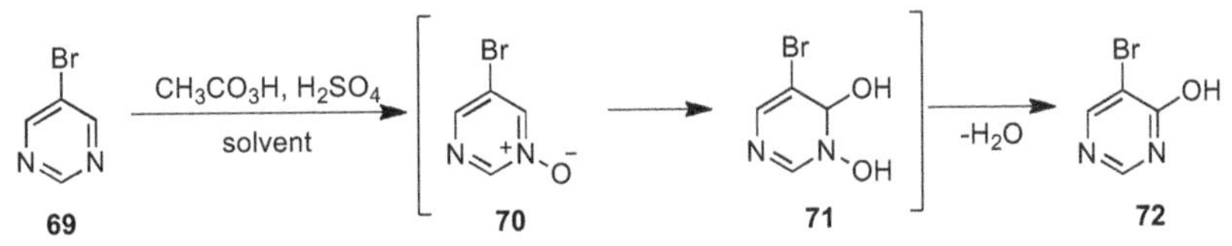

SCHEME 8.5

Procedure

- Charge slowly 255 g (2.6 mol) of H$_2$SO$_4$ into a 5-L flask containing 1.5 L of water.
- Cool the resulting solution to 20 °C.
- Charge 410 g (2.58 mol) of **69**.
- Charge 628 g (2.64 mol) of 32% peracetic acid via an addition funnel at <70 °C.
- Stir the batch at 70 °C for 2 h.
- Cool to 50 °C.
- Add a solution of Na$_2$SO$_3$ (110 g, 0.87 mol) in water (0.5 L) at <70 °C (starch–iodide test should be negative).
- Cool to 50 °C.
- Add 580 g (7.25 mol) of 50 wt% NaOH solution at <70 °C.
- Cool the resulting slurry to 20 °C.
- Filter and wash with water (2×0.5 L).
- Transfer the wet cake to a 1-L flask.
- Add toluene (750 mL) and heat to reflux.
- Distill azeotropically to remove ca. 250 mL of toluene.
- Cool to room temperature.
- Filter to give the product **72** as a wet cake (415 g of 11% toluene wet cake, 83%).

8.1.6.3   Oxidation of Phosphite

The preparation of di-*tert*-buty potassium phosphate **75** could be accomplished by oxidation of phosphite **73** with iodine. As shown in Scheme 8.6,[59] the reaction proceeded through a phosphoryl iodide intermediate **74** that hydrolyzed in situ to give the di-*tert*-butyl potassium phosphate **75** after treatment with KO*t*Bu.

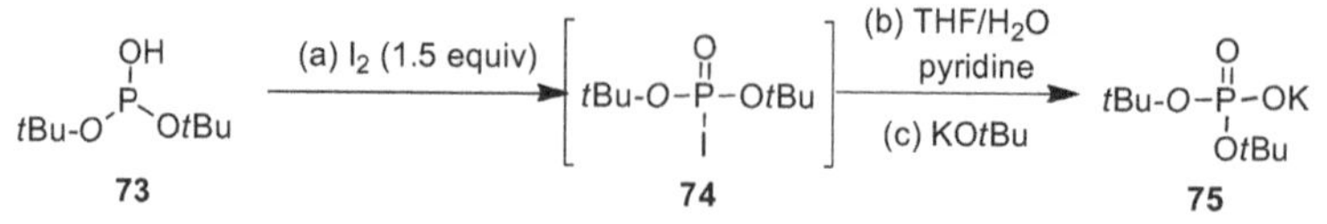

SCHEME 8.6

(I) Problems

- The reaction protocol required a stoichiometric amount of iodine, which is a controlled substance.
- The process provided a rather large amount of waste (>2 kg of pyridine hydroiodide per kg of the isolated product **75**).

(II) Solutions

To obviate these aforementioned issues, a catalytic reaction protocol was developed, in which the iodine was generated in situ by a reaction of KI (0.15 equiv) with 30% H_2O_2. The reaction was carried out in a mixture of toluene and acetonitrile (3:1) under biphasic conditions.

8.2 REDUCTION REACTIONS

Among organic transformations, reduction reactions are used much more frequently than oxidative transformations due to process safety, toxicity, and waste disposal concerns related to oxidants.[60] Catalytic hydrogenation is the most frequently used approach, followed by hydride and borane reduction.

Boron-based reductants are overwhelmingly the preferred choice for reductions of aldehydes, ketones, carboxylic acids, esters, and amides because of their diversity, chemoselectivity, and commercial availability (in many cases as solutions, which facilitates their use in the kilo lab or pilot plant facilities).

8.2.1 REDUCTION WITH NaBH₄-BASED AGENTS

Sodium borohydride ($NaBH_4$) is a preferred reducing reagent for the reduction of various aldehydes and ketones to alcohols. $NaBH_4$ is commercially available in various forms (powder, pellets, caustic solution). $NaBH_4$ reductions are typically carried out in THF, alcohols (MeOH, EtOH), or combinations thereof. Other solvent combinations include toluene/MeOH and MTBE/H_2O (biphasic mixture with nBu₄NCl as phase-transfer agent). $NaBH_4$ reductions may be performed under aqueous or anhydrous conditions. In addition, the reducing power of $NaBH_4$ can be adjusted by the choice of solvents and additives.

Although the safety and ease of handling $NaBH_4$ is better than that of other reducing agents such as aluminum hydrides, its reduction reactions generate large amounts of heat and hydrogen gas which need to be controlled. To avoid runaway reactions, semibatch operations in multi-purpose reactors are frequently used by pharmaceutical companies. The addition of $NaBH_4$ solution (preferred over solid $NaBH_4$) to the reactor is recommended. Safety evaluation showed that both DMF and DMAc as reaction solvents can cause violent runaway reactions.[61]

8.2.1.1 Reduction of Acids

Reduction of carboxylic acid **76** was accomplished by simply treating a mixture of **76** with 1.5 equiv of anhydrous HCl in THF, followed by slow addition of $NaBH_4$ (1.5 equiv)/diglyme (Equation 8.35).[62] The slow addition of $NaBH_4$ was required to prevent lactam reduction.

$$(8.35)$$

Procedure

- Charge 4.70 kg (124 mol, 1.5 equiv) of $NaBH_4$ into an inerted 800-L reactor (reactor 1) followed by adding 90 kg of diglyme.
- Stir the mixture at 25 °C for 3 h.
- Charge 27 kg of THF into a separate inerted 800-L reactor (reactor 2) and cool to 0 °C.
- Add 4.60 kg (126 mol) of HCl (gas).
- Charge 26.0 kg (83.2 mol) of **76** into another inerted 800-L reactor (reactor 3) followed by adding 90 kg of THF and cooling the resulting slurry to 0 °C.
- Charge HCl/THF in reactor 2 to the slurry (in reactor 3) over 30 min and stir for 30 min.
- Charge $NaBH_4$/diglyme in reactor 1 into reactor 3 over 2 h.
- Warm the contents in reactor 3 to 25 °C and hold at that temperature for 2.5 h.
- Cool the batch to 0 °C.
- Add MeOH (206 kg) over 30 min.
- Distill under reduced pressure to remove solvents.
- Add water (437 kg) and azeotropic distillation at 50 °C under full vacuum until a minimum stirring volume is reached.
- Cool the batch to 25 °C.
- Add 39 kg of 0.5 M NaOH and heptane (57 kg).
- Stir the resulting two-phase slurry for 10 h.
- Filter and rinse with water (40 kg).
- Dry at 35 °C/10–50 mbar for 6 h.
- Yield: 20.5 kg (80%).

REMARKS

(a) Borane can be generated in situ under a variety of conditions, involving reactions of sodium borohydride with Lewis or protic acid.[63]

(b) Reactions (Equations 8.36 through 8.38) are examples of the reduction of acids and amides by using the in situ generated borane.[64]

$$(8.36)$$

$$(8.37)$$

$$(8.38)$$

8.2.1.2   Reduction of Esters

8.2.1.2.1   NaBH₄/AlCl₃

Carboxylic acid **80** is a prostaglandin D$_2$ (PGD$_2$) receptor antagonist that is a promising alternative anti-allergic drug candidate. As a key intermediate, the benzoic acid salt of amino alcohol **79** was prepared by diastereoselective reduction of **78**. Although lithium aluminum hydride (LAH) with sulfuric acid gave **79** in good yield, LAH is difficult to handle in large-scale manufacturing. Bouveault–Blanc reduction rendered the resulting product in poor diastereoselectivity due to epimerization of the alkyl group at the 3-C position caused by α-proton abstraction of O-methyl oxime. Eventually, the use of NaBH$_4$/AlCl$_3$ reduced ester and O-methyl oxime to afford **79** in 75% yield without epimerization of the alkyl group at the 3-position (Equation 8.39).[65]

(8.39)

(I) Reaction problems

The safety of this diastereoselective reduction was of considerable concern when scaling up as the total heat of the reaction was 437.3 kJ/mol.

(II) Solutions

The exothermic reduction was controlled by adding a solution of NaBH$_4$ in diglyme to the reaction mixture. In addition, the reduction heat was further controlled by conducting the reaction at two stages: reduction of ester at 35 °C, followed by diastereoselective reduction of O-methyl oxime at 65 °C. Excess borohydride in the reaction mixture was destroyed by dropwise addition of acetone after the reaction. The process was safely applied to pilot manufacturing of **79** on a 160 kg/lot basis.

Procedure

Stage (a)
- Charge a solution of NaBH$_4$ (26.3 g, 0.69 mol) in diglyme (468 mL) to a mixture of **78** (127.6 g, 0.50 mol) and AlCl$_3$ (34.7 g, 0.26 mol) in diglyme (264 mL) dropwise at 35 °C.
- Stir the resulting mixture at 35 °C for 1 h.
- Heat to 65 °C, followed by stirring for 2 h.
- Cool to 10 °C.
- Add acetone (39.0 g) dropwise at 10 °C.
- Transfer the resulting mixture to 18% aqueous HCl solution (271 g, 1.34 mol) at 10–30 °C.
- Charge 291.6 g (2.48 mol) of 34% NaOH dropwise to the above mixture, followed by adding toluene (840 mL).
- Separate layers.

Stage (b)
- Wash the organic layer with aqueous 10% NaCl (2×240 g).
- Distill azeotropically to give 702 g of residue.
- Charge a solution of benzoic acid (47.7 g, 0.39 mol) in acetone (175 mL) at 30 °C.
- Cool the resulting slurry to 5 °C.
- Filter, wash with cold acetone (300 mL), and dry.
- Yield of benzoic acid of **79**: 94.23 g (72.1%).

REMARK

Reduction of Boc-amino ester **81** was most favorably performed with $LiBH_4$ in the presence of methanol as activating agent (Equation 8.40),[66] in which methanol acts probably as a proton donor.[67]

$$\underset{\textbf{81}}{\text{MeO}\overset{\text{NHBoc}}{\diagdown}\text{CO}_2\text{Me}} \xrightarrow[\text{MeOH, THF}]{\text{LiBH}_4} \underset{\textbf{82}}{\text{MeO}\overset{\text{NHBoc}}{\diagdown}\text{OH}} \tag{8.40}$$

8.2.1.2.2 $NaBH_4/ZnCl_2$

The combination of sodium borohydride with zinc chloride in the presence of a tertiary amine showed powerful reducing properties for the reduction of carboxylic esters to their corresponding alcohols.[68] This combination demonstrated well in the reduction of a carboxylic ester such as **83** to the corresponding amino alcohol **84** (in 81% yield after crystallization from toluene) (Equation 8.41).[69] Compared to flammable, corrosive, and moisture-sensitive LAH, this reduction system is easy to handle on large scale.

$$\underset{\textbf{83}}{\text{(structure with CO}_2\text{Et and CN)}} \xrightarrow[\text{DME}]{\text{NaBH}_4/\text{ZnCl}_2} \underset{\textbf{84}}{\text{(structure with OH and NH}_2\text{)}} \tag{8.41}$$

Procedure

- Charge 58.6 g (1.55 mol, 4.0 equiv) of $NaBH_4$ to a solution of **83** (100 g, 0.39 mol) in DME (1.5 L) followed by adding 105 g (0.78 mol, 2.0 equiv) of $ZnCl_2$.
- Stir the mixture under reflux for 7 h.
- Cool to room temperature.
- Add DME (215 mL), water (85 mL), and saturated NH_4Cl (700 mL).
- Separate layers,
- Add 6 M HCl to the aqueous layer to pH 1 and stir at room temperature for 2 h.
- Add toluene (1.0 L) followed by aqueous 10 M NaOH to pH 14.
- Separate layers.
- Distill the organic layer under reduced pressure.
- Crysallize the resulting residue from toluene (200 mL).
- Filter, wash with cold toluene (200 mL), and dry.
- Yield: 69 g (81%).

Note

Under the reaction conditions, both the ester and nitrile groups in **83** were reduced.

8.2.1.2.3   NaBH₄/CaCl₂

Reduction of ester **85** using in situ generated $Ca(BH_4)_2$ from a mixture of $NaBH_4$ and $CaCl_2$ in EtOH afforded a clean reduction to the corresponding primary alcohol **86** (Equation 8.42).[70]

$$(8.42)$$

Procedure

- Charge, into a 140 L reaction vessel, 2.81 kg (16.4 mol) of **85**, followed by adding EtOH (84 L).
- Charge, in three portions over 15 min, 3.65 kg (32.9 mol, 2.0 equiv) of $CaCl_2$.
- Stir at below 22 °C to form a solution.
- Charge, in three portions over 20 min, 2.49 kg (65.7 mol, 4.0 equiv) of $NaBH_4$.
- Stir the batch for 20 h.
- Cool to 5 °C.
- Add 11.2 L of 6 N HCl over 30 min.
- Warm the batch to room temperature and stir for 2 h.
- Filter over Solka Floc and rinse with EtOH.
- Distill the combined filtrates (together with another batch with 14.4 mol of **85**) under reduced pressure.
- Switch the solvent to water (32 L).
- Extract the aqueous solution with 1-butanol (53 L, then, 2×26.5 L).
- Wash the combined organic layers with brine (10.5 L).
- Distill under reduced pressure.
- Switch the solvent to water, then to IPA to give a suspension (40 L).
- Filter over Solka Floc and wash with IPA (2×10 L),
- Solution yield of **86**: 4.13 kg (94% assay yield).

8.2.1.3   Reduction of Amides

Generally, amides are less reactive under reductive conditions. Amides can be reduced to the corresponding amines with LAH, borane, or by catalytic hydrogenation, though high temperatures and pressures are usually required for the latter. LAH is a highly reactive and flammable solid, and difficult in handling on a production scale.

8.2.1.3.1   NaBH₄/BF₃·THF

Amide reduction with in situ generated borane from $NaBH_4/BF_3$·THF was realized by adding the solution of amide mandelate salt **87** dropwise to a suspension of $NaBH_4$ in THF at 0 °C, followed by a slow addition of an excess of BF_3·THF at 15 °C (Equation 8.43).[71]

$$(8.43)$$

After the completion of the reaction, methanol was added at 0 °C to quench the borane complexes and a small amount of unreacted $NaBH_4$. The mixture was then distilled to 6 residual volumes at atmospheric pressure.

Procedure

Stage (a)
- Dissolve 1 kg of **87** in 6.5 L of THF containing 20 g of TBAB at 40 °C.
- Charge the solution to a suspension of 0.38 kg of $NaBH_4$ in THF (1.5 L) at 25 °C.
- Stir the resulting suspension at 25 °C for 2 h.
- Charge 2.3 L of BF_3·THF over ca. 50 min.
- Stir the resulting mixture at 35 °C for 18 h and Stage (b).

NOTES

(a) The use of TBAB helped the dissolution of the mandelate salt **87** in THF at room temperature by increasing the ionic character of the solvent.

(b) As inorganic fluoride can damage the glass-lined reactor, confirmation of this absence is extremely important when a fluorine-containing reagent is used in the process. ^{19}F NMR method was implemented during the amide reduction by $NaBH_4$ with BF_3·THF complex as inorganic F^- was expected in the region −110/−130 ppm.

(c) Borane is not compatible with some aprotic solvents such as acetone, *N,N*-dimethylacetamide, and acetonitrile.

8.2.1.3.2 NaBH₄/AcOH

Reduction of amide **89** to amine **90** was achieved using either in situ generated sodium acetoxyborohydride or diborane (Equation 8.44).[72]

$$(8.44)$$

Procedure (a) 100 g scale

- Charge 58.2 g (1.54 mol) of $NaBH_4$ to a solution of **89** (100 g, 0.31 mol) in MeTHF (700 mL).
- Cool the resulting suspension to 2 °C.
- Charge a solution of AcOH (88 mL, 1.5 mol) in MeTHF (300 mL) over 1 h.
- Heat to reflux and stir for 22 h.
- Cool to 10 °C.
- Transfer the reaction mixture to ice/water (1.4 L).
- Add 50 wt% NaOH (120 g, 1.5 mol).
- Separate layers.
- Extract the aqueous layer with MeTHF.
- Wash with combined organic layers of brine (1 L).
- Distill until ~300 mL.
- Adde 2 N aqueous HCl (1 L).
- Reflux for 2 h.
- Cool to 20 °C.
- Filter and rinse with water (200 mL) and MeTHF (2×200 mL).
- Dry at 40 °C under vacuum.
- Yield: 88.6 g (83% yield).

Procedure (b) 11.3 kg scale

- Charge 12.3 kg (86.4 mol) of $BF_3 \cdot OEt_2$ to a solution of **89** (11.3 kg, 34.6 mol) in THF (33.9 L).
- Transfer the solution slowly to a hot (55 °C) suspension of $NaBH_4$ (2.45 kg, 64.8 mol) in THF (28.2 L) over 3 h at 50–60 °C.
- Stir the batch at 60 °C for 3 h.
- Cool to 5 °C.
- Add a mixture of 10:1 (w/w) water and THF (20.6 kg) at 5–20 °C.
- Stir at 20 °C for 2 h.
- Add water (45.2 L).
- Extract with MeTHF (45.2 L).
- Wash with an organic layer with water (56.5 L) and brine (65 kg).
- Add MeOH (11.3 L) and concentrated HCl (6.81 kg, 60.1 mol).
- Heat to 60 °C over 2 h and stir for 1 h.
- Distill at 35 °C under reduced pressure (0.1 bar) to 25 L.
- Add MeTHF (113 L).
- Distill at atmospheric pressure at 80 °C to 65 L.
- Cool to 10 °C and stir for 2 h.
- Filter and wash with MeTHF (11 L).
- Yield: 8.36 kg (69% yield).

8.2.1.3.3 NaBH₄/Iodine

Reduction of amide **91** could also be achieved with $NaBH_4$/iodine system. The reaction was carried out by adding a solution of I_2 (2.0 equiv) in THF to a mixture of the amide starting material and $NaBH_4$ (4.2 equiv) in THF (Equation 8.45).[73]

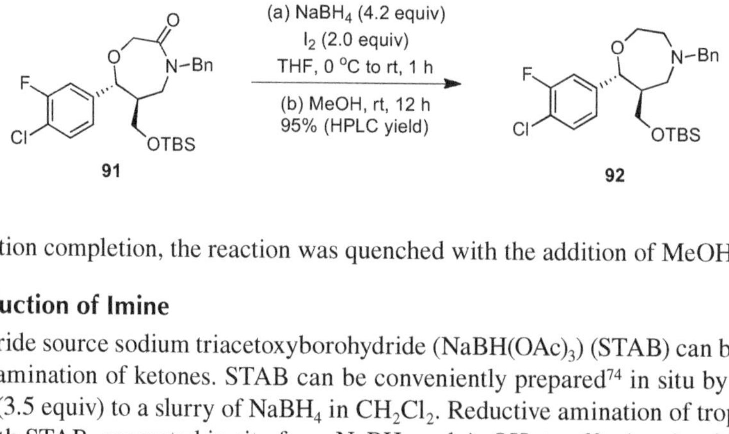

$$(8.45)$$

Upon the reaction completion, the reaction was quenched with the addition of MeOH.

8.2.1.4 Reduction of Imine

As a mild hydride source sodium triacetoxyborohydride ($NaBH(OAc)_3$) (STAB) can be employed in the reductive amination of ketones. STAB can be conveniently prepared[74] in situ by slow addition of acetic acid (3.5 equiv) to a slurry of $NaBH_4$ in CH_2Cl_2. Reductive amination of tropinone **93** was carried out with STAB, generated in situ from $NaBH_4$ and AcOH, to afford amine **95** via imine **94** (Scheme 8.7).[74]

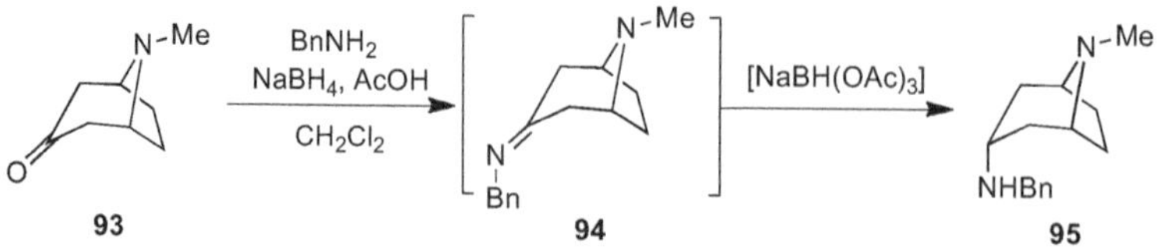

SCHEME 8.7

REMARKS

(a) In situ generated STAB is the reagent of choice in large-scale preparation in order to reduce the cost of the reagent and increase the robustness of the process, though STAB is commercially available.

(b) Due to its enhanced solubility in most organic solvents, sodium tripropionoxyborohydride [NaBH(OCOEt)$_3$] could replace STAB to reduce imine.[75]

(c) After the reaction completion, the acidic reaction mixture is typically quenched with bases.

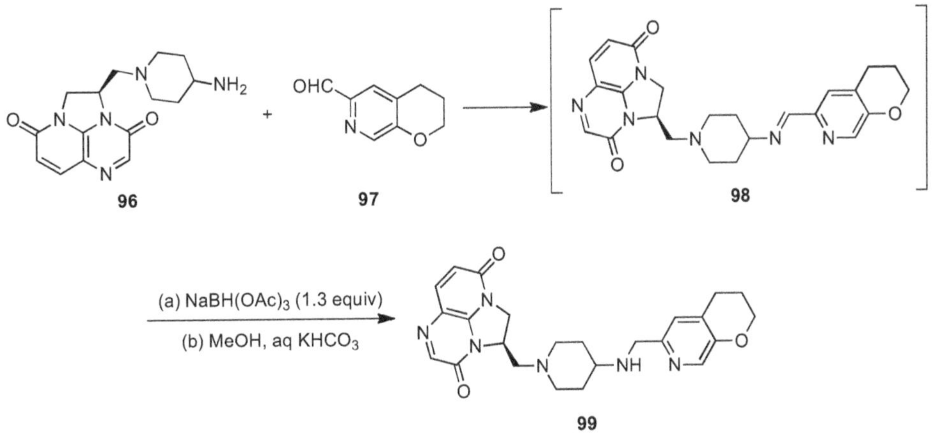

SCHEME 8.8

A near-miss incident was reported[76] from a reductive amination of aldehyde **97** with amine **96** in a pilot plant operation (Scheme 8.8), where the drummed, aqueous waste stream from the STAB reduction of imine **98** generated unexpected internal pressure, causing significant deformation of the waste drum during the temporary storage. It was found that the gas was carbon dioxide generated slowly during the storage from a reaction of a boric acid by-product with excess potassium bicarbonate (Equation 8.46).

$$H_3BO_3 + KHCO_3 \longrightarrow K[H_4BO_4] + CO_2 \qquad (8.46)$$

8.2.2 REDUCTION WITH BORANE

Borane (BH$_3$) is a colorless gas that exists as a gaseous dimer. Diborane has a very low auto ignition temperature of about 38–52 °C and a wide explosive range in air (0.8–90% vol).[77] Diborane is a useful reagent with many applications. When scaling up borane reactions, a scrubbing system is recommended in order to trap the escaping diborane. Methanol (or aqueous sodium hydroxide solution[71]) is commonly used to destroy the diborane to hydrogen and methyl borate.

Dialkylborane, trialkylborane, and trialkylborohydride derivatives are pyrophoric in pure form or at higher concentrations (>15 wt%). Trialkylboranes are resistant to hydrolysis under acidic or basic aqueous conditions and will remain in the organic layer. Therefore, the trialkylborane or other alkylboranes in a reaction mixture must be oxidized prior to air exposure to avoid possible spontaneous combustion caused by rapid exothermic air oxidations. Hydrogen peroxide is usually employed to oxidize alkylboranes. In practice, the addition of hydrogen peroxide should be slow while maintaining nitrogen flow. Normally, oxidation is carried out in the presence of sodium hydroxide. Alternatively, the use of hydrogen peroxide under acidic conditions allows the oxidation of organoboranes at ambient temperatures.

Borane–Lewis base complexes are widely used as borane sources in organic synthesis due to their ease to handle compared to diborane. Such complexes include $BH_3 \cdot THF$ (by dissolving diborane in THF), $BH_3 \cdot DMS$ (dimethyl sulfide), etc. Both $BH_3 \cdot THF$ and $BH_3 \cdot DMS$ are commercially available in solution. $BH_3 \cdot DMS$ is more stable than $BH_3 \cdot THF$ but has an unpleasant odor. Borane can also form complexes with amines, that is, $BH_3 \cdot tert$-butylamine and $BH_3 \cdot N,N$-diethylaniline ($BH_3 \cdot DEAN$).

8.2.2.1 $BH_3 \cdot THF$ Complex

$BH_3 \cdot THF$ complex is a versatile reagent with a multitude of applications in organic and inorganic syntheses.[78] However, the $BH_3 \cdot THF$ complex is unstable and must be transported and stored below 8 °C. The thermal decomposition of $BH_3 \cdot THF$ complex occurs at 10–50 °C to generate butyl borate via cleavage of the THF ring (Equation 8.47).[79]

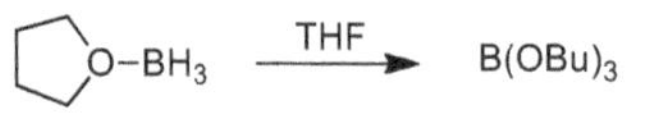

$$\text{(structure)} \quad O-BH_3 \xrightarrow{\text{THF}} B(OBu)_3$$

(8.47)

The $BH_3 \cdot THF$ complex can liberate diborane from the solution at temperatures above 50 °C (Equation 8.48).

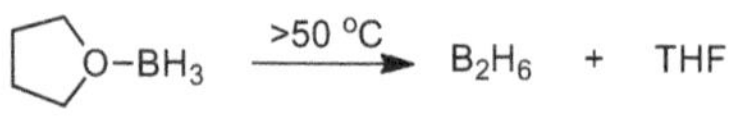

$$\text{(structure)} \quad O-BH_3 \xrightarrow{>50\ ^\circ C} B_2H_6 \ + \ THF$$

(8.48)

Because of these decomposition reactions, it is recommended to perform the $BH_3 \cdot THF$ complex reaction at temperatures below 35 °C. Furthermore, the addition sequence is equally important: the addition of $BH_3 \cdot THF$ complex to the substrate, instead of vice versa, is recommended to decrease the thermal exposure of $BH_3 \cdot THF$ complex and eliminate over-reduction side products. When employing an excess of $BH_3 \cdot THF$ complex, the subsequent workup will generate hydrogen gas from the quenching reaction with methanol, which is a safety concern during scale-up. To address the safety concerns during scale-up, additives are frequently used to minimize the equivalents of the $BH_3 \cdot THF$ complex required for a given reaction.

8.2.2.1.1   Hydroboration

Due to the chelating ability of amine with borane, excess of $BH_3 \cdot THF$ is required when conducting borane reduction of amino group-containing compounds. For instance, hydroboration of alkene **100** required 2.2 equiv of borane to achieve consistent results (Scheme 8.9).[80] In order to reduce the quantity of borane, boron trifluoride diethyl etherate ($BF_3 \cdot EE$) was utilized to complex the amino group in the molecule **100**. The addition of 1.2 equiv of $BF_3 \cdot EE$ allowed the hydroboration to be carried out with 1.4 equiv of borane, resulting in less hydrogen evolution during the quenching.

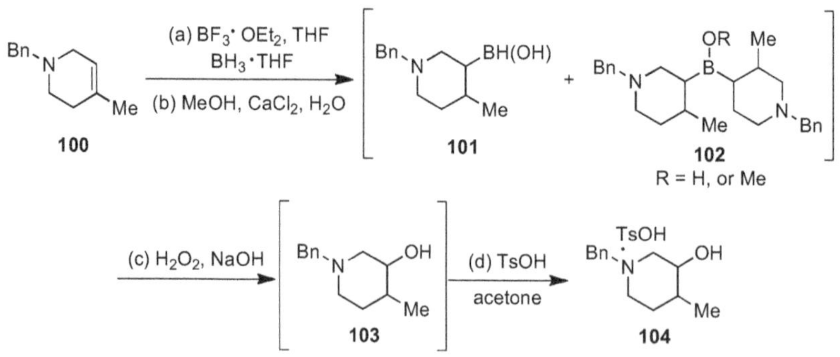

SCHEME 8.9

Procedure

Stage (a) Hydroboration
- Charge, to a 640-L reactor, 27.78 kg (123 mol) of **100** and 62 L of THF.
- Cool to 5 °C.
- Charge 22.4 kg (158 mol, 1.3 equiv) of $BF_3 \cdot OEt_2$ at 5–10 °C.
- Charge 171 L (1 M in THF, 1.4 equiv) of $BH_3 \cdot THF$ at 8–13 °C.
- Stir the batch at 20 °C for 14.5 h.

Stage (b) Quenching
- Cool to 7 °C.
- Add 37 L of MeOH over 2 h at 7–12 °C.
- Warm the reaction mixture to 19 °C.
- Add 40 L of 10% $CaCl_2$ in 0.2 N HCl (pH was 2 then).
- Stir at 20 °C for 1 h and then at 41 °C for 1 h.
- Distill off 135 L of solvent at 40 °C.

Stage (c)
- Cool to 7 °C.
- Charge 28.4 kg (146 mol, 1.2 equiv) of 17.5% H_2O_2 at 7–12 °C.
- Stir at 19–22 °C for 68 h.
- Add 13 L of 10% Na_2SO_3 and 50 L of MTBE.
- Add 37 L of 30% NaOH to pH 12 at 9–18 °C.
- Separate layers and extract the aqueous layer with MTBE (2×50 L).
- Wash the organic layer with 25 L of water and 25 L of saturated NaCl solution.
- Distill off 155 L of solvent at 45 °C.

Stage (d)
- Prepare, in a 100-L reactor, a solution of TsOH by dissolving $TsOH \cdot H_2O$ (25.5 kg, 134 mol, 1.1 equiv) in 61 L of acetone.
- Charge seed crystals (0.1 g) in acetone (0.5 L).
- Charge this solution to the solution of a free base at 44–47 °C.
- Stir the resulting white suspension at 5 to –2 °C for 1.5 h.
- Filter and wash with acetone (2×18 L).
- Dry at 40 °C.
- Yield: 41.37 kg (88%).

REMARKS

(a) The calcium chloride was added to trap any fluoride liberated from the quench of BF_3.
(b) $BH_3 \cdot THF$ has poor selectivity for amide versus nitrile reduction.
(c) An explosion from a 400-L cylinder of 2 M borane-THF during storage at Pfizer's Groton, CT, research campus (June 25, 2002) resulted in injuries and property damage. As a result of this incident, 2 M borane-THF is classified as a hazard class 4.1 self-reactive hazardous material with a self-accelerating decomposition temperature (SADT) of 40 °C. The primary decomposition pathway of borane-THF complexes at ambient temperatures proceeds via THF ring-opening mechanism, ultimately forming tributylborate.

8.2.2.1.2 Reduction of Imide

The formation of borane–amine complexes during the reduction of amides, imides, and imines not only requires excess $BH_3 \cdot THF$ complex but also needs an extra step to release the amine product.

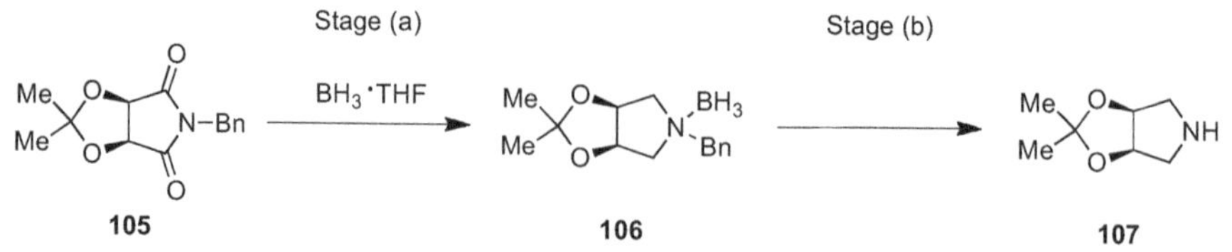

Stage (a)   Stage (b)

105   106   107

SCHEME 8.10

For instance, the borane reduction of imide **105** using commercial BH$_3$·THF complex led to a stable borane–amine adduct **106** (Scheme 8.10).[81] The cleavage of such complex often requires harsh reaction conditions, most commonly performed by refluxing in aqueous HCl.

It was found that treatment of **106** with Pd (or Raney Ni) not only affected the hydrogenolysis of the *N*-benzyl moiety but also catalyzed the decomplexation of the borane–amine adduct.

Procedure

Stage (a)
- Charge 87.5 mL (87.5 mmol) of 1 M BH$_3$·THF solution to an ice-cold solution of **105** (6.53 g, 25.0 mmol) in anhydrous THF (25 mL).
- Warm the resulting mixture to room temperature and stir for 2.5 h.
- Add MeOH (25 mL).
- Distill to dryness.
- Slurry the residual crystalline solid in a mixture of EtOAc/hexane (1:1, 20 mL).
- Cool to 0 °C and filter to give **106** (5.01 g, 83%).

Stage (b)
- Charge a solution of **106** (3.0 g, 12.0 mmol) in MeOH (15 mL) to a suspension of 10% Pd/C (0.6 g, 50% wet) in MeOH (15 mL).
- Seal the reactor and stir at room temperature for 12 h.
- Filter through cotton.
- Concentrate under a slight vacuum (200 mmHg).
- Purify the residue by flash chromatography.
- Yield of **107**: 1.63 g (94%).

REMARKS

(a) It is noteworthy that the borane–amine complex **106** is a crystalline material, which provides a purification opportunity.
(b) This palladium (or Raney nickel)-catalyzed debenzylation was conducted without external hydrogen source as the methanolysis of borane-amine adducts produced hydrogen (Equation 8.49):

$$R_3N \cdot BH_3 + MeOH \longrightarrow R_3N + B(OMe)_3 + H_2 \qquad (8.49)$$

8.2.2.2 BH$_3$·DMS Complex

Borane·dimethyl sulfide (BH$_3$·DMS) complex is another extremely useful borane reagent. Although the reactivity of this complex falls slightly behind BH$_3$·THF complex, it offers higher concentration (~10 M) and thermal stability. Reduction with BH$_3$·DMS may be performed at a temperature of up to 70 °C. BH$_3$·DMS complex reduces the amide selectively with minimal impact on the nitrile by charging the reagent at 0–5 °C and warming it to room temperature. The drawbacks of BH$_3$·DMS complex are the toxicity and volatility (bp 38 °C) of DMS, which limits its applications on large scale.

REMARK
Generally, two scrubbing methods are available for safe and simple ways to mitigate DMS odor: the use of activated carbon in a drum to absorb DMS and the oxidation with sodium hypochlorite. If activated carbon is chosen, any diborane in the vent stream must be destroyed before venting through the carbon. Due to the high surface area and significant water content of activated carbon, any vented diborane will react, generating localized hot spots and potentially igniting the evolving hydrogen.

8.2.2.3 BH₃·Amine Complex

8.2.2.3.1 BH₃·tBuNH₂ Complex

Borane-*tert*-butylamine (BH$_3$·*t*BuNH$_2$) complex was used in the reduction of enamine **109** to eliminate potentially hazardous hydrogen evolution when using hydride reducing agent. In addition to stabilizing borane, the *tert*-butylamine in the BH$_3$·*t*BuNH$_2$ complex moderated the acidity of the reaction mixture (Scheme 8.11).[82]

After dilution with water (exothermic) the reaction mixture was quenched with 25% aqueous KOH solution.

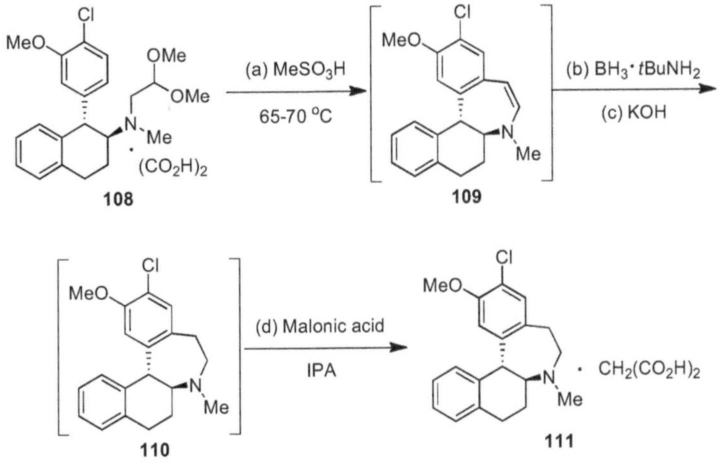

SCHEME 8.11

Procedure

Stage (a)
- Charge 100 g (0.208 mol) of **108** in portions to methanesulfonic acid (160 mL, 2.47 mol, 11.9 equiv) at 45–50 °C.
- Heat the resulting suspension at 55–60 °C for 2 h.

Stage (b)
- Cool the dark red-brown reaction mixture to 0 °C.
- Add MTBE (160 mL).
- Charge a slurry of BH$_3$·*t*BuNH$_2$ (16 g, 0.813 mol, 3.9 equiv) in MTBE (80 mL) at 10–15 °C.

Stage (c)
- Cool to 0 °C.
- Add water (320 mL), followed by adding 25% aqueous KOH solution (640 mL).

- Separate layers.
- Extract the aqueous layer with MTBE (200 mL).
- Wash the combined MTBE layers with water (160 mL).
- Distill atmospherically under N_2 to 200 mL.
- Add IPA (320 mL).
- Distill to 300 mL.

Stage (d)
- Charge a warm solution (45 °C) of malonic acid (23 g, 0.22 mol) in IPA (200 mL) to the stirred free base solution of **110**.
- Cool the resulting slurry to 0 °C and stir for 1 h.
- Filter, wash with cold IPA (120 mL) and cold MTBE (160 mL), and dry at 50 °C under vacuum.
- Yield of **111**: 83 g (92%).

8.2.2.3.2   BH₃·DEAN Complex

Borane·*N,N*-diethylaniline (BH₃·DEAN)[83] complex, derived from a less basic amine, engages in the same types of reaction as BH₃·THF and BH₃·DMS. It can be hydrolyzed by water or methanol after the reaction. This compound offers excellent long-term stability at ambient temperatures. Reduction with BH₃·DEAN can be performed at temperatures up to 100 °C.

Enantioselective reduction of ketone **112** at 40 °C with 1 equiv of BH₃·DEAN in the presence of (*S*)-(–)-2-Methyl-CBS-oxazaborolidine (Corey–Bakshi–Shibata) (0.5 mol%) catalyst in MTBE gave alcohol **113** in 98% yield and 98.9% ee on a multi-kg scale (Equation 8.50).[84]

$$(8.50)$$

Procedure

- Charge a solution of **112** (5.042 kg, 26.46 mol) in MTBE (16 L) to a solution of (S)-MeCBS (131 mL of 1.0 M solutions in toluene, 0.131 mol, 0.5 mol%) and BH₃·DEAN (4.26 kg, 26.1 mol) in MTBE (10 L) at 40 °C over 10 h.
- Stir the resulting solution at 40 °C for 1 h.
- Cool to 18 °C and stir overnight.
- Add MeOH (2.3 L) at <20 °C over 1 h.
- Stir the resulting solution for 30 min.
- Add 2.5 N aqueous HCl (31.5 L) at 22–25 °C over 30 min.
- Stir the resulting mixture for 30 min before layer separation.
- Wash the organic layer with saturated NaCl.
- Distill under reduced pressure.
- Yield of **113**: 5.040 kg (99%).

As a practical and safe borane reagent, BH₃·DEAN was employed in the enantioselective reduction of ketone **114** on a kg scale (Equation 8.51).[85]

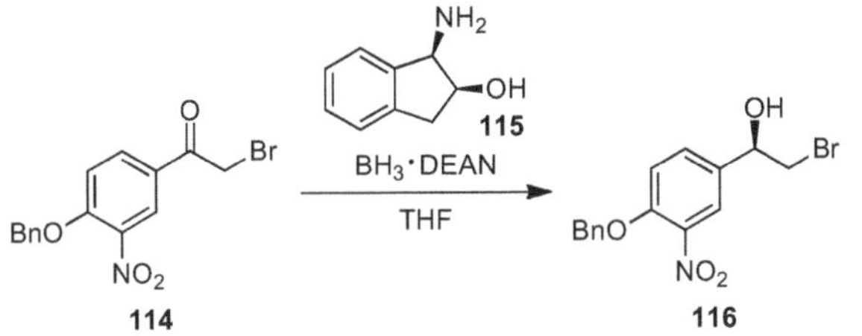

(8.51)

Procedure

- Charge 50 g of **115** followed by 5.75 kg of THF to a 50-L reactor.
- Charge 0.81 kg of BH$_3$·DEAN at 5–20 °C.
- Stir at 5–15 °C for 30 min, then cool to 0–5 °C.
- Charge a pre-prepared solution of **114** (2.3 kg) in 18.4 kg of THF over 2 h.
- Charge 1.8 kg of acetone at 0–15 °C.
- Warm the solution to 20–25 °C and stir for 30 min.
- Distill under reduced pressure to approximately 7 L.
- Add toluene (13.8 kg).
- Wash with 2.8 kg of DI water.
- Distill under reduced pressure to approximately 7 L.
- Cool to 25 °C.
- Add 5.3 g of seeds of **116**.
- Cool to 19 °C.
- Add 3.2 kg of heptane.
- Filter and wash with heptane.
- Dry under vacuum.
- Yield: 1.84 kg (80% yield, 99:1 er).

8.2.2.3.3 BH$_3$·Pyridine Complex

Methylation of pyrrole ring in **117** was accomplished through a three-step sequence involving a Mannich reaction, *N*-methylation, and reduction of quaternary salts **118** of Mannich base with sodium cyanoborohydride.[86] Despite the usefulness of sodium cyanoborohydride, it is not acceptable for industrial synthesis because of the evolution of hydrogen cyanide gas during the reaction.

As an alternative reagent, BH$_3$·pyridine complex was used to reduce successfully quaternary ammonium **118** to the corresponding methyl group in **119** (Scheme 8.12)[87] on large scale.

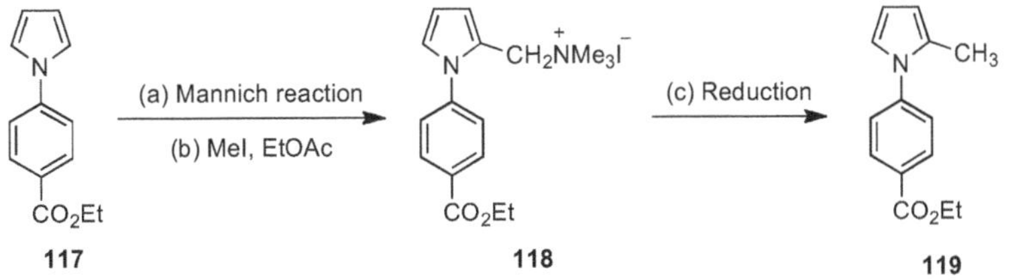

SCHEME 8.12

Procedure

Stage (c)

- Mix 130 kg (314 mol) of **118** with 64.2 kg (691 mol) of BH$_3$–pyridine complex in 400 L of *N,N*'-dimethylimidazolidinone (DMI).

- Heat at 104–113 °C for 2 h.
- Cool to 30 °C.
- Add *n*-heptane (650 L) followed by water (390 L).
- Separate the layers and extract the aqueous layer with *n*-heptane (390 L).
- Wash the combined organic layers with 12% formalin (390 L) to remove (or destroy) BH_3–pyridine complex followed by water (390 L).
- Distill under reduced pressure to give **119** as an oil.
- Yield: 46.1 kg (64%).

REMARKS

(a) BH_3·pyridine complex can be prepared from the reaction of pyridinium chloride, which was easily prepared from pyridine and hydrogen chloride, with sodium borohydride.
(b) A safe and scalable procedure for the preparation of α-picoline·borane from sodium mono-acyloxyborohydrides and α-picoline was reported.[88]

8.2.3 REDUCTION WITH LITHIUM ALUMINUM HYDRIDE

Lithium aluminum hydride (LAH) is used for the reduction of esters, amides, carboxylic acids, and ketones. Compared to $NaBH_4$, LAH is more reactive than $NaBH_4$ and reductions with LAH are less selective. Occasionally, amide reduction with LAH may accompany the cleavage of the amide bond. Therefore, in order to suppress side reactions, additives, such as LiI, are used together with LAH.

In addition, LiI was used in the selective LAH reduction of cyclopentenone **120** (Scheme 8.13).[89] The addition of LiI reduced the 1,4-hydride addition product **123** whose subsequent tautomerization/1,2-reduction led to over-reduction side product **122**. As a result, the reduction of **120** with LAH/LiI (2.1 equiv) was carried out at –30 °C furnishing the desired alcohol **121** in a 39:1 ratio of *cis:trans* with only trace amounts of **122**.

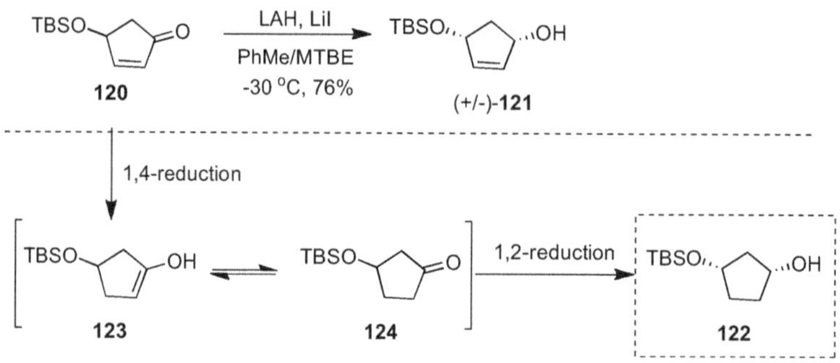

SCHEME 8.13

Procedure

- Mix 8 kg (59.7 mol) of LiI with 600 g (15.8 mol) of LAH in 48 L of toluene under N_2 and cool to –30 °C.
- Charge a solution of **120** (6 kg, 28.25 mol) in MTBE (28 L) while maintaining the batch temperature <–20 °C.
- Stir at –30 °C for 3 h.
- Quench with saturated NH_4Cl (20 L) while maintaining the batch temperature <10 °C.
- Filter off the aluminum salts and wash with toluene (6 L).
- Separate layers.

- Extract the aqueous layer with toluene (6 L).
- Distill the combined organic layers at 50 °C/100 Torr.
- Dissolve the resulting residue in ethyl acetate (8 L) and dry over $MgSO_4$ (500 g).
- Filter and distill the filtrate at 50 °C/100 Torr.
- Yield: 6.4 kg (76%).

REMARKS

(a) To minimize exposure of the flammable LAH and anhydrous LiI to air during the handling, these raw materials were purchased in pre-weighed toluene-soluble bags.
(b) Commercial LAH solutions in various solvents (for example, THF, 2-methoxyethyl ether, DME) are safer and more practical alternatives for large-scale manufacturing.

NOTES

1. United Nations; *Recommendations on the Transport of Dangerous Goods Manual of Tests and Criteria*, 5th ed.; United Nations, New York, 2009; pp 29–30.
2. Lothrop, W. C.; Handrick, G. R. *Chem. Rev.* **1949**, *44*, 419.
3. Peer, M. *Spec. Chem.* **1998**, *18*, 256.
4. Sperry, J. B.; Azuma, M.; Stone, S. *Org. Process Res. Dev.* **2021**, *25*, 212.
5. Prasad, J. S.; Vu, T.; Totleben, M. J.; Crispino, G. A.; Kacsur, D. J.; Swaminathan, S.; Thornton, J. E.; Fritz, A.; Singh, A. K. *Org. Process Res. Dev.* **2003**, *7*, 821.
6. Brenek, S. J.; Caron, S.; Chisowa, E.; Delude, M. P.; Drexler, M. T.; Ewing, M. D.; Handfield, R. E.; Ide, N. D.; Nadkarni, D. V.; Nelson, J. D.; Olivier, M.; Perfect, H. H.; Phillips, J. E.; Teixeira, J. J.; Weekly, R. M.; Zelina, J. P. *Org. Process Res. Dev.* **2012**, *16*, 1348.
7. Kubota, A.; Takeuchi, H. *Org. Process Res. Dev.* **2004**, *8*, 1076.
8. Zhang, X.; Hu, A.; Pan, C.; Zhao, Q.; Wang, X.; Lu, J. *Org. Process Res. Dev.* **2013**, *17*, 1591.
9. Williams, D. R.; Klein, J. C.; Kopel, L. C.; Nguyen, N.; Tantillo, D. J. *Org. Lett.* **2016**, *18*, 424.
10. (a) Reed, K. L.; Gupton, J. T.; Solarz, T. L. *Synth. Commun.* **1989**, *19*, 3579. (b) Straub, T. S. *Tetrahedron Lett.* **1995**, *36*, 663.
11. Lehmann, T. E.; Kuhn, O.; Krüger, J. *Org. Process Res. Dev.* **2003**, *7*, 913.
12. Roydhouse, M. D.; Ghaini, A.; Constantinou, A.; Cantu-Perez, A.; Motherwell, W. B.; Gavriilidis, A. *Org. Process Res. Dev.* **2011**, *15*, 989.
13. Caille, J.-C.; Govindan, C. K.; Junga, H.; Lalonde, J.; Yao, Y. *Org. Process Res. Dev.* **2002**, *6*, 471.
14. Coutts, L. D.; Geiss, W. B.; Gregg, B. T.; Helle, M. A.; King, C.-H. R.; Itov, Z.; Mateo, M. E.; Meckler, H.; Zettler, M. W.; Knutson, J. C. *Org. Process Res. Dev.* **2002**, *6*, 246.
15. Cabaj, J. E.; Kairys, D.; Benson, T. R. *Org. Process Res. Dev.* **2007**, *11*, 378.
16. Roth, G. P.; Landi, J. J., Jr.; Salvagno, A. M.; Müller-Bötticher, H. *Org. Process Res. Dev.* **1997**, *1*, 331.
17. Karlsson, S.; Gardelli, C.; Lindhagen, M.; Nikitidis, G.; Svensson, T. *Org. Process Res. Dev.* **2018**, *22*, 1174.
18. Leanna, M. R.; Sowin, T. J.; Morton, H. E. *Tetrahedron Lett.* **1992**, *33*, 5029.
19. March, J. *Advanced Organic Chemistry*, 4th ed.; John Wiley & Sons, Inc., New York, 1992; pp 1193–1195.
20. Luca, L. D.; Giacomelli, G.; Porcheddu, A. *J. Org. Chem.* **2001**, *66*, 7907.
21. Nguyen, T. V.; Hall, M. *Tetrahedron Lett.* **2014**, *55*, 6895.
22. (a) Ohsugi, S.; Nishide, K.; Oono, K.; Okuyama, K.; Fudesaka, M.; Kodama, S.; Node, M. *Tetrahedron* **2003**, *59*, 8393. (b) Nishide, K.; Ohsugi, S.; Fudesaka, M.; Kodama, S.; Node, M. *Tetrahedron Lett.* **2002**, *43*, 5177.
23. Liu, C.; Ng, J. S.; Behling, J. R.; Yen, C. H.; Campbell, A. L.; Fuzail, K. S.; Yonan, E. E.; Mehrotra, D. V. *Org. Process Res. Dev.* **1997**, *1*, 45.
24. (a) Nicolaou, K. C.; Peng, X.-S.; Sun, Y.-P.; Polet, D.; Zou, B.; Lim, C. S.; David Y.-K. Chen, D. Y.-K. *J. Am. Chem. Soc.* **2009**, *131*, 10587. (b) Berliner, M. A.; Dubant, S. P. A.; Makowski, T.; Ng, Karl, Sitter, B.; Wager, C.; Zhang, Y. *Org. Process Res. Dev.* **2011**, *15*, 1052. (c) Evans, P. A.; Murthy, V. S.; Roseman, J. D.; Rheingold, A. L. *Angew. Chem. Int. Ed.* **1999**, *38*, 3175. (d) Waser, M.; Moher, E. D.; Borders, S. S. K.; Hansen, M. M.; Hoard, D. W.; Laurila, M. E.; LeTourneau, M. E.; Miller, R. D.;

Phillips, M. L.; Sullivan, K. A.; Ward, J. A.; Xie, C.; Bye, C. A.; Leitner, T.; Herzog-Krimbacher, B.; Kordian, M.; Müllner, M. *Org. Process Res. Dev.* **2011**, *15*, 1266. (e) Zanka, A.; Itoh, N.; Kuroda, S. *Org. Process Res. Dev.* **1999**, *3*, 394.

25. Chen, L.; Lee, S.; Renner, M.; Tian, Q.; Nayyar, N. *Org. Process Res. Dev.* **2006**, *10*, 163.

26. Bernhardson, D.; Brandt, T. A.; Hulford, C. A.; Lehner, R. S.; Preston, B. R.; Price, K.; Sagal, J. F.; St. Pierre, M. J.; Thompson, P. H.; Thuma, B. *Org. Process Res. Dev.* **2014**, *18*, 57.

27. Tidwell, T. T. *Synthesis* **1990**, *10*, 857.

28. Bowles, P.; Brenek, S. J.; Caron, S.; Do, N. M.; Drexler, M. T.; Duan, S.; Dubé, P.; Hansen, E. C.; Jones, B. P.; Jones, K. N.; Ljubicic, T. A.; Makowski, T. W.; Mustakis, J.; Nelson, J. D.; Olivier, M.; Peng, Z.; Perfect, H. H.; Place, D. W.; Ragan, J. A.; Salisbury, J. J.; Stanchina, C. L.; Vanderplas, B. C.; Webster, M. E.; Weekly, R. M. *Org. Process Res. Dev.* **2014**, *18*, 66.

29. Appell, R. B.; Duguid, R. J. *Org. Process Res. Dev.* **2000**, *4*, 172.

30. (a) Barnett, C.; Wilson, T. M.; Kobierski, M. E. *Org. Process Res. Dev.* **1999**, *3*, 184. (b) Abele, S.; Inauen, R.; Funel, J.-A.; Weller, T. *Org. Process Res. Dev.* **2012**, *16*, 129.

31. (a) Varie, D. L.; Beck, C.; Borders, S. K.; Brady, M. D.; Cronin, J. S.; Ditsworth, T. K.; Hay, D. A.; Hoard, D. W.; Hoying, R. C.; Linder, R. J.; Miller, R. D.; Mohner, E. D.; Remacle, J. R.; Rieck, J. A. III, *Org. Process Res. Dev.* **2007**, *11*, 546. (b) Grongsaard, P.; Bulger, P. G.; Wallace, D. J.; Tan, L.; Chen, Q.; Dolman, S. J.; Nyrop, J.; Hoerrner, R. S.; Weisel, M.; Arredondo, J.; Itoh, T.; Xie, C.; Wen, X.; Zhao, D.; Muzzio, D. J.; Bassan, E. M.; Shultz, C. S. *Org. Process Res. Dev.* **2012**, *16*, 1069.

32. Wennekes, T.; Lang, B.; Leeman, M.; van der Marel, G. A.; Smits, E.; Weber, M.; van Wiltenburg, J.; Wolberg, M.; Aerts, J. M. F. G.; Overkleeft, H. S. *Org. Process Res. Dev.* **2008**, *12*, 414.

33. Kopach, M. E.; Kobierski, M. E.; Coffey, D. S.; Alt, C. A.; Zhang, T.; Borghese, A.; Trankle, W. G.; Roberts, D. J. *Process Res. Dev.* **2010**, *14*, 1229.

34. Goodman, S. N.; Dai, Q.; Wang, J.; Clark, W. M., Jr. *Org. Process Res. Dev.* **2011**, *15*, 123.

35. Hobson, L. A.; Akiti, O.; Deshmukh, S. S.; Harper, S.; Katipally, K.; Lai, C. J.; Livingston, R. C.; Lo, E.; Miller, M. M.; Ramakrishnan, S.; Shen, L.; Spink, J.; Tummala, S.; Wei, C.; Yamamoto, K.; Young, J.; Parsons, R. L.; Jr. *Org. Process Res. Dev.* **2010**, *14*, 441.

36. Davies, A. J.; Scott, J. P.; Bishop, B. C.; Brands, K. M.; Brewer, S. E.; DaSilva, J. O.; Dormer, P. G.; Dolling, U.-H.; Gibb, A. D.; Hammond, D. C.; Lieberman, D. R.; Palucki, M.; Payack, J. F. *J. Org. Chem.* **2007**, *72*, 4864.

37. Lima, H. M.; Sivappa, R.; Yousufuddin, M.; Lovely, C. J. *J. Org. Chem.* **2014**, *79*, 2481.

38. Sasano, Y.; Kogure, N.; Nagasawa, S.; Kasabata, K.; Iwabuchi, Y. *Org. Lett.* **2018**, *20*, 6104.

39. Lopez, F. C.; Shankar, A.; Thompson, M.; Shealy, B.; Locklear, D.; Rawalpally, T.; Cleary, T.; Gagliardi, C. *Org. Process Res. Dev.* **2005**, *9*, 1003.

40. Hasegawa, T.; Kawanaka, Y.; Kasamatsu, E.; Ohta, C.; Nakabayashi, K.; Okamoto, M.; Hamano, M.; Takahashi, K.; Ohuchida, S.; Hamada, Y. *Org. Process Res. Dev.* **2005**, *9*, 774.

41. Cooper, M. S.; Heaney, H.; Newbold, A. J.; Sanderson, W. R. *Synlett* **1990**, *9*, 533.

42. (a) Ravikumar, K. S.; Zhang, Y. M.; Bégué, J.-P.; Bonnet-Delpon, D. *Eur. J. Org. Chem.* **1998**, *12*, 2937. (b) Ravikumar, K. S.; Bégué, J.-P.; Bonnet-Delpon, D. *Tetrahedron Lett.* **1998**, *39*, 3141.

43. (a) Shokrolahi, A.; Zali, A.; Pouretedal, H. R.; Mahdavi, M. *Catal. Commun.* **2008**, *9*, 859. (b) Bahrami, K. In *Regio- and Stereocontrolled Oxidations and Reductions*; Roberts, S. M., Whitall, J., Eds.; Catalysts for Fine Chemical Synthesis, Vol. 5; Wiley, Chichester, England; Hoboken, NJ, 2007; pp 283–287. (c) Shaabani, A.; Rezayan, A. H. *Catal. Commun.* **2007**, *8*, 1112. (d) Velusamy, S.; Kumar, A.; Saini, R.; Punniyamurthy, T. *Tetrahedron Lett.* **2005**, *46*, 3819.

44. (a) Ward, R. S.; Roberts, D. W.; Diaper, R. L.; Richard, L. *Sulfur Lett.* **2000**, *23*, 139. (b) Kropp, P. J.; Breton, G. W.; Fields, J. D.; Tung, J. C.; Loomis, B. R. *J. Am. Chem. Soc.* **2000**, *122*, 4280. (c) Llauger, L.; He, H.; Chiosis, G. *Tetrahedron Lett.* **2004**, *5*, 9549. (d) Magnus, N. A.; Braden, T. M.; Buser, J. Y.; DeBaillie, A. C.; Heath, P. C.; Ley, C. P.; Remacle, J. R.; Varie, D. L.; Wilson, T. M.; *Org. Process Res. Dev.* **2012**, *16*, 830.

45. Candiani, I.; D'Arasmo, G.; Heidempergher, F.; Tomasi, A. *Org. Process Res. Dev.* **2009**, *13*, 456.

46. Yamagami, T.; Moriyama, N.; Kyuhara, M.; Moroda, A.; Uemura, T.; Matsumae, H.; Moritani, Y.; Inoue, I. *Org. Process Res. Dev.* **2014**, *18*, 437.

47. Davies, I. W.; Marcoux, J.-F.; Corley, E. G.; Journet, M.; Cai, D.-W.; Palucki, M.; Wu, J.; Larsen, R. D.; Rossen, K.; Pye, P. J.; DiMichele, L.; Dormer, P.; Reider, P. J. *J. Org. Chem.* **2000**, *65*, 8415.

48. Grosjean, C.; Henderson, A. P.; Hérault, D.; Ilyashenko, G.; Knowles, J. P.; Whiting, A.; Wright, A. R. *Org. Process Res. Dev.* **2009**, *13*, 434.

49. Naganathan, S.; Andersen, D. L.; Andersen, N. G.; Lau, S.; Lohse, A.; Sørensen, D. *Org. Process Res. Dev.* **2015**, *19*, 721.

50. Meckler, H.; Herr, R. J. *Org. Process Res. Dev.* **2012**, *16*, 550.

51. Trankle, W. G.; Kopach, M. E. *Org. Process Res. Dev.* **2007**, *11*, 913.

52. Fox, R. J.; Markwalter, C. E.; Lawler, M.; Zhu, K.; Albrecht, J.; Payack, J.; Eastgate, M. D. *Org. Process Res. Dev.* **2017**, *21*, 754.

53. Linghu, X.; Wong, N.; Iding, H.; Jost, V.; Zhang, H.; Koenig, S. G.; Sowell, C. G.; Gosselin, F. *Org. Process Res. Dev.* **2017**, *21*, 387.

54. Prakash, G. K. S.; Mathew, T.; Panja, C.; Olah, G. A. *J. Org. Chem.* **2007**, *72*, 5847.

55. Sammons, M.; Jennings, S. M.; Herr, M.; Hulford, C. A.; Wei, L.; Hallissey, J. F.; Kiser, E. J.; Wright, S. W.; Piotrowski, D. W. *Org. Process Res. Dev.* **2013**, *17*, 934.

56. Kress, T. J. *J. Org. Chem.* **1985**, *50*, 3073.

57. Dietsche, T. J.; Gorman, D. B.; Orvik, J. A.; Roth, G. A.; Shiang, W. R. *Org. Process Res. Dev.* **2000**, *4*, 275.

58. Dietsche, T. J.; Powers, B. *Chem. Eng. News* **1995**, *73*, 4 (July 10, 1995).

59. Zheng, B.; Fox, R. J.; Sugiyama, M.; Fritz, A.; Eastgate, M. D. *Org. Process Res. Dev.* **2014**, *18*, 636.

60. Dugger, R. W.; Ragan, J. A.; Ripin, D. H. B. *Org. Process Res. Dev.* **2005**, *9*, 253.

61. Shimizu, S.; Osata, H.; Imamura, Y.; Satou, Y. *Org. Process Res. Dev.* **2010**, *14*, 1518.

62. Busacca, C. A.; Cerreta, M.; Dong, Y.; Eriksson, M. C.; Farina, V.; Feng, X.; Kim, J.-Y.; Lorenz, J. C.; Sarvestani, M.; Simpson, R.; Varsolona, R.; Vitous, J.; Campbell, S. J.; Davis, M. S.; Jones, P.-J.; Norwood, D.; Qiu, F.; Beaulieu, P. L.; Duceppe, J.-S.; Haché, B.; Brong, J.; Chiu, F.-T.; Curtis, T.; Kelley, J.; Lo, Y. S.; Powner, T. H. *Org. Process Res. Dev.* **2008**, *12*, 603.

63. (a) Kanth, J. V. B.; Brown, H. C. *Inorg. Chem.* **2000**, *39*, 1795. (b) Wang, Y.; Papamichelakis, M.; Chew, W.; Sellstedt, J.; Noureldin, R.; Tadayon, S.; Daigneault, S. *Org. Process Res. Dev.* **2008**, *12*, 1253. (c) Vo, L.; Ciula, J.; Cooding, O. W. *Org. Process Res. Dev.* **2003**, *7*, 514. (d) Abiko, A.; Masamune, S. *Tetrahedron Lett.* **1992**, *33*, 5717. (e) Giannis, A.; Sandhoff, K. *Angew. Chem., Int. Ed. Eng.* **1989**, *28*, 2, 218. (f) Freeguard, G. F.; Long, L. H. *Chem. Ind.* **1965**, 471. (g) Kanth, J. V. B.; Periasamy, M. *J. Org. Chem.* **1991**, *56*, 5964. (h) McKennon, M. J.; Meyers, A. I. *J. Org. Chem.* **1993**, *58*, 3568.

64. (a) Braish, T. F. *Org. Process Res. Dev.* **2009**, *13*, 336. (b) Kamath, V. P.; Juarez-Brambila, J. J.; Morris, C. B.; Winslow, C. D.; Morris, P. E. Jr. *Org. Process Res. Dev.* **2009**, *13*, 928. (c) Campeau, L.-C.; Dolman, S. J.; Gauvreau, D.; Corley, E.; Liu, J.; Guidry, E. N.; Ouellet, S. G.; Steinhuebel, Weisel, M.; O'Shea, P. D. *Org. Process Res. Dev.* **2011**, *15*, 1138. (d) Molinaro, C.; Shultz, S.; Roy, A.; Lau, S.; Trinh, T.; Angelaud, R.; O'Shea, P. D.; Abele, S.; Cameron, M.; Corley, E.; Funel. J.-A.; Steinhuebel, D.; Weisel, M.; Krska, S. *J. Org. Chem.* **2011**, *76*, 1062.

65. Hida, T.; Mitsumori, S.; Honma, T.; Hiramatsu, Y.; Hashizume, H.; Okada, T.; Kakinuma, M.; Kawata, K.; Oda, K.; Hasegawa, A.; Masui, T.; Nogusa, H. *Org. Process Res. Dev.* **2009**, *13*, 1413.

66. Mattei, P.; Moine, G.; Püntener, K.; Schmid, R. *Org. Process Res. Dev.* **2011**, *15*, 353.

67. Soai, K.; Ookawa, A. *J. Org. Chem.* **1986**, *51*, 4000.

68. Yamakawa, T.; Masaki, M.; Nohira, H. *Bull. Chem. Soc. Jpn.* **1991**, *64*, 2730.

69. Fujino, K.; Takami, H.; Atsumi, T.; Ogasa, T.; Mohri, S.; Kasai, M. *Org. Process Res. Dev.* **2001**, *5*, 426.

70. Girardin, M.; Ouellet, S. G.; Gauvreau, D.; Moore, J. C.; Hughes, G.; Devine, P. N.; O'Shea, P. D.; Campeau, L.-C. *Org. Process Res. Dev.* **2013**, *17*, 61.

71. Guercio, G.; Manzo, A. M.; Goodyear, M.; Bacchi, S.; Curti, S.; Provera, S. *Org. Process Res. Dev.* **2009**, *13*, 489.

72. Tao, Y.; Widlicka, D. W.; Hill, P. D.; Couturier, M.; Young, G. R. *Org. Process Res. Dev.* **2012**, *16*, 1805.

73. Ishimoto, K.; Yamaguchi, K.; Nishimoto, A.; Murabayashi, M.; Ikemoto, T. *Org. Process Res. Dev.* **2017**, *21*, 2001.

74. Burks, J. E., Jr.; Espinosa, L.; Labell, E. S.; McGill, J. M.; Ritter, A. R.; Speakman, J. L.; Williams, M.; Bradley, D. A.; Haehl, M. G.; Schmid, C. R. *Org. Process Res. Dev.* **1997**, *1*, 198.

75. Golden, M.; Legg, D.; Milne, D.; Bharadwaj M., A.; Deepthi, K.; Gopal, M.; Dokka, N.; Nambiar, S.; Ramachandra, P.; Santhosh, U.; Sharma, P.; Sridharan, R.; Sulur, M.; Linderberg, M.; Nilsson, A.; Sohlberg, R.; Kremers, J.; Oliver, S.; Patra, D. *Org. Process Res. Dev.* **2016**, *20*, 675.

76. Dixon, F., Jr.; Flanagan, R. C. *Org. Process Res. Dev.* **2020**, *24*, 1063.

77. (a) Urben, P. G., Ed. *Bretherick's Handbook of Reactive Chemical Hazards*, 6th ed.; Butterworth-Heinemann, Boston, MA, 1999; p 1937. (b) Chung, C. K.; Bulger, P. G.; Kosjek, B.; Belyk, K. M.; Rivera, N.; Scott, M. E.; Humphrey, G. R.; Limanto, J.; Bachert, D. C.; Emerson, K. M. *Org. Process Res. Dev.* **2014**, *18*, 215. (c) A recent research showed that the auto ignition temperature of diborane is 136 °C-139 °C. Monteriro, A. M.; Flanagan, R. C. *Org. Process Res. Dev.* **2017**, *21*, 241.

78. Potyen, M.; Josyula, K. V. B.; Schuck, M.; Lu, S.; Gao, P.; Hewett, C. *Org. Process Res. Dev.* **2007**, *11*, 210.

79. am Ende, D. J.; Vogt, P. F. *Org. Process Res. Dev.* **2003**, *7*, 1029.

80. Ripin, D. H. B.; Abele, S.; Cai, W.; Blumenkopf, T.; Casavant, J. M.; Doty, J. L.; Flanagan, M.; Koecher, C.; Laue, K. W.; McCarthy, K.; Meltz, C.; Munchhoff, M.; Pouwer, K.; Shah, B.; Sun, J.; Teixeira, J.; Vries, T.; Whipple, D. A.; Wilcox, G. *Org. Process Res. Dev.* **2003**, *7*, 115.

81. Couturier, M.; Tucker, J. L.; Andresen, B. M.; Dubé, P.; Megri, J. T. *Org. Lett.* **2001**, *3*, 465.

82. Gala, D.; Dahanukar, V. H.; Eckert, J. M.; Lucas, B. S.; Schumacher, D. P.; Zavialov, I. A.; Buholzer, P.; Kubisch, P.; Mergelsberg, I.; Scherer, D. *Org. Process Res. Dev.* **2004**, *8*, 754.

83. Burkhardt, E. R.; Matos, K. *Chem. Rev.* **2006**, *106*, 2617.

84. Chung, J. Y. L.; Cvetovich, R.; Amato, J.; McWilliams, J. C.; Reamer, R.; DiMichele, L. *J. Org. Chem.* **2005**, *70*, 3592.

85. Wilkinson, H. S.; Tanoury, G. J.; Wald, S. A.; Senanayake, C. H. *Org. Process Res. Dev.* **2002**, *6*, 146.

86. Yamada, K.; Itoh, N.; Iwakuma, T. *J. Chem. Soc., Chem. Commun.* **1978**, *24*, 1089.

87. Zanka, A.; Nishiwaki, M.; Morinaga, Y.; Inoue, T. *Org. Process Res. Dev.* **1998**, *2*, 230.

88. Kawase, Y.; Yamagishi, T.; Kutsuma, T.; Huo, Z.; Yamamoto, Y.; Kimura, T.; Nakata, T.; Kataoka, T.; Yokomatsu, T. *Org. Process Res. Dev.* **2012**, *16*, 495.

89. Watson, T. J. N.; Curran, T. T.; Hay, D. A.; Shah, R. S.; Wenstrup, D. L.; Webster, M. E. *Org. Process Res. Dev.* **1998**, *2*, 357.

9 Other Hazardous Reactions

Chemical reactions may produce different types of hazards. One of the most dangerous hazards is an uncontrolled exothermic reaction which may generate high temperatures and pressure, and result in the bursting of the vessel. Many exothermic runaway incidents were caused by under-rated controls and safety backup systems, as in large reactors, the heat transfer is much less sufficient due to the diminished ratio of surface area/volume. As a consequence, compared to the laboratory reactor, the heat removal rates can be 10 times slower per unit volume at the pilot scale, and as much as 30 times slower on a commercial scale.[1] In addition, poor understanding of the reaction chemistry and kinetics, and inadequate process training are also the source of runaway reactions.

Reaction calorimeter (RC-1) (Mettler Toledo) gives much information in a single experiment and is considered an integrated method in process development,[2] especially in early phase process development. It is very useful to perform the initial "familiarization" reactions with an RC-1 to collect basic information that will benefit the process development.

To reduce the hazards and avoid the potential runaway reaction, an inherently safer process design is necessary. The following are process development opportunities:

- Replace hazardous materials with safer ones.
- Use a semi-batch process to charge the hazardous material portionwise to prevent its accumulation.
- Introduce process controls, including the use of sensors, alarms, and other control systems.
- Set up process safety parameters for a given exothermic reaction.
- Run the reaction under dilute conditions to ensure the presence of thermal mass to absorb the reaction heat.
- Use a continuous flow reactor instead of a batch reactor for exothermic reactions.

9.1 CATALYTIC CROSS-COUPLING REACTIONS

9.1.1 HECK REACTION

Heck reaction of aryl bromide **1** with alkene **2** was used to prepare ketone **3** (Equation 9.1).[3] It was found that the Heck reaction under conditions (Pd(OAc)$_2$, LiCl, LiOAc, and TEA, 90 °C) proceeded rapidly with a noticeable temperature rise, which posed a safety concern at scale.

$$(9.1)$$

Consequently, the exotherm Heck reaction could be controlled by dosing triethylamine (TEA) at 75 °C.

Procedure

- Charge dimethyl acetamide (DMAc) (26 gal) into a 200-gal reactor containing **1** (39.59 kg, 163 mol), followed by adding LiCl (13.9 kg, 327 mol), LiOAc (5.4 kg, 81.8 mol), and water (3 gal).

DOI: 10.1201/9781003288411-9

- Purge the resulting solution with N$_2$ for 30 min.
- Charge 24.8 kg (196.5 mol) of **2** and rinse with DMAc (2 gal).
- Charge 1800 g (8.03 mol) of Pd(OAc)$_2$ and 1.7 kg (16.8 mol, 10%) of TEA.
- Purge the headspace with nitrogen.
- Heat to 75 °C.
- Charge 3.3 kg (32.61 mol, 20%) of TEA and stir for 15 min.
- Charge 5.8 kg (57.21 mol, 70%) of TEA and stir for 2 h.
- Cool to 28 °C over 1 h.
- Add water (20 gal) and MTBE (20 gal).
- Add DARCO (10 kg) into a separate, 500-gal reactor, followed by Celite (10 kg).
- Transfer the content in the 200-gal reactor into the 500-gal reactor, followed by adding water (26 gal) and MTBE (26 gal).
- Stir for 2 h.
- Filter and rinse with MTBE (35 gal).
- Separate layers.
- Wash the organic layer with water (2×18 gal).
- Distill the organic layer under atmospheric pressure until ca 21 gal volume.
- The MTBE solution of **3** (KF 0.09%) was used in the subsequent step.

REMARK

The use of LiCl and LiOAc combined with ligating base (TEA) enhanced the reactivity and reduced the catalyst loading.

9.1.2   Negishi Cross-Coupling Reaction

The Negishi cross-coupling reaction was used to synthesize pyrazolopyrazine **6**, a drug candidate for the treatment of depression, anxiety, addiction, etc. The cross-coupling was initially carried out by heating a mixture of **4**, **5**, and the catalyst in THF (Equation 9.2).[4]

(9.2)

(I) Reaction problems

When heating the reaction mixture, an exothermic event was observed and the temperature increased from 50 °C to reflux (~66 °C) over a span of 5 min despite maintaining a jacket temperature of 55 °C.

(II) Solutions

A semi-batch process was developed to address the risk of the uncontrolled exotherm reaction. Ultimately, in the revised process, **5** was dosed gradually into the mixture solution of **4** and the catalyst at 55–60 °C.

9.2   BLAISE REACTION

Employing the Blaise reaction of α-bromoester and nitrile in the presence of zinc provides an efficient way to prepare β-ketoester. For example, the synthesis of 2,6-dichloro-5-fluoronicotinoyl acetate **8** was accomplished in a single step using the Blaise reaction of 3-cyano-2,6-dichloro-5-fluoropyridine

7 with ethyl bromoacetate (Equation 9.3).[5] However, an induction period was observed for the Reformatsky reagent-related reaction, which was a safety concern. To address the safety issue, the use of methanesulfonic acid as an in situ activator of zinc removed the induction period, rendering the Blaise reaction safe and viable for a large-scale operation.

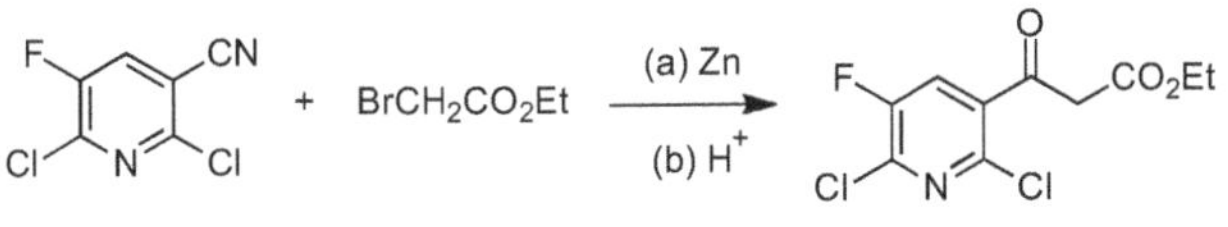

$$(9.3)$$

Procedure

- Charge 250.0 g (2.6 mol, 5 mol%) of methanesulfonic acid to a stirred suspension of **7** (10.0 kg, 52.4 mol) and zinc dust (5.1 kg, 78.5 mol, 1.5 equiv) in THF (50 L).
- Heat the resulting mixture to reflux and stir for 10 min.
- Charge 11.5 kg (68.1 mol, 1.3 equiv) of ethyl bromoacetate dropwise over 2.5 h.
- Reflux after the addition for 0.5 h.
- Cool to 0–5 °C.
- Add 25 L of 6 N HCl and water (5 L).
- Warm the mixture to room temperature and stir for 1 h.
- Cool to 5 °C.
- Filter and wash with EtOH/water (30 L, 7:3, v:v).
- Dry by nitrogen purge.
- Yield: 10.6 kg (72%).

9.3   RITTER REACTION

Ritter reaction is a reaction between nitriles and alcohols or olefins to give carboxylic amides. Although the Ritter reaction is employed in various large-scale applications,[6] a problem associated with the Ritter reaction is the necessity of a strong acid catalyst in order to produce carbon cation intermediates. Besides the environmental risk associated with using such a strong acid, a thermal runaway reaction would occur when scaling up the Ritter reaction. For instance, the Ritter reaction of acrylonitrile with cyclohexanol was conducted by adding concentrated sulfuric acid to a mixture of **9** and **10** (Scheme 9.1).[7]

(I) Ritter reaction incident

An incident occurred when scaling up the Ritter reaction on a 2 mole scale. Several hours after all of the reagents had been mixed at room temperature, a violent reaction occurred, ejecting most of the flask contents.

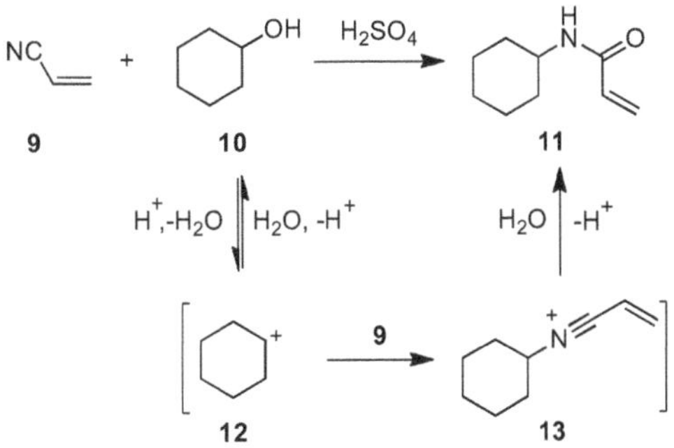

SCHEME 9.1

(II) Solutions

To address the safety issue, it is necessary to run the reaction by adding the concentrated sulfuric acid to a premixed solution of **9** and **10** at relatively high temperatures (45–50 °C) to prevent buildup of reactive species.

9.4 HYDROGEN–LITHIUM EXCHANGE

Although the preparation of aldehyde **15** was achieved through lithiation of Bromo benzotriazole **14** with LDA followed by quenching the organolithium intermediate with DMF, performing this hydrogen-lithium exchange chemistry on large scale posed safety concerns associated with potential benzyne formation (Equation 9.4).[8]

$$ \text{(9.4)} $$

To obviate this safety issue, magnesium ate complex, prepared in situ by reaction of hexyllithium with butylmagnesium chloride and 2,2,6,6-tetramethylpiperidine (TMP) at –40 °C, was used to replace LDA.

Procedure

- Cool the solution of 22.8 kg (161 mol) of TMP in THF (90.2 kg) to –20 °C.
- Charge 45.4 kg (165 mol) of 2.5 M hexyllithium in hexane at <–10 °C over 45 min.
- Age the resulting slurry at –15 °C for 5 min followed by adding 41 kg of 1.93 M butylmagnesium chloride in THF at <–10 °C over 10 min.
- Age the batch at –15 °C for 5 min.
- Cool the batch to –45 °C.
- Charge a solution of 20.5 kg (80.7 mol) of **14** in THF (54.1 kg) at <–39 °C over 30 min followed by adding 41.3 kg (565 mol) of DMF at <–35 °C over 55 min.
- Warm the batch to –25 °C and age for 2 h.
- The resulting **15** in THF was directly used in the subsequent reaction.

9.5 HALOGENATION REACTIONS

9.5.1 CHLORINATION REACTION

N-Chlorosuccinimide (NCS) is one of the commonly used chlorination reagents in organic synthesis. As most of the chlorination reactions with NCS are exothermic, control of the exothermic events is critical in process development. Generally, a controlled addition approach is adopted by adding the chlorination reagent NCS to a warm reaction mixture in order to mitigate the exotherm and avoid the accumulation of NCS in the reaction mixture. For example, dichlorination of benzodioxan **16** with NCS was carried out by adding the NCS slurry in AcOH into a hot reaction mixture (between 70 and 100 °C), affording 75% yield of the dichlorobenzodioxan **17** (Equation 9.5).[9]

$$ \text{(9.5)} $$

REMARKS

(a) It was observed that upon dissolving *N*-bromosuccinimide (NBS) at a concentration of 15% w/v in cold THF an extremely exothermic reaction ensued, resulting in a highly colored boiling solution. This NBS/THF incompatibility is probably due to a radical chain reaction initiated by peroxides in THF.[10] An article[11] reveals that NBS is not compatible with solvents, such as DMF, DMAc, NMP, THF, and toluene. Acetonitrile, dichloromethane, and ethyl acetate are the recommended solvents when using NBS for bromination reactions.

(b) Bromination with bromine should not use MTBE as the solvent as mixing these two materials would lead to a spontaneous and strongly exothermic reaction.[12]

9.5.2 FLUORINATION REACTIONS

9.5.2.1 Deoxyfluorination

The increasing demand for organofluorine compounds prompts extensive development of fluorination reagents that are capable of incorporating fluorine atoms in a safe, efficient, and controlled fashion. Among the currently available fluorination reagents, aminodifluorosulfinium salts are widely used as fluorination reagents. For example, diethylaminosulfur trifluoride (DAST) and bis(2-methoxyethyl)aminosulfur trifluoride (Deoxo-Fluor) are commonly utilized in the deoxyfluorination of alcohols (Figure 9.1).

However, both DAST and Deoxo-Fluor[13] are thermally unstable and highly explosive, which limits their application on a large scale. Recently, a powerful deoxyfluorination system[14] was developed by a conjunction of diethylaminodifluorosulfinium tetrafluoroborate (XtalFluor-E) or difluoro(morpholino)sulfonium tetrafluoroborate (XtalFluor-M) with promoters, such as Et$_3$N·3HF, Et$_3$N·2HF, or DBU. The dialkylamino(difluoro)sulfinium tetrafloroborates, XtalFluor-E and XtalFluor-M, are easily handled crystalline solids which do not generate highly toxic and corrosive-free HF, and are therefore compatible with borosilicate glassware. Unlike DAST and Deoxo-Fluor, XtalFluor-E and XtalFluor-M are thermally stable. The transformation of alcohols or carbonyl groups into alkyl fluorides or *gem*-difluorides was realized by a combination of the XtalFluor reagents with promoters such as Et$_3$N·3HF, Et$_3$N·2HF, or DBU (Equations 9.6 and 9.7).

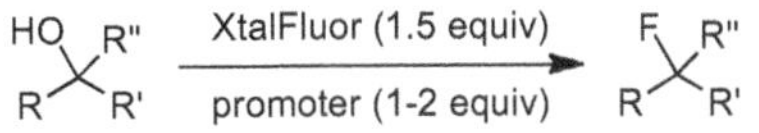

$$\text{(9.6)}$$

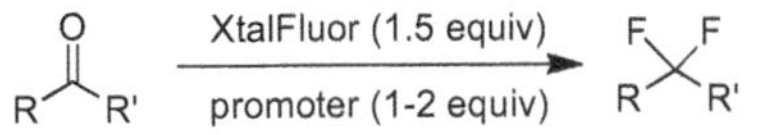

$$\text{(9.7)}$$

2,2-Difluoro-1,3-dimethylimidazolidine (DFI) is a fuming liquid used to react with primary, secondary, and tertiary alcohols to afford the corresponding alkyl fluorides in good yields under mild

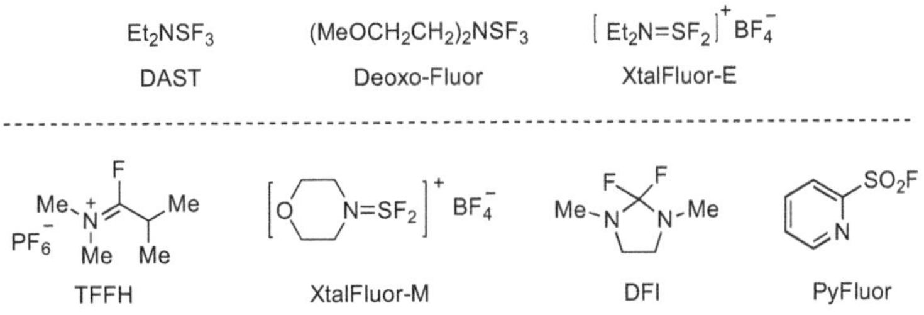

FIGURE 9.1 Structures of some fluorination reagents.

conditions.[15] *N,N,N′,N′*-tetramethylfluoroformamidinium hexafluorophosphate (TFFH) is a non-hygroscopic, crystalline solid used for peptide synthesis and deoxyfluorination reactions.[16]

2-Pyridinesulfonyl fluoride (PyFluor)[17] is a colorless crystalline solid, being able to be stored at room temperature for 30 days. PyFluor could deoxyfluorinate alcohols to give the corresponding fluoro products in good yields.

9.5.2.2   Hydrofluorination of Aziridines

Scheme 9.2[18] outlines a hydrofluorination reaction of aziridines **18** with amine·HF, generated in situ, providing β-fluoroamines **19**. Experimentally, the reaction was conducted by the combination of PhCOF (2.0 equiv) and 1,1,1,3,3,3-hexafluoroisopropanol (HFIP) in the presence of a catalytic amount of 1,5-diazabicyclo[4.3.0]non-5-ene (DBN) in MTBE at 50 °C.

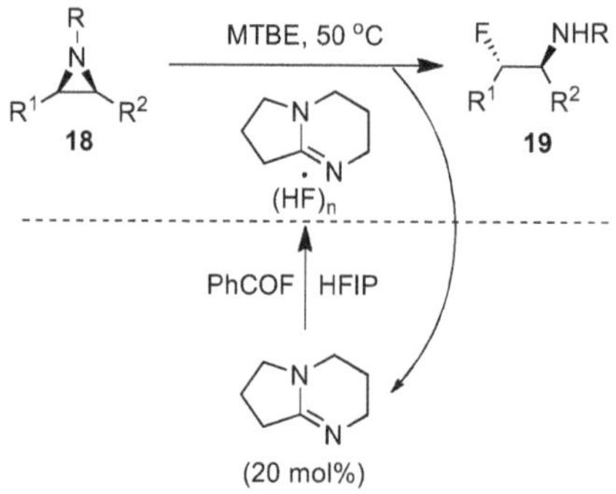

SCHEME 9.2

This protocol displays a broad scope with respect to aziridine substitution and *N*-protecting groups. Commonly employed protecting groups are tolerated, including Boc, Cbz, and Bn. *N*-Alkyl aziridines, such as R=Bn, are sufficiently basic that hydrofluorination occurs in the absence of DBN. However, *N*-tosyl aziridine reacted sluggishly, providing only 46% product along with 48% unreacted aziridine. Acyl aziridines, including those derived from picolinic and 4-thiazolecarboxylic acid, provided β-fluoroamides in good yields.

A review paper describes the use of various continuous gas–liquid flow reactors for processes involving fluorine gas for the selective fluorination of a range of aromatic substrates, benzaldehyde, and diester substrates.[19]

9.5.2.3   Electrophilic Fluorination

Fluorinated enol ether **21** could be prepared by electrophilic fluorination of enol ether **20** with fluorinating agent *N*-fluorobenzenesulfonimide (NFSI) (Equation 9.8).[20] The reaction was carried out by adding NFSI to a solution of the anion of **20**, prepared by a reaction of **20** with LiHMDS, at –20 °C to 0 °C.

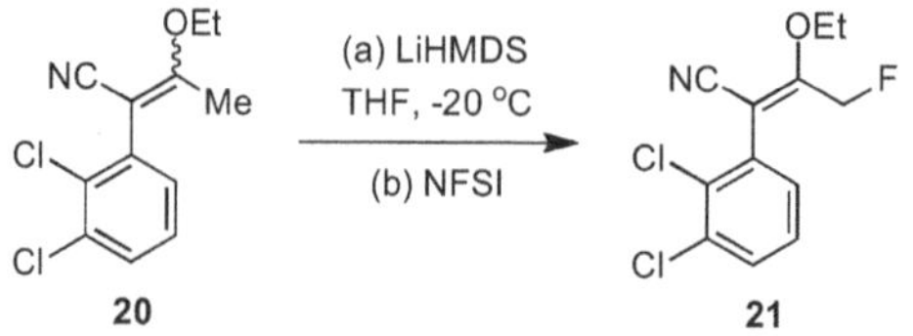

$$(9.8)$$

(I) Reaction problems

The rate of adding NFSI is critical for reducing the potential of anion exchange, thereby reducing levels of under- and over-fluorinated impurities, and fast addition of NFSI is required. However, such fast addition renders an unsafe process as the reaction is very exothermic.

(II) Solutions

To address the safety and impurity issues, a reverse addition protocol was adopted, wherein the solution of the anion of **20** was added to the solution of NFSI.

9.5.3 DEOXYCHLORINATION

Deoxychlorination of hydroxyl heteroaromatic derivatives with phosphoryl chloride ($POCl_3$) is a well-established reaction for the preparation of chlorinated heteroarenes. Reactions usually proceed through a dichlorophosphate intermediate and subsequent displacement of the dichlorophosphonate with chloride. In general, two reaction protocols have been developed:

(i) Neat $POCl_3$:
Using large excess $POCl_3$ combined with base, usually diisopropylethylamine (DIPEA) under reflux presents a significant scale-up challenge from both a safety and environmental standpoint.

(ii) $POCl_3$/solvent:
Deoxychlorination using $POCl_3$ (1–2 equiv) in an organic solvent, such as toluene, CH_2Cl_2, or PhCl addresses both process safety and environmental issues.

REMARKS

(a) Chlorinated products are often unstable during workup and storage due to their susceptibility to hydrolysis. Therefore, in order to prevent the hydrolysis of the chlorinated products back to the starting materials, anhydrous workup conditions can be applied.

(b) Alternatively, a telescoping method may be used to avoid the isolations.

9.5.3.1 Deoxychlorination of Triazinone

Deoxychlorination of hydroxypyrimidine **22** in neat $POCl_3$ required distillation to remove excess reagent (Condition A, Equation 9.9)[21]. The deoxychlorination with $POCl_3$ and triethylamine in ethyl acetate showed, however, exothermic decomposition, with the onset of decomposition at 35 °C (Condition B, Equation 9.9). To address these process safety issues, the deoxychlorination of **22** was carried out readily with $POCl_3$ (1.1 equiv) and DIPEA (1.1 equiv) in toluene on a 35 kg scale, producing **23** in 85–87% yields (Condition C, Equation 9.9).

$$(9.9)$$

It was, however, noticed that when the $POCl_3$/DIPEA/toluene system was applied to chlorinate pyrazolotriazinone **24** excess of $POCl_3$ (4.0 equiv) was required in order to drive the reaction to completion. It was found that using a catalytic amount of *N*-methylmorpholine (NMM) (0.5 mol%) was able to reduce the amount of $POCl_3$ from 4.0 equiv to 1.3 equiv. Thus, a high-yielding, safe, and more environmentally friendly process for the deoxychlorination step was developed (Equation 9.10).[22]

$$(9.10)$$

Procedure[22]

- Charge 83 mL (740 mmol, 0.5 mol%) of NMM to a slurry of **24** (42.2 kg, 148 mol) in dry MeCN (232 L), followed by adding DIPEA (51.5 L, 296 mol, 2.0 equiv).
- Cool the content to 0 °C.
- Charge 17.9 L (192 mol, 1.3 equiv) of $POCl_3$.
- Warm the mixture to 25 °C and stir for 1 h.
- Cool the batch to –5 °C.
- Add MeCN (32 L).
- Proceed the subsequent step without isolation of **25**.

REMARKS

(a) $POCl_3$-promoted Friedel–Crafts cyclization and subsequent deoxychlorination afforded **27** efficiently in one-pot (Equation 9.11).[23] The use of base (TEA) is essential in terms of suppressing the formation of impurity **28** whose formation is presumably catalyzed by HCl.

$$(9.11)$$

Procedure

- Charge 8.9 kg (58.0 mol) of $POCl_3$ slowly to a suspension of **26** (4.0 kg, 14.3 mol) and TEA (6.4 kg, 63.4 mol) in 20 L of toluene at 8–10 °C.
- Heat at reflux for 4 h.
- Cool the batch to 10 °C.
- Transfer the batch to ice water (10 L) at <10 °C.
- Add 25% NaOH at 10 °C to pH 8–9.
- Filter to give a crude product.
- Slurry the crude product in toluene (5 L) at 70 °C.
- Cool to 10 °C.
- Filter and dry at 65 °C.
- Yield of **27**: 3.8 kg (95%).

(b) Using $POCl_3$ in the deoxychlorination is usually combined with TEA or DIPEA to trap the HCl by-product. However, introducing such tertiary organic base, such as TEA, led to some side products, due to the reaction of TEA with the chlorinated product (Equation 9.12):[21]

$$(9.12)$$

(c) An incident[24] that resulted in an explosion at Pfizer's research facility in Groton, Connecticut, USA involved the combination of phosphorus oxychloride and acetone. Apparently, the incompatibility between POCl₃ and acetone results in heat generation and significant gas evolution when they are combined.

9.5.3.2 Deoxychlorination of Quinazolone

Research on the deoxychlorination of quinazolones with $POCl_3$ demonstrated that the reaction occurs in two distinct stages, which can be separated through appropriate temperature control (Scheme 9.3).[25] An initial phosphorylation reaction occurs readily under basic conditions (R_3N, aq $pKa > 9$) at $<25\,°C$ to give the corresponding phosphorylated intermediates **30**.

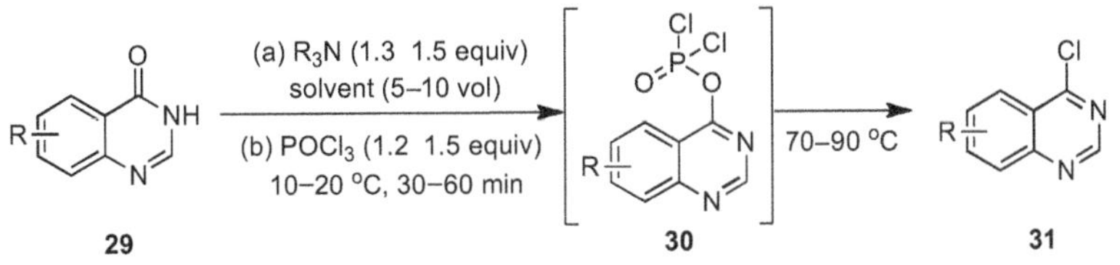

SCHEME 9.3

Under these conditions, pseudo-dimer formation, arising from a reaction between phosphorylated intermediates and unreacted quinazolones **29**, is completely suppressed. Clean turnover of phosphorylated quinazolones to the corresponding chloroquinazolines **31** is then achieved by heating to $70–90\,°C$. The hold time following $POCl_3$ addition is necessary to ensure complete phosphorylation of **29** prior to heating. Furthermore, aromatic solvents are preferred, particularly those of higher polarity, such as anisole and chlorobenzene. Acetonitrile should be avoided as the reaction solvent as a significant quantity of dimer was observed when CH_3CN was used.

9.5.3.3 Deoxychlorination of Triazine

As described in Scheme 9.4,[26] deoxychlorination of triazine **32** proceeds through the chlorophosphate intermediate **33**, followed by displacement of dichlorophosphonate by chloride.

Compared with the formation of **33**, the dichlorophosphonate displacement is relatively slow. In order to drive the reaction to completion in a reasonable time frame, external chlorides are sometimes required.[27] It was found, however, that the reaction in the presence of a substoichiometric amount of DIPEA (0.9 equiv) was faster than with 1.1 equiv of DIPEA. It was hypothesized that the in situ generated free HCl is the catalyst for the displacement step.

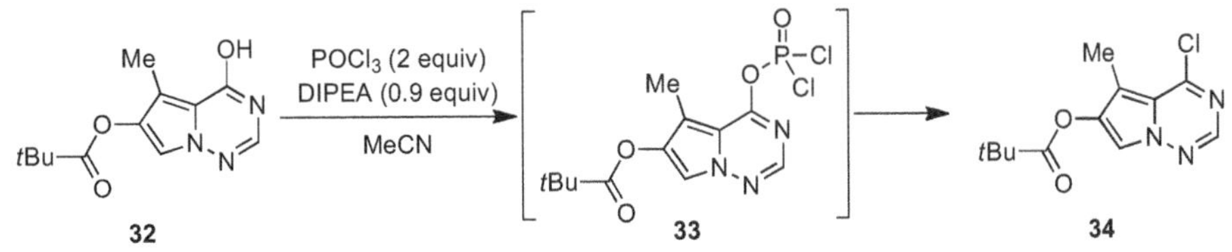

SCHEME 9.4

Procedure

- Charge 66 L (706 mol) of $POCl_3$ to a slurry of **32** (89.0 kg, 357 mol) in dry MeCN (156 kg, <0.05% water) at 23–24 °C over 8 min.
- Rinse with MeCN (45 kg).

- Charge 56 L (321 mol, 0.9 equiv) of DIPEA at 22–29 °C over 20 min.
- Rinse with MeCN (45 kg).
- Heat to reflux over 57 min and stir at reflux for 11.5 h.
- Cool the batch to 30–35 °C.
- Workup.

NOTE

Preparation of 2,6-dichloropurine via deoxychlorination of xanthine with phosphorus oxychloride in the presence of weak nucleophilic organic bases, such as amidine, guanidine base, or Proton-Sponge, was described.[28]

9.5.3.4   Deoxychlorination of 4,6-Dihydroxypyrimidine

A significant exotherm was observed in the conversion of 4,6-dihydroxypyrimidine derivative **35** to the corresponding dichloropyrimidine **36** by treatment of **35** with POCl$_3$ (Equation 9.13).[29]

$$(9.13)$$

Fortunately, the exotherm was successfully controlled by adding POCl$_3$ to a solution of **35** in the presence of 2,6-lutidine as a promoter at an elevated temperature in toluene where the accumulation of reactive intermediates could be avoided.

Procedure

- Charge 5.0 kg (23.6 mol, KF=0.2%) of **35** into a 50-gal reactor, followed by adding toluene (20 L) and 2.7 L (23.6 mol) of 2,6-lutidine.
- Stir the resulting mixture at room temperature.
- Warm the mixture to 50 °C.
- Charge 4.85 L (53.02 mol, mol, 2.25 equiv) of POCl$_3$ slowly over 1 h.
- Stir the batch at 70 °C for 24 h.
- Cool to 0 °C.
- Add 20% aqueous NaOH (5.5 kg/28 L H$_2$O) slowly to pH ~5.5 at ≤30 °C.
- Add EtOAc (13 L) and stir for 30 min.
- Filter through Celite and rinse with EtOAc (2×5 L).
- Separate layers.
- Extract the aqueous layer with EtOAc (3×12 L).
- Wash the combined organic layers with saturated aq NH$_4$Cl (3×12 L) and brine (12 L).
- Dry over MgSO$_4$.
- Filter and distill the filtrate under reduced pressure.
- Yield: 5.63 kg as a brown oil (which crystallized overnight).

DISCUSSIONS

(a) The use of triethylamine instead of 2,6-lutidine gave similar results.
(b) Compared to acetonitrile, the use of toluene improved the process safety.
(c) n additional advantage of using toluene was that it allowed for a more efficient extractive workup.

9.5.3.5   Deoxychlorination of 6,7-Dihydrothieno[3.2-d]pyrimidine-2,4-diol

Similarly, the control-addition approach was successfully applied in the deoxychlorination of 6,7-dihydrothieno[3.2-*d*]pyrimidine-2,4-diol **37** by adding POCl$_3$ portionwise into a mixture of **37** and *N,N*-diethylaniline (Equation 9.14).[30] *N,N*-Diethylaniline was used as both the base and the solvent. The product **38** could be conveniently isolated by simply quenching into the water followed by filtration.

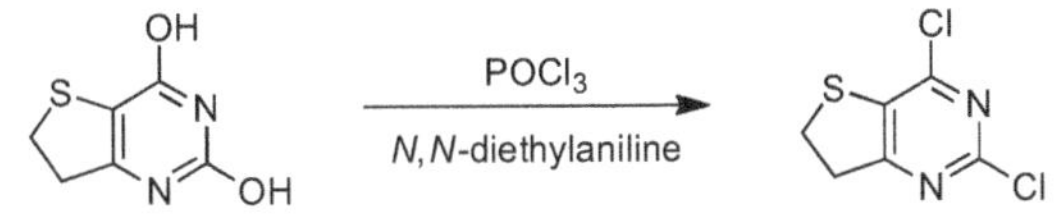

(9.14)

Procedure

- Charge 800 g (4.66 mol) of **37** into an inert and dry jacketed reactor (Reactor 1) as a solid followed by adding 1.5 L (9.31 mol) of *N,N*-diethylaniline at <25 °C over 0.5 to 1 h.
- Heat the content to 105 to 110 °C.
- Charge 0.68 equiv (868 mL, 34% of the total) of POCl$_3$ over 5 to 10 min and stir until the internal temperature began to decrease.
- Charge the rest of POCl$_3$ (1.32 equiv, or 68% of the total) at 110 °C over 30 to 40 min.
- Stir the batch at 105 to 110 °C for 18 to 24 h.
- Cool to 45 °C and add 400 mL of THF.
- Transfer the content to Reactor 2.
- Add water (4.8 L) into Reactor 1 and keep it at 5 to 10 °C.
- Transfer the reaction mixture from Reactor 2 to Reactor 1 at 5 to 10 °C.
- Stir the quenching mixture at 5 °C for 0.5 to 1 h.
- Filter and wash with water (2×1.6 L).
- Air-dry in the funnel for 6 to 8 h.
- Yield of crude product: 964 g (88% yield and 92 wt% purity).

9.5.3.6   Deoxychlorination of 6-Chlorophthalazin-1-ol

The preparation of 4-(6-chlorophthalazin-1-yl)morpholine **41** involved deoxychlorination of 6-chlorophthalazin-1-ol **39** followed by an S$_N$Ar amination with morpholine (Scheme 9.5).[31] The transformation of 6-chlorophthalazin-1-ol **39** to 1,6-dichlorophthalazine **40** was achieved by utilizing 2.0 equivalents of POCl$_3$ in PhCl at 90 °C.

Like most chlorinated heteroaromatic compounds, the crude **40** presented stability issues due mainly to its sensitivity to moisture. The crude **40** was unstable and developed a colored insoluble precipitate. In order to address the stability issues, a telescoped process was developed and the deoxychlorination and S$_N$Ar amination steps were conducted in one-pot.

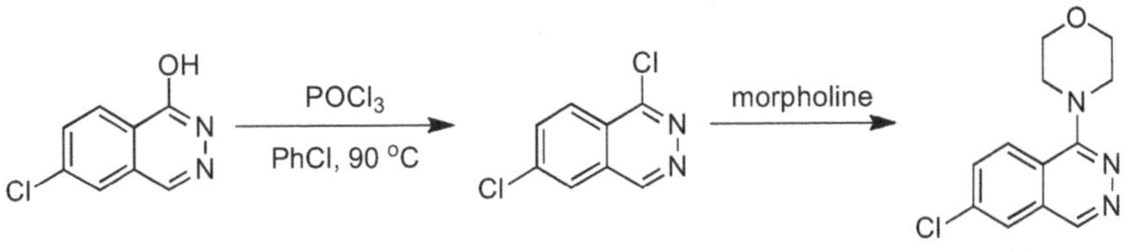

SCHEME 9.5

9.6   THIOCYANATION

A thiocyanation of propargyl bromdie was carried out by an S_N2 reaction of propargyl bromide with potassium thocyanate; the resulting crude product was purified by vacuum distillation (Equation 9.15).[32]

$$\text{Br} + \text{KSCN} \longrightarrow \text{SCN} \tag{9.15}$$

During the course of distillation, an explosion occurred. The culprit for the accident is assumed to be the accumulation of highly reactive isothiocyanatopropa-1,2-diene, generated in situ via an [3,3]-sigmatropic rearrangement under thermal conditions (Equation 9.16). The isothiocyanato-propa-1,2-diene would undergo exothermic polymerization which leads to this explosion.

$$\text{SCN} \longrightarrow \tag{9.16}$$

Thereby it was suggested that the distillation shall be carried out with efficient vacuum at a relatively low bath temperature ($\leq 80\,°C$) to suppress the undesired rearrangement. The operation shall wear safety glasses with a protective shield.

Thiocyanogen, $(SCN)_2$, a pseudo-halide, reacts with electron-rich aniline through a standard electrophilic aromatic substitution reaction. Thiocyanation of 2-chloroaniline **42** with thiocyanogen, generated in situ from a reaction of sodium thiocyanate and bromine, afforded 2-chloro-4-thiocyanatoaniline **43** (Equations 9.17 and 9.18).[33]

$$\tag{9.17}$$

$$2\,NaSCN + Br_2 \longrightarrow (NCS)_2 + 2\,NaBr \tag{9.18}$$

Conducting the reaction in methanol presented a problem due to the interaction of bromine with methanol which can produce unstable methyl hypobromite (Equation 9.19). To avoid the hazardous interaction of bromine with methanol, acetonitrile was used to dilute the bromine.

$$CH_3OH + Br_2 \longrightarrow CH_3OBr + HBr \tag{9.19}$$

9.7   GAS-INVOLVED REACTIONS

Besides exothermic runaway reactions, the uncontrolled evolution of gaseous by-products is also a safety concern. The gaseous by-products can be carbon dioxide, hydrogen, isobutene, acetylene, etc. These gaseous by-products, except carbon dioxide, are flammable. Acetylene is very sensitive to pressure, temperature, static electricity, and mechanical shock.

9.7.1   Boc Protection

Di-*tert*-butyl dicarbonate (Boc_2O) is frequently used to protect various functional groups, such as the amino group, hydroxy group, and carboxylate group. The protection of carboxylate, such as **44**, will potentially generate 4.5 equiv of CO_2 gas (if 2.25 equiv of Boc_2O is used) (Equation 9.20).[34] Notably, the gas evolution consisted of two major phases: non-dose-controlled release at ~0.6 L/

min with an induction period and a dose-controlled release at ~1.2 L/min. To address the non-dose-controlled release of CO_2 at the beginning of the reaction, an initial charge of ~10% of the Boc_2O/THF solution was performed. After the initial gas evolution phase was over, the remainder of the Boc_2O/THF solution was added in an addition-controlled fashion. The higher temperature of the reaction (55–60 °C) combined with the subsequent aqueous alkaline washes ensured very little dissolved CO_2 would be present after the workup.

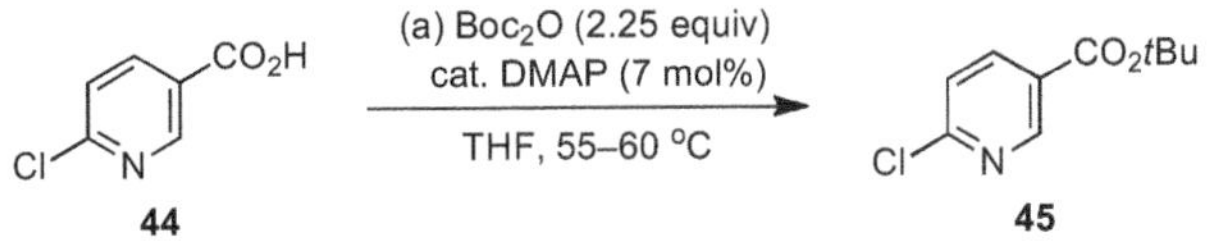

(9.20)

REMARK

The hypothesis regarding the induction period is that at the beginning of the reaction the majority of the DMAP is not initially available to enter the catalytic cycle as it forms a salt with the nicotinic acid.

9.7.2 N-ACETYLATION

2-Butanone **47** is an intermediate for the synthesis of an anti-tumor agent **51**. The preparation of **47** was carried out by heating a mixture of amino acid **46** and acetic anhydride in pyridine at 120 °C (Scheme 9.6).[35]

(I) Problems

An induction period was observed, followed by a strong evolution of carbon dioxide.

(II) Chemistry diagnosis

Investigations revealed that the strong evolution of CO_2 is caused by the rapid decarboxylative hydrolysis of the accumulated **50** (or **49**) with water (Scheme 9.6). The generation of water is the result of the dehydration of acetic acid, a by-product during the formation of **48** and **49**.

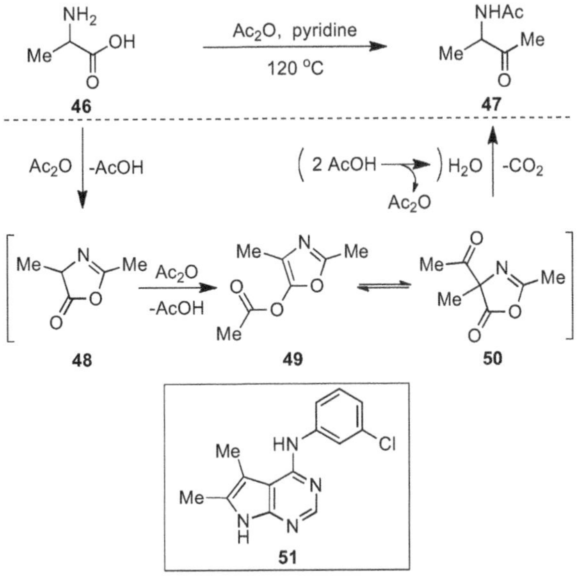

SCHEME 9.6

(III) Solutions

To diminish the accumulation of the intermediate **50** (or **49**), 1 equiv of acetic acid was added to the initial mixture of acetic anhydride and pyridine, in the event, the reaction could be performed in an addition-controlled mode. This modified procedure was run successfully and with well-controlled carbon dioxide evolution on a 0.5 kmol-scale in a 600-L reactor, whereby acetic anhydride, acetic acid, and pyridine were charged into the reactor and heated to ca. 120 °C, and alanine **46** (45 kg) was then added as a slurry in pyridine over 2–3 h.

REMARK

A safer and more economical protocol was developed by the addition of **46** to a mixture of acetic acid (0.5 equiv), acetic anhydride (2.2 equiv), triethylamine (2.5 equiv), and 4-(dimethylamino)pyridine (1.0 mol%) at 50 °C. Using this protocol, over 0.5 kmoles of alanine (50 kg) were successfully converted to acetylaminobutanone **47** in a 400-L reactor with over 90% yield.

9.7.3 BOC DEPROTECTION

Deprotection of the Boc group is typically achieved under acidic conditions,[36] generating CO_2 and isobutylene as gaseous by-products. Isobutylene is a volatile organic compound and is subject to regulation by Environmental Protection Agency (EPA). Isobutylene (boiling point of –6.9 °C) can pass through most process condensers to the atmosphere in the absence of suitable end-of-line devices; thus, emissions of isobutylene must be dealt with accordingly before a process can be run on manufacturing scales. Isobutylene is generated from *t*-butyl cation intermediate by loss of proton during the removal of Boc group. Anisole or thioanisole are frequently used as the *t*-butyl cation scavenger to prevent the formation of isobutylene. Besides, to reduce isobutylene emissions, a mixture of toluene/$MeSO_3H$ (8 g of $MeSO_3H$ in 225 g of toluene) is used as an external scrubber to convert isobutylene to non-volatile oligomers (Scheme 9.7). This system was employed successfully on the pilot plant-scale production.[37]

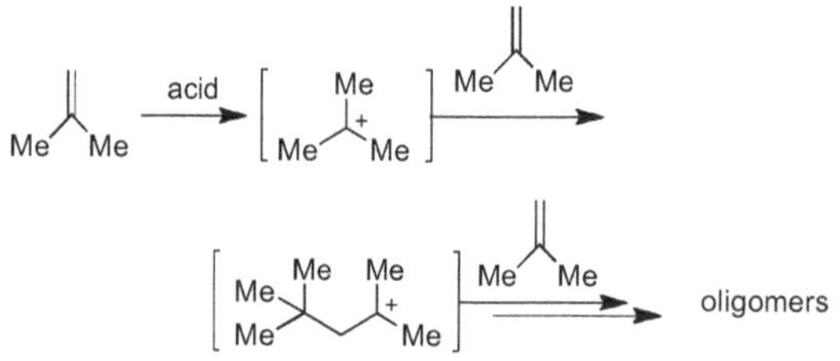

SCHEME 9.7

REMARKS

(a) The extent of isobutylene off-gas is solvent-dependent, for example, in alcohol solvents, the isobutylene concentration is reduced due to the fact that the *tert*-butyl cation is partially trapped as the corresponding alkyl *tert*-butyl ethers. Methanol is the most efficient solvent for this purpose showing a 90% trapping efficiency, while ethanol and 2-propanol are not efficient, likely due to steric effects. Surprisingly, isobutylene emissions from both dichloromethane and toluene were found to be relatively low as well, due to the propensity of isobutylene to polymerize under the reaction conditions to produce non-volatile oligomers. However, all of the isobutylene is liberated as a reaction off-gas in THF.

(b) The scrubbing efficiency can be obtained from equation (Equation 9.21)[37] by simply integrating the area under both profiles of inlet and outlet:

$$\text{Efficiency} = 1 - \text{Area (outlet)/Area (inlet)} \qquad (9.21)$$

9.7.3.1 Selective Deprotection with TFA

Aqueous trifluoroacetic acid (TFA) was demonstrated to be the reagent of choice in the deprotection of the Boc group from **52** (Equation 9.22).[38] Under the reaction conditions, *tert*-butyl cation was converted by TFA and water to *t*-butyl trifluoroacetate and *t*-butanol, respectively. As a result, the deprotection evolved only <10% of isobutene gas.

$$(9.22)$$

REMARKS

(a) The use of sulfuric acid led to a foaming problem in the pilot plant.
(b) The use of HCl in EtOH was less selective, leading to significant amounts of demesylation side product.

9.7.3.2 Deprotection with NaOH

Although acidic conditions were effective at removing the Boc-protecting group on the indole nitrogen in Boc-protected indole derivative **54**, the CO_2 off-gassing was a concern, as it was not easily controlled (Condition A, Equation 9.23).[39] In order to address the process safety issue associated with the uncontrolled CO_2 off-gassing, basic conditions were employed at an elevated temperature (65 °C) to remove the Boc group and the CO_2 by-product was released safely via a controlled addition of HCl during the subsequent neutralization (Condition B, Equation 9.23).

$$(9.23)$$

9.7.4 *N-TERT*-BUTYLAMIDE FORMATION

Isobutylene could also be generated from *tert*-butyl alcohol under acidic conditions. For instance, a process for the preparation of amide **57** from nitrile **56** under Ritter reaction conditions using *tert*-butyl alcohol and sulfuric acid generated isobutylene side product (Equation 9.24).[40] The difficulty in controlling the isobutylene evolution rate was a process safety concern.

$$(9.24)$$

To overcome this problem, *tert*-butyl alcohol was replaced with *tert*-butyl acetate and the reaction was carried out by a controlled addition of *t*BuOAc into a solution of **56** in a mixture of sulfuric acid and acetic acid.

Procedure

- Charge 96.0 g (1.8 mol, 2.25 equiv) of concentrated H_2SO_4 to a mixture of 94.4 g (0.8 mol) of **56** and 150 mL (2.62 mol, 3.275 equiv) of AcOH at ≤30 °C over 0.5 h.
- Charge 215.6 mL (1.6 mol, 2.0 equiv) of tBuOAc over 45 min at 25±4 °C
- Stir the resulting solution at room temperature for 12 h.
- Quench the batch into 9.0% aqueous NaOH solution (360 g in 3.64 kg of water) at 8±4 °C.
- Stir the mixture at room temperature for 1.5 h.
- Filter to give a solid.
- Suspend the solid in water (600 g) and stir for 0.5 h.
- Filter and dry at 44±5 °C under 25 mbar for 14 h.
- Yield: 138.44 g (90%).

9.7.5 DEPROTECTION OF N-TERT-BUTYL GROUP

Tetrahydrofluorenone **59** was identified by Merck as a potent selective estrogen receptor beta agonist. The final step synthesis of **59** on large scale involved deprotection of the t-butyl group.

Two reaction protocols were developed for the removal of the t-butyl group. Under Condition A (Scheme 9.8),[8] sulfuric acid (100%, 7 volumes), generated by mixing fuming sulfuric acid with concentrated sulfuric acid, was necessary to achieve a clean reaction at an acceptable rate.

Procedure

- Charge 14.08 kg (7.22 L) of 20% fuming sulfuric acid to 96% sulfuric acid (18.1 kg, 9.84 L) at ≤40 °C.
- Charge 2.58 kg (6.52 mol) of **58** portionwise to the 100% H_2SO_4 at ≤29 °C.
- Add additional 96% H_2SO_4 (1.84 kg, 1.0 L).
- Heat the resulting mixture to 75 °C and age for 4 h.
- Cool to 10 °C.
- Add MeCN (18 L).
- Add a solution of 5 N NaOH (40 L) slowly.
- Filter, wash with water (2×13 L), and dry.
- Yield: 2.08 kg (89%) as a light brown solid.

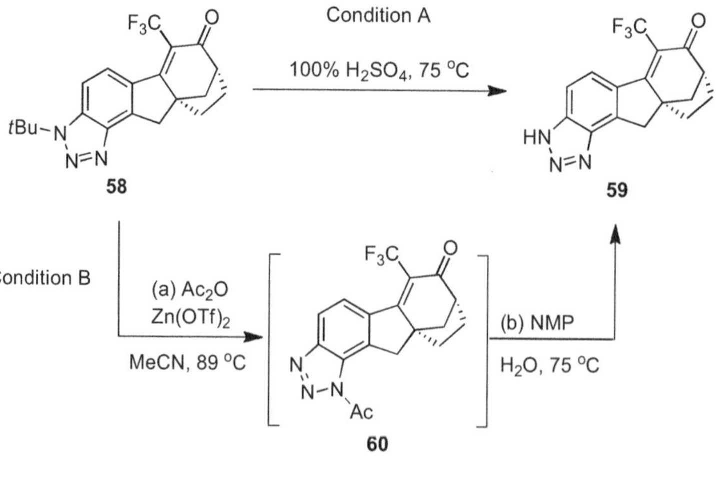

SCHEME 9.8

Note

In this protocol, the formation of isobutene was only addressed during the workup by using aceto-nitrile to trap the *tert*-butyl cation intermediate.

In the second protocol (Condition B, Scheme 9.8), the removal of the *t*-butyl group could be realized via *N*-acetylation with acetic anhydride in the presence of zinc triflate to give intermediate **60** whose subsequent hydrolysis afforded **59** in good yield.

Procedure

Stage (a)
- Charge 61.8 kg (605 mol) of acetic acid to a slurry of **58** (45.4 kg, 121 mol) and zinc triflate (8.80 kg, 48.4 mol) in MeCN (61.5 kg).
- Reflux the mixture for 48 h.
- Charge 8.80 kg (48.4 mol) of zinc triflate and continue to reflux for 12 h.

Stage (b)
- Add MeCN (71.5 kg) followed by water (91 kg) and NMP (46.8 kg).
- Stir at 75 °C for 3 h.
- Add water (113 kg) and cool to 25 °C.
- Add water (424 kg).
- Filter and wash with water (2×50 kg).
- Yield: 31.6 kg (82%) as a light brown solid.

REMARK

The reaction condition B eliminated the unfavorite strong sulfuric acid and the formation of isobu-tene off-gas by trapping the *t*-butyl cation with MeCN.

9.7.6 DECARBOXYLATIVE ETHOXIDE ELIMINATION

Ketone **65** was synthesized by a reaction of ethyl 4-iodobenzoate **61** with dianion **63**, generated in situ from a reaction of 2-phenylbutanoic acid **62** with LDA (Scheme 9.9).[41] The preparation was performed by mixing **62** with LDA at room temperature, followed by adding ester **61** THF solution at 0 °C.

(I) Problems

No significant CO_2 evolution was observed during the reaction until the aqueous workup. This delayed evolution of CO_2 raises safety concerns and must be addressed.

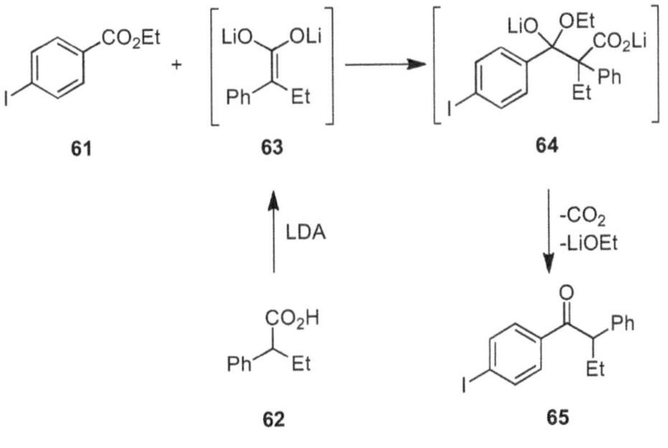

SCHEME 9.9

(II) Solutions

Presumably, CO_2 could be trapped in solution as adducts with bases such as LDA (and ethoxide) at the reaction temperature. To address the delayed CO_2 evolution, the additional protocol was modified wherein the dianion **63** solution was added to the solution of **61** at room temperature. Therefore, carbon dioxide evolution was safely controlled by the addition rate.

Procedure

- Charge dry THF (11.0 kg) and 183.5 kg of 25.2% (432 mol) of LDA in heptane/ethylbenzene/THF to a reaction vessel.
- Rinse with dry THF (4.0 kg).
- Charge a solution of **62** (35.9 kg, 218.7 mol) in THF (65 kg) at 23–24 °C over 85 min.
- Rinse with dry THF (30 kg).
- Stir the resulting mixture at 10–20 °C for 1.5 h to form a solution of **63**.
- Charge the solution of **63** to the solution of **61** (48.4 kg, 175.3 mol) in toluene (156.4 kg) over 30 min.
- Rinse with dry THF (30 kg).
- Stir the reaction mixture at 20–23 °C for 1 h.
- Add the resulting reaction mixture to an aqueous HCl solution (110 kg of 37% HCl in 140 kg of H_2O) at 17–27 °C over 1 h.
- Rinse with toluene (30 kg) and stir for 5 min.
- Separate layers.
- Wash the organic layer with water (170 L).
- Distill under reduced pressure to collect solvents (450 L).
- Add IPA (192 kg).
- Heat to 59 °C.
- Cool to 0 °C over 3.5 h and stir at 0 °C for 1 h.
- Filter, wash with cold IPA (77 kg), and dry at 25–30 °C for 24 h.
- Yield of **65**: 57.5 kg (94%).

9.7.7 DEPROTONATION

Sodium hydride reacts with proton source exothermically and can be hazardous on a large scale as explosive hydrogen gas is generated as a by-product. For example, the heat generated from the reaction of NaH with water can ignite the hydrogen gas. Thus, handling the flammable NaH solid on large scale poses a safety challenge. In practice, NaH in deprotonation can be replaced with alkaline alkoxide. For the synthesis of α,β-unsaturated ester **67**, NaOEt replaced the flammable NaH, which avoided dangerous foaming during the deprotonation (Scheme 9.10).[42]

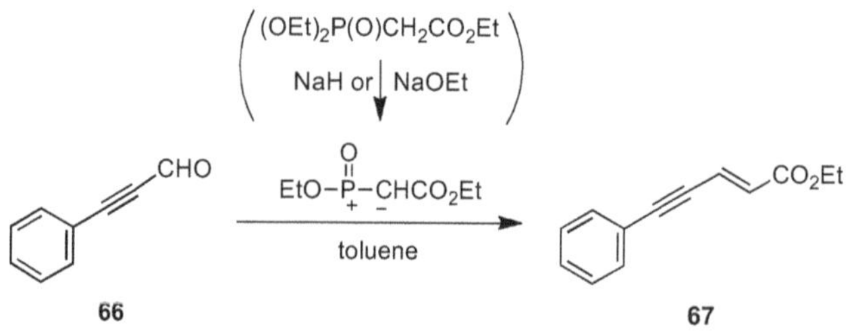

SCHEME 9.10

9.7.8 S$_N$2 REACTION

Acetylene is a colorless and extremely flammable gas and very sensitive to pressure, temperature, static electricity, and/or mechanical shock. It readily forms explosive mixtures with air over an unusually broad range of concentrations. The preparation of hydroxyl acetylene **70** via a reaction of chloride **69** with 2.1 equivalents of acetylene–ethylene diamine complex **68** in 1,3-dimethyl-3,4,5,6-tetrahydro-2(1H)pyrimidone (DMPU) generated 1 equivalent of the dangerous acetylene gas.

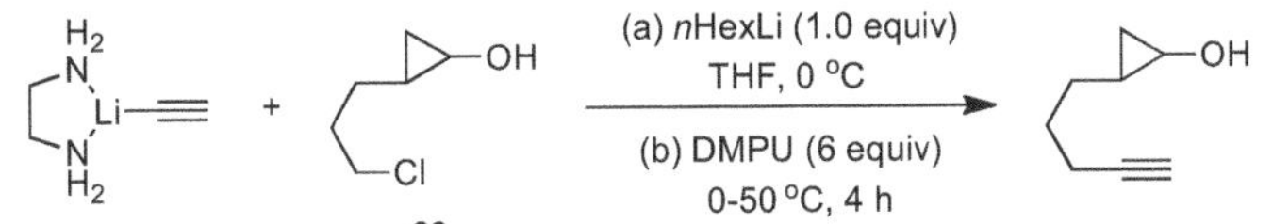

$$(9.25)$$

In order to limit the formation of acetylene, hexyllithium was identified as a sacrificial base to replace the extra equivalent of **68** (Equation 9.25).[43]

REMARK
In order to monitor the evolution of acetylene in the process, ReactIR was used to measure the acetylene gas (absorption at 3300 cm^{-1}) in the headspace.

9.7.9 CHLORINATION REACTION

To set a stage for safely manufacturing 2-chloroethyl phenyl ether **73**, a controlled addition of thionyl chloride was utilized during the chlorination of 2-phenoxyethanol **71** with thionyl chloride to control the evolution of toxic sulfur dioxide (Scheme 9.11).[41] This protocol was to add 1 equiv of thionyl chloride to a mixture of **71** and 1 equiv of pyridine at reflux in toluene, regulating effectively the evolution of sulfur dioxide.

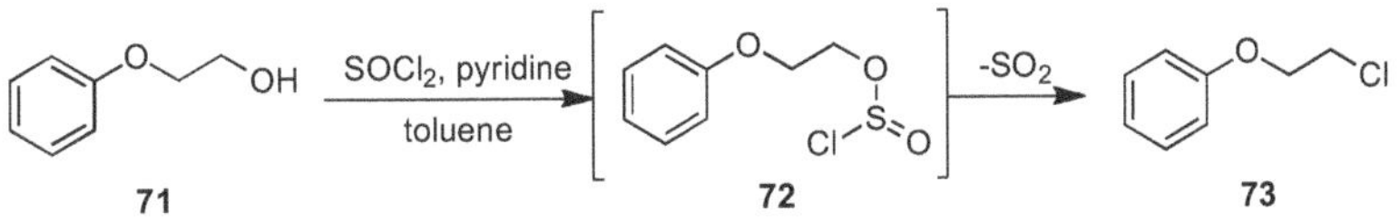

SCHEME 9.11

Procedure
* Charge toluene (395 kg) into a clean, dry, nitrogen-flushed vessel with a Dean–Stark water separator.
* Heat to reflux.
* Cool to 100–105 °C.
* Charge 70 kg (506.6 mol) of **71**, followed by adding 43.0 kg (554.9 mol, 1.1 equiv) of pyridine and rinsing with toluene (21 kg).
* Heat the resulting mixture to reflux.
* Charge 63.5 kg (533.9 mol, 1.05 equiv) of thionyl chloride as a solution in toluene (42 kg) over 90 min, followed by rinsing with toluene (21 kg).
* Reflux the batch continuously for 1 h.
* Cool to 20–25 °C.
* Add water (490 L) and stir at 20–25 °C for 10 min.
* Separate layers.
* Add water (490 L) to repeat the water wash.
* Distill at <60 °C under reduced pressure.

- Dry at 110–120 °C under vacuum (10–15 mbar)
- Yield of **73**: 69.7 kg (88%).

9.8 DARZENS REACTION

Accumulation of unreacted reagents may create unexpected exotherm during scale-up. The methyl glycidate intermediate **75** was prepared via the Darzens reaction by treating benzophenone **74** with methyl chloroacetate in the presence of NaOMe in methyl *tert*-butyl ether (MTBE) (Equation 9.26).[44]

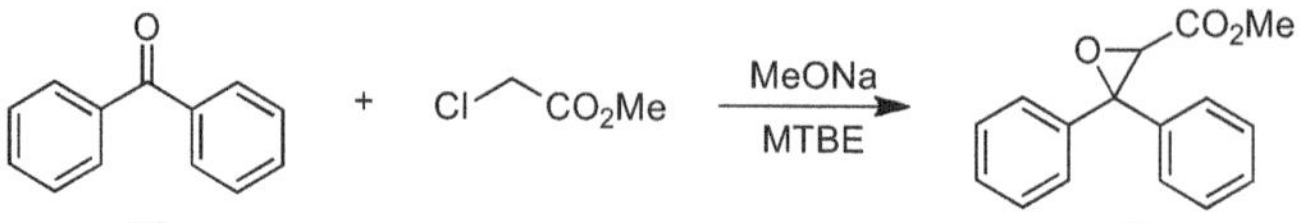

$$\text{(9.26)}$$

Although the reaction appeared to be no problem in the laboratory experiments and the calorimetric experiments, the first pilot scale synthesis was accompanied by an unwelcome surprise. The inefficient agitation and poor solubility of sodium methoxide in MTBE resulted in the accumulation of methyl chloroacetate, which, in turn, led to uncontrollable exotherm with a result that most of the MTBE evaporated within seconds.

9.9 HOFMANN REARRANGEMENT

Hofmann rearrangement presents exothermic hazards, which is a concern for the process safety. In order to address the safety issues, the Hofmann rearrangement for the conversion of pyridyl amide **76** to the corresponding amine **78** (as shown in Scheme 9.12)[45] was conducted in two stages: the formation of *N*-chloropyridyl amide **77** at 20 °C followed by a rearrangement with a controlled addition of the solution of **77** into a hot sodium hydroxide solution (80 °C). Using this protocol, the rearrangement was instantaneous without heat accumulation.

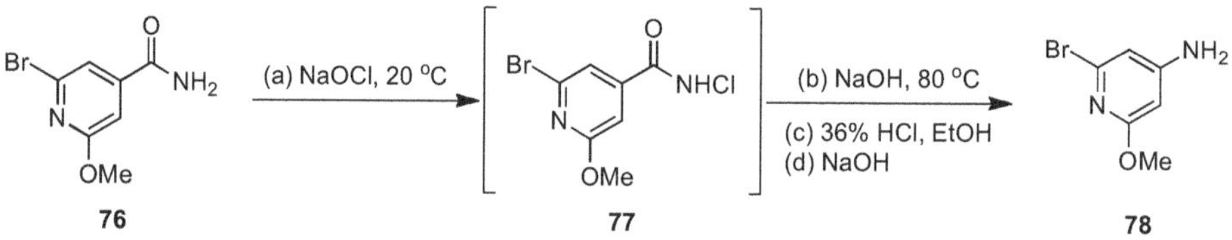

SCHEME 9.12

9.10 FRIEDEL–CRAFTS REACTION

Nitromethane is frequently used as a cosolvent in Friedel–Crafts reactions. However, the use of nitromethane solvent on at scale is a concern as nitromethane is a highly energetic compound. A scalable protocol was developed for the Friedel–Crafts acylation of azaindole **79** (Equation 9.27).[46] Instead of using nitrometane as the reaction solvent, the reaction was carried out in methylene chloride with tetra-*n*-butylammonium bisulfate as an additive.

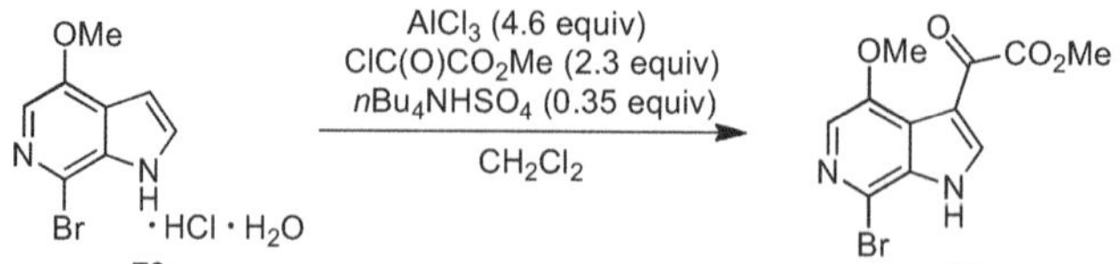

$$\text{(9.27)}$$

NOTES

1. McConville, F. X. *The Pilot Plant Real Book: A Unique Handbook for the Chemical Process Industry,* FXM Engineering and Design, Worcester, MA, 2002, pp 1–3.

2. Roberge, D. M. *Org. Process Res. Dev.* **2004**, *8*, 1049.

3. Camp, D.; Matthews, C. F.; Neville, S. T.; Rouns, M.; Scott, R. W.; Truong, Y. *Org. Process Res. Dev.* **2006**, *10*, 814.

4. Contcharov, A.; Dunetz, J. R. *Org. Process Res. Dev.* **2014**, *18*, 1145.

5. Choi, B. S.; Chang, J. H.; Choi, H.; Kim, Y. K.; Lee, K. K.; Lee, K. W.; Lee, J. H.; Heo, T.; Nam, D. H.; Shin, H. *Org. Process Res. Dev.* **2005**, *9*, 311.

6. Those applications include: (a) Merck's industrial-scale synthesis of anti-HIV drug indinavir (Crixivan), Thompson, W. J.; Fitzgerald, P. M. D.; Holloway, M. K.; Emini, E. A.; Darke, P. L.; McKeever, B. M.; Schlief, W. A.; Quintero, J. C.; Zugay, J. A.; Tucker, T. J.; Schwering, J. E.; Homnick, C. F.; Nunberg, J.; Springer, J. P.; Huff, J. R. *J. Med. Chem.* **1992**, *35*, 1685. (b) Synthesis of the alkaloid aristotelone, Kurti, L.; Czako, B. *Strategic Applications of Named in Organic Synthesis*, Elsevier Academic Press, Burlington, MA, 2005. (c) Synthesis of Amantadine, an antiviral and antiparkinsonian drug, Vardanyan, R.; Hruby, V.J. *Synthesis of Essential Drugs*, 1st ed. Elsevier, Amsterdam, 2006, p 137.

7. Chang, S.-J. *Org. Process Res. Dev.* **1999**, *3*, 232.

8. Maddess, M. L.; Scott, J. P.; Alorati, A.; Baxter, C.; Bremeyer, N.; Brewer, S.; Campos, K.; Cleator, E.; Dieguez-Vazquez, A.; Gibb, A.; Gibson, A.; Howard, M.; Keen, S.; Klapars, A.; Lee, J.; Li, J.; Lynch, J.; Mullens, P.; Wallace, D.; Wilson, R. *Org. Process Res. Dev.* **2014**, *18*, 528.

9. Kowalczyk, B. A.; Robinson, J., III; Gardner, J. O. *Org. Process Res. Dev.* **2001**, *5*, 116.

10. Mauragis, M. A.; Veley, M. F.; Lipton, M. F. *Org. Process Res. Dev.* **1997**, *1*, 39.

11. Shimizu, S.; Imamura, Y.; Ueki, T. *Org. Process Res. Dev.* **2014**, *18*, 354.

12. Chung, J. Y. L.; Steinhuebel, D.; Krska, S. W.; Hartner, F. W.; Cai, C.; Rosen, J.; Mancheno, D. E.; Pei, T.; DiMichele, L.; Ball, R. G.; Chen, C.; Tan, L.; Alorati, A. D.; Brewer, S. E.; Scott, J. P. *Org. Process Res. Dev.* **2012**, *16*, 1832.

13. Deoxo Fluor was employed to flourinate methyl 2,3,6-tri-*O*-benzoyl-r-D-galactopyranoside: Weiberth, F. J.; Gill, H. S.; Jiang, Y.; Lee, G. E.; Lienard, P.; Pemberton, C.; Powers, M. R.; Subotkowski, W.; Tomasik, W.; Vanasse, B. J.; Yu, Y. *Org. Process Res. Dev.* **2010**, *14*, 623.

14. L'Heureux, A.; Beaulieu, F.; Bennett, C.; Bill, D. R.; Clayton, S.; LaFlamme, F.; Mirmehrabi, M.; Tadayon, S.; Tovell, D.; Coururier, M. *J. Org. Chem.* **2010**, *75*, 3401.

15. Hayashi, H.; Sonoda, H.; Fukumura, K.; Nagata, T. *Chem. Commun.* **2002**, *15*, 1618.

16. (a) Carpino, L. A.; El-Faham, A. *J. Am. Chem. Soc.* **1995**, *117*, 5401. (b) El-Faham, A.; Khattab, S. N. *Synlett* **2009**, *6*, 886. (c) Bellavance, G.; Dubé, P.; Nguyen, B. *Synlett* **2012**, *23*, 569.

17. Nielsen, M. K.; Ugaz, C. R.; Li, W.; Doyle, A. G. *J. Am. Chem. Soc.* **2015**, *137*, 9571.

18. Kalow, J. A.; Schmitt, D. E.; Doyle, A. G. *J. Org. Chem.* **2012**, *77*, 4177.

19. McPake, C. B.; Sandford, G. *Org. Process Res. Dev.* **2012**, *16*, 844.

20. Walker, M. D.; Albinson, F. D.; Clark, H. F.; Clark, S.; Henley, N. P.; Horan, R. A. J.; Jones, C. W.; Wade, C. E.; Ward, R. A. *Org. Process Res. Dev.* **2014**, *18*, 82.

21. Anderson, N. G.; Ary, T. D.; Berg, J. L.; Bernot, P. J.; Chan, Y. Y.; Chen, C.-K.; Davies, M. L.; DiMarco, J. D.; Dennis, R. D.; Deshpande, R. P.; Do, H. D.; Droghini, R.; Early, W. A.; Gougoutas, J. Z.; Grosso, J. A.; Harris, J. C.; Haas, O. W.; Jass, P. A.; Kim, D. H.; Kodersha, G. A.; Kotnis, A. S.; LaJeunesse, J.; Lust, D. A.; Madding, G. D.; Modi, S. P.; Moniot, J. L.; Nguyen, A.; Palaniswamy, V.; Phillipson, D. W.; Simpson, J. H.; Thoraval, D.; Thurston, D. A.; Tse, K.; Polomski, R. E.; Wedding, D. L.; Winter, W. J. *Org. Process Res. Dev.* **1997**, *1*, 300.

22. Broxer, S.; Fitzgerald, M. A.; Sfouggatakis, C.; Defreese, J. L.; Barlow, E.; Powers, G. L.; Peddicord, M.; Su, B.-N.; Tai-Yuen, Y.; Pathirana, C.; Sherbine, J. P. *Org. Process Res. Dev.* **2011**, *15*, 343.

23. Liu, Y.; Zhang, Z.; Wu, A.; Yang, X.; Zhu, Y.; Zhao, N. *Org. Process Res. Dev.* **2014**, *18*, 349.

24. Brenek, S. J.; am Ende, D. J.; Clifford, P. J. *Org. Process Res. Dev.* **2000**, *4*, 585.

25. Arnott, E. A.; Chan, L. C.; Cox, B. G.; Meyrick, B.; Phillips, A. *J. Org. Chem.* **2011**, *76*, 1653.

26. Pesti, J. A.; LaPorte, T.; Thornton, J. E.; Spangler, L.; Buono, F.; Crispino, G.; Gibson, F.; Lobben, P.; Papaioannou, C. G. *Org. Process Res. Dev.* **2014**, *18*, 89.

27. Cleator, E.; Scott, J. P.; Avalle, P.; Bio, M. M.; Brewer, S. E.; Davies, A. J.; Gibb, A. D.; Sheen, F. J.; Stewart, G. W.; Wallace, D. J.; Wilson, R. D. *Org. Process Res. Dev.* **2013**, *17*, 1561.

28. Zeng, Q.; Huang, B.; Danielsen, K.; Shukla, R.; Nagy, T. *Org. Process Res. Dev.* **2004**, *8*, 962.

29. (a) Lane, J. W.; Spencer, K. L.; Shakya, S. R.; Kallan, N. C.; Stengel, P. J.; Remarchuk, T. *Org. Process Res. Dev.* **2014**, *18*, 1641. (b) Remarchuk, T.; St-Jean, F.; Carrera, D.; Savage, S.; Yajima, H.; Wong, B.;

Babu, S.; Deese, A.; Stults, J.; Dong, M. W.; Askin, D.; Lane, J. W.; Spencer, K. L. *Org. Process Res. Dev.* **2014**, *18*, 1652.

30. Frutos, R. P.; Tampone, T. G.; Mulder, J. A; Rodriguez, S.; Yee, N. K.; Yang, B.-S.; Senanayake, C. H. *Org. Process Res. Dev.* **2016**, *20*, 982.

31. Achmatowica, M.; Thiel, O. R.; Wheeler, P.; Bernard, C.; Huang, J.; Larsen, R. D.; Faul, M. M. *J. Org. Chem.* **2009**, *74*, 795.

32. Banert, K.; Richter, F.; Hagedorn, M. *Org. Process Res. Dev.* **2015**, *19*, 1068.

33. Patel, B.; Firkin, C. R.; Snape, E. W.; Jenkin, S. L.; Brown, D.; Chaffey, J. G. K.; Hopes, P. A.; Reens, C. D.; Butters, M.; Moseley, J. D. *Org. Process Res. Dev.* **2012**, *16*, 447.

34. Delmonte, A. J.; Fan, Y.; Girard, K. P.; Jones, G. S.; Waltermire, R. E.; Rosso, V.; Wang, X. *Org. Process Res. Dev.* **2011**, *15*, 64.

35. Fisher, R. W.; Misum, M. *Org. Process Res. Dev.* **2001**, *5*, 581.

36. (a) Greene, T. W.; Wuts, P. G. M. *Protective Groups in Organic Synthesis*, 3rd ed.; Wiley, New York, 1999; pp 518–525. (b) Lawrence, S. A. *Chimica Oggi* **1999**, 15. (c) Carpino, L. A. *Acc. Chem. Res.* **1973**, *6*, 191. (d) Carpino, L. A. *J. Am. Chem. Soc.* **1957**, *79*, 98. (3) Strazzolini, P.; Melloni, T.; Giumanini, A. G. *Tetrahedron* **2001**, *57*, 9033.

37. Dias, E. L.; Hettenbach, K. W.; am Ende, D. J. *Org. Process Res. Dev.* **2005**, *9*, 39.

38. Chung, J. Y. L.; Scott, J. P.; Anderson, C.; Bishop, B.; Bremeyer, N.; Cao, Y.; Chen, Q.; Dunn, R.; Kassim, A.; Lieberman, D.; Moment, A. J.; Sheen, F.; Zacuto, M. *Org. Process Res. Dev.* **2015**, *19*, 1760.

39. Bien, J.; Davulcu, A.; DelMonte, A. J.; Fraunhoffer, K. J.; Gao, Z.; Hang, C.; Hsiao, Y.; Hu, W.; Katipally, K.; Littke, A.; Pedro, A.; Qiu, Y.; Sandoval, M.; Schild, R.; Soltani, M.; Tedesco, A.; Vanyo, D.; Vemishetti, P. *Org. Process Res. Dev.* **2018**, *22*, 1393.

40. Jiang, X.; Lee, G. T.; Villhauer, E. B.; Prasad, K.; Prashad, M. *Org. Process Res. Dev.* **2010**, *14*, 883.

41. Ace, K. K.; Armitage, M. A.; Bellingham, R. K.; Blackler, P. D.; Ennis, D. S.; Hussain, N.; Lathbury, D. C.; Morgan, D. O.; O'Connor, N.; Oakes, G. H.; Passey, S. C.; Powling, L. C. *Org. Process Res. Dev.* **2001**, *5*, 479.

42. Deussen, H.-J.; Jeppesen, L.; Schärer, N.; Junager, F.; Bentzen, B.; Weber, B.; Weil, V.; Mozer, S. J.; Sauerberg, P. *Org. Process Res. Dev.* **2004**, *8*, 363.

43. Bassan, E. M.; Baxter, C. A.; Beutner, G. L.; Emerson, K. M.; Fleitz, F. J.; Johnson, S.; Keen, S.; Kim, M. M.; Kuethe, J. T.; Leonard, W. R.; Mullens, P. R.; Muzzio, D. J.; Roberge, C.; Yasuda, N. *Org. Process Res. Dev.* **2012**, *16*, 87.

44. Jansen, R.; Knopp, M.; Amberg, W.; Bernard, H.; Koser, S.; Müller, S.; Münster, I.; Pfeiffer, T.; Riechers, H. *Org. Process Res. Dev.* **2001**, *5*, 16.

45. Daver, S.; Rodeville, N.; Pineau, F.; Arlabosse, J.-M.; Moureou, C.; Muller, F.; Pierre, R.; Bouquet, K.; Dumais, L.; Boiteau, J.-G.; Cardinaud, I. *Org. Process Res. Dev.* **2017**, *21*, 231.

46. Zheng, B.; Silverman, S. M.; Steinhardt, S. E.; Kolotuchin, S.; Iyer, V.; Fan, J.; Skliar, D.; McLeod, D. D.; Bultman, M.; Tripp, J. C.; Murugesan, S.; La Cruz, T. E.; Sweeney, J. T.; Eastgate, M. D.; Conlon, D. A. *Org. Process Res. Dev.* **2017**, *21*, 1145.

10 Hazardous Reagents

Exothermic chemical processes often release energy, in the form of heat and pressure, under certain conditions, which may create a serious incident with risk to people's life and damage to property and the environment. Such chemical reactions may be caused by using hazardous starting materials and reagents, such as diazonium salts, azide compounds, hydrazine, etc.

Chemical reagents are classified, according to their hazardous nature, as flammable, harmful or toxic to humans, corrosive to the production equipment, hazardous to the environment, and explosive when exposed to heat or under friction. Some reagents bear a combination of such hazards. To prevent possible accidents, it is important during handling hazardous materials to refer to the materials safety data sheet (MSDS) and related regulations and use safety measures appropriate to the characteristics of the reagent.

10.1 DIAZONIUM SALTS

Some reactions, such as diazotization chemistry (e.g., the Sandmeyer reaction) and aromatic nitrations, require significant efforts to generate the necessary supporting calorimetric and safety data prior to running the chemistry at scale.

The major hazards of diazonium salts arise from their inherent instability which can lead to explosion with the release of nitrogen gas. Thus, the accumulation of the diazonium salt intermediates during the course of diazotization poses a safety concern. Most of the incidents[1] were triggered by friction during the cleaning of the equipment due to the sensitive diazonium salts precipitated in valves or in the reactor.[2] To avoid the accumulation of diazonium salt and maintain a low concentration of the hazardous intermediate in the reaction system, a common practice in the process development is the slow addition of sodium nitrite to a warm aqueous substrate solution. This dose-controlled protocol is well accepted by process chemists.

10.1.1 Hydrolysis of Diazonium Salt

A process for the preparation of phenol **3** from aniline **1** via diazonium salt intermediate **2** was devised (Scheme 10.1).[3]

To avoid the accumulation of unstable **2**, a solution of $NaNO_2$ in water was added to a solution of **1** in aqueous H_2SO_4 and acetonitrile at 45–55 °C over a period of 35 min. The in situ hydrolytic decomposition of the diazonium salt **2** and the gas release (practically no foaming) were controlled by the addition rate. The reaction was completed approximately 10 min after the end of dosage to afford the desired product **3** in 81% yield.

Procedure

- Charge a freshly prepared solution of $NaNO_2$ (137.5 g, 1.99 mol, 1.2 equiv) in water (270 mL) to a mixture of 30% H_2SO_4 (5 kg, 15.3 mol) and **1** (266.6 g, 1.66 mol) in MeCN (270 mL) at 50–55 °C over 35 min.
- Cool to 20 °C.
- Extract with MTBE (1.7 L, 2×1 L).
- Wash the combined organic layers with water (1 L).

DOI: 10.1201/9781003288411-10

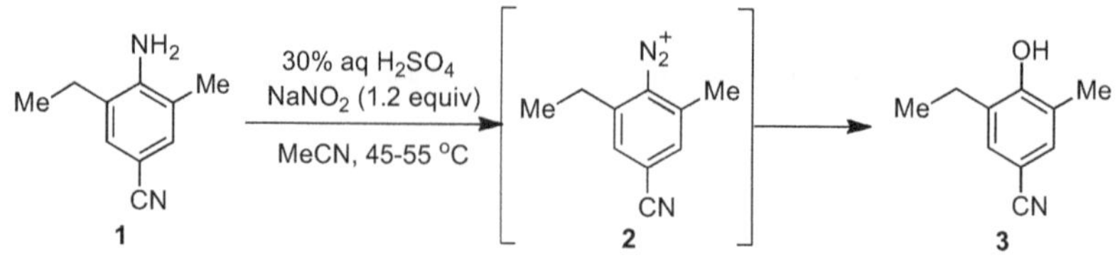

SCHEME 10.1

- Distill under reduced pressure at 70 °C to dryness to give 271 g crude solid.
- Crystallize the crude **3** from EtOAc (120 mL)/cyclohexane (1.2 L).
- Yield of **3**: 218 g (81%).

REMARK

The residual diazonium salt can be decomposed by treating the reaction mixture with urea (0.3 equiv) for 10 min at 50 °C.

10.1.2   Diazonium Salt-Involved Cyclization

The synthesis of 3-hydroxypyrazole **6** involves a diazonium intermediate **5** (Scheme 10.2).[4] Accumulations of the sticky solid of **5** are a big concern at scale-up.

In order to address these safety issues, the reaction was carried out by charging an aqueous solution of NaNO$_2$ slowly to the solution of hydrazide **4** in a mixture of 1 N HCl and ethanol at reflux. The in situ generated diazonium salt **5** immediately cyclized and the resulting indazolone product **6** precipitated out from the reaction mixture.

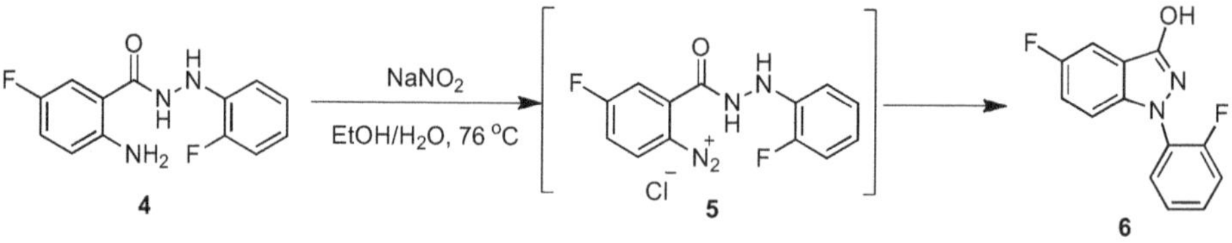

SCHEME 10.2

Procedure

- Heat a suspension of **4** (300 g, 1.14 mol) and 1 N HCl (4 L) in EtOH (4 L) to 76 °C to form a light-red solution
- Charge a solution of NaNO$_2$ (236 g, 3.42 mol) in water (600 mL) over 1 h.
- Stir the resulting mixture at 76 °C for an additional 1 h.
- Cool to 20 °C.
- Filter, wash with water (3×500 mL), and dry at 50 °C under vacuum for 18 h.
- Yield of **6**: 196 g (70%).

10.1.3   Nitroindazole Formation

According to the literature procedure,[5] the preparation of nitroindazole **9** was conducted by a fast, one-portion addition of sodium nitrite solution to a mixture of nitrotoluidine **7** and acetic acid (Scheme 10.3).[6]

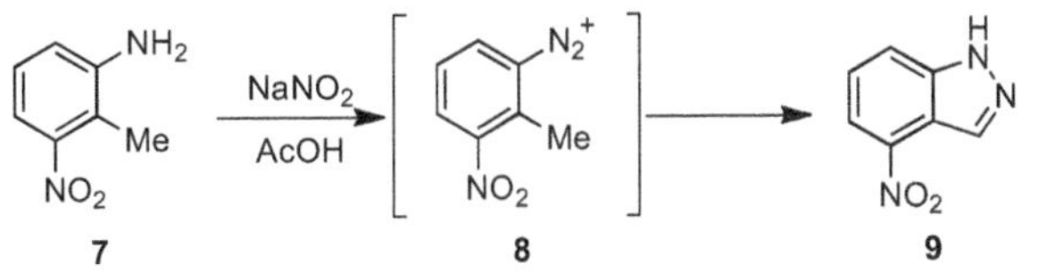

SCHEME 10.3

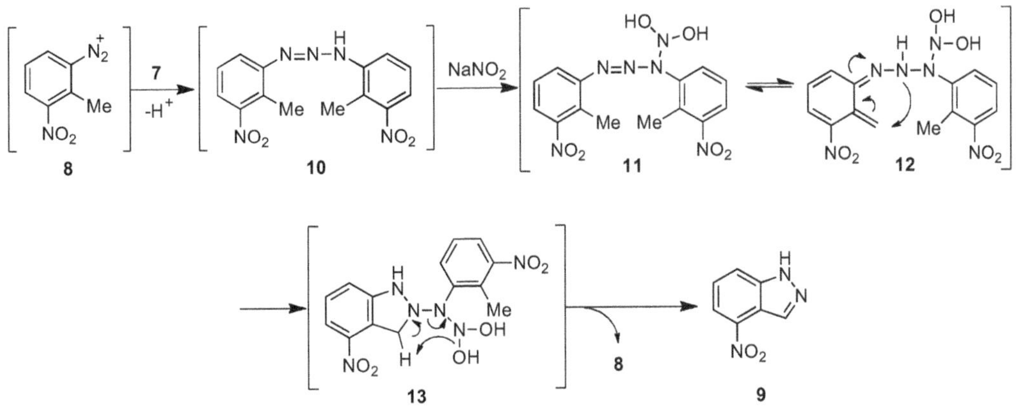

SCHEME 10.4

(I) Reaction problems

Such fast addition of $NaNO_2$ led to a large exotherm and a slow addition of the $NaNO_2$ solution, however, resulted in massive precipitation of an unknown compound and incomplete reaction.

(II) Chemistry diagnosis

Being aware of the incomplete reaction associated with the unknown precipitation, efforts were made to identify the structure of the unknown precipitate and to investigate its chemical properties. As a result, the unknown compound was identified as triazene **10** which, fortunately, could be converted to the desired nitroindazole **9** by adding excess sodium nitrite.

The sodium nitrite-promoted transformation of **10** to **9** was hypothesized to proceed through intermediates **11–13** as illustrated in Scheme 10.4.

(III) Solutions

With the hypothesized reaction mechanism in mind, a new process was developed which included a slow addition (over 1 h period) of the sodium nitrite solution to the preheated reaction mixture at 50 °C. This process provides good control of the exothermic event and prevents intermediate triazene precipitation.

Procedure

- Charge slowly an aqueous solution of $NaNO_2$ (2.37 kg, 34.3 mol, 2.09 equiv) in 6.0 kg of water to a warm solution (50 °C) of **7** (2.5 kg, 16.4 mol) in 53 kg of acetic acid at 45–55 °C.
- Stir the mixture at 50 °C for an additional 1 h.
- Distill under reduced pressure to ca 10 L.
- Add 30 kg of water.
- Distill continuously to ca. 10 L.
- Add 30 kg of water and cool to 20 °C.

- Filter, wash with 30 kg of water, and dry.
- Yield: 2.6 kg (95%).

10.1.4   SYNTHESIS OF TRIFLUOROMETHYL-SUBSTITUTED CYCLOPROPANES

The slow addition approach was adopted by scientists in academia as well. For example, a mild and environmentally benign protocol for the preparation of trifluoromethyl-substituted cyclopropanes involved two in situ generated reactive intermediates, the diazonium salt **17** and diazoalkane **18** (Scheme 10.5).[7]

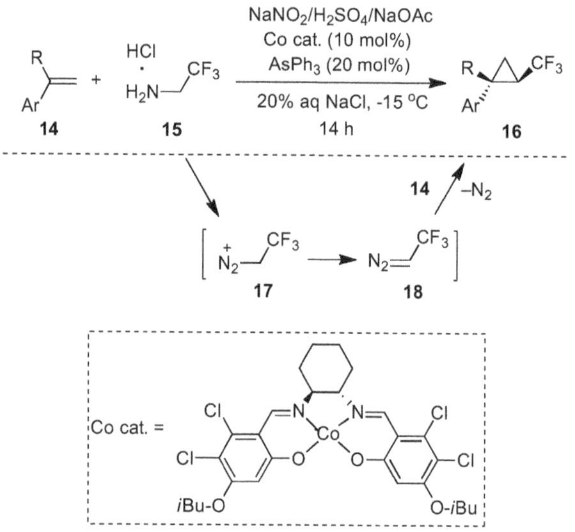

SCHEME 10.5

The addition of sodium nitrite to an aqueous solution of substrates **14** and **15** in the presence of a cobalt catalyst allowed this tandem process to proceed smoothly, in which the cyclization of **18** concomitantly with **14**, avoiding the accumulation of **17** or **18** in the reaction system.

10.1.5   SANDMEYER REACTION

Due to the nature of the Sandmeyer reaction, diazonium salts have to be prepared prior to mixing with sodium cyanide/copper cyanide. Given the unpredictable nature of the diazonium functionality, a safety evaluation of the Sandmeyer reaction of 2-chloro-5-trifluoromethylaniline **19** was carried out, leading to an improved process for the production of 2-chloro-5-trifluoromethyl-benzonitrile **21**.

In the new process as described in Scheme 10.6,[8] the diazonium salt **20** was prepared at a temperature below 3 °C, which was subsequently transferred into a hot aqueous mixture (80 °C) of sodium carbonate, copper cyanide, and sodium cyanide.

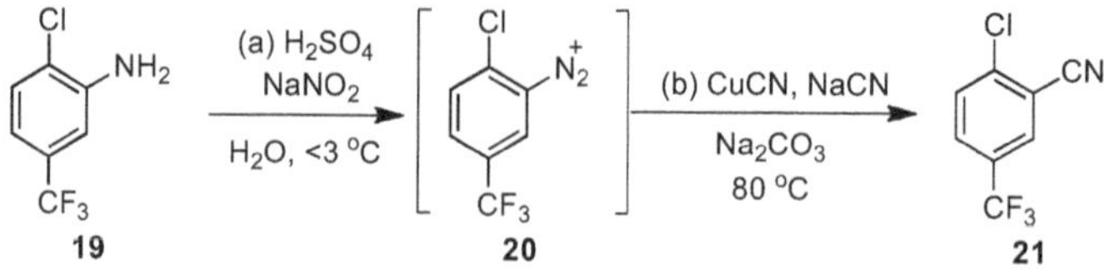

SCHEME 10.6

Procedure

Stage (a)
- Prepare an aqueous sulfuric acid solution by adding 29 kg (245 mol) H_2SO_4 into 80 L of H_2O.
- Cool the solution to <14 °C.
- Charge 24 kg (123 mol) of **19** at <15 °C.
- Add toluene (50 L) and cool to 3 °C.
- Charge a solution of $NaNO_2$ (9.6 kg, 139 mol) in H_2O (20 L) at <3 °C.
- Stir at <3 °C for 45 min.
- Separate layers.

Stage (b)
- Charge the aqueous layer to a hot (80 °C) slurry of Na_2CO_3 (36 kg, 340 mol), CuCN (11.3 kg, 402 mol), and NaCN (9.6 kg, 196 mol) in H_2O (87 L).
- Stir at 65 °C for 2 h.
- Add heptane (175 L).
- Separate layers.
- Distill the heptane layer under reduced pressure.
- Yield: 12.5 kg (50% yield and 94.4% purity).

REMARKS

(a) The temperature of the bulk diazonium ion solution never exceeds 15 °C.
(b) To avoid any mechanical stress to the diazonium salt and the risk of having diazonium salt deposits in the pump chambers, the transfer of the diazonium salt solution is performed using a vacuum.
(c) The excess cyanide can be destroyed with sodium hypochlorite.[9]

10.2 AZIDE COMPOUNDS

Sodium azide[10] is used in the literature to construct nitrogen-containing heterocyclic compounds. Albeit present in the literature, the use of sodium azide on large scale poses major safety concerns, especially under elevated temperatures. The formation of hydrazoic acid (HN_3) is inherent to many azide-involved processes due to the presence of small amounts of protic components in the reaction mixture. Hydrazoic acid is an unstable compound that may decompose violently, forming nitrogen and hydrogen (Equation 10.1).[11]

$$2\,HN_3 \longrightarrow H_2 + 3\,N_2 \tag{10.1}$$

Therefore, it is important that any azide-involved reactions have to maintain basic conditions in order to suppress the formation of hydrazoic acid. Any risk of the potential accumulation of hyrazoic acid in the headspace must be reduced by applying a constant sweep of nitrogen over the head space during plant operation. In addition, to avoid any opportunity for explosive liquid hydrazoic acid to condense on the condenser, it is suggested not to use the condensing system during processing. ICP analysis for metal residues in input materials and diligent cleaning of the reactors are necessary to avoid any presence of heavy metals, which could catalyze azide decomposition.

Hydrazoic acid, as well as sodium azide, is highly toxic and extremely shock sensitive. The recommended airborne exposure limit is 0.11 ppm for HN_3 or 0.3 mg/m^3 for NaN_3.

Organic azides are well-known for their poor thermal stability and must be treated with great care. A thorough risk assessment study must be conducted prior to scale-up of any organic azide reactions. In addition, any potential in situ formation of organic azides should be carefully evaluated.

For example, reactions of azides that involve phase transfer catalysts, such as methyl trioctyl ammonium, ethyl tribenzyl ammonium, or methyl tributyl ammonium, may produce in situ trialkylamine and alkyl azides such as methyl or ethyl azides via nucleophilic displacement. Methyl azide and ethyl azide are extremely hazardous and explosive.

10.2.1 Nucleophilic Displacement

Most azide-involved chemical transformations are accomplished via nucleophilic substitution processes.

10.2.1.1 Synthesis of 3,4-Dihydropyrrole

The synthesis of 2,3-dihydro-pyrrole **24** was realized by employing a large excess (5 equiv) of NaN$_3$ at an elevated temperature (80 °C) (Scheme 10.7).[12] Apparently, these reaction conditions are not suitable for scale-up.

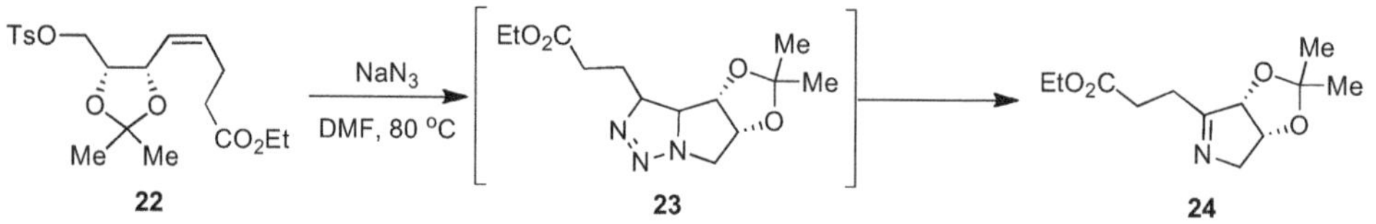

SCHEME 10.7

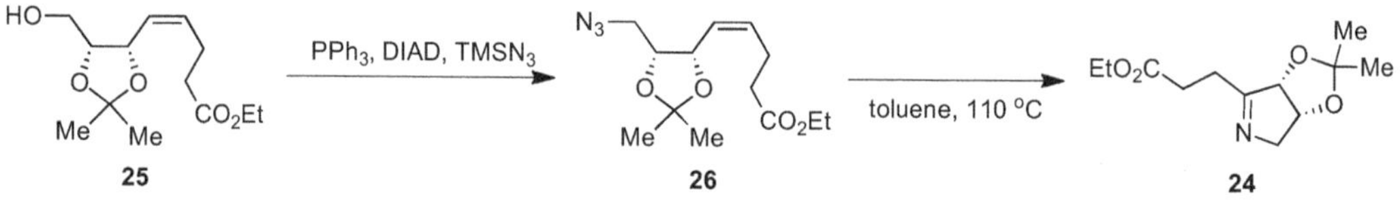

SCHEME 10.8

Scheme 10.8[12] depicts the optimized process which was scaled up to make a kilograms quantity of **24**. The new process was performed at 0 °C under Mitsunobu reaction[13] conditions using TMSN$_3$ as the source of azide.

REMARK

The use of a polymeric azidation reagent for the preparation of explosive azidomethanes at ambient temperature is described (Equation 10.2).[14] Thus, methyl iodide was converted into methyl azide, methylene bromide, and methylene chloride into diazido methane and bromoform into triazido methane.

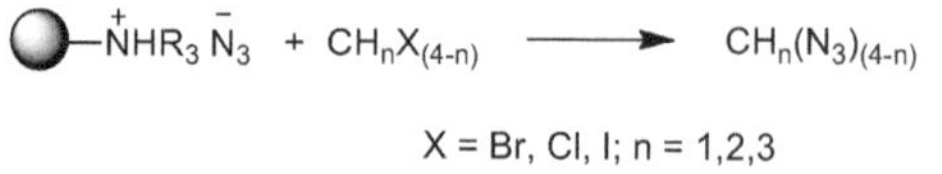

$$X = Br, Cl, I; n = 1,2,3$$

10.2.1.2 Synthesis of 3-(1,2-Diarylbutyl) Azide

During a displacement reaction of mesylate **27** with sodium azide (Equation 10.3),[15] a small amount of hydrazoic acid was observed as per online FTIR.

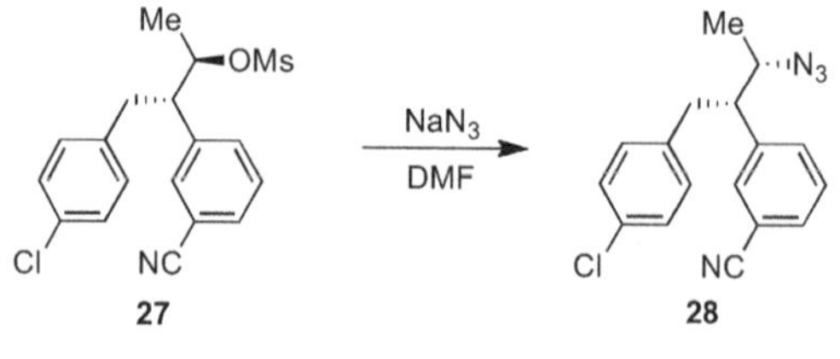

To eliminate the potential safety hazard associated with HN_3, the use of DIPEA suppressed completely the formation of HN_3.

Procedure

- Charge 2.03 L (11.64 mol) of DIPEA to a solution of mesylate **27** (3.53 kg, 9.70 mol) in DMF.
- Charge 757 g (11.64 mol) of NaN_3.
- Heat slowly to 70 °C and stir for 8–11 h.
- Cool to 25 °C.
- Add 34.4 L of 5% $NaHCO_3$ and 13.6 L of toluene.
- Separate layers, wash the organic layer with water, and then with 1 N HCl (the resulting solution was used in the subsequent step).

REMARK

The use of an inorganic base such as K_2CO_3 was less effective. A blast shield should be used for this reaction.

10.2.1.3 Preparation of Aryl (Alkyl) Azides

The risk assessment study (by DSC, drop-weight testing, and advanced reactive system screening tool) of aryl azide **31** proved that the neat **31** is highly shock sensitive and has a very large exothermic energy with a potential of explosion upon heating, whereas a diluted solution of **31** is safe enough to handle (Scheme 10.9).[16]

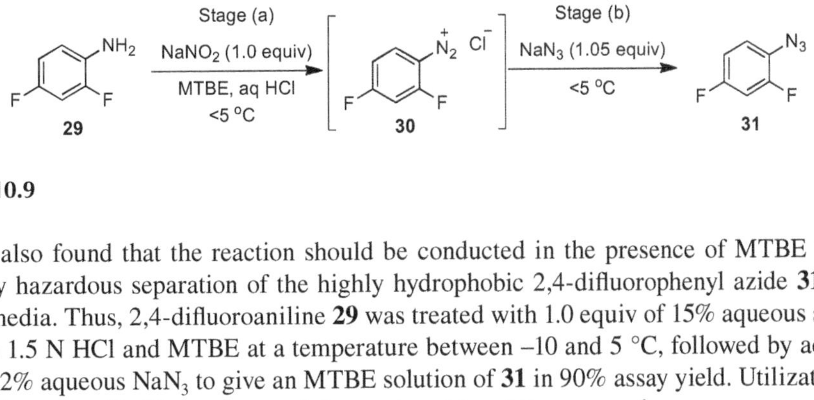

SCHEME 10.9

It was also found that the reaction should be conducted in the presence of MTBE to prevent potentially hazardous separation of the highly hydrophobic 2,4-difluorophenyl azide **31** from the aqueous media. Thus, 2,4-difluoroaniline **29** was treated with 1.0 equiv of 15% aqueous solution of $NaNO_2$ in 1.5 N HCl and MTBE at a temperature between –10 and 5 °C, followed by adding 1.05 equiv of 12% aqueous NaN_3 to give an MTBE solution of **31** in 90% assay yield. Utilization of this protocol, the aryl azide **31** was successfully produced in a kilogram scale.

Procedure

Stage (a)
- Charge 14 L of 1.5 N HCl into a 50-L vessel, followed by adding MTBE (3.5 L) and 900 g (6.97 mol) of **29**.
- Cool to –10 °C.
- Charge an aqueous solution of $NaNO_2$ (480 g, 6.97 mol in 2.7 L of water) at –10 °C to 5 °C.
- Continue to stir at 0–5 °C for 15 min.

Stage (b)
- Charge a 12 wt% aqueous solution of NaN_3 (480 g, 7.32 mol in 3.6 L of water) at –10 to 5 °C.
- Continue to stir at 0–5 °C for 30 min.

- Add MTBE (6.98 L).
- Separate layers.
- Wash the organic layer with 1 N NaOH (4.5 L) and water (4.5 L) successively.
- Azeotrope with MTBE (ca 22.5 L) at <40 °C until KF <500 ppm.
- Yield: 970 g (90%) (MTBE solution assay yield).

NOTES

(1) Due to the high shock sensitivity and thermal instability, aryl azide **31** must not be isolated.

(2) Considering the explosive nature of aryl azide **31** with metal and heat, the reaction temperature was controlled by an oil bath, not an internal heating/cooling pipe.

(3) The reaction vessel was covered with aluminum foil to obscure light since the aryl azide **31** is decomposed by light.

(4) Aqueous solutions greater than 15 wt% NaN$_3$ could pose a thermal hazard. Therefore, it is recommended to prepare a more dilute sodium azide aqueous solution, the temperature of the aqueous sodium azide solution must be maintained cold (–5 °C) to avoid volatilization of hydrogen azide (the boiling point of hydrogen azide is 36–38 °C).

(5) Distillation of alkyl azides is extremely hazardous and can result in explosions.[17] A laboratory-scale preparation of benzyl azide **33** was carried out via a reaction of benzyl chloride **32** with sodium azide (Equation 10.4).[18] The purification of **33** with wiped film evaporation (WFE) technology was demonstrated in 94% yield with 98% purity.

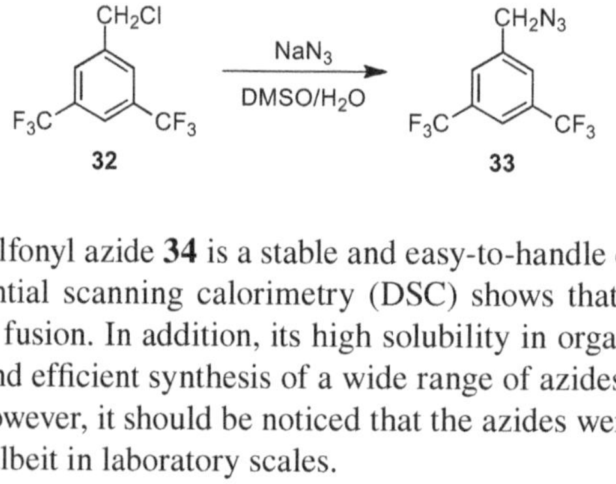

(10.4)

(6) Benzotriazol-1-yl-sulfonyl azide **34** is a stable and easy-to-handle crystalline diazo-transfer reagent. Differential scanning calorimetry (DSC) shows that **34** is stable below 95 °C with 166.7 J/g of fusion. In addition, its high solubility in organic and aqueous media allows convenient and efficient synthesis of a wide range of azides and diazo compounds (Scheme 10.10).[19] However, it should be noticed that the azides were prepared in the presence of copper salt albeit in laboratory scales.

Safety Notes

(a) Azide reactions shall be free of heavy metals, such as lead, copper, silver, gold, and mercury, due to the ready formation of insoluble explosive heavy-metal azide salts.[20] These salts can settle to low points in the processing equipment and detonate when

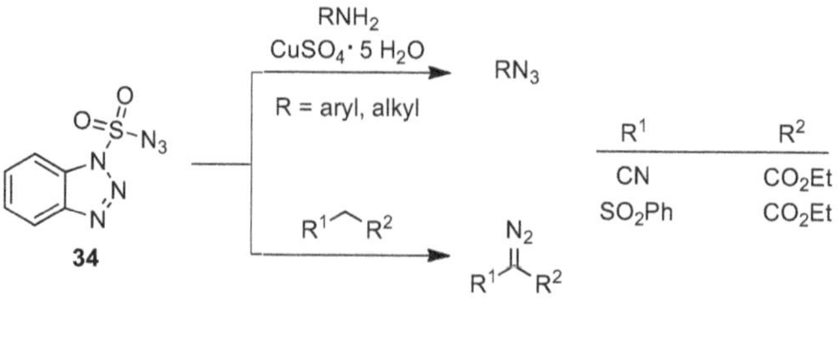

SCHEME 10.10

they are subjected to mechanical or thermal stress. When preparation of alkyl azide, dichloromethane shall not be used as the reaction solvent as a reaction of dichloromethane with sodium azide can form diazidomethane, $CH_2(N_3)_2$, which exhibits a strong explosive propensity.[14,21]

(b) An explosion occurred during a 200 g scale synthesis of azidotrimethylsilane from chlorotrimethylsilane and sodium azide in poly(ethylene glycol) (PEG) as solvent (Equation 10.5).[22]

$$TMSCl + NaN_3 \longrightarrow TMSN_3 + NaCl \qquad (10.5)$$

It was postulated that the explosion may be caused by hydrazoic acid that may have been generated in the presence of a proton source (PEG) or the failure of the magnetic stirring during the distillation purification, causing overheat of unreacted azide salts at the bottom of the flask, which eventually led to an explosion.

10.2.2 NUCLEOPHILIC ADDITION

10.2.2.1 Synthesis of Tetrazole

Tetrazoles are generally regarded as energy-rich, thermo-dynamically unstable compounds. Sodium azide is widely used as an azide source in organic synthesis. However, the use of sodium azide can potentially generate hazardous hydrazoic acid, especially in acidic reaction media.

10.2.2.1.1 Use of Sodium Azide

To suppress the formation of hydrazoic acid from sodium azide during the synthesis of tetrazole **36**, a catalytic amount of ZnO was used to facilitate the cyclization reaction (Equation 10.6).[23] This modification ensured the safety of the reaction as only ca. 2 ppm hydrazoic acid was detected in the head space, which is far below the detonation threshold (ca. 15000 ppm).

$$(10.6)$$

Procedure

- Charge 10.96 kg (56 wt%, 16.8 mol) of **35** into a reaction vessel that was connected to a scrubber containing 3 N NaOH.
- Charge 137 g (1.68 mol) of ZnO followed by THF (6.2 L) and H_2O (31 L).
- Charge an aqueous solution of 1.2 kg (18.5 mol) of NaN_3 in 3.1 L of H_2O via an additional funnel over 20 min.
- Heat the mixture to 78 °C for 15 h.
- Cool to ambient temperature.
- Add saturated aqueous $NaHCO_3$ (16 L), brine (16 L), and 2-MeTHF (31 L).
- Separate the organic product layer.
- Yield: 6.88 kg (100%) (a solution assay yield).

Another cycloaddition of nitrile **37** with sodium azide was developed for the synthesis of tetrazole **38** (Equation 10.7).[24]

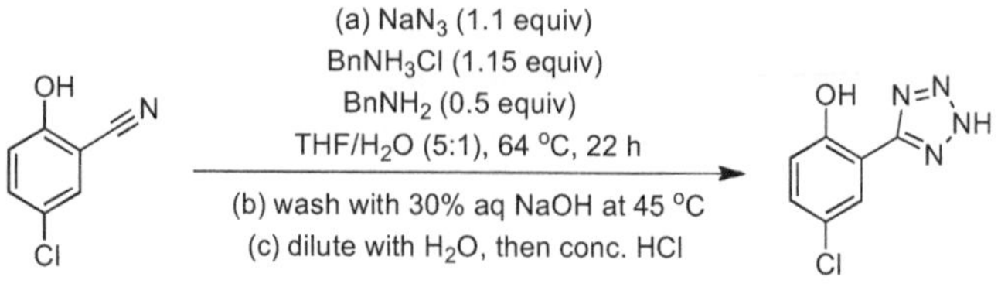

(10.7)

Benzylamine hydrochloride was employed to replace ammonium chloride used in the literature[25] to prevent sublimation of explosive NH_4N_3 (vapor pressure at 80 °C is 38 mmHg). Water was introduced to improve the reaction rate. In addition, buffering the reaction mixture with benzylamine (0.5 equiv) reduced the level of HN_3 in the headspace.

REMARK

All aqueous solutions of sodium azide shall be kept above pH 7, all equipment shall be washed with copious aqueous base post-use, and all azide-containing waste shall be collected separately, kept basic, and disposed of separately.

10.2.2.1.2 Use of Sodium Tetrazole-5-Carboxylate

The synthesis of tetrazole **40** required a reaction of nitrile **39** with sodium azide at 75 °C (Equation 10.8).[26] ReactIR analysis of the head space showed the presence of significant amounts of hydrazoic acid (HN_3), which was a serious safety concern.[27]

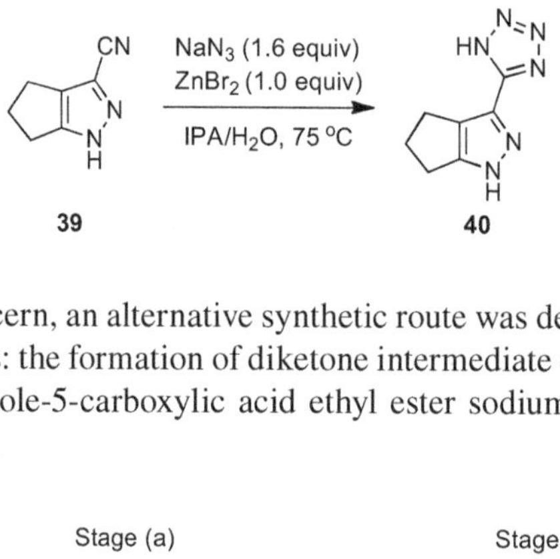

(10.8)

To circumvent the safety concern, an alternative synthetic route was developed (Scheme 10.11). The new route contains two stages: the formation of diketone intermediate **43** via condensation of cyclopentanone **41** with 1*H*-tetrazole-5-carboxylic acid ethyl ester sodium salt **42** and the subsequent cycloaddition with hydrazine.

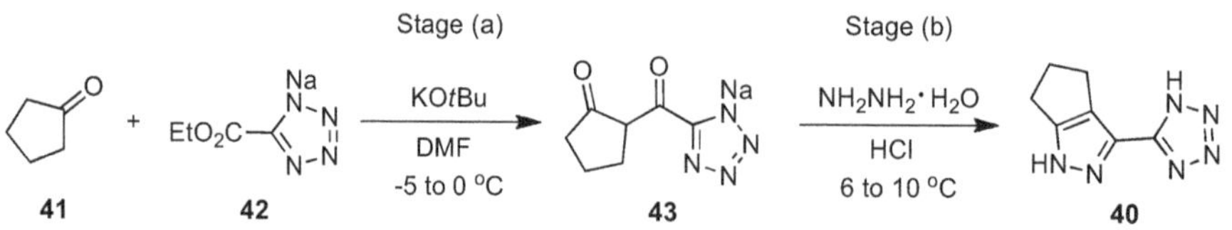

SCHEME 10.11

Procedure

Stage (a)
- Charge 3.36 kg (40 mol) of **41** to a solution of **42** (7.87 kg, 48 mol) and KO*t*Bu (6.74 kg, 60 mol) in DMF (75 L) at −2.5 °C to −5 °C over 1 h.
- After the addition, age the batch at 0 °C for 1 h.
- Charge HCl (9.88 L, 11 N) over 85 min at <6 °C.

Stage (b)

- Charge 8 L (88.8 mol) of 35.5 wt% solution of hydrazine in water over 35 min at <10 °C.
- Age the batch for 18 h.
- Distill the reaction mixture under reduced pressure to remove 65 L of solvent.
- Add water (25 L).
- Distill off 31 L of solvent under reduced pressure.
- Add water (125 L) and age overnight.
- Add conc HCl (5.74 kg) to pH 1.
- Age at 20–25 °C for 2 h.
- Filter and rinse with water (3×25 L).
- Dry at 45 °C under vacuum for 72 h.
- Yield: 6.33 kg (90%).

10.2.2.1.3 Use of Hydrazine

Synthesis of BMS-317180, a potent, orally active agonist of the human growth hormone secreta-gogue (GHS) receptor, involves tetrazole intermediate **45**. Conversion of amide **44** to tetrazole **45** was carried out according to the literature method,[28] giving **45** in 85% yield (Scheme 10.12).[29]

(I) Safety concerns

Trimethylsilyl azide is a flammable and highly toxic liquid. Its vapors may form an explosive mixture with air. Thereby the safe storage and handling of trimethylsilyl azide are a big concern at scale.

(II) Solutions

An alternative route was developed to avoid the use of trimethylsilyl azide. The application of amidrazone formation/diazotization methodology eliminated the need for azide salts, allowing the development of a safe, scalable synthesis of tetrazole core **50**. Scheme 10.13 showed a four-step,

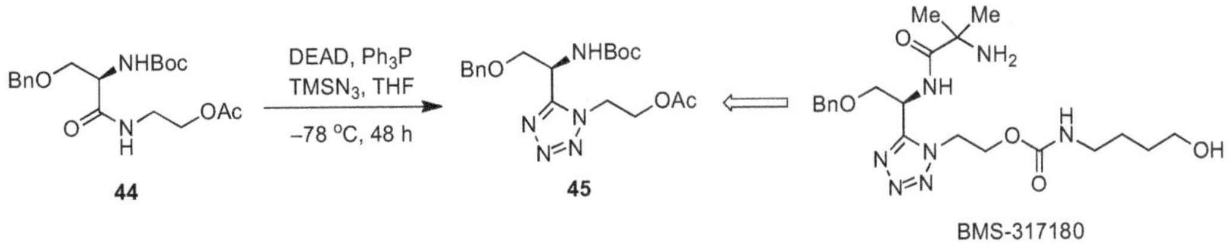

SCHEME 10.12

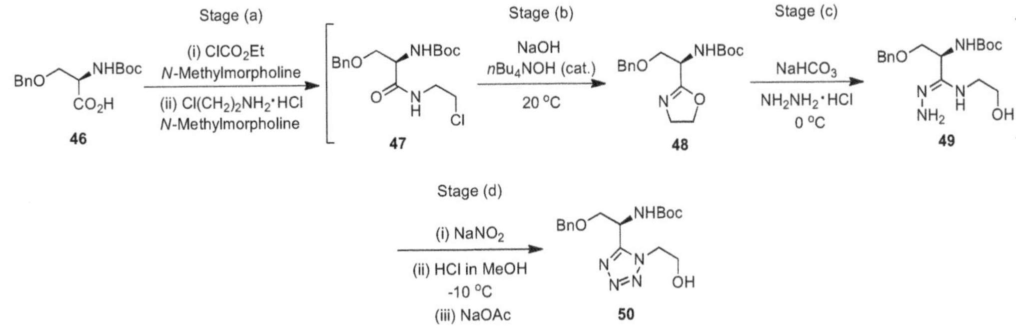

SCHEME 10.13

telescoped sequence which enabled the preparation of **50** from *N*-Boc-*O*-benzyl-D-serine **46** in 75% yield with only one isolation.

Procedure

Stage (a)_(i)
- Charge 40.0 kg (135 mol, 1.0 equiv) of **46** into an 800-L Hastelloy reactor, followed by adding CH$_2$Cl$_2$ (200 L).
- Cool the resulting solution to –35 °C.
- Charge 15.0 kg (148 mol, 1.1 equiv) of NMM at <–30 °C.
- Rinse with CH$_2$Cl$_2$ (10 L).
- Charge 16.3 kg (150 mol, 1.1 equiv) of ethyl chloroformate at <–30 °C.
- Rinse with CH$_2$Cl$_2$ (200 L).
- Age the batch at –30±3 °C for 50 min.

Stage (a)_(ii)
- Charge 26.0 kg (257 mol, 1.9 equiv) of NMM at –30 °C.
- Cool the mixture to –40 °C.
- Charge 17.3 kg (149 mol, 1.1 equiv) of milled chloroethylamine hydrochloride in one portion via the reactor mainway at –40 to –28 °C.
- Cool to –30 °C and age for 45 min.
- Warm the batch to 0 °C over 40 min.
- Age at 0 °C until complete consumption of **46** (≤2.0 A%).
- Add water (140 L) at 0 °C.
- Warm the mixture to 20 °C over 30 min.
- Separate layers and discard the top aqueous layer.
- Wash the organic layer with 0.5 N HCl (140 L).
- Separate layers and discard the top aqueous layer.
- Wash the organic layer with water (140 L).
- Separate layers and discard the top aqueous layer.

Stage (b)
- Charge, to the CH$_2$Cl$_2$ solution, water (97 L), 105 L of 15% NaOH, and 4.4 kg (17 mol, 0.05 equiv) of 40 % of tetrabutylammonium hydroxide.
- Stir the resulting biphasic mixture at 20 °C for 10.5 h.
- Separate layers and discard the top aqueous layer.
- Wash the organic layer with 5% NaHCO$_3$ (224 L), then with 5% NaCl (224 L).
- Add 7.9 kg of NaHCO$_3$.
- Distill at ≤35 °C under 200–500 mbar to a final volume of 40–60 L.
- Add MeOH (100 L) and continue to distill at ≤35 °C under 50–500 mbar until ≤1.0% of CH$_2$Cl$_2$.
- Add MeOH (80 L) and filter for use in the next step.

Stage (c)
- Charge 12.1 kg (176.6 mol, 1.3 equiv) of hydrazine monohydrochloride to an 800-L Hastelloy reactor, followed by adding MeOH (520 L).
- Cool to 0 °C.
- Transfer the methanolic solution from Stage (b) into the reactor while maintaining the temperature below 5 °C.
- Age the batch at 0 °C for 1.5 h.

Stage (d)

- Charge MeOH (400 L) into a 2000-L Hastelloy reactor.
- Cool to 0 °C.
- Charge 21.3 kg (271 mol, 2.0 equiv) of AcCl at 0–5 °C.
- Age the resulting solution at 0 °C for 30 min.
- Cool to –50 °C.
- Charge 16.4 kg (238 mol, 1.75 equiv) of NaNO$_2$ to a solution of **49** (obtained from Stage (c)) at 0 °C in the 800-L reactor.
- Stir the content at 0 °C for 15 min prior to cooling to –30 °C.
- Transfer the mixture into the 2000-L Hastelloy reactor at ≤–10 °C.
- Stir the resulting batch at –10 °C for 20 min.
- Add 26.0 kg (316 mol, 2.3 equiv) of NaOAc.
- Warm the mixture to 20 °C (pH 5–6).
- Add toluene (600 L).
- Distill the batch at ≤40 °C to ca. 530 L under reduced pressure.
- Add toluene (600 L).
- Distill the batch at ≤40 °C to ca. 530 L under reduced pressure.
- Filter the batch and adjust the temperature of the filtration to 20 °C.
- Add water (320 L) and stir for 15 min.
- Separate layers and discard the aqueous layer.
- Add 5% NaHCO$_3$ (320 L) and stir for 15 min.
- Separate layers and discard the aqueous layer.
- Add 5% NaCl (320 L) and stir for 15 min.
- Separate layers and discard the aqueous layer.
- Distill the resulting toluene solution at ≤35 °C under reduced pressure to ca. 266 L (KF≤500 ppm).
- Filter into an 800-L Hastelloy reactor.
- Warm to 50 °C.
- Add heptane (180 L) at 45–50 °C (56:44, toluene/heptane).
- Cool the resulting solution to 20 °C over 3 h.
- Cool the resulting slurry to 0 °C over 1 h and age for 1 h.
- Filter and wash with heptane (2×60 L).
- Dry at 35 °C/25 mmHg.
- Yield: 36.9 kg (75%).

10.2.2.2 Synthesis of Triazole

Compound **53** was identified, by Merck, as a potent, orally bioavailable NK-1 receptor antagonist. The first generation of synthesis of **53** (Condition (a), Scheme 10.14)[30] involved a triazole intermediate **52a**, obtained via cycloaddition of **51a** (X=CONMe$_2$) with azide. This cycloaddition was sluggish requiring a temperature of 75 °C and 2 equiv of acetic acid to avoid the formation of the Michael-addition side products, such as **54**. Consequently, the potential formation of highly explosive HN$_3$ in the presence of AcOH rendered this process unsafe.

Solutions

The ease of azide–alkyne cycloaddition to triazole ring system is governed by the polarization the acetylene. It was surmised that the replacement of the dimethylcarbamoyl with a stronger electron-withdrawing formyl group would facilitate the triazole-ring formation. In the event, under optimized conditions (Condition (b), Scheme 10.14) the cycloaddition of **51b** with sodium azide occurred at room temperature in the absence of AcOH.

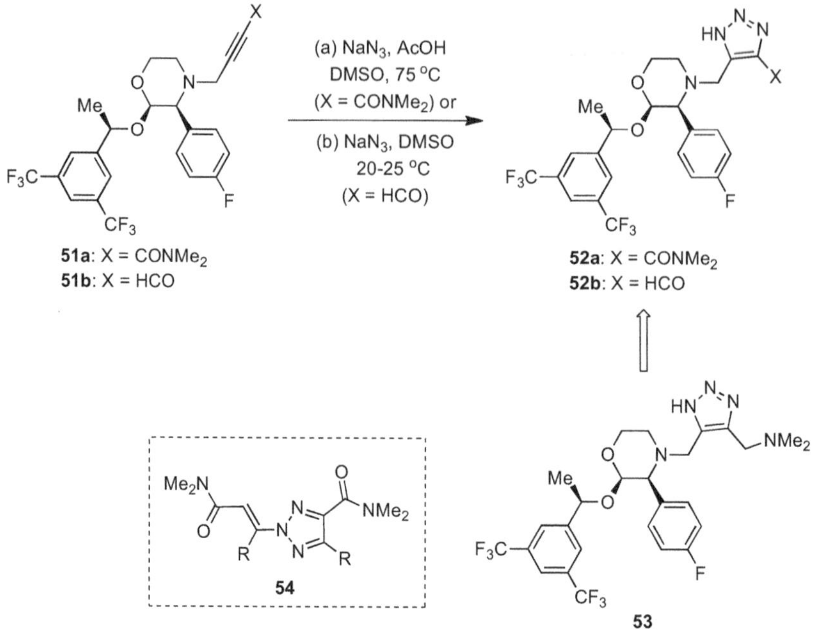

51a: X = CONMe$_2$
51b: X = HCO

(a) NaN$_3$, AcOH
DMSO, 75 °C
(X = CONMe$_2$) or

(b) NaN$_3$, DMSO
20-25 °C
(X = HCO)

52a: X = CONMe$_2$
52b: X = HCO

54

53

SCHEME 10.14

REMARK

The combination of p-toluenesulfonyl hydrazines and 1-aminopyridinium iodide was employed as an azide surrogate in the synthesis of 4-aryl-1,2,3-triazoles under metal- and azide-free conditions.[31] As shown in Equation 10.9, this iodine-mediated multi-component reaction appears to have a broad substrate scope tolerating various functional groups in the aryl methyl aryl ketones.

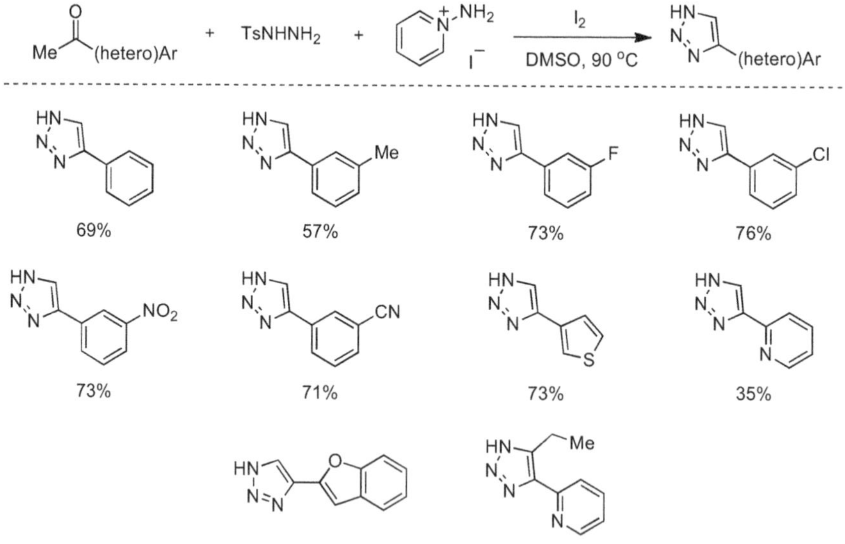

$$\text{Me}\overset{O}{\underset{}{\|}}\text{(hetero)Ar} + \text{TsNHNH}_2 + \text{Py}^+\text{-NH}_2 \; \text{I}^- \xrightarrow[\text{DMSO, 90 °C}]{I_2} \text{HN-triazole-(hetero)Ar}$$ (10.9)

69%　　　　57%　　　　73%　　　　76%

73%　　　　71%　　　　73%　　　　35%

74%　　　　trace

10.2.2.3 Synthesis of Carbamate

Carbamate **58** is a drug candidate that was under investigation as an inhibitor of glucosylceramide synthase for potential use in Fabry disease. The initial small-scale synthesis of **58** involved acyl

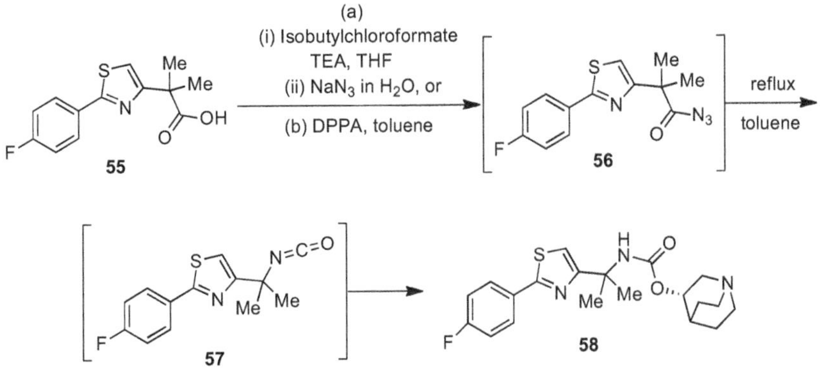

SCHEME 10.15

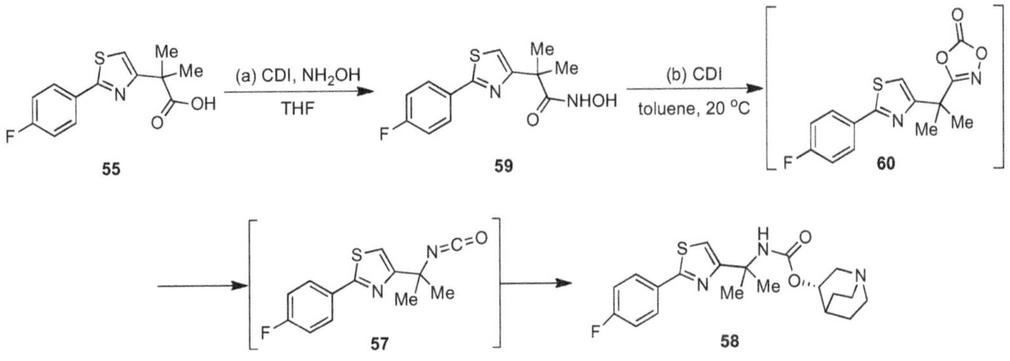

SCHEME 10.16

azide formation followed by Curtius rearrangement and quenching with (S)-(+)-quinuclidinol (Scheme 10.15).[32] Although Curtius rearrangement has been used in the preparation of amines and the amine derivatives from carboxylic acids, the hazards of azide chemistry are a serious concern.

Therefore, an alternative and safer process for the synthesis of **58** was developed. Compared with Curtius and Hofmann rearrangements, Lossen rearrangement involves the conversion of hydroxamic acid **59** to the corresponding isocyanate **57** with less hazardous reagents and intermediates under mild conditions. Thereby a CDI-mediated Lossen rearrangement protocol was utilized in the synthesis of **58** to mitigate the process hazards. The new process was demonstrated at a kilogram scale successfully (Scheme 10.16).

Procedure

Stage (b)

- Charge 1.34 kg (8.3 mol, 1.1 equiv) of CDI into a solution of **59** (2.11 kg, 7.5 mol) in toluene (17.3 kg) at 20 °C.
- Stir at 20 °C for 3 h, then at 60 °C for 16 h.
- Charge 1.13 kg (8.9 mol, 1.2 equiv) of (S)-(+)-quinuclidinol and reflux for 16 h.
- Cool to room temperature and partition with toluene (10 L) and water (20 L).
- Separate layers.
- Wash the organic layer with water (20 L).
- Cool the organic layer to 0 °C.
- Extract with aqueous HCl (1 N, 2×20 L).
- Partition the combined aqueous layers with isopropyl acetate (36.9 kg).

- Add 5.3 kg of aqueous NaOH (10 N) to the biphasic mixture at 5 °C.
- Separate layers.
- Distill the organic layer under reduced pressure to ca. 16 L.
- Add heptane (13.7 kg) and filter.
- Add heptane (7.7 kg) to the filtrate.
- Distill under reduced pressure to ca. 16 L.
- Stir the resulting mixture at 20 °C for 4 h (during which time, the product crystallized).
- Filter, wash with heptane (13.7 kg), and dry at 40 °C under vacuum.
- Yield of **58**: 2.09 kg (71%) as an off-white solid.

10.2.2.4   Curtius Rearrangement

Diphenyl phosphoryl azide (DPPA) (bp 157 °C/0.17 mmHg, d=1.223 g/mL, 25 °C was selected as a reagent for the Curtius rearrangement. From a safety perspective, DPPA was added slowly into the solution of starting material **61** in *t*BuOH/toluene at 80 °C, allowing rapid consumption of hazardous acyl azide intermediate, thus avoiding accumulation. The in situ formed isocyanate intermediate **62** was trapped by *tert*-butanol to afford the desired Boc-protected amine product **63** (Scheme 10.17).[33]

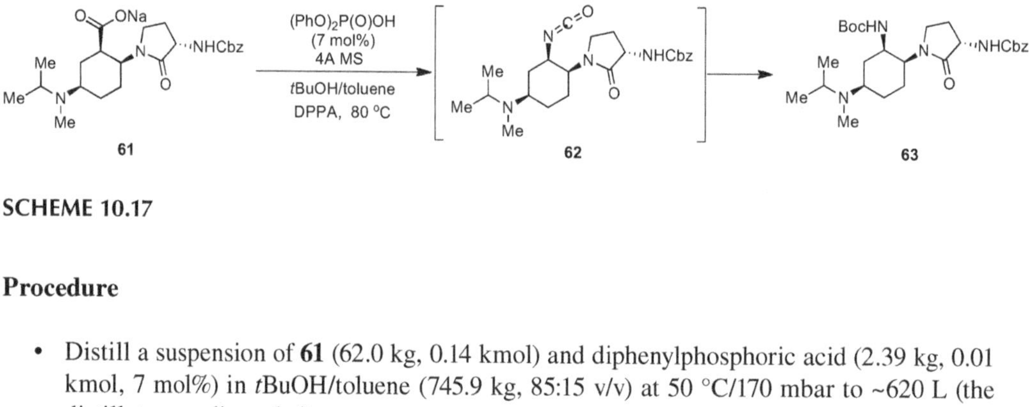

SCHEME 10.17

Procedure

- Distill a suspension of **61** (62.0 kg, 0.14 kmol) and diphenylphosphoric acid (2.39 kg, 0.01 kmol, 7 mol%) in *t*BuOH/toluene (745.9 kg, 85:15 v/v) at 50 °C/170 mbar to ~620 L (the distillate was discarded).
- Add 248.6 kg of *t*BuOH/toluene (85:15 v/v).
- Heat the batch to reflux (90–92 °C) at ambient pressure and recirculate the distillate through molecular sieves (55 kg, 4 Å) with an inline polish filter (5 µm) over a 2 h period (KF ≤150 ppm, methanol ≤0.1 mg/mL by GC).
- Adjust the batch temperature to 75–82 °C.
- Add 41.4 kg (0.15 kmol, 1.1 equiv) of DPPA slowly over 75 min.
- Rinse with toluene (10 kg).
- Monitor the reaction progress with HPLC (sodium carboxylate ≤5%).
- Concentrate at 110 °C/ambient temperature to remove ~200 L of distillate (with was replaced with toluene).
- Cool the batch to 17–25 °C.
- Add water (168 kg) and separate layers.
- Wash the organic layer with 15 wt% KH_2PO_4 aqueous solution (335 kg and 506 kg) and water (496 kg) (the aqueous washing was disposed into non-metallic containers after adjusting pH ≥12 with 30% aqueous NaOH
- Distill the organic phase to ~190 L at 60–65 °C/200 mbar.
- Heat the batch to 75–85 °C.
- Add heptane (424 kg) followed by a slurry of seeds (0.3 kg) in toluene (20 kg).

- Cool to 15–25 °C over 3 h and hold for 4 h.
- Filter and rinse with toluene/heptane (222 kg, 1:5 v/v).
- Yield: ~55 kg (~80% yield) as a wet white solid (that was used directly in the subsequent step).

REMARKS

(1) The presence of azides requires dry, non-metallic glass-lined equipment with inert transfer lines (PTFE).
(2) The completion of the reaction was determined by monitoring the isocyanate intermediate via quenching the in-process sample with benzylamine
(3) Diphenyl phosphoric acid (7 mol%) was used to trap the residual NaOMe (or NaOH) present in the sodium carboxylate starting material, as a strong alkali base can catalyze the oligomerization of isocyanates.
(4) The N-Boc-carbamate product was mutagenic and discharge of this product from the filter shall be avoided.

The in situ formed isocyanate intermediate **65** in the Curtius rearrangement could also be trapped by benzyl alcohol to furnish the Cbz-protected amine product **66** (Scheme 10.18).[34]

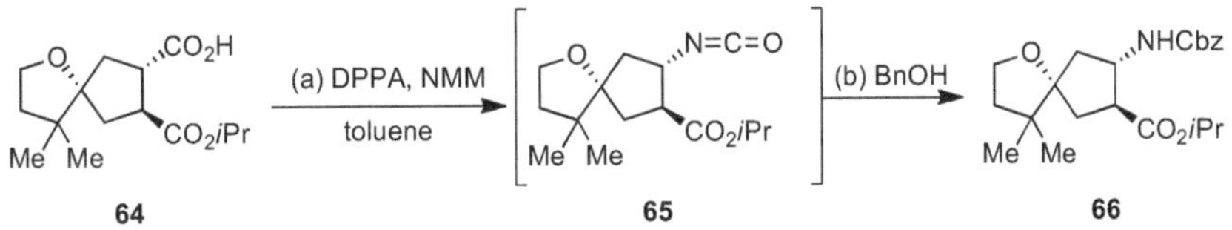

SCHEME 10.18

Procedure

Stage (a)
- Charge 793 g (89% w/w, 2.48 mol) of **64** into a 25-L jacketed vessel followed by adding dry toluene (7 L) and NMM (0.68 L, 6.21 mol).
- Add DPPA (0.642 L, 2.98 mol) at 104 °C over 2 h.
- After the addition, stir the batch at 104 °C for 15 min.

Stage (b)
- Add BnOH (403 g, 3.72 mol) at 100 °C.
- Stir the batch at 100 °C for 3 h.
- Cool to 20 °C.
- Add an aqueous saturated solution of Na_2CO_3 (3 L) and stir for 5 min.
- Separate layers.
- Wash the organic layer with water (3×1 L).
- Concentrate the organic layer under reduced pressure.
- Yield: 1.15 kg (81% w/w, 2.39 mol, 96% yield) as a viscous oil.

REMARKS
Multiple ^{1}H NMR samples were taken and recorded during the course of the reaction to ensure no accumulation of the potentially hazardous acyl azide intermediate.

10.3   HYDRAZINE

Besides extensive use as a rocket propellant, hydrazine, a colorless flammable liquid, is an important reagent in fine chemical and pharmaceutical industries. Hydrazine is highly toxic and dangerously unstable, especially in its anhydrous form. For safety reasons, hydrazine is usually handled in
an aqueous solution. An aqueous solution of hydrazine is considered to be safe for the application
of chemical transformations. As long as a ratio of 1.86 mol of water to 1.0 mol of hydrazine (35
mol%) is maintained in the liquid reaction mixture, the vapors in the head space of the reaction
cannot become flammable due to the fact that water is more volatile and will be enriched in the
vapor phase. Aqueous hydrazine containing >40 mol% hydrazine is flammable and can support
combustion even in the absence of oxygen. While the ignition temperature of hydrazine vapors is
270 °C in the presence of air, a nitrogen atmosphere should be maintained during the heat-up, age,
and cool-down of reactions.

The anhydrous condition should be avoided in the presence of hydrazine. Thus, heating of hydrazine hydrate in organic solvents, especially solvents which form an azeotropic mixture with water,
will lead to explosive anhydrous hydrazine in the vapor phase. A reduction in pressure will result
in higher hydrazine concentrations above liquid hydrazine and its solutions. Therefore, hydrazine
should not be distilled under a vacuum without careful consideration of the hazards.[35]

Depending on reaction solvents, two approaches are generally used to remove excess hydrazine
after completion of a reaction: extraction of the desired product from an aqueous reaction mixture
with organic solvent (for reaction under aqueous conditions) or washing the organic reaction mixture with water (for reaction in organic solvent). Aqueous waste of hydrazine should be treated with
alkali and bleach prior to disposal.[36] Alternatively, hydrazine can be diluted with water and neutralized with dilute sulfuric acid prior to disposal.

10.3.1   WOLFF–KISHNER REDUCTION

10.3.1.1   Reduction of Asymmetric Ketone

Hydrazine is used in the deoxygenation of aldehydes and ketones to methyl or methylene derivatives in Wolff–Kishner reduction. A practical, kilogram-scale implementation of the Wolff–Kishner
reduction was realized for the reaction of sterically demanding ketone **67** (Equation 10.10).[37] Key
to the successful implementation of this reaction was to identify two safety process parameters as
the process safety margins: temperature and water contents in the distillates and in the reaction
mixture.

(10.10)

Procedure

- Charge 2.49 kg (5.28 mol) of **67** into a 72-L round-bottom flask equipped with a mechanical stirrer, thermocouple, and reflux condenser.
- Charge 1.18 kg of powdered KOH (~85%).
- Charge 25 L of diethylene glycol (DEG) followed by adding 2.1 L (43.2 mol) of hydrazine
hydrate and 0.5 L (27.8 mol) of H_2O.
- Heat the mixture to 143 °C over 2 h and maintain at this temperature of reflux for ca.
30 min.

- Continue to reflux and begin to collect H_2O with Dean–Stark apparatus (then the temperature reached 155 °C) (total of 790 mL of water collected).
- Cool to 70–75 °C.
- Rinse the sides of the flask with 7.5 L of MeCN.
- Cool further to 55 °C.
- Add 17.5 L of H_2O dropwise.
- Cool to room temperature and stir overnight.
- Filter and wash with 35 L of 1:1.5 MeCN:H_2O followed by H_2O (25 L).
- Dry under vacuum.
- Yield: 2.0 kg (85%).

REMARK
While this reaction was safely and efficiently conducted at kilogram scale, further scale-up of this reaction, or any Wolff–Kishner reduction, would require further testing of all the reaction parameters prior to execution.

10.3.1.2 Reduction of Symmetric Ketone
Wolff–Kishner reaction lent itself readily to the preparation of 2,2,6,6-tetramethylpiperidine **72** in a continuous reactor via the reduction of 2,2,6,6-tetramethyl-4-piperidinone **70** (Scheme 10.19).[38] The reaction was conducted in two stages: (a) the formation of hydrazone **71** at 60–90 °C and (b) the conversion of **71** into **72** at 210 °C. The resulting product **72** and the released H_2O were continuously removed by distillation. This protocol was safely demonstrated on a >10-mole laboratory scale.

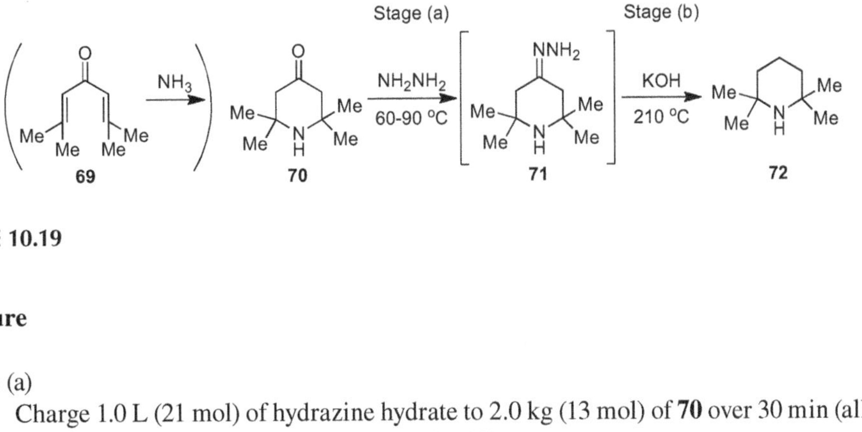

SCHEME 10.19

Procedure

Stage (a)
- Charge 1.0 L (21 mol) of hydrazine hydrate to 2.0 kg (13 mol) of **70** over 30 min (allowing the temperature rose temporarily to 90 °C).
- Maintain the reaction in a 60 °C bath.

Stage (b)
- Prepare a solution of KOH (48 g) (0.85 mol) in diethylene glycol (0.25 L) and paraffin oil (60 mL) at 210 °C in a reactor equipped with a 50 cm long Vigreus column.
- Transfer the reaction mixture from Stage (a) into the reactor at a rate of 2.5 g/min.
- Distill through the Vigreus column into a separation funnel.
- Yield: 1.51 kg (83%).

10.3.1.3 Synthesis of Indazole
Aminoindazole **74** was prepared via a reaction of commercially available 2,6-dichloro-benzonitrile **73** with hydrazine monohydrate (Equation 10.11).[39] Initially, the reaction was conducted in NMP

at 70 °C. However, a safety evaluation by adiabatic calorimetry tests revealed that a high risk of runaway reactions may occur in NMP.

$$(10.11)$$

Thus, switching the reaction solvent from the high boiling NMP (bp 202 °C) to the lower boiling pyridine (bp 115 °C) would provide a greater heat sink between the reaction temperature and the runaway reaction temperature (T_{ONSET}=130 °C). Importantly, the use of sodium acetate (or another weak base such as potassium acetate, potassium carbonate, or sodium carbonate) as an additive in combination with hydrazine hydrate and pyridine as the solvent at reflux (104 °C) raised the T_{ONSET} for exothermic decomposition to 200 °C.

Procedure

- Heat a suspension of 50.0 kg (267 mol) of **73** and 26.2 kg (319 mol) of NaOAc in pyridine (216 kg) to 100 °C.
- Charge 150 kg (3000 mol) of hydrazine monohydrate over 30 min.
- Stir the batch at 105 °C for 16 h.
- Cool to 30 °C.
- Add water (595 kg) and filter.
- Add water (184 kg).
- Add a seed slurry (280 g of **74** in 6.5 kg of water), followed by adding more water (1746 kg).
- Cool to 0 °C.
- Filter, wash with water (2×100 kg), and dry at 60 °C under vacuum for 16 h.
- Yield of **74**: 38.0 kg (74%) as a white solid.

10.3.1.4 Synthesis of Pyrazole

5-Amino-3-(2-chlorophenyl)-1H-pyrazole **76** was prepared by condensation of 2-chlorobenzoyl acetonitrile **75** with hydrazine in ethanol (Equation 10.12).[40] Upon the reaction completion, ethanol was removed by vacuum displacement with ethyl acetate, and this solution was washed with water to remove residual hydrazine. This workup protocol is considered unsafe and not suitable for scaling up.

$$(10.12)$$

A safer process was developed by using toluene as a reaction solvent to replace ethanol. The excess hydrazine was removed simply by aqueous extraction.

10.3.1.5 Preparation of Alkylamine

Cleavage of phthalimide **77** in the preparation of amine **78** was usually achieved using hydrazine. To avoid handling toxic hydrazine at scale, methylamine (40%) aqueous solution was selected as a hydrazine surrogate (Equation 10.13).[41]

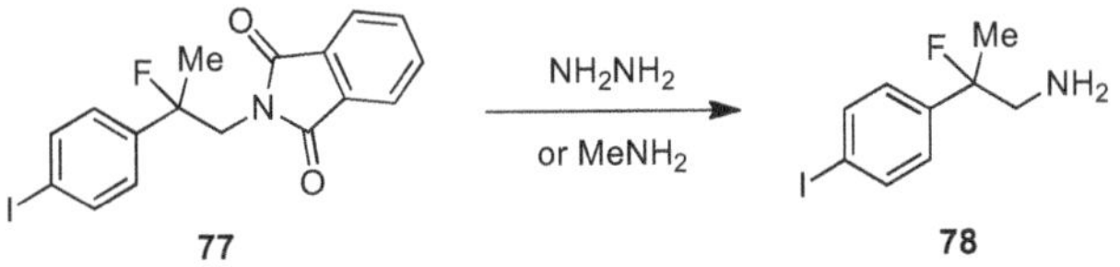

(10.13)

10.3.2   PREPARATION OF ARYL (OR ALKYL) HYDRAZINES AND RELATED REACTIONS

Aryl (or alkyl) hydrazines are important synthetic intermediates for access to pharmaceuticals, such as the anti-tuberculosis medication isoniazid and the anti-fungal fluconazole. Diazonium salts can be reduced with stannous chloride ($SnCl_2$) to the corresponding hydrazine derivatives. The use of ascorbic acid (or sodium dithionite) is an improvement over stannous chloride since it is a non-toxic and environmentally benign reducing agent. Ascorbic acid is commercially available and less expensive.

10.3.2.1   Preparation of 5-Hydrazinoquinoline

5-Hydrazinoquinoline **81** was prepared originally from 5-aminoquinoline **79** by diazotization followed by reduction with tin and hydrochloric acid or stannous chloride and mineral acid (Scheme 10.20).[42]

The product obtained from the tin reduction protocol was not suitable for pharmaceutical use as it was typically contaminated with tin residues and colored side products.

As shown in Scheme 10.21,[43] ascorbic acid was used as the reducing agent in the synthesis of **81**. The proposed reaction pathway involves a labile lactone intermediate **84**.

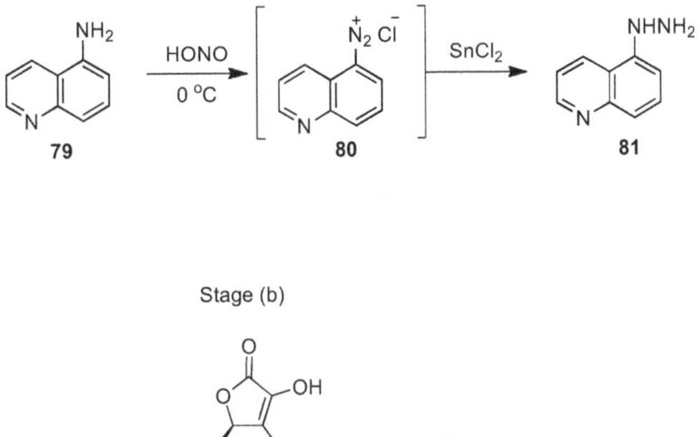

SCHEME 10.20

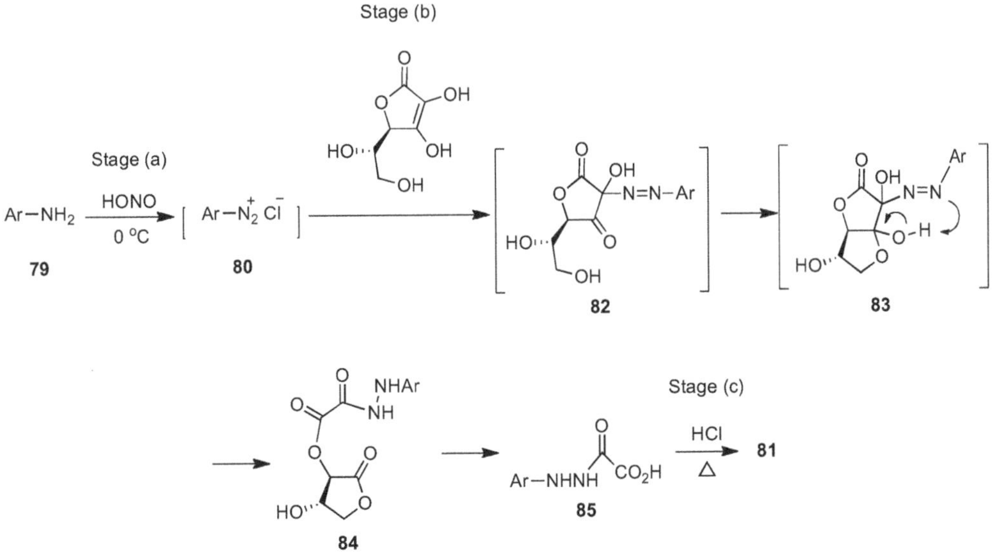

SCHEME 10.21

Procedure

Stage (a)

- Charge 32% aqueous HCl (75 L) into vessel A and cool to 0.5 °C over 1.75 h.
- Charge 15 kg (104 mol) of **79** in four portions every 15 min.
- Cool the resulting mixture to 0 °C.
- Charge 8 kg (116 mol, 1.1 equiv) of sodium nitrite and water (15 L) into vessel B.
- Stir at 15–25 °C to form a clear yellow-green solution.
- Charge the solution in vessel B into vessel A over 2 h while maintaining the internal temperature in vessel A at ≤2 °C.
- Stir at –2 to 2 °C for 1 h.
- Purge vessel A with nitrogen for 15 min to remove NO_x red fumes from the reactor headspace to the scrubber system.

Stage (b)

- Charge 19 kg (108 mol, 1.04 equiv) of L-ascorbic acid followed by adding 32% aqueous HCl (30 L) into vessel B.
- Stir at 15–20 °C to form a homogeneous suspension.
- Transfer the cold diazonium salt suspension in vessel A into vessel B at 15–20 °C over 1 h.
- Rinse vessel A with 32% aqueous HCl (4 L) into vessel B.
- Stir the mixture in vessel B at 15–20 °C for 20 min, then at 35–45 °C for 3 h.
- Cool the batch to 15–20 °C.
- Add water (41 L) to vessel B, cool to –5 to 0 °C, and stir for 1.5 h.
- Filter and rinse with MeOH (22.5 L).
- Dry at 40 °C.
- Yield of **85**: 21.5 kg;

Stage (c)

- Charge 21.5 kg of **85** into vessel B, followed by adding water (30 L) and 32% aqueous HCl (90 L).
- Stir the resulting suspension at 88–91 °C for 2 h.
- Cool to 18–22 °C and age for 2 h.
- Filter, wash with MeOH (22.5 L), and dry at 40–43 °C.
- Yield of **81**: 16 kg as a yellow solid.

10.3.2.2 Synthesis of Aminopyrazole

To prepare aminopyrazole **90**, a telescoped process was used from diazonium salt **87** (Scheme 10.22).[44] This one-pot sequence involves the reduction of diazonium salt **87** with ascorbic acid to oxalate derivative **88** followed by transesterfication to **89** as a stable crystalline solid. Without isolation, this mixture was hydrolyzed under basic conditions and then treated with 2-ethoxymethylene malononitrile to give **90**.

10.3.2.3 Fischer Indole Synthesis

Fischer reaction[45] proceeds via condensation of aryl hydrazine with a carbonyl compound, followed by a 1,3-sigmatropic rearrangement and subsequent elimination of ammonia. Attempts to synthesize indole **95** via Fischer indole synthesis failed due to the instability of aryl hydrazine intermediate **96**. Even the hydrochloride salt of **96** degraded rapidly at ambient temperature with the best-isolated yield of 13% obtained by using tin(I) chloride as the reductant.

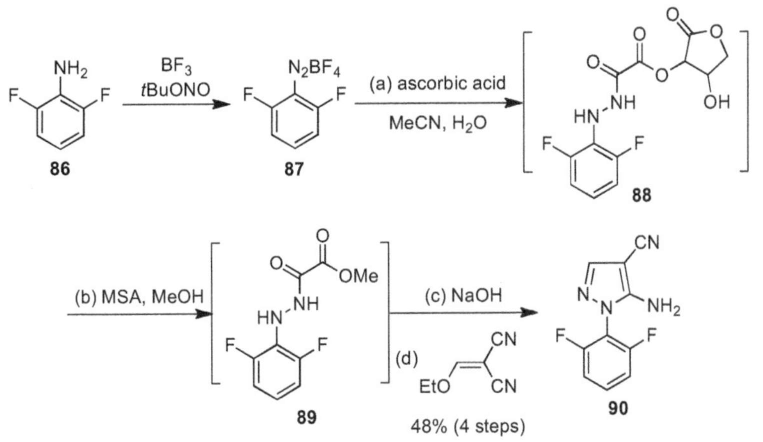

SCHEME 10.22

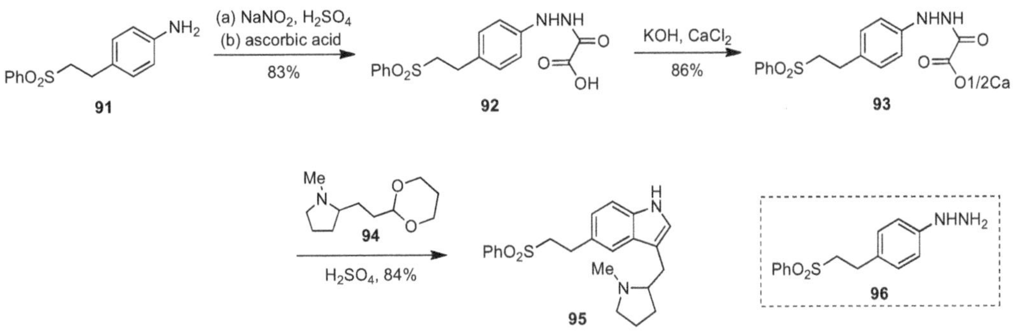

SCHEME 10.23

A practical approach for the synthesis of **95** was developed. In order to mitigate problems associated with the instability of **96**, the reduction of diazo intermediate, derived from **91**, with ascorbic acid converted the resulting unstable hydrazine **96** to a crystalline calcium salt **93** via oxalyl hydrazide **92** in high yields (Scheme 10.23).[46]

10.4 HYDROXYLAMINE

Hydroxylamine is unstable under heat and may decompose during storage via N–O bond cleavage. It is known that ferrous and ferric salts accelerate the decomposition of 50% NH_2OH solutions. It is recommended that hydroxylamine and its derivatives should be converted to their corresponding salts during isolation or storage,[47] as their salts are relatively stable (the decomposition reaction of the hydroxylamine hydrochloride salt–35 wt% hydroxylamine hydrochloride water solution–has an onset temperature of ~145 °C[48]).

Several approaches in the literature for syntheses of alkyl hydroxylamines include partial oxidation of primary amines using Oxone[49] or dimethyl dioxirane.[50] Dibenzoyl peroxide is used to prepare hydroxylamines of secondary amines.[51] In addition, partial reduction of a nitro compound to access hydroxylamine is also reported.[52]

Hydroxylamines and their salts are commonly used as reducing agents in numerous organic and inorganic reactions. For instance, a three-step synthesis of chiral alkyl hydroxylamine **99** involves a reductive cleavage of nitrone **98** with hydroxylamine (Scheme 10.24).[53]

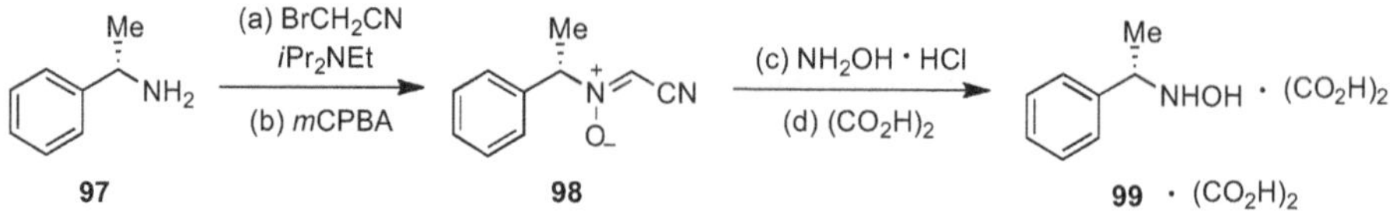

SCHEME 10.24

As hydroxylamine and its hydrochloride salt are hazardous especially when heating at 60 °C, the process described in Scheme 10.24 is not the process of scale-up choice.

A practical protocol was developed (Equation 10.14), in which p-toluenesulfonic acid monohydrate was chosen to replace hydroxylamine hydrochloride.

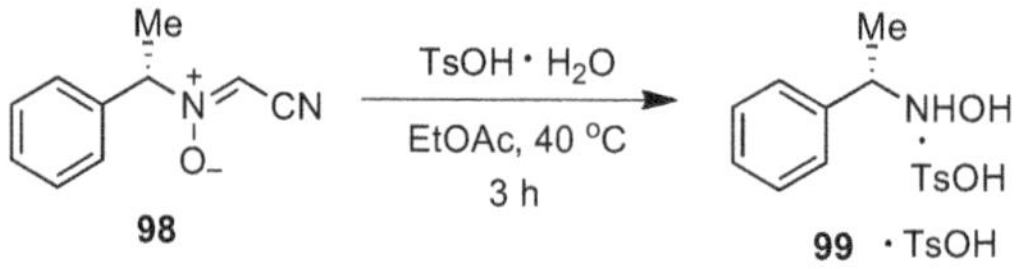

$$(10.14)$$

The reaction was conducted in ethyl acetate at 40 °C for 3 h and the resulting p-toluenesulfonic acid salt of **99** was obtained as a colorless solid (mp 126–128 °C) in 71% overall yield.

REMARKS

(a) The commercial grade (70–77 wt%) mCPBA is stabilized by chlorobenzoic acid and water shall be stored at 4 °C. mCPBA is not compatible with certain solvents such as DMF and DMSO.

(b) It is recommended to use glass-lined reactors, as aqueous hydroxylamine is not compatible with most metals.[54]

(c) O-(diphenylphosphinyl) hydroxylamine (DPPH) is useful in the electrophilic amination for the installation of N–N bonds. DPPH is prepared via a reaction of diphenylphosphinic chloride with hydroxylamine hydrochloride in the presence of sodium hydroxide (Equation 10.15).[55]

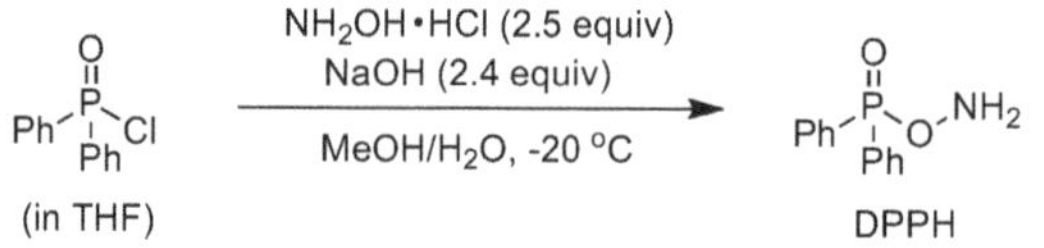

$$(10.15)$$

The decomposition energy of DPPH is >500 J/g and the onset temperature is <150 °C (determined by DSC). During the isolation at scale using an agitated pressure filter/dryer, an unexpected thermal decomposition of DPPH occurred. The decomposition propagated outside of the confines of the filter and glovebox used for discharging, resulting in the further decomposition of previously discharged DPPH contained in a bag below the glovebox. The investigation suggests that this incident was most likely caused by an electrostatic discharge during heel removal from the filter.

10.5   OXIME

The process for converting oxime **100** into isatin **101** is exothermic (Equation 10.16).[56] To control exothermic events, a semi-batch approach is frequently used to prevent the accumulation of reactive materials. Ultimately, the process was carried out safely by adding the oxime **100** in ten equal portions to the sulfuric acid at 60 °C.

(10.16)

Procedure

- Charge 316.3 kg (3225 mol) of sulfuric acid to a precooled (5 °C) 50 wt% of sulfuric acid (28 kg, 143 mol) at 5 °C.
- Heat the content to 60 °C.
- Charge 42.81 kg (214 mol) of **100** in 10 equal portions at 58–62 °C over 5 h.
- Stir at 58–62 °C for 1.5 h before cooling to 10 °C.
- Add pre-cooled water (5 °C) (800 L) and wash the line with water (170 L).
- Stir for 2 h.
- Filter and wash with water (0–6 °C) (3×210 L).
- Dry at 45 °C under reduced pressure.
- Yield: 35.35 kg (75%).

10.6 *N*-OXIDE

An incident occurred unexpectedly during the production of 2,3-dimethyl-4-(2,2,2-trifluoroetho xy)pyridine *N*-oxide **103**, from a reaction of 2,3-dimethyl-4-nitropyridine *N*-oxide **102** with trifluo-roethanol in a mixture of methyl ethyl ketone (MEK) and water using triethylbenzylammonium chloride (TEBA) as a phase transfer catalyst (Equation 10.17).[57]

(10.17)

The incident happened as follows:

> *The vessel, at room temperature, was first charged with 450 kg of **102**, then 720 kg of K_2CO_3 and 15 kg of TEBA were added without stirring, and finally 15 L of water was added. The incident started 10 min after the vessel was closed for alternate vacuum/nitrogen purge. A high increase of the internal temperature associated with fast gas evolution was detected, generating a sudden pressure rise that ended up with the explosive breakage of the reactor.*

Quoted from *Organic Process Research & Development*, 2007, *11*, 1131 with permission from the American Chemical Society.

Evaluation of the *N*-oxide **102** by differential scanning calorimetry (DSC) revealed that compound **102** (melts at 101 °C) begins a severe decomposition at 237 °C, with an associated heat of 1797 J/g. It was observed via an accelerating rate calorimeter (ARC) that a slow exotherm reaction of the mixture of **102** with TEBA and potassium carbonate starts at 72 °C, which became extremely fast at 89 °C with an adiabatic temperature rise higher than 364 °C and an internal pressure rise higher than 21 bar. The exothermic dissolution of potassium carbonate in water generated enough heat to locally increase the temperature of the reaction mass and trigger the decomposition of the nitro compound **102**, causing the observed thermal runaway.

Thereby a safer process was developed by adding successively at room temperature all of the reagents except **102** into the reaction vessel and letting the system release all of the evolved heat, followed by addition of the nitro compound **102** and then heating to 85 °C.

REMARKS

(a) A recommended maximum operating temperature (T_{MR}) (in °C) may be obtained from the evaluation thermal stability of a given reaction using differential scanning calorimetry (DSC) (Equation 10.18):[58]

$$T_{MR} = 0.7 \times T_{on} - 46$$

(10.18)

T_{on} (in °C) is the temperature of exothermic onset.

(b) The stability of *N*-oxides **104** and **106** (655 J/g with a 101 °C onset temperature) is also a safety concern for scale-up (Equation 10.19).[59]

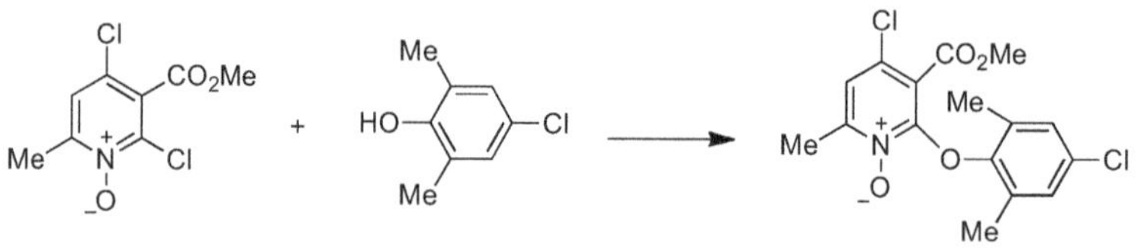

104   105   106

(10.19)

10.7   NITRO COMPOUNDS

Aromatic nitro compounds are important intermediates in synthetic organic chemistry. Nitro compounds are often highly explosive, especially when the compound contains more than one nitro group and is not pure. In general, the thermal stability of nitro compounds decreases when they are mixed with other chemicals or contaminated by impurities. Contaminated nitro compounds or their solutions may decompose at much lower temperatures than pure products. Pure organic nitro compounds decompose at high temperatures, exhibiting large decomposition exotherms. In most cases, the decomposition is violent or explosive.

10.7.1   PREPARATION OF NITRO COMPOUNDS BY NITRATION

Generally, nitration is conducted in a nitric acid/sulfuric acid (or mixed acid), in which the sulfuric acid acts as a catalyst as well as an absorbent for water by-product. The use of organic solvents in nitration reactions has several advantages:[60] (i) to make the reaction mixture uniform, (ii) to reduce the consumption of nitric acid, and (iii) to improve the product yield. Typical nitration solvents are 1,2-dichloroethane, dichloromethane, toluene, acetic acid, trifluoroacetic acid, and sulfolane.[61] The selection of nitration solvents should be cautious, however, as the presence of organic solvents and concentrated nitric acid in the nitration reaction mixture allows possible secondary redox reactions between the solvents and nitric acid.

The formation and separation of heavies of nitro organic compounds may be a hazard. Unless the heavies are well-identified, the heavies should not be allowed to accumulate. A process hazard review should be prompted on nitration and washing steps to reduce or suppress the generation of heavies.

Workup of nitration reactions is usually performed by quenching reaction mixtures into a large amount of water. Contamination of nitro compounds with caustic soda, salts such as sodium sulfate, sodium nitrate, and other metal salts may decrease the thermal stability of the desired nitro compounds. Drying nitro products may concentrate inorganic impurities in the organic phase thereby lowering its thermal stability.

REMARK

The continuous process is increasingly playing a pivotal role in the pharmaceutical industry due to its operational simplicity, shorter processing cycle time, and inherently safer nature. With the implementation of emerging technologies like microfluidics, microreactors, and continuous flow chemistry,[62] synthetic chemists can develop fast, efficient, and safe organic processes. Due to their high surface-to-volume ratio, microreactors provide efficient heat transfer to prevent run-away reactions. In addition, by using microreactors reaction temperatures can be accurately tuned and thus avoid hot spots that are often responsible for the formation of by-products. Microreactors were implemented in a number of potential scale-up hazardous reactions including nitration and bromination reactions,[63] Curtius rearrangement,[64] Swern oxidation,[65] etc.[66] Using flow reactors for scaling up of exothermic nitration reactions is the current interest in both chemical and pharmaceutical industries.[67]

10.7.1.1 NaNO₃/TFA/TFAA Nitration System

Selective *meta*-nitration of 1-aminoindan **107** under classic nitration conditions afforded nitrate ester intermediate **108**, isolated as the nitrate salt (Scheme 10.25).[68] However, conversion of the nitrate ester **108** to the desired product **109** proved to be challenging. In addition, the isolation of **108** as its nitrate salt with three nitro groups represented an unacceptable scale-up hazard.

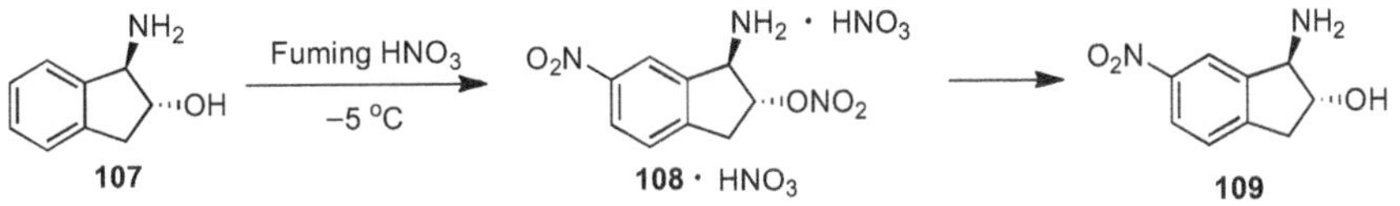

SCHEME 10.25

To address these safety issues, mild nitration conditions were developed, whereby $NaNO_3$ was activated with TFAA at $-20\,^\circ C$ (Equation 10.20).

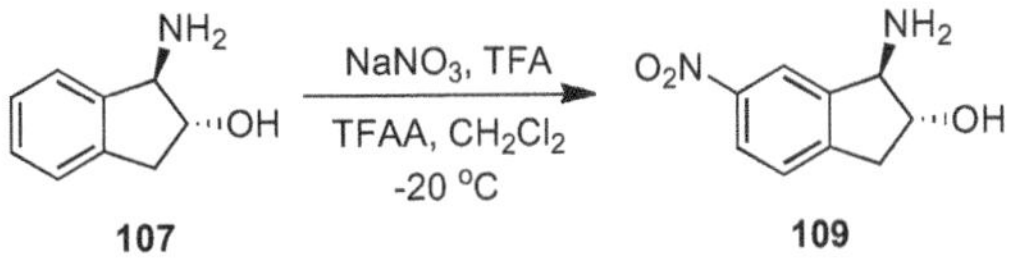

(10.20)

The reaction was conducted in a mixture of CH_2Cl_2 and TFA, affording the nitro-aminoindanol **109** in 55–60% yield and high purity (>99 area% by HPLC).

Procedure

- Charge 15.8 kg (105.9 mol) of **107** to CH_2Cl_2 (160 L), followed by adding 10.6 kg (124.7 mol, 1.18 equiv) of $NaNO_3$.
- Cool the resulting slurry to $-15\,^\circ C$.
- Charge 98.7 kg (865.6 mol, 8.2 equiv) of TFA over 90 min at $-20\,^\circ C$.
- Charge 60.8 kg (289.5 mol, 2.7 equiv) of TFAA over 90 min at $-20\,^\circ C$.
- Rinse with CH_2Cl_2 (15 L).
- Stir the resulting slurry at $-20\,^\circ C$ for 5 h.
- Transfer the batch into 95 L of water at $0–5\,^\circ C$ over 60 min.
- Rinse the vessel with CH_2Cl_2 (150 L).

- Add 260 L of 5 N NaOH at 0–5 °C.
- Add an additional 6 L of 5 N NaOH until pH 13.
- Stir the resulting mixture for 1 h.
- Filter, wash with cold water (210 L), and dry at 35–40 °C under vacuum.
- Yield: 12.3 kg (60%).

REMARK

Trifluoroacety-protected amine **110** could be nitrated with nitronium trifluoromethylsulfonate salt $(NO_2^+CF_3SO_3^-)$ (Equation 10.21).[69]

$$\text{110} \xrightarrow[\substack{CH_2Cl_2,\ 20\ °C \\ 20\ h,\ 85\%}]{HNO_3,\ CF_3SO_3H} \text{111} \tag{10.21}$$

The nitronium trifluoromethylsulfonate salt, readily prepared in high yield by mixing trifluoro-methanesulfonic acid and nitric acid,[70] is a stable white crystalline solid at 0 °C under a nitrogen atmosphere.

10.7.1.2   KNO₃/TFA Nitration System

The traditional nitration system (HNO_3/H_2SO_4) suffers from volume inefficiency due to the water quenching and the following neutralization. An alternative nitration system (KNO_3/TFA) was iden-tified for the nitration of anisole derivative **112**, which provided a cleaner reaction profile and higher yield. The product could be crystallized directly upon the addition of water thus avoiding the dif-ficult aqueous workup (Equation 10.22).[71]

$$\text{112} \xrightarrow[TFA]{KNO_3} \text{113} \tag{10.22}$$

Procedure

- Charge 9.0 kg (40.1 mol) of **112** followed by adding 4.25 kg (42.0 mol) of KNO₃.
- Charge 45.4 L of TFA at –10 °C.
- Warm the batch slowly to 22 °C over 3.5 h (A rapid exotherm to 36 °C was observed as the reaction went to completion.).
- Cool to 22 °C.
- Add H₂O (135.1 L) at 17 °C.
- Stir the resulting tan slurry for 2 h.
- Filter, wash with H₂O (45.0 L), and dry.
- Yield: 9.96 kg (91.4%).

10.7.1.3   Use of Acetyl Nitrate

The nitration reaction of *N*-[(4-methoxyphenyl)ethyl]acetamide **114** was initially carried out with nitric acid (10 equiv) in acetic acid (Equation 10.23).[72] However, the HNO₃/AcOH nitration system is considered unsafe and a dramatic temperature increase from 40 °C to 103 °C over 10 min was observed which was accompanied by heavy gas evolution.

$$(10.23)$$

Note

The calorimetry showed that the nitration with nitric acid occurred as soon as the addition of substrate solution to the nitric acid was started but could stall if the temperature drops below 15 °C.

Replacing nitric acid with acetyl nitrate, generated in situ, as the nitration agent offers some advantages. As a result, the nitration required fewer equiv of the nitration agent and was also shown to be safe using an alternative addition mode with fuming nitric acid (1.46 equiv) being added in small aliquots while keeping the reaction mixture between 15 and 25 °C. These modifications resulted in significant reduction in side products derived from over-nitration and hence enhanced yield.

Procedure

- Charge 460 kg of AcOH, followed by adding 115 kg (595 mol) of **114**, 102 kg (999 mol) of Ac_2O, and 11 kg of 98% H_2SO_4.
- Charge, to the mixture, 56 kg (871 mol, 1.46 equiv) of 98% HNO_3 in 5 kg aliquots maintaining the reaction temperature within 15 to 20 °C during the addition.
- Stir for 1 h.
- Transfer the reaction mixture to 1150 L of water and stir for 16 h.
- Add 2200 L of 20% NaOH at <25 °C.
- Filter, wash with water, and dry.
- Yield: 99.1 kg (70%).

10.7.1.4 NaNO₃/Py·SO₃ System

Nitration of electron-rich heteroaromatics presents particular challenges due to the inherent instability of these compounds under strongly acidic and oxidizing conditions. Because of its less acidic conditions, the acetyl nitrate system is commonly employed for the nitration of electron-rich heterocycles. However, acetyl nitrate undergoes a facile thermal decomposition, which raises a major safety concern for its application.

Activation of sodium nitrate by pyridine·SO₃ complex in acetonitrile leads to the formation of an insoluble nitronium sulfate intermediate (Equation 10.24). The nitronium sulfate salt was considered to be the active nitrating species and the nitration rate was controlled by the dissolution rate of the salt.

High-yielding nitration of pyrrole **116** with a mixture of sodium nitrate and pyridine·SO₃ complex was realized (Equation 10.25).[73]

$$(10.24)$$

$$(10.25)$$

10.7.1.5   Stepwise Nitration

Nitration of 4-(4-methoxyphenyl)morpholine requires tight control of the substrate to nitric acid in a 1:1 molar ratio in order to avoid under/or over nitration situations. Under/or over nitration led not only to a yield loss but also to a big burden on the purification of the product. A step-wise nitration strategy was utilized, which provides an ease and reliable way to control the 1:1 molar ratio of the substrate to nitric acid (Equation 10.26).[74]

$$ (10.26) $$

The nitration was carried out by using the isolated nitric acid salt **118**, which was conveniently prepared by the addition of 1.0 equiv of 70% nitric acid to the TBME/ethyl acetate solution of 4-(4-methoxyphenyl)morpholine. The subsequent nitration was conducted by adding a solution of **118** in CH_2Cl_2 to concentrated sulfuric acid while maintaining the batch temperature at 0–5 °C.

10.7.1.6   Use of Chlorotrimethylsilane and Nitrate Salt

Electrophilic nitration of arenes often results in mixtures of isomeric nitrated products. A mixture of nitrate salt and chlorotrimethylsilane is found to be an efficient regioselective nitrating agent for *ipso*-nitration of arylboronic acids to produce the corresponding nitroarenes in moderate to excellent yields (Equation 10.27).[75] This reaction provides an alternative approach for access to nitro compounds with high regioselectivity in good purity. Another significant feature of this method is the complete avoidance of dinitro by-product formation.

$$ (10.27) $$

10.7.1.7   Metal-Catalyzed Nitration

10.7.1.7.1   Palladium-Catalyzed Nitration

The recently developed palladium-catalyzed *ipso*-nitration reaction provides a mild and regioselective method for the nitration of variety of arenes (Equation 10.28).[76]

$$ Ar-X \ + \ NaNO_2 \xrightarrow[\substack{TDA\ (5\ mol\%),\ tBuOH \\ 130\ ^\circ C,\ 24\ h}]{\substack{Pd_2(dba)_3\ (0.5\ mol\%) \\ tBuBrettPhos\ (1.2\ mol)}} Ar-NO_2 \qquad (10.28) $$

X   Cl, OTf, ONf
TDA = tris(3,6-dioxaheptyl)amine

The addition of tris(3,6-dioxaheptyl)amine (TDA), as a phase-transfer catalyst, increased the solubility of sodium nitrite in *t*BuOH. Among chloro-, Bromo-, and iodoarenes, the rate of transmetalation follows the order of Cl>Br>I, and the chloro derivatives are the best substrates.

10.7.1.7.2   Cu(II)-Catalyzed Nitration

Copper-catalyzed nitration allows nitration of aminotetrazoles in a chemo- and regio-selective manner (Equation 10.29).[77]

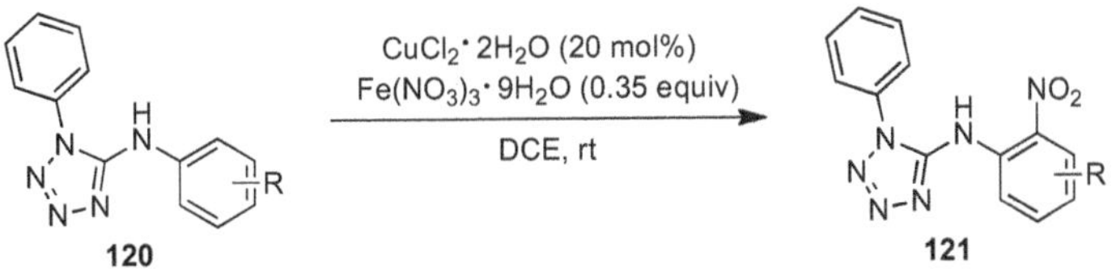

(10.29)

This reaction features mild reaction conditions with good functional group compatibility. The tetrazole directing group from the resulting nitration products **121** could be removed under basic conditions to afford nitroanilines (Equation 10.30).

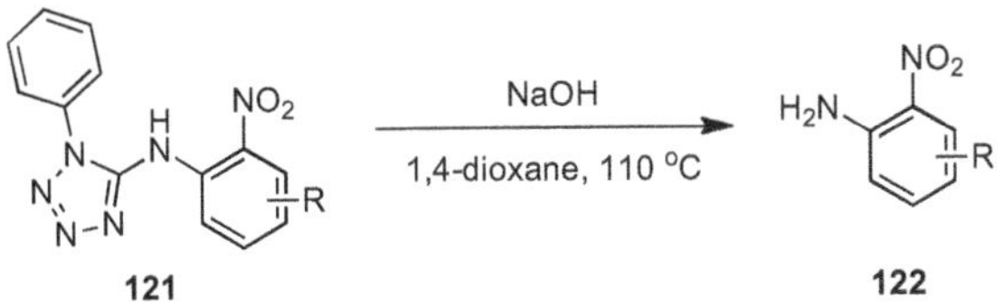

(10.30)

10.7.1.8 *N*-Nitro Intermediate Rearrangement

Traditional nitration of 2-amino-4-picoline **123** could produce a mixture of 3- and 5-nitro isomers in an unfavorable 1:4 ratio. The following bromination of the isolated desired minor isomer **124** afforded **125** in less than 10% overall yield (Scheme 10.26).[78]

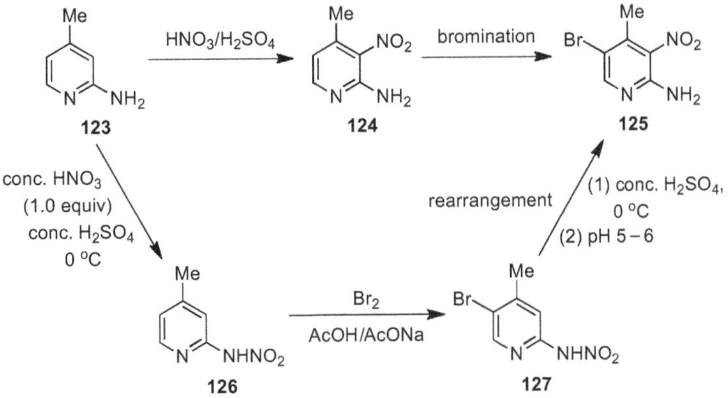

SCHEME 10.26

Process research and development led to an efficient, atom-economic, and regioselective synthesis of **125** via rearrangement of 2-nitramino-picoline intermediate **127**. Conceivably, 2-nitramino functionality[79] in the intermediate 2-nitramino-4-picoline **126** plays a unique role in deactivating and providing the necessary steric environment for the regioselective bromination, leading to the desired 5-Bromo intermediate **127**. The subsequent intramolecular acid-catalyzed rearrangement delivered the nitro functionality to the C3 position in a selective and atom-economic manner.

10.7.2 HAZARDOUS REACTIONS OF NITRO COMPOUNDS

Due to the electron-deficiency in nitro-containing aromatic rings, aromatic nitro compounds can participate in nucleophilic aromatic substitution (S_NAr) reactions with appropriate nucleophiles under basic conditions. To conduct these S_NAr reactions, care must be taken due to the instability of the nitro group under alkaline conditions.

10.7.2.1   Unstable Nitrophenate

Alkaline nitro compounds are unstable, for example, sodium or potassium nitrophenates exhibit shock sensitivity and autocatalytic decomposition. Hydrolysis of p-chloronitrobenzene using NaOH resulted in unstable nitrophenols and nitrophenates, which led to an explosion caused by the auto-catalytic decomposition of sodium nitrophenate (Equation 10.31).[80] This incident occurred in a process where p-chloronitrobenzene was hydrolyzed with 50% caustic soda, in a batch operation, to produce p-nitrophenol. During cooling, an explosion occurred due to the decomposition of the nitrophenate (crystallized on the vessel wall) that exhibits autocatalytic behavior.

$$(10.31)$$

10.7.2.2   Accumulation of Reactant

Another incident occurred in a semi-batch process to manufacture o-nitroanisole by reacting meth-anolic caustic soda with o-chloronitrobenzene (Equation 10.32).[80]

$$(10.32)$$

Due to the absence of agitation, un-reacted caustic soda could be accumulated in the reaction mix-ture where two separate liquid phases were present, one containing o-chloronitrobenzene and the other methanolic caustic soda. When agitation was restarted, the accumulated reactants led to the runaway reaction.

REMARKS

(a) As impurities may lower a solution's thermal stability, recovery of nitro compounds by evaporation of solvent may be a risky operation.

(b) The presence of active carbon used to de-colorize products may also contribute to the ini-tiation of decomposition because active carbon or graphite lowers the thermal stability of organic nitro compounds.

10.8   VOLATILE ORGANIC COMPOUNDS

Volatile organic compounds (VOCs) are organic chemicals that have a high vapor pressure at room temperature. Normally, such compounds have low boiling and flash points which pose high risk of fire and explosion. In addition, some VOCs are dangerous to human health or cause harm to the environment.

10.8.1   ETHYLENE OXIDE

Ethylene oxide is a colorless gas at 25 °C with a boiling point of 10.7 °C. Although it is a raw mate-rial with diverse applications, ethylene oxide itself is a very hazardous substance: at room tempera-ture it is flammable, with a flash point of –24 °C, carcinogenic, mutagenic, irritating, and anesthetic gas with a misleadingly pleasant aroma.

Ethylene oxide was used for introducing the hydroxyethyl chain for the laboratory synthesis of a drug intermediate **129** as shown in the reaction (Equation 10.33).[81]

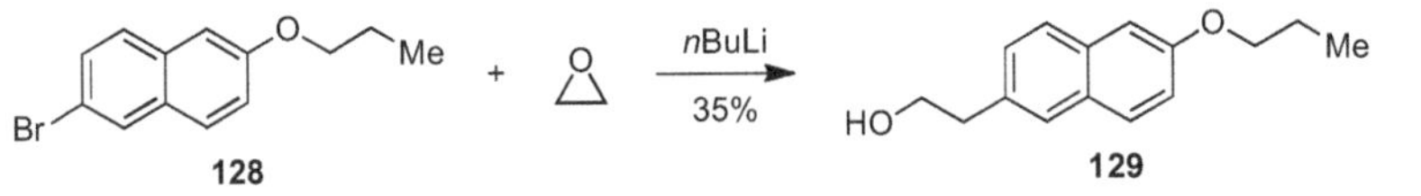

(10.33)

To avoid using ethylene oxide on large scale, the preparation of alcohol **129** was replaced by a two-step process, involving palladium-catalyzed alkylation followed by reduction (Scheme 10.27).

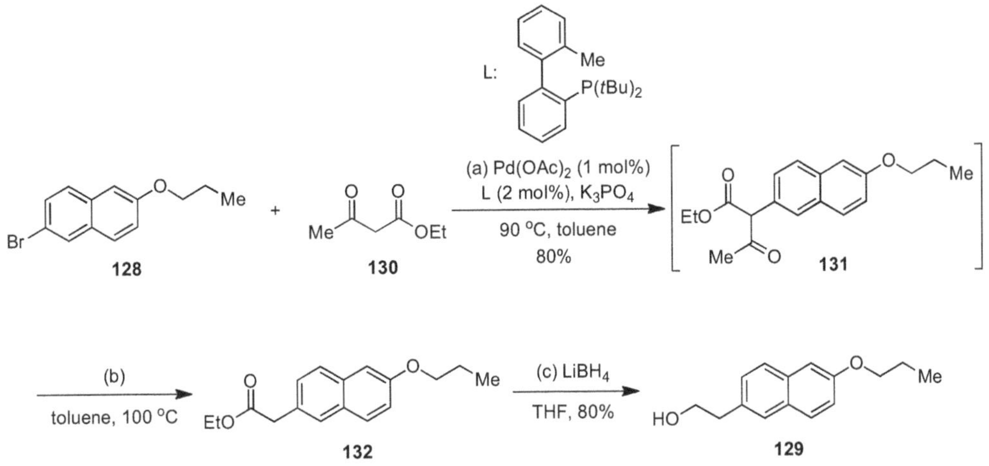

SCHEME 10.27

Procedure

Stages (a) and (b)
- Charge 8.2 kg (63.1 mol) of **130** to slurry of **128** (12.8 kg, 48.2 mol), potassium phosphate monohydrate (36.1 kg, 170.0 mol), and ligand (0.30 kg, 0.96 mol) and Pd(OAc)$_2$ (0.11 kg, 0.48 mol) in toluene (92 L) at rt.
- Stir the mixture at 90 °C for 2 h, then at 100 °C for 5 h.
- Cool to 0 °C.
- Add water (50 L) at ≤10 °C.
- Warm the resulting mixture to 20 °C and hold for 1 h.
- Separate layers.
- Wash the organic layer with 5% aqueous NaCl solution (2×25 kg), sat. aqueous NaCl solution (25 kg).
- Filter through Celite and rinse with toluene (8 kg).
- Distill the combined filtrates under reduced pressure.
- Yield of **132**: 14.2 kg (82%).

REMARK

The use of K$_3$PO$_4$ monohydrate in the deacylation caused slight etching of the glassware, and a metal reactor is recommended for this purpose.

10.8.2 ACETYLENE

Acetylene is a colorless and extremely flammable gas and very sensitive to pressure, temperature, static electricity, and/or mechanical shock. It readily forms explosive mixtures with air over an unusually broad range of concentrations. Calcium carbide is inexpensive and abundant solid and commercially available. A one-pot, three-component reaction was developed to access useful 3-methyl-2-arylimidazo[1,2-a]pyridine derivatives **135** utilizing calcium carbide as acetylene surrogate (Equation 10.34).[82]

Calcium carbide generates cuprous acetylide intermediate **136** in the presence of water and CuI. The reaction of the cuprous acetylide **136** with an in situ formed imines from amines and aldehydes would furnish the desired imidazopyridine products **135**. Studies found that the reaction is sensitive to steric effect and aldehydes **134** with substituents such as F, Cl, or Br at the *ortho* position failed to provide the desired products. In addition, reactions of aliphatic aldehydes, for example, *n*-butanal, did not produce the desired product either.

10.8.3 HALOGENATED METHYL ETHERS

10.8.3.1 Bis(bromomethyl)ether

Albeit being effective alkylation reagents, bis(chloromethyl)ether and bis(bromomethyl)ether are carcinogen and acutely hazardous materials.[83]

In order to avoid using bis(bromomethyl)ether, the Bromomethylation[84] of mesitylene **137** was carried out with paraformaldehyde and anhydrous HBr in acetic acid (Equation 10.35).[85] However, the harmful bis(bromomethyl)ether can be generated under the reaction conditions. Therefore, the reaction should be carefully evaluated prior to scaling up.

10.8.3.2 Chloromethyl Ethyl Ether

To reduce the risk associated with exposure to toxic chemicals, such as chloromethyl ethyl ether, during handling and storage, an approach of in situ generation of chloromethyl ethyl ether was adopted in the final *N*-alkylation step for the synthesis of anxiolytic agent **141** (Scheme 10.28).[86]

   The reaction was carried out by adding a solution of **139** in the presence of DBU (1.0 equiv) in DMF to a solution of chloromethyl ethyl ether, generated in situ from a reaction of diethyoxymethane with acetyl chloride, followed by adding diisopropylethylamine.

Note

An attempt to avoid the addition of amide **139** as a solid, DBU (1.0 equiv) was used to increase the solubility of **139** in DMF.

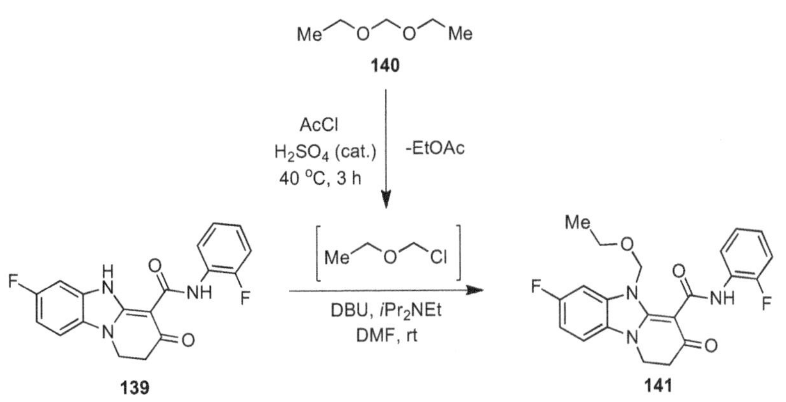

SCHEME 10.28

10.8.4 METHYL IODIDE

Methyl iodide is commonly used as a methylation agent. However, methyl iodide (bp 42.5 °C) is a toxic volatile material and should be avoided at large-scale production. Alternatives include methyl tosylate, dimethyl carbonate, dimethyl sulfate,[87] and trimethylsulfoxonium iodide.[88]

10.8.4.1 *N*-Methylation of Heterocycles

The *N*-methylation of heterocycles such as indoles and pyrroles has traditionally been accomplished using reagents such as methyl iodide in the presence of a strong base.[89] Dimethyl carbonate (DMC) as a viable alternative to toxic alkylating agents is considered as a "green reagent".[90] General *N*-methylation reaction conditions with DMC, DMF, and catalytic 1,4-diazabicyclo[2.2.2]octane (DABCO) have been developed and applied to a variety of electron-deficient pyrroles (Equation 10.36)[91] and indoles (Equation 10.37),[92] affording the corresponding *N*-methyl products in good to excellent yields.

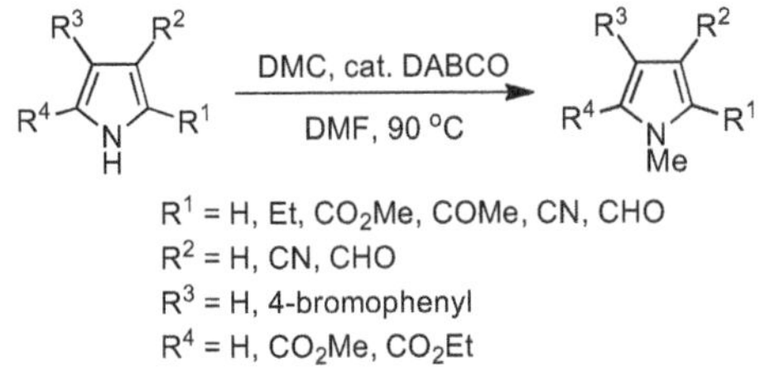

$$R^1 = H, Et, CO_2Me, COMe, CN, CHO$$
$$R^2 = H, CN, CHO$$
$$R^3 = H, \text{4-bromophenyl}$$
$$R^4 = H, CO_2Me, CO_2Et$$

(10.36)

$$R = CN, NO_2, Cl, Br, CHO, CO_2Me$$

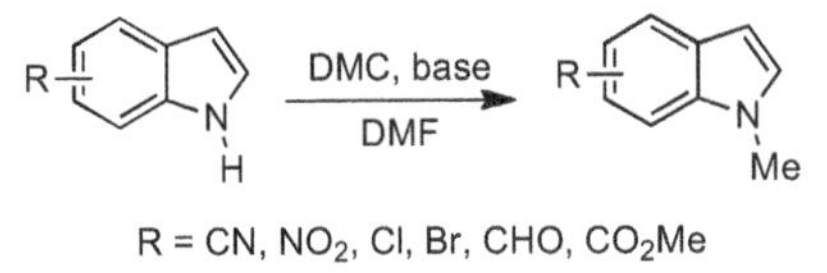

(10.37)

10.8.4.2 *N*-Methylation of Amide

Although methyl iodide poses potential environmental problems, the *N*-methylation of amide **142** was achieved by using excess of methyl iodide (Equation 10.38).[93]

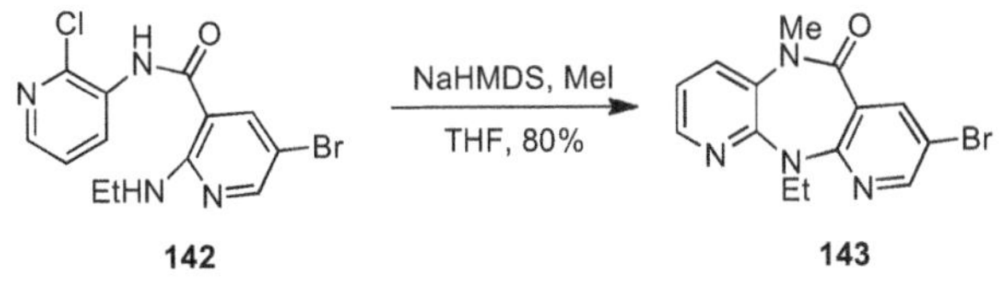

(10.38)

Removal of excess of methyl iodide from the waste stream was achieved by distilling the volatiles into a solution of 4-picoline. The contents were heated at 50 °C for 5 h to complete destruction of MeI.

Procedure

- Charge 114 kg (321 mol) of **142**, followed by adding 323 kg of THF.
- Warm to 50 °C.
- Charge 650 kg (719 mol, 2.2 equiv) of 1 M NaHMDS in THF over 2 h.
- Stir at 50 °C for 1 h.
- Charge 59.2 kg (66 mol, 0.2 equiv) of 1 M NaHMDS in THF.
- Stir at 50 °C for 1 h.
- Cool to 25 °C.
- Charge 119 kg (836 mol, 2.6 equiv) of MeI over 2 h at <30 °C.
- Stir at 25 °C for 10 h.
- Distill at 5 kPa the volatiles into a solution of 4-picoline (168 kg, 1803 mol, 5.6 equiv) in MeOH (88 kg) containing 2,6-di-*tert*-butyl-4-methylphenol (600 g).
- Heat the distillates at 50 °C for 5 h.
- Workup.

REMARKS[94]

(a) Methyl iodide is not compatible with certain reducing agents such as $LiAlH_4$, $LiEt_3BH$, and $LiBH_3NR_2$. Reactions of methyl iodide with those reagents are violent and exothermic, generating flammable methane gas (Equations 10.39–10.41).

$$LiAlH_4 \ + \ CH_3I \ \longrightarrow \ CH_4 \ + \ LiAlH_3I \qquad\qquad (10.39)$$

$$LiEt_3BH \ + \ CH_3I \ \longrightarrow \ CH_4 \ + \ LiI \ + \ Et_3B \qquad\qquad (10.40)$$

$$LiBH_3NR_2 \ + \ CH_3I \ \longrightarrow \ CH_4 \ + \ LiI \ + \ H_2B{-}NR_2 \qquad\qquad (10.41)$$

(b) Alkaline scrubber is used in plant to trap the volatile CH_3I.[95]

10.8.4.3 *N*-Methylation of 2-Methoxypyridine Derivative

Instead of using methyl iodide, a combination of trimethylsulfoxonium iodide with magnesium hydroxide was utilized to methylate 2-methoxypyridine derivative **144** in DMF at 88 °C, affording a clean conversion within 2 h (Equation 10.42).[88]

$$(10.42)$$

REMARK

Reactions under anhydrous conditions were very slow and led to incomplete conversion of pyridine to pyridone.

10.8.5 METHYL BROMIDE

As a volatile organic compound, methyl bromide (bp 3 °C) is a colorless, odorless, and non-flammable gas. Methyl bromide is an ozone-depleting and potentially occupational carcinogen chemical. As a result, methyl bromide, generated as a by-product during demethylation of **146** (Equation 10.43),[96] poses environmental and safety concerns and, therefore, must be controlled upon scaling up.

$$\text{(10.43)}$$

Accordingly, a methyl bromide scrubber system was developed by using aqueous ethanolamine (20%). This scrubber medium effectively converted the methyl bromide off-gas to amine·HBr salts (Equations 10.44–10.46) which are ready for aqueous waste disposal.

$$RNH_2 \ + \ MeBr \ \longrightarrow \ R\overset{+}{N}H_2Me \ \overset{-}{Br} \tag{10.44}$$

$$RNH_2 \ + \ R\overset{+}{N}H_2Me \ \overset{-}{Br} \ \rightleftharpoons \ R\overset{+}{N}H_3 \ \overset{-}{Br} \ + \ RNHMe \tag{10.45}$$

$$RNHMe \ + \ MeBr \ \longrightarrow \ R\overset{+}{N}HMe_2 \ \overset{-}{Br} \tag{10.46}$$

REMARK

Methyl bromide off gas was also effectively trapped using a PPh_3/DMF scrubber to avoid any environmental release.[97]

10.9 HIGH ENERGY COMPOUNDS

10.9.1 5-HYDROXYBENZOFURAZAN

Benzofurazan derivatives are high-energy compounds, having relatively low onset temperatures of decomposition with large energy release (e.g., 2664 J/g, onset at 133 °C for 5-hydroxybenzofurazan **148**). The original procedure employed by medicinal chemists for the synthesis of **150** was shown in the reaction (Equation 10.47),[98] involving an S_NAr reaction of pyridyl chloride **149** with 5-hydroxy-benzofurazan **148**. Compared to **148**, the product, 5-aryloxybenzofurazan **150**, has an improved safety margin with regard to decomposition onset (1872 J/g, onset at 250 °C).

$$\text{(10.47)}$$

Concerning the potentially hazardous decomposition of 5-hydroxybenzofurazan **148**, the one-step synthesis of **150** was replaced with a safer, three-step sequence, including a S$_N$Ar reaction of **149** with nitrophenol **151**, oxidative cyclization, and *N*-oxide reduction (Scheme 10.29).[98] The optimized synthesis avoided the use of **148** and produced **150** on a 32 kg scale.

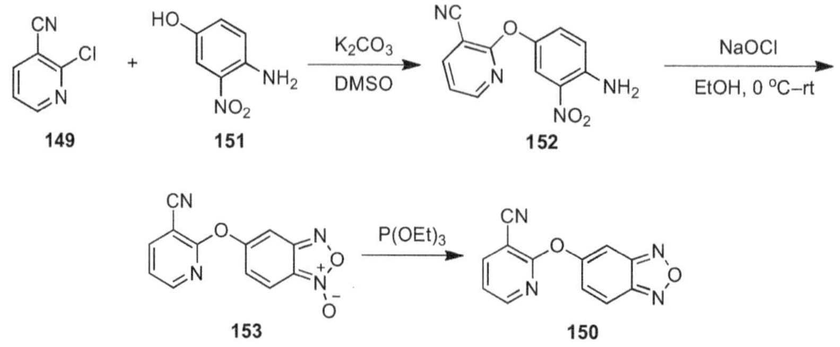

SCHEME 10.29

10.9.2 1-Hydroxybenzotriazole (HOBt)

One of the popular carbonyl activating agents for amide bond formation is 1-hydroxybenzotriazole (HOBt). However, HOBt is known to explode with high decomposition energy when heated beyond its melting point (156 °C).[99] Several safe, effective, and potential alternatives to HOBt[100] are available, such as 2-hydroxypyridine and 2-hydroxy-5-nitropyridine.

10.9.2.1 Ethyl Acetate/Water Two-Phase System

As the anhydrous HOBt dust is explosive in nature, HOBt-mediated preparation of amide **156** was conducted in a two-phase mixture of ethyl acetate and water (2.5/1, v/v) (Equation 10.48).[47]

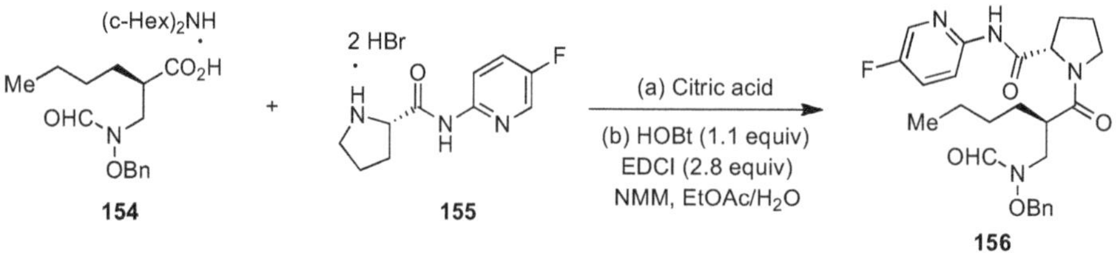

$$(10.48)$$

Procedure

Stage (a)
* Treat a solution of **154** (41.7 kg, 90.5 mol) in EtOAc (360 L) with 360 kg of 10% citric acid at room temperature.
* Separate layers.
* Wash the organic layer with water (2×270 L).
* Charge, to the EtOAc solution of the free acid of **154**, 40.1 kg (108.1 mol, 1.2 equiv) of **155**.
* Cool the resulting mixture to 5 °C.

Stage (b)
* Add 72 L of water, followed by adding HOBt (15.4 kg, 100.6 mol) and 1-ethyl-3-(3-di methylaminopropyl)-carbodiimide hydrochloride (EDCI) (48.4 kg, 252.2 mol).

- Add 72 L of water, followed by adding *N*-methylmorpholine (57.5 kg, 568 mol) slowly at <20 °C.
- Stir the mixture at 21 °C for 10 h.
- Separate layers.
- Wash the organic layer with water (4×270 L) to remove HOBt.
- Split the EtOAc solution into two equal portions (85.7 kg each).
- Filter each portion through a bed of silica gel (50 kg) using EtOAc as the eluent.
- Distill the combined fractions at 45 °C under reduced pressure to a final volume of 260 L.
- Heat the batch to 55 °C.
- Add heptane (815 L).
- Cool to 45 °C.
- Add seeds.
- Cool to 35 °C and hold for 2 h.
- Cool to 20 °C and hold for 8 h.
- Cool to −10 °C and stir fro 2 h.
- Filter and wash with heptane (94 L), EtOAc (48 L), EtOH (2.7 L), and water (0.14 L).
- Dry at 42 °C under vacuum for 8 h.
- Recrystallize the crude product from EtOAc (173 L)/heptane (695 L).
- Yield of **156**: 28.9 kg (68%) as an off-white crystalline solid.

REMARKS

(a) The residual HOBt was removed by water wash.
(b) Dicyclohexylamine is a good candidate for the transformation of low-melting chiral acid into the corresponding crystalline salt. Another example of using dicyclohexylamine was reported in the literature.[101]
(c) HOBt is sold as a monohydrate and drying of the hydrated material shall be avoided.

10.9.2.2   Replacement of HOBt with 2-Hydroxypyridine

For coupling under anhydrous conditions, the use of HOBt at a production scale is not safe. Thus, 2-hydroxypyridine was employed to replace HOBt during the synthesis of cathepsin S inhibitor **159** (Equation 10.49).[101]

(10.49)

Procedure

- Charge 10.8 g (114 mmol, 1.3 equiv) of 2-hydroxypyridine to a solution of **157** (87.3 mmol) (~12% in MeTHF).

- Charge 24.23 g (96 mmol, 1.1 equiv) of aminonitrile **158**.
- Charge 21.76 g (114 mmol, 1.3 equiv) of EDCI.
- Stir for 16 h.
- Workup.

REMARK

This laboratory process was successfully demonstrated on an 11 kg scale with **159** in 88% yield.

10.9.2.3 Replacement of HOBt with 2-Chloro-4,6-dimethoxy-1,3,5-triazine

2-Chloro-4,6-dimethoxy-1,3,5-triazine (CDMT) was chosen as the coupling reagent in the amide formation between acid **160** and amine **161** to minimize safety concerns arising from the use of HOBt on the manufacturing scale (Equation 10.50).[97]

$$(10.50)$$

10.10 TOXIC COMPOUNDS

Many cyanides are highly toxic by halting cellular respiration via inhibiting an enzyme in the mitochondria called cytochrome c oxidase. Hydrocyanic acid, also known as hydrogen cyanide (HCN), is a highly volatile and toxic liquid (boiling point: 25.6 °C).

10.10.1 ZINC CYANIDE

Palladium-catalyzed cyanation of aryl halides is a common approach to introducing the cyano group in an aromatic system (Equation 10.51).[15] Besides the toxicity, cyanide potentially deactivates the palladium catalyst. Thus, heterogeneous conditions in which the cyanide source is insoluble are generally required. Zinc cyanide is frequently utilized in palladium-catalyzed cyanation because mainly of its poor solubility in most solvents (its solubility in water is 0.05 mg/100 mL). Precaution shall be taken not only during the reaction, but in the workup of the reaction mixture as well. It is important to maintain the system under basic conditions to ensure that no HCN is generated. The addition of concentrated ammonium hydroxide to the reaction mixture during the workup sequestered free cyanide ions completely.

$$(10.51)$$

Procedure

- Charge 11.3 L of DMF into a reactor.
- Sparge the DMF with nitrogen.
- Charge 49.2 g (0.219 mol) of Pd(OAc)$_2$ and 267 g (0.876 mol) of P(*o*-Tol)$_3$.
- Heat to 56 °C and stir the catalyst solution for 15 min while sparging with nitrogen.
- Charge 301 mL (0.331 mol) of Et$_2$Zn over 15 min and age at 56 °C for 1 h.
- Charge 778 g (6.62 mol) of Zn(CN)$_2$ to a solution of **163** (3.75 kg, 11.04 mol) in 15 L of DMF in a separate flask.
- Sparge the mixture with nitrogen for 20 min.
- Heat to 56 °C and transfer the mixture to the reactor.
- Stir the batch at 56 °C for 2 h.
- Cool to 5–10 °C.
- Add 3.75 L of concentrated NH$_4$OH at <30 °C.
- Filter the batch after aging for 1 h.
- Rinse with 20 L of toluene.
- Add 26 L of 20% aqueous NH$_4$OH and 17.5 L of toluene to the filtrate.
- Separate layer after well-mixing.
- Wash the organic layer with 26 L of 15% NaCl and 26 L of water.
- Yield: 2.90 kg (92.0%, a solution assay).

10.10.2 POTASSIUM FERROCYANIDE

Potassium ferrocyanide (K$_4$[Fe(CN)$_6$]) is non-toxic, although upon contact with strong acid it can release extremely toxic HCN gas. It is not decomposed to cyanide in the body. Therefore, K$_4$[Fe(CN)$_6$] was identified as an inexpensive cyanide source for the cyanation of aryl bromide **165** (Equation 10.52).[102]

$$\text{(10.52)}$$

Procedure

- Charge a deoxygenated solution of Pd(OAc)$_2$ (2.1 kg, 9.2 mol) and P(*o*-Tol)$_3$ in 302.7 kg of DMAc to a deoxygenated suspension of **165** (160.4 kg, 463.2 mol), potassium ferrocyanide (78.1 kg, 184.9 mol), Na$_2$CO$_3$ (49.0 kg, 463.2 mol) in a mixture of toluene (277.1 kg) and DMAc (305.0 kg) at 25 °C.
- Heat the resulting suspension to 120–130 °C over 1 h and stir at 125 °C for 9 h.
- Filter, after cooling, and rinse with toluene (277.2 kg).
- Wash the combined filtrates with water (640 kg).
- Extract the aqueous layer with toluene (277.0 kg).
- Wash the combined organic layers with water (320 kg).
- Distill the organic layer to 480 L.
- Add IPA (753.0 kg) to the toluene solution.
- Distill the resulting organic mixture to 400 L.
- Add IPA (768.4 kg).
- Heat the resulting organic mixture to 60 °C and treat with 16.1 kg of activated charcoal.

- Filter through a Celite pad and rinse with IPA (251.3 kg).
- Distill the combined filtrates to 400 L.
- Add IPA (260 kg).
- Treat the resulting organic mixture with 6.4 kg of *N*-methyl thiourea at 55 °C until the solid dissolved.
- Cool to 20 °C.
- Add water (800 kg) and age for 1 h.
- Filter, wash with a mixture of IPA (125.8 kg) and water (320 kg), and dry at 60 °C under reduced pressure.
- Yield: 116.6 kg (86%) as a white solid.

10.10.3 Acetone Cyanohydrin

Acetone cyanohydrin is a useful cyanide surrogate that has been utilized in various chemical transformations. Acetone cyanohydrin is extremely hazardous and decomposes into highly toxic hydrogen cyanide when making contact with water under acidic conditions.

10.10.3.1 Epoxide Opening

As a cyanide surrogate, acetone cyanohydrin was treated with lithium hexamethyldisilazane (LiHMDS) to give lithium cyanide. The in situ generated LiCN was used in an epoxide ring-opening reaction (Equation 10.53).[103]

$$(10.53)$$

Procedure

- Charge 37.9 kg (441.1 mol) of acetone cyanohydrin into a reactor which contains 190.6 kg (352.5 mol) of lithium bis(trimethylsilyl)amide (LiHMDS) in THF (pre-concentrated by distillation at <30 °C under reduced pressure to a volume of 140 L) at 0–22 °C.
- Prepare a solution of **167** by mixing 44.2 kg (98.1 wt%, 173 mol) of **167** in 40 kg of THF at 39 °C.
- Charge the solution of **167** into the reactor.
- Heat the batch to 65 °C at reflux and hold at reflux for 12 h.
- Cool to 1 °C.
- Add 243 kg of 13.7 wt% aqueous HCl to pH 9–10.
- Add 51 kg of water and 166 kg of isopropanol.
- Distill to 550 L at 40 ° under reduced pressure.
- Cool to –11 °C and hold for 15 h.
- Isolate the product by centrifuge.
- Wash with 25% aqueous isopropanol (3×15 kg) at 15 °C.
- Dry at 45 °C under vacuum.
- Yield: 43.5 kg (90.2%) with 99.6% purity.

REMARK

Safety requirements in the plant involve chemical-resistant gloves, safety goggles, a supplied air respirator, and chemical resistant suit, such as Saranax, for any charging or sampling operations. All reaction streams were kept caustic (pH >8) to minimize the possibility of the formation of HCN.

10.10.3.2 Michael Addition

Acetone cyanohydrin as cyanide substituent was applied toward the Michael addition to prepare ketonitrile **170** (Equation 10.54).[104] The reaction was carried out in aqueous acetone by treatment of α,β-unsaturated ketone **169** with 1.25 equiv of acetone cyanohydrin in the presence of 6 mol% of tetramethylammonium hydroxide as a catalyst.

$$(10.54)$$

Procedure

- Charge 9.3 kg (32.9 mol) of **169** into 32 L of acetone at room temperature to form a slurry.
- Charge 3.5 kg (41.1 mol, 1.25 equiv) acetone cyanohydrin, followed by adding 0.75 L (2.08 mol, 6 mol%) of 25% aqueous solution of tetramethylammonium hydroxide.
- Heat the resulting mixture at reflux for 10 h.
- Cool the mixture.
- Add H_2O (18 L) and cool the resulting slurry to 0–10 °C.
- Isolate by centrifugation.
- Dry at 40–50 °C under vacuum.
- Yield: 8.6 kg (85%).

REMARK

The process waste could be distilled to remove acetone (which does not azeotrope with water) and the resulting aqueous waste was treated with sodium hypochlorite to destroy the excess cyanide under basic conditions (Equation 10.55).[105] The resulting cyanate (OCN^-) is not toxic.

$$CN^- \ + \ OCl^- \ \longrightarrow \ OCN^- \ + \ Cl^- \qquad (10.55)$$

10.11 CORROSIVE REAGENT

TBAF is often used to deprotect TIPS group-protected terminal alkynes such as **171** (Equation 10.56). However, TBAF is corrosive to a glass-lined reactor, thereby the removal of the TIPS group using corrosive TBAF is undesired on large scale.

$$(10.56)$$

As an alternative method to the removal of the TIPS group at a large scale, a mixture of tetrabutyl-ammonium hydroxide (20 mol%) and K_2HPO_4 buffer was found to be optimum in providing good product yield and purity.[106]

NOTES

1. An incident occurred during the isolation of diazonium salt on laboratory scale: Bassan, E.; Ruck, R. T.; Dienemann, E.; Emerson, K. M.; Humphrey, G. R.; Raheem, I. T.; Tschaen, D. M.; Vickery, T. P.; Wood, H. B.; Yasuda, N. *Org. Process Res. Dev.* **2013**, *17*, 1611.

2. (a) Ullrich, R.; Grewer, Th. *Thermochim. Acta* **1993**, *225*, 201. (b) Urben, P. G. *Bretherick's Handbook of Reactive Chemical Hazards*, 6th ed.; Butterworth Heinemann, Oxford, 1999; Vol. 2, pp 96–97. (c) Anon.; *Sichere Chemiearbeit* **1993**, *45*, 8.

3. Schmidt, G.; Reber, S.; Bolli, M. H.; Abele, S. *Org. Process Res. Dev.* **2012**, *16*, 595.

4. Magano, J.; Waldo, M.; Geene, D.; Nord, E. *Org. Process Res. Dev.* **2008**, *12*, 877.

5. (a) Porter, H. D.; Peterson, W. D. *Org. Synth.* **1940**, *20*, 73. (b) Benchidmi, M.; Bouchet, P.; Lazaro, R. *J. Heterocycl. Chem.* **1979**, *16*, 1599.

6. Lukin, K.; Hsu, M. C.; Chambournier, G.; Kotecki, B.; Venkatramani, C. J.; Leanna, M. R. *Org. Process Res. Dev.* **2007**, *11*, 578.

7. Morandi, B.; Mariampillai, Carreira, E. M. *Angew. Chem. Int. Ed.* **2011**, *50*, 1101.

8. Nielsen, M. A.; Nielsen, M. K.; Pittelkow, T. *Org. Process Res. Dev.* **2004**, *8*, 1059.

9. Ellis, J. E.; Davis, E. M.; Brower, P. L. *Org. Process Res. Dev.* **1997**, *1*, 250.

10. Bennett, R. B.; Choi, J. R.; Montgomery, W. D.; Cha, J. K. *J. Am. Chem. Soc.* **1989**, *111*, 2580.

11. Wiss, J.; Fleury, C.; Heuberger, C.; Onken, U. *Org. Process Res. Dev.* **2007**, *11*, 1096.

12. Sharma, P. K.; Shah, R. N.; Carver, J. P. *Org. Process Res. Dev.* **2008**, *12*, 831.

13. He, L.; Wanunu, M.; Byun, H.; Bittman, R. *J. Org. Chem.* **1999**, *64*, 6049.

14. Hassner, A.; Stern, M.; Gottlieb, H. E.; Frolow, F. *J. Org. Chem.* **1990**, *55*, 2304.

15. Chen, C.; Frey, L. F.; Shultz, S.; Wallace, D. J.; Marcantonio, K.; Payack, J. F.; Vazquez, E.; Springfield, S. A.; Zhou, G.; Liu, P.; Kieczykowski, G. R.; Chen, A. M.; Phenix, B. D.; Singh, U.; Strine, J.; Izzo, B.; Krska, S. W. *Org. Process Res. Dev.* **2007**, *11*, 616.

16. Tsuritani, T.; Mizuno, H.; Nonoyama, N.; Kii, S.; Akao, A.; Sato, K.; Yasuda, N.; Mase, T. *Org. Process Res. Dev.* **2009**, *13*, 1407.

17. Marsh, F. D. *J. Org. Chem.* **1972**, *37*, 2966.

18. Kopach, M. E.; Murray, M. M.; Braden, T. M.; Kobierski, M. E.; Williams, O. L. *Org. Process Res. Dev.* **2009**, *13*, 152.

19. Katritzky, A. R.; Khatib, M. E.; Bol'shakov, O.; Khelashvili, L.; Steel, P. J. *J. Org. Chem.* **2010**, *75*, 6532.

20. (a) Tomlinson, W. R.; Ottoson, K. G.; Audrieth, L. F. *J. Am. Chem. Soc.* **1949**, *71*, 375. (b) Egghart, H. C. *Inorg. Chem.* **1968**, *7*, 1225.

21. Conrow, R. E.; Dean, W. D. *Org. Process Res. Dev.* **2008**, *12*, 1285.

22. Dale, D.; Ironside, M. D.; Shaw, S. M. *Org. Process Res. Dev.* **2014**, *18*, 1778.

23. Girardin, M.; Dolman, S. J.; Lauzon, S.; Ouellet, S. G.; Hughes, G.; Fernandez, P.; Zhou, G.; O'Shea, P. D. *Org. Process Res. Dev.* **2011**, *15*, 1073.

24. Treitler, D. S.; Leung, S.; Lindrud, M. *Org. Process Res. Dev.* **2017**, *21*, 460.

25. Finnegan, W. G.; Henry, R. A.; Lofquist, R. *J. Am. Chem. Soc.* **1958**, *80*, 3908.

26. Wilson, R. D.; Cleator, E.; Ashwood, M. S.; Bio, M. M.; Brands, K. M. J.; Davies, A. J.; Dolling, U.-H.; Emerson, K. M.; Gibb, A. D.; Hands, D.; McKeown, A. E.; Oliver, S. F.; Reamer, R. A.; Sheen, F. J.; Stewart, G. W.; Zhou, G. X. *Org. Process Res. Dev.* **2009**, *13*, 543.

27. Demko, Z. P.; Sharpless, K. B. *J. Org. Chem.* **2001**, *66*, 7945.

28. Duncia, J. V.; Pierce, M. E.; Santella, J. B., III. *J. Org. Chem.* **1991**, *56*, 2395.

29. Davulcu, A. H.; McLeod, D. D.; Li, J.; Katipally, K.; Littke, A.; Doubleday, W.; Xu, Z.; McConlogue, C. W.; Lai, C. J.; Gleeson, M.; Schwinden, M.; Parsons, R. L., Jr. *J. Org. Chem.* **2009**, *74*, 4068.

30. Journet, M.; Cai, D.; Hughes, D. L.; Kowal, J. J.; Larsen, R. D.; Reider, P. J. *Org. Process Res. Dev.* **2005**, *9*, 490.

31. Huang, C.; Geng, X.; Zhao, P.; Zhou, Y.; Yu, X.-X.; Wang, L.-S.; Wu, Y.-D.; Wu, A.-X. *J. Org. Chem.* **2021**, *86*, 13664.

32. Zhao, J.; Gimi, R.; Katti, S.; Reardon, M.; Nivorozhkin, V.; Konowicz, P.; Lee, E.; Sole, L.; Green, J.; Siegel, C. S. *Org. Process Res. Dev.* **2015**, *19*, 576.

33. (a) Deerberg, J.; Prasad, S. J.; Sfouggatakis, C.; Eastgate, M. D.; Fan, Y.; Chidambaram, R.; Sharma, P.; Li, L.; Schild, R.; Müslehiddinoğlu, J.; Chung, H.-J.; Leung, S.; Rosso, V. *Org. Process Res. Dev.* **2016**, *20*, 1949. (b) For examples of large-scale preparation of 2-amino-5-methylpyrazine, see: Steven, A.; Hopes, P. *Org. Process Res. Dev.* **2018**, *22*, 77.

34. Karlsson, S.; Gardelli, C.; Lindhagen, M.; Nikitidis, G.; Svensson, T. *Org. Process Res. Dev.* **2018**, *22*, 1174.

35. Niemeier, J. K.; Kjell, D. P. *Org. Process Res. Dev.* **2013**, *17*, 1580.

36. Eisenbraun, E. J.; Payne, K. W.; Bymaster, J. S. *Ind. Eng. Chem. Res.* **2000**, *39*, 1119.

37. Kuethe, J. T.; Childers, K. G.; Peng, Z.; Journet, M.; Humphrey, G. R.; Vickery, T.; Bachert, D.; Lam, T. T. *Org. Process Res. Dev.* **2009**, *13*, 576.

38. Kampmann, D.; Stuhlmüller, G.; Simon, R.; Cottet, F.; Leroux, F.; Schlosser, M. *Synthesis* **2005**, *6*, 1028.
39. (a) Wang, Z.; Richter, S. M.; Gandarilla, J.; Kruger, A. W.; Rozema, M. J. *Org. Process Res. Dev.* **2013**, *17*, 1603. (b) Kruger, A. W.; Rozema, M. J.; Chu-Kung, A.; Ganderilla, J.; Haight, A. R.; Kotecki, B. J.; Richter, S. M.; Schwartz, A. M.; Wang, Z. *Org. Process Res. Dev.* **2009**, *13*, 1419.
40. Brandt, T. A.; Caron, S.; Damon, D. B.; DiBrino, J.; Ghosh, A.; Griffith, D. A.; Kedia, S.; Ragan, J. A.; Rose, P. R.; Vanderplas, B. C.; Wei, L. *Tetrahedron* **2009**, *65*, 3292.
41. Magnus, N. A.; Aikins, J. A.; Cronin, J. S.; Diseroad, W. D.; Hargis, A. D.; LeTourneau, M. E.; Parker, B. E.; Reutzel-Edens, S. M.; Schafer, J. P.; Staszak, M. A.; Stephenson, G. A.; Tameze, S. L.; Zollars, L. M. H. *Org. Process Res. Dev.* **2005**, *9*, 621.
42. Dufton, S. F. *J. Chem. Soc.* **1892**, 782.
43. Norris, T.; Bezze, C.; Franz, S. Z.; Stivanello, M. *Org. Process Res. Dev.* **2009**, *13*, 354.
44. Milburn, R. R.; Thiel, O. R.; Achmatowicz, M.; Wang, X.; Zigterman, J.; Bernard, C.; Colyer, J. T.; DiVirgilio, E.; Crockett, R.; Correll, T. L.; Nagapudi, K.; Ranganathan, K.; Hedley, S. J.; Allgeier, A.; Larsen, R. D. *Org. Process Res. Dev.* **2011**, *15*, 31.
45. Robinson, B. *The Fischer Indole Synthesis*; John Wiley and Sons, Chichester, 1982.
46. Ashcroft, C.; Hellier, P.; Pettman, A.; Watkinson, S. *Org. Process Res. Dev.* **2011**, *15*, 98.
47. Slade, J.; Parker, D.; Girgis, M.; Mueller, M.; Vivelo, J.; Liu, H.; Bajwa, J.; Chen, G.-P.; Carosi, J.; Lee, P.; Chaudhary, A.; Wambser, D.; Prasad, K.; Bracken, K.; Dean, K.; Boehnke, H.; Repič, O.; Blacklock, T. J. *Org. Process Res. Dev.* **2006**, *10*, 78.
48. Cisneros, L. O.; Rogers, W. J.; Mannan, M. S. *Can. J. Chem. Eng.* **2004**, *82*, 1307.
49. Fields, J. D.; Kropp, P. J. *J. Org. Chem.* **2000**, *65*, 5937.
50. Wittman, M. D.; Halcomb, R. L.; Danishefsky, S. J. *J. Org. Chem.* **1990**, *55*, 1981.
51. Biloski, A. J.; Ganem, B. *Synthesis* **1983**, 537.
52. Zhang, Y.; Xu, G. Z. *Naturforsch., B: Chem. Sci.* **1989**, *44*, 1475.
53. Patel, I.; Smith, N. A.; Tyler, S. N. G. *Org. Process Res Dev.* **2009**, *13*, 49.
54. Santhosh, U.; Kshirsagar, Y. M.; Venkatesan, K.; Hazra, D.; Kindel, J.; Sridharan, R.; Ennis, D.; Howells, G. E.; Stefinovic, M.; Manjunatha, S. G.; Nambiar, S. *Org. Process Res. Dev.* **2014**, *18*, 1802.
55. Rieder, C. J.; Smith, M. V. *Org. Process Res. Dev.* **2021**, *25*, 2308.
56. Ford, J. G.; Pointon, S. M.; Powell, L.; Siedlecki, P. S. *Org. Process Res. Dev.* **2010**, *14*, 1078.
57. Barbas, R.; Botija, M.; Camps, H.; Portell, A.; Prohens, R.; Puigjaner, C.; *Org. Process Res. Dev.* **2007**, *11*, 1131.
58. Colleville, A. P.; Horan, R. A. J.; Tomkinson, N. C. O. *Org. Process Res. Dev.* **2014**, *18*, 1128.
59. (a) Caron, S.; Do, N. M.; Sieser, J. E.; Whritenour, D. C.; Hill, P. D. *Org. Process Res. Dev.* **2009**, *13*, 324. (b) Caron, S.; Do, N. M.; Sieser, J. E. *Tetrahedron Lett.* **2000**, *41*, 2299.
60. Davis, G.; Cook, N. *CHEMTECH*, **1977**, *7*, 626.
61. Walker, M. D.; Andrews, B. I.; Burton, A. J.; Humphreys, L. D.; Kelly, G.; Schilling, M. B.; Scott, P. W. *Org. Process Res. Dev.* **2010**, *14*, 108.
62. (a) Jähnisch, K.; Hessel, V.; Lowe, H.; Baerns, M. *Angew. Chem. Int. Ed.* **2004**, *43*, 406. (b) Ehrfeld, W.; Hessel, V.; Löwe, H. In *Microreactors in Organic Synthesis and Catalysis*; Wirth, T., Ed. Wiley-VCH, Weinheim, 2008. (c) Wirth, T., Ed. *Microreactors in Organic Synthesis and Catalysis*; Wiley-VCH: Weinheim, 2008. (d) Brian, P.; Mason, K.; Price, J.; Bogdan, A.; Tyler McQuade, D. *Chem. Rev.* **2007**, *107*, 2300. (e) Anderson, N. G. *Org. Process Res. Dev.* **2012**, *16*, 852.
63. Pelleter, J.; Renaud, F. *Org. Process Res. Dev.* **2009**, *13*, 698.
64. Rumi, L.; Pfleger, C.; Spurr, P.; Klinkhamer, U.; Bannwarth, W. *Org. Process Res. Dev.* **2009**, *13*, 747.
65. McConnell, J. R.; Hitt, J. E.; Daugs, E. D.; Rey, T. A. *Org. Process Res. Dev.* **2008**, *12*, 940.
66. Grongsaard, P.; Bulger, P. G.; Wallace, D. J.; Tan, L.; Chen, Q.; Dolman, S. J.; Nyrop, J.; Hoerrner, R. S.; Weisel, M.; Arredondo, J.; Itoh, T.; Xie, C.; Wen, X.; Zhao, D.; Muzzio, D. J.; Bassan, E. M.; Shultz, C. S. *Org. Process Res. Dev.* **2012**, *16*, 1069.
67. A continous nitration of substituted pyridine substrates on large scale was described: Gage, J. R.; Guo, X.; Tao, J.; Zheng, C. *Org. Process Res. Dev.* **2012**, *16*, 930.
68. Hansen, M. M.; Borders, S. S. K.; Clayton, M. T.; Heath, P. C.; Kolis, S. P.; Larsen, S. D.; Linder, R. J.; Reutzel-Edens, S. M.; Smith, J. C.; Tameze, S. L.; Ward, J. A.; Weigel, L. O. *Org. Process Res. Dev.* **2009**, *13*, 198.
69. Li, X.; Ng, R. A.; Zhang, Y.; Russell, R. K.; Sui, Z. *Org. Process Res. Dev.* **2009**, *13*, 652.
70. Coon, C. L.; Blucher, W. G.; Hill, M. E. *J. Org. Chem.* **1973**, *38*, 4243.
71. Allwein, S. P.; Roemmele, R. C.; Haley, J. J. Jr.; Mowrey, D. R.; Petrillo, D. E.; Reif, J. J.; Gingrich, D. E.; Bakale, R. P. *Org. Process Res. Dev.* **2012**, *16*, 148.

72. Giles, M. E.; Thomson, C.; Eyley, S. C.; Cole, A. J.; Goodwin, C. J.; Hurved, P. A.; Morlin, A. J. G.; Tornos, J.; Atkinson, S.; Just, C.; Dean, J. C.; Singleton, J. T.; Longton, A. J.; Woodland, I.; Teasdale, A.; Gregertsen, B.; Else, H.; Athwal, M. S.; Tatterton, S.; Knott, J. M.; Thompson, N.; Smith, S. J. *Org. Process Res. Dev.* **2004**, *8*, 628.

73. Beutner, G. L.; Desai, L.; Fanfair, D.; Lobben, P.; Anderson, E.; Leung, S. W.; Eastgate, M. D. *Org. Process Res. Dev.* **2014**, *18*, 1812.

74. Zhang, P.; Shankar, A.; Cleary, T. P.; Cedilote, M.; Locklear, D.; Pierce, M. E. *Org. Process Res. Dev.* **2007**, *11*, 861.

75. Prakash, G. K. S.; Panja, C.; Mathew, T.; Surampudi, V.; Petasis, N. A.; Olah, G. A. *Org. Lett.* **2004**, *6*, 2205.

76. (a) Prakash, G. K. S.; Mathrew, T. *Angew. Chem. Int. Ed.* **2009**, *49*, 1726. (b) Fors, B. P.; Buchwald, S. L. *J. Am. Chem. Soc.* **2009**, *131*, 12898.

77. Sadhu, P.; Alla, S. K.; Punniyamurthy, T. *J. Org. Chem.* **2014**, *79*, 8541.

78. Bhattacharya, A.; Purohit, V. C.; Deshpande, P.; Pullockaran, Grosso, J. A.; DiMarco, J. D.; Gougoutas, J. Z. *Org. Process Res. Dev.* **2007**, *11*, 885.

79. For example of nitration via rearrangement of *N*-nitramino intermediate, see: Porcs-Makkay, M.; Mezei, T.; Simig, G. *Org. Process Res. Dev.* **2007**, *11*, 490.

80. Gustin, J.-L. *Org. Process Res. Dev.* **1998**, *2*, 27.

81. Jiang, X.; Gong, B.; Prasad, K.; Repič, O. *Org. Process Res. Dev.* **2008**, *12*, 1164.

82. Chen, W.; Li, Z. *J. Org. Chem.* **2022**, *87*, 76.

83. (a) Sittig, M. *Handbook of Toxic and Hazardous Chemical and Carcinogens,* 2nd ed.; Noyes, Park Ridge, NJ, 1985; pp 133–135. (b) Lewis, R. J. *Sax's Dangerous Properties of Industrial Materials,* 10th ed.; Wiley, New York, 2000; Vol. II, p 478.

84. van der Made, A. W.; van der Made, R. H. *J. Org. Chem.* **1993**, *58*, 1262.

85. Clair, J. D.; Valentine, J. R. *Org. Process Res. Dev.* **2005**, *9*, 1013.

86. Cohen, J. H.; Maryanoff, C. A.; Stefanick, S. M.; Sorgi, K. L.; Villani, F. J., Jr.; *Org. Process Res. Dev.* **1999**, *3*, 260.

87. Prashad, M.; Har, D.; Hu, B.; Kim, H.-Y.; Girgis, M. J.; Chaudhary, A.; Repič, O.; Blacklock, T. J. *Org. Process Res. Dev.* **2004**, *8*, 330.

88. Campeau, L.-C.; Dolman, S. J.; Gauvreau, D.; Corley, E.; Liu, J.; Guidry, E. N.; Ouellet, S. G.; Steinhuebel, Weisel, M.; O'Shea, P. D. *Org. Process Res. Dev.* **2011**, *15*, 1138.

89. For examples of the *N*-methylation of pyrroles and indoles see: (a) Baltazzi, E.; Krimen, L. I. *Chem. Rev.* **1963**, *63*, 511. (b) Reinecke, M. G.; Sebastian, J. F.; Johnson, H. W.; Pyun, C. *J. Org. Chem.* **1972**, *37*, 3066. (c) Santaniello, E.; Farachi, C.; Ponit, F. *Synthesis* **1979**, *8*, 617.

90. Rekha, V. V.; Ramani, M. V.; Ratnamala, A.; Rupakalpana, V.; Subbaraju, G. V.; Satyanarayana, C.; Rao, C. S. *Org. Process Res. Dev.* **2009**, *13*, 769.

91. Laurila, M. L.; Magnus, N. A.; Staszak, M. A. *Org. Process Res. Dev.* **2009**, *13*, 1199.

92. Jiang, X.; Tiwari, A.; Thompson, M.; Chen, Z.; Cleary, T. P.; Lee, T. B. K. *Org. Process Res. Dev.* **2001**, *5*, 604.

93. Busacca, C. A.; Cerreta, M.; Dong, Y.; Eriksson, M. C.; Farina, V.; Feng, X.; Kim, J.-Y.; Lorenz, J. C.; Sarvestani, M.; Simpson, R.; Varsolona, R.; Vitous, J.; Campbell, S. J.; Davis, M. S.; Jones, P.-J.; Norwood, D.; Qiu, F.; Beaulieu, P. L.; Duceppe, J.-S.; Haché, B.; Brong, J.; Chiu, F.-T.; Curtis, T.; Kelley, J.; Lo, Y. S.; Powner, T. H. *Org. Process Res. Dev.* **2008**, *12*, 603.

94. Pasumansky, L.; Goralski, C. T.; Singaram, B. *Org. Process Res. Dev.* **2006**, *10*, 959.

95. Yanagisawa, A.; Taga, M.; Atsumi, T.; Nishimura, K.; Ando, K.; Taguchi, T.; Tshumuki, H.; Chujo, I.; Mohri, S. *Org. Process Res. Dev.* **2011**, *15*, 376.

96. Hettenbach, K.; am Ende, D. J.; Leeman, K.; Dias, E.; Kasthurikrishnan, N.; Brenek, S. J.; Ahlijanian, P. *Org. Process Res. Dev.* **2002**, *6*, 407.

97. Kallman, N. J.; Liu, C.; Yates, M. H.; Linder, R. J.; Ruble, J. C.; Kogut, E. F.; Patterson, L. E.; Laird, D. L. T.; Hansen, M. M. *Org. Process Res. Dev.* **2014**, *18*, 501.

98. Caron, S.; Dugger, R. W.; Ruggeri, S. G.; Ragan, J. A.; Ripin, D. H. B. *Chem. Rev.* **2006**, *106*, 2943.

99. For a review, see: Ray, T. In *Encyclopedia of Reagents for Organic Synthesis*; Paquette, L., Ed.; John Wiley and Sons, New York; Vol. 4, p 2752. In the DSC studies, an exotherm was noted beginning at 150 °C [$\Delta H \approx 1129$ J/g].

100. Dunn, P. J.; Hoffmann, W.; Kang, Y.; Mitchell, J. C.; Snowden, M. J. *Org. Process Res. Dev.* **2005**, *9*, 956.

101. Lorenz, J. C.; Busacca, C. A.; Feng, X.; Grinberg, N.; Haddad, N.; Johnson, J.; Kapadia, S.; Lee, H.; Saha, A.; Sarvestani, M.; Spinelli, E. M.; Versolona, R.; Wei, X.; Zeng, X.; Senanayake, C. H. *J. Org. Chem.* **2010**, *75*, 1155.

102. Utsugi, M.; Ozawa, H.; Toyofuku, E.; Hatsuda, M. *Org. Process Res. Dev.* **2014**, *18*, 693.
103. Pesti, J.; Chen, C-K.; Spangler, L.; Delmonte, A. J.; Benoit, S.; Berglund, D.; Bien, J.; Brodfuehrer, P.; Chan, Y.; Corbett, E.; Costello, C.; DeMena, P.; Discordia, R. P.; Doubleday, W.; Gao, Z.; Gingras, S.; Grosso, J.; Haas, O.; Kacsur, D.; Lai, C.; Leung, S.; Miller, M.; Muslehiddinoglu, J.; Nguyen, N.; Qiu, J.; Olzog, M.; Reiff, E.; Thoraval, D.; Totleben, M.; Vanyo, D.; Vemishetti, P.; Wasylak, J.; Wei, C. *Org. Process Res. Dev.* **2009**, *13*, 716.
104. Ellis, J. E.; Davis, E. M.; Dozeman, G. J.; Lenoir, E. A.; Belmont, D. T.; Brower, P. L. *Org. Process Res. Dev.* **2001**, *5*, 226.
105. Gerritsen, C. M.; Margerum, D. W. *Inorg. Chem.* **1990**, *29*, 2757.
106. Sharif, E. U.; Miles, D. H.; Rosen, B. R.; Jeffrey, J. L.; Debien, L. P. P.; Powers, J. P.; Leleti, M. R. *Org. Process Res. Dev.* **2020**, *24*, 1254.

11 Grignard Reagent and Related Reactions

The Grignard reagent has played a significant role in organic synthesis since its discovery in 1900 by a French chemist, Victor Grignard. The Grignard reagent is such a diverse reagent that it can participate in a wide range of reactions, such as coupling reactions, oxidations, nucleophilic aliphatic substitutions, and eliminations. The Grignard reaction is the addition of organomagnesium halides (Grignard reagents) to ketones or aldehydes, to form tertiary or secondary alcohols. The reaction with formaldehyde leads to primary alcohols.

11.1 PREPARATION OF GRIGNARD REAGENTS

In general, the Grignard reagent can be made through reactions of alkyl or aryl halides with magnesium. Magnesium can take three basic physical forms: turnings, powder chips, and finely divided metal. Magnesium turnings are easy to handle but can cause abrasiveness to glass-lined reaction vessels if stirred for a long period of time. Powdered magnesium is more reactive but can be pyrophoric. Magnesium chips have a smaller surface area than magnesium turnings, which results in less reactivity.

The main hazard associated with the preparation of the Grignard reagent is the induction period because of the oxide layer on the metal surface. In addition, the reaction is sensitive to water and impurities which can result in delayed initiation periods. As the insertion of magnesium into C–X bond is an exothermic process (380 kj/mol Mg), accumulation of the halide in the presence of magnesium metal should be avoided or a runaway reaction may occur. Real-time monitoring[1] of the reaction progress by FTIR can provide valuable information including initiation and subsequent reactions.[2]

Therefore, the development of safe and robust initiation methods becomes critical for large-scale production. The preparation of the Grignard reagent from halide with magnesium metal turnings is a heterogeneous reaction and shall be performed in a flat-bottomed reactor with efficient agitation to provide a large surface area. With conical-bottomed reactor Mg turnings accumulated at the bottom valve, resulting in a slow or incomplete reaction.

11.1.1 USE OF CHLOROTRIMETHYLSILANE

A number of approaches are developed to activate the magnesium metal. The activation of magnesium can be done by employing 1–5 mol% of chlorotrimethylsilane (TMSCl) and the reaction can be initiated under mild reaction conditions, usually at room temperature to 35 °C.

11.1.1.1 Preparation of 4-Fluoro-2-methylphenylmagnesium Bromide

TMSCl could be employed to activate magnesium in the preparation of 4-fluoro-2-methylphenylmagnesium bromide **2** (Equation 11.1).[3]

$$(11.1)$$

Procedure

- Charge 0.75 mL (5.91 mmol) of TMSCl to a suspension of Mg (16.5 g, 0.68 mol) in anhydrous THF (450 mL) at 25 °C under N_2.
- Charge a portion (~5 mL) of **1** (134 g, 0.71 mol).
- Stir at 35 °C until the reaction was initiated.
- Charge the remaining **1** at 35–45 °C while monitoring the heat released during the addition.
- Continue to stir for 3 h prior to cooling to 25 °C.
- The resulting Grignard mixture was used directly in the subsequent step.

Observations

During the addition, heat evolution (ca. 10 °C increase in the temperature) and discoloration were observed.

11.1.1.2   Preparation of (4-(2-(Pyrrolidin-1-yl)ethoxy)phenyl)magnesium Bromide

Magnesium metal was activated by the addition of 2.3 mol% of TMSCl, which resulted in a smooth initiation with a 10% charge of bromide **3**. In the event, the Grignard **4** was prepared on a multi-kilogram scale by adding the remainder of the bromide **3** below 50 °C followed by a 1 hour of reflux (Equation 11.2).[4]

$$ \tag{11.2} $$

Procedure

- Charge 100 kg of THF into a nitrogen-purged vessel, followed by adding 4.8 kg (197.5 mol) of magnesium.
- Stir the mixture at 25–28 °C.
- Charge 3.6 kg (4.1 mol, 2.5 mol%) of 1.0 M TMSCl in THF and wash with THF (2 kg).
- Charge 45 kg (166.6 mol) of **3** into a separate nitrogen-purged vessel, followed by adding THF (100 kg) to form a solution.
- Charge 15 L of the solution of **3** to the suspension of Mg.
- Stir until initiation occurred (5 to 40 min and the temperature rise to ca. 50 °C).
- Charge the remaining solution of **3** over 45 min at 38–51 °C and wash with THF (25 kg).
- Stir the resulting mixture at 45–50 °C for 30 min.
- Cool to 20–25 °C (and used in the subsequent step after filtration).

NOTE

Alternatively, the Grignard formation can be initiated with 1,2-dibromoethane (1.0 equiv).

11.1.2   USE OF DIISOBUTYLALUMINUM HYDRIDE

For the preparation of aryl Grignard reagents, diisobutylaluminum hydride (DIBAL-H) can be utilized for the initiation at or below 20 °C. As the temperature is well below the boiling point of THF, it is possible to detect the initiation of the reaction through the temperature rise. Reactions (Equation 11.3) and (Equation 11.4)[5] demonstrated that employing 1 mol% of DIBAL-H as the activation reagent, both the initiation and formation of the aryl Grignard reagents could be performed at or below 20 °C.

$$(11.3)$$

$$(11.4)$$

Procedure[5]

- Prepare a solution of **5** by dissolving 68.6 kg (282.2 mol) of **5** in 280 L of THF (<0.1% H_2O).
- Charge 6.54 kg (269.0 mol, 0.95 equiv) of magnesium turnings and 66 L of THF (<0.1% H_2O) into a 630-L glass-lined reactor.
- Charge 27 kg of the pre-prepared solution of **5** and 2 kg of 20% DIBAL-H (2.81 mol, 1 mol%) toluene solution.
- Charge the rest of the solution of **5**, after the initiation has faded, under a dose-controlled manner.

REMARKS

Analogously, sodium bis(2-methoxyethoxy)aluminum hydride (Red-Al) (65 wt% in toluene) is also able to activate magnesium.[6]

A higher initiation temperature was also employed (Equation 11.5).[7] Furthermore, the initiation with DIBAL-H was able to be extended to alkyl bromides, such as 1-chloropentylbromide (Equation 11.6).

$$(11.5)$$

$$(11.6)$$

11.1.3 Use of Diisobutylaluminum Hydride/Iodine

The use of the DIBAH/iodine system was proved to be efficient in the initiation of the reaction of magnesium with 4-chlorotetrahydropyran **7** during the preparation of 4-tetrahydropyranylmagnesium chloride **8** (Equation 11.7).[8] Compared with the cases where DIBAL-H or iodine was used separately, this initiation was fast and insensitive to moisture.

$$(11.7)$$

Procedure

- Charge magnesium (3.31 g, 138 mmol) and 180 mL of THF.
- Charge iodine (118.2 mg, 0.465 mmol).

- Charge DIBAL-H (3.6 mL, 1 M solution in toluene, 3.6 mmol).
- Heat to 60 °C and stir at that temperature for 45 min.
- Charge 20% of **7** (16.52 g, 137 mmol).
- Stir at 60 °C for 2 h.
- Charge the remaining **7** in four equal portions.

REMARK

The preparation of **8** was also initiated by iodine at 60 °C, however, it is exceptionally sensitive to residual water and grades of magnesium.

11.1.4 USE OF GRIGNARD REAGENTS

Grignard reagents are able to initiate reactions of alkyl or aryl halides with magnesium.

Advantages
- Indigenous to the process and no new chemical entities were induced.
- The Grignard reagents can act as drying agents.

11.1.4.1 Use of MeMgCl

3,4-Dimethoxy-*o*-toluic acid **10** was prepared through a two-step, one-pot process, involving the formation of the Grignard reagent **11**, followed by a reaction with dry ice (Scheme 11.1).[9] The addition of catalytic amounts of methylmagnesium chloride (in THF) to the mixture of magnesium and aryl bromide **9** led to more consistent initiation. The exotherm of the reaction was then easily controlled by adjusting the addition rate of the remaining aryl bromide solution.

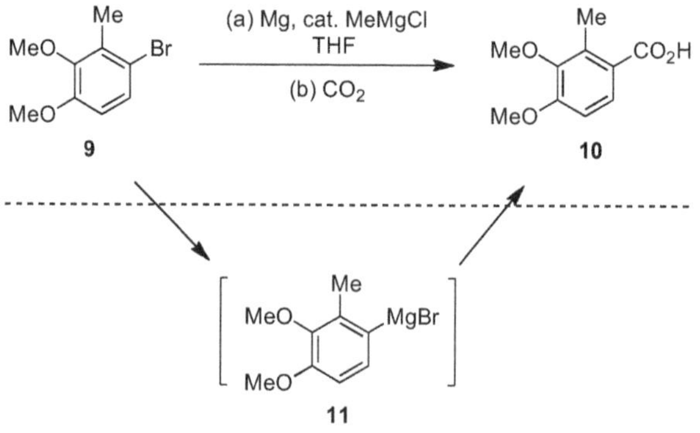

SCHEME 11.1

Procedure

Stage (a)
- Charge 2.78 kg (5%) of pre-prepared solution of **9** (20 kg, 86.5 mol in 35 kg of THF) into a 200-L reactor at 25 °C containing a mixture of 2.2 kg (90.5 mol) of magnesium turnings in THF (17.7 kg).
- Charge 0.8 kg (2.34 mol) of MeMgCl (3 M solution in THF).
- Charge slowly, after the initial exotherm subsided, the remaining solution of **9** over 4 h while maintaining the temperature below 45 °C.
- Keep the resulting solution at 25 °C overnight under N_2 to give the solution of **11**.

Stage (b)

- Transfer the solution of **11** into a 400-L reactor containing a pre-chilled (−60 °C) mixture of 38.2 kg (868 mol) of dry ice and THF (75 kg) over 50 min while maintaining the temperature below −10 °C.
- Warm the mixture to 20–25 °C.
- Add water (140 kg).
- Distill under reduced pressure (60 mbar) until 60% of the solvents (~83 kg) is distilled out.
- Add 14.6 kg of 4.5 M H_2SO_4.
- Stir at 30 °C for 1 h.
- Cool to 5 °C and stir for 1 h.
- Filter and wash with water (2×30 kg).
- Dry at 65 °C under vacuum.
- Yield: 14.76 kg (86.2%) as a white solid.

11.1.4.2 Use of EtMgBr

Analogously, ethylmagnesium bromide was used to initiate the reaction of aryl bromide **12** with magnesium (Scheme 11.2).[2a] In the event, the magnesium insertion reaction reproducibly initiated at 30 °C in <15 min with 1.16 equiv of ethylmagnesium bromide (relative to the water content in the solution).

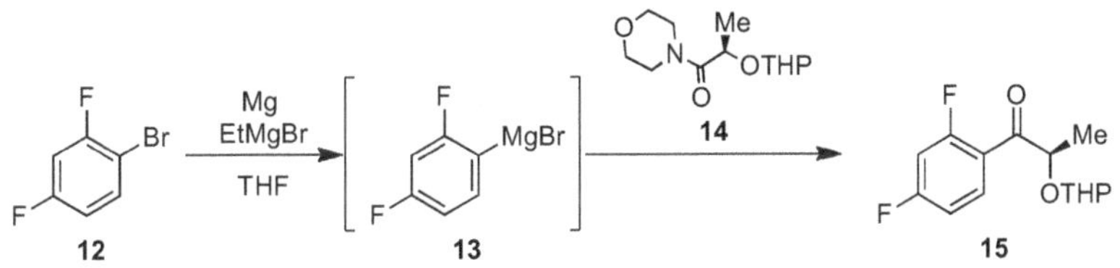

SCHEME 11.2

Procedure

- Charge THF (303 kg) into a stainless steel reactor which was equipped with a Mettler FTIR followed by adding 9.30 kg (383 mol, 1.0 equiv) of Mg via a solid addition funnel.
- Charge 2.25 kg of 1 N solution of EtMgBr (1.16 equiv versus water in solution) followed by rinsing with THF (10 kg).
- Stir the mixture for 1 h.
- Charge 9.15 kg (47.4 mol) of **12** at 29–32 °C over 5 min followed by rinsing with THF (1 kg) (initiation was observed during the addition).
- Charge the additional 69.7 kg (361.2 mol, total 408.6 mol, 1.07 equiv) of **12** at <32 °C over 2.5 h followed by rinsing with THF (1 kg).
- Stir the batch at 25–27 °C until the reaction is complete (by FTIR).
- Charge a 30.0 wt% solution of **14** (257.3 kg, 317.3 mol, 0.83 equiv) at 35 °C over 20 min.
- Stir the resulting mixture at 35 °C for 18 min.
- Transfer the batch into a glass-lined reactor containing acetic acid (41.5 kg, 691 mol) in water (207 kg) at <18 °C over 40 min followed by rinsing with THF (35 kg).
- Warm the mixture to 25 °C over 30 min.
- Add EtOAc (207 kg).
- Separate layers.
- Add heptane (85 kg) to the organic layer.

- Wash with water (206 kg).
- Distill at 15–20 °C under 0.1 bar to <0.1% water by KF.
- Add additional EtOAc (398 kg).
- Distill continuously to 125 L (final water content = 0.08%).
- Add DMF (90 kg).
- Distill at 22–50 °C under 0.5 bar until EtOAc <0.004% (by GC).
- Yield of **15**: 122.7 kg as a DMF solution (56.7 wt%, 81.1% yield, and 98.0% ee).

REMARKS

(a) Ethylmagnesium chloride functions as a drying agent as well, as azeotropic distillation was ineffective.

(b) The excess of aryl bromide (1.07 equiv) led to more consistent reactions and ensured the consumption of all of the metal, minimizing the risk of residual highly pyrophoric magnesium remaining in the workup.

11.1.4.3   Use of Heel

11.1.4.3.1   Preparation of Alkylmagnesium Bromide

A small amount of heel (or residue) from previous Grignard preparation was used as an initiator for the subsequent formation of the Grignard **17** (Equation 11.8).[10]

$$F_3CF_2C\diagdown\diagup S\diagdown(\ )_7 Br \xrightarrow[\text{THF}]{\text{Mg}} F_3CF_2C\diagdown\diagup S\diagdown(\ )_7 MgBr \qquad (11.8)$$

$$\textbf{16} \qquad\qquad\qquad\qquad \textbf{17}$$

Procedure

- Charge magnesium (1.17 equiv) and 8.4 volumes of THF to a heel mixture (~0.05 equiv) obtained from a previous batch.
- Heat the mixture to 45 °C.
- Charge 0.14 equiv of **16** and 2.2 volumes of THF.
- Charge the rest of **16** (0.86 equiv) in several portions.
- Cool and decant ca. 83% of the Grignard solution (17% of the retained mixture is used for the subsequent Grignard formation).

11.1.4.3.2   Preparation of Arylmagnesium Chloride

During the preparation of the Grignard reagent **19**, the initiation could be carried out in the presence of 1–2 mol% of the product Grignard **19** by adding 5–10% of aryl chloride **18** (Equation 11.9).[11]

$$\underset{\textbf{18}}{\overset{\text{Cl}}{\diagdown}\text{SMe}} \xrightarrow[\text{THF, reflux}]{\text{Mg}} \underset{\textbf{19}}{\overset{\text{MgCl}}{\diagdown}\text{SMe}} \qquad (11.9)$$

NOTE

The initiation of the reaction of aryl chloride **18** with magnesium required charging 20–60% of 4-chlorothioanisole **18** in the presence of 1,2-dibromoethane (2.5 mol%) under reflux in THF. This lack-of-controlling initiation was a safety issue in large-scale production of the Grignard reagent **19**.

11.1.5   Use of Alkyl Halides

11.1.5.1   Use of Iodomethane

Aryl magnesium bromide **21** could be prepared from the reaction of aryl bromide **20** with magnesium turnings in the presence of catalytic amounts of iodomethane in anhydrous THF (Equation 11.10).[12]

$$(11.10)$$

20   **21**

11.1.5.2   Use of 1,2-Dibromoethane

A method using 1,2-dibromoethane as an activator emerged around 1990, which was found to be more reliable than the activation with iodine. The use of 1,2-dibromoethane is particularly advantageous as its action can be monitored by the observation of bubbles of ethylene. Furthermore, the side products are innocuous (Equation 11.11).

$$BrCH_2CH_2Br \;+\; Mg \longrightarrow C_2H_4 \;+\; MgBr_2 \qquad\qquad (11.11)$$

The preparation of Grignard reagents with 1,2-dibromoethane[13] as a magnesium activator is no longer necessary to reflux the reaction mixture. For example, the activation in the preparation of *p*-(dimethylamino)phenylmagnesium bromide **23** was performed at 50–55 °C by adding 5 mol% of 1,2-dibromoethane to a mixture of magnesium and 10% of *p*-(dimethylamino)phenyl bromide **22** in THF (Equation 11.12).[5]

$$(11.12)$$

22   **23**

REMARKS

(a) On several occasions, the reaction did not initiate even with stirring at high temperatures for several hours.

(b) Another drawback of this method is that 1,2-dibromoethane is a suspected carcinogen, which poses potential hazardous issues for operators.

11.1.6   Halogen–Magnesium Exchange

An attractive halogen–magnesium exchange method is developed in order to avoid the problematic exothermic reaction during the Grignard preparation.[14] This approach has received much attention due to its generality, mild reaction conditions, and excellent functional group compatibility. Employing alkyl Grignard reagents, such as *i*PrMgCl, promotes the metal–halogen exchange under mild reaction conditions, which prevents potential runaway reactions during the preparation of aryl Grignard reagents. However, with less reactive aryl bromides, the Br–Mg exchange requires long reaction times with only moderate conversion. This sluggish Br–Mg exchange can be accelerated dramatically by using *i*PrMgCl·LiCl complex (Turbo Grignard) (Equation 11.13).[15]

$$(11.13)$$

The application either of *i*PrMgCl/LiCl or *s*BuMgCl/LiCl in halogen-magnesium exchange reactions results in higher conversions and also in less gaseous side products.

11.1.6.1 Preparation of Trifluoromethyl Substituted Aryl Grignard Reagents

While trifluoromethyl-substituted aromatic compounds are widely used as pharmaceutical intermediates, trifluoromethyl-substituted aryl Grignard reagents are less stable and care must be taken when handling them. Figure 11.1[16] shows the four most frequently used trifluoromethyl substituted aryl Grignard reagents **24–27** that have safety concerns at scale.

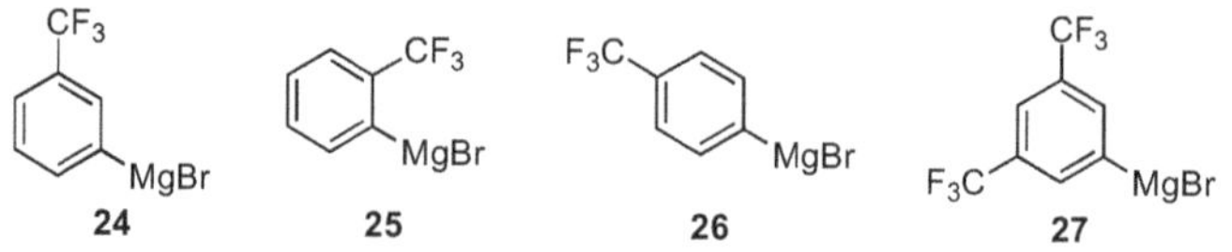

FIGURE 11.1   Trifluoromethyl substituted aryl Grignard reagents.

The Grignard reagents **24–27** exhibited significant exothermic activity on Reactive System Screening Tool (RSST) and Differential Thermal Analysis (DTA) analysis. The RSST and DTA results suggest that the initiation of exothermic reactions of these compounds may lead to uncontrollable runaway reactions, potentially resulting in an explosion. In addition, these trifluoromethylphenyl Grignard reagents can detonate on the loss of solvent contact or upon moderate heating. Therefore, the preparation of these Grignard reagents by the traditional method using magnesium metal is extremely dangerous,[17] which hampered their application in large-scale organic synthesis.

Knochel's halogen–magnesium exchange protocols[15] proved to be a method of choice in the preparation of trifluoromethyl-substituted phenyl Grignard reagents. Equation 11.14[18] illustrates the preparation of 2-trifluoromethylphenyl magnesium chloride **29** from the reaction of 2-bromobenzotrifluoride **28** with *i*PrMgCl.

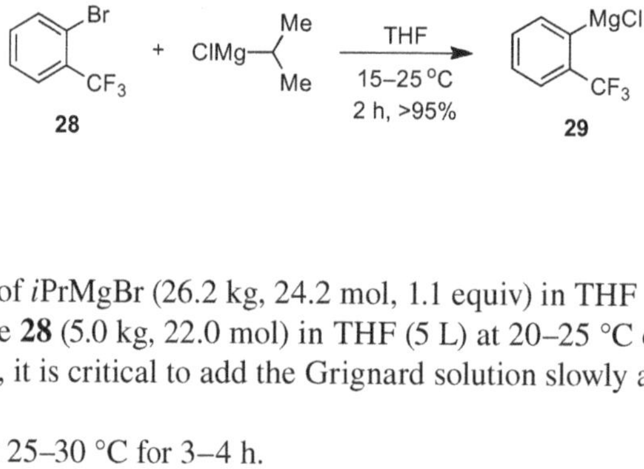

(11.14)

Procedure[18c]

- Charge a solution of *i*PrMgBr (26.2 kg, 24.2 mol, 1.1 equiv) in THF to a solution of 2-bromobenzotrifluoride **28** (5.0 kg, 22.0 mol) in THF (5 L) at 20–25 °C over 15 min (Caution: For safety reasons, it is critical to add the Grignard solution slowly and keep the temperature under 30 °C)
- Stir the mixture at 25–30 °C for 3–4 h.
- The resulting aryl Grignard solution of **29** was used directly in the subsequent reaction.

REMARKS

(a) This Grignard chemistry was successfully scaled up producing 83 L(50 mol) of 2-trifluoromethylphenyl magnesium bromide solution in THF at 0.6 M concentration.[18a]

(b) An adiabatic calorimetry study of the solution of 2-trifluoromethylphenyl magnesium chloride (ca. 1.1 M) in an Advanced Reactive Systems Screening Tool (ARSST) showed a highly exothermic event with onset temperature at ca. 85 °C. Thus, it was recommended that the Grignard should be prepared at a low concentration (<0.6 M).

Similarly, 3,5-di(trifluoromethyl)phenyl magnesium chloride **31** was prepared via bromine–magnesium exchange between aryl bromide **30** and isopropylmagnesium chloride (Equation 11.15).[16]

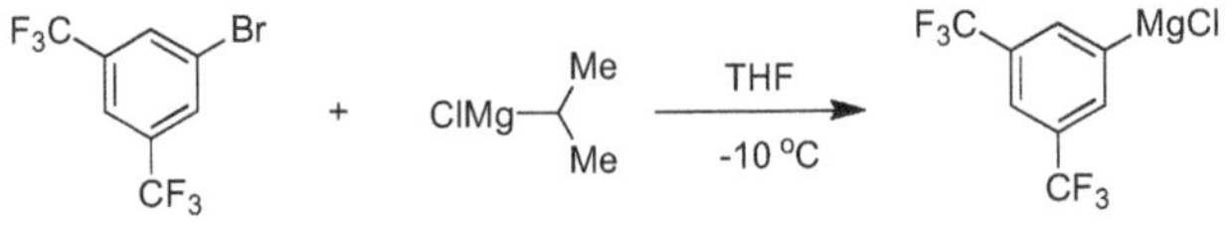

$$(11.15)$$

Advantages

- Mild reaction conditions (–10–25 °C) in THF.
- No induction period.

11.1.6.2 Preparation of *N*-Methylpyrazole Grignard Reagent

Analogous to aryl bromides, the halogen–magnesium exchange between iodo-substituted pyrazole **32** and isopropylmagnesium chloride was utilized for the preparation of *N*-methylpyrazole the Grignard reagent **33** during the synthesis of pyrazole oxadiazole ketone **35** (Scheme 11.3).[19]

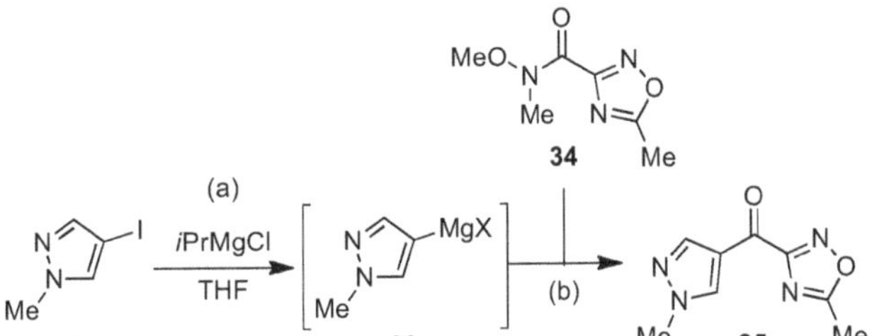

SCHEME 11.3

Procedure

Stage (a)
- Charge 4.49 kg (21.2 mol) of **32** and THF (9.0 L) into a 100-L vessel under N_2.
- Cool the content to –19 °C.
- Charge 11.1 L of 2.0 M isopropylmagnesium chloride (22.1 mol) over 1.75 h at <5 °C.

Stage (b)
- Charge, to the solution of **33**, a solution of **34** (3.15 kg, 18.4 mol) in THF (5.33 L) at –9 °C to 21 °C over 25 min.
- Stir the reaction mixture for 10 min at room temperature followed by workup.

11.1.6.3 Preparation of (4-Bromonaphthalen-1-yl)magnesium Chloride

The Br–Mg exchange with isopropylmagnesium chloride may be plagued by side reactions, such as protonation and double Br–Mg exchange when the substrate has more than one bromine substituent. For instance, the preparation of 4-Bromo-1-naphthoic acid **38** through a mono-Br–Mg exchange of dibromonaphthalene with isopropylmagnesium chloride, followed by reaction with solid carbon dioxide led to, besides the desired product **38**, high levels of desbromo **39**, monoacid **40**, and diacid **41** (Condition (a), Scheme 11.4).[20]

In order to mitigate the formation of side products, the Br–Mg exchange was accomplished using Turbo Grignard (*i*PrMgCl/LiCl, 1.5 equiv) (Condition (b), Scheme 11.4), giving a better reaction profile with lower levels of the three impurities.

Procedure Condition (b)

- Charge 100 L of a solution of *i*PrMgCl/LiCl (1.35 M in THF, 135 mol) to a solution of **36** (25 kg, 87 mol) in THF (175 L) at 0 °C.

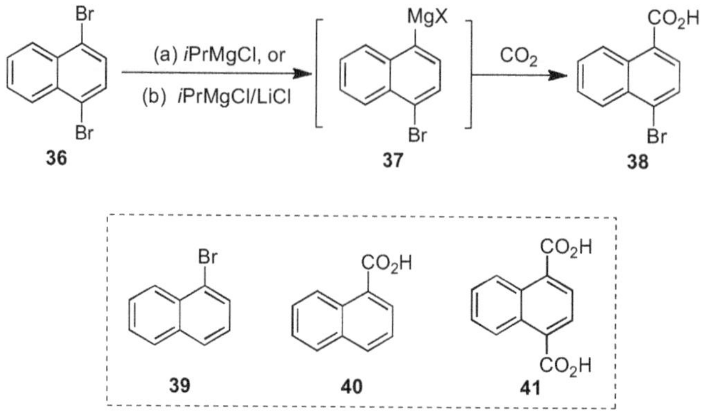

SCHEME 11.4

- Stir at 0 °C for 0.5 h.
- Warm to 20 °C and stir until the complete formation of the Grignard reagent (<2% of **36**).
- Cool to –10 °C and sparge with CO_2 gas for 1 h.
- Workup.

This magnesium–halogen exchange protocol lent itself readily to the preparation of other aryl Grignard reagents such as pyridyl magnesium halide **43** (Equation 11.16).[21]

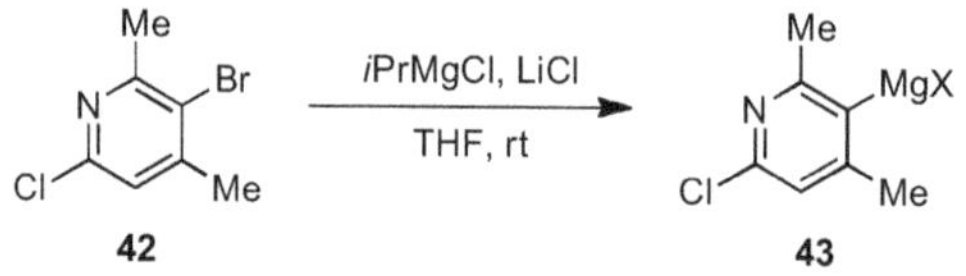

$$(11.16)$$

11.1.6.4 Magnesium-ate Complex

In addition to alkylmagnesium halides (RMgX), a magnesium-ate complex (R_3MgLi)[22] would also be utilized in the halogen–magnesium exchange reaction (Scheme 11.5).[23]

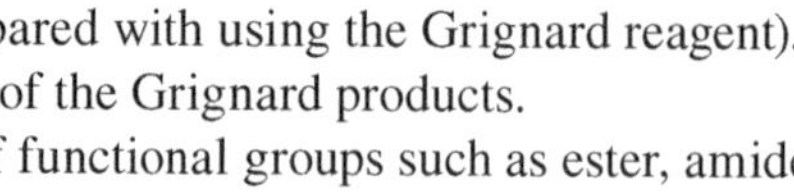

SCHEME 11.5

Advantages:

- Fast exchange rate (compared with using the Grignard reagent).
- Good to excellent yields of the Grignard products.
- Tolerance of a number of functional groups such as ester, amide, and the cyano group.

Limitations:

- Low or cryogenic temperature is required.
- Possible exothermic decomposition may be a scale-up concern.[24]
- More development work is needed prior to scale-up.

The synthesis of drug candidate **44**, a muscarinic receptor antagonist, involves the preparation of 6-Bromo-2-formylpyridine **47**. The Br–Mg exchange reaction with isopropylmagnesium chloride was relatively sluggish and required excess *i*PrMgCl for completion. In addition, the formation of a side product 2-formyl-6-isopropylpyridine after treatment with DMF could be more than 5%.

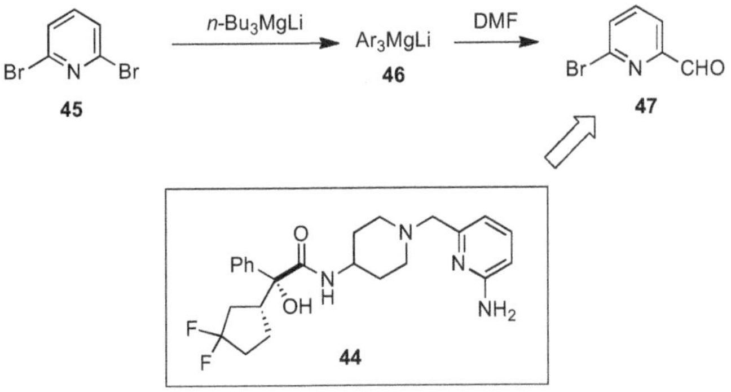

SCHEME 11.6

Thereby Br–metal (nBu_3MgLi) exchange protocol was developed and demonstrated to be robust, and scaleable for selective functionlization of 2,6-dibromopyridine (Scheme 11.6).[25]

This new and practical metalation protocol obviated cryogenic conditions and upon quenching with DMF provided 6-Bromo-2-formylpyridine 47 in excellent yield.

11.1.6.5 Alkylmagnesium Alkoxides

Knochel's group developed a new halogen-magnesium exchange reagent, alkylmagnesium alkoxide ($sBuMgOR·LiOR$) (R = 2-ethylhexyl), for Br–Mg exchange of electron-rich aryl bromides to generate the desired aryl and heteroaryl magnesium alkoxides (ArMgOR·LiOR) in toluene (Scheme 11.7).[26]

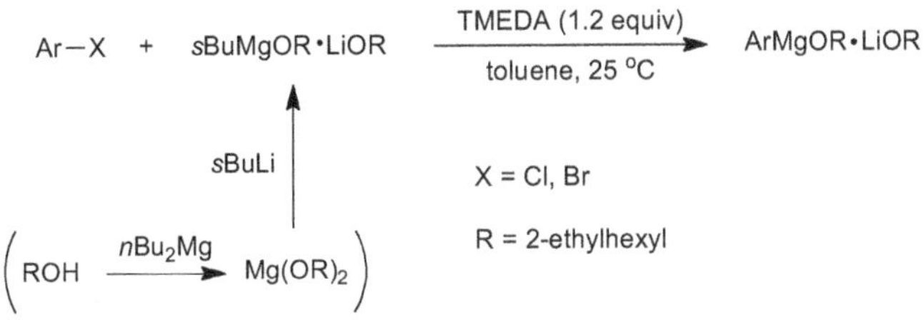

SCHEME 11.7

The resulting arylmagnesium alkoxides can be utilized in various reactions, including reactions with ketones, acyl chlorides, and Pd-catalyzed Kumada cross-coupling reactions. Furthermore, the related reagent $sBu_2Mg·2LiOR$ allows a Cl–Mg exchange with various electron-rich aryl chlorides in toluene to produce organomagnesium species of type $Ar_2Mg·2LiOR$, which react well with aldehydes and allyl bromides. The $sBuMgOR·LiOR$ (or $sBu_2Mg·2LiOR$) can be conveniently prepared in two steps: a reaction of 2-ethylhexanol (2.0 equiv) with nBu_2Mg to form magnesium alkoxide and subsequent treating the magnesium alkoxide with one (or two) equivalent(s) of s-BuLi.

11.2   REACTIONS OF GRIGNARD REAGENTS

11.2.1   REACTIONS WITH KETONES

11.2.1.1   Vinyl Grignard Reagents

The Grignard reaction of 2-indanone 48 with vinyl Grignard reagent in toluene led to tertiary alcohol 49 in good yield (Equation 11.17).[27]

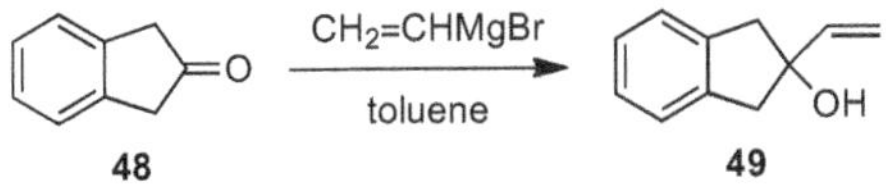

(11.17)

Procedure

- Charge 250 mL of vinyl Grignard in THF (1.0 M, 250 mmol) into a reactor.
- Dilute the Grignard with 120 mL of toluene.
- Concentrate in vacuo to remove THF.
- Cool to 0 °C.
- Charge a solution of 2-indanone **48** (13.2 g, 100 mmol) in 65 mL of MTBE.
- Warm to room temperature upon completion of the reaction.
- Charge 200 mL of 1 N HCl to quench the reaction.
- Separate the organic layer from the aqueous layer and dry over Na_2SO_4.
- Concentrate to give a dark-brown oil.

REMARK

Due to competitive enolization of **48** using commercial $CH_2=CHMgBr$ in tetrahydrofuran (THF) provided only ca. 40% conversion to the tertiary allylic alcohol **49**.

Solvent effects were observed during the Grignard reaction of ketone **52** with vinyl Grignard reagent **51**, obtained in situ from Mg–I exchange between vinyl iodide **50** and *i*PrMgCl·LiCl. When THF was used as a solvent, **51** reacted with itself upon formation, thus 4–5 equiv of **50** were required to achieve full conversion (Scheme 11.8).[28]

When the reaction was carried out in toluene, the self-reaction of the Grignard reagent was not observed perhaps due to greater aggregation in the non-coordinating solvent.

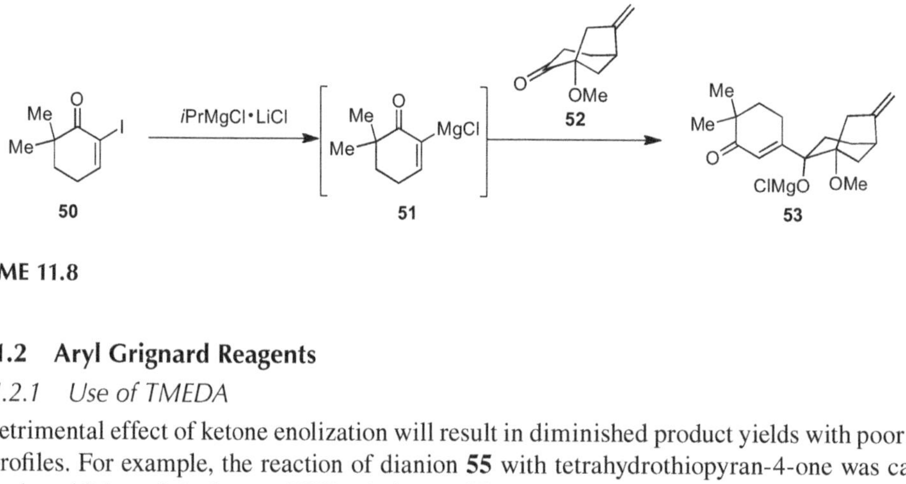

SCHEME 11.8

11.2.1.2 Aryl Grignard Reagents

11.2.1.2.1 Use of TMEDA

The detrimental effect of ketone enolization will result in diminished product yields with poor reaction profiles. For example, the reaction of dianion **55** with tetrahydrothiopyran-4-one was carried out by the addition of the ketone THF solution to **55**, producing the desired tertiary alcohol **56** in only 63% isolated yield along with a side product **57**, which was formed via the proton transfer from the ketone (Scheme 11.9).[29]

In order to suppress the ketone enolization, TMEDA (3 equiv) was used as an additive which restrained effectively the ketone enolization, resulting in higher product yields (80–90%).

Procedure

- Charge 10.0 g (29.6 mmol) of **54** into a 250-mL, 3-necked flask, followed by adding THF (60 mL) and TMEDA (10.3 g, 88.9 mmol).
- Cool the resulting solution to –25 °C.
- Charge 31.4 mL (62.8 mmol) of 2 M *i*PrMgCl over 15 min at <–25 °C.
- Stir the batch at –25 °C for 20 min.

- Charge a solution of tetrahydrothiopyran-4-one (4.8 g, 41.5 mmol) in 12 mL of THF drop-wise at <–20 °C over 10 min.
- Stir the resulting mixture at –20 °C for 1 h.
- Quench the reaction by transferring the batch into a mixture of glacial acetic acid (13 mL) and water (66 mL) at 0–27 °C.
- Add MTBE (25 mL) and stir the resulting mixture for 20 min.
- Separate layers.
- Extract the aqueous layer with MTBE (25 mL) and octane (15 mL to help the layer separation).
- Wash the combined organic layers with sat. NH$_4$Cl (25 mL).
- Extract the aqueous NH$_4$Cl layer with MTBE (25 mL).
- Distill the combined organic layers under reduced pressure to 50 mL.
- Add octane (50 mL).
- Distill under reduced pressure to 50 mL (the product oiled out).
- Cool to 23 °C.
- Filter, wash with octane (2×25 mL), and dry at 40 °C under vacuum.
- Yield of **56**: 9.4 g (96%) as an off-white solid.

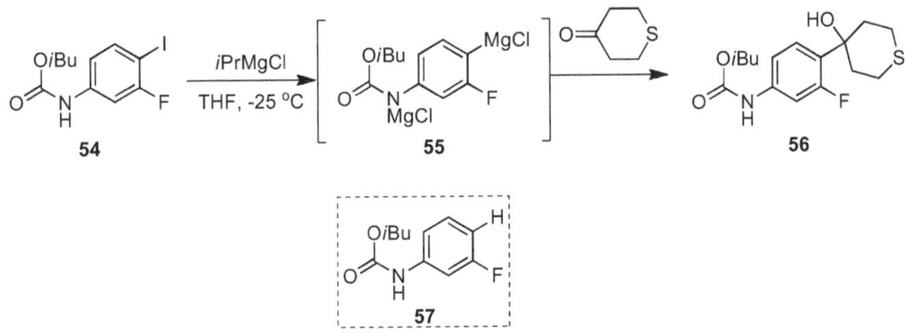

SCHEME 11.9

11.2.1.2.2   Use of TEA/MgCl$_2$

The synthesis of carbinol **62** was accomplished by a chemoselective addition of Grignard **59** to ketone moiety in **61** (Scheme 11.10).[30] The magnesium salt of carboxylic acid **60** was prepared by removing carboxylic acid proton with sacrificing the Grignard reagent (1 equiv). During this preparation, the temperature for the addition of *i*PrMgCl is critical; above –50 °C, isopropyl addition to the ketone moiety was observed to give diastereomeric alcohols **63**.

To avoid the undesired reaction during the preparation of **61**, a practical and convenient method was to treat **60** with equimolar amounts of triethylamine and anhydrous MgCl$_2$ in THF at ambient temperature. Filtration of the insoluble Et$_3$N·HCl salt yielded a solution of the magnesium salt **61**, which could be directly used in the subsequent Grignard reaction.

REMARK

The limitation of isopropylmagnesium chloride is the generation of highly flammable propane on quenching excess of *i*PrMgX. This drawback can be overcome by using cyclohexylmagnesium chloride as an alternative to *i*PrMgX.

11.2.1.3 Methylmagnesium Bromide

The Grignard reaction of 4-acetylbenzonitrile **64** with methylmagnesium bromide produced tertiary carbinol product **65** along with 5–15% of adol side product **66** due to the enolization of **64** by the basic methylmagnesium bromide (Equation 11.18).[31]

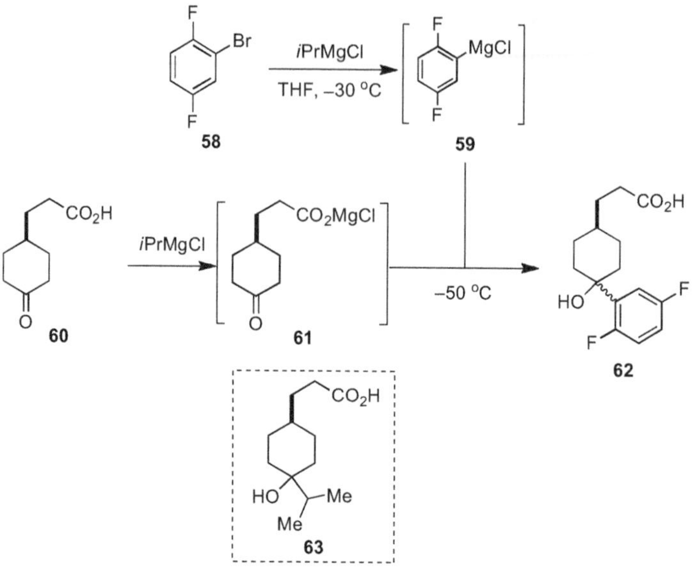

SCHEME 11.10

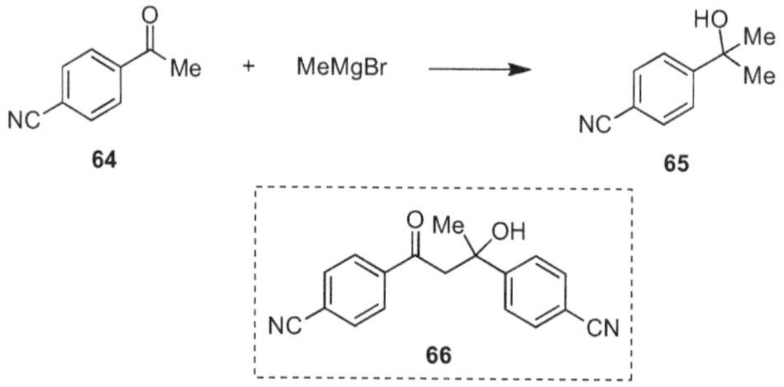

(11.18)

In order to mitigate such undesired enolization, a less basic methylation reagent (MeTiCl$_3$) was identified for the Grignard reaction. MeTiCl$_3$ was prepared from a reaction of MeLi with TiCl$_4$, and employing the in situ generated MeTiCl$_3$, **65** was produced in 81–85% yields on a multi-kilogram scale.

Procedure

- Charge 85 kg of anisole into a reactor and cool to –10 to –5 °C.
- Charge 17 kg (89.6 mol, 1.5 equiv) of TiCl$_4$ –10 to –5 °C over 1 h.
- Stir the resulting mixture at –10 to –5 °C for 15 min (an orange suspension was obtained then).
- Charge a solution of MeLi in cumene/MeTHF (61 kg, 3.2 wt%, 88.9 mol, 1.5 equiv) at –10 to –5 °C over 100 min.
- Stir the resulting reddish solution at –10 to –5 °C for 35 min.
- Charge a solution of **64** (8.5 kg, 58.6 mol) in anisole (43 kg) at –10 to –2 °C over 80 min.
- Stir the batch at –5 to –2 °C for 1 h.
- Add water (4.3 kg) at –10 to 0 °C over 0.5 h, followed by 38 kg of additional water.
- Warm the batch to 18 °C.

- Separate layers and wash the organic layer with water (2×43 kg), aqueous $NaHCO_3$ (8% w/v, 2×110 kg), and water (2×43 kg).
- Distill azeotropically at <55 °C/45–50 Torr to remove anisole with water (a total of 255 kg of water was added over five portions) until <0.5% anisole by GC.
- Mix the resulting mixture with cumene (22 kg), MTBE (31 kg), and aqueous NaCl (25 wt%, 20 kg).
- Separate layers.
- Distill the organic layer at <60 °C/70–75 Torr to remove MTBE until 30 L of volume was achieved.
- Filter through 0.8 μm cartridge.
- Heat the filtrate to 65–70 °C.
- Add heptane (5.8 kg) and cool the resulting solution to 20 °C.
- Charge 8.5 g of **65** as the seed (crystallization commenced within 5 min).
- Stir the resulting suspension at 20–22 °C for 1 h.
- Cool the batch to 0–5 °C and age for 3 h.
- Filter and wash with cumene/heptane (9 kg, 17:1) and heptane (8.5 kg).
- Dry at 35 °C under reduced pressure.
- Yield: 8.0 kg (84% yield, 100% purity) as a white crystalline solid.

11.2.2 REACTION WITH ACID CHLORIDE

The coupling reaction between acyl compounds with organometallic reagents is an approach that is frequently used in ketone synthesis. One of the problems associated with this approach is the over-addition reaction. An efficient method of avoiding the over-addition side reaction was demonstrated in the synthesis of diaryl ketone **58** (Equation 11.19).[18b] As shown in the reaction (Equation 11.19), the treatment of acid chloride **67** in THF with a preformed complex of Grignard reagent **68** with bis[2-(N,N-dimethylamino)ethyl] ether (BDMAEE) at −10 to 0 °C for 1 h produced the benzophenone **69** in 80% isolated yield. The complexation of **68** with BDMAEE moderated the reactivity of the aromatic magnesium species and, thereby suppressed the over-addition side reaction.[32]

$$(11.19)$$

NOTE

A protocol that was designed to circumvent the over-addition issue is to use Weinreb amide[33] as the coupling partner.

11.2.3 REACTION WITH DIMETHYLFORMAMIDE

The formylation of acetylene **70** with DMF in the presence of n-BuLi provided the desired aldehyde **71a** (Equation 11.20).[34] Due to the poor chemoselectivity of the highly reactive n-BuLi, the quantity of n-BuLi had to be controlled to exactly 1 equiv, less than 1 equiv will lead to low product yield wasting expensive acetylene **70** and more than 1 equiv will generate a side product **71b**. The use of less reactive ethylmagnesium chloride (1.2 equiv) allowed the reaction to run at room temperature and **71b** was controlled below the detectable level.

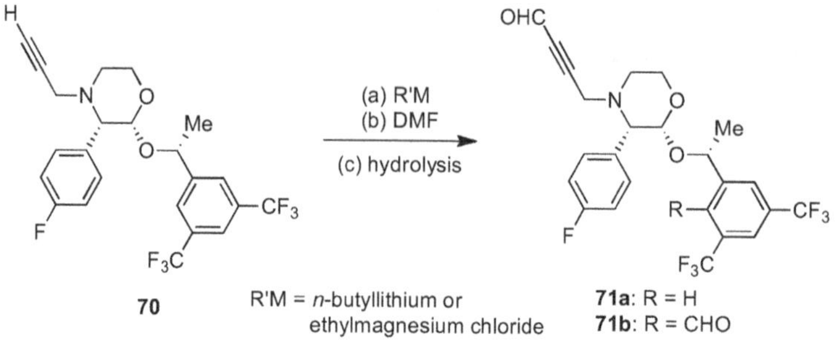

$$(11.20)$$

70   R'M = n-butyllithium or   **71a**: R = H
             ethylmagnesium chloride   **71b**: R = CHO

REMARKS

(a) Ketones are commonly prepared by acylation of Grignard or organozinc reagents with acid fluorides, chlorides, anhydrides, esters, thioesters, acyl pyrrazolides,[35] Weinreb amides,[31] or acyl hemiacetals.[36] Furthermore, the acylation of Grignard and heteroaryllithium reagents with stable and easily accessible N-acylbenzotriazoles, derived from a variety of aliphatic, unsaturated, (hetero)aromatic, and N-protected α-amino carboxylic acids, provides an alternative tool for the synthesis of ketones. This method proved to be successful with a wide variety of organometallic reagents and N-acylbenzotriazoles, affording ketones in high yields (Scheme 11.11).[37]

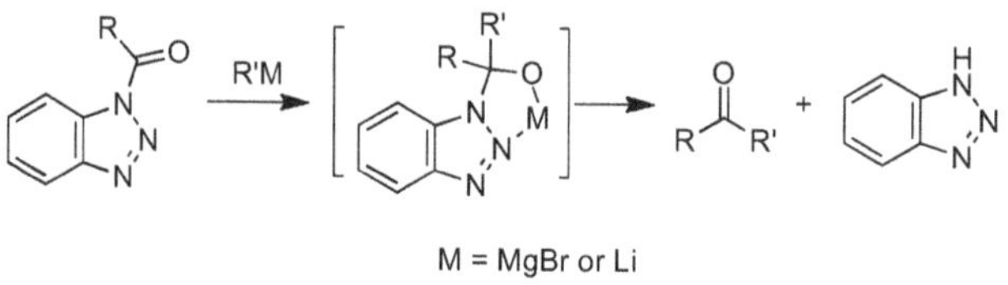

M = MgBr or Li

SCHEME 11.11

(b) The stoichiometry and quality of the organolithium reagent used in the deprotonation was of critical importance in controlling the formation of impurities. In order to control the stoichiometry in the deprotonation of (S)-**72** with n-butyllithium, catalytic amounts of 1,10-phenanthroline (0.1–0.4 mol%) was used as an internal indicator during the preparation of the lithium alkoxide. The colorless solution became light yellow during the addition of n-butyllithium and eventually turned dark brown at the equivalence point (Equation 11.21).[38]

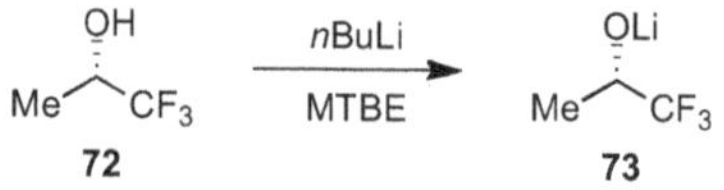

$$(11.21)$$

However, the use of an internal colorimetric indicator (1,10-phenanthroline) poses concerns regarding contamination with phenanthroline decomposition products, which led to the pursuit of an in situ spectroscopic method. In situ Raman spectroscopy was ultimately used to monitor this deprotonation reaction.

11.2.4 REACTION WITH WEINREB AMIDE

Compared with organolithium reagents, reactions of Grignard reagents with Weinreb amides provide better reaction profiles mainly due to their relatively low reactivity. For instance, the reaction of the Grignard reagent **75** with Weinreb amide **76** provided the desired alkyl aryl ketone **77** in 77–82% yields (Scheme 11.12).[39]

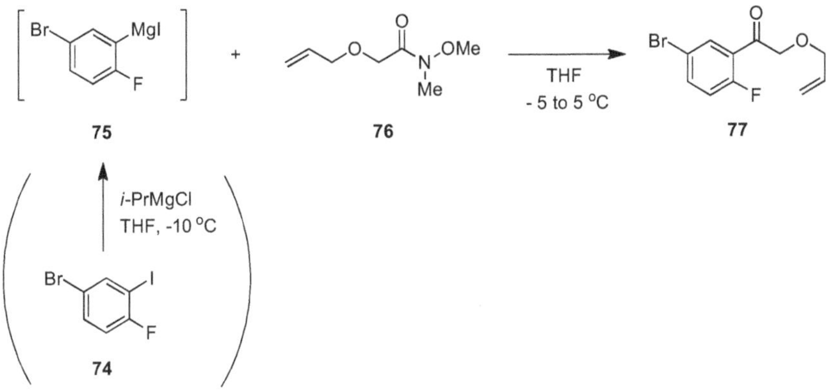

SCHEME 11.12

The reaction conditions overcome limitations associated with the use of the organolithium reagents, such as cryogenic temperature and side reactions (proton abstraction and benzyne formation).

11.2.5 MICHAEL ADDITION

Syntheses of certain key impurities are frequently required to assist in analytical method development, drug registration, and confirmation of impurity identity. A dimer **81** is an impurity during the synthesis of 7α-[9-(4,4,5,5,5-pentafluoropentylsulfinyl)-nonyl]estra-1,3,5-(10)-triene-3,17β-diol (Fulvestrant, Faslodex), a drug molecule for the treatment of breast cancers. The synthesis of the dimer impurity **81** involved 1,6-conjugative addition of the Grignard reagent onto ketone **78** (Scheme 11.13).[40] Dimethyl sulfide was found to have an important effect in stabilizing the active organo-dicuprate species formed by the reaction of Cu(I)Cl with di(bromomagnesio)nonane. In the absence of dimethyl sulfide, a 1,2-addition of the Grignard and loss of acetate are the dominant processes.

The preparation of *trans*-2,6-dimethylpyranone **83** via copper-catalyzed Grignard 1,4-addition was problematic, generating two aldol dimerization products (**86** and **87**) in 15% combined yield (Equation 11.22),[41] in addition to *cis*-isomer **84** (<2%) and ring-opening side product **85** (<5%).

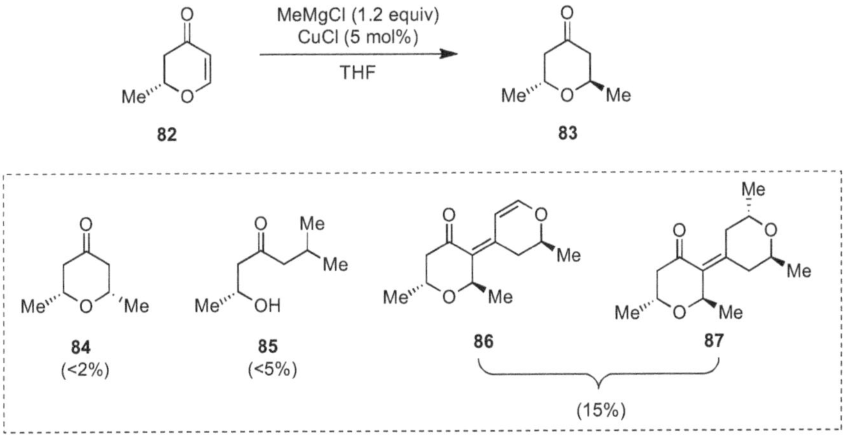

(11.22)

Apparently, these two aldol dimerization side products were formed via enolization of the product.

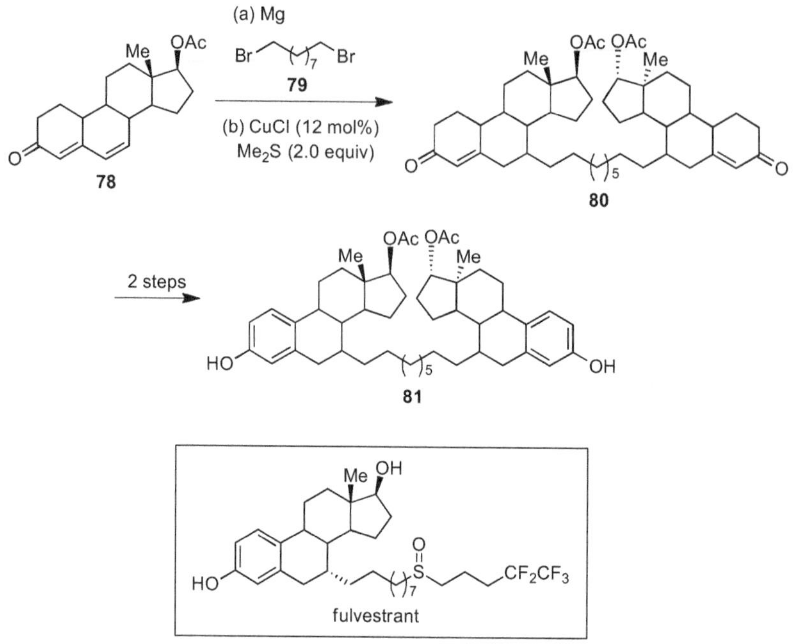

SCHEME 11.13

Solutions

In order to mitigate the enolization problems, TMSCl was used as an additive to trap the enolate as the silyl enol ether. Ultimately, the reaction was performed in the presence of TMSCl (1.5 equiv) and 1,3-bis(diphenylphosphino)propane (DPPP) (7 mol%), followed by hydrolysis of the silyl enol ether with citric acid.

11.2.6 REACTION WITH EPOXIDE

The R group in the Grignard reagent (RMgX) is often used as a nucleophile or a base, and little attention is paid to the halide X (= Cl or Br). However, it was recognized during a three-step synthesis of (R)-propyloctanenitrile **91** that the selection of the Grignard reagent (EtMgBr or EtMgCl) in an epoxide ring-opening step affected the optical purity of the nitrile product **91** (Scheme 11.14).[42] The synthetic sequence started with a ring-opening of an optical pure (R)-1,2-epoxyoctane starting material **88** (98.5% ee) with EtMgBr, followed by mesylation of the resulting (S)-decan-4-ol **89**. The crude mesylate **90** underwent a displacement reaction with lithium cyanide (generated in situ from lithium hydride and acetone cyanohydrin) to obtain the nitrile **91**. Besides the desired product **89**, the ethylation reaction of (R)-1,2-epoxyoctane **88** also produced bromohydrin **92** (0.4–1.1 area% by GC analysis), that could not be separated from alcohol **89** by distillation.

(I) Chemistry diagnosis

The impurity **92** was formed by the side reaction of the epoxide **88** with $MgBr_2$, generated from a reaction of $Et_2CuMgBr$ complex (Equation 11.23).

$$2 \; Et_2CuMgBr \; \rightleftharpoons \; MgBr_2 \; + \; (Et_2Cu)_2Mg \tag{11.23}$$

A two-step transformation of the bromohydrin **92** to dinitrile **93** generated lithium bromide (Equation 11.24) whose subsequent reactions with mesylate **90** led to the enatiomer **94** in two-S_N2

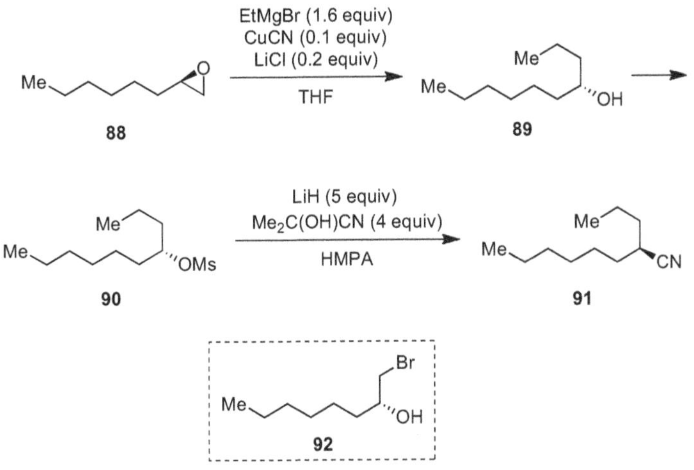

SCHEME 11.14

steps (Equation 11.25). In the event, a decrease in the optical purity of the product **91** (82.7% ee) was observed.

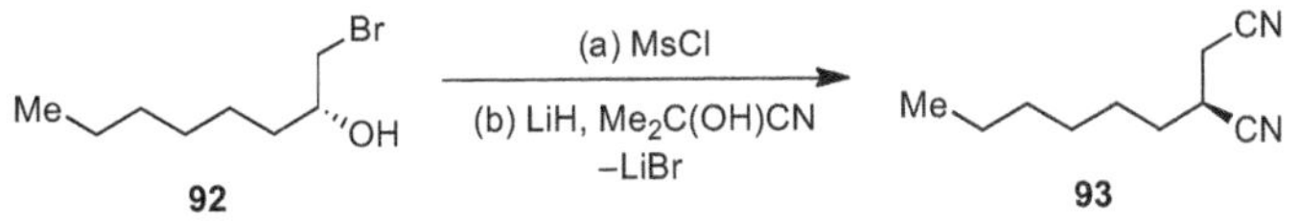

$$(11.24)$$

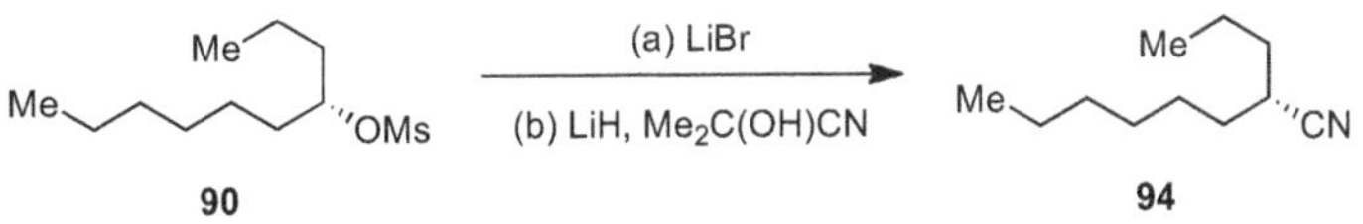

$$(11.25)$$

(II) Solutions

To avoid this chiral erosion, a less nucleophilic chloride source would be effective for reducing the corresponding halohydrin formation during the alkylation step. Accordingly, the use of EtMgCl (1.4 equiv)/CuCl (0.02 equiv) to replace EtMgBr/CuCN worked well without the chlorohydrin formation. As a result, the crude **89** was converted to the nitrile **91** in high enantiomeric excess and no loss of optical purity was observed.

11.2.7 OTHER GRIGNARD REAGENT-INVOLVED REACTIONS

11.2.7.1 Ramberg–Bäcklund Reaction

Traditionally, Ramberg–Bäcklund reaction is used to convert α-halosulfones into alkenes via epi-sulfone intermediates and subsequent extrusion of sulfur dioxide. A Ramberg–Bäcklund reaction of alkyl triflones **95** with cyclohexylmagnesium bromide was developed to synthesize highly value-added *gem*-difluoroalkenes **97** (Scheme 11.15).[43]

This streamlined protocol allows for the rapid synthesis of *gem*-difluoroalkenes **97** in good yields. The success of the transformation relies on the use of Grignard reagent (i.e., cyclohexylmagnesium bromide) that played dual roles: as a base in the deprotonation of the alkyl triflone substrates **95** and as an activator for the C–F bond by complexation of fluoride leaving group. The alkyl triflone substrates, readily accessible via α-alkylation, could be utilized with a variety of useful functional groups such as chlorine, alkenyl, alkynyl, acetal, siloxy, and amino groups.

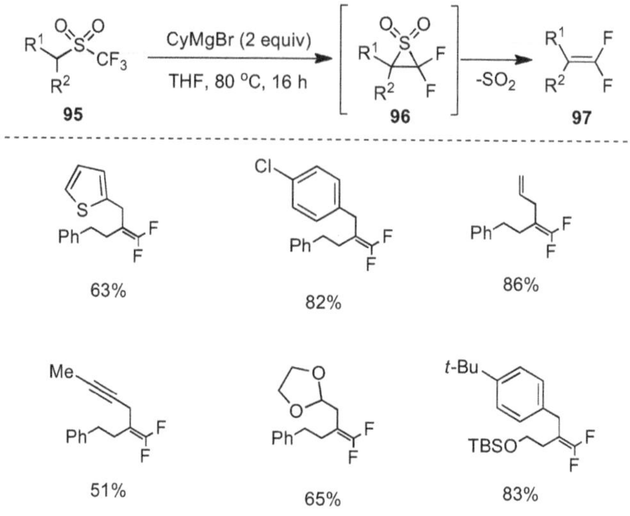

SCHEME 11.15

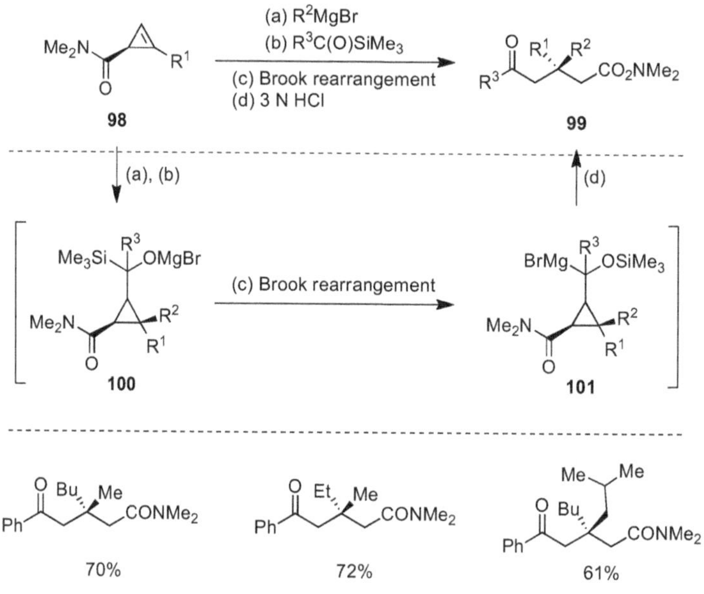

SCHEME 11.16

11.2.7.2   Synthesis of Chiral δ-Ketoamides

Access to chiral acyclic compounds can be achieved via stereoselective ring-opening of strained carbocycles. This strategy was utilized to synthesize δ-ketoamides **99** containing a quaternary stereogenic center from chiral cyclopropenes **98** (Scheme 11.16).[44]

A key step in this approach is the Brook rearrangement that serves as a trigger for the opening of cyclopropane intermediates **101**. This stereoselective carbometallation of cyclopropenyl amides **98** with Grignard reagents (R²MgBr) followed by a reaction with acyl silanes established the stage for the Brook rearrangement. Hydrolysis of the corresponding silyl enol ethers **101** under aqueous acidic conditions provided the desired δ-ketoamides **99** in a one-pot process. Using this method, several δ-ketoamides were prepared bearing alkyl substituents (R¹ and R²) on the quaternary carbons.

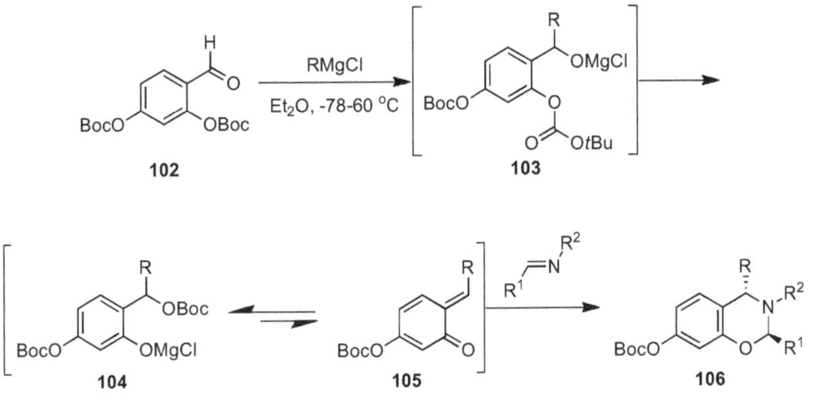

SCHEME 11.17

11.2.7.3 Access to *o*-Quinone Methide Intermediates

Quinone methides are reactive intermediates that are commonly encountered in many areas of chemistry and biology. Both reductive and oxidative metabolisms from quinine methides have become a very important topic in drug design as well as drug safety. *o*-Quinone methides can readily engage in [4+2] cycloadditions with electron-rich dieneophiles, leading to re-aromatization of the ring. To access the *o*-quinone methides, a common method involves heating a Mannich base of a corresponding polycyclic *o*-phenolic compound to yield a stable polyaryl *o*-quinone methide. *o*-Quinone methides such as **105** could be generated in situ in low concentration via the addition of Grignard reagents onto salicyladehyde followed by intramolecular Boc-group migration (Scheme 11.17).[45]

Albeit in low concentrations, the electrophilic *o*-quinone methides **105** would undergo 1,4-conjugate addition with imines followed by an intramolecular cyclization to provide an assortment of *N*-substituted 3,4-dihydro-2*H*-1,3-benzoxines **106** in a single pot.

NOTES

1. Kryk, H.; Hessel, G.; Schmitt, W. *Org. Process Res. Dev.* **2007**, *11*, 1135.
2. (a) Pasti, J.; Chen, C.-K.; Spangler, L.; DelMonte, A. J.; Benoit, S.; Berglund, D.; Bien, J.; Brodfuehrer, P.; Chan, Y.; Corbett, E.; Costello, C.; DeMena, P.; Discordia, R. P.; Doubleday, W.; Gao, Z.; Gingras, S.; Grosso, J.; Haas, O.; Kacsur, D.; Lai, C.; Leung, S.; Miller, M.; Muslehiddinoglu, J.; Nguyen, N.; Qiu, J.; Olzog, M.; Reiff, E.; Thoraval, D.; Totleben, M.; Vanyo, D.; Vemishetti, P.; Wasylak, J.; Wei, C.; *Org. Process Res. Dev.* **2009**, *13*, 716. (b) am Ende, D. J.; Clifford, P. J.; DeAntonis, D. M.; SantaMaria, C.; Brenek, S. J. *Org. Process Res. Dev.* **1999**, *3*, 319.
3. (a) Guercio, G.; Bacchi, S.; Goodyear, M.; Carangio, A.; Tinazzi, F.; Curti, S. *Org. Process Res. Dev.* **2008**, *12*, 1188. (b) For safety concerns in Grignard reaction, see, Pines, S. *Org. Process Res. Dev.* **2008**, *12*, 1283.
4. Ace, K. W.; Armitage, M. A.; Bellingham, R. K.; Blackler, P. D.; Enns, D. S.; Hussain, N.; Lathbury, D. C.; Morgan, D. O.; O'Connor, N.; Oakes, G. H.; Passey, S. C.; Powling, L. C. *Org. Process Res. Dev.* **2001**, *5*, 479.
5. Tilstam, U.; Weinmann, H. *Org. Process Res. Dev.* **2002**, *6*, 906.
6. Vit, J. U.S. Patent 1,332,718, 1973.
7. Alorati, A. D.; Gibb, A. D.; Mullens, P. R.; Stewart, G. W. *Org. Process Res. Dev.* **2012**, *16*, 1947.
8. Kopach, M. E.; Singh, U. K.; Kobierski, M. E.; Trankle, W. G.; Murray, M. M.; Pietz, M. A.; Forst, M. B.; Stephenson, G. A. Mancuso, V.; Giard, T.; Vanmarsenille, M.; DeFrance, T. *Org. Process Res. Dev.* **2009**, *13*, 209.
9. Connolly, T. J.; Matchett, M.; McGarry, P.; Sukhtankar, S.; Zhu, J. *Org. Process Res. Dev.* **2004**, *8*, 624.
10. Hogan, P. J.; Powell, L.; Robinson, G. E. *Org. Process Res. Dev.* **2010**, *14*, 1188.
11. Engelhardt, F. C.; Shi, Y.-J.; Cowden, C. J.; Conlon, D. A.; Pipik, B.; Zhou, G.; McNamara, J. M.; Dolling, U.-H. *J. Org. Chem.* **2006**, *71*, 480.

12. Li, X.; Reuman, M.; Russell, R. K.; Adams, R.; Ma, R.; Beish, S.; Branum, S.; Youells, S.; Roberts, J.; Jain, N.; Kanojia, R.; Sui, Z. *Org. Process Res. Dev.* **2007**, *11*, 414.

13. Li, X.; Wells, K. M.; Branum, S.; Damon, S.; Youells, S.; Beauchamp, D. A.; Palmer, D.; Stefanick, S.; Russell, R. K.; Murray, W. *Org. Process Res. Dev.* **2012**, *16*, 1727.

14. (a) Abarbri, M.; Dehmel, F.; Knochel, P. *Tetrahedron Lett.* **1999**, *40*, 7449. (b) Boymond, L.; Rottländer, M.; Cahiez, G.; Knochel, P. *Angew. Chem., Int. Ed.* **1998**, *37*, 1701. (c) Jensen, A. E.; Dohle, W.; Sapountzis, I.; Lindsay, D. M.; Vu, V. A.; Knochel, P. *Synthesis* **2002**, *4*, 565. (d) Knochel, P.; Dohle, W.; Gommermann, N.; Kneisel, F. F.; Kopp, F.; Korn, T.; Sapountzis, I.; Vu, V. A. *Angew. Chem., Int. Ed.* **2003**, *42*, 4302.

15. (a) Knochel, P.; Krasovskiy, A. EP 1 582 523 A1, 2005. (b) Knochel, P.; Krasovskiy, A. *Angew. Chem. Int. Ed.* **2004**, *43*, 3333. (c) Krasovskiy, A.; Straub, B. F.; Knochel, P. *Angew. Chem. Int. Ed.* **2006**, *45*, 159.

16. Leazer, J. L., Jr.; Cvetovich, R.; Tsay, F.-R.; Dolling, U.; Vickery, T.; Bachert, D. *J. Org. Chem.* **2003**, *68*, 3695.

17. (a) Waymouth, R.; Moore, E. *J. Chem. Eng. News* **1997**, *75*, 6. (b) Ashby, E. C.; Al-Fekri, D. M. *J. Organomet. Chem.* **1990**, *390*, 275. (c) Appleby, I. C. *Chem. Ind.* **1971**, *35*, 120.

18. (a) Tang, W.; Sarvestani, M.; Wei, X.; Nummy, L. J.; Patel, N.; Narayanan, B.; Byme, D.; Lee, H.; Yee, N. K.; Senanayake, C. H. *Org. Process Res. Dev.* **2009**, *13*, 1426. (b) for another aryl Grignard preparation using Knochel's protocol, see: Wang, X.; Zhang, L.; Sun, X.; Lee, H.; Krishnamurthy, D.; O'Meara, J. A.; Landry, S.; Yoakim, C.; Simoneau, B.; Yee, N. K.; Senanayake, C. H. *Org. Process Res. Dev.* **2012**, *16*, 561. (c) Tang, W.; Patel, N. D.; Wei, X.; Byrne, D.; Chitroda, A.; Narayanan, B.; Sienkiewicz, A.; Nummy, L. J.; Sarvestani, M.; Ma, N.; Senanayake, C. H. *Org. Process Res. Dev.* **2013**, *17*, 382.

19. Ruck, R. T.; Huffman, M. A.; Stewart, G. W.; Cleator, E.; Kandur, W. V.; Kim, M. M.; Zhao, D. *Org. Process Res. Dev.* **2012**, *16*, 1329.

20. Bret, G.; Harling, S. J.; Herbal, K.; Langlade, N.; Loft, M.; Negus, A.; Sanganee, M.; Shanahan, S.; Strachan, J. B.; Turner, P. G.; Whiting, M. P. *Org. Process Res. Dev.* **2011**, *15*, 112.

21. Crawford, J. B.; Chen, G.; Carpenter, B.; Wilson, T.; Ji, J.; Skerlj, R. T.; Bridger, G. *J. Org. Process Res. Dev.* **2012**, *16*, 109.

22. Marinetti, A.; Mathey, F.; Richard, L. *Organometallics* **1993**, *12*, 1207.

23. Kitagawa, K.; Inoue, A.; Shinokubo, H.; Oshima, K. *Angew. Chem. Int. Ed.* **2000**, *39*, 2481.

24. Reeves, J. T.; Sarvestani, M.; Song, J. J.; Tan, Z.; Nummy, L. J.; Lee, H.; Yee, N.; Senanayake, C. H. *Org. Process Res. Dev.* **2006**, *10*, 1258.

25. Mase, T.; Houpis, I. N.; Akao, A.; Dorziotis, I.; Emerson, K.; Hoang, T.; Iida, T.; Itoh, T.; Kamei, K.; Kato, S.; Kato, Y.; Kawasaki, M.; Lang, F.; Lee, J.; Lynch, J.; Maligres, P.; Molina, A.; Nemoto, T.; Okada, S.; Reamer, R.; Song, J. Z.; Tschaen, D.; Wada, T.; Zewge, D.; Volante, R. P.; Reider, P. J.; Tominoto, K. *J. Org. Chem.* **2001**, *66*, 6775.

26. Ziegler, D. S.; Karaghiosoff, K.; Knochel, P. *Angew. Chem. Int. Ed.* **2018**, *57*, 6701.

27. Ragan, J. A.; Ende, D. J.; Brenek, S. J.; Eisenbeis, S. A.; Singer, R. A.; Tickner, D. L.; Teixeira, J. J.; Vanderplas, B. C.; Weston, N. *Org. Process Res. Dev.* **2003**, *7*, 155.

28. Cernijenko, A.; Risgaard, R.; Baran, P. S. *J. Am. Chem. Soc.* **2016**, *138*, 9425.

29. Herrinton, P. M.; Owen, C. E.; Gage, J. R. *Org. Process Res. Dev.* **2001**, *5*, 80.

30. (a) Davies, A. J.; Scott, J. P.; Bishop, B. C.; Brands, K. M.; Brewer, S. E.; Dasilva, J. O.; Dormer, P. G.; Dolling, U.-H.; Gibb, A. D.; Hammond, D. C.; Lieberman, D. R.; Palucki, M.; Payack, J. F. *J. Org. Chem.* **2007**, *72*, 4864. (b) For the chemoselective addition of an allyl organometallic reagent to a ketone in the presence of an acid moiety, see: Kumar, S.; Kaur, P.; Chimni, S. S. *Synlett.* **2002**, 573.

31. Weiberth, F. J.; Gill, H. S.; Lee, G. E.; Ngo, D. P.; Shrimp, F. L., II; Chen, X.; D'Netto, G.; Jackson, B. R.; Jiang, Y.; Kumar, N.; Roberts, F.; Zlotnikov, E. *Org. Process Res. Dev.* **2015**, *19*, 806.

32. For BDMAEE mediated Grignard reaction with nitrile, see: Li, W.; Gao, J. J.; Lorenz, J. C.; Xu, J.; Johnson, J.; Ma, S.; Lee, H.; Grinberg, N.; Busacca, C. A.; Lu, B.; Senanayake, C. H. *Org. Process Res. Dev.* **2012**, *16*, 836.

33. Nahm, S.; Weinreb, S. M. *Tetrahedron Lett.* **1981**, *22*, 3815.

34. Journet, M.; Cai, D.; Hughes, D. L.; Kowal, J. J.; Larsen, R. D.; Reider, P. J. *Org. Process Res. Dev.* **2005**, *9*, 490.

35. (a) Zhang, Y.; Rovis, T. *J. Am. Chem. Soc.* **2004**, *126*, 15964. (b) Hansford, K. A.; Dettwiler, J. E.; Lubell, W. D. *Org. Lett.* **2003**, *5*, 4887. (c) Dieter, R. K. *Tetrahedron* **1999**, *55*, 4177. (d) Shirley, D. A. In *Organic Reactions*; Wiley, New York, 1954; Vol. 8, p 28. (e) Lauer, D.; Staab, H. A. *Chem. Ber.* **1969**, *102*, 1631. (f) Staab, H.; Jost, E. *Ann.* **1962**, *655*, 90. (g) Kuhn, R.; Staab, H. A. *Chem. Ber.* **1954**, *87*, 262.

36. Mattson, M. N.; Rapoport, H. *J. Org. Chem.* **1996**, *61*, 6071.

37. Katritzky, A. R.; Le, K. N. B.; Khelashvili, L.; Mohapatra, P. P. *J. Org. Chem.* **2006**, *71*, 9861.
38. Thiel, O. R.; Achmatowicz, M.; Bernard, C.; Wheeler, P.; Savarin, C.; Correll, T. L.; Kasparian, A.; Allgeier, A.; Bartberger, M. D.; Tan, H.; Larsen, R. D. *Org. Process Res Dev.* **2009**, *13*, 230.
39. Kolis, S. P.; Hansen, M. M.; Arslantas, E.; Brändli, L.; Buser, J.; DeBaillie, A. C.; Frederick, A. L.; Hoard, D. W.; Hollister, A.; Huber, D.; Kull, T.; Linder, R. J.; Martin, T. J.; Richey, R. N.; Stutz, A.; Waibel, M.; Ward, J. A.; Zamfir, A. *Org. Process Res. Dev.* **2015**, *19*, 1203.
40. Powell, L.; Mahmood, A.; Robinson, G. E. *Org. Process Res. Dev.* **2011**, *15*, 49.
41. Young, I.; Haley, M. W.; Tam, A.; Tymonko, S. A.; Xu, Z.; Hanson, R. L.; Goswami, A. *Org. Process Res. Dev.* **2015**, *19*, 1360.
42. Hasegawa, T.; Kawanaka, Y.; Kasamatsu, E.; Ohta, C.; Nakabayashi, K.; Okamoto, M.; Hamano, M.; Takahashi, K.; Ohuchida, S.; Hamada, Y. *Org. Process Res. Dev.* **2005**, *9*, 774.
43. Maekawa, Y.; Nambo, M.; Yokogawa, D.; Crudden, C. M. *J. Am. Chem. Soc.* **2020**, *142*, 15667.
44. Zhang, F.-G.; Eppe, G.; Marek, I. *Angew. Chem. Int. Ed.* **2016**, *55*, 714.
45. Chen, P. K. C.; Wong, Y. F.; Yang, D.; Pettus, T. R. R. *Org. Lett.* **2019**, *21*, 7746.

12 Challenging Reaction Intermediates

To facilitate the optimization of chemical processes, the most important and effective approach is mechanism-guided process development (MPD). Methods that are frequently used to investigate the reaction mechanism include (but are not limited to) the identification of intermediates and products (the desired product, by-products, and side products), the effect of catalysts or isotopes, stereochemicals and kinetic evidence, and the effect of solvents. Usually, the combination of two or three approaches will provide sufficient reaction information if one method does not work well. Practically, a reaction mechanism can be hypothesized based on, in most cases, limited information on intermediates and side products. The process optimization would be conveniently guided by the hypothesized reaction mechanism.

12.1 EFFECT OF INTERMEDIATES

12.1.1 IN TELESCOPING STEPS

Isolation of reaction intermediates and study of their chemical properties can assist in developing telescoped processes. Epoxidation of **1** with hydrogen peroxide catalyzed by hexachloroacetone gave a mixture of α-epoxide **2a** and β-epoxide **2b** (2:1) (Scheme 12.1).[1] The desired **2a** was obtained by crystallization in 49% yield. Arylation of **3a**, obtained from the protection of **2a** with TMS group, with Grignard reagent **4**, followed by acidic hydrolysis to cleave the ketal and silyl groups afforded **5a** in yields of 73–79% along with **6** (5–10%). Aromatization of dienone **5a** was carried out using a mixture of acetyl bromide and acetic anhydride, furnishing **7**, after saponification, as HCl salt.

(I) Reaction problems
- The epoxidation of **1** provided a moderate selectivity (2:1 ratio of **2a**:**2b**) with an unsatisfactory isolated yield (49%) of **2a**.
- The addition of **3a** with Grignard reagent afforded a 73–79% yield of **5a** along with 5–10% side product **6**.

(II) Observations
During the aromatization of **5a**, an isolable intermediate **9**, obtained from **6**, could be transformed into β-aryl product **10** via a common intermediate **8** within 4 h in a highly stereoselective way (Scheme 12.2). These observations opened a way to the synthetic use of the "undesired" β-epoxide **2b**. The dienone **6** was the major product of silylation–arylation of the epoxide **2b**, and the trienol acetate **9** was expected to be the first intermediate in the aromatization of dienone **6**.

(III) Solutions
These observations led to a highly efficient telescoped process by submitting the crude mixture of epoxides **2a/2b** (α/β) to the sequence silylation, arylation, aromatization, saponification, and salt formation to furnish the aryl estrone hydrochloride **7** (Scheme 12.3).

DOI: 10.1201/9781003288411-12

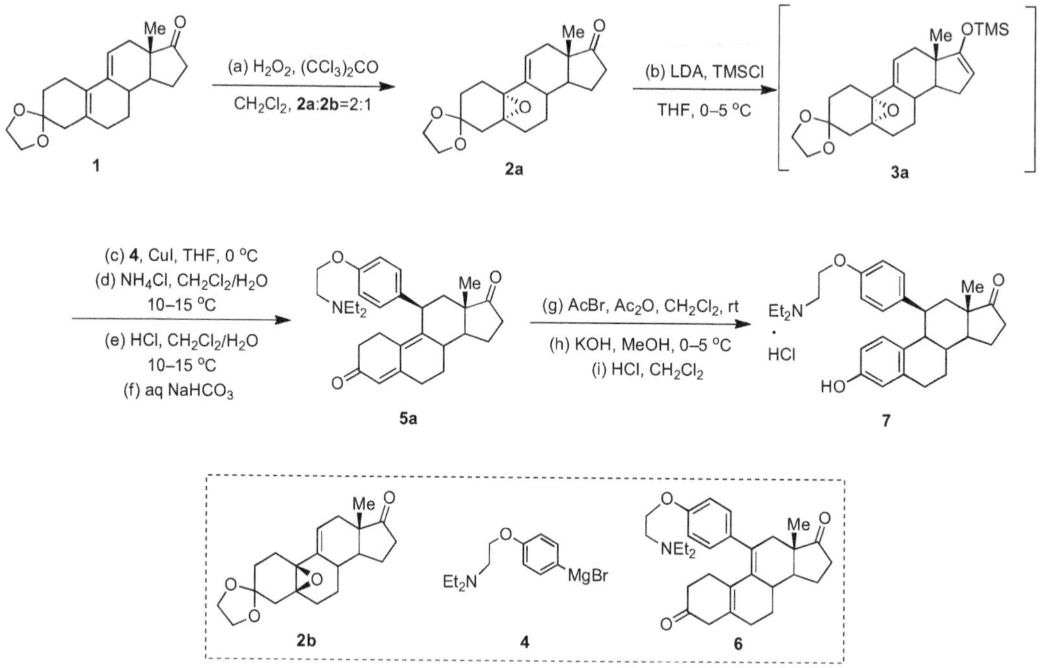

SCHEME 12.1

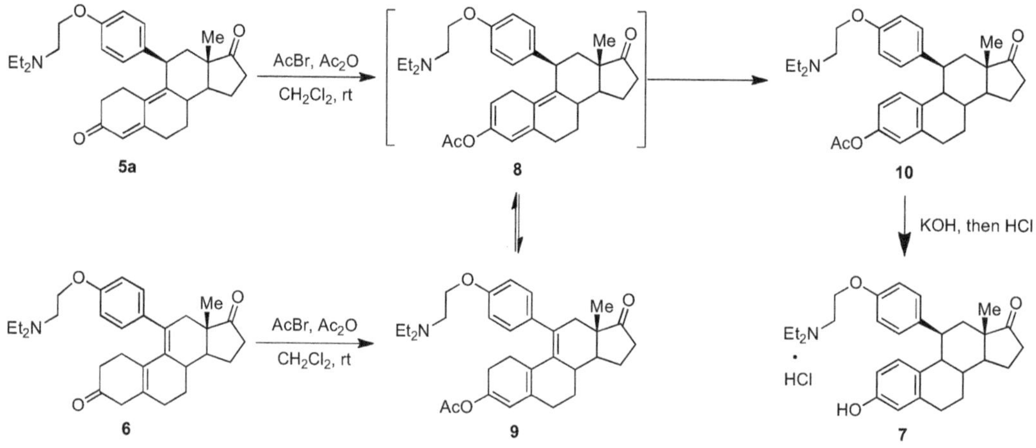

SCHEME 12.2

12.1.2 IN DESIGNING SYNTHETIC ROUTE

A rapid assembly of bisanilino-1*H*-pyrrolo[2,3-*d*]pyrimidines **14** by means of S$_N$Ar reaction of **11a** with **12** furnished **14a** in 40% yield over three steps. In sharp contrast, a reaction of **11b** did not provide product **14b** under the same reaction conditions (Scheme 12.4).[2]

It was found that the reaction of aryl chloride **11a** and amine **12** constitutes an intramolecular activation of the pyrimidine ring by a C(1′)-carboxamide moiety to give tetracyclic intermediate **15**. Subsequent selective S$_N$Ar at C(6) leads to **16** which upon ring-opening displacement by ammonia generated amide **17** (Scheme 12.5).

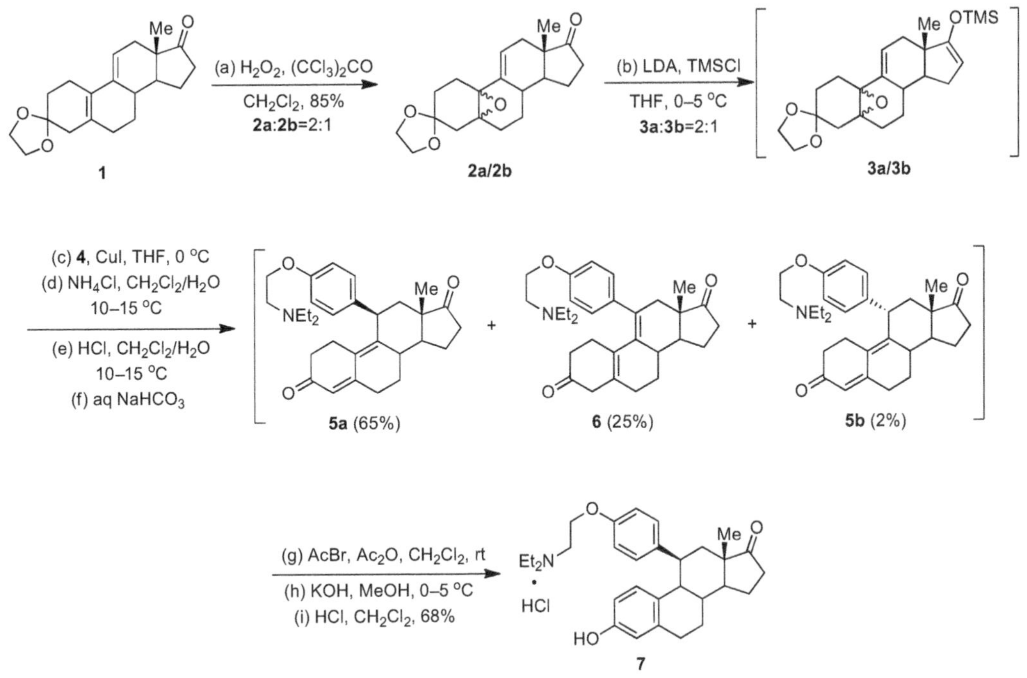

SCHEME 12.3

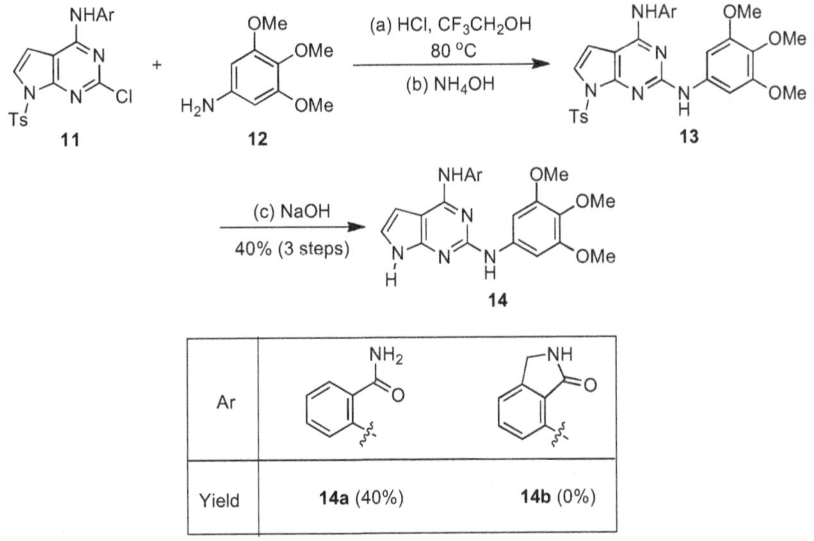

SCHEME 12.4

With this mechanistic information in hand, scientists at GlaxoSmithKline were able to develop a practical one-pot synthesis for compound **14b** by using the intramolecular activation strategy (Scheme 12.6).[2]

12.1.3 IN AGITATION

The undesired physical property of the by-product can sometimes create unexpected processing problems, such as poor stirring. The preparation of amine **26** involves mesylation of alcohols

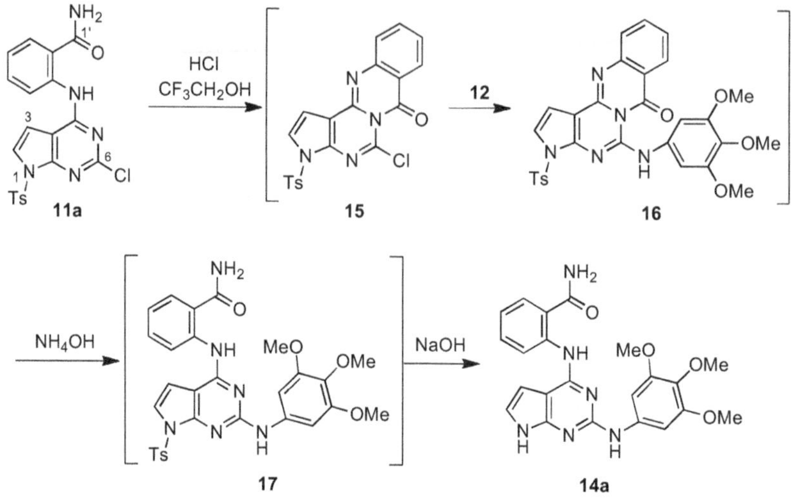

SCHEME 12.5

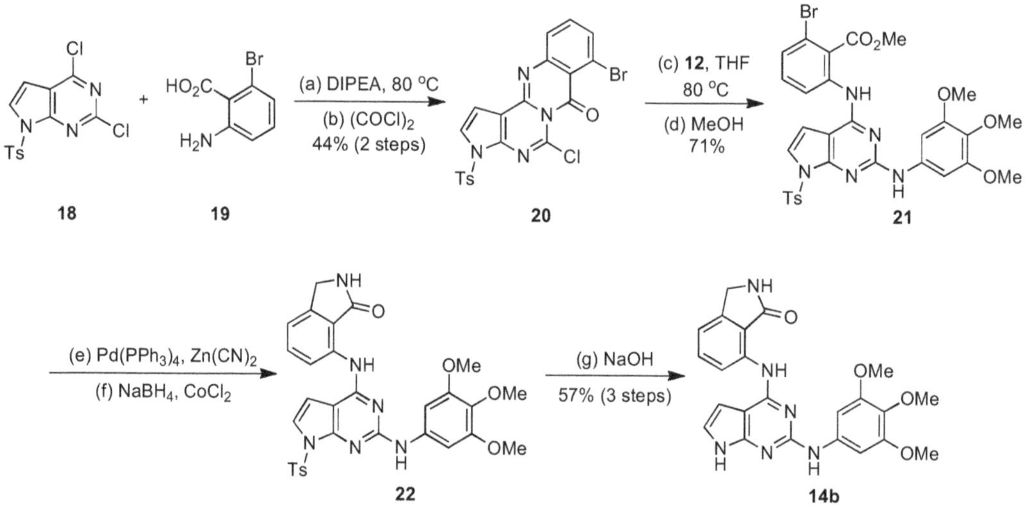

SCHEME 12.6

23 and **24** at −10 °C, followed by warming up to 10 °C prior to adding 1,3-propanediamine (Scheme 12.7).[3]

(I) Problems

During this preparation, a white, flocculent solid (triethylamine hydrochloride salt) changed to a thick, gummy material that quickly settled to the bottom of the reactor and seized the agitator. The gummy material was identified as Et$_3$NH$^+$OMs$^-$ and formed as the result of counterion exchange between the mesylate anion in the aziridinium intermediate **25a** and the chloride ion in by-product Et$_3$NH$^+$Cl$^-$ (Scheme 12.8). ^{1}H NMR studies suggest that the counterion exchange occurred during the warm-up.

(II) Solutions

Accordingly, in an optimized process, water was introduced after the complete mesylation to dissolve the gummy material generated during the warm-up.

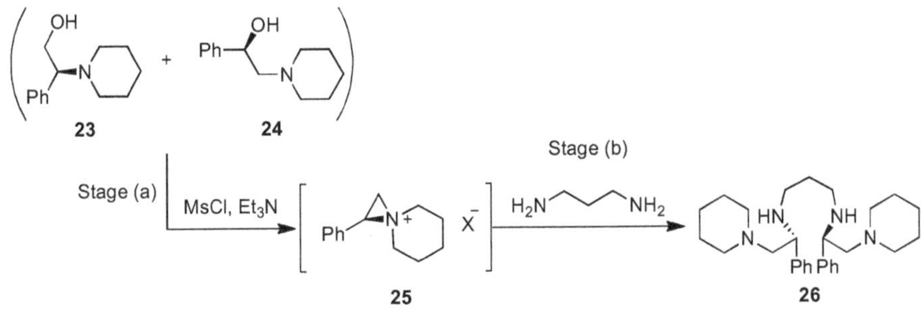

SCHEME 12.7

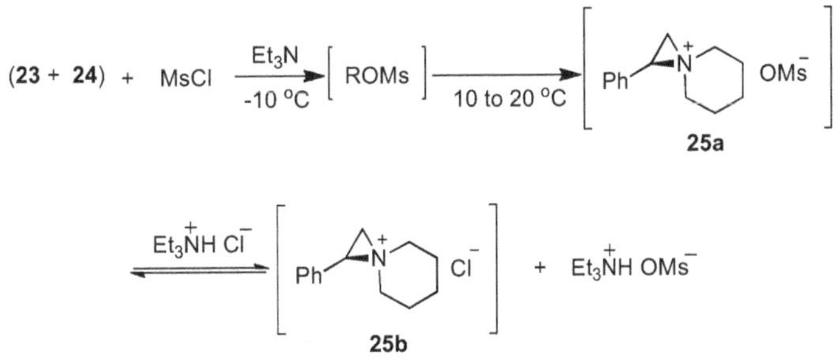

SCHEME 12.8

Procedure

Stage (a)
- Charge 4.37 L (31.4 mol, 3.7 equiv) of Et$_3$N to a mixture solution of **23** and **24** (8.48 mol) in MTBE (9.0 L).
- Cool the contents to –10 °C.
- Charge 790 mL (10.2 mol, 1.2 equiv) of MsCl at a rate of 3.3 mL/min at –8 °C to –10 °C.
- Stir at –10 °C for 30 min.

Stage (b)
- Charge 1,3-diaminopropane (354 mL, 4.2 mol, 0.5 equiv), followed by adding water (4.1 L).
- Warm the batch to 20 °C over 1 h (during which time the precipitated triethylamine salt dissolved).
- Stir at 20 °C for 18 h.
- Separate layers.
- Wash the organic layer with sat. ammonium chloride solution (3.5 L) and brine (3.5 L).
- Distill under reduced pressure (ca. 20 Torr) to ca. 4.0 L.
- Add IPA (5.0 L).
- Distill under reduced pressure (ca. 20 Torr) at 50 °C to ca. 3.5 L.
- Cool to 22 °C.
- Add IPA (5.0 L).
- Cool 0 °C.
- Add 2.0 g of **26** as seeds.
- Charge water (4.5 L) portionwise and hold the batch at 0 °C for 30 min.

- Warm to 22 °C and stir for 22 h.
- Filter, rinse with IPA/water (1.1:1.0, 2×850 mL), and dry.
- Yield: 1.33 kg (70%).

12.1.4 IN PRODUCT ISOLATIONS/PURIFICATIONS

12.1.4.1 Pictet–Spengler Reaction

A solid reaction intermediate can provide a convenient way for purification. For instance, during the preparation of 6-methoxytetrahydroisoquinoline·HCl **30** via Pictet–Spengler reaction, a solid intermediate (bis(6-methoxy-3,4-dihydroisoquinolin-2(1*H*)-yl)-methane **29**) was isolated following the basification of the reaction mixture (Scheme 12.9).[4] Treatment of the isolated solid intermediate **29** with concentrated HCl was able to produce the desired product as HCl salt **30** in high purity.

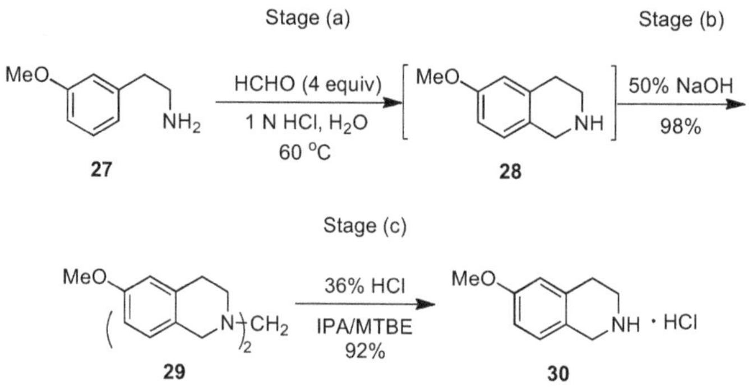

SCHEME 12.9

Procedure

Stage (a)
- Charge an aq solution of formaldehyde (416.5 g, 37 wt%, 5.13 mol) to a solution of **27** (200.0 g, 1.28 mol) in 1 N HCl aqueous solution (1.92 L, 1.92 mol).
- Heat the mixture at 60 °C for 1 h.
- Cool room temperature.

Stage (b)
- Add 69.76 g of 50% NaOH solution at 25–30 °C over 10 min with cooling in an ice/water bath.
- Stir the resulting light suspension at 25–30 °C for 1 h.
- Add 26.16 g of 50% NaOH solution at 25–30 °C over 5 min.
- Stir the resulting light suspension at 25–30 °C for 1 h.
- Add 26.16 g of 50% NaOH solution at 25–30 °C over 5 min.
- Stir the resulting light suspension at 25–30 °C for 1 h.
- Add 53.32 g of 50% NaOH solution at 25–30 °C over 5 min.
- Stir the resulting suspension at 25–30 °C for 2 h.
- Filter, rinse with water (200 mL), and dry.
- Yield of **29**: 213.4 g (98%).

Stage (c)
- Suspend 213.4 g (0.63 mol) of **29** in IPA (1.06 L).
- Charge 140.4 g of 36% HCl (1.38 mol) slowly.

- Stir the resulting suspension at room temperature for 18 h.
- Add MTBE (0.53 L) and stir at room temperature for 4 h.
- Filter, wash with a mixture of MTBE/IPA (1:1, 100 mL), and dry.
- Yield of **30**: 230.3 g (92%).

12.1.4.2 Imide Reduction

Borane reagents[5] are effective for the reductions of amides and nitrile groups to their corresponding amines. The synthesis of pyrrolidine **33** was achieved via a two-step process involving the reduction of 1-benzylpyrrolidine-2,5-dione **31** to 1-benzylpyrrilidine **32** followed by hydrogenolysis. The reduction using borane was undermined by the formation of a side product **34** in 15% that had to be removed prior to catalytic debenzylation (Scheme 12.10).[6]

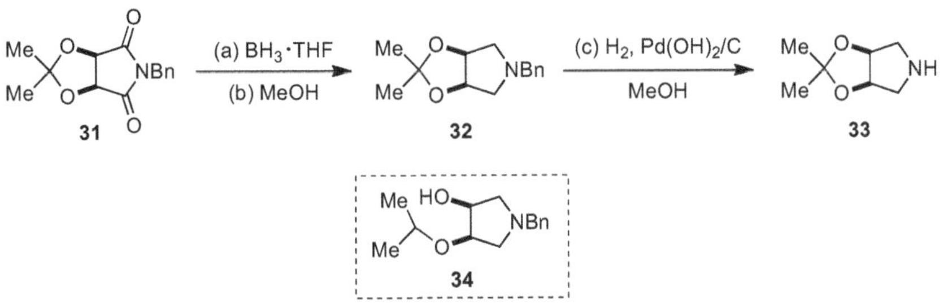

SCHEME 12.10

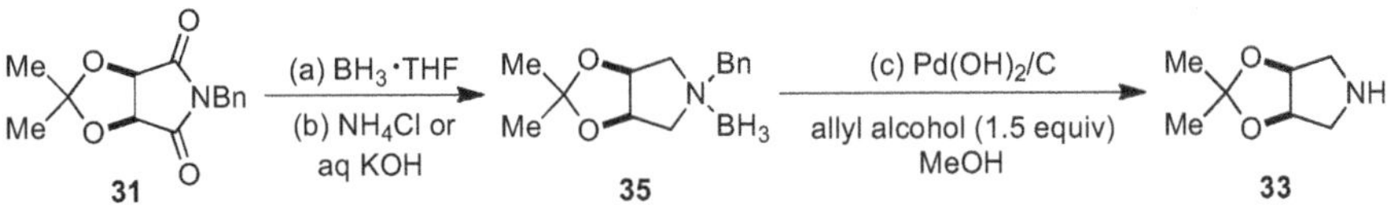

SCHEME 12.11

Solutions

In order to address the issues associated with the side product **34**, a strategy was developed by converting the benzyl-protected pyrrolidine intermediate **32** into a solid borane–amine complex **35** (Scheme 12.11).[6] It was found that quenching the reaction mixture with either ammonium chloride (pH 1) or 6 M potassium hydroxide (pH 14) could give a crystalline borane–amine complex **35**. Therefore, the isolation of **35** could readily purge the side product **34**.

REMARKS

(a) The isolation of this stable, crystalline intermediate **35** offers an additional benefit: **35** could be directly used in the subsequent catalytic debenzylation without an external hydrogen source as **35** was able to serve as its own hydrogen source.

(b) Allyl alcohol served as a hydrogen scavenger to consume the extra hydrogen.

12.1.5 IN IMPROVING PRODUCT YIELDS

12.1.5.1 Amide Formation

Amides can be prepared directly from an addition/elimination reaction by mixing esters with amines. Therefore, the synthesis of amide **37** was designed through the reaction of ethyl benzofuran-3-carboxylate **36** with methylamine (Equation 12.1).[7]

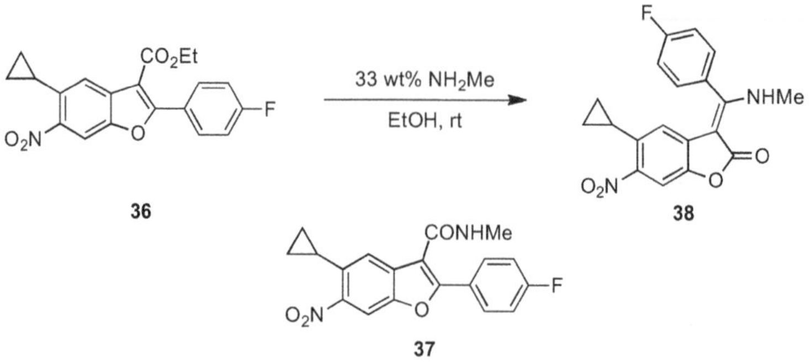

(12.1)

However, this simple reaction led to a surprising result; instead of the desired amide **37**, enamine lactone **38** was obtained as a crystalline solid in 75% yield.

Solutions

The formation of **38** was proposed, as shown in Scheme 12.12, to proceed via a Michael addition of methylamine to benzofuran ring to give a ring-opening intermediate **39** followed by intramolecular condensation.

Fortunately, the intermediate **38**, without isolation, could be converted to the desired product **37** by heating the reaction mixture at 57 °C. Ultimately, the amide product **37** could be isolated in 71% yield.

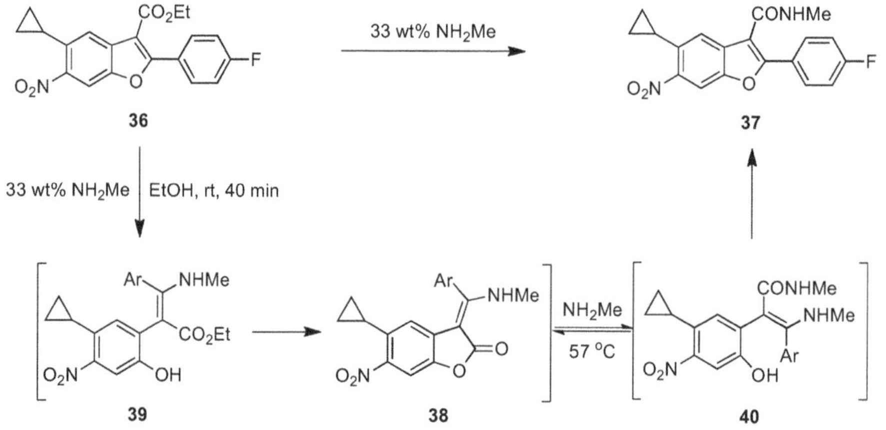

SCHEME 12.12

12.1.5.2   Synthesis of Hydroxybenzisoxazole

Often, a reactive intermediate can participate, in addition to the desired reaction path, in competitive side reactions. Compound **44** was a drug candidate for the treatment of Alzheimer's disease and cognitive impairment. As an intermediate for the synthesis of **44**, hydroxybenzisoxazole **43** was accessed via a reaction of hydroxamic acid **41** with 1,1-carbonyldiimidazole (CDI) (Scheme 12.13).[8] The formation of **43** presumably occurs via the formation of **42** as an intermediate followed by the opening of the cyclic carbonate by the adjacent phenolic oxygen with concomitant generation of carbon dioxide.

(I) Reaction problems

Although straightforward, the reaction was lack of reproducibility due to the competitive side reaction of the intermediate **42**. In the undesired process, the cyclic carbonate **42** may undergo a Lossen

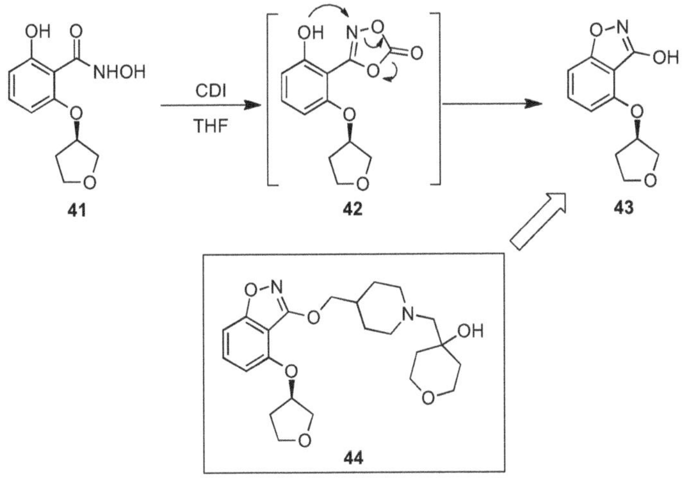

SCHEME 12.13

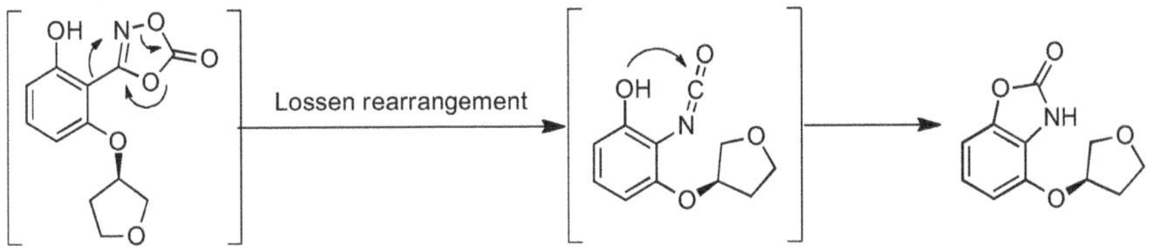

SCHEME 12.14

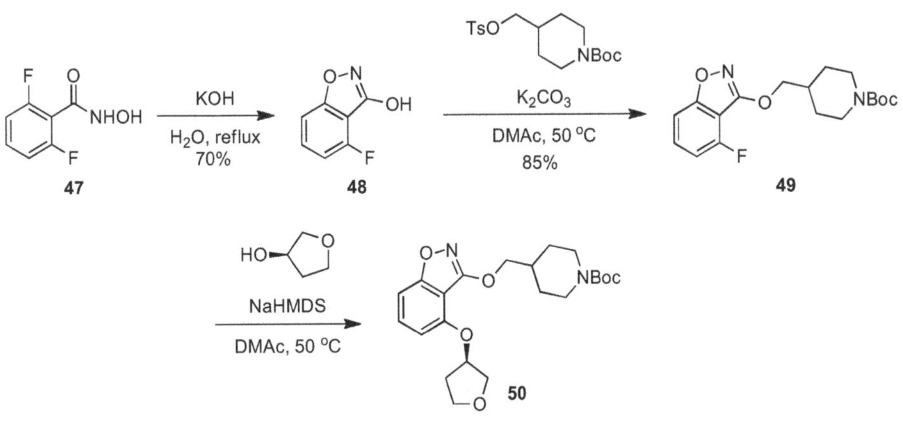

SCHEME 12.15

rearrangement to give isocyanate **45** (Scheme 12.14). This rearrangement was likely facilitated by the *ortho* electron-donating groups. The adjacent phenolic oxygen can then close on the isocyanate to form the undesired benzoxazolinone **46**.

(II) Solutions

In order to address this issue, 2,6-difluorohydroxamic acid **47** was used to replace hydroxamic acid **41**, and the intramolecular cyclization of **47** was achieved by refluxing in water in the presence of KOH to furnish hydroxybenzisoxazole **48** (Scheme 12.15).

The S_N2 substitution of fluoride in **49** with (R)-tetrahydrofuran-3-ol in the presence of NaHMDS furnished intermediate **50**. This new route avoided the formation of cyclic carbonate such as **42**.

12.1.5.3   Synthesis of Vinyl Bromide

In the first-generation process, vinyl bromide **53** was prepared by chemists at Bristol Myers Squibb following the chemistry as shown in Scheme 12.16.[9]

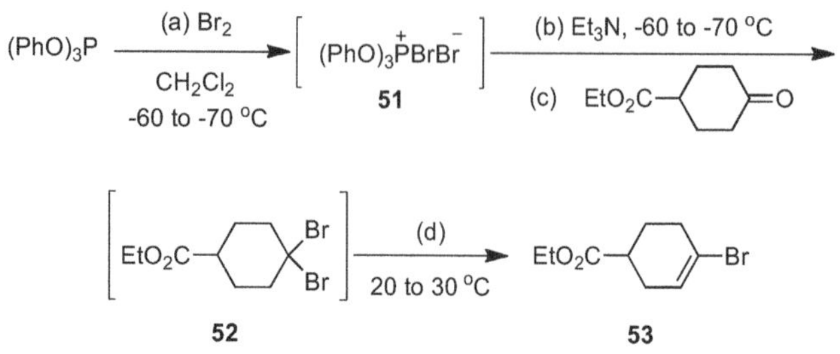

SCHEME 12.16

The yield of **53** (on a plant-scale batch of 17 kg input) dropped significantly from ~70% to 55%, which was presumably caused by instability of the reactive ionic species **51** which converted to the inactive covalent form **54** (Equation 12.2) during the processing (8 h for the addition of bromine and 10 h for warming to ambient temperature).

$$(PhO)_3\overset{+}{P}Br\,Br^{-} \longrightarrow (PhO)_3PBr_2 \qquad\qquad (12.2)$$
$$\mathbf{51} \qquad\qquad\qquad \mathbf{54}$$

Studies showed that the reaction of the ionic species **51** with ethyl 4-oxocyclohexanecarboxylate is kinetically fast. Thereby a robust and high-yielding process (the second generation process) was developed by adding bromine to a pre-cooled solution (–20 to –15 °C) of ethyl 4-oxocyclohexane-carboxylate, triethylamine, and triphenyl phosphite in dichloromethane at <–5 °C. This new process eliminated the cryogenic conditions, reduced the reaction time from 18–24 h to 1 h, and yielded 85–90% product **53**.

Procedure

The first-generation process
- Charge Br_2 (6.2 L, 121.0 mol) to a solution of $(PhO)_3P$ (37.2 kg, 119.9 mol) in dichloromethane (347 L) at –65 to –75 °C.
- Stir the mixture at –65 to –75 °C for 30 min.
- Charge triethylamine (14.2 kg, 139.8 mol) followed by adding ethyl 4-oxocyclohexanecarboxylate (17 kg, 99.8 mol) at –65 to –75 °C.
- Stir the batch at –65 to –75 °C for 1 h.
- Warm the batch to ambient temperature over 10 h and stir for 23 h.
- Add 0.5 N HCl (85 L).
- Separate layers.
- Wash the organic layer with water (85 L).
- Pass the organic layer through a silica gel bed (60 120 mesh, 51 kg) and rinse with dichloromethane (87 L).
- Distill the combined filtrates under reduced pressure.
- Yield: 59.2 kg (21.7 wt%) (55% yield).

The second-generation process
- Charge Br_2 (180 mL, 3.49 mol) dropwise to a solution of ethyl 4-oxocyclohexanecar-boxylate (500 g, 2.94 mol), triethylamine (580 mL, 4.16 mol), and $(PhO)_3P$ (1100 g, 3.50 mol) in dichloromethane (10 L) at <–5 °C.
- After the addition, warm the batch to ambient temperature and stir for 1 h.
- Add 0.5 N HCl (2.5 L).
- Separate layers.
- Wash the organic layer with water (2.5 L).
- Pass the organic layer through a silica gel pad (60–120 mesh, 900 g).
- Distill the filtrate under reduced pressure.
- Yield: 1950 g (32 wt%) (90% yield).

12.1.6 In Improving Operational Profile

The difference in reactivity of reaction intermediates necessitates the installation of multiple operation stages in a single transformation. The synthesis of pyridazinone 56 from hydroxy ketone 55 involves four intermediates (57–60) with different inherent reactivity, which renders a complicated operational profile as shown in Scheme 12.17.[10] The cycloaddition of 55 with hydrazine could furnish dihydropyridazinone 58 at –10 °C (Stage (a), Scheme 12.17); conversion of 58 to the desired product 56 is a slow process requiring 40 °C (Stage (b), Scheme 12.17). In the event, a competitive side reaction of 58 led to the formation of hydrazone intermediate 60 via intermediate 59. The intermediate 60 remained in the aqueous layer during the workup. The intermediate 60, however, could be converted to 56 by heating at 70 °C for 24 h (Stage (c), Scheme 12.17).

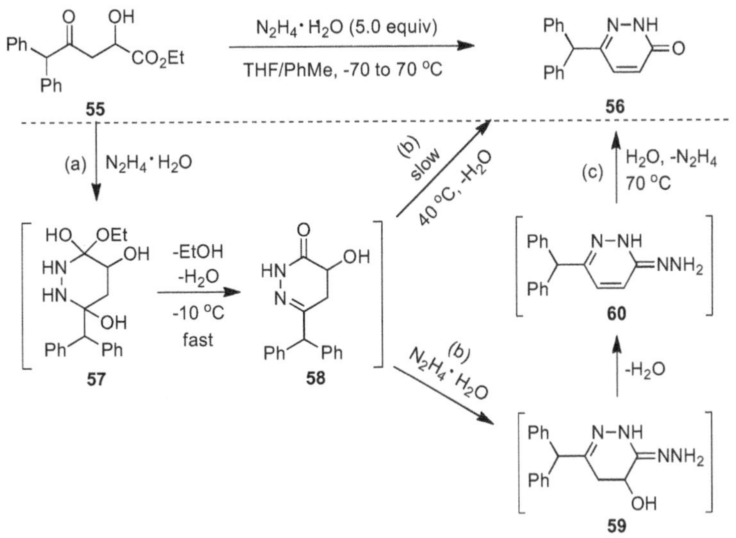

SCHEME 12.17

Procedure

Stage (a) Formation of 58
- Charge 33.1 kg (661.2 mol) of hydrazine hydrate at <–70 °C to a solution of 55 (132.2 mol, used directly as an aldol reaction mixture).
- Transfer the mixture into toluene (20.1 kg).

- Rinse with THF (13.1 kg) and toluene (12.0 kg).
- Stir at –10 °C for 12 h.

Stage (b) Conversion of **58** to **56** (along with **60**)
- Warm the batch to 40 °C and age for 3 h.
- Add water (139 kg) and toluene (40 kg).
- Add 6 N HCl (32.7 kg) to pH 10.1.
- Separate layers.
- Wash the organic layer twice with aqueous NaCl (28 L, 20% w/w).

Stage (c) Conversion of **60** to **56**
- Heat the combined aq layers at 70 °C for 24 h.
- Add THF (25.0 kg), toluene (96.0 kg), and aqueous NaCl (28 L, 20% w/w).
- Separate layers.
- Combine the organic layers (including the organic layers from Stage (b)).
- Distill under vacuum to 82 L.
- Add heptane (57.0 kg) and age for 2 h.
- Filter and wash with a mixture of toluene (29.0 kg) and heptane (34.0 kg).
- Dry at 40 °C under vacuum.
- Yield: 26.3 kg (75.8%) of **56**.

12.2 MONITORING INTERMEDIATES

Knowing the information of reaction intermediates is important for the chemical process development. The reaction progress can be monitored by measuring concentrations of the starting materials or reagents remaining in the reaction mixture; depending on the individual reaction, the reaction intermediate information may be obtained either directly or indirectly.

12.2.1 DIRECT MONITORING INTERMEDIATE

Normally, the reaction information can be obtained directly by means of classical methods such as GC, HPLC, or NMR. For reactions involving labile intermediates, the real-time monitoring of a reaction's progress affords opportunities to improve controls of a process. ReactIR provides a direct way to monitor the formation of unstable reaction intermediates. For example, the progress of the one-pot reaction to convert amide **61** to acylurea **63** via aroylisocyanate intermediate **62** was conveniently scrutinized with ReactIR by monitoring stretches of **61** (at 1686 cm^{-1}), **62** (at 2250 cm^{-1}), and **63** (at 1737 cm^{-1}) (Scheme 12.18).[11]

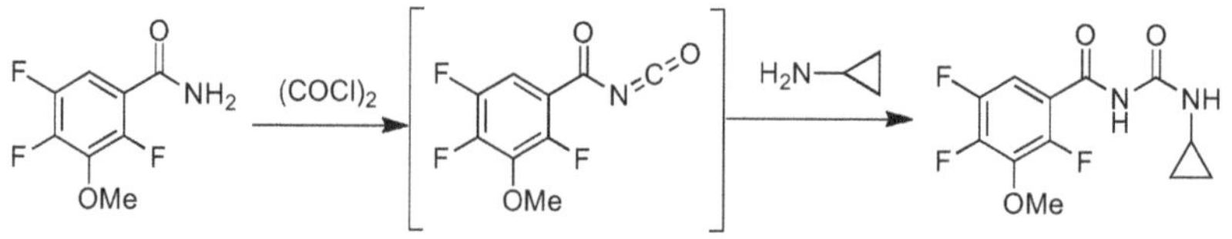

SCHEME 12.18

12.2.2 INDIRECT MONITORING INTERMEDIATES

For reactions with labile reaction intermediates, it is not possible to isolate the intermediates or to obtain the reaction information directly by means of classical methods such as GC, HPLC, or NMR. Therefore, as an indirect approach, the derivatization of the intermediates is frequently used.

12.2.2.1 Derivatization of Acylimidazolide

It is not practical to monitor directly the formation of the reactive reaction intermediate, such as acylimidazolide **65**, by means of classical methods (GC, HPLC, or NMR) without derivatization. The intermediate, obtained from the activation of acid **64** with 1,1-carbonyldiimidazole (CDI), was hydrolytically unstable under HPLC conditions (0.1% TFA in acetonitrile–water). Accordingly, an indirect analytical method for the activation step was developed by monitoring the formation of benzylamide **67** formed by quenching an aliquot of the reaction mixture with an excess of benzyl-amine (Scheme 12.19).[12]

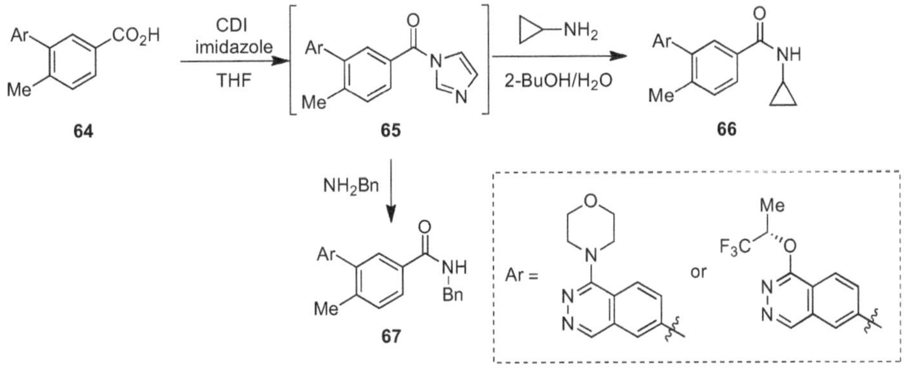

SCHEME 12.19

REMARK

Although this approach usually provides consistent analyses during laboratory optimization, the extended times involved in withdrawing a sample from pilot-plant equipment and transferring it to the lab for sample preparation can create artifacts in the data. Extended times for off-line assays can also allow significant decomposition of the product within processing equipment.

12.2.2.2 Derivatization of N-Methylene Bridged Dimer

The synthesis of spirocyclic hydantoin **70** proceeded through N-methylene bridged dimer interme-diate **69** formed via cycloaddition of **68** with hexamethylenetetramine (HMTA) in the presence of glycine (Scheme 12.20).[13]

The levels of **69** could not be directly determined by HPLC as it displayed a broad peak on a reverse phase HPLC due to its hydrolytic instability. Ultimately, a method was developed by con-verting **69** to **70a** and **71** by treating a sample of the reaction mixture with sodium triacetoxyboro-hydride (Equation 12.3). Therefore, the levels of **69** could be determined indirectly by measuring the levels of **70a** and/or **71**.

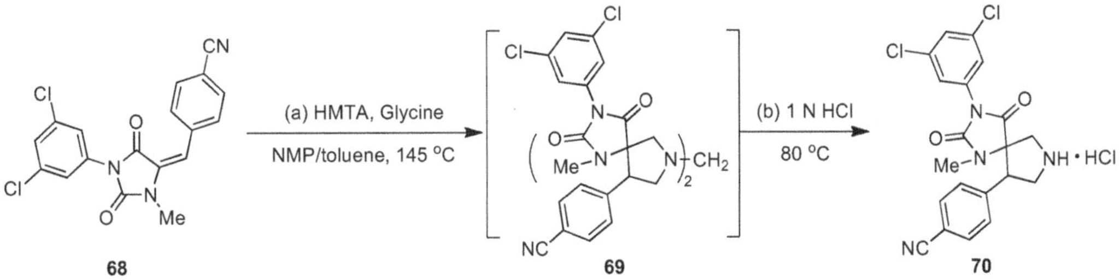

SCHEME 12.20

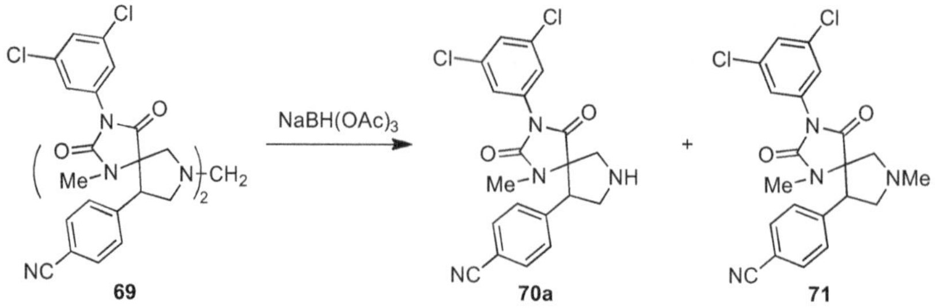

(12.3)

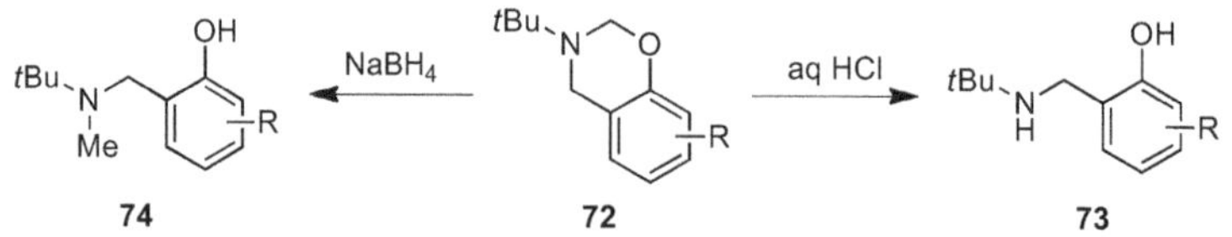

SCHEME 12.21

DISCUSSIONS

Analogously, direct monitoring of the progress of the hydrolysis of 3,4-dihydrobenzo[3,4]oxazine **72** to phenol **73** by a reverse phase LC-MS is not possible due to the instability of **72** under aqueous acidic conditions (Scheme 12.21).[14] Thus, an indirect method was developed by treating in-process samples with $NaBH_4$ and monitoring the levels of tertiary amine **74** as any unreacted **72** will be readily converted to **74** with $NaBH_4$ quantitatively.

NOTES

1. Prat, D.; Benedetti, F.; Girard, G. F.; Bouda, L. N.; Larkin, J.; Wehrey, C.; Lenay, J. *Org. Process Res. Dev.* **2004**, *8*, 219.
2. Chamberlain, S. D.; Gerding, R. M.; Lei, H.; Moorthy, G.; Patnaik, S.; Redman, A. M.; Stevens, K. L.; Wilson, J. W.; Yang, B.; Shorwell, J. B. *J. Org. Chem.* **2008**, *73*, 9511.
3. Frizzle, M.; Caille, S.; Marshall, T. L.; McRae, K.; Nadeau, K.; Guo, G.; Wu, S.; Martinelli, M. J.; Moniz, G. A. *Org. Process Res. Dev.* **2007**, *11*, 215.
4. Zhong, H. M.; Villani, F. J.; Marzouq, R. *Org. Process Res. Dev.* **2007**, *11*, 463.
5. Burkhardt, E. R.; Matos, K. *Chem. Rev.* **2006**, *106*, 2617.
6. Couturier, M.; Andresen, B. M.; Jorgensen, J. B.; Tucker, J. L.; Busch, F. R.; Brenek, S. J.; Dubé, P.; Ende, D. J.; Negri, J. T. *Org. Process Res. Dev.* **2002**, *6*, 42.
7. Bowman, R. K.; Bullock, K. M.; Copley, R. C. B.; Deschamps, N. M.; McClure, M. S.; Powers, J. D.; Wolters, A. M.; Wu, L.; Xie, S. *J. Org. Chem.* **2015**, *80*, 9610.
8. Widlicka, D. W.; Murray, J. C.; Coffman, K. J.; Xiao, C.; Brodney, M. A.; Rainville, J. P.; Samas, B. *Org. Process Res. Dev.* **2016**, *20*, 233.
9. Likhite, N.; Ramasamy, S.; Tendulkar, S.; Sathasivam, S.; Lusung, M.; Zhu, Y.; Strotman, N.; Nye, J.; Ortiz, A.; Kiau, S.; Eastgate, M. D.; Vaidyanathan, R. *Org. Process Res. Dev.* **2016**, *20*, 977.
10. Yoshida, S.; Yishino, T.; Miyafuji, A.; Yasuda, H.; Kimura, T.; Takahashi, T.; Mukuta, T. *Org. Process Res. Dev.* **2012**, *16*, 1544.
11. Beylin, V.; Boyles, D. C.; Curran, T. T.; Macikenas, D.; Parlett, R. V. IV; Vrieze, D. *Org. Process Res. Dev.* **2007**, *11*, 441.
12. (a) Thiel, O. R.; Achmatowicz, M.; Bernard, C.; Wheeler, P.; Savarin, C.; Correll, T. L.; Kasparian, A.; Allgeier, A.; Bartberger, M. D.; Tan, H.; Larsen, R. D. *Org. Process Res. Dev.* **2009**, *13*, 230. (b) Achmatowicz, M.; Thiel, O. R.; Wheeler, P.; Bernard, C.; Huang, J.; Larsen, R. D.; Faul, M. M. *J. Org. Chem.* **2009**, *74*, 795.
13. Delmonte, A. J.; Waltermire, R. E.; Fan, Y.; McLeod, D. D.; Gao, Z.; Gesenberg, K. D.; Girard, K. P.; Rosingana, M.; Wang, X.; Kuehne-Willmore, J.; Braem, A. D.; Castoro, J. A. *Org. Process Res. Dev.* **2010**, *14*, 553.
14. The author's unpublished work.

13 Protecting Groups

Protection and deprotection of functional groups have been playing a great role in organic synthesis. Protections fall predominately into three categories: protection of amino groups, protection of carboxylic acids as esters, and protection of hydroxyl groups. The most common protecting group[1] for the amino group is *tert*-butyloxycarbonyl (Boc), followed by benzyl and carbobenzyloxy (Cbz) groups (these two groups can be removed by catalytic hydrogenolysis). Frequently, carboxylic acids are protected as their corresponding alkyl esters, such as methyl ester or ethyl ester. A wide variety of protecting groups are used for the protection of the hydroxyl group. The most common protecting groups are benzyl, silicon-containing groups, and acetyl and benzoyl groups. In general, the benzyl-protecting group works well and tolerates acidic or basic conditions. There are concerns, however, regarding product contamination over the use of palladium-catalyzed deprotection of protecting groups, especially at the late stage in the synthesis. Both acetyl and benzoyl groups proved to be labile under basic conditions and extensive protecting group migration was observed.

13.1 PROTECTION OF HYDROXYL GROUP

13.1.1 PREVENTION OF SIDE REACTIONS

13.1.1.1 Friedel–Crafts Alkylation

Selection of the right protecting group is critical for the success of a reaction. For example, attempts to make indeno-β-lactam **2a** under Friedel–Crafts alkylation conditions by treatment of benzyl-protected phenol **1a** with TfOH failed, giving, instead of **2a**, chroman **3** in 44% yield (Equation 13.1).[2]

(13.1)

1a: R = Bn
1b: R = TIPS

2a: R = Bn
2b: R = TIPS

3

To circumvent the undesired cyclization, a bulky triisopropylsilyl (TIPS) group was used to reduce the nucleophilicity of the oxygen atom. The reaction was conducted in dichloromethane in the presence of Sc(OTf)$_3$ and 2,6-di-*tert*-butylpyridine (DTBP) affording the desired indeno-β-lactam **2b** in 80% yield.

DOI: 10.1201/9781003288411-13

13.1.1.2 Removal of Trifluoromethanesulfonyl Group

Attempts to remove the trifluoromethanesulfonyl group in phenol **4a** by treatment of **4a** with LiAlH$_4$ at reflux in toluene failed to give the desired deprotection product **5a**, but a ring-opening product **7** was obtained instead in a quantitative yield (Scheme 13.1).[3]

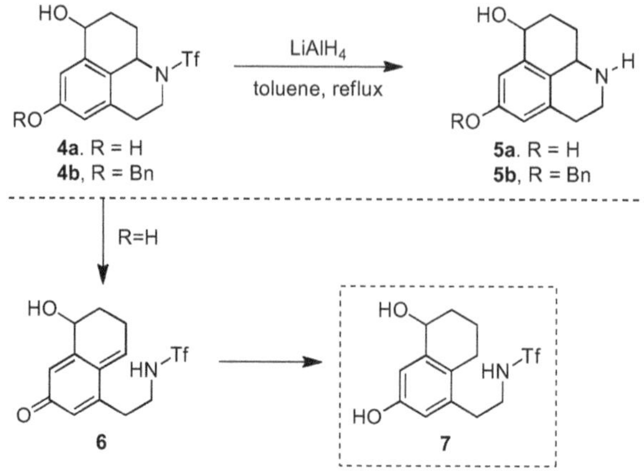

SCHEME 13.1

The reaction of **4a** with LiAlH$_4$ may take place via a methaquinone intermediate **6** since trifluoromethanesulfonamide is a good leaving group and the presence of the hydroxyl group would allow retro-Michael reaction occurs to form **6**. Therefore, in order to remove the triflate-protecting group, the hydroxyl group in phenol **4a** was protected as its benzyl ether, and the secondary amine in the resulting **4b** was successfully unmasked under the same conditions (LiAlH$_4$, refluxing toluene), affording **5b** in an excellent yield.

REMARK

Selective deprotection is critical in chemical transformations. Ammonia, pyridine, and ammonium acetate were found to be extremely effective as inhibitors for palladium-catalyzed benzyl ether hydrogenolysis. Thus, in the presence of ammonia, pyridine, or ammonium acetate (0.1–0.5 equiv), reactions, such as palladium-catalyzed deprotection of Cbz group (**8c**) and benzyl ester (**8d**) and palladium-catalyzed hydrogenation of olefin (**8b**) and azide (**8a**), proceeded smoothly without deprotection of benzyl ethers (Equation 13.2).[4]

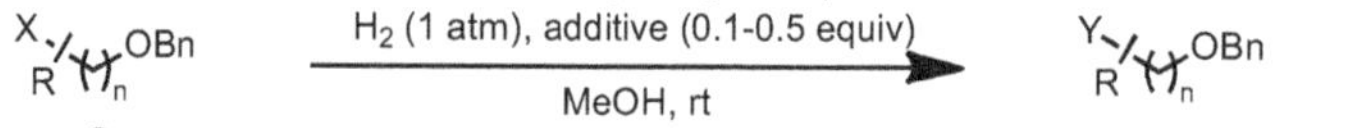

$$(13.2)$$

8a, X = N$_3$	**9a**, Y = NH$_2$
8b, X = PhCHCH	**9b**, Y = PhCH$_2$CH$_2$
8c, X = R'NCbz	**9c**, Y = R'NH
8d, X = CO$_2$Bn	**9d**, Y = CO$_2$H

13.1.1.3 S$_N$2 Reaction

The outcome of a reaction can be influenced by the stability of the protecting group. For instance, an *O*-alkylation of phenol **10** with TBS-protected mesylate **11** suffered from poor reproducibility along with the formation of several troublesome impurities (**13–15**) (Equation 13.3).[5] The cleavage of the *tert*-butyldimethylsilyl (TBS)-ether in product **12** by a small amount of H$_2$O, which could exist under the alkylation conditions, resulted in the formation of des-TBS **13** and **14** as well as ester dimer **15**.

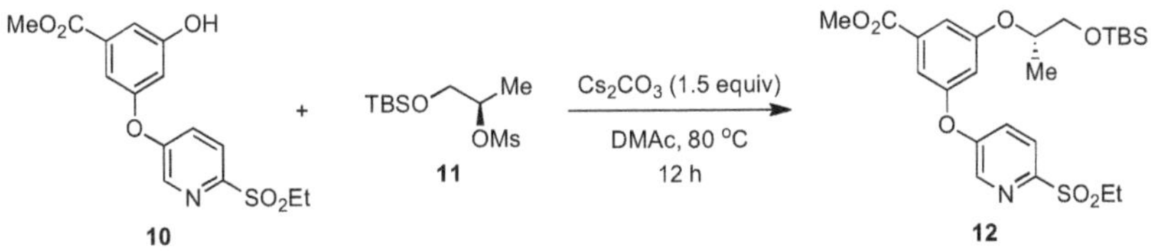

(13.3)

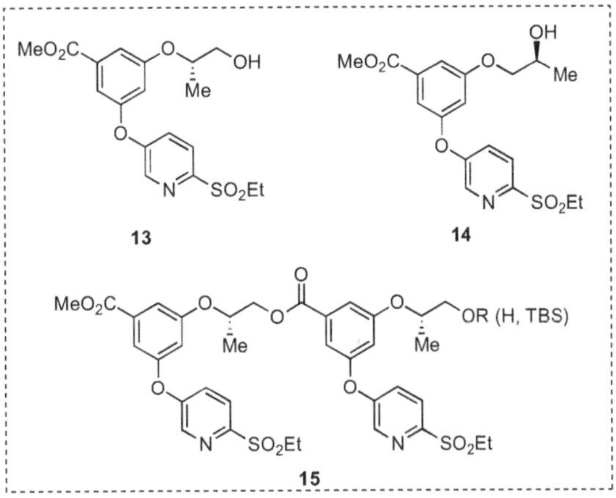

To address the impurity issues, the more stable protecting group, TIPS, was used in the S_N2-alkylation resulting in an improved reaction profile with only a negligible amount of dimeric impurity **15**.

DISCUSSIONS

Although the functional group protection enhanced reaction selectivity, using such an approach on large production scales is not an ideal option. Driven by reducing production costs, developing protection-free synthesis has become a relentless effort for process chemists. A protection-free process for a selective mono-*O*-alkylation of a triol intermediate **16** was developed based on the following inherent reactivity order: OH (C24)>OH (C15)>OH (C25) (Equation 13.4).[6]

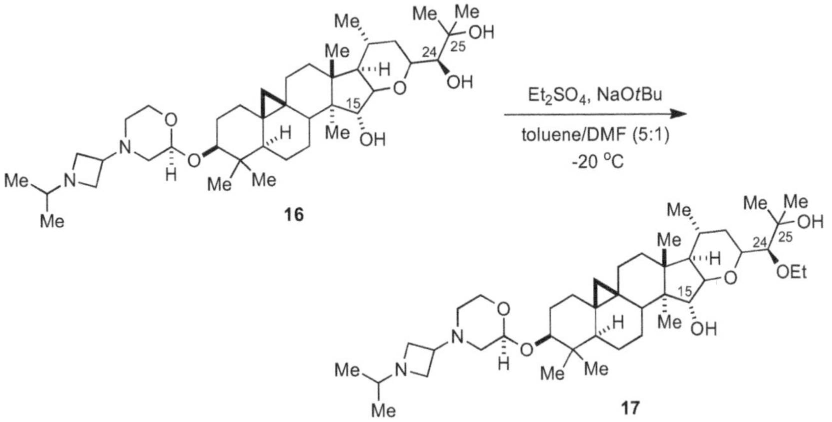

(13.4)

13.1.2 INCREASING CATALYST ACTIVITY

As one of the process development strategies, functional group protection was utilized to improve the enantioselectivity during the asymmetric reduction of ketone **18** (Equation 13.5).[7] The catalytic asymmetric reduction of ketone **18** (with R=H) required high loading of an expensive chiral catalyst, which was presumably attributed to the low catalyst activity as a result of a chelating property of the hydroxyl group.

$$(13.5)$$

R = H: 80%, 94.0% ee
        SM/Cat: 200:1

R = Bn: 81%, 98.5% ee
        SM/Cat.: 2500:1

With the goal of reducing the hydrogenation catalyst loading without sacrificing the enantioselectivity, an economically viable process was developed by the protection of the hydroxyl group with the benzyl group. As a result, the asymmetric ketone reduction was improved by lowering the catalyst loading from 200:1 to 2500:1 (the ratio of the starting material to the catalyst) with an increased enantiopurity of 98.5% ee (from 94.0% ee).

REMARK

Protection does not always render positive results as expected. For instance, the protection of the hydroxyl group in phenol **20** displays a negative effect on the selectivity of C-arylation of oxindole enolate. With the protection of phenol functionality in **20** (R^1=OMOM, R^2=MOM, R^3=Br), the cyclization underwent not at the carbon center, but at the oxygen leading to O-arylated side product **22** (Equation 13.6 and Entry 1 of Table 13.1).[8] Notably, the compound **20** with a free hydroxyl group (R^1=OH) cyclized to **21** in 46 % yield (Entry 2 of Table 13.1). It appears that the "naked" hydroxyl group helps to keep co-planarity of the tyrosine arene and the oxindole enolate by taking advantage of hydrogen bonding between the phenol of the tyrosine and the enolate alkoxide, which favors the C-arylation of oxindole enolate.

$$(13.6)$$

13.1.3   SEPARATION OF DIOL DIASTEREOMERS

Ketones and aldehydes are often utilized to protect diols by forming ketals or acetals. An approach was developed for the separation of *anti-* and *syn-*2,4-pentanediol by using acetophenone **24** as the protecting group (Scheme 13.2).[9]

TABLE 13.1

C-Arylation of 20

Entry	R^1	R^2	R^3	R	Base	Temp	Yield of 21, %
1	OMOM	MOM	Br	Cbz	NaH	rt	0
2	OH	MOM	H	Boc	Na$_2$CO$_3$	65 °C	46
3	OH	MOM	Br	Boc	Na$_2$CO$_3$	65 °C	0
4	OH	H	Br	Boc	Na$_2$CO$_3$	65 °C	56

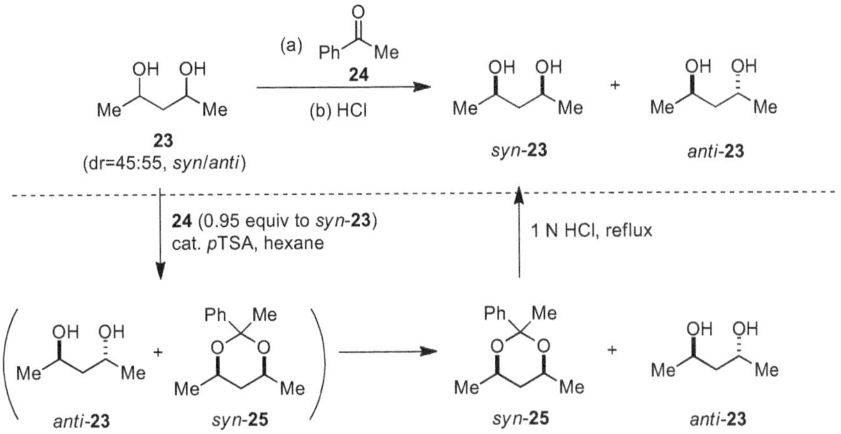

SCHEME 13.2

Selective protection of a mixture of *anti*- and *syn*-2,4-pentanediol with **24** was realized, giving the *syn*-enriched ketal (*syn*-**25**, dr=6:1) that was isolated by a simple phase separation from the *anti*-enriched diol (*anti*-**23**) aqueous mixture. The subsequent selective hydrolysis of *anti*-ketal (*anti*-**25**) in the crude *syn*-**25** (dr=6:1) with 1.0 N HCl at ambient temperature afforded a pure *syn*-**25** in a dr >99:1. Hydrolysis of the *syn*-**25** in aqueous methanol under reflux gave the isolated *syn*-2,4-pentanediol in 75–79% yield with dr >99:1. The resulting *anti*-diol-enriched mixture was further treated with **24** to remove residual *syn*-**23**, providing the *anti*-**23** in 79–85% yield with dr >98:2.

Procedure

Stage (a) Selective protection and hydrolysis
- Charge 115.77 g of 2,4-pentanediol (500 mmol of *syn*-**23** and 611 mmol of *anti*-**23**) into a 1-L three-necked flask, followed by adding 0.38 g (2.0 mmol) of pTSA, 57.07 g (475 mmol, 0.95 equiv relative to *syn*-**23**) of **24**, and hexane (650 mL).
- Heat the mixture at reflux (68–70 °C) under N$_2$ for 18 h (9.56 g of water was removed by Dean–Stark trap (dr (*syn*-**25**/*anti*-**25**)=6.3:1 and dr (*syn*-**23**/*anti*-**23**)=1:7 by NMR).
- Cool to 25 °C.
- Add 25 mL of 1.0 N HCl to the batch followed by adding the contents of the Dean–Stark trap (9.56 g of water and a small amount of hexane).
- Stir the batch at ambient temperature for 17–20 h (until all *anti*-**25** disappeared).
- Separate layers.
- Keep the top hexane layer containing *syn*-**25** and **24** in vessel A and the bottom aqueous layer containing *anti*-enriched **23** in vessel B.

- Wash the content in vessel A with water (3×25 mL) and transfer the water washings to vessel B.

Stage (b) Hydrolysis of *syn*-**25**
- Distill the volatiles in vessel A under reduced pressure to give 92.36 g of residue.
- Add MeOH (250 mL), water (10 mL), and 1.0 N HCl (25 mL).
- Heat at reflux for 4–8 h.
- Cool to ambient temperature.
- Add $NaHCO_3$ powder (2.5 g, 30 mmol) and stir for 1 h.
- Distill under reduced pressure.
- Add water (50 mL) and hexane (300 mL).
- Stir for 10–15 min.
- Separate layers.
- Extract the top hexane layer with water (2×50 mL).
- Distill the hexane layer under reduced pressure to give **24** (50.57 g, 88.6% recovery),
- Distill the combined aqueous layers.
- Add toluene (100 mL) and distill azeotropically under reduced pressure.
- Add CH_2Cl_2 (100 mL) to triturate the residue.
- Filter through a pad of Celite (5 g) to remove insoluble salts and rinse with CH_2Cl_2 (50 mL).
- Distill the combined filtrates under reduced pressure and dry the product in a vacuum oven.
- Yield of *syn*-**23**: 41.47 g (78.9%, dr=100:0–1).

Stage (c) Purification and isolation of *anti*-**23**
- Charge hexane (650 mL) to vessel B followed by adding 17.0 g (141 mmol, 1.4 equiv relative to *syn*-**23**) of **24**
- Reflux the mixture for 36 h (water byproduct was removed by Dean–Stark trap).
- Cool to ambient temperature.
- Separate layers.
- Transfer the bottom *anti*-**23** layer into vessel C.
- Wash the hexane layer with water (50 mL) and add the water washing to vessel C.
- Wash the combined water layers in vessel C with hexane (100 mL).
- Add solid $NaHCO_3$ (3.0 g) to vessel C.
- Distill to remove water under reduced pressure at 45 °C.
- Dry the residue by azeotropic distillation with toluene (100 mL).
- Add CH_2Cl_2 (50 mL) to the residue.
- Filter to remove insoluble salts and rinse with CH_2Cl_2 (50 mL).
- Distill and dry under vacuum.
- Yield of *anti*-**23**: 52.42 g (81.6%, dr=100:2).

13.1.4 DEVELOPING TELESCOPED PROCESS

A telescoped process was developed for a four-component reaction using a highly efficient in situ protection strategy during the total synthesis of (–)-mandelalide A. Nucleophilic attack of epoxide **28** with 2-lithio-2-TBS-1,3-dithiane **27** generated alkoxide anion **29**. The in situ protection of **29** was realized via a solvent-triggered Brook rearrangement (Scheme 13.3).[10] The resulting carbon nucleophile **30**, in the same flask, reacted with (S)-epichlorohydrin giving an electrophilic terminal epoxide **31**. The addition of vinylmagnesium bromide and copper iodide completed the construction of the requisite advanced intermediate homoallylic alcohol (–)-**32**.

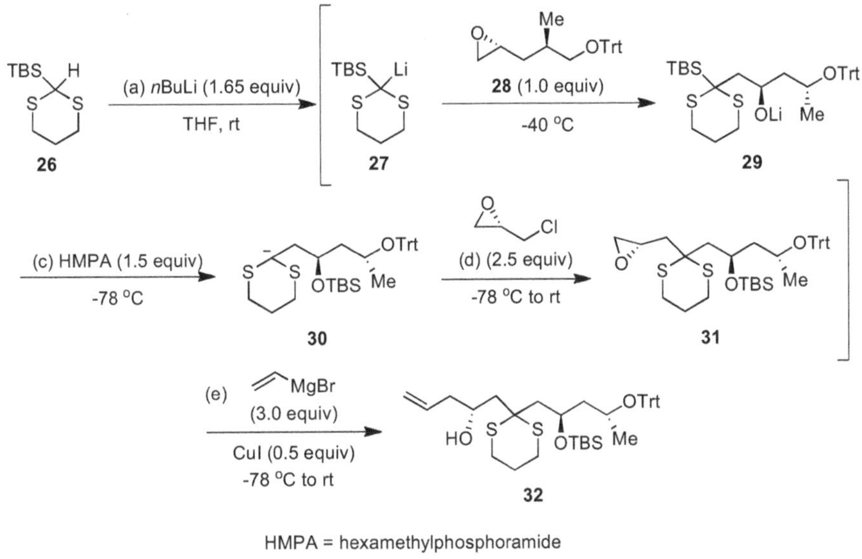

HMPA = hexamethylphosphoramide

SCHEME 13.3

13.2 PROTECTION OF AMINO GROUP

The presence of amino groups usually impedes reactions of interest due to their inherent basic and nucleophilic properties. Therefore, amino groups are often required to be protected by various protecting groups including Boc, acetyl, and 2,5-hexanedione. In addition, the formation of imines by reactions with aldehydes or ketones is also used as a protection approach.[11]

13.2.1 PROTECTION WITH (4-NITROPHENYL)SULFONYL GROUP

Friedel–Crafts alkylation of unprotected indole **33** (Equation 13.7)[12] rendered three major side products, dimer **36**, trimer **37**, and acid **38**, due primarily to the high reactivity and poor selectivity of **33**.

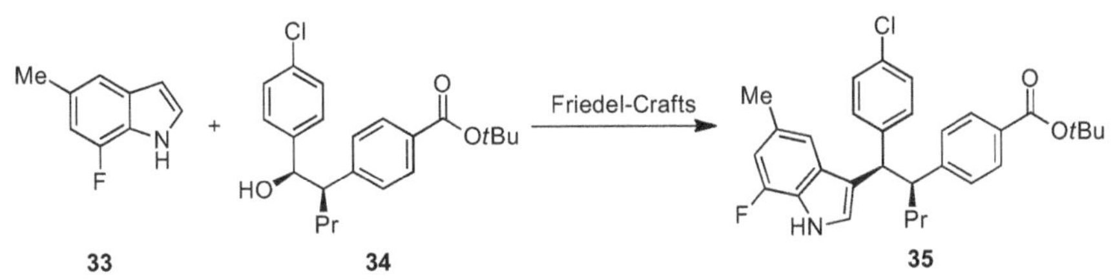

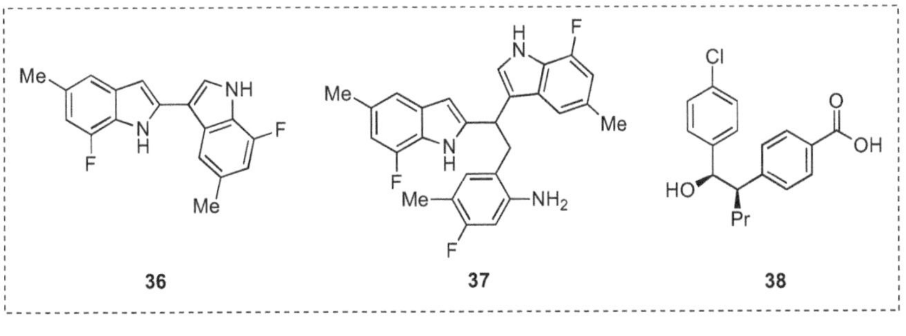

To modify the reactivity of **33**, the nitrogen in **33** was masked with (4-nitrophenyl)sulfonyl (nosyl) protecting group. In the event, the Friedel–Crafts alkylation of **39** with methanesulfonic acid catalyst using TFA as the solvent gave optimal diastereoselectivity and yield (Equation 13.8).

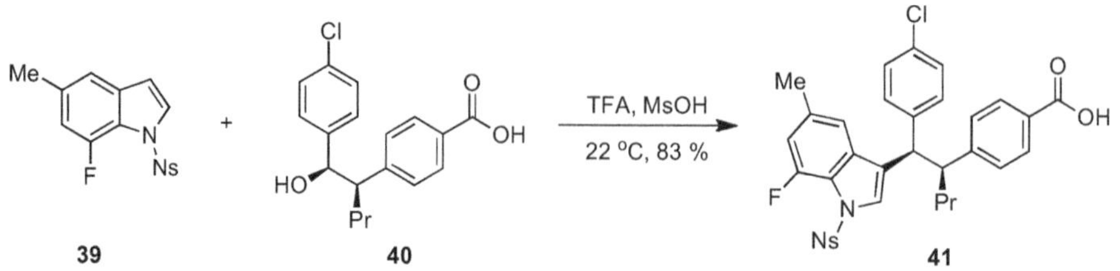

13.2.2 PROTECTION WITH 2,5-HEXANEDIONE

During process research and development, the selection of an appropriate protecting group to mask reactive sites can render the success of the downstream chemistry. The impact of the amino group protection of (*R*)-2-amino-5-oxiranylpyridine **46** on the subsequent epoxide ring-opening was significant in terms of the regioselectivity and the optical purity of the ring-opening product (Scheme 13.4).[13] The coupling reaction (Route a, Scheme 13.4) of **42** (with acetyl protecting group) with alkylamine **44** produced the desired β-opening product that was contaminated with a significant amount of the undesired α-opening side product plus extensive racemization.

Switching the protecting group from acetyl to 2,5-hexanedione suppressed side reactions, providing the desired ring-opening product **45** (Route b, Scheme 13.4) without erosion of the chirality.

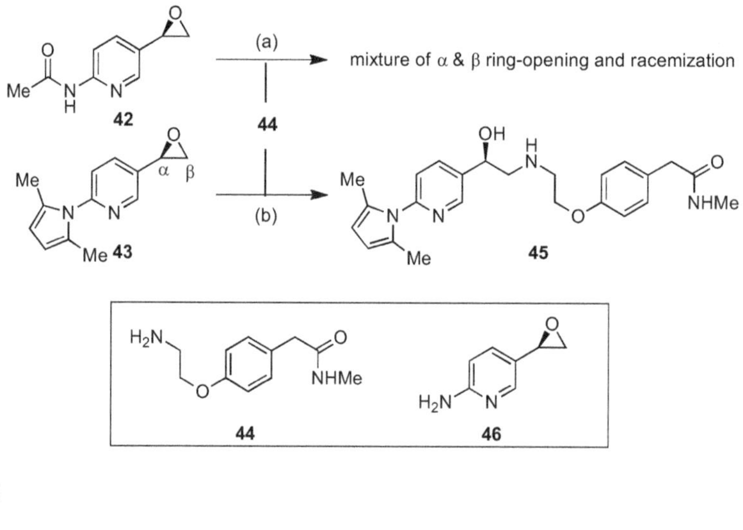

SCHEME 13.4

13.2.3 PROTECTION WITH ALDEHYDE

Selective protection of the amino group in the presence of the hydroxyl group can be achieved by a reaction with aldehydes to form the corresponding imine. For instance, the amino group in tyramine **48** could be conveniently protected by condensation with aldehyde **47** to form Schiff base **49**. Deprotection was accomplished by treatment of **50** with aqueous HCl to obtain the amine product as the hydrochloride salt in 91% yield (Scheme 13.5).[14]

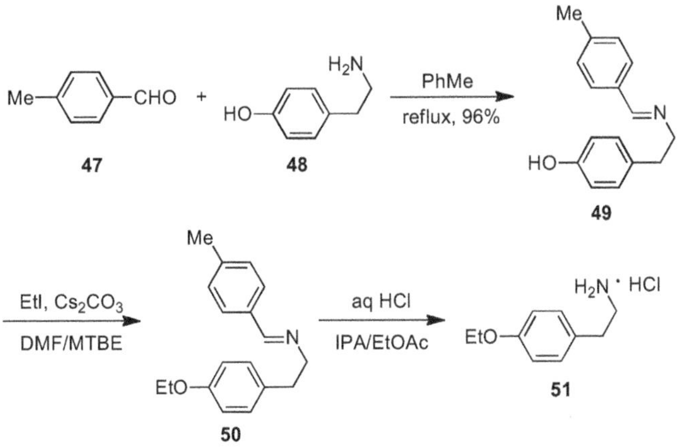

SCHEME 13.5

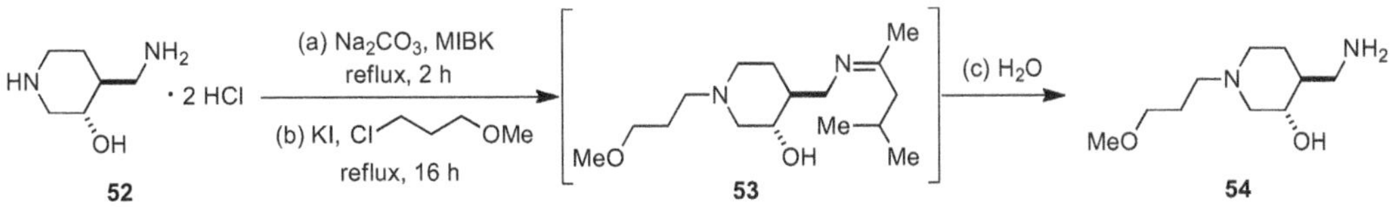

SCHEME 13.6

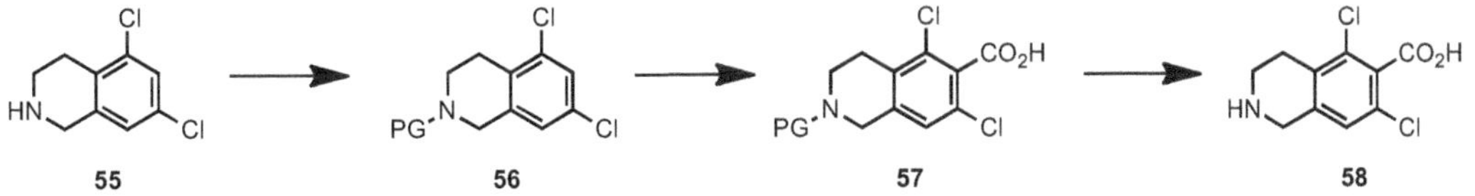

SCHEME 13.7

13.2.4 PROTECTION WITH 4-METHYL-2-PENTANONE

Selective deactivation of the primary amino group in the presence of secondary amine functionality via the formation of imine provides an efficient and cost-effective protection method in organic synthesis. Such an approach lent itself readily in the selective alkylation of the secondary amino group in the presence of a protected primary amino group (Scheme 13.6).[15] In this approach, methyl isobutyl ketone (MIBK), as an inexpensive solvent, was used not only as the reaction medium but also as a temporary protecting group for the protection of the primary amino group in **52**. Thus, heating **52** in MIBK with sodium carbonate in the presence of 1-chloro-3-methoxypropane led selectively to the N-alkylated imine **53**. Hydrolysis of **53**, in turn, provided the free primary amine **54** in good yield and high purity.

Such in situ protection strategy has been nicely extended to the selective alkylation and acylation of other secondary amines in the presence of primary amino groups.[15]

13.2.5 PROTECTION WITH METHYLENE GROUP

The synthesis of 5,7-dichlorotetrahydroisoquinoline-6-carboxylic acid **58** from tetrahydroisoquinoline **55** generally requires multiple stages involving the protection of the amino group in **55** followed by carboxylation and deprotection (Scheme 13.7), which makes the process inefficient.

To avoid the undesired protection/deprotection reaction sequence, a practical approach was developed by converting the starting material **55a** into aminoacetal **59** whose methylene bridge serves as the protecting group. Subsequent lithiation, quenching with CO_2, and hydrolysis provided the desired product **58a** in 88.5% yield (Scheme 13.8).[16]

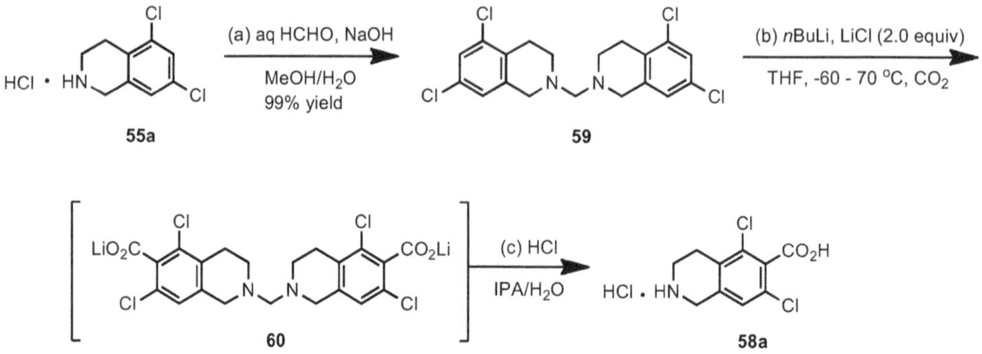

SCHEME 13.8

13.2.6   Protection with *tert*-Butyloxycarbonyl Group

The selection of an appropriate protection group can improve a reaction's efficiency. *tert*-Butyloxycarbonyl (Boc) protecting group is ubiquitous in organic synthesis. A step-economic process was demonstrated in the transformation of Boc-protected amine **61** to benzimidazolone **63** (Scheme 13.9).[17]

Upon the reduction of the nitro group, the resulting Boc-protected diamine intermediate **62** was cyclized without using CDI as the carbonyl source to give the desired benzimidazolone **63** in 85% yield (over two steps).

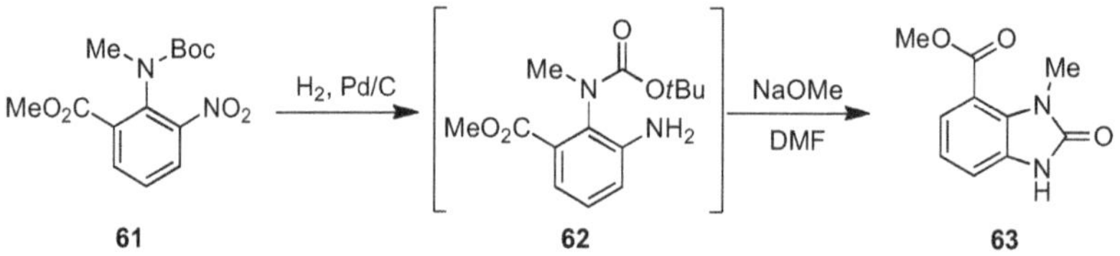

SCHEME 13.9

13.2.7   Protection with 2,2,6,6-Tetramethylpiperidin-1-yloxycarbonyl Group

2,2,6,6-Tetramethylpiperidin-1-yloxycarbonyl (Tempoc) was identified as an effective protecting group for primary, secondary, and heterocyclic amines.[18] As shown in Scheme 13.10, the Tempoc protecting group has excellent orthogonal properties compared to Boc- and Cbz-carbamates, being resistant to acidic and hydrogenolysis conditions.

Tempoc can be selectively removed under Cu-catalyzed conditions (10 mol% of $CuCl_2$, rt, 12 h).

The *p*-nitrophenol tempo carbonate (NPTC) can be conveniently prepared in a two-step process: reduction of Tempo with sodium ascorbate and *O*-acylation of the resulting hydroxylamine intermediate with *p*-nitrophenol chloroformate (Equation 13.9).

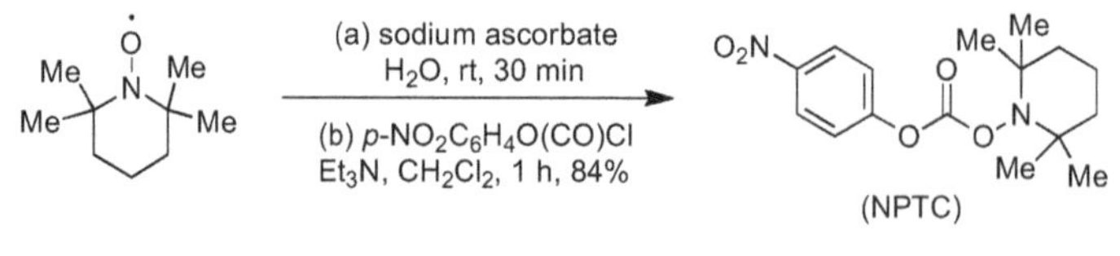

(13.9)

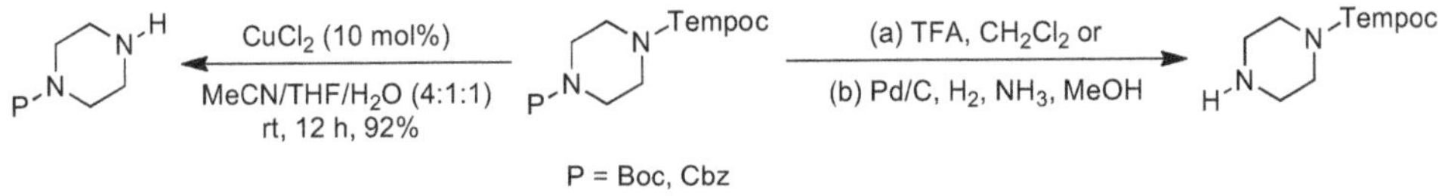

SCHEME 13.10

13.2.8 PROTECTION WITH TRIMETHYLSILYL GROUP

For a compound with a resonance structure, an indirect protection approach may be feasible by masking the associated functional group to suppress the side product formation. Two side products **66** and **67** were observed during the *N*-alkylation of unprotected quinazolinedione **64** with ethyl bromoacetate in the presence of potassium carbonate as a base in DMF (Equation 13.10).[19]

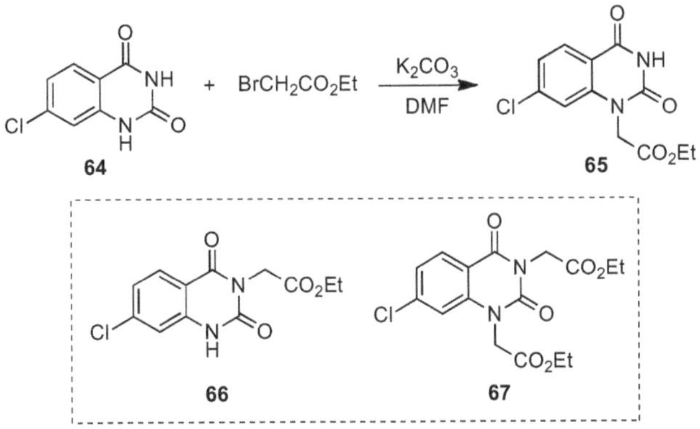

$$(13.10)$$

Efforts in limiting the side product formation led to an indirect protection approach by converting **64** to bis(trimethylsilyloxy)-pyrimidine intermediate **68** whose reaction with ethyl bromoacetate on heating to afford selectively the desired 1-ethoxycarbonylmethyl derivative **65** upon removal of the TMS group (Scheme 13.11).

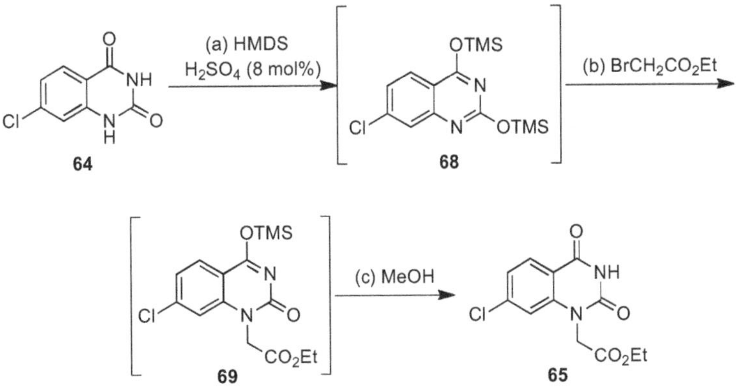

SCHEME 13.11

Procedure

Stage (a)

- Charge 0.73 kg of sulfuric acid carefully to a suspension of **64** (18.25 kg) in toluene (54.8 L) and hexamethyldisilazane (HMDS, 36.0 kg).

- Heat at reflux for 8 h until a clear solution was obtained.
- Distill under reduced pressure to remove toluene and excess HMDS.

Stage (b)
- Charge 36.5 kg of ethyl bromoacetate to the residue.
- Heat the resulting mixture at 115–130 °C for 3 h.

Stage (c)
- Add 36.4 L of 1,4-dioxane at 100 °C.
- Add MeOH (54.8 L) at 70 °C.
- Cool the resulting suspension below 5 °C.
- Filter, wash with MeOH (20 L), water (90 L), and dry under vacuum.
- Yield of **65**: 25.4 kg (96.8%) as a white powder.

REMARK

Silylation with trimethylchlorosilane and triethylamine in dichloromethane proceeded well, but the removal of triethylamine hydrochloride by filtration was unfavorable at scale.

13.3 PROTECTION OF CARBOXYLIC ACID

Because of the acidity of carboxylic acids, the acid functionality may have detrimental effects on reactions, causing various reaction problems. Therefore, the acid functionality is often converted to neutral species, such as ester. In addition, the use of protecting group can improve the reaction performance, such as selectivity and yield.

As described in Scheme 13.12,[20] the initial bromoacetic acid protection strategy employed *tert*-butyl protecting group to mask the carboxylic acid functionality as the *tert*-butyl ester in order to perform the S_N2 reaction of the bromoacetic acid **70** with allyl alcohol. The resulting ether **71** was used to synthesize alkyl aryl ketone **74** via a sequence involving ester hydrolysis, Weinreb amide formation, and reaction with aryl Grignard reagent.

Albeit straightforward, the synthesis required two additional steps: *t*-butyl ester formation and deprotection, which renders this process not efficient.

Thus, a step-economic synthesis was developed and Scheme 13.13 illustrates the process for the synthesis of **74**. This new process employed the Weinreb amide functional group in **76** acting as the protecting group for the S_N2 ether formation and the resulting Weinreb amide **73** allowed direct coupling with the Grignard reagent. This process furnished **74** in a three-step process, avoiding the protection/deprotection steps.

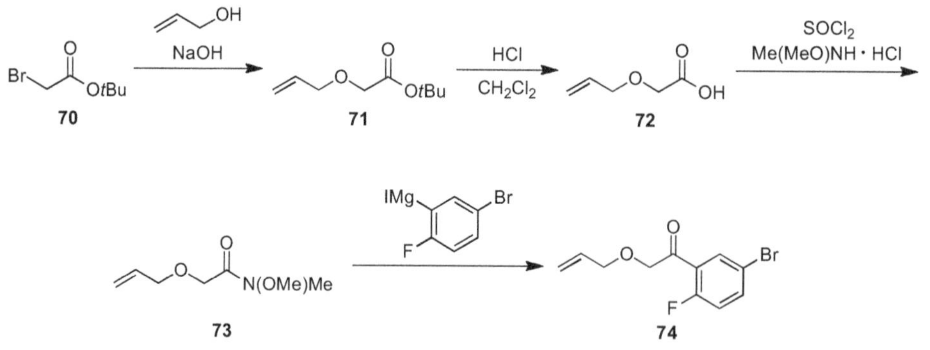

SCHEME 13.12

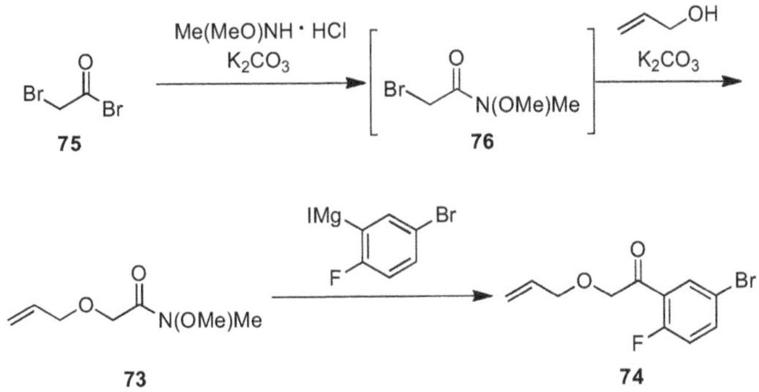

SCHEME 13.13

REMARKS

(a) Carboxylic acids could be protected as ally ester and the deprotection was carried out under Pd-catalyzed conditions.[21]

(b) Saponification of bulky *tert*-butyl ester can be ineffective, while acid-mediated conditions may give low product yields due to decarboxylation. "Anhydrous hydroxide", generated from the reaction of KOtBu with water, could be used to carry out the saponification of a variety of bulky esters in an organic solvent.[22]

13.4 PROTECTION OF ALDEHYDES AND KETONES

Aldehydes and ketones can participate in various chemical reactions, such as addition/elimination with nucleophiles and deprotonation with a base to form enolates. Due to the relatively high reactivity, the carbonyl functionality in aldehydes and ketones has to be converted to less reactive species with an appropriate protecting group during the course of organic synthesis.

13.4.1 PROTECTION OF ALDEHYDE

During the enantioselective cyclopropanation of enals, the carbonyl group in the enals **77** was protected in situ with chiral catalyst, diphenylprolinol silyl ether **79**, to form iminium salt **81** and/or hemiaminal **84** (Scheme 13.14).[23]

In the proposed reaction mechanism (Route I, Scheme 13.14),[23] the benzyl chloride **78**, in the presence of base DIPEA, would act as a nucleophile to react with iminium salt **81** followed by intramolecular cyclopropanation. The reaction may also proceed via an alternative pathway (Route II, Scheme 13.14) in which **78** serves as an electrophile to reaction with hemiaminal **84**. The resulting aziridinium **85** would undergo 1,5-hydrogen shift followed by intramolecular cyclopropanation to give intermediate **87**. The diphenylprolinol silyl ether **79** plays a dual role: as a chiral catalyst and as an in situ protecting group.

13.4.2 PROTECTION OF KETONES

13.4.2.1 Protection with Methanol

In order to obviate side reactions associated with the carbonyl group in cyclobutanone **88** during the cross-coupling reaction with pyridazinone **90**, the cyclobutanone **88** was effectively protected as the dimethyl ketal by the addition of MeOH to the acidic reaction mixture of **88** (Scheme 13.15).[24]

The subsequent copper-catalyzed C–N cross-coupling reaction of the resulting ketal **89** with **90** in the presence of N,N'-dimethylethylene diamine (DMEDA) afforded a crystalline product **91** in 84% yield with >98% purity.

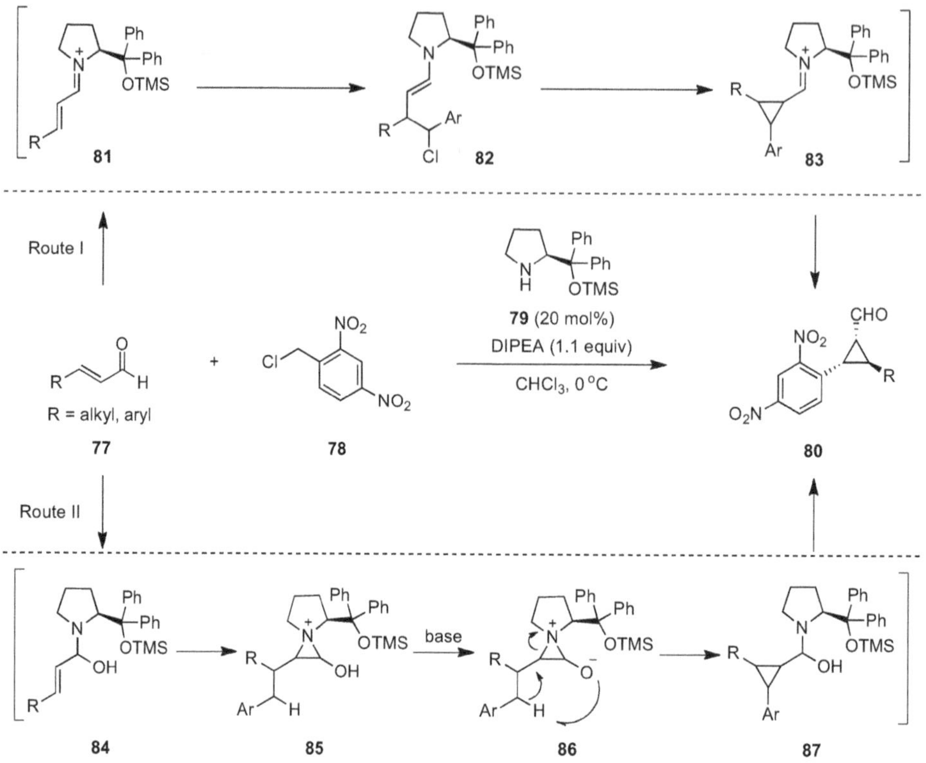

SCHEME 13.14

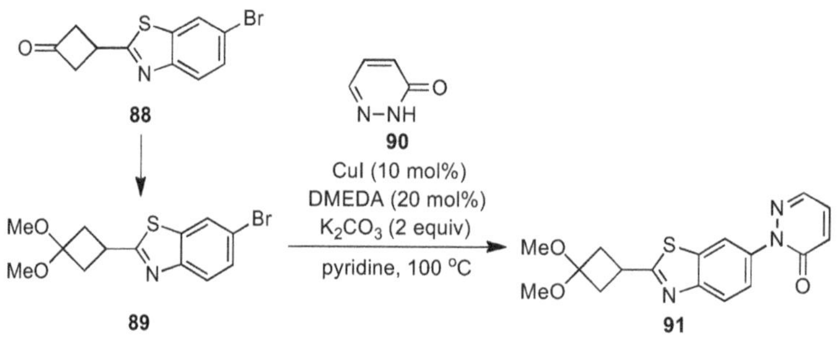

SCHEME 13.15

13.4.2.2   Protection with Ethylene Glycol

In addition to improvements in reaction selectivity, the protecting group approach is able to solve challenging reaction problems. For example, the synthesis of tetrahydro-azaindole **94** via Pictet–Spengler reaction of ketone **92** failed, due presumably to the involvement of a putative intermediate **96** (Scheme 13.16).[25]

Thus, conversion of the carbonyl carbon (sp^2) to sp^3 hybridization would obviate the formation of the intermediate **96**. Indeed, the protection of **92** to dioxolane **93** with ethylene glycol enabled a smooth cyclization under the standard Pictet–Spengler condition to give **94**.

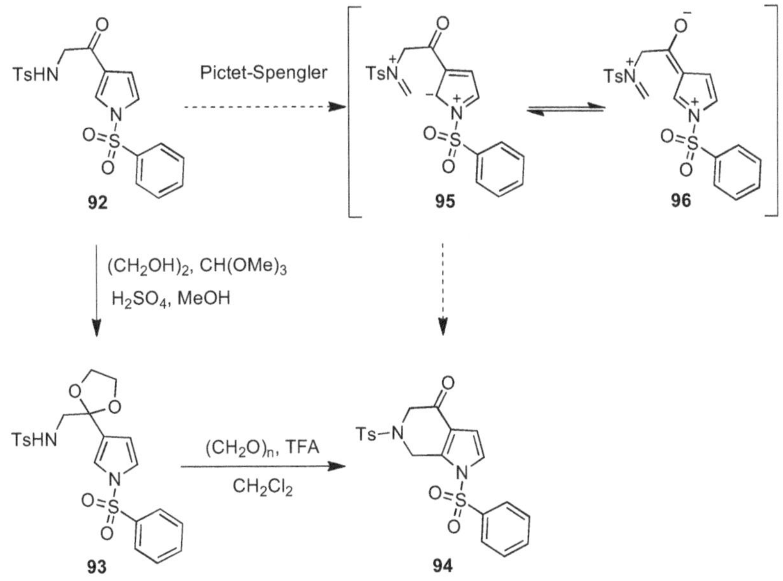

SCHEME 13.16

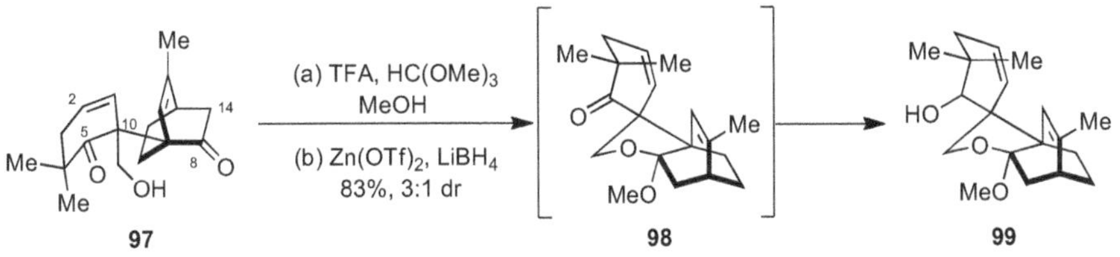

SCHEME 13.17

13.4.2.3 Using Internal Protection Approach

The selective reduction of the more sterically hindered C-5 carbonyl group in the presence of the C-8 ketone in hydroxyl diketone **97** was adopted as an internal protection approach. The reaction was accomplished by a selective in situ formation of ketal intermediate **98** (TFA, CH(OMe)$_3$), followed by reduction with LiBH$_4$/Zn(OTf)$_2$ to furnish the desired alcohol **99** in 83% yield (Scheme 13.17).[26]

13.5 PROTECTION OF ACETYLENE

Acetylene is often protected because of the acidity of the acetylenic hydrogen and its gaseous nature. 2-Hydroxypropyl and TMS groups are usually used to mask the acetylene. Using the former protecting group required 4.0 equiv of the Grignard reagent in the Grignard reaction step, while the use of TMS as the protecting group allowed reducing the amount of the Grignard reagent from 4.0 equiv to 1.5 equiv (Scheme 13.18).[27]

In addition, switching the acetylene protecting group from tertiary alcohol in **101a** to TMS-acetylene **101b** reduced the palladium catalyst loading from 6% to 2%.

NOTE

It was difficult to cleave the tertiary alcohol-protecting group in **103a** (P=Me$_2$COH), while the TMS-deprotection could be realized under mild conditions (1 equiv of K$_2$CO$_3$ in MeOH).

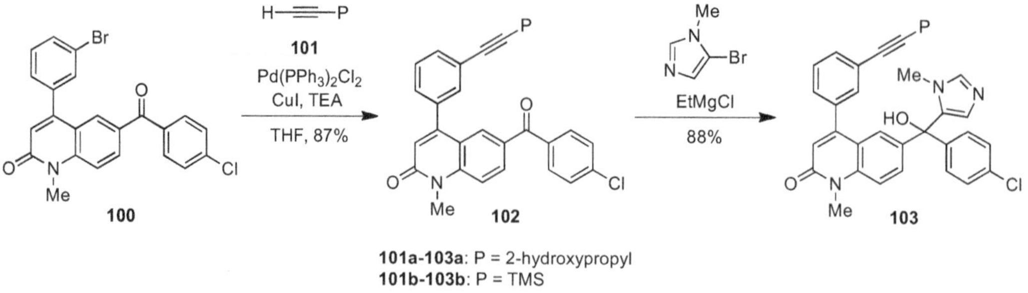

101a-103a: P = 2-hydroxypropyl
101b-103b: P = TMS

SCHEME 13.18

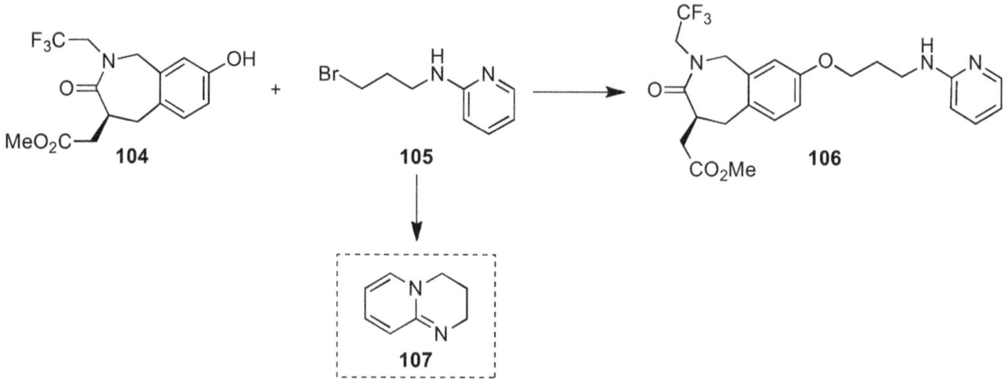

SCHEME 13.19

13.6 UNUSUAL PROTECTING GROUPS

13.6.1 BORON-CONTAINING PROTECTING GROUPS

13.6.1.1 Borane Complex

13.6.1.1.1 Protection of Pyridine Nitrogen

The synthesis of vitronectin inhibitor **106** would require an *O*-alkylation of phenol **104** with bromide **105** (Scheme 13.19).[28] However, an intramolecular cyclization of the bromide **105** would produce 3,4-dihydro-2*H*-pyrido[1,2-*a*]pyrimidine side product **107**.

In order to address the undesired side reaction, the synthesis with a commercially available but expensive 2-chloropyridine *N*-oxide **108** led to bromide **109**, followed by a reaction with **104** and subsequent reduction to provide the desired product **106** (Scheme 13.20). The bromide **109**, however, was thermally unstable, which was a safety concern when scaling up.

Alternatively, the pyridine nitrogen was blocked by converting **105** to its corresponding pyridine borane complex **111**. The *O*-alkylation of **104** with **111** afforded **106** upon removal of the borane under acidic conditions (Scheme13.21).

13.6.1.1.2 Protection of Phosphorus

The boranato group acts as a protecting group to prevent the phosphine group from reactions with oxygen or other electrophiles such as alkyl halides. Phosphine–borane adduct **114** was used as an isolable intermediate and applied in the synthesis of phosphine oxazoline ligands **115** (Scheme 13.22).[29]

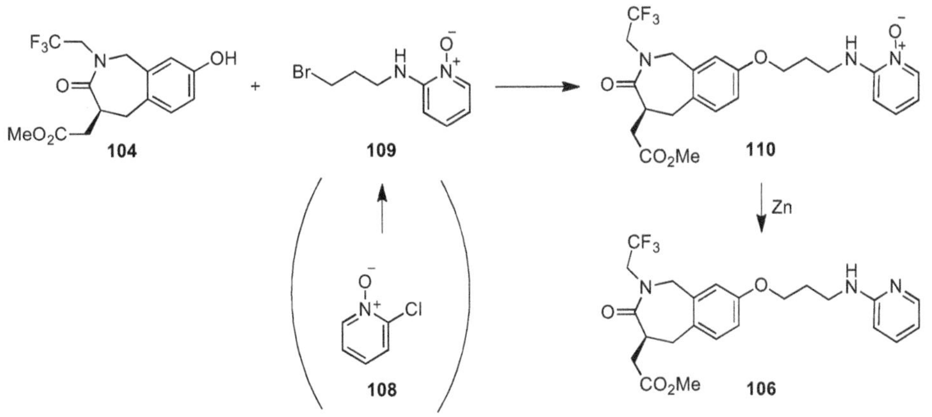

SCHEME 13.20

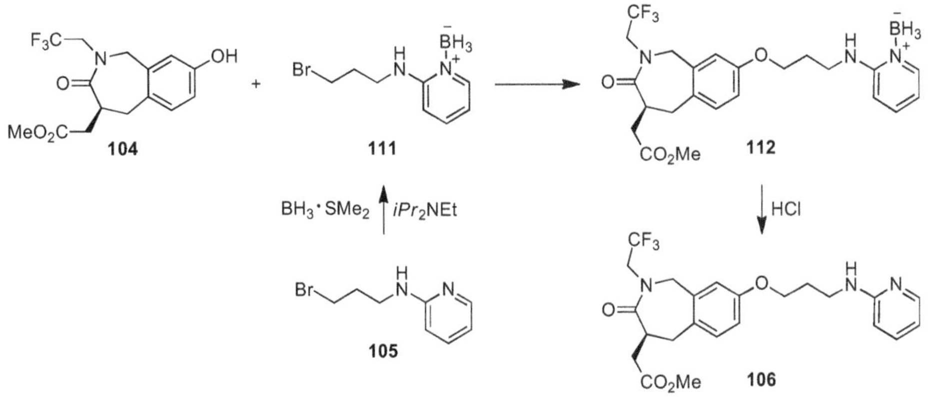

SCHEME 13.21

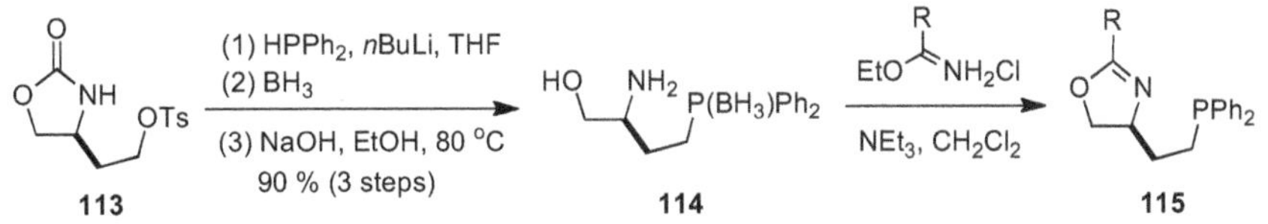

SCHEME 13.22

13.6.1.2 Boronic Acids

13.6.1.2.1 Protection of Diol

Phenylboronic acid was utilized to protect 1,2-diol in the synthesis of nucleoside (Scheme 13.23).[30] The phenylboronic acid in **119** could be conveniently removed by treatment with aqueous citric acid.

13.6.1.2.2 As Directing/Blocking Group for Regioselective Halogenations

Aryl boronic acids can undergo thermal protodeboronation[31] which is one of the most common side reactions in metal-catalyzed cross-coupling reactions. Application of this protodeboronation, however, led to a synthetic protocol, in which the boronic acid and its derivative were used as directing groups or blocking groups to access *ortho-* and *meta-*functionalized phenols.[31]

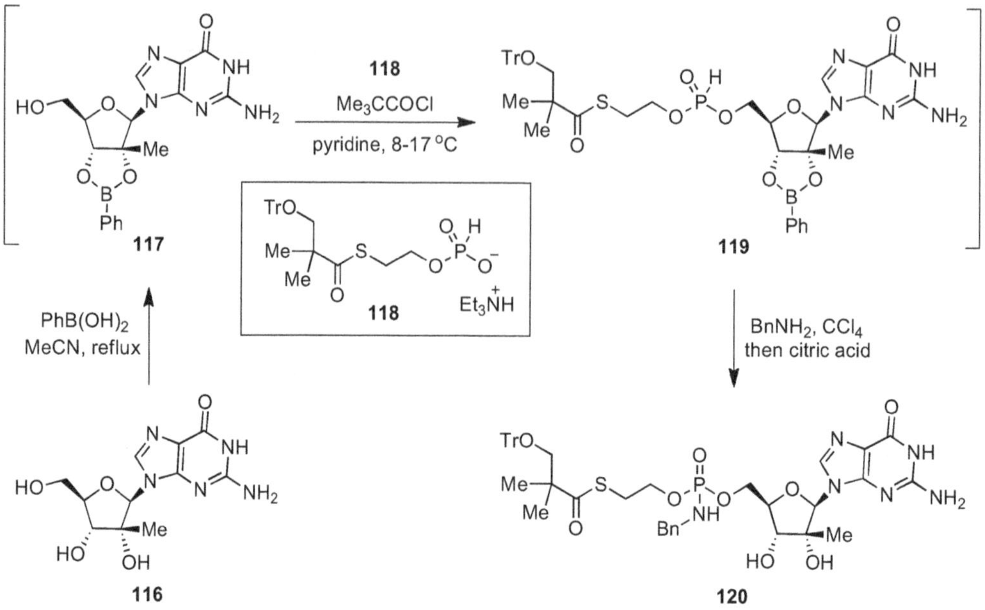

SCHEME 13.23

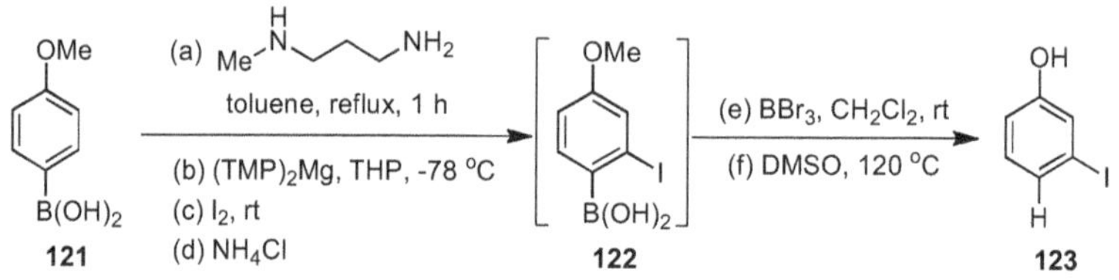

SCHEME 13.24

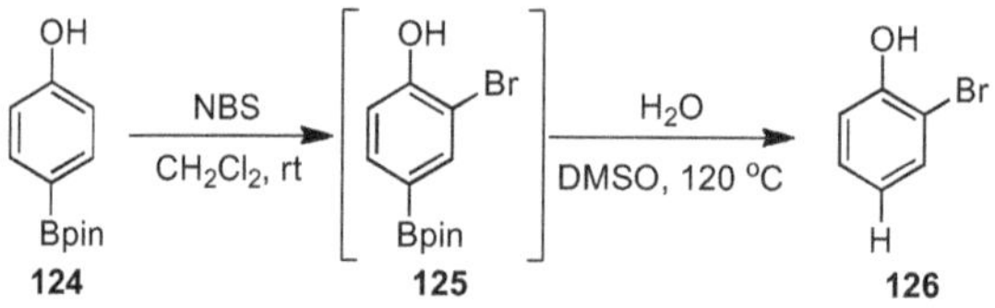

SCHEME 13.25

For instance, the use of boronic acid as a directing group to access *meta*-iodophenol **123** could be realized (Scheme 13.24).

Furthermore, a selective bromination of 4-hydroxyphenyl pinacol boronate **124** with NBS (1 equiv) gave 2-bromophenol **126** in 87% overall yield upon protodeboronation by heating at 120 °C in DMSO (Scheme 13.25). The pinacol boronate serves as a blocking group.

13.6.2   *N*-NITRO PROTECTING GROUP

13.6.2.1   Regioselective Nitration

Traditionally, the synthesis of 2-amino-5-bromo-3-nitro-4-picoline **128** would involve nitration of 2-amino-4-picoline **127** producing a mixture of 3- and 5-nitro isomers in an unfavorable 1:4 ratio.

The following bromination of the isolated desired minor isomer afforded **128** in less than 10% overall yield (Equation 13.11).

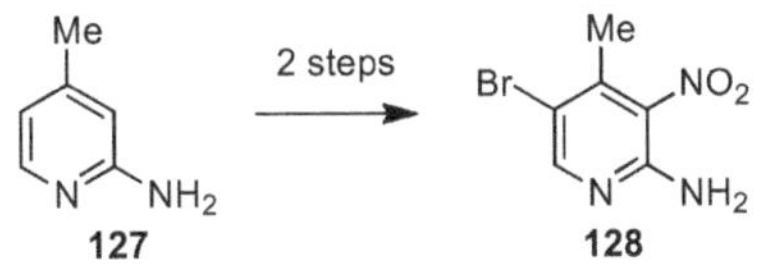

$$\text{(13.11)}$$

An efficient, atom-economic, and regioselective process was developed, involving a rearrangement of 2-nitramino-picoline intermediate **130** (Scheme 13.26).[32]

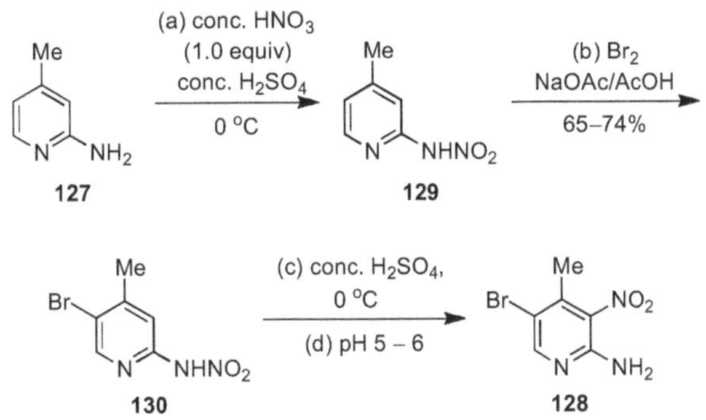

SCHEME 13.26

Protection of the amino group in **127** with the nitro group provides the necessary electronic and steric environment for the subsequent bromination. As expected, the regioselective bromination of 2-nitramino-4-picoline **129** furnished the desired brominated intermediate **130** in good yield. The following acid-catalyzed rearrangement installed the nitro group at the C3 position, leading to the desired 2-amino-5-bromo-3-nitro-4-picoline product **128**.

13.6.2.2 Activation of Aniline

A direct *N*-alkylation of the amino group of compound **131** with bromide **132** failed to provide the desired *N*-alkylated product **133** (Scheme 13.27).[33] Attempts to activate the aniline nitrogen via acetylation proved to be unsuccessful. The nitro-protecting group was applied to activate the aniline nitrogen, giving an intermediate **134** whose subsequent *N*-alkylation afforded the *N*-alkylated derivative **135**.

The reduction of the nitro group in **135** was accompanied by the removal of the *N*-nitro group, providing the desired product **136** in 81% yield after isolation as its hydrochloride salt.

13.6.3 Halogen as Protecting Group

13.6.3.1 Bromine Protecting Group

13.6.3.1.1 Selective Halogenation

Owing to the reversibility of aromatic bromination, bromine was used as a strategic protecting group in the synthesis of compounds that are normally considered difficult to prepare. Schemes (13.28 and 13.29)[34] show two examples of using bromine as the protecting group in the preparation of *N*-(3-bromophenyl)acetamide **139** and 2,6-dichlorophenylamine **142**.

13.6.3.1.2 Nitration

Installation of an amino group onto 1,4-benzodioxan-5-carboxylic acid **143** was achieved in three steps involving bromination, nitration, and hydrogenation/debromination (Scheme 13.30).[35]

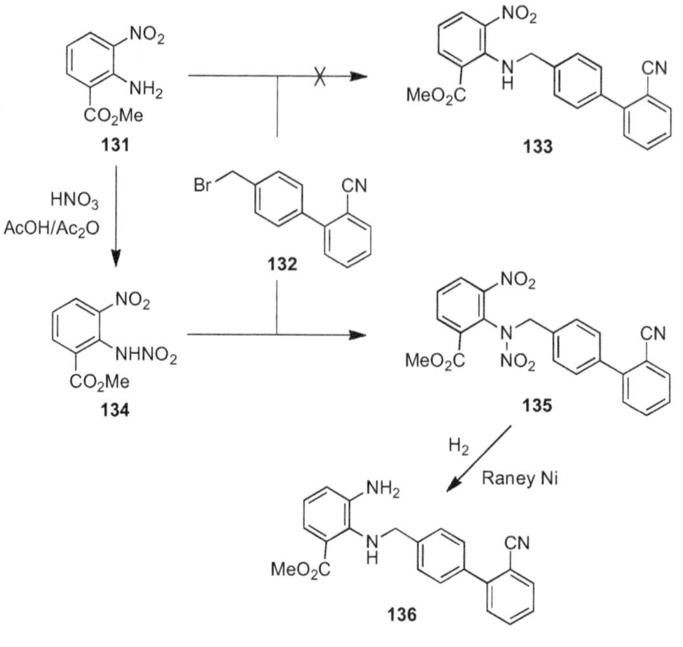

SCHEME 13.27

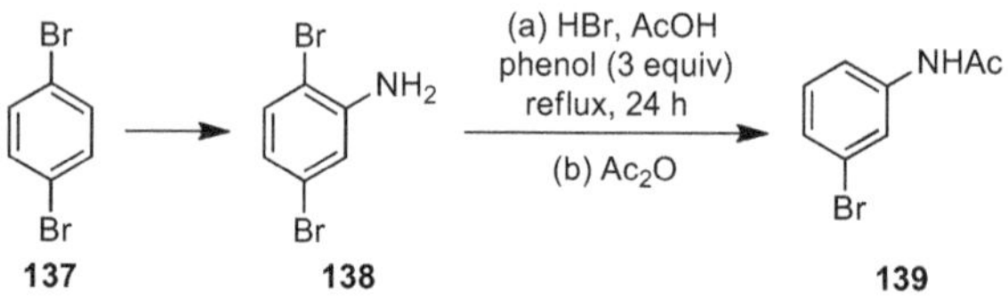

SCHEME 13.28

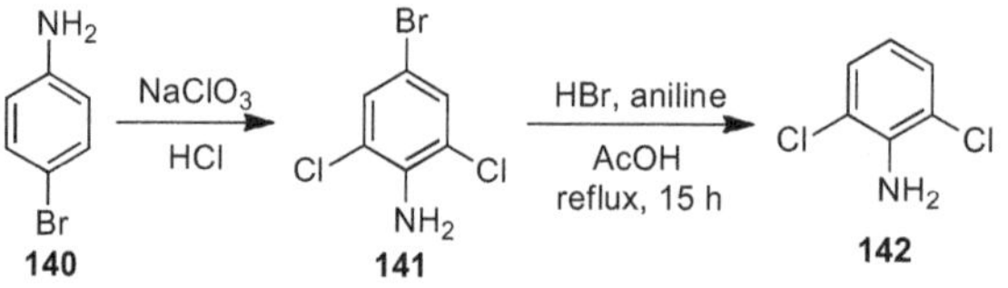

SCHEME 13.29

The Bromo substituents were utilized as the protecting group which set the stage for the subsequent nitration at 8-position (Step (b), Scheme 13.30). The following catalytic reduction of the nitro group and catalytic debromination were carried out in a one-pot fashion furnishing the desired product **144**.

13.6.3.1.3 Regioselective Alkylation

Direct alkylation of commercially available indazole derivative **147** with 4-chloro-2-(trifluoromethyl) benzyl bromide **149** suffers from poor regioselectivity, resulting in a mixture of N1/N2 alkylated products in a ratio of 1:1.

To address this issue, bromine was selected as the protecting group and installed on the 3-position in the indazole ring by a regioselective bromination with NBS (Scheme 13.31).[36] The subsequent N-alkylation with **149** provided the desired product **150** in a good yield. Under these conditions,

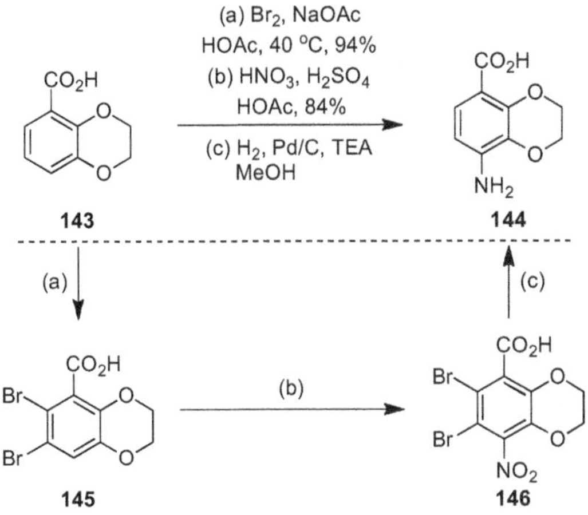

SCHEME 13.30

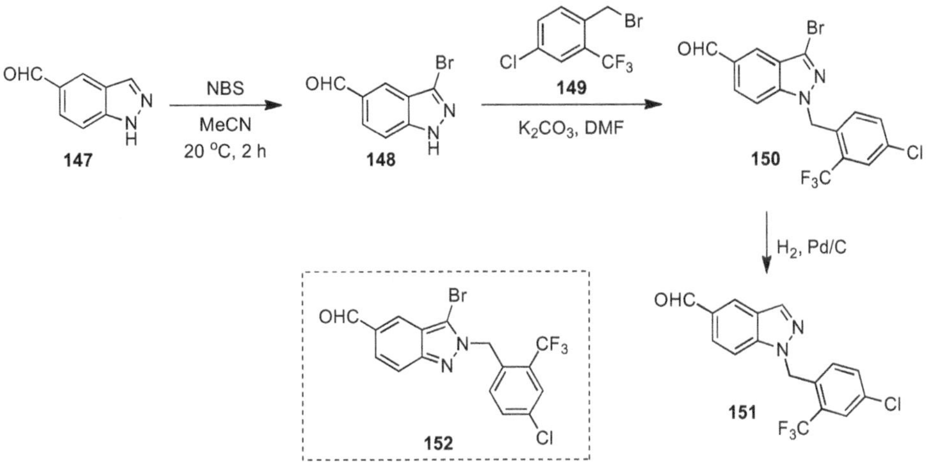

SCHEME 13.31

only 6–8% of the undesired side product **152** was generated, which was removed completely by crystallization. Ultimately, the bromine-protecting group was readily removed by catalytic debromination with Pd/C as the catalyst.

13.6.3.2   Chlorine Protecting Group

Analogously, the less reactive chlorine was selected as the protecting group in the regioselective nitration of **153** (Scheme 13.32).[37]

Procedure

Stage (a)
- Charge 9.0 kg (40.1 mol) of **153** into an inverted vessel, followed by adding 4.250 kg (42.0 mol) of KNO_3.
- Charge 45.4 L of TFA at –10 °C over 1 min.
- Warm slowly the batch to 22 °C over 3.5 h (A rapid exotherm to 36 °C was observed).
- Cool to 22 °C.

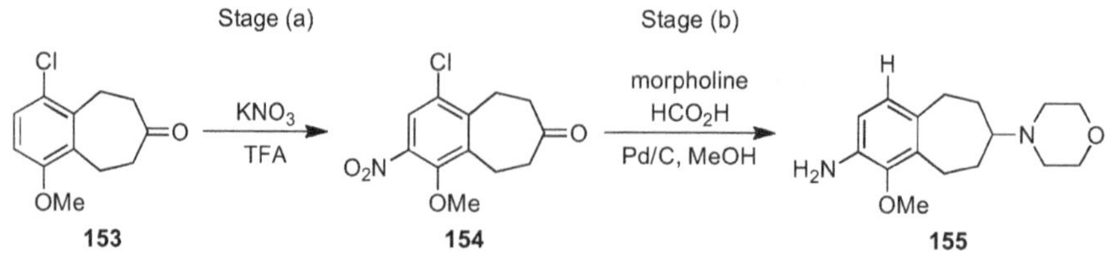

Stage (a)   Stage (b)

153   **154**   **155**

SCHEME 13.32

- Add water (135.1 L) and stir the resulting slurry for 2 h.
- Filter, wash with water (45.0 L), and dry.
- Yield of **154**: 9.87 kg (91.4%).

Stage (b)

- Charge 5.40 kg (117 mol, 8.8 equiv) of formic acid to a solution of morpholine (12.10 kg, 139 mol, 10.5 equiv) in MeOH (24.8 kg) at <40 °C over 10 min.
- Charge this solution to a slurry of **154** (3.90 kg, 91 wt%, 13.2 mol) and 10% Pd/C (4.45 kg, 50% wet) in MeOH (91.2 kg) at 65 °C over 1.5 h.
- Stir at 60 °C for 1.5 h.
- Cool to 25 °C.
- Filter through Celite 545 (0.8 kg) and wash with EtOAc (30 kg).
- Distill at 23–29 °C under reduced pressure (73–144 mmHg).
- Add EtOAc (60.3 kg) and 8% aqueous solution of $NaHCO_3$ (7.6 gal).
- Separate layers.
- Extract the aqueous layer with EtOAc (30 kg).
- Dry the combined EtOAc layers over Na_2SO_4 (4 kg) and filter.
- Distill to dryness and dry at 50 °C/50 mmHg.
- Yield of **155**: 3.04 kg (85.4%).

NOTE

When the bromine-protecting group was used for the nitration of the anisole derivative, a 2:1:1 mixture of *ortho*, *para*, and *ortho/para* over-nitration products (resulting from the *ipso* substitution) was obtained.

NOTES

1. Greene, T. W.; Wuts, P. G. M. *Protective Groups in Organic Synthesis*, 3rd ed.; John Wiley & Sons, New York, 1999.
2. Momoi, Y.; Okuyama, K.; Toya, H.; Sugimoto, K.; Okano, K.; Tokuyama, H. *Angew. Chem. Int. Ed.* **2014**, *53*, 13215.
3. Ho, W.-B.; Broka, C. *J. Org. Chem.* **2000**, *65*, 6743.
4. Sajiki, H. *Tetrahedron Lett.* **1995**, *36*, 3465.
5. Yoshikawa, N.; Xu, F.; Arredondo, J. D.; Itoh, T. *Org. Process Res. Dev.* **2011**, *15*, 824.
6. Fuller, N. O.; Hubbs, J. L.; Austin, W. F.; Shen, R.; Ives, J.; Osswald, G.; Bronk, B. S. *Org. Process Res. Dev.* **2014**, *18*, 683.
7. Palmer, A. M.; Webel, M.; Scheufler, C.; Haag, D.; Müller, B. *Org. Process Res. Dev.* **2008**, *12*, 1170.
8. Mai, C.-K.; Sammons, M. F.; Sammakia, T. *Angew. Chem. Int. Ed.* **2010**, *49*, 2397.
9. Pan, H.; Tu, S.; Zhang, C. *Org. Process Res. Dev.* **2015**, *19*, 463.
10. Nguyen, M. H.; Imanishi, M.; Kurogi, T.; Smith, Amos, B. III, *J. Am. Chem. Soc.* **2016**, *138*, 3675.
11. Savage, S. A.; Waltermire, R. E.; Campagna, S.; Bordewekar, S.; Toma, J. D. R. *Org. Process Res. Dev.* **2009**, *13*, 510.

12. Chung, J. Y. L.; Steinhuebel, D.; Krska, S. W.; Hartner, F. W.; Cai, C.; Rosen, J.; Mancheno, D. E.; Pei, T.; DiMichele, L.; Ball, R. G.; Chen, C.; Tan, L.; Alorati, A. D.; Brewer, S. E.; Scott, J. P. *Org. Process Res. Dev.* **2012**, *16*, 1832.

13. Scott, R. W.; Fox, D. E.; Wong, J. W.; Burns, M. P. *Org. Process Res. Dev.* **2004**, *8*, 587.

14. Magnus, N. A.; Astleford, B. A.; Brennan, J.; Stout, J. R.; Tharp-Taylor, R. W. *Org. Process Res. Dev.* **2009**, *13*, 280.

15. Laduron, F.; Tamborowski, V.; Moens, L.; Horváth, A.; Smaele, D. D.; Leurs, S. *Org. Process Res. Dev.* **2005**, *9*, 102.

16. Xu, W.; Gong, X.; Odilov, A.; Hu, T.; Jiang, X.; Zhu, F.; Guo, S.; Jiang, D.; Wu, M.; Shen, J. *Org. Process Res. Dev.* **2021**, *25*, 2447.

17. Sawai, Y.; Yabe, O.; Nakaoka, K.; Ikemoto, T. *Org. Process Res. Dev.* **2017**, *21*, 222.

18. Lizza, J. R.; Bremerich, M.; McCabe, S. R.; Wipf, P. *Org. Lett.* **2018**, *20*, 6760.

19. Goto, S.; Tsuboi, H.; Kanoda, M.; Mukai, K.; Kagara, K. *Org. Process Res. Dev.* **2003**, *7*, 700.

20. Kolis, S. P.; Hansen, M. M.; Arslantas, E.; Brändli, L.; Buser, J.; DeBaillie, A. C.; Frederick, A. L.; Hoard, D. W.; Hollister, A.; Huber, D.; Kull, T.; Linder, R. J.; Martin, T. J.; Richey, R. N.; Stutz, A.; Waibel, M.; Ward, J. A.; Zamfir, A. *Org. Process Res. Dev.* **2015**, *19*, 1203.

21. Takao, K.; Noguchi, S.; Sakamoto, S.; Kimura, M.; Yoshida, K.; Tadano, K. *J. Am. Chem. Soc.* **2015**, *137*, 15971.

22. (a) Gassman, P. G.; Schenk, W. N. *J. Org. Chem.* **1977**, *42*, 918. (b) Ku, Y.-Y.; Chan, V. S.; Christesen, A.; Grieme, T.; Mulhern, M.; Pu, Y.-M.; Wendt, M. D. *J. Org. Chem.* **2019**, *84*, 4814.

23. Meazza, M.; Ashe, M.; Shin, H. Y.; Yang, H. S.; Mazzanti, A.; Yang, J. W.; Rios, R. *J. Org. Chem.* **2016**, *81*, 3488.

24. Kallemeyn, J. M.; Ku, Y.-Y.; Mulhern, M. M.; Bishop, R.; Pal, A.; Jacob, L. *Org. Process Res. Dev.* **2014**, *18*, 191.

25. Chen, K.; Risatti, C.; Bultman, M.; Soumeillant, M.; Simpson, J.; Zheng, B.; Fanfair, D.; Mahoney, M.; Mudryk, B.; Fox, R. J.; Hsiao, Y.; Murugesan, S.; Conlon, D. A.; Buono, F. G.; Eastgate, M. D. *J. Org. Chem.* **2014**, *79*, 8757.

26. Cernijenko, A.; Risgaard, R.; Baran, P. S. *J. Am. Chem. Soc.* **2016**, *138*, 9425.

27. Andresen, B. M.; Couturier, M.; Cronin, B.; D'Occhio, M.; Ewing, M. D.; Guinn, M.; Hawkins, J. M.; Jasys, V. J.; LaGreca, S. D.; Lyssikatos, J. P.; Moraski, G.; Ng, K.; Raggon, J. W.; Stewart, A. M.; Tickner, D. L.; Tucker, J. L.; Urban, F. J.; Vazquez, E.; Wei, L. *Org. Process Res. Dev.* **2004**, *8*, 643.

28. Zajac, M. A. *J. Org. Chem.* **2008**, *73*, 6899.

29. (a) Hou, D.-R.; Reibenspies, J. H.; Burgess, K. *J. Org. Chem.* **2001**, *66*, 206. (b) Porte, A. M.; Reibenspies, J.; Burgess, K. *J. Am. Chem. Soc.* **1998**, *120*, 9180.

30. (a) Mayes, B. A.; Arumugasamy, J.; Baloglu, E.; Bauer, D.; Becker, A.; Chaudhuri, N.; Latham, G. M.; Li, J.; Mathieu, S.; McGarry, F. P.; Rosinovsky, E.; Stewart, A.; Trochet, C.; Wang, J.; Moussa, A. *Org. Process Res. Dev.* **2014**, *18*, 717. (b) Mayes, B. A.; Wang, J.; Arumugasamy, J.; Arunachalam, K.; Baloglu, E.; Bauer, D.; Becker, A.; Chaudhuri, N.; Glynn, R.; Latham, G. M.; Li, J.; Lim, J.; Liu, J.; Mathieu, S.; McGarry, F. P.; Rosinovsky, E.; Soret, A. F.; Stewart, A.; Moussa, A. *Org. Process Res. Dev.* **2015**, *19*, 520.

31. (a) Lee, C.-Y.; Ahn, S.-J.; Cheon, C.-H. *J. Org. Chem.* **2013**, *78*, 12154. (b) Ahn, S.-J.; Kim, N.-K.; Lee, C.-Y.; Cheon, C.-H. *J. Org. Chem.* **2014**, *79*, 7277.

32. Bhattacharya, A.; Purohit, V. C.; Deshpande, P.; Pullockaran, Grosso, J. A.; DiMarco, J. D.; Gougoutas, J. Z. *Org. Process Res. Dev.* **2007**, *11*, 885.

33. Porcs-Makkay, M.; Mezei, T.; Simig, G. *Org. Process Res. Dev.* **2007**, *11*, 490.

34. (a) Choi, H. Y.; Chi, D. Y. *J. Am. Chem. Soc.* **2001**, *123*, 9202. (b) Effenberger, F. *Angew. Chem. Int. Ed.* **2002**, *41*, 1699.

35. Lienard, P.; Gradoz, P.; Greciet, H.; Jegham, S.; Legroux, D. *Org. Process Res. Dev.* **2017**, *21*, 18.

36. Li, X.; Russell, R. K.; Spink, J.; Ballentine, S.; Teleha, C.; Branum, S.; Wells, K.; Beauchamp, D.; Patch, R.; Huang, H.; Player, M.; Murray, W. *Org. Process Res. Dev.* **2014**, *18*, 321.

37. Allwein, S. P.; Roemmele, R. C.; Haley, J. J. Jr.; Mowrey, D. R.; Petrillo, D. E.; Reif, J. J.; Gingrich, D. E.; Bakale, R. P. *Org. Process Res. Dev.* **2012**, *16*, 148.

14 Telescope Approach

Traditional laboratory chemical synthesis is operated in a stepwise manner to accomplish multi-step sequences. The sequence is interrupted after each chemical transformation for necessary isolation and purification prior to the subsequent transformation. This "stop-and-go" approach[1] is plagued with poor step economy. Telescoping two or more steps into a one-pot operation is one of the most common tools in process development used by process chemists. A telescoped process has several advantages including:

- To reduce the number of processing solvents.
- To reduce the number of unit operations.
- To improve process efficiency.
- To limit exposure to toxic reaction intermediate(s).
- To avoid isolation of oil intermediate(s).
- To reduce the amount of waste.

A review article[2] highlights the current telescoping techniques including multi-component reaction (MCR), cascade reaction, and tandem reaction.

14.1 IMPROVING PROCESS SAFETY

The telescoping process is a powerful method in chemical process development that enables the integration of two or more reactions into a one-pot operation without the isolation of intermediate products. Handling hazardous or toxic intermediates during isolation/purification on a production scale may pose process safety concerns. Therefore, a telescoping process appears to be an ideal approach to address these issues by limiting exposure to hazardous and toxic materials.

14.1.1 CHLOROKETONE INTERMEDIATE

The preparation of 6-chloro-5-(2-chloroethyl)oxindole 4 in the synthesis for ziprasidone involves a hazardous chloroketone intermediate 3. Chloroketone 3 is a skin sensitizer, manufacturing and handling of this intermediate on a large scale is potentially hazardous. In order to mitigate the exposure to 3, a telescoped process was developed (Scheme 14.1).[3] Using the telescoped process, clean and complete conversion of the ketone 3 to the desired product 4 was observed by the addition of tetramethyl disiloxane (TMDS) (2.0 equiv) to the mixture of 3 with only <1% dehalogenated side product 5.

14.1.2 LACHRYMATORY CHLOROMETHACRYATE INTERMEDIATE

Analogously, the telescoping approach was employed to avoid exposure to lachrymatory chloromethacrylate 7 during the synthesis of chiral amine 8, an intermediate for the synthesis of sampatrilat (Scheme 14.2).[4] Sampartrilat was a drug candidate for the treatment of chronic heart failure.

14.1.3 CHLOROMETHYL BENZOIMIDAZOLE

Scheme 14.3[5] described a telescoped process in order to avoid exposure to mutagenic intermediate 11.

DOI: 10.1201/9781003288411-14

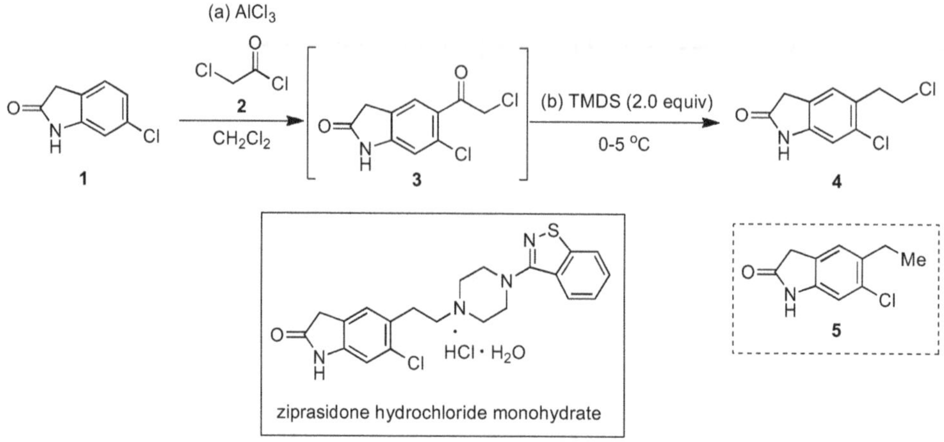

SCHEME 14.1

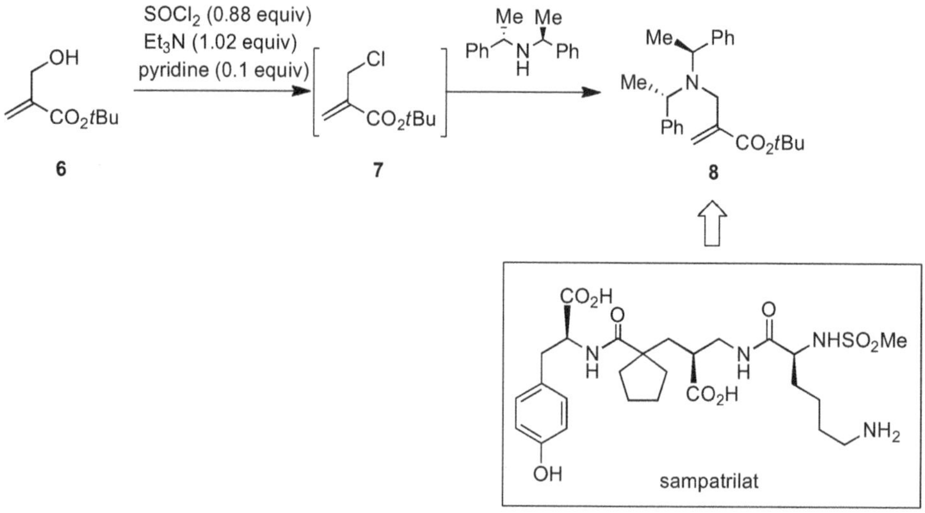

SCHEME 14.2

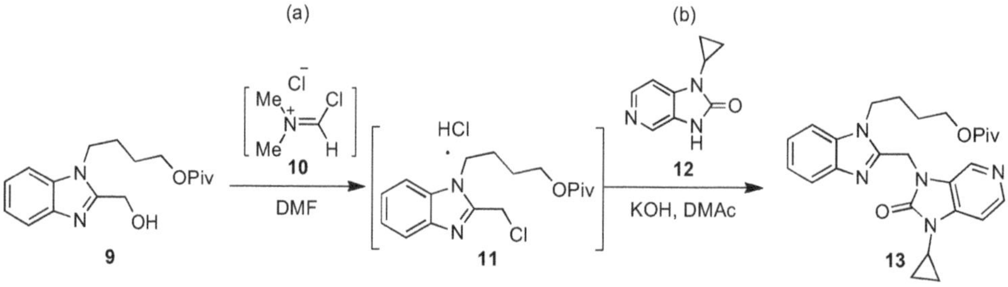

SCHEME 14.3

Procedure

Stage (a)
- Charge a solution of **9** (1.54 kg, 5.06 mol) in DMF (6.1 L) to a suspension of **10** (741 g, 5.80 mol) in DMF (1.3 L) over 30 min.
- Upon stirring for 1 h, charge 17.0 g (0.133 mol) of additional **10**.
- Stir for 20 min.

Stage (b)
- In a separate vessel, charge 980 mL (12.4 mol) of 12.7 N KOH aq solution to a suspension of **12** (795 g, 4.54 mol) in DMAc (6.58 L).
- Charge the solution of **11**, prepared in Stage (a), over 10 min at <35 °C.
- Stir for 2 h at room temperature.
- Filter to remove the salt.
- Cool the organic layer in a 15–20 °C bath.
- Add a solution of NH_4Cl (2.8 M, 7.2 L) at <25 °C.
- Cool to 18 °C and add seeds (5.34 g).
- Add water (20 L).
- Cool to 10 °C and stir for 2 h.
- Filter and wash with water (18.6 L) and heptane (16.8 L).
- Dry at 38 °C under vacuum.
- Yield: 16.3 kg (78%).

14.1.4 PYRIDINE *N*-OXIDE

Hydroxymethylpyridine **16** is an intermediate for the synthesis of compound **17**, identified by Bristol Myers Squibb Company, as a development candidate for the treatment of depression. The synthesis of **16** was accomplished in a two-step process involving a pyridine *N*-oxide intermediate **15**. The pyridine *N*-oxide **15** was a thermal hazard; handling such a high-energy compound in a solvent-free form or a concentrated solution would be dangerous. Therefore, a telescoped process (Scheme 14.4)[6] was developed to avoid the isolation of **15**.

Following the oxidation with *m*CPBA, the reaction mixture was quenched with an aqueous sodium thiosulfate solution followed by washing and vacuum distillation to a concentration of ~4

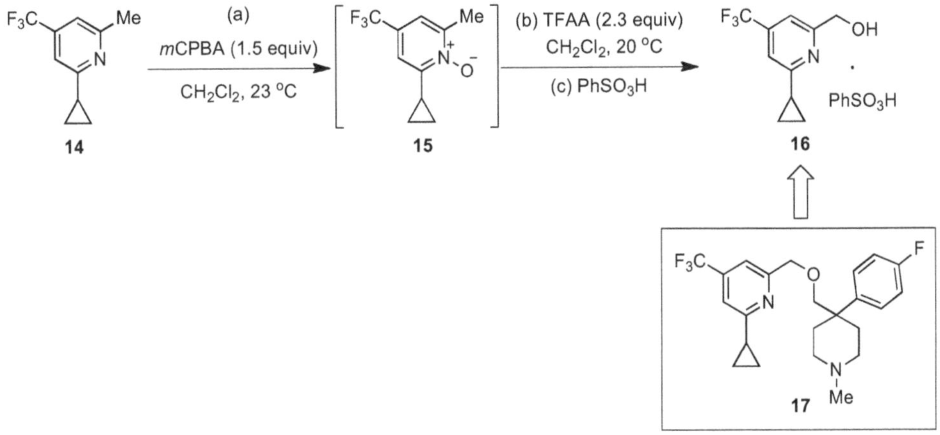

SCHEME 14.4

mL/g. The *N*-oxide stream was adjusted to a concentration of 10 mL/g with methylene chloride and then used immediately in the subsequent Boekelheide rearrangement.

14.1.5 BENZYL BROMIDE

The telescoping method has been utilized to avoid exposure to toxic reagents during product isolation.[7] LY2497282 is a potent and selective DPP IV inhibitor for the potential treatment of diabetes and the synthesis of LY2497282 involves amino alcohol **19** as a key intermediate. As depicted in Scheme 14.5,[8] a mutagenic benzyl bromide was used in the preparation of **19** during the dibenzylation of amino ester **20**. In order to mitigate exposure to the mutagen benzyl bromide, the three-step process was telescoped into a one-pot operation.

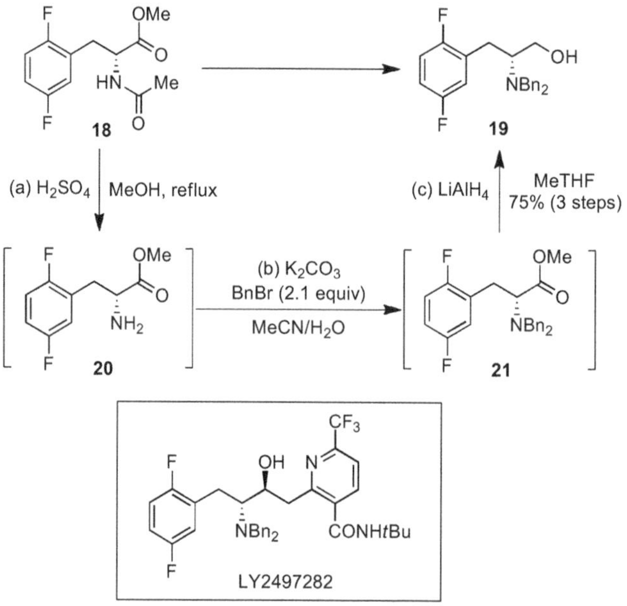

SCHEME 14.5

Upon completion of the dibenzylation, the resulting amino ester **21** was directly reduced to amino alcohol **19** with LiAlH$_4$ without isolation, in which the excess of benzyl bromide could be completely consumed. This three-step telescoped process was carried out at a 10 kg scale, affording the desired amino alcohol **19** in an overall 75% yield with 99% ee.

14.1.6 METHYL IODIDE

Generally, *N*-methylation of amines can be achieved via either methylation with toxic reagents, such as methyl iodide, dimethyl sulfate, and dimethyl carbonate, or reductive methylation with formaldehyde. The use of methanol as an inexpensive reductive methylation reagent was able to address this issue.

A new phosphine-based ligand containing a pyridyl moiety was designed to activate the O–H bond in methanol (shown in structure **24**, Scheme 14.6).[9] Such activation would allow the oxidative addition of Pd(0) to the O–H bond leading to palladium hydride complex **25**. The subsequent β-hydride elimination would generate formaldehyde and hydrogen in situ. Thus, a telescoped process was developed involving catalytic hydrogenation with the in situ formed hydrogen and subsequent *N*-methylation of the resulting aniline with the in situ formed formaldehyde. Methanol in this reaction acts as a solvent, hydrogen source, and methylation agent.

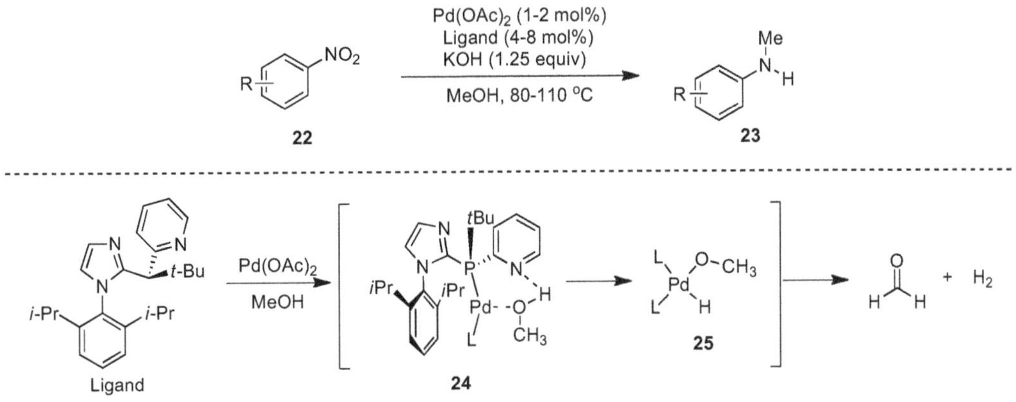

SCHEME 14.6

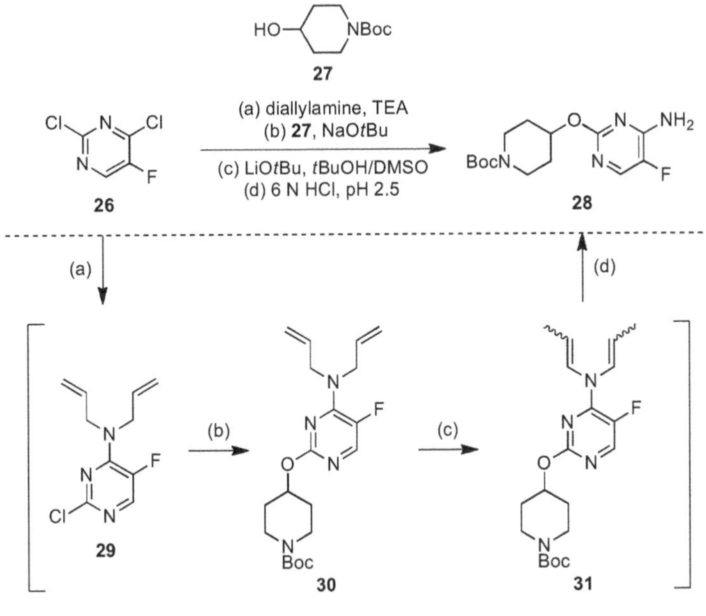

SCHEME 14.7

14.2   PROCESSING PROBLEMATIC INTERMEDIATES

Isolation of reaction intermediates usually serves as the purity control point (PCP) during multistep organic synthesis. Isolation of low melting and hygroscopic solids, however, is problematic, particularly at scale. Purification of such non-crystalline intermediates often requires column chromatography. In order to maximize process throughput, eliminating the requirement for chromatographic purification would be a key objective. Achieving this objective would rely on the identification of crystalline intermediates in the sequential synthesis. Once the PCP is identified, bypassing the isolation of such hygroscopic intermediates is feasible by implementing the telescoping technique.

14.2.1   OILY INTERMEDIATES

The synthesis of 5-fluoro-2-(piperidin-4-yloxy)pyrimidin-4-amine **28** involves three oily intermediates (**29**, **30**, **31**) (Scheme 14.7),[10] which makes the isolation and purification inconvenient.

Ultimately, telescoping the steps through intermediates **29**, **30**, and **31** provided directly **28** as a crystalline solid.

Note

Hydrolysis of **31** was carried out at pH 2.0–3.0 without affecting the Boc group.

14.2.2 HYGROSCOPIC INTERMEDIATE SOLID

Imine intermediate **34** is a hygroscopic and low-melting solid and isolation of such an intermediate is not feasible on a large scale. Thereby, a telescoped process (Scheme 14.8)[11] was developed, avoiding the isolation of the hygroscopic imine **34**.

   Without any purification, the crude **34** was directly used for the subsequent cycloaddition. Furthermore, the resulting cycloadduct **35** was submitted to the subsequent salt formation. Ultimately, using this one-pot process in 500-gal reactors, the desired product **36** was obtained with 57% overall yield.

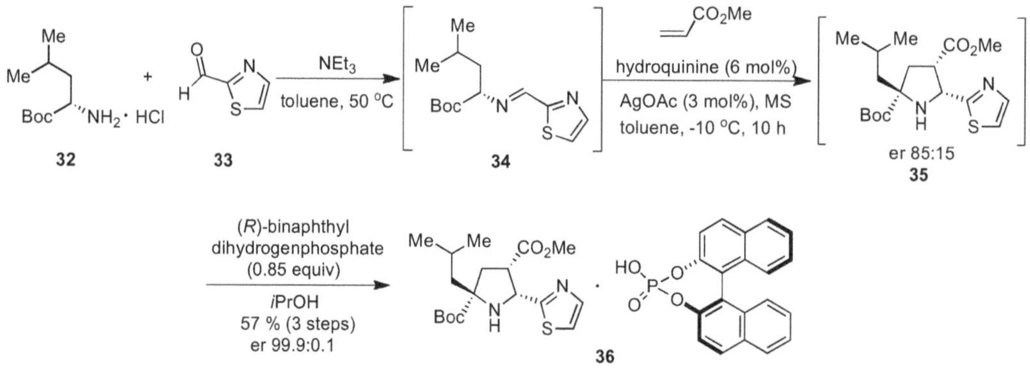

SCHEME 14.8

14.2.3 HYGROSCOPIC AMINE SALT

A four-step preparation of lactam **39** involving a reaction of 3-nitro-1*H*-pyrazole-5-carboxylic acid **38** with highly hygroscopic 2-chloro-*N*-methylethanamine hydrochloride salt **40** led to inconsistent product yields and purity. In order to avoid handling the highly hygroscopic **40**, a one-pot process was developed, in which **40** was prepared in situ via chlorination of 2-methylaminoethanol **37** and used directly without isolation (Scheme 14.9).[12]

   This one-pot process consists of four chemical transformations that occurred at three process temperatures: chlorination of **37** to **40** at 54 °C (Stage (a), Scheme 14.9), activation of acid **38** to dimeric intermediate **42** with SOCl₂ at 70 °C (Stage (b), Scheme 14.9), and ring opening of **42** with **40** (Stage (c), Scheme 14.9) and cyclization of **43** (Stage (d), Scheme 14.9) at room temperature to 45 °C.

Procedure

   Stage (a)
   - Charge 220 g (2.92 mol, 1.18 equiv) of **37** into a reactor followed by adding toluene (2.68 L) and 383 g (2.47 mol) of **38**.
   - Cool the contents to 15 °C.
   - Charge 498 mL (6.83 mol) of thionyl chloride at ≤25 °C over 10 min followed by adding DMF (18.8 mL).
   - Heat the batch to 54 °C and stir at that temperature for 10 min.

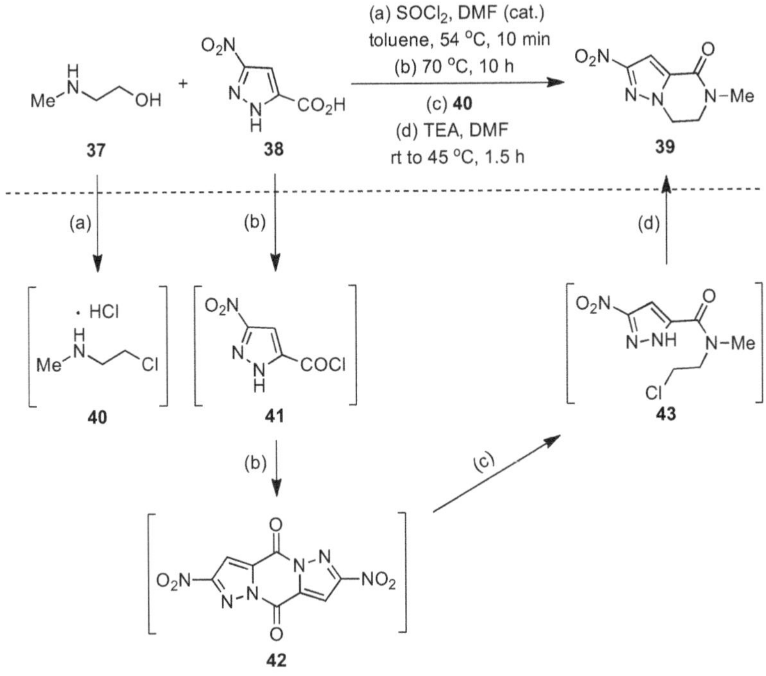

SCHEME 14.9

Stage (b)
- Heat to 70 °C and stir at 70 °C for 10 h.
- Concentrate at 60 °C/100 mmHg to remove excess SOCl$_2$ and a portion of toluene (1.72 L).
- Cool to room temperature.

Stage (c, d)
- Add DMF (1.9 L) followed by adding TEA (1.53 L, 12.0 mol) at <45 °C.
- Stir the resulting mixture at room temperature for 1.5 h.
- Distill at 60 °C/100 mmHg to remove solvent (~1.5 L).
- Cool the batch to room temperature.
- Add water (4.6 L) and heptane (0.5 L).
- Stir at ambient temperature overnight and at 5 °C for 2 h.
- Filter and wash with water (2×1.1 L) and dry.
- Yield: 390 g (81.5%) as a tan solid.

REMARK
The formation of HCl salt **40** prevented the oligomerization of the free base of **40**.

14.2.4 HIGH WATER-SOLUBLE INTERMEDIATE

Isolation of water-soluble organic compounds may result in a significant loss of products during the isolation. In order to avoid the isolation of high-water soluble alcohol **45**, a thorough process was developed for the transformation of epoxide **44** to tosylate **46** (Scheme 14.10).[13] After workup and a solvent switch from methylene chloride to toluene, the tosylate **46** could be crystallized by simply adding heptane, affording a 64% yield of **46** over 2 steps.

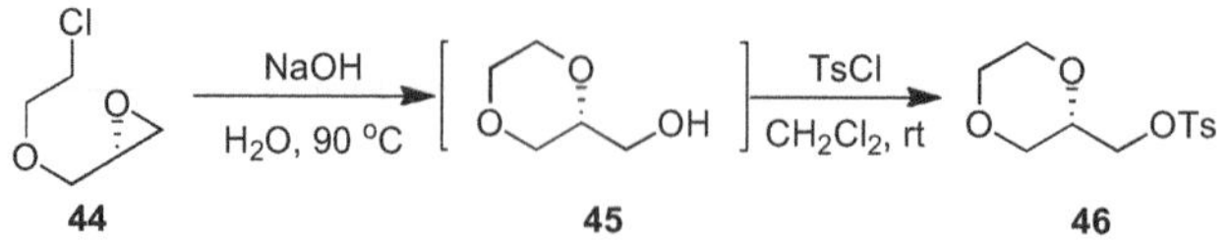

SCHEME 14.10

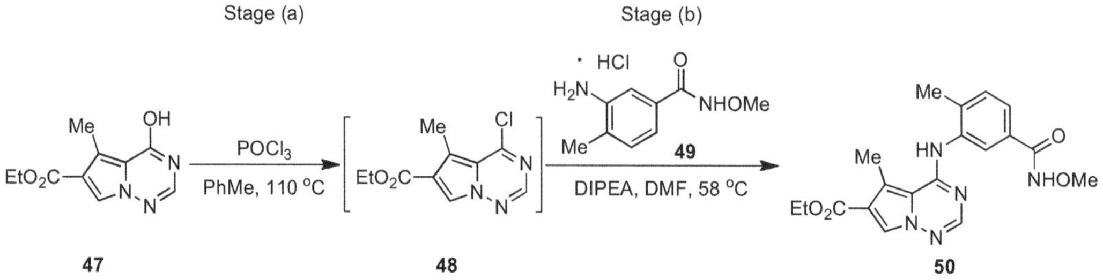

SCHEME 14.11

The telescoping strategy has been further demonstrated in a number of other applications[14] in chemical development including various types of hard-to-handle process intermediates.

14.2.5 Unstable Intermediates

Isolation of unstable reaction products may lead not only to loss of product, but also to a high level of impurities. Application of telescoped process allows preclusion of unstable process intermediates from isolation.

14.2.5.1 Heteroaryl Chloride

Heteroaryl halides are valuable organic intermediates and can be converted into a variety of products. However, most of the heteroaryl halides have limited stability because they are inherently susceptible to hydrolysis. For instance, hydrolytic decomposition of chloride **48** was observed during the isolation of **48** (Scheme 14.11).[15] This stability issue was addressed by telescoping the chlorination step with the subsequent amine coupling, which was done by simply performing a solvent exchange from toluene to DMF after the workup of the chlorination reaction.

14.2.5.2 Toluenesulfonate Intermediate

A telescoped process was incorporated into the synthetic sequence for the preparation of DG-051B, a small molecule inhibitor for the prevention of heart attack. The two-step transformation was carried out in a telescoped fashion to prepare alkyl aryl ether **52**. Thus, the initial tosylation yielded the unstable tosylate **53**, which was submitted to S_N2 etherification without isolation of **53**. This telescoped process was successfully performed at a multi-kilogram scale (Scheme 14.12).[16]

REMARK

Unstable intermediates can sometimes be converted into their corresponding salts to enhance stability.[17]

14.2.5.3 Aldehyde Intermediates

14.2.5.3.1 Reduction/Grignard-Type Reaction

Compared with acids and esters, aldehydes exhibit higher reactivity toward electrophiles. The presence of the electron-withdrawing trifluoromethyl group in aldehydes such as **55** further increases the electrophilic ability of the carbonyl group.[18] Such high reactivity renders 3,3,3-trifluoropropanal

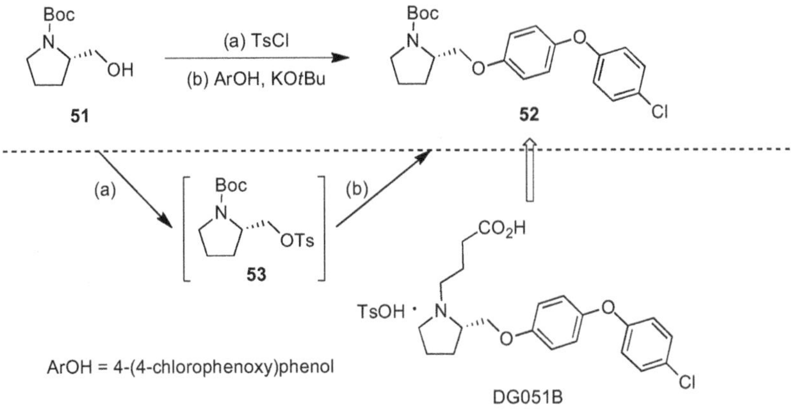

ArOH = 4-(4-chlorophenoxy)phenol

SCHEME 14.12

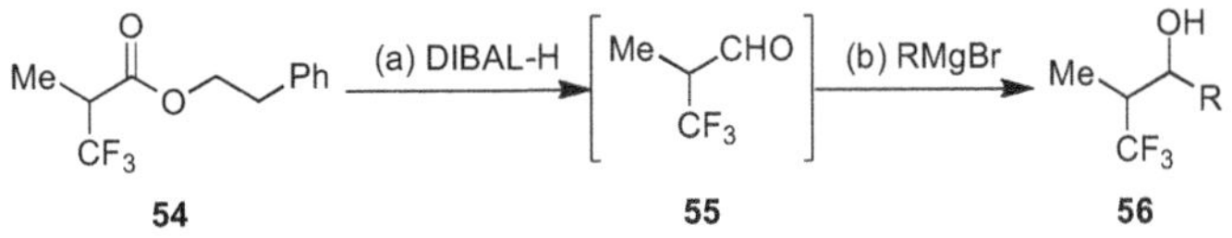

SCHEME 14.13

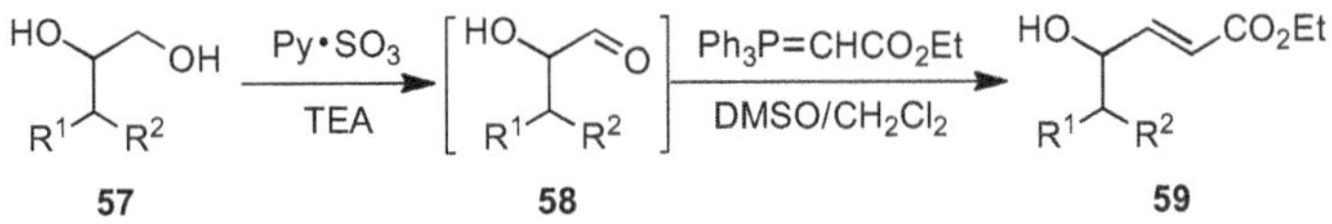

SCHEME 14.14

55 not feasible for isolation from the reaction mixture. Thus, a one-pot protocol was developed (Scheme 14.13)[19] to accommodate the highly reactive **55** that was formed in situ via the reduction of 2-phenylethyl 2-(3,3,3-trifluoromethyl)propionate **54** with DIBAL-H. The resulting aldehyde **55** smoothly reacted with Grignard reagents (or lithium enolates) in a one-pot to afford the desired alcohol products **56** in good yields.

14.2.5.3.2   Oxidation/Wittig Reaction

Analogously, an aldehyde intermediate **58**, generated in situ from chemoselective oxidation of diol **57** with pyridine·SO$_3$ complex, was exposed to the Witting alkenylation conditions affording olefin **59** in a simple and convenient one-pot protocol (Scheme 14.14).[20] Pyridine·SO$_3$ complex is a white solid and commercially available.

14.2.5.4   Unstable Alkene Intermediates

Diels–Alder reaction is a powerful tool for a rapid build-up of complex structures of defined geometry with minimal waste. However, the number of industrial-scale Diels–Alder reactions is limited,[21] which is mainly attributed to safety issues. Those safety concerns are often linked to the instability of the dienes and/or dienophiles and the intrinsic full-batch mode operations. Therefore, the telescoping approach is often employed to cope with these issues.

14.2.5.4.1   Dehydration/Diels–Alder Reaction

In order to avoid isolation of the unstable diene intermediate **61**, the preparation of cyclohexene derivative **62** was realized by employing a dehydration/Diels–Alder tandem sequence and conducted in a one-pot manner as illustrated in Scheme 14.15.[22]

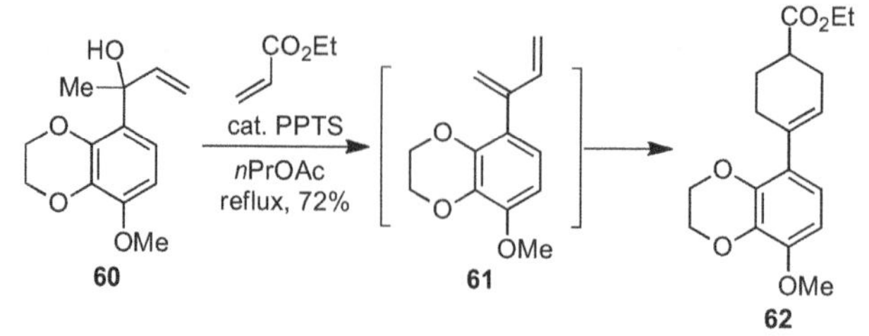

SCHEME 14.15

14.2.5.4.2   Pd-Catalyzed Dehydrogenation/Diels–Alder Reaction

A tandem Pd-catalyzed dehydrogenation/Diels–Alder reaction would enable rapid construction of diverse molecular skeletons from simple starting materials. The tandem sequence (Equation 14.1)[23] involves a dehydrogenation via palladium-catalyzed C–H bond activation and subsequent β-hydride elimination followed by the Diels–Alder reaction.

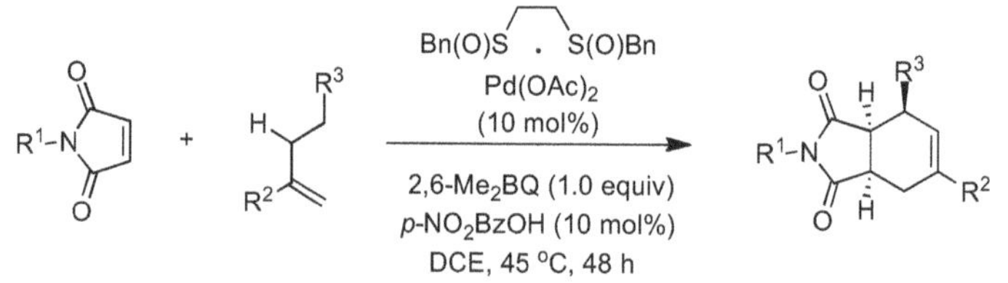

(14.1)

R[1] = Ph, *p*-OMeC$_6$H$_4$, *p*-AcC$_6$H$_4$, *p*-FC$_6$H$_4$, *p*-BrC$_6$H$_4$, etc.
R[2] = H, Me,
R[3] = CH$_2$OTBS, NPhth, (CH$_2$)$_3$NO$_2$, etc.

The versatility of the tandem reaction enables rapid access to complex skeletons found in biologically active molecules. For example, the tandem dehydrogenation/Diels–Alder reaction was utilized to generate an adduct **64** (Scheme 14.16),[23] an intermediate for the synthesis of polycycle **65**. This isoindoloquinoline skeleton in **65** is found in several alkaloids, such as jamtine, that exhibit significant anti-hyperglycemic activity.

14.2.5.4.3   Acrylate Formation/Heck Coupling

Owning to the fact that acrylates are prone to polymerization, especially under elevated temperatures, preparation, and isolation of acrylate derivatives, such as **68**, would be challenging. Ultimately, a telescope strategy was applied toward the synthesis of olefin **70** by telescoping the acrylate formation and Heck coupling reaction (Scheme 14.17)[24] into a one-pot process.

Procedure

Stage (a)
- Charge 12.5 L of MeCN into a 30-L jacketed reactor followed by adding 1.477 kg (9.49 mol) of **66**.
- Charge 5.3 L (38.0 mol) of TEA.
- Cool the resulting mixture to 0 °C.
- Charge 2.26 L (23.0 mol) of acetic anhydride slowly at <15 °C.
- Warm the mixture to 20 °C and stir for 30 min.

Stages (b) and (c)
- Charge 2.5 kg (8.62 mol) of **69** followed by adding 2.65 L (19.0 mol) of TEA, 131.8 g (0.43 mol, 5.0 mol%) of tri-*o*-tolylphosphine, and 38.9 g (0.17 mol, 2.0 mol%) Pd(OAc)$_2$.

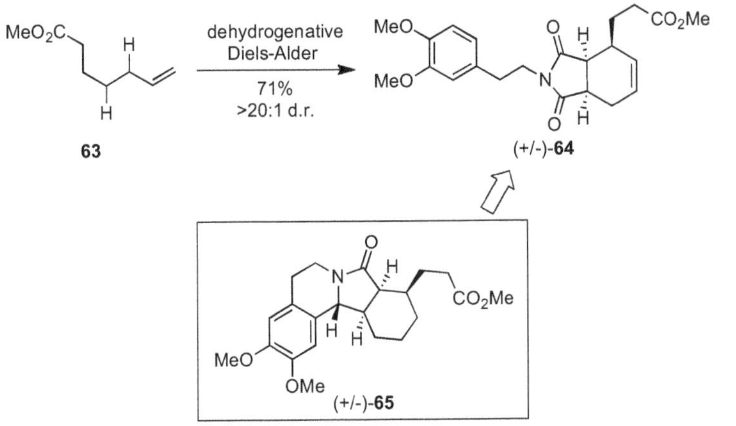

SCHEME 14.16

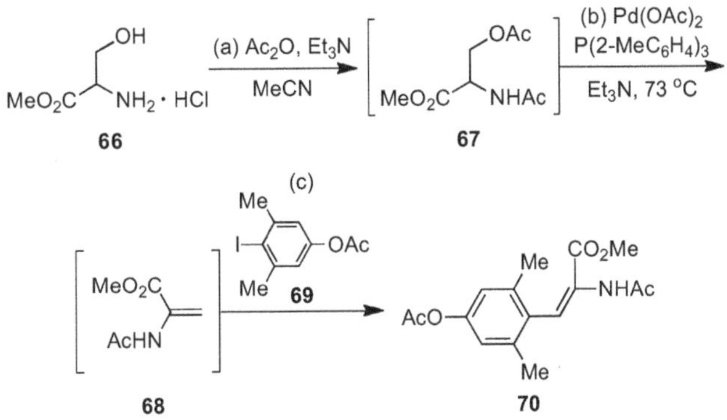

SCHEME 14.17

- Heat the reaction mixture to 73 °C and stir for 13 h.
- Cool to room temperature.
- Filter through a bed of Arbocel and rinse with MeCN (7.5 L).
- Distill the combined filtrates under reduced pressure to remove ca 20 L of solvent.
- Add EtOAc (25 L) and sat NaHCO$_3$ (10 L).
- Separate layers.
- Add EtOAc (2.5 L) to the organic layer.
- Wash the diluted organic layer with sat NaHCO$_3$ (10 L) followed by water (10 L).
- Dry the organic layer over MgSO$_4$ (2 kg), filter, and rinse with EtOAc.
- Stir the combined filtrates with Carbon Darko KB (wet powder, 100 mesh, 1 kg) at room temperature for 2.5 h.
- Filter through a pad of Arbocel and silica, and rinse with EtOAc (5 L).
- Separate the filtrates into roughly two equal portions that were distilled to dryness under reduced pressure to give a yellow solid.
- Slurry the solid in a mixture of EtOAc (6.5 L) and heptane (13 L) at reflux for 2 h.
- Cool to 20 °C and stir for 16 h.
- Filter and wash with a mixture of EtOAc (0.8 L) and heptane (1.6 L), then with heptane (1.4 L).
- Yield: 1721.8 g (848.7 g + 873.1 g) (65%).

14.2.5.4.4 Protection/Heck Reaction/Deprotection

Approved in early 2012 by FDA, axitinib **74** was developed by Pfizer for the treatment of advanced renal cell carcinoma. To set the final stage for the synthesis of **74**, the Heck reaction was chosen for installing the vinyl functionality. Scheme 14.18[25] illustrates the telescoped protocol, including acylation, the Heck reaction, and deprotection.

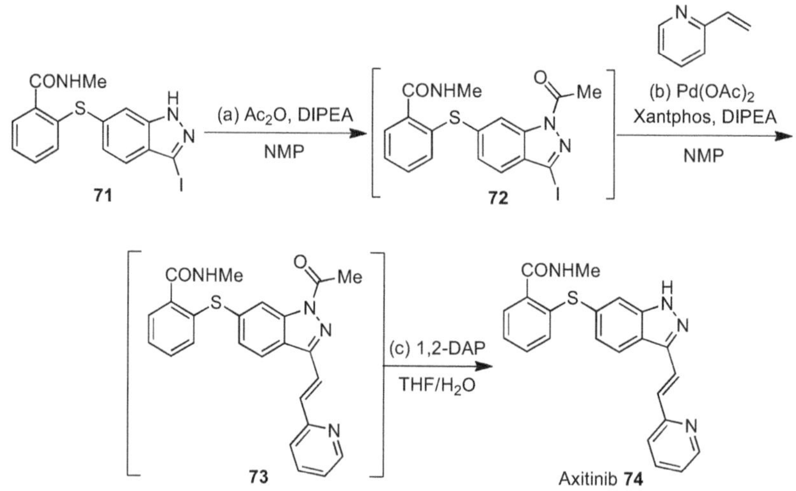

SCHEME 14.18

Typically, the three-step telescoped process was conducted at 50 °C in the presence of Pd(OAc)$_2$, Xantphos, iodide **71**, Hunig's base, and acetic anhydride in NMP to complete the *N*-acyl protection step. The subsequent Heck reaction proceeded at 90 °C following the addition of 2-vinyl pyridine. After completing the Heck reaction, the acylated product **73** was deprotected in situ with 1,2-diaminopropane (1,2-DAP), followed by recrystallization to afford axitinib **74** in 75% yield over three steps.

REMARK

1,2-Diaminopropane was chosen not only for the deprotection of the acyl group, but also for the removal of Pd.

14.2.5.5 Unstable β-Hydroxyketone

During the synthesis of 6-benzhydryl-2*H*-pyridazin-3-one **80**, isolation of unstable β-hydroxyketone intermediate **77** with quenching and concentration operations proved to be challenging, leading to impurities **81** and **82** (Scheme 14.19).[26]

In order to address the instability of **77**, a telescoped process was developed by performing the preparation in one-pot without isolation of **77**, affording pyridazinone **80** in 76% yield (Scheme 14.20).

Hydrazone **84** was observed in the aqueous layer after quenching. Therefore, heating the combined aqueous layers at 70 °C for 24 hours (Stage (c), Scheme 14.20) was able to convert **84** to the product **80**, leading to an 11% increase in the product yield.

Procedure

Stage (a)
- Charge, to a solution of diisopropylamine (14.7 kg, 145.3 mol) in THF (135 kg), 62.2 kg (15% in hexane, 145.3 mol) of *n*BuLi at −30 °C to −5 °C.

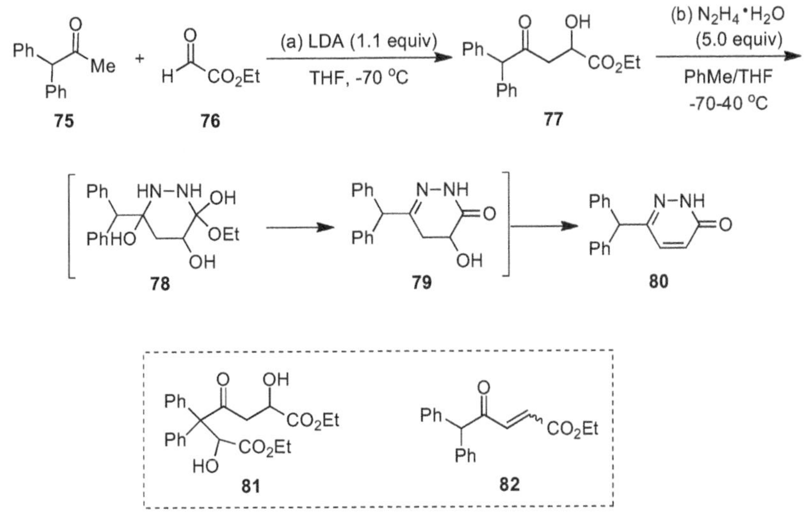

SCHEME 14.19

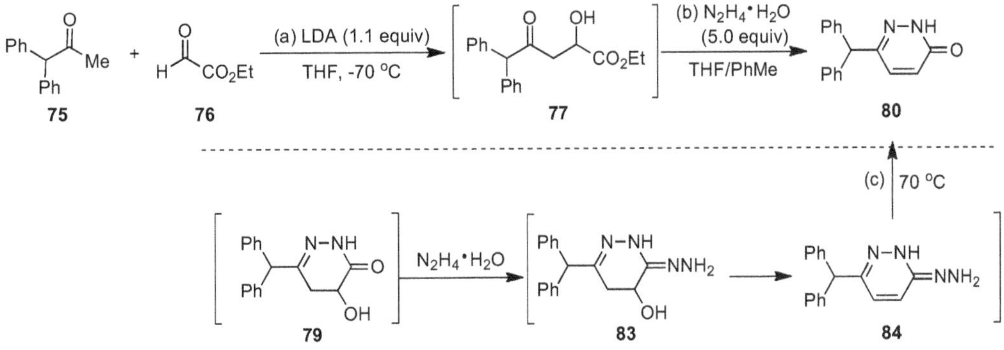

SCHEME 14.20

- Stir for 1 h, then cool to −78 °C.
- Charge, to the LDA solution, a solution of **75** (27.8 kg, 132.2 mol) in THF (70.1 kg) at <−70 °C.
- Stir for 1 h.
- Charge 54.1 kg (265.0 mol) of **76** (50% solution in toluene) at <−70 °C.
- Stir for 0.5 h.

Stage (b)
- Charge 33.1 kg (661.2 mol) of hydrazine hydrate at <−70 °C.
- Transfer the mixture into toluene (20.1 kg).
- Rinse with THF (13.1 kg) and toluene (12.0 kg).
- Stir for 12 h.
- Warm the batch to 40 °C and age for 3 h (<1% of **79**).
- Add water (139 kg) and toluene (40 kg).
- Add 6 N HCl (32.7 kg) to pH 10.1.
- Separate layers.
- Wash the organic layer twice with aqueous NaCl (20% w/w, 28 L).

Stage (c)

- Heat the combined aq layers to 70 °C and age for 24 h (<1% of **84**).
- Add THF (25.0 kg), toluene (96.0 kg), and 20% (w/w) aqueous solution of NaCl (28 L).
- Combine the organic layers (including the organic layer from Stage (b)).
- Distill the combined organic layers under reduced pressure to 82 L.
- Add heptane (57.0 kg) over 1 h and age for 2 h.
- Filter, rinse with a mixture of toluene (29.0 kg) and heptane (34.0 kg), and dry at 40 °C under vacuum.
- Yield of **80**: 26.3 kg (75.8%).

14.3   IMPROVING FILTRATION

Filtration is one of the most frequently used unit operations to achieve solid–liquid separations, such as the isolation of solid products from slurry reaction mixtures or the removal of insoluble catalysts and impurities. For the isolation of solid products with a large quantity of fine particles, the filtration can be unacceptably slow. To mitigate such a slow filtration problem which could be a process bottleneck, the telescope approach is frequently chosen in the process development to avoid the isolation of such solid products.

14.3.1   PREPARATION OF AMIDE

The initial step-by-step synthesis of drug intermediate **88** involved the isolation of intermediate **87**, which required 14 days for filtering and drying of this material on a 76 kg-scale campaign. To address problems associated with filtration and drying, a telescoped process was developed (Scheme 14.21).[27]

Upon completion of the amide formation, the reaction mixture was quenched by adding water to hydrolyze excess acid chloride **86**. The subsequent step to access **88** was directly carried out by adding tetramethylguanidine (TMG) to the resulting reaction mixture followed by a reaction with piperidine. This telescoped process was demonstrated in the pilot plant campaign, and offered a dramatic improvement in processing times. Ultimately, the synthesis of **88** was completed in 2 days on an 80-kg scale.

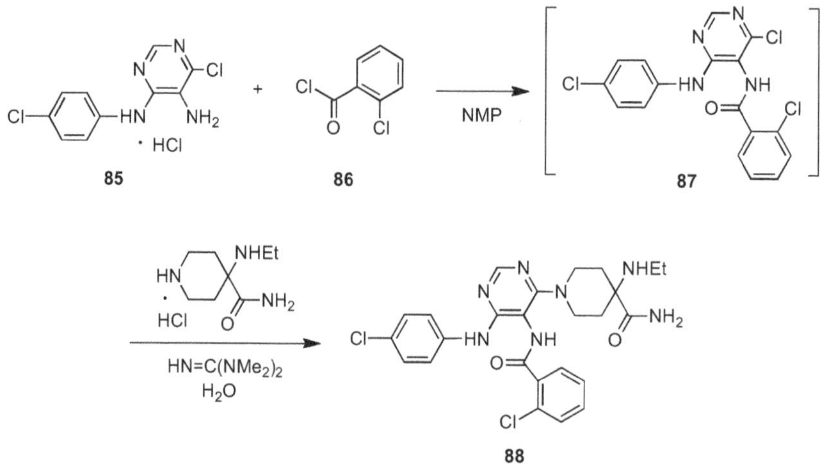

SCHEME 14.21

14.3.2   SYNTHESIS OF β-NITROSTYRENE

Because of the instability and reactivity, aldehydes are frequently converted into their corresponding aldehyde–sodium bisulfite adducts for isolation/purification and storage. The synthesis of β-nitrostyrene **90** from isocroman **89** was realized via a four-step process: (a) ring-opening of **89**, (b) formation of hexaminium salt **92** from a reaction between benzyl chloride **91** and hexamethylenetetramine (HMTA), (c) hydrolysis of **92** to give benzaldehyde **93**, and (d) condensation of the purified benzaldehyde **93** with nitromethane (Scheme 14.22).[28] The purification of **93** via its crystalline sodium bisulfite addition complex **94** proved time-consuming due to problems in filtering the microcrystalline complex.

Therefore, a telescoped process was developed, which eliminated the problematic filtration during the purification of **93** via the bisulfite addition complex **94**.

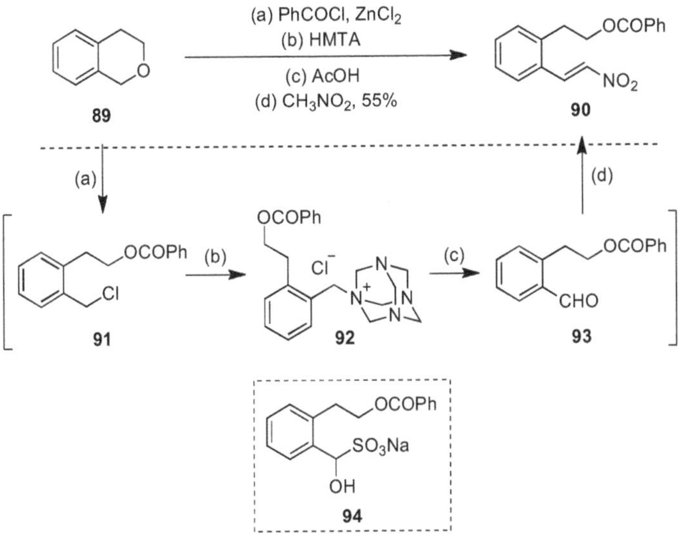

SCHEME 14.22

Procedure

Step (a)
- Charge 10.5 kg (77 mol, 10 mol%) of ZnCl$_2$ at 25 °C into a stirred solution of benzoyl chloride (115 kg, 822 mol, 1.05 equiv) in CH$_2$Cl$_2$ (635 kg).
- Charge slowly 105 kg (786 mol) of **89** into the above mixture under reflux.
- Continue to reflux for 1 h.
- Cool to 20–30 °C.

Step (b)
- Charge 150 mL of industrial methylated spirit (IMS) followed by adding 166 kg (1186 mol, 1.5 equiv) of HMTA.
- Distill at atmospheric pressure until 600 L of distillate is collected.

Step (c)
- Charge a mixture of AcOH (315 L) and water (315 L).
- Distill at atmospheric pressure until 525 L of distillate is collected.
- Cool the batch.

- Extract with MTBE (810 L).
- Wash the organic layer with 2 M H_2SO_4 (574 L), 10% w/v aqueous Na_2CO_3 (2×525 L), 26% w/v aqueous NaCl (525 L).
- Distill to give a mobile oil **93**.

Step (d)
- Dissolve **93** in MeOH (360 L).
- Charge, to the solution of **93**, 52 kg (852 mol) of nitromethane, 16.9 kg (282 mol) of AcOH, and 20.6 kg (282 mol) of *n*-butylamine.
- Stir the resulting mixture at 22 °C for 18 h.
- Isolate the product **90** with a centrifuge.
- Wash with IPA (520 mL).
- Dry to give the product **90**.
- Yield: 128 kg (55% from **89**).

14.4   TELESCOPING CATALYTIC REACTIONS

Application of the telescope strategy toward reactions that involve expensive catalysts would potentially allow these steps to share the expensive catalysts, thereby reducing manufacturing costs. If imine reduction is immediately followed by debenzylation deprotection, utilization of a palladium-based catalyst in both steps is more desirable due to the possibility of combining the reduction with the debenzylation into a one-pot operation. Analogously, catalytic dehalogenation can share the same palladium catalyst with the subsequent Suzuki cross-coupling reaction without isolation of the dehalogenated product.

14.4.1   IMINE REDUCTION/DEBENZYLATION

A telescoped process, involving two palladium-mediated steps (imine reduction and debenzylation), was developed for the synthesis of chiral amine **99** (Scheme 14.23).[29]

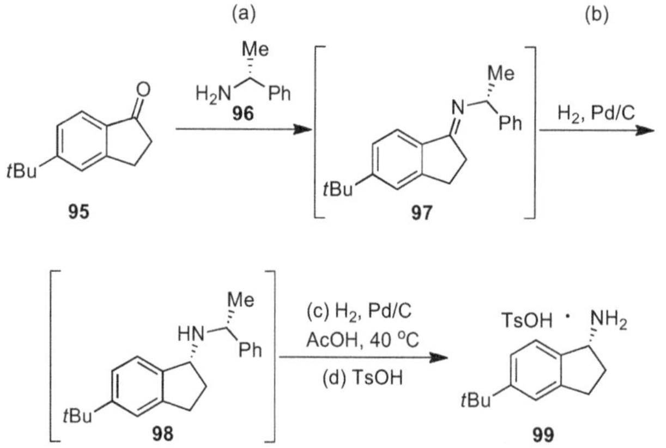

SCHEME 14.23

Procedure

Stage (a) The imine **97** formation
- Charge **96** (4.8 kg, 40 mol) to a solution of **95** (3.66 kg, 19.5 mol) in 26 kg of toluene.

- Distill under atmospheric pressure until ~12 L of toluene is collected.
- Cool the resulting solution to room temperature.

Stage (b) Pd-catalyzed hydrogenation
- Transfer the toluene solution of **97** into a hydrogenator.
- Charge 30 kg of methanol.
- Charge 1.15 kg of 5% Pd/C under N_2 atmosphere.
- Conduct the reaction at 10 °C under 40 psi of hydrogen.

Stage (c) Pd-catalyzed debenzylation
- Charge 2.92 kg of acetic acid at 20±5 °C.
- Conduct the reaction at 40 °C until ~96% conversion was achieved.
- Cool to room temperature and filter.

Stage (d) Salt formation
- Charge the filtrate to TsOH (3.4 kg, 17.8 mol) in 5.1 kg of MeOH.
- Distill the resulting solution under reduced pressure to ~20 L volume.
- Add 32 L of water at 60–65 °C.
- Age at 65 °C for 1 h.
- Cool to 20 °C.
- Filter and rinse with 5 kg of water.
- Charge the resulting crude salt back into the reactor.
- Add 21 kg of toluene and 2.3 kg of MeOH.
- Heat to ~65 °C.
- Cool to ~20 °C.
- Filter and wash with toluene.
- Dry under vacuum.
- Yield: 4.9 kg (70%) with 97% ee.

REMARK

In Stage (b), an alternative (also a safer) way of charging Pd/C catalyst is to charge the catalyst slurry.

14.4.2 DEBROMINATION/SUZUKI CROSS-COUPLING REACTION

Scheme 14.24[30] presents an efficient and high-yielding one-pot process for the synthesis of 2-amino-5-phenylpyrazine **103**. This telescoped process contains a sequential palladium-catalyzed regioselective debromination of dibromopyrazine **100** and the subsequent Suzuki cross-coupling reaction.

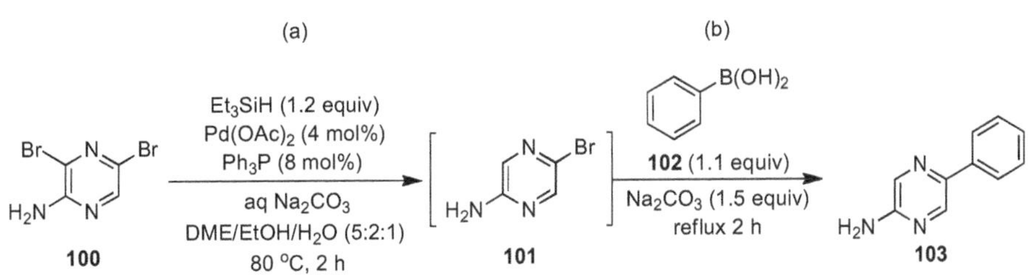

SCHEME 14.24

Procedure

Stage (a)

- Degas a mixture of **100** (100 g, 395 mmol), Et$_3$SiH (75.7 mL, 474 mmol, 1.2 equiv), 1.5 M aqueous Na$_2$CO$_3$ (132 mL, 198 mmol, 0.5 equiv), and DME/EtOH/H$_2$O (5:2:1, 1.5 L) by vacuum/N$_2$-fill cycles.
- Charge Pd(OAc)$_2$ (4 mol%) and PPh$_3$ (8 mol%).
- Degas the above mixture twice.
- Heat to 80 °C and stir for 2 h.
- Cool to 30–40 °C.

Stage (b)

- Charge **102** (53.0 g, 435 mmol, 1.1 equiv) and solid Na$_2$CO$_3$ (62.8 g, 593 mmol, 1.5 equiv).
- Degas the above mixture.
- Heat at reflux for 2 h.
- Workup.

14.5   IMPROVING OVERALL PRODUCT YIELDS

14.5.1   SYNTHESIS OF SPIROCYCLIC HYDANTOIN

Isolation of reaction intermediates as salts is an approach frequently used by process chemists to remove undesired side products. However, a drawback of this approach is the additional step plus salt-break operations, which renders the process less efficient. In some cases, using such a salt-formation approach gave relatively low overall yields. For example, the synthesis of spirocyclic hydantoin **107** employed the salt-formation strategy by forming HCl salt **106**, followed by salt-break and diastereomeric salt resolution (Scheme 14.25).[31]

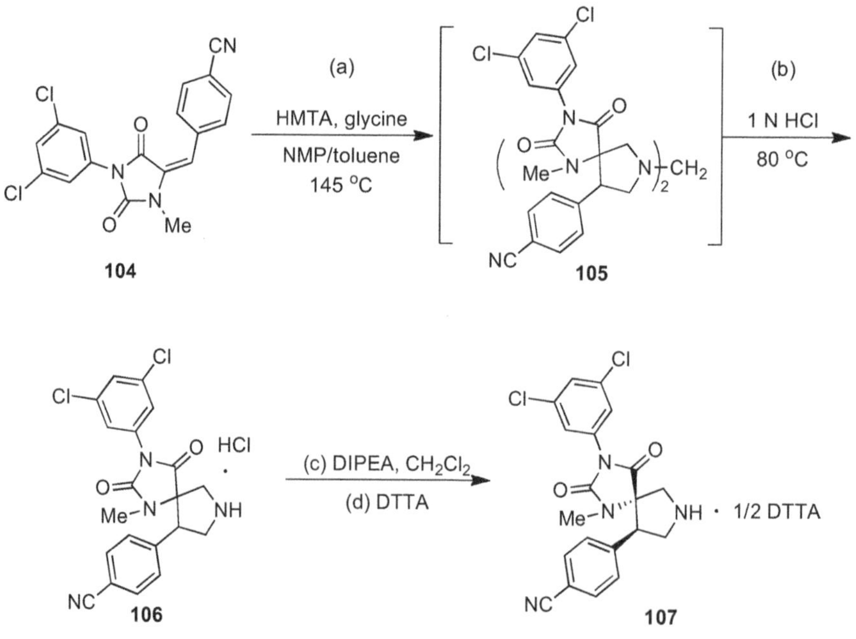

SCHEME 14.25

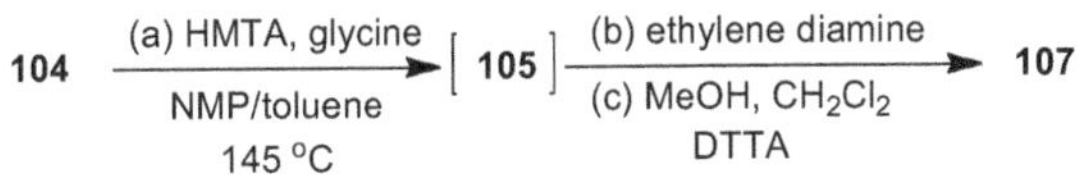

$$104 \xrightarrow[\substack{\text{NMP/toluene} \\ 145\,°C}]{\text{(a) HMTA, glycine}} \Big[\;105\;\Big] \xrightarrow[\substack{\text{(c) MeOH, CH}_2\text{Cl}_2 \\ \text{DTTA}}]{\text{(b) ethylene diamine}} 107$$

SCHEME 14.26

Although this approach for producing **107** was effective, the slow filtration of **106** and relatively low overall yield of **107** (23.6%) suggested the process needs optimization.

Scheme 14.26 illustrates a telescoped process involving the formation of dimer **105**, cleavage of the *N*-methylene bridge in **105**, and diastereomeric salt resolution. Implementation of this telescoped process provided a 32.2% overall yield with 98.7% ee and 99.9% purity.

Procedure

Stage (a)
- Heat a mixture of **104** (68.8 kg, 184.4 mol), glycine (26.4 kg, 351.7 mol), hexamethylenetetramine (HMTA, 8.5 kg, 60.6 mol) in NMP (355 kg), and toluene (150 kg) at 144 °C for 6 h.
- Cool to 55 °C.
- Add THF (355 kg) and toluene (10 kg).

Stage (b)
- Charge 18.6 kg (309.5 mol) of ethylene diamine and THF (12.2 kg).
- Hold the resulting solution at 55 °C for 1 h.
- Cool to 20 °C.

Stage (c)
- Add 20% NaCl aq solution (1022 kg) and toluene (20 kg).
- Stir for 15 min.
- Separate layers.
- Add toluene (150 kg) to the organic layer.
- Distill at 55 °C under reduced pressure (<50 mbar).
- Cool to below 25 °C.
- Add CH_2Cl_2 (840 kg).
- Sample for KF (target <0.2%) and toluene (target <0.09 g/mL).
- Polish filtration.
- Add CH_2Cl_2 (272 kg), water (4.2 kg), and MeOH (26.8 kg).
- Stir for 15 min.
- Charge 175 g of seeds and 16.1 kg of (+)-DTTA.
- Hold the batch for 20 h.
- Cool to 0–5 °C and hold for 3.5 h.
- Filter and wash with CH_2Cl_2 (120 kg, 270 kg, and 120 kg).
- Dry for 37 h.
- Yield: 36.2 kg (32.2%).

DISCUSSIONS

The dimer **105** displayed a broad peak on a reverse phase HPLC due to its hydrolytic instability. Therefore, direct monitoring **105** using the reverse phase HPLC was not possible. Ultimately, the levels of **105** could be monitored indirectly by treating a sample of the reaction mixture with sodium triacetoxyborohydride (Equation 14.2):

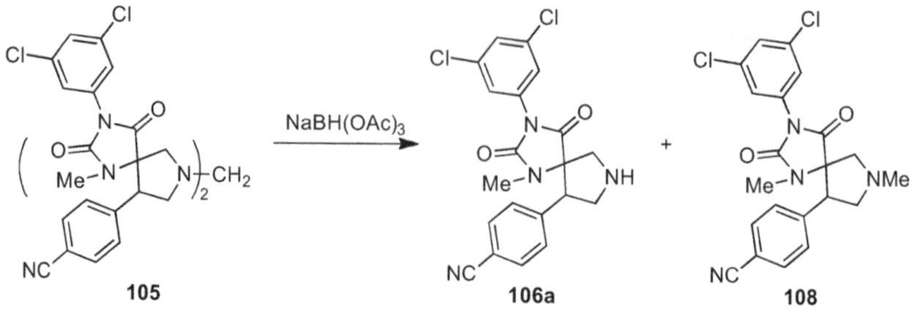

$$(14.2)$$

14.5.2 TRANSITION METAL-CATALYZED CROSS-COUPLING REACTION

In addition to a relatively long overall processing time, the isolation of process intermediates often results in a yield loss. Compound **112** possesses an anti-inflammatory activity with an improved side effect profile in vivo. Diaryl **111** was identified as an intermediate for the synthesis of **112**. Initially, the preparation of **111** was realized via Suzuki cross-coupling of aryl bromide **110** with arylboronic acid, generated from lithiation of 1,3-dimethoxybenzene **109** followed by trapping with triisopropyl borate and hydrolysis. This three-step, two-pot process furnished the product **111** with an overall yield of 58% (Condition A, Scheme 14.27).[32]

Compared with the two-pot Suzuki coupling approach, Negishi coupling of **110** with organozinc reagent, generated in situ from lithiation of 1,3-dimethoxybenzene **109** with nBuLi at 0 °C in THF followed by transmetalation with $ZnCl_2$, provided **111** in one-pot operation (Condition B, Scheme 14.27). Consequently, this highly efficient one-pot protocol improved the product yield from 58% to 90%.

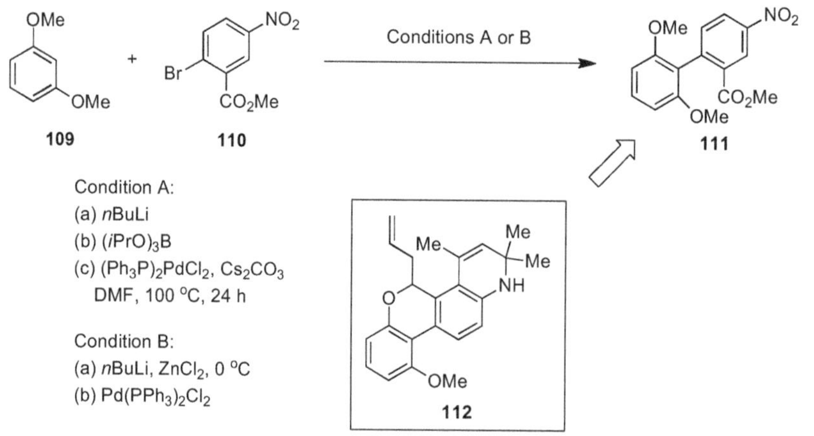

SCHEME 14.27

14.6 REDUCTION IN PROCESSING SOLVENTS

14.6.1 TOLUENE AS THE COMMON SOLVENT

A preferred and commonly used telescope technique is to conduct several reaction steps in the same solvent.[33] Scheme 14.28[34] demonstrates a two-step, one-pot process for the preparation of carboxylic acid **115** from diester **113**. A selective hydrolysis strategy was developed to access **115** via the formation of δ-lactone **114** followed by hydrogenolysis of the pseudobenzylic C−O bond. Toluene was selected as the solvent for the two-step, one-pot transformation. Upon the completion of the intramolecular condensation, toluene was able to remove acetic acid by azeotropic distillation prior

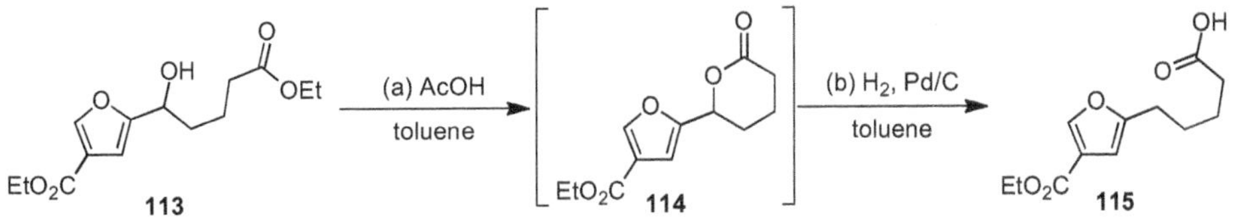

SCHEME 14.28

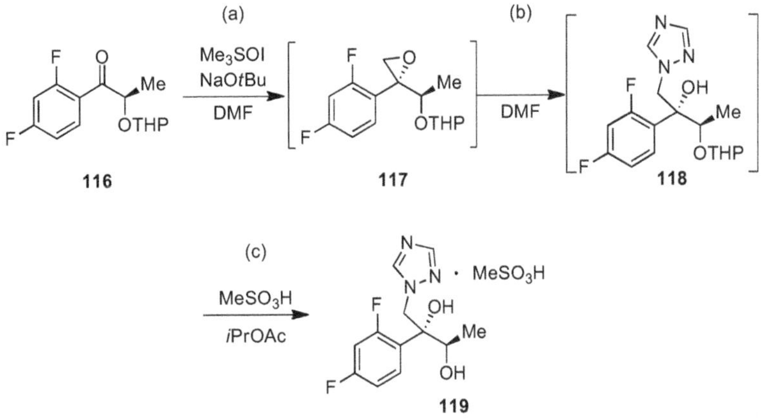

SCHEME 14.29

·to the hydrogenolysis. This one-pot process was readily scaled up to multi-kilogram production in good yields.

14.6.2   DMF as the Common Solvent

The ability to dissolve most organic compounds makes dimethylformamide (DMF) a useful solvent for numerous reactions. Telescoping steps with DMF as the solvent would prevent solvent-switch between steps, making the process more efficient. For instance, DMF was chosen as one of the solvents in a telescoped process for the synthesis of triazolium methanesulfonate salt **119** (Scheme 14.29).[35]

This telescoped process involved epoxidation, epoxide-opening with 1,2,4-triazole, and THP-deprotection/salt formation. The final salt formation was carried out in isopropyl acetate.

Procedure

Stage (a)
- Charge a solution of 30 wt% sodium *tert*-butoxide in THF (197.3 kg, 615.9 mol) to a solution of trimethylsulfoxonium iodide (70.8 kg, 321.7 mol), 1,2,4-triazole (22.2 kg, 321.4 mol) in DMF (498 kg) at <35 °C.
- Stir the resulting mixture at 20–25 °C for 30 min.
- Charge a 53.1 wt% solution of **116** in DMF (142.2 kg, 279.8 mol) at <30 °C followed by DMF (26 kg) rinse.
- Stir the resulting mixture for 30 min (HPLC showed <0.1% of **116** relative to **117**).

Stage (b)
- Heat the batch to 85–90 °C for 12 h (2.9% of **117** relative to **118**).
- Cool to 20 °C.

- Add water (381 kg).
- Extract with *i*PrOAc (2×393 kg).
- Wash the combined extracts with 12 wt% NaCl solution (378 kg) followed by water (2×378 kg).
- Distill the organic solution to 150 L at 55 °C under 0.08 bar.
- Add *i*PrOAc (298 kg).
- Distill to 150 L (with <0.3% water).
- Cool to 25 °C.

Stage (c)
- Charge 34.9 kg (363.1 mol) of MeSO$_3$H at <35 °C.
- Stir the batch at 20–25 °C for 30 min and then for an additional 12 h.
- Cool to 0–5 °C and stir for 3 h.
- Filter, wash with *i*PrOAc (2×90 kg) and heptane (77 kg), and dry at 45–50 °C under vacuum.
- Yield of **119**: 52.48 kg (51.3%).

14.6.3   EtOAc as the Common Solvent

Ethyl acetate was used as the common solvent in a telescoped process in the synthesis of amide **123** for Sildenafil (Scheme 14.30).[36] Sildenafil is an agent for the treatment of male erectile dysfunction.

This one-pot synthesis included nitro reduction, acid activation, and amidation. The use of ethyl acetate as a single solvent allowed for easy solvent recovery with no aqueous waste stream.

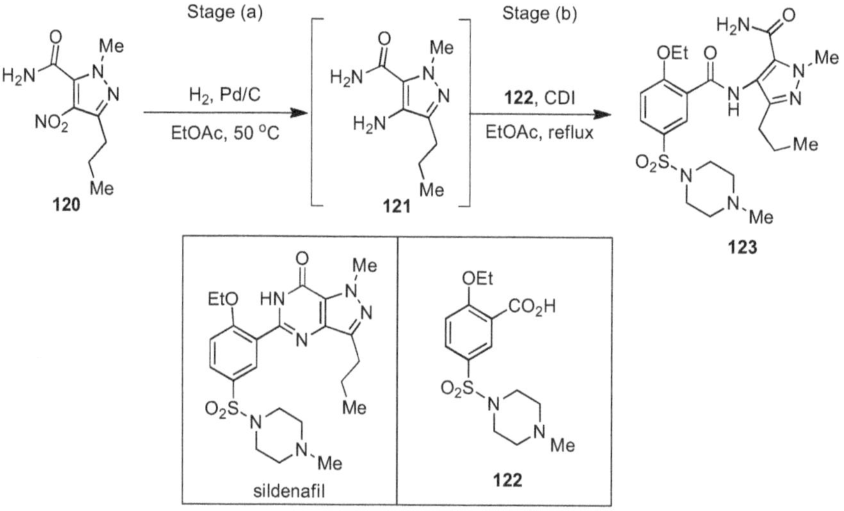

SCHEME 14.30

14.6.4   THF as the Common Solvent

Protection/deprotection is a common approach in organic synthesis. However, these operations on large scale are not favorable because it makes the process less economic in terms of number of steps. Under certain conditions, a deprotection step is able to be telescoped into a one-pot operation. The preparation of hexahydropyridazine derivative **126** was realized in a two-step, one-pot process, involving cyclization of bromide **124** and in situ selective deprotection of Cbz group (Scheme 14.31).[37]

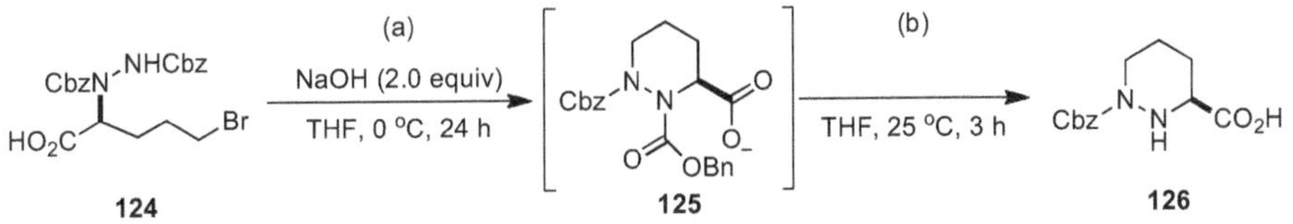

SCHEME 14.31

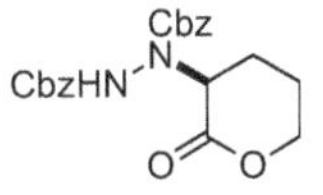

FIGURE 14.1 The structure of δ-lactone **127**.

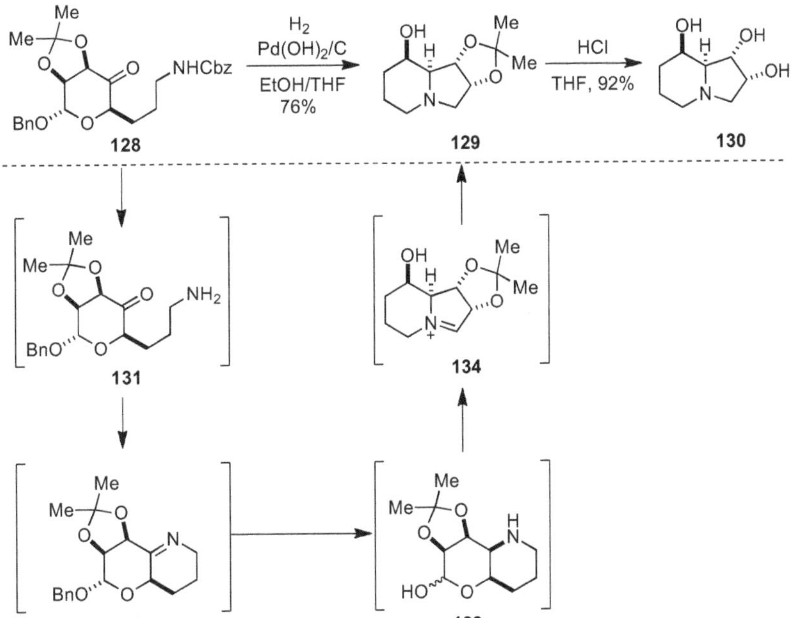

SCHEME 14.32

The first cyclization step was carried out at 0 °C with 2.0 equiv of NaOH to give intermediate **125**. The subsequent removal of the 2-benzyloxycarbonyl group was accomplished when warming the reaction mixture to 25 °C for 3 h. This selective Cbz-deprotection is presumably promoted by the neighboring carboxylate group.

REMARK

It was interesting that δ-Lactone **127** (Figure 14.1) was obtained when using alkoxides such as NaOEt and NaOtBu:

14.6.5   EtOH/THF as the Common Solvent

A one-pot reductive cyclization was used to install the indolizidine ring in 8a-*epi*-swainsonine **130** (Scheme 14.32).[38] This cascade transformation was carried out in a mixture of EtOH and THF involving hydrogenolytic deprotection and double reductive amination with a catalytic amount of palladium hydroxide and at 1 atm of hydrogen. Hydrogenolysis of the Cbz-protecting group in **128** should provide primary amine **131**, which would be converted to the Schiff base **132**. Hydrogenolytic deprotection of the benzyl protecting group and imine reduction in **132** should give hemi-acetal **133**, which can rearrange to a bicyclic iminium ion intermediate **134** whose reduction should afford the protected 8a-*epi*-swainsonine **129**. This cascade process provided a 76% yield of **129**. Finally, the 8a-*epi*-swainsonine **130** was obtained by acidic hydrolysis of acetonide **129** in an excellent yield (92%).

DISCUSSIONS

Solvent plays an important role in organic transformations. Some reactions are not compatible with certain solvents, which demand specific solvents in order to achieve desired results. Thereby solvent exchange is necessary when telescoping reactions which have distinct solvent requirements. For instance, the synthesis of chiral hydroxylamine **139** from amino alcohol **135** was completed in three steps in three different solvents (Scheme 14.33).[39] A condensation of amino alcohol **135** with aldehyde **136** requires reflux in toluene, the subsequent oxidation of the resulting imine **137** with *m*-chloroperbenzoic acid has to be in methylene chloride, and the final reaction of **138** with hydroxylamine hydrochloride demands alcohol solvent. In the event, this three-step telescoped process had to switch the reaction solvent two times (toluene→CH$_2$Cl$_2$→MeOH). Such frequent solvent switch renders the process less efficient.

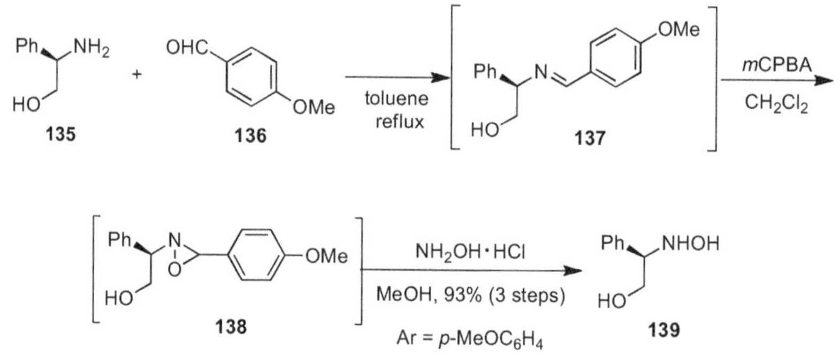

SCHEME 14.33

14.7   OTHER TELESCOPE PROCESSES

To shine a light on a telescoped process, the incorporation of a cascade or tandem reaction sequence into the process would enhance greatly the efficiency of the telescope approach. A number of outstanding examples have been reported to access structurally diverse chemicals in a step-economic manner.[40]

14.7.1   BROMINATION/ISOMERIZATION REACTIONS

Bromination of diketone **140** with dibromodimethylhydantoin (DBDMH) occurred under either base or acid-catalyzed conditions, producing bromide **141** (Scheme 14.34).[41] The subsequent isomerization of **141** to the bromide **144** required acidic conditions. As a result, an acid-catalyzed one-pot bromination/isomerization process was developed by Abbvie Laboratories. This telescoped process was conducted at 60 °C employing DBDMH and a catalytic amount of trifluoroacetic acid (TFA) in acetonitrile, providing cleanly the desired secondary amine **144**.

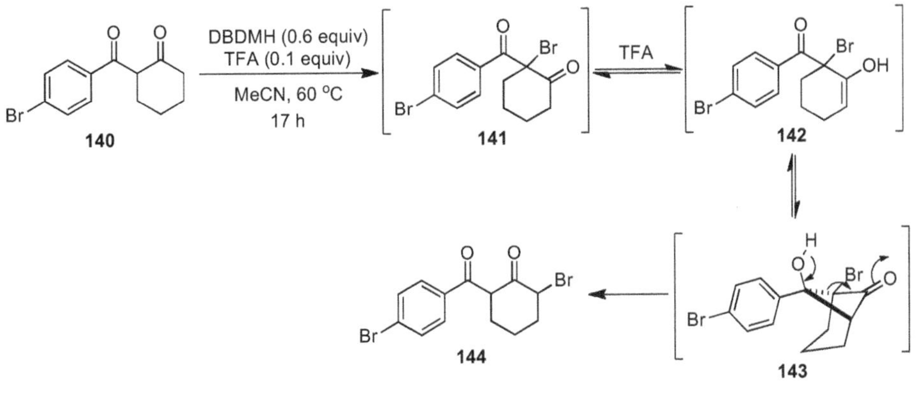

SCHEME 14.34

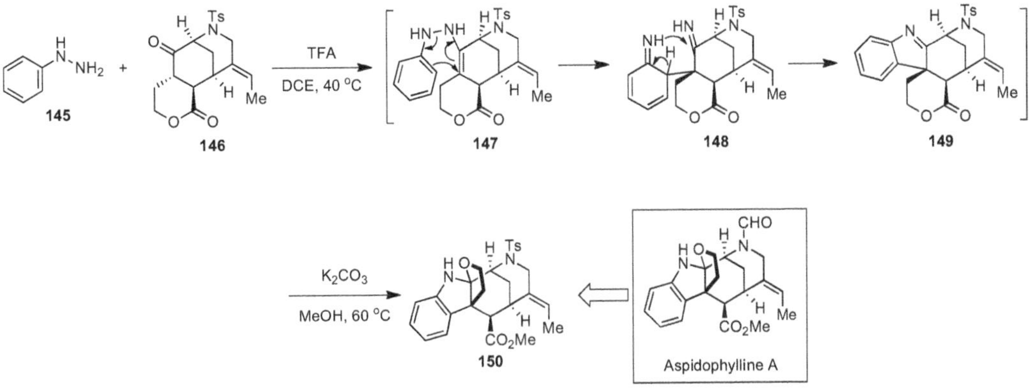

SCHEME 14.35

14.7.2 Fisher Indole Synthesis/Ring Rearrangemet

Application of the telescope approach toward the total synthesis of (±)-Aspidophylline A, a three-ring system was efficiently constructed (Scheme 14.35).[42]

The Fisher indole synthesis was realized via a reaction of cyclic ketone **146** with phenylhydrazine **145** in the presence of TFA in dichloroethane, the subsequent ring rearrangement of indole **149** (not isolated) afforded, after the solvent swap to methanol and addition of potassium carbonate, the methyl ester **150** in 70% overall yield. The rearrangement involves lactone methanolysis followed by cycloaddition. This one-pot protocol led to the introduction of three rings by the assembly of one C–C bond and three C–heteroatom bonds, all with complete diastereoselectivity.

14.7.3 Ylide Formation/Wittig Reaction/Cycloaddition

A three-step telescoped process was designed to access racemic pagoclone **155**, including the formation of ylide **152**, Wittig olefination, and cycloaddition. Pagoclone (+)-**155** is a drug candidate for the treatment of general anxiety disorder and panic disorder. The approach was based on the assumption that the hydroxy-isoindolinone **153** would be in equilibrium with the aldehyde **154**, whose Wittig reaction with **152** would produce the racemic **155** upon cycloaddition (Scheme 14.36).[43]

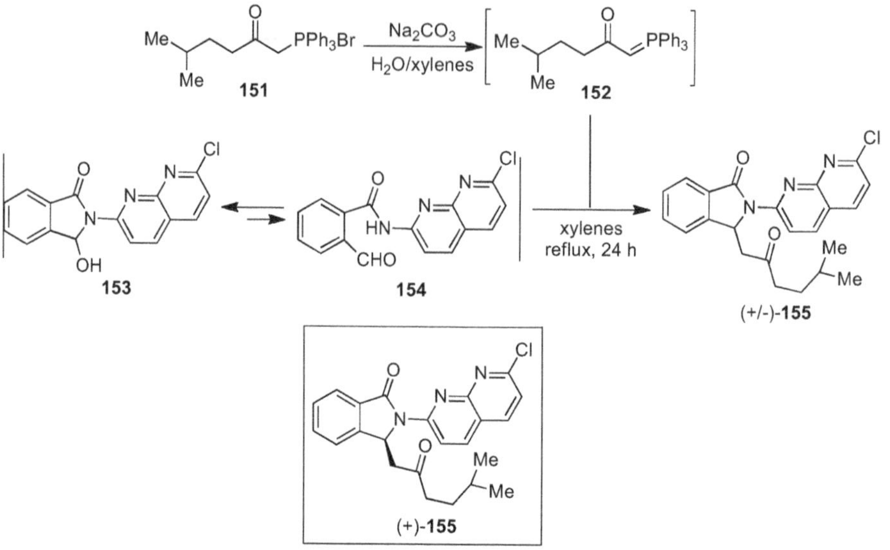

SCHEME 14.36

Procedure

- Charge H_2O (1185 L), followed by the addition of 120 kg (1700 mol) of Na_2CO_3, 370 kg (813 mol) of **151**, and xylenes (1350 kg).
- Stir for 30 min (a clear reaction mixture is obtained).
- Remove the lower aqueous layer.
- Wash the organic layer with aq Na_2CO_3 (60 kg in 1185 L of H_2O).
- Charge, to the organic layer, 158 kg (507 mol) of **153**.
- Heat at 136 °C for 24 h (initially, azeotropic distillation to remove residual water).
- Cool to 80 °C and distill under reduced pressure.
- Add IPA (2284 L).
- Heat the resulting mixture to reflux.
- Cool to 20 °C.
- Filter and wash with IPA (600 L) and MeOH (300 L).
- Dry at 60 °C under vaccum.
- Yield: 170 kg (82%).

14.7.4   OVERMAN REARRANGEMENT

Although Overman rearrangement has been widely applied in academic institutes,[44] its applications on large scale proved challenging due to mainly the relatively harsh reaction conditions and complex molecular structure. A report[45] presented nicely an enantioselective Overman rearrangement for access to an intermediate **159** in the synthesis of quinuclidine **160**, a drug molecule for the treatment of schizophrenia and acute manic disorders (Scheme 14.37). A total of 156.6 kg of **159** was produced in six pilot-plant batches.

14.7.5   ACYLATION/REDUCTION/O-ALKYLATION/BROMINATION

A telescoped process was developed by scientists at Eli Lilly for access to aryl bromide **163**, an intermediate for the synthesis of anti-cancer agent **164**. Scheme 14.38[46] illustrates the four-step, one-pot process which involves acylation, reduction, O-alkylation, and bromination. This telescoped process furnished **163** in 78% overall yield along with two impurities, **165** and **166**.

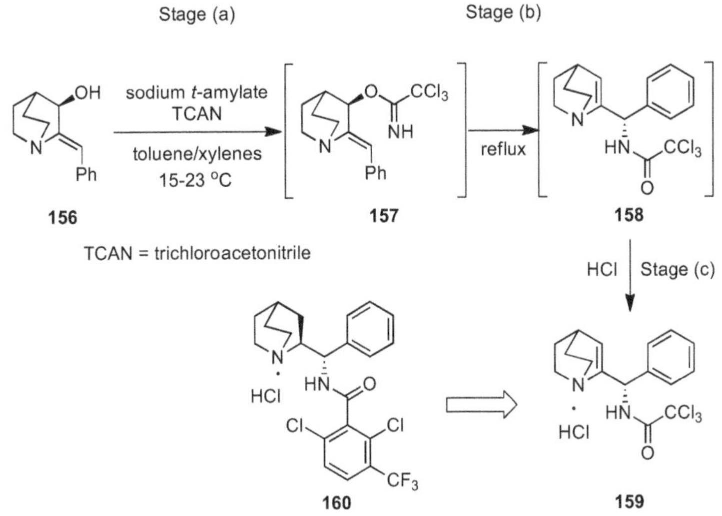

SCHEME 14.37

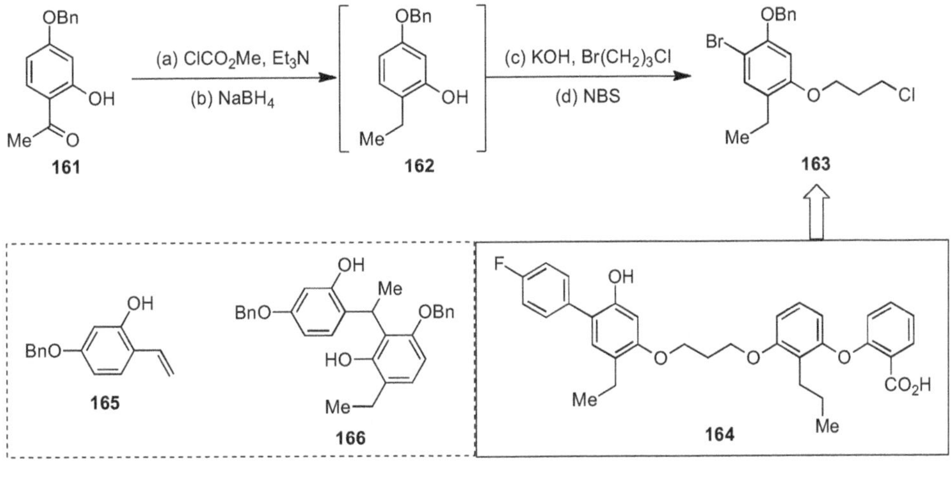

SCHEME 14.38

Utilization of the mechanism-guided process development approach revealed that quinone methide intermediate **169** is responsible for the formation of the two impurities (Scheme 14.39).[46] Thus, maintaining a low concentration of **169** in the reaction mixture was critical to mitigate the impurity formation. Ultimately, levels of **165** and **166** could be suppressed by the controlled addition of the intermediate **167** to NaBH₄ solution.

14.7.6 SYNTHESIS OF (–)-OSELTAMIVIR

In order to achieve a good overall yield for a multi-step synthesis, the yield of each individual reaction must be high. For example, an overall yield of 35% for a ten-step synthesis requires each individual reaction of the sequence proceeds with a 90% yield. The best yield yet achieved for the total synthesis of Tamiflu is approximately 35%.[47] The use of the telescope approach can avoid product loss during the isolation of intermediates, thereby, high overall yields are feasible.

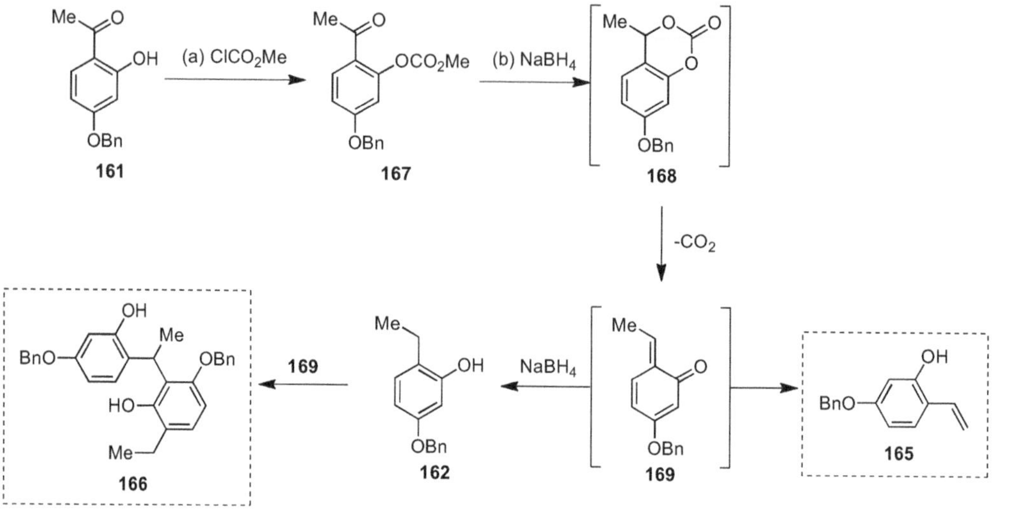

SCHEME 14.39

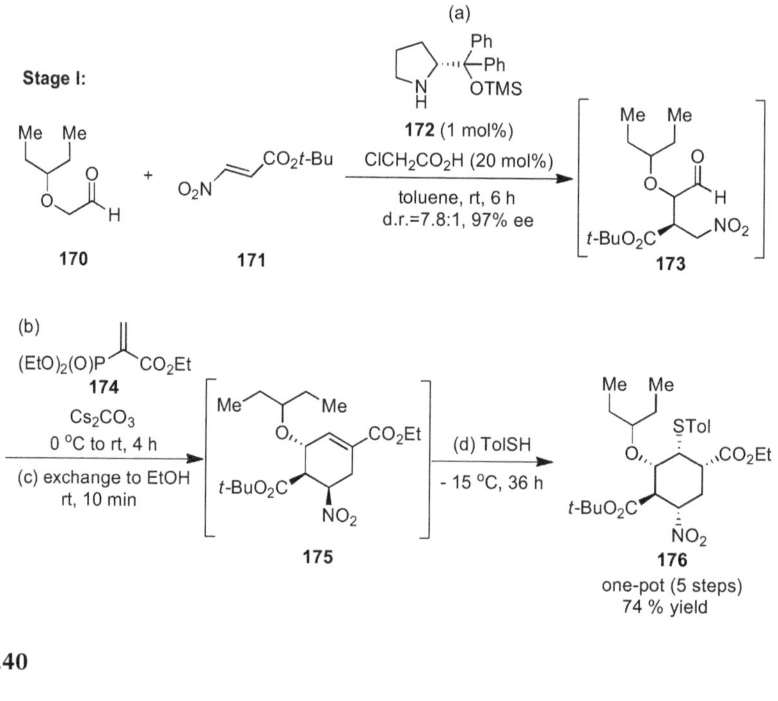

SCHEME 14.40

An efficient asymmetric total synthesis of (–)-oseltamivir **179** (a free base of Tamiflu), an anti-viral reagent, was accomplished through two "one-pot" reaction sequences with excellent overall yield (60%).[48] Scheme 14.40 shows the first one-pot reaction sequence (Stage I) which consists of diphenylprolinol silyl ether mediated asymmetric Michael reaction, domino Michael reaction/ Horner–Wadsworth–Emmons reaction, thiol Michael reaction, and base-catalyzed isomerization.

Two major side products, **177** and **178**, were generated from the cascade Michael reaction/ Horner–Wadsworth–Emmons reaction of **173** with vinyl phosphonate derivative **174** (Scheme 14.41). It was found that these two side products could be converted to the desired intermediate **175** via retro-aldol/Horner–Wadsworth–Emmons reaction and retro Michael reactions by simply switching toluene to ethanol in the presence of Cs_2CO_3.

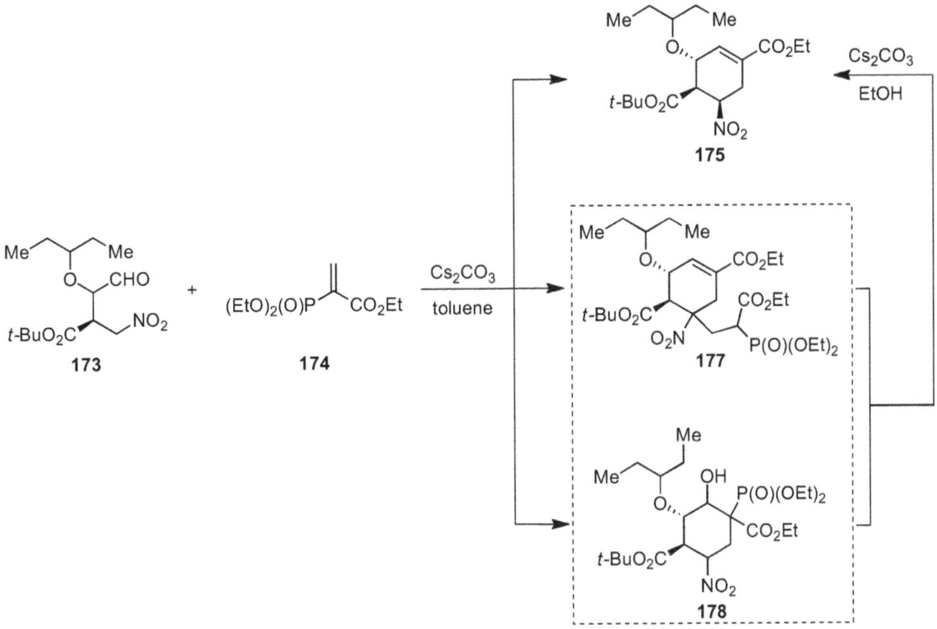

SCHEME 14.41

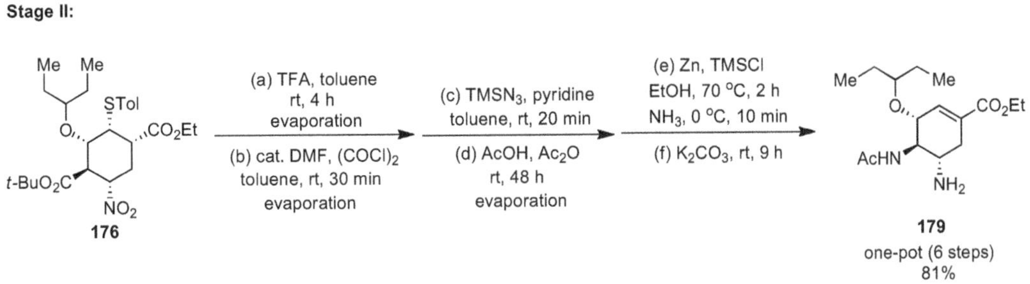

Stage II:

SCHEME 14.42

Scheme 14.42 outlines six reactions (in Stage II) including deprotection of the *tert*-butyl ester and conversion into acyl chloride then acyl azide, Curtius rearrangement, amide formation, reduction of the nitro group into an amine, and retro Michael reaction of thiol moiety. A total yield of 81% was achieved.

14.8 LIMITATION OF THE TELESCOPE APPROACH

While the telescoped process can offer numerous advantages over the classic step-wise process, extensive telescoping operations may provide little control over potential impurities, possibly leading to unacceptable levels of impurities in the isolated product. Such limitations become a concern especially when a telescoped process is installed at the end of the synthetic sequence.

14.8.1 LACK OF PURITY CONTROL

As an illustrated example in the development of LFA-1 (leukocyte function-associated antigen-1) inhibitor **184**, the initial synthesis was accomplished in 89% overall yield via a three-step, one-pot

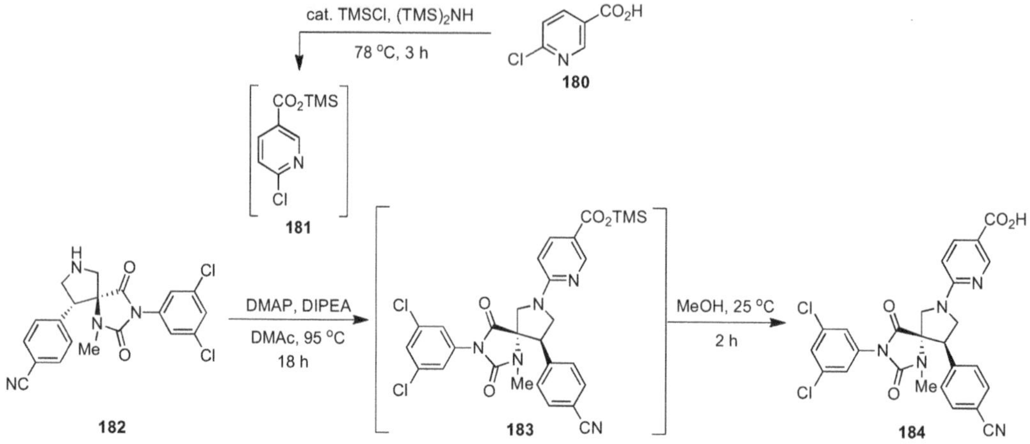

SCHEME 14.43

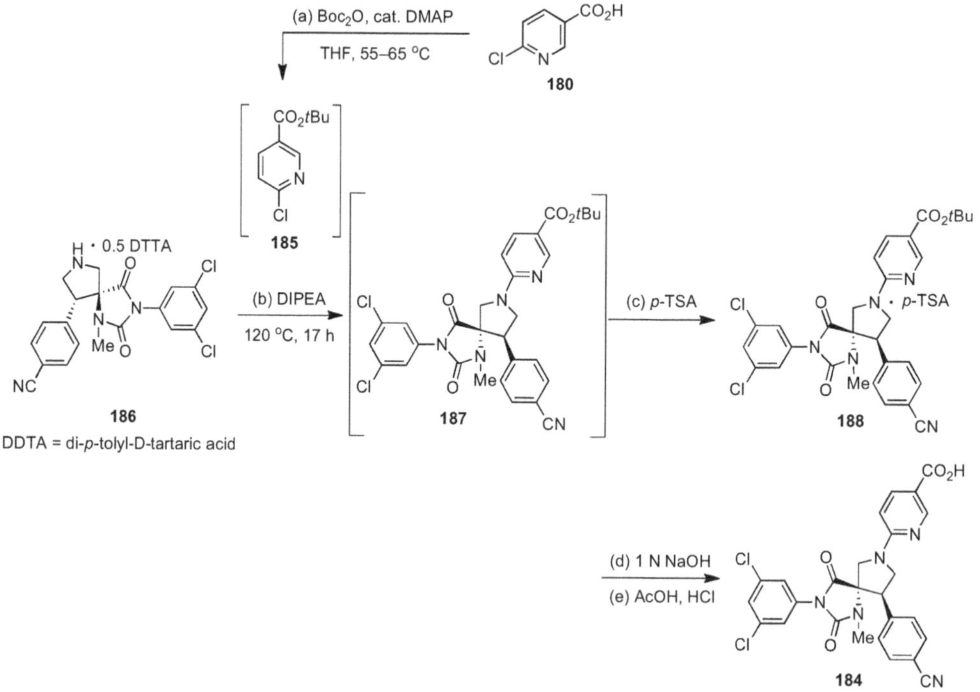

SCHEME 14.44

synthesis, involving intermediates **181** and **183** (Scheme 14.43).[49] Due to the lack of controlling impurity formation, this one-pot synthesis was abandoned.

Consequently, in searching for purity control point (PCP), a crystalline intermediate **188** was identified as the necessary PCP in the modified synthesis. The isolated **188** was in turn submitted to the final hydrolysis step, affording **184** in good purity (Scheme 14.44).

14.8.2   POOR PRODUCT YIELD

The presence of reaction by-products from the previous step may have a detrimental effect on the downstream steps. For example, crystallization of amide **192**, obtained from a telescoped process

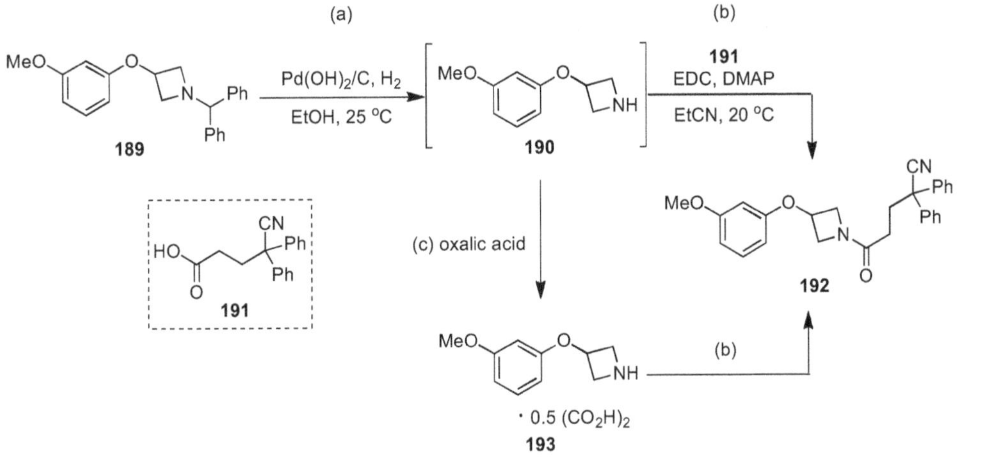

SCHEME 14.45

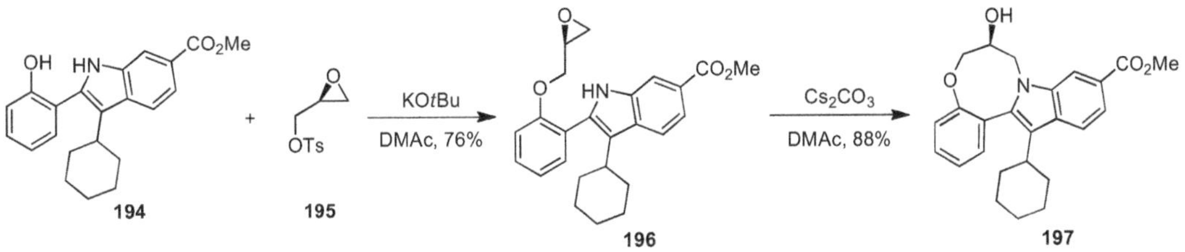

SCHEME 14.46

(**189**→ **190**→ **192**), proved not robust, resulting in **192** in poor yields (<40%). It was postulated that the diphenylmethane by-product, liberated from the benzhydryl protecting group under hydrogenolysis conditions, inhibited the crystallization of product **192** (Scheme 14.45).[50]

Thus, the one-pot protocol was replaced with a step-wise process and the amine **190** was isolated as a hemi-oxalate salt **193** in 84% yield (5.4 kg over two batches) with excellent purity.

14.8.3 LACK OF COMPATIBILITY

Certain types of reaction cannot be telescoped together or a significant amount of side product may be generated. For instance, a S_N2 ether formation and subsequent cycloaddition had to be carried out as discrete steps (Scheme 14.46),[51] albeit both steps in the same solvent (DMAc). In the event, the S_N2 product **196** was isolated to remove potassium *tert*-butoxide, as the formation of eight-membered ring **197** required a base catalyst and a template for the cyclization. Cesium carbonate was the base of choice.

NOTE

The reaction was conducted using a reverse addition protocol by adding **196** to the slurry of Cs_2CO_3 in order to limit the intermolecular reaction.

NOTES

1. Walji, A. M.; MacMillan, D. W. C. *Synlett* **2007**, *10*, 1477.
2. Zhao, W.; Chen, F.-E. *Current Organic Synthesis* **2012**, *9*, 873.

3. Nadkarni, D. V.; Hallissey, J. F. *Org. Process Res. Dev.* **2008**, *12*, 1142.
4. Dunn, P. J.; Hughes, M. L.; Searle, P. M.; Wood, A. S. *Org. Process Res. Dev.* **2003**, *7*, 244.
5. Provencal, D. P.; Gesenberg, K. D.; Wang, H.; Escobar, C.; Wong, H.; Brown, M. A.; Staab, A. J.; Pendri, Y. R. *Org. Process Res. Dev.* **2004**, *8*, 903.
6. Risatti, C.; Natalie, K. J., Jr.; Shi, Z.; Conlon, D. A. *Org. Process Res. Dev.* **2013**, *17*, 257.
7. Young, I. S.; Ortiz, A.; Sawyer, J. R.; Conlon, D. A.; Buono, F. G.; Leung, S. W.; Burt, J. L.; Sortore, E. W. *Org. Process Res. Dev.* **2012**, *16*, 1558.
8. Yu, H.; Richey, R. N.; Stout, J. R.; LaPack, M. A.; Gu, R.; Khau, V. V.; Frank, S. A.; Ott, J. P.; Miller, R. D.; Carr, M. A.; Zhang, T. Y. *Org. Process Res. Dev.* **2008**, *12*, 218.
9. Wang, L.; Neumann, H.; Beller, M. *Angew. Chem. Int. Ed.* **2019**, *58*, 5417.
10. Zhang, H.; Yan, J.; Kanamarlaudi, R. C.; Wu, W.; Keyes, P. *Org. Process Res. Dev.* **2009**, *13*, 807.
11. Agbodjan, A. A.; Cooley, B. E.; Copley, R. C. B.; Corfield, J. A.; Flanagan, R. C.; Glover, B. N.; Guidetti, R.; Haigh, D.; Howes, P. D.; Jackson, M. M.; Matsuoka, R. T.; Medhurst, K. J.; Millar, A.; Sharp, M. J.; Slater, M. J.; Toczko, J. F.; Xie, S. *J. Org. Chem.* **2008**, *73*, 3094.
12. Shu, L.; Wang, P.; Gu, C.; Garofalo, L.; Alabanza, L. M.; Dong, Y. *Org. Process Res. Dev.* **2012**, *16*, 1870.
13. Stewart, G. W.; Brands, K. M. J.; Brewer, S. E.; Cowden, C. J.; Davies, A. J.; Edwards, J. S.; Gibson, A. W.; Hamilton, S. E.; Katz, J. D.; Keen, S. P.; Mullens, P. R.; Scott, J. P.; Wallace, D. J.; Wise, C. S. *Org. Process Res. Dev.* **2010**, *14*, 849.
14. (a) Peters, R.; Waldmeier, P.; Joncour, A. *Org. Process Res. Dev.* **2005**, *9*, 508. (b) Alcaraz, M.-L.; Atkinson, S.; Cornwall, P.; Foster, A. C.; Gill, D. M.; Humphries, L. A.; Keegan, P. S.; Kemp, R.; Merifield, E.; Nixon, R. A.; Noble, A. J.; O'Beirne, D.; Patel, Z. M.; Perkins, J.; Rowan, P.; Sadler, P.; Singleton, J. T.; Tornos, J.; Watts, A. J.; Woodland, I. A. *Org. Process Res. Dev.* **2005**, *9*, 555. (c) Pu, Y.-M.; Grieme, T.; Gupta, A.; Plata, D.; Bhatia, A. V.; Cowart, M.; Ku, Y.-Y. *Org. Process Res. Dev.* **2005**, *9*, 45. (d) Ikunaka, M.; Matsumoto, J.; Fujima, Y.; Hirayama, Y. *Org. Process Res. Dev.* **2002**, *6*, 49. (e) de Koning, P. D.; Castro, N.; Gladwell, I. R.; Morrison, N. A.; Moses, I. B.; Panesar, M. S.; Pettman, A. J.; Thomson, N. M. *Org. Process Res. Dev.* **2011**, *15*, 1256. (f) Savage, S. A.; Waltermire, R. E.; Campagna, S.; Bordawekar, S.; Toma, J. D. R. *Org. Process Res. Dev.* **2009**, *13*, 510. (g) Xu, D. D.; Waykole, L.; Calienni, J. V.; Ciszewski, L.; Lee, G. T.; Liu, W.; Szewczyk, J.; Vargas, K.; Prasad, K.; Repič, O.; Blacklock, T. J. *Org. Process Res. Dev.* **2003**, *7*, 856.
15. Shi, Z.; Kiau, S.; Lobben, P.; Hynes, J., Jr.; Wu, H.; Parlanti, L.; Discordia, R.; Doubleday, W. W.; Leftheris, K.; Dyckman, A. J.; Wrobleski, S. T.; Dambalas, K.; Tummala, S.; Leung, S.; Lo, E. *Org. Process Res. Dev.* **2012**, *16*, 1618.
16. Enache, L. A.; Kennedy, I.; Sullins, D. W.; Chen, W.; Ristic, D.; Stahl, G. L.; Dzekhtser, S.; Erickson, R. A.; Yan, C.; Muellner, F. W.; Krohn, M. D.; Winger, J.; Sandanayaka, V.; Singh, J.; Zembower, D. E.; Kiselyov, A. S.; *Org. Process Res. Dev.* **2009**, *13*, 1177.
17. (a) Scalone, M.; Waldmeter, P. *Org. Process Res. Dev.* **2003**, *7*, 418. (b) Jong, R. L. D.; Davidson, J. G.; Dozeman, G. J.; Fiore, P. J.; Giri, P.; Kelly, M. E.; Puls, T. P.; Seamans, R. E. *Org. Process Res. Dev.* **2001**, *5*, 216.
18. (a) Lanier, M.; Haddach, M.; Pastor, R.; Riess, J. G. *Tetrahedron Lett.* **1993**, *34*, 2469. (b) Ishihara, T.; Hayashi, H.; Yamanaka, H. *Tetrahedron Lett.* **1993**, *34*, 5777. (c) Haas, A. M.; Hägele, G. *J. Fluorine Chem.* **1996**, *78*, 75. (d) Ishihara, T.; Takahashi, A.; Hayashi, H.; Yamanaka, H.; Kubota, T. *Tetrahedron Lett.* **1998**, *39*, 4691.
19. Yamazaki, T.; Kobayashi, R.; Kitazume, T.; Kubota, *J. Org. Chem.* **2006**, *71*, 2499.
20. Crisóstomo, F. R. P.; Carrillo, R.; Martín, T.; García-Tellado, F.; Martín, V. S. *J. Org. Chem.* **2005**, *70*, 10099.
21. Funel, J.-A.; Abele, S. *Angew. Chem. Int. Ed.* **2013**, *52*, 3822.
22. Yanagisawa, A.; Nishimura, K.; Ando, K.; Nezu, T.; Maki, A.; Kato, S.; Tamaki, W.; Imai, E.; Mohri, S. *Org. Process Res. Dev.* **2010**, *14*, 1182.
23. Stang, E. M.; White, C. *J. Am. Chem. Soc.* **2011**, *133*, 14892.
24. Praquin, C. F. B.; de Koning, P. D.; Peach, P. J.; Howard, R. M.; Spencer, S. L. *Org. Process Res. Dev.* **2011**, *15*, 1124.
25. Chekal, B. P.; Guinness, S. M.; Lillie, B. M.; McLaughlin, R. W.; Palmer, C. W.; Post, R. J.; Sieser, J. E.; Singer, R. A.; Sluggett, G. W.; Vaidyanathan, R.; Withbroe, G. J. *Org. Process Res. Dev.* **2014**, *18*, 266.
26. Yoshida, S.; Yishino, T.; Miyafuji, A.; Yasuda, H.; Kimura, T.; Takahashi, T.; Mukuta, T. *Org. Process Res. Dev.* **2012**, *16*, 1544.

27. Ragan, J. A.; Bourassa, D. E.; Blunt, J.; Breen, D.; Busch, F. R.; Cordi, E. M.; Damon, D. B.; Do, N.; Engtrakul, A.; Lynch, D.; McDermott, R. E.; Mongillo, J. A.; O'Sullivan, M. M.; Rose, P. R.; Vanderplas, B. C. *Org. Process Res. Dev.* **2009**, *13*, 186.

28. Hayler, J. D.; Howie, S. L. B.; Giles, R. G.; Negus, A.; Oxley, P. W.; Walsgrove, T. C.; Whiter, M. *Org. Process Res. Dev.* **1998**, *2*, 3.

29. Lukin, K.; Hsu, M. C.; Chambournier, G.; Kotecki, B.; Venkatramani, C. J.; Leanna, M. R. *Org. Process Res. Dev.* **2007**, *11*, 578.

30. Itoh, T.; Kato, S.; Nonoyama, N.; Wada, T.; Maeda, K.; Mase, T. *Org. Process Res. Dev.* **2006**, *10*, 822.

31. Delmonte, A. J.; Waltermire, R. E.; Fan, Y.; McLeod, D. D.; Gao, Z.; Gesenberg, K. D.; Girard, K. P.; Rosingana, M.; Wang, X.; Kuehne-Willmore, J.; Braem, A. D.; Castoro, J. A. *Org. Process Res. Dev.* **2010**, *14*, 553.

32. Ku, Y.-Y.; Grieme, T.; Raje, P.; Sharma, P.; Morton, H. E.; Rozema, M.; King, S. A. *J. Org. Chem.* **2003**, *68*, 3238.

33. For telescoping steps in one common solvent, see: (a) Leroy, V.; Lee, G. E.; Lin, J.; Herman, S. H.; Lee, T. B. *Org. Process Res. Dev.* **2001**, *5*, 179.

34. Ragan, J. A.; Murry, J. A.; Castaldi, M. J.; Conrad, A. K.; Jones, B. P.; Li, B.; Makowski, T. W.; McDermott, R.; Sitter, B. J.; White, T. D.; Young, G. R. *Org. Process Res. Dev.* **2001**, *5*, 498.

35. Pasti, J.; Chen, C.-K.; Spangler, L.; DelMonte, A. J.; Benoit, S.; Berglund, D.; Bien, J.; Brodfuehrer, P.; Chan, Y.; Corbett, E.; Costello, C.; DeMena, P.; Discordia, R. P.; Doubleday, W.; Gao, Z.; Gingras, S.; Grosso, J.; Haas, O.; Kacsur, D.; Lai, C.; Leung, S.; Miller, M.; Muslehiddinoglu, J.; Nguyen, N.; Qiu, J.; Olzog, M.; Reiff, E.; Thoraval, D.; Totleben, M.; Vanyo, D.; Vemishetti, P.; Wasylak, J.; Wei, C.; *Org. Process Res. Dev.* **2009**, *13*, 716.

36. Dale, D. J.; Dunn, P. J.; Golightly, C.; Hughes, M. L.; Levett, P. C.; Pearce, A. K.; Searle, P. M.; Ward, G.; Wood, A. S. *Org. Process Res. Dev.* **2000**, *4*, 17.

37. Chen, Y.; Lu, Y.; Zou, Q.; Chen, H.; Ma, D. *Org. Process Res. Dev.* **2013**, *17*, 1209.

38. Abrams, J. N.; Babu, R. S.; Guo, H.; Le, D.; Le, J.; Osbourn, J. M.; O'Doherty, G. A. *J. Org. Chem.* **2008**, *73*, 1935.

39. Tamura, O.; Gotanda, K.; Yoshino, J.; Morita, Y.; Terashima, R.; Kikuchi, M.; Miyawaki, T.; Mita, N.; Yamashita, M.; Ishibashi, H.; Sakamoto, M. *J. Org. Chem.* **2000**, *65*, 8544.

40. (a) Wu, X.; Lin, H.-C.; Li, M.-L.; Li, L.-L,; Han, Z.-Y.; Gong, L.-Z. *J. Am. Chem. Soc.* **2015**, *137*, 13476. (b) Zhang, F.-G.; Eppe, G.; Marek, I. *Angew. Chem. Int. Ed.* **2016**, *55*, 714. (c) Liu, Z.; Zeng, T.; Yang, K. S.; Engle, K. M. *J. Am. Chem. Soc.* **2016**, *138*, 15122. (d) de Gracia Retamosa, M.; Ruiz-Olalla, A.; Bello, T.; de Cózar, A.; Cossío, F. P. *Angew. Chem. Int. Ed.* **2018**, *57*, 668. (e) Chen, J.-Q.; Yu, W.-L.; Wei, Y.-L.; Li, T.-H.; Xu, P.-F. *J. Org. Chem.* **2017**, *82*, 243. (f) Sun, J.; Qiu, J.-K.; Jiang, B.; Hao, W.-J.; Guo, C.; Tu, S.-J. *J. Org. Chem.* **2016**, *81*, 3321. (g) Wang, L.; Neumann, H.; Beller, M. *Angew. Chem. Int. Ed.* **2019**, *58*, 5417. (h) Veliu, R.; Schneider, C. *J. Org. Chem.* **2021**, *86*, 11960.

41. Ravn, M. M.; Wagaw, S. H.; Engstrom, K. M.; Mei, J.; Kotecki, B.; Souers, A. J.; Kym. P. R.; Judd, A. S.; Zhao, G. *Org. Process Res. Dev.* **2010**, *14*, 417.

42. Zu, L.; Boal, B. W.; Garg, N. K. *J. Am. Chem. Soc.* **2011**, *133*, 8877.

43. Stuk, T. L.; Assink, B. K.; Bates, Jr., R. C.; Erdman, D. T.; Fedij, V.; Jennings, S. M.; Lassig, J. A.; Smith, R. J.; Smith, T. L. *Org. Process Res. Dev.* **2003**, *7*, 851.

44. For leading references to the Overman rearrangement, see: (a) Overman, L. E.; Carpenter, N. E. The Allylic Trihaloacetimidate Rearrangement. In *Organic Reactions*; Overman, L. E., Ed.; John Wiley & Sons, Inc., New York, 2005; Vol. 66, pp 3–107. (b) Doherty, A. M.; Kornberg, B. E.; Reily, M. D. *J. Org. Chem.* **1993**, *58*, 795. (c) Overman, L. E. *Acc. Chem. Res.* **1980**, *13*, 218. (d) Anderson, C. E.; Overman, L. E. *J. Am. Chem. Soc.* **2003**, *125*, 12412. (e) Chen, B.; Mapp, A. K. *J. Am. Chem. Soc.* **2004**, *126*, 5364. (f) Lee, E. E.; Batey, R. A. *Angew. Chem. Int. Ed.* **2004**, *43*, 1865. (g) Fischer, D. F.; Xin, Z.-Q.; Peters, R. *Angew. Chem. Int. Ed.* **2007**, *46*, 7704. (h) Nishikawa, T.; Asai, M.; Ohyabu, N.; Isobe, M. *J. Org. Chem.* **1998**, *63*, 188. (i) Chen, Y. K.; Lurain, A. E.; Walsh, P. J. *J. Am. Chem. Soc.* **2002**, *124*, 12225. (j) Marion, N.; Gealageas, R.; Nolan, S. P. *Org. Lett.* **2007**, *9*, 2653.

45. Chandramouli, S. V.; Ayers, T. A.; Wu, X.-D.; Tran, L. T.; Peers, J. H.; Disanto, R.; Roberts, F.; Kumar, N.; Jiang, Y.; Choy, N.; Pemberton, C.; Powers, M. R.; Gardetto, A. J.; D'Netto, G. A.; Chen, X.; Gamboa, J.; Ngo, D.; Copeland, W.; Rudisill, D. E.; Bridge, A. W.; Vanasse, B. J.; Lythgoe, D. J. *Org. Process Res. Dev.* **2012**, *16*, 484.

46. Yates, M. H.; Koenig, T. M.; Kallman, N. J.; Ley, C. P.; Mitchell, D. *Org. Process Res. Dev.* **2009**, *13*, 268.

47. Federspiel, M.; Fischer, R.; Hennig, M.; Mair, H.-J.; Oberhauser, T.; Rimmler, G.; Albiez, T.; Bruhin, J.; Estermann, J.; Gandert, C.; Göckel, V.; Götzö, S.; Hoffmann, U.; Huber, G.; Janatsch, G.; Lauper, S.; Röckel-Stäbler, O.; Trussardi, S.; Zwahlen, A. G. *Org. Process Res. Dev.* **1999**, *3*, 266.
48. Ishikawa, H.; Suzuki, T.; Orita, H.; Uchimaru, T.; Hayashi, Y. *Chem. Eur. J.* **2010**, *16*, 12616.
49. Delmonte, A. J.; Fan, Y.; Girard, K. P.; Jones, G. S.; Waltermire, R. E.; Rosso, V.; Wang, X. *Org. Process Res. Dev.* **2011**, *15*, 64.
50. Dillon, B. R.; Roberts, D. F.; Entwistle, D. A.; Glossop, P. A.; Knight, C. J.; Laity, D. A.; James, K.; Praquin, C. F.; Stang, R. S.; Watson, C. A. L. *Org. Process Res. Dev.* **2012**, *16*, 195.
51. Scott, J. P.; Alam, M.; Bremeyer, N.; Goodyear, A.; Lam, T.; Wilson, R. D.; Zhou, G. *Org. Process Res. Dev.* **2011**, *15*, 1116.

15 Design of New Synthetic Route

Evaluation of existing synthetic routes and design of new routes for scale-up need to consider a number of factors. In addition to the established six SELECT (safety, environmental impact, legality, economy, control, and throughput) criteria, other factors, such as the number of synthetic steps, type of reactions, reaction sequence, and product purification, also play important roles during the design and development of a chemical process. Furthermore, the instability of starting materials/intermediates pose concerns during transportation and storage.

Efficient synthetic methods require the assembly of complex molecules with reactions that are both selective (chemo-, regio-, diastereo-, and enantio-) and economical in atom count (maximum number of atoms of reactants appearing in the products). The addition reaction is preferred over the elimination reaction as the addition will build up the molecular skeleton while elimination will lose fragment(s) of the molecule which, in most cases, become waste. Substitution reaction is inherently a more balanced transformation given that the replacement occurs between two compatible pieces in terms of mass. A good rule of thumb when designing a new synthetic route is:

(a) use of a catalysis system.
(b) use of a convergent over linear approach.
(c) use of a tandem or cascade process.
(d) use of a multicomponent reaction (MCR).
(e) use of a telescope strategy to minimize isolation steps.

15.1 IMPROVING PROCESS SAFETY

Safety is of primary importance for any chemical process. Process safety can be attributed to several factors including, but not limited to, the use of toxic reagents, exothermic reactions, and harsh reaction conditions.

15.1.1 TOXIC REAGENTS AND INTERMEDIATES

15.1.1.1 Trimethylsilyl Cyanide as Cyanide Source

Trimethylsilyl cyanide (TMSCN) is frequently used in organic synthesis as the source of cyanide. For example, the synthesis of 2-aminomethyl-3-fluoropyridine **3** (as the hydrochloride salt) involved the oxidation of 3-fluoropyridine **1**, a subsequent reaction of the resulting *N*-oxide with TMSCN, and hydrogenation of the nitrile **2** under aqueous acidic conditions to give **3** in 55% overall yield (Scheme 15.1).[1]

However, TMSCN may hydrolyze to give hydrogen cyanide. In addition, oxidation with hydrogen peroxide was also a safety concern.

In order to address these process issues, a new synthetic route was developed (Scheme 15.2). In the new synthesis, the use of TMSCN and the potentially hazardous peroxide oxidation were eliminated. This synthetic route has been demonstrated to be viable for the preparation of multi-kilogram quantities of **3**.

15.1.1.2 CuCN in Sandmeyer Reaction

Compound **8** is a fragment for the synthesis of potent bradykinin 1 antagonist **11**, developed by Merck (Scheme 15.3).[2] The preparation of **8** involved the installation of a 1,2,4-oxadiazole ring. The original synthesis (Route I, Scheme 15.3) relied on the preparation of a highly functionalized benzonitrile

DOI: 10.1201/9781003288411-15

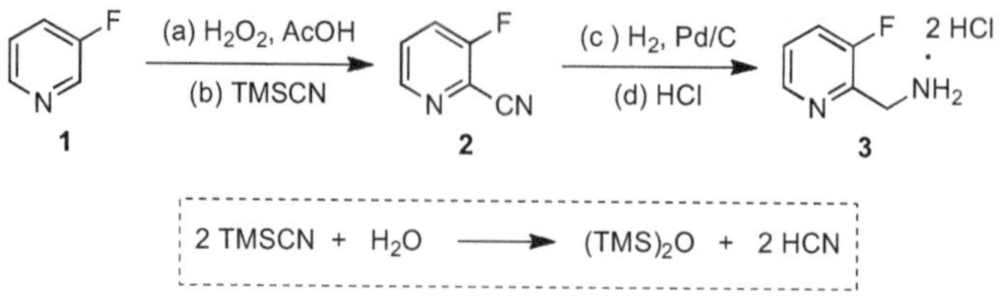

SCHEME 15.1

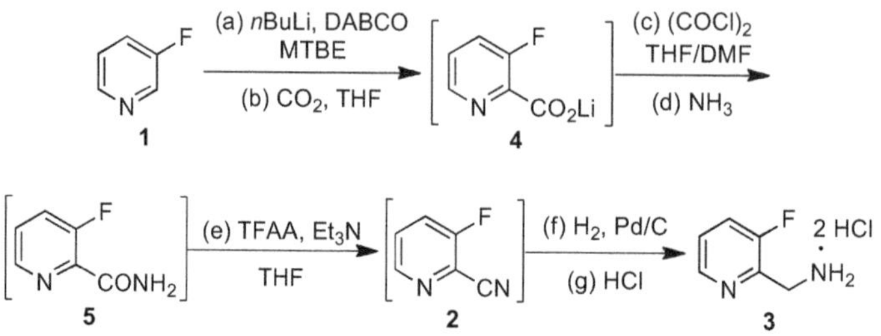

SCHEME 15.2

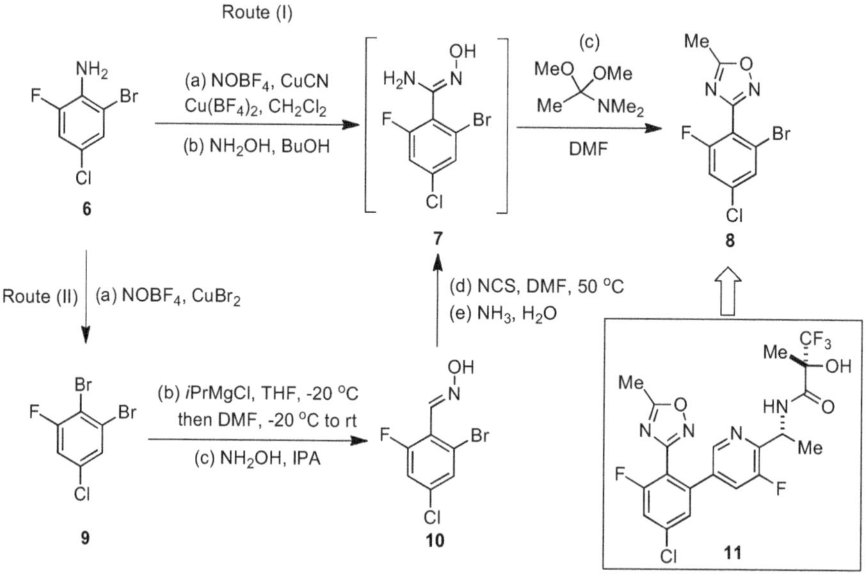

Route (I)

Route (II)

SCHEME 15.3

using Sandmeyer reaction with cyanide, which suffered from one major safety issue: evolution of HCN under the reaction conditions (pH=1). Thus, a new synthetic route (Route II, Scheme 15.3) was developed using a novel regioselective metal-halogen exchange reaction of 1,2-dibromo-5-chloro-3-fluorobenzene **9** with isopropyl magnesium chloride to install the 1,2,4-oxadiazole ring structure.

15.1.1.3 Toxic Reagent (HF)

Sevoflurane, 1,1,1,3,3,3-hexafluoro-2-(fluoromethoxy)propane, is one of the most widely used inhalation anesthetics. Sevoflurane is currently manufactured by a reaction of commercially available 1,1,1,3,3,3-hexafluoro-2-propanol **12** with paraformaldehyde in the presence of large molar excess of sulfuric acid and hydrogen fluoride gas (Route I, Scheme 15.4).[3]

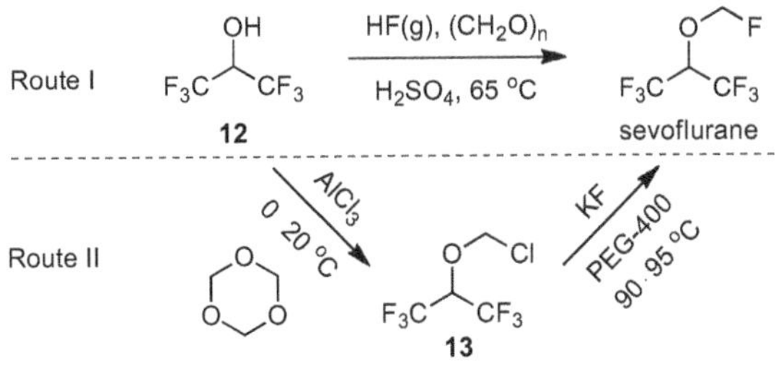

SCHEME 15.4

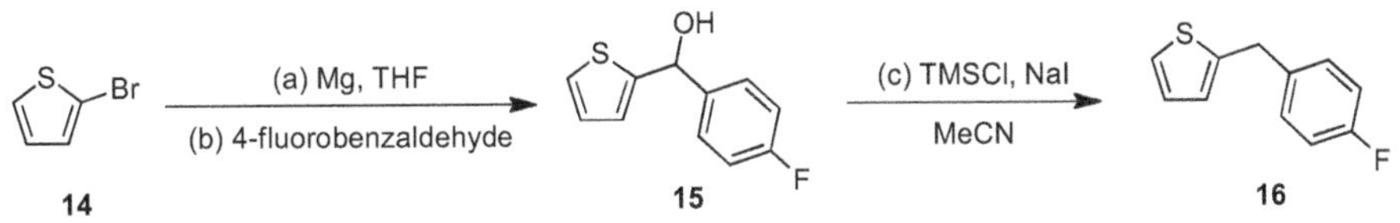

SCHEME 15.5

Although this one-step process reportedly produces the crude sevoflurance in yields of up to 71%, HF gas is extremely corrosive and highly toxic and requires special precautions for its handling and use. Due to the caustic nature of HF, special reactors must be used that resist degradation by the action of this reagent.

In order to avoid using HF, a safe and efficient process for the synthesis of the inhalation anesthetic sevoflurane was developed by Abbott Laboratories (as shown in Route II, Scheme 15.4).[4] This two-step protocol was conducted in one-vessel and addressed many of the hazards and complications of the current process for sevoflurane manufacture. The new method was easily scaled up to produce 10-kg batches of 99.4% pure sevoflurane.

15.1.1.4   Toxic Benzyl Halides

15.1.1.4.1   Benzylation of 2-Bromothiohene

Direct benzylation of 2-bromothiophene **14** with benzyl halides is limited on large scale due to the toxicity of benzylation agents. Therefore, an alternative benzylation method was developed[5] via a tandem Grignard reaction and iodotrimethylsilane (TMSI)-mediated reduction (Scheme 15.5).

The deoxygenation of biaryl methanol **15**, formed from the Grignard reaction with 4-fluorobenzaldehyde, was realized with in situ generated TMSI to afford an excellent yield of **16**.

Procedure

Stage (c)
- Charge 4189 g (28.0 mol, 4.36 equiv) of NaI into a 50-L, three-necked reactor, followed by adding MeCN (1.5 L) and 3554 mL (3042 g, 28.0 mol, 4.36 equiv) of TMSCl.
- Stir the resulting mixture at room temperature for 15 min.
- Cool to 0 °C.
- Charge slowly a solution of crude **15** (1470 g, 91% pure, 6.42 mol) in MeCN (1.5 L) via an addition funnel over 3.5 h at <10 °C.
- Warm the batch to room temperature and stir overnight (12 h).
- Cool the batch to 5 °C.
- Add an aqueous solution of NaOH (760 g in 5 L of water, 2.96 equiv) at ≤40 °C over 10 min.
- Cool the resulting mixture to room temperature and stir for 45 min.

- Add EtOAc (3.5 L) and a solution of $Na_2S_2O_3 \cdot 5H_2O$ (3598 g, 14.50 mol, 2.25 equiv) in water (3.5 L) and stir for 1 h.
- Separate layers.
- Wash the organic layer with a solution of NaOH (140 g, 0.50 equiv) and $Na_2S_2O_3 \cdot 5H_2O$ (553.4 g, 0.35 equiv) in water (4 L).
- Stir the resulting organic layer with water (3 L) containing 6.5 mL of TEA for 1 h.
- Add 500 g of NaCl prior to layer separation.
- Dry the organic layer over $MgSO_4$, filter, and distill the filtrate under reduced pressure.
- Yield of **16**: 1319 g (97%, 93% pure) as a pale-brown oil.

15.1.1.4.2 Benzylation of Indole Derivative

To prepare racemic acid **20**, the initial pilot plant campaign used the chemistry as shown in Scheme 15.6,[6] producing 22 kg of the material.

(I) Problems

- Bromoacid **19** showed mild exothermic activity by DSC above 70 °C.
- **19** was lachrymatory and showed hydrolytic sensitivity as well.

(II) Solutions

To address these issues associated with the bromoacid **19**, a more efficient approach was developed by an acid-promoted direct coupling of 3,4-methylenedioxymandelic acid in the presence of trifluoroacetic acid in acetonitrile (Equation 15.1).

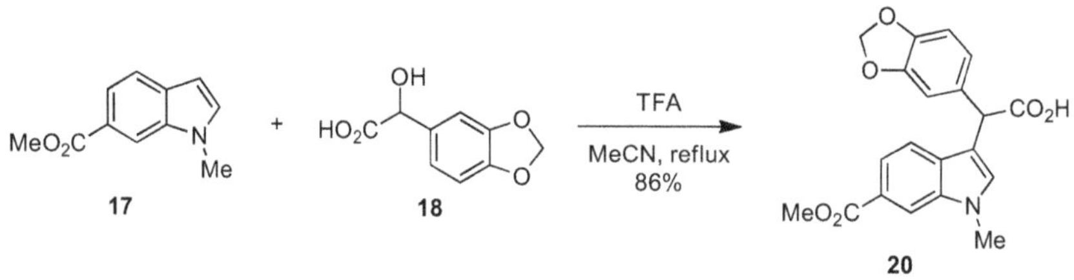

$$(15.1)$$

This new protocol was successfully employed in the pilot plant campaign on a maximum scale of 24.9 kg of **18**, avoiding the use of the lachrymatory and thermal labile bromoacid **19**.

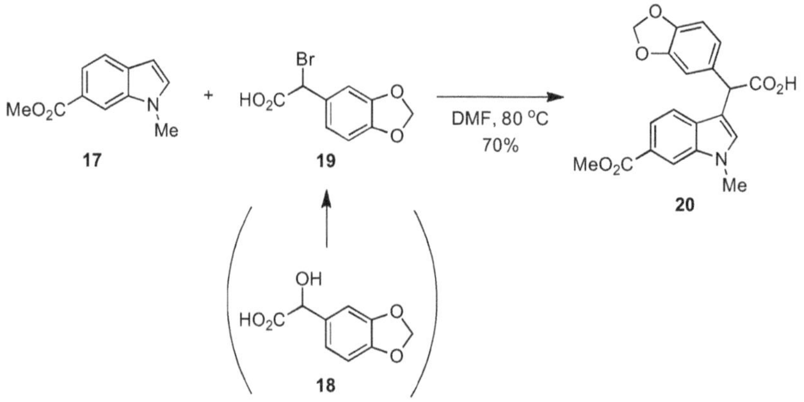

SCHEME 15.6

Procedure

- Charge 28.9 kg (253.6 mol) of TFA to a stirred suspension of **17** (24 kg, 126.8 mol) and **18** (24.9 kg, 126.8 mol) in MeCN (240 L).
- Stir under reflux for 24 h.
- Cool to room temperature and granulate for 16 h.
- Filter, wash with MeCN (2×24 L), and dry at 45 °C under vacuum.
- Yield: 40.2 kg (86%) as an off-white crystalline solid.

15.1.1.5 Phosphorus Oxychloride

The use of phosphorus oxychloride in the preparation of intermediate **23** presented handling and potential safety challenges (Scheme 15.7).[7]

To obviate the process safety issue, a convenient approach was to re-organize the bond-formation sequence. Scheme 15.8 demonstrated that phosphorus oxychloride could be eliminated via re-ordering the bond-formation steps.

15.1.1.6 Sulfonyl Chloride Intermediate

Sildenafil is the first agent for the treatment of male erectile dysfunction (a disease more commonly known as male impotence). This new prescription drug was approved by the United States and the European Union in 1998. The final step for the synthesis of sildenafil in the medicinal chemistry route (a linear nine-step synthesis) is a coupling reaction of sulphonyl chloride **27** with N-methylpiperazine (Scheme 15.9).[8] Due to the potentially toxic material **27**, multiple recrystallizations of the final product were required to reduce the toxic impurity to accepted low levels.

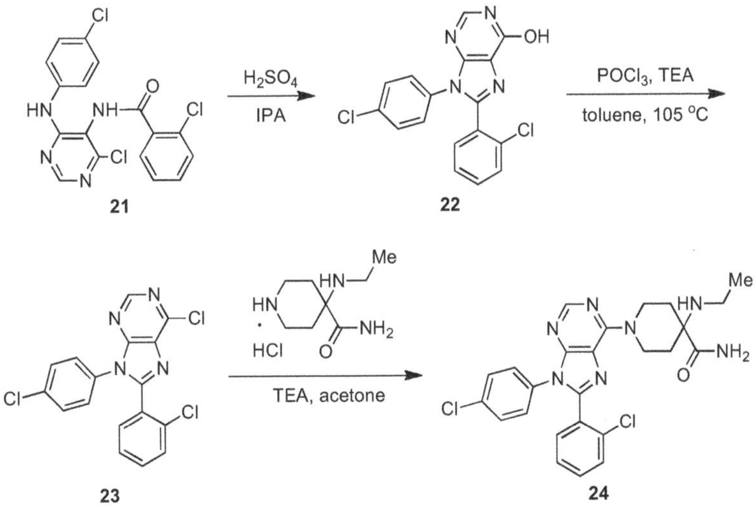

SCHEME 15.7

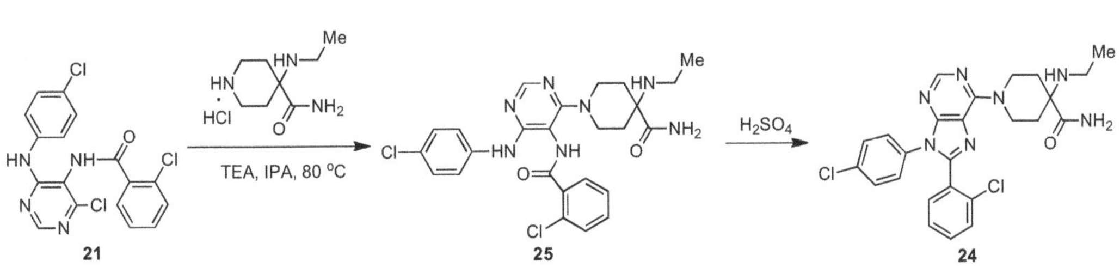

SCHEME 15.8

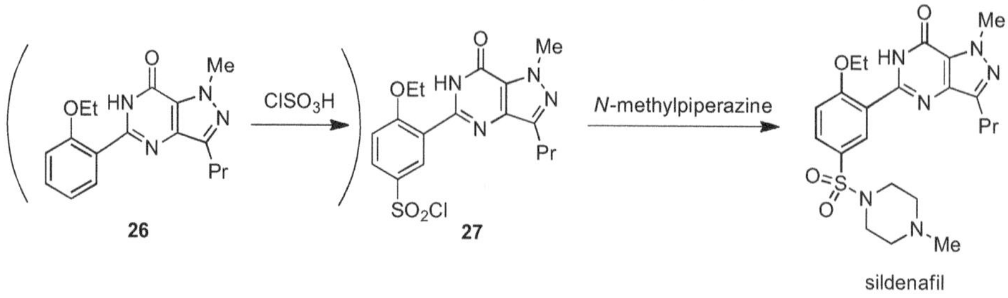

SCHEME 15.9

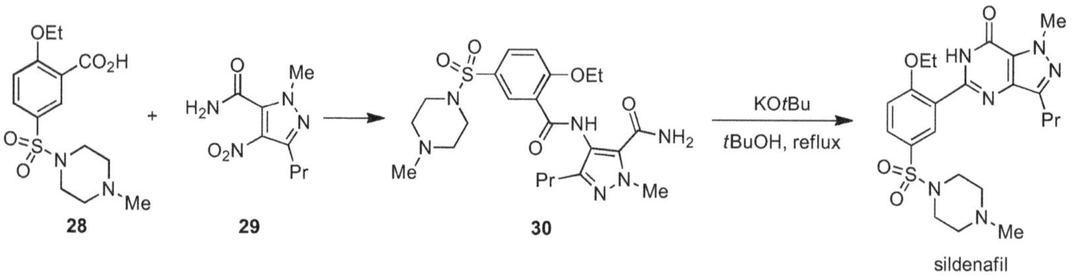

SCHEME 15.10

To circumvent the drawbacks of the medicinal route, a convergent synthetic route was developed consisting of a clean cyclization as the final step, which avoided using the toxic **27** (Scheme 15.10).

15.1.2 HIGH ENERGY REAGENTS

15.1.2.1 Azide-Involved Cycloaddition

Compound **33** displays promising antibacterial activity, and Equation 15.2[9] outlines the final step for the synthesis of **33** in the discovery synthesis.

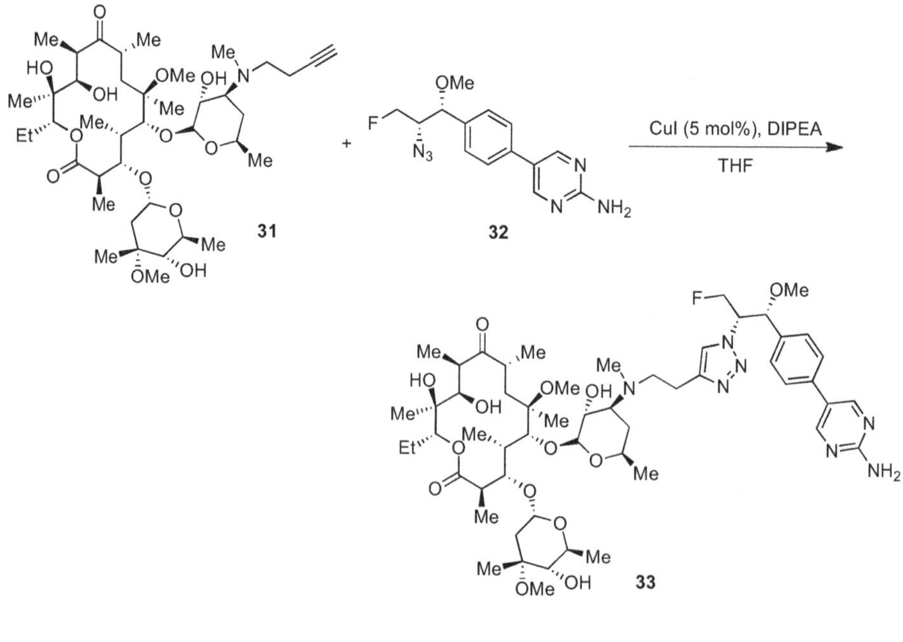

(15.2)

However, the discovery synthesis has a possibility of forming highly explosive copper azide during the course of the copper-catalyzed cycloaddition reaction. In order to address the process safety issue, the triazole intermediate was assembled via a reaction of hydrazone **34** with amine **35** (Equation 15.3).

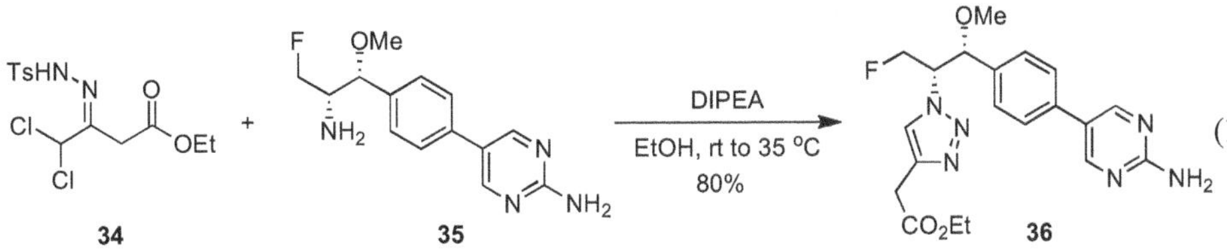

$$(15.3)$$

15.1.2.2 Diazonium Salt-Involved Indazole Formation

The synthesis of indazole **42**, as shown in Scheme 15.11,[10] had two major issues prior to scaling up: long synthetic sequence and unstable diazonium salt-involved indazole ring formation.

To address these issues, especially the safety associated with diazonium intermediate **38**, a new synthetic route was developed (Scheme 15.12) and **42** was prepared in four steps starting from commercially available 3-hydroxybenzaldehyde **43**.

The key indazole ring formation was accomplished by a copper-catalyzed cyclization of *N*-methylhydrazone **46**, which was formed by the reaction of aldehyde **45** with aqueous methylhydrazine. Overall, **42** was produced in 31% yield from **43**.

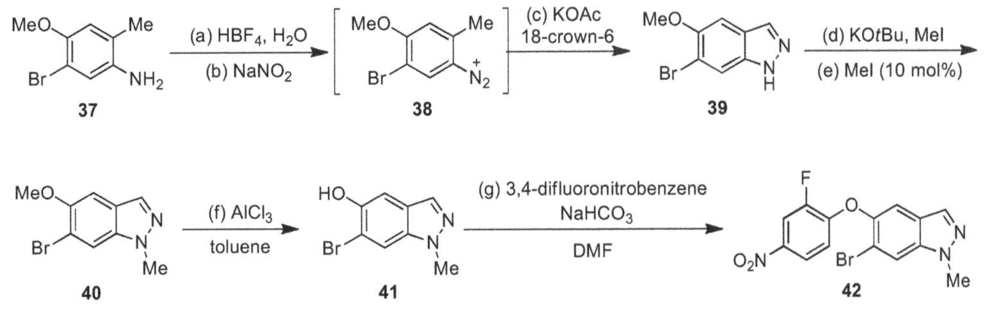

SCHEME 15.11

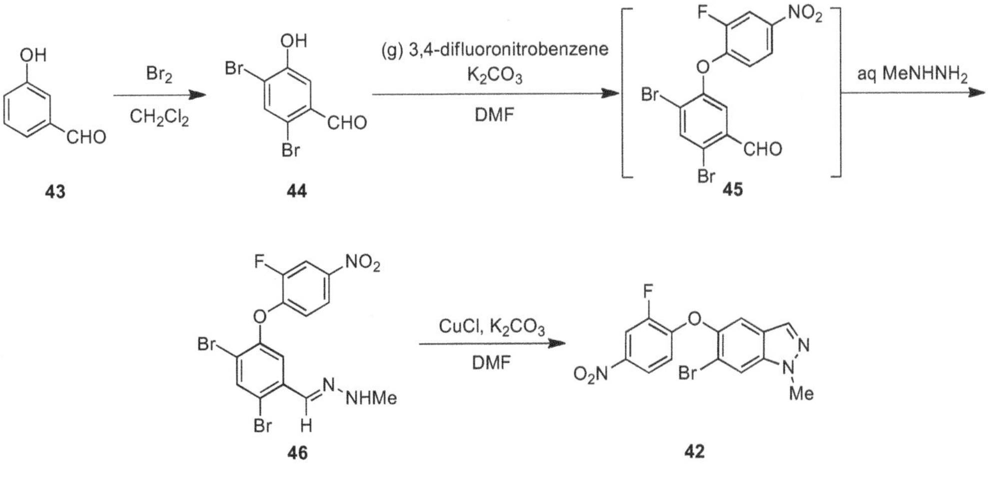

SCHEME 15.12

15.1.2.3   Lithium Aluminum Hydride Reduction

Lithium aluminum hydride (LAH) is a powerful reducing agent and is frequently used in the reduction of carboxylic acid derivatives to their corresponding alcohols. Due to its high reactivity, LAH is less discriminating and in addition to acid derivatives, it readily attacks other functionalities. Sodium borohydride ($NaBH_4$) is a widely used reagent for the reduction of aldehydes and ketones to the corresponding alcohols. However, carboxylic acid derivatives are usually quite resistant to borohydride reduction.

Although the conversion of acid **47** to alcohol **48** was effected using LAH in the discovery chemistry route (Route I, Scheme 15.13),[6] an alternative method[11] was used to reduce **47** to the alcohol **48**, avoiding using LAH. The new protocol involves the in-situ formation of an acyl imidazole intermediate **49**, followed by reduction with $NaBH_4$ in aqueous THF.

Procedure

Route II

- Charge 271 g (1.67 mol) of CDI to a slurry of **47** (748.3 g, 1.39 mol, 90% ee) in anhydrous THF (2.25 L) at room temperature.
- Stir the resulting mixture for 1 h.
- Transfer the reaction mixture to a solution of $NaBH_4$ (158.8 g, 4.19 mol, 3.0 equiv) in a mixture of THF (1.5 L) and water (0.94 L) at 0–18 °C over 30 min.
- Stir the resulting mixture at room temperature for an additional 1 h.
- Add the mixture into a mixture of EtOAc (3.75 L) and aqueous citric acid (2.15 kg in 3.75 L of water).
- Separate layers.
- Wash the organic layer with water (2×1.88 L).
- Distill under reduced pressure at 45 °C.
- Add MeOH (3.0 L).
- Cool to room temperature and stir for 16 h.
- Filter, wash with MeOH (360 mL), and dry at 50 °C under vacuum.
- Yield: 560.2 g (76%) as a white crystalline solid.

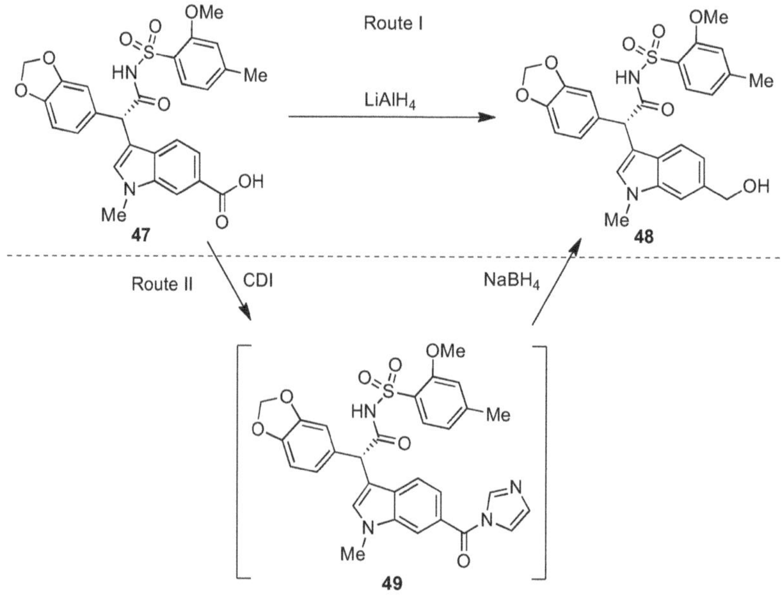

SCHEME 15.13

REMARK

This convenient protocol for selective reduction of carboxylic acid imidazolides to primary alcohols tolerates Bromo, nitro, cyano, ester, and amide functional groups.

15.1.3   UNDESIRED REACTION CONDITIONS

15.1.3.1   Acylation Reaction

A neat reaction at high temperatures would pose safety hazards and operational difficulty on large scale. The preparation of amide **52**, an intermediate for the synthesis of thiazolo[5,4-*d*]pyrimidine **53**, required acylation of 5-aminopyrimidine **50** with acid chloride **51** at 125 °C under neat conditions due to the poor nucleophilicity of the amino group in **50** (Scheme 15.14).[12]

Process optimization for the synthesis of **53** led to a much safer process. The sluggish acylation of **50** was circumvented by the installation of the aniline moiety first followed by acylation (Scheme 15.15). Consequently, the acylation of **55** with **51** was completed in dimethylacetamide in only 1 h at room temperature, owing to the enhanced reactivity by the electron-donating effect of the aniline substituent.

15.1.3.2   S$_N$Ar Reaction

The eight-step process for the synthesis of a clinical candidate **62** for the treatment of asthma raised a toxic hazard concern in addition to the unsafe harsh reaction conditions (Scheme 15.16).[13]

(I) Problems

- Albeit operationally simple, the S$_N$Ar chlorine replacement with ethanolamine became vigorously exothermic at temperatures near 200 °C by calorimetry experiments.

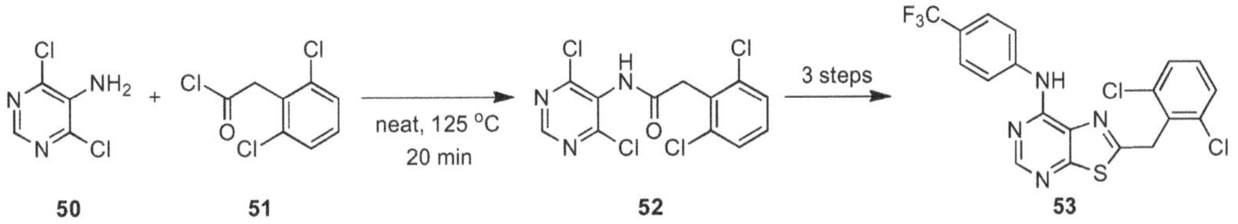

SCHEME 15.14

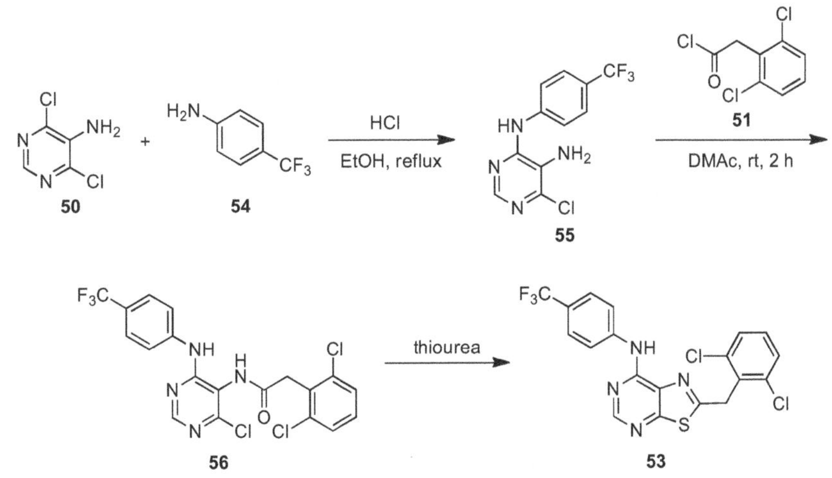

SCHEME 15.15

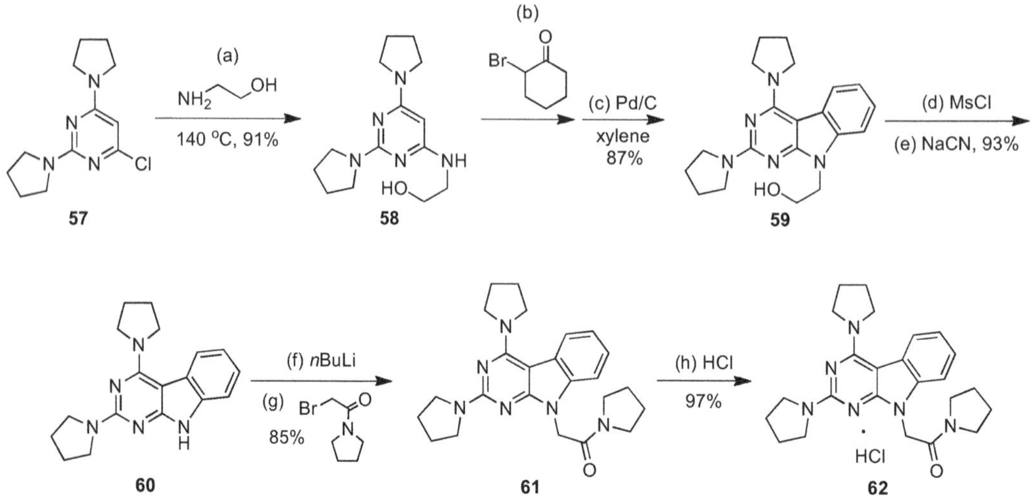

SCHEME 15.16

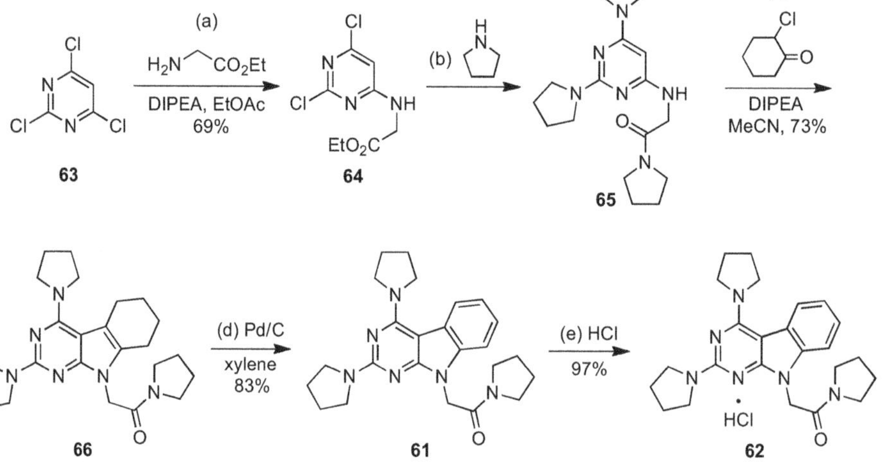

SCHEME 15.17

- The **59→60** conversion required large quantities of cyanide.
- The intermediate **60** exhibited some genotoxicity (the Ames test is positive).[14]

(II) Solutions

To obviate the aforementioned issues, Scheme 15.17 describes a safe, short (five steps), operationally simple, and readily scalable process that was demonstrated for the preparation of **62** in 50 kg scale in 35% overall yield.

15.2   IMPROVING PRODUCT YIELD

Low product yields may be caused by a number of factors, such as incomplete conversions, side reactions, and isolation/purification losses. Most of these issues can be fixed via process optimization, for example, alternation of a reaction temperature in order to achieve complete conversion or suppress side reactions. In addition to optimizing the existing process, a strategy of designing new synthetic routes is also frequently utilized to improve product yields.

15.2.1 CYCLOADDITION REACTION

5-Chloro-3-(4-methylsulfonyl)phenyl-2-(2-methyl-5-pyridinyl)pyridine **69** was identified as a very potent and specific COX-2 inhibitor. Synthetic efforts led to three routes for access of **69** by annulation of ketone **67** (Schemes 15.18).[15] Route I involves an acid-promoted annulation/Friedlander condensation leading to a 47% assay yield. Routes II employed base-promoted annulations of ketone **67** with dichloroacrolein **70**, furnishing **69** in 58% yield.

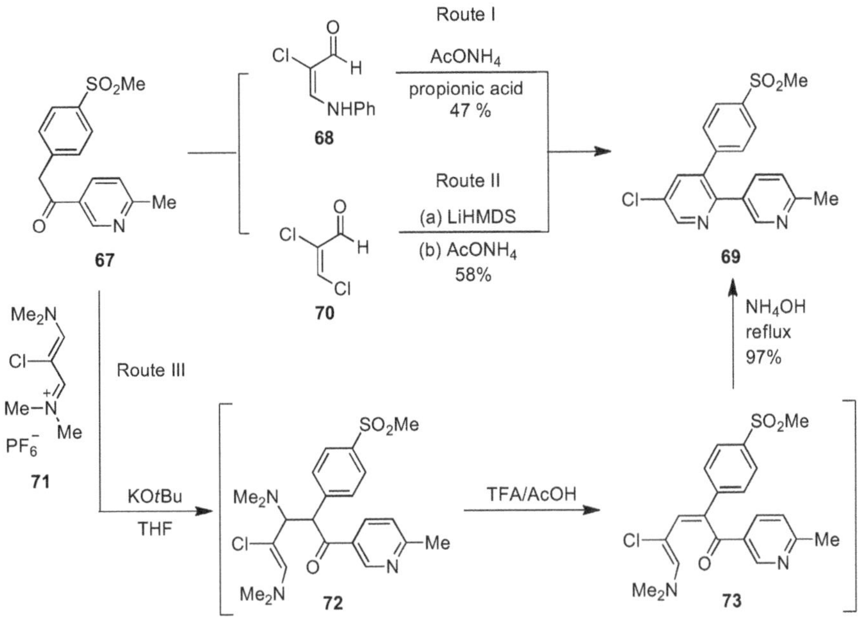

SCHEME 15.18

Route III constitutes the reaction of ketone **67** (1.0 equiv) with 2-chloro-*N,N*-dimethylaminotrimethinium hexafluorophosphate salt **71** (1.05 equiv) in the presence of an equimolar amount of KO*t*Bu in THF at ambient temperature. The resulting adduct **72** was converted to a putative intermediate **73** by inverse quench into a mixture of HOAc (7 equiv) and TFA (0.75 equiv) in THF at <30 °C. The final ring closure occurred upon heating at reflux in the presence of an excess of aqueous ammonium hydroxide to give the **69** in 97% assay yield. Apparently, this telescoped process became the route of choice.

15.2.2 RESOLUTION/AMIDE FORMATION/CYCLIZATION

A strategy of conducting racemic separation at an early stage of the synthetic sequence is frequently adopted by process chemists in order to enhance a process throughput.[16] For instance, an early resolution approach (Route II, Scheme 15.19) was developed to replace the original route (Route I, Scheme 15.19) during the process development for the synthesis of (*S,S*)-reboxetine succinate **78**, a drug candidate for the treatment of fibromyalgia (Scheme 15.19).[17] The new route offered a much more efficient process with a threefold increase in throughput and an improved overall yield.

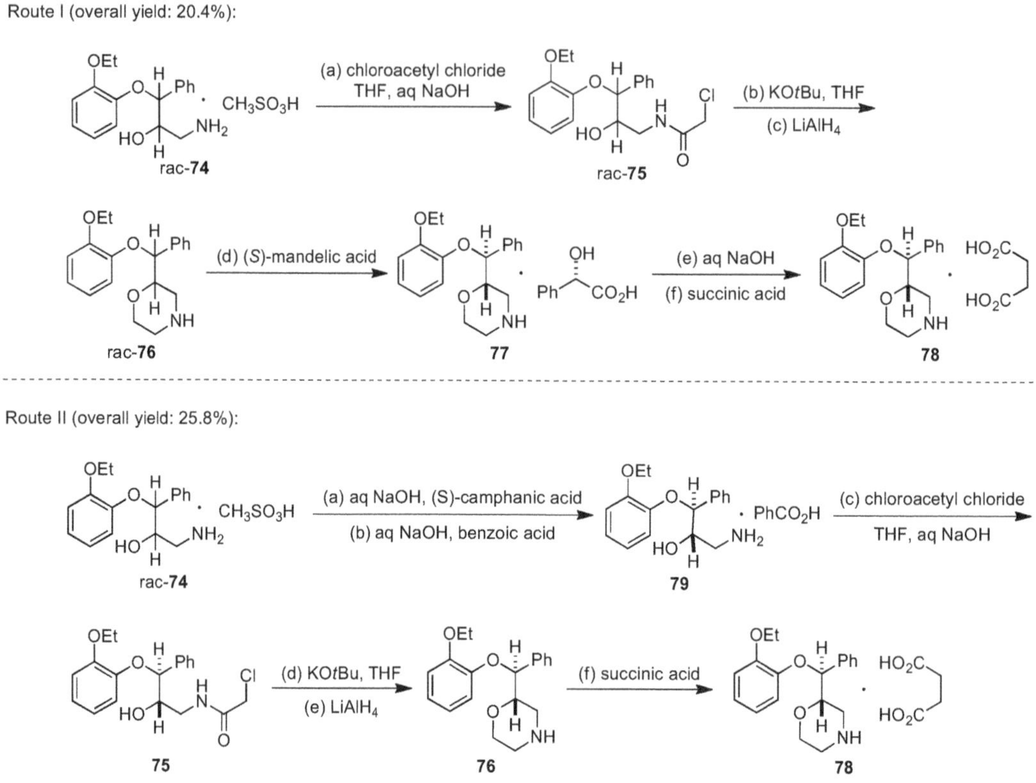

Route I (overall yield: 20.4%):

Route II (overall yield: 25.8%):

SCHEME 15.19

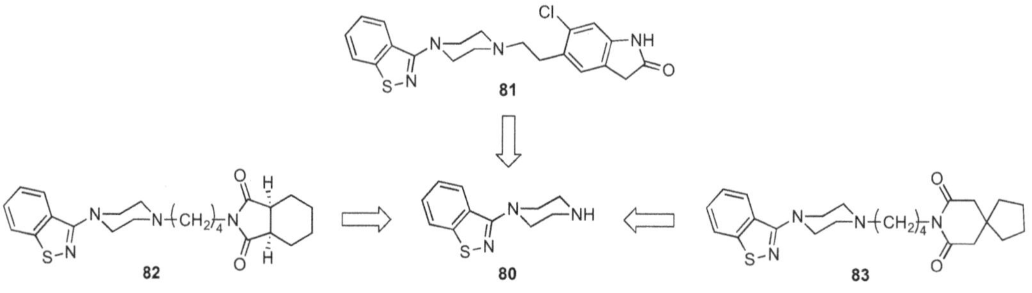

SCHEME 15.20

15.2.3 Chlorine Replacement

3-(1-Piperazinyl)-1,2-benzisothiazole **80** represents a key intermediate for the synthesis of a new class of "atypical anti-psychotic" drugs, such as ziprasidone **81**, Sumitomo's SM-9018 **82**, and tiospirone **83**, for the treatment of schizophrenia (Scheme 15.20).

A reaction between 3-chloro-1,2-benzisothiazole **84** and piperazine was complex involving intermediates such as **85** and the yields of the desired product **80** depend the reactant stoichiometry, concentration, and temperature (Route I, Scheme 15.21).[18] For example, the reaction of **84** with anhydrous piperazine (4.4 equiv) in THF for 17 h at 60–65 °C cleanly cleaved the 1,2-benzisothiazole ring to produce sulfenamide **85** in 85% yield with <0.1% of the desired **80**. However, when anhydrous piperazine (10 equiv) reacted with **84** at higher temperatures (11 h at 140–145 °C) in

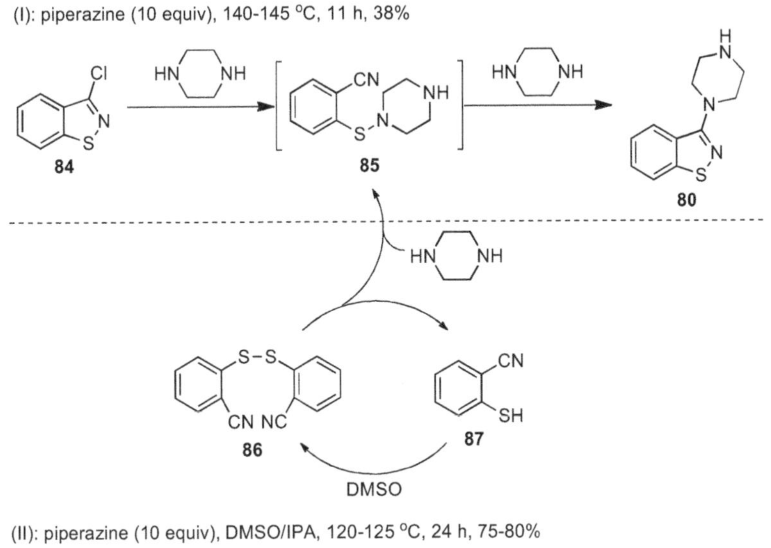

(I): piperazine (10 equiv), 140-145 °C, 11 h, 38%

(II): piperazine (10 equiv), DMSO/IPA, 120-125 °C, 24 h, 75-80%

SCHEME 15.21

diglyme, **80** was formed in 38% yield. The unexpected ring-opening/ring-closing reaction pathway (Route I, Scheme 15.21) rendered the process not robust.

In order to address the aforementioned issues, a new, robust commercial process was developed. As shown in Route II (Scheme 15.21), the reaction of **86** with excess piperazine in the presence of DMSO at 120–125 °C afforded **80** in 75–80% yield. DMSO was used to re-oxidize the reaction-liberated 2-mercaptobenzonitrile **87** to **86**, thereby utilizing both halves of the symmetrical disulfide to generate the product.

Procedure

- Mix 53.3 kg (198.7 mol) of **86**, 171 kg (1987 mol) of piperazine, 34 kg of DMSO (435 mol), and 64 L of IPA at 80 °C under nitrogen.
- Heat at reflux (120–125 °C) for 24 h.
- Cool the solution to 85–90 °C.
- Add H_2O (350 L) and cool the resulting slurry to 30–35 °C.
- Distill at 50–60 °C under reduced pressure (25–30 mmHg) until ~580 L of batch volume was reached by pulling a vacuum through the scrubber to remove residual dimethyl sulfide.
- Add H_2O (190 L) and IPA (75 L) and cool the resulting slurry to 25–30 °C.
- Filter and rinse with a mixture of IPA (75 L)/H_2O (75 L).
- Extract the combined filtrates and wash with toluene (800 L).
- Extract the aqueous layer with toluene (400 L).
- Wash the combined toluene layers with H_2O (320 L).
- Treat the toluene layer with Darco KB-B (5.3 kg).
- Filter through Celite.
- Distill and cool to 20 °C.
- Add IPA (580 L).
- Add concentrated HCl to pH 4–5.
- Cool to 0–5 °C.
- Filter and dry the resulting solid product at 40–50 °C/25 mmHg for 48 h.
- Yield: 73.1 kg (72.0%).

REMARK

Vapors from the reflux condenser were vented through a scrubber solution containing 15% (w/w) NaOCl (320 L) to destroy dimethyl sulfide.

15.2.4 WITTIG REACTION

Skepinone-L is a drug candidate under development for the treatment of sorafenib-resistant liver cancer. The discovery synthesis of skepinone-L involved a key intermediate **92** which was prepared in 8 steps with an overall yield of 12% (Scheme 15.22).[19] The Wittig reaction performed poorly, generating only a 40% yield of the olefin intermediate **91** along with triphenylphosphine oxide by-product. The formation of triphenylphosphine oxide complicated the workup, requiring aqueous quench and extraction with dichloromethane.

Modification of the discovery route led to a more efficient, streamlined 5-step process, furnishing the intermediate **92** in 46% overall yield (an almost fourfold increase in the product yield) (Scheme 15.23). The problematic Wittig reaction was replaced with the Heck cross-coupling reaction, producing consistently high yields (88%) of the olefin intermediate **96**.

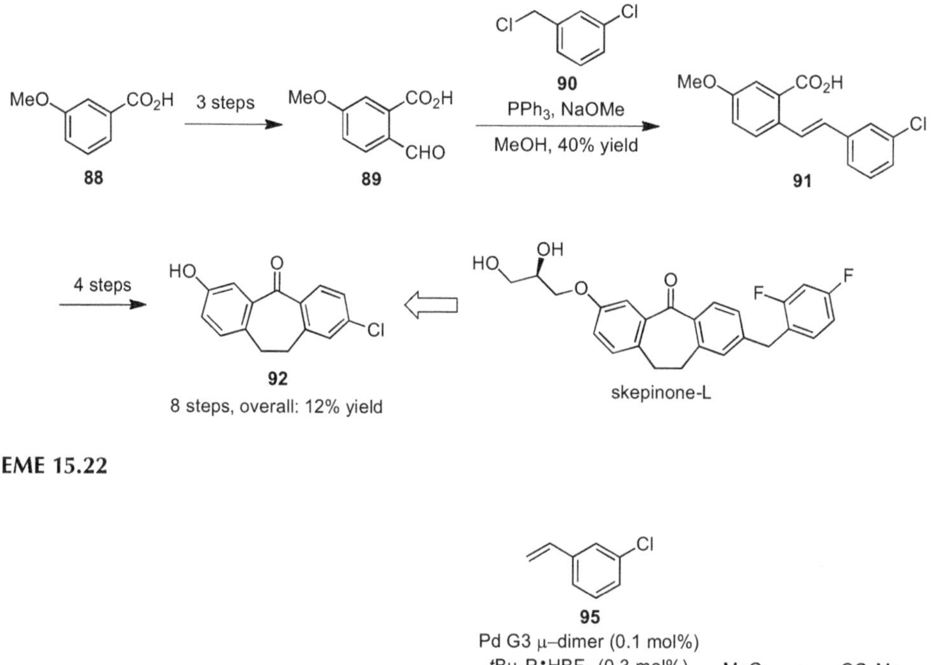

SCHEME 15.22

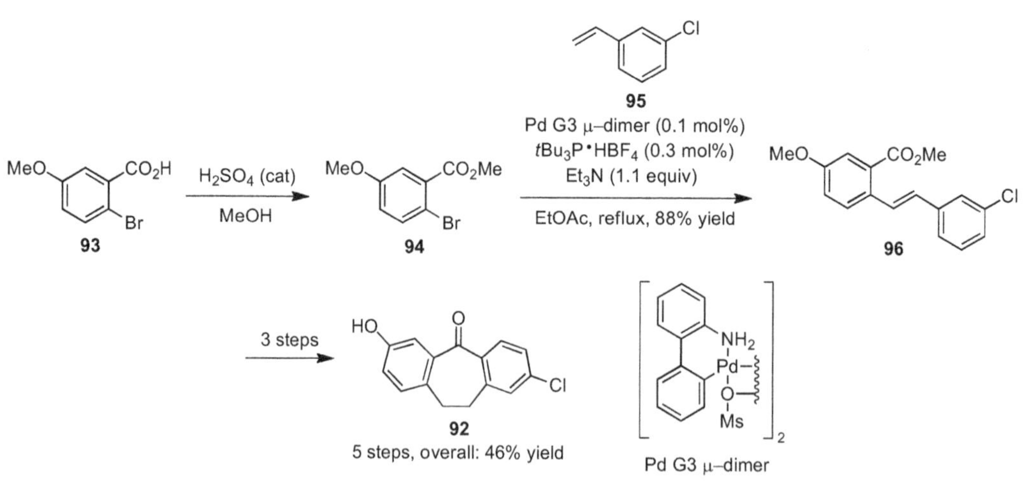

SCHEME 15.23

15.3 IMPROVING REACTION SELECTIVITY

15.3.1 CHLORINATION

N-Chlorosuccinimide (NCS) and cyanuric chloride are frequently used to chlorinate aromatic compounds. However, developing a practical regioselective chlorination reaction can be challenging. To achieve regioselective chlorination of heteroaromatics, such as 7-azaindole 97, a two-step strategy was developed to convert 97 into 7-azaindole N-oxide 99 followed by a reaction with methanesulfonyl chloride to provide the desired 4-chloro-7-azaindole 98 in good yield (Scheme 15.24).[20]

15.3.2 IODINATION

Iodination of compound 100 with N-iodosuccinimide (NIS) at ambient temperature gave a 1:4 mixture of 101 and 102 (Route I, Scheme 15.25).[21] Removal of diiodide 101 from 102 required chromatography which was highly undesired on large scale.

To circumvent the over-iodination problems, a new synthetic route was devleoped, in which regiospecific iodination of diethyl phosphate intermediate 105 was installed (Route II, Scheme 15.25). This chromatography-free process was demonstrated in the pilot plant affording a multi-kilogram quantity of 102 in >98% purity.

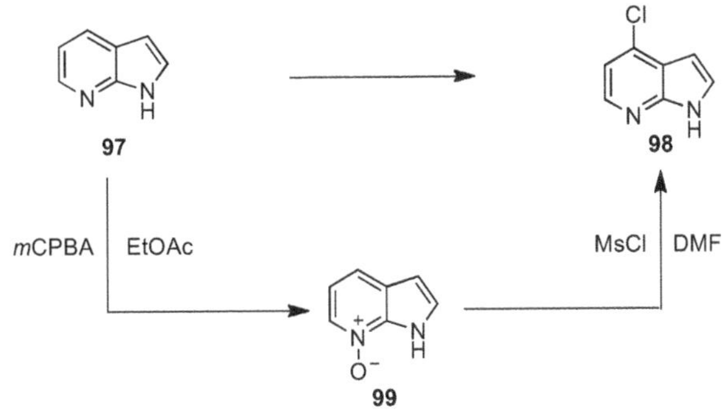

SCHEME 15.24

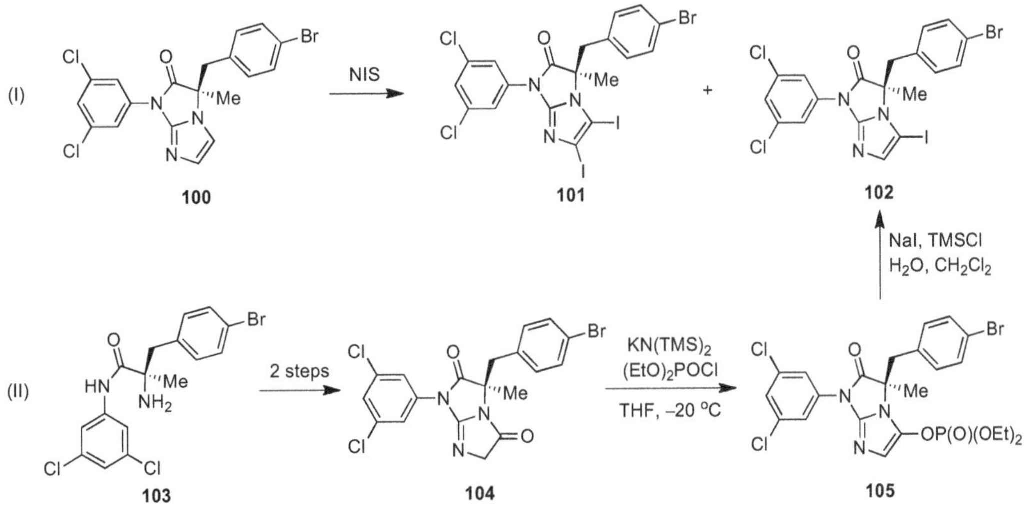

SCHEME 15.25

15.3.3 N-ALKYLATION REACTION

The synthesis of tetrahydropyrazolopyrazine **109** involved *N*-alkylation of nitropyrazole **106** with 1,2-dibromoethane, giving the desired product **107** in a modest yield due to the formation of regioisomer **108** (Scheme 15.26).[22]

To overcome the poor selectivity, a four-step, one-pot process was developed for the preparation of lactam **112**, a precursor of **109** (Scheme 15.27). This one-pot process consists of the amidation of acid **110** with 2-chloro-*N*-methylethanamide followed by regiospecific intramolecular *N*-alkylation.

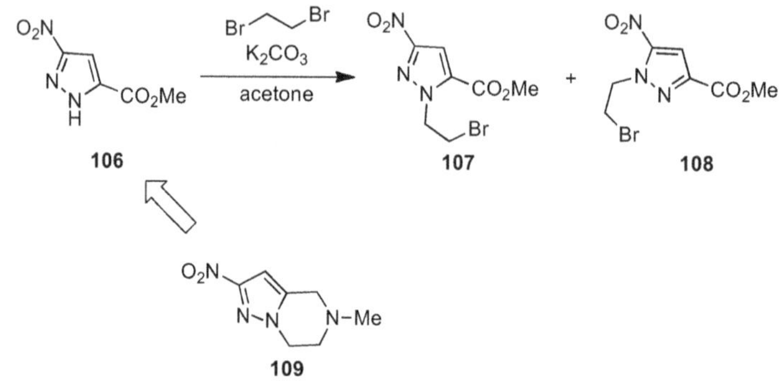

SCHEME 15.26

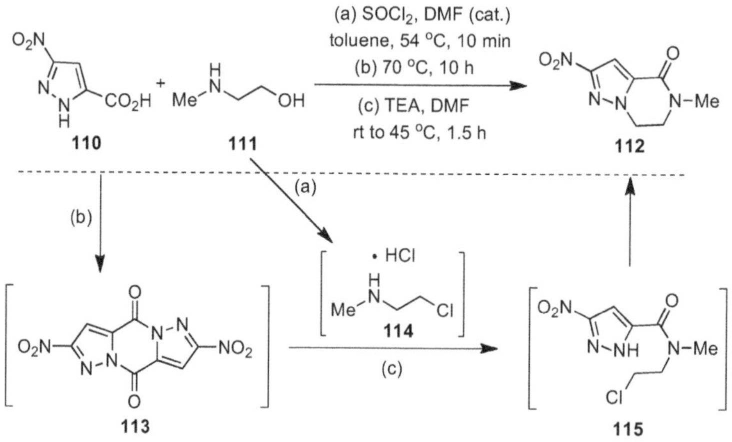

SCHEME 15.27

15.3.4 FORMATION OF SEVEN-MEMBERED RING

The synthesis of 4-oxo-5,6,7,8-tetrahydro-4H-cyclohepta[*b*]furan-3-carboxylic acid **119** could be achieved at a small scale using the chemistry as shown in Scheme 15.28.[23] Activation of one of the two non-equivalent carboxylates in the intermediate **118** led to an intramolecular Friedel-Crafts cyclization to furnish the desired cyclic product **119**. However, under such conditions, **118** could also undergo intermolecular oligomerization.

In order to circumvent this side reaction, a new synthetic route was developed to differentiate the two carboxylic acid groups by the formation of δ-lactone **122** upon treatment of alcohol **121** with AcOH in toluene (Scheme 15.29).[23] Hydrogenolysis of the pseudobenzylic C–O bond then provided carboxylic acid **123**, in which the desired carboxylate was now differentiated from the ethyl ester. This new route afforded the desired keto acid **119** in 26% yield over five steps.

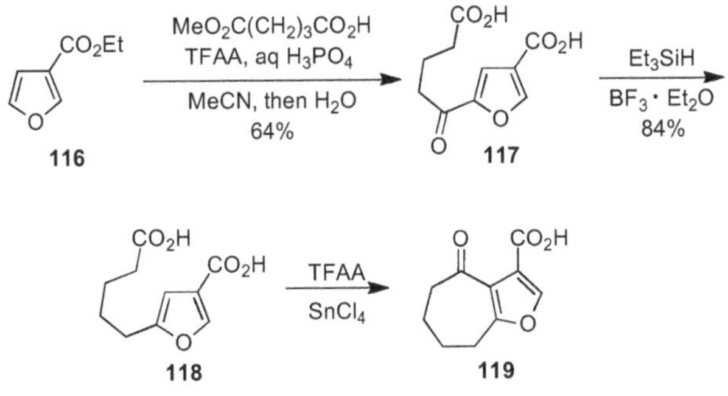

SCHEME 15.28

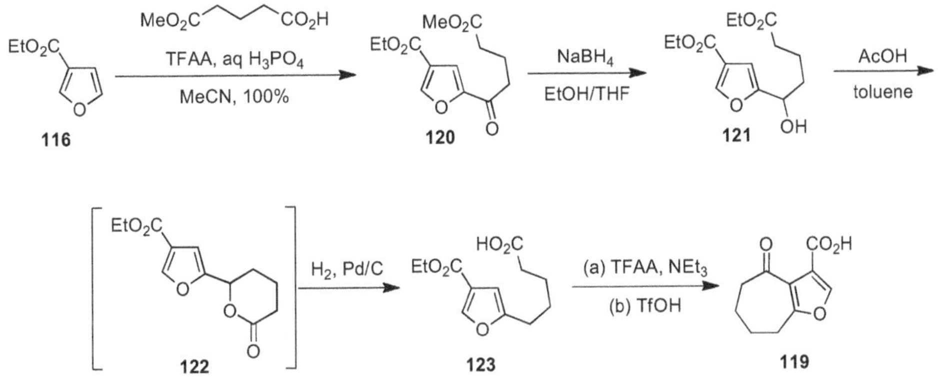

SCHEME 15.29

15.4 OTHER ROUTE DESIGN STRATEGIES

15.4.1 Using Less Expensive Starting Material

Besides process safety, the costs of starting materials[24] and reagents should be considered seriously as they are one of the major cost contributors in a given process.

MN-447 is a potent antagonist of the integrins $\alpha_v\beta_3$ and $\alpha_{IIb}\beta_3$ for the treatment of acute ischemic diseases. The original synthesis of MN-447 developed by Japanese scientists consisted of ten steps with an overall yield of 10% (Scheme 15.30, Route I).[25] Besides, the expensive 4-fluorobenzoic acid starting material (¥14,800/g) renders this synthetic route less favorable.

Accordingly, a new synthetic protocol was developed that produced MN-447 in an overall yield of 45% (Scheme 15.30, Route II). In this protocol, the starting material, 3-hydroxy-4-nitrobenzoic acid **126**, was less expensive ($2.08/g), which made the new synthesis more cost-effective. Furthermore, this API synthesis required a total of eight steps and avoided chromatography purification.

15.4.2 Using Convergent Approach

A convergent synthesis is a strategy that aims to improve the efficiency of multi-step chemical synthesis. In contrast to a linear synthesis, the convergent synthesis can rapidly build the target molecule by coupling the two reaction partners that are prepared beforehand simultaneously.

15.4.2.1 Decarboxylative Cross-Coupling Reaction

In addition to Suzuki–Miyaura, Negishi, Sanogashira, and Kumada cross-coupling reactions, transition metal-catalyzed decarboxylative cross-coupling[26] of arene carboxylic acids with aryl halides

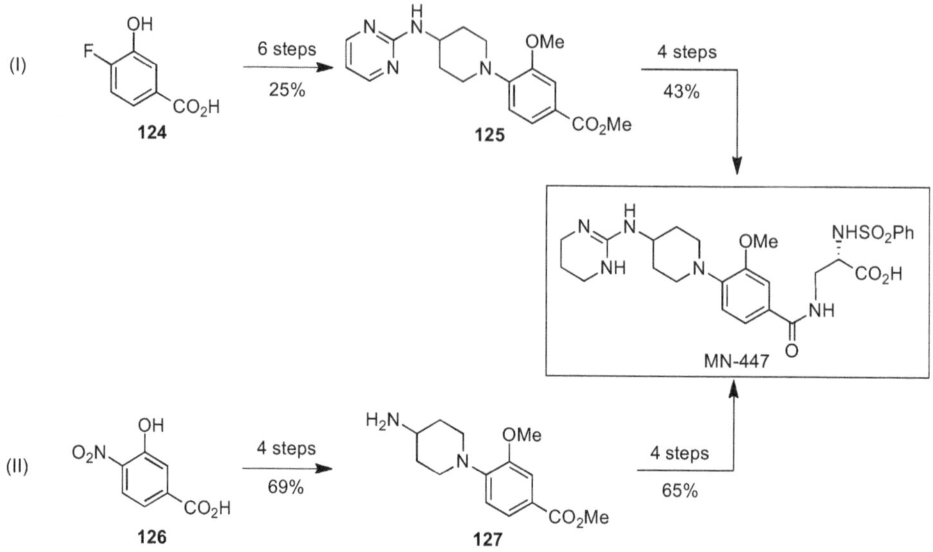

SCHEME 15.30

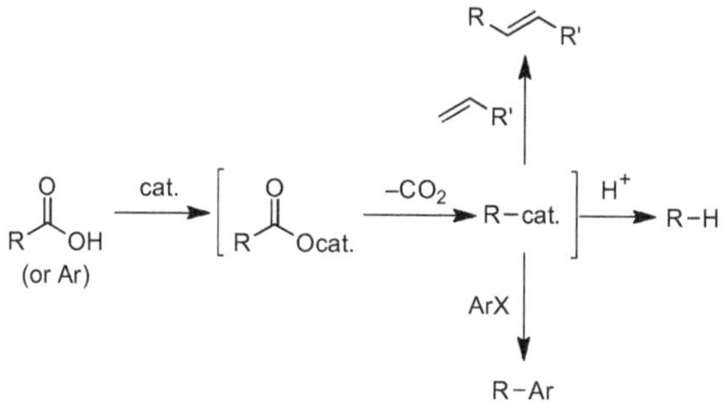

SCHEME 15.31

and triflates has recently received considerable attention as an attractive tool for biaryl formation. Scheme 15.31 outlines the three types of decarboxylative reactions including protodecarboxylation, decarboxylative cross-coupling, and decarboxylative Heck reaction. Thus, carboxylic acids can be used as synthetic equivalents of aryl, or alkyl halides, as well as organometallic reagents. The ready availability of carboxylic acids makes them extremely promising raw materials for chemical synthesis. Furthermore, this synthetic approach offers ecological advantages with respect to a number of traditional arene functionalizations such as halogenation, nitration/reduction/diazotization, or Friedel–Crafts reactions, which are often waste-intensive.

In decarboxylative cross-coupling reactions, the initially formed metal carboxylates are converted into organometallic compounds with the extrusion of CO_2. This decarboxylation step is usually highly endothermic and requires a transition metal mediator, such as mercury, silver, or copper, at high temperatures. Due to the toxicity and non-catalytic nature the application of those metals is severely limited at an industrial scale. Recently, palladium-catalyzed decarboxylative cross-coupling reactions have been becoming one of the new synthetic tools for the construction of pharmaceutical intermediates and APIs. The diversity of the decarboxylative cross-coupling reactions is highlighted in a number of reports.[27]

One of the recent applications of the decarboxylative cross-coupling reaction toward the synthesis of 3-amino-2-(4-chlorophenyl)thiophene **132** led to a process improvement to a great extent.

Originally, the synthesis of **132** involves a radical bromination, S_N2 reaction, Fisher esterification, Dieckmann condensation, and hydrolysis/decarboxylation, followed by a reaction with hydroxylamine with overall yields of 36–45% (Scheme 15.32).[28] Although telescoping of steps minimized isolated intermediates, the six-step transformation was deemed less efficient when consideration of commercial production.

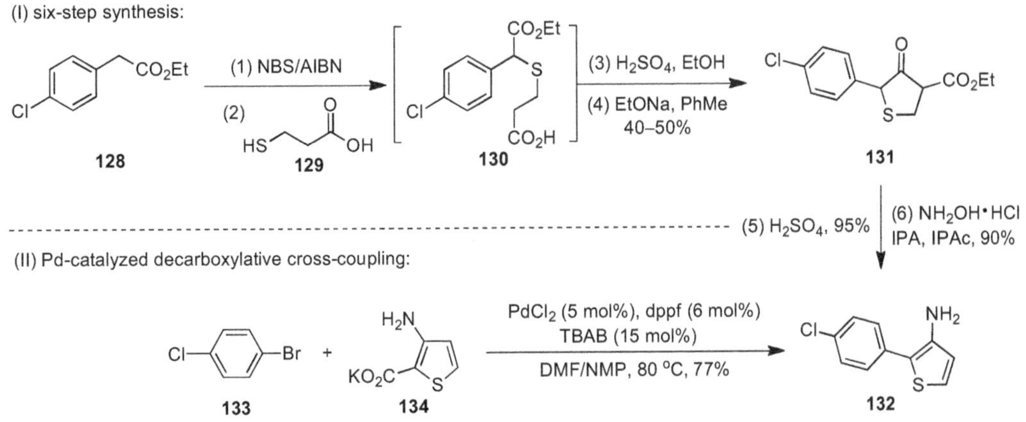

SCHEME 15.32

A Pd-catalyzed decarboxylative cross-coupling of 3-aminothiophene-2-carboxylate **134** with 4-chlorophenylbromide **133** offers a practical synthetic route, producing **132** in a single transformation in 77% isolated yield.

15.4.2.2 Synthesis of Chiral Amide

The synthesis of chiral compounds via the traditional kinetic resolution (KR) will produce products with a maximun yield of 50%, which is a major drawback when the KR is performed at the end of linear syntheses. This is the case for the synthesis of (*S*)-(4-benzylmorpholin-2-yl)(morpholino)methanone methanesulfonate **138**, a key regulatory API starting material for the preparation of Edivoxetine hydrochloride, via the original linear route (Scheme 15.33).[29] The five-step, linear synthesis involves a kinetic resolution at the penultimate step, which led to poor throughput. In addition, 2-chloroacrylonitrile **135**, as the raw material, is highly toxic and corrosive.

In order to address these issues, a convergent synthesis was developed at Eli Lilly & Company. As illustrated in Scheme 15.34, the new process utilized *D*-serine **139** as the charility source to

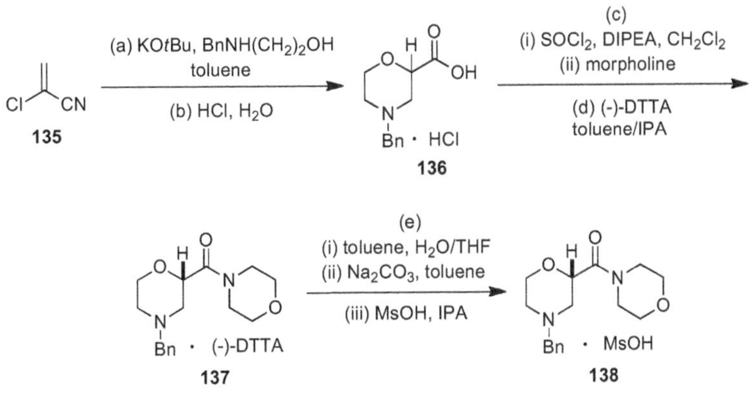

SCHEME 15.33

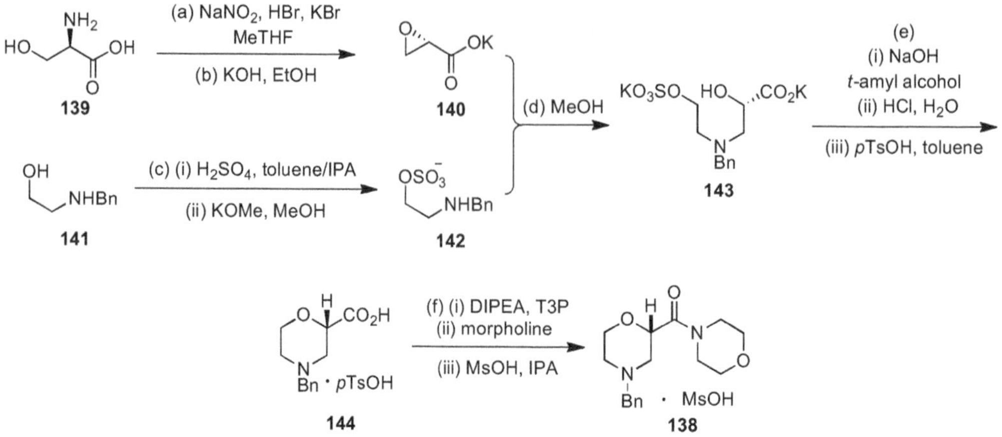

SCHEME 15.34

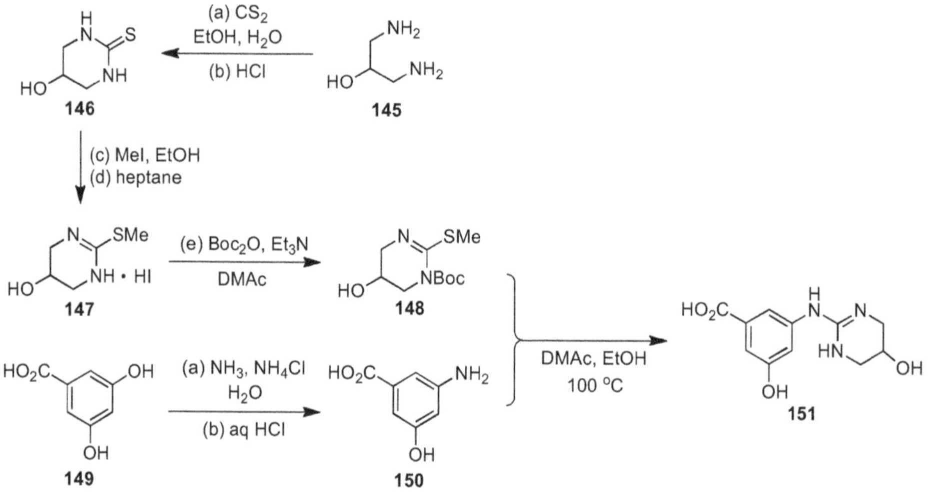

SCHEME 15.35

generate one of the convergent coupling partners, a chiral epoxide **140**. The subsequent convergent coupling with 2-(benzylamino)ethyl hydrogen sulfate **142** furnished the desired product **138** following cyclization and salt formation in 28% overall yield (from **139**) with 99.9% purity and >99.9% ee.

DISCUSSIONS

However, convergent synthesis is not always the route of choice, for instance, a convergent synthesis of drug intermediate **151** suffers a number of concerns associated with safety, technical, and operational issues (Scheme 15.35).[30]

To address these issues, a linear approach, involving a one-pot (three-step) process as described in Scheme 15.36, avoided the use of toxic carbon disulfide, eliminating protection/deprotection. Employing this efficient linear process, a total of 85.6 kg of **151** was prepared in the production campaign with an overall yield of 74% in 96.6 wt% purity.

REMARK

The gases of methyl mercaptan and methylamine by-products were scrubbed with solutions of caustic and sulfuric acid, respectively.

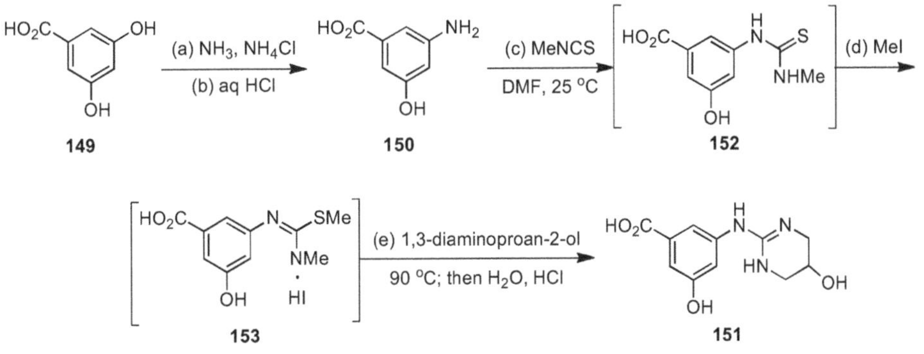
(a) NH_3, NH_4Cl
(b) aq HCl
(c) MeNCS
DMF, 25 °C
(d) MeI
(e) 1,3-diaminoproan-2-ol
90 °C; then H_2O, HCl
149
150
152
153
151

SCHEME 15.36

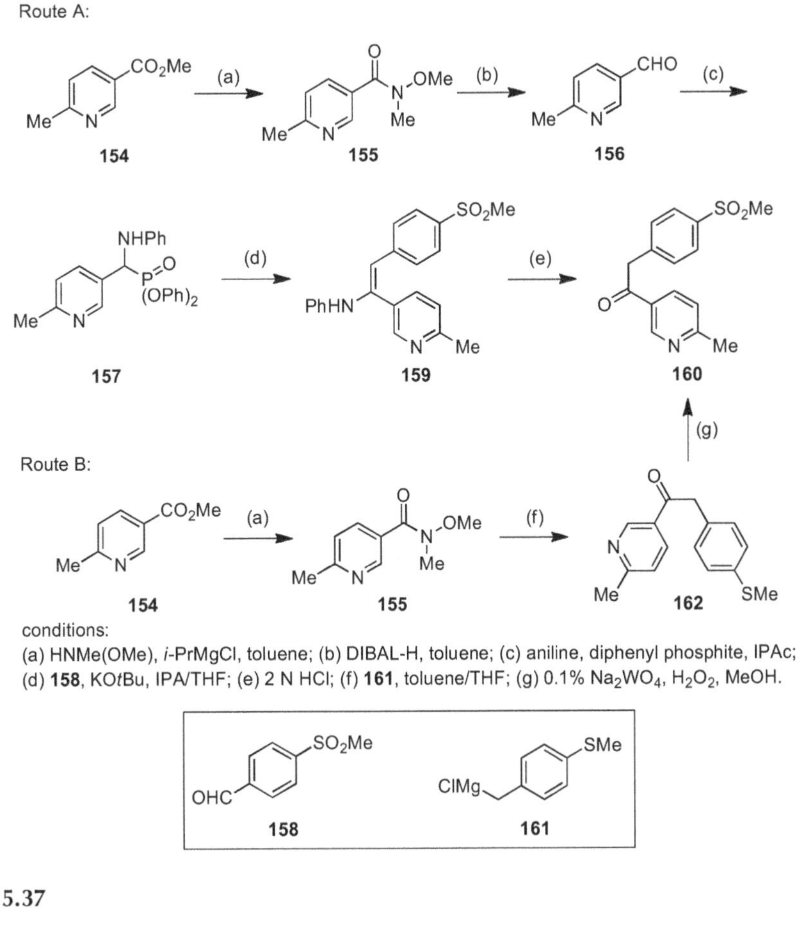
Route A:
Route B:
154
155
156
157
159
160
162
158
161
conditions:
(a) HNMe(OMe), *i*-PrMgCl, toluene; (b) DIBAL-H, toluene; (c) aniline, diphenyl phosphite, IPAc;
(d) **158**, KO*t*Bu, IPA/THF; (e) 2 N HCl; (f) **161**, toluene/THF; (g) 0.1% Na_2WO_4, H_2O_2, MeOH.

SCHEME 15.37

15.4.3 Step-Economy Synthesis

15.4.3.1 Synthesis of Keto–Sulfone Intermediate

Step-economy synthesis presents a number of advantages such as fewer synthetic steps, shorter processing times, and less wastes. Bearing this in mind, scientists at Merck designed two synthetic routes for the preparation of keto–sulfone **160** (Scheme 15.37).[15] The initial route A utilized five steps producing **160** in 65% overall yield.

Compared to Route A, Route B represents a significant improvement in terms of synthetic steps. The Grignard reaction/oxidation in Route B afforded the keto–sulfone **160** in 68% yield in only three steps.

15.4.3.2 Synthesis of Bendamustine

Efforts for developing sustainable chemical processes includes the application of new technologies and reactions toward not only the on-going drug programs, but the existing commercial APIs as well. For instance, the implementation of a reductive alkylation reaction into a new process for the manufacture of bendamustine, a drug approved by the FDA in 2009 for the treatment of chronic lymphocytic leukemia (CLL), addressed the process safety, robustness, and efficiency. Compared with the current commercial process, the newly designed route reduced the number of synthetic steps from eight to five and improved overall yield from 12% to 45% (Scheme 15.38).[31]

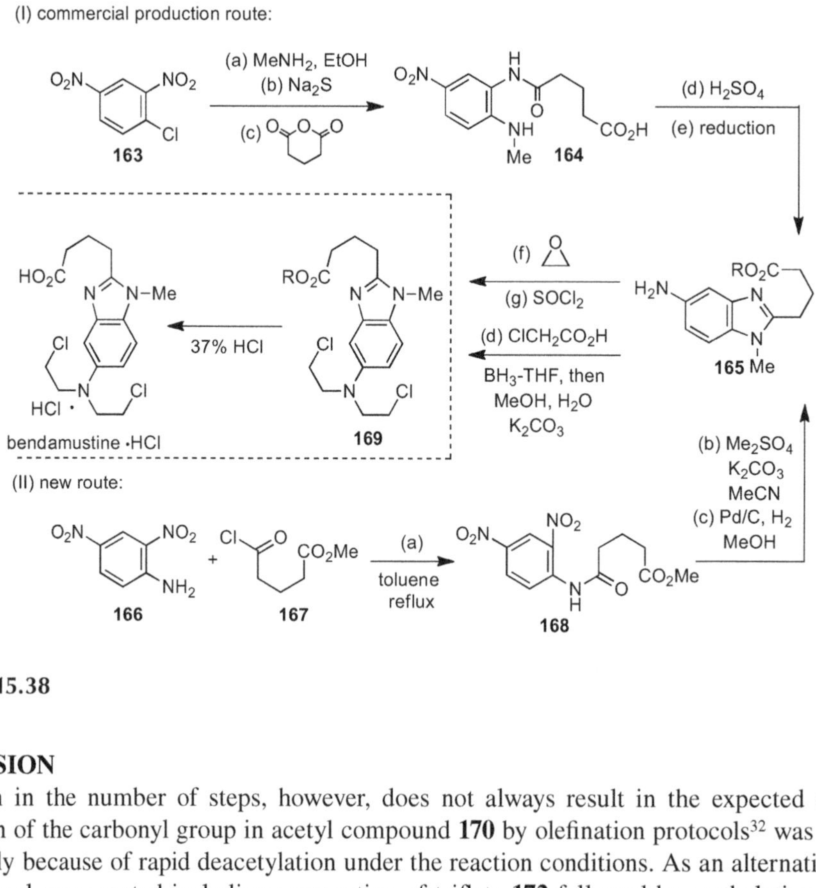

SCHEME 15.38

DISCUSSION

Reduction in the number of steps, however, does not always result in the expected outcomes. Vinylation of the carbonyl group in acetyl compound **170** by olefination protocols[32] was not effective mainly because of rapid deacetylation under the reaction conditions. As an alternative, a two-step protocol was created including preparation of triflate **172** followed by methylation of triflate **172** by Negishi-type Pd(0)-catalyzed alkylation,[33] affording the desired isopropenyl compound **171** (Scheme 15.39).[34]

15.4.4 ATOM-ECONOMY SYNTHESIS

It is always challenging to develop a process that meets all of the 12 Green Chemistry principles.[35] Atom-economic synthesis has been one of the most important subjects for process chemists on account of its direct impact on the limited availability of raw materials, process cost, wastes, and environments.

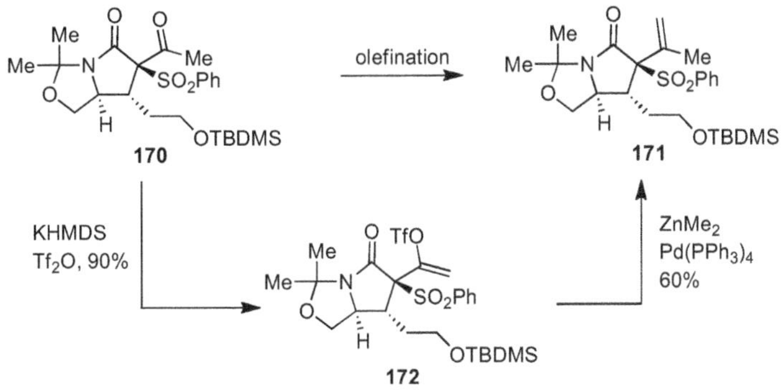

SCHEME 15.39

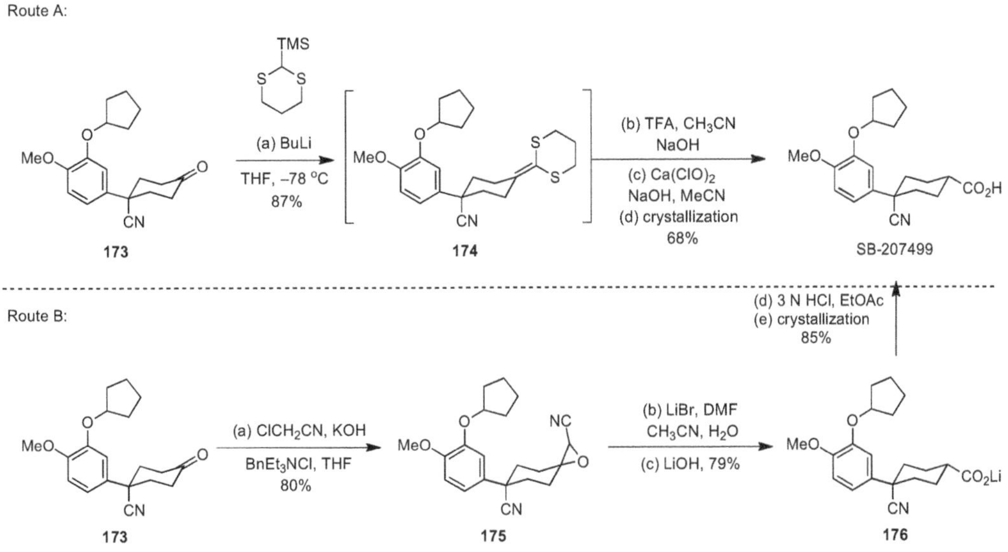

Route A:

Route B:

SCHEME 15.40

15.4.4.1 Synthesis of Carboxylic Acid

Scheme 15.40[36] presents two synthetic approaches for the synthesis of SB-207499, a potent second-generation inhibitor of PDE4 (phosphodiesterase-4) in clinical development for asthma and chronic obstructive pulmonary diseases. In addition to unfavorable cryogenic reaction conditions, the use of 2-(trimethylsilyl)-1,3-dithiane in Route A (Scheme 15.40) was not efficient in terms of atom economy.

In contrast, an atom-economic synthesis was demonstrated in Route B (Scheme 15.40), in which Darzens reaction was employed to access intermediate epoxynitrile **175** followed by rearrangement. This route also avoided cryogenic reaction conditions.

15.4.4.2 Stereoselective Synthesis of Diol

One of the research synthesis routes developed by Novartis during the synthesis of fluvastatin employing stereoselective synthesis, starting with phloroglucinol **177** (Scheme 15.41).[37]

Although the designed route delivered the side-chain intermediate **181** with the desired *syn*-configuration, this synthesis employed unacceptable oxidizing reagent pyridinium chlorochromate (PCC) and expensive *tert*-butyldiphenylsilyl chloride, rendering the process too costly with poor atom-economy.

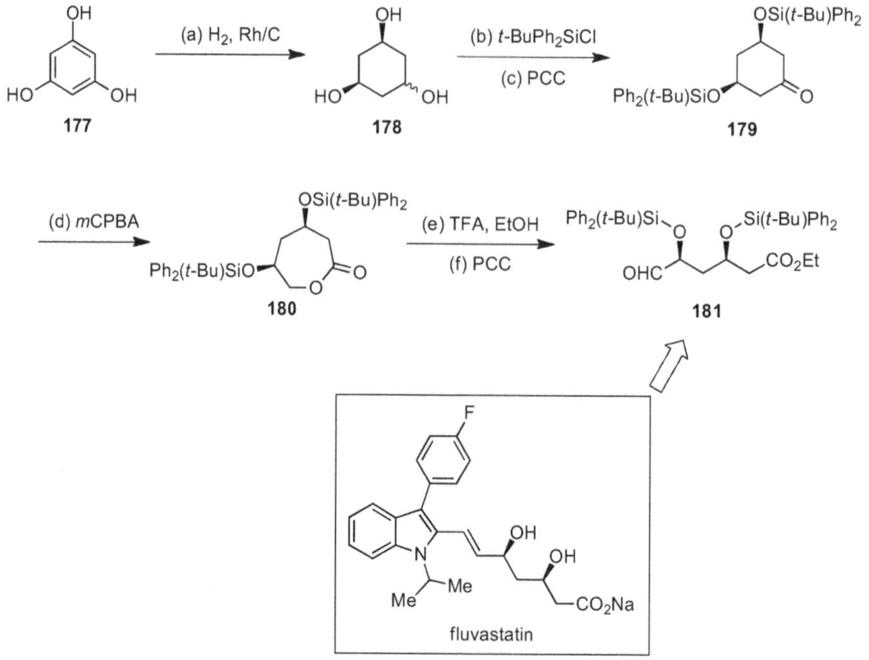

SCHEME 15.41

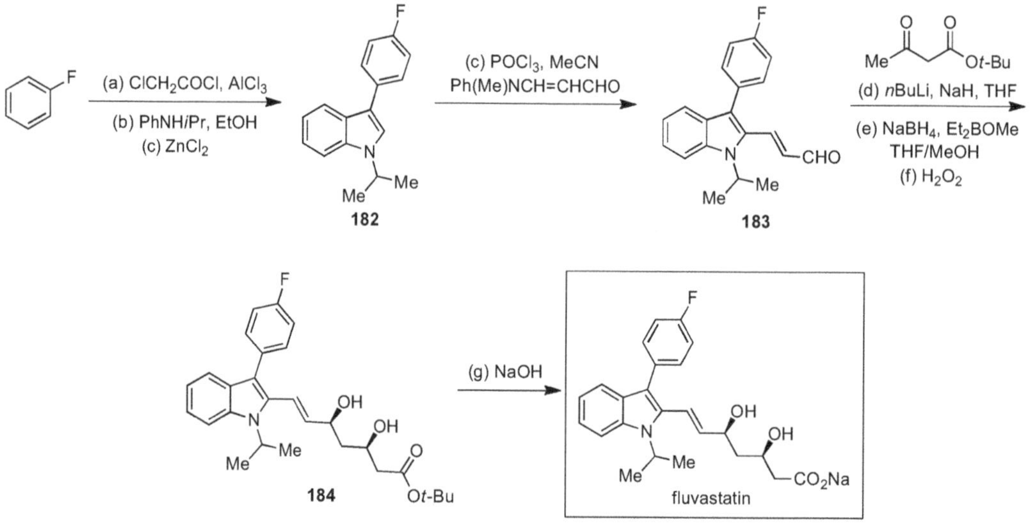

SCHEME 15.42

An atom-economic commercial process was developed (Scheme 15.42). The synthesis was only six steps long, entirely stereoselective in the E-olefin and *syn*-diol formation, and it requires no chromatography. The *syn*-diol intermediate **184**, obtained from stereoselective reduction with $NaBH_4/Et_2BOMe$, is a crystalline solid, which could be purified by recrystallization.

15.4.5   Alternating Bond-Formation Order

An oil or moisture, air, or light-sensitive process intermediate poses a significant purification challenge; in this event, the design of a new synthetic route to avoid such an intermediate becomes

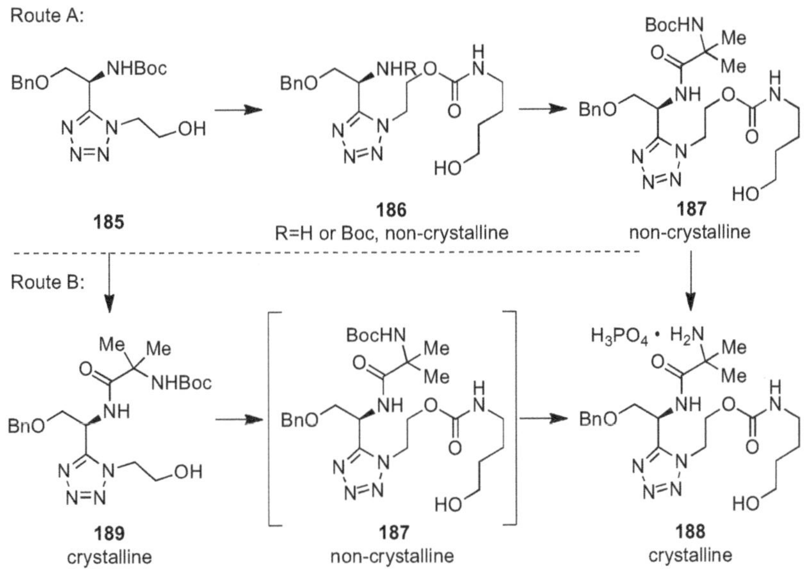

Route A:

185

186
R=H or Boc, non-crystalline

187
non-crystalline

Route B:

189
crystalline

187
non-crystalline

188
crystalline

SCHEME 15.43

necessary. A viable manufacturing process is preferred to have a number of crystalline intermediates in order to have a recourse for purity control by recrystallization. Tetrazole-based drug candidate **188** was developed by Bristol Myers Squibb Company as a potent, orally active antagonist of the human growth hormone secretagogue receptor. The synthetic route, as outlined in Route A of Scheme 15.43,[38] presents a major concern because it was in lack of crystalline intermediates until the final stage of the synthesis (i.e., crystallization of **188** as its phosphate salt).

Therefore, a process was developed by alternating the bond-connection order as illustrated in Route B of Scheme 15.43. This synthetic route was able to form a crystalline intermediate **189** which was telescoped through a non-crystalline **187** to the final API.

15.4.6 Minimizing Oxidation Stage Change

15.4.6.1 Minimizing Nitrogen Oxidation Stage Adjustment

Redox economy[39] addresses the minimization of changes in oxidation states throughout a synthesis. For this reason, there is a strong preference to design synthetic routes in which the functional groups in the starting materials have the same oxidation states in the products. The synthesis of a novel anti-inflammatory agent **194** required a Ullmann reaction to form diaryl ether bond. Route I in Scheme 15.44[40] describes a five-step synthesis involving an inefficient oxidation stage adjustment. In contrast, this inefficient transformation was eliminated in Route II of Scheme 15.44 and the synthesis of **194** was accomplished in three steps.

15.4.6.2 Minimizing Carbon Oxidation Stage Adjustment

15.4.6.2.1 Synthesis of Carboxylate Ester

Ifetroban sodium (BMS-180291, **197**, a single enantiomer), under development in the Bristol Myers Squibb Pharmaceutical Research Institute, is a long-acting, orally bioavailable, highly selective thromboxane A2 receptor antagonist with potent anti-thrombotic and anti-ischemic activity. One of the notable inefficiencies present in the original synthesis (Route I, Scheme 15.45)[41] is caused by the oxidation-state adjustments during the synthesis of a key intermediate **198**. The access to the carboxylate ester **198** underwent a redundant process involving reduction/protection/oxidation sequence, which renders the process step inefficient.

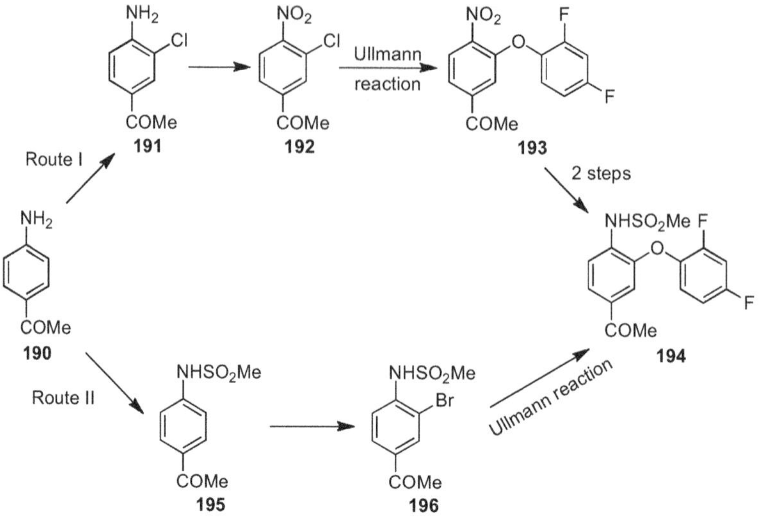

SCHEME 15.44

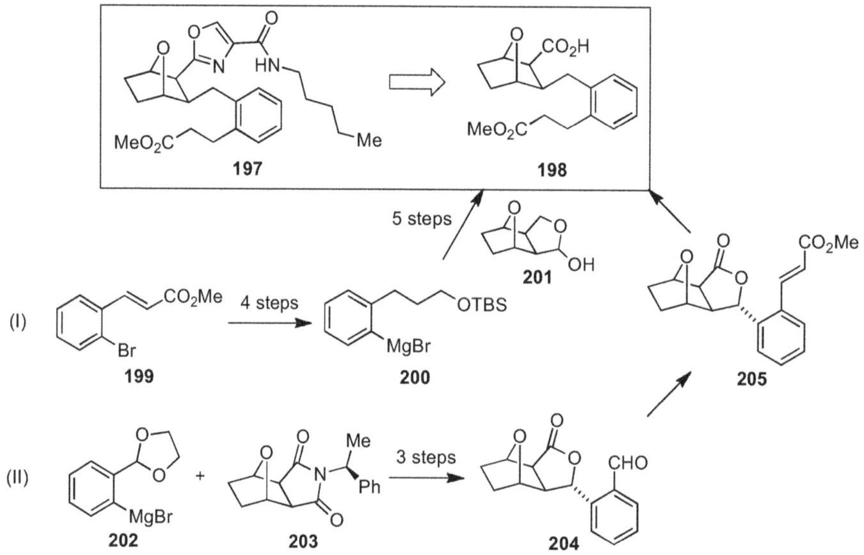

SCHEME 15.45

Thus, the newly developed synthesis (Route II, Scheme 15.45) avoided the step inefficiency by maintaining the correct oxidation state throughout the process.

15.4.6.2.2   Synthesis of Alkyl Chloride

The same strategy was applied to the synthesis of vinyl methyl chloride **206** (Scheme 15.46).[42] The synthesis of **206** involved five steps starting with acetylene **210** in 18% overall yield, which was characterized by the transformation of ester **208** to alcohol **207** (the oxidation state adjustment with DIBAL-H). Substitution of the five-step synthesis with a two-step procedure through the secondary alcohol **211** without oxidation/reduction operation allowed the preparation of **206** in a much more efficient manner.

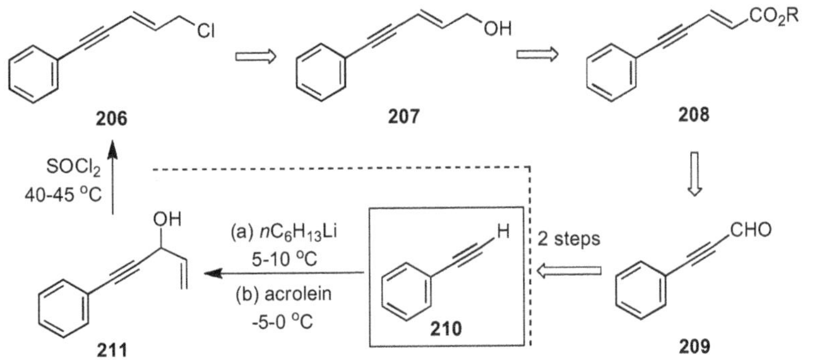

SCHEME 15.46

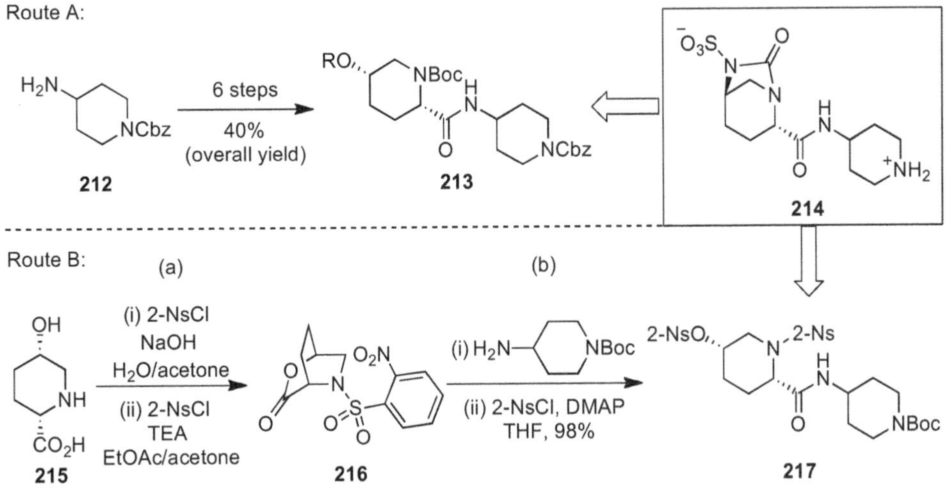

SCHEME 15.47

15.4.7 COUPLING REAGENT-FREE AMIDE FORMATION

Compound **214** is a drug candidate, developed by Merck for the treatment of bacterial infections. The initial synthetic route involved the preparation of *cis*-pipecolic amide intermediate **213** from Cbz-protected aminopiperidine **212** in six steps with 40% overall yield (Route A, Scheme 15.47).[43]

An optimized synthetic route involved lactone **216**, obtained in two-step, one-pot operations, starting from *cis*-5-hydroxypipecolic acid **215** (Route B, Scheme 15.47). The subsequent conversion of lactone **216** to **217** was accomplished by treating with the commercially available Boc-protected aminopiperidine in THF, followed by a reaction with 2-NsCl and DMAP in one-pot. Using this novel amide formation approach avoided the need for typical coupling reagents (e.g., EDC, CDI, T3P), allowing the preparation of **217** with a cleaner reaction profile and in a cost-effective and environmentally benign manner.

15.4.8 PREVENTING ETCHING OF GLASS REACTOR

Generally, it is recommended not to use glass-lined reactors if a reaction is conducted in concentrated caustic base solutions, such as NaOH and KOH, or acidic fluoride ion-containing solutions, because of the possibility of etching the glass-lined reactors under those two conditions. In order to minimize glass erosion, sodium bases are generally preferred over potassium bases.

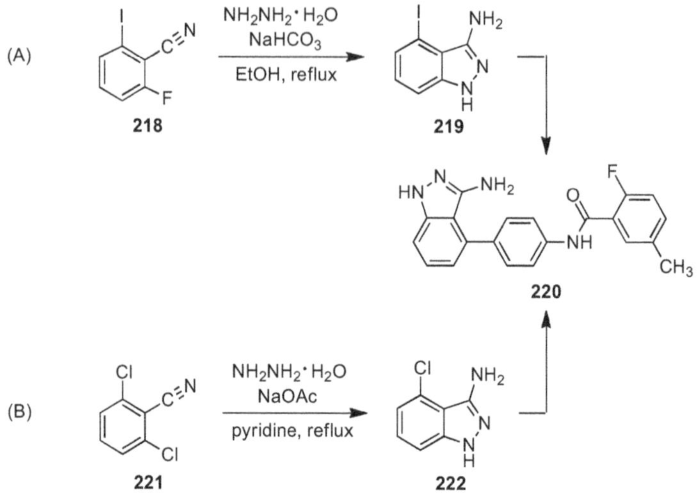

SCHEME 15.48

Among halogenated aromatics, fluorine-containing aromatic compounds are the best substrates for aromatic nucleophilic substitution reactions (S_NAr). However, the fluoride by-products from the S_NAr reactions may potentially cause etching of glass-lined reactors. For example, the glass etching problem in the production of indazole **220** from 2-fluoro-6-iodobenzonitrile **218** limited the scalability of the chemistry (Route A, Scheme 15.48).[44] In an effort to avoid etching of glass vessels, Route B (Scheme 15.48) was developed by using 2,6-dichlorobenzonitrile **221** as an inexpensive, readily available starting material to replace 2-fluoro-6-iodobenzonitrile **218**.

Procedure (Route B, production of **222**)

- Suspend 50.0 kg (267 mol) of **221** and 26.2 kg (319 mol) of NaOAc in pyridine (216 kg).
- Heat the mixture to 100 °C.
- Charge 150 kg (3000 mol) of hydrazine monohydrate over 30 min.
- Stir the batch at 105 °C for 16 h.
- Cool to 30 °C.
- Add water (595 kg) and filter.
- Add water (184 kg) to the filtrate followed by seeding with a slurry of **222** (280 g in 6.5 kg of water).
- Add an additional 1746 kg of water and cool to 0 °C.
- Filter, wash with water (2×100 kg), and dry at 60 °C under vacuum.
- Yield of **222**: 38.0 kg (74%) as a white solid.

NOTES

1. Ashwood, M. S.; Alabaster, R. J.; Cottrell, I. F.; Cowden, C. J.; Davies, A. J.; Dolling, U. H.; Emerson, K. M.; Gibb, A. D.; Hands, D.; Wallace, D. J.; Wilson, R. D. *Org. Process Res. Dev.* **2004**, *8*, 192.
2. Menzel, K.; Machrouhi, F.; Bodenstein, M.; Alorati, A.; Cowden, C.; Gibson, A. W.; Bishop, B.; Ikemoto, N.; Nelson, T. D.; Kress, M. H.; Frantz, D. E. *Org. Process Res. Dev.* **2009**, *13*, 519.
3. Coon, C. L.; Simon, R. L. U.S. Patent 4,469,898, 1984.
4. Ramakrishna, K.; Behme, C.; Schure, R. M.; Bieniarz, C. *Org. Process Res. Dev.* **2000**, *4*, 581.
5. Stoner, E. J.; Cothron, D. A.; Balmer, M. K.; Roden, B. A. *Tetrahedron* **1995**, *51*, 11043.
6. Ashcroft, C. P.; Challenger, S.; Clifford, D.; Derrick, A. M.; Hajikarimian, Y.; Slucock, K.; Silk, T. V.; Thomson, N. M.; Williams, J. R. *Org. Process Res. Dev.* **2005**, *9*, 663.

7. Ragan, J. A.; Bourassa, D. E.; Blunt, J.; Breen, D.; Busch, F. R.; Cordi, E. M.; Damon, D. B.; Do, N.; Engtrakul, A.; Lynch, D.; McDermott, R. E.; Mongillo, J. A.; O'Sullivan, M. M.; Rose, P. R.; Vanderplas, B. C. *Org. Process Res. Dev.* **2009**, *13*, 186.

8. Dale, D. J.; Dunn, P. J.; Golightly, C.; Hughes, M. L.; Levett, P. C.; Pearce, A. K.; Searle, P. M.; Ward, G.; Wood, A. S. *Org. Process Res. Dev.* **2000**, *4*, 17.

9. Hanselmann, R.; Job, G. E.; Johnson, G.; Lou, R.; Martynow, J. G.; Reeve, M. M. *Org. Process Res. Dev.* **2010**, *14*, 152.

10. Kallman, N. J.; Liu, C.; Yates, M. H.; Linder, R. J.; Ruble, J. C.; Kogut, E. F.; Patterson, L. E.; Laird, D. L. T.; Hansen, M. M. *Org. Process Res. Dev.* **2014**, *18*, 501.

11. Sharma, R.; Voynov, G. H.; Ovaska, T. V.; Marquez, V. E. *Synlett* **1995**, *8*, 839.

12. Liu, J.; Fitzgerald, A. E.; Lebsack, A. D.; Mani, N. S. *Org. Process Res. Dev.* **2011**, *15*, 382.

13. Bundy, G. L.; Banitt, L. S.; Dobrowolski, P. J.; Palmer, J. R.; Schwartz, T. M.; Zimmermann, D. C.; Lipton, M. F.; Mauragis, M. A.; Veley, M. F.; Appell, R. B.; Clouse, R. C.; Daugs, E. D. *Org. Process Res. Dev.* **2001**, *5*, 144.

14. Ames, B. N.; McCann, J.; Yamasaki, E. *Mutation Res.* **1975**, *31*, 347.

15. Davies, I. W.; Marcoux, J.-F.; Corley, E. G.; Journet, M.; Cai, D.-W.; Palucki, M.; Wu, J.; Larsen, R. D.; Rossen, K.; Pye, P. J.; DiMichele, L.; Dormer, P.; Reider, P. J. *J. Org. Chem.* **2000**, *65*, 8415.

16. For early resolution examples, see: (a) Jansen, R.; Knopp, M.; Amberg, W.; Bernard, H.; Koser, S.; Müller, S.; Münster, I.; Pfeiffer, T.; Riechers, H. *Org. Process Res. Dev.* **2001**, *5*, 16.

17. Hayes, S. T.; Assaf, G.; Checksfield, G.; Cheung, C.; Critcher, D.; Harris, L.; Howard, R.; Mathew, S.; Regius, C.; Scotney, G.; Scott, A. *Org. Process Res. Dev.* **2011**, *15*, 1305.

18. Walinsky, S. W.; Fox, D. E.; Lambert, J. F.; Sinay, T. G. *Org. Process Res. Dev.* **1999**, *3*, 126.

19. Forster, M.; Wentsch-Teltschik, H. K.; Laufer, S. A. *Org. Process Res. Dev.* **2021**, *25*, 1831.

20. Han, C.; Green, K.; Pfeifer, E.; Gosselin, F. *Org. Process Res. Dev.* **2017**, *21*, 664.

21. Frutos, R. P.; Eriksson, M.; Wang, X.-J.; Byrne, D.; Varsolona, R.; Johnson, M. D.; Nummy, L.; Krishnamurthy, D.; Senanayake, C. H. *Org. Process Res. Dev.* **2005**, *9*, 137.

22. Shu, L.; Wang, P.; Gu, C.; Garofalo, L.; Alabanza, L. M.; Dong, Y. *Org. Process Res. Dev.* **2012**, *16*, 1870.

23. Ragan, J. A.; Murry, J. A.; Castaldi, M. J.; Conrad, A. K.; Jones, B. P.; Li, B.; Makowski, T. W.; McDermott, R.; Sitter, B. J.; White, T. D.; Young, G. R. *Org. Process Res. Dev.* **2001**, *5*, 498.

24. Starting materials are different from the API starting material. For the designation and justification of API SMs, see: (a) Faul, M. M.; Kiesman, W. F.; Smullkowski, M.; Pfeiffer, S.; Busacca, C. A.; Eriksson, M. C.; Hicks, F.; Orr, J. D. *Org. Process Res. Dev.* **2014**, *18*, 587. (b) Faul, M. M.; Busacca, C. A.; Eriksson, M. C.; Hicks, F.; Kiesman, W. F.; Smullkowski, M.; Orr, J. D.; Pfeiffer, S. *Org. Process Res. Dev.* **2014**, *18*, 594.

25. Ishikawa, M.; Tsushima, M.; Kubota, D.; Yanagisawa, Y.; Hiraiwa, Y.; Kojima, Y.; Ajito, K.; Anzai, N. *Org. Process Res. Dev.* **2008**, *12*, 596.

26. (a) Goossen, L. J.; Zimmermann, B.; Linder, C.; Rodriguez, N.; Lange, P. P.; Hartung, J. *Adv. Synth. Catal.* **2009**, *351*, 2667. (b) Wang, Z.; Ding, Q.; He, X.; Wu, J. *Tetrahedron* **2009**, *65*, 4635. (c) Goossen, L. J.; Deng, G.; Levy, L. M. *Science* **2006**, *313*, 662. (d) Forgione, P.; Brochu, M. C.; St Onge, M.; Thesen, K. H.; Bailey, M. D.; Bilodeau, F. *J. Am. Chem. Soc.* **2006**, *128*, 11350. (e) Zhang, F.; Greaney, M. F. *Org. Lett.* **2010**, *12*, 4745. (f) Becht, J. M.; Le Drian, C. *Org. Lett.* **2008**, *10*, 3161. (g) Goossen, L. J.; Rodriguez, N.; Linder, C. *J. Am. Chem. Soc.* **2008**, *130*, 15248. (h) Baudoin, O. *Angew. Chem., Int. Ed.* **2007**, *46*, 1373. (i) Goossen, L. J.; Rodriguez, N.; Melzer, B.; Linder, C.; Deng, G.; Levy, L. M. *J. Am. Chem. Soc.* **2007**, *129*, 4824. (j) Becht, J. M.; Catala, C.; Le Drian, C.; Wagner, A. *Org. Lett.* **2007**, *9*, 1781. (k) Myers, A. G.; Tanaka, D.; Mannion, M. R. *J. Am. Chem. Soc.* **2002**, *124*, 11250.

27. (a) Chou, C.-M.; Chatterjee, I.; Studer, A. *Angew. Chem. Int. Ed.* **2011**, *50*, 8614. (b) Gooßen, L. J.; Rodríguez, N.; Gooßen, K. *Angew. Chem. Int. Ed.* **2008**, *47*, 3100. (c) Gooßen, L. J.; Zimmermann, B.; Knauber, T. *Angew. Chem. Int. Ed.* **2008**, *47*, 7103. (d) Lou, S.; Westbrook, J. A.; Schaus, S. E. *J. Am. Chem. Soc.* **2004**, *126*, 11440. (e) Wang, C.; Piel, I.; Glorius, F. *J. Am. Chem. Soc.* **2009**, *131*, 4194. (f) Grenning, A. J.; Tunge, J. A. *Org. Lett.* **2010**, *12*, 740. (g) Shang, R.; Xu, Q.; Jiang, Y.-Y.; Wang, Y.; Liu, L. *Org. Lett.* **2010**, *12*, 1000. (h) Xie, K.; Yang, Z.; Zhou, X.; Li, X.; Wang, S.; Tan, Z.; An, X.; Guo, C.-C. *Org. Lett.* **2010**, *12*, 1564. (i) Miyasaka, M.; Fukushima, A.; Satoh, T.; Hirano, K.; Miura, M. *Chem. Eur. J.* **2009**, *15*, 3674.

28. Mitchell, D.; Coppert, D. M.; Moynihan, H. A.; Lorenz, K. T.; Kissane, M.; McNamara, O. A.; Maguire, A. R. *Org. Process Res. Dev.* **2011**, *15*, 981.

29. Kopach, M. E.; Heath, P. C.; Scherer, R. B.; Pietz, M. A.; Astleford, B. A.; McCauley, M. K.; Singh, U. K.; Wong, S. W.; Coppert, D. M.; Kerr, M. S.; Houghton, P. G.; Rhodes, G. A.; Tharp, G. A. *Org. Process Res. Dev.* **2015**, *19*, 543.

30. Clark, J. D.; Collins, J. T.; Kleine, H. P.; Weisenburger, G. A.; Anderson, D. K. *Org. Process Res. Dev.* **2004**, *8*, 571.
31. Chen, J.; Przyuski, K.; Roemmele, R.; Bakale, R. P. *Org. Process Res. Dev.* **2011**, *15*, 1063.
32. (a) Corey, E. J.; Clark, D. A.; Goto, G.; Marfat, A.; Mioskowski, C.; Samuelsson, B.; Hammerstro¨m, S. *J. Am. Chem. Soc.* **1980**, *102*, 1436. (b) (i) Tebbe, F. N. *J. Am. Chem. Soc.* **1978**, *100*, 3611. (ii) Pines, S. H. *Synthesis* **1991**, *2*, 165. (c) Takai, K. *J. Am. Chem. Soc.* **1986**, *108*, 7408.
33. Hadei, E. A. B.; Kantchev, C. J.; O'Brien, M. G. *Org. Lett.* **2005**, *7*, 3805.
34. Jung, Y. C.; Yoon, C. H.; Turos, E.; Yoo, K. S.; Jung, K. W. *J. Org. Chem.* **2007**, *72*, 10114.
35. Ahluwalia, V. K. *Strategies for Green Organic Synthesis*, Boca Raton, CRC Press, 2012, P2.
36. Badham, N. F.; Chen, J.-H.; Cummings, P. G.; Dell'Orco, P. C.; Diederich, A. M.; Eldridge, A. M.; Mendelson, W. L.; Mills, R. J.; Novack, V. J.; Olsen, M. A.; Rustum, A. M.; Webb, K. S.; Yang, S. *Org. Process Res. Dev.* **2003**, *7*, 101.
37. Repič, O.; Prasad, K.; Lee, G. T. *Org. Process Res. Dev.* **2001**, *5*, 519.
38. Davulcu, A. H.; McLeod, D. D.; Li, J.; Katipally, K.; Littke, A.; Doubleday, W.; Xu, Z.; McConlogue, C. W.; Lai, C. J.; Gleeson, M.; Schwinden, M.; Parsons, R. L., Jr. *J. Org. Chem.* **2009**, *74*, 4068.
39. For reviews of redox economy, see: (a) Gaich, T.; Baran, P. S. *J. Org. Chem.* **2010**, *75*, 4657. (b) Burns, N. Z.; Baran, P. S.; Hoffmann, R. W. *Angew. Chem. Int. Ed.* **2009**, *48*, 2854. (c) Newhouse, T.; Baran, P. S.; Hoffmann, R. W. *Chem. Soc. Rev.* **2009**, *38*, 3010.
40. Zanka, A.; Kubota, A.; Hirabayashi, S.; Nakamura, H. *Org. Process Res. Dev.* **1997**, *1*, 71.
41. Mueller, R. H.; Wang, S.; Pansegrau, P. D.; Jannotti, J. Q.; Poss, M. A.; Thottathil, J. K.; Singh, J.; Humora, M. J.; Kissick, T. P.; Boyhan, B. *Org. Process Res. Dev.* **1997**, *1*, 14.
42. Deussen, H.-J.; Jeppesen, L.; Schärer, N.; Junager, F.; Bentzen, B.; Weber, B.; Weil, V.; Mozer, S. J.; Sauerberg, P. *Org. Process Res. Dev.* **2004**, *8*, 363.
43. Miller, S. P.; Zhong, Y.-L.; Liu, Z.; Simeone, M.; Yasuda, N.; Limanto, J.; Chen, Z.; Lynch, J.; Capodanno, *Org. Lett.* **2014**, *16*, 174.
44. Kruger, A. W.; Rozema, M. J.; Chu-Kung, A.; Ganderilla, J.; Haight, A. R.; Kotecki, B. J.; Richter, S. M.; Schwartz, A. M.; Wang, Z. *Org. Process Res. Dev.* **2009**, *13*, 1419.

16 Stereochemistry

Some enantiomeric drugs display an improved pharmacological profile compared to the race-mates or the enantiomeric isomers.[1] For example, racemic bupivacaine hydrochloride (Marcaine) (Figure 16.1)[2] is currently used as an epidural anesthetic during labor and for ulna nerve block for local anesthesia in minor operations. At high doses, however, it is potentially hazardous due to car-diotoxicity problems. A clinical study found that the myocardial contractility index and the stroke index were reduced by a greater degree with racemic bupivacaine than with levobupivacaine. This evidence indicates that the use of levobupivacaine is significantly safer than racemic bupivacaine due to its lower cardiotoxicity.

Tramadol [2-(dimethylaminomethyl)-1-(3-methoxyphenyl)cyclohexanol] (Figure 16.2) is a mild, non-addictive, centrally acting binary analgesic agent, introduced in Germany in the later 1970s by Grunenthal.[3] It was approved for use in the United States in 1995 and is currently marketed as Ultram by Ortho-McNeil Pharmaceuticals, Inc. While the marketed form is the racemic hydrochlo-ride salt of the *cis*-isomer, (+)-enantiomer of the *cis*-isomer exhibits analgesic activity 10-fold higher than that of the (–)-enantiomer.[4]

On average, 56% of contemporary pharmaceuticals contain one or multiple chiral centers. Generally, two approaches are now available to address absolute stereochemistry, including (1) asymmetric synthesis and (2) chiral resolution and chromatographic separation. Despite tremen-dous development in recent years, asymmetric synthesis accounts for only 20% of the stereogenic centers generated during the synthesis of drug molecules.[5] The resolution methods remain the major tool to prepare chiral drug molecules in the pharmaceutical industry.

16.1 ASYMMETRIC SYNTHESIS

In the past two decades, asymmetric synthesis has been one of the most interesting areas in syn-thetic chemistry. Utilization of asymmetric catalysis in industrial processes can provide superior yield and reduce the waste streams during the manufacture of pharmaceutical products. The asym-metric reaction may be characterized into three categories: (a) the use of a chiral catalyst (asym-metric catalysis), (b) the use of chiral substrates or reagents (chiral pool synthesis), and (c) the use of chiral auxiliaries.

16.1.1 Asymmetric Catalysis

16.1.1.1 Desymmetrization of Anhydride

3,4-Dihydropyridin-2-one derivative **6** is a core structure in the synthesis of potent $P2X_7$ receptor antagonists (Scheme 16.1).[6]

Scientists at Roche developed a practical asymmetric synthesis mediated by sulfonamide **7** or thiourea-based organocatalyst **8**. A stereogenic center in the dihydropyridinone ring **6** was gener-ated by catalytic desymmetrization of anhydride **3** using either catalyst **7** or **8**. Under this protocol, dihydropyridinone **6** was prepared in 82% ee (using **7**) or 80% ee (using **8**), which upon treatment in toluene/hexane gave 97% ee.

16.1.1.2 Asymmetric Reduction of Enone

The reduction of ketones to alcohols in both non-asymmetric and asymmetric fashion is a very general practice in organic transformation.[7] In particular, the preparation of chiral, secondary alco-hols from prochiral ketones is of paramount importance. Asymmetric reduction of enone **9** with

DOI: 10.1201/9781003288411-16

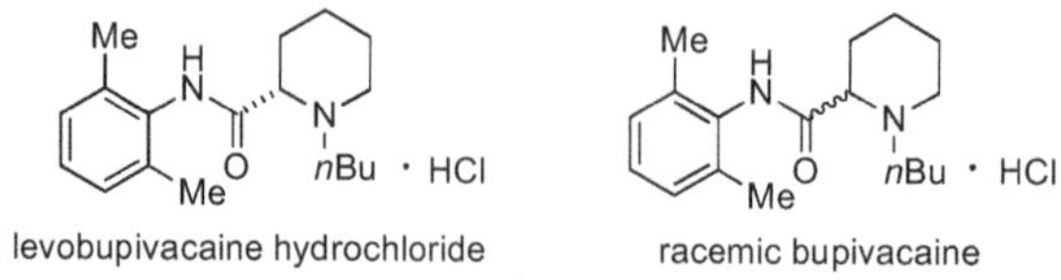

FIGURE 16.1   Structures of Levobupivacaine and Bupivacaine.

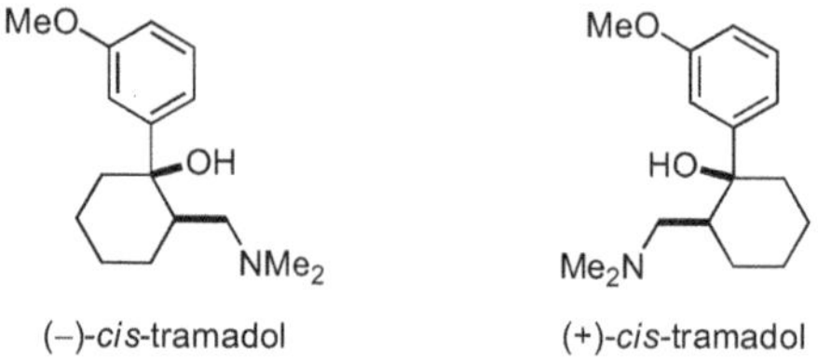

FIGURE 16.2   Structures of (−)-*cis*-Tramadol and (+)-*cis*-Tramadol.

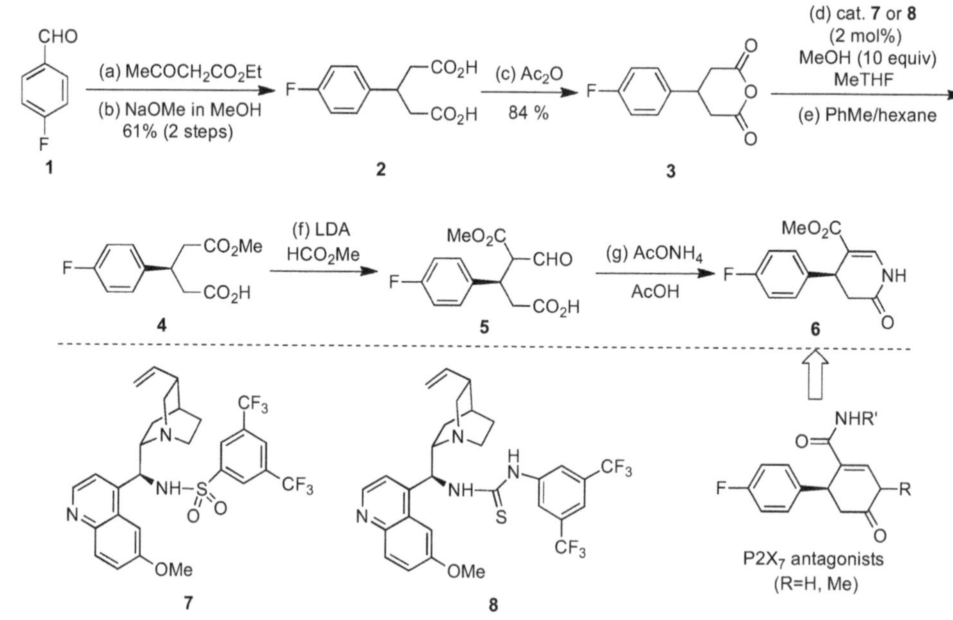

SCHEME 16.1

a well-defined chiral Ru catalyst, [(R,R)-TsDPEN]Ru(p-cymene)Cl,[8] allows access to the enantio-merically enriched allylic alcohol **10** (Equation 16.1).[9]

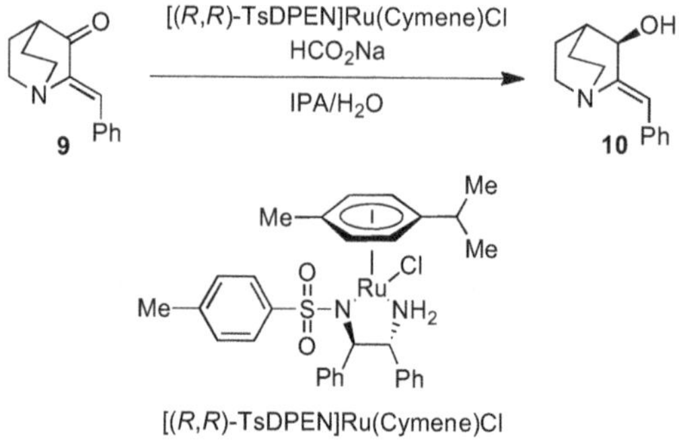

$$(16.1)$$

[(R,R)-TsDPEN]Ru(Cymene)Cl

This asymmetric transfer hydrogenation reaction was carried out in the presence of [(R,R)-TsDPEN] Ru(cymene)Cl catalyst with sodium formate as the hydrogen source.

Procedure

- Charge 0.22 kg (0.35 mol) of [(R,R)-TsDPEN]Ru(p-cymene)Cl, H_2O (30.0 kg), and 2-propanol (24.0 kg) into a 100-gal stainless steel reactor.
- Adjust the temperature to 40 °C and hold for 1 h.
- Charge, into a separate 100-gal Hastelloy reactor, 48 kg (706 mol) of HCO_2Na, H_2O (120.0 kg), 30.0 kg (141 mol) of **9**, and 2-propanol (48 kg).
- Adjust the temperature to 40 °C while sparging with N_2.
- Charge the catalyst solution into the heterogeneous solution of **9**.
- Heat the resulting mixture to 67 °C and hold for 5 h.
- Cool to 37 °C.
- Separate layers.
- Add water (180 kg) and adjust the temperature to 22 °C.
- Filter and wash with water (2×90 kg).
- Dry at 60 °C under vacuum.
- Yield: 26.6 kg (88%).

16.1.1.3 Sharpless Asymmetric Dihydroxylation

Chiral alcohols are key intermediates in the synthesis of many biologically active compounds. Sharpless asymmetric dihydroxylation of terminal olefins is one of the most reliable and general reactions to date for preparing chiral alcohols with high stereoselectivity. The reaction is typically performed using a catalytic amount of osmium oxide, which after the reaction is regenerated with either potassium ferricyanide or N-methylmorpholine N-oxide (NMO), in the presence of hydroquinine 1,4-phthalazinediyl diether, $(DHQ)_2$-PHAL or $(DHQD)_2$-PHAL, as the chiral ligand. The use of catalytic amounts of osmium oxide is favorable because it dramatically reduces the amount of the highly toxic and very expensive osmium oxide needed. At large scales, the use of NMO has several advantages. First, no large amounts of salts are required in the reaction (3 molar equiv of each salt is used in the AD mix). Unlike the iron by-products formed from the use of potassium ferricyanide, the by-product from NMO, N-methylmorpholine, is easily removed and recycled if necessary. Second, NMO is economical, by a factor of 5 when compared to the amounts of ferricyanide/carbonate salts required per mole of olefin. Third, the NMO-based reactions can be run in high concentration.

The asymmetric dihydroxylation of 2-isopropoxy-3-methoxystyrene **11** was demonstrated on a 2.5-kg scale using NMO as the oxidant to give both high yield and good stereoselectivity (Equation 16.2).[10] The reaction was run in a mixture of t-butanol and water at ambient temperature. The chiral ligand, which is the most expensive reagent, was used catalytically and can be recovered by simple acid/base extraction. The osmium was also used catalytically in a reduced form which is less toxic. Being water soluble, the osmium was easily separated from the organic product and can also be recovered in the fully oxidized form for reuse.

$$(16.2)$$

Procedure

- Charge 34.5 g (0.09 mol) of $K_2OsO_2(OH)_4$ into a 50-L, three-necked reactor, followed by hydroquinine 1,4-phthalazinediyl diether (81.6 g, 0.10 mol), N-methylmorpholine N-oxide (60 wt% in water, 3.0 L, 17.4 mol), tBuOH (14 L), and H_2O (10 L).
- Stir until a clear solution is obtained.
- Charge 2.5 kg (13.0 mol) of **11** at a rate of 5.6 mL/min using a peristaltic pump while maintaining the batch temperature at 20±5 °C.
- Stir the resulting orange solution until the olefin content was <0.5%.

Workup/isolation

- Add toluene (12 L) and a solution of Na_2SO_3 (1.9 kg) in H_2O (4.7 L).
- Stir the resulting mixture overnight.
- Separate layers.
- Wash the organic layer with an aqueous solution of Na_2SO_4 (0.8 kg) in H_2O (5 L).
- Dry the organic phase over K_2CO_3 (1.0 kg).
- Distill at 60 °C under vacuum.
- Yield: 2.5 kg in 94% of GC purity and 95% of enantiomeric purity.

Recovery of the ligand

- Extract the organic phase with aqueous H_2SO_4 (0.38 L) in aqueous Na_2SO_4 (1.6 kg) in H_2O (10 L).
- Basify the resulting acidic solution with NaOH and extract with toluene (0.7 L).
- Distill to give 61 g of the ligand (98%) as a white powder.

16.1.1.4   Enantioselective Alkylation

In addition to the established methods[11] for the construction of chiral carbon centers by the asymmetric α-alkylation of carbonyl systems, enantioselective catalytic alkylation of malonyl systems was achieved via asymmetric phase-transfer catalysis. Direct mono-alkylation of 1,3-dicarbonyl systems is challenging, mainly due to their ease of racemization under basic or acidic conditions.

Phase-transfer catalytic reaction, as one of the economical and environmentally benign synthetic methods, is commonly used in chemical and pharmaceutical industries. The phase-transfer catalytic alkylation of N,N-diarylmalonamic tert-butyl ester **13** in the presence of (S,S)-3,4,5-trifluorophenyl-NAS bromide **14** (1 mol%) afforded a (S)-mono-α-alkylated product **15** (up to 96% ee) (Equation 16.3).[12]

Application of this enantioselective alkylation was extended toward the synthesis of quaternary amino ester **17** from imine **16** using the chiral phase-transfer catalyst **14**, providing 61% product yield in 97% ee on multi-gram scales (Equation 16.4).[13]

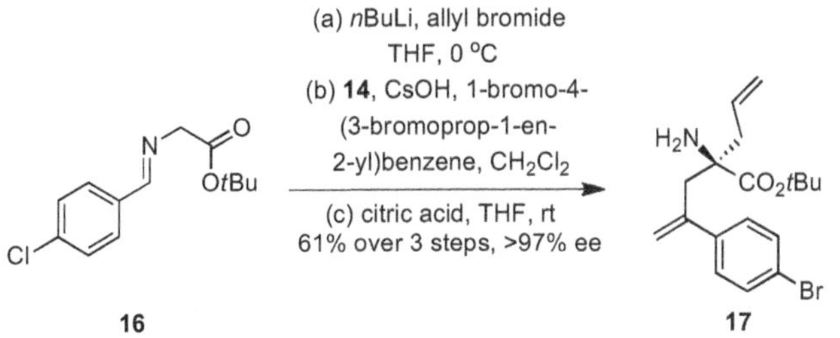

$$(16.4)$$

REMARK

Direct asymmetric functionalization of the γ-position in α,β-unsaturated carbonyl derivatives is formidably challenging because of the lack of a vicinal chelating group. Thereby an indirect method for an asymmetric γ-alkylation of α,β-unsaturated malonates was developed (Scheme 16.2).[14]

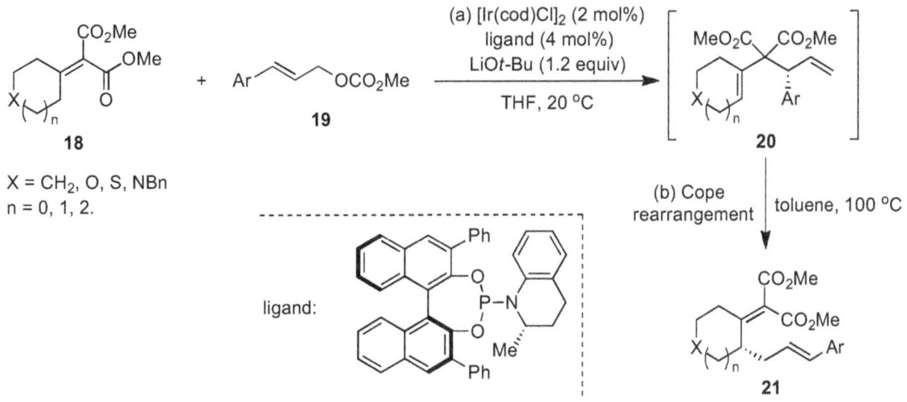

SCHEME 16.2

The initial reaction was catalyzed by an iridium catalyst in the presence of a chiral ligand to alkylate extended enolates of the α,β-unsaturated malonates **18** asymmetrically at the α-position to give 1,5-dienes **20** in good yields and excellent enantiopurity (>99% ee). The subsequent thermal Cope rearrangement of **20** allowed an efficient chirality transfer to occur intramolecularly, providing the desired γ-alkylated products **21**.

16.1.1.5 Enantioselective Protonation of Enamines

Enantioselective protonation of enamines was designed in the synthesis of chiral α-alkylated ketones. As shown in Scheme 16.3,[15] the reaction of an in situ formed o-quinone methide (not shown) from **22** with 1,3-diketones would deliver chroman intermediates **24** whose retro-Claisen reaction may form racemic α-alkylated ketones **25**. Subsequently, a reaction between the racemic ketones **25** and chiral amine catalyst would occur to afford enamine intermediates **26**. A stereospecific protonation of the enamines furnishes the desired chiral α-alkylated ketones **27** in good yields and high enantioselectivities.

16.1.1.6 CuH-Catalyzed Synthesis of 2,3-Disubstituted Indolines

A CuH-catalyzed stereoselective synthesis of 2,3-disubstituted indolines was realized in the presence of a chiral bisphosphine ligand (Scheme 16.4).[16] The reaction was initiated by an olefinic insertion into Cu–H bond of L*CuH, generated in situ from a reaction of $Cu(OAc)_2$ with diethoxy(methyl) silane (DEMS) in the presence of chiral ligand, to form alkylcopper intermediate **29**.

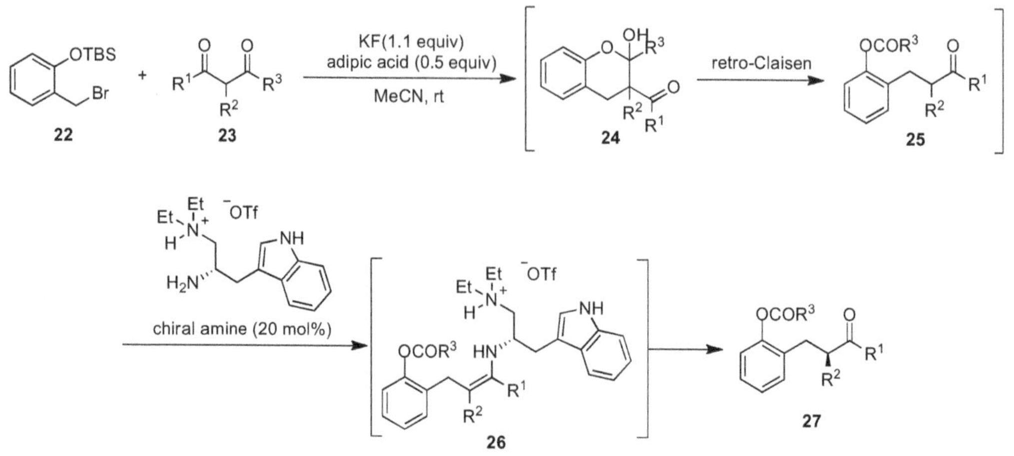

SCHEME 16.3

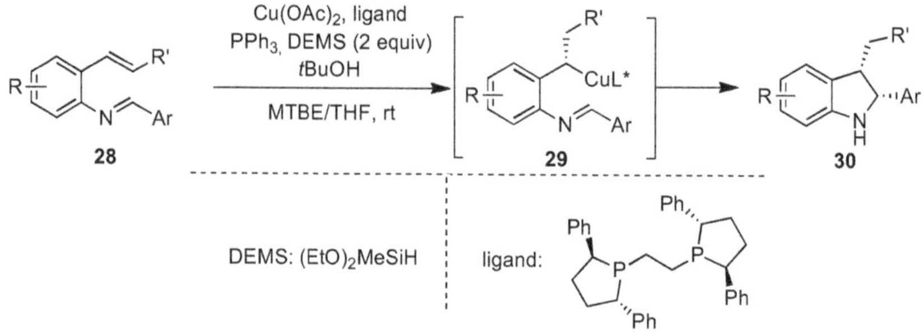

SCHEME 16.4

The Markovnikov organocopper adduct **29** was trapped intramolecularly by tethered imine to give the desired product **30** in a high diastereo- and enantioselective manner.

16.1.1.7 CuH-Catalyzed Synthesis of Chiral Amines

Furthermore, the CuH-catalyzed reaction was utilized to synthesize chiral branched amine **35** using 1,2-benzisoxazole **32** as an electrophilic ammonia equivalent (Scheme 16.5).[17] This hydroamination of 2-(methylthio)-5-vinylpyrimidine **31** occurred through an intermolecular reaction of alkylcopper intermediate **33** with **32** to give a chiral Schiff base intermediate **34**.

An application of this approach toward the synthesis of the anti-retroviral drug Maraviroc was illustrated in Scheme 16.6.[17]

16.1.2 CHIRAL POOL SYNTHESIS

16.1.2.1 Condensation of Indoline with Benzaldehyde

A diastereoselective synthesis of indoline aminal **40** was realized from a reaction of (*R*)-5-Bromo-2-(5-bromoindolin-2-yl)phenol **38** with benzaldehyde. This condensation was catalyzed by TFA (5 mol%) in acetonitrile at 35 °C (Scheme 16.7).[18] The reaction gave indoline aminal product **40** in a yield of 93% with a dr ratio of 99:1. It was suggested[18] that a crystallization-induced dynamic resolution (CIDR) attributed to the high diastereoselectivity.

The aminal **40** was identified as a key intermediate for the synthesis of MK-8742, a potent and selective NS5a inhibitor developed by Merck & Co for the treatment of HCV.

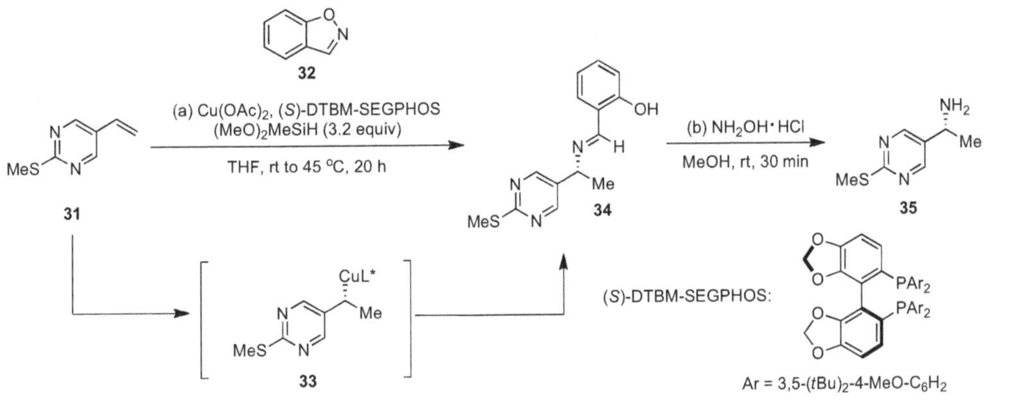

SCHEME 16.5

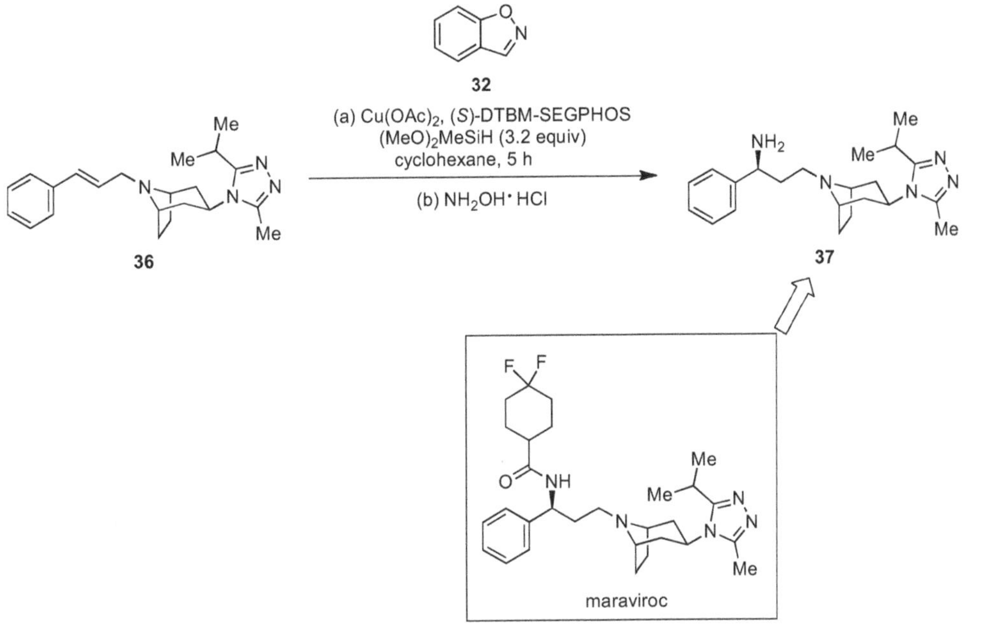

maraviroc

SCHEME 16.6

16.1.2.2 Claisen Rearrangement

A chiral pool synthesis was devised to prepare a common intermediate **45** in the total synthesis of (–)-fructigenine A and (–)-5-*N*-acetylareemin (Scheme 16.8).[19]

The synthetic sequence involves olefination/isomerization, acetyl deprotection, and Claisen rearrangement, furnishing the desired oxindole **45** in 89% yield and 99% ee. This redox economy synthetic sequence accomplished two goals: the oxidation state transformation from C3 to C2 and the stereogenic center transformation from the tethered chiral alkoxy to C3 in a one-pot operation.

16.1.3 USE OF CHIRAL AUXILIARIES

16.1.3.1 Diastereoselective Diels–Alder Reaction

The use of a chiral auxiliary approach in the synthesis of chiral benzyl ether was started with a diastereoselective Diels–Alder reaction employing diene (+)-**46** and methyl crotonate to deliver an

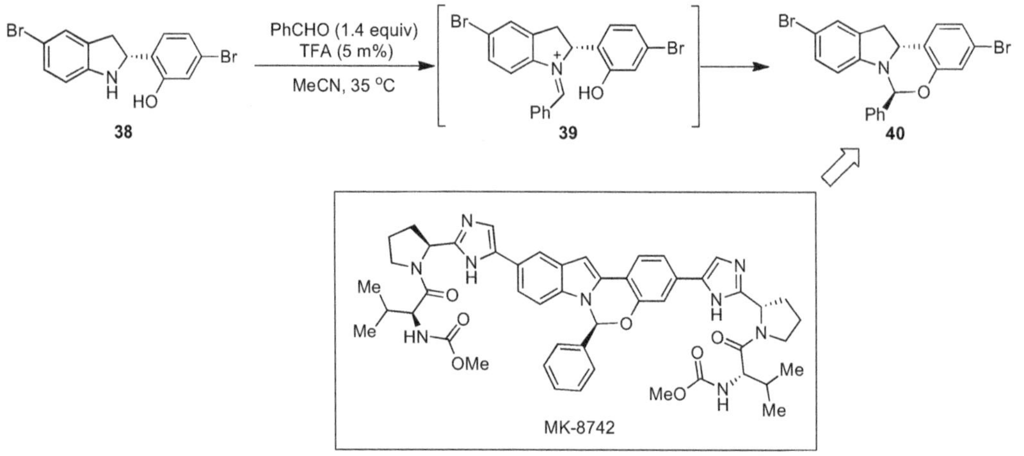

SCHEME 16.7

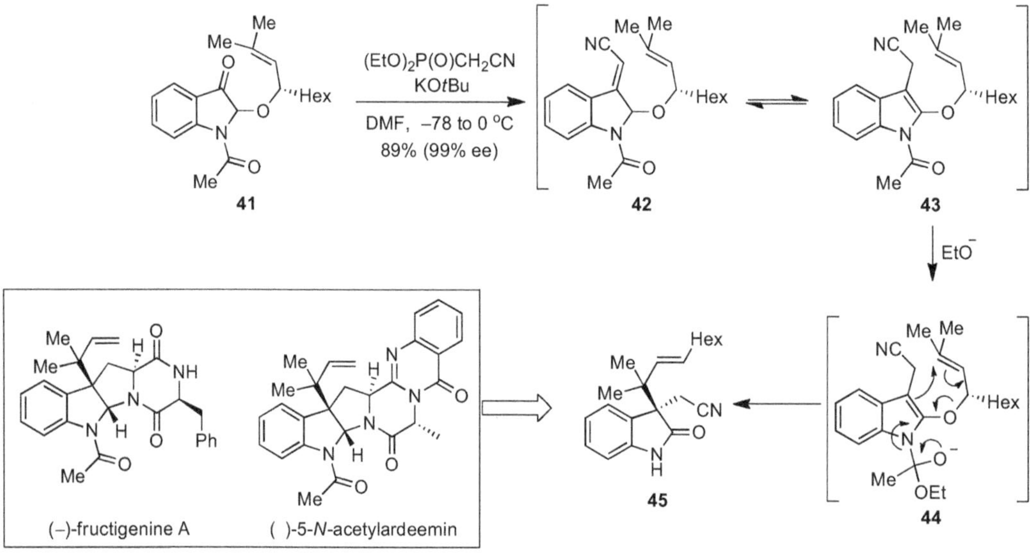

SCHEME 16.8

intermediate ester, which upon reduction with LiAlH$_4$ furnished primary alcohol (+)-**47**. Following the protection of the hydroxyl group as the benzyl ether, removal of the pyrrolidine auxiliary by HF in acetonitrile furnished (+)-**48** (Scheme 16.9).[20]

16.1.3.2 Diastereoselective Synthesis of Boronic Acid

L-Valyl-pyrrolidine-(2R)-boronic acid **54** was synthesized using (+)-pinanediol as chiral auxiliary group. Removal of the chiral auxiliary pinanediol from valyl-pyrrolidine-(2R)-boronic acid pinanediol ester **53** was realized successful via boronic acid exchange with phenylboronic acid (Scheme 16.10).[21]

REMARK

Attempts to recycle the expensive (+)-pinanediol led to a practical boronic acid exchange approach (Equation 16.5).[21] The optimized boronic acid exchange between **49** and **53** furnished the product

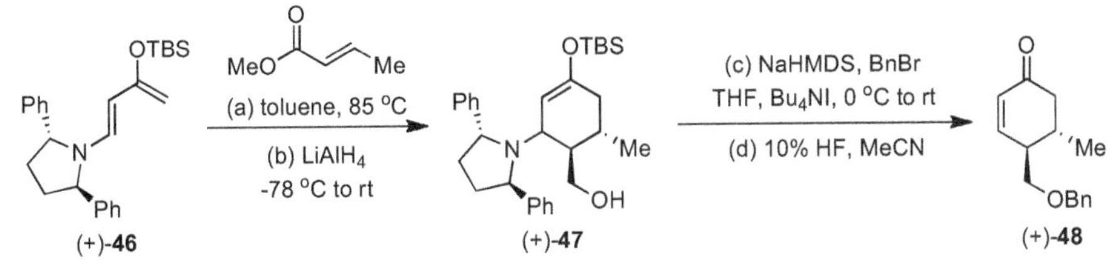

SCHEME 16.9

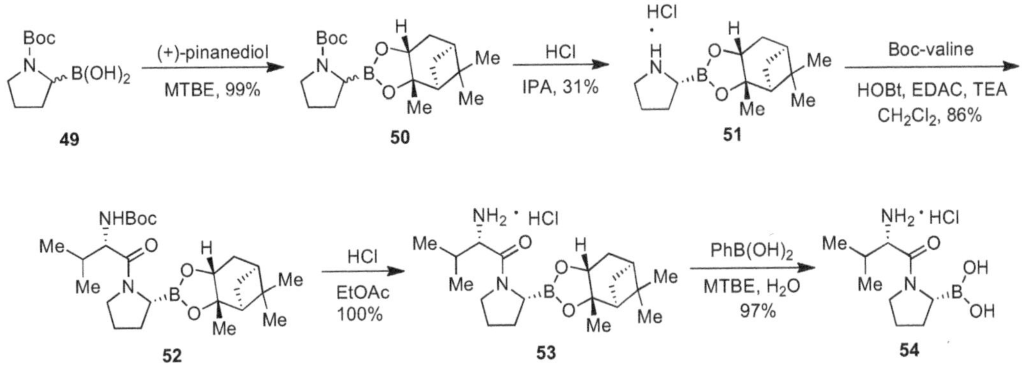

SCHEME 16.10

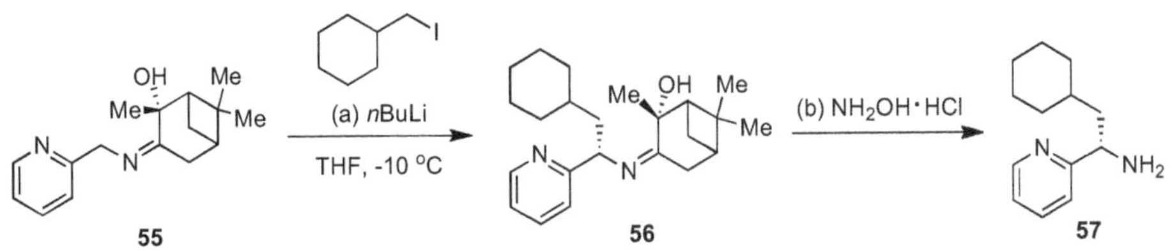

SCHEME 16.11

54 along with **50**. The efficiency of these exchanges is due to the fact that **50** is completely insoluble in water and completely partitions to the MTBE layer, driving the process to completion. The overall route was more environmentally friendly by eliminating phenylboronic acid.

$$
\begin{array}{ccc}
\underset{\text{(in water)}}{49\ +\ 53} & \xrightarrow[\substack{\text{MTBE}\\99\%}]{} & \underset{\text{(in water)}}{54}\ +\ \underset{\text{(in MTBE)}}{50}
\end{array} \qquad (16.5)
$$

16.1.3.3 Synthesis of Chiral (*S*)-Pyridyl Amine

A chiral (*S*)-pyridyl amine was synthesized from a two-step process, including α-alkylation of chiral imine **55** followed by removal of the chiral auxiliary via imine-exchange with hydroxylamine hydrochloride without detectable racemization (Scheme 16.11).[22] The subsequent acid/base extraction allowed for the efficient removal of the oxime by-product.

REMARK

In general, ketimines can be hydrolyzed by either acids or bases. However, hydrolysis of **56** with a citric acid solution was sluggish, while using 1–2 N HCl led to decomposition as well as racemization of the imine.

16.1.3.4   Synthesis of L-Carnitine

L-Carnitine ((R)-3-hydroxy-4-(trimethylammonio)butanoate) is a medicine for improving myocardial function.

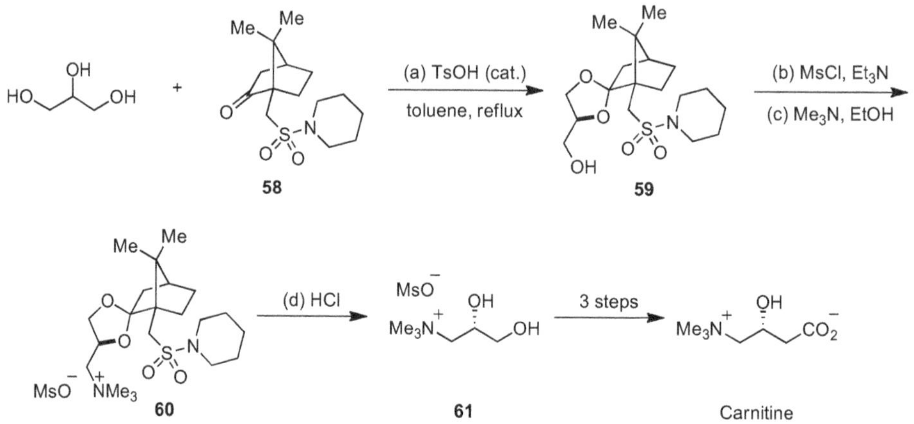

SCHEME 16.12

Scheme 16.12[23] presents a seven-step synthesis of carnitine starting with a reaction of glycerol with (1R)-10-camphorsulfonamide **58** to give a spiro-acetal **59** in 60% yield. **59** was then converted, in almost quantitative yield, into the corresponding mesylate whose reaction with trimethylamine gave **60**. The chiral auxiliary in **60** was then cleaved off by HCl treatment, and thoroughly recovered from the organic layer, while the dihydroxypropyltrimethylammonium mesylate **61** was obtained by evaporation of the aqueous solution. The conversion of **61** into carnitine was achieved in three steps involving bromination, S_N2 substitution, and nitrile hydrolysis.

16.2   KINETIC RESOLUTION

Although great progress has been made in both chemocatalytic and biocatalytic asymmetric syntheses, optical resolution of racemic compounds still represents a valuable method and continues to play an important role in producing enantiopure materials on industrial scales. These optical resolutions are generally characterized as kinetic resolution and dynamic kinetic resolution.[24] In general, enantiomers have identical physical and chemical properties. However, their chemical properties may differ in unsymmetrical environments; for example, enantiomers react with chiral compounds at different reaction rates. By making use of the difference in reactivity under asymmetric environments, the kinetic resolution is the most common way to obtain enantiopure compounds.[25] Like normal chemical reactions, the outcome of kinetic resolution may depend on various factors such as solvent, temperature, resolution reaction time and concentration, and the ratio of resolution agent.

A significant disadvantage of the kinetic resolution process is a 50% theoretical yield of the desired enantiomer with a 50% loss of the undesired enantiomeric substrate. In order for the kinetic resolution approach to be viable, the substrates must be inexpensive or easily accessible from inexpensive commercial starting materials.

16.2.1   HYDROLYTIC KINETIC RESOLUTION OF EPOXIDE

The hydrolytic kinetic resolution (HKR) has been extensively adopted as the method of choice for the preparation of a variety of terminal epoxides and/or diols in enantioenriched form from inexpensive racemic epoxides (Equation 16.6).[26]

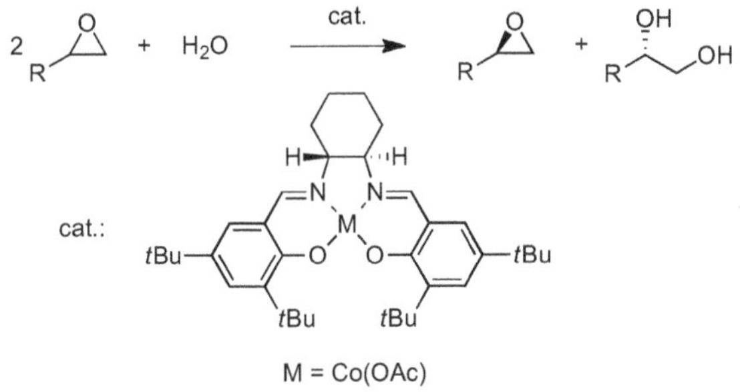

(16.6)

In this HKR process, water is used as the reagent for effecting the resolution reaction and the rate of the ring-opening reaction can be controlled simply by modulating the water addition rate to the epoxide–catalyst mixture.

16.2.2 Resolution of Diol via Stereoselective Esterification

Resolution via the formation diastereomeric salts is not applicable for compounds that lack acidic or basic functional groups. Thus, a derivatization protocol was developed for the resolution of racemic diol (±)-**62**. Diastereoselective esterification of (±)-**62** with (–)-camphanic acid allowed conversion of (±)-**62** into the corresponding diastereomeric ester **63**, followed by recrystallization and hydrolysis, furnishing the desired enantiopure diol (R)-**62** (Scheme 16.13).[27]

Procedure

Stage (a)
- Charge, at 5–10 °C, 14.0 kg (73.0 mol, 2.2 equiv) of EDC·HCl to a stirred solution containing (±)-**62** (9.44 kg, 33.2 mol), (–)-CpOH (11.8 kg, 59.5 mol, 1.8 equiv), and DMAP (1.22 kg, 10.0 mol, 0.3 equiv) in DMF (47 L).
- Stir the resulting mixture at 5–10 °C for 2 h prior to cooling to <10 °C.
- Transfer the batch to a stirred mixture of EtOAc (472 L) and water (94 L).
- Separate layers and wash the organic layer with 5% aqueous NaHCO$_3$ (2×99 kg), 16.7% aqueous NaCl.
- Distill under reduced pressure.
- Add heptane (189 L).
- Distill under reduced pressure until 142 L.
- Add heptane (47 L) and stir at 20–25 °C for 1.5 h.
- Filter and wash with heptane (76 L).
- Dry at 40–45 °C under vacuum.
- Yield of **63**: 14.8 kg (96%).

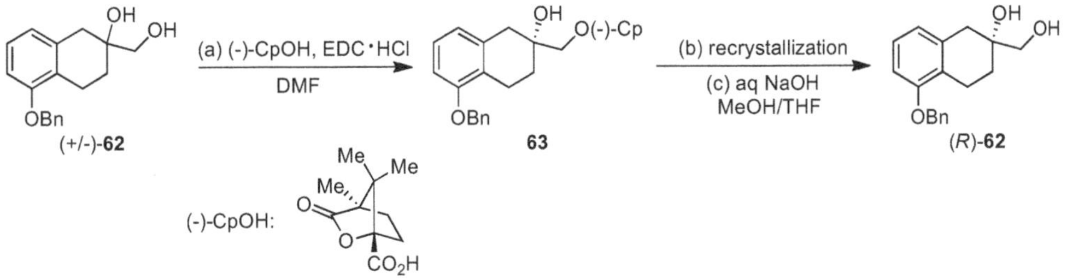

SCHEME 16.13

Stage (b)

- Dissolve 14.7 kg of **63** in 588 L of EtOH at >65 °C.
- Cool the resulting solution to 50–55 °C.
- Charge 5.6 g of seed crystals.
- Stir the batch at 50–55 °C for 0.5 h.
- Cool and age at 15–20 °C for 1 h.
- Filter and rinse with EtOH (118 L).
- Dry at 40–45 °C under vacuum.
- Yield of crude **63**: 5.93 kg (78% de).
- Dissolve the crude **63** (5.93 kg) in EtOH (237 L) at 75–80 °C prior to cooling to 25 °C.
- Age at 25 °C for 1 h.
- Filter and wash with EtOH (47 L).
- Dry at 40–45 °C under vacuum.
- Yield of **63**: 5.05 kg (>99% de).

Stage (c)

- Charge water (5.7 kg) and 48% NaOH (2.9 kg) at 20–25 °C to a solution of **63** (5.05 kg) in a mixture of MeOH (51 L) and THF (51 L).
- Stir at 20–25 °C for 1 h.
- Add water (51 L) followed by adding aqueous HCl (2.1 kg of 35% HCl and 25 L of water).
- Distill to 30 L.
- Add 5% aqueous $NaHCO_3$ (53.5 kg).
- Stir the resulting slurry at 20–25 °C for 1 h.
- Filter and wash with water (51 L).
- Dry at 40–45 °C under vacuum.
- Yield of ®-**62**: 2.81 kg (>99% ee).

16.2.3 RESOLUTION OF PHOSPHINE LIGAND VIA STEREOSELECTIVE LIGAND EXCHANGE

Chiral ligands have been playing a significant role in contemporary asymmetric organic synthesis. Atropisomerically chiral ligands, such as 1'(2'-diphenylphosphin'-1'-naphthyl)isoquinoline **64**, have been applied to a number of asymmetric syntheses.[28] A practical resolution protocol was developed usi®(R)-Pd dimer **65** as the resolving agent to separate the R enant®er (R)-**64** from (S)-**64** (Equation 16.7).[29] The (R)-phosphine ligand **64** was obtained in 46% yield from the reaction mixture by filtration, and the isolated complex was treated with 1,2-bis(diphenylphosphino)ethane (dppe) to give pure (S)-enantiomer.

(16.7)

64:65 = 4:1

acetone
55 °C, 2 h

64   **65**

(R)-**64**
46%

(R,S)-**66**
45%

Procedure

Stage (a)

- Charge 17.4 g (25.6 mmol) of **65** in acetone (0.6 L) to a suspension of **64** (45 g, 100 mmol) in acetone (1 L).
- Stir the resulting mixture at 55 °C for 2 h.
- Add potassium hexafluorophosphate (9.42 g, 51.2 mmol) in acetone (0.3 L).
- Stir the resulting mixture at 55 °C for 2 h.
- Cool to room temperature and filter (the filtrate was further processed in Stage (b)).
- Dissolve the solid in CHCl$_3$.
- Filter through Celite and wash with acetone (0.1 L).
- Dry® give (R)-**64**.
- Yield: 20.8 G (46%).

Stage (b)

- Distill the filtrate to give an orange solid.
- Wash the solid with toluene (2×0.1 L) and CHCl$_3$ (2×0.1 L).
- Mix the resulting pale-yellow solid (40 g, 45 mmol) with 19.7 g (49.4 mmol) of dppe in CH$_2$Cl$_2$ (0.4 L).
- Stir at room temperature for 3 h.
- Distill under reduced pressure to give a pale-yellow solid.
- Slurry the solid in acetone (0.2 L) for 30 min.
- Filter and wash with acetone.
- Yield of (S)-**64**: 19.2 g (97%).

16.2.4 Resolution of Diastereomeric Mixture via Salt Formation

The resolution of the diastereomeric mixture of the foam-like carboxylic acid **67** was realized by the formation of ammonium salt with adamantylamine (Equation 16.8).[30] After crystallization, a yield of 40% of **68** was obtained with a diastereomeric ratio of 92:8.

$$(16.8)$$

16.3 ENZYMATIC RESOLUTION

Despite the dramatic improvements in enantioselective synthesis, bioresolution remains one of the most inexpensive and operationally simple methods for producing pure enantiomers on large scales in the industry. Enzymes are highly enantioselective catalysts, and enzymatic resolution often yields enantiomeric excesses of more than 99%.

16.3.1 Resolution of Esters

16.3.1.1 Resolution of Methyl Piperidine-4-Carboxylate

Enzymatic resolution[31] renders a process with less environmental impact as the reaction uses water as the solvent. The enzymatic resolution was proved to be practical on the large scale for the resolution of the racemic piperidine hydrochloride salt (±)-**69** (Equation 16.9).[85]

(16.9)

The resolution was carried out with lipase from *Candida antarctica* (immozyme CALB-T3-150) and 1.57 equiv of dipotassium phosphate at pH 8. The resolved (2R,4S)-**69** was used directly in the subsequent step.

Procedure

- Charge 13.1 kg (95.7 wt%, 50.2 mol) of (±)-**69** into a reactor, followed by adding 13.7 kg (78.66 mol) of K$_2$HPO$_4$ and 80 L of water.
- Stir to form a pH 8 solution.
- Charge 3.95 kg of immobilized lipase.
- Stir the batch at 35 °C for 40 h (99% conversion).
- Cool to 5 °C.
- Add MeTHF (39.5 L) and 25 wt% aqueous KOH (4.33 kg of KOH in 13.1 L of water).
- Filter and wash with MeTHF (35 L).
- Separate layers.
- The product organic layer was directly used in the subsequent step.

16.3.1.2   Resolution of Ethyl α-Amino Acetate

Enzymes, which are composed of chiral amino acids, catalyze chemical reactions with high stereoselectivity. Specifically, esterase enzymes catalyze the hydrolysis of esters to carboxylic acids. Pig liver esterase (PLE) is a widely used enzyme for asymmetric ester hydrolysis. Although it was originally used for the desymmetrizing hydrolysis of glutarate esters,[32] PLE can hydrolyze malonates, cyclic diesters, monoesters, and other substrates.

The enzymatic resolution of racemic ester **70** was affected by PLE leaving the acid as waste (Equation 16.10).[33]

(16.10)

16.3.1.3   Resolution of Diazepane Acetate

The maximum yield for a given kinetic resolution is 50%, resulting in the loss of the other enantiomer. Therefore, recycling of the other enantiomer by racemization is important, especially on a commercial manufacturing scale. Racemization of (R)-ester [(R))-**71**] was realized in dimethyl carbonate as a cosolvent with MeOH and in the presence of NaOMe (Scheme 16.14).[34]

REMARK

If a specialty enzyme is used in an economically feasible industrial process, enzyme recycling is essential. Therefore, the stability of the enzyme is important. Immobilization on solid supports such as EUPERGIT can improve the operational stability of enzymes, making enzyme recycling amenable.[35]

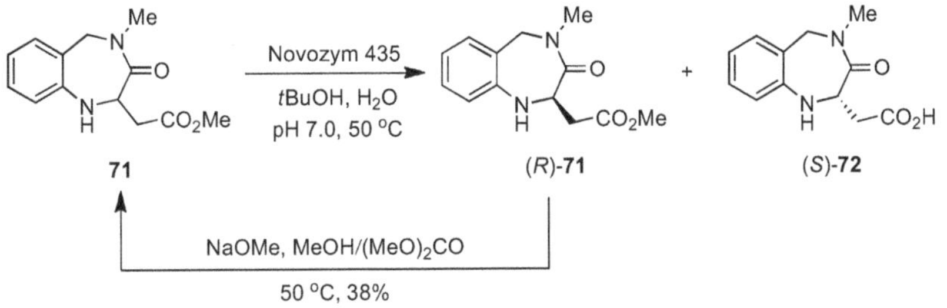

SCHEME 16.14

16.3.2 RESOLUTION OF AMINO ACIDS

Acylase I enzymes isolated from the porcine kidney (PKA) and the fungus *Aspergillus* sp. (AA) are commercially available, inexpensive, stable in aqueous solution, and possess high specific activity.

Enzymatic resolution of amino acids **73** and **75** in the presence of enzyme acylase I was realized at 35–37 °C giving **74** and **76** in yields of 45% (with >99% ee) and 43% (97%ee), respectively (Equations 16.11 and 16.12).[36]

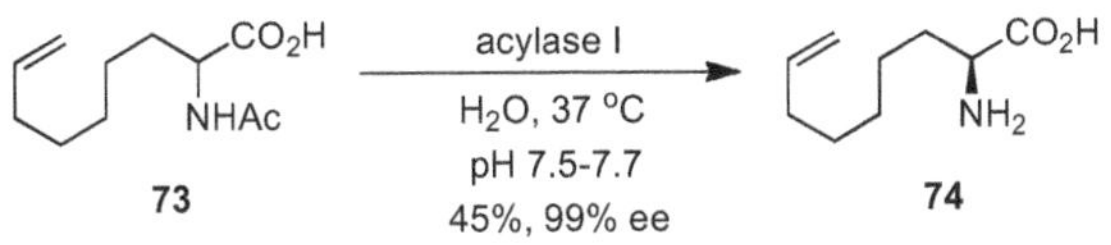

$$(16.11)$$

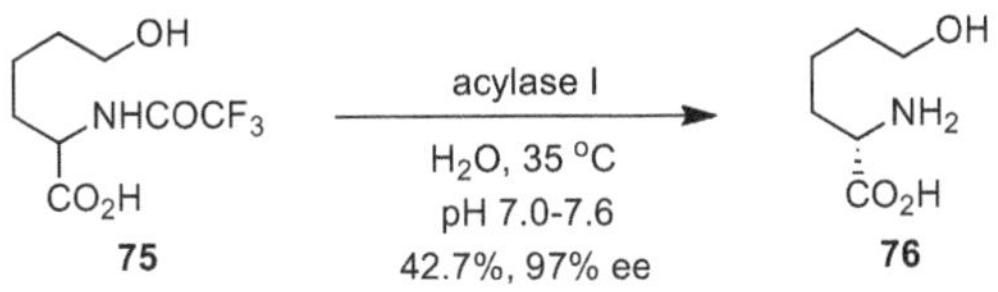

$$(16.12)$$

16.3.3 RESOLUTION SECONDARY ALCOHOLS

Generally, an enzymatic resolution requires no protection/deprotection steps, and the reactions are typically carried out under ambient and neutral conditions. For example, hydrolase-catalyzed kinetic resolutions have been widely used for the synthesis of enantiopure secondary alcohols such as **79** as shown in Equation 16.13.[37]

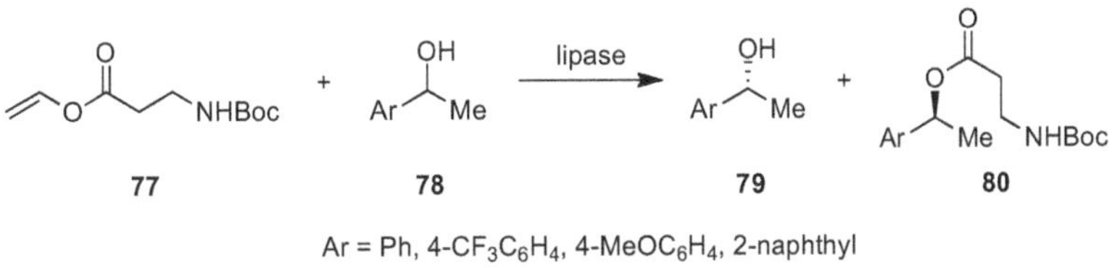

$$(16.13)$$

Ar = Ph, 4-CF$_3$C$_6$H$_4$, 4-MeOC$_6$H$_4$, 2-naphthyl

16.4 SEPARATION WITH CHIRAL CHROMATOGRAPHY

The optical pure chromene (*R*)-**81** was the most active compound with the desired selective estrogen receptor modulator (SERM) activity, while the enantiomer (*S*)-**81** showed weak SERM activity. The synthesis of **81** was achieved in eight linear synthetic steps. Attempts to introduce the required chiral center via either enantioselective asymmetric synthesis of (*R*)-**81** or resolution of the racemate **81** with chiral acids were unfruitful. Therefore, the optically pure (>99% ee) (*R*)-**81** was obtained on a kilogram scale by chiral chromatographic separation of the racemate **81** (Scheme 16.15).[38]

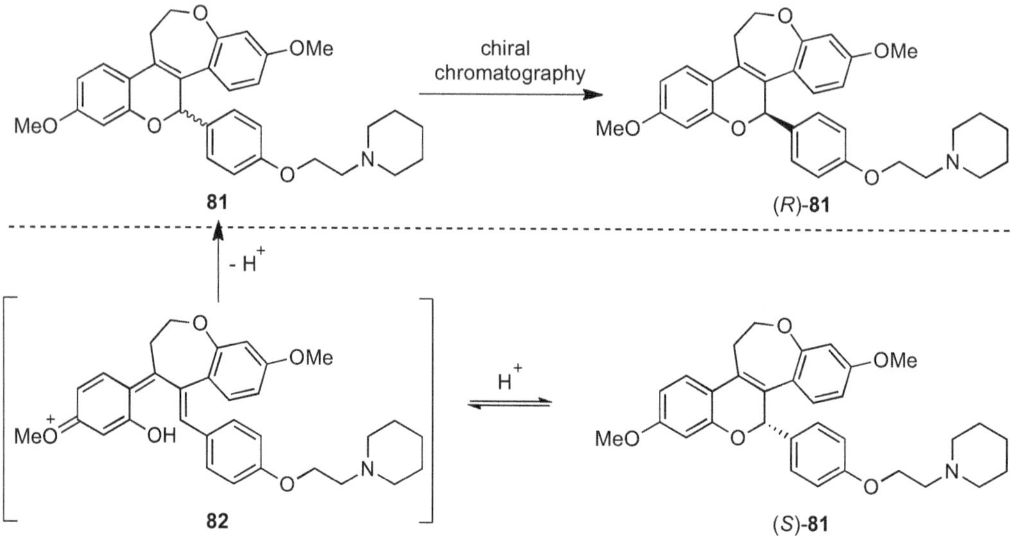

SCHEME 16.15

Efforts in converting the undesired (S)-**81** into the racemate **81** by adopting the literature conditions[39] failed due to problems caused by the 8-methoxy group. An acid-catalyzed racemization process was developed to convert (S)-**81** back to the racemate **81** via intermediate **82** (Scheme 16.15). Refluxing **81** with either HCl (4.0 equiv) in EtOH for 76 h or H_2SO_4 (2.0 equiv) in water for 68 h afforded a near racemic mixture in >96% isolated yield and good chemical purity (87–95%).

16.5   CLASSICAL RESOLUTION

Classical resolution is to convert the enantiomers to their corresponding diastereomers. The resulting diastereomeric products possess different physical and chemical properties, which provide an excellent opportunity to manipulate and separate chiral molecules. The classical resolution of racemic mixtures via diastereomeric salt pairs plays an important role in the preparation of optically active compounds. Theoretically, only 0.5 equiv of the resolving agent is required to form the diastereomeric salt with one enantiomer from the racemic mixture. However, usually, 1.0 equiv of the resolving agent is used to resolve a racemic compound in order to achieve better resolution. Interestingly, the resolution of (±)-*threo*-methylphenidate hydrochloride (**83**, Ritalin hydrochloride) with R-(–)-binaphthyl-2,2'-diyl hydrogen phosphate **84** showed that the use of 1 equiv of the resolving agent gave poorer resolution compared to 0.5 equiv (Figure 16.3).[1d] (±)-*Threo*-methylphenidate hydrochloride is a mild nervous system stimulant, marketed for the treatment of children with attention deficit hyperactivity disorder (ADHD). (2R,2'R)-(+)-*Threo*-methylphenidate hydrochloride (+)-**83** has been reported to be 5 to 38 times more active than the corresponding (2S,2'S)-(–)-*threo*-methylphenidate hydrochloride.[40]

The major drawback of this method is the sacrifice of the undesired enantiomer.

16.5.1   RESOLUTION OF RACEMIC ACID

(S)-3-Methyl-2-phenylbutylamine **85** is a versatile resolving agent to resolute a number of racemic acids. Scheme 16.16[41] lists four optical carboxylic acids, resolved by using (S)-3-methyl-2-phenylbutylamine **85**.

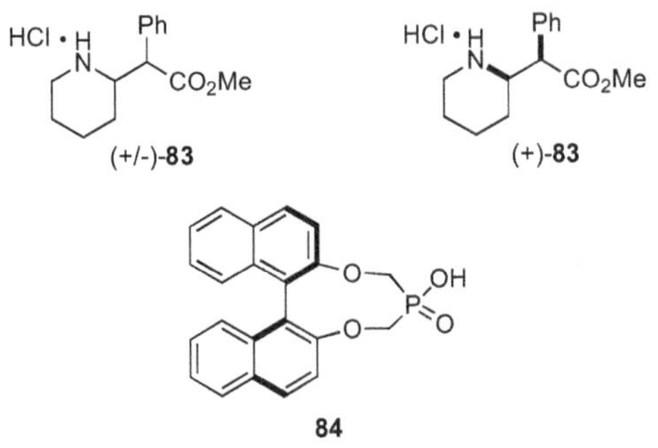

FIGURE 16.3 Structures of *threo*-methylphenidate hydrochlorides **83** and *R*-(–)-binaphthyl-2,2'-diyl hydrogen phosphate **84**.

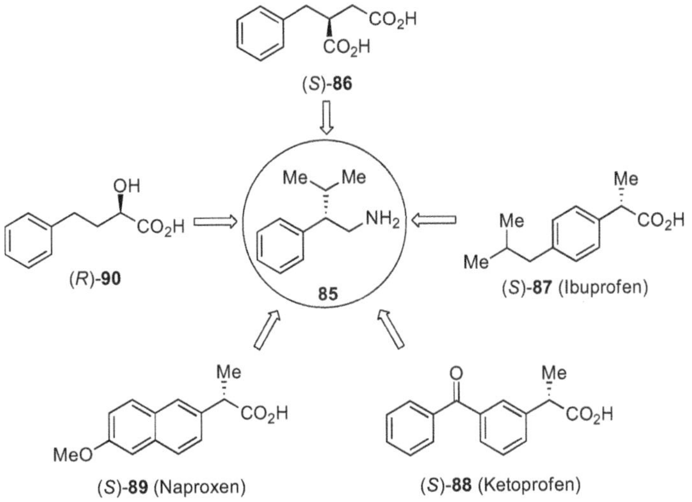

SCHEME 16.16

16.5.2 RESOLUTION OF RACEMIC BASES

16.5.2.1 Use of Optical Pure *tert*-Leucine Derivative

The racemic *tert*-leucinol **92** was resolved by using optically active *tert*-leucine derivative **91** with a sterically demanding *tert*-butyl group as the resolving agent, affording the (*R*)-*tert*-leucinol as its corresponding salt **93** (Equation 16.14).[42]

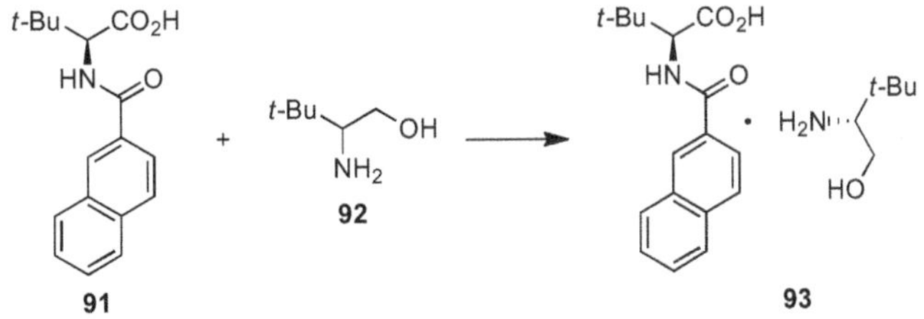

$$(16.14)$$

The remaining *tert*-leucinol **92**, predominantly (*S*)-configurated, was isolated from the mother liquor via a reaction with (*S*)-mandelic acid **94**, giving the salt **95** in 63% yield after double recrystallization (Equation 16.15).[42]

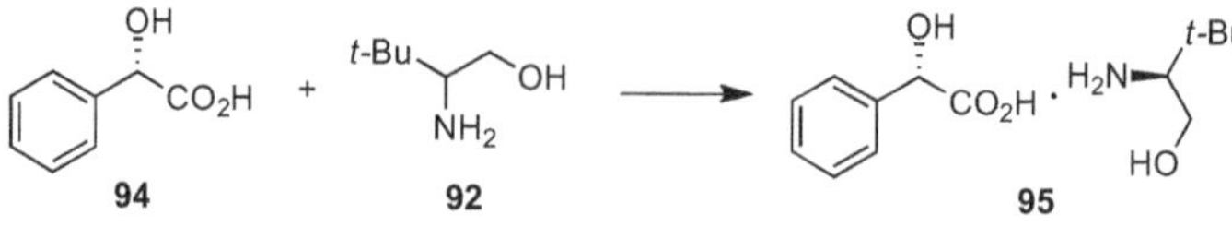

(16.15)

16.5.2.2 Use of di-p-Toluoyl-D-Tartaric Acid

A viable classical resolution process was developed for the preparation of (3R,4S)-4-(4-bromophe nyl)-1-methyl-pyrrolidine-3-carboxylic acid ethyl ester **98** (Scheme 16.17).[43] Treatment of the race mic mixture of *trans-N*-methylpyrroline *trans-rac*-**96** with di-p-toluoyl-D-tartaric acid (*D*-DTTA) in ethanol at 76 °C afforded the corresponding diastereomeric salt **97** in 42% yield. A pure ester **98** was obtained following treatment with aqueous NaOH in 99.6% ee.

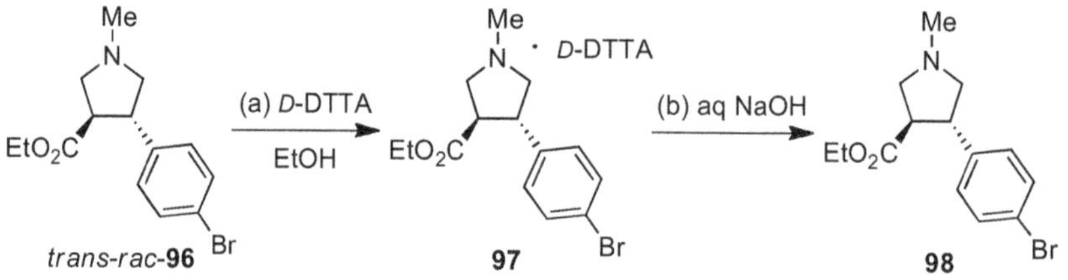

SCHEME 16.17

Procedure

Stage (a)
- Charge 31 kg (80.2 mol) of *D*-DTTA into a 500-L reactor, followed by adding a solu tion of *trans-rac*-**96** (49.2 kg, 51.1 wt%, 80 mol) in ethanol and ethanol (165.0 kg).
- Heat the batch to 76 °C over 1 h and maintain at that temperature for 10 min.
- Cool to 15 °C over 2 h and maintain at 10–15 °C for 1 h.
- Centrifuge and wash with ethanol (26.0 kg).
- Charge the cake back into the reactor, followed by adding EtOH (188 kg).
- Heat to 75 °C and stir at that temperature for 10 min.
- Cool to 10–15 °C and maintain at that temperature for 1 h.
- Centrifuge and wash with EtOH (14 kg).
- Yield of **97**: 23.4 kg (42.0%).

Stage (b)
- Charge 8.0 kg (11.5 mol) of **97** into a 30-L reactor, followed by adding water (10 kg).
- Charge 9 kg 10 wt% NaOH to pH 7.5–8.0 slowly maintaining the batch temperature at 10–25 °C.
- Stir the resulting mixture at 20–25 °C for 40 min.
- Filter and extract the filtrate with EtOAc (3×6 L).
- Dry the combined EtOAc layers over Na_2SO_4.
- Filter and concentrate.
- Yield of **98**: 3.71 kg (97.2%) as a light-yellow oil.

Similarly, the resolution of a racemic amine **99** with *D*-DTTA was achieved, giving the tartaric acid salt **100** in quantitative yield (Equation 16.16).[44]

(16.16)

16.5.2.3 Use of bis((S)-Mandelic Acid)-3-Nitrophthalate

Analogously, separation of the (S)-(−)-isomer of amlodipine from the racemate amlodipine was achieved by using bis((S)-mandelic acid)-3-nitrophthalate (Equation 16.17).[45]

(16.17)

REMARK

Besides the resolution of racemic mixtures, the classical resolution lent itself readily to enantiomeric enrichment. For instance, the (S)-pyridyl amine **101** (90% ee), obtained from asymmetric synthesis, required enantiomeric enrichment. A classical resolution was utilized to purify **101** via a diastereomeric salt formation with L-tartaric acid (1.0 equiv). As shown in Equation 16.18,[22] treatment of wet 2-propanol solution of the crude **101** with 1 equiv of L-tartaric acid, after stirring overnight, furnished the corresponding salt, which could be isolated as a stable white solid in greater than 99% ee.

(16.18)

16.5.3 RESOLUTION OF KETONE

The resolution approach that based on the formation of diastereomeric salts cannot be directly applied to substrates that are lack suitable functional groups.[46] Accordingly, development effort led to a novel resolution approach by introducing an appropriate functional group temporarily into a molecule followed by forming diastereomeric salts with a resolution agent. Using this approach the resolution of racemic bicyclo[3.2.0]hept-2-en-6-one **103** was realized by converting **103** into the corresponding α-hydroxysulfonic acid **105** that would react spontaneously with (+)-α-methylbenzylamine to lead to the diastereomeric salt **104** (Scheme 16.18).[47]

16.5.4 RESOLUTION OF RACEMIC AMMONIUM SALT

Besides the desymmetrization of glycerol with a chiral auxiliary,[23] carnitine was also synthesized via classical resolution of the racemic trimethylammonium chloride intermediate with (+)-L-tartaric acid.[48]

The synthesis commenced with an epichlorohydrin ring-opening reaction with trimethylamine followed by resolution of the resulting racemic trimethylammonium chloride with L-tartaric acid

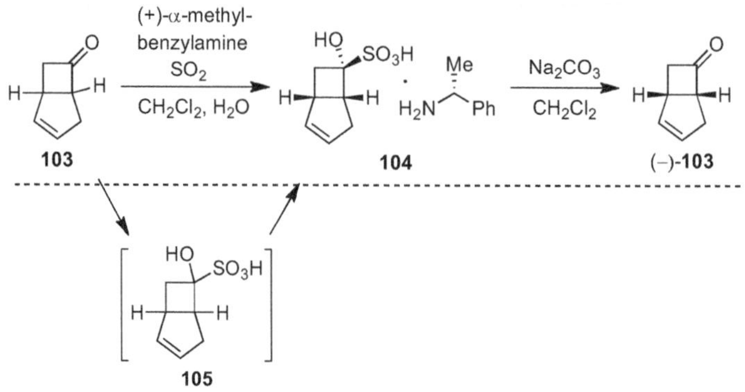

SCHEME 16.18

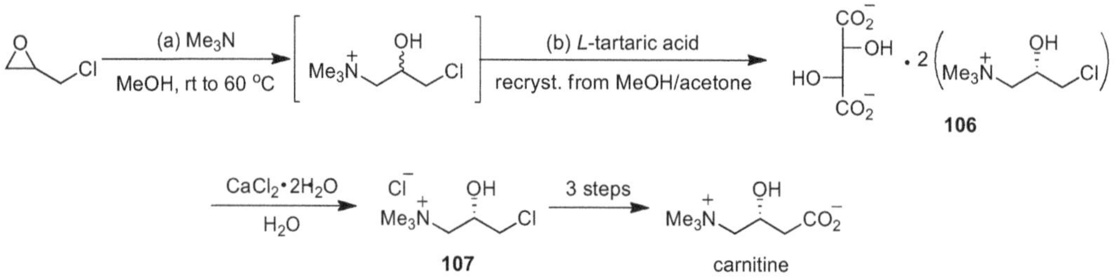

SCHEME 16.19

and recrystallization from MeOH/acetone to afford the tartrate salt **106** in one-pot operation (Scheme 16.19). The subsequent salt exchange with calcium chloride in water was able to separate the key chiral trimethylammonium chloride intermediate **107** from the solid calcium tartrate salt by-product by simple filtration.

16.5.5   DIASTEREOMER SALT BREAK

Generally, standard acid or base treatments are employed to break diastereomeric salts to afford the corresponding free acids or bases. However, the salt-break of diastereomeric salts of amino acids can be achieved under neutral conditions without external acids or bases due to the fact that the internal amino or carboxylic groups may serve as a base or an acid to break the salts intramolecularly. For example, a diastereomeric salt **108** could be broken with wet THF under neutral conditions to give **109** with an $S{:}R$ ratio of 99.1:0.1 (Equation 16.19).[1h]

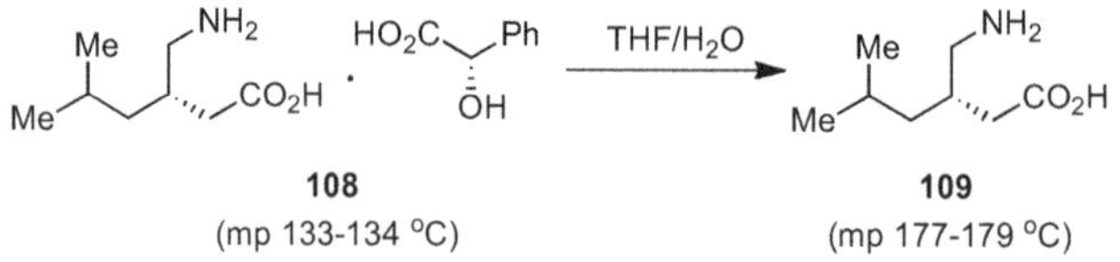

(16.19)

108
(mp 133-134 °C)

109
(mp 177-179 °C)

Procedure

- Transfer **108** (obtained from a resolution of 29.7 kg (168 mol) of racemic mixture of **109**) to a 400-L reactor with THF (195 L) and H_2O (10 kg).
- Heat the batch to 60–65 °C.
- Cool to 0–5 °C.

- Centrifuge and rinse with a mixture of THF (28 L)/H$_2$O (1 kg).
- Transfer the damp solid to a 200 L still with IPA (113 L) and H$_2$O (38 kg).
- Heat until a solution is obtained (75–80 °C).
- Filter.
- Cool to 0–5 °C.
- Centrifuge and wash with IPA (25 L).
- Dry at 35–45 °C under vacuum.
- Yield: 7.4 kg (55.6%).

16.5.6 EXAMPLES OF DIASTEREOMERIC SALTS

Research on the classical resolutions has resulted in numerous publications; it is beyond the scope of this book to collect all these works. Tables 16.1 and 16.2 highlight various useful classical resolution examples that appeared in the literature.

16.6 DYNAMIC KINETIC RESOLUTION

Although the kinetic resolution of racemic compounds is increasing in synthetic significance, a disadvantage of the resolution process is that a maximum 50% yield of the desired product is obtained based on the racemic starting material. Compared with the kinetic resolution, dynamic kinetic resolution (DKR) is preferred in which a racemization reaction takes place concurrently, leading to the desired product in higher yield. The applications of the dynamic kinetic resolution have increased dramatically during the past few years by making use of a combination of (chemo-catalytic) in situ racemization of a substrate and (bio-catalytic) enantioselective transformations. DKRs are often achieved through the equilibration of stereoisomers by proton transfer,[49] addition–elimination reactions,[50] oxidation/reduction reactions,[51] or isomerization of configurationally labile carbanions.[52] On the other hand, equilibration between stereoisomers containing all-carbon quaternary stereocenters can only be achieved through reversible C–C bond-forming reactions. Crystallization-induced dynamic resolution (CIDR)[53] via in situ racemization and selective crystallization is a frequently used DKR approaches.

16.6.1 DKR VIA IMINE INTERMEDIATES

Amine racemizations rely mainly on chiral centers that have acidic α-protons with low pKa values, for example, Schiff bases of some amino acids can allow racemization of unwanted isomers. This type of DKRs involves an equilibrium mixture of three iminium salts (113, 114, and 115) as illustrated in Scheme 16.20.

16.6.1.1 3,5-Dichlorosalicylaldehyde Catalyst

Crystallization-induced dynamic resolution (CIDR) processes were demonstrated in a number of industrial examples, such as Merck's asymmetric transformation of an aminodiazepam 116 with (S)-camphorsulfonic acid 117 and 3,5-dichlorosalicylaldehyde 118 (Equation 16.20).[54] Racemization (6.1 kg scale) with a catalytic amount of 118 (3 mol%) was carried out at room temperature in the presence of 0.92 equiv of 117. Excess of amine 116 probably deprotonated the imine methines, leading to the racemization of the undesired enantiomeric amine.

$$(16.20)$$

TABLE 16.1
Diastereomeric Salts for Resolution of Racemic Acids

Entry	Salt	Appearance	Mp (°C)	Salt break
1		Solid[a]	–	NaOH
2		Solid[b]	165–168	–
3		White solid[c]	–	–
4		Solid[d]	–	32% HCl
5		White crystals[e]	177	1 N aqueous HCl
6		Solid[f]	145–147	2 N HCl
7		Solid[g]	99 (Form I) 107 (Form II)	37% HCl
8		Crystalline solid[h]	-	NaOH
9		Crystalline solid[i]	–	2 N aq HCl

(Continued)

TABLE 16.1 (CONTINUED)
Diastereomeric Salts for Resolution of Racemic Acids

Entry	Salt	Appearance	Mp (°C)	Salt break
10		Solid[57]	–	1 N aq HCl
11		Solid[j]	–	3.8 N aqueous HCl
12		Solid[k]	–	–
13		White solid[l]	–	Aqueous NaOH
14		Solid[m]	–	NaHSO$_4$
15		Solid[n]	176–178	Aqueous AcOH
16		Solid[o]	–	1 N HCl

(Continued)

TABLE 16.1 (CONTINUED)
Diastereomeric Salts for Resolution of Racemic Acids

Entry	Salt	Appearance	Mp (°C)	Salt break
17		Solid[p]	141–143	–

[a] Linderberg, M. T.; Moge, M.; Sivadasan, S. *Org. Process Res. Dev.* **2004**, *8*, 838.

[b] Rasmy, O. M.; Vaid, R. K.; Semo, M. J.; Chelius, E. C.; Robey, R. L.; Alt, C. A.; Rhodes, G. A.; Vicenzi, J. T. *Org. Process Res. Dev.* **2006**, *10*, 28.

[c] Shu, L.; Li, Z.; Gu, C.; Fishlock, D. *Org. Process Res. Dev.* **2013**, *17*, 247.

[d] Waser, M.; Moher, E. D.; Borders, S. S. K.; Hansen, M. M.; Hoard, D. W.; Laurila, M. E.; LeTourneau, M. E.; Miller, R. D.; Phillips, M. L.; Sullivan, K. A.; Ward, J. A.; Xie, C.; Bye, C. A.; Leitner, T.; Herzog-Krimbacher, B.; Kordian, M.; Müllner, M. *Org. Process Res. Dev.* **2011**, *15*, 1266.

[e] Koch, G.; Kottirsch, G.; Wietfeld, B.; Küsters, E. *Org. Process Res. Dev.* **2002**, *6*, 652.

[f] Hu, B.; Prashad, M.; Har, D.; Prasad, K.; Repič, O.; Blacklock, T. J. *Org. Process Res. Dev.* **2007**, *11*, 90.

[g] Yue, T.-Y.; Mcleod, D. D.; Albertson, K. B.; Beck, S. R.; Deerberg, J.; Fortunak, J. M.; Nugent, W. A.; Radesca, L. A.; Tang, L.; Xiang, C. D. *Org. Process Res. Dev.* **2006**, *10*, 262.

[h] Cai, S.; Dimitroff, M.; McKennon, T.; Reider, M.; Robarge, L.; Ryckman, D.; Shang, X.; Therrien, J. *Org. Process Res. Dev.* **2004**, *8*, 353.

[i] Ashcroft, C. P.; Challenger, S.; Clifford, D.; Derrick, A. M.; Hajikarimian, Y.; Slucock, K.; Silk, T. V.; Thomson, N. M.; Williams, J. R. *Org. Process Res. Dev.* **2005**, *9*, 663.

[j] Karlsson, S.; Brånalt, J.; Halvarsson, M. Ö.; Bergman, J. *Org. Process Res. Dev.* **2014**, *18*, 969.

[k] Jansen, R.; Knopp, M.; Amberg, W.; Bernard, H.; Koser, S.; Müller, S.; Münster, I.; Pfeiffer, T.; Riechers, H. *Org. Process Res. Dev.* **2001**, *5*, 16.

[l] Boesch, H.; Cesco-Cancian, S.; Hecker, L. R.; Hoekstra, W. J.; Justus, M.; Maryanoff, C. A.; Scott, L.; Shah, R. D.; Solms, G.; Sorgi, K. L.; Stefanick, S. M.; Thurnheer, U.; Villani, F. J., Jr.; Walker, D. G. *Org. Process Res. Dev.* **2001**, *5*, 23.

[m] Walker, S. D.; Borths, C. J.; DiVirgilio, E.; Huang, L.; Liu, P.; Morrison, H.; Sugi, K.; Tanaka, M.; Woo, J. C. S.; Faul, M. M. *Org. Process Res. Dev.* **2011**, *15*, 570.

[n] Liu, J.; Wang, Z.; Levin, A.; Emge, T. J.; Rablen, P. R.; Floyd, D. M.; Knapp, S. *J. Org. Chem.* **2014**, *79*, 7593.

[o] Bunegar, M. J.; Dyer, U. C.; Evans, G. R.; Hewitt, R. P.; Jones, S. W.; Henderson, N.; Richards, C. J.; Sivaprasad, S.; Skead, B. M.; Stark, M. A.; Teale, E. *Org. Process Res. Dev.* **1999**, *3*, 442.

[p] Woods, M.; Dyer, U. C.; Andrews, J. F.; Morfitt, C. N.; Valentine, R.; Sanderson, J. *Org. Process Res. Dev.* **2000**, *4*, 418.

Procedure

Stage (a)
- Mix a solution of racemic amine **116** (6.1 kg, 23 mol) in *i*PrOAc (188 L) with a solution of **117** (4.9 kg, 21.16 mol, 0.92 equiv) in MeCN (47 L) at 20–25 °C.
- Seed the mixture with 10 g of **119**.
- Stir for 4 h.

Stage (b)
- Treat the resulting white slurry with **118** (132 g, 0.69 mol, 3 mol%).
- Stir for 12 h.
- Filter and wash with *i*PrOAc.
- Dry under vacuum.
- Yield: 10.4 kg (91%).

TABLE 16.2
Diastereomeric Salts for Resolution of Racemic Bases

Entry	Salt	Appearance	Mp (°C)	Salt break
1	(tetrahydronaphthalene with NHMe, OH) · 0.5 (HO2C–OH / HO2C–OH, tartaric acid)	Solid[a]	214–216	K_2CO_3
2	(quinuclidine–phenyl–NH_2) · (HO2C–OH / HO2C–OH, tartaric acid)	Solid[9]	–	NaOH
3	(benzothiophene–tropane N–Me) · (HO2C–OH / HO2C–OH, tartaric acid)	Solid[b]	–	NaOH
4	(F_3C–tetrahydroquinoline, NH_2, Et, CO_2Et) · 0.5 (HO2C–OBz / HO2C–OBz, dibenzoyltartaric acid)	White crystalline solid[c]	189.5–191.5	NaOH
5	(Cl–benzoxazole–diazepane NH, Me) · (HO2C–OBz / HO2C–OBz, dibenzoyltartaric acid)	White solid[d]	164.2	4 N NaOH
6	(Cl–benzoxazole–diazepane NH, Me) · (HO2C–OBz / HO2C–OBz, dibenzoyltartaric acid)	White crystalline solid[e]	–	10 N NaOH
7	(benzothiazole–NH–phenyl ketone, Me2N–Me) · (HO2C–OC(O)-p-Tol / HO2C–OC(O)-p-Tol, di-p-toluoyltartaric acid)	White solid[f]	–	NaOH
8	(piperidine–Me_3C–OH) · (HO2C–OC(O)-p-Tol / HO2C–OC(O)-p-Tol, di-p-toluoyltartaric acid)	Crystalline colorless solid[g]	–	NaOH
9	(indole–Me–NH_2, OBn) · 0.5 (HO2C–OC(O)-p-Tol / HO2C–OC(O)-p-Tol, di-p-toluoyltartaric acid)	Solid[h]	–	NaOH
10	(phenyl–OH–NH_2) · 0.5 (HO2C–OC(O)-p-Tol / HO2C–OC(O)-p-Tol, di-p-toluoyltartaric acid)	Solid[i]	–	NaOH
11	(Br–tetrahydropyridinium, Me, N^+Bn H) · (HO2C–OC(O)-p-Tol / HO2C–OC(O)-p-Tol, di-p-toluoyltartaric acid)	White solid[j]	–	$NaHCO_3$
12	(tetrahydro-β-carboline NH, benzofuran) · (HO2C–NHAc, Me_2CH, N-acetyl leucine)	Off-white solid[k]	–	K_2CO_3

(Continued)

TABLE 16.2 (CONTINUED)

Diastereomeric Salts for Resolution of Racemic Bases

Entry	Salt	Appearance	Mp (°C)	Salt break
13	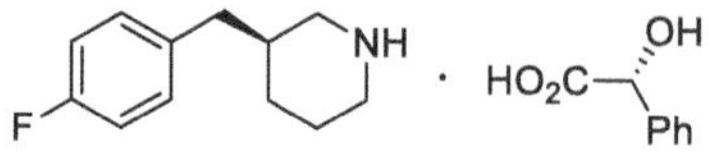	Solid[l]	172	–
14	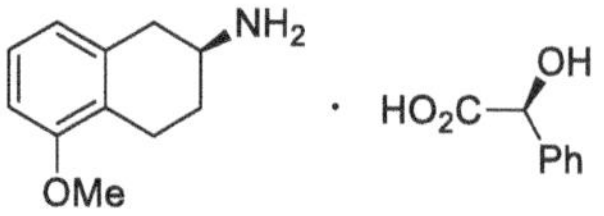	Solid[m]	200–201	NaOH
15	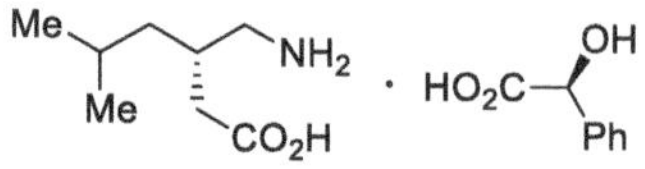	White solid[mh]	133–134	60–65 °C THF/H_2O
16	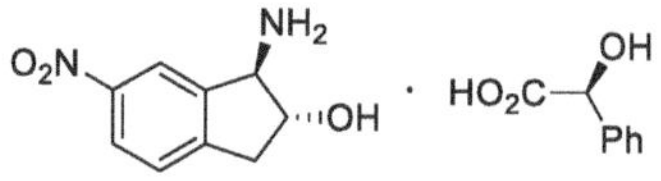	Solid[n]	–	NaOH
17	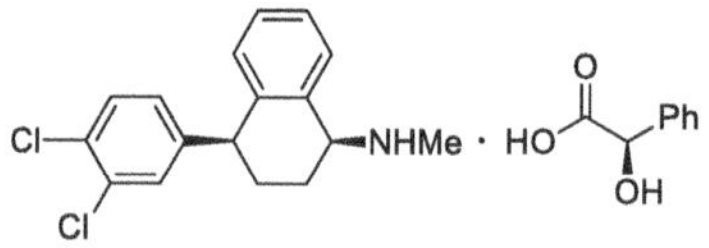	Solid[o]	–	–
18	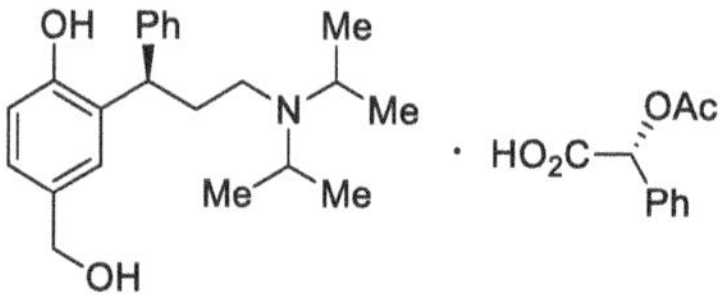	Solid[p]	–	K_2CO_3
19	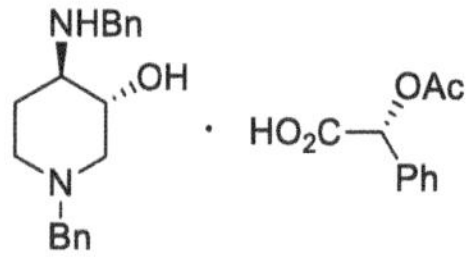	Solid[q]	–	–
20	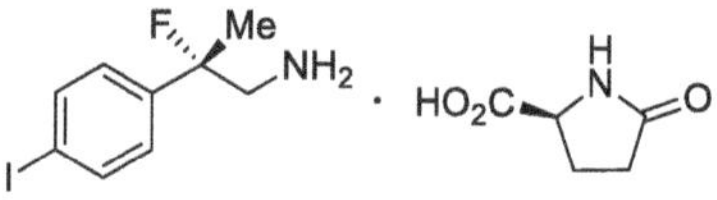	Crystalline solid[r]	–	NaOH
21	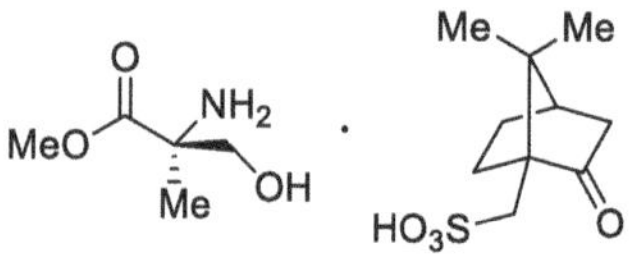	White solid[s]	–	Et_3N
22	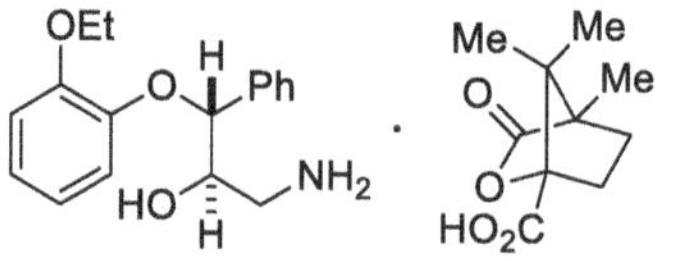	White solid[m]	194–196	–
23	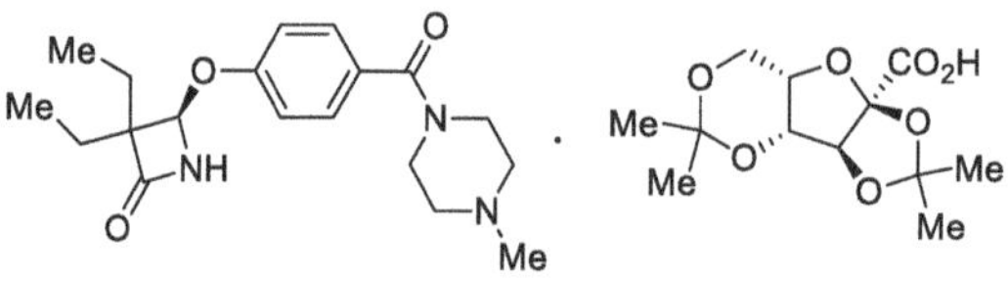	Solid[t]	–	K_2CO_3
24	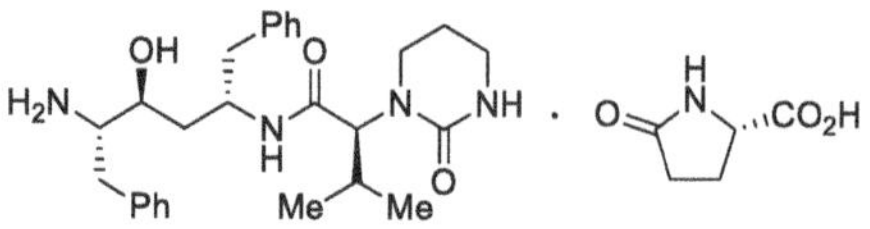	Solid[u]	–	$NaHCO_3$

(*Continued*)

TABLE 16.2 (CONTINUED)
Diastereomeric Salts for Resolution of Racemic Bases

Entry	Salt	Appearance	Mp (°C)	Salt break
25		Solid[v]	–	–

(structure: piperidine bearing CO$_2$Me and a C(Me)$_2$ group, N–H; paired with a cyclic phosphate bearing OMe, P(=O)OH, O, O, and a benzo-fused ring with Me Me)

[a] (a) Draper, R. W.; Hou, D.; Iyer, R.; Lee, G. M.; Liang, J. T.; Mas, J. L.; Tormos, W.; Vater, E. J.; Günter, F.; Mergelsberg, I.; Scherer, D. *Org. Process Res. Dev.* **1998**, *2*, 175. (b) Gala, D.; Dahanukar, V. H.; Eckert, J. M.; Lucas, B. S.; Schumacher, D. P.; Zavialov, I. A.; Buholzer, P.; Kubisch, P.; Mergelsberg, I.; Scherer, D. *Org. Process Res. Dev.* **2004**, *8*, 754.

[b] Malmgren, H.; Cotton, H.; Frøstrup, B.; Jones, D. S.; Loke, M.-L.; Peters, D.; Schultz, S.; Sölver, E.; Thomsen, T.; Wennerberg, J. *Org. Process Res. Dev.* **2011**, *15*, 408.

[c] Damon, D. B.; Dugger, R. W.; Magnus-Aryitey, G.; Ruggeri, R. B.; Wester, R. T.; Tu, M.; Abramov, Y. *Org. Process Res. Dev.* **2006**, *10*, 464.

[d] Baxter, C. A.; Cleator, E.; Brands, K. M. J.; Edwards, J. S.; Reamer, R. A.; Sheen, F. J.; Stewart, G. V.; Strotman, N. A.; Wallace, D. J. *Org. Process Res. Dev.* **2011**, *15*, 367.

[e] Goswami, A.; Howell, J. M.; Hua, E. Y.; Mirfakhrae, K. D.; Soumeillant, M. C.; Swaminathan, S.; Qian, X.; Quiroz, F. A.; Vu, T. C.; Wang, X.; Zheng, B.; Kronenthal, D. R.; Patel, R. N. *Org. Process Res. Dev.* **2001**, *5*, 415.

[f] Aelterman, W.; Lang, Y.; Willemsens, B.; Vervest, I.; Leurs, S.; Knaep, F. D. *Org. Process Res. Dev.* **2001**, *5*, 467.

[g] Hoard, D. W.; Moher, E. D.; Turpin, J. A. *Org. Process Res. Dev.* **1999**, *3*, 64.

[h] Ikunaka, M.; Kato, S.; Sugimori, D.; Yamada, Y. *Org. Process Res. Dev.* **2007**, *11*, 73.

[i] Lohse, O.; Spöndlin, C. *Org. Process Res. Dev.* **1997**, *1*, 247.

[j] Lombardo, V. M.; Bernier, L.; Chen, M. Z.; Farrell, W.; Flick, A.; Nuhant, P.; Sach, N. W.; Tao, Y.; Trujillo, J. I. *Org. Process Res. Dev.* **2021**, *25*, 2315.

[k] Li, X.; Branum, S.; Russell, R. K.; Jiang, W.; Sui, Z. *Org. Process Res. Dev.* **2005**, *9*, 640.

[l] Yue, T.-Y.; Mcleod, D. D.; Albertson, K. B.; Beck, S. R.; Deerberg, J.; Fortunak, J. M.; Nugent, W. A.; Radesca, L. A.; Tang, L.; Xiang, C. D. *Org. Process Res. Dev.* **2006**, *10*, 262.

[m] Hirayama, Y.; Ikunaka, M.; Matsumoto, J. *Org. Process Res. Dev.* **2005**, *9*, 30.

[n] Hansen, M. M.; Borders, S. S. K.; Clayton, M. T.; Heath, P. C.; Kolis, S. P.; Larsen, S. D.; Linder, R. J.; Reutzel-Edens, S. M.; Smith, J. C.; Tameze, S. L.; Ward, J. A.; Weigel, L. O. *Org. Process Res. Dev.* **2009**, *13*, 198.

[o] Taber, G. P.; Pfisterer, D. M.; Colberg, J. C. *Org. Process Res. Dev.* **2004**, *8*, 385.

[p] Dirat, O.; Bibb, A. J.; Burns, C. M.; Checksfield, G. D.; Dillon, B. R.; Field, S. E.; Fussell, S. J.; Green, S. P.; Mason, C.; Mathew, J.; Mathew, S.; Moses, I. B.; Nikiforov, P. I.; Pettman, A. J.; Susanne, F. *Org. Process Res. Dev.* **2011**, *15*, 1010.

[q] Young, I. S.; Ortiz, A.; Sawyer, J. R.; Conlon, D. A.; Buono, F. G.; Leung, S. W.; Burt, J. L.; Sortore, E. W. *Org. Process Res. Dev.* **2012**, *16*, 1558.

[r] Magnus, N. A.; Aikins, J. A.; Cronin, J. S.; Diseroad, W. D.; Hargis, A. D.; LeTourneau, M. E.; Parker, B. E.; Reutzel-Edens, S. M.; Schafer, J. P.; Staszak, M. A.; Stephenson, G. A. Tameze, S. L.; Zollars, L. M. H. *Org. Process Res. Dev.* **2005**, *9*, 621.

[s] Anson, M. S.; Clark, H. F.; Evans, P.; Fox, M. E.; Graham, J. P.; Griffiths, N. N.; Meek, G.; Ramsden, J. A.; Roberts, A. J.; Simmonds, S.; Walker, M. D.; Willets, M. *Org. Process Res. Dev.* **2011**, *15*, 389.

[t] Storace, L.; Anzalone, L.; Confalone, P. N.; Davis, W. P.; Fortunak, J. M.; Giangiordano, M.; Haley, J. J.; Kamholz, K.; Li, H.-Y.; Ma, P.; Nugent, W. A.; Parsons, Jr., R. L.; Sheeran, P. J.; Silverman, C. E.; Waltermire, R. E.; Wood, C. C. *Org. Process Res. Dev.* **2002**, *6*, 54.

[u] Stoner, E. J.; Cooper, A. J.; Dickman, D. A.; Kolaczkowski, L.; Lallaman, J. E.; Liu, J.-H.; Oliver-Shaffer, P. A.; Patel, K. M.; Paterson, J. B., Jr.; Plata, D. J.; Riley, D. A.; Sham, H. L.; Stengel, P. J.; Tien, J.-H. J. *Org. Process Res. Dev.* **2000**, *4*, 264.

[v] Andersen, S. M.; Bollmark, M.; Berg, R.; Fredriksson, C.; Karlsson, S.; Liljeholm, C.; Sörensen, H. *Org. Process Res. Dev.* **2014**, *18*, 952.

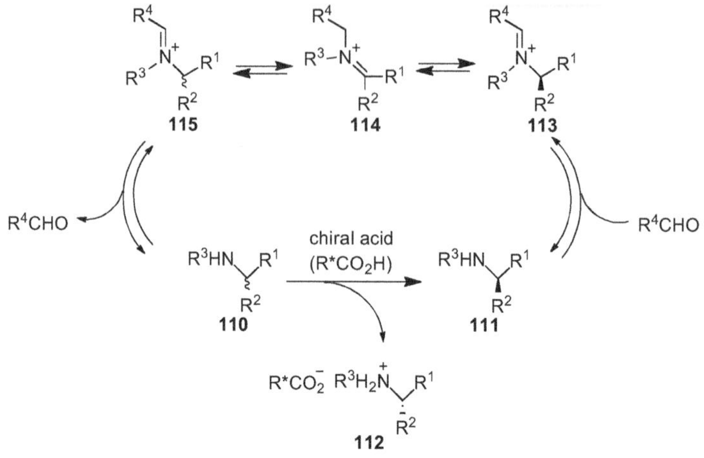

SCHEME 16.20

3,5-Dichlorosalicylaldehyde **118** was also utilized in the dynamic kinetic resolution of the racemic secondary amine **120** to prepare piperazinone **122** (as carboxylic salt) (Equation 16.21).[55]

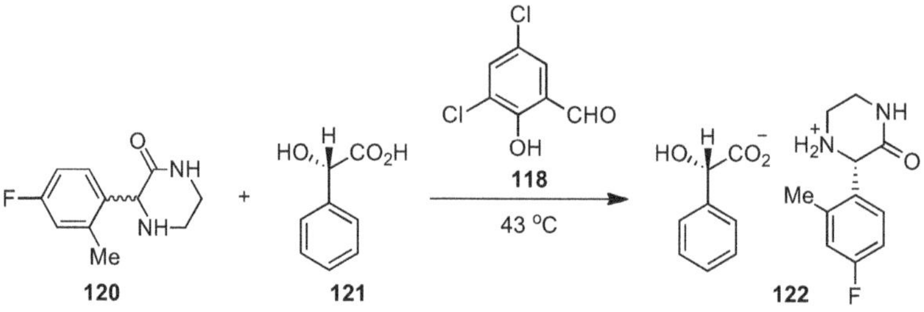

(16.21)

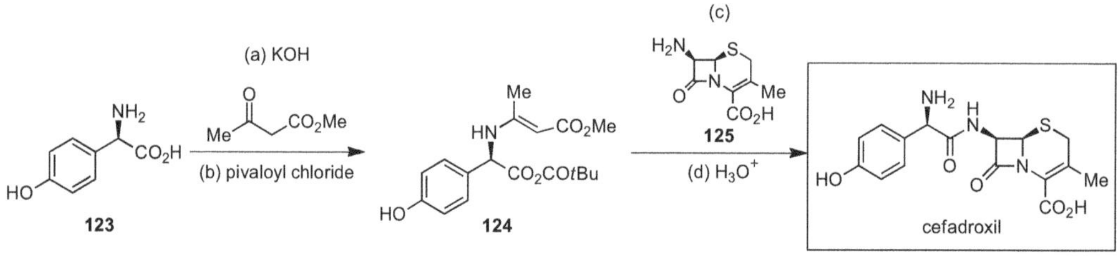

SCHEME 16.21

REMARK

In contrast, resolutions of racemic aldehydes or ketones were also realized by converting the aldehydes or ketones to their corresponding diastereomeric imines with chiral amines.[56]

16.6.1.2 2-Hydroxy-6-(hydroxymethyl)benzaldehyde Catalyst

Cefadroxil is an important member of the cephalosporin anti-biotics having a global market of approximately 300 tons per year. Cefadroxil is produced by coupling of Dane salt-protected (D)-*p*-hydroxyphenylglycine[57] **124** with β-lactam nucleus **125** (Scheme 16.21).[58]

In contrast to the current industrial synthesis that employs the D-amino acid in enantiopure form, Cefadroxil could be synthesized starting from racemic *p*-hydroxyphenylglycine. The subsequent CIDR process was realized by clathration with 2,7-dihydroxynaphthalene **126**, in which the desired diastereomer can be selectively withdrawn from the equilibrating mixture of epimers (Equation 16.22).[95]

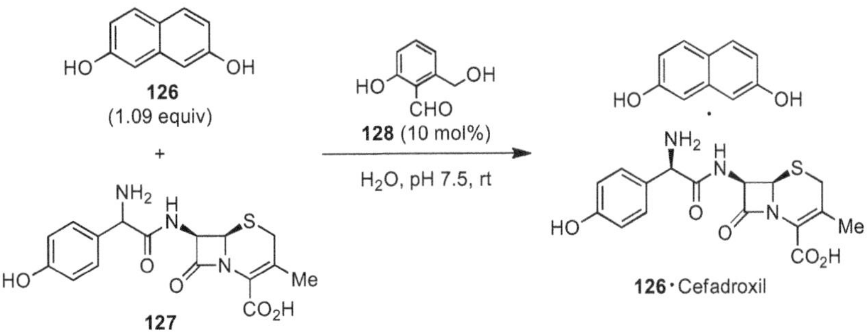

(16.22)

As the solubility difference in water between Cefadroxil and *epi*-Cefadroxil is rather small, selective crystallization of the desired diastereomer could not be achieved. However, 2-hydroxy-6-(hydroxymethyl)benzaldehyde (**128**)-mediated clathration of Cefadroxil with **126** was an efficient method, through which Cefadroxil precipitated as clathrate, while its epimer remained in the solution. After filtration, the precipitated clathrate was obtained with a yield of 86%.

16.6.1.3 Picolinaldehyde Catalyst

Biomimetic catalysis is an attractive technology for dynamic kinetic resolutions because of low-cost and environmentally benign conditions. Scheme 16.22[59] demonstrates a picolinaldehyde-involved DKR process for the resolution of amino acids, wherein the in situ formed Zn-picolinaldehyde complex could catalyze the racemization via equilibrium between **129** and **130**. Reactions were carried out at room temperature in the presence of Li_2CO_3 (0.6 equiv), aldehyde (0.1 equiv), and $Zn(OTf)_2$ (0.05 equiv), furnishing the desired amino acids in fairly good yields with high chiral purity.

16.6.1.4 DRK without Catalyst

[3,3]-Sigmatropic rearrangement has been widely used in organic synthesis. Scheme 16.23[60] describes a tandem process for enantioselective synthesis of angularly substituted 1-azabicyclic rings.

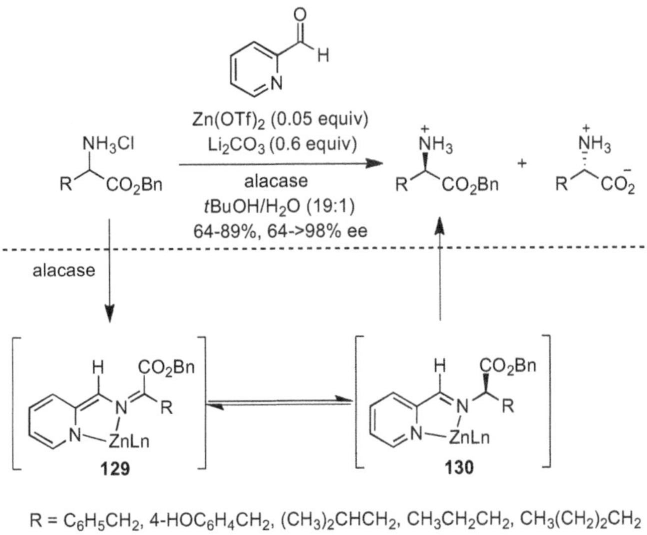

$R = C_6H_5CH_2$, 4-$HOC_6H_4CH_2$, $(CH_3)_2CHCH_2$, $CH_3CH_2CH_2$, $CH_3(CH_2)_2CH_2$

SCHEME 16.22

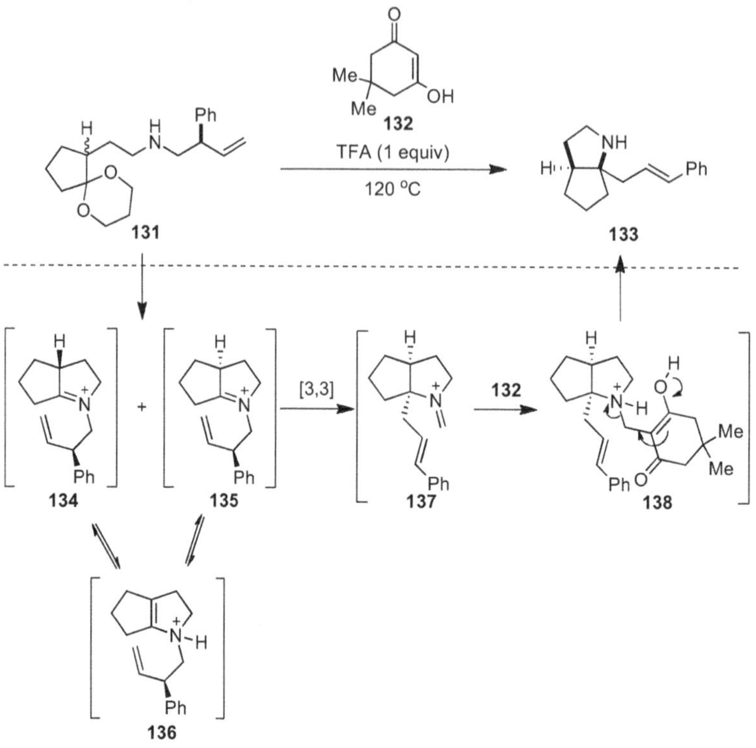

SCHEME 16.23

In this synthesis, treatment of a mixture of aminoketal **131** and dimedone **132** (2.5 equiv) with TFA (1.0 equiv) at 120 °C afforded the desired azabicyclic product **133** in good yield and enantiomeric purity. The synthetic methodology was based on a rapid dynamic kinetic epimerization via iminium/enamonium tautomerization (**134**⇌**136**⇌**135**) and a subsequent slow aza-Cope rearrangement. The resulting iminium intermediate **137** was trapped by **132** followed by C–N bond cleavage to provide the desired product **133**.

16.6.1.5   Iridium-Involved DKR

An application of the DKR strategy toward asymmetric amination was able to access chiral amines in diastereo- and enantioselective manner (Scheme 16.24).[61]

The protocol involves Ir-catalyzed oxidation of α-branched alcohols **139** followed by asymmetric reductive amination. The DKR process occurred during the reductive amination in which the racemization of the undesired iminium intermediates **143** was achieved via enamine intermediates **144**.

16.6.2   DKR via Enolate Intermediates

The acid- or base-catalyzed enolization process is widely utilized in the dynamic kinetic resolution of ketones with α-hydrogen(s) to the carbonyl group. The ease of such proton transfer is based on the relative acidity of the α-hydrogen.

16.6.2.1   Enolization with Base

Bromo alcohol **148** was identified as a key intermediate for the synthesis of taranabant, a drug candidate developed by Merck for the treatment of obesity. In October 2008, Merck has stopped its phase III clinical trials due to side effects, mainly depression and anxiety. The initial route to

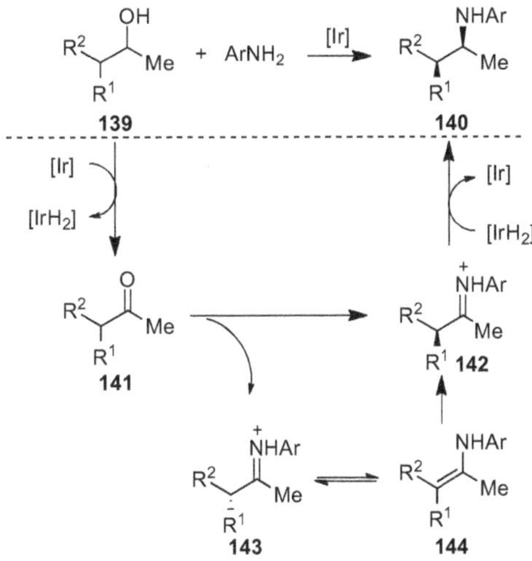

SCHEME 16.24

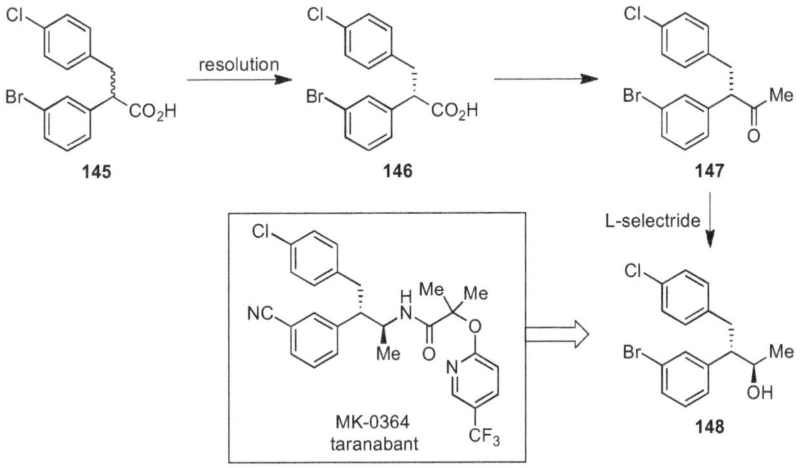

SCHEME 16.25

148 involved the resolution of racemic acid 145 and reduction of the resulting chiral acid 146 to ketone 147, followed by diastereoselective reduction of 147 using L-selectride at –78 °C (Scheme 16.25).[62]

Based on the relative acidity of the α-H to the carbonyl of ketone 147, a telescoped process was developed. This one-pot process (Scheme 16.26)[99] involved an enantioselective reduction of ketone 147 with Noyori's catalyst combining with DKR in the presence of a catalytic amount of potassium *tert*-butoxide. The desired product 148 was obtained in a dr ratio of 8:1 and 94% ee.

16.6.2.2 Enolization without Base

Since the proton at the α-position to a carbonyl group makes the stereogenic center rather sensitive to the acidity or basicity of the reaction media, it is not surprising that a DKR can be realized under mild or base-free conditions. As the racemization of enantiomers (151α and 151β) via enolization was rapid enough (k_{INV}/k_{RR}=92) with respect to the hydrogenation and rates of hydrogenation of 151α and 151β were substantially different (k_{SR}/k_{RR}=15), the DKR of the racemic β-ketoester 151 was achieved with Noyori's catalyst under hydrogenation conditions (Scheme 16.27).[63] This process

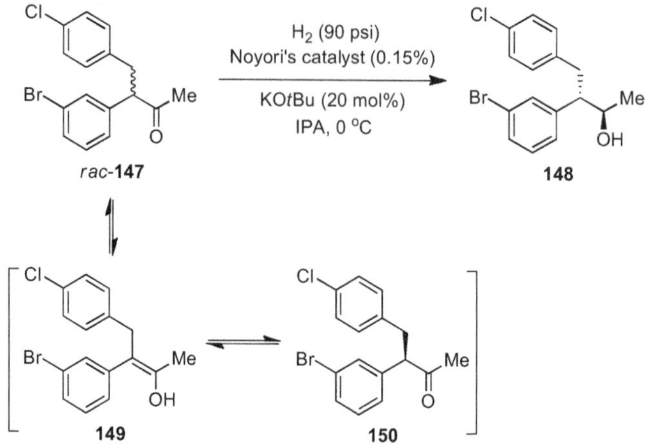

SCHEME 16.26

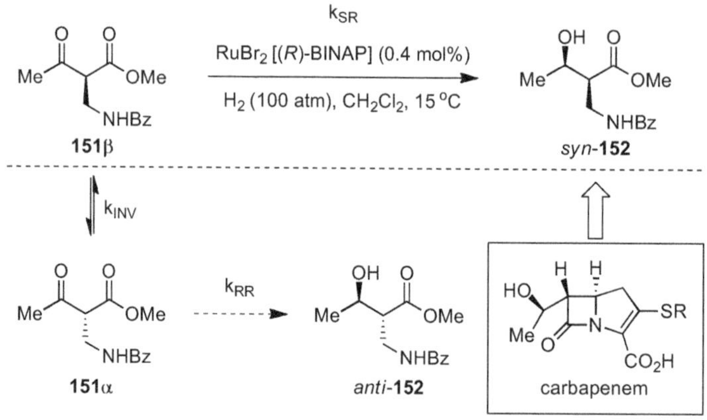

SCHEME 16.27

was licensed by Takasago International Corporation for the multi-kilogram scale synthesis of car-bapenem antibiotics.

16.6.3 DKR via Diastereomeric Salt Formation

A dynamic resolution of racemic acid **153** was effected using (S)-(−)-1-phenylethylamine to afford the desired salt in 84% yield and acceptable optical purity (94% de). Excess of chiral amine was utilized as a base catalyst for the base-catalyzed epimerization. Furthermore, the optical purity was able to upgrade to greater than 98% by crystallization of disodium salt of the free acid from aqueous ethanol (Scheme 16.28).[57]

Procedure

Stage (a) Resolution
- Charge 2.94 kg (24.3 mol, 2.0 equiv) of (S)-(−)-1-phenethylamine to a stirred suspension of **153** (6.52 kg, 12.15 mol) in THF (35.8 L, 5.5 mL/g) and 1,2-dimethoxyethane (35.8 L, 5.5 mL/g) at 60 °C over 15 min.
- Cool the resulting mixture to 55 °C over 1 h.

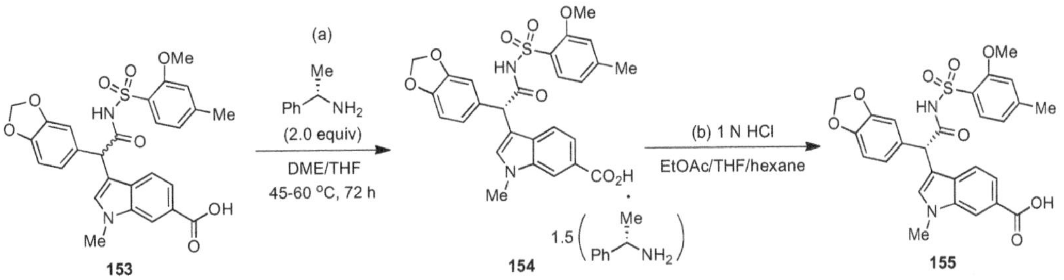

SCHEME 16.28

- Seed with 65 g of **154**.
- Stir for 24 h.
- Cool the resulting suspension to 50 °C over 5 h.
- Stir for 24 h.
- Cool the suspension to 45 °C.
- Stir for 24 h.
- Cool the suspension to room temperature.
- Filter and wash with DME (13.0 L).
- Dry at 45 °C under vacuum.
- Yield: 7.2 kg (82%).

Stage (b) Formation of free acid
- Partition **154** (7.2 kg, 10.02 mol) between EtOAc (36 L), THF (7.2 L), and aqueous HCl (1 N, 36 L).
- Separate layers.
- Wash the organic layer with 1 N HCl (3×4.46 L) and water (3.6 L).
- Azeotropically distill with EtOAc at constant volume.
- Distill to a volume of approximately 18 L.
- Add hexane (18 L).
- Filter and wash with hexane (7.2 L).
- Dry at 45 °C under vacuum.
- Yield: 4.88 kg (91%).

16.6.4 DKR OF SIX-MEMBERED RING SYSTEMS

16.6.4.1 Epimerization of *cis*-Isomer to *trans*-Isomer

Crystallization-induced dynamic resolution (CIDR) is an adaptation of dynamic resolution, a process that can afford in principle a quantitative yield of chiral product from a racemic starting material through in situ racemization.

Drug candidate **160** was developed by Merck for the treatment of type 2 diabetes mellitus (T2DM). The synthesis of lactam intermediate *trans*-**159** involved decarboxylation of carboxylic acid **158**, giving a mixture of *cis*-**159**/*trans*-**159** in a ratio of 97:3. However, the desired *trans*-**159** isomer was the minor product (Scheme 16.29).[64] Epimerization of the *cis* isomer **159** to its corresponding thermodynamically favored *trans* isomer could be realized by treatment with weak base. Therefore, upon exposure of the mixture of *cis*-**159**/*trans*-**159** to NaHCO$_3$ at 45 °C, the ratio of *cis*-**159**/*trans*-**159** was improved from 97:3 to 1:2. Furthermore, employing a base-catalyzed, dynamic crystallization-driven protocol was able to isolate the desired *trans*-**159** in 76% overall yield (3 steps).

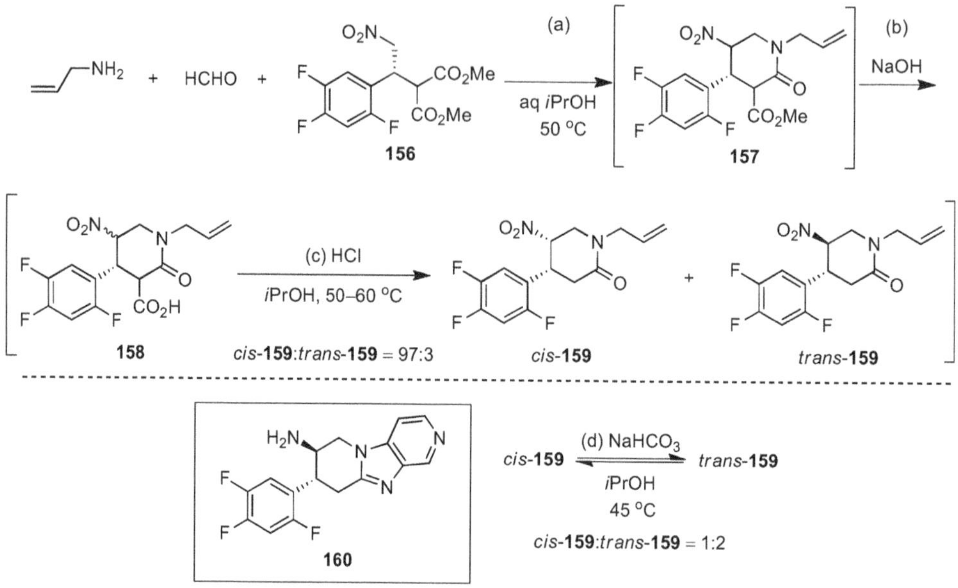

SCHEME 16.29

Procedure

Stage (a)
- Charge 25.42 g (0.31 mol) of 37% HCHO to a solution of **156** (100 g, 0.30 mol) and allylamine (22.14 g, 0.39 mol) in IPA (500 mL) and water (100 mL) at 50 °C over 1–2 h.
- Stir the resulting mixture at 50 °C for 2–3 h.

Stage (b)
- Charge 104.4 mL of 5 N NaOH (0.52 mol) at 50 °C in one portion.
- Stir the reaction mixture at 50 °C for 1–2 h.

Stage (c)
- Charge 54.2 mL (0.66 mol) of concentrated HCl dropwise at 50–60 °C over 30 min.
- Stir the reaction mixture at 50–60 °C for 2–3 h.

Stage (d)
- Cool to 45 °C.
- Charge 100 mL of 5% NaHCO$_3$ dropwise to pH 7–8 followed by adding water (100 mL).
- Stir the resulting slurry at 45 °C for 1–2 h.
- Cool to ambient temperature and age overnight.
- Filter, wash with aqueous 50% IPA (2×150 mL), and dry at ambient temperature.
- Yield: 72 g (76%) as solid.

16.6.4.2   Isomerization of Cyclohexane Derivative

It is not feasible to enrich the desired cyclohexenyl derivative (S)-**161** by direct isomerization, because of the small energy difference between the two diastereoisomers (the difference in energy between these two conformers (R)-**161** and (S)-**161** is 0.11 kcal/mol) (Equation 16.23).[65]

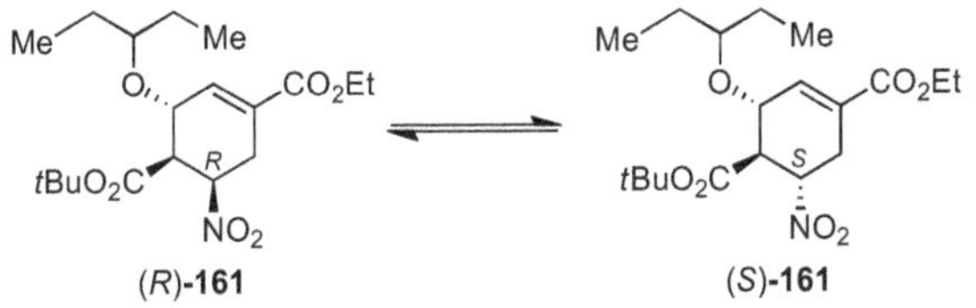

(16.23)

(R)-**161** (S)-**161**

Based on the fact that the 1,3 steric repulsion in a cyclohexane ring system is larger than in a cyclohexene ring system, a strategy was developed to convert the cyclohexene ring system into a cyclohexane framework by thiol Michael reaction (Equation 16.24). As a result, the anticipated isomerization of **161** occurred, affording **162** in the desired 5S stereochemistry.

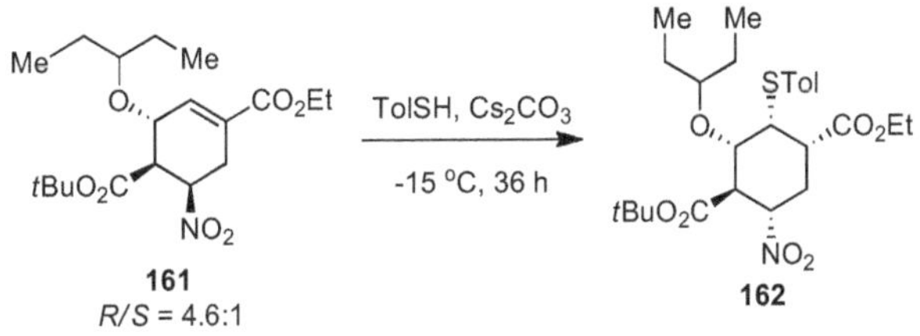

(16.24)

161
R/S = 4.6:1

162

16.6.4.3 Fischer Indole Synthesis

An asymmetric Fischer indole synthesis was developed with a spirocyclic chiral phosphoric acid catalyst (5 mol%) in the presence of amberlite CG50 as an additive (Scheme 16.30).[66] The reaction involved a DKR process which proceeded via imines **165** and enamines **166** and **167** in a highly enantioselective manner. Amberlite CG50 was required to scavenge ammonia by-products that otherwise would poison the catalyst.

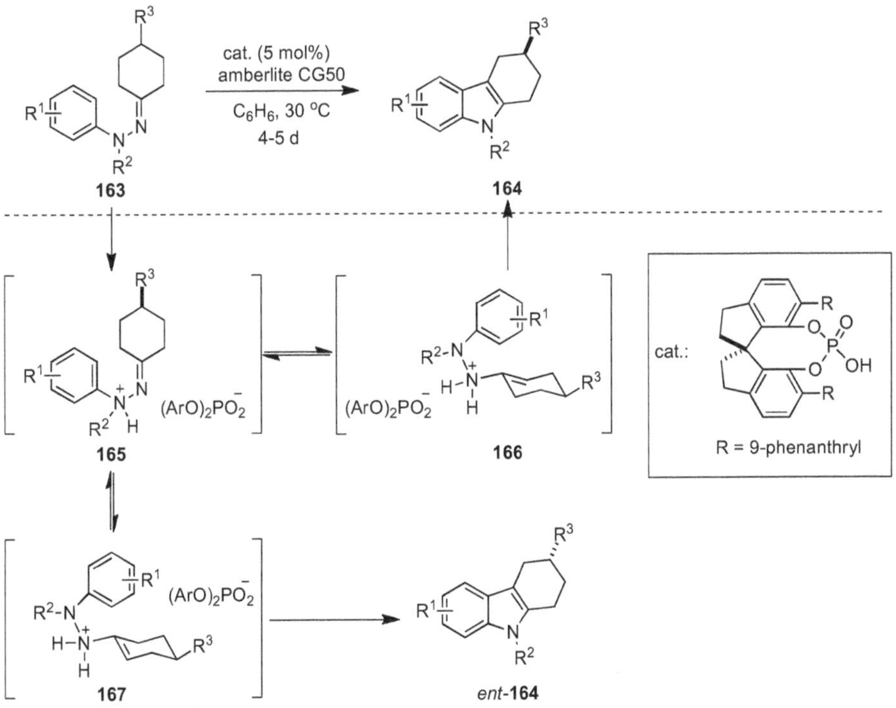

SCHEME 16.30

16.6.5 DKR via Reversible Bond Formation

The combination of reversible bond formation with selective recrystallization permits dynamic kinetic resolution.

16.6.5.1 Reversible C–C Bond Formation

16.6.5.1.1 Retro-Aldol/Aldol Equilibration

Attractive cascade reactions, as illustrated in Scheme 16.31,[67] involve a sequence initiated by nucleophilic addition of Grignard reagent, generated in situ, followed by anion-promoted oxy-Cope and Dieckmann cyclization. Noteworthy also is the integration of the DKR process into the reaction sequence, which allowed the establishment of multiple stereogenic centers in a one-pot operation. This dynamic kinetic resolution occurred via the participation of the retro-aldol/aldol equilibrium[68] between intermediates **172** and **173**.

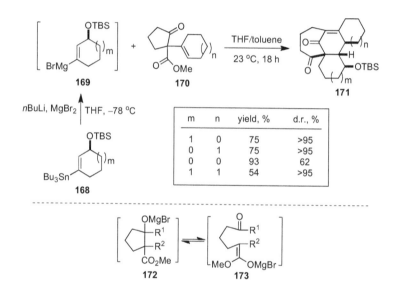

m	n	yield, %	d.r., %
1	0	75	>95
0	1	75	>95
0	0	93	62
1	1	54	>95

SCHEME 16.31

16.6.5.1.2 Strecker Reaction

Diastereoselective Strecker reaction of (R)-phenylglycine amide **175** with pivaldehyde **174** (1.05 equiv) and NaCN is accompanied by an in-situ crystallization-induced asymmetric transformation, whereby one diastereomer **176** selectively precipitates and could be isolated in 93% yield and dr >99:1 (Scheme 16.32).[69]

16.6.5.2 Reversible C–N Bond Formation

16.6.5.2.1 Reversible Cycloaddition Reaction

The combination of reversible C–N bond formation with CIDR was utilized to prepare *trans*-imidazolidinone **181** (Scheme 16.33).[70] The condensation of the amine **179** with isobutyraldehyde gave an imine intermediate **180** whose subsequent cycloaddition afforded a mixture of the *cis*- and *trans*-isomers. However, the pure *trans*-isomer **181** was selectively crystallized out of the mixture in heptane. The undesired *cis*-**181** was in equilibrium with the imine **180** in solution, which permitted the production of high yields of the *trans*-**181**. This CIDR was applied in the preparation of *trans*-isomer of imidazolidinone **181** in multikilogram quantities.

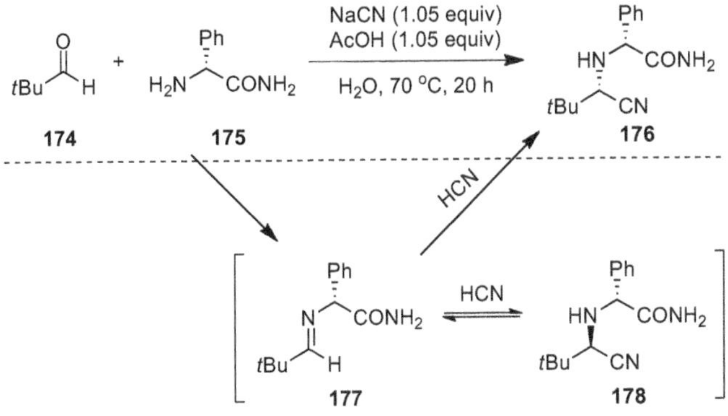

SCHEME 16.32

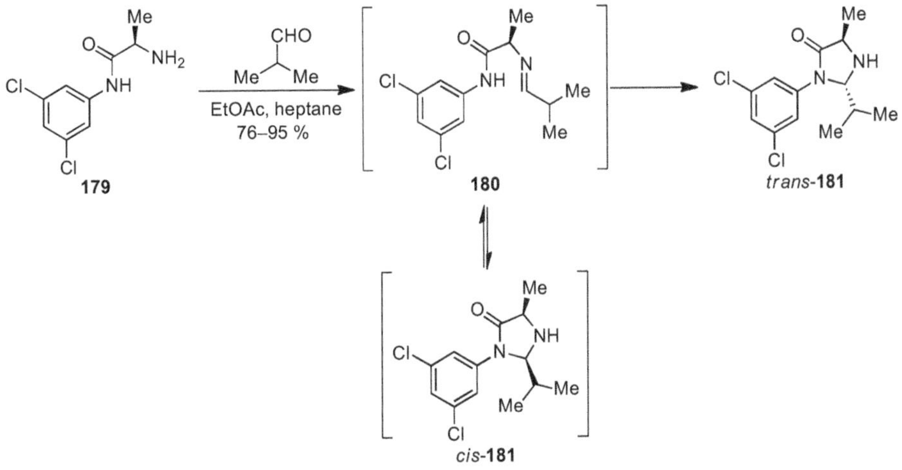

SCHEME 16.33

Procedure

- Charge 571 mL (6.00 mol) of isobutyraldehyde to a stirred EtOAc solution of **179** (5.10 mol) at 17–23 °C over 30 min.
- Distill azeotropically with a Dean–Stark apparatus until no more water was collected (ca 10 h).
- Distill the batch to ca 1.5 L.
- Add heptane (6.0 L).
- Distill at 50–60 °C under reduced pressure to ca 1.5 L.
- Add heptane (6.0 L) and stir the resulting mixture at 50 °C for 12 h while the crystallization occurred.
- Cool to 20–25 °C.
- Filter and wash with heptane (1.0 L).
- Yield: 1113 g (76%) as a light-yellow solid.

16.6.5.2.2 Retro-Michael Addition

CIDR can be realized via a retro-Michael addition reaction. Taking advantage of the solubility difference between Michael addition adducts **184a** and **184b**, the desired diastereomeric acid **184a** was selectively crystallized leaving **184b** in the solution, which would rapidly revert into **182** and **183** via retro-Michael reaction in warm ethanol solution (Scheme 16.34).[71]

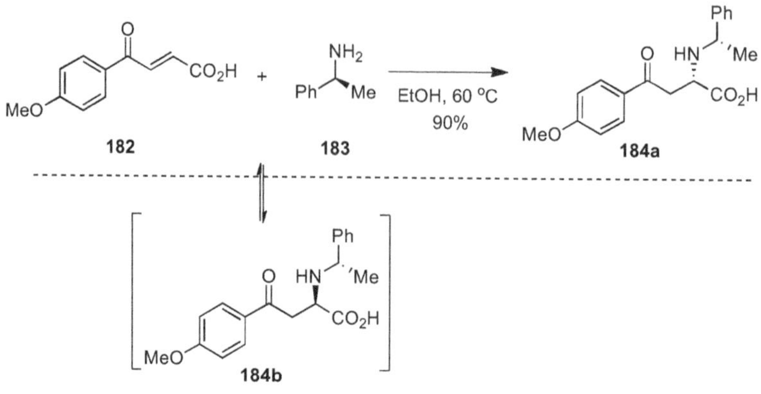

SCHEME 16.34

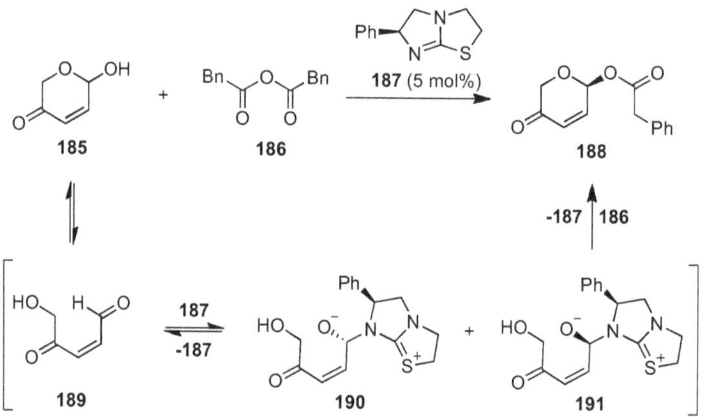

SCHEME 16.35

16.6.5.2.3 Formation of Enantiopure Ester

The dynamic kinetic resolution of lactol **185** was developed by scientists of Bristol Myers Squibb. Levamisole **187**, as an inexpensive and commercially available material, was chosen as the DKR catalyst. Although the detailed reaction mechanism has not been released yet, the key step would involve a reversible formation of **190** (Scheme 16.35).[72]

The reaction of (Z)-5-hydroxy-4-oxopent-2-enal **189** with levamisole **187** would form a mixture of **190** and **191**. Apparently, **191** would readily cyclize to generate chiral lactol **185** whose reaction with **186** furnishes the desired phenylacetate **188**. In contrast, because of the steric hindrance, the lactol ring formation of **190** would be slow; instead, **190** would dissociate back to **189**.

Using this protocol, **188** was obtained in 74% ee, which could be enriched to 99% ee through recrystallization.

16.6.5.3 Reversible C–O Bond Formation

*16.6.5.3.1 Synthesis of **trans**-Fluorinated Oxazolidine Chiral Auxiliary*

Chiral fluorinated oxazolidine (Fox) is a valuable chiral auxiliary used in stereoselective synthesis. The condensation of fluoral ethyl hemiacetal **192** with (R)-phenylglycinol **193** gave a diastereomeric mixture of Fox (*trans:cis* = 62:38) (Scheme 16.36).[73]

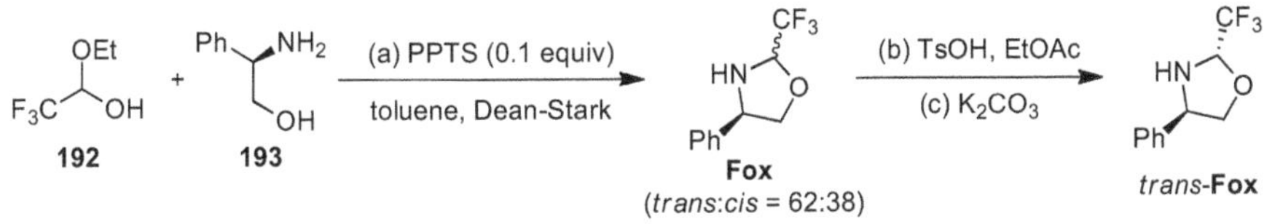

SCHEME 16.36

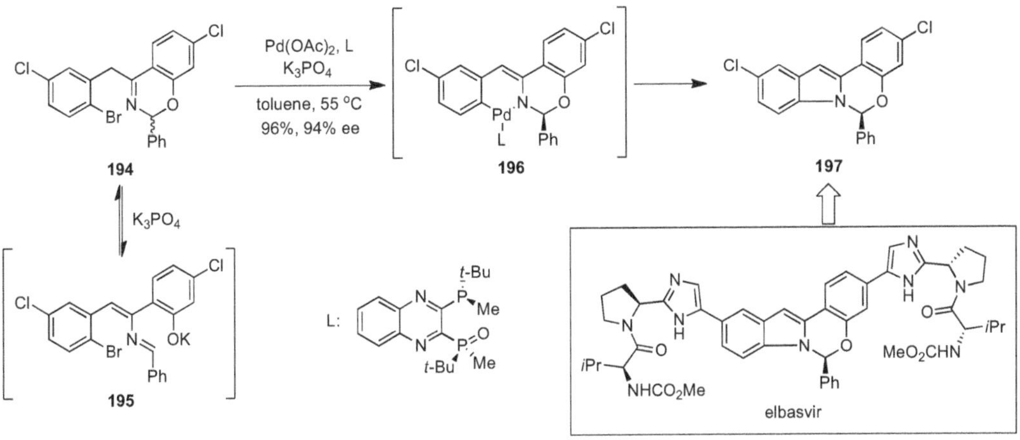

SCHEME 16.37

In order to resolve the desired *trans*-Fox, a CIDR was carried out with TsOH in ethyl acetate to effect the epimerization at the C-2 center. Ultimately, the *trans*-Fox diastereomer crystallized as the pure *p*-toluenesulfonate salt in 79% yield.

16.6.5.3.2 Asymmetric Synthesis of Hemiaminal

DKR process using reversible C–O bond formation was demonstrated in the synthesis of hemiaminal **197** (Scheme 16.37), an intermediate for access of the HCV drug candidate elbasvir.[74]

As shown in Scheme 16.37, the reversible C–O bond cleavage/formation led to the stereo-defined imido–Pd complex **196** whose reductive elimination afforded the desired hemiaminal product **197**.

16.6.5.4 Reversible C–S Bond Formation

A crystallization-induced dynamic resolution of a diastereomeric mixture of tertiary benzylic alcohol **198**(*cis:trans*=1:1) was developed[75] by a process group from Merck. Based on the solubility difference between *cis*-**199** and *trans*-**199** and the reversibility of the carbon–sulfur bond formation, the desired *cis*-**199** product was obtained from the reaction of **198** with 4-chlorothiophenol (4-CTP) and isolated directly from the reaction mixture by a simple filtration (Equation 16.25).

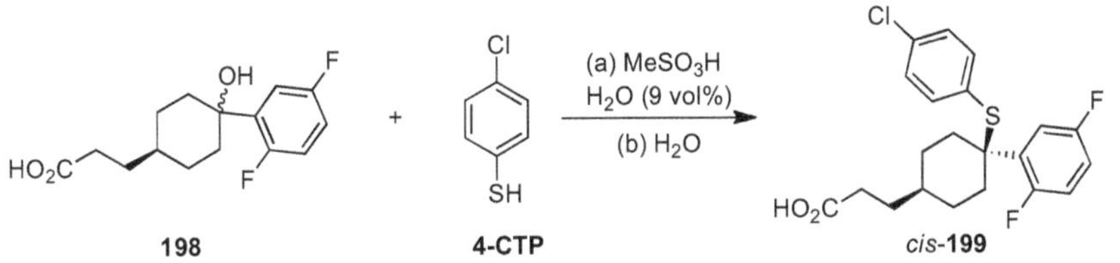

$$(16.25)$$

Procedure

Stage (a)
- Charge water (118 mL) to a solution of 4-CTP (59.8 g, 0.414 mol, 1.1 equiv) in MeSO$_3$H (1070 mL) at below 40 °C.
- Charge 106.8 g (0.376 mol) of **198** in one portion.
- After 10 min, charge 1 g of *cis*-**199** and stir at 36–42 °C overnight.
- Cool to ambient temperature.

Stage (b)
- Add water (1.07 L) at below 40 °C over 50 min.
- Cool the resulting suspension to ambient temperature and age for 30 min.
- Filter, wash with water (5 L), and dry at 40 °C overnight.
- Yield of *cis*-**199**: 152 g (95%, containing 6% of *trans*-**199**).

16.6.6 OTHER DKR METHODS

16.6.6.1 Bromide-Catalyzed DKR

16.6.6.1.1 Racemization of α-Bromo Carboxylic Acid

Tetrabutylammonium bromide (TBAB) could be employed as the catalyst for the CIDR of racemic α-Bromo carboxylic acid **200** via racemization of the undesired diastereomeric salt **203** (Scheme 16.38).[76] The racemization may proceed through a reaction of **203** with TBAB to lead to bromine and ion pair **204** whose subsequent reaction with bromine generates racemic **200**.

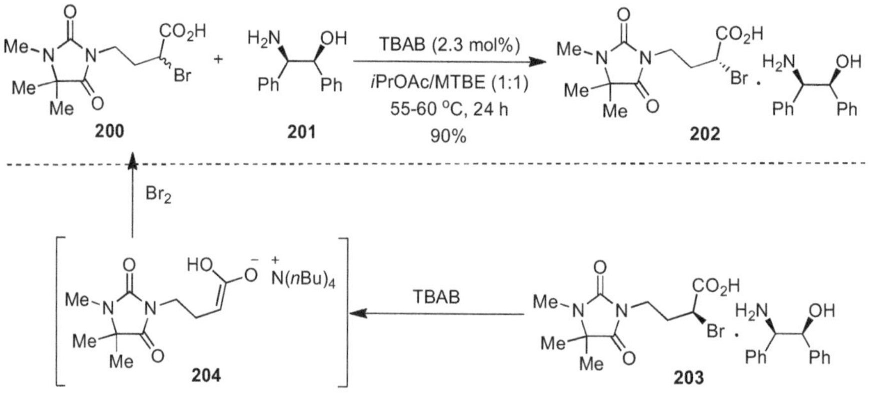

SCHEME 16.38

16.6.6.1.2 Epimerization of Ester

Alternatively, a method for in-situ racemization is based on a reversible nucleophilic substitution of halogen, in which the halogen is situated in an electronically activated position. For instance, a TBAB-mediated epimerization was integrated into a CIDR process, developed for the resolution of racemic dibromo alcohol via transesterification. In the event, the equilibration of solid esters *rac*-**206** with a deficiency of *rac*-**205** would lead to enantiomerically pure free alcohol (*R*)-**205** in solution and a solid mixture of stereoisomeric esters (**206a**:**206b** = <1) (Scheme 16.39).[77] The resulting mixture of esters could be epimerized to a 1:1 mixture by bromide exchange and used to treat another batch of racemic alcohol.

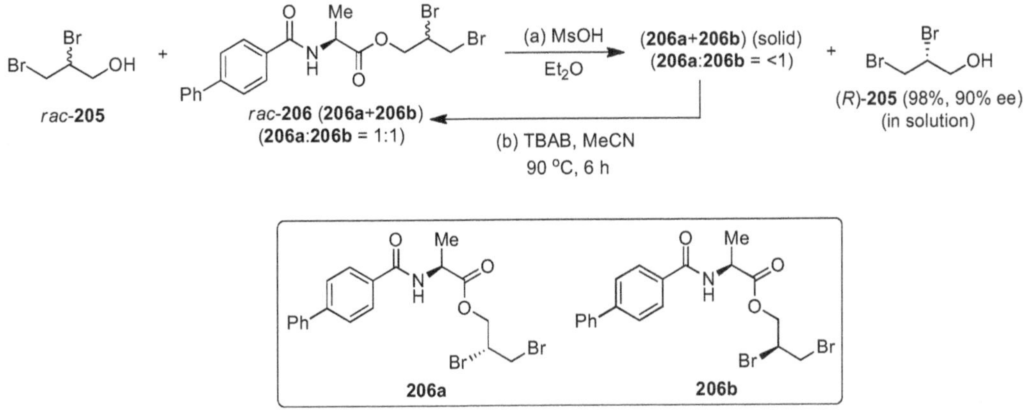

SCHEME 16.39

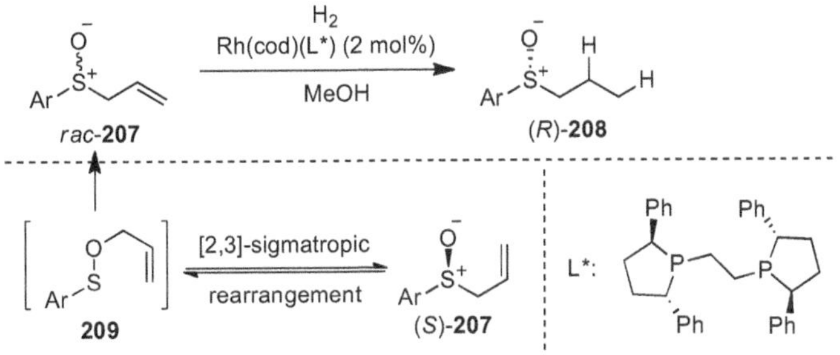

SCHEME 16.40

One interesting feature of this approach is that the enantiomerically pure alcohol is produced directly free from any derivatization.

16.6.6.2 Resolution of Sulfoxide

Chiral sulfoxides have been utilized in asymmetric synthesis to introduce chiral centers. The sulfoxides are usually synthesized from the oxidation of sulfides in the presence of a chiral catalyst.[78] A synthetic method for the synthesis of chiral sulfoxides (*R*)-**208** was developed through DKR of allylic sulfoxides (Scheme 16.40).[79]

As illustrated in Scheme 16.40, the DKR process was achieved by combination of Mislow [2,3]-sigmatropic rearrangement of (*S*)-**207** with catalytic asymmetric hydrogenation. To achieve optimal results, controlling the relative rates of hydrogenation and racemization is critical. The optimized conditions were using low pressure of hydrogen gas to decrease the rate of hydrogenation relative to the rate of sigmatropic rearrangement.

16.6.6.3 Dynamic Kinetic Isomerization via Ir-Catalyzed Internal Redox Transfer Hydrogenation

A dynamic kinetic isomerization of dihydropyranones **211** was achieved via an Ir-catalyzed stereoselective internal redox transfer hydrogenation (Scheme 16.41).[80]

It was demonstrated that 2,6-dichlorobenzoic acid, as an additive, could accelerate the equilibration rate between complexes **213** and **214**, leading to a high *cis:trans* diastereomeric ratio via dynamic kinetic isomerization. Dihydropyranones **211** could be accessed conveniently through Achmatowicz rearrangement of feedstock furans **210**.

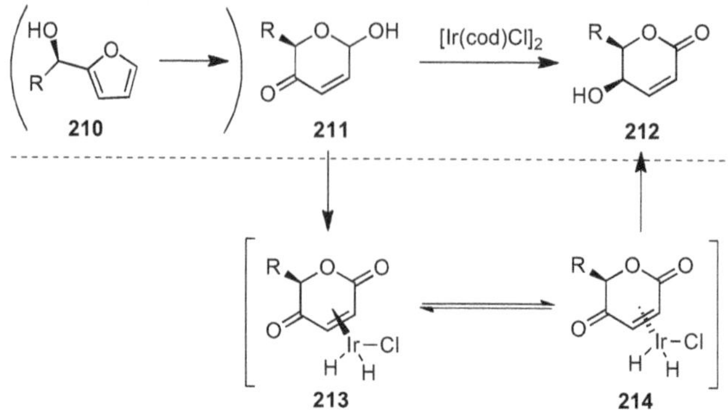

SCHEME 16.41

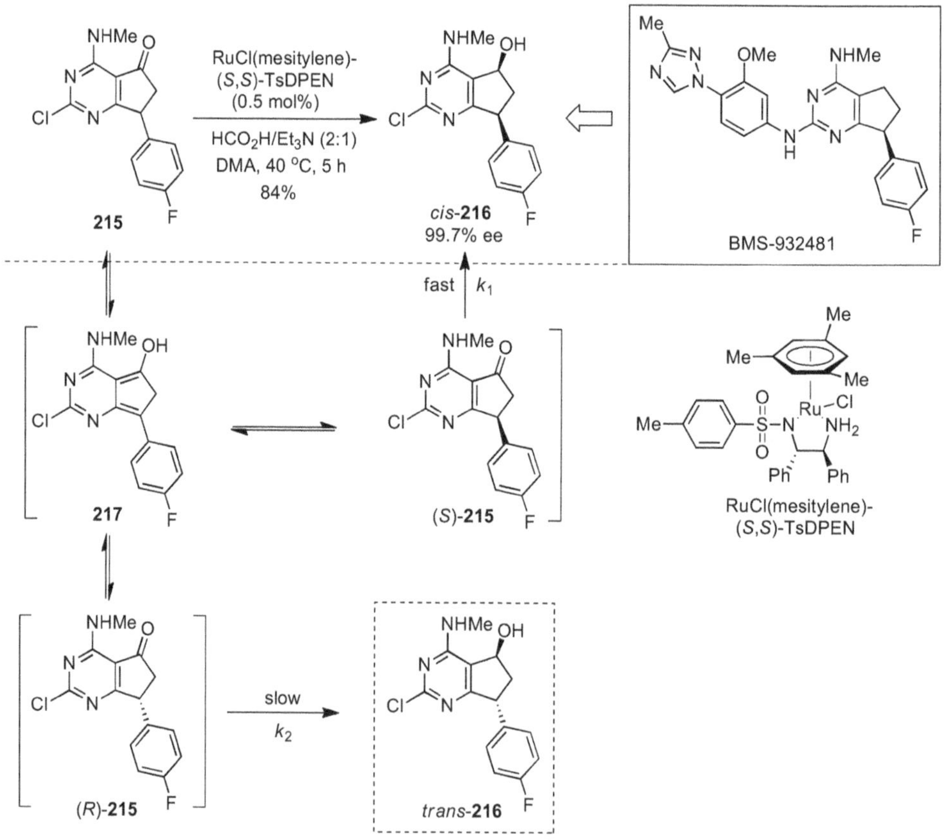

SCHEME 16.42

16.6.6.4 Vinylogous Dynamic Kinetic Resolution

Scientists from Bristol Myers Squibb demonstrated the first example of vinylogous dynamic kinetic resolution (VDKR) (Scheme 16.42).[81] during the synthesis of BMS-932481, a drug candidate for the treatment of Alzheimer's disease.

Unlike the normal DKR of ketone, VDKR is a process in which the racemizable stereocenter and the carbonyl are not contiguous. The key to the success of this VDKR hinged on the racemization

TABLE 16.3
List of Dynamic Kinetic Resolution Examples

Entry	DKR Examples	Comments
1

HNB: 2-hydroxy-5-nitrobenzaldehyde

| Racemization of the unwanted enantiomer was via enolization of the azomethine formed with 2-hydroxy-5-nitrobenzaldehyde (HNB) as a catalyst. The employment of (R)-CSA (0.95 equiv) in toluene gave the (R)-CSA salt in 79% yield (99.8% ee) on a multi-kilogram scale.[a] |
2		Salicylaldehyde was used as the catalyst in this process. Addition of a small amount of water was necessary.[b]
3		Catalytic amounts of water (1.5-2.0 mol%) was required for this CIDR process.[ca]
4		The CIDR was carried out by heating at 70–75 °C for 6–8 h followed by cooling to 20 °C and filtration.[d]
5		The CIDR was accomplished with an excess of 2-methyoxyethylamine in acetonitrile.[e]

(Continued)

TABLE 16.3 (CONTINUED)
List of Dynamic Kinetic Resolution Examples

Entry	DKR Examples	Comments
6		The acid-catalyzed CIDR was conducted in MeOH/EtOAc at 50 °C in 92% yield.[70]
7		The acid-catalyzed CIDR was accomplished via aldehyde intermediate in 98% yield by equilibration of a solution of *cis*-epimer and slow precipitation of the *trans*-lactol as the HCl salt.[f]
8		The acid-catalyzed anomerization occurred, and the product crystallized selectively as the salt with nBu_3N. However, the product was isolated as the bis-cyclohexylamine salt.[g]
9		The CIDR process was carried out in the presence of 1,5-diazobicyclo[4.3.0] non-5-nene (DBN) (0.1 w/w) with seeds (0.5 wt%) at 65 °C, affording 73% product yield with 99.4% ee.[72]

(Continued)

TABLE 16.3 (CONTINUED)
List of Dynamic Kinetic Resolution Examples

Entry	DKR Examples	Comments
10	Me Me OK Me / Me (0.3 equiv) heptane -12 to -7 °C, 5 h [CF3 ... Me,,, O N O Bn] → [CF3 ... product] intermediate: [Me,,, CF3 O N OH Bn]	The CIDR was catalyzed by potassium 3,7-dimethyloct-3-oxide (0.3 equiv), leading to the desired product via an epimerization of the undesired diastereomer.[h]
11	Me O Ph N Cl H Ph + NH4OH, MeOH, rt, 1 week → Me O Ph NH Cl Ph (97:3 dr) intermediate: [Me OH Ph N Cl Ph]	The base (NH4OH)-catalyzed CIDR process enabled to convert a mixture of diastereomeric α-chloroamides to the desired diastereomer with a dr ratio of 97:3.[i]
12	F ... CN N t-Bu (trans:cis=80:20) + NaOH, EtOH, 83 °C → F ... CO2H N t-Bu (>99.9% ee)	NaOH-mediated epimerization/saponification of nitrile was able to produce the corresponding acid in 95% yield and >99.9% ee after crystallization.[j]
13	MeO—C6H4—C(O)—CH=CH—CO2H + Me Me NH2 OH (1.1 equiv) CH2Cl2 25-30 °C, 7d (DKR via retro-Michael addition) → product.	The CIDR proceeded via a retro-Michael addition, in which the chiral amine was used as both the reagent (1.0 equiv) and the base catalyst (10 mol%).[k]
14	EtO2C ... CO2Et Me N CHO H (thiophene) + HCl·H2N SH, EtO2C, NaOAc, EtOH, rt, 3 h, 79% → product with thiazolidine CO2Et; intermediate: imine form	The CIDR is facilitated by NaOAc via imine intermediate.[l]

(Continued)

TABLE 16.3 (CONTINUED)
List of Dynamic Kinetic Resolution Examples

Entry	DKR Examples	Comments
15		The racemization took place in the presence of a catalytic amount of tetrabutylammonium bromide (TBAB) (20 mol%) in THF. Evaporation of THF gave the desired diastereomer in 91% yield and >98% de.[m]
16		The CIDR could be mediated with $ZnCl_2$.[n]
17		This resolution was carried out without base. The CIDR process involved a diastereoselective Strecker reaction of (R)-phenylglycine amide with ketone and NaCN.[106]
18		This resolution was carried out without base. 1:1 or 2:1 salt was obtained from CIDR reactions in ethanol.[o]
19		This resolution was carried out without base. The CIDR was conducted in MeOH at 40 °C, followed by cooling to 0 °C.[p]

(Continued)

TABLE 16.3 (CONTINUED)
List of Dynamic Kinetic Resolution Examples

[a] Singh, J.; Kronenthal, D. R.; Schwinden, M.; Godfrey, J. D.; Fox, R.; Vawter, E. J.; Zhang, B.; Kissick, T. P.; Patel, B.; Mneimne, O.; Humora, M.; Papaioannou, C. G.; Szymanski, W.; Wong, M. K. Y.; Chen, C. K.; Heikes, J. E.; DiMarco, J. D.; Qiu, J.; Deshpande, R. P.; Gougoutas, J. Z.; Mueller, R. H. *Org. Lett.* **2003**, *5*, 3155.

[b] Colson, P.-J.; Przybyla, C. A.; Wise, B. E.; Babiak, K. A.; Seaney, L. M.; Korte, D. E. *Tetrahedron: Asymmetry* **1998**, *9*, 2587.

[c] (a) Karmakar, S.; Byri, V.; Gavai, A. V.; Rampulla, R.; Mathur, A.; Gupta, A. *Org. Process Res. Dev.* **2016**, *20*, 1717. (b) Yasuda, N.; Cleator, E.; Kosjek, B.; Yin, J.; Xiang, B.; Chen, F.; Kuo, S.-C.; Belyk, K.; Mullens, P. R.; Goodyear, A.; Edwards, J. S.; Bishop, B.; Ceglia, S.; Belardi, J.; Tan, L.; Song, Z. J.; DiMichele, L.; Reamer, R.; Cabirol, F. L.; Tang, W. L.; Liu, G. *Org. Process Res. Dev.* **2017**, *21*, 1851.

[d] Zhao, M. M.; McNamara, J. M.; Ho, G.-J.; Emerson, K. M.; Song, Z. J.; Tschaen, D. M.; Brands, K. M. J.; Dolling, U.-H.; Grabowski, E. J. J.; Reider, P. J.; Cottrell, I. F.; Ashwood, M. S.; Bishop, B. C. *J. Org. Chem.* **2002**, *67*, 6743.

[e] Chung, J. Y. L.; Marcune, B.; Strotman, H. R.; Petrova, R. I.; Moore, J. C.; Dormer, P. G. *Org. Process Res. Dev.* **2015**, *19*, 1418.

[f] Pye, P. J.; Rossen, K.; Weissman, S. A.; Maliakal, A.; Reamer, R. A.; Ball, R.; Tsou, N. N.; Volante, R. P.; Reider, P. J. *Chem. Eur. J.* **2002**, *8*, 1372.

[g] Komatsu, H.; Awano, H. *J. Org. Chem.* **2002**, *67*, 5419.

[h] Brands, K. M. J.; Payack, J. F.; Rosen, J. D.; Nelson, T. D.; Candelario, A.; Huffman, M. A.; Zhao, M. M.; Li, J.; Craig, B.; Song, Z. J.; Tschaen, D. M.; Hansen, K.; Devine, P. N.; Pye, P. J.; Rossen, K.; Dormer, P. G.; Reamer, R. A.; Welch, C. J.; Mathre, D. J.; Tsou, N. N.; McNamara, J. M.; Reider, P. J. *J. Am. Chem. Soc.* **2003**, *125*, 2129.

[i] Lee, S.-K.; Lee, S. Y.; Park, Y. S. *Synlett* **2001**, *12*, 1941.

[j] Chung, J. Y. L.; Cvetovich, R.; Amato, J.; McWilliams, J. C.; Reamer, R.; DiMichele, L. *J. Org. Chem.* **2005**, *70*, 3592.

[k] (a) Kolarovic, A.; Berkeš, D.; Baran, P.; Povazanec, F. *Tetrahedron Lett.* **2001**, *42*, 2579. (b) For more CIDR processes involving a base-catalyzed retro-Michael addition, see: Kolarovic, A.; Berkeš, D.; Baran, P.; Povazanec, F. *Tetrahedron Lett.* **2005**, *46*, 975.

[l] Marchalín, S.; Cvopová, K.; Križ, M.; Baran, P.; Oulyadi, H.; Daïch, A. *J. Org. Chem.* **2004**, *69*, 4227.

[m] Caddick, S.; Jenkins, K. *Tetrahedron Lett.* **1996**, *37*, 1301.

[n] Napolitano, E.; Farina, V. *Tetrahedron Lett.* **2001**, *42*, 3231.

[o] Chaplin, D. A.; Johnson, N. B.; Paul, J. M.; Potter, G. A. *Tetrahedron Lett.* **1998**, *39*, 6777.

[p] Konoike, T.; Matsumura, K.; Yorifuji, T.; Shinomoto, S.; Ide, Y.; Ohya, T. *J. Org. Chem.* **2002**, *67*, 7741.

---

of the undesired (R)-**215** conjugation through a pyrimidine ring; also, the reduction rate of k_1 shall be greater than k_2 in order to achieve high diastereoselectivity.

16.6.7 VARIOUS DKR EXAMPLES

Research on DKRs has resulted in numerous DKR-related publications; it is beyond the scope of this book to collect all this literature work. Table 16.3 highlights various useful DKR approaches.

NOTES

1. (a) Hayes, S. T.; Assaf, G.; Checksfield, G.; Cheung, C.; Critcher, D.; Harris, L.; Howard, R.; Mathew, S.; Regius, C.; Scotney, G.; Scott, A. *Org. Process Res. Dev.* **2011**, *15*, 1305. (b) Pflum, D. A.; Wilkinson, H. S.; Tanoury, G. J.; Kessler, D. W.; Kraus, H. B.; Senanayake, C. H.; Wald, S. A. *Org. Process Res. Dev.* **2001**, *5*, 110. (c) Vartak, A. P.; Crooks, P. A. *Org. Process Res. Dev.* **2009**, *13*, 415. (d) Prashad, M.; Hu, B.; Repič, O.; Blacklock, T. J.; Giannousis, P. *Org. Process Res. Dev.* **2000**, *4*, 55. (e) Graves, P. E.; Salhanick, H. A. *Endocrinology* **1979**, *105*, 52. (f) Hett, R.; Fang, Q. K.; Gao, Y.; Wald, S. A.; Senanayake, C. H. *Org. Process Res. Dev.* **1998**, *2*, 96. (g) Bannister, R. M.; Brookes, M. H.; Evans, G.; Katz, R. B.; Tyrrell, N. D. *Org. Process Res. Dev.* **2000**, *4*, 467. (h) Hoekstra, M. S.; Sobieray, D. M.; Schwindt, M. A.; Mulhern, T. A.; Grote, T. M.; Huckabee, B. K.; Hendrickson, V. S.; Franklin, L. C.; Granger, E. J.; Karrick, G. L. *Org. Process Res. Dev.* **1997**, *1*, 26. (i) However, thalidomide has

been developed as a racemate as the optical pure drug molecule proved unstable. See: Muller, G. W.; Konnecke, W. E.; Smith, A. M.; Khetani, V. D. *Org. Process Res. Dev.* **1999**, *3*, 139. (j) Kammermeier, B.; Beck, G.; Reitz, J.; Sommer, C.; Wollmann, T.; Jendralla, H. *Org. Process Res. Dev.* **1997**, *1*, 121.

2. Langston, M.; Dyer, U. C.; Frampton, G. A. C.; Hutton, G.; Lock, C. J.; Skead, B. M.; Woods, M.; Zavareh, H. S. *Org. Process Res. Dev.* **2000**, *4*, 530.

3. (a) Flick, K.; Frankus, E. U.S. Patent 3,652,589, March 28, 1972; *Chem. Abstr.* **1972**, *76*, 153321. (b) Flick, K.; Frankus, E. U.S. 3,830,934, August 20, 1974; *Chem. Abstr.* **1974**, *82*, 21817.

4. (a) Frankus, E. V.; Friedrichs, E.; Kim, S. M.; Osterloh, G. *Arzneim.-Forsch./Drug Res.* **1978**, *28*, 114. (b) Goeringer, K. E.; Logan, B. K.; Christian, G. D. *J. Anal. Toxicol.* **1997**, *21*, 529.

5. Laird, T. *Org. Process Res. Dev.* **2006**, *10*, 851.

6. Huang, X.; Broadbent, S.; Dvorak, C.; Zhao, S.-H. *Org. Process Res. Dev.* **2010**, *14*, 613.

7. A review on asymmetric homogeneous hydrogenations at scale, *Chem. Soc. Rev.* **2012**, *41*, 3340.

8. (a) Noyori, R.; Ohkuma, T. *Angew. Chem., Int. Ed.* **2001**, *40*, 40. (b) Watanabe, M.; Murata, K.; Ikariya, T. *J. Org. Chem.* **2002**, *67*, 1713.

9. Chandramouli, S. V.; Ayers, T. A.; Wu, X.-D.; Tran, L. T.; Peers, J. H.; Disanto, R.; Roberts, F.; Kumar, N.; Jiang, Y.; Choy, N.; Pemberton, C.; Powers, M. R.; Gardetto, A. J.; D'Netto, G. A.; Chen, X.; Gamboa, J.; Ngo, D.; Copeland, W.; Rudisill, D. E.; Bridge, A. W.; Vanasse, B. J.; Lythgoe, D. J. *Org. Process Res. Dev.* **2012**, *16*, 484.

10. Sutin, L.; Ahrgren, L. *Org. Process Res. Dev.* **1997**, *1*, 425.

11. For reviews, see: (a) Evans, D. A. in *Asymmetric Synthesis*, ed. Morrison, J. D. Academic Press, New York, 1983, vol. 3, ch. 1, pp 83–110. (b) Meyers, A. I. in *Asymmetric Synthesis*, ed. Morrison, J. D. Academic Press, New York, 1983, vol. 3, ch. 3, pp 213–274. (c) Hayashi, T. in *Catalytic Asymmetric Synthesis*, ed. Ojima, I. VCH, New York, 1993, ch. 7, pp 323–331, ch. 8, pp 389–398. (d) Trost B. M.; Van Vranken, D. L. *Chem. Rev.* **1996**, *96*, 395. For recent advances, see: (e) Murphy, K. E.; Hoveyda, A. H. *J. Am. Chem. Soc.* **2003**, *125*, 4690. (f) Doyle A. G.; Jacobsen, E. N, *J. Am. Chem. Soc.* **2005**, *127*, 62.

12. Kim, M.; Choi, S.; Lee, Y.–J.; Lee, J.; Nahm, K.; Jeong, B.-S.; Park, H.; Jew, S. *Chem. Commun.* **2009**, *7*, 782.

13. Fix–Stenzel, S. R.; Hayes, M. E.; Zhang, X.; Wallace, G. A.; Grongsaard, P.; Schaffter, L. M.; Hannick, S. M.; Franczyk, T. S.; Stoffel, R. H.; Cusack, K. P. *Tetrahedron Lett.* **2009**, *50*, 4081.

14. Liu, W.-B.; Okamoto, N.; Alexy, E. J.; Hong, A. Y.; Tran, K.; Stoltz, B. M. *J. Am. Chem. Soc.* **2016**, *138*, 5234.

15. Zhu, Y.; Zhang, L.; Luo, S. *J. Am. Chem. Soc.* **2016**, *138*, 3978.

16. Ascic, E.; Buchwald, S. L. *J. Am. Chem. Soc.* **2015**, *137*, 4666.

17. Guo, S.; Yang, J. C.; Buchwald, S. L. *J. Am. Chem. Soc.* **2018**, *140*, 15976.

18. Li, H.; Chen, C.; Nguyen, H.; Cohen, R.; Maligres, P. E.; Yasuda, N.; Mangion, I.; Zavialov, I.; Reibarkh, M.; Chung, Y. L. *J. Org. Chem.* **2014**, *79*, 8533.

19. Takiguchi S.; Iizuka, T.; Kumakura, Y.; Murasaki, K.; Ban, N.; Higuchi, K.; Kawasaki, T. *J. Org. Chem.* **2010**, *75*, 1126.

20. Smith, A. B., III.; Bosanac, T.; Basu, K. *J. Am. Chem. Soc.* **2009**, *131*, 2348.

21. Gibson, F. S.; Singh, A. K.; Soumeillant, M. C.; Manchand, P. S.; Humora, M.; Kronenthal, D. R. *Org. Process Res. Dev.* **2002**, *6*, 814.

22. Roth, G. P.; Landi, J. J., Jr.; Salvagno, A. M.; Müller-Bötticher, H. *Org. Process Res. Dev.* **1997**, *1*, 331.

23. Marzi, M.; Minetti, R.; Moretti, G.; Tinti, M. O.; de Angelis, F. *J. Org. Chem.* **2000**, *65*, 6766.

24. Beak, P.; Anderson, D. R.; Curtis, M. D.; Laumer, J. M.; Pippel, D. J.; Weisengurger, G. A. *Acc. Chem. Res.* **2000**, *33*, 715.

25. (a) Breuer, M.; Ditrich, K.; Habicher, T.; Hauer, B.; Kesseler, M.; Stuermer, R.; Zelinski, T.; *Angew. Chem. Int. Ed.* **2004**, *43*, 788. (b) Fogassy, E.; Nogradi, M.; Dozma, D.; Egri, G.; Palovics, E.; Kiss, V. *Org. Biomol. Chem.* **2006**, *4*, 3011.

26. Schaus, S. E.; Brandes, B. D.; Larrow, J. F.; Tokunaga, M.; Hansen, K. B.; Gould, A. E.; Furrow, M. E.; Jacobsen, E. N. *J. Am. Chem. Soc.* **2002**, *124*, 1307.

27. Ohigashi, A.; Kanda, A.; Moriki, S.; Baba, Y.; Hashimoto, N.; Okada, M. *Org. Process Res. Dev.* **2013**, *17*, 658.

28. (a) Ghosh, A. K.; Matsuda, H. *Org. Lett.* **1999**, *1*, 2157. (b) Hii, K. K.; Claridge, T. D. W.; Brown, J. M.; Smith, A.; Deeth, R. J. *Helv. Chim. Acta* **2001**, *84*, 3043. (c) Lloyd-Jones, G. C.; Stephen, S. C. *Chem. Eur. J.* **1998**, *4*, 2539. (d) Namyslo, J. C.; Kaufmann, D. E. *Synlett* **1999**, *6*, 804. (e) Rabeyrin, C.; Nguefack, C.; Sinou, D. *Tetrahedron Lett.* **2000**, *41*, 7461. (f) Tillack, A.; Koy, C.; Michalik, D.; Fischer, C. *J. Organomet. Chem.* **2000**, *603*, 116. (g) Yamashita, M.; Nagata, T. (Takeda Chemical Industries, Ltd., Japan). PCT Int. Appl. Wo 0119777, 2001.

29. Lim, C. W.; Tissot, O.; Mattison, A.; Hooper, M. W.; Brown, J. M.; Cowley, A. R.; Hulmes, D. I.; Blacker, A. J. *Org. Process Res. Dev.* **2003**, *7*, 379.

30. Karlsson, S.; Sörensen, J. H. *Org. Process Res. Dev.* **2012**, *16*, 586.

31. (a) Murtagh, L.; Dunne, C.; Gabellone, G.; Panesar, N. J.; Field, S.; Reeder, L. M.; Saenz, J.; Smith, G. P.; Kissick, K.; Martinez, C.; Alsten, J. G. V.; Evans, M. C.; Franklin, L. C.; Nanninga, T.; Wong, J. *Org. Process Res. Dev.* **2011**, *15*, 1315. (b) Bassan, E. M.; Baxter, C. A.; Beutner, G. L.; Emerson, K. M.; Fleitz, F. J.; Johnson, S.; Keen, S.; Kim, M. M.; Kuethe, J. T.; Leonard, W. R.; Mullens, P. R.; Muzzio, D. J.; Roberge, C.; Yasuda, N. *Org. Process Res. Dev.* **2012**, *16*, 87.

32. Cohen, S.; Khedouri, E. *J. Am. Chem. Soc.* **1961**, *83*, 1093.

33. Savage, S. A.; Waltermire, R. E.; Campagna, S.; Bordawekar, S.; Toma, J. D. R. *Org. Process Res. Dev.* **2009**, *13*, 510.

34. Atkins, R. J.; Banks, A.; Bellingham, R. K.; Breen, G. F.; Carey, J. S.; Etridge, S. K.; Hayes, J. F.; Hussain, N.; Morgan, D. O.; Oxley, P.; Passey, S. C.; Walsgrove, T. C.; Wells, A. S. *Org. Process Res. Dev.* **2003**, *7*, 663.

35. Boller, T.; Meier, C.; Menzler, S. *Org. Process Res. Dev.* **2002**, *6*, 509.

36. (a) Wang, X.-J.; Zhang, L.; Smith-Keenan, L. L.; Houpis, I. N.; Farina, V. *Org. Process Res. Dev.* **2007**, *11*, 60. (b) Horgan, S. W.; Burkhouse, D. W.; Cregge, R. J.; Freund, D. W.; LeTourneau, M.; Margolin, A.; Webster, M. E. *Org. Process Res. Dev.* **1999**, *3*, 241.

37. Brossat, M.; Moody, T. S.; de Nanteuil, F.; Taylor, S. J. C.; Vaughan, F. *Org. Process Res. Dev.* **2009**, *13*, 706.

38. Li, X.; Russell, R. K.; Horváth, A.; Jain, N.; Depré, D.; Ormerod, D.; Aelterman, W.; Sui, Z. *Org. Process Res. Dev.* **2009**, *13*, 102.

39. Gauthier, S.; Caron, B.; Cloutier, J.; Dory, Y. L.; Favre, A.; Larouche, D.; Mailhot, J.; Ouellet, C.; Schwerdtfeger, A.; Leblanc, G.; Martel, C.; Simard, J.; Merand, Y.; Belanger, A.; Labrie, C.; Labrie, F. *J. Med. Chem.* **1997**, *40*, 2117.

40. (a) Rometsch, R. U.S. Patent 2,957,880, October 25, 1960. (b) Maxwell, R. A.; Chaplin, E.; Eckhardt, S. B.; Soares, J. R.; Hite, G. *J. Pharmacol. Exp. Ther.* **1970**, *173*, 158.

41. Chikusa, Y.; Fujimoto, T.; Ikunaka, M.; Inoue, T.; Kamiyama, S.; Maruo, K.; Matsumoto, J.; Matsuyama, K.; Moriwaki, M.; Nohira, H.; Saijo, S.; Yamanishi, M.; Yoshida, K. *Org. Process Res. Dev.* **2002**, *6*, 291.

42. Drauz, K.; Jahn, W.; Schwarm, M. *Chem. Eur. J.* **1995**, *1*, 538.

43. Chen, J.; Chen, T.; Hu, Q.; Püntener, K.; Ren, Y.; She, J.; Du, Z.; Scalone, M. *Org. Process Res. Dev.* **2014**, *18*, 1702.

44. Zhong, Y.-L.; Pipik, B.; Lee, J.; Kohmura, Y.; Okada, S.; Igawa, K.; Kadowaki, C.; Takezawa, A.; Kato, S.; Conlon, D. A.; Zhou, H.; King, A. O.; Reamer, R. A.; Gauthier, Jr., D. R.; Askin, D. *Org. Process Res. Dev.* **2008**, *12*, 1245.

45. Lee, H. W.; Shin, S. J.; Yu, H.; Kang, S. K.; Yoo, C. L. *Org. Process Res. Dev.* **2009**, *13*, 1382.

46. Stuk, T. L.; Assink, B. K.; Bates, Jr., R. C.; Erdman, D. T.; Fedij, V.; Jennings, S. M.; Lassig, J. A.; Smith, R. J.; Smith, T. L. *Org. Process Res. Dev.* **2003**, *7*, 851.

47. Boulton, L. T.; Brick, D.; Fox, M. E.; Jackson, M.; Lennon, I. C.; McCague, R.; Parkin, N.; Rhodes, D.; Ruecroft, G. *Org. Process Res. Dev.* **2002**, *6*, 138.

48. Voeffray, R.; Perlberger, J.-C.; Tenud, L.; Gosteli, J. *Helv. Chim. Acta* **1987**, *70*, 2058.

49. (a) Tang, L.; Deng, L. *J. Am. Chem. Soc.* **2002**, *124*, 2870. (b) Jurkauskas, V.; Buchwald, S. L. *J. Am. Chem. Soc.* **2002**, *124*, 2892. (c) Nunami, K.-I.; Kubota, H.; Kubo, A. *Tetrahedron Lett.* **1994**, *35*, 8639. (d) Ward, R. S. *Tetrahedron: Asymmetry* **1995**, *6*, 1475.

50. (a) Node, M.; Nishide, K.; Shigeta, Y.; Shiraki, H.; Obata, K. *J. Am. Chem. Soc.* **2000**, *122*, 1927. (b) Brand, S.; Jones, M. F.; Rayner, C. M. *Tetrahedron Lett.* **1995**, *36*, 8493.

51. (a) Kitamura, M.; Tokunaga, M.; Noyori, R. J. *J. Am. Chem. Soc.* **1993**, *115*, 144. (b) Allen, J. V.; Williams, J. M. J. *Tetrahedron Lett.* **1996**, *37*, 1859. (c) Persson, B. A.; Larsson, A. L. E.; Le Ray, M.; Baeckvall, J.-E. *J. Am. Chem. Soc.* **1999**, *121*, 1645. (d) Pamies, O.; Baeckvall, J.-E. *Chem. Rev.* **2003**, *103*, 3247.

52. (a) Hayashi, T.; Konishi, M.; Fukushima, M.; Mise, T.; Kagotani, M.; Tajika, M.; Kumada, M. *J. Am. Chem. Soc.* **1982**, *104*, 180. (b) Mikami, K.; Yoshida, A. *Tetrahedron* **2001**, *57*, 889.

53. Anderson, N. G. *Org. Process Res. Dev.* **2005**, *9*, 800.

54. Reider, P. J.; Davis, P.; Hughes, L. D.; Grabowski, E. J. S. *J. Org. Chem.* **1987**, *52*, 955.

55. Guercio, G.; Bacchi, S.; Goodyear, M.; Carangio, A.; Tinazzi, F.; Curti, S. *Org. Process Res. Dev.* **2008**, *12*, 1188.

56. Košmrlj, J.; Weigel, L. O.; Evans, D. A.; Downey, C. W.; W. J. *J. Am. Chem. Soc.* **2003**, *125*, 3208.

57. (a) Bhattacharya, Araullo-Mcadams, C.; Meier, M. B. *Synth. Commun.* **1994**, *24*, 2449. (b) Yokozeki, K.; Nakamori, S.; Yamanaka, S.; Eguchi, C.; Mitsugi, K.; Yoshinaga, F. *Agric. Biol. Chem.* **1987**, *51*, 715. (c) Yokozeki, K.; Kubota, K. *Agric. Biol. Chem.* **1987**, *51*, 721.

58. Kemperman, G. J.; Zhu, J.; Klunder, A. J. H.; Zwanenburg, B. *Org. Lett.* **2000**, *2*, 2829.

59. Felten, A. E.; Zhu, G.; Aron, Z. D. *Org. Lett.* **2010**, *12*, 1916.

60. Aron, Z. D.; Ito, T.; May, T. L.; Overman, L. E.; Wang, J. *J. Org. Chem.* **2013**, *78*, 9929.

61. Rong, Z.-Q.; Zhang, Y.; Chua, R. H. B.; Pan, H.-J.; Zhao, Y. *J. Am. Chem. Soc.* **2015**, *137*, 4944.

62. Chen, C.; Frey, L. F.; Shultz, S.; Wallace, D. J.; Marcantonio, K.; Payack, J. F.; Vazquez, E.; Springfield, S. A.; Zhou, G.; Liu, P.; Kieczykowski, G. R.; Chen, A. M.; Phenix, B. D.; Singh, U.; Strine, J.; Izzo, B.; Krska, S. W. *Org. Process Res. Dev.* **2007**, *11*, 616.

63. Noyori, R.; Ikeda, T.; Ohkuma, T.; Widhalm, M.; Kitamura, M.; Takaya, H.; Akutagawa, S.; Sayo, N.; Saito, T.; Taketomi, T.; Kumobayashi, H. *J. Am. Chem. Soc.* **1989**, *111*, 9134.

64. Xu, F.; Corley, E.; Zacuto, M.; Conlon, D. A.; Pipik, B.; Humphrey, G.; Murry, J.; Tschaen, D. *J. Org. Chem.* **2010**, *75*, 1343.

65. Ishikawa, H.; Suzuki, T.; Orita, H.; Uchimaru, T.; Hayashi, Y. *Chem. Eur. J.* **2010**, *16*, 12616.

66. Müller, S.; Webber, M. J.; List, B. *J. Am. Chem. Soc.* **2011**, *133*, 18534.

67. Xu, K.; Lalic, G.; Sheehan, S. M.; Shair, M. D. *Angew. Chem. Int. Ed.* **2005**, *44*, 2259.

68. Tice, C. M.; Heathcock, C. H. *J. Org. Chem.* **1981**, *46*, 9.

69. Boesten, W. H. J.; Seerden, J.-P.; de Lange, B.; Dielemans, H. J. A.; Eisenberg, H. L.; Kaptein, B.; Moody, H. M.; Kellogg, R. M.; Broxterman, Q. B. *Org. Lett.* **2001**, *3*, 1121.

70. Wang, X.-J.; Frutos, R. P.; Zhang, L.; Sun, X.; Xu, Y.; Wirth, T.; Nicola, T.; Nummy, L. J.; Krishnamurthy, D.; Busacca, C. A.; Yee, N.; Senanayake, C. H. *Org. Process Res. Dev.* **2011**, *15*, 1185.

71. Yamada, M.; Nagashima, N.; Hasegawa, J.; Takahashi, S. *Tetrahedron Lett.* **1998**, *39*, 9019.

72. Ortiz, A.; Benkovics, T.; Beutner, G. L.; Shi, Z.; Bultman, M.; Nye, J.; Sfouggatakis, C.; Kronenthal, D. R. *Angew. Chem. Int. Ed.* **2015**, *54*, 7185.

73. Lubin, H.; Dupuis, C.; Pytkowicz, J.; Brigaud, T. *J. Org. Chem.* **2013**, *78*, 3487.

74. Li, H.; Belyk, K. M.; Yin, J.; Chen, Q.; Hyde, A.; Ji, Y.; Oliver, S.; Tudge, M. T.; Campeau, L.-C.; Campos, K. R. *J. Am. Chem. Soc.* **2015**, *137*, 13728.

75. Davies, A. J.; Scott, J. P.; Bishop, B. C.; Brands, K. M.; Brewer, S. E.; Dasilva, J. O.; Dormer, P. G.; Dolling, U.-H.; Gibb, A. D.; Hammond, D. C.; Lieberman, D. R.; Palucki, M.; Payack, J. F. *J. Org. Chem.* **2007**, *72*, 4864.

76. Kiau, S.; Discordia, R. P.; Madding, G.; Okuniewicz, F. J.; Rosso, V.; Venit, J. J. *J. Org. Chem.* **2004**, *69*, 4256.

77. Brunetto, G.; Gori, S.; Fiaschi, R.; Napolitano, E. *Helvetica Chimica Acta* **2002**, *85*, 3785.

78. For literature examples, see: (a) Drago, C.; Caggiano, L.; Jackson, R. R. W. *Angew. Chem. Int. Ed.* **2005**, *44*, 7221. (b) Egami, H.; Katsuki, T. *J. Am. Chem. Soc.* **2007**, *129*, 8940. (c) Liao, S.; Čorić, I.; Wang, Q.; List, B. *J. Am. Chem. Soc.* **2012**, *134*, 10765. (d) Sun, J.; Zhu, C.; Dai, Z.; Yang, M.; Pan, Y.; Hu, H. *J. Org. Chem.* **2004**, *69*, 8500.

79. Dornan, P. K.; Kou, K. G. M.; Houk, K. N.; Dong, V. M. *J. Am. Chem. Soc.* **2014**, *136*, 291.

80. Wang, H.-Y.; Yang, K.; Bennett, S. R.; Guo, S.; Tang, W. *Angew. Chem. Int. Ed.* **2015**, *54*, 8756.

81. Strotman, N. A.; Ramirez, A.; Simmons, E. M.; Soltani, O.; Parsons, A. T.; Fan, Y.; Sawyer, J. R.; Rosner, T.; Janey, J. M.; Tran, K.; Li, J.; La Cruz, T. E.; Pathirana, C.; Ng, A. T.; Deerberg, J. *J. Org. Chem.* **2018**, *83*, 11133.

17 Various Quenching Strategies

Reaction workup involves a series of manipulations, including quenching, extraction/washing, distillation, filtration, and crystallization. After completion of the reaction, the reaction mixture is normally quenched with an aqueous solution. The quenching operation serves mainly two purposes: to discontinue the reactions including both the desired and undesired reactions and to allow the products to partition into either an organic or aqueous layer, which sets the stage for the subsequent separation/purification. Occasionally, reactions have to be stopped/or decelerated by diluting with an organic solvent prior to the product isolation, especially when products are labile under standard quenching conditions. For direct isolation, the quenching, in addition to the discontinuation of the reaction, is utilized to precipitate the desired product followed by the filtration.

17.1 ACIDIC QUENCHING

Quenching with an acidic aqueous solution is used to stop a reaction by bringing the basic reaction media to neutral or acidic ones. In most cases, such quenchings help to facilitate the removal of inorganic by-products.

17.1.1 REMOVAL OF MAGNESIUM SALTS

17.1.1.1 Reaction of Grignard Reagent with Weinreb Amide

α-Chloroketone **4** was prepared via a reaction of Grignard reagent **2** with Weinreb amide **3** (Scheme 17.1).[1] Based on the fact that magnesium acetate salt $Mg(OAc)_2$ is inherently more soluble in water (390 g/L) than the complex $(NH_4)MgCl_3$, acidic quenching ($AcOH/H_2O$) was developed to remove magnesium as $Mg(OAc)_2$ salt in the aqueous layer.

This acid quenching protocol also mitigated the formation of a dimer impurity **5** that was generated via dimerization of **4** under basic conditions.

17.1.1.2 Weinreb Amide Formation

A similar workup protocol was used in the preparation of Weinreb amide **7** (Equation 17.1).[2]

$$(17.1)$$

Upon completion of the reaction, the reaction mixture was quenched into a 10% aqueous acetic acid (1.0 equiv) solution to remove magnesium salts by converting the magnesium into highly water-soluble salt $Mg(OAc)_2$ that was readily rejected into the aqueous layer.

17.1.2 REMOVAL OF ZINC

The enantioselective preparation of diarylmethanol **11** was realized via a nucleophilic addition of organozinc reagent **9** to aryl aldehyde **8** in the presence of piperidine-based chiral ligand **10** as shown in Equation 17.2.[3]

DOI: 10.1201/9781003288411-17

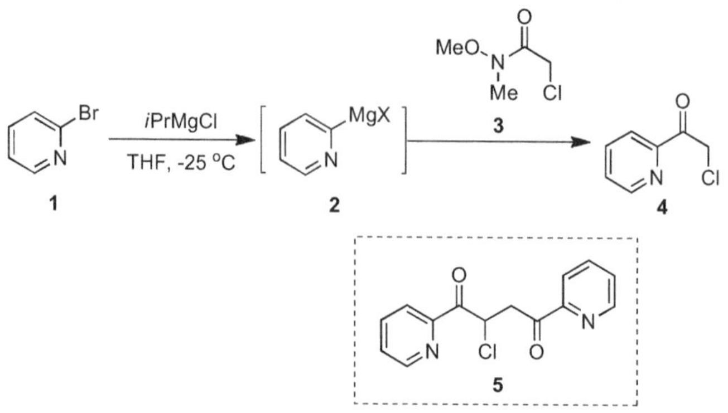

SCHEME 17.1

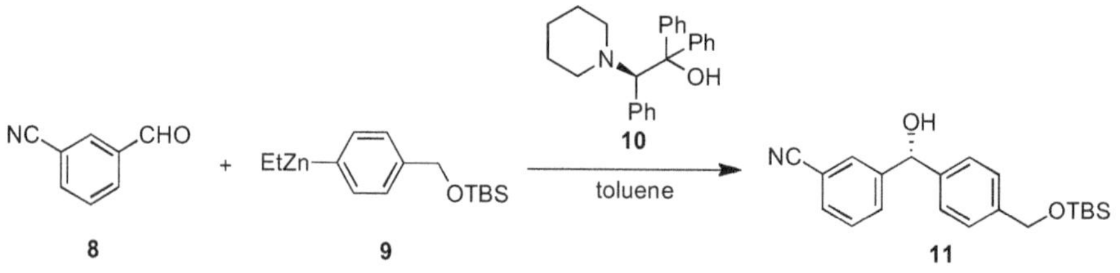

$$(17.2)$$

To cope with the zinc waste produced from this chemistry, a precipitation/filtration approach was employed by treatment of the reaction mixture with wet acetic acid. The resulting precipitated $Zn(OAc)_2 \cdot 2H_2O$ as a granular solid was removed by simple filtration.

Procedure

- Charge a solution of **10** (6.7 g, 18.7 mmol, 15.4 mol%) in toluene (70 mL) to a cold mixture (–10 °C) of **9** (121.2 mmol) in toluene (312 mL) via syringe.
- Stir the resulting mixture for 30 min.
- Charge a solution of **8** (12.2 g, 93.3 mmol) in toluene (40 mL) dropwise via an additional funnel at <–5 °C.
- Stir the batch at –10 to –5 °C for 4 h.
- Add a mixture of AcOH (59 mL) and water (14 mL) dropwise.
- Filter the resulting slurry and rinse the solid (zinc acetate dihydrate) with toluene (50 mL).
- Wash the combined filtrates with 0.5 N HCl (2×200 mL), water (2×100 mL), 0.5 N NaOH (100 mL), and water (100 mL).
- Distill at 45 °C under reduced pressure to give the crude **11** (40.4 g).
- Dissolve the crude **11** in MeCN (330 mL).
- Wash the MeCN solution with heptane (66 mL, 5×33 mL).
- The resulting MeCN solution of **11** (29.1 g, 88.3%) was used directly in the following step.

REMARK

The use of dry AcOH produced slimy $Zn(OAc)_2$ which was difficult to remove by filtration.

17.2 BASIC QUENCHING

Analogously, reactions under acidic conditions are usually quenched with bases. In addition to stopping reactions and controlling side reactions, basic quenching is also utilized to prevent glass-lined reactors from etching by HF.

17.2.1 SUPPRESSING THIADIAZOLE ISOMERIZATION

It was found that thiadiazole **13** is not stable and gradually isomerized to the *cis*-isomer **14** (Equation 17.3).[4] To minimize the undesired isomerization, a basic quenching of the reaction mixture with an aqueous solution of sodium carbonate was used.

$$(17.3)$$

12

13

14

Procedure

- Charge 34.1 kg (68.3 mol) of **12** to 424 kg of THF followed by adding 19.7 kg (88.8 mol) of P_2S_5 at 30 °C.
- Age the resulting mixture at 30 °C for 2 h.
- Add an aqueous solution of Na_2CO_3 (21.7 kg, 205 mol) in water (512 kg) at 7 °C.
- Stir the mixture for 17 h.
- Filter and rinse with water (68.0 kg) to give a product wet cake.
- Mix the wet cake with a mixture of water (205 kg) and THF (121 kg) and stir at 10 °C for 1 h.
- Filter, wash with water (68.0 kg), and dry at 60 °C.
- Yield: 31.9 kg (93.8%) with 0.5% of *cis*-isomer **14**.

17.2.2 PREVENTION OF ETCHING GLASS REACTOR

Glass and glass-lined reactors shall avoid long-time exposure to aqueous sodium hydroxide (and potassium hydroxide) solutions because the slow reactions between the glass and the corrosive caustic solutions to form soluble silicates would damage the reactors. In addition, to prevent etching, a reaction mixture in a glass reactor shall be free of fluoride under acidic conditions because hydrofluoric acid is very corrosive and can etch glass when hydrated.

17.2.2.1 Quenching with Sodium Bicarbonate

Fluorinated enol ether **16** was prepared via electrophilic fluorination of enol ether **15** with *N*-fluorobenzenesulfonimide (NFSI) (Equation 17.4).[5] To avoid the potential etching of the glass reactor by HF, a basic workup was desired and the reaction was quenched by adding an aqueous sodium bicarbonate solution to the reaction mixture.

(a) LiHMDS
THF, -20 °C

(b) NFSI
(c) NaHCO₃

$$(17.4)$$

15

16

17.2.2.2   Quenching with Sodium Hydroxide

Trifluoromethylated compounds have found a large number of industrial uses from dyes and polymers to pharmaceuticals and agrochemicals. Methyl fluorosulfonyldifluoroacetate **17** is a good difluorocarbene precursor and an excellent trifluoromethylating agent (Scheme 17.2).[6]

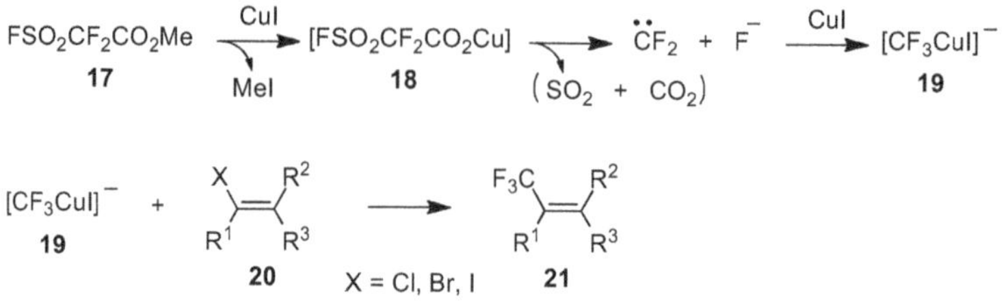

SCHEME 17.2

As shown in Scheme 17.2, the interaction between **17** and CuI would generate copper complex **18** whose dissociation will form difluorocarbene and fluoride along with sulfur dioxide and carbon dioxide. Ultimately, a reaction of difluorocarbene and fluoride in the presence of CuI would generate nucleophilic trifluoromethylation species **19**.

As fluoride anion is generated in situ, it is necessary to maintain the reaction medium under basic conditions during the copper-mediated trifluoromethylation of vinyl iodide **22** with **17** as small amounts of HF may be formed from the reaction (Equation 17.5).[7]

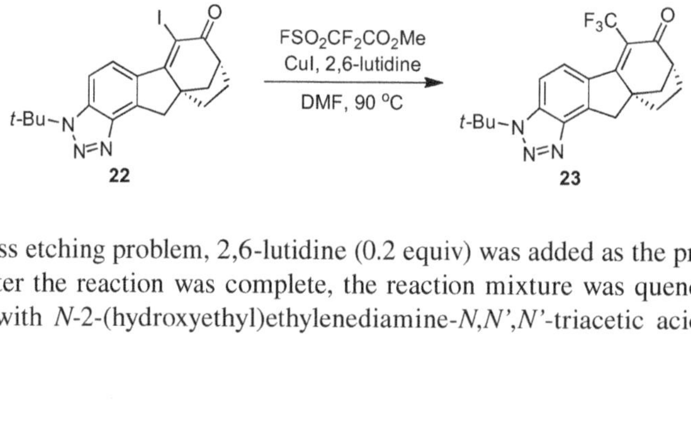

(17.5)

To obviate the glass etching problem, 2,6-lutidine (0.2 equiv) was added as the proton scavenger.

In addition, after the reaction was complete, the reaction mixture was quenched with sodium hydroxide along with N-2-(hydroxyethyl)ethylenediamine-N,N',N'-triacetic acid to facilitate the copper removal.

Procedure

- Mix 35.6 kg (80.0 mol) of **22** with 3.05 kg (16.0 mol) of CuI in DMF (505 kg).
- Charge 1.7 kg (16.0 mol) of 2,6-lutidine to the above mixture.
- Heat to 90–92 °C.
- Charge 31.2 kg (162 mol) of methyl fluorosulfonyldifluoroacetate over 1 h.
- Continue to heat at this temperature for 1 h.
- Cool to 20 °C.
- Add 26.8 kg (240 mol) of 12 N NaOH followed by adding a solution of N-2-(hydroxyethyl) ethylenediamine-N,N',N'-triacetic acid (29 kg, 80 mol) in water (290 kg).
- Add water (450 kg) and age the resulting slurry for 14 h.
- Filter, wash with water (2×150 kg), and dry at 50 °C.
- Yield: 25.6 kg (77%).

17.3   ANHYDROUS QUENCHING

Destroying excess reagents during workup is important in controlling impurity formation. Quenching the reaction mixture with an aqueous solution is frequently used to stop the reaction and at the same time set up the stage for subsequent product isolation by either filtration or extraction. Occasionally, aqueous quenching leads to difficulty in the subsequent layer separations or degradation of the products. Therefore, an anhydrous workup approach (by quenching the reaction mixture with organic solution) is developed to overcome these processing issues.

17.3.1 REMOVAL OF ZINC BY-PRODUCT

Aqueous quenching of organometallic reactions with either saturated ammonium chloride or dilute HCl solutions led to the precipitation of metal by-products, thereby complicating ensuing layer separations. This situation was observed during enantioselective cyclopropanation[8] of allylic alcohol **24**. Employing Charette's conditions[9] by exposure of alcohol **24** to a mixture of zinc reagent [(Zn(CH_2I)_2] and stoichiometric (R,R)-dioxaborolane **25** gave cyclopropylmethanol **26** in 83% yield (Equation 17.6).[10] The subsequent workup by quenching the reaction mixture with aqueous ammonium chloride or diluted aqueous HCl solution made the layer separation difficult.

$$(17.6)$$

To circumvent this problem, anhydrous quenching conditions were employed. Thus, upon completion of the reaction, the addition of 2-propanol and acetic acid at 35 °C followed by cooling to 0–5 °C resulted in the precipitation of zinc acetate. After filtration, two subsequent washes of the filtrate with 3 N NaOH removed butylboronic acid and N,N,N',N'-tetramethyltartaric amide by-products from the product stream (See 17.1.2 Removal of Zinc).

17.3.2 AVOIDING INSOLUBLE ORGANIC MASS

Aqueous quenching may result in insoluble organic mass due to the poor solubility of the product in the resulting aqueous–organic mixtures. For instance, insoluble organic "balls" were observed during the aqueous workup of the palladium-mediated cross-coupling reaction of pyridyl bromide **27** with isopropyl 2-cyanoacetate (Equation 17.7).[11] The "balls", formed from the aqueous quenching of residual NaH, were difficult to break up. In order to avoid the formation of such "balls", an anhydrous quenching protocol was developed, wherein 2-propanol was employed to quench the reaction mixture, resulting in readily stirrable slurries.

$$(17.7)$$

Procedure

- Charge 12.0 kg of 60% NaH/oil (300 mol, 2.5 equiv) into an inerted 800-L glass-lined reactor followed by adding 0.27 kg (1.2 mol, 1 mol%) of Pd(OAc)_2, 1.26 kg (4.8 mol, 4 mol%) of PPh_3, and 40 kg (120 mol) of **27**.
- Charge toluene (70 kg).

- Heat the mixture to 60 °C.
- Charge a solution of isopropyl 2-cyanoacetate (17.4 L, 138 mol, 1.15 equiv) in toluene (20 L) over 2 h.
- Heat the batch to 100 °C and maintain at that temperature for 1.5 h.
- Cool the mixture to 35 °C.
- Add 19 kg of IPA (308 mol, 2.6 equiv) over 20 min.
- Transfer the batch into a stainless steel reactor.
- Add 438 kg of 1 N NaOH solution for the subsequent hydrolysis/decarboxylation.

17.3.3 AVOIDING DEGRADATION OF PRODUCT

Unstable products can degrade during manufacturing, transportation, and storage. Not surprisingly, it is challenging to handle a reaction mixture that contains such unstable products during the workup.

17.3.3.1 Use of Ethyl Acetate

As shown in Reaction (17.8),[12] triisopropylsilyl (TIPS) protected thiol ether **30** was prepared from a Pd-catalyzed coupling reaction between bromide **29** and TIPS thiol (1.2 equiv). The reaction was carried out in the presence of NaOtBu (1.2 equiv) and PdCl$_2$(dppf) (1.0 mol %) under refluxing in toluene. After aqueous workup, **30** was crystallized in 80% yield with excellent purity.

(17.8)

(I) Workup problems

The main concern on scale-up was the stability of the product **30** toward the aqueous workup conditions. Investigations indicated that longer processing times may lead to loss of the TIPS protecting group in **30**, resulting in side reactions (e.g., formation of disulfide).

(II) Solutions

The degradation of **30** was caused by the hydrolytic instability of **30** under the aqueous workup conditions. Accordingly, a non-aqueous workup protocol was developed to avoid such degradations. Upon completion of the reaction, ethyl acetate was added and the resulting reaction mixture was filtered through celite to remove sodium bromide. The resulting filtrate was then diluted with heptane to induce crystallization of **30**.

Procedure

- Charge, under N$_2$, 349.6 L of toluene, 26 kg (91.5 mol) of aryl bromide **29**, and 10.55 kg (109.8 mol, 1.2 equiv) of NaOtBu.
- Charge 20.91 kg (109.8 mol, 1.2 equiv) of TIPS thiol.
- Heat the mixture to 75 °C.
- Charge 0.67 kg (0.92 mol, 1.0 mol%) of [1,1'-bis(diphenylphosphino)ferrocene]-dichloropalladium (II)
- Stir the resulting mixture at reflux for 2 h (the reaction is complete then).
- Cool the mixture to 22 °C.
- Add EtOAc (780 L) and stir at rt for 30 min.

- Filter the resulting slurry through Celite (39 kg).
- Rinse the cake with EtOAc (260 L).
- Distill under reduced pressure to one-fifth of the volume.
- Add heptane (520 L).
- Cool to 5 °C and stir for 1 h.
- Filter and wash the product cake with heptane (260 L).
- Dry at 50 °C under vacuum.
- Yield: 28.45 kg (79%).

REMARKS

(a) This mode of addition was considered to be safe as the coupling process was not exothermic.

(b) This addition sequence is crucial for the complete conversion. The addition of the catalyst to the reaction mixture at room temperature, followed by a slow heating ramp resulted in an incomplete conversion, which was interpreted as resulting from the catalyst degradation by the TIPS thiolate prior to the mixture reaching a temperature at which the coupling occurred. An alternative explanation was that the heating of the reaction mixture at 75 °C prior to adding the catalyst could expel dissolved oxygen gas, thus avoiding oxidation of the catalyst.

17.3.3.2 Use of Diisopropylethylamine

Anhydrous workup is necessary when a reaction product is unstable in the presence of moisture. The enol ether **32** was prepared in quantitative yield via protection of oxyindole disulfide **31** with *tert*-butyldimethylsilyl (TBS) triflate in toluene along with the liberation of triethylammonium triflate by-product (Equation 17.9).[13]

$$\left(\underset{\underset{\text{Me}}{\overset{}{\text{N}}}}{\overset{\text{S}}{\bigcirc}} = O \right)_2 \quad \xrightarrow[\text{(b) } i\text{Pr}_2\text{NEt}]{\text{(a) TBSOTf, Et}_3\text{N}} \quad \left(\underset{\underset{\text{Me}}{\overset{}{\text{N}}}}{\overset{\text{S}}{\bigcirc}} - \text{OTBS} \right)_2 \qquad (17.9)$$

31   **32**

(I) Problems

- The product bis-*O*-silyl enol ether **32** was unstable in the presence of moisture.
- After completion of the reaction, the majority of triethylammonium triflate by-product in the lower liquid layer was separated from the upper product layer. Further reduction of the triethylammonium triflate to below 2 mol% proved to be challenging.

(II) Solutions

An anhydrous workup protocol was adopted by treatment of the product toluene solution with DIPEA, which avoided the decomposition of the product **32**. This protocol was also able to reduce the residual triethylammonium triflate level by converting it to a solid diisopropylethylammonium triflate that was removed by filtration.

Procedure

Stage (a)
- Charge 23.3 kg (230.3 mol, 2.3 equiv) of Et$_3$N to a slurry of 34.8 kg (97.6 mol) of **31** in toluene (241.5 kg), followed by rinsing with toluene (16.0 kg).
- Cool to 0 °C.
- Charge 57.8 kg (218.7 mol, 2.2 equiv) of TBSOTf at <10 °C, followed by rinsing with toluene (15.5 kg).

- Stir the resulting mixture at 0–10 °C for 15 min.
- Warm the batch to 19 °C.
- Allow the biphasic mixture to settle.
- Remove the lower triethylammonium triflate (in toluene) layer.

Stage (b)
- Add a mixture of DIPEA (12.9 kg) and toluene (15.3 kg) to the upper layer.
- Distill the resulting mixture under reduced pressure at ≤32.3 °C.
- Cool it down to 20 °C.
- Filter to remove diisopropylethylammonium triflate.
- Add a mixture of DIPEA (6.3 kg) and toluene (142.2 kg).
- Distill and then filter.
- Yield: 56.4 kg (99%) of **32** as a toluene solution.

17.3.4 DECOMPOSITION OF EXCESS REAGENT

17.3.4.1 Use of Methanol

Chemoselective reduction of lactam **33** was carried out using borane–THF complex (Scheme 17.3).[14] Methanol proves to be the reagent of choice for quenching the borane reduction mixture. In addition to the decomposition of excess borane, MeOH was capable of breaking the complex **35** at 55–60 °C.

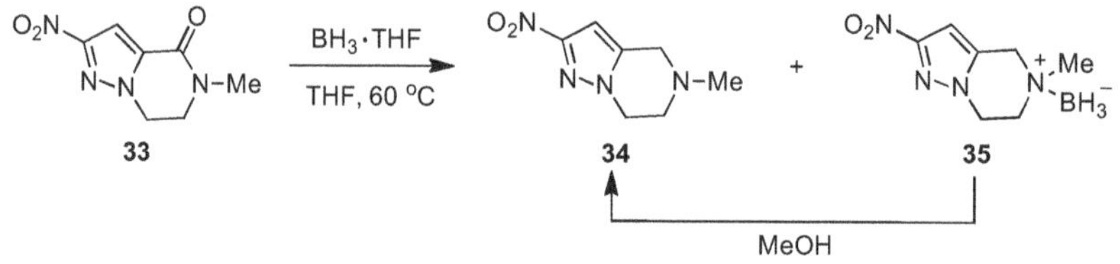

SCHEME 17.3

Procedure

- Charge 5.82 L (5.82 mol, 2.06 equiv) of a solution of $BH_3 \cdot THF$ (1M in THF) to a solution of **33** (554 g, 2.82 mol) at 60 °C over 1 h.
- Stir at 60 °C for 5 h.
- Add MeOH (1.0 L, 24.7 mol) at 55 °C and stir at that temperature for 1 h.
- Concentrate the resulting solution under reduced pressure to ~4 L.
- Add heptane (3.32 L).
- Concentrate to 4 L followed by adding 3.32 L of heptane.
- Cool to rt and stir for 30 min.
- Filter and wash with heptane (2.2 L).
- Yield: 579 g.

17.3.4.2 Use of Silicon Dioxide

Aqueous workup had to be avoided during the workup of the reaction mixture containing moisture-sensitive trimethylsilyl ketal product **37** (Equation 17.10).[15] Thus, silica gel was used to quench the excess trimethyl(trifluoromethyl)silane.

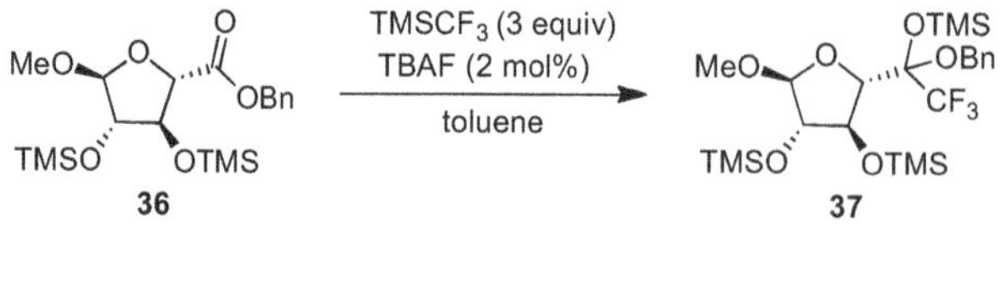

(17.10)

Procedure

- Charge a solution of TBAF in THF (1 M, 87 mL, 2 mol%) to a solution of **36** (1.8 kg, 4.4 mol) in toluene (11.4 kg), immediately followed by adding trimethyl(trifluoromethyl)silane (1.94 L, 13.1 mol, 3 equiv) at 20 ± 5 °C within 30 min.
- After the addition, stir the mixture at that temperature for 2.5 h.
- Charge 1.8 kg of SiO_2 and stir for 1 h.
- Filter and rinse with toluene (3.6 L).
- Distill the combined toluene solution under reduced pressure at <40 °C of batch temperature.
- Yield: 2.28 kg (93.0% yield) as a 2:1 diastereomeric mixture.

17.4 OXIDATIVE QUENCHING

Oxidative quenching can be utilized to convert problematic by-products to species that are readily removed during the workup.

17.4.1 OXIDATION OF HYDROGEN IODIDE

The reaction of chloroiodomethane **38** with chlorosulfonic acid could be employed to prepare chloromethyl chlorosulfate **39** (Scheme 17.4).[16] It was found that the removal of the by-product HI from the reaction system was critical in order to drive the reaction to completion. An approach was developed using 2 equiv of chorosulfonic acid, in which 1 equiv was for the conversion of **38** to **39** and the other equiv for the oxidation of HI to iodine.

(I) Problematic iodine

Solid iodine settled at the bottom of a reactor, complicating transfers of the reaction stream.

(II) Solutions

Oxidative quenching with bleach converted iodine to iodate.

Procedure

- Charge methylene chloride (1.44 L) to a 5-L reactor, followed by adding 170.9 g (1.47 mol, 1.65 equiv) of chlorosulfonic acid and 84.6 g (0.711 mol, 0.8 equiv) of thionyl chloride at 20 °C.
- Charge 160 g (0.889 mol, 1.0 equiv) of **38**.
- Stir the resulting mixture at 20–25 °C for 1 h.
- Heat to 35 °C over 30 min and stir at that temperature until **38** was <5%.
- Cool to 0 °C.
- Charge a cold 10 wt% bleach solution (1.72 kg, 0 °C, 2.6 equiv) in portionwise (1.5%, 3.5%, 5%, 10%, 30%, 50%) to the cold (0 °C) reaction mixture.
- Stir for 30 min.
- Separate layers.
- Wash the CH_2Cl_2 product layer with 0.25% bleach solution (800 mL).
- Distill to give **39** as a dense colorless liquid.

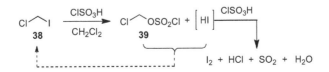

SCHEME 17.4

DISCUSSIONS

(a) The general reductive quenching with aqueous sodium thiosulfate led to the decomposition of **39** resulting in only 47% product yield.

(b) Thionyl chloride was used as a water scavenger to deplete the in situ generated water by-product.

(c) Thionyl chloride is not compatible with solvents such as alcohols, water, DMF, THF, and MTBE. Quenching the thionyl chloride in a toluene solution with aqueous sodium hydroxide proved ineffective.[17] Therefore, *n*-butanol was identified as an effective reagent in destroying thionyl chloride in toluene (Equation 17.11).[17]

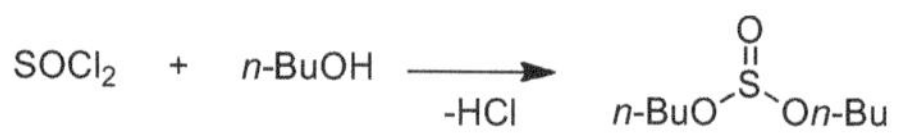

$$SOCl_2 \ + \ n\text{-BuOH} \ \xrightarrow[-HCl]{} \ n\text{-BuO}\overset{O}{\underset{O}{\overset{\|}{\underset{\|}{S}}}}On\text{-Bu} \tag{17.11}$$

17.4.2 OXIDATION OF PINACOL

Preparation of arylboronic acid **42** via palladium-catalyzed borylation of aryl bromide **40** followed by hydrolysis of boronic ester **41**. However, the hydrolysis of **41** suffered incomplete conversion due to the reverse reaction of **42** with the pinacol by-product. In order to address this undesired reaction caused by the pinacol by-product, sodium periodate was used to oxidize the pinacol to acetone. Thus, the implementation of this oxidation step into the hydrolysis process leading to a complete conversion (Scheme 17.5).[18]

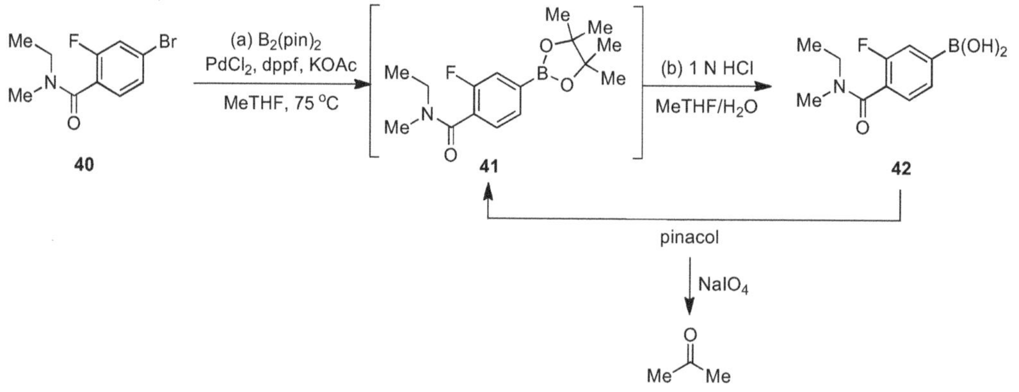

SCHEME 17.5

Procedure

Stage (b)

- Cool the MeTHF product solution obtained from Stage (a) (starting with 3.17 moles of **40**) to 6.5 °C.
- Charge 960 g (4.49 mol) of $NaIO_4$ and water (5.1 L).
- Stir the resulting mixture at 6.5 °C for 1 h.
- Warm to 20 °C followed by adding 1 N HCl (4.46 L).
- Stir the biphasic mixture at room temperature overnight.
- Separate layers.
- Extract the aqueous layer with MeTHF (2.1 L).
- Wash the combined organic layers with 20% sodium thiosulfate (3 L) and half-saturated brine (3 L).
- Filter the organic layer into a clean reactor.

- Distill to 3.8 L at 70 °C/90 mmHg (25 °C batch temperature).
- Charge heptane (5.7 kg) and 26 g of seed crystals into a clean crystallizer.
- Charge, to the crystallizer, 3.8 L of the product MeTHF solution and heptane at 20 °C simultaneously over 2 h.
- Stir the resulting thick white slurry at 20 °C overnight.
- Filter and wash with MeTHF/heptane (1:1, 3 L) then heptane (1 L).
- Dry at 45 °C for 72 h.
- Yield: 486 g (69%).

17.5 REDUCTIVE QUENCHING

Analogously, during the workup of oxidation reactions, reductive quenching is frequently employed to destroy excess oxidants which may, otherwise, have detrimental effects.

17.5.1 Restroying tert-Butyl Hydroperoxide

Epoxidation of cinnamyl alcohol **43** was accomplished under Sharpless epoxidation conditions, providing epoxide product **44** in good yield with 92% ee (Equation 17.12).[19] After completion of the reaction, using an appropriate quenching protocol to destroy the excess of tert-butyl hydroperoxide (TBHP) was critical as **44** is acid-sensitive and unstable under standard TBHP quenching conditions.

$$
\begin{array}{c}
\text{L-DIPT (0.15 equiv)} \\
\text{Ti(}i\text{PrO)}_4 \text{ (0.098 equiv)} \\
\hline
\text{TBHP (2.0 equiv)} \\
\text{4 A MS, CH}_2\text{Cl}_2
\end{array}
$$

(cinnamyl alcohol **43** → epoxide **44**)   (17.12)

L-DIPT: L-(+)-diisopropyl tartrate

(I) Problems

The initial quenching with aqueous $FeSO_4$ solution (pH 3.5 in the beginning and pH 1.5 after quenching) led to the decomposition of the epoxide **44**.

(II) Solutions

Triethylphosphite was chosen as the TBHP quenching reagent. Thus, $P(OEt)_3$ (1.1 equiv) was added slowly after the completion of the reaction and the resulting solution, after filtering through Celite 545, was used directly in the subsequent step.

REMARK

The use of standard peroxide test strips or iodometric titration is not effective in determining the TBHP concentration. 4-Aminophenol appears to be a very sensitive TBHP color indicator at ppm level; after reaction mixtures are resolved on TLC plates, it stains the TBHP spot as a pink color.

17.5.2 Destroying Hydrogen Peroxide

Oxidation of methylsulfanylpyrimidine **45** to sulfone **47** was originally carried out in full batch mode with hydrogen peroxide (2.5 equiv) and sodium tungstate dihydrate (5 mol%) at 20–30 °C for 30 h, then additional 0.5 equiv H_2O_2 were added with further 48 h aging at 55–62 °C. The reaction was quenched with aqueous $Na_2S_2O_3$, and 90% of sulfone **47** was isolated by filtration (Condition A, Scheme 17.6).[20] This process suffered safety concerns because of the accumulation of H_2O_2.

Changing from a full batch to a semibatch operation by dosing 2.3 equiv H_2O_2 into a mixture of **45** and sodium tungstate dihydrate (1 mol%) at 60 °C made the process safer (Condition B, Scheme 17.6). After completion of the reaction, sodium bisulfate was used to destroy excess H_2O_2.

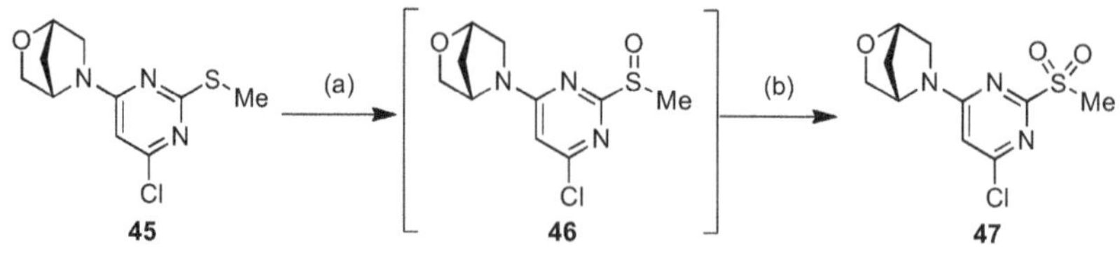

Condition A: (a) H_2O_2 (3.0 equiv), $Na_2WO_4 \cdot H_2O$ (5 mol%)
MeOH, 20 to 62 °C.
(b) aq $Na_2S_2O_3$, 90% yield.
Condition B: (a) H_2O_2 (2.3 equiv), $Na_2WO_4 \cdot H_2O$ (1 mol%)
MeOH, 60 °C.
(b) aq $NaHSO_3$, 89.3% yield.

SCHEME 17.6

Procedure

- Add 40.8 kg 35% H_2O_2 (420 mol) to a suspension of **45** (47.3 kg, 184.0 mol) and sodium tungstate dehydrate (0.61 kg, 1.85 mol) in MeOH (368 kg) and water (241 kg) at 60 °C over 4 h.
- Rinse with water (10.6 kg).
- Stir the batch at 60 °C for 3 h (O_2 level was controlled <5%).
- Cool to 22 °C within 75 min.
- Add 38% aq $NaHSO_3$ (47.8 kg) within 20 min and rinse with water (5.4 kg) (residual H_2O_2 <10 ppm).
- Stir 22 °C for 3 h.
- Isolate the precipitated product by centrifugation.
- Dry at 45 °C under reduced pressure (45–47 mbar) for 5 h.
- Yield: 47.6 kg (89.3% yield).

17.5.3 DESTROYING OXONE

Besides hydrogen peroxide, other oxidants such as oxone can be used to oxidize sulfide to sulfone. For instance, oxidation of sulfide **48** with oxone provided sulfone **49** in 80% yield (Scheme 17.7).[21] A reductive workup using $Na_2S_2O_3$ allowed the excess oxone to be destroyed. However, $Na_2S_2O_3$ also decomposed to elemental sulfur during the quenching, which would affect the subsequent alkylation reaction and product isolation.

Therefore, sodium bisulfite ($NaHSO_3$) was identified as an alternative reductant to quench the oxidation reaction mixture.

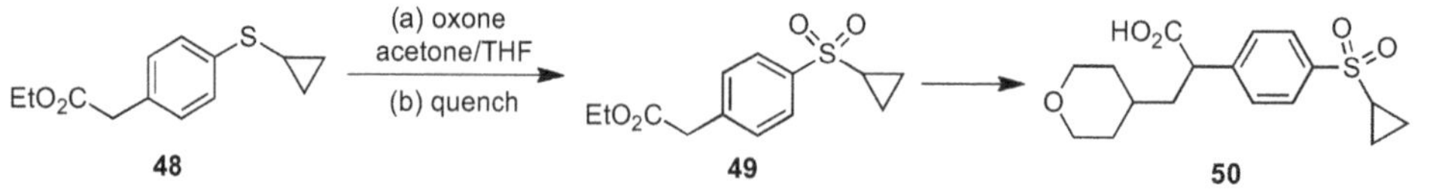

SCHEME 17.7

17.5.4 DESTROYING HALOGENS

Quenching halogenation reactions with sodium bisulfite or thiosulfate is usually used to destroy the excess of bromine or iodine. Ascorbic acid has also been used to quench halogenation reactions, and using the ascorbic acid protocol is able to obviate the potential product contamination by sulfur when using sodium thiosulfate.

17.5.4.1 Use of Ascorbic Acid to Destroy Bromine

Bromination of alkene **51** with bromine in methylene chloride provided dibromide **52** (Equation 17.13).[22]

$$(17.13)$$

51   **52**

(I) Problems

Initially, quenching the bromination reaction mixture with thiosulfate to destroy the excess of bromine had two issues:

- The resulting quenching suspension was difficult to filter.
- The product contained sulfur impurities which had to be removed by column chromatography in order to avoid poisoning the palladium catalyst in the following Suzuki reaction.

(II) Solutions

Ascorbic acid was selected to quench the reaction mixture. As a result, the product **52** was isolated in 93% yield and 88% purity on a 7 kg scale, obviating the aforementioned problems.

Procedure

- Charge 7.0 kg (35.8 mol) of **51** and CH_2Cl_2 (49 L) into a reactor.
- Cool to 10–15 °C.
- Charge 42% solution of bromine (6.07 kg, 38.0 mol) in CH_2Cl_2 (11.9 L) at ca. 15 °C over 30 min.
- Warm the batch to 20 °C and hold at 20 °C for 2 h.
- Add 9% solution of ascorbic acid (31.0 kg) at <25 °C over 5 min.
- Stir the resulting mixture for 10 min.
- Separate layers.
- Wash the organic layer with water (28.0 L).
- Distill the organic layer at 30–35 °C under reduced pressure to dryness.
- Dissolve the crude product in THF (139.9 L).
- Disill at 30–35 °C under reduced pressure.
- Yield: 11.84 kg (92.7%).

17.5.4.2 Use of Ascorbic Acid to Destroy Iodine

Ascorbic acid was also utilized to destroy the excess iodine during the iodination of indazole **53** (Equation 17.14).[23]

$$(17.14)$$

53   **54**

The iodination was carried out with 2 equiv of iodine in the presence of aqueous potassium hydroxide. Upon completion of the reaction, the reaction mixture is transferred into an aqueous, methanolic ascorbic acid solution to destroy the excess iodine.

17.6 DISPROPORTIONATION QUENCHING

Disproportionation quenching has been employed to decompose excess hydrogen peroxide into oxygen and water. Oxidation of thioacetate **55** with performic acid, generated in situ from the reaction of formic acid with hydrogen peroxide, produced the corresponding sulfonic acid **56** (Equation 17.15).[24] The workup included the addition of activated carbon (Darco G-60) followed by filtration. The use of activated carbon provides two advantages: one is to decompose the excess peroxide into oxygen and water by residual metals in the activated carbon and the other is to remove color impurities.

$$(17.15)$$

Procedure

Stage (a)
- Charge 34.2 kg (71.3 mol) of **55** as solids into a nitrogen-purged reactor at 20 °C, followed by adding 246.2 L of 96% formic acid.
- Charge 3.77 L (0.467 equiv) of 30% H_2O_2.
- Stir the resulting mixture at 20 °C for 2 h.
- Repeat the H_2O_2 addition at 2 h stirring eight more times.

Stage (b)
- Upon completion of the addition, transfer the batch to a suspension of Darco G-60 (1.70 kg) in 96% formic acid (39 L) at 20 °C under a high-volume nitrogen sweep over 2 h.
- Stir the resulting suspension at 20 °C for 12 h.
- Filter through Celite.
- Add glacial acetic acid (316 L) to the filtrate.
- Distill at ≤50 °C under reduced pressure to 30 L.
- Add MeOH (140 L) to the residue followed by adding isopropyl ether (1400 L) at 20 °C over 1 h.
- Stir the resulting slurry at 20 °C for 2 h.
- Filter, wash, and dry.
- Yield: 91%.

REMARK[25]

As the decomposition of peroxides can be catalyzed by trace metal impurities to generate heat and O_2 which potentially leads to a foaming or over-pressurization incident. Therefore, the reaction vessel shall be carefully cleaned to reduce the risk of rapid metal-catalyzed decomposition of peroxides. An efficient nitrogen purge is necessary throughout the quenching process in order to prevent the accumulation of oxygen.

17.7 REVERSE QUENCHING

The mode of quenching a reaction mixture is important and an improper quenching mode may diminish product yields and generate undesired side products. Direct quenching is a unit operation by adding the quenching solution into the reaction mixture; while reverse quenching is adding the reaction mixture into the quenching solution.

17.7.1   CONTROL OF IMPURITY FORMATION

17.7.1.1   Preparation of Ketone

Initially, isolation of alkyl aryl ketone **59** from the reaction mixture by direct quenching method with aqueous acetic acid led to a double addition side product **60** in 10–15% (Equation 17.16).[26]

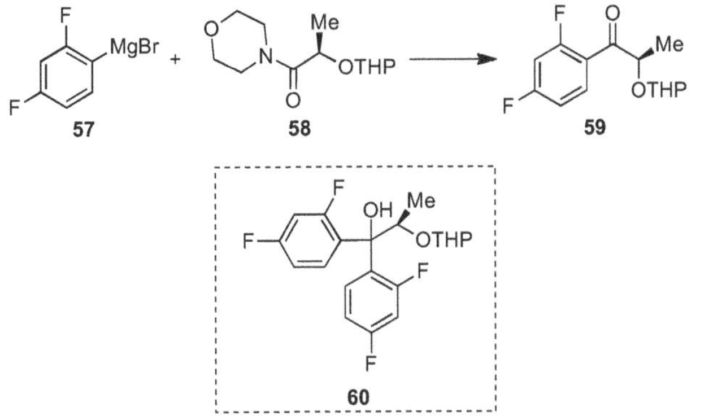

$$(17.16)$$

In order to mitigate this side product formation, reverse quenching by adding the reaction mixture into a cold aqueous acetic acid solution reduced the level of **60** from 10–15% to 2–7%.

17.7.1.2   Preparation of Aldehyde

It is always challenging handling an aldehyde reaction mixture because most aldehydes are susceptible to oxidation or nucleophilic attack. For example, the nucleophilic dimethylamine, released from the intermediate **62** during the preparation of alkynal **63**, reacted with the product **63** to generate side products **64–66** (Scheme 17.8).[27]

To prevent the formation of **64–66**, initially, a reverse quenching protocol (for R'M=BuLi) was developed by adding the α-aminoalkoxide mixture into a buffered biphasic mixture of 10% aqueous monobasic potassium phosphate (4 equiv) and MTBE. Further optimization led to a modified

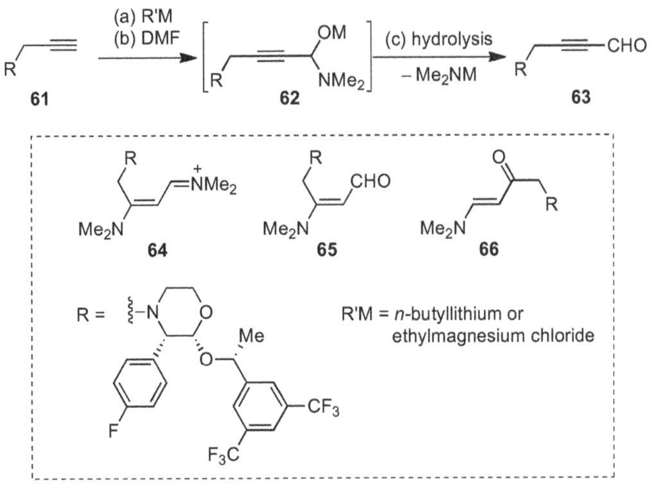

SCHEME 17.8

reverse quenching protocol (for R'M=EtMgCl) by addition of the reaction mixture into a biphasic solution of 7% aqueous sodium dihydrogen citrate (2 equiv, pH=3.9–4.5) and toluene. In this case, the acetylenic aldehyde **63** was completely stable and the yield of **63** was improved from 40–50% to 99%.

17.7.1.3 Grignard Reaction

A reverse quenching protocol was employed during the workup of the Grignard reaction of the in situ generated Grignard reagent **68** with tetrahydrothiopyran-4-one **69**. It has been observed that the reverse quenching by adding the cold reaction mixture into an aqueous acetic acid solution was critical in controlling the side reaction of the aldol condensation of the excess **69** (Scheme 17.9).[28]

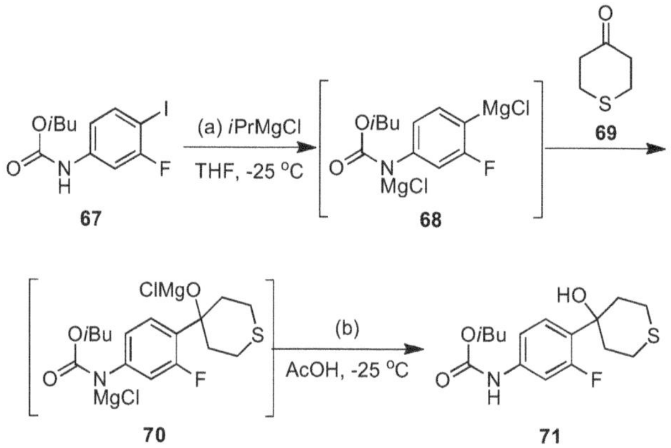

SCHEME 17.9

17.7.2 Removal of Excess Reagent

The direct isolation approach allows the isolation of the product directly from the reaction mixture via simple filtration. In the protection of 3-bromo-5-chloro-2-hydroxybenzaldehyde **72** with 2-methoxye-thoxymethyl chloride (MEMCl), water was used as both an anti-solvent and quenching reagent during the workup (Equation 17.17).[29] The addition of water to the reaction mixture resulted in immediate precipitation of the product **73**. However, the isolated product **73** was contaminated MEMCl.

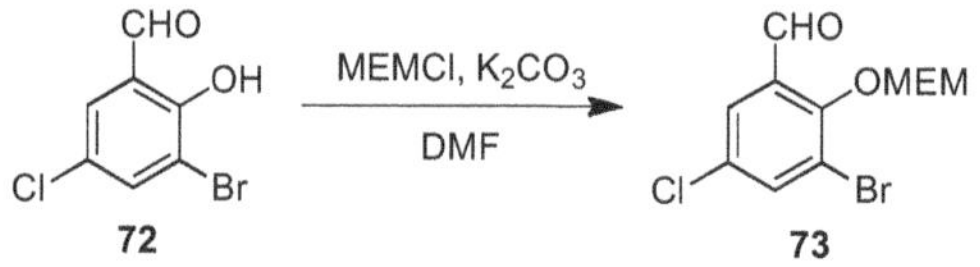

$$(17.17)$$

Ultimately, utilization of the reverse quenching approach by charging the reaction mixture into water could effectively remove the excess of MEMCl and no MEMCl was detected in the isolated product **73**.

17.7.3 Increase in Conversion

The synthesis of aldehyde **75** was achieved by lithiation of 4-bromo-2-fluorochlorobenzene **74** with LDA in THF at –50 °C and treatment of the resulting anion with DMF, followed by quenching with water and extracting into MTBE, to give a solution of the aldehyde product **75** (Scheme 17.10).[30]

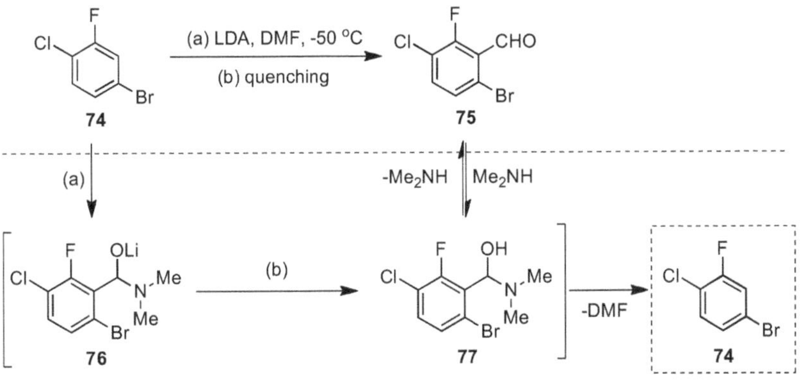

SCHEME 17.10

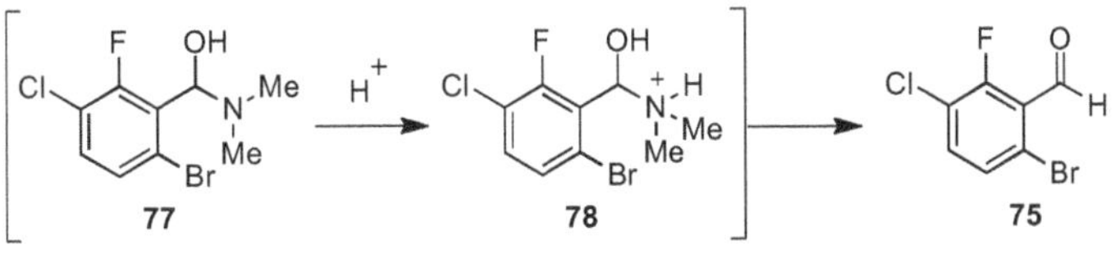

SCHEME 17.11

(I) Problems

When the MTBE solution of **75** (post water quenching) was left to stand at 20 °C overnight, 30% of **74** was reformed.

(II) Solutions

As illustrated in Scheme 17.10, the water quenching protocol may result in an intermediate **77** that may decompose to produce the desired product **75** along with dimethylamine. **77** would also decompose to reform **74** with losing DMF. In addition, the aldehyde product **75** would further react with dimethylamine to reform **77**.

Based on the hypothesis, quenching with strong acid would lead to protonation of the nitrogen in **77**, thus favoring the elimination of the dimethylamine fragment from the resulting **78** to generate the desired product **75** (Scheme 17.11). Therefore, an optimized reverse quenching protocol was developed, in which the reaction mixture was added into a biphasic mixture of 4 M hydrochloric acid and MTBE at 0 °C. As expected, using this protocol the reformation of **74** was suppressed, affording the complete conversion to the desired product **75**.

Procedure

Stage (a)
- Charge 10.7 kg (38.4 mol) of *n*-butyllithium (23% in hexane) to a mixture of diisopropylamine (4.0 kg, 40.1 mol) and THF (35 L) at 0–10in a 400-L vessel.
- Charge the resulting LDA solution to a solution of **74** (7.0 kg, 33.4 mol) in THF (35 L) at −60 °C over 1 h.
- Age at −60 °C for 1 h.
- Charge 40 kg (100.0 mol) of DMF at −50 to −55 °C.
- Age for 1 h.

Stage (b)

- Transfer the batch into a vigorously stirred mixture of water (45 kg), concentrated HCl (33 kg), and MTBE (70 L) at 0–5 °C.
- Warm the batch to 15 °C and stop the agitation.
- Separate the layers.
- Extract the aqueous layer with MTBE (35 L).
- Concentrate the combined organic layers under reduced pressure.
- Yield: 54 kg (7.5 kg of **75** by assay, 95% yield).

17.7.4   SUPPRESSING PRODUCT HYDROLYSIS

Isolation of a hydrolytically labile chloride product **80** from a deoxychlorination reaction mixture proved to be challenging (Equation 17.18).[31]

$$\text{(17.18)}$$

It was conceived that reverse quenching and isolation from a buffered aqueous media could mitigate the stability issues. Thus, upon completion of the reaction, the reaction mixture was transferred into an aqueous phosphate buffer solution with concomitant adjustment pH to 6.8–8.0 using 2 N aqueous NaOH while maintaining the temperature below 30 °C.

Procedure

- Charge 2.50 kg (13.9 mol) of **79** and 5.8 L of anhydrous MeCN (<300 ppm water) into a 100-L reactor.
- Set the jacket temperature to 20 °C.
- Charge 1.94 L (20.8 mol, 1.5 equiv) of $POCl_3$.
- Rinse with MeCN (1.7 L).
- Heat at 80 ± 5 °C for 2 h.
- Cool to 20 °C over 1.5 h.
- Prepare a quenching solution by dissolving $NaH_2PO_4 \cdot H_2O$ (8.0 kg) and Na_2HPO_4 (21.1 kg) in water (15.5 L).
- Transfer the reaction mixture into the quenching solution at ≤30 °C while maintaining the pH at 6.8–8.0 with 2 N NaOH (37.0 L).
- Rinse with MeCN (5.0 L).
- Add 2 N NaOH (2.5 L) to adjust pH to 7.5.
- Cool the resulting mixture to 20 °C.
- Filter and wash with water (2×25 L).
- Dry at 25 °C under vacuum with dry nitrogen sweep.
- Yield: 2.59 kg (78%).

REMARK

Alternatively, the excess $POCl_3$ was destroyed by anhydrous quenching with 2-propanol during deoxychlorination of 7-bromo-6-methoxy-2*H*-isoquinolin-1-one **81** (Equation 17.19):[32]

$$\text{(17.19)}$$

17.7.5 PREVENTION OF PRODUCT DECOMPOSITION

Dimethyltitanocene **84** is a useful methylenation reagent and can be used for methylenation of esters, ketones, amides, and other carbonyl compounds (Scheme 17.12).[33] Compound **84** could be prepared by treating titanocene dichloride **83** with the Grignard reagent. However, the workup under aqueous conditions proved problematic. The addition of the water to the reaction mixture led to the decomposition of **84**, resulting in substantial amounts of **83** and methyltitanocene chloride.

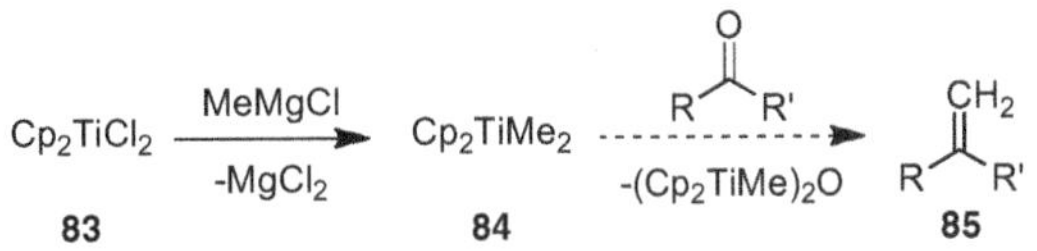

SCHEME 17.12

Eventually, the product decomposition was mitigated using a reverse quenching approach by adding the reaction mixture into an aqueous solution buffered with 6% ammonium chloride.

17.7.6 PREVENTION OF EMULSION

Emulsion is a frequently encountered problem during extractive workup, which may lead to loss of product and long processing times.

17.7.6.1 Copper-Catalyzed Amination

During a copper-catalyzed amination of aryl bromide **86** with morpholine, an extractive workup with acid or base washes led to emulsions (Equation 17.20).[34]

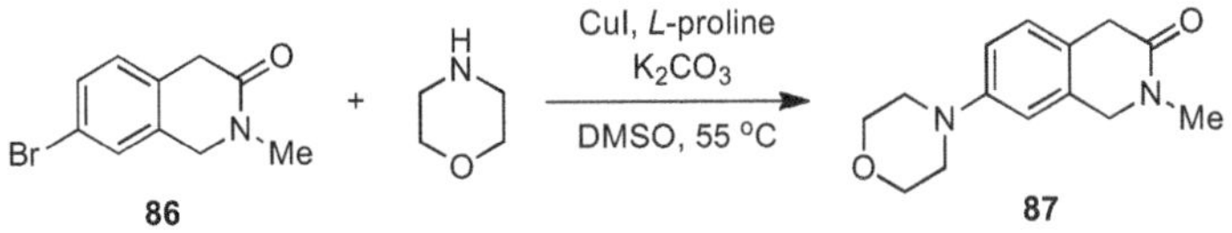

(17.20)

The reverse quenching by charging the reaction mixture into a brine solution efficiently avoided such emulsion during the subsequent methylene chloride extraction.

17.7.6.2 Lithium Aluminum Hydride Reduction

Aqueous quenching of aluminum hydride reduction often ends up with the formation of massive emulsions caused by the formation of aluminum hydroxide gels, resulting in not only difficulties in extraction/filtration, but also loss of product. Rochelle salt (potassium sodium tartrate) could be used to break up the emulsion.[27] Rochelle salt was proved to be the quenching of choice for the reduction of benzyllactam **88** with lithium aluminum hydride (LAH) in terms of ease of phase separation and control of heat evolution. Reverse quenching was employed by transferring the reaction mixture into Rochelle salt solution (Equation 17.21).[35]

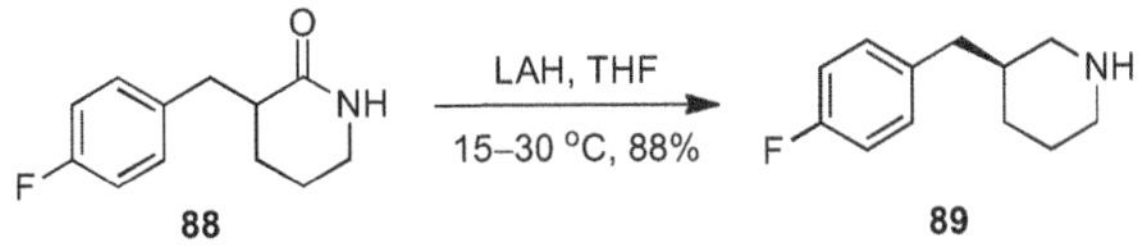

(17.21)

Procedure

- Charge 27.5 kg (133 mol) of **88** and THF (254 kg) into a 200-gallon glass-lined reactor.
- Cool to 15–20 °C.

- Charge a solution of LAH (50.3 kg, 133 mol, 10 wt% in THF) at <20 °C.
- Heat to 30 °C upon complete addition and stir for 2 h.
- Transfer slowly the reaction mixture to Rochelle's salt solution (112 kg, 397 mol) in water (358 L).
- Stir for 15 min and separate layers.
- Extract the aqueous layer with EtOAc (189 kg).
- Combine EtOAc layer with THF layer.
- Distill the combined layers to remove THF.
- Wash with brine.
- Yield: 22.5 kg (88%).

17.7.7 PREVENTION OF EXOTHERMIC RUNAWAY

Phosphorus oxychloride ($POCl_3$) is a powerful deoxychlorination reagent and is commonly used to convert hydroxyheteroaromatics into the corresponding chloroheteroaromatics or to generate the Vilsmeier-Haack reagent in situ. Albeit three chlorine atoms in $POCl_3$, most of these transformations require more than 1 equiv of $POCl_3$ in order to drive reactions to completion. As a consequence, the excess of $POCl_3$ has to be destroyed, in most cases, via hydrolysis during the workup. Generally, reverse quenching is the preferred approach by adding the postreaction mixture into ice-water or an aqueous base solution.

However, quenching the Vilsmeier reaction mixture into ice-water may result in incomplete hydrolysis of $POCl_3$ and a delayed exotherm. In order to address this safety concern, a reverse quenching protocol was developed for the preparation of aldehyde **91** via formylation of indole compound **90** with $POCl_3$ (2.5 equiv)/DMF (1.4 equiv) (Equation 17.22).[36] Thus, the reaction was quenched safely by the slow addition of the postreaction mixture into a sodium acetate aqueous solution at 35–40 °C. This reverse quenching hydrolyzed instantaneously the unreacted $POCl_3$ completely, preventing an accumulation of unstable intermediates and, in turn, the possibility of a runaway reaction from occurring. These conditions afforded **91** as a solid in high purity and quantitative yield.

(17.22)

REMARK
Biphasic workup protocol is not recommended due to the potential delay of the exothermic event.

17.8 CONCURRENT QUENCHING

The preparation of 2-chloroquinoline-3-carbaldehyde **93** was realized via a reaction of acetamide **92** with phosphorus oxychloride at 70 °C under Vilsmeier reaction conditions (6.0 equiv) (Equation 17.23).[37] During the workup, the resulting viscous postreaction mixture containing a large quantity of unreacted $POCl_3$ was quenched reversely into a mixture of 50:50 (v/v) acetonitrile/water (20 vol) at below 5 °C. The desired product **93** gradually crystallized out from the solution and was subsequently separated from the resulting highly acidic medium by filtration.

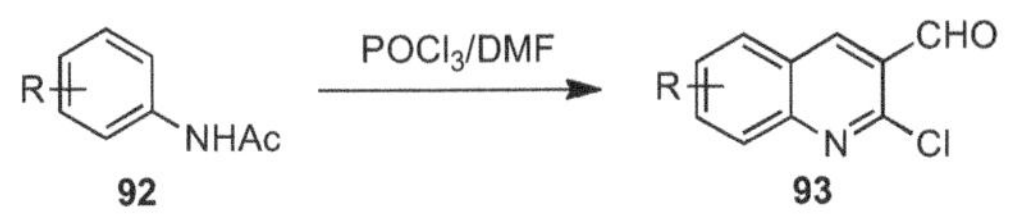

$$(17.23)$$

(I) Problems

It was noted, after isolation, that an exothermic reaction occurred in the aqueous filtrates. The exotherm was due to the accumulation of metastable intermediates, phosphorodichloridic acid, and phosphoro-chloridic acid, resulting from the incomplete hydrolysis of phosphorus oxychloride (Scheme 17.13).

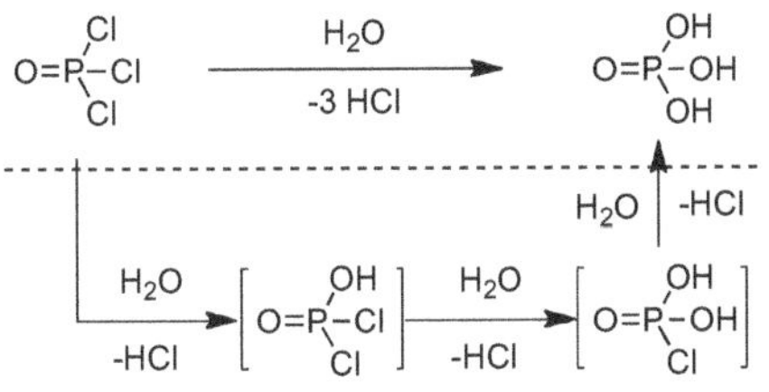

SCHEME 17.13

(II) Solutions

Accordingly, a concurrent quenching protocol was implemented by co-dosing the reaction solution with aqueous NaOH (1.5 equiv) solution in order to avoid the accumulation of P–Cl bond-containing intermediates.

Procedure

- Charge **92** (0.33 mol) into a 2-L reactor followed by adding 183 mL (1.96 mol, 6.0 equiv) of $POCl_3$ at 0 ± 5 °C.
- Stir until homogeneous.
- Charge 63.5 mL (2.5 equiv) of DMF at 0 ± 5 °C.
- Heat the resulting mixture to 70 ± 5 °C and hold at the temperature for >16 h.
- Cool to 20 ± 5 °C.
- Add MeCN (5 vol) and stir until homogeneous.
- Add MeCN (10 vol) into a 5-L reactor followed by adding water (10 vol) and stir at 20 ± 5 °C.
- Charge the contents of the 2-L reactor into the 5-L reactor simultaneously with 5 N NaOH (2.94 mol, 1.5 equiv relative to $POCl_3$) at 20 ± 5 °C with pH<7.
- Stir at 20 ± 5 °C for 1 h.
- Filter and wash with 50:50 v/v MeCN/water (5 vol) and water (5 vol).
- Analyze the filtrates by ^{31}P NMR for completion of hydrolysis of $POCl_3$ prior to disposal.

17.9   DOUBLE QUENCHING

17.9.1   Acetone/HCl Combination

17.9.1.1   Ketone Reduction

Residual $NaBH_4$ from ketone reduction reactions is typically destroyed by aqueous quenching using HCl or H_2SO_4. If the product is unstable under aqueous acidic conditions, an anhydrous quenching normally is an alternative approach by employing organic acids such as acetic acid and citric acid. However, hydrogen gas evolution and concurrent exotherm associated with the quenching must be addressed. In the event, a double-quenching approach has been developed by using ketone, such as acetone, as an alternative quenching agent,[38] to destroy excess $NaBH_4$ followed by adding 37% HCl. This protocol avoided off-gassing and minimized heat generation. For example, in the preparation of indanol **95** via the reduction of indanone **94** with $NaBH_4$, quenching the reaction mixture with acetone prevented the evolution of hydrogen gas (Equation 17.24).[39]

(17.24)

Procedure

Stage (a)

- Prepare a solution of indanone by dissolving 17.4 kg of **94** (98.2 mol, containing approximately 20% of 4-nitroisomer) in 35 L of THF.
- Charge 1.02 kg (27.0 mol) of $NaBH_4$ into a separate tank containing 155 L of EtOH (denatured with toluene and precooled to –15 °C).
- Charge the solution of **94** to the cold $NaBH_4$ slurry over 30 min.
- Stir for 1 h.

Stage (b)

- Charge 3.5 L of acetone.
- Add 4.0 kg of concentrated HCl to pH 3.
- Distill under reduced pressure to 35 L.
- Filter and rinse with MTBE (140 L).
- Wash the combined filtrates with H_2O (3×175 L).
- Treat the organic layer with 7 kg of Darco.
- Filter and rinse with MTBE (65 L).
- Distill azeotropically under reduced pressure to remove water.
- Add 5 kg of molecular sieves (4Å, powdered) and stir for 6 h.
- Filter and rinse with MTBE (35 L).
- Distill under reduced pressure to 50 L.
- Add heptane (100 L) over 6 h.
- Stir the slurry for 6 days.
- Filter and wash with a cold mixture of heptane (25 L)/MTBE (15 L).
- Dry at 45 °C in a vacuum for 16 h.
- Yield: 15.6 kg (84%).

17.9.1.2 S_NAr Reaction

An excess quantity of methylmagnesium chloride (2 equiv) was required for the nucleophilic aromatic substitution (S_NAr) of aryl fluoride **96** to achieve a complete conversion (Equation 17.25).[40] In an effort to develop a workup protocol that minimizes side reactions associated with the excess of Grignard reagent (MeMgCl), a double-quenching approach was utilized by adding acetone to destroy the excess of MeMgCl, followed by hydrochloride acid.

(17.25)

17.9.2 ACETONE/CITRIC ACID COMBINATION

An inappropriate quenching protocol may not be able to stop the reaction completely and the excess reducing agent left in the reaction mixture leads to the generation of over-reduction side products. For instance, quenching the ethyl ester reduction reaction with an aqueous solution of citric acid resulted in an over-reduction impurity **100** in 20% (Condition A, Equation 17.26).[41]

Thus, a double-quenching method was developed and the reaction was stopped by adding acetone followed by an aqueous solution of citric acid (Condition B, Equation 1.26). In the event, the reduction was accomplished with NaBH$_4$/MgCl$_2$, giving cleanly the alcohol product **99** in 76% yield and >98% purity with only 5% of the over-reduction impurity **100**.

17.9.3 ACETONE/MeOH/H$_2$O COMBINATION

The double-quenching strategy was also applied to the asymmetric reduction of ketimine **101** (Equation 17.27).[42] After the completion of the reaction, acetone was added to destroy the excess NaBH$_4$, followed by the addition of methanol and water.

17.9.4 ETHYL ACETATE/WATER COMBINATION

Lithium aluminum hydride (LAH) is highly reactive toward water, releasing hydrogen gas. Consequently, in order to avoid the evolution of hydrogen gas, generated during the workup, the double-quenching strategy was used: first treating the reaction mixture with ester (i.e., ethyl acetate) to destroy the excess of LAH, followed by adding aqueous quenching solution. For example, the reduction of ester **103** with LAH to the corresponding alcohol **104** was quenched by adding ethyl acetate, followed by water (Equation 17.28).[43]

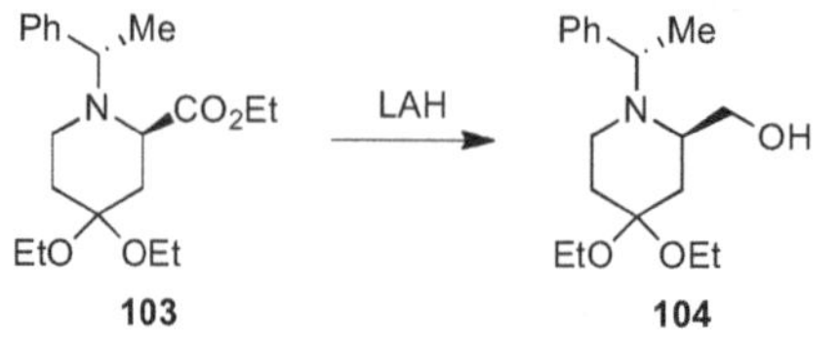

(17.28)

17.9.5 Ethyl Acetate/Tartaric Acid Combination

The double-quenching approach could also be used in the reduction of Weinreb amide **105** with diisobutyl aluminum hydride (DIBAL-H) (Equation 17.29).[2] Upon the completion of the reaction, the excess reducing agent was destroyed by adding ethyl acetate, followed by reverse quenching into a 20% tartaric acid aqueous solution.

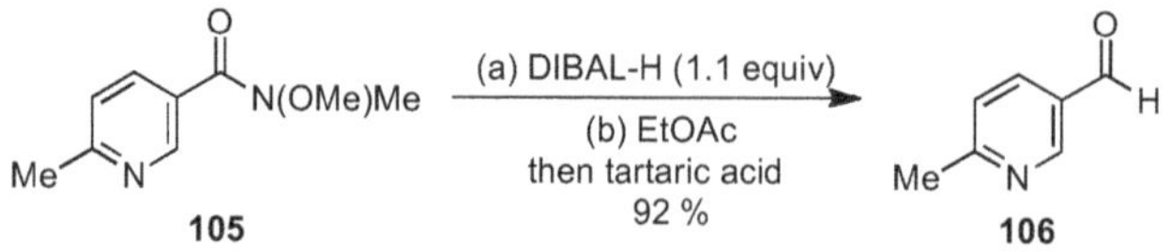

(17.29)

17.9.6 Ethyl Acetate/Aqueous Sodium Bicarbonate Combination

Reduction of optically pure lactam **107** en route to bicyclic amine **109**, a core intermediate for the synthesis of an anti-tumor compound, was achieved with DIBAL-H (Scheme 17.14).[44] After the reduction was complete, the postreaction mixture was quenched with ethyl acetate to consume the unreacted DIBAL-H, followed by a second quenching with saturated aqueous $NaHCO_3$ solution to precipitate the aluminum salts as granular solids which were easily separated by filtration.

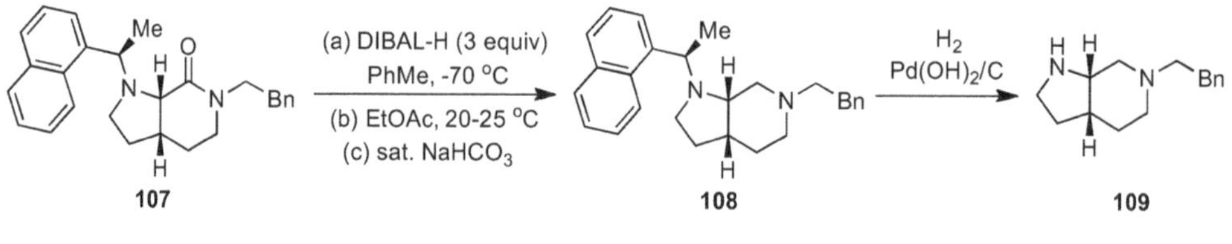

SCHEME 17.14

17.9.7 Isopropanol/Citric Acid Combination

The reduction of ethyl ester **110** with DIBAL-H in toluene/methylcyclohexane at −90 °C afforded the desired aminoaldehyde **111** (Equation 17.30).[35] Quenching the reaction employed a double-quenching approach by adding IPA, followed by treating the resulting mixture with a citric acid aqueous solution. This double-quenching protocol allowed the effective prevention of potential emulsion during the subsequent extraction.

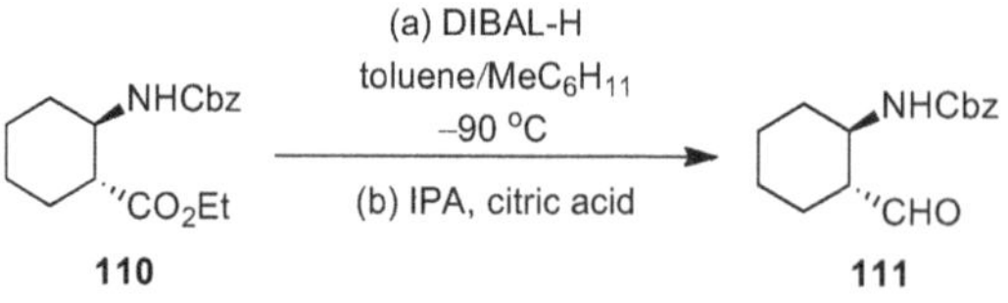

(17.30)

Procedure

Stage (a)
- Charge 17.5 kg (57.3 mol) of **110** into a 600-L reactor, followed by adding toluene (167 kg) and methylcyclohexane (27.0 kg).
- Cool the content to −90 to −96 °C.
- Charge 90.1 kg (127 mol) of 20 wt% solution of DIBAL-H at <−90 °C and stir for 30 min.

Stage (b)
- Add a mixture of IPA (13.8 kg) and heptane (23.0 kg).
- Transfer the resulting mixture into a 1000-L reactor containing a solution of citric acid (54.3 kg, 258 mol) in water (112 L) while maintaining the temperature at <30 °C.
- Separate layers.
- Wash the organic layer with 5% aqueous $NaHCO_3$ (147 kg) and brine (150 kg).
- Add IPA (27.5 kg) to the organic layer, followed by adding $NaHSO_3$ (11.9 kg, 62.6 mol) in water (24.5 L).
- Filter to give **111** as a solid bisulfite adduct.

17.9.8 METHYL FORMATE/AQUEOUS HCL COMBINATION

A double-quenching approach was used during the workup of the reaction of 2,6-dichloroisonicotinic acid **112** with ethylmagnesium chloride (3 equiv) in THF. The reaction was quenched by adding methyl formate to destroy the excess Grignard reagent, followed by adding aqueous HCl (Equation 17.31).[38] Using this protocol essentially eliminated the formation of tertiary alcohol side products.

$$(17.31)$$

17.10 REACTIVE QUENCHING

Glecaprevir was identified as a potent hepatitis C virus (HCV) NS3/4A protease inhibitor, active against all major HCV genotypes. The synthesis of glecaprevir involved macrocycle intermdiate **116** that was obtained from an intramolecular cyclization of an allylic bromide **115** (Scheme 17.15).[45]

As allyl bromide **115** is potentially mutagenic, therefore, a reactive quenching protocol was developed to convert unreacted **115** into non-mutagenic **117** by replacing the allyl bromide with 2-mercaptoethylamine (Equation 17.32).

$$(17.32)$$

This reactive quenching with 2-mercaptoethylamine hydrochloride salt (2 equiv relative to remaining allylic bromide) is very fast with a half-life of less than 1 min; a 30 min stir time ensures complete destroy all of the remaining allylic bromide.

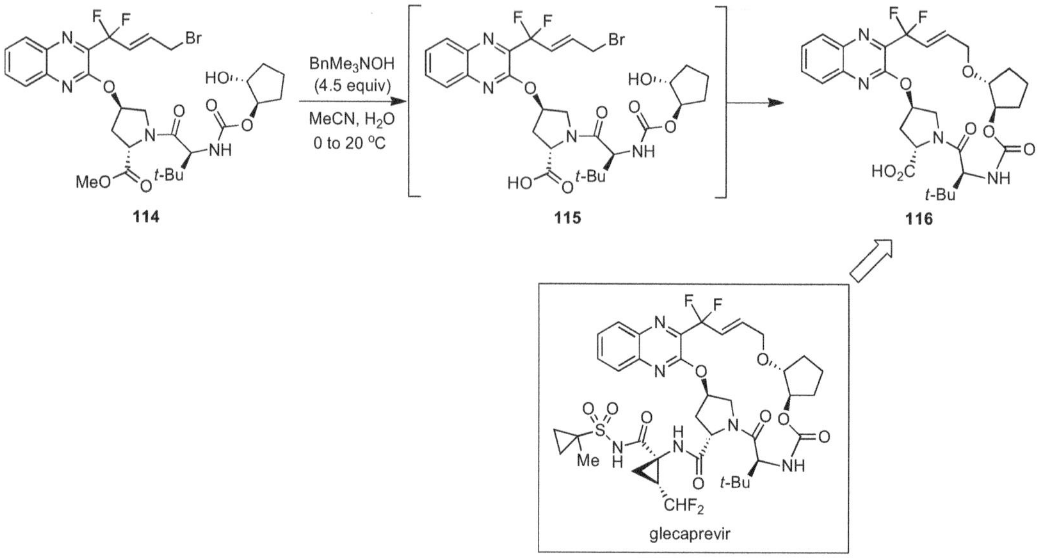

SCHEME 17.15

NOTES

1. Bradley, P. A.; Carroll, R. J.; Lecouturier, Y. C.; Moore, R.; Noeureuil, P.; Patel, B.; Snow, J.; Wheeler, S. *Org. Process Res. Dev.* **2010**, *14*, 1326.
2. Davies, I. W.; Marcoux, J.-F.; Corley, E. G.; Journet, M.; Cai, D.-W.; Palucki, M.; Wu, J.; Larsen, R. D.; Rossen, K.; Pye, P. J.; DiMichele, L.; Dormer, P.; Reider, P. J. *J. Org. Chem.* **2000**, *65*, 8415.
3. Magnus, N. A.; Anzeveno, P. B.; Coffey, D. S.; Hay, D. A.; Laurila, M. E.; Schkeryantz, J. M.; Shaw, B. W.; Staszak, M. A. *Org. Process Res. Dev.* **2007**, *11*, 560.
4. Yoshida, S.; Ohigashi, A.; Morinaga, Y.; Hashimoto, N.; Takahashi, T.; Ieda, S.; Okada, M. *Org. Process Res. Dev.* **2013**, *17*, 1252.
5. Walker, M. D.; Albinson, F. D.; Clark, H. F.; Clark, S.; Henley, N. P.; Horan, R. A. J.; Jones, C. W.; Wade, C. E.; Ward, R. A. *Org. Process Res. Dev.* **2014**, *18*, 82.
6. Chen, Q.-Y. *J. Fluorine Chem.* **1995**, *72*, 241.
7. Maddess, M. L.; Scott, J. P.; Alorati, A.; Baxter, C.; Bremeyer, N.; Brewer, S.; Campos, K.; Cleator, E.; Dieguez-Vazquez, A.; Gibb, A.; Gibson, A.; Howard, M.; Keen, S.; Klapars, A.; Lee, J.; Li, J.; Lynch, J.; Mullens, P.; Wallace, D.; Wilson, R. *Org. Process Res. Dev.* **2014**, *18*, 528.
8. For reviews of enantioselective Simmons-Smith reactions, see: (a) Singh, V. K.; DattaGupta, A.; Sekar, G. *Synthesis* **1997**, *2*, 137. (b) Lebel, H.; Marcoux, J.-F.; Molinaro, C.; Charette, A. B. *Chem. Rev.* **2003**, *103*, 977.
9. (a) Charette, A. B.; Molinaro, C.; Brochu, C. *J. Am. Chem. Soc.* **2001**, *123*, 12168. (b) Charette, A. B.; Juteau, H. *J. Am. Chem. Soc.* **1994**, *116*, 2651. (c) Charette, A. B.; Prescott, S.; Brochu, C. *J. Org. Chem.* **1995**, *60*, 1081. (d) Charette, A. B.; Juteau, H.; Lebel, H.; Molinaro, C. *J. Am. Chem. Soc.* **1998**, *120*, 11943.
10. Anthes, R.; Benoit, S.; Chen, C.-K.; Corbett, E. A.; Corbett, R. M.; DelMonte, A. J.; Gingras, S.; Livingston, R. C.; Pendri, Y.; Sansker, J.; Soumeillant, M. *Org. Process Res. Dev.* **2008**, *12*, 178.
11. Busacca, C. A.; Cerreta, M.; Dong, Y.; Eriksson, M. C.; Farina, V.; Feng, X.; Kim, J.-Y.; Lorenz, J. C.; Sarvestani, M.; Simpson, R.; Varsolona, R.; Vitous, J.; Campbell, S. J.; Davis, M. S.; Jones, P.-J.; Norwood, D.; Qiu, F.; Beaulieu, P. L.; Duceppe, J.-S.; Haché, B.; Brong, J.; Chiu, F.-T.; Curtis, T.; Kelley, J.; Lo, Y. S.; Powner, T. H. *Org. Process Res. Dev.* **2008**, *12*, 603.
12. de Koning, P. D.; Murtagh, L.; Lawson, J. P.; Embse, R. A. V.; Kunda, S. A.; Kong, W. *Org. Process Res. Dev.* **2011**, *15*, 1046.
13. Alcaraz, M.-L.; Atkinson, S.; Cornwall, P.; Foster, A. C.; Gill, D. M.; Humphries, L. A.; Keegan, P. S.; Kemp, R.; Merifield, E.; Nixon, R. A.; Noble, A. J.; O'Beirne, D.; Patel, Z. M.; Perkins, J.; Rowan, P.; Sadler, P.; Singleton, J. T.; Tornos, J.; Watts, A. J.; Woodland, I. A. *Org. Process Res. Dev.* **2005**, *9*, 555.

14. Shu, L.; Wang, P.; Gu, C.; Garofalo, L.; Alabanza, L. M.; Dong, Y. *Org. Process Res. Dev.* **2012**, *16*, 1870.
15. Achmatowicz, M. M.; Allen, J. G.; Bio, M. M.; Bartberger, M. D.; Borths, C. J.; Colyer, J. T.; Crockett, R. D.; Hwang, T.-L.; Koek, J. N.; Osgood, S. A.; Roberts, S. W.; Swietlow, A.; Thiel, O. R.; Caille, S. *J. Org. Chem.* **2016**, *81*, 4736.
16. Zheng, B.; Sugiyama, M.; Eastgate, M. D.; Fritz, A.; Murugesan, S.; Conlon, D. A. *Org. Process Res. Dev.* **2012**, *16*, 1827.
17. Durrwachter, J. R. *Org. Process Res. Dev.* **2017**, *21*, 1423.
18. Conlon, D. A.; Natalie, Jr., K. J.; Cuniere, N.; Razler, T. M.; Zhu, J.; de Mas, N.; Tymonko, S.; Fraunhoffer, K. J.; Sortore, E.; Rosso, V. W.; Xu, Z.; Adams, M. L.; Patel, A.; Huang, J.; Gong, H.; Weinstein, D. S.; Quiroz, F.; Chen, D. C. *Org. Process Res. Dev.* **2016**, *20*, 921.
19. Henegar, K. E.; Cebula, M. *Org. Process Res. Dev.* **2007**, *11*, 354.
20. Hoffmann-Emery, F.; Niedermann, K.; Rege, P. D.; Konrath, M.; Lautz, C.; Kraft, A. K.; Steiner, C.; Bliss, F.; Hell, A.; Fischer, R.; Carrera, D. E.; Beaudry, D.; Angelaud, R.; Malhotra, S.; Gosselin, F. *Org. Process Res. Dev.* **2022**, *26*, 313.
21. Yamagami, T.; Moriyama, N.; Kyuhara, M.; Moroda, A.; Uemura, T.; Matsumae, H.; Moritani, Y.; Inoue, I. *Org. Process Res. Dev.* **2014**, *18*, 437.
22. Breining, S. R.; Genus, J. F.; Mitchener, J. P.; Cuthbertson, T. J.; Heemstra, R.; Melvin, M. S.; Dull, G. M.; Yohannes, D. *Org. Process Res. Dev.* **2013**, *17*, 413.
23. Chekal, B. P.; Guinness, S. M.; Lillie, B. M.; McLaughlin, R. W.; Palmer, C. W.; Post, R. J.; Sieser, J. E.; Singer, R. A.; Sluggett, G. W.; Vaidyanathan, R.; Withbroe, G. J. *Org. Process Res. Dev.* **2014**, *18*, 266.
24. Belecki, K.; Berlinger, M.; Bibart, R. T.; Meltz, C.; Ng, K.; Phillips, J.; Ripin, D. H. B.; Vetelino, M. *Org. Process Res. Dev.* **2007**, *11*, 754.
25. Ripin, D. H. B.; Weisenburger, G. A.; am Ende, D. J.; Bill, D. R.; Clifford, P. J.; Meltz, C. N.; Phillips, J. E. *Org. Process Res. Dev.* **2007**, *11*, 762.
26. Pasti, J.; Chen, C.-K.; Spangler, L.; DelMonte, A. J.; Benoit, S.; Berglund, D.; Bien, J.; Brodfuehrer, P.; Chan, Y.; Corbett, E.; Costello, C.; DeMena, P.; Discordia, R. P.; Doubleday, W.; Gao, Z.; Gingras, S.; Grosso, J.; Haas, O.; Kacsur, D.; Lai, C.; Leung, S.; Miller, M.; Muslehiddinoglu, J.; Nguyen, N.; Qiu, J.; Olzog, M.; Reiff, E.; Thoraval, D.; Totleben, M.; Vanyo, D.; Vemishetti, P.; Wasylak, J.; Wei, C.; *Org. Process Res. Dev.* **2009**, *13*, 716.
27. Journet, M.; Cai, D.; Hughes, D. L.; Kowal, J. J.; Larsen, R. D.; Reider, P. J. *Org. Process Res. Dev.* **2005**, *9*, 490.
28. Herrinton, P. M.; Owen, C. E.; Gage, J. R. *Org. Process Res. Dev.* **2001**, *5*, 80.
29. Clark, J. D.; Weisenburger, G. A.; Anderson, D. K.; Colson, P.-J.; Edney, A. D.; Gallagher, D. J.; Kleine, H. P.; Knable, C. M.; Lantz, M. K.; Moore, C. M. V.; Murphy, J. B.; Rogers, T. E.; Ruminski, P. G.; Shah, A. S.; Storer, N.; Wise, B. E. *Org. Process Res. Dev.* **2004**, *8*, 51.
30. Goodyear, A.; Xin, L.; Bishop, B.; Chen, C.; Cleator, E.; McLaughlin, M.; Sheen, F. J.; Stewart, G. W.; Xu, Y.; Yin, J. *Org. Process Res. Dev.* **2012**, *16*, 605.
31. Thiel, O. R.; Achmatowicz, M.; Bernard, C.; Wheeler, P.; Savarin, C.; Correll, T. L.; Kasparian, A.; Allgeier, A.; Bartberger, M. D.; Tan, H.; Larsen, R. D. *Org. Process Res. Dev.* **2009**, *13*, 230.
32. Song, Z. J.; Tellers, D. M.; Dormer, P. G.; Zewge, D.; Janey, J. M.; Nolting, A.; Steinhuebel, D.; Oliver, S.; Devine, P. N.; Tschaen, D. M. *Org. Process Res. Dev.* **2014**, *18*, 423.
33. (a) Petasis, N. A.; Lu, S.-P. *Tetrahedron Lett.* **1995**, *36*, 2393. (b) Hughes, D. L.; Payack, J. F.; Cai, D.; Verhoeven, T. R.; Reider, P. J. *Organometallics* **1996**, *15*, 663. (c) Payack, J. F.; Huffman, M. A.; Cai, D.; Hughes, D. L.; Collins, P. C.; Johnson, B. K.; Cottrell, I. F.; Tuma, L. D. *Org. Process Res. Dev.* **2004**, *8*, 256.
34. Yang, Q.; Ulysse, L. G.; McLaws, M. D.; Keefe, D. K.; Haney, B. P.; Zha, C.; Guzzo, P. R.; Liu, S. *Org. Process Res. Dev.* **2012**, *16*, 499.
35. Yue, T.-Y.; Mcleod, D. D.; Albertson, K. B.; Beck, S. R.; Deerberg, J.; Fortunak, J. M.; Nugent, W. A.; Radesca, L. A.; Tang, L.; Xiang, C. D. *Org. Process Res. Dev.* **2006**, *10*, 262.
36. Li, X.; Wells, K. M.; Branum, S.; Damon, S.; Youells, S.; Beauchamp, D. A.; Palmer, D.; Stefanick, S.; Russell, R. K.; Murray, W. *Org. Process Res. Dev.* **2012**, *16*, 1727.
37. Achmatowicz, M. M.; Thiel, O. R.; Colyer, J. T.; Hu, J.; Elipe, M. V. S.; Tomaskevitch, J.; Tedrow, J. S.; Larsen, R. D. *Org. Process Res. Dev.* **2010**, *14*, 1490.
38. Henegar, K. E.; Ashford, S. W.; Baughman, T. A.; Sih, J. C.; Gu, R.-L. *J. Org. Chem.* **1997**, *62*, 6588.
39. Hansen, M. M.; Borders, S. S. K.; Clayton, M. T.; Heath, P. C.; Kolis, S. P.; Larsen, S. D.; Linder, R. J.; Reutzel-Edens, S. M.; Smith, J. C.; Tameze, S. L.; Ward, J. A.; Weigel, L. O. *Org. Process Res. Dev.* **2009**, *13*, 198.

40. Daniewski, A. R.; Liu, W.; Püntener, K.; Scalone, M. *Org. Process Res. Dev.* **2002**, *6*, 220.

41. Hanselmann, R.; Job, G. E.; Johnson, G.; Lou, R.; Martynow, J. G.; Reeve, M. M. *Org. Process Res. Dev.* **2010**, *14*, 152.

42. Butcher, K. J.; Denton, S. M.; Field, S. E.; Gillmore, A. T.; Harbottle, G. W.; Howard, R. M. H.; Laity, D. A.; Ngono, C. J.; Pibworth, B. A. *Org. Process Res. Dev.* **2011**, *15*, 1192.

43. Lau, J. F.; Hansen, T. K.; Kilburn, J. P.; Frydenvang, K.; Holsworth, D. D.; Ge, Y.; Uyeda, R. T.; Judge, L. M.; Andersen, H. S. *Tetrahedron* **2002**, *58*, 7339.

44. Shieh, W.-C.; Chen, G.-P.; Xue, S.; McKenna, J.; Jiang, X.; Prasad, K.; Repič, O.; Straub, C.; Sharma, S. K. *Org. Process Res. Dev.* **2007**, *11*, 711.

45. Kallemeyn, J. M.; Engstrom, K. M.; Pelc, M. J.; Lukin, K. A.; Morrill, W. H.; Wei, H.; Towne, T. B.; Henle, J.; Nere, N. K.; Welch, D. S.; Shekhar, S.; Ravn, M. M.; Zhao, G.; Fickes, M. G.; Ding, C.; Vinci, J. C.; Marren, J.; Cink, R. D. *Org. Process Res. Dev.* **2020**, *24*, 1373.

18 Various Isolation and Purification Strategies

Isolation (and/or purification) of liquid (or low-melting solid) chemical products may involve extraction and/or distillation; isolation (and/or purification) of solid products would require filtration and crystallization. In some cases, column chromatography is needed especially for certain high-boiling liquids and low-melting or non-crystalline solids.

18.1 EXTRACTION

Extraction is a unit operation that is frequently used to separate (or recover) the product from an organic or aqueous mixture. The extraction operation is usually followed by washing to remove traces of by-products/impurities from the extractant. One of the problems encountered during the extraction operation is the emulsion which leads to poor layer separation, loss of product, and long processing time.

18.1.1 Aqueous Extractions

Aqueous extractions are frequently employed to separate products from aqueous mixtures with organic solvents. This operation requires that the organic solvents are immiscible with water; thus, tetrahydrofuran and alcoholic solvents, such as methanol, ethanol, and isopropanol, cannot be utilized as the extraction solvents. However, water-immiscible alcohols with carbon chains greater than 3, such as *n*-butanol, are sometimes chosen as the extraction solvents. The ideal extraction solvents (Class 3) should have good solubility for organic compounds with relatively low boiling points for easy removal/or recycling and low toxicity.

18.1.1.1 Use of Methyl *tert*-Butyl Ether

During extractive workup, layer separation can be the bottleneck that determines the processing cycle times. Reaction (18.1)[1] displays the preparation of disulfide **3** via the coupling reaction of acid chloride **1** with *L*-isoleucine **2**. The workup of the coupling reaction mixture involves multiple extractions with methyl *tert*-butyl ether (MTBE), followed by washing with water to remove excess *L*-isoleucine.

$$(18.1)$$

(I) Problems

It was noted that the layer separation between MTBE and aqueous phases became longer with each successive extraction, requiring several hours for the fourth extraction. This problem would have greatly increased cycle times on a pilot scale.

DOI: 10.1201/9781003288411-18

(II) Solutions

These layer separation problems were resolved by adding dilute hydrochloric acid. When the aqueous phase was kept acidic, the two layers separated almost instantaneously. The crude product **3** was precipitated by the addition of hexane.

Procedure

- Charge 46.2 kg (135 mol) of **1** followed by adding 38.8 kg (296 mol, 2.2 equiv) of **2**, 34.2 kg (407 mol, 3.0 equiv) of $NaHCO_3$, and THF (462 L).
- Heat the batch to 60–65 °C and stir at that temperature for 2 h.
- Add the resulting slurry slowly to a rapidly stirred mixture of HCl (37%, 37.6 kg, 381 mol), water (353 L), and MTBE (496 L).
- Rinse the reactor with THF (94 L), MTBE (95 L), and water (94 L).
- Stir the combined quenching mixture for 15 min.
- Separate layers.
- Wash the upper organic layer with water (235 L) followed by washing with dilute HCl (0.06%, 235 L, then 118 L).
- Add hexane (663 L) and stir for 3 h.
- Filter and wash with hexane (120 L).
- Dry at 60–65 °C under vacuum.
- Yield: 59.7 kg of crude **3**.

REMARK

This reverse quenching avoided excessive foaming.

18.1.1.2   Use of 2-Methyltetrahydrofuran

Analogous to most inorganic salts, small-molecule organic salts, such as triethylammonium sulfate **5**, may have a high-water solubility. In the context, the normal aqueous workup followed by extraction with 2-methyltetrahydrofuran (2-MeTHF) will result in low product recovery due to the loss of the product in the aqueous phase. Thus, a salt exchange protocol was developed by converting **5** into a more lipophilic tetrabutylammonium sulfate **6** that was readily extracted into the organic phase, while the other salts were rejected into the aqueous phase (Scheme 18.1).[2]

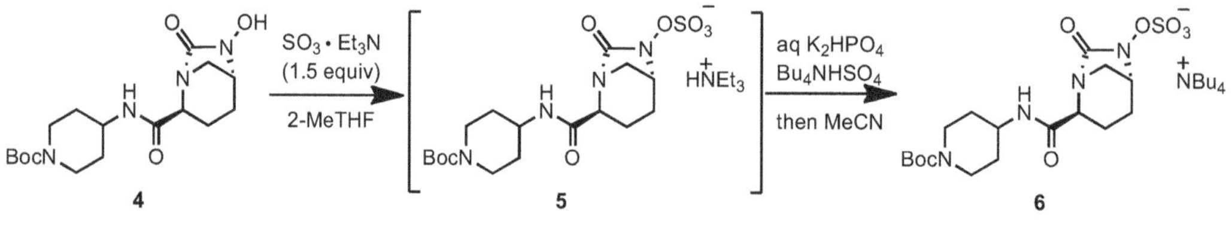

SCHEME 18.1

Procedure

- Add 400 g (1.09 mol) of hydroxy urea **4** to a 10-L glass cylindrical vessel followed by adding 311 g (94.8 wt%, 1.63 mol, 1.5 equiv) of $SO_3 \cdot Et_3N$, and 2-MeTHF (4 L).
- Heat the batch to 38–40 °C and age for 17 h.
- Cool the batch to 20 °C.
- Add a solution of K_2HPO_4 (405 g, 1.95 mol) in water (5.2 L) over 2 h and age for 20 min.
- Add 405 g (1.19 mol) of Bu_4NHSO_4 and stir the mixture for 1 h.
- Separate layers.
- Extract the aqueous layer with 2-MeTHF (2.4 L).

- Wash the combined organic layers with 20% aqueous NaCl (2.0 L).
- Distill the organic layer to 2.2 L under reduced pressure at <40 °C.
- Switch the solvent to MeCN and adjust the batch volume to 4 L.
- Filter through 80 g of activated carbon, and wash with MeCN.
- Distill azeotropically to get a solution of **6** in MeCN (3.6 L, 471 g of **6**, 97% assay yield).

18.1.1.3   Use of Ethyl Acetate

In order to enhance the separation power, a combination of the extractive workup with semicarbazide was demonstrated to be a unique tool to remove unreacted ketone starting material from
the reaction mixture. The conversion of (*R*)-(+)-pulegone **7** to **11** was accomplished via a two-step
process involving bromination and a subsequent Favorskii rearrangement (Scheme 18.2).[3] The crude
product **11** thus obtained contained several impurities including the unreacted ketone **7**. To remove
7 from the oily product **11** without resorting to chromatography would be a great challenge.

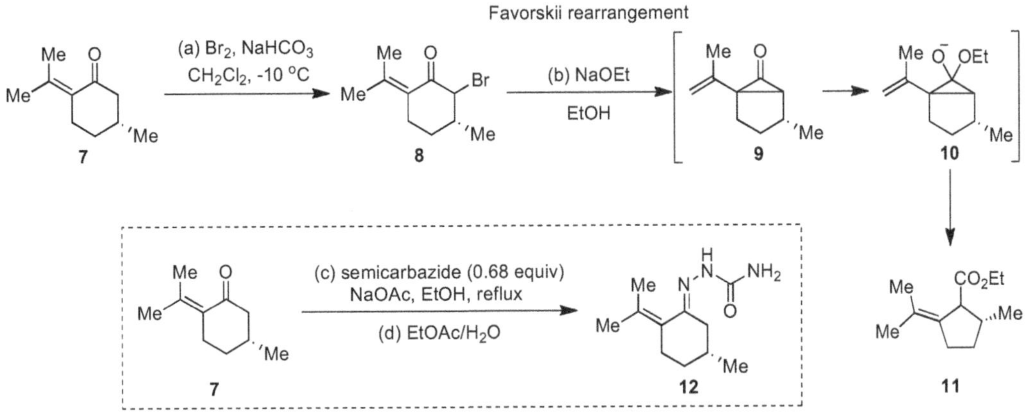

SCHEME 18.2

A team of Array Biopharma and Genentech developed a purification approach by converting the
ketone **7** into a water-soluble semicarbazone **12**, followed by ethyl acetate extraction. The cyclopentyl ester **11** could then be selectively extracted into EtOAc and isolated as oil, which was of acceptable quality for further processing.

18.1.1.4   Use of Dodecane

Enone **15** was prepared from the reaction of amide **13** with vinyl Grignard **14** (Equation 18.2).[4]
Isolation of **15** from the reaction mixture was challenging due to its low boiling point.

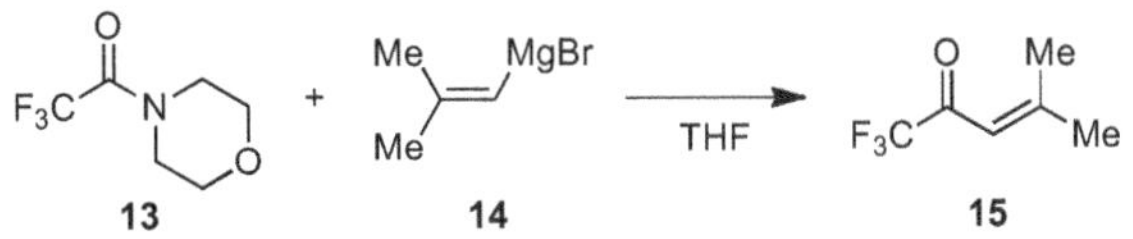

(18.2)

To address this isolation issue, an extractive workup was developed using dodecane as the extractive
solvent. After quenching with concentrated HCl, extracting with dodecane, and washing with water
to remove THF, the enone **15** was collected by distillation (bp 108 °C at 1 atm) from the high boiling
dodecane (bp 217 °C at 1 atm) solution.

Procedure

- Charge 34.75 kg (18.25 mol, 1.1 equiv) of the solution of **14** (0.5 M in THF) into a reactor.
- Cool the content to −10 °C.

- Charge 3.038 kg (16.59 mol) of **13** at ≤5 °C.
- Warm the batch to ca. 15 °C and hold at this temperature for 1 h.
- Cool the batch to ca. −10 °C.
- Add 4.65 L of conc. HCl at ≤20 °C.
- Add water (13.7 L) and dodecane (7.6 L).
- Stir for 10 min.
- Separate layers.
- Wash the organic layer with a mixture of water (15.8 L) and MeOH (3.6 L), then with water (15.2 L).
- Distill using a thin film evaporator.
- Yield: 1.64 kg (65%) as a light orange liquid.

REMARK

Isolation of volatile products by extraction and distillation usually suffers low product yields due to the loss of products during the distillation. For instance, the isolation of 5-chloro-3-fluoro-2-(trifluoromethyl)pyridine **17** through extraction followed by careful distillation to remove the extraction solvent gave low product yields (<20%) (Equation 18.3).[5]

$$(18.3)$$

In order to address the product loss issue, a steam distillation was implemented during the workup and the product **17** was isolated readily in 98% average yield and used directly in the subsequent Suzuki cross-coupling reaction. Another advantage of using steam distillation was that the spent copper catalyst was left in the distillation pot and no copper contamination in the isolated product was observed.

18.1.1.5   Use of *n*-Butanol

n-Butanol was employed to extract amine products from an aqueous reaction mixture. Benzimidazole **19** was prepared from condensation of 3,6-dimethoxybenzene-1,2-diamine with *N*-methylpyrrolidone (NMP) (Equation 18.4).[6] The product **19** was extracted into *n*-butanol organic layer from the aqueous reaction mixture.

$$(18.4)$$

REMARK

It was found that the separation of the black *n*-butanol–aqueous layers was extremely difficult. Ultimately, using a conductivity meter facilitated the challenging layer separation.

18.1.2   Anhydrous Extraction

Selective anhydrous extraction is a unit operation in which the organic reaction mixture is extracted with an organic solvent. Apparently, the extractive solvent has to be immiscible in the organic reaction mixture. This approach has been applied to remove undesired products or unreacted reagents.

18.1.2.1 Heptane/Acetonitrile System

As acetonitrile is not miscible with heptane, less polar side product **24** was effectively removed from the product acetonitrile solution by selective extraction with heptane (Equation 18.5).[7]

$$(18.5)$$

Procedure

- Dissolve the crude **23** (40.4 g) in MeCN (330 mL).
- Wash the MeCN solution with heptane (66 mL, 5×33 mL).
- The resulting MeCN solution of **23** (29.1 g, 88.3%) was used directly in the subsequent step.

18.1.2.2 Heptane–Cyclohexane/N-Methyl-2-pyrrolidone System

The anhydrous extractive workup is also used to remove excess less polar reagents from the polar reaction mixture. Due to the fact that NMP is immiscible with heptane and cyclohexane, extraction of the S_NAr reaction mixture in NMP with heptane/cyclohexane allowed to remove the excess of *tert*-butyl nicotinic ester **26** readily (Equation 18.6).[8]

$$(18.6)$$

18.1.3 DOUBLE EXTRACTION

A double extraction strategy was developed to isolate the reaction product from the reaction mixture. Due to the high water solubility of aldehyde **30** and the presence of DMSO in the reaction mixture, extraction of the product **30** out of the reaction media was not feasible. Therefore, in order to isolate **30**, a double extraction approach was utilized (Scheme 18.3).[9]

First, extraction with aliquat 336 in toluene allowed to transport of sodium salt **29** from the reaction mixture into toluene, which separated **29** from DMSO reaction mixture. Subsequently, extraction of the resulting toluene solution with aqueous acetic acid allowed to release **30** into the aqueous solution. Finally, the removal of water and acetic acid by solvent exchange with 2-butanol gave the triazole aldehyde **30** in 67% overall yield.

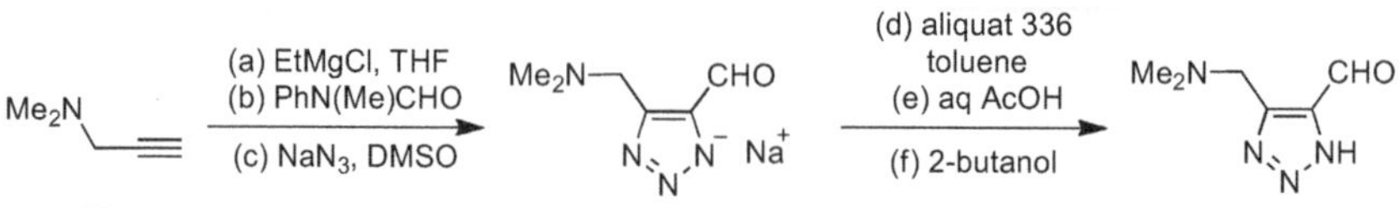

SCHEME 18.3

Procedure

Stage (d)
- Add aliquat 336 (807 g, 2.0 mol) and toluene (2.0 L) to the DMSO solution of **29** (1.0 mole scale, pH≈9.7).
- Stir for 0.5 h, then add water (1.1 L).
- Separate layers.
- Extract the aqueous layer twice with aliquat 336 (200 g, 0.5 mol) in toluene (2.0 L).
- Wash the combined organic layers with water (820 mL) to remove DMSO.

Stage (e)
- Wash the organic layer with 12.5% aqueous AcOH solution (930 mL, 2.0 mol) to release **30** into the aqueous layer.
- Wash the organic layer with water (365 mL).

Stage (f)
- Azeotropic distillation of the combined aqueous solution with 2-butanol until water <0.5 vol%.
- Adjust the volume to 750 mL and cool.
- Filter, wash with 2-butanol, and dry at 40 °C under vacuum.
- Yield: 102.5 g (67%).

18.2 DIRECT ISOLATION

Anderson defines direct isolation as a process that is designed to crystallize or precipitate the product directly from the reaction mixture without resorting to an extractive workup.[10] To achieve this, four approaches are frequently used by process chemists: cooling, adding anti-solvent, a combination of the cooling/anti-solvent, and salt formation. The selection of reaction solvent is key in the application of direct isolation. Compared to extractive isolation, direct isolation provides a high product yield with less waste and avoids the formation of impurities on account of limited contact between the product and the reaction mixture.

18.2.1 USE OF COOLING

Direct isolation via cooling reaction mixture offers an easy and convenient way to separate the product from the reaction mixture. The solubility of a given solute in a solvent typically depends on temperature. For many solid organic compounds, solubility increases as the temperature goes up and vice versa. When using this approach, the reaction is usually performed under reflux, and the product crystallizes upon cooling. Ethyl acetate, methanol, ethanol, isopropanol, tetrahydrofuran, acetonitrile, dichloromethane, heptane, and toluene are frequently used process solvents.

18.2.1.1 Direct Isolation from Isopropanol

Direct isolation of dihydropyridazinone **32** was accomplished by cooling the reaction mixture in isopropanol from 80 °C to <5 °C, followed by filtration (Equation 18.7).[11]

$$(18.7)$$

Procedure

- Mix 375 g (0.864 mol, 1.0 equiv) of **31** with 64.9 g (1.3 mol, 1.5 equiv) of hydrazine mono-hydrate in IPA (3.2 L) in a 6-L, four-necked flask.
- Heat at 80 °C with stirring for 18 h.
- Cool the batch to <5 °C and age for 1.5 h.
- Filter and wash with IPA (400 mL).
- Dry at 60 °C under vacuum.
- Yield: 314 g (84%).

18.2.1.2 Direct Isolation from Ethyl acetate

The preparation of amide **35** was accomplished via the reaction of acid **33** with amine **34** in the presence of *N,N*-carbonyldiimidazole (CDI) in ethyl acetate under reflux (Equation 18.8).[12]

$$(18.8)$$

The product **35** was conveniently isolated in 91% yield by filtration upon cooling the reaction mixture, rejecting the imidazole by-product in the mother liquor.

Procedure

- Distill a solution of **33** (0.875 kg, 2.55 mol) in EtOAc (7 L) at atmospheric pressure to remove 1.75 L of EtOAc.
- Cool the resulting slurry to room temperature.
- Charge 0.43 kg (2.65 mol) of CDI in one portion.
- Heat the mixture at 35 °C for 30 min, at 45–50 °C for 30 min, and at reflux for 1 h.
- Cool to 45–50 °C.
- Charge 0.59 kg (2.42 mol) of **34** in one portion.
- Heat to reflux and distill off 0.875 L of EtOAc at atmospheric pressure.
- Heat at reflux for 16 h.
- Cool to 10–15 °C and age for 1 h.
- Filter, wash with EtOAc, and dry at 50 °C under vacuum.
- Yield: 1.252 kg (90.7%).

18.2.1.3 Direct Isolation from Isopropyl Acetate

It is rather challenging to handle acid-labile compounds during the workup of reactions under acidic conditions. The selective deprotection of the terminal acetonide group in **36** was initially utilized acetic acid in water (1:1) (Equation 18.9).[13]

$$(18.9)$$

Although this method was effective, an over-hydrolysis by-product was observed due to the distillation at ambient temperature (under high vacuum) in order to remove water and acetic acid during the workup. To avoid this problem, a direct isolation approach was adopted. The reaction was conducted in isopropyl acetate with propionic acid, a controlled amount of water (5 equiv), and a catalytic amount of p-toluene sulfonic acid. Under these reaction conditions, the product **37** was precipitated out directly from the reaction mixture, thus avoiding vacuum distillation and extractive workup. With this modification, the selective hydrolysis step was reproducible and complete within 2–3 h, offering the product **37** in an 83% yield.

Procedure

- Charge a solution of p-toluenesulfonic acid (0.308 g, 1.6 mmol) in water (5.6 mL) to a stirred solution of **36** (98 g, 0.324 mol) in iPrOAc (490 mL), propionic acid (243 g, 3.284 mol), and water (29.2 g).
- Heat the resulting solution at 40 °C for 2 h.
- Cool to 4 °C and hold for 1 h.
- Filter and wash with cold iPrOAc (2×225 mL).
- Dry at 50 °C.
- Yield: 71 g (83%) as a white crystalline solid.

18.2.1.4   Direct Isolation from Acetonitrile

Acetonitrile was selected as the solvent for the reaction of indole **38** and alcohol **39** (1.0 equiv) in the presence of trifluoroacetic acid (Equation 18.10).[14] The product **40** was insoluble in the reaction mixture and was isolated in 86% yield directly by filtration upon cooling the reaction mixture to room temperature.

$$(18.10)$$

Procedure

- Charge 28.9 kg (253.6 mol) of TFA into a stirred suspension of **38** (24 kg, 126.8 mol) and **39** (24.9 kg, 126.8 mol) in MeCN (240 L).
- Stir under reflux for 24 h.
- Cool to room temperature and granulate for 16 h.
- Filter, wash with MeCN (2×24 L), and dry at 45 °C under vacuum.
- Yield: 40.2 kg (86%) as an off-white crystalline solid.

18.2.2 Use of Anti-Solvent

Using the anti-solvent approach in direct isolation is frequently applied in reactions wherein polar solvents are used as the reaction solvents, such as DMAc, DMF, DMSO, and AcOH. In general, the anti-solvent shall be miscible with the reaction solvent.

18.2.2.1 Adding Water to Acetic Acid

Scheme 18.4[15] shows two different workups for the bromination of methylcoumalate **41** with pyridinium bromide perbromide (PBPB). The initial extractive isolation of the desired product **42** required washing the extraction solution with aqueous bisulfite to destroy excess bromine, which resulted in the formation of impurity **43**.

The direct isolation allowed the isolation of **42** by adding water to the reaction mixture, followed by filtration and crystallization. No impurity **43** was observed using this isolation protocol.

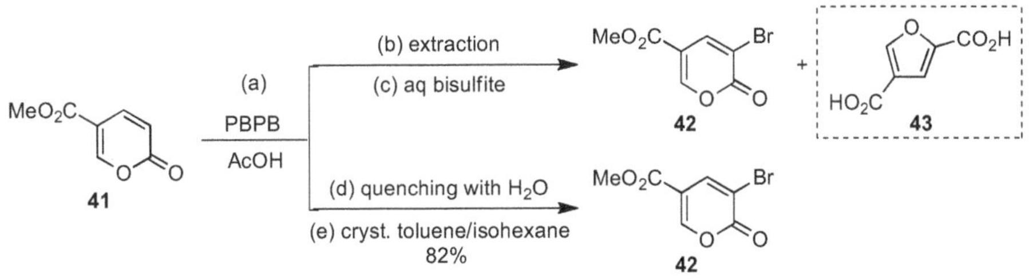

SCHEME 18.4

18.2.2.2 Addition of Water to Dimethylformamide

Although direct isolation provides an efficient approach to product isolation, the application of this strategy is not always problem-free. 3-Amino-5-Bromo-1-methyl-2H-pyridone **46** was prepared from Bromo nitropyridone **44** via N-methylation, followed by iron reduction of the nitro group to the amino group (Scheme 18.5).[16] Initially, the N-methylated intermediate **45** was isolated directly by adding water to the reaction mixture, followed by filtration and slurrying the crude **45** in EtOAc/heptane to remove O-methylated side product **47**.

(I) Problems

The subsequent iron reduction of **45**, however, showed only a 60% yield.

(II) Solutions

Examination of the isolated **45** by elemental analysis found that it contained 15 wt% of potassium salt. Furthermore, the UV spectrum revealed that the conjugated π-system in **45** was pH-dependent.

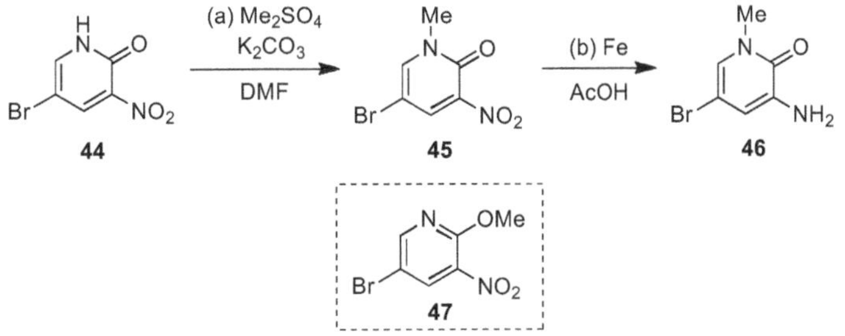

SCHEME 18.5

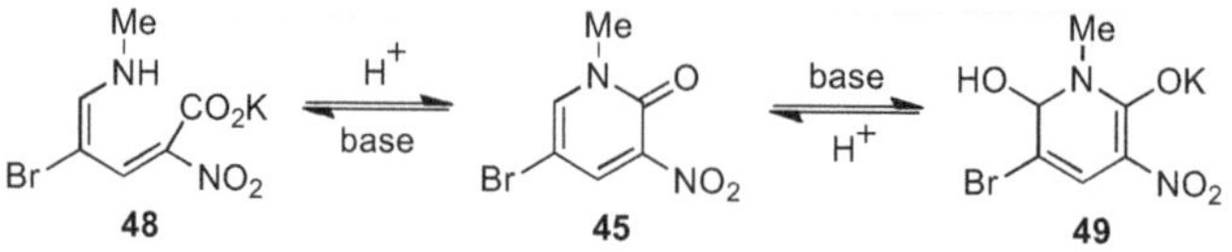

SCHEME 18.6

Therefore, the intermediate **45** would equilibrate with two hypothesized species, **48** and **49** and basic conditions appear to favor the formation of **48** and **49** as shown in Scheme 18.6.

Accordingly, a new isolation protocol was developed by adding concentrated sulfuric acid to the reaction mixture prior to the addition of water.

Procedure

- Charge 20 kg (91.3 mol) of **44** into a reactor, followed by adding 18.9 kg (137 mol) of K_2CO_3 and 80 L of DMF.
- Cool the resulting slurry to –5 °C.
- Charge 13.8 kg (109.6 mol) of Me_2SO_4 at <15 °C over 1.5 h.
- Warm the batch to 25 °C and stir for 12 h.
- Filter to remove inorganic salts.
- Add toluene (21 L) to the filtrate, followed by adding 5 kg of H_2SO_4.
- Add water (100 L).
- Filter and wash with water (20 L) and toluene (20 L).
- Yield: 16.3 kg (76%) as a yellow solid.

18.2.2.3 Addition of Water to Dimethylacetamide

The diaryl ether **52**, prepared by the reaction of phenol **50** with 4,6-dichloropyrimidine **51** in dimethylacetamide (DMAc), was isolated directly from the reaction mixture by a simple filtration following quenching with water (Equation 18.11).[17]

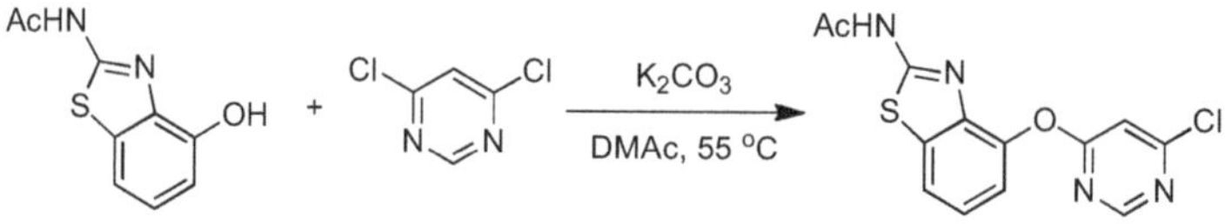

$$(18.11)$$

Procedure

- Charge DMAc (8.8 L), 1.377 kg (9.24 mol, 1.05 equiv) of **51**, 1.832 kg (8.8 mol) of **50**, and 1.338 kg (9.68 mol, 1.1 equiv) of K_2CO_3 into a glass-lined reactor.
- Heat the batch to 55 °C and stir for 2 h (<0.5% of **50** by HPLC).
- Cool to 10 °C.
- Add water (8.8 L) at <20 °C.
- Cool the resulting mixture to 10 °C and age for 2 h.
- Filter, wash with water (1×4.4 L, 4×2.2 L), and dry at 60 °C under vacuum.
- Yield of **52**: 2.54 kg (90%).

18.2.2.4 Addition of Water to Dimethylsulfoxide

To prepare *N*-aryl piperidine **55**, the S_NAr reaction of ethyl 4-fluorobenzoate **53** was selected to arylate piperidine **54** in DMSO (Equation 18.12).[18] Direct isolation of the product **55** was realized by adding water to the reaction mixture, followed by filtration.

$$(18.12)$$

Procedure

- Charge 15.4 kg (91.6 mol, 1.05 equiv) of **53**, 20.4 kg (87.3 mol) of **54**, and 24.1 kg (174.5 mol, 2.0 equiv) of K_2CO_3 to 112 kg of DMSO.
- Heat the mixture to 110 °C and age for 24 h.
- Cool the batch to 65 °C.
- Add water (143 kg).
- Cool the resulting slurry to 25 °C and age for 2 h.
- Filter, wash with water (143 kg), and dry at 40 °C.
- Yield: 27.3 kg (90.7%).

18.2.2.5 Addition of Methanol to Dimethylsulfoxide

Depending on product solubility, methanol or other solvents can be employed as the anti-solvent in direction isolation. The S_NAr reaction of 7-chloro-6-nitro-2,4(1H,3H)-quinazolinedione **56** with ethylamine (70% aqueous solution) in DMSO at 80 °C was utilized to prepare amine **57**. The product **57** was isolated directly via crystallization by adding MeOH (Equation 18.13).[19]

$$(18.13)$$

Procedure

- Charge dropwise a 70% aqueous solution of NH_2Et (194 mL, 2.4 mol, 2.7 equiv) to a solution of **56** (200 g, 0.88 mol) in DMSO (2.0 L) at 80 °C over a period of 1 h.
- Stir at 80 °C for 3 h.
- Cool the batch to 30 °C.
- Add 2.0 L of MeOH and stir at 30 °C for 2 h.
- Filter and rinse with MeOH (600 mL).
- Suspend the resulting solid in MeOH (1.0 L) at 30 °C for 3 h.
- Filter and wash with MeOH (200 mL).
- Dry under vacuum.
- Yield: 180 g (86%).

18.2.3 USE OF COOLING AND ANTI-SOLVENT

18.2.3.1 Isolation of Sonogashira Product

Isolation of transition metal-catalyzed reaction products usually requires column chromatography to remove metal catalysts and impurities. Standard conditions of Sonogashira cross-coupling reaction require palladium catalyst (1 to 10 mol%), copper iodide (10 to 20 mol%), and triethylamine (10 to 20 equiv) as a base. It is rather challenging to develop a direct isolation process to remove such large quantities of metals and bases. Attempts in the process development of Sonogashira reaction of aryl iodide **58** with alkyne **59** identified ethanol as the reaction solvent. The use of ethanol solvent was able to reduce the TEA amount to 4 equiv and the product **60** was isolated directly by cooling and adding water and toluene (Equation 18.14).[20]

(18.14)

Procedure

- Charge 800 mL (4 equiv) of TEA into a solution of **59** (501.5 g, 82% purity, 1.36 equiv) in 2.5 L of ethanol, followed by adding CuI (2.77 g, 1 mol%), **58** (500.0 g, 1.46 mol), and Pd(PPh$_3$)$_2$Cl$_2$ (10.2 g, 1 mol%).
- Heat the batch to 57 °C and stir at that temperature for 80 min.
- Concentrate to 2.5 L under reduced pressure.
- Cool to 10 °C.
- Add 2.5 L of water and 2.0 L of toluene.
- Adjust pH from 9.4 to 8.3 with 220 mL of 6 M HCl.
- Stir the resulting suspension at 10 °C for 20 min prior to filtration.
- Wash with a cake with toluene (100 mL), water (500 mL), and EtOH (600 mL).
- Dry at 65 °C.
- Yield: 493.0 g (83.5%) in 99.2% purity.

18.2.3.2   Isolation of 6-Chlorophthalazin-1-ol

The preparation of 6-chlorophthalazin-1-ol **62** was realized via a ring expansion of hydroxyisoinolinone **61** using hydrazine (Equation 18.15).[21]

(18.15)

The following workup involved the addition of water as an anti-solvent, followed by cooling to provide the product **62** by simple filtration.

Procedure

Stage (a)
- Charge 2.06 kg (8.59 mol) of **61** into a 30-L jacketed reactor, equipped with a mechanical stirrer, reflux condenser, addition funnel, and temperature probe, followed by adding AcOH (6.4 L).
- Heat to 90 °C to form a homogeneous solution.
- Charge dropwise at 90–93 °C, 0.53 kg (9.0 mol, 1.05 equiv) hydrazine hydrate (54 wt%) (*caution: exotherm*) over 3 h.
- Continue to heat at 90 °C for 1.5 h.

Stage (b)
- Add pre-heated water (80 °C) (12.8 L) at 80–90 °C.
- Cool to 20 °C over 4 h.
- Filter and rinse with water (3 L).

- Dry the wet cake in the air overnight.
- Charge 3.59 kg of the crude **62** (obtained from two batches) into a 30-L reactor.
- Add CH_2Cl_2 (18.0 L) and stir at 20 °C for 1 h.
- Filter and wash with CH_2Cl_2 (4.0 L).
- Dry at 50 °C.
- Yield: 2.76 kg (89.0%) as pale yellow crystals.

NOTE

The addition of hydrazine hydrate at 90 °C avoided the accumulation of hydrazine in the reaction mixture.

18.2.3.3 Isolation of S_NAr Product

Direct isolation of alkyl aryl ether **65** from an S_NAr reaction mixture in THF was achieved by diluting the mixture with water and cooling from 55 °C to 20 °C; the precipitated product **65** was collected by filtration (Equation 18.16).[22]

$$(18.16)$$

Procedure

- Charge 0.96 kg of 50% aqueous NaOH solution (12 mol, 1.5 equiv) to a solution of **63** (2.35 kg, 7.98 mol) and **64** (1.05 kg, 9.62 mol, 1.2 equiv) in THF (16 L) at 20 °C.
- Heat the resulting mixture at 50–55 °C for 10 h.
- Charge 0.090 kg (0.82 mol, 0.10 equiv) of **64** and stir at 50–55 °C for 12 h.
- Charge 0.045 kg (0.41 mol, 0.050 equiv) of **64** and stir at 50–55 °C for 5 h.
- Add water (23 L) and cool to 20 °C over 2 h.
- Filter, wash with water (28 L), and dry at 45 °C under vacuum with nitrogen sweep.
- Yield of **65**: 2.63 kg (90%).

18.2.4 USE OF NEUTRALIZATION

Direct isolation can also be achieved by neutralization of an acidic or basic reaction mixture with base or acid to precipitate the desired product. Acid **67** is a potent antagonist of the integrins $\alpha_v\beta_3$ and $\alpha_{IIb}\beta_3$ for the treatment of acute ischemic diseases. To isolate **67** from the acidic reaction mixture, a direct isolation protocol was developed by a step-wise neutralization with an aqueous NaOH solution (Equation 18.17).[23]

$$(18.17)$$

Procedure

Stage (a)
- Heat a solution of **66** (200 g, 94.4 %, 0.309 mol) in 0.5 N HCl (4.0 L) at 60–61 °C for 5.5 h.
- Cool the mixture to 35 °C.

- Add water (4.0 L) followed by 226 mL of 5 N aqueous NaOH solution to pH 1.6–1.7.

Stage (b)
- Charge a slurry of 10% Pd/C (10.0 g) in water (400 mL).
- Hydrogenate at 30–36 °C using a hydrogen balloon for 7.5 h.
- Filter and rinse with 0.1 N HCl (4.7 L).

Stage ©
- Add 148 mL of aqueous 5 N NaOH to pH 2.8–2.9.
- Add MeOH (2.0 L).
- Add 96 mL of aqueous 5 N NaOH to pH 5.0–5.2.
- Add 22.5 mL of aqueous 1 N NaOH to pH 7.4–7.6.
- Add seeds.
- Stir the batch for 20 h.
- Distill to remove ca. 1.6 L of solvents.
- Filter and dry at 35 °C under reduced pressure for 18 h.
- Yield: 149 g (83.5%).

DISCUSSIONS

(a) The use of HCl allowed not only the removal of the *tert*-butyl group, but also the formation of the pyrimidine HCl salt, which is required for the subsequent hydrogenation.
(b) As the solubility of most water-soluble organic compounds is pH-dependent, the crystallization yields will fluctuate. For example, crystallization of **67** from a pH 5 solution gave a 56% yield, while a greater than 80% yield was achieved from pH 7.
(c) Initial attempts of crystallization by increasing pH resulted in a gummy form of **67**. To circumvent this problem, methanol was added prior to adjusting the pH (from pH 2.8) to pH 7 and **67** was isolated in 84% yield with 96.5% purity.

18.2.5 Use of Salt Formation

One of the common approaches in direct isolation is to convert the product into its insoluble salt followed by filtration. A demonstrative example was presented in Reaction (18.18),[24] in which triol **68** was converted to acetonide **69** via selective protection of the two secondary hydroxyl groups in **68**. This transformation was accomplished with 0.95 equiv of sulfuric acid (rather than a catalytic amount) in a mixture of acetone and 2 equiv of 2,2-dimethoxypropane (2,2-DMP). The product **69** thus formed precipitated out directly from the reaction mixture (direct drop process) as the hydrogen sulfate salt and was isolated by filtration.

$$(18.18)$$

Procedure

- Charge 5.73 kg (58.5 mol, 0.95 equiv) of H_2SO_4 (98%) to a suspension of triol **68** (32.0 kg, 61.6 mol) in acetone (149 L) and 12.83 kg (123 mol, 2 equiv) of 2,2-DMP.

- Stir the resultant solution for 50 min at 20 °C to lead to a thick slurry.
- Filter and wash with acetone (32 L).
- Dry at 50 °C under vacuum.
- Yield: 35.6 kg (88%).

REMARK

Care should be taken when practicing direct isolation via the salt formation approach. For example, impurities **72** were observed in the dried salt product **71** (Equation 18.19).[25]

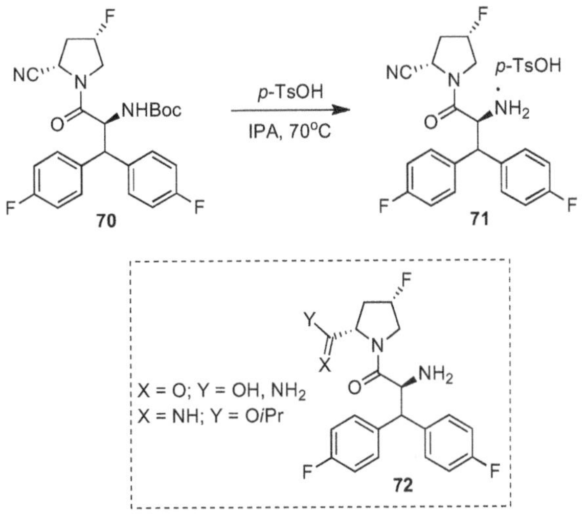

(18.19)

Upon investigations, it was found that the excess of *p*-TsOH in the product wet cake led to the degradation of the product during drying. Thus, the implementation of sufficient wash with IPA (2×2 vol) solved the product degradation issue.

18.2.6   OTHER DIRECT ISOLATION APPROACHES

18.2.6.1   Direct Drop Process

An ideal workup situation is that the product is insoluble in the reaction media and precipitates without cooling or adding anti-solvent. Thus, the product can be readily isolated by simple filtration. Occasionally, a by-product may interfere with the isolation, resulting in diminished product yields. Thereby the direct drop process has to be modified.

18.2.6.1.1   Removal of MeOH By-product

Condensation of urea **73** with methyl glyoxylate hemiacetal **74** in toluene at 23 °C furnished hydroxyl urea ester **75** as a solid precipitate (Equation 18.20).[26] Prior to filtration, the removal of methanol by-product from the reaction mixture proved critical for full precipitation of the product to ensure good and consistent yields.

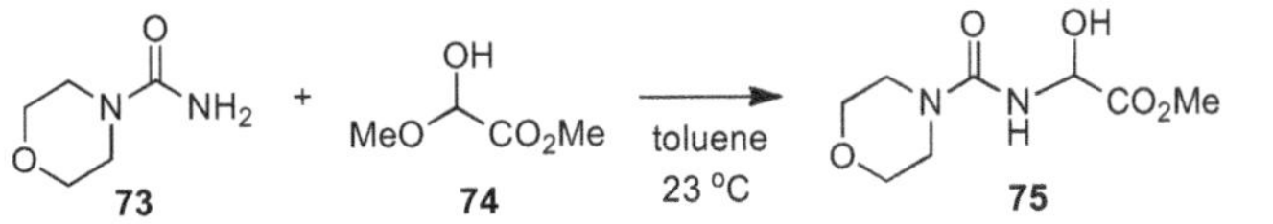

(18.20)

Procedure:

- Charge 33 kg (195 mol) of **73** into a 500-L reactor, followed by adding toluene (260 kg).
- Charge 25.0 kg (208 mol, 1.07 equiv) of **74**.

- Stir the batch at 20 °C for 20 h.
- Distill at 50 °C under reduced pressure to remove MeOH.
- Cool to 20 °C over 1 h.
- Filter and wash with toluene (32 kg).
- Dry at 50 °C under vacuum for 6 h.
- Yield of the crude **75**: 38.7 kg (91%).

18.2.6.1.2   Removal of Triethylamine Hydrochloride By-product

In addition, it is also important to remove the co-precipitated solid by-product completely prior to filtration or it will contaminate the isolated product as an impurity. A reaction of cyclopropylamine **76** and sulfonyl chloride **77** in the presence of triethylamine (TEA) in toluene generated the desired *N*-cyclopropyl sulfonamide **78** as solid precipitate along with co-precipitated triethylamine hydrochloride salt (Equation 18.21).[27]

$$(18.21)$$

To obviate the potential product contamination by triethylamine hydrochloride, water was introduced into the reaction mixture to dissolve the triethylamine hydrochloride, followed by filtration. This process was successfully demonstrated on a 110-kg scale.

18.2.6.2   Removal of By-Product by Direct Drop Approach

In contrast to direct isolation, the direct removal approach is used to remove the solid by-product from the reaction mixture by filtration without resorting to extractive workup.

Oseltamivir phosphate **79**[28] (Tamiflu, Figure 18.1), the water-soluble and orally bioavailable prodrug of the corresponding pharmacologically active acid **80**, was first introduced to the market in November 1999 by Gilead Sciences and Hoffmann-La Roche for the treatment and prevention of seasonal influenza virus infections. It is active against the currently spreading avian H5N1 and H1N1 swine influenza strains.

Development efforts[29] led to the presently used technical synthesis of **79** from shikimic acid **81**. Mesylate **83**, an intermediate in the synthesis of **79**, was obtained from mesylation of **82** (Equation 18.22).[29b] Due to the formation of emulsions after quenching with water, the normal extractive aqueous workup was replaced with operations including filtration of the reaction mixture to remove insoluble triethylamine HCl salt and concentration of the filtrate followed by crystallization, which led to 82 % yield of **83**.

$$(18.22)$$

In comparison among the direct isolation approaches, Table 18.1 summarizes each approach with advantages and limitations.

18.3   FILTRATION PROBLEMS

Filtration is a unit operation that is used to separate solids from liquids and is accomplished by passing the slurry mixture through a filtration paper or cloth. A common filtration problem is the slow

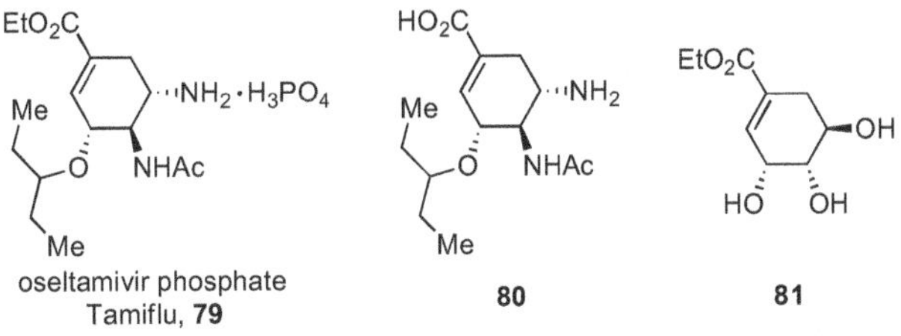

oseltamivir phosphate
Tamiflu, 79

80

81

FIGURE 18.1 Structures of compounds **79–81**.

TABLE 18.1
Comparion of the Direct Isolation Approaches

Isolation Approach	Advantage	Limitation
Cooling	• Easy operation. • Reaction temperature is higher than the isolation temperature.	• The isolated yield may be low if the temperature difference between reaction temperature and isolation temperature is small. • The solubility of undesired products should be greater than the desired product.
Anti-solvent	Easy operation.	• The product may oil out if strong anti-solvent is used. • Extra solvents are required, thereby leading to more wastes.
Neutralization	Relatively simple operation.	The product may oil out.
Salt formation	The salt product is usually isolated as crystalline solid.	• This method requires acidic or basic functional groups in the product. • The salt product sometimes is hygroscopic.
Direct drop approach	This is the most economic process as no cooling or anti-solvent is required.	This method requires the starting material and reagent(s) are soluble, while the product is not.
Removal of by-product by direct drop approach	Removal of by-product from the process stream prevents contamination of the product.	This method adds an additional filtration step.

filtration rate caused by the clog of the filter cloth by small particles. In addition, small particles can also lead to low product yields, due to the loss of small particle product into the mother liquor.

18.3.1 METAL-RELATED FILTRATION PROBLEMS

18.3.1.1 Copper-Related Filtration Problems

Copper (I/II) salts are inexpensive, stable, and extensively used as catalysts in organic synthesis. The spent copper catalysts intend to chelate with compounds bearing amino and thiol groups forming copper complexes. These copper complexes often have poor solubility and precipitate as fine insoluble solids which may create filtration problems during the reaction workup.

18.3.1.1.1 Addition of Thioacetamide

The reaction of (methylamino)acetaldehyde dimethyl acetal **84** and cyclopropane carbonitrile **85** at 85 °C in the presence of CuCl provided an amidine **86**, an intermediate for the synthesis of imidazole **87** (Scheme 18.7).[30]

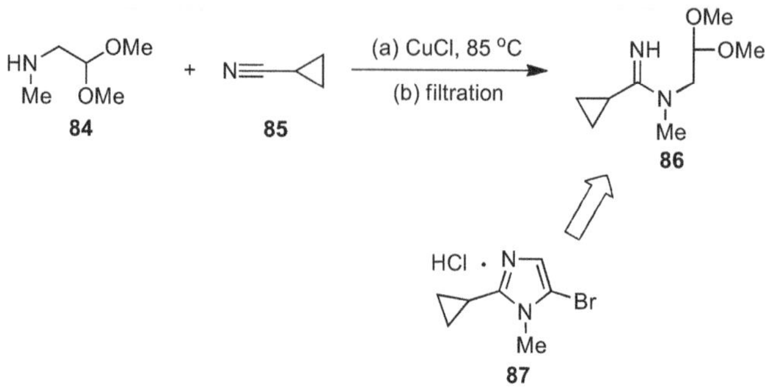

SCHEME 18.7

(I) Slow filtration problem

Filtration of the reaction mixture turned out to be very slow and laborious, and the use of filtering agents such as celite or cellulose filter aid resulted in no improvement.

(II) Solutions

The slow filtration was presumably caused by the formation of a viscous copper complex with **86**. It was envisioned that the addition of a chelating ligand could improve the filtration rate. Thus, the addition of thioacetamide to the reaction mixture enhanced the filtration rate significantly, due to the possible ligand displacement by thioacetamide.

18.3.1.1.2   Addition of Zinc Dust

A copper-catalyzed C–N cross-coupling of 5-ethylthio-1*H*-tetrazole **89** with 2-iodobenzoic acid **88** provided the desired 2-(5-(ethylthio)-2*H*-tetrazol-2-yl)benzoic acid (Equation 18.23).[31] The aqueous acidic workup resulted in the precipitation of insoluble copper salts, which made the filtration difficult.

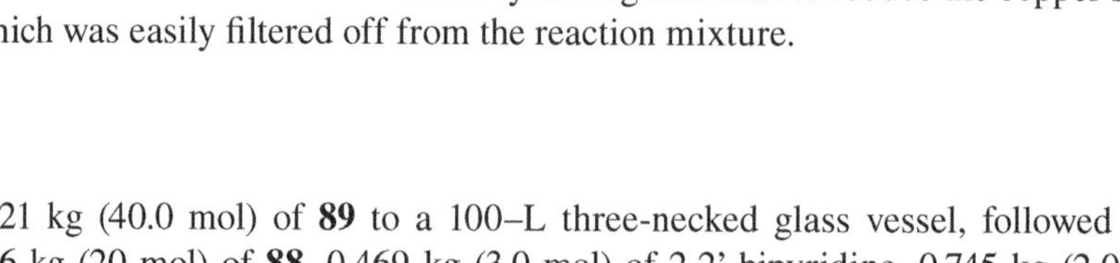

(18.23)

Ultimately, this filtration issue was addressed by adding zinc dust to reduce the copper salts to copper metal, which was easily filtered off from the reaction mixture.

Procedure

- Add 5.21 kg (40.0 mol) of **89** to a 100–L three-necked glass vessel, followed by adding 4.96 kg (20 mol) of **88**, 0.469 kg (3.0 mol) of 2,2'-bipyridine, 0.745 kg (2.0 mol) of Cu(MeCN)$_4$PF$_6$, 12.74 kg (60.0 mol) of K$_3$PO$_4$, and 1,4-dioxane (24.80 L).
- Heat the resulting dark red mixture at 86–88 °C for 14 h.
- Cool to 42 °C.
- Add 0.497 kg (7.6 mol) zinc dust, followed by adding 5 N aqueous HCl (24 L, 120 mol, precooled to 5 °C) and water (9.92 L) at <63 °C.
- Add Solka-Floc (500 g), Darco KB charcoal (0.5 kg), and CPME (12.4 L).
- Stir the mixture for 45 min.

- Filter through a Solka-Floc pad and wash the cake with CPME (37.2 L).
- Heat the combined filtrates to 40 °C, separate layers, and wash the organic layer with water (2×9.92 L).
- Concentrate to 37 L volume at 70–80 °C under vacuum.
- Add diisopropylamine (6.27 L, 44.0 mol) over 30 min at 76–90 °C to give a tan slurry.
- Cool to 25 °C over 1 h and age for 1 h.
- Filter and wash with CPME (25 L).
- Slurry the filtration cake with water twice (25 L) to remove residual tetrazole.
- Dry under a nitrogen stream.
- Yield: 5.22 kg.

18.3.1.2 TiCl$_4$-Related Problems

A difficult filtration was encountered during the isolation of enamine **92**, prepared from the reaction of 4-*tert*-butylacetophenone **91** with morpholine in the presence of TiCl$_4$ as a water scavenger and Hünig base (DIPEA) in toluene (Equation 18.24).[32] This slow filtration was presumably caused by the sticky agglomerates formed from the preformed TiCl$_4$-morpholine complex and reaction by-products (TiO$_2$ and the DIPEA·HCl salt).

$$(18.24)$$

It was found that the addition of sodium sulfate at the beginning of the reaction gave a mixture slurry which could be easily filtered.

Procedure

- Mix sodium sulfate (87.5 kg) in 175.0 L of toluene to form a slurry.
- Charge 52.0 kg (597 mol, 6.0 equiv) of morpholine to the above slurry at 5–8 °C.
- Charge 18.8 kg (99 mol, 1.0 equiv) of TiCl$_4$ at <15 °C.
- Charge 64.2 kg (497 mol, 5.1 equiv) of DIPEA.
- Charge 17.5 kg (99 mol) of **91**.
- Heat at 70–73 °C for 3 h (the reaction was complete then).
- Cool to 20–23 °C and centrifuge.
- Wash the cake with 35.0 L of toluene.
- Distill the combined filtrates at reduced pressure until morpholine <10% and DIPEA <10%.
- Add 73.5 L of ethyl acetate and 58.3 kg of triethylamine.
- Yield: quantitative (used directly in the subsequent step).

18.3.2 Small Particle Size

18.3.2.1 Addition of Acetic Acid

The reactive crystallization by adding acid to the basic hydrolysis mixture will lead to a high local supersaturation and a large degree of uncontrolled nucleation and growth, resulting in excessive fines and wide particle size distribution.[119] Fine particle solids may either block the filtration cloth or pass through the cloth, resulting in poor filtration or loss of product in the mother liquor. For example, the neutralization of a basic hydrolysis mixture by adding 1 N HCl produced carboxylic acid **94** as a very fine particle that led to a slow filtration (Equation 18.25).[33]

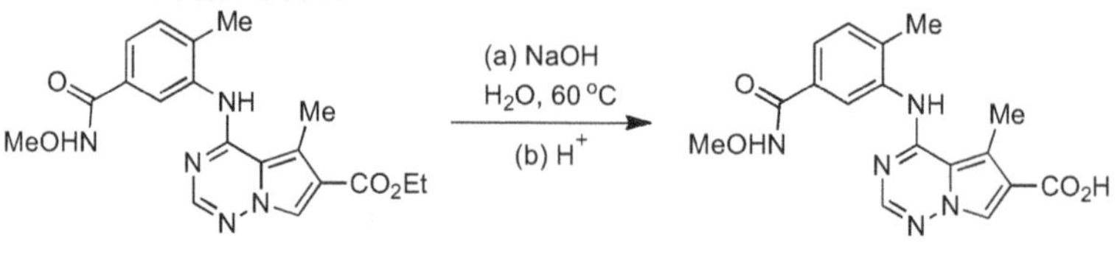

$$(18.25)$$

This issue was successfully addressed by replacing HCl with acetic acid during the neutralization. The use of weak acid (AcOH) makes the neutralization an equilibrium process. As a result, small particles of **94** can, due to their higher solubility,[34] be dissolved via the salt exchange equilibrium back to its corresponding salt (Equation 18.26).

$$R\text{-}CO_2Na + AcOH \rightleftharpoons R\text{-}CO_2H + AcONa \qquad (18.26)$$

Procedure

- Heat a mixture of **93** (2.06 kg, 5.37 mol), water (6.0 kg), and 50% NaOH (1.15 kg, 15.375 mol, 2.7 equiv) at 60–65 °C for 5 h.
- Add water (7.2 kg) at 60–65 °C.
- Charge 6 wt% of aqueous solution (6.6 kg, 109.9 mol, 20.5 equiv) of AcOH over 2 h.
- Charge 4.9 kg of additional AcOH solution over 2 h.
- Cool the resulting slurry to 25 °C over 2 h.
- Filter and wash with water (10 kg) and heptane (8.0 kg).
- Dry at 70–75 °C under vacuum.
- Yield: 1.68 kg (88%).

REMARK

During the centrifuge isolation, the buildup of static charge is possible especially when rinse with hydrocarbon solvent such as heptane. To prevent the buildup of the static charge, heptane blended with 5% ethanol is used as the final cake wash.[35]

Reaction (18.27)[36] presents another example of the application of AcOH toward the neutralization of a basic saponated reaction mixture.

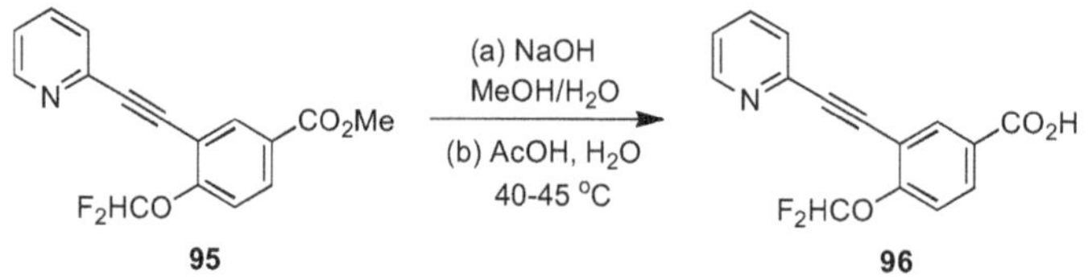

$$(18.27)$$

Procedure

- Charge water (31.5 kg) and MeOH (27.8 kg) at 20 °C followed by adding 6.33 kg (21.8 mol) of **95** as solid.
- Heat to 35–40 °C.
- Charge 50% NaOH (2.10 kg, 26.2 mol, 1.2 equiv) while maintaining the temperature at 35–40 °C.
- Stir for 1 h until the reaction completes.
- Add water (32.5 kg) and warm to 45 °C.

- Add a solution of AcOH (6.80 kg, 113.2 mol, 5.2 equiv) in water (13.7 kg) over 3 h.
- Cool to 20 °C and stir for 1 h.
- Filter and wash with water (78.0 kg).
- Yield: 5.86 kg (99.1%).

The filtration of 4-amino-6-hydroxyisophthalic acid **98**, prepared by adding hydrochloride acid to the aqueous solution of trisodium salt **97**, took a long time (ca. 4 h) on a 10-g scale batch (Equation 18.28)[37] due to the fine particles. Again, the use of acetic acid to replace HCl addressed successfully the slow filtration issue.

$$(18.28)$$

However, for compound **98** being highly soluble in acetic acid, the quantity of AcOH used has a significant impact on the isolated product yield. For instance, the yields of the isolated **98** increased from 60% to 95% with a reduction of the AcOH amount from 10.0 to 3.6 equiv.

More examples were reported in the literature (Equations 18.29 and 18.30).[38]

$$(18.29)$$

$$(18.30)$$

18.3.2.2 Addition of Isopropanol

A filtration problem was encountered during the isolation of indole **105**, obtained from the Fischer indole reaction (Equation 18.31).[39] Treatment of hydrazine **103** with 4-piperidone monohydrate hydrochloride **104** in isopropanol under reflux furnished indole **105** in a good yield with a purity of >99.7 wt%. The isolated crude product was suspended in water to remove ammonium chloride, which gave rise to a fine particulate solid that was comprised of agglomerated needles with a particle size distribution in the 5–30 μm range, causing poor filtration.

$$(18.31)$$

The use of a mixture of solvent systems could improve the physical properties of a solid material by breaking the agglomerated solid mass. In the event, a controlled crystallization was realized by adding IPA into the batch (IPA/water=1:9), which provided a solid in larger particle size (50–100 μm) with faster filtration.

Procedure

- Charge 21.2 kg (138 mol) of **104** into a pre-prepared 12 wt% IPA solution of **103** (ca. 129 mol obtained directly from the previous step).
- Heat the batch to 40 °C.
- Charge a solution of HCl in IPA (5.5 N, 27.8 kg).
- Heat the batch to reflux until complete conversion (>98.5%).
- Cool to <20 °C.
- Filter and rinse with 67.9 kg of IPA (in three portions).
- Charge 175 L of water into a 100-gallon glass-lined reactor.
- Charge the wet cake (97.5 kg which contains ca. 19 kg of IPA).
- Heat to 105 °C until a clear solution forms.
- Cool and hold at 90 to 80 °C for 20 min, 80 to 60 °C for 40 min.
- Cool to 5 °C and hold for 1 h.
- Filter and wash with 60 L of precooled water (3–5 °C) in three portions.
- Dry at 70 °C/50 mmHg.
- Yield: 24.6 kg with 100 wt% purity.

18.3.2.3 Temperature Control

Initially, the isolation of Suzuki coupling product **108** (Equation 18.32)[40] by filtration was problematic due to the small particles caused by a nucleation-dominated crystallization under supersaturated conditions.

$$(18.32)$$

In order to address this issue, a cooling-crystallization approach was used. The aqueous product solution was heated to 70 °C followed by pH adjustment and controlled cooling. Utilizing the optimized crystallization conditions, the particle size was increased by greater than 100-fold.

Procedure

- Charge 2.0 kg (93.8 wt%, 5.52 mol) of **106** into a 100-L jacketed reactor, followed by adding 1.33 kg (87.2 wt%, 6.45 mol) of **107**, and 5.0 L of IPA.
- Charge a solution of K_3PO_4 (3.52 kg, 16.6 mol) in water (10.0 L) followed by adding a slurry of palladium bis(4-dimethylaminophenyl-di-*tert*-butylphosphine) dichloride (4.2 g, 6 mmol) in IPA (20 mL).
- Inert the content with nitrogen.
- Heat the jacket to 85 °C and stir the batch at that temperature for 1 h.
- Cool the batch to 20 °C.
- Separate layers.
- Add water (1.5 L) to the aqueous layer.
- Wash the resulting aqueous solution with toluene (2.5 L).
- Polish filter the aqueous layer.

- Add IPA (8.0 L).
- Heat the resulting solution to 70 °C.
- Add 2 N aqueous HCl (4 L) over 45 min to pH 3.
- Cool the batch to 20 °C over 1 h and age for 12 h.
- Filter, wash, and dry.
- Yield of **108**: 2.05 kg (93.4%).

18.3.2.4 Polymorph Transformation

Because various polymorphic forms have different arrangements in the crystal lattice, the transformation of one polymorphic form to another can change the crystalline particle size. Therefore, the slow filtration problem can be resolved by the polymorph transformation. For instance, a slow filtration was observed during the isolation of aldehyde product **111** from the postformylation reaction mixture (Scheme 18.8).[41]

To address this filtration problem, the crystallization batch was held for 1 h prior to filtration, in which a polymorph transformation occurred.

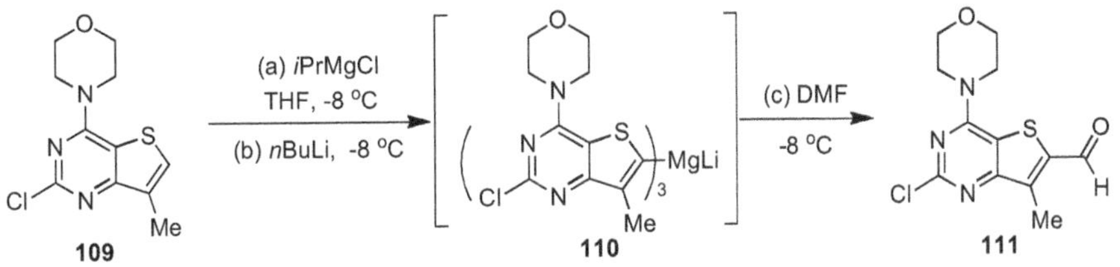

SCHEME 18.8

Procedure

- Charge 9.45 kg (35.0 mol) of **109** into a reactor followed by adding anhydrous THF (94.5 L).
- Cool the mixture to <−8 °C.
- Charge 8.64 kg (16.8 mol) of 20 wt% solution of *i*PrMgCl in THF followed by adding 9.84 kg (38.4 mol) of 25 wt% solution of *n*BuLi in heptane at <−8 °C.
- Stir the mixture at <−8 °C for 2 h.
- Add 5.11 kg (69.9 mol) of anhydrous DMF at <−8 °C.
- Stir the batch for 1 h.
- Transfer the content to a mixture of AcOH (23.6 kg), 37% HCl (9.10 kg), IPA (94.4 L), and water (91.7 L).
- Stir the resulting slurry at 50–55 °C for 1 h.
- Distill the suspension under reduced pressure to remove THF.
- Cool to room temperature.
- Filter, wash with water (2×20 L), and dry at 40–60 °C under reduced pressure.
- Yield: 10.0 kg (96%) as a yellow solid.

18.3.3 LOW-MELTING SOLID

Low-melting solids may cause filtration problems due to clogging of the filtration cloth by the gummy/paste material, which poses a challenging issue at scale. Not surprisingly, filtration of the low-melting solid product **114** (mp 33.5–35.5 °C), prepared via the alkylation of resorcinol monobenzoate **112**, was very slow (Equation 18.33).[42] It was found that the addition of α-cellulose as a solid support to the suspension of crude product improved effectively the filtration.

$$(18.33)$$

TDA = tris(3,6-dioxaheptyl)amine or tris[2-(2-methoxyethoxy)ethyl]amine

Procedure

- Charge heptane (0.4 L) into a 5 L, four-necked round-bottomed flask, followed by adding 64.7 g (0.2 mol) of TDA, 428.4 g (1.9 mol) of **112**, 414 g (3.0 mol) of K_2CO_3, and 429.2 g (2.6 mol) of **113**.
- Heat the content to 110 °C for 10 h.
- Cool to room temperature.
- Add H_2O (1 L) and heptane (1 L).
- Wash the organic layer with 1 N NaOH (0.5 L) followed by water.
- Distill at 50 °C/40 mbar.
- Add MeOH (0.2 L).
- Distill at 50 °C/40 mbar.
- Add MeOH (0.2 L).
- Distill at 50 °C/40 mbar.
- Add MeOH (1.5 L).
- Add 21 g of α-cellulose.
- Cool to –10 °C and stir for 2 h.
- Filter and wash with chilled MeOH (–5 °C).
- Dry to give 490 g of **114** with cellulose as an off-white solid.

18.4 PURIFICATION STRATEGIES

In process chemistry, workup refers to a series of manipulations including quenching of reaction mixture, extraction, distillation, filtration, and drying. Purification is a dedicated operation with attempts to remove by-products/side products from the desired product. Removal of impurities and reducing them to process acceptable levels are critical because their presence in the active pharmaceutical ingredient (API), even in small amounts, may affect drug safety and efficacy. Common methods to remove impurities include extraction, distillation, crystallization, and chromatography. Besides, other purification methods, such as salt formation and derivatization, are also frequently utilized.

18.4.1 USE OF SALT FORMATION

Crystallization often accompanies a loss of product in the mother liquor, which is a concern, especially at the late stage of synthesis. Given a product that contains acidic or basic functionality, the salt formation can be a valuable approach to improve the product purity without the compromising on the isolation yield. For the formation of a stable salt, the p*Ka* difference between the acid and base partners should be ≥2.[43]

A number of acids including both inorganic and organic acids are able to convert basic amines into their corresponding salts. For the purpose of purification, salts have to be stable and non-hygroscopic crystalline solids. Similar to the purification of organic amines, organic acids can also be purified by converting to their corresponding salts via reactions with requisite quantities of bases. The commonly used inorganic bases include LiOH, NaOH, KOH, CsOH, $Mg(OH)_2$, $Ca(OH)_2$, and $Ba(OH)_2$.[44] A number of organic amines are available for salt formation with organic acids as well.

18.4.1.1 Hydrochloric Acid Salts

Hydrochloric acid salts are widely used to improve the purity of basic organic products (usually contain amino groups) and can be prepared by mixing the organic amine solutions with either concentrated aqueous hydrochloric acid or anhydrous hydrogen chloride solution.

18.4.1.1.1 Purification of Oily Product

When reaction products are oil, isolation, and purification are challenging if distillation is not feasible. Therefore, the salt formation becomes a method of choice without resorting to chromatography. For example, the oily 2,3,4,6-tetra-*O*-benzyl-1-deoxynojirimycin **117** was purified by converting **117** into its hydrochloride salt (Scheme 18.9).[45] The hydrochloride salt of **117** was precipitated from acetone at 0 °C and isolated by centrifugation to provide [**117**·HCl] as an off-white solid in 56% yield over three reaction steps. Minor impurities were removed by means of an additional reslurrying step in acetone that provided [**117**·HCl] with 93% recovery.

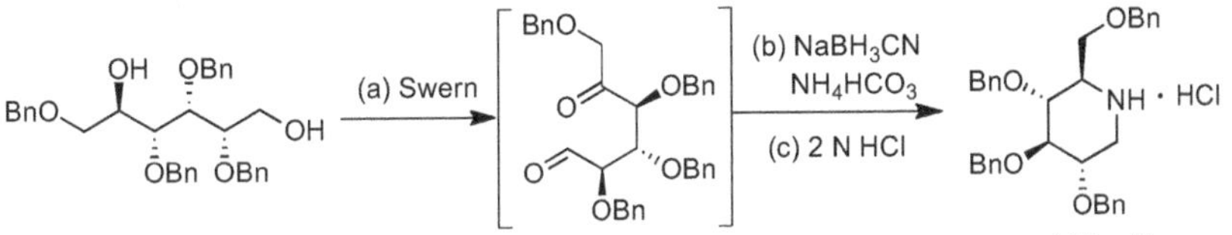

SCHEME 18.9

Procedure

Stage (c) Formation of hydrochloride salt
- Charge aqueous 2 N HCl into a solution of **117** in CH_2Cl_2.
- Distill CH_2Cl_2 until residue temperature reached 55 °C.
- Add acetone with vigorous stirring.
- Distill out ~1/3 solvent to remove residual CH_2Cl_2.
- Add additional acetone.
- Stir for 1 h at room temperature.
- Cool to 0–5 °C over 1 h.
- Filter and wash with precooled acetone.
- Dry at 35–40 °C.
- Yield: 56%.

REMARKS

(a) The salt break was achieved by adding 1 N NaOH into a suspension of the salt.
(b) In addition, the salt formation strategy can also be used to remove impurities by converting the impurity into its appropriate water-soluble salt.
(c) Care should be taken when strong acids are utilized in the salt preparation, as the acid, such as sulfuric acid, may promote product degradation. For example, using sulfuric acid to convert **118** to salt **119** led to an impurity **120** (Equation 18.34).[46]

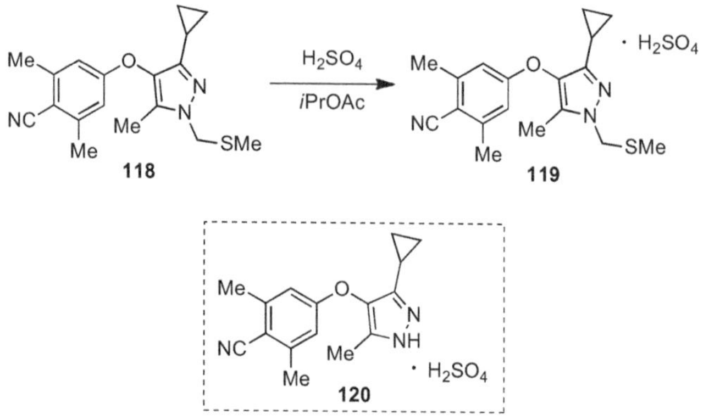

(18.34)

18.4.1.1.2 Purification of Acid-Sensitive Compounds

For a compound with an acid-sensitive group such as the formamide group in hydroxy amine **121**, the hydrochloric acid salt formation by direct addition of strong acid HCl will lead to problematic decomposition. With the aim of avoiding such decomposition, a salt-exchange approach was developed by forming a weak acetic acid salt followed by converting the weak acetic acid salt to the desired hydrochloride salt (Equation 18.35).[47]

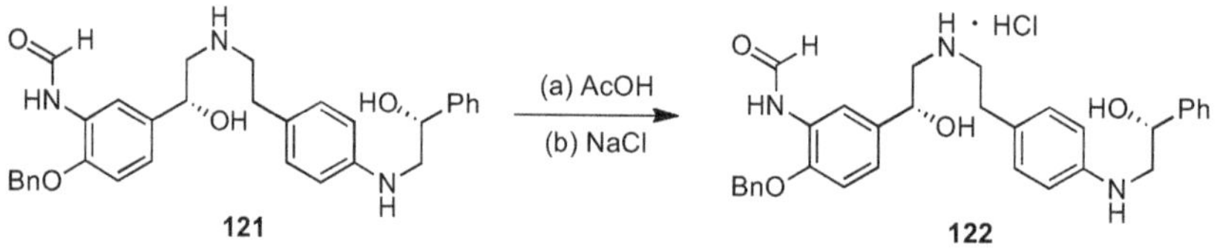

(18.35)

Thus, treatment of the solution of the free base **121** in 1-pentanol with acetic acid, followed by washing with aqueous sodium chloride solution, afforded the corresponding HCl salt **122** in good yield and purity.

Procedure

- Treat a solution of **121** (≤215 mmol) in 1-pentanol (500 mL) with 18.5 mL (324 mmol, 1.5 equiv) of AcOH.
- Wash the resulting mixture with 10 w/v% aqueous NaCl solution (2×500 mL).
- Dilute the organic layer with 1 L of 1-pentanol.
- Add 0.2 g of **122** and age for 30 min.
- Distill under reduced pressure (100 mbar) to ca. 1.4 L.
- Cool to room temperature.
- Filter, wash with 1-pentanol (2×300 mL) and EtOAc (300 mL), and dry.
- Yield: 86.09 g.

REMARK

This salt formation protocol offered an additional advantage, in which the stoichiometry of the hydrochloric acid was controlled to produce the desired mono HCl salt **122**.

Analogously, the preparation of acid-sensitive 2-(1,1-dimethoxymethyl)-piperidine hydrochloride salt **124** was achieved by using ammonium chloride as the HCl source via a one-pot process (Equation 18.36).[48]

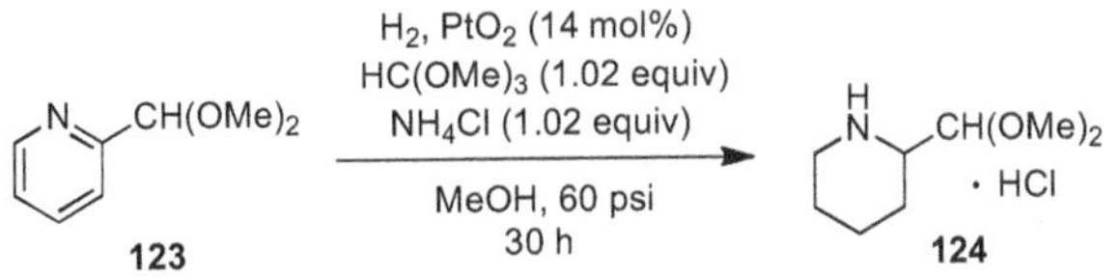

(18.36)

Procedure

- Charge 7 g (45 mmol) of **123**, followed by adding 1.45 g (6.4 mmol) of PtO$_2$, 4.85 g (46 mmol) of trimethylorthoformate, 2.45 g (46 mmol) of NH$_4$Cl, and MeOH (200 mL).
- Hydrogenate at 60 psig for 30 h.
- Filter and distill under vacuum to 50 mL.
- Add cold Et$_2$O (200 mL) and filter to remove a white precipitate.
- Distill the filtrate under reduced pressure to 20 mL.
- Add Et$_2$O (200 mL) and swirl the flask vigorously.
- Filter to give 7.1 g (81%) of **124**.

18.4.1.1.3 Purification of Olefin

Removal of a minor olefin isomer from the *E/Z* olefin mixture without resorting to chromatography poses a significant challenge. An olefin purification approach was developed for the purification of compound **126**, a potential drug candidate for the treatment of stroke. The crude **126** (*E/Z*=85:15), obtained from hydrolysis of ester amide **125** (Scheme 18.10),[49] was purified by converting to Schiff base hydrochloride salt **127**. Accordingly, refluxing the crude **126** in acetone, followed by adding concentrated HCl, afforded **127** with an *E/Z* ratio is greater than 99.9:0.1.

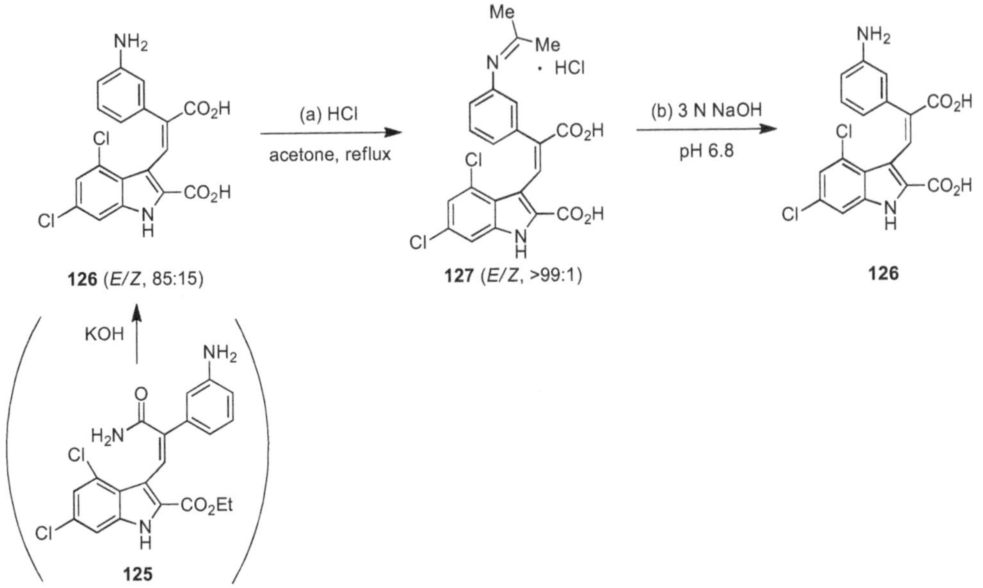

SCHEME 18.10

Procedure

Stage (a)
- Charge 1.017 kg (2.6 mol) of **126** and 14 L of acetone into a 22 L, three-necked reactor.
- Heat to reflux and hold for 30 min.

- Filter the resulting turbid solution into another 22 L, three-necked reactor.
- Rinse with acetone (2 L).
- Charge 200 mL of conc. HCl at 50 °C over 5 min.
- Cool to room temperature over 2 h.
- Filter, wash with acetone (2×1 L), and dry at 25 °C/50 torr.
- Yield of **127**: 989.2 g (78%).

Stage (b)

- Charge 2.66 kg (5.68 mol) of **127** and 26.6 L of water into a 35-L flask.
- Charge 3.1 L of 3 N NaOH to pH 6.8 (from pH 1.3).
- Filter the resulting solution through sintered glass and rinse with water (1 L).
- Charge 3.0 L of 3 N HCl to the combined filtrate to pH 3.1.
- Filter the resulting slurry, wash with water (4×5 L), and dry at 60 °C/50 torr.
- Yield of **126**: 1.97 kg (89%).

18.4.1.2   Acetic Acid Salt

It is challenging to isolate/purify unstable reaction products. One of the frequently used approaches is to convert the unstable reaction product to its corresponding salt. Acetic acid was selected to transfer the unstable amine-free base **129** to the mono-acetic acid salt **130** (Scheme 18.11).[50]

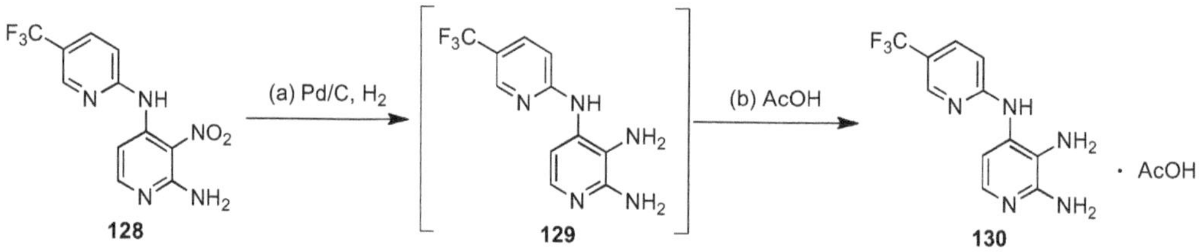

SCHEME 18.11

Procedure

- Charge 6.7 kg (22.39 mol) of **128** into a vessel, followed by adding EtOAc (95.5 L).
- Charge 670 g of 5 wt% Pd/C as a suspension in EtOAc (5 L), followed by adding EtOAc (20.6 L).
- Stir the mixture at 28 °C under H_2 (50 psi) for 18 h.
- Filter and rinse with EtOAc (15.9 L).
- Transfer the combined filtrates into a second vessel.
- Add AcOH (1.5 L, 26.20 mol, 1.17 equiv) over 30 min.
- Filter the resulting slurry, wash with EtOAc (15.9 L), and dry at 40 °C under vacuum.
- Yield: 6.03 kg (91%).

18.4.1.3   (*R*)-Mandelate Salt

Since the isolation of non-crystalline carbamate **132** is not convenient, a telescoping approach was proposed to convert **131** to **133** directly without isolation of **132** (Scheme 18.12).[51] However, the formation of phosphate salt **133** allowed only a minimal purity improvement, leaving impurities in **133** at an unacceptable level.

Thus, a "bridging salt" strategy was employed to purge impurities, and the crystalline (*R*)-mandelate salt **136** was identified as the "bridging salt". As shown in Scheme 18.13, an elegant telescope process furnished **136** in 83% overall isolated yield for the four-step sequence. The expected purity upgrade was realized, evidenced by assay values of 99.5% (HPLC area), 99.2% (wt/wt), and

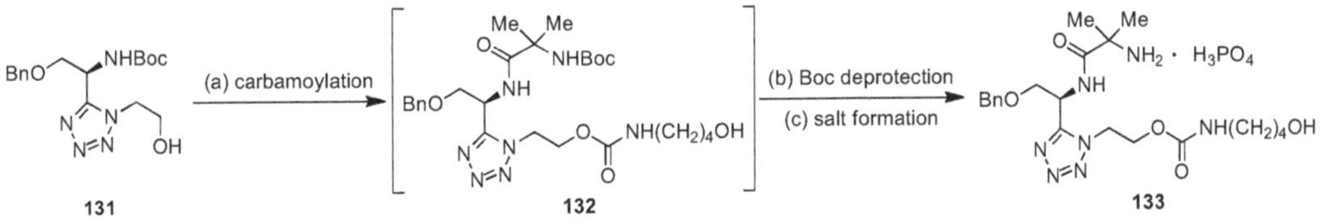

SCHEME 18.12

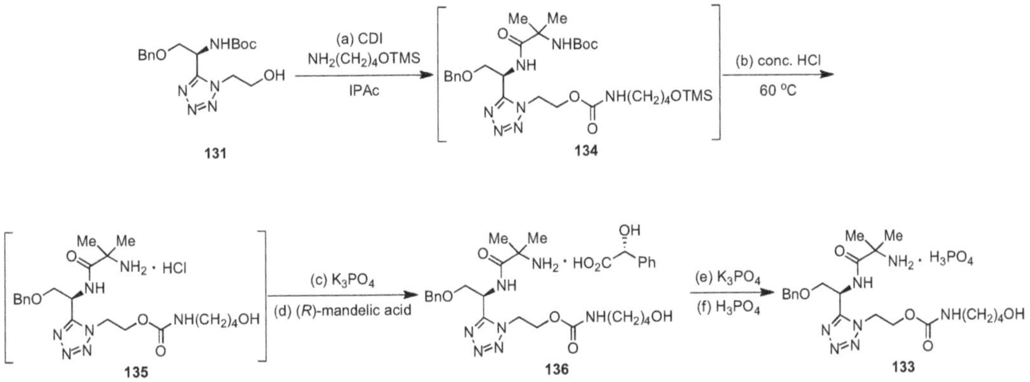

SCHEME 18.13

>99.9% ee. Finally, using the high-purity (*R*)-mandelate salt **136**, the desired phosphate salt **133** was isolated in an 81% yield with greater than 99.5% purity.

18.4.1.4 L-Tartaric Acid Salt

Analogously, L-tartaric acid salt was also utilized to improve the optical purity of chiral sulfoxide. Recrystallization of sulfoxide **137**, a drug candidate as a neurokinin antagonist, with L-tartaric acid removed the dimeric impurity **139** and improved the product ee (Equation 18.37).[52]

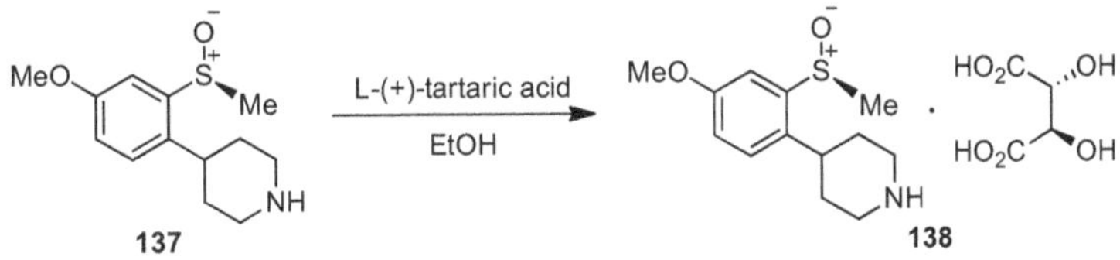

$$(18.37)$$

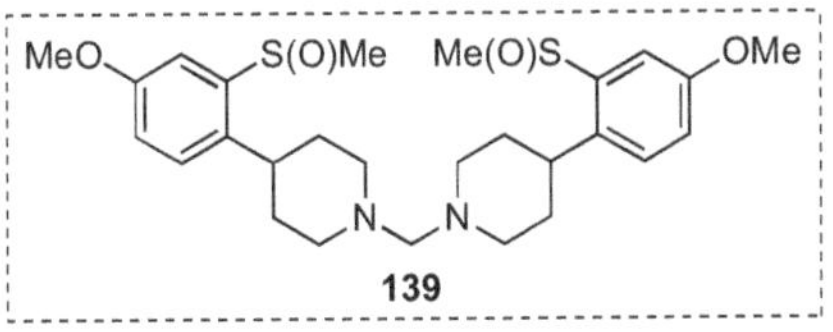

18.4.1.5 2-Picolinic Acid Salt

Tetrahydrocarbazole **142** was identified as a drug candidate for the treatment of human papillomavirus infections. The synthesis of **142** involved the deprotection of **140**, giving the corresponding free amine intermediate. However, isolation of the free amine could not achieve satisfactory purity and yield. Thus, a salt formation approach was developed to isolate the amine intermediate as a 1:1 salt with 2-picolinic acid in 81% yield and 99.2% ee on scale-up preparation (Scheme 18.14).[53]

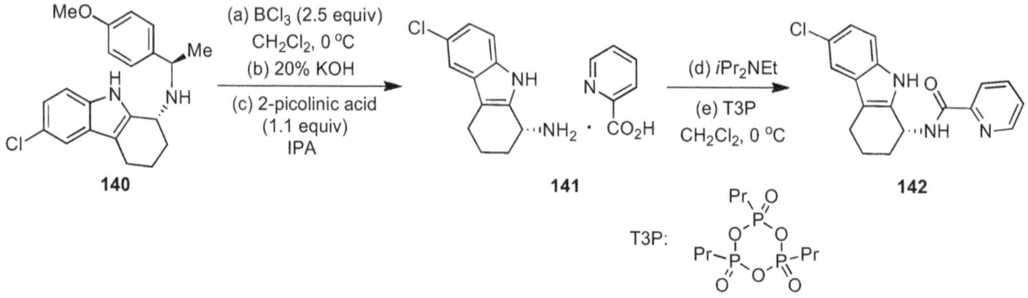

SCHEME 18.14

Procedure

Stage (a)
- Charge 13.5 kg (34.5 mol) of **140** into a 200-gallon reactor, followed by adding CH_2Cl_2 (67.5 L).
- Charge 86.3 L (86.3 mol) of 1 M BCl_3 in CH_2Cl_2 at ca. 25 °C over 30 min.
- Stir at 25 °C for 3 h and cool to 10 °C and stir overnight.

Stage (b)
- Add 135 kg of 20 wt% aqueous KOH over 1 h while maintaining the temperature at ca. 10 °C.
- Warm the resulting mixture to 25 °C.
- Separate layers.
- Extract the aqueous layer with 135 L of a mixture of CH_2Cl_2/MeOH (9:1).
- Wash the combined organic layers with 135 L of 10 wt% NaCl.

Stage (c)
- Charge 4.73 kg (38.3 mol, 1.1 equiv) of 2-picolinic acid to the organic layer.
- Distill at atmospheric pressure to ca. 95 L.
- Add IPA (95 L).
- Distill at atmospheric pressure to ca. 95 L.
- Cool to 10 °C and stir for 1 h.
- Filter, wash with 27 L of a mixture of IPA/CH_2Cl_2 (2:1), and dry at 60 °C under vacuum.
- Yield of **141**: 9.6 kg (81%).

Stage (d)
- Charge 12.9 kg (37.5 mol) of **141** into a 200-gallon reactor, followed by adding CH_2Cl_2 (90 L) and 16.0 kg (7.68 mol) of DIPEA.
- Charge 464 g (3.76 mol, 0.1 equiv) of 2-picolinic acid.

Stage (e)
- Cool the resulting mixture to 0 °C.
- Charge 33.4 kg (52.5 mol) of 50 wt% solution of T3P in EtOAc over 30 min at 0–10 °C.
- Stir the resulting solution at 0 °C for 2 h.
- Add water (90 L) at <25 °C and stir for 30 min.
- Separate layers.
- Extract the aqueous layer with CH_2Cl_2 (65 L).
- Wash the combined organic layers with water (90 L) and 15 wt% NaCl (90 L).

- Distill the organic layer to ca. 52 L at atmospheric pressure (until KF<0.21 wt% of water content).
- Add EtOH (65 L).
- Distill at reduced pressure to ca. 65 L.
- Add EtOH (206 L) and heat the mixture at 76–80 °C to form a solution.
- Filter and cool the filtrate to 20–25 °C.
- Add water (13 L) over 1 h.
- Cool to 0 °C and stir for 2 h.
- Filter, wash with a mixture of EtOH/water (65 L, 1:1), and dry at 80 °C under vacuum.
- Yield of **142**: 9.7 kg (79%).

REMARK
The 2-picolinic acid used in the salt **141** is the same acid that was to be incorporated into the target **142** in the last step, which adds an additional bonus in the final amide formation.

18.4.1.6 Toluenesulfonic Acid Salts

18.4.1.6.1 Improvement of Product Yield and Purity

A common problem in purification via crystallization is the loss of product in the mother liquor, primarily due to the relatively high solubility of the product in the given system. Recovery of the product from the mother liquor, however, suffers, in most cases, unsatisfactory purity and long cycle times. For instance, crystallization of Horner–Wadsworth–Emmons (HWE) olefination product **145** from MeOH/H_2O gave a yield of 70% (50:1, E/Z) with 10% of the product loss in the mother liquor.

$$(18.38)$$

In order to address this issue, the free base **145** was converted to p-toluenesulfonic acid (p-TSA) salt (**145**·p-TSA), which was isolated as a highly crystalline solid in a yield of 77% (Equation 18.38).[54] The free base was obtained, by treatment of the salt with triethylamine, in 98% yield and high purity (E/Z ratio, 1000:1).

Procedure

- Charge 1.8 kg (16.1 mol, 1.5 equiv) of sodium amylate in one portion to a solution of **144** (3.9 kg, 21.4 mol, 2.0 equiv) in THF (30 L) at 17–32 °C.
- Stir the resulting slurry for 2 h and cool to –2 °C.
- Charge 2.3 kg (10.7 mol) of **143** in two equal portions.
- Age the batch at –3 to 2 °C for 23 h.
- Add 85% H_3PO_4 (1.1 L, 16.2 mol) in water (9.3 L) followed by adding a solution of p-TSA monohydrate (2.1 kg, 10.9 mol) in water (1.2 L).
- Age the resulting mixture for 1 h.
- Distill under reduced pressure to ca. 27 L.
- Add water (23 L) and stir at 20 °C for 2 h.
- Filter, wash with water (15 L), and dry under vacuum for 48 h.
- Yield: 3.7 kg (77%) of salt **145**·p-TSA as a yellow crystalline solid.

18.4.1.6.2 Improvement of Physical Properties

Trifluoroacetic acid (TFA)-promoted cyclization of imine **146** was achieved, furnishing the desired azaindole as TFA salt **147** (Equation 18.39).[54] Albeit as a crystalline solid, **147** is hygroscopic, causing problems in the isolation.

(18.39)

In order to avoid isolation of such hygroscopic solids on a large scale, a salt-exchange method was developed by converting TFA salt **147** to p-TSA salt. The conversion of **147** to the p-TSA salt was easily achieved by diluting the TFA salt slurry with water (1.2 vol) followed by adding a solution of p-TSA (2.7 equiv) in IPAc/water (2:1, 15 vol).

Procedure

- Charge TFA (50 L) to a solution of **146** (38.4 mol) in xylene.
- Age the batch at 60 °C for 5 h.
- Cool to room temperature.
- Distill at 10–25 °C under reduced pressure to remove excess TFA (45 L).
- Switch solvent to IPAc and adjust to 60 L.
- Prepare, in a separate container, a homogeneous biphasic solution of p-TSA by mixing 6.5 kg (34.1 mol) of p-TSA, IPAc (25 L), and water (12.5 L).
- Add water (3 L) to the TFA salt slurry.
- Charge the p-TSA solution at 30 °C over 40 min.
- Age the resulting thick slurry at 30 °C for 30 min.
- Cool to room temperature.
- Filter, wash with a mixture of IPAc/MeOH/H$_2$O (100:5:2.5; 10 L) then IPAc (20 L), and dry.
- Yield: 10.8 kg.

REMARK

The p-TSA salt was converted to the free base using aqueous triethylamine to minimize potential ester hydrolysis.

Table 18.2 lists some organic amine salts with acids including both inorganic and organic acids.

18.4.1.7 Sodium Salt

Triazole acid **150** was prepared by coupling aryl iodide **148** with triazole **149** in the presence of potassium carbonate and copper iodide in THF/DMF (Equation 18.40).[55] Under these conditions, the reaction typically reached >98% conversion and afforded a mixture of product **150** and regioisomer **151** in an 81:19 ratio. Attempts to reject the regioisomer **151** by crystallization were not successful due to the lower solubility of this compound compared to that of **150**. Fortunately, the formation of the sodium salt of **150** in THF was able to reject the undesired isomer **151** albeit at the expense of around 15% of the desired isomer. In this way the sodium salt of **150** could be isolated in 65% yield, typically containing approximately 1% of the isomer **151**.

TABLE 18.2
Various Organic Amine Salts for the Isolation/Purification

Acids	Amine Salts	Comments
HCl	(structure: BnO-substituted piperidine, OBn, NH · HCl, BnO, OBn)	The salt was isolated by precipitation in acetone; the salt-break was achieved with 1 N NaOH.[45]
	(structure: formamide/BnO aryl amino-alcohol, HO,,,Ph · HCl)	The salt was prepared by treating the amine with AcOH followed by adding aq NaCl.[47]
	(structure: piperidine CH(OMe)$_2$ · HCl)	The salt was prepared by treating the amine with NH$_4$Cl.[48]
	(structure: triazole F$_3$C, NH · HCl)	The salt was prepared by reaction of the free base with 1.01 equiv of HCl in 2-propanol.[a]
	(structure: Et$_2$N–O–aryl steroid, Me, O, HO · HCl)	The salt was prepared by treating the free base CH$_2$Cl$_2$ solution with aq HCl.[b]
HBr	(structure: benzisothiazolone, S, N, CO$_2$H, Me · HBr, Me)	The salt was isolated from AcOH via direct-drop process. The salt-break was realized by mixing in MTBE/water biphasic system without base.[1]
Sulfuric acid (H$_2$SO$_4$)	(structure: Me, O, cyclopropyl pyrazole, NC, Me, Me, N, SMe · H$_2$SO$_4$)	The salt was prepared by adding sulfuric acid to a solution of the free base in isopropyl acetate at <25 °C.[46]
	(structure: F$_3$C aryl, O, OH, Me, N, Me, Me, Me · H$_2$SO$_4$)	The salt was prepared by adding sulfuric acid to a solution of the free base in THF.[c]
Formic acid	(structure: NHBoc, tBuO$_2$C, S, NH$_2$, Me · HCO$_2$H)	The salt was prepared by adding formic acid (1.0 equiv) to a solution of the amine in ethyl acetate at 5 °C.[d]
Acetic acid	(structure: Cl aryl, N–NH, HN, pyrazole, Me · AcOH)	The salt was prepared by adding glacial acetic acid (1.5 equiv) to the free base solution in toluene/EtOAc. The salt-break was done by adding 2 N NaOH to a mixture of the salt in MeTHF.[e]

(*Continued*)

TABLE 18.2 (CONTINUED)
Various Organic Amine Salts for the Isolation/Purification

Acids	Amine Salts	Comments
Trifluoroacetic acid		The salt was obtained directly from catalytic hydrogenation in a mixture of acetic acid and trifluoroacetic acid.[f]
Oxalic acid		The salt was prepared by adding oxalic acid (1.0 equiv) to EtOAc solution of the amine.[g]
		The salt was prepared by adding oxalic acid to the amine toluene solution at room temperature.[h]
		This salt was made from the oil free base and isolated as a crystalline oxalate salt from MTBE and IPA (1:1).[i]
		This salt was prepared from ethyl acetate; the salt was broken back to the free base with aqueous NaOH.[j]
		L-2-Amino-3-hydroxy-N-pentylpropanamide, an intermediate in the synthesis of the long-acting thromboxane receptor antagonist developed by BMS, was readily prepared and isolated as its oxalate salt in 81% yield and >99% ee through a "one-pot" process from L-serine methyl ester.[k]
		Compared with the salt of dibenzoyl tartaric acid, oxalic acid salt was preferred due to its ease of isolation (reasonably fast filtrations), good chemical stability, the low cost of oxalic acid, and the high potency of the resulting salt (i.e., the low molecular weight of oxalic acid relative to the free base).[l]
		The salt was prepared from CPME and MTBE with oxalic acid (1.1 equiv). Isolation of the oxalate salt allowed for the effective purge of process-related impurities.[m] Neutralization of the salt with aqueous NaOH provided a crystalline free base (5R diastereomer).

(*Continued*)

TABLE 18.2 (CONTINUED)
Various Organic Amine Salts for the Isolation/Purification

Acids	Amine Salts	Comments
Methanesulfonic acid		The salt was prepared in methylene chloride under reflux for 20 h.[n]
Benzenesulfonic acid		The benzenesulfonic acid salt was prepared in EtOAc at 45 °C; the salt-break was effected with aq 1 N NaOH in MeTHF.[o]
p-Toluenesulfonic acid (p-TSA)		The salt was prepared in IPAc. K_3PO_4 was used to break the salt.[69]
		The tosylate salt is a white crystalline solid that allowed isolation of the amine in high purity.[p]
		This bis-tosylate salt was obtained from deprotection of Boc-amine in acetonitrile at 60 °C with 3 equiv of p-TSA·H_2O in 85% yield (29.4 kg).[q]
		Pd-catalyzed hydrogenolysis to remove N-benzyl group was carried out in the presence of p-TSA and the resulting amine product was isolated as the p-TSA salt by adding MTBE.[r]
		Removal of the Boc group from Boc-protected amine with p-TSA resulted in precipitation of the amine tosylate salt that could be isolated by filtration in high yield.[25]
		The acyloxazolidinone, obtained from a telescoped, three-step sequence, was isolated as p-TSA salt.[s]
Malonic acid		The malonate salt was isolated from IPA and MTBE.[63]
Fumaric acid		The fumaric acid salt was isolated from ethanol as a white solid. The salt-break was affected with aqueous K_2CO_3.[t]

(Continued)

TABLE 18.2 (CONTINUED)
Various Organic Amine Salts for the Isolation/Purification

Acids	Amine Salts	Comments
4-Nitrobenzoic acid		Using 4-nitrobenzoic acid permits isolation of Boc-protected pyrrolidine derivative as the 4-nitrobenzoic acid salt.[u]
L-Tartaric acid		The salt was prepared by adding L-tartaric acid into a solution of the amine in ethanol; the salt-break was achieved with 1 N NaOH.[v]
		The salt was prepared by adding L-(+)-tartaric acid ethanol solution to a solution of the amine in ethanol; the salt-break was achieved with aq NaOH.[52]
		The salt was prepared by adding L-(+)-tartaric acid to the free base solution in 1-butanol/MeOH under reflux.[w]
(R)-2-Hydroxy-2-phenylacetic acid ((R)-(−)-mandelic acid)		The salt was prepared by mixing the amine with (R)-(−)-mandelic acid in toluene/acetonitrile at 70 °C.[x]
(R)-(-)-Camphorsulfonic acid		The crude intermediate was crystallized via salt formation with (R)-camphorsulfonic acid to remove residual impurities and also upgrade the ee (from ~98% to >99%).[64]
(S)-(+)-Camphorsulfonic acid		The diastereomeric purity of piperidine could be greatly improved by forming the (S)-(+)-camphorsulfonic acid salt.[y]

[a] Hansen, K. B.; Balsells, J.; Dreher, S.; Hsiao, Y.; Kubryk, M.; Palucki, M.; Rivera, N.; Steinhuebel, D.; Armstrong III, J. D.; Askin, D.; Grabowski, E. J. *J. Org. Process Res. Dev.* **2005**, *9*, 634.

[b] Prat, D.; Benedetti, F.; Girard, G. F.; Bouda, L. N.; Larkin, J.; Wehrey, C.; Lenay, J. *Org. Process Res. Dev.* **2004**, *8*, 219.

[c] Han, Z. S.; Xu, Y.; Fandrick, D. R.; Rodriguez, S.; Li, Z.; Qu, B.; Gonnella, N. C.; Sanyal, S.; Reeves, J. T.; Ma, S.; Grinberg, N.; Haddad, N.; Krishnamurthy, D.; Song, J. J.; Yee, N. K.; Pfrengle, W.; Ostermeier, M.; Schnaubelt, J.; Leuter, Z.; Steigmiller, S.; Däubler, J.; Stehle, E.; Neumann, L.; Trieselmann, T.; Tielmann, P.; Buba, A.; Hamm, R.; Koch, G.; Renner, S.; Dehli, J. R.; Schmelcher, F.; Stange, C.; Mack, J.; Soyka, R.; Senanayake, C. H. *Org. Lett.* **2014**, *16*, 4142.

[d] Rassias, G.; Hermitage, S. A.; Sanganee, M. J.; Kincey, P. M.; Smith, N. M.; Andrews, I. P.; Borrett, G. T.; Slater, G. R. *Org. Process Res. Dev.* **2009**, *13*, 774.

[e] Brandt, T. A.; Caron, S.; Damon, D. B.; DiBrino, J.; Ghosh, A.; Griffith, D. A.; Kedia, S.; Ragan, J. A.; Rose, P. R.; Vanderplas, B. C.; Wei, L. *Tetrahedron* **2009**, *65*, 3292.

[f] Armitage, M.; Bret, G.; Choudary, B. M.; Kingswood, M.; Loft, M.; Moore, S.; Smith, S.; Urquhart, M. W. J. *Org. Process Res. Dev.* **2012**, *16*, 1626.

(Continued)

TABLE 18.2 (CONTINUED)
Various Organic Amine Salts for the Isolation/Purification

[g] Ikunaka, M.; Kato, S.; Sugimori, D.; Yamada, Y. *Org. Process Res. Dev.* **2007**, *11*, 73.

[h] Ikunaka, M.; Maruoka, K.; Okuda, Y.; Ooi, T. *Org. Process Res. Dev.* **2003**, *7*, 644.

[i] Gala, D.; Dahanukar, V. H.; Eckert, J. M.; Lucas, B. S.; Schumacher, D. P.; Zavialov, I. A.; Buholzer, P.; Kubisch, P.; Mergelsberg, I.; Scherer, D. *Org. Process Res. Dev.* **2004**, *8*, 754.

[j] de Koning, P. D.; McAndrew, D.; Moore, R.; Moses, I. B.; Boyles, D. C.; Kissick, K.; Stanchina, C. L.; Cuthberton, T.; Kamatani, A.; Rahman, L.; Rodriguez, R.; Urbina, A.; Sandoval, A.; Rose, P. R. *Org. Process Res. Dev.* **2011**, *15*, 1018.

[k] Reid, J. G.; Chen, B.-C.; Njegovan, J. F.; Quinlan, S. L.; Stark, D. R.; Reynolds, A.-M.; Pickup, B.; Harrison, B. *Org. Process Res. Dev.* **1997**, *1*, 174.

[l] Peng, Z.; Ragan, J. A.; Colon-Cruz, R.; Conway, B. G.; Cordi, E. M.; Leeman, K.; Letendre, L. J.; Ping, L.-J.; Sieser, J. E.; Singer, R. A.; Sluggett, G. W.; Strohmeyer, H.; Vanderplas, B. C. *Org. Process Res. Dev.* **2014**, *18*, 36.

[m] Bowles, P.; Brenek, S. J.; Caron, S.; Do, N. M.; Drexler, M. T.; Duan, S.; Dubé, P.; Hansen, E. C.; Jones, B. P.; Jones, K. N.; Ljubicic, T. A.; Makowski, T. W.; Mustakis, J.; Nelson, J. D.; Olivier, M.; Peng, Z.; Perfect, H. H.; Place, D. W.; Ragan, J. A.; Salisbury, J. J.; Stanchina, C. L.; Vanderplas, B. C.; Webster, M. E.; Weekly, R. M. *Org. Process Res. Dev.* **2014**, *18*, 66.

[n] Braden, T. M.; Coffey, D. S.; Doecke, C. W.; LeTourneau, M. E.; Martinelli, M. J.; Meyer, C. L.; Miller, R. D.; Pawlak, J. M.; Pedersen, S. W.; Schmid, C. R.; Shaw, B. W.; Staszak, M. A.; Vicenzi, J. T. *Org. Process Res. Dev.* **2007**, *11*, 431.

[o] Risatti, C.; Natalie, K. J., Jr.; Shi, Z.; Conlon, D. A. *Org. Process Res. Dev.* **2013**, *17*, 257.

[p] Ripin, D. H. B.; Abele, S.; Cai, W.; Blumenkopf, T.; Casavant, J. M.; Doty, J. L.; Flanagan, M.; Koecher, C.; Laue, K. W.; McCarthy, K.; Meltz, C.; Munchhoff, M.; Pouwer, K.; Shah, B.; Sun, J.; Teixeira, J.; Vries, T.; Whipple, D. A.; Wilcox, G. *Org. Process Res. Dev.* **2003**, *7*, 115.

[q] Ruck, R. T.; Huffman, M. A.; Stewart, G. W.; Cleator, E.; Kandur, W. V.; Kim, M. M.; Zhao, D. *Org. Process Res. Dev.* **2012**, *16*, 1329.

[r] Bret, G.; Harling, S. J.; Herbal, K.; Langlade, N.; Loft, M.; Negus, A.; Sanganee, M.; Shanahan, S.; Strachan, J. B.; Turner, P. G.; Whiting, M. P. *Org. Process Res. Dev.* **2011**, *15*, 112.

[s] Singer, R. A.; Ragan, J. A.; Bowles, P.; Chisowa, E.; Conway, B. G.; Cordi, E. M.; Leeman, K. R.; Letendre, L. J.; Sieser, J. E.; Sluggett, G. W.; Stanchina, C. L.; Strohmeyer, H. *Org. Process Res. Dev.* **2014**, *18*, 26.

[t] Content, S.; Dupont, T.; Fédou, N. M.; Penchev, R.; Smith, J. D.; Susanne, F.; Stoneley, C.; Twiddle, S. J. R. *Org. Process Res. Dev.* **2013**, *17*, 202.

[u] Mangion, I. K.; Chen, C.; Li, H.; Maligres, P.; Chen, Y.; Christensen, M.; Cohen, R.; Jeon, I.; Klapars, A.; Krska, S.; Nguyen, H.; Reamer, R. A.; Sherry, B. D. Zavialov, I. *Org. Lett.* **2014**, *16*, 2310.

[v] Pippel, D. J.; Mills, J. E.; Pandit, C. R.; Young, L. K.; Zhong, H. M.; Villani, F. J.; Mani, N. S. *Org. Process Res. Dev.* **2011**, *15*, 638.

[w] Larrow, J. F.; Roberts, E.; Verhoeven, T. R.; Ryan, K. M.; Senanayake, C. H.; Reider, P. J.; Jacobsen, E. N. *Org. Synth.* **2004**, *10*, 29.

[x] Yue, T.-Y.; Mcleod, D. D.; Albertson, K. B.; Beck, S. R.; Deerberg, J.; Fortunak, J. M.; Nugent, W. A.; Radesca, L. A.; Tang, L.; Xiang, C. D. *Org. Process Res. Dev.* **2006**, *10*, 262.

[y] Girardin, M.; Ouellet, S. G.; Gauvreau, D.; Moore, J. C.; Hughes, G.; Devine, P. N.; O'Shea, P. D.; Campeau, L.-C. *Org. Process Res. Dev.* **2013**, *17*, 61.

$$(18.40)$$

Procedure

- Charge 6.04 kg (23.0 mol) of **148**, followed by adding THF (45 L) and DMF (9 L).
- Charge 218 g (1.15 mol) of CuI and 7.94 kg (57.4 mol) of K_2CO_3.
- Heat the mixture to 40 °C.
- Charge 3.16 kg (4.60 mol) of **149** as a solution in THF (6 L).
- Heat the mixture to 65 °C.
- Add N,N-dimethylethylenediamine (244 mL, 2.30 mol) once the reaction is complete.
- Cool to room temperature.
- Add 3.6 N HCl (36 L).
- Extract with EtOAc (2×30 L).
- Wash the combined organic layers with LiCl solution (2×20 L).
- Distill to exchange solvent from EtOAc to THF (90 L).
- Charge 2.42 kg (25.2 mol) of NaO*t*Bu and stir at room temperature overnight.
- Filter and wash with THF (2×20 L).
- Dry at 40 °C under vacuum to afford 4.22 kg of crude sodium salt.

REMARK

N,N-Dimethylethylenediamine was added to aid in copper removal.

18.4.1.8   Potassium Salt

The preparation of acid **154** was accomplished via ammonium salt **153**, formed by the slow addition of anhydride **152** to an excess of ammonium hydroxide, followed by pH adjustment with aqueous HCl (Scheme 18.15).[56] As a result, the free acid **154** was obtained in 80% yield along with approximately 4% of phthalic acid **155**.

(I) Problems

The acid **154** was found to be unstable upon standing in acidic media even at 0 °C, rapidly hydrolyzed to diacid **155**.

(II) Solutions

To circumvent the instability of **154** in the presence of an acid, the salt-exchange strategy was developed to convert the ammonium salt **153** to potassium salt **156** by the addition of potassium hydroxide. The potassium salt had the favorable properties of being a crystalline, non-hygroscopic solid that could be used directly in the subsequent Hofmann rearrangement. Thus, **156** could be

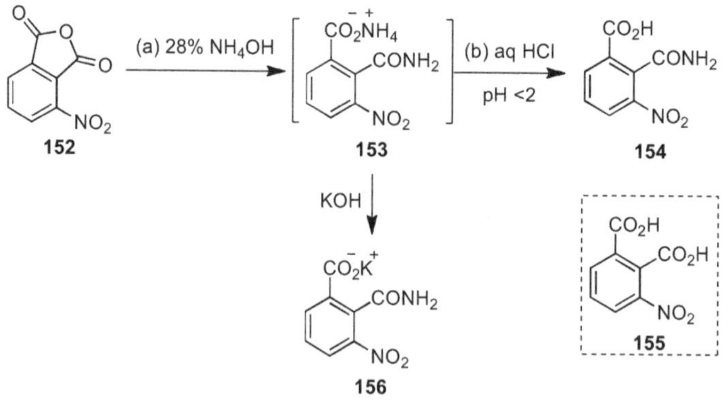

SCHEME 18.15

readily obtained from a reaction of **153** with KOH and isolated by crystallization from 2-propanol in 86% isolated yield with 99.8% purity.

Procedure

- Charge 65.6 kg (525 mol) of 28% NH_4OH into an N_2-purged 115-L glass-lined reactor.
- Cool the content to 0±5 °C.
- Charge 38.4 kg (200.9 mol) of **152** as solid over 2.5 h at <15 °C.
- Charge 12.4 kg (221.4 mol, 1.1 equiv) of KOH in H_2O (19.4 kg) at 20±5 °C.
- Distill under vacuum at <50 °C to reduce the volume to approximately 90 L.
- Add isopropanol (76 kg) at <30 °C.
- Distill under vacuum at <50 °C to reduce the volume to approximately 90 L.
- Add 2-propanol (76 kg) at <30 °C and stir the resulting slurry for <1 h.
- Cool to 0 °C and stir for <1 h.
- Filter and wash with isopropanol (94 kg).
- Dry at 45 °C.
- Yield: 40.7 kg.

18.4.1.9 Magnesium Salt

Scheme 18.16[57] illustrates a one-pot synthesis of sulfide acid **161** starting with propargyl alcohol **157**.

The process involved carbometalllation of propargyl alcohol **157** with aryl Grignard reagent **158**, in which methylmagnesium chloride was employed as a sacrificial alkyl Grignard reagent to deprotonate the propargyl alcohol **157** because methylmagnesium chloride was less reactive toward carbometallation in the absence of CuI. Quenching the resulting vinylmagnesium chloride **159** with CO_2 gave **160** whose reaction with acetic anhydride produced the desired product **161** along with side product **162**.

As shown in Scheme 18.17, the intermediate **160** would react further with residual CO_2 to form **163** that was not able to react with Ac_2O. Instead, **163** was protonated to lead to the side product **162** through **164**. Therefore, in order to suppress the formation of **162**, potassium tertiary butoxide (0.5 equiv) was introduced to destroy residual CO_2 prior to the addition of Ac_2O.

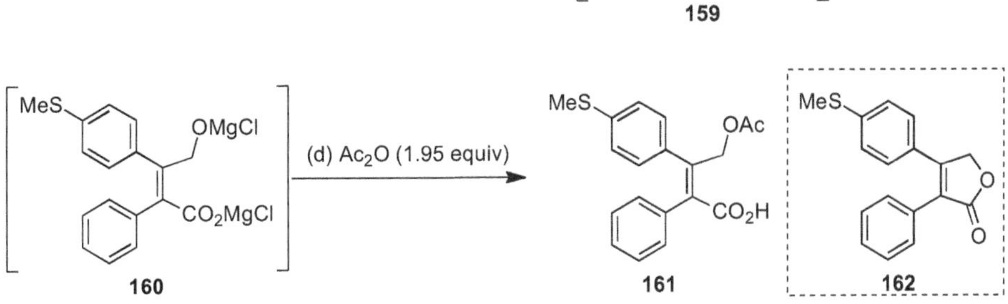

SCHEME 18.16

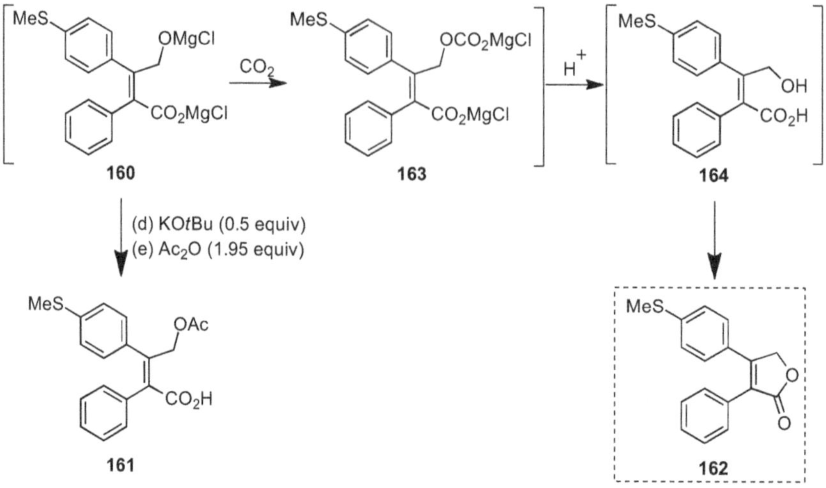

SCHEME 18.17

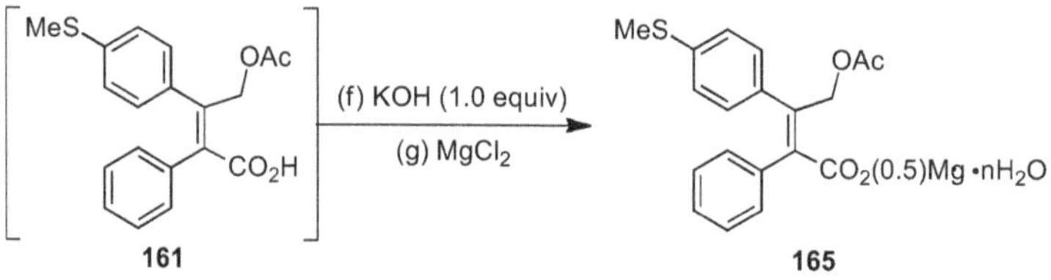

$$(18.41)$$

The isolation/purification of sulfide acid **161** from a reaction mixture without resorting to chromatography posed a considerable challenge. To aid in purification, magnesium salt **165** was prepared by first buffering the crude reaction mixture with KOH (1.0 equiv) within pH 6.4–7.2 range followed by washing with an aqueous magnesium chloride solution (Equation 18.41). This salt **165** was isolated as a pure crystalline solid.

REMARK

The salt-break was easily achieved by dissolving **165** in DMF followed by adding 2 N aqueous AcOH. The precipitated free acid **161** was isolated by filtration.

18.4.1.10 Dicyclohexylamine Salts

Dicyclohexylamine is a popular organic amine that has been used to convert organic acids into their corresponding salts. The salt-break can be accomplished by employing an appropriate acid such as sulfuric acid and citric acid.

18.4.1.10.1 Isolation of Oily Acid

An oily material such as chiral free acid **167** (with an mp of <−40 °C) was converted into its corresponding dicyclohexylamine salt **168** and isolated as a crystalline solid (Scheme 18.18).[26]

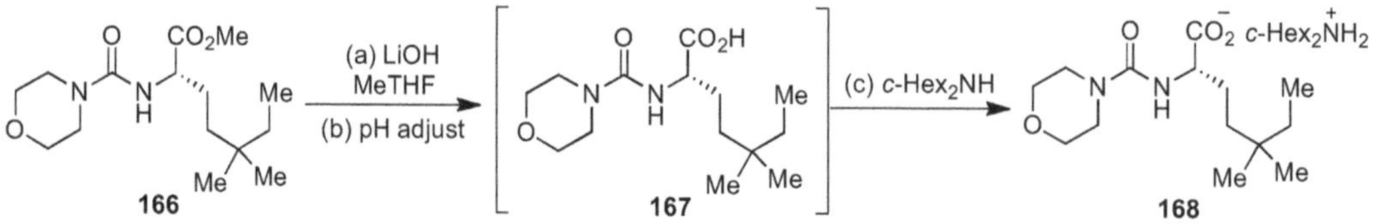

SCHEME 18.18

Procedure

Stage (b) pH adjustment
- Charge 5.57 L of 6 N HCl (33.4 mol, 1.1 equiv) into the reaction mixture from stage (a).
- Separate the organic layer from the aqueous layer.
- Extract the aqueous layer with IPAc (4.6 L).
- Wash the combined organic layers with 10% NaCl (5 L).
- Distill under reduced pressure until KF below 0.3%.
- Dilute with IPAc to ~11 wt% to precipitate a small amount of the racemate followed by filtration.

Stage (c)
- Charge 5.82 kg (31.9 mol, 1.05 equiv) of dicyclohexylamine followed by adding 10 g seed.
- Filter after 4 h of aging.
- Wash with heptane.
- Dry at 45 °C under vacuum.
- Yield of **168**: 15.07 kg (86%).

REMARK
The salt-break was achieved using 1.2 equiv of 0.5 M H_2SO_4 solution.

18.4.1.10.2 Isolation of β-Hydroxyacid

To aid in the isolation of thick oily β-hydroxyl acid **169**, a robust isolation protocol was developed by precipitation of **169** as the dicyclohexylamine salt **170** (Equation 18.42).[58]

(18.42)

Procedure

- Charge 59 kg (325 mol) of dicyclohexylamine into a solution of **169** (163 mol) in diisopropyl ether (IPE).
- Stir the mixture at room temperature for 2.5 h.
- Cool to 5 °C.
- Filter and wash with cold IPE (25 gal).
- Dry at 46 °C.
- Yield: 59.0 kg (68.6%).

REMARK
The use of citric acid to break the salt **170** made the workup easy with a single extraction due to the high water solubility of the dicyclohexylammonium citrate salt.

18.4.1.10.3 Removal of Undesired Diastereomer

A convenient method was developed to isolate the desired diastereomer **172** via the dicyclohexylamine salt formation (Equation 18.43).[59] Treatment of the hydrogenation product mixture (dr=67:33) with dicyclohexylamine allowed to precipitate **172** as the dicyclohexylamine salt, followed by filtration, rejecting the undesired isomer in the mother liquor.

$$(18.43)$$

Procedure

Stage (a)

- Charge 27.7 kg (438.5 mol) of ammonium formate to a solution of **171** (40.7 kg, 146.25 mol) and 5% Pd/C (50% wet, 12.2 kg) in EtOH (407 L), followed by adding 15.5 kg (146.25 mol) of cyclohexylamine.
- Stir the resulting mixture at 40 °C.
- Charge 23.0 kg (292.5 mol) of ammonium formate at below 45 °C.
- Stir the mixture at 40–45 °C for 10 h.
- Filter and rinse with EtOH (122 L).
- Add water (204 L) to the combined filtrates.
- Distill under reduced pressure to ca. 204 L.
- Add EtOAc (285 L).
- Distill again under reduced pressure to ca. 204 L.
- Add EtOAc (407 L) and 10% aqueous NaCl (436 kg)
- Adjust pH to 4.0 with 5 N HCl (34.9 kg).
- Separate layers.
- Wash the organic layer with 10% aqueous NaCl (349 kg) twice, with 20% aqueous NaCl (139 kg).
- Distill to 204 L.
- Add EtOAc (407 L) and distill to 204 L.
- Add acetone (488 L) and distill to 204 L.
- Add acetone (366 L).

Stage (b)

- Charge 26.5 kg (146.25 mol) of dicyclohexylamine.
- Warm to 35 °C and add 20.4 g of seeds.
- Warm to 50 °C and stir for 2 h.
- Cool to 0–5 °C at a rate of 1 °C/min.
- Stir for 2 h.
- Filter, wash with acetone (204 L), and dry at 40 °C under vacuum.
- Yield: 40.5 kg (59.8%, dr=95.0:5.0).

In addition to dicyclohexylamine, other organic amines are also used in the isolation/purification of reaction products. Table 18.3 lists some examples that utilize organic amine salts to purify organic acid products.

18.4.1.11 Quaternary Salt

Among various approaches in process development, chromatography-free synthesis is one of the objectives for large-scale production. An S_N2 reaction of bromide **173** with $1H$-1,2,3-triazole **174** produced a mixture of desired product **175** and side product **176** in 1:1 ratio (Scheme 18.19).[71]

TABLE 18.3

Examples of Salts between Organic Acids and Organic Bases

Organic Amines	Chemical Structures	Comments
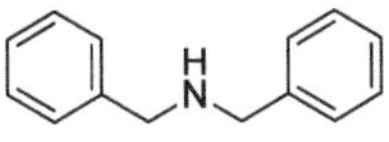 Dibenzylamine (Bn$_2$NH)	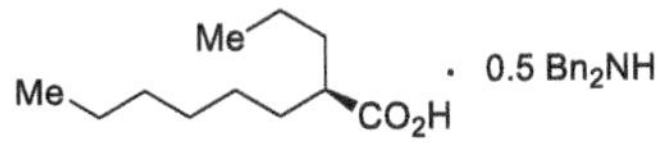 $\cdot$ 0.5 Bn$_2$NH	NaOH was used to break the salt.[60a]
	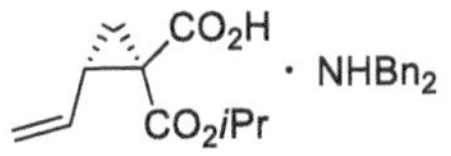 $\cdot$ NHBn$_2$	The optical purity could be increased from 94% ee to >99% ee by recrystallization from IPA/water; 15% (w/w) H$_3$PO$_4$ solution was used to break the salt.[61b]
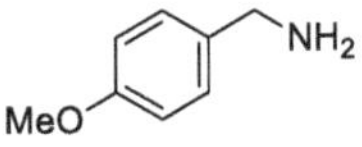 4-Methoxy-benzylamine	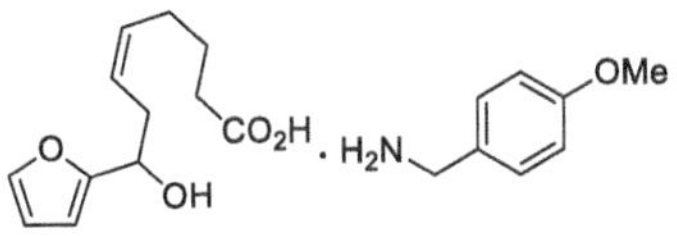 $\cdot$ H$_2$N	Off-white solid, mp 90.6–91.1 °C.[62c]
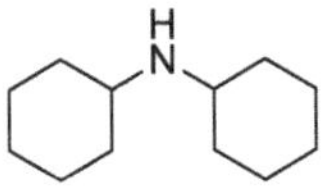 Dicyclohexylamine (c-Hex$_2$NH)	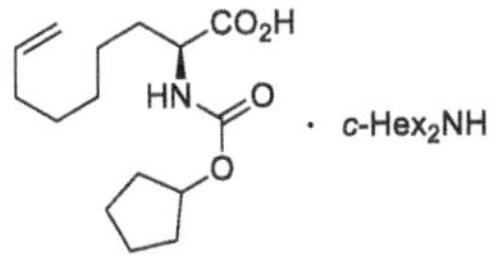 $\cdot$ c-Hex$_2$NH	White solid, mp 90–93 °C.[63d]
	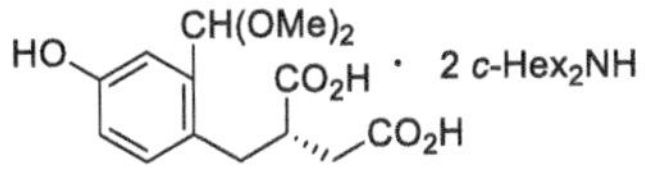 $\cdot$ 2 c-Hex$_2$NH	H$_2$SO$_4$ was used to break the salt.[64e]
	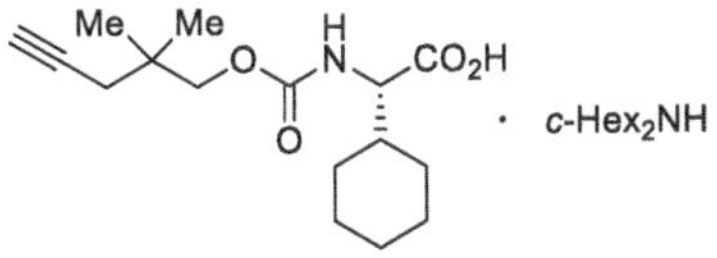 $\cdot$ c-Hex$_2$NH	Mp 128–130 °C.[65f]
	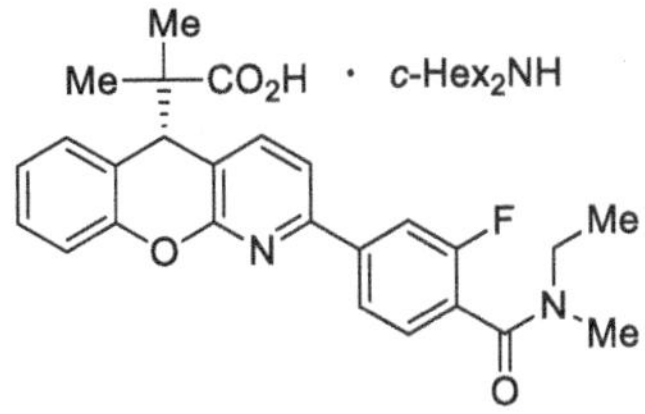 $\cdot$ c-Hex$_2$NH	Solid.[66g]
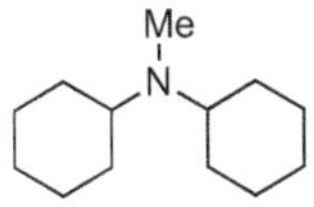 N,N-Dicyclohexylmethylamine (c-Hex$_2$NMe)	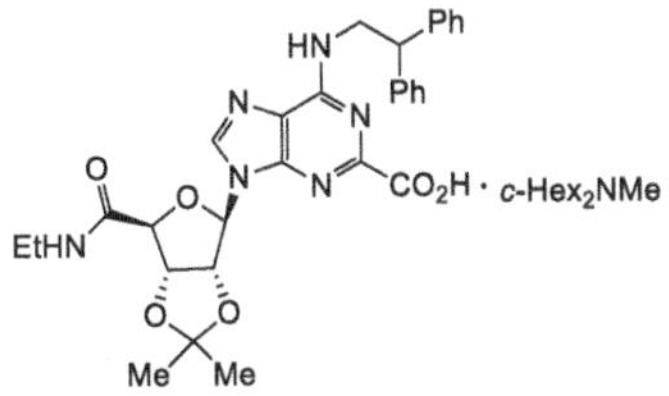 $\cdot$ c-Hex$_2$NMe	HCl was used to break the salt.[24]
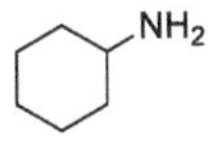 Cyclohexylamine (c-HexNH$_2$)	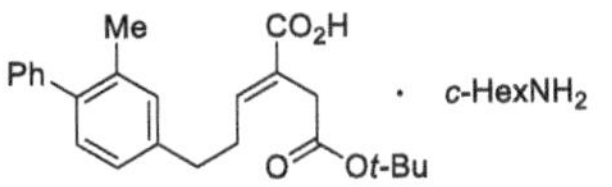 $\cdot$ c-HexNH$_2$	Colorless solid, mp 133–135 °C.[67h]
	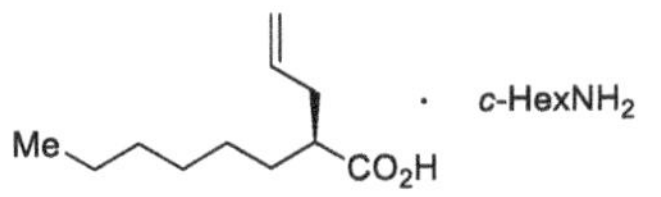 $\cdot$ c-HexNH$_2$	Aqueous HCl was used to break the salt.[68i]

(Continued)

TABLE 18.3 (CONTINUED)

Examples of Salts between Organic Acids and Organic Bases

Colorless solid, mp 168–169 °C dec.[69j]

(R)-(+)-1-Phenylethylamine

Aqueous HCl was used to break the salt.[78]

(R)-(+)-Bornylamine

MsOH was used to break the salt.[70k]

[a] Hasegawa, T.; Kawanaka, Y.; Kasamatsu, E.; Ohta, C.; Nakabayashi, K.; Okamoto, M.; Hamano, M.; Takahashi, K.; Ohuchida, S.; Hamada, Y. *Org. Process Res. Dev.* **2005**, *9*, 774.

[b] Tang, W.; Wei, X.; Yee, N. K.; Patel, N.; Lee, H.; Savoie, J.; Senanayake, C. H. *Org. Process Res. Dev.* **2011**, *15*, 1207.

[c] Henschke, J. P.; Liu, Y.; Huang, X.; Chen, Y.; Meng, D.; Xia, L.; Wei, X.; Xie, A.; Li, D.; Huang, Q.; Sun, T.; Wang, J.; Gu, X.; Huang, X.; Wang, L.; Xiao, J.; Qiu, S. *Org. Process Res. Dev.* **2012**, *16*, 1905.

[d] Wang, X.-J.; Zhang, L.; Smith-Keenan, L. L.; Houpis, I. N.; Farina, V. *Org. Process Res. Dev.* **2007**, *11*, 60.

[e] Wallace, M. D.; McGuire, M. A.; Yu, M. S.; Goldfinger, L.; Liu, L.; Dai, W.; Shilcrat, S.; *Org. Process Res. Dev.* **2004**, *8*, 738.

[f] Song, Z. J.; Tellers, D. M.; Dormer, P. G.; Zewge, D.; Janey, J. M.; Nolting, A.; Steinhuebel, D.; Oliver, S.; Devine, P. N.; Tschaen, D. M. *Org. Process Res. Dev.* **2014**, *18*, 423.

[g] Conlon, D. A.; Natalie, Jr., K. J.; Cuniere, N.; Razler, T. M.; Zhu, J.; de Mas, N.; Tymonko, S.; Fraunhoffer, K. J.; Sortore, E.; Rosso, V. W.; Xu, Z.; Adams, M. L.; Patel, A.; Huang, J.; Gong, H.; Weinstein, D. S.; Quiroz, F.; Chen, D. C. *Org. Process Res. Dev.* **2016**, *20*, 921.

[h] Ashcroft, C. P.; Challenger, S.; Derrick, A. M.; Storey, R.; Thomson, N. M. *Org. Process Res. Dev.* **2003**, *7*, 362.

[i] Hasegawa, T.; Kawanaka, Y.; Kasamatsu, E.; Iguchi, Y.; Yonekawa, Y.; Okamoto, M.; Ohta, C.; Hashimoto, S.; Ohuchida, S. *Org. Process Res. Dev.* **2003**, *7*, 168.

[j] Komatsu, H.; Awano, H. *J. Org. Chem.* **2002**, *67*, 5419.

[k] Chen, J. G.; Zhu, J.; Skonezny, P. M.; Rosso, V.; Venit, J. J. *Org. Lett.* **2004**, *6*, 3233.

Isolation of triazole **175** from the mixture without resorting to column chromatography posed a significant challenge. Based on the fundamental difference in nucleophilicity of isomeric triazoles, a non-chromatographic isolation approach was developed to remove the undesired regioisomer **176**. Accordingly, submitting the mixture of **175** and **176** to a selective *N*-methylation converted **176** into the corresponding water-soluble quaternary salt **177**, which was rejected readily by simple water wash.

18.4.2 DERIVATIZATION

As a purification approach, derivatization has been utilized to isolate unstable, low-melting/oily, or non-crystalline products by converting those products into crystalline solid derivatives.

18.4.2.1 Isolation/Purification of Aldehydes

18.4.2.1.1 Use of Sodium Metabisulfite

Reactions of aldehydes with sodium metabisulfite ($Na_2S_2O_5$) or sodium bisulfite ($NaHSO_3$) have been extensively utilized in the purification of aldehydes, albeit with limited synthetic values.

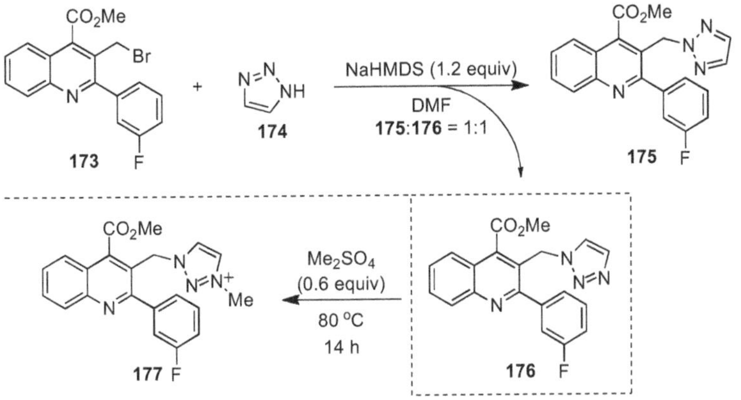

SCHEME 18.19

Treatment of contaminated aldehydes with sodium metabisulfite or sodium bisulfite can lead to aldehyde–bisulfite adducts which precipitated out from the reaction mixture and can be isolated by filtration. The aldehydes can be released from the adducts by treating them with a base such as sodium bicarbonate or sodium hydroxide and the bisulfite is liberated as sulfur dioxide. More information regarding applications of the bisulfite adduct can be found in Chapter 3.

18.4.2.1.1.1 Isolation of Substituted Benzaldehyde

Substituted benzaldehyde **180**, obtained from S_NAr reaction of 4-fluorobenzaldehyde **179**, is an oil, vulnerable to oxidation, and is difficult to handle during isolation/purification. To circumvent these issues, **180** was converted to the bisulfite adduct **181** as a stable solid (Scheme 18.20).[72] Thus, following aqueous workup and solvent exchange to acetonitrile, treatment of the acetonitrile solution of **180** with aqueous sodium metabisulfite gave the solid bisulfite adduct **181**, which was collected by filtration.

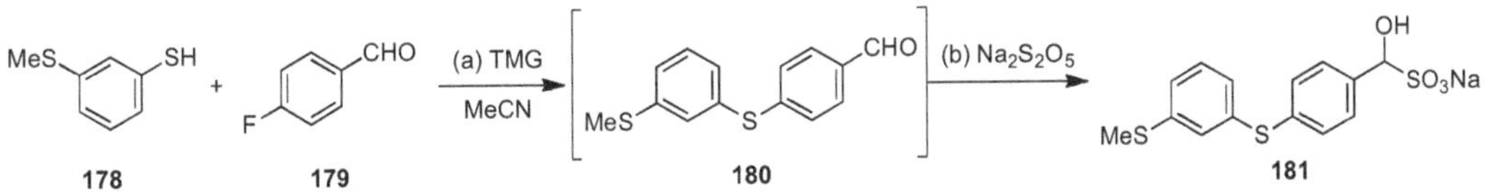

SCHEME 18.20

Procedure

Stage (a)
- Charge 1.06 kg (8.55 mol) of **179** into a solution of **178** (1.33 kg, 8.51 mol) in acetonitrile at 10 °C.
- Charge slowly 1.08 kg'(9.3' mol) of 1,1',3,3'-tetramethyl guanidine (TMG).
- Heat the mixture to 50 °C for 16 h.
- Cool to 20 °C.
- Add ethyl acetate (12 kg) followed by 2 N HCl (6.68 kg).
- Stir for 20 min.
- Separate layers.
- Wash the organic layer with 1 N $NaHCO_3$ (15.42 kg) and brine (1.73 kg NaCl in 6.7 kg of water).
- Add acetonitrile (3.9 kg).

- Distill at atmospheric pressure to 10 L volume.
- Dilute with acetonitrile (11.8 kg) and distill to a volume of 10 L.
- Dilute with acetonitrile (11.8 kg) and distill to a volume of 15 L.

Stage (b)
- Cool to 22 °C.
- Charge 1.88 kg (9.89 mol) of sodium metabisulfite in water (15 kg).
- Stir at 22 °C for 48 h.
- Filter and wash with water (2×10.1 kg).
- Dry at 50 °C under vacuum.
- Yield: 2.18 kg (70%).

18.4.2.1.1.2   Isolation of Biaryl Propanal
Biaryl propanal was obtained from a reaction of biaryl bromide **182** with allyl alcohol **183** (Equation 18.44)[92] under Jeffery's conditions[73] and was conveniently isolated as solid sodium bisulfite adduct **184**.

$$(18.44)$$

Procedure

Stage (a)
- Heat a mixture of 20 g (81 mmol) of **182**, 14 mL (200 mmol) of **183**, 22 g (81 mmol) of TBACl, 17 g (200 mmol) of NaHCO$_3$, 0.91 g (4.0 mmol) of Pd(OAc)$_2$, 2.5 g (8.1 mmol) of tri-*o*-tolylphosphine in MeCN (200 mL) at reflux for 1 h.
- Cool the mixture.
- Add EtOAc (200 mL).
- Separate layers.
- Wash the the organic layer with H$_2$O (2×200 mL), aqueous citric acid (10 %, 100 mL), brine (100 mL).
- Add MgSO$_4$ (20 g) and charcoal (2 g).
- Filter and concentrate.

Stage (b)
- Add MeOH (100 mL) and the aqueous solution of 11.2 g of Na$_2$S$_2$O$_5$ in H$_2$O (20 mL).
- Stir the mixture at room temperature for 16 h.
- Filter and wash with EtOAc (3×20 mL).
- Dry the product **184** under vacuum.
- Yield: 15.9 g (64%), mp 170–178 °C.

18.4.2.1.2   Use of Sodium Bisulfite

18.4.2.1.2.1   Isolation of Aldehyde from Alcohol
Reduction of amino ester **185** with DIBAL-H generated not only the desired aldehyde **186** but also the over-reduction product **187** (in 4.4%) even at −90 °C (Scheme 18.21).[78]

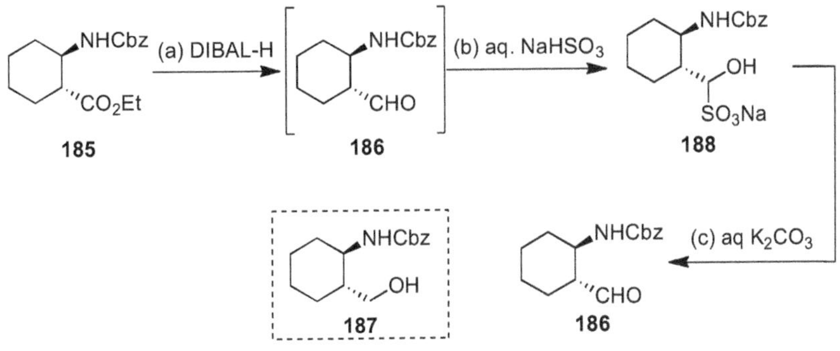

SCHEME 18.21

In order to remove this side product **187** and unreacted starting material **185**, the isolation of **186** was carried out by treating the postreaction mixture with aqueous $NaHSO_3$ solution to selectively precipitate the aldehyde **186** as the bisulfite adduct **188**, followed by filtration. After the liberation of the amino aldehyde **186** from the bisulfite adduct **188** with aqueous potassium carbonate, an 85% yield of **186** was obtained.

Procedure

Stage (a)
- Charge 17.5 kg (57.3 mol) of **185** into a 600-L reactor, followed by adding toluene (167 kg) and methylcyclohexane (27.0 kg).
- Stir to form a solution.
- Cool to −90 to −96 °C.
- Charge 20 wt% solution of DIBAL-H (90.1 kg, 127 mol) in hexanes at <−90 °C.
- Hold for an additional 30 min after the addition.
- Add a mixture of IPA (13.8 kg) and heptane (23.0 kg).
- Transfer the batch into a 1000 L reactor containing a solution of citric acid (54.3 kg, 258 mol) in water (112 L) at 30 °C.
- Separate layers.
- Wash the organic layer with 5% $NaHCO_3$ (147 kg) and 10% brine (150 kg).
- Add IPA (27.5 kg) to the organic layer.

Stage (b)
- Charge 11.9 kg (62.6 mol) of $NaHSO_3$ in water (24.5 L).
- Filter to give **188**.

Stage (c)
- Charge the wet cake of **188** back into the 1000 L reactor.
- Add water (140 L), followed by MTBE (90.9 kg) and 50% aqueous K_2CO_3 (40.8 kg).
- Stir to dissolve the solids.
- Separate layers.
- Wash the organic layer with 15% brine (180 kg) twice.
- Distill under reduced pressure to complete solvent exchange to heptane (final volume ca. 88 L).
- Filter and dry at 30 °C under vacuum.
- Yield of **186**: 12.75 kg (85%).

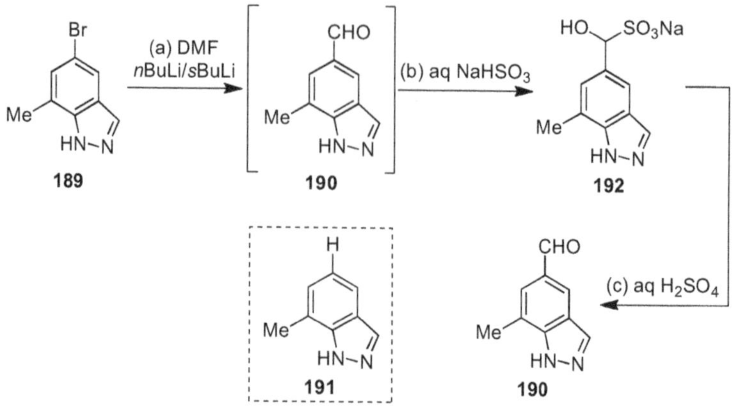

SCHEME 18.22

18.4.2.1.2.2 Isolation of Aldehyde from Indazole

Instead of filtration, isolation of water-soluble aldehyde–bisulfite adduct requires extraction. For instance, the water-insoluble bisulfite adduct **192** was separated from the unreacted bromoindazole **189** and debromo indazole **191** by extracting **192** into the aqueous layer (Scheme 18.22).[74]

Aldehyde **190** was regenerated by treatment of **192** with aqueous sulfuric acid.

Procedure

Stage (a)

- Charge 8.6 kg (40.5 mol) of **189** into a reactor, followed by adding THF (128 L).
- Cool the solution to –70 °C.
- Charge 16.9 kg (60.8 mol) of 2.5 M *n*-BuLi solution in hexanes at <–60 °C.
- Stir at –70 °C for 1 h.
- Charge 54.1 kg (101.3 mol) of 2.0 M *sec*-BuLi solution in cyclohexane at <–60 °C.
- Stir the batch at –70 °C for 1 h.
- Charge 11.8 kg (162.0 mol) of DMF at –55 °C.
- Warm the batch to 5 °C over 2 h.

Stage (b)

- Add water (86 L) at <20 °C and stir for 15 min.
- Separate layers.
- Distill the organic layer at 40 °C under reduced pressure until a final volume of 55 L.
- Extract the concentrated solution with aqueous sodium bisulfite (10%, 5×26 L).
- Add EtOAc (128 L) to the combined aqueous extracts.

Stage (c)

- Add aqueous H_2SO_4 (2 M, 17 L) to pH 8–9 followed by adding NaCl (43 kg).
- Stir the resulting mixture for 30 min.
- Separate layers.
- Distill the organic layer at 45 °C under reduced pressure to a volume of 26 L.
- Add heptane (60 L).
- Continue distillation to <5% v/v of EtOAc.
- Add heptane (34 L) to ajust the volume to 60 L.
- Cool to 5 °C over 2 h and stir at 5 °C for 2 h.
- Filter and wash with heptane (9 L).
- Dry at 40 °C/25 mmHg.
- Yield: 5.84 kg (90.3%).

18.4.2.2 Isolation/Purification of Amine

A protocol was developed for the purification of crude amine **195** by derivatization of **195** with a Boc-protecting group (Scheme 18.23).[75] The resulting Boc-protected amine **196** was a crystalline solid and readily purified by recrystallization. The purified **196** was treated with TFA to remove the Boc group and isolated as the bis-trifluoroacetate salt of **195** in good overall yield (80–85% from the aldehyde) and purity (>98%). The bis-trifluoroacetate salt of **195** was not hygroscopic and could be kept at room temperature for at least six months.

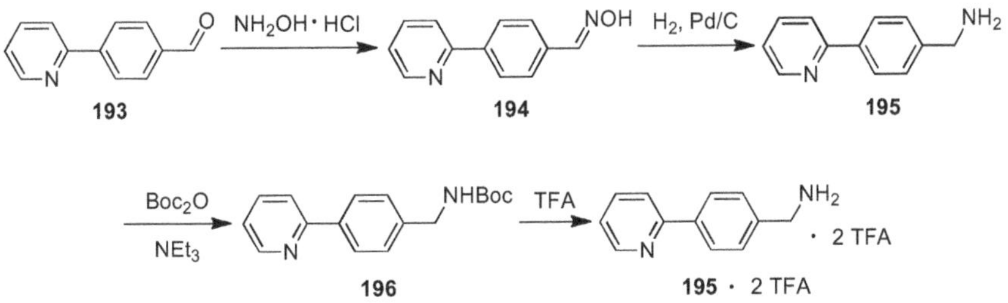

SCHEME 18.23

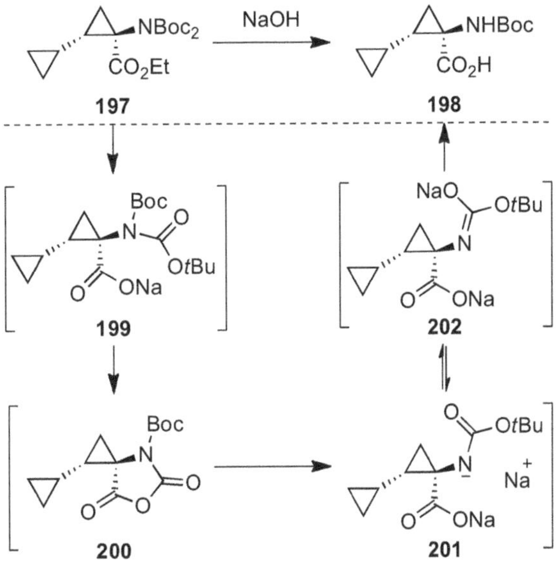

SCHEME 18.24

REMARK

Generally, a Boc-protective group is stable under basic conditions and its deprotection is usually carried out under acidic conditions. However, a selective mono-Boc removal in cyclopropane derivative **197** was achieved under basic ester hydrolysis conditions (Scheme 18.24).[76]

This base-promoted Boc-deprotection is presumably via a spirocarbamate intermediate **200** formed by an intramolecular nucleophilic cycloaddition. The remaining mono-Boc protecting group survived under the basic conditions due to the formation of di-sodium salts **202**.

18.4.2.3 Isolation/Purification of Diol

Utilization of the derivatization approach was nicely demonstrated in the isolation of diol. Separation of the desired *syn*-diol **203** from the *anti*-isomer was realized by converting **203** into its corresponding cyclic borate **204** with boric acid in isopropanol (Equation 18.45),[77] followed by recrystallization of **204**, rejecting the *anti*-isomer into the mother liquor.

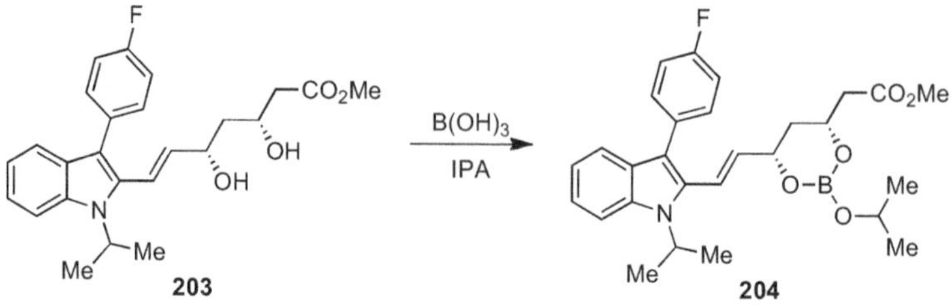

(18.45)

The boron could then be removed by treatment of **204** with methanol; the resulting methyl borate formed an azeotrope with methanol and could be removed by distillation.

18.4.2.4   Isolation/Purification of Amino Diol

The salt formation approach for the isolation/purification of amino diols, such as **206**, may not be applicable due to the potential high-water solubility and hygroscopicity of the corresponding salts. Scheme 18.25[78] describes a practical protocol for the purification of water-soluble amino diol **206**. Conversion of **206** into a mixture of Schiff base **207** and oxazolidines **208** and **209** with benzaldehyde, followed by extraction and acidic workup, furnished the amino diol as its hydrochloride salt **210**.

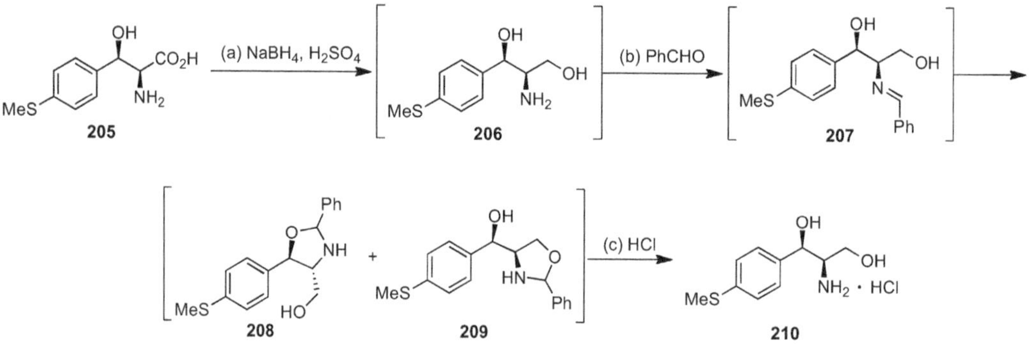

SCHEME 18.25

18.4.3   Various Approaches for Impurity Removal

18.4.3.1   Removal of Ammonium Chloride

A convergent process for the preparation of amide **215** was developed by coupling amine hydrochloride salt **213** with acid **214** as shown in Scheme 18.26.[79]

(I) Problems

The amine hydrochloride salt **213** was contaminated with residual ammonium chloride that was carried in from reductive amination and esterification. The presence of ammonium chloride led to the formation of amide impurity **216** in the amide formation step. Removal of **216** from **215** was challenging.

(II) Solutions

Normally, the inorganic salt can be removed by dissolving the organic compound in a solvent and washing the organic solution with water. However, this approach could not be applied in the purification of a highly water-soluble compound such as **212**. Alternatively, the ammonium chloride was removed under anhydrous conditions by treating the methanol solution of the hydrochloride salt of

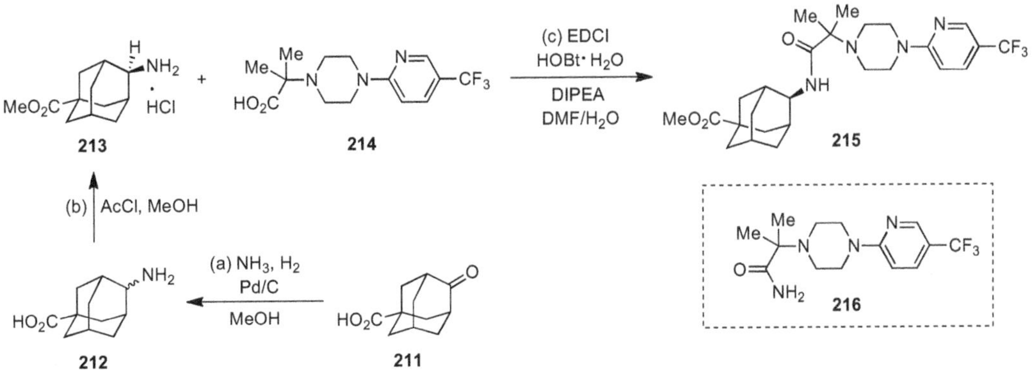

SCHEME 18.26

213 with diisopropylethylamine (DIPEA) to convert ammonium chloride to ammonia. The subsequent evaporation under reduced pressure removed efficiently the liberated ammonia.

Procedure

Stage (b)

- Charge MeOH (7.2 kg) into a 25 L glass flask.
- Cool to 0 °C.
- Charge 2.30 kg (29.30 mol) of AcCl at <30 °C.
- Stir the resulting solution at 25 °C for 30 min.
- Charge 0.90 kg (4.61 mol) of 212 into a 50-L reactor, followed by adding the HCl/MeOH solution.
- Heat the content to 45 °C and stir at that temperature for 16 h.
- Cool to 30 °C.
- Add MeCN (7.2 kg) and distill to near dryness.
- Add MeCN (7.2 kg) and distill to near dryness.
- Add MeCN (7.2 kg) to form a slurry.
- Filter, wash with MeCN (7.2 kg), and dry at 50 °C under vacuum for 12 h to give crude 213.
- Charge the crude 213 into a 25-L reactor.
- Add MeOH (7.5 kg) and 0.9 kg (6.97 mol) of DIPEA.
- Stir for 2 h.
- Distill to dryness to remove the ammonia.
- Add MeOH (7.5 kg) and stir to dissolve the solids.
- Repeat the distillation.
- Add MeOH (7.5 kg) to a separate container and cool to 0 °C.
- Charge 0.5 kg (6.37 mol) of AcCl at <30 °C.
- Stir at 25 °C for 30 min.
- Add HCl/MeOH solution to the 25-L reactor and stir to dissolve all solids.
- Distill to near dryness.
- Add EtOAc (32 kg) and IPA (3.4 kg) into a separate container.
- Add ca. 1/3 of the EtOAc/IPA to the 25-L reactor and stir to form a slurry.
- Filter, wash in two portions with the remaining EtOAc/IPA, then with EtOAc.
- Dry at 55 °C.
- Yield of 213·HCl salt: 0.90 kg (74.2%).

NOTE

The DIPEA hydrochloride was removed by reslurry of the solids in ethyl acetate/IPA followed by filtration.

18.4.3.2 Removal of 9-BBN

A one-pot preparation of silane **219** was outlined in Reaction (18.46)[80] involving a regiospecific hydroboration of vinyl silane **218** with an excess of 9-BBN and Suzuki coupling with bromide **217**.

$$(18.46)$$

A convenient approach was developed to remove the unreacted 9-BBN. It was found that 9-BBN can form a 1:1 adduct with ethanolamine, which was insoluble in hexane or toluene. Thus, the reaction mixture was treated with ethanolamine and the resulting 9-BBN-ethanolamine adduct was removed by simple filtration.

Procedure

Stage (a)
- Charge 9.1 kg of (74.6 mol, 1.4 equiv) of 9-BBN to an inerted 400-L reactor.
- Reinert the reactor.
- Charge toluene (66 kg) and 11.9 kg (62 mol, 1.17 equiv) of **218**.
- Heat the resulting mixture at 70 °C for 2 h.
- Cool to 22 °C.

Stage (b)
- Charge 46.9 kg of 3.2 M NaOH (134 mol, 2.5 equiv).
- Agitate at 22 °C for 45 min.

Stage (c)
- Charge 17.8 kg (53 mol) of **217**, followed by adding 4.22 kg (16 mol, 30 mol%) of PPh$_3$, and 178 g (1 mol, 1.9 mol%) of Pd(OAc)$_2$.
- Stir at 40 °C with 140 rpm for 30 min, heat to 70 °C over 3 h, and maintain at 70 °C for 10 h.
- Cool to 22 °C.
- Charge 9.1 kg (149 mol, 2.8 equiv) of ethanolamine at <50 °C over 20 min.
- Stir at 22 °C for 1 h.
- Filter off the 9-BBN–ethanolamine complex and rinse with toluene (31 kg).
- Separate layers.
- Extract the aqueous layer with toluene (20 kg).
- Wash the combined organic layers with water (50 kg).
- Distill the organic layer to minimal stirrable volume (127 L of distillate collected).
- Cool to 22 °C to give a crude **219**.

18.4.3.3 Removal of Acetic Acid

A one-pot synthesis of 2-aminoindan was realized by oximation of ninhydrin **220** followed by catalytic reduction of the resulting oxime intermediate **221** (Scheme 18.27).[81] During the isolation of the HCl salt **222**, the main difficulty encountered was the formation of a strong emulsion upon basification with sodium hydroxide. This emulsion was caused by acetic acid carried over from the oxime formaton step. To avoid this problem, acetic acid was removed by azeotropic distillation with xylene; as a result, a clean layer separation was realized upon basification. After the layer separation, 2-aminoindan hydrochloride salt **222** was precipitated by adding HCl and isolated in 66% yield.

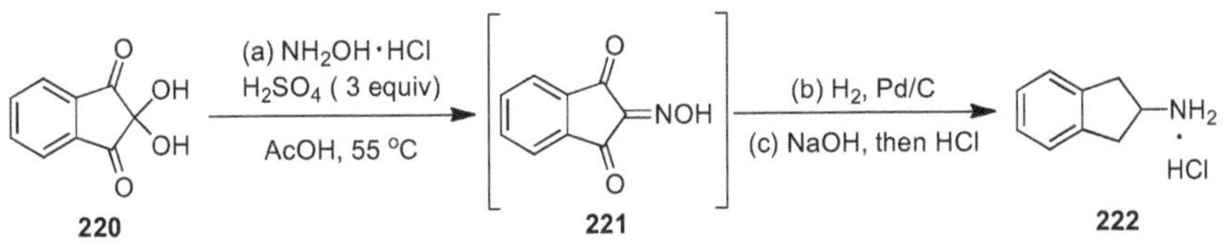

SCHEME 18.27

REMARK

Azeotropic removal of acetic acid with toluene was not efficient (i.e., the toluene/acetic acid azeotrope contains 28% acetic acid), giving 2-aminoindan hydrochloride **222** in only 44% yield, while xylene forms azeotrope with 71% of acetic acid, leading to a more thorough removal of acetic acid.

18.4.3.4 Removal of (1*E*,3*E*)-Dienol Phosphate

Isolation of (1*E*,3*Z*)-dienol phosphate **223a** from the mixture of **223a** and (1*E*,3*E*)-dienol phosphate **223b** was rather challenging. Because of steric hindrance, the Diels–Alder reaction of the mixture of **223a** and **223b** (70/30) with maleic anhydride produced tetrahydroisobenzofuran-1,3-dione **224** with **223a** intact. Therefore, a purification approach was developed utilizing the Diels–Alder reaction of maleic anhydride (Scheme 18.28).[82] The resulting Diels–Alder adduct tetrahydroisobenzofuran-1,3-dione **224** was in turn hydrolyzed to water-soluble diacid sodium salt **225** by treatment of the reaction mixture with aqueous NaOH and rejected into the aqueous layer during extractive workup.

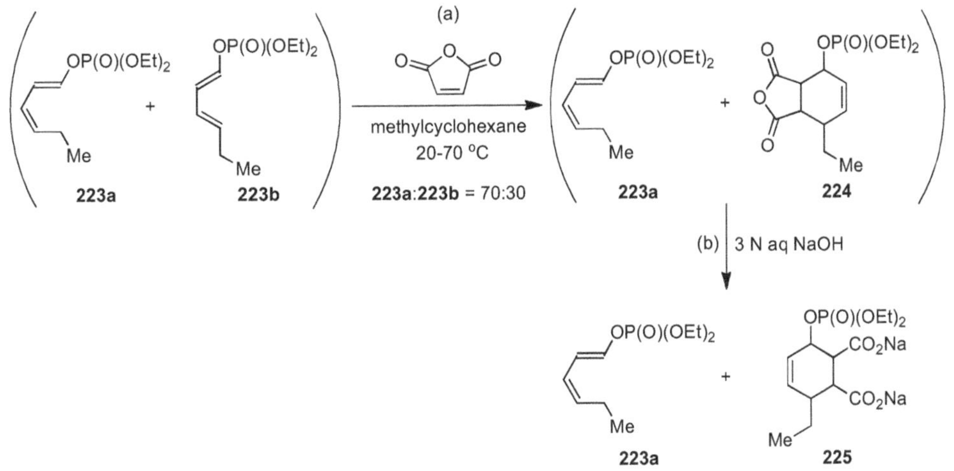

SCHEME 18.28

Procedure

- Add a mixture of isomers **223a** and **223b** (310 g, 1.25 mol, 70/30) into a 2-L reactor, followed by adding maleic anhydride (122 g, 1.25 mol, 1.0 equiv), and methylcyclohexane (1 L).
- Heat the batch at 70 °C for 3 h.
- Cool to 20 °C and stir at that temperature overnight.
- Add 2 N aqueous NaOH (1 L) at 40 °C (pH was then 14).
- Separate layers.

- Extract the aqueous layer with heptane (2×600 mL) and wash the combined organic layers with saturated aq NaCl (2×600 mL).
- Concentrate the organic layer under reduced pressure.
- Dissolve the residue in MTBE (200 mL), filter, and concentrate the filtrate under reduced pressure.
- Yield of **223a**: 210.1 g (68%) as an orange liquid.

18.5   CRYSTALLIZATION

Finishing technologies, such as crystallization, filtration, and drying, are crucial elements in process development for manufacturing active pharmaceutical ingredients (APIs). Crystallization is a major unit operation and is well established as an essential separation and purification technique in the pharmaceutical industry. Although a continuous crystallization using a continuous oscillatory baffled crystallizer (COBC) was recently developed,[83] which offers more control over the crystallization process, most crystallizations are carried out batchwise. Problems associated with batch crystallization include (1) inconsistencies of batch-to-batch in terms of the crystal size and crystal size distribution (CSD) and (2) the purity profile (residual impurities in crystals, wrong polymorph, or poor chiral purity). These issues can have a significant impact on both product quality and the performance of downstream processes such as filtration, drying, milling, and the product formulation and biopharmaceutical characteristics of the drug product. Therefore, considerable time and effort are invested in the development of the crystallization process in order to obtain desired and consistent crystal properties.

Monitoring and control of the crystallization process are critical to meet special requirements for products such as desired crystal morphology and proper crystal size distribution. Focused beam reflectance measurement (FBRM) and particle vision measurement (PVM) are powerful tools developed by Lasentec for in-line real-time measurement of particle size and morphology.[84] FBRM is a probe-based high solid concentration particle characterization tool with measurements of particle size in the range of 0.25–1000 μm.[85] PVM is a high-resolution video microscope, which is typically used for in-process high-resolution imaging of particles within the process environment.[86]

In general, a crystallization process contains operations such as heating, cooling, and filtering. Black's rule[87] states that solubility doubles every 20 °C; thus, the yield of cooling crystallizations can be expected according to Equation 18.47:

$$\text{Yield} = 1 - \frac{1}{2^{\Delta T/20}} \tag{18.47}$$

For example, if a temperature in the crystallization is 60 °C greater than the isolation temperature, according to Black's rule, on average, the cooling crystallizations may be expected to give a yield of 87.5% (Equation 18.48):

$$\text{Yield} = 1 - \frac{1}{2^{\Delta T/20}} = 1 - \frac{1}{2^{60/20}} = 87.5\% \tag{18.48}$$

Usually, a non-linear cooling protocol is preferred. The rule of thumb for "optimum" cooling of a crystallization process would be to cool slowly at first, while the concentration of solids, and hence surface area for growth, is low. As the concentration of solids increases and the surface area is increased, an increase in cooling rate is applied to maintain the supersaturation.

In most cases, crystallization needs to combine cooling with the addition of an anti-solvent. Basically, there are two ways of performing anti-solvent additions, adding it at the beginning of the process to help induce nucleation or adding it at the end of the cooling period to help enhance yield. Evaporation of solvents can also facilitate the crystallization process.

REMARK

A "good" solvent should have a solubility of ~100 mg/mL for a given material, while a "poor" solvent (or anti-solvent) may dissolve the material in less than 20 mg/mL at room temperature.

Although several techniques have been proposed for the in situ generations of seeds via controlled nucleation/dissolution events,[88] seeded crystallization is still predominantly applied in the chemical and pharmaceutical industries. The method of seeding plays an important role in defining the properties of crystals produced, such as polymorphism,[89] and in controlling crystal size distribution.[90] Development of a seed-induced crystallization process requires the optimization of several parameters, for example, temperature, saturation conditions at the point of seed addition, the amount of seed, and the cooling profile. Seed loading varies from as low as 0.5% to as high as 10%. Seed can be added in dry or slurry form. Adding the seed in the slurry form (either in anti-solvent or mother liquor) may help the seed disperse to its primary crystal size. The addition of dry seed may induce unwanted solvent entrapment and agglomeration, as the dry seed may have aggregated during storage/preparation. Furthermore, the use of seed in slurry or dry form may impact the crystallization kinetics. For example, during the seed-induced crystallization of Boc-protected piperazine derivative, a shorter desaturation time (2.6 h) was observed with the dry seed when comparing >5 h of desaturation time with seed in the slurry form (Figure 18.2).[91]

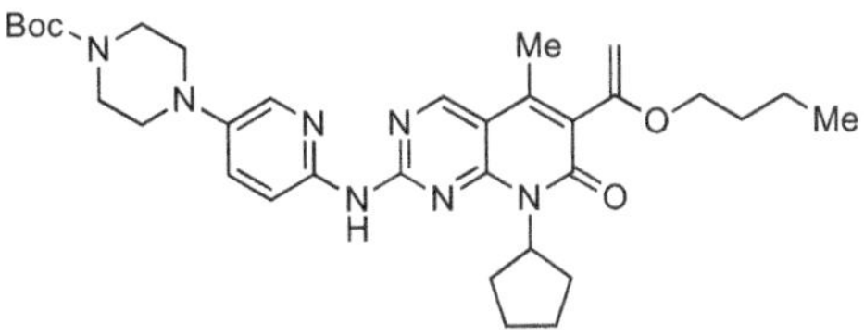

FIGURE 18.2 The structure of Boc-protected piperazine derivative.

Generally, several steps are involved during the preparation of seeds, such as milling, blending, grinding, sieving, and washing.[92] Fine particles in the seeds should be avoided, since fine particles can clog filters and agglomerate, causing solvent inclusion in the product.

One of the main difficulties in batch crystallization is accomplishing a uniform and reproducible CSD. Seed-induced crystallization allows crystallization to occur in a controlled fashion, resulting in fast filtration at scale. In order to determine the region in which seed addition should lead to a controlled crystallization and good filtration rates, kinetic and thermodynamic solubility data are required. Figure 18.3 describes the relationship between kinetic and thermodynamic solubilities. Identifying the meta-stable zone is usually the first step in crystallization development as it provides the operating envelope to favor crystal growth over nucleation. Unless automation is applied, precise determining the meta-stable zone can be a tedious process. Determining the meta-stable zone for cooling crystallization requires some attention, as the meta-stable zone width is typically a function of the cooling rate employed.[93] A wide meta-stable zone is favorable since it will provide a flexible opportunity to run polish filtration.

18.5.1 SEED-INDUCED CRYSTALLIZATION

Crystallization is the most general practice for process chemists to purify process intermediates and final drug substances. The development of practical and robust crystallization processes is challenging due to various potential processing issues. Commonly encountered challenges include low product purity and yield, oiling out, polymorph control, and compound stability during processing. Other issues related to particle size and particle size distribution (PSD), crystal habit, filterability, crystal attrition during the crystallization and processing, crystal agglomeration, and the bulk powder properties associated with formulation, such as flowability and bulk density.

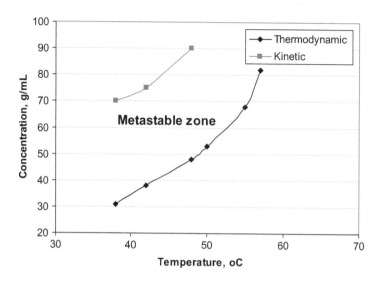

FIGURE 18.3 Kinetic versus thermodynamic solubilities.

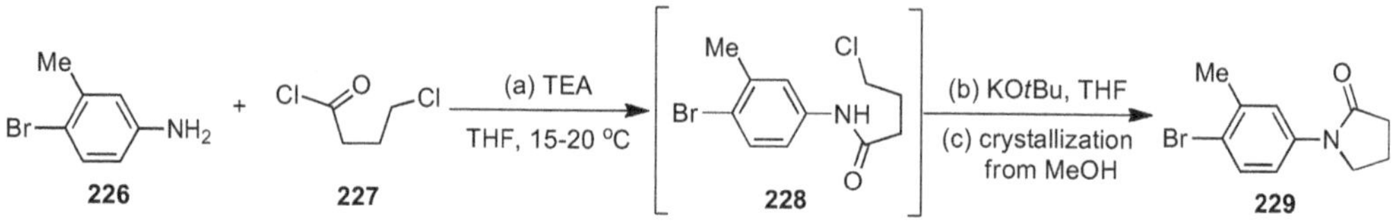

SCHEME 18.29

18.5.1.1 Avoiding Uncontrolled Crystallization

Rapid crystallization may occur during a reactive crystallization such as salt formation by adding an acid to the free base, which leads to a high supersaturation of the salt in the reaction system and a large degree of uncontrolled nucleation and growth. As a result, excessive fines and wide particle size distribution[94] will result in slow filtration and inefficient washing. A sudden crystallization would make the crystallization process lose of control, potentially leading to damage to the agitator.

18.5.1.1.1 Immobile Suspension

Extremely rapid crystallization of aryl bromide **229** occurred at 20–25 °C, resulting in an immobile suspension. In order to mitigate the risk of the uncontrolled-crystallization, seeding the solution of **229** at 50–55 °C led to a gradual crystallization of the product, and the resulting suspension was filtered to give the product **229** in reproducible yield and purity (Scheme 18.29).[95]

Procedure

Stage (a)
- Charge 4.25 kg (42.04 mol, 1.03 equiv) of TEA to a mixture of 7.75 kg (98 %, 40.82 mol) of **226** in THF (116 L) at 20–25 °C.
- Cool the resulting solution to 15 °C.
- Charge 5.98 kg (99.2%, 42.04 mol, 1.03 equiv) of **227** at 15–20 °C over 1 h.
- Stir the resulting suspension at 15–20 °C for 1 h.

Stage (b)
- Cool the batch down to 0–5 °C.
- Charge a solution of KOtBu (11.05 kg, 99.5 %, 98.01 mol, 2.40 equiv) in THF (93 L) at 0–5 °C over 1 h.
- Stir the batch at 0–5 °C for 15 min.

Stage (c)
- Add water (51 L) and EtOAc (67 L).
- Separate layers.
- Wash the organic layer with 2 N HCl (51 L) and brine (39 L).
- Distill under atmospheric pressure to remove ca. 230 L of solvents.
- Replace EtOAc with MeOH.
- Cool the resulting methanolic solution to 50–55 °C.
- Add seeds and cool to 15–20 °C (crystallization occurred at 44 °C).
- Filter and wash with cold (0–5 °C) MeOH (9 L).
- Dry at 40 °C under vacuum (1 mmHg).
- Yield: 7.5 kg (74 %).

18.5.1.1.2   Crystallization on Reactor Walls

Conversion of *R*-mandelate salt **230** to the desired phosphate salt **133** was accomplished in two stages: salt-break with potassium phosphate and treatment of the resulting free base with phosphoric acid. The phosphate salt formation proved to be problematic due to the exceedingly low solubility of **133** in the crystallization medium. As a result, the addition of phosphoric acid resulted in uncontrolled crystallization on the reactor walls. Therefore, using the seed-induced crystallization by charging the seed crystals of **133** prior to the controlled introduction of phosphoric acid provided controllable crystallization, giving 81% isolated yield of **133** with greater than 99.5% purity (Equation 18.49).[51]

$$(18.49)$$

Procedure

Stage (a)
- Charge 2.37 kg of **230** into a glass-lined 100-L reactor followed by adding water (40.17 kg).
- Stir to effect dissolution.
- Add EtOAc (15.5 kg).
- Adjust pH to 8.5–9.0 by adding 2.0 M K_3PO_4 solution (ca. 2.5 kg).
- Stir for 15 min and stop agitation.
- Separate layers.
- Extract the aq layer with EtOAc (2×15.5 kg).
- Wash the combined organic layers with 0.075 M K_3PO_4 (10.2 kg) and 5.0% aqueous NaCl (10.5 kg).
- Distill the organic layer at 35–50 °C under reduced pressure to ca. 30 L.
- Filter the batch through a 5 μm cartridge filter and transfer the filtrate into another glass-lined 100-L reactor.
- Adjust the volume to 36 L by adding filtered EtOAc.

Stage (b)
- Charge 0.36 kg (1.0 % v/v) of water into the reactor.
- Heat the batch to 50 °C.
- Charge, with vigorous agitation, a small portion (ca. 0.5 L) of a pre-prepared aqueous solution of H_3PO_4 (0.365 kg, 0.97 equiv), water (0.36 kg), and EtOAc (32.0 kg) at a rate of ≤100 mL/min.

- Charge 34 g (1.9%) of seed.
- Charge the rest of the acid solution over 3 h.
- Cool the batch to 20 °C over ca. 1 h and age at the temperature for 1 h.
- Filter and wash with EtOAc (18.1 kg).
- Dry at 50 °C under 25 mmHg.
- Yield: 1.8 kg (81.2 %).

18.5.1.2 Avoiding Oiling Out

Common issues during crystallization are oiling out or forming a gumlike solid. Oiling out usually occurs when a mixture of solvents is used, and in this case, the solute is generally not evenly distributed between the two phases. The large difference in affinity of the target compound for each solvent can be the driving force for the liquid–liquid phase separation. As the oil phase intends to trap impurities, the isolated product is usually in poor purity. Besides, the oil phase can pass through the filter into the waste receiver. As a result, low isolated product yields are expected due to incomplete crystallization. For instance, the workup of the deamination of **231** (Equation 18.50)[96] by adding aqueous ammonia gave the crude free base **232** in low isolated yields. The free base formation with aqueous ammonia led initially to oily product **232**, and during the profile cooling **232** eventually crystallized. However, observation of the change from oil to solid was difficult due to the very similar visual appearance of both phases. As a result, the un-solidified oil passed through the filter into the waste receiver.

$$(18.50)$$

NOTES

(a) Sulfamic acid was used to destroy excess nitrous acid (Equation 18.51).

$$NH_2SO_3H + HNO_2 \longrightarrow H_2SO_4 + N_2 + H_2O \qquad (18.51)$$

(b) Alkyl aryl ketone API compound **233** could only be obtained as an amorphous solid. One possible reason for the disinclination to crystallization is the dynamic equilibrium between the open-chain ketone **233** and the ring-closed lactam **234** (Equation 18.52):[97]

$$(18.52)$$

18.5.1.2.1 Amide Formation

The use of anti-solvent in crystallization is a common approach to increase the recovery yield. The selection of the anti-solvent can be critical for the success of the crystallization. For instance, the isolation/crystallization of drug candidate **235** by adding ethanol anti-solvent to the reaction mixture in 2-butanone made the product **236** be oiled out (Equation 18.53).[98]

(18.53)

Changing the anti-solvent from ethanol to isopropanol avoided the product oiling out and under optimal conditions (a ratio of 65:35 of 2-butanone/IPA) **236** was precipitated out from this solvent mixture upon cooling.

Procedure

Stage (a)
- Charge 21.4 kg (94.1 wt%, 63.5 mol) of **235** and 2-butanone (114 kg) at 25 °C.
- Charge a mixture of 12.4 kg (1.2 equiv) of CDI and 2-butanone (33 kg) at 20 °C over 30 min.
- Rinse with 2-butanone (16 L) and stir for 1 h.
- Charge 1.72 kg (0.17 equiv) of CDI and 2-butanone (4 kg).
- Stir for 7.5 h.

Stage (b)
- Charge a solution of benzylsulfonamide (13.2 kg, 1.21 equiv) in 2-butanone (33 kg) over 10 min.
- Rinse with 2-butanone (8 kg).
- Distill at 90 °C to distill off 2-butanone until the batch reached 170 L.
- Reflux for 17 h.

Stage (c)
- Add 2-butanone (35 kg) at >80 °C, followed by adding IPA (64 kg) at >70 °C.
- Filter at 80 °C and rinse with 2-butanone (12 kg) and IPA (7 kg).
- Cool the filtrate to 60 °C.

Stage (d)
- Add a solution of 37% HCl (9.8 kg) in IPA (18.1 kg) over 40 min at 60 °C.
- Add seed (0.24 kg) and stir for 1 h.
- Add a solution of 37% HCl (16.0 kg) in IPA (29.6 kg) over 2 h at 60 °C.
- Cool to 0 °C over 10 h and stir 7 h.
- Filter, wash with cold 2-butanone/IPA (1:1, w/w. 32 kg) and cold water (2 °C, 3×40 kg), and dry.
- Yield: 24.7 kg (83%).

NOTE
Addition of HCl/IPA rejected the cocrystallized imidazole into the mother liquor.

18.5.1.2.2 Isolation of Phenol
Nucleation during crystallization may be inhibited by high levels of either organic impurities or residual metal, such as Pd(II),[99] which may lead to product oiling out. Utilization of Saegusa oxidation, phenol **238** was obtained from enone **237** (Equation 18.54).[100]

(18.54)

(I) Problems

Initially, triethylamine was used to remove the unreacted Pd(II) by reduction of Pd(II) to Pd(0) followed by filtration. However, this protocol was less efficient and the residual Pd(II) in the solution caused oiling out of the product **238**.

(II) Solutions

Replacing triethylamine with potassium formate gave a more efficient reduction resulting in large particles of the precipitated Pd(0) that were easily removed by filtration. Therefore, with low levels (<80 ppm) of palladium, phenol **238** could be smoothly crystallized by adding seeds.

REMARKS

(a) This reduction protocol allowed an efficient recovery of palladium from the process.
(b) A viable alternative to the Saegusa oxidation was to use $CuBr_2$ in acetonitrile with LiBr as an additive. The residual copper and lithium levels were efficiently reduced to <1 ppm by washing with acetonitrile.

18.5.1.2.3   Isolation of 1,7-Diazacarbazole

The presence of polymeric impurities can also cause product oiling out during crystallization. For instance, *N*-protection of 1,7-diazacarbazole **239** with propyl vinyl ether (PVE) gave the desired product **240**. The direct isolation of **240**, however, led to the product oiling out due to the contamination of polymeric impurities generated from PVE (Equation 18.55).[101]

(18.55)

To remove these polymeric impurities, the product THF solution was swapped to EtOAc and then treated with a mixture of charcoal and celite followed by filtration.

18.5.1.2.4   Isolation of Purine Derivative

Although water is the desired solvent for the cyclization of 2-chloropurine derivative **241** with 1,2-dibromoethane, the oil product layer was contaminated with significant amounts of unreacted 1,2-dibromoethane, which inhibited the product crystallization (Equation 18.56).[102]

(18.56)

The addition of 2-propanol and seeding did not prevent the product from oiling out. In order to address the product oiling out, a mixture of ethanol/water (1.3:1) was used to replace isopropanol/water, leading to a robust crystallization process.

Procedure

- Charge 45.10 kg of water into a 100 L reactor, followed by adding 5.40 kg (96.24 mol, 3 equiv) of KOH, 9.40 kg (31.57 mol) of **241**, 2.92 kg (9.06 mol, 0.29 mol%) of TBAB, and 17.70 kg (94.22 mol, 2.98 equiv) of 1,2-dibromoethane.
- Stir the resulting mixture at 47 °C for 20 h.
- Charge EtOH (46.2 kg) and seed slurry (96.6 g) in EtOH (0.60 kg).
- Age the batch for 2 h.
- Cool to 5 °C over 2 h and age for 1 h.
- Transfer the mixture into a filter dryer, wash with water (27.4 kg), and dry at 60 °C/vacuum with N_2 purge.
- Yield: 6.85 kg (67%) as a light yellow solid.

18.5.1.3 Control of Exothermic Crystallization

As most crystallization processes are exothermic, seeding has been used to control exothermic events in order to avoid runaway situations. Biginelli three-component coupling was utilized to prepare 3,4-dihydropyrimidinone **245**, which crystallized directly from the reaction mixture upon cooling to room temperature (Equation 18.57).[103] As expected, an exotherm was observed upon crystallization. To circumvent this exotherm, seed-induced crystallization was employed prior to the onset of crystallization.

$$(18.57)$$

18.5.1.4 Control of Polymorph

Compound **247**, developed by Bristol Myers Squibb Company, is a member of the drug class known as statins and is used for lowering blood cholesterol. This calcium salt formation was affected by adopting the salt exchange approach as illustrated in Reaction (18.58).[128]

$$(18.58)$$

Although this process was maintained as crystalline slurry, seeding was employed to ensure that the desired polymorph was obtained during the crystallization.

18.5.2 REACTIVE CRYSTALLIZATION

Crystallization is a major unit operation for the separation and purification of pharmaceutical inter-mediates and products. Traditionally, crystallization is implemented after completion of the reaction. Compared to the traditional crystallization approach, reactive crystallization, by conducting the crystallization during the course of the reaction, can improve the process efficiency and reduce the number of unit operations.

18.5.2.1 Deprotection/Salt Formation

Removal of the Boc-protecting group is generally under acidic conditions, using either TFA or HCl, and the resulting free base may conveniently be converted to its hydrochloride salt. For instance, Boc deprotection of amine **248** with HCl in ethyl acetate gave HCl salt **249** that was crystallized in situ and readily isolated by filtration (Equation 18.59).[104]

$$(18.59)$$

Procedure

- Charge 100.6 g (0.218 mol) of **248** portionwise to HCl solution in EtOAc (HCl, 79.2 g, 2.17 mol, 10.0 equiv; EtOAc, 420 mL) at –30 to –20 °C.
- Stir the resulting mixture at –15 to –10 °C for 2 h and –10 to 0 °C for 1.5 h.
- Warm the batch to 25 °C over 1.5 h and stir at that temperature for 4 h.
- Add EtOAc (700 mL) and stir for 4 h at 25 °C.
- Filter and wash with EtOAc (200 mL) and heptane (200 mL).
- Dry under vacuum with nitrogen sweep.
- Yield: 86.7 g (100%).

REMARK

Care should be taken when performing heating/cooling recrystallization protocols to upgrade the purity of product salts, especially for salts with strong acids. For example, at 80 °C recrystallization of API tosylate **71** resulted in the decomposition of **71** into **250** (Equation 18.60).[25]

$$(18.60)$$

In order to avoid problematic product decomposition, the recrystallization was conducted at a relatively low temperature (70 °C).

18.5.2.2 Enamine Preparation

Preparation of enamine **252** (Equation 18.61)[105,55] was achieved via amination of β-keto amide **251** with ammonium acetate. A seed-induced reactive crystallization was employed for the isolation/

purification of **252**. In the presence of seed (3.2%), the crystallization of the desired product **252** occurred at 45 °C during the reaction. Upon completion of the reaction and cooling to 0–5 °C, **252** was directly isolated as a white crystalline solid through a simple filtration.

(18.61)

Procedure

- Charge 10% (1.18 mol) of the solution of ketoamide **251** in MeCN to a solution of NH_4OAc (0.91 kg, 11.81 mol) in MeOH (27 L) at 45 °C dropwise over 30 min.
- Stir the resulting mixture at 45 °C for 1.5 h.
- Add 140 g of seed.
- After 30 min at 45 °C, charge the remaining **251** (10.66 mol) solution dropwise over 3–6 h.
- Age the batch for 3 h.
- Add MeOH (12 L) at 40–45 °C over 2 h.
- Cool the slurry to 0–5 °C over 3–4 h and age for 1 h.
- Filter and wash with cold (0 °C) MeOH (29 L).
- Dry at ambient temperature under vacuum.
- Yield: 4.37 kg (82 %).

18.5.2.3 Free Acid Formation

Conversion of carboxylate salt to its corresponding free acid is usually done by adding acid to the salt solution. The success of reactive crystallization of the resulting free acid hinges upon the selection of solvent, in which the free acid crystallizes while adding the acid. Generally, the solvent should be able to dissolve the salt by dissociating the ionic pair. Accordingly, polar aprotic solvents such as DMF and DMSO, or protic solvents, for example, methanol, ethanol, 2-propanol, and water are frequently the solvents of choice. For instance, ethanol was chosen as the solvent for the dissolution of sodium salt **253**. Once the dissolution was complete, the conversion of **253** to the corresponding free acid **254** was straightforward by adding hydrochloric acid (1.1 equiv) (Equation 18.62).[106]

(18.62)

Dilute hydrochloric acid (0.5 N) was used for a purpose to obtain a final solvent ratio that ensured high recovery (>99% yield, >99.8% purity).

Procedure

- Charge 4.98 kg of crude **253** and 25.0 L of EtOH.
- Stir at 100 rpm until complete dissolution (3.34 kg, 8.38 mol of **253** based on solution assay).

- Filter the resulting solution through a 5 μm polypropylene filter.
- Set the jacket temperature to 20 °C and stir at 158 rpm.
- Charge 0.5 N HCl (18.4 L, 9.20 mol, 1.1 equiv) at ≤30 °C.
- Cool the batch to 20±5 °C and age for 1 h.
- Filter and wash with a mixture of EtOH/water (1:1, 13.0 L).
- Dry at 55 °C under vacuum.
- Yield: 3.18 kg (100%).

18.5.2.4 Boc Protection

A seed-induced reactive crystallization was applied to prepare Boc-protected amine **256**. As **256** has limited solubility in a mixture of water and EtOH, the reaction was seeded with **256** (0.17%) prior to the addition of Boc anhydride to facilitate the formation of large particles, thereby easing the subsequent filtration (Equation 18.63).[107]

$$ (18.63) $$

Procedure

- Charge 16.5 kg (62.4 mol) of **255** and 24.5 kg of EtOH into a glass-lined 400-L reactor to form a solution at 20 °C.
- Charge 61.8 kg of 16 wt% aq K$_2$CO$_3$ (71.5 mol), followed by seeding with 25 g of crystals of **256**.
- Charge 14.9 kg (68.3 mol) of Boc$_2$O as a solid in portions at <30 °C.
- Stir the mixture overnight.
- Filter and wash with 29 kg of a mixture of water/EtOH (3:1), followed by 30 kg of water.
- Dry under a stream of nitrogen.
- Yield: 15.0 kg (90%).

18.5.2.5 Limitations of Reactive Crystallization

Albeit effective in product isolation, the reactive crystallization shows limitations in the purification of 4-substituted phthalazin-1(2H)-one **258** (Equation 18.64).[108]

$$ (18.64) $$

The reaction was carried out by refluxing the mixture of 2-acylbenzoic acid **257** and 5.0 equiv of aqueous hydrazine in 5 volumes of ethanol. The low solubility of **258** in the reaction mixture (1.5 mg/mL at 80 °C) led to a rapid uncontrolled crystallization and inconsistent levels of hydrazine retained in the isolated crystalline product **258** (529–3108 ppm). Since hydrazine is genotoxic, the level of hydrazine in the isolated **258** had to be controlled to ≤110 ppm.

In order to reduce the hydrazine level in the isolated **258**, a CDI-mediated cyclization was developed (Scheme 18.30), which allowed using only 1.1 equiv of hydrazine. However, the product **258**, obtained from the reaction mixture in acetonitrile via the reactive crystallization, was contaminated with hydrazine in levels ranging between 719 and 4275 ppm.

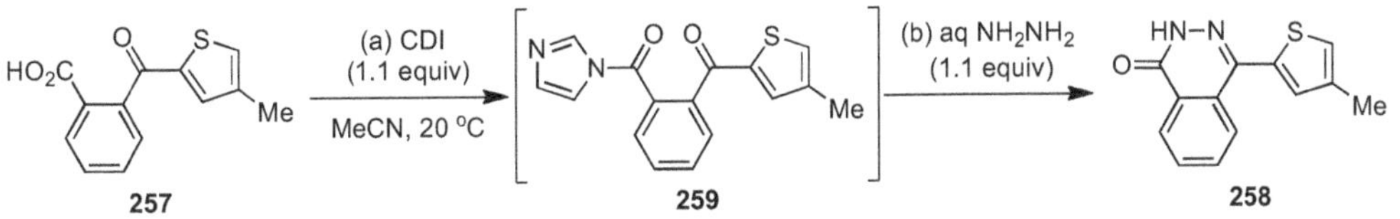

SCHEME 18.30

Compared with acetonitrile which has low solubility for **258** (0.42 mg/mL at 20 °C), DMF showed improved solubility (25 mg/mL at 20 °C) and was chosen as the reaction solvent. In the event, the product **258** was crystallized by adding water as the anti-solvent and isolated in good yields and <110 ppm of hydrazine.

18.5.3 OTHER CRYSTALLIZATION APPROACHES

18.5.3.1 Addition of Water

18.5.3.1.1 Avoiding Ball Formation

Generally, salts are less soluble in organic solvents, which may crystallize out during the course of salt formation. Occasionally, agglomeration is an issue during salt preparation. Such agglomeration can form large balls that may lead to difficult agitation or break the agitator. For example, a large ball was formed during the preparation of hydrochloride salt **261** using anhydrous HCl (Equation 18.65).[109] To address this issue, the addition of a small amount (2.2 wt%) of water improved the solid property.

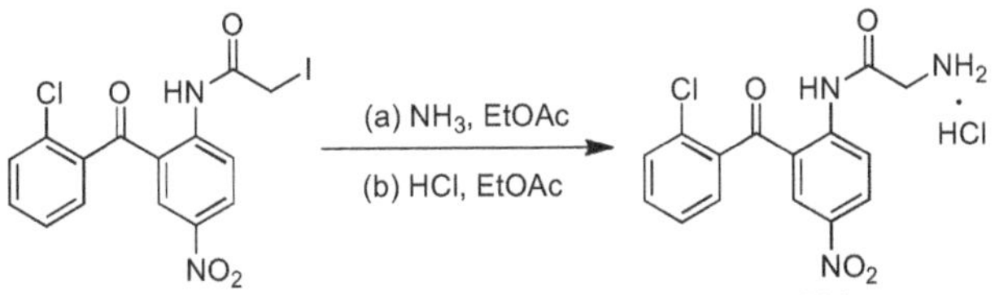

(18.65)

Fortunately, the window for water amount is sufficiently wide that allowed the use of aqueous concentrated hydrochloride acid instead of anhydrous HCl.

Procedure

Stage (a)
- Charge 1.132 kg (2.55 mol, based on 90% yield from the previous S_N2 iodination reaction of **260** into a glass reactor followed by adding EtOAc (25 kg).
- Bubble ammonia gas (0.4–0.5 kg, 25–30 mol) at 25 °C over 3.5 h.
- Displace excess ammonia by bubbling N_2 for 1 h followed by degassing at ca. 30 °C under vacuum (25 in. gauge) for 16 h.

Stage (b)
- Charge 0.57 kg (5.8 mol) of 37% aqueous HCl at ≤30 °C.
- Add water (0.57 kg).
- Cool to 6 °C and hold for 9 h.
- Filter and wash with cold (5 °C) EtOAc (2×0.8 kg) to give **261** in ca. 60% yield.

REMARK

Care should be taken when isolation of salts that are prepared under anhydrous conditions as some salts are hygroscopic. Those hygroscopic salts may cause problems during filtration. Sometimes, such a hygroscopic property is due to the formation of hydrate. Thus, the addition of stoichiometric amounts of water during the course of salt formation can solve the problem.

18.5.3.1.2   Reducing Residual MIBK

Depending on the API chemical structure and its crystalline morphology, the presence of residual crystallization solvent in the crystallized product can be at unacceptably high levels even under aggressive drying conditions. Vismodegib is a drug that was approved by the FDA in 2012 for the treatment of skin cancer (basal cell carcinoma). Recrystallization of crude vismodegib from methyl isobutyl ketone (MIBK) resulted in the isolated vismodegib contaminated with MIBK in levels of >10000 ppm (Equation 18.66).[110]

(18.66)

The residual MIBK was hypothesized to exist as a channel solvate and the addition of water (0.5 wt%) as an additive in the recrystallization reduced the residual MIBK in the isolated API to levels of <2000 ppm.

Procedure

- Dissolve 12.5 kg (29.7 mol) of crude vismodegib in MIBK (162.5 kg).
- Distill out azeotropically ca. 40% of MIBK at atmospheric pressure (115 °C).
- Cool to 105 °C and add ca. 65 kg MIBK.
- Filter through a 1 μm filter and rinse with MIBK (18.8 kg).
- Distill to ca. 9.5 volumes of MIBK.
- Cool to 95 °C.
- Add ca. 0.5% water (relative to total volume).
- Add seed (85 g).
- Age the resulting suspension at 95 °C for 2 h.
- Cool to 5 °C at a rate of ca. 18 °C/h.
- Age at 5 °C for 3–24 h before centrifugation.
- Wash with MIBK (ca. 2.7 vol), dry at 65 °C under vacuum for ca. 5 h, then at ca. 90 °C for ≥6 h.
- Yield of pure vismodegib: 10.7 kg (85.5%).

18.5.3.2   Addition of Polymer

Occasionally, using conventional crystallization to remove a low-level impurity with poor solubility is less effective because the impurity intends to cocrystallize with the product. Considering the energy barrier to nucleation and growth kinetics of impurity crystals, scientists at AbbVie Inc. developed a strategy of using non-adsorbing polymer (copovidone) as a crystallization inhibitor during the purification of the Sonogashira coupling product **264** (Equation 18.67).[111]

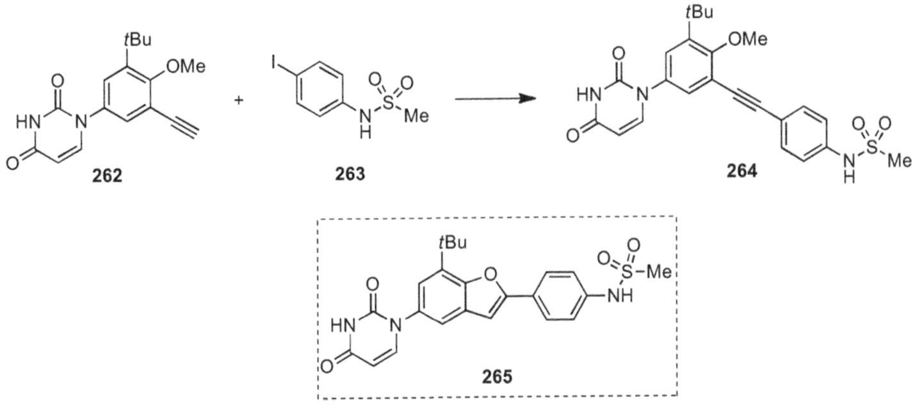

(18.67)

Thus, using this approach, the nucleation of the impurity **265** was significantly delayed, until after the isolation of the product, resulting in a successful production of a cGMP scale-up batch with the required purity within a tight time frame.

18.5.3.3 Crystallization from Extraction Solvent

The ability to extract product from the aqueous phase into the organic phase depends on the partition difference between the aqueous phase and the organic phase. Normally, an extraction solvent should possess high solubility for the given compound. Thus, such extraction solvent will not be suitable for crystallization or great yield loss will result.

A profound effect of small amounts of water on the solubility of organic compounds in ethyl acetate was noticed during the purification of compound **267** by crystallization (Equation 18.68).[68] Taking advantage of the solubility difference of **267** in wet ethyl acetate (38.1 mg/mL at 23 °C) and anhydrous ethyl acetate (3.7 mg/mL at 23 °C), ethyl acetate was utilized in both extractive workup and crystallization.

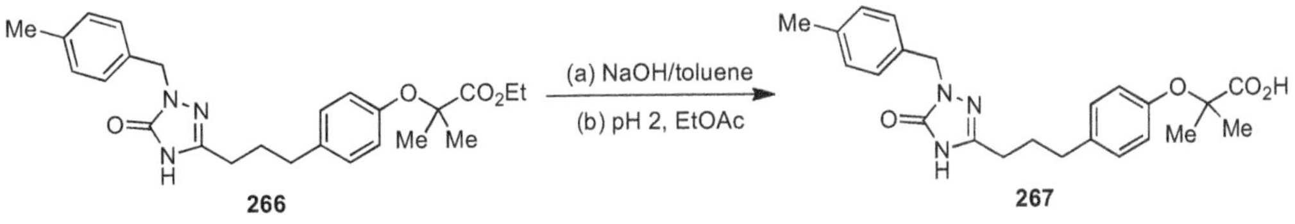

(18.68)

Procedure

Stage (a)
- Charge 0.8 kg (1.828 mol) of **266**, followed by adding 4 L of toluene and 4.023 L (4.023 mol, 2.2 equiv) of 1 N NaOH.
- Stir the resulting biphasic mixture at room temperature for 5 h.

Stage (b)
- Add 0.337 L of concentrated HCl to the aqueous layer to pH 2.
- Add 8 L of EtOAc under stirring.
- Separate layers.
- Distill the organic layer to remove ca. 4 L of EtOAc.
- Add 3.6 L of EtOAc.

- Continue to distill to remove ca. 3.6 L of EtOAc.
- Cool to 60–65 °C.
- Add 0.8 g (0.1%) of seed crystals of **267**.
- Cool to 0–5 °C.
- Filter and wash with cold EtOAc.
- Dry at 55 °C.
- Yield: 713.6 g (95.3%).

18.5.3.4   Three-Solvent System

A two-solvent system (ethyl acetate/heptane) for crystallization of **270** proved unreliable, and oiling out of the product was periodically observed after charging heptane or during the cooling process (Equation 18.69).[112]

$$ (18.69) $$

One of the solutions is to adjust the system polarity by adding a solvent with dissolution ability between good solvent and poor solvent. For example, toluene is frequently introduced as a third solvent to ethyl acetate/heptane crystallization system. Because the solubility of **270** is very low in heptane and very high in ethyl acetate; in such a two-solvent system **270** has the propensity to become highly supersaturated resulting in oiling out during the crystallization. Therefore, a three-solvent system, ethyl acetate/toluene/heptane (1:1:3), for the crystallization of **270** was developed.

Procedure

Stage (a)
- Charge a solution of **269** (81.2 kg, 464 mol, 1.13 equiv) in EtOAc (428 L) to a slurry of CDI (75.2 kg, 464 mol, 1.13 equiv) in EtOAc (325 L) at 25 °C and rinse with EtOAc (50 L).
- Stir the resulting yellow solution at 25 °C for 21 h.
- Charge the reaction mixture to a solution of **268** (221.6 kg, 410.3 mol) in NMP (239 L) over 19 min and rinse with EtOAc (50 L).
- Stir the batch at 25 °C for 4 h.
- Add EtOAc (828 L).
- Wash with aq 1 N HCl (1188 L) followed by water (3×1170 L).

Stage (b)
- Distill the organic solution at 71–84 °C under atmospheric pressure to remove EtOAc (1090 L).
- Cool the solution to 55 °C.
- Add toluene (275 L) at >50 °C.
- Add EtOAc (169 L) to achieve the desired EtOAc:toluene (1.03:1) weight ratio.
- Add heptane (550 L) at >45 °C.
- Heat the resulting solution to 50 °C.
- Add 1.7 kg (1.0%) of **270** as seed.
- Stir the resulting slurry at 50 °C for 4 h.
- Add heptane (137 L) and stir at 50 °C for 100 min.
- Add heptane (137 L) and cool to 45 °C over 50 min and hold for 75 min.

- Cool to 30 °C over 3.25 h and hold for 1 h.
- Cool to –5 °C over 3.42 h and hold for 37.25 h.
- Filter and wash with a mixture of cold toluene/heptane (217 L/412 L).
- Dry in a tumble dryer.
- Yield: 170.6 kg (84.8 %).

REMARK

The following solvent combinations are frequently chosen in crystallization:

(a) EtOAc/toluene/heptane
(b) water/MeCN
(c) water/MeOH
(d) water/IPA

18.5.3.5 Derivatization

Purification of low melting solids is always challenging as column chromatography is not desired option at scale. Besides salt formation, derivatization is one of the purification strategies that are frequently used by process chemists to convert the low-melting solids to their corresponding solid derivatives followed by crystallization. For example, in order to achieve high purity of aldehyde **272** by crystallization, **272** was converted to the corresponding hydrazone intermediate **273**, which was then crystallized from MTBE in 54% yield as a highly crystalline solid. The subsequent hydrolysis of **273** with CuCl$_2$ in aqueous tBuOH provided **272** in a yield of 95% (Scheme 18.31).[113]

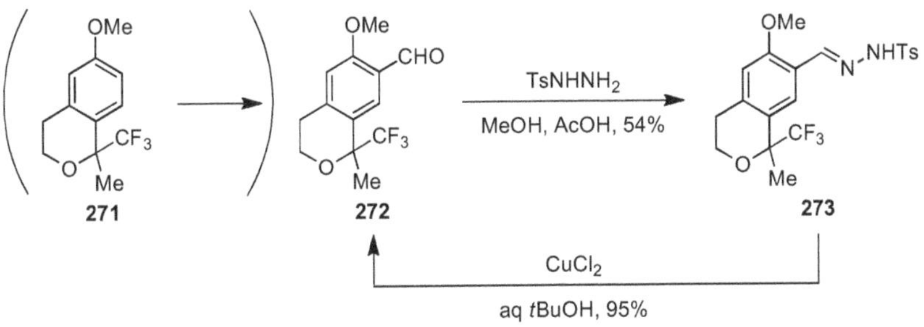

SCHEME 18.31

REMARKS

(a) The moderate yield in the formation of tosyl hydrazone indicates a further optimization may be necessary.
(b) At scale, **272** crystallized directly from heptane after the formylation in a sufficient chemical purity for the subsequent transformation.

Analogously, an oily chiral aldehyde **274** was purified by converting **274** to a crystalline aldehyde–benzotriazole adduct **275** (Equation 18.70).[114]

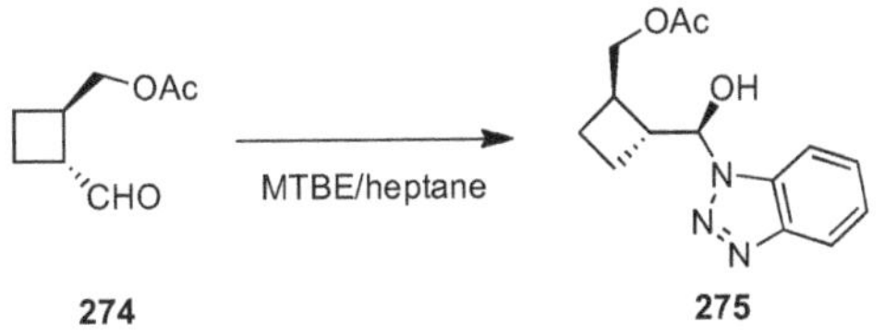

(18.70)

18.5.3.6 Control of Crystal Size Distribution

Morphology or crystal habit refers to the appearance of crystals. The shape of crystals is catego-rized into three types: needle-like, plate-like, and rod-like. The needle-like or plate-like crystals are less favorable because they are prone to breakage to produce fines during handling, which can potentially lead to operational complications such as poor cake filtration and washing and lower product yield. In general, rod-like crystals are preferred. Various factors, such as crystal structure, solvents, additives, and impurities, can affect the crystal morphology. Additionally, heat/cool cycles can be employed to modify the crystal morphology as well.

Pharmaceutical products with small particles with narrow particle size distribution are desirable for enhanced oral bioavailability or for inhalation therapy. In many cases, these particle sizes can be achieved by different milling devices. Another way to generate small particles is through crystal-lization, that is, nucleation, under high supersaturation conditions. To accomplish it, the simplest approach is the reverse addition by charging a highly concentrated feed solution directly into an anti-solvent. Very high supersaturation is generated locally, and small nuclei/crystals are rapidly formed.

Fine crystals are defined to be crystals with chord lengths of <20 μm, whereas coarse are defined to be crystals with chord lengths of >100 μm.[115] In many cases, the formation of fine crystals is undesired, as the fine crystals can clog the filtration cloth resulting in slow filtration or pass through the filtration cloth leading to loss of product. For example, during the workup of the Suzuki coupling reaction, the acidification step produced a thick suspension of fine materials that was extremely slow to filter (Equation 18.71).[120]

$$(18.71)$$

It was found that the acidification temperature of the reaction mixture was crucial and acidification at 80 °C in MeOH/H_2O (1:3) reduced the fines significantly, which led to rapid filtration and good yields.

Another useful technique for getting rid of unwanted fine crystals without losing yield is tem-perature cycling. Since smaller particles have a slightly higher solubility than larger ones and the temperature fluctuations increase the kinetics of Ostwald ripening, warming the slurry of crystals to dissolve the fine particles followed by cooling will effectively solve the slow filtration and product loss issues.

Note

Special attention should be paid when the application of the Ostwald ripening toward heat-sensitive compounds.

18.5.3.7 Cocrystallization

18.5.3.7.1 Water-Soluble Product

An aqueous workup is usually used to isolate the water-insoluble product from the reaction mixture. For water-soluble products, such an approach is not amenable on large scale as the recovery of the product from the aqueous mixture is a tedious task. For instance, the amine product **277**, obtained from the deprotection of amide **276**, was highly water soluble, and the aqueous workup resulted in poor yields. Its hydrochloride salt was hygroscopic and not amenable to recrystallization. In order to address these isolation issues, a two-step, anhydrous protocol was developed by converting **276** to imidoyl chloride **278** followed by alcoholysis (Scheme 18.32).[116]

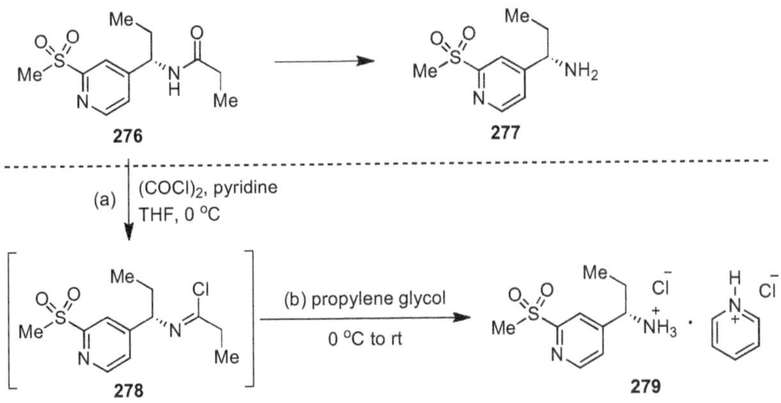

SCHEME 18.32

The amine product thus obtained was isolated in 86% yield via direct filtration as a 1:1 cocrystal with pyridinium hydrochloride **279** which was a non-hygroscopic white solid.

Procedure

Stage (a)
- Cool a solution of 10.0 g (37.0 mmol) of **276** and pyridine (3.60 mL, 44.5 mmol, 1.2 equiv) in THF (60 mL) to 0 °C.
- Charge 3.60 mL (40.4 mmol, 1.09 equiv) of oxalyl chloride at <5 °C.
- After the addition, stir the resulting yellow slurry at 0 °C for 20 min.

Stage (b)
- Charge 5.50 mL (74.9 mmol, 2.0 equiv) of propylene glycol at 0 °C.
- Warm the resulting mixture to rt.
- Add THF (30 mL) and stir for 1 h.
- Filter, wash with EtOAc followed by heptane, and dry at room temperature under vacuum.
- Yield: 11.90 g (86%).

REMARK

The deprotection of the propionyl group using acids, such as HCl, HBr, and H_3PO_4, proved to be inferior to the $(COCl)_2$-pyridine/propylene glycol system.

18.5.3.7.2 Amorphous Solid

Although salt formation or derivatization are successfully utilized in the isolation and purification of oily or amorphous products, these methods are less effective for compounds without any basic nitrogen or necessary functionality. A report demonstrated a cocrystallization approach for the purification of amorphous API **281** (Scheme 18.33).[117] Practically, the amorphous **281** was converted to the L-phenylalanine complex and crystallized as monohydrate **282**. After filtration and drying, **282** was produced in 80–88% yields (58 kg) and with 99.9 A% purity. This material has been used for clinical investigations.

Procedure

- Mix 64.40 kg of **280** in water (68.70 kg) and EtOH (62.44 kg) at 20 °C with minimum agitation.

- Charge 57.02 kg of 0.1 N NaOH.
- Heat to 40–50 °C and stir for 1–2 h.
- Cool to 20 °C.
- Add water (154.56 kg).
- Stir at 18–25 °C for 1 h.
- Add 37% HCl (~123.44 kg) to pH 6.3.
- Charge 20.20 kg of L-phenylalanine and water (141.68 kg).
- Heat the resulting slurry to 75 °C.
- Pass the solution through a polish filter.
- Cool to 55–60 °C.
- Add seed (322 g).
- Cool to 40 °C over 1 h and maintain at 40 °C for 4 h.
- Cool to 18–25 °C over 2 h and stir at this temperature for 12–16 h.
- Filter and wash with 10 °C water (322 kg) until the conductivity of the wash was <0.001 σ siemens.
- Wash the cake further with EtOAc (290 kg) to remove any excess compound **281**.
- Dry at 18–25 °C under vacuum for 4 h then at 40 °C for 12 h.
- Yield: 54–58 kg (80–88%).

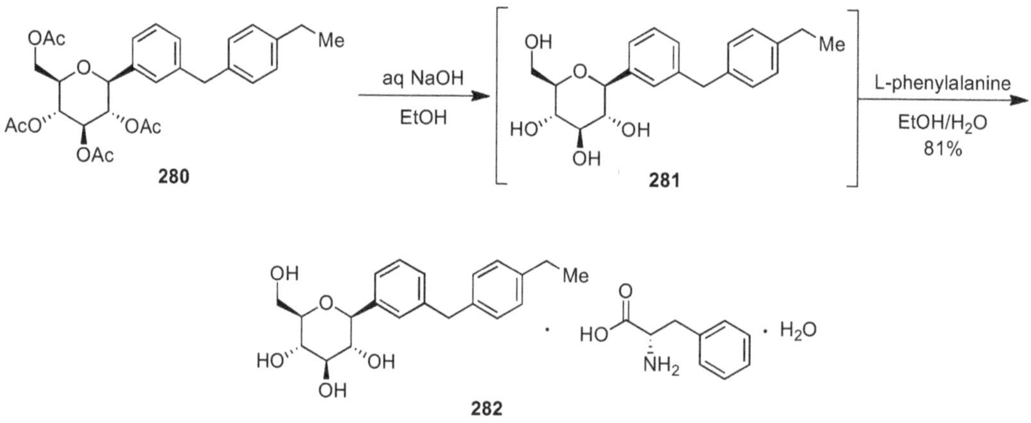

SCHEME 18.33

18.5.3.7.3   Zwitterion Cocrystal

Zwitterion molecules, such as betaine, can also be selected as cocrystal partners and crystallization of betaine cocrystal has been employed to purge critical impurities in the late-stage synthesis. For example, the crystallization of the cocrystal **284** of the selective estrogen receptor degrader (SERD) drug candidate **285** with betaine was successfully utilized to purify the drug substance (Scheme 18.34).[118]

The conversion of the betaine cocrystal **284** to the crystalline **285** could be achieved in an acetic acid/H_2O system providing a high-purity crystalline product **285** in 76% yield.

18.5.3.7.4   Low Melting Solid

Similarly, a cocrystallization approach was developed to purify 4-Bromo-2-isopropylphenol **287**. Compound **287** is a low melting solid (mp 43–44 °C) and was prepared from bromination of 2-isopropylphenol **286**. To purify **287** without resorting to chromatography, 1,4-diazabicyclo[2.2.2] octane (DABCO) was chosen as the cocrystal partner and the resulting cocrystal (**287**·0.5 DABCO) (mp 95–97 °C) was conveniently isolated by simple filtration (Equation 18.72).[119]

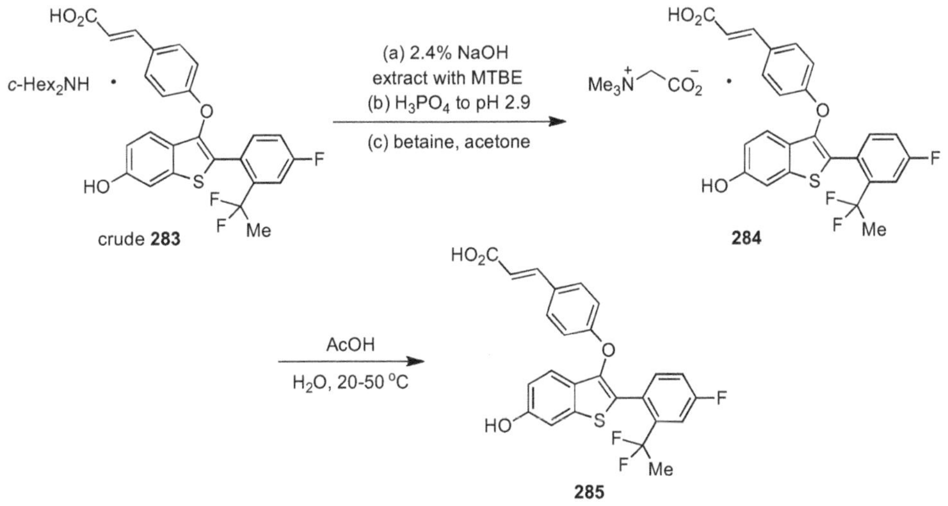

SCHEME 18.34

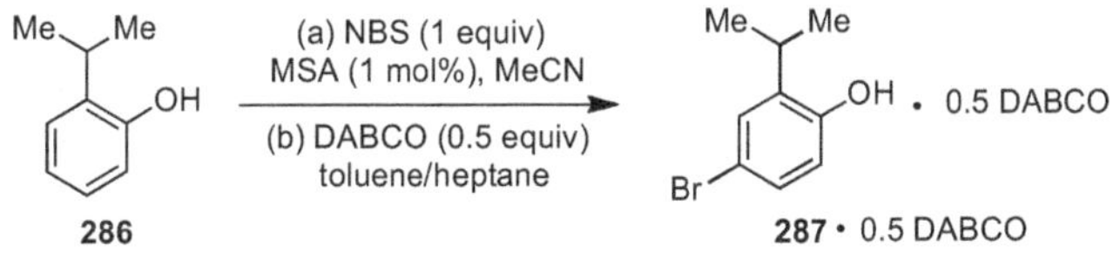

(18.72)

Procedure

- Charge 90 kg (661 mol) of **286** and 270 L of acetonitrile into a 1500 L vessel.
- Charge methanesulfonic acid (429 mL, 6.6 mol).
- Cool the batch to −10 °C.
- Add solid 118 kg (661 mol) NBS in four portions (4×29.5 kg) at <−5 °C.
- After aging for 1 h, warm the batch to 0 °C.
- Quench the batch with 0.5 wt% NaHSO$_3$ (540 L) at <−5 °C.
- Extract with toluene (270 L) at 45 °C and wash the toluene layer with 5% v/v H$_3$PO$_4$ (180 L) and 5% brine (180 L).
- Concentrate the toluene solution to remove water (5000 ppm) and acetonitrile (2%).
- Heat the residual stream to 50 °C.
- Add 36.7 kg (330 mol) DABCO to form a homogeneous solution.
- Add 225 L heptanes, lecithin (90 g), and seed (90 g).
- Agitate for 1 h.
- Add heptanes (225 L) over 1 h.
- Cool the resulting slurry to 20 °C over 2 h.
- Filter, wash with heptanes (90 L), and dry under vacuum.
- Yield: 165 kg (92%).

NOTES

1. Fiore, P. J.; Puls, T. P.; Walker, J. C. *Org. Process Res. Dev.* **1998**, *2*, 151.
2. Kim, J.; Itoh, T.; Xu, F.; Dance, Z. F. X.; Waldman, J. H.; Wallace, D. J.; Wu, F.; Kats-Kagan, R.; Ekkati, A. R.; Brunskill, A. P. J.; Peng, F.; Fier, P. S.; Obligacion, J. V.; Sherry, B. D.; Liu, Z.; Emerson, K. M.; Fine, A. J.; Jenks, A. V.; Armenante, M. E. *Org. Process Res. Dev.* **2021**, *25*, 2249.

3. Lane, J. W.; Spencer, K. L.; Shakya, S. R.; Kallan, N. C.; Stengel, P. J.; Remarchuk, T. *Org. Process Res. Dev.* **2014**, *18*, 1641.

4. Reeves, J. T.; Fandrick, D. R.; Tan, Z.; Song, J. J.; Rodriguez, S.; Qu, B.; Kim, S.; Niemeier, O.; Li, Z.; Byrne, D.; Campbell, S.; Chitroda, A.; DeCroos, P.; Fachinger, T.; Fuchs, V.; Gonnella, N. C.; Grinberg, N.; Haddad, N.; Jäger, B.; Lee, H.; Lorenz, J. C.; Ma, S.; Narayanan, B. A.; Nummy, L. J.; Premasiri, A.; Roschangar, F.; Sarvestani, M.; Shen, S.; Spinelli, E.; Sun, X.; Varsolona, R. J.; Yee, N.; Brenner, M.; Senanayake, C. H. *J. Org. Chem.* **2013**, *78*, 3616.

5. 5-Chloro-3-fluoro-2-(trifluoromethyl)pyridine is expensive with limited commercial suppliers. This work was conducted by the author in Jacobus Pharmaceutical Company, Inc. and had not been published.

6. Funel, J.-A.; Brodbeck, S.; Guggisberg, Y.; Litjens, R.; Seidel, T.; Struijk, M.; Abele, S. *Org. Process Res. Dev.* **2014**, *18*, 1674.

7. Magnus, N. A.; Anzeveno, P. B.; Coffey, D. S.; Hay, D. A.; Laurila, M. E.; Schkeryantz, J. M.; Shaw, B. W.; Staszak, M. A. *Org. Process Res. Dev.* **2007**, *11*, 560.

8. Delmonte, A. J.; Fan, Y.; Girard, K. P.; Jones, G. S.; Waltermire, R. E.; Rosso, V.; Wang, X. *Org. Process Res. Dev.* **2011**, *15*, 64.

9. Journet, M.; Cai, D.; Hughes, D. L.; Kowal, J. J.; Larsen, R. D.; Reider, P. J. *Org. Process Res. Dev.* **2005**, *9*, 490.

10. Anderson, N. G. *Org. Process Res. Dev.* **2004**, *8*, 260.

11. Wang, Y.; Przyuski, K.; Roemmele, R. C.; Hudkins, R. L.; Bakale, R. P. *Org. Process Res. Dev.* **2013**, *17*, 846.

12. Dale, D. J.; Draper, J.; Dunn, P. J.; Hughes, M. L.; Hussain, F.; Levett, P. C.; Ward, G. B.; Wood, A. S. *Org. Process Res. Dev.* **2002**, *6*, 767.

13. Xu, D. D.; Waykole, L.; Calienni, J. V.; Ciszewski, L.; Lee, G. T.; Liu, W.; Szewczyk, J.; Vargas, K.; Prasad, K.; Repič, O.; Blacklock, T. J. *Org. Process Res. Dev.* **2003**, *7*, 856.

14. Ashcroft, C. P.; Challenger, S.; Clifford, D.; Derrick, A. M.; Hajikarimian, Y.; Slucock, K.; Silk, T. V.; Thomson, N. M.; Williams, J. R. *Org. Process Res. Dev.* **2005**, *9*, 663.

15. Ashworth, I. W.; Bowden, M. C.; Dembofsky, B.; Levin, D.; Moss, W.; Robinson, E.; Szczur, N.; Virica, J. *Org. Process Res. Dev.* **2003**, *7*, 74.

16. Hong, J.-B.; Davidson, J. P.; Jin, Q.; Lee, G. R.; Matchett, M.; O'Brien, E.; Welch, M.; Bingenheimer, B.; Sarma, K. *Org. Process Res. Dev.* **2014**, *18*, 228.

17. Thiel, O. R.; Bernard, C.; King, T.; Dilmeghani-Seran, M.; Bostick, T.; Larsen, R. D.; Faul, M. M. *J. Org. Chem.* **2008**, *73*, 3508.

18. Yoshida, S.; Ohigashi, A.; Morinaga, Y.; Hashimoto, N.; Takahashi, T.; Ieda, S.; Okada, M. *Org. Process Res. Dev.* **2013**, *17*, 1252.

19. Fujino, K.; Takami, H.; Atsumi, T.; Ogasa, T.; Mohri, S.; Kasai, M. *Org. Process Res. Dev.* **2001**, *5*, 426.

20. Štimac, V.; Škugor, M. M.; Jakopović, I. P.; Vinter, A.; Ilijaš, M.; Alihodžić, S.; Mutak, S. *Org. Process Res. Dev.* **2010**, *14*, 1393.

21. Achmatowicz, M.; Thiel, O. R.; Wheeler, P.; Bernard, C.; Huang, J.; Larsen, R. D.; Faul, M. M. *J. Org. Chem.* **2009**, *74*, 795.

22. Gontcharov, A.; Dunetz, J. R. *Org. Process Res. Dev.* **2014**, *18*, 1145.

23. Ishikawa, M.; Tsushima, M.; Kubota, D.; Yanagisawa, Y.; Hiraiwa, Y.; Kojima, Y.; Ajito, K.; Anzai, N. *Org. Process Res. Dev.* **2008**, *12*, 596.

24. Ashcroft, C. P.; Dessi, Y.; Entwistle, D. A.; Hesmondhalgh, L. C.; Longstaff, A.; Smith, J. D. *Org. Process Res. Dev.* **2012**, *16*, 470.

25. Patterson, D. E.; Powers, J. D.; LeBlanc, M.; Sharkey, T.; Boehler, E.; Irdam, E.; Osterhout, M. H. *Org. Process Res. Dev.* **2009**, *13*, 900.

26. Lorenz, J. C.; Busacca, C. A.; Feng, X.; Grinberg, N.; Haddad, N.; Johnson, J.; Kapadia, S.; Lee, H.; Saha, A.; Sarvestani, M.; Spinelli, E. M.; Varsolona, R.; Wei, X.; Zeng, X.; Senanayake, C. H. *J. Org. Chem.* **2010**, *75*, 1155.

27. Mudryk, B.; Zheng, B.; Chen, K.; Eastgate, M. D. *Org. Process Res. Dev.* **2014**, *18*, 520.

28. (a) Kim, C. U.; Lew, W.; Williams, M. A.; Liu, H.; Zhang, L.; Swaminathan, S.; Bischofberger, N.; Chen, M. S.; Mendel, D. B.; Tai, C. Y.; Laver, W. G.; Stevens, R. C. *J. Am. Chem. Soc.* **1997**, *119*, 681. (b) Kim, C. U.; Lew, W.; Williams, M. A.; Wu, H.; Zhang, L.; Chen, X.; Escarpe, P. A.; Mendel, D. B.; Laver, W. G.; Stevens, R. C. *J. Med. Chem.* **1998**, *41*, 2451.

29. (a) Rohloff, J. C.; Kent, K. M.; Postich, M. J.; Becker, M. W.; Chapman, H. H.; Kelly, D. E.; Lew, W.; Louie, M. S.; McGee, L. R.; Prisbe, E. J.; Schultze, L. M.; Yu, R. H.; Zhang, L. *J. Org. Chem.* **1998**, *63*, 4545. (b) Federspiel, M.; Fischer, R.; Hennig, M.; Mair, H. –J.; Oberhauser, T.; Rimmler, G.; Albiez, T.

Bruhin, J.; Estermann, H.; Gandert, C.; Göckel, V.; Götzö, S.; Hoffmann, U.; Huber, G.; Janatsch, G.; Lauper, S.; Röckel-Stäbler, O.; Trussardi, R.; Zwahlen, A. G. *Org. Process Res. Dev.* **1999**, *3*, 266.

30. Frutos, R. P.; Rodríguez, S.; Patel, N.; Johnson, J.; Saha, A.; Krishnamurthy, D.; Senanayake, C. H. *Org. Process Res. Dev.* **2007**, *11*, 1076.

31. Song, Z. J.; Maligres, P.; Molinaro, C.; Humphrey, G.; Fritzen, J.; Wilson, J.; Chen, Y. *Org. Process Res. Dev.* **2019**, *23*, 2354.

32. Oh, L. M.; Wang, H.; Shilcrat, S. C.; Herrmann, R. E.; Patience, D. B.; Spoors, P. G.; Sisko, J. *Org. Process Res. Dev.* **2007**, *11*, 1032.

33. Shi, Z.; Kiau, S.; Lobben, P.; Hynes, J., Jr.; Wu, H.; Parlanti, L.; Discordia, R.; Doubleday, W. W.; Leftheris, K.; Dyckman, A. J.; Wrobleski, S. T.; Dambalas, K.; Tummala, S.; Leung, S.; Lo, E. *Org. Process Res. Dev.* **2012**, *16*, 1618.

34. For Ostwald ripening, see: Noorduin, W. L.; Vlieg, E.; Kellog, R. M.; Kaptein, B. *Angew. Chem. Int. Ed.* **2009**, *48*, 9600.

35. Slade, J.; Parker, D.; Girgis, M.; Muelller, M.; Vivelo, J.; Liu, H.; Bajwa, J.; Chen, G.-P.; Carosi, J.; Lee, P.; Chaudhary, A.; Wambser, D.; Prasad, K.; Bracken, K.; Dean, K.; Boehnke, H.; Repič, O.; Blacklock, T. J. *Org. Process Res. Dev.* **2006**, *10*, 78.

36. Sperry, J. B.; Farr, R. M.; Levent, M.; Ghosh, M.; Hoagland, S. M.; Varsolona, R. J.; Sutherland, K. *Org. Process Res. Dev.* **2012**, *16*, 1854.

37. The author's unpublished work.

38. (a) Magnus, N. A.; Aikins, J. A.; Cronin, J. S.; Diseroad, W. D.; Hargis, A. D.; LeTourneau, M. E.; Parker, B. E.; Reutzel-Edens, S. M.; Schafer, J. P.; Staszak, M. A.; Stephenson, G. A. Tameze, S. L.; Zollars, L. M. H. *Org. Process Res. Dev.* **2005**, *9*, 621. (b) Bänziger, M.; Cercus, J.; Stampfer, W.; Sunay, U. *Org. Process Res. Dev.* **2000**, *4*, 460.

39. Hobson, L. A.; Nugent, W. A.; Anderson, S. R.; Deshmukh, S. S.; Haley, J. J. III; Liu, P.; Magnus, N. A.; Sheeran, P.; Sherbine, J. P.; Stone, B. R. P.; Zhu, J. *Org. Process Res. Dev.* **2007**, *11*, 985.

40. Milburn, R. R.; Thiel, O. R.; Achmatowicz, M.; Wang, X.; Zigterman, J.; Bernard, C.; Colyer, J. T.; DiVirgilio, E.; Crocketttt, R.; Correll, T. L.; Nagapudi, K.; Ranganathan, K.; Hedley, S. J.; Allgeier, A.; Larsen, R. D. *Org. Process Res. Dev.* **2011**, *15*, 31.

41. Tian, Q.; Hoffmann, U.; Humphries, T.; Cheng, Z.; Hidber, P.; Yajima, H.; Guillemot-Plass, M.; Li, J.; Bromberger, U.; Babu, S.; Askin, D.; Gosselin, F. *Org. Process Res. Dev.* **2015**, *19*, 416.

42. Prasad, K.; Xu, D. D.; Tempkin, O.; Villhauer, E. B.; Repič, O. *Org. Process Res. Dev.* **2003**, *7*, 743.

43. Bastin, R. J.; Bowker, M. J.; Slater, B. J. *Org. Process Res. Dev.* **2000**, *4*, 427.

44. Aalla, S.; Gilla, G.; Bojja, Y.; Anumula, R. R.; Vummenthala, P. R.; Padi, P. R. *Org. Process Res. Dev.* **2012**, *16*, 682.

45. Wennekes, T.; Lang, B.; Leeman, M.; van der Marel, G. A.; Smits, E.; Weber, M.; van Wiltenburg, J.; Wolberg, M.; Aerts, J. M. F. G.; Overkleeft, H. S. *Org. Process Res. Dev.* **2008**, *12*, 415.

46. de Koning, P. D.; McManus, D. J.; Bandurek, G. R. *Org. Process Res. Dev.* **2011**, *15*, 1081.

47. Caine, D. M.; Paternoster, I. L.; Sedehizadeh, S.; Shapland, P. D. P. *Org. Process Res. Dev.* **2012**, *16*, 518.

48. Jones, K.; Newton, R. F.; Yarnold, C. J. *Tetrahedron* **1996**, *52*, 4133.

49. Watson, T. J. N.; Horgan, S. W.; Shah, R. S.; Farr, R. A.; Schnettler, R. A.; Nevill, C. R., Jr.; Weiberth, F. J.; Huber, E. W.; Baron, B. M.; Webster, M. E.; Mishra, R. K.; Harrison, B. L.; Nyce, P. L.; Rand, C. L.; Goralski, C. T. *Org. Process Res. Dev.* **2000**, *4*, 477.

50. Cleator, E.; Scott, J. P.; Avalle, P.; Bio, M. M.; Brewer, S. E.; Davies, A. J.; Gibb, A. D.; Sheen, F. J.; Stewart, G. W.; Wallace, D. J.; Wilson, R. D. *Org. Process Res. Dev.* **2013**, *17*, 1561.

51. Davulcu, A. H.; McLeod, D. D.; Li, J.; Katipally, K.; Little, A.; Doubleday, W.; Xu, Z.; McConlogue, C. W.; Lai, C. J.; Gleeson, M.; Schwinden, M.; Parsons, R. L., Jr. *J. Org. Chem.* **2009**, *74*, 4068.

52. Bowden, S. A.; Burke, J. N.; Gray, F.; McKown, S.; Moseley, J. D.; Moss, W. O.; Murray, P. M.; Welham, M. J.; Young, M. J. *Org. Process Res. Dev.* **2004**, *8*, 33.

53. Boggs, S. D.; Cobb, J. D.; Gudmundsson, K. S.; Jones, L. A.; Matsuoka, R. T.; Millar, A.; Patterson, D. E.; Samano, V.; Trone, M. D.; Xie, S.; Zhou, X. *Org. Process Res. Dev.* **2007**, *17*, 539.

54. Tudge, M.; Savarin, C. G.; DiFelice, K.; Maligres, P.; Humphrey, G.; Reamer, B.; Tellers, D. M.; Hughes, D. *Org. Process Res. Dev.* **2010**, *14*, 787.

55. Baxter, C. A.; Cleator, E.; Brands, K. M. J.; Edwards, J. S.; Reamer, R. A.; Sheen, F. J.; Stewart, G. W.; Strotman, N. A.; Wallace, D. J. *Org. Process Res. Dev.* **2011**, *15*, 367.

56. Barkalow, J. H.; Breting, J.; Gaede, B. J.; Haight, A. R.; Henry, R.; Kotecki, B.; Mei, J.; Pearl, K. B.; Tedrow, J. S.; Viswanath, S. K. *Org. Process Res. Dev.* **2007**, *11*, 693

57. Engelhardt, F. C.; Shi, Y.-J.; Cowden, C. J.; Conlon, D. A.; Pipik, B.; Zhou, G.; McNamara, J. M.; Dolling, U.-H. *J. Org. Chem.* **2006**, *71*, 480.

58. Camp, D.; Matthews, C. F.; Neville, S. T.; Rouns, M.; Scott, R. W.; Truong, Y. *Org. Process Res. Dev.* **2006**, *10*, 815.

59. Michida, M.; Takayanagi, Y.; Imai, M.; Furuya, Y.; Kimura, K.; Kitawaki, T.; Tomori, H.; Kajino, H. *Org. Process Res. Dev.* **2013**, *17*, 1430.

60. [a] Hasegawa, T.; Kawanaka, Y.; Kasamatsu, E.; Ohta, C.; Nakabayashi, K.; Okamoto, M.; Hamano, M.; Takahashi, K.; Ohuchida, S.; Hamada, Y. *Org. Process Res. Dev.* **2005**, *9*, 774.

61. [b] Tang, W.; Wei, X.; Yee, N. K.; Patel, N.; Lee, H.; Savoie, J.; Senanayake, C. H. *Org. Process Res. Dev.* **2011**, *15*, 1207.

62. [c] Henschke, J. P.; Liu, Y.; Huang, X.; Chen, Y.; Meng, D.; Xia, L.; Wei, X.; Xie, A.; Li, D.; Huang, Q.; Sun, T.; Wang, J.; Gu, X.; Huang, X.; Wang, L.; Xiao, J.; Qiu, S. *Org. Process Res. Dev.* **2012**, *16*, 1905.

63. [d] Wang, X.-J.; Zhang, L.; Smith-Keenan, L. L.; Houpis, I. N.; Farina, V. *Org. Process Res. Dev.* **2007**, *11*, 60.

64. [e] Wallace, M. D.; McGuire, M. A.; Yu, M. S.; Goldfinger, L.; Liu, L.; Dai, W.; Shilcrat, S.; *Org. Process Res. Dev.* **2004**, *8*, 738.

65. [f] Song, Z. J.; Tellers, D. M.; Dormer, P. G.; Zewge, D.; Janey, J. M.; Nolting, A.; Steinhuebel, D.; Oliver, S.; Devine, P. N.; Tschaen, D. M. *Org. Process Res. Dev.* **2014**, *18*, 423.

66. [g] Conlon, D. A.; Natalie, Jr., K. J.; Cuniere, N.; Razler, T. M.; Zhu, J.; de Mas, N.; Tymonko, S.; Fraunhoffer, K. J.; Sortore, E.; Rosso, V. W.; Xu, Z.; Adams, M. L.; Patel, A.; Huang, J.; Gong, H.; Weinstein, D. S.; Quiroz, F.; Chen, D. C. *Org. Process Res. Dev.* **2016**, *20*, 921.

67. [h] Ashcroft, C. P.; Challenger, S.; Derrick, A. M.; Storey, R.; Thomson, N. M. *Org. Process Res. Dev.* **2003**, *7*, 362.

68. [i] Hasegawa, T.; Kawanaka, Y.; Kasamatsu, E.; Iguchi, Y.; Yonekawa, Y.; Okamoto, M.; Ohta, C.; Hashimoto, S.; Ohuchida, S. *Org. Process Res. Dev.* **2003**, *7*, 168.

69. [j] Komatsu, H.; Awano, H. *J. Org. Chem.* **2002**, *67*, 5419.

70. [k] Chen, J. G.; Zhu, J.; Skonezny, P. M.; Rosso, V.; Venit, J. J. *Org. Lett.* **2004**, *6*, 3233.

71. Wang, H.; Yin, H. *Org. Process Res. Dev.* **2010**, *14*, 474.

72. Knight, C. J.; Millan, D. S.; Moses, I. B.; Robin, A. A.; Selby, M. D. *Org. Process Res. Dev.* **2012**, *16*, 697.

73. Jeffery, T. *Tetrahedron Lett.* **1991**, *32*, 2121.

74. Cann, R. O.; Chen, C.-P. H.; Gao, Q.; Hanson, R. L.; Hsieh, D.; Li, J.; Lin, D.; Parsons, R. L.; Pendri, Y.; Nielsen, R. B.; Nugent, W. A.; Parker, W. L.; Quinlan, S.; Reising, N. P.; Remy, B.; Sausker, J.; Wang, X. *Org. Process Res. Dev.* **2012**, *16*, 1953.

75. Oumata, N.; Ferandin, Y.; Meijer, L.; Galons, H. *Org. Process Res. Dev.* **2009**, *13*, 641.

76. Young, I. S.; Qiu, Y.; Smith, M. J.; Hay, M. B.; Doubleday, W. W. *Org. Process Res. Dev.* **2016**, *20*, 2108.

77. (a) Repič, O.; Prasad, K.; Lee, G. T. *Org. Process Res. Dev.* **2001**, *5*, 519. (b) Dale, J. *J. Chem. Soc.* **1961**, 910.

78. Kaptein, B.; van Dooren, T. J. G. M.; Boesten, W. H. J.; Sonke, T.; Duchateau, A. L. L.; Broxterman, Q. B.; Kamphuis, J. *Org. Process Res. Dev.* **1998**, *2*, 10.

79. Becker, C. L.; Engstrom, K. M.; Kerdesky, F. A.; Tolle, J. C.; Wagaw, S. H.; Wang, W. *Org. Process Res. Dev.* **2008**, *12*, 1114.

80. Busacca, C. A.; Cerreta, M.; Dong, Y.; Eriksson, M. C.; Farina, V.; Feng, X.; Kim, J.-Y.; Lorenz, J. C.; Sarvestani, M.; Simpson, R.; Varsolona, R.; Vitous, J.; Campbell, S. J.; Davis, M. S.; Jones, P.-J.; Norwood, D.; Qiu, F.; Beaulieu, P. L.; Duceppe, J.-S.; Haché, B.; Brong, J.; Chiu, F.-T.; Curtis, T.; Kelley, J.; Lo, Y. S.; Powner, T. H. *Org. Process Res. Dev.* **2008**, *12*, 603.

81. Roche, D.; Sans, D.; Girgis, M. J.; Prasad, K.; Repic, O.; Blacklock, T. J. *Org. Process Res. Dev.* **2001**, *5*, 167.

82. Chourreu, P.; Guerret, O.; Guillonneau, L.; Gayon, E.; Lefevre, G. *Org. Process Res. Dev.* **2020**, *24*, 1335.

83. Lawton, S.; Steele, G.; Shering, P.; Zhao, L.; Laird, I. *Org. Process Res. Dev.* **2009**, *13*, 1357.

84. (a) Nere, N. K.; Ramkrishna, D.; Parker, B. E. *Ind. Eng. Chem. Res.* **2007**, *46*, 3041. (b) Wang, Z. Z.; Wang, J. K.; Dang, L. P. *Org. Process Res. Dev.* **2006**, *10*, 450.

85. Kougoulos, E.; G. Jones, A.; Jennings, K. H.; Wood-Kaczmar, M. W. *J. Cryst. Growth* **2005**, *273*, 529.

86. O'Sullivan, B.; Barrett, P.; Hsiao, G.; Carr, A.; Glennon, B. *Org. Process Res. Dev.* **2003**, *7*, 977.

87. Muller, F. L.; Fielding, M.; Black, S. *Org. Process Res. Dev.* **2009**, *13*, 1315.

88. (a) Abu Bakar, M. R.; Nagy, Z. K.; Saleemi, A. N.; Rielly, C. D. *Cryst. Growth Des.* **2009**, *9*, 1378. (b) Woo, X. Y.; Nagy, Z. K.; Tan, R. B. H.; Braatz, R. D. *Cryst. Growth Des.* **2009**, *9*, 182.

89. (a) Beckmann, W.; Otto, W.; Budde, U. *Org. Process Res. Dev.* **2001**, *5*, 387. (b) Beckmann, W. *Org. Process Res. Dev.* **2000**, *4*, 372. (c) Beckmann, W.; Nickisch, K.; Budde, U. *Org. Process Res. Dev.* **1998**, *2*, 298.

90. (a) Patience, D. B.; Dell'Orco, P. C.; Rawlings, J. B. *Org. Process Res. Dev.* **2004**, *8*, 609. (b) Warstat, A.; Ulrich, J. *Chem. Eng. Technol.* **2006**, *29*, 187. (c) Yu, Z. Q.; Chow, P. S.; Tan, R. B. H. *Org. Process Res. Dev.* **2006**, *10*, 717. (d) Kubota, N.; Noihito, D. *Powder Technol.* **2001**, *121*, 31. (e) Loi Mi Lung-Somarribaa, B.; Moscosa-Santillanb, M.; Portea, C.; Delacroix, A. *J. Cryst. Growth* **2004**, *270*, 624.

91. Maloney, M. T.; Jones, B. P.; Olivier, M. A.; Magano, J.; Wang, K.; Ide, N. D.; Palm, A. S.; Bill, D. R.; Leeman, K. R.; Sutherland, K.; Draper, J.; Daly, A. M.; Keane, J.; Lynch, D.; O'Brien, M.; Tuohy, J. *Org. Process Res. Dev.* **2016**, *20*, 1203.

92. (a) Kubota, N.; Doki, N.; Yokota, M.; Jagadesh, D. *J. Chem. Eng. Jpn.* **2002**, *35*, 1036. (b) Doki, N.; Kubota, N.; Sato, A.; Yokota, M. *Chem. Eng. Jpn.* **2001**, *81*, 313. (c) Adi, H.; Larson, I.; Stewart, P. *Powder Technol.* **2007**, *179*, 95. (d) Ludwick, J. C.; Henderson, P. L. *Sedimentology* **1968**, *11*, 197.

93. (a) Parsons, A. R.; Black, S. N.; Colling, R. *Chem. Eng. Res. Des.* **2003**, *81*, 700. (b) Fujiwara, M.; Chow, P. S.; Ma, L. M.; Braatz, R. D. *Cryst. Growth Des.* **2002**, *2*, 363.

94. Kim, S.; Lotz, B.; Lindrud, M.; Girard, K.; Moore, T.; Nagarajan, K.; Alvarez, M.; Lee, T.; Nikfar, F.; Davidovich, M.; Srivastava, S.; Kiang, S. *Org. Process Res. Dev.* **2005**, *9*, 894.

95. Ennis, D. S.; McManus, J.; Wood-Kaczmar, W.; Richardson, J.; Smith, G. E.; Carstairs, A. *Org. Process Res. Dev.* **1999**, *3*, 248.

96. Adlington, N. K.; Black, S. N.; Adshead, D. L. *Org. Process Res. Dev.* **2013**, *17*, 557.

97. Karlsson, S.; Bergman, R.; Broddefalk, J.; Löfberg, C.; Moore, P. R.; Stark, A.; Emtenäs, H. *Org. Process Res. Dev.* **2018**, *22*, 618.

98. Andersen, S. M.; Aurell, C.-J.; Zetterberg, F.; Bollmark, M.; Ehrl, R.; Schuisky, P.; Witt, A. *Org. Process Res. Dev.* **2013**, *17*, 1543.

99. Chekal, B. P.; Campeta, A. M.; Abramov, Y. A.; Feeder, N.; Glynn, P. P.; McLaughlin, R. W.; Meenan, P. A.; Singer, R. A. *Org. Process Res. Dev.* **2009**, *13*, 1327.

100. Harris, R. M.; Andrews, B. I.; Clark, S.; Cooke, J. W. B.; Gray, J. C. S.; Ng, S. Q. Q. *Org. Process Res. Dev.* **2013**, *17*, 1239.

101. Stumpf, A.; Cheng, Z. K.; Wong, B.; Reynolds, M.; Angelaud, R.; Girotti, J.; Deese, A.; Gu, C.; Gazzard, L. *Org. Process Res. Dev.* **2015**, *19*, 661.

102. Stumpf, A.; McClory, A.; Yajima, H.; Segraves, N.; Angelaud, R.; Gosselin, F. *Org. Process Res. Dev.* **2016**, *20*, 751.

103. Hobson, L. A.; Akiti, O.; Deshmukh, S. S.; Harper, S.; Katipally, K.; Lai, C. J.; Livingston, R. C.; Lo, E.; Miller, M. M.; Ramakrishnan, S.; Shen, L.; Spink, J.; Tummala, S.; Wei, C.; Yamamoto, K.; Young, J.; Parsons, R. L.; Jr. *Org. Process Res. Dev.* **2010**, *14*, 441.

104. Zhong, Y.-L.; Pipik, B.; Lee, J.; Kohmura, Y.; Okada, S.; Igawa, K.; Kadowaki, C.; Takezawa, A.; Kato, S.; Conlon, D. A.; Zhou, H.; King, A. O.; Reamer, R. A.; Gauthier, D. R.; Askin, D. *Org. Process Res. Dev.* **2008**, *12*, 1245.

105. Hansen, K. B.; Hsiao, Y.; Xu, F.; Rivera, N.; Clausen, A.; Kubryk, M.; Krska, S.; Rosner, T.; Simmons, B.; Balsells, J.; Ikemoto, N.; Sun, Y.; Spindler, F.; Malan, C.; Grabowski, E. J. J.; Armstrong III, J. D. *J. Am. Chem. Soc.* **2009**, *131*, 8798.

106. Thiel, O. R.; Achmatowicz, M.; Bernard, C.; Wheeler, P.; Savarin, C.; Correll, T. L.; Kasparian, A.; Allgeier, A.; Bartberger, M. D.; Tan, H.; Larsen, R. D. *Org. Process Res. Dev.* **2009**, *13*, 230.

107. Naganathan, S.; Andersen, D. L.; Andersen, N. G.; Lau, S.; Lohse, A.; Sørensen, D. *Org. Process Res. Dev.* **2015**, *19*, 721.

108. Mennen, S. M.; Mak-Jurkauskas, M. L.; Bio, M. M.; Hollis, S.; Nadeau, K. A.; Clausen, A. M.; Hansen, K. B. *Org. Process Res. Dev.* **2015**, *19*, 884.

109. McDonough, J. A.; Durrwachter, J. R. *Org. Process Res. Dev.* **1997**, *1*, 268.

110. Angelaud, R.; Reynolds, M.; Venkatramani, C.; Savage, S.; Trafelet, H.; Landmesser, T.; Demel, P.; Levis, M.; Ruha, O.; Ruechert, B.; Jaeggi, H. *Org. Process Res. Dev.* **2016**, *20*, 1509.

111. Czyzewski, A. M.; Chen, S.; Bhamidi, V.; Yu, S.; Marsden, I.; Ding, C.; Becker, C.; Napier, J. *Org. Process Res. Dev.* **2017**, *21*, 1493.

112. Weisenburger, G. A.; Anderso, D. K.; Clark, J. D.; Edney, A. D.; Karbin, P. S.; Gallagher, D. J.; Knable, C. M.; Pietz, M. A. *Org. Process Res. Dev.* **2009**, *13*, 60.

113. Caron, S.; Do, N. M.; Sieser, J. E.; Arpin, P.; Vazquez, E. *Org. Process Res. Dev.* **2007**, *11*, 1015.

114. Caille, S.; Cui, S.; Faul, M. M.; Mennen, S. M.; Tedrow, J. S.; Walker, S. D. *J. Org. Chem.* **2019**, *84*, 4583.

115. Bakar, M. R. A.; Nagy, Z. K.; Rielly, C. D. *Org. Process Res. Dev.* **2009**, *13*, 1343.

116. Reeves, J. T.; Tan, Z.; Reeves, D. C.; Song, J. J.; Han, Z. S.; Xu, Y.; Tang, W.; Yang, B.-S.; Razavi, H.; Harcken, C.; Kuzmich, D.; Mahaney, P. E.; Lee, H.; Busacca, C. A.; Senanayake, C. H. *Org. Process Res. Dev.* **2014**, *18*, 904.

117. Deshpande, P. P.; Singh, J.; Pullockaran, A.; Kissick, T.; Ellsworth, B. A.; Gougoutas, J. Z.; Dimarco, J.; Fakes, M.; Reyes, M.; Lai, C.; Lobinger, H.; Denzel, T.; Ermann, P.; Crispino, G.; Randazzo, M.; Gao, Z.; Randazzo, R.; Lindrud, M.; Rosso, V.; Buono, F.; Doubleday, W. W.; Leung, S.; Richberg, P.; Hughes, D.; Washburn, W. N.; Meng, W.; Volk, K. J.; Mueller, R. H. *Org. Process Res. Dev.* **2012**, *16*, 577.

118. Baenziger, M.; Baierl, M.; Devanathan, K.; Eswaran, S.; Fu, P.; Gschwend, B.; Haller, M.; Kasinathan, G.; Kovacic, N.; Langlois, A.; Li, Y.; Schuerch, F.; Shen, X.; Wan, Y.; Wickendick, R.; Xie, S.; Zhang, K.; *Org. Process Res. Dev.* **2020**, *24*, 1405.

119. Peng, F.; Humphrey, G. R.; Maloney, K. M.; Lehnherr, D.; Weisel, M.; Lévesque, F.; Naber, J. R.; Brunskill, A. P.; Larpent, P.; Zhang, S-W.; Lee, A. Y.; Arvary, R. A.; Lee, C. H.; Bishara, D.; Narsimhan, K.; Sirota, E.; Whittington, M. *Org. Process Res. Dev.* **2020**, *24*, 2453.

19 Methods for Residual Metal Removal

Precious metal catalysts are extensively used in catalytic reactions to produce a wide range of products across a variety of industries. Many organic syntheses rely on metal-catalyzed reactions to prepare advanced intermediates or final products, with palladium chemistry dominating this arena.[1] Typical precious metal-promoted reactions include carbon–carbon bond formation, carbon–nitrogen bond formation, and catalytic hydrogenation and deprotection. Among many variants of palladium-catalyzed C–C coupling reactions, Suzuki, Negishi, Sonogashira, and Heck reactions have received the most significant attention. One of the major benefits provided by cross-coupling technologies is the increased convergence when compared to classical methods, as two highly functionalized molecules can be coupled together under relatively mild conditions. As a result, the metal-catalyzed reaction is often situated late in the synthetic sequence. The convergent coupling approach, however, limits the number of downstream opportunities for metal residue removal. Considering the therapeutical safety, acceptable limits for residual heavy metals in the drug substance are quite stringent,[2] and reducing residual metals to meet the ever more stringent regulatory restrictions on the level of metals (a few parts per million)[3] in active pharmaceutical ingredients remains a major challenge.

Numerous strategies developed during the last decades are presented in this Chapter for the removal of precious and transition metals, mainly palladium.[4]

19.1 REMOVAL OF RESIDUAL PALLADIUM

Depending on the chemistry of Pd participated, palladium exhibits two main states, Pd(0) and Pd(II). Several techniques have been developed to reduce residual Pd to acceptable levels, including adsorption, crystallization, distillation, extraction, and precipitation/filtration. The insoluble palladium such as carbon-supported palladium used in hydrogenation reaction can be removed by simple filtration of the reaction mixture prior to the product purification. Crystallization and extraction are the preferred industrial-scale approaches for the removal of soluble palladium impurities. In most cases, a combination of two or three Pd-removal approaches is necessary in order to achieve ideal results.

19.1.1 CRYSTALLIZATION

Diminishing the palladium content by means of crystallization is a commonly used technology in the pharmaceutical industry. Through crystallization, the soluble palladium impurity can be rejected into the mother liquor upon filtration. For instance, typical levels of palladium observed in sodium salt (S)-3 obtained from the Suzuki cross-coupling reaction were 300–600 ppm, whereas after crystallization from ethanol and water, the corresponding acid (S)-4 was isolated with 40–80 ppm Pd (Scheme 19.1).[5]

In many cases, however, crystallization failed to reduce the residual palladium to the desired level especially when the product bears chelating functional groups. In order to address these challenges, the crystallization was modified by adding various chelating agents, such as cysteine, N-acetylcysteine, ethanolamine, etc. Table 19.1 lists some chelating agents that are frequently utilized as additives in crystallization operations.

DOI: 10.1201/9781003288411-19

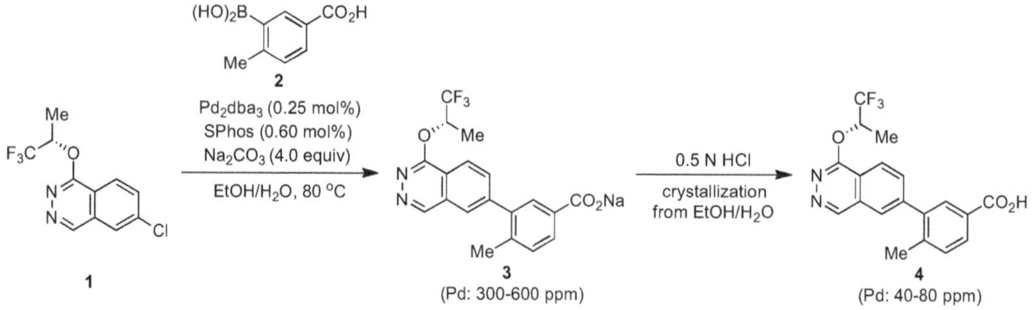

SCHEME 19.1

TABLE 19.1
Common Crystallization Additives

Name	Structure	Mw	Mp (Bp)
Ethanolamine	 HO–CH₂CH₂–NH₂	61.8	10–11 °C (170 °C)
Maleic acid	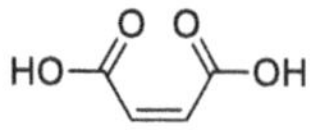 HOOC–CH=CH–COOH	116.07	137–140 °C
N,N,N',N'-Tetramethylethylenediamine	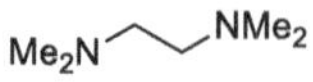 Me₂N–CH₂CH₂–NMe₂	116.20	–50 °C (120–122 °C)
Triethanolamine	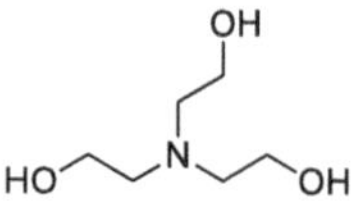 N(CH₂CH₂OH)₃	149.19	18–21 °C (190–193 °C/5 mmHg)
1,2-Diaminopropane[a]	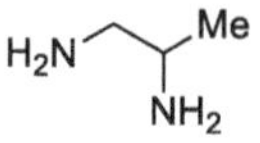 H₂N–CH₂–CH(Me)–NH₂	74.12	–37.1 °C (119.6 °C)
Thiourea	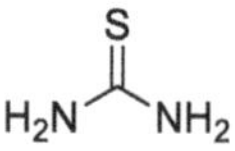 H₂N–C(=S)–NH₂	76.12	170–176 °C
Cysteine	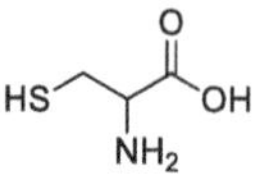 HS–CH₂–CH(NH₂)–COOH	121.16	225 °C (dec)
N-Acetylcysteine	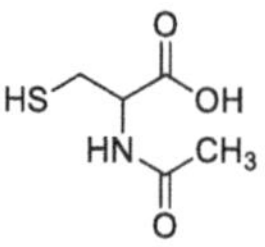 HS–CH₂–CH(NHC(=O)CH₃)–COOH	163.19	106–108 °C
Imidazole	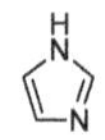 (imidazole ring)	68.08	89–91 °C (256 °C)

[a] Chekal, B. P.; Guinness, S. M.; Lillie, B. M.; McLaughlin, R. W.; Palmer, C. W.; Post, R. J.; Sieser, J. E.; Singer, R. A.; Sluggett, G. W.; Vaidyanathan, R.; Withbroe, G. J. *Org. Process Res. Dev.* **2014**, *18*, 266.
[b] Achmatowica, M.; Thiel, O. R.; Wheeler, P.; Bernard, C.; Huang, J.; Larsen, R. D.; Faul, M. M. *J. Org. Chem.* **2009**, *74*, 795.

19.1.1.1 Crystallization in the Presence of Cysteine

Because of the high affinity for the transition metal of the sulfhydryl group, cysteine was used as an additive in the crystallization of Sonogashira cross-coupling product **7** to aid in reducing residual palladium (Equation 19.1).[6]

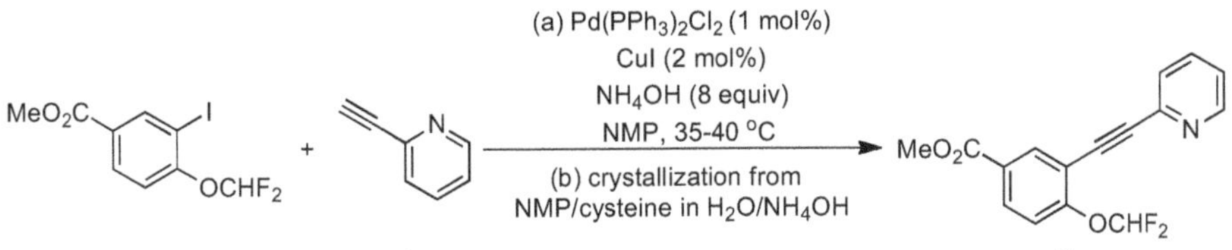

$$\text{(19.1)}$$

Procedure

- Charge 2.30 kg (22.3 mol, 1.01 equiv) of **6** to a solution of **5** (7.25 kg, 22.1 mol) in NMP (30.7 kg).
- Charge 84 g (0.442 mol, 2 mol%) of CuI.
- Degas by evacuating and refilling via nitrogen bleed three times.
- Charge 0.16 kg (0.221 mol, 1 mol%) of $Pd(PPh_3)_2Cl_2$.
- Heat the mixture to 35 °C.
- Charge 12.6 kg (8 equiv) of 28% NH_4OH solution over ca. 45 min while maintaining the temperature at 35–40 °C.
- Stir the batch for 1 h.
- Add a solution of L-cysteine (0.5 kg, 3.69 mol, 0.167 equiv) in water (3.0 kg) and 28% NH_4OH (0.7 kg).
- Stir for 2 h.
- Cool to 20 °C to effect crystallization.
- Add water (47.3 kg) over 45 min.
- Cool the resulting suspension to 5 °C and stir for 1 h at 5 °C.
- Filter and wash with water (30.9 kg), 28% NH_4OH (2×14.2 kg), and water (23.5 kg).
- Wash the cake with a 2:1 MeOH/water solution (28.6 kg).
- Yield: 6.34 kg (86.1%).

REMARK

The use of NH_4OH allowed for the reaction to be run at a lower temperature (35–40 °C); some additional benefits of using NH_4OH were to increase the water solubility of cysteine and facilitate copper removal.

19.1.1.2 Crystallization in the Presence of *N*-Acetylcysteine

N-Acetylcysteine is a versatile chelating agent to remove residual metal impurities and can be utilized in aqueous or organic media.

19.1.1.2.1 Heck Reaction

Late-stage Heck coupling reaction in the synthesis of drug candidate **9** was carried out with high loading of palladium acetate (8 mol%), leading to 1411 ppm Pd in the isolated product **9** (Equation 19.2).[7]

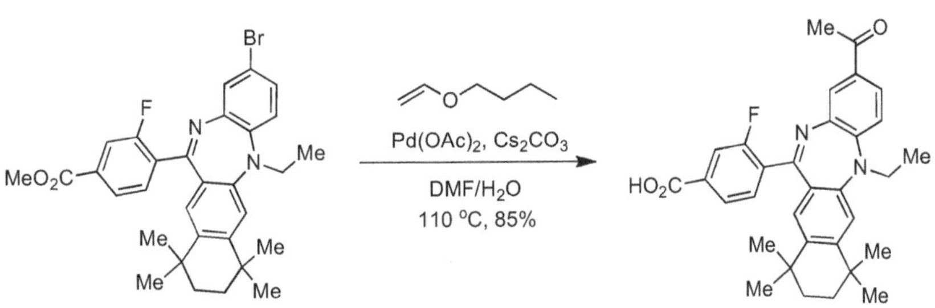

$$\text{(19.2)}$$

To reduce the residual Pd in **9**, a modified crystallization was developed by adding a basic solution of *N*-acetylcysteine (in 3 N NaOH) into the suspension of the drug substance in EtOH at 40 °C, followed by cooling and filtration, cutting the residual Pd from 1411 ppm to 8 ppm.

NOTE

Using this protocol, the drug molecule must be stable under basic conditions.

19.1.1.2.2 Negishi Cross-Coupling Reaction

The Negishi coupling product **12** (crude) contains Pd at a level of 20–50 ppm along with 200–1000 ppm Zn after treatment with ethylenediamine (Scheme 19.2).[8] The subsequent S_NAr reaction of **12** with aminopyrimidine **13** furnished the product **14** with Pd in <10–30 ppm and trace amounts of Zn (<10 ppm) upon diluting with toluene and washing with an aqueous solution of *N*-acetylcysteine. These metal impurities could be further reduced by recrystallization from ethanol/H_2O in the presence of *N*-acetylcysteine, giving **14** in excellent purity (99.4 area% in HPLC) and very low Pd and Zn (<1 ppm).

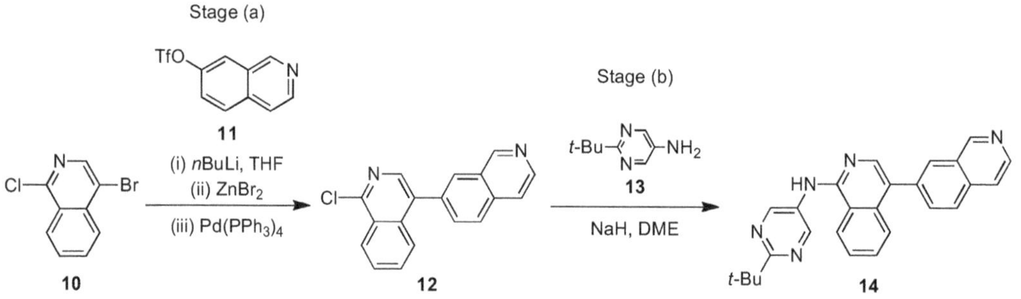

SCHEME 19.2

Procedure

Stage (b)

- Charge 1.074 kg (26.9 mol, 2.5 equiv) of NaH (60% dispersion in mineral oil) to a solution of **13** (1.615 kg, 10.7 mmol) in DME (40 L) within 10 min.
- Charge 3.42 kg (11.76 mol, 1.1 equiv) of **12** within 20 min.
- Stir at reflux for 1 h.
- Cool the batch to 20 °C.
- Add toluene (40 L).
- Add slowly a solution of *N*-acetylcysteine (0.2 kg) in water (30 L).
- Separate layers.
- Extract the aqueous layer with toluene (20 L).
- Wash the combined organic layers with water (20 L).
- Distill to 20 L.
- Add toluene (60 L) and distill to 19 L.
- Cool to 0 °C and stir overnight.
- Filter, wash with toluene (3×5 L), and dry at 55 °C under reduced pressure.
- Yield of **14**: 4.78 kg (27 ppm Pd; 17 ppm Zn).
- Suspend 4.73 kg of the crude product **14** in a mixture of EtOH (86 L) and water (4.5 L) in the presence of activated charcoal (240 g).
- Stir at reflux for 1 h.
- Filter and wash with hot EtOH (5 L).
- Add a solution of *N*-acetylcysteine (240 g) and ethylenediamine tetraacetic acid disodium salt (430 g) in water (9.5 L) at 70 °C followed by adding water (94 L).

- Cool the batch to 0 °C within 5 h.
- Stir at 0 °C overnight.
- Filter and wash with water (95 L).
- Dry at 60 °C under reduced pressure.
- Yield: 3.7 kg (85 %; Pd and Zn <1 ppm).

19.1.1.2.3 Sonogashira Reaction

A key step for making the unsymmetrical diaryl ethane **21** involves one-pot double Sonogashira cross-coupling reactions, followed by hydrogenation of ethyne **20** (Scheme 19.3).[9]

(I) Problems

Upon workup of the Sonogashira reactions, the Pd level in the isolated product **20** varies from batch to batch ranging from 19 ppm to 700 ppm. In addition, the palladium level in the isolated hydrogenation product **21** was not acceptable (6 ppm) after crystallization (required Pd <4 ppm). It was found that Pd(0)/PPh$_3$ complex, derived from Sonogashira coupling, decomposed to metallic, insoluble Pd during the solvent switch distillation.

(II) Solutions

To alleviate the precipitation of palladium black, an oxidative workup procedure was implemented to convert Pd(0) into Pd(II) by adding H$_2$O$_2$.[10] As shown in Table 19.2, the crystallization in the presence of N-acetylcysteine, in turn, reduced the Pd level down to an acceptable level.

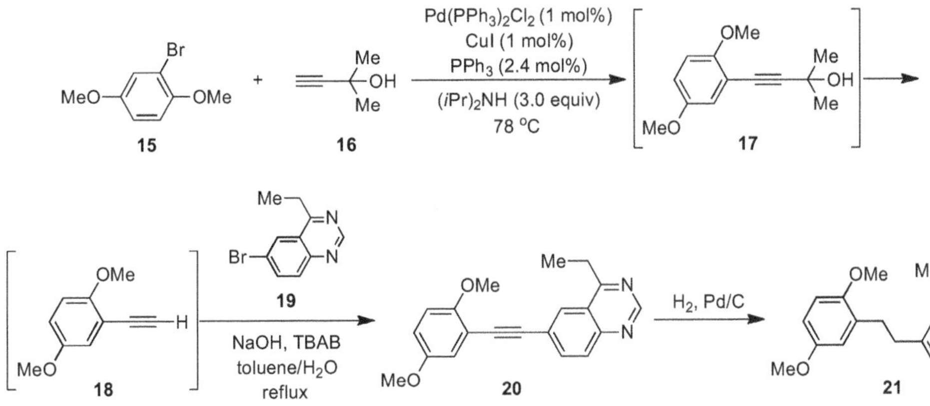

SCHEME 19.3

TABLE 19.2
Results of Pd Removal by Crystallization in the Presence of N-Acetylcysteine[11]

N-Acytyl-cysteine (g/mL)	Pd content, ppm		
	20	21, before crystallization	21, after crystallization
-	17	10	6
0.5%	17	10	1
0.5%	30	11	<2
1.0%	120	21	<1

REMARK

The *N*-acetylcysteine-involved crystallization was applied also in the removal of residual Pd from the Pd-catalyzed carbonylation product.[11]

19.1.1.2.4   Suzuki Cross-Coupling Reaction

GDC-0134 is a small molecule DLK inhibitor, which was developed for the potential treatment of amyotrophic lateral sclerosis. As the penultimate intermediate for the synthesis of GDC-0134, compound **24** was prepared via Suzuki coupling reaction of chloropyrimidine **23** with boronate **22,** and the isolated product **24** was contaminated with Pd at levels of 100–150 ppm (Scheme 19.4).[12]

To reduce the residual Pd, a crystallization protocol was developed by treatment of the post-reaction mixture with aqueous *N*-acetylcysteine (0.1 equiv) at 60 °C for 30 min before adding *n*-heptane. At a kilogram scale, crystallization did not happen spontaneously, and seeding after *N*-acetylcysteine addition was implemented. Pyrido-pyrimidine **24** was isolated in 84% yield and 99.3% w/w assay purity with Pd levels $\leq$5 ppm.

Procedure

- Add 0.60 kg (0.81 mol) of PdCl$_2$(dppf) to a suspension of 47.5 kg (163.9 mol) of **23**, 54.0 kg (188.8 mol) of **22,** and 68.0 kg (492.0 mol) of K$_2$CO$_3$ in THF (510 kg) and water (165 kg) at 20 °C.
- Heat the batch to 63 °C and stir for 3 h (the reaction was complete then).
- Cool to 58 °C.
- Add a solution of *N*-acetylcysteine (2.8 kg, 17.2 mol) in water (20.2 kg) within 15 min and rinse with water (9.6 kg).
- Stir for 2 h.
- Add 160 g of seed **24** and stir for 75 min.
- Add *n*-heptane (97.0 kg) within 40 min.
- Cool the suspension to 22 °C within 3 h and stir at 22 °C for 6 h.
- Isolate the product by centrifugation, wash with a mixture of THF (189.2 kg) and water (191.2 kg), and dry at 45 °C under reduced pressure for 9.5 h.
- Yield of **24**: 57.0 kg (84%).

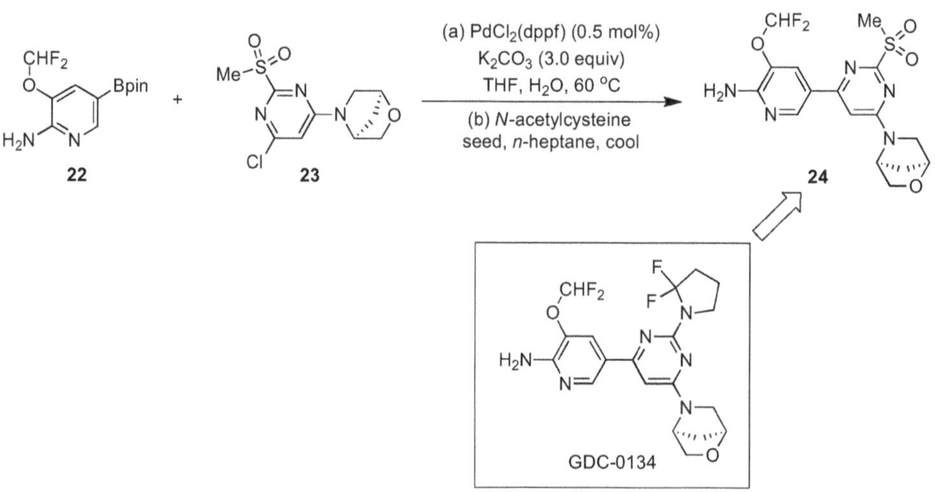

SCHEME 19.4

19.1.2 EXTRACTION

The extraction method can be effectively utilized to reduce palladium levels, providing that the distribution of the drug molecule between the aqueous and organic phases is different than that of the palladium complex. However, the palladium impurity tends, in some circumstances, to have a similar distribution pattern to the drug molecule, which renders the extraction less effective. Thus, a modified extraction technology can be generated by introducing additives into the extraction system.[13] In practice, two experimental scenarios have to consider: one is to migrate the organic soluble palladium impurity from the organic phase into the aqueous phase (liquid–liquid transportation) and vice versa, and the other is to precipitate the palladium complex out of the organic phase and remove it by filtration (extractive precipitation).

19.1.2.1 Liquid–Liquid Transportation

Chelating metal impurities with water-soluble chelating agents can transfer metal impurities from the organic phase into the aqueous phase. In contrast, palladium impurity can be rejected into the organic phase by using organic ligands. In practice, using trialkylphosphine (or triarylphosphine) ligand can generate organic soluble Pd complexes to reject the Pd impurities into the organic phase, while keeping the desired product in the aqueous phase.

19.1.2.1.1 Use of Triethylamine

The convergent synthesis of 4,5-disubstituted oxazole **27**, a potent and selective inhibitor of the stress-activated kinase p38α, was accomplished via Suzuki coupling of aryl boronic acid **25** with 4-Bromo-oxazole **26** (Equation 19.3).[14] The isolated crude product **27** was plagued with a high level of metal residues (2100 ppm of Pd and 3200 ppm of Fe), which is presumably due to the chelating effect of the weakly basic triazole moiety in **27**. An extraction method was utilized to purge palladium and iron by boiling the solution in the presence of triethylamine followed by the addition of water to crystallize the product. After a single repetition, the metal residues in the isolated product **27** were reduced to 9.0 ppm for Pd and 8.85 ppm for Fe, respectively.

(19.3)

Procedure

Stage (a)
- Mix 1.80 kg (10.2 mol, 1.09 equiv) of **25** with 3.0 kg (9.34 mol, 1.0 equiv) of **26** in a mixture of water (4.5 L) and MeTHF (45 L).
- Heat the batch in the presence of 374 g of Pd(dppf)Cl$_2$·CH$_2$Cl$_2$, 141 g (0.93 mol) of CsF, and 1.55 kg (11.2 mol) of K$_2$CO$_3$ at 70 °C for 1 h.
- Distill under reduced pressure to remove MeTHF.
- Partition the resulting residue between aqueous 2 N HCl (30 L) and CH$_2$Cl$_2$ (48 L).
- Distill the resulting organic layer at atmosphere pressure and displace with toluene (45 L).
- Add 4 N HCl (30 L) to transfer the product to the aqueous phase.
- Add 50 wt% NaOH to the aqueous layer to pH 10.
- Filter to give the crude product **27**.

Stage (b)

- Treat the crude product **27**, obtained in Stage (a), with IPA (6 L) and TEA (9 L) at reflux for 3 h.
- Cool to 35 °C.
- Add water (75 L).
- Cool the batch to 15 °C.
- Filter to give a wet cake product.
- Charge the wet cake back to the reactor.
- Repeat the treatment by mixing the wet cake with IPA (6 L) and TEA (9 L) at reflux for 3 h, cooling to 35 °C, adding water (75 L), and cooling to 15 °C.
- Filter and dry under vacuum to give 3.08 kg (88.7%) of **27**.
- Slurry the product in IPA (76 L).
- Distill azeotropically to remove 5 L of solvent.
- Cool to 50 °C.
- Filter through a 0.2 μm filter.
- Distill to 6 L volume.
- Cool to 22 °C and stir for 48 h.
- Filter and dry.
- Yield: 2.83 kg.

REMARKS

Alternative strategy to alleviate the Pd contamination is the use of water-soluble sulfonylated phosphine ligands such as 3,3′,3″-phosphanetriyltris(benzenesulfonic acid) trisodium salt (TPPTS)[15] and sodium 3-(diphenylphosphino)benzenesulfonate (TPPMS).[16] Application of the water-soluble phosphine ligands was demonstrated in the Sonogashira cross-coupling reaction of 2,5-dimethoxyphenyl bromide **28** with **16** in aqueous acetonitrile (Equation 19.4).[11]

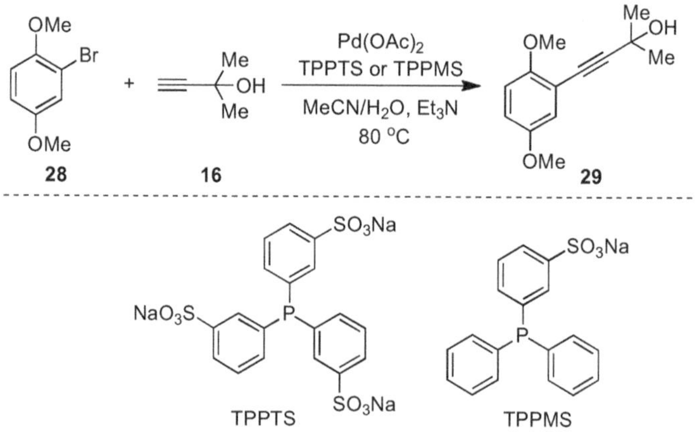

$$(19.4)$$

However, these ligands are limited to laboratory scale due to the high price and limited commercial availability.

19.1.2.1.2 Use of Tri(n-butyl)phosphine

The convergent synthesis via Suzuki coupling reaction gave the isolated crude **32** with ca. 8000 ppm of Pd. Using the liquid–liquid transportation approach was able to reduce levels of Pd/Fe in the isolated crude product **32** to less than 50 ppm (Equation 19.5).[4b]

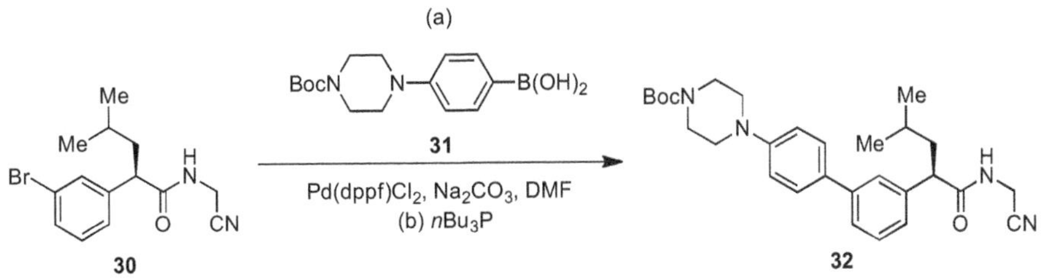

(19.5)

The workup protocol involved treatment of the crude **32** solution in ethyl acetate with 20 mol% of nBu$_3$P and aqueous lactic acid to transfer **32** from the organic layer into the aqueous layer, followed by salt-break with Na$_2$CO$_3$ and extracting the product back into ethyl acetate. The levels of Pd and Fe in **32**, after crystallization from toluene, were less than 50 ppm.

Procedure

Stage (b)

- Charge 56 g (0.277 mol) of nBu$_3$P to a solution of crude **32** (1.01 kg) in EtOAc (20 L).
- Stir the resulting mixture at room temperature for 1 h.
- Charge water (15 L) and 805 g of lactic acid (85%, 7.6 mol).
- Separate layers.
- Extract the aqueous layer with EtOAc (3×5 L).
- Wash the combined organic layers with water (5 L).
- Add a suspension of Na$_2$CO$_3$ (805 g, 7.6 mol) in EtOAc (15 L) to the combined aqueous layers, and stir the mixture for ca. 15 min until the carbonate dissolved.
- Separate layers.
- Wash the organic layer with water (5 L).
- Add DarkoKB (200 g) and Solka Floc 40 NF (200 g).
- Age the mixture at room temperature overnight.
- Filter and wash with EtOAc (4 L).
- Distill under reduced pressure and switch solvent to toluene.
- Adjust the concentration to 10 mL/g.
- Heat the batch to 75 °C to form a solution.
- Cool to 65 °C and add seeds.
- Cool to room temperature over 8 h.
- Filter, wash with toluene (2 L), and dry under vacuum.
- Yield: 640 g (89%).

More examples of using tri(n-butyl)phosphine to remove Pd can be found in the literature.[17]

19.1.2.1.3   Use of Triphenylphosphine

Sulopenem is a broad-spectrum anti-bacterial candidate developed by Pfizer. Scheme 19.5[18] shows a two-step transformation of allyl ester **33** into sulopenem involving TBS deprotection followed by

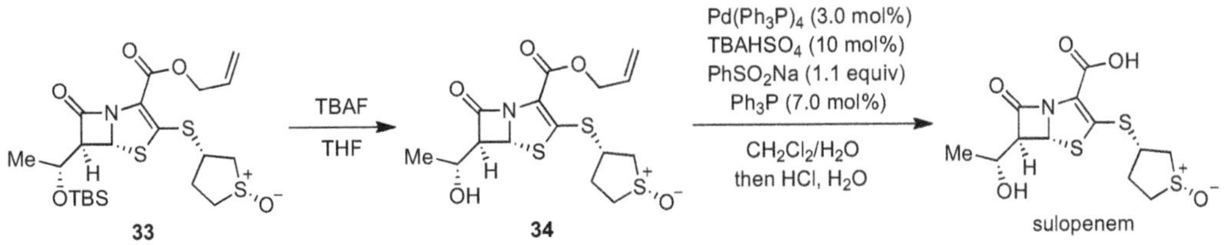

SCHEME 19.5

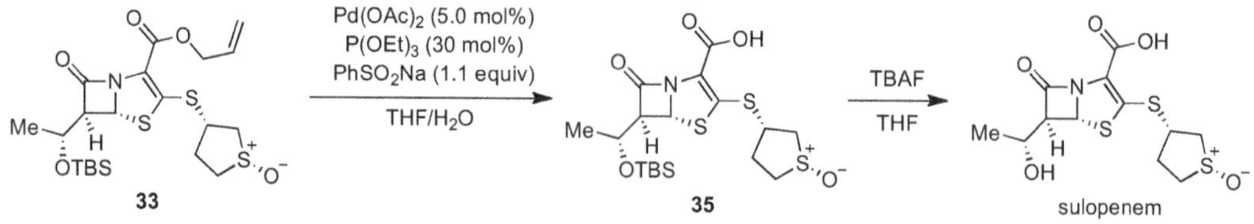

SCHEME 19.6

palladium-catalyzed deallylation. Triphenylphosphine was utilized to prevent Pd(0) from precipitation during the palladium-catalyzed deallylation.

The use of triphenylphosphine ligand (7 mol%) retained the Pd(0) in the methylene chloride organic phase, while the product sodium salt was taken into the aqueous phase prior to acidification and isolation.

REMARKS

(a) Sulfinic acids or their salts are the most effective allyl scavengers in the presence of palladium catalysts.[19]

(b) Process optimization led to reordered deprotection sequence and improved deallylation conditions using more stable Pd(Oac)$_2$ as a precatalyst with an excess of triethylphosphite ligand as illustrated in Scheme 19.6.[20]

19.1.2.2 Extractive Precipitation

The extractive precipitation approach is used to extract the product into the organic (or aqueous) phase while leaving the metal impurity as solid precipitates.

19.1.2.2.1 Use of Sodium Bisulfite

The extractive precipitation strategy was applied toward the purification of the Suzuki–Miyaura coupling product as illustrated in Reaction (19.6).[21] Direct isolation, by adding water to precipitate the coupling product **38**, resulted in palladium levels of 12000 ppm. Accordingly, the isolation of the disubstituted indole product **38** by washing with toluene/NaHSO$_3$ and filtration reduced the palladium level to 100 ppm.

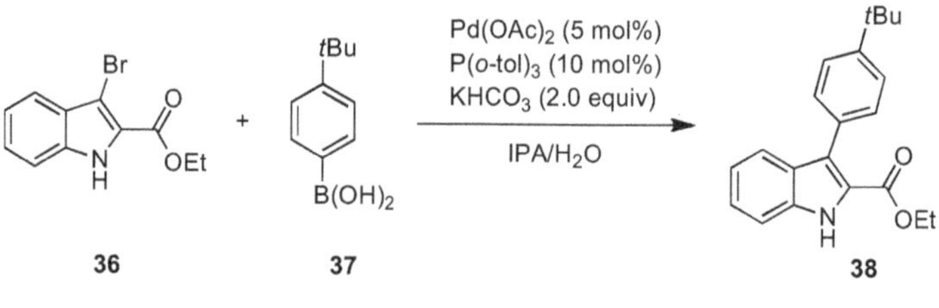

(19.6)

Procedure

Stage (a)
- Charge 1.0 kg of **36** (1.0 equiv) followed by adding 0.8 kg (1.2 equiv) of **37**, 0.75 kg (2.0 equiv) of KHCO$_3$, 110 g (10 mol%)) of P(o-tol)$_3$, and 40 g (5 mol%) of Pd(Oac)$_2$.
- Charge water (1.0 L) and IPA (4 L).
- Heat the mixture at 60 °C for 2 h under N$_2$.

Stage (b)
- Add toluene (6 L) and 20% NaHSO$_3$ (6 L).
- Stir at 60 °C for 1 h.

- Filter to remove insoluble Pd and separate layers.
- Distill the organic layer to 3 L volume.
- Add heptane (5 L).
- Cool to 0 °C and stir for overnight.
- Filter and wash with heptane.
- Dry under vacuum.
- Yield: 983 g (82%).

19.1.2.2.2 Use of 2,4,6-Trimercaptotriazine

2,4,6-Trimercaptotriazine (TMT) can be employed to reduce the residual Pd level via the formation of insoluble TMT–Pd complex.[4a] For instance, the addition of TMT (6 mol%) to the Heck cyclization reaction mixture and heating at reflux reduced the levels of residual palladium in ester **40** from ca. 2000 ppm to <15 ppm (Equation 19.7).[22]

$$(19.7)$$

Procedure

Stage (a)
- Charge 87 kg (166 mol) of **39** followed by adding 2.177 kg (8.3 mol) of Ph$_3$P, 0.589 kg (3.32 mol) of PdCl$_2$, and toluene (1060 kg).
- Charge 22.3 kg (220 mol) of TEA.
- Heat at reflux for 4 h.

Stage (b)
- Charge 1.74 kg (9.81 mol) of TMT.
- Continue to reflux for 1 h.
- Cool to 45 °C.
- Filter and rinse with toluene (152 kg) to remove the solids.
- Wash, at 40–45 °C, the filtrate with H$_2$O (2×696 L).
- Distill under reduced pressure until 680 L volume.
- Add isooctane (602 kg) at 41–44 °C over 100 min.
- Cool to 18 °C and age for 2 h.
- Filter and wash with heptane/isooctane (1:1, v/v, 2×177 L).
- Dry at 45–50 °C under vacuum.
- Yield: 48.7 kg (74%).

REMARKS

(a) The low level of residual palladium is an important criterion for pharmaceuticals, and the use of a heterogeneous catalyst is motivated by the ease of separation of the palladium from the product. Recent development of a heterogeneous palladium catalyst allows the recovery of the expensive palladium and ligand after cross-coupling.[23] The Pd residue level in the isolated product tends to be reduced significantly simply by filtration.[25a] Besides hydrogenation reactions, Suzuki cross-couplings can be possibly carried out by employing

724

Handbook for Chemical Process Research and Development

heterogeneous catalysts to replace homogeneous catalysts. Two examples[24] of using heterogeneous Pd catalyst in Suzuki reactions were successfully demonstrated, in which palladium residues were readily reduced to accepted levels simply by filtration.

(b) In addition, a recently developed polymer fiber-supported catalyst[25] provides stability benefits toward mechanical agitation. The application of the fiber catalyst was successfully demonstrated in hydrogenations at both pilot plant (250 L) and industrial (4000 L) scales.[26]

In general, nitrogen- or sulfur-containing soluble chelating agents and polyacids are frequently used in the extractive workup to achieve the Pd-removal goal. Table 19.3 lists some commonly used chelating additives.

19.1.3 ADSORPTION

Adsorption is a process to reduce metal levels by adsorbing soluble metals onto a solid phase. Activated carbon is an inexpensive and widely used solid material to remove residual palladium. A polymer-supported trimercaptotriazine (TMT) palladium scavenger was demonstrated to effectively reduce palladium acetate concentration in THF solution by a factor of 1000. The macroporous polystyrene–2,4,6-trimercaptotriazine (MP–TMT) resin (Table 19.4, Entry 1) was prepared by covalently attaching TMT to an insoluble gel-type polystyrene support.[27] MP–TMT resin has a TMT loading of 0.7–1.1 mmol/g and is commercially available from Argonaut Technologies. Being not dependent on solvent swelling, it can function effectively in both organic and aqueous conditions with 2–5 equiv of resin relative to the palladium content. Smopex-111, a thiol functionalized fiber, is commonly used as an efficient precious metal scavenger.

Table 19.4 lists a number of commercially available polymer-supported reagents applied as metal scavengers.[28]

19.1.3.1 Activated Carbon

The reduction of the residual palladium in fosfluconazole was effected by treating the Pd-catalyzed debenzylation mixture with activated carbon (Equation 19.8).[29] Fosfluconazole is a water-soluble prodrug of Diflucan, developed by Pfizer as a broad-spectrum anti-fungal agent and launched in 1988.

(19.8)

Procedure

Stage (a)
- Mix 30.1 kg (53.13 mol) of **41** with 1.5 kg 5% Pd/C (50% wet) and 4.36 kg (108.9 mol) of NaOH in 75.7 L of low-endotoxin water.
- Conduct the reaction at room temperature under 60 psi of H_2 for 12 h.

TABLE 19.3
Frequently Used Additives during the Extractive Workup

Entry	Additive	Structure	Mw	Mp (Bp)	Treatment
1	Trimercaptotriazine[a] (TMT)	SH (triazine with three SH groups)	177.27	>300 °C	Aqueous wash or in combination with NaOH
2	TMT-Na • nH$_2$O[b]	SNa (triazine with SNa groups) · n H$_2$O	243.22 (anhydrous basis)	82–84 °C	Aqueous wash or in combination with NaOH
3	2-Mercaptobenzoic acid[c]	(benzoic acid with SH group)	154.19	162–165 °C	The sodium 2-mercaptobenzoate is prepared by treatment with NaOH.
4	Sodium sulfide[d]	Na$_2$S	78.04	950 °C	Aqueous wash
5	Sodium bisulfite	NaHSO$_3$	104.06	-	Aqueous wash
6	DL-Lactic acid	(lactic acid structure)	90.08	(122 °C/15 mmHg)	Aqueous wash
7	Triethylamine	Et$_3$N	110.19	89.7 °C	Wash and filtration
8	Tris(2-aminoethyl)amine[30]	(tris(2-aminoethyl)amine structure)	146.23	(114 °C/15 mmHg)	In combination with silica gel.
9	Tri(n-butyl)phosphine	(tributylphosphine structure)	202.32	(150 °C/50 mmHg)	–
10	2-(Dimethylamino)- ethanethiol hydrochloride[30]	(structure) · HCl	141.66	150–160 °C	In combination with silica gel.
11	1,2-Ethanedithiol[30]	HS—CH$_2$CH$_2$—SH	94.20	–41 °C (144–146 °C)	In combination with silica gel.
12	L-Cysteine[30]	(cysteine structure)	121.16	240 °C (dec)	In combination with silica gel.
13	N-Acetyl-L-cysteine[30]	(N-acetyl cysteine structure)	163.19	106–108 °C	-
14	2-Mercaptoethanol[30]	HS—CH$_2$CH$_2$—OH	78.13	(157 °C)	In combination with silica gel.
15	Ethylenediamine[30]	H$_2$N—CH$_2$CH$_2$—NH$_2$	60.10	8.5 °C (118 °C)	In combination with silica gel.

[a] Jacks, T. E.; Belmont, D. T.; Briggs, C. A.; Horne, N. M.; Kanter, G. D.; Karrick, G. L.; Krikke, J. J.; McCabe, R. J.; Mustakis, J. G.; Nanninga, T. N.; Risedorph, G. S.; Seamans, R. E.; Skeean, R.; Winkle, D. D.; Zennie, T. M. *Org. Process Res. Dev.* **2004**, *8*, 201.

[b] Wang, L.; Green, L.; Li, Z.; Dunn, J. M.; Bu, X.; Welch, C. J.; Li, C.; Wang, T.; Tu, Q.; Bekos, E.; Richardson, D.; Eckert, J.; Cui, J. *Org. Process Res. Dev.* **2011**, *15*, 1371.

[c] Chung, J. Y. L.; Steinhuebel, D.; Krska, S. W.; Hartner, F. W.; Cai, C.; Rosen, J.; Mancheno, D. E.; Pei, T.; DiMichele, L.; Ball, R. G.; Chen, C.; Tan, L.; Alorati, A. D.; Brewer, S. E.; Scott, J. P. *Org. Process Res. Dev.* **2012**, *16*, 1832.

[d] Brodszki, M.; Bäckström, B.; Horvath, K.; Larsson, T.; Malmgren, H.; Pelcman, M.; Wähling, H.; Wallberg, H.; Wennerberg, J. *Org. Process Res. Dev.* **2011**, *15*, 1027.

TABLE 19.4
Some Polymer-Supported Reagents

Entry	Name	Structure	Application	Comments
1	Macro-porous polystyrene–2,4,6-trimercaptotriazine (MP–TMT).[33]		Removal of Pd(II) in both aqueous and organic media.	Available from Argonaut Technologies.
2	Macro-porous polystyrene-based resin beads (QuadraPure).[a]	QuadraPure-TU, QuadraPure-IDA, QuadraPure-AMPA	QuadraPure scavengers, having functionality loading of 1–7 mmol/g, are stable up to 60 °C and can be used in aqueous or organic systems under a wide range of pH (2–14). The beads are highly cross-linked (up to 20%) having relatively low swell in organic solvents.	Available from Sigma-Aldrich.
3	Macro-porous thiourea functionalized polysiloxane (Deloxan THP II) Macro-porous thiol (mercapto) functionalized polysiloxane (Deloxan MP)	Deloxan THP II Deloxan MP	Application in aqueous media.[44]	Available from Meryer Chemical Technology Shanghai Company.
4	Polymer-supported ethylenediamine (P-EDA) Polymer-supported bis(2-aminoethyl)-amine (P-BiAA) Polymer-supported tris(2-aminoethyl)-amine (P-TriAA)	P-EDA P-BiAA P-TriAA	DiAION CR20 chelate resin is a cross-linked styrene-divinyl benzene copolymer resin having polyamine functional chelating groups in the copolymer and has an average of 1.2 micron particle diameter.[b]	P-EDA, P-BiAA, P-TriAA are available from Aldrich, DiAION CR20 is available from Mitsubishi Kasei of Tokyo, Japan.

(Continued)

TABLE 19.4 (CONTINUED)
Some Polymer-Supported Reagents

Entry	Name	Structure	Application	Comments
5	Polyamine-type chelate resin (DiAION CR20)	DiAION CR20	These multidentate scavengers are highly effective adsorbents for the removal of precious metals, specifically palladium.	These metal scavengers, provided by Phosphonics Technology company,[c] can be applied toward a wide range of metals in different oxidation states.[d]
	Silica-supported sulfur-based metal scavengers.			
6	SiliaMets Cysteine (Si-Cys) is a silica bound equivalent, wherein the thiol group remains free and accessible for high metal scavenging efficiency.[e]		It is a versatile metal scavenger for a variety of metals, including Pd, Sn, Ru, Pt, Cu, Rh, Cd, and Sc under a wide range of conditions. Si-Cys is the preferred metal scavenger for tin residues.	Silica-based materials offer good and broad solvent compatibility, no swelling characteristics, and excellent stability even at temperatures above 100 °C. Use of SiliaBond Imidazole combined with L-cysteine reduced residual Fe and Pd.[f]
7	SiliaBond Imidazole		SiliaBond® Imidazole (Si-IMI) is a versatile scavenger for metals including Cd, Co, Cu, Fe, Ni, Pd, and Rh under a wide range of conditions.	

[a] (a) Hinchcliffe, A.; Hughes, C.; Pears, D. A.; Pitts, M. R. *Org. Process Res. Dev.* **2007**, *11*, 477. (b) Girgis, M. J.; Kuczynski, L. E.; Berberena, S. M.; Boyd, C. A.; Kubinski, P. L.; Scherholz, M. L.; Drinkwater, D. E.; Shen, X.; Babiak, S.; Lefebvre, B. *Org. Process Res. Dev.* **2008**, *12*, 1209.
[b] Urawa, Y.; Miyazawa, M.; Ozeki, N.; Ogura, K. *Org. Process Res. Dev.* **2003**, *7*, 191.
[c] The company website address: www.phosphonics.com.
[d] Galaffu, N.; Man, S. P.; Wilkes, R. D.; Wilson, J. R. H. *Org. Process Res. Dev.* **2007**, *11*, 406.
[e] (a) de Koning, P. D.; McAndrew, D.; Moore, R.; Moses, I. B.; Boyles, D. C.; Kissick, K.; Stanchina, C. L.; Cuthberton, T.; Kamatani, A.; Rahman, L.; Rodriguez, R.; Urbina, A.; Sandoval, A.; Rose, P. R. *Org. Process Res. Dev.* **2011**, *15*, 1018. (b) de Koning, P. D.; Murtagh, L.; Lawson, J. P.; Embse, R. A. V.; Kunda, S. A.; Kong, W. *Org. Process Res. Dev.* **2011**, *15*, 1046.
[f] Hicks, F.; Hou, Y.; Langston, M.; McCarron, A.; O'Brien, E.; Ito, T.; Ma, C.; Matthew, C.; O'Bryan, C.; Provencal, D.; Zhao, Y.; Huang, J.; Yang, J.; Yang, Q.; Li, H.; Johnson, M.; Yan, S.; Liu, Y. *Org. Process Res. Dev.* **2013**, *17*, 829.

- Filter and wash with low-endotoxin water (9.8 L).
- Separate the toluene by-product from the aqueous filtrate.
- Add 3.1 kg of carbon to the aqueous solution.
- Stir for 30 min and filter.

Stage (b)
- Add aqueous solution of H_2SO_4 (6.69 kg) in low-endotoxin water (25 L) to pH 1.45 over 2 h.
- Filter and wash with low-endotoxin water (103 L) and acetone (103 L).
- Dry at 50 °C for 12 h.
- Yield: 18.1 kg (88%).

REMARK

The use of a loose scavenger, that is, charcoal, poses processing issues such as manual handling of the loose scavenger and contamination of reaction vessels by the loose scavenger.

19.1.3.2 MP–TMT

The use of MP–TMT scavenger at a ratio of 11 g/mmol of MP–TMT to Pd was able to reduce residual palladium to 37 ppm in the amination product **45** (Equation 19.9).[30]

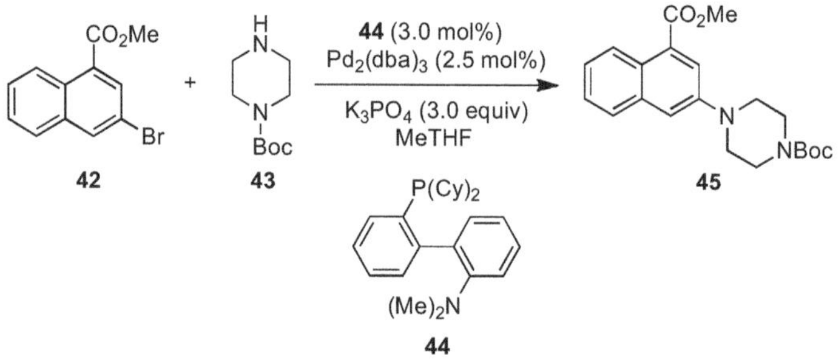

(19.9)

The combination of MP–TMT with activated charcoal was used to reduce the level of palladium impurity from 450 ppm to <40 ppm in the API candidate **48** obtained from Suzuki cross-coupling reaction (Equation 19.10).[31]

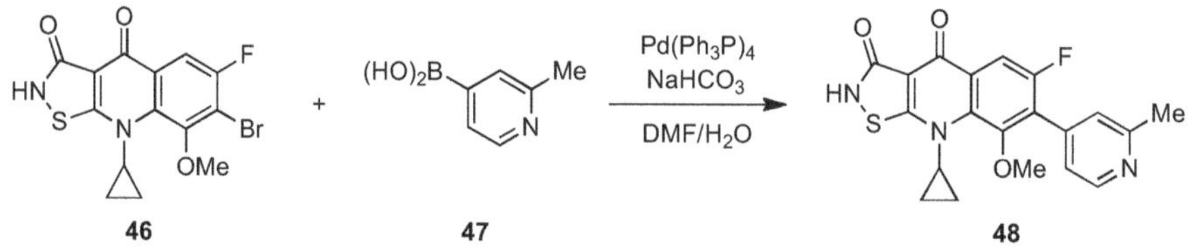

(19.10)

19.1.3.3 Deloxan THP-II

The removal of palladium residue from Suzuki coupling product **51** (Equation 19.11)[32a] was achieved by using Deloxan THP-II resin (see Table 19.4, Entry 3). Following deprotection treatment of the aqueous methanolic solution of **51** with Deloxan THP-II resin reduced the palladium level from ca. 240 ppm to <2 ppm.

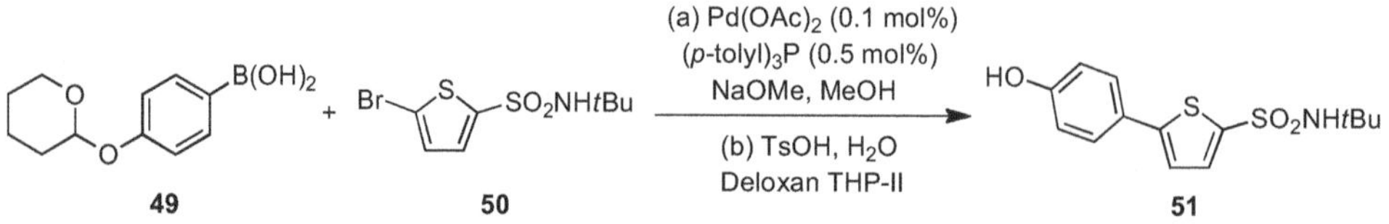

(19.11)

19.1.3.4 Smopex 110

The use of Smopex 110 to reduce palladium content under aqueous acidic conditions was demonstrated in the Suzuki coupling reaction of **52** with **53** (Equation 19.12).[33] Thus, treatment of the crude **54** with Smopex 110 at 70 °C reduced the residual palladium from 1300 ppm to <1 ppm.

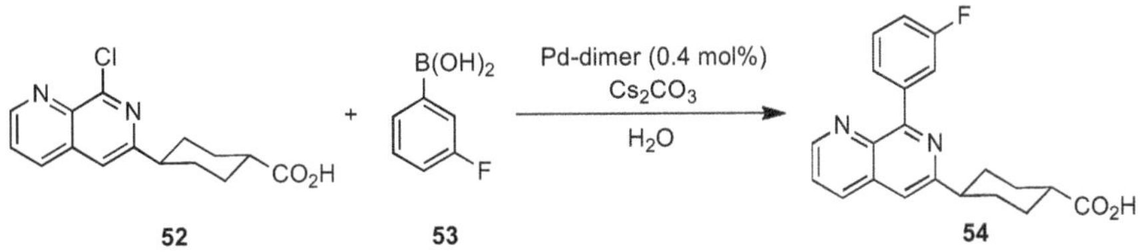

$$(19.12)$$

52   **53**   **54**

NOTES

(a) Although effective, the use of chelating polymer resins suffers drawbacks such as relatively high cost, potential product losses during the treatment, and leaching of impurities from the adsorbent.

(b) The utilization of cartridges, packed with solid adsorbent, was demonstrated[34] on a large-scale for scavenging various metals from process streams. Benefits of the cartridge format include (1) a reduced number of unit operations, (2) a reduced batch cycle time with increased productivity, and (3) reduced postbatch cleaning activities.

(c) A report revealed that a Pd-Smopex-111 complex with a palladium loading of 4.4-4.7 wt%, prepared by treating Smopex-111 with palladium acetate, was a highly reactive catalyst for Heck and Suzuki coupling reactions (Equations 19.13 and 19.14).[35] This catalyst is reusable with negligible leaching of palladium.

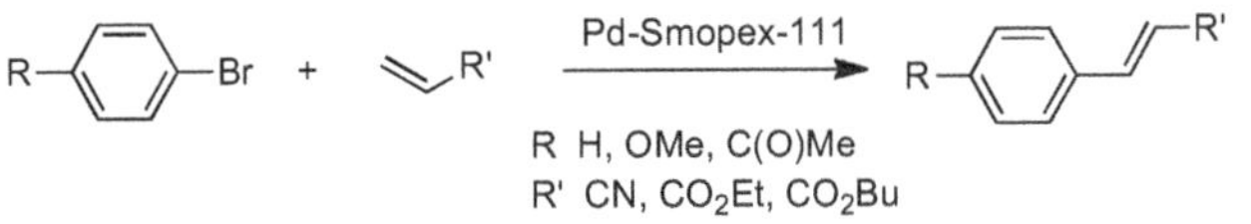

$$(19.13)$$

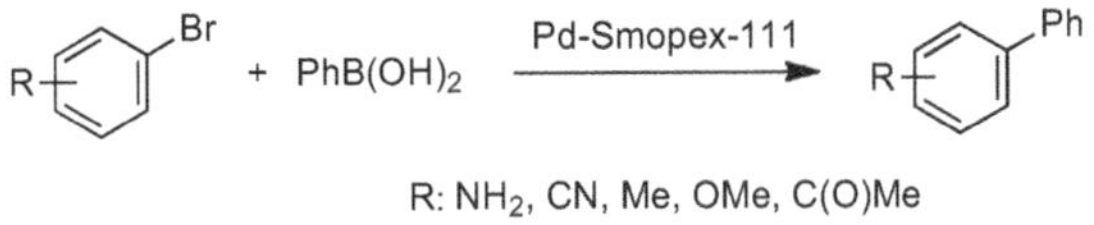

R H, OMe, C(O)Me
R' CN, CO$_2$Et, CO$_2$Bu

$$(19.14)$$

R: NH$_2$, CN, Me, OMe, C(O)Me

19.1.4 DISTILLATION

Compared with organic compounds, palladium and other heavy metals are not volatile. Therefore, it is possible to purify the desired organic product by means of distillation, leaving the heavy metals in the distillation residue. For instance, steam distillation separated the indole product **58** from the by-products and residual palladium (Scheme 19.7).[31]

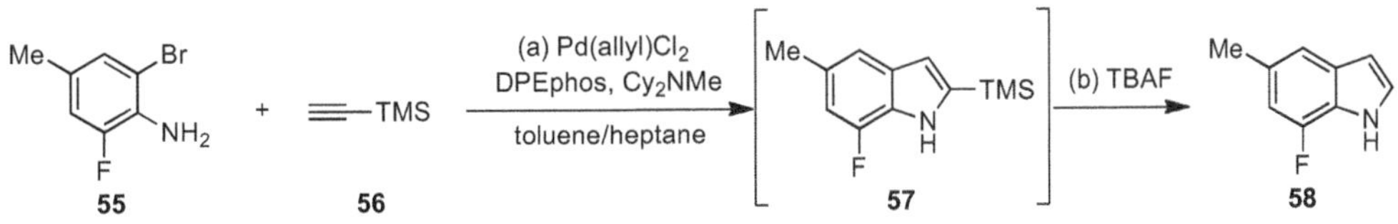

55   **56**   **57**   **58**

Cy$_2$NMe: *N,N*-dicyclohexylmethylamine

SCHEME 19.7

Procedure

Stage (b)

- Charge 282 kg (895 mol, 1.1 equiv) of tetrabutylammonium fluoride (TBAF) trihydrate, MeOH (41 L), and THF (878 L) into **57** (177.5 kg, 801.9 mol, 1.0 equiv) toluene solution (24.7 wt%) over 2 h at <30 °C.
- Heat the mixture to 80–85 °C and stir for 3 h.
- Distill at 50–55 °C (jacket) at reduced pressure (to remove THF).
- Wash the resulting mixture with 10% aqueous NaCl (2×3600 kg).
- Separate layers.
- Extract the aqueous layer with toluene (513 L).
- Filter the combined organic layers.
- Distill (steam distillation) using a ratio of 58 L DI water to 1 kg of the expected final product.
- Add 68 kg of NaCl to the distillate containing a mixture of the product, toluene, and water.
- Separate layers.
- Extract the aqueous layer with toluene (92 L).
- Use the combined organic layers for the next step.
- Yield: 87% (solution yield).

19.1.5   OTHER METHODS

19.1.5.1   Adsorption–Crystallization

In some cases, the removal of residual metals needs a combination of two or more methods in order to achieve ideal results. For instance, the removal of residual Pd from **60**, obtained from Sonogashira cross-coupling reaction/intramolecular cycloaddition (Equation 19.15),[36] proved unsuccessful via either adsorption with solid phase (such as Amberlite IRC resins, silica gel, or alumina); extraction with TMT, ammonium formate, 2-hydroxypyridine, or dimethylaminoethanethiol; or crystallization.

(19.15)

In the event, an approach was developed by a combination of adsorption with crystallization. The protocol involved dissolving **60** in acidic aqueous media, followed by treatment with Deloxan THP and crystallization from acetonitrile/MTBE, giving the product **60** containing Pd of 20 ppm (from 242 ppm) and Cu of 2 ppm (from 105 ppm).

Procedure

Stage (a)

- Charge 985 g (2.23 mol) of **59** into a 22 L reactor, followed by adding 42.4 g (0.22 mol, 10 mol%) of CuI and 15.7 g (0.0266 mol, 1 mol%) of Pd(Ph$_3$P)$_2$Cl$_2$.
- Charge EtOH (10 L), TEA (630 mL, 4.52 mol, 2 equiv), and 257 mL (3.39 mol, 1.5 equiv) of 3-butyn-1-ol.

- Distill off 5 L of EtOH (the reaction was complete then).
- Add water (7 L).
- Distill under reduced pressure to remove the remaining EtOH.
- Cool to room temperature.
- Add Celite (450 g) followed by adding 1 N HCl (3 L, 1.3 equiv) to pH 1.2.
- Filter and rinse with water (3 L).

Stage (b)
- Transfer the combined filtrates into a 22-L flask and add 726 g of Deloxan THP.
- Stir the resulting mixture overnight.
- Filter and rinse with water (0.7 L).
- Transfer the combined filtrates into a 20 L wash tank.
- Add Na$_2$CO$_3$ (287 g, 2.71 mol, 1.2 equiv) portionwise to pH 9.
- Extract the resulting solution with CH$_2$Cl$_2$ (3×11 L).
- Transfer the combined organic layers into 22 L flask.
- Distill under reduced pressure to 6 L volume.
- Add MeCN (1.3 L) and MTBE (7 L).
- Distill to 6 L volume.
- Add MTBE (7 L) and distill to 4 L volume.
- Cool the resulting yellow slurry in ice bath.
- Filter and wash with MTBE.
- Dry at 50 °C under vacuum.
- Yield: 519 g (60%, Cu: 2 ppm and Pd: 20 ppm).

19.1.5.2 Adsorption–TMT Wash

A binary system,[30] consisting of chelating agents and carbon or silica gel adsorbents, proved to be a cost-effective method to remove palladium from pharmaceutical intermediates and APIs. It was demonstrated that many chelating agents can synergize with carbon or silica gel and improve significantly the efficiency of palladium removal. For example, the combination of TMT with activated charcoal proved to be effective in removing Pd from amide **64** (Scheme 19.8).[37]

19.1.5.3 Protecting Group

Apart from the improved regioselectivity, the use of protecting group can also offer an opportunity to reject the residual palladium that otherwise would be difficult to remove due to the chelating effect of heteroatoms in the product such as LY2784544. LY2784544 was a drug candidate for the treatment of several myeloproliferative disorders. As the penultimate intermediate for the synthesis of LY2784544, the *tert*-butyl protected compound **67** was prepared from palladium-catalyzed amination of chloro-imidazopyridizine **65** with aminopyrazole **66** (Scheme 19.9).[44c] The presence of *tert*-butyl group in **67** alleviated the burden of removing residual palladium.

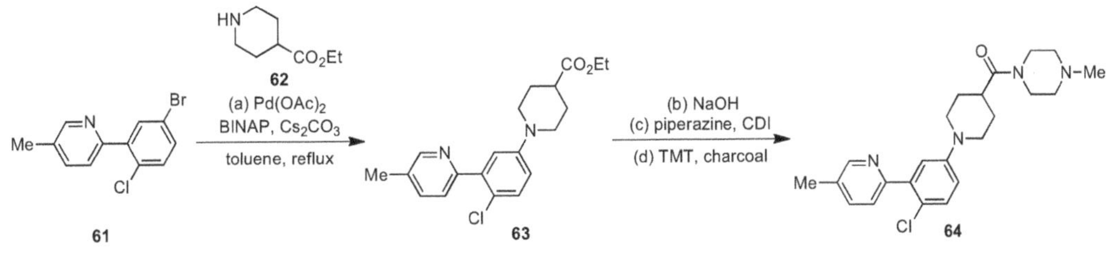

SCHEME 19.8

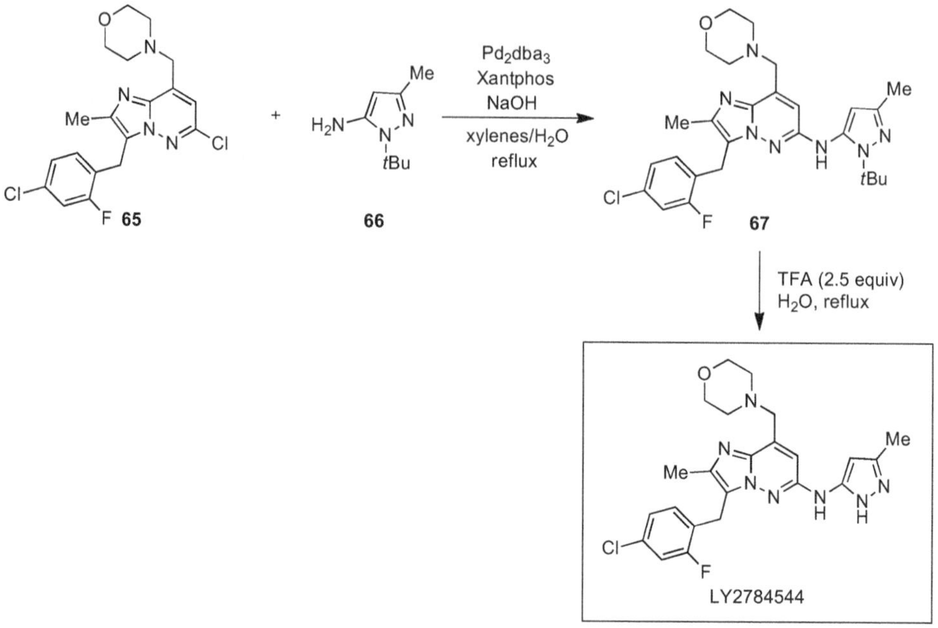

SCHEME 19.9

19.1.5.4 Salt Formation

The late-stage Pd-catalyzed convergent cross-coupling reaction usually leads to a high palladium content in the drug substance, especially when the drug molecule contains good chelating groups. For example, the final Negishi cross-coupling (Equation 19.16)[38] for the synthesis of 5-[2-metho xy-5-(4-pyridinyl)phenyl]-2,1,3-benzoxadiazole **70**, a drug candidate for the treatment of asthma, posed a serious palladium removal issue. Normal crystallization of the free base could reduce the residual palladium only to a level of 100–800 ppm.

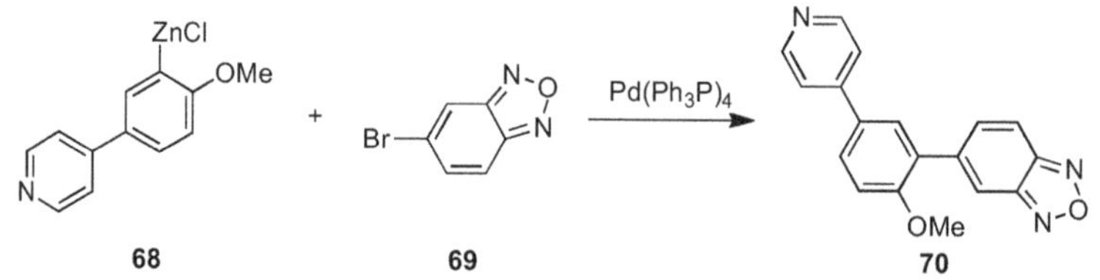

$$(19.16)$$

Ultimately, an efficient approach was developed by preparing its hemi-maleate salt, followed by converting the salt back to the free base. By using this method the residual Pd in the drug substance was reduced to <0.5 ppm.

19.1.6 Conclusion

Palladium-catalyzed reactions have become one of the fundamentals in modern organic chemistry and have been extensively utilized in the synthesis of organic compounds during the last two decades. Palladium-catalyzed processes contribute a large proportion of chemical transformations that are employed in the synthesis of drug substances. Palladium is a Class 1 metal and its residual levels in the APIs are regulated with strict guidelines. Therefore, a number of practical methods have been developed for the efficient removal of palladium metal from reaction products and waste streams. Table 19.5 summarizes the Pd-removal methods discussed in this chapter.

TABLE 19.5
Summary of Methods in Removing Residual Pd

	Pd source	Principle	Reagent	Comments
Filtration	Heterogeneous and homogeneous palladium catalysts	Removal of Pd along with supports	Filtration aids, such as celite	Separation of solids from liquids
Adsorption	Soluble palladium catalysts	Binding or adsorbing Pd onto non-drug substance phase and removing by filtration	Silica- and polymer-based scavengers with chelating functional groups or adsorbents	Easy separation from process stream; volume efficient (compared to the extraction method); good product recovery (compared to recrystallization). However, polymer-based solid scavengers are expensive. Another drawback is the possibility of leaching that may lead to product contamination.
Extraction	Soluble palladium catalysts	Removal of Pd is based on solubility difference between aqueous and organic phases.	Water immiscible organic solvents with chelating agents	The process may result in volume inefficiency and long processing cycle times.
Crystallization	Soluble palladium catalysts	The solubility difference between the Pd complex and the API	Crystallization solvent(s) and/or chelating additive(s)	Pd, as an impurity, is rejected into the crystallization mother liquor.
Distillation	Heterogeneous and homogeneous palladium catalysts	Separation of volatile products	–	The product should be stable under distillation conditions.

19.2   REMOVAL OF RESIDUAL COPPER

Removal of residual copper from organic products can be achieved with the aid of various agents, including TMT, N,N-dimethylethylenediamine,[39] aqueous ammonia, ammonium chloride,[40] and thiourea.

19.2.1   Use of Aqueous Ammonia

Copper-catalyzed Ullmann coupling of indazolone **71** with aryl bromide **72** in the presence of N,N-dimethyl ethylenediamine (DMEDA) furnished the product **73** (Scheme 19.10),[41] a key intermediate for the synthesis of heat shock protein 90 (Hsp90) inhibitor **74** as a therapeutic target for cancer treatment. It was demonstrated that aqueous ammonia was able to remove copper by converting copper to water-soluble copper salts.

Procedure

- Purge a solution of **71** (2.48 kg, 10.68 mol), **72** (3.50 kg, 11.86 mol, 1.11 equiv), and K_2CO_3 (3.45 kg, 24.96 mol, 2.34 equiv) in anhydrous 1,4-dioxane (28 L) with N_2 for 30 min.
- Charge, to the above solution, a degassed solution of CuI (0.45 kg, 2.37 mol, 0.22 equiv) and DMEDA (0.56 L, 5.15 mol) in 1,4-dioxane (8.41 L).

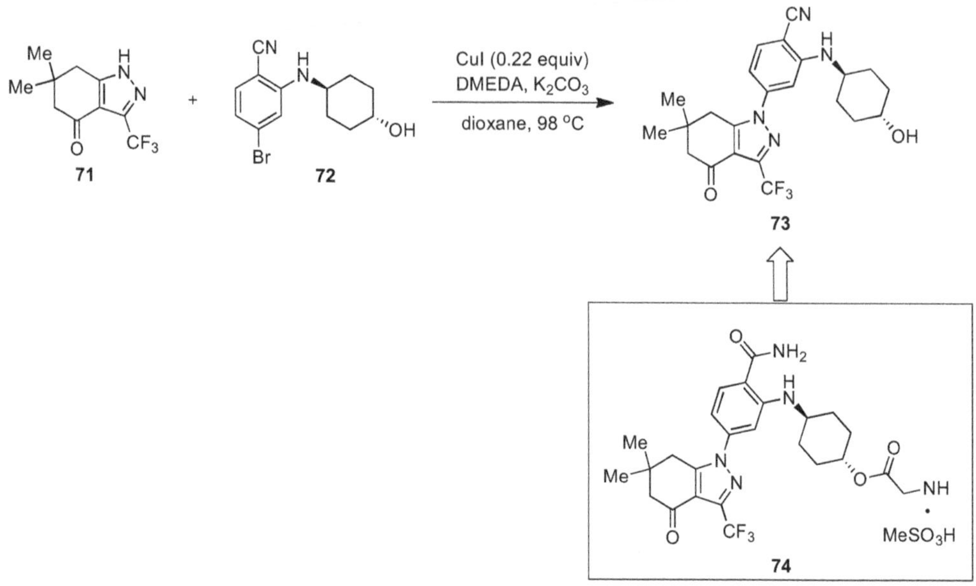

SCHEME 19.10

- Heat the batch to 98 °C over 2 h and stir at the temperature for 65 h.
- Cool to ≤30 °C.
- Add concentrated NH₄OH (4.74 L), water (11 L), and sat. NH₄Cl (15.8 L).
- Separate layers.
- Distill at 40 °C under reduced pressure to 12 L.
- Heat to 60 °C and add heptane (13.2 L).
- Cool to 20 °C.
- Filter and wash with ethyl acetate/heptane (21 L, 1:2).
- Purify the crude product by crystallization from ethyl acetate/heptane.
- Yield: 3.49 kg (73%).

Analogously, the Cu content in the Sonogashira product **77** was reduced to <20 ppm with aqueous ammonia during the workup (Equation 19.17).[42]

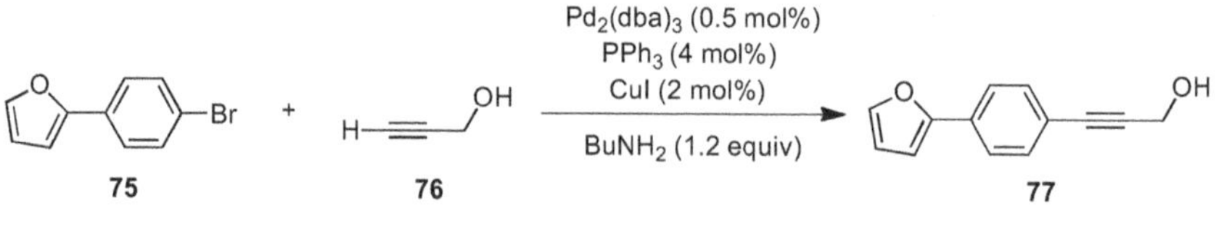

Aqueous ammonia wash was also applied to remove residual copper from the copper-catalyzed amination product **79** (Equation 19.18).[43]

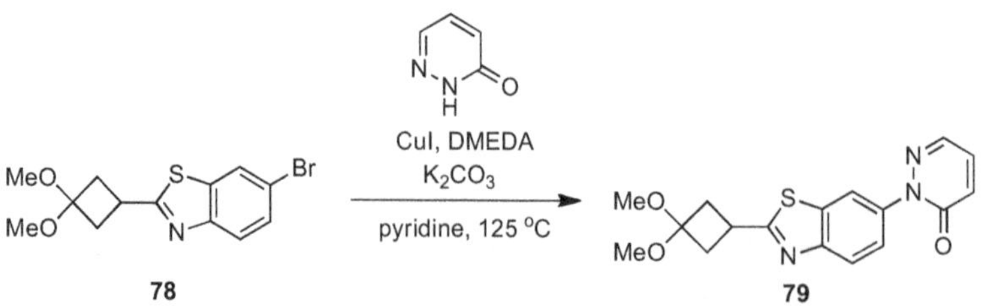

19.2.2 USE OF THIOUREA

Removal of copper from compounds containing sulfur, such as **81**, by using aqueous ammonium chloride or potassium chloride was less effective. It was found, however, that the addition of thiourea, upon completion of the aromatization, to the reaction mixture could remove copper effectively by the formation of a highly insoluble copper–thiourea complex (Equation 19.19).[44]

$$
\text{(19.19)}
$$

REMARKS

(a) The use of thiourea or thioacetamide[45] could lead to a stench in the waste stream.

(b) Oxalic acid is also used to remove copper.[57]

(c) It was reported[46] that the use of thioacetamide could facilitate the filtration of a copper-containing reaction mixture by forming a less viscous copper-thioacetamide complex.

19.2.3 USE OF 2,4,6-TRIMERCAPTOTRIAZINE

A method was developed to reduce the level of Cu(I) residue in copper-mediated reaction product **84** via oxidizing Cu(I) to Cu(II) by air, followed by precipitating Cu(II) with TMT, trimming the copper level down to 15-25 ppm (Equation 19.20).[47]

$$
\text{(19.20)}
$$

Procedure

Stage (a)

- Charge 8.56 kg (214 mol) of NaOH into a glass-lined reactor, followed by adding DMAc (169 kg).
- Stir for 15 min prior to adding 25.5 kg (229 mol) of **83**.
- Stir the mixture at <27 °C for 1 h.
- Charge 36.0 kg (142 mol) of **82** and 11.5 kg (80.4 mol) of copper (I) oxide.
- Heat the batch at 80–90 °C for 4 h.
- Cool to <30 °C over 1 h.
- Add a mixture of HCl (76 kg, 36% w/w) and water (858 kg) to pH <1 and stir for 15 min (keep the mixture overnight without stirring).
- Pass the mixture through a series of filters: 10 μm bag filter, 1 μm bag filter, and 0.45 μm cartridge filter.
- Wash the filters with water (142 kg).
- Transfer the combined filtrates into a glass-lined reactor.
- Add NH₄OH (67 kg, 27% w/w) and stir for 30 min (the pH was then 9.5).
- Isolate the precipitated product via centrifugation and wash with water.
- Charge the wet cake into the reactor, followed by adding water (558 kg).

- Add 76 kg (36% w/w) HCl to dissolve the product.
- Add NH_4OH (67 kg, 27% w/w) and stir for 35 min to precipitate the product.
- Filter and wash with water (141 kg) to give a product wet cake.

Stage (b)
- Charge the wet cake into the reactor with water (558 kg).
- Add 14 kg of TMT aqueous solution (15%) and stir for 1 h.
- Add 61 kg (36% w/w) HCl to lower the pH to <1.
- Bubble compressed air into the stirred mixture for 1 h.
- Add TMT (7×7.5 kg) at 20 min intervals while keeping the pH low by adding 2.1 kg of HCl.
- Add 4.5 kg of activated charcoal, water (15 kg), and Celite 521 (1.5 kg).
- Stir for 1 h.
- Pass the mixture through a series of filters: 25 μm bag filter, 1 μm bag filter, and 0.45 μm cartridge filter.
- Wash the filters with water (94 kg).
- Transfer the combined filtrates into a glass-lined reactor.
- Add aqueous NH_4OH (53 kg, 27% w/w).
- Isolate the product via centrifugation and wash with water (180 kg) and EtOH (44 kg).
- Dry at 60–70 °C/9 mbar for 48 h.
- Yield: 34.0 kg (84%) as an off-white solid.

19.3   REMOVAL OF RESIDUAL RHODIUM

Rhodium has been frequently used as a catalyst in organic transformations, such as hydrogenation,[48] C–H activation,[49] C–H insertion,[50] and Michael addition. Like palladium, Rh is toxic and the residual Rh in the drug substances and excipients has to be restricted on the foundation of safety- and quality-based criteria. To address these toxicity and safety concerns associated with residual Rh and reduce the residual Rh to acceptable levels, methods that have been employed include crystallization, extraction, distillation, and adsorption with absorbents such as silica gel, activated carbon,[51] and polymer-supported metal scavengers. Johnson Matthey developed Smopex® fibers as metal scavengers for aqueous and organic media. Owing to their unique structure, thin fibers with grafted functional side chains, Smopex® fibers exhibit fast kinetics in metal extraction, making them especially advantageous in scavenging metals from solution.

19.3.1   USE OF SMOPEX-234

The Smopex-234 was selected to remove residual Rh from the product **87** obtained from asymmetric Michael addition (Equation 19.21).[52]

Treatment of the postreaction mixture with Oxone, followed by adding Smopex-234 showed that 97% of Rh was removed. Smopex-234 can be returned to the supplier for rhodium recovery.

REMARK

It is postulated that exposure of the reaction mixture to Oxone may cause oxidation of rhodium or the BINAP-rhodium complex to an appropriate oxidation state, thus facilitating its removal from the solution.

19.3.2 Use of Ecosorb C-941

An enantioselective hydrogenation of **88** under optimized conditions [NH_4Cl (0.15 mol%), [Rh(COD)Cl]$_2$ (0.15 mol%), tBu JOSIPHOS (0.155 mol%), 250 psig of hydrogen at 50 °C] in MeOH, as shown in Reaction (19.22),[53] afforded Sitagliptin **89** in 98% yield and 95% ee. Sitagliptin **89** received FDA approval for the treatment of type 2 diabetes.

$$(19.22)$$

Ecosorb C-941, a polymer impregnated with activated carbon, was found to be efficient in removing rhodium from the crude hydrogenation stream, and with as little as 10 wt% of Ecosorb resulted in near complete removal of the dissolved rhodium.

19.4 REMOVAL OF RESIDUAL RUTHENIUM

Ruthenium-based catalysts have been frequently utilized in various synthetic applications, especially in the synthesis of macrocyclic compounds via ruthenium-catalyzed ring-closing metathesis (RCM). Examples of API synthesis using RCM can be found in the literature.[54] Therefore, developing practical methods to remove ruthenium from the product is of great importance. Methods in residual ruthenium removal were discussed in a review paper.[55]

19.4.1 Use of Activated Carbon

Besides applications for the removal of color impurities, activated carbon is widely utilized in removing transition metals.[56] The reductive transformation of aryl ketone **90** to the corresponding chiral alcohol **91** was achieved via ruthenium-catalyzed enantioselective hydrogenation (Equation 19.23).[57]

$$(19.23)$$

Selective adsorption was chosen to remove the ruthenium from the product. Activated carbon Norit CA 1 is the adsorbent of choice and most of the ruthenium was removed upon treating the asymmetric reduction mixture with activated carbon Norit CA 1.

19.4.2 Use of Supercritical Carbon Dioxide

Efforts in the synthesis of the drug candidate **94** for the treatment of hepatitis C virus identified a ruthenium-catalyzed ring-closing metathesis (RCM) for the preparation of the core macrocycle **93**, a key intermediate for the synthesis of **94** (Scheme 19.11).[58]

Supercritical CO_2 was used to remove the ruthenium catalyst from a crude ring-closing metathesis product. The method was implemented in a semi-continuous fashion and allowed for the efficient removal of the toxic metal impurities to meet the specifications for the final drug substance.

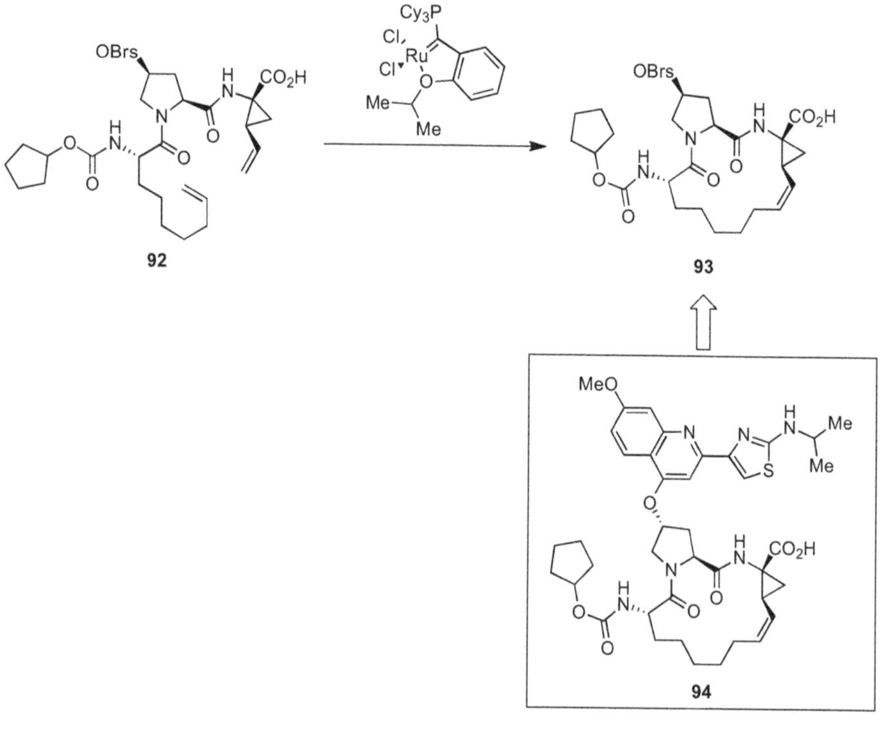

SCHEME 19.11

19.5 REMOVAL OF ZINC

19.5.1 Extraction with Trisodium Salt of EDTA

Negishi coupling reaction is less popular at large-scale production due to the use of zinc reagent. Despite this disadvantage, Negishi cross-coupling was successfully applied in the synthesis of 2-chloro-5-(pyridin-2-yl) pyrimidine **98** (Scheme 19.12).[59] As the product **98** has a strong metal chelating ability, the removal of zinc and Pd required a combination of two approaches, including extraction with trisodium salt of ethylenediamine tetraacetic acid (EDTA), and treatments with Silica gel and thiol-modified silica gel, and activated charcoal. As a result, the Pd and Zn contents were reduced to in the range of 10–20 ppm and 25–50 ppm, respectively.

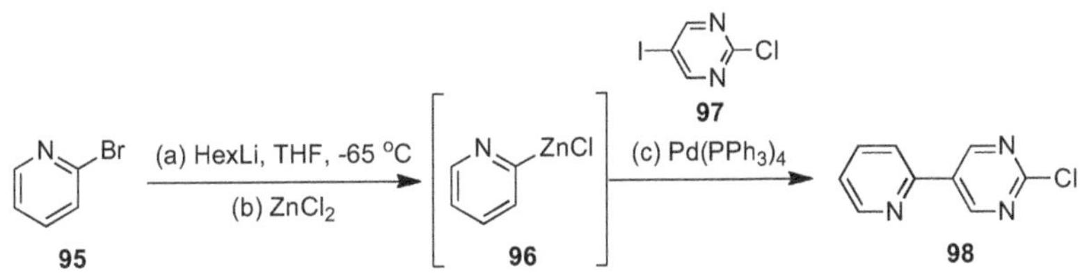

SCHEME 19.12

Procedure

Stage (a) Lithiation
- Charge 130 mL of THF into a 2 L flask.
- Purge with N_2.
- Cool to -70 °C.
- Charge a solution (175 mL, 2.5 M, 0.436 mol, 1.05 equiv) of hexyllithium in hexanes at <-50 °C.
- Cool to -65 °C.
- Charge 65.7 g (0.416 mol) of **95** at <-65 °C.

Stage (b) Transmetallation
- Charge a solution of $ZnCl_2$ (57.8 g, 0.416 mol, 1.0 equiv) in 268 mL of THF at <-55 °C.
- Warm to ~22 °C and stir for 2 h.

Stage (c) Negishi coupling
- Charge 4.8 g (0.004 mol, 0.96 mol%) of $Pd(PPh_3)_4$.
- Charge a solution of **97** (50 g, 0.2 mol, 0.93 equiv) in 160 mL of THF at <30 °C.
- Stir for 1 h.
- Add 1 L of 0.39 M aqueous solution of EDTA·3Na and 100 mL of CH_2Cl_2.
- Stir vigorously for 15 min.
- Separate layers.
- Distill the organic layer under reduced pressure.
- Add 300 mL of CH_2Cl_2 to the residue (a black oil).
- Add 66 g of Kiesegel 60 and 6.1 g of thiol-modified silica.
- Stir for 20 min and filter and rinse with CH_2Cl_2.
- Distill the filtrate to ~350 mL.
- Cool to ~22 °C.
- Wash with 2 N HCl (2×350 mL).
- Treat the combined aqueous layers with Norit A Supra (1.4 g) and filter.
- Add aqueous ammonia to pH 6–7.
- Filter and dry under vacuum at 40 °C.
- Yield: 25.6 g (67%).

19.5.2 Use of Ethylenediamine

The use of ethylenediamine can also effectively reduce the Zn content. For example, the reduction of zinc content in the crude Negishi coupling product **12** was carried out by suspending the crude **12** in the presence of ethylenediamine in a mixture of water and THF. Due to the excellent water solubility of the zinc–ethylenediamine complex, this treatment reduced the Zn level from 7–14 wt% to 4500 ppm (Pd at a level of 200 ppm) (Equation 19.24).[10,11]

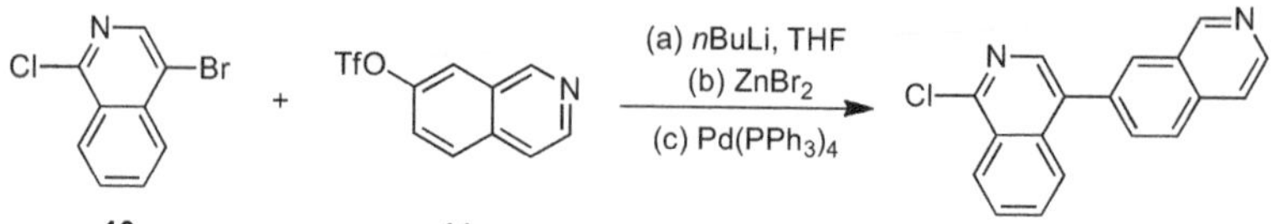

$$(19.24)$$

Procedure

Stage (a) Lithiation

- Charge n-BuLi (1.6 M in hexane, 13.37 kg, 30.6 mol) to a solution of **10** (7.05 kg, 29 mol) in THF (80.8 L) at <-70 °C.
- Stir for 10 min.

Stage (b) Transmetallation

- Charge a solution of $ZnBr_2$ (6.5 kg, 28.9 mol) in 17.3 L of THF at <-70 °C.
- Stir at -70 °C for 10 min.
- Warm the batch to 0 °C.

Stage (c) Negishi coupling

- Charge a solution of **11** (7.62 kg, 27.5 mol) and $Pd(PPh_3)_4$ (0.46 kg, 0.4 mol) in THF (17.3 L).
- Warm the batch to 25 °C and stir for 1 h.
- Add 25 L of 10% aqueous NH_4Cl and separate layers.
- Wash the organic layer with 10% NaCl.
- Distill until ~147 L distillate was collected.
- Cool to 10 °C and stir at 10 °C for 2 h.
- Isolate the crude **12** by centrifugation.
- Yield of **12**: 6.39 kg (78.7% and 11.5% Zn).
- Suspend crude **12** (6.0 kg) in 65.5 L of water and 16.4 L of THF.
- Charge 6.0 kg of ethylenediamine.
- Heat to 45 °C.
- Filter and wash with water (50 L).
- Dry at 50 °C under vacuum.
- Yield: 3.99 kg (4500 ppm Zn and 200 ppm Pd).

19.6   REMOVAL OF MAGNESIUM

The presence of metal ions in the reaction mixture can create problems during the isolation, for instance, magnesium salt in the postreaction mixture caused severe emulsions during the isolation of the tetrahydroquinoline derivative **101** (Scheme 19.13).[60]

To circumvent this emulsion problem, the addition of citric acid in the quenching solution allowed the removal of magnesium by the formation of a magnesium–citric acid complex.[61]

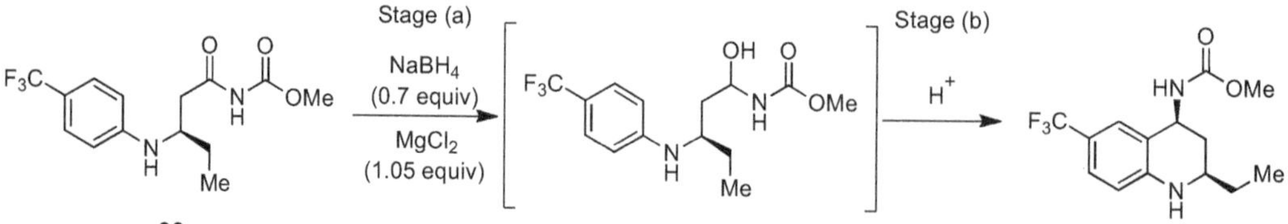

SCHEME 19.13

Procedure
Stage (b)
- Transfer the reduction mixture of **100** into a 200-L reactor containing a mixture of CH_2Cl_2 (70 L), conc. HCl (5.8 L), citric acid (10.5 kg), and water (64 L).
- Stir at room temperature for 2 h.
- Remove the aqueous layer.
- Extract the organic layer with aqueous citric acid solution (6.3 kg of citric acid in 34 L of water).
- Add Darco-activated carbon (G-60 grade, 700 g).
- Stir for 30 min.
- Filter through Celite and rinse with CH_2Cl_2.
- Exchange solvent to hexanes via distillation.
- Cool to room temperature.
- Filter and wash with hexanes (14 L).
- Dry at 40 °C under vacuum.
- Yield: 5.291 kg (80%).

REMARK
Magnesium salt can also be removed by converting it into water-soluble magnesium acetate.[62]

19.7 REMOVAL OF ALUMINUM

19.7.1 USE OF TRIETHANOLAMINE

Lithium aluminum hydride is widely used in organic synthesis as a reducing agent. However, the use of LAH on large scale poses significant challenges, due not only to process safety, but also the removal of aluminum and lithium waste. To address the workup issues, an aqueous disodium tartrate wash was initially developed. However, this suffered from mass transfer issues upon scaleup. Hence an alternative method was developed by utilizing triethanolamine (Equation 19.25).[63]

$$(19.25)$$

Procedure

Stage (a)
- Dissolve 50 kg (152.7 mol) of **102** in 250 L of anhydrous THF.
- Charge 2.4 M solution of LAH (63.5 L, 152.7 mol, 1.0 equiv) in THF at 20–25 °C over 4 h.
- Stir the resulting mixture for 2 h.

Stage (b)
- Add a solution of triethanolamine (53.6 kg, 360.0 mol, 2.4 equiv) in THF (100 L) over 2 h and stir for 0.5 h.
- Add aqueous NaOH (25.7 kg in 250 L of H_2O) solution and stir for 15 min.
- Add cyclohexane (150 L).
- Separate layers and wash the organic phase with aqueous NaOH (2×12.8 kg NaOH in 125 L H_2O) followed by H_2O (50 L).

- Distill azeotropically to ~125 L (~2.5 L/kg).
- Add ethanol and distill (solvent exchange from cyclohexane to ethanol and used directly in the subsequent step).

19.7.2 Use of Crystallization

Crystallization is also an efficient way to eliminate aluminum from the product. Friedel–Crafts acylation of indole **104** with acid chloride **105** in the presence of diethylaluminum chloride (Et$_2$AlCl) provided 3-acylated indole **106** (Equation 19.26).[68]

$$(19.26)$$

Et$_2$AlCl plays a dual role in this acylation, as Lewis acid activates acid chloride and as an HCl scavenger.[64] The crude product **106** was isolated by the addition of MeOH (5 equiv) to the reaction mixture. The isolated solids contained the product **106** as well as the aluminum salts. Crystallization of the crude solid product **106** from boiling MeOH/H$_2$O (4:1) gave **106** in 91% isolated yield and 99.8 wt% purity with <30 ppm of Al.

Procedure

- Mix a solution of **104** (2.56 kg, 11.9 mol) in MeOH (26 L) with toluene (36 L).
- Distill the resulting solution under reduced pressure to a volume of 13 L (to remove MeOH).
- Cool the mixture to 0 °C.
- Charge 7.9 L (15.3 mol, 1.2 equiv) of 1.8 M Et$_2$AlCl at 0 °C to 5 °C over 20 min.
- Charge a solution of **105** (2.50 kg, 15.28 mol, 1.2 equiv) in heptane (3.7 L) at 0–13 °C over 1.5 h.
- Stir at room temperature for 16 h.
- Cool to 0 °C.
- Add MeOH (2.3 L, 57.12 mol), followed by adding heptane (22 L).
- Filter and wash with heptane (28 L) to give the crude **106**.
- Recrystallize the crude **106** from boiling MeOH/H$_2$O (4:1, 90 L).
- Yield: 3.84 kg (91%) with 99.8 wt% purity.

19.8 REMOVAL OF IRON AND NICKEL

19.8.1 Removal of Iron

Employing the Leimgruber–Batcho indole synthesis protocol, indole **108** was obtained from (E)-2-nitropyrrolidinostyrene **107** in 86% yield with 99.9% purity (Equation 19.27).[65] In addition to the desired indole **108**, debenzylated side product **109** was also generated in various amounts. In order to mitigate this side product formation, iron acetate (2.5 mol%) was used to poison the catalyst.

$$(19.27)$$

After the reaction, the level of iron in the product **108** was effectively reduced by washing with aqueous ammonia.

19.8.2 REMOVAL OF NICKEL

Analogously, aqueous ammonia was also employed to remove the residual nickel from Ni-catalyzed cross-coupling reaction product **112**. As shown in Reaction (19.28),[66] the nickel could be easily removed through an aqueous ammonia wash, followed by crystallization.

$$(19.28)$$

Procedure

- Charge 0.6 kg of water to a mixture of **110** (47.6 kg, 110 mol) in MeCN (477 L), followed by adding 70.1 kg (330 mol, 3.0 equiv) of K_3PO_4 and 40.7 kg (165 mol, 1.5 equiv) of **111**.
- Degas the mixture with subsurface nitrogen sparge for 30 min.
- Charge 9.05 g (31.1 mmol, 0.03 mol%) of nickel (II) nitrate hexahydrate and 16.4 g (62.5 mmol, 0.06 mol%) of Ph_3P.
- Degas the mixture for 1 h.
- Heat the batch at 60 °C for 25 h.
- Add a mixture of water (390 L) and 28% aqueous NH_4OH (85 kg) over 2 h.
- Stir the resulting mixture for 2 h.
- Cool the batch to 10–20 °C over 4 h.
- Separate layers.
- Add MeCN (190 L) to the organic layer.
- Heat the resulting organic layer to 40–50 °C and age at this temperature for 1 h.
- Cool the batch to 20–30 °C over 4 h and age for 1 h.
- Filter, wash with MeCN (237 L), then water (238 L) and MeCN (237 L).
- Dry at 50 °C under vacuum.
- Yield: 54.4 kg (79%).

NOTES

1. (a) Heck, R. F. *Palladium Reagents in Organic Syntheses*, Academic Press, New York, 1985. (b) Negishi, E.; de Meijere, A. *Handbook of Organopalladium Chemistry for Organic Synthesis*, Wiley, New York, 2002.
2. "Guideline on the specification limits for residues of metals catalysts or metal reagents", EMEA/CHMP/SWP/4446/2000.
3. Magano, J. In *Transition Metal-Catalyzed Couplings in Process Chemistry: Case Studies from the Pharmaceutical Industry*; Magano, J., Dunetz, J. R., Eds.; Wiley-VCH, Weinheim, 2013; Chapter 2.
4. (a) Rosso, R. W.; Lust, D. A.; Bernot, P. J.; Grosso, J. A.; Modi, S. P.; Rusowicz, A.; Sedergran, T. C.; Simpson, J. H.; Srivastava, S. K.; Humora, M. J.; Anderson, N. G. *Org. Process Res. Dev.* **1997**, *1*, 311. (b) Chen, C.; Dagneau, P.; Grabowski, E. J. J.; Oballa, R.; O'Shea, P.; Prasit, P.; Robichaud, J.; Tillyer, R.; Wang, X. *J. Org. Chem.* **2003**, *68*, 2633. (c) For reviews, see Garrett, C. E.; Prasad, K. *Adv. Synth. Catal.* **2004**, *346*, 889.
5. Thiel, O. R.; Achmatowicz, M.; Bernard, C.; Wheeler, P.; Savarin, C.; Correll, T. L.; Kasparian, A.; Allgeier, A.; Bartberger, M. D.; Tan, H.; Larsen, R. D. *Org. Process Res. Dev.* **2009**, *13*, 230.

6. Sperry, J. B.; Farr, R. M.; Levent, M.; Ghosh, M.; Hoagland, S. M.; Varsolona, R. J.; Sutherland, K. *Org. Process Res. Dev.* **2012**, *16*, 1854.

7. Jiang, X.; Lee, G. T.; Prasad, K.; Repič, O. *Org. Process Res. Dev.* **2008**, *12*, 1137.

8. Denni-Dischert, D.; Marterer, W.; Bänziger, M.; Yusuff, N.; Batt, D.; Ramsey, T.; Geng, P.; Michael, W.; Wang, R-M. B.; Taplin, F., Jr.; Versace, R.; Cesarz, D.; Perez, L. B. *Org. Process Res. Dev.* **2006**, *10*, 70.

9. Königsberger, K.; Chen, G.-P.; Wu, R.; Girpis, M. J.; Prasad, K.; Repič, O.; Blacklock, T. J. *Org. Process Res. Dev.* **2003**, *7*, 733.

10. (a) Rönn, M.; Bäckvall, J.-E.; Anderson, P. G. *Tetrahedron Lett.* **1995**, *36*, 7749. (b) Larock, R. C.; Hightower, T. R. *J. Org. Chem.* **1993**, *58*, 5298.

11. Tang, W.; Patel, N. D.; Wei, X.; Byrne, D.; Chitroda, A.; Narayanan, B.; Sienkiewicz, A.; Nummy, L. J.; Sarvestani, M.; Ma, N.; Senanayake, C. H. *Org. Process Res. Dev.* **2013**, *17*, 382.

12. Hoffmann-Emery, F.; Niedermann, K.; Rege, P. D.; Konrath, M.; Lautz, C.; Kraft, A. K.; Steiner, C.; Bliss, F.; Hell, A.; Fischer, R.; Carrera, D. E.; Beaudry, D.; Angelaud, R.; Malhotra, S.; Gosselin, F. *Org. Process Res. Dev.* **2022**, *26*, 313.

13. Villa, M.; Cannata, V.; Rosi, A.; Allegrini, P.; World Patent WO98/51646, 1998.

14. Li, B.; Buzon, R.; Zhang, Z. *Org. Process Res. Dev.* **2007**, *11*, 951.

15. (a) Kuntz, E. G. U.S. Patent 4,248,802, 1981. (b) Sinou, D. *Bull. Soc. Chim. Fr.* **1987**, *3*, 480.

16. Ahrland, S.; Chatt, J.; Davies, N. R.; Williams, A. A. *J. Chem. Soc.* **1958**, 276.

17. (a) Daïri, K.; Yao, Y.; Faley, M.; Tripathy, S.; Rioux, E.; Billot, X.; Rabouin, D.; Gonzalez, G.; Lavallée, J.-F.; Attardo, G. *Org. Process Res. Dev.* **2007**, *11*, 1051. (b) Larsen, R. D.; King, A. O.; Chen, C.-Y.; Corley, E. G.; Foster, B. S.; Roberts, F. E.; Yang, C.; Lieberman, D. R.; Reamer, R. A.; Tschaen, D. M.; Verhoeven, T. R.; Reider, P. J.; Lo, Y. S.; Rossano, L. T.; Brookes, S.; Meloni, D.; Moore, J. R.; Arnett, J. F. *J. Org. Chem.* **1994**, *54*, 6391.

18. Brenek, S. J.; Caron, S.; Chisowa, E.; Colon-Cruz, R.; Delude, M. P.; Drexler, M. T.; Handfield, R. E.; Jones, B. P.; Nadkarni, D. V.; Nelson, J. D.; Oliver, M.; Weekly, R. M.; Bellinger, G. C. A.; Brkic, Z.; Choi, N.; Desneves, J.; Lee, M. A.-P.; Pearce, W.; Watson, J. K. *Org. Process Res. Dev.* **2012**, *16*, 1338.

19. Honda, M.; Morita, H.; Nagakura, I. *J. Org. Chem.* **1997**, *62*, 8932.

20. Brenek, S. J.; Caron, S.; Chisowa, E.; Delude, M. P.; Drexler, M. T.; Ewing, M. D.; Handfield, R. E.; Ide, N. D.; Nadkarni, D. V.; Nelson, J. D.; Oliver, M.; Perfect, H. H.; Phillips, J. E.; Teixeira, J. J.; Weekly, R. M.; Zelina, J. P. *Org. Process Res. Dev.* **2012**, *16*, 1348.

21. Bullock, K. M.; Mitchell, M. B.; Toczko, J. F. *Org. Process Res. Dev.* **2008**, *12*, 896.

22. Banks, A.; Breen, G. F.; Caine, D.; Carey, J. S.; Drake, C.; Forth, M. A.; Gladwin, A.; Guelfi, S.; Hayes, J. F.; Maragni, P.; Morgan, D. O.; Oxley, P.; Perboni, A.; Popkin, M. E.; Rawlinson, F.; Roux, G. *Org. Process Res. Dev.* **2009**, *13*, 1130.

23. (a) Shieh, W. C.; Shekhar, R.; Blacklock, T.; Tedesco, A. *Syn. Comm.* **2002**, *32*, 1059. (b) A review of Heck reaction: Bhanage, B. M.; Arai, M.; *Catalysis Reviews* **2001**, *43*, 315.

24. (a) Conlon, D. A.; Drahus-Paone, A.; Ho, G.-J.; Pipik, B.; Helmy, R.; McNamara, J. M.; Shi, Y.-J.; Williams, J. M.; Macdonald, D.; Deschênes, D.; Gallant, M.; Mastracchio, A.; Roy, B.; Scheigetz, J. *Org. Process Res. Dev.* **2006**, *10*, 36. (b) Miller, W. D.; Fray, A. H.; Quatroche, J. T.; Sturgill, C. D. *Org. Process Res. Dev.* **2007**, *11*, 359.

25. Näsman, J. H.; Ekman, K. B.; Sundell, M. J. U.S. Patent 5,326,825, 1993.

26. Helminen, J.; Paatero, E.; Hotanen, U. *Org. Process Res. Dev.* **2006**, *10*, 51.

27. Ishihara, K.; Nakayama, M.; Kurihara, H.; Itoh, A.; Haraguchi, H. *Chem. Lett.* **2000**, *29*, 1218.

28. A micro-tube screening approach has been developed for metal impurity removal: Welch, C. J.; Albaneze-Walker, J.; Leonard, W. R.; Biba, M.; DaShilva, J.; Henderson, D.; Laing, B.; Mathre, D. J.; Spencer, S.; Bu, X.; Wang, T. *Org. Process Res. Dev.* **2005**, *9*, 198.

29. Bentley, A.; Butters, M.; Green, S. P.; Learmonth, W. J.; MacRae, J. A.; Morland, M. C.; O'Connor, G.; Skuse, J. *Org. Process Res. Dev.* **2002**, *6*, 109.

30. Kuethe, J. T.; Childers, K. G.; Humphrey, G. R.; Journet, M.; Peng, Z. *Org. Process Res. Dev.* **2008**, *12*, 1201.

31. Hashimoto, A.; Pais, G. C. G.; Wang, Q.; Lucien, E.; Incarvito C. D.; Deshpande, M.; Bradbury, B. J.; Wiles, J. A. *Org. Process Res. Dev.* **2007**, *11*, 389.

32. (a) Allsop, G. L.; Cole, A. J.; Giles, M. E.; Merifield, E.; Noble, A. J.; Pritchett, M. A.; Purdie, L. A.; Singleton, J. T. *Org. Process Res. Dev.* **2009**, *13*, 751. (b) Kerdesky, F. A. J.; Leanna, M. R.; Zhang, J.; Li, W.; Lallaman, J. E.; Ji, J.; Morton, H. E. *Org. Process Res. Dev.* **2006**, *10*, 512. (c) Mitchell, D.; Cole, K. P.; Pollock, P. M.; Coppert, D. M.; Burkholder, T. P.; Clayton, J. R. *Org. Process Res. Dev.* **2012**, *16*, 70.

33. Jiang, X.; Lee, G. T.; Villhauer, E. B.; Prasad, K.; Prasad, M. *Org. Process Res. Dev.* **2010**, *14*, 883.

34. Reginato, G.; Sadler, P.; Wilkes, R. D. *Org. Process Res. Dev.* **2011**, *15*, 1396.

35. Jiang, X.; Sclafani, J.; Prasad, K.; Repič, O.; Blacklock, T. *J. Org. Process Res. Dev.* **2007**, *11*, 769.

36. Dorow, R. L.; Herrinton, P. M.; Hohler, R. A.; Maloney, M. T.; Mauragis, M. A.; McGhee, W. E.; Moeslein, J. A.; Strohbach, J. W.; Veley, M. F. *Org. Process Res. Dev.* **2006**, *10*, 493.

37. Betti, M.; Castagnoli, G.; Panico, A.; Coccone, S. S.; Wiedenau, P. *Org. Process Res. Dev.* **2012**, *16*, 1739.

38. Manley, P. W.; Acemoglu, M.; Marterer, W.; Pachinger, W. *Org. Process Res. Dev.* **2003**, *7*, 436.

39. Baxter, C. A.; Cleator, E.; Brands, K. M. J.; Edwards, J. S.; Reamer, R. A.; Sheen, F. J.; Stewart, G. W.; Strotman, N. A.; Wallace, D. J. *Org. Process Res. Dev.* **2011**, *15*, 367.

40. Gala, D.; Dahanukar, V. H.; Eckert, J. M.; Lucas, B. S.; Schumacher, D. P.; Zavialov, I. A.; Buholzer, P.; Kubisch, P.; Mergelsberg, I.; Scherer, D. *Org. Process Res. Dev.* **2004**, *8*, 754.

41. Duan, S.; Venkatraman, S.; Hong, X.; Huang, K.; Ulysse, L.; Mobele, B. I.; Smith, A.; Lawless, L.; Locke, A.; Garigipati, R. *Org. Process Res. Dev.* **2012**, *16*, 1787.

42. Houpis, I. N.; Shilds, D.; Nettekoven, U.; Schnyder, A.; Bappert, E.; Weerts, K.; Canters, M.; Vermuelen, W. *Org. Process Res. Dev.* **2009**, *13*, 598.

43. Kallemeyn, J. M.; Ku, Y.-Y.; Mulhern, M. M.; Bishop, R.; Pal, A.; Jacob, L. *Org. Process Res. Dev.* **2014**, *18*, 191.

44. Brazier, E. J.; Hogan, P. J.; Leung, C. W.; O'Kearney-McMullan, A.; Norton, A. K.; Powell, L.; Robinson, G.; Williams, E. G. *Org. Process Res. Dev.* **2010**, *14*, 544.

45. Mulder, J. A.; Frutos, R. P.; Patel, N. D.; Qu, B.; Sun, X.; Tampone, T. G.; Gao, J.; Sarvestani, M.; Eriksson, M. C.; Haddad, N.; Shen, S.; Song, J. J.; Senanayake, C. H. *Org. Process Res. Dev.* **2013**, *17*, 940.

46. Frutos, R. P.; Rodríguez, S.; Patel, N.; Johnson, J.; Saha, A.; Krishnamurthy, D.; Senanayake, C. H. *Org. Process Res. Dev.* **2007**, *11*, 1076.

47. Malmgren, H.; Bäckström, B.; Søver, E.; Wennerberg, J. *Org. Process Res. Dev.* **2008**, *12*, 1195.

48. Thompson, H. W.; McPherson, E. *J. Am. Chem. Soc.* **1974**, *96*, 6232.

49. Bengali, A. A.; Schultz, R. H.; Moore, B.; Bergman, *J. Am. Chem. Soc.* **1994**, *116*, 9585.

50. (a) Taber, D. F.; Petty, E. H.; Raman, K. *J. Am. Chem. Soc.* **1985**, *107*, 196. (b) Wenkert, E.; Mylari, B. L.; Davis, L. L. *J. Am. Chem. Soc.* **1968**, *90*, 3870. (c) Taber, D. F.; Petty, E. H. *J. Org. Chem.* **1982**, *47*, 4808. (d) Wenkert, E.; Davis, L. L.; Mylari, B. L.; Solomon, M. F.; da Silva, R. R.; Shulman, S.; Warnet, R. J.; Ceccherelli, P.; Curini, M.; Pellicciari, R. *J. Org. Chem.* **1982**, *47*, 3242.

51. (a) O'Shea, P. D.; Gauvreau, D.; Gosselin, F.; Hughes, G.; Nedeau, C.; Roy, A.; Shultz, S. *J. Org. Chem.* **2009**, *74*, 4547. (b) Magnus, N. A.; Braden, T. M.; Buser, J. Y.; DeBaillie, A. C.; Heath, P. C.; Ley, C. P.; Remacle, J. R.; Varie, D. L.; Wilson, T. M.; *Org. Process Res. Dev.* **2012**, *16*, 830. (c) DeBaillie, A. C.; Magnus, N. A.; Laurila, M. E.; Wepsiec, J. P.; Ruble, J. C.; Petkus, J. J.; Vaid, R. K.; Niemeier, J. K.; Mick, J. F.; Gunter, T. Z. *Org. Process Res. Dev.* **2012**, *16*, 1538.

52. Brock, S.; Hose, D. R. J.; Moseley, J. D.; Parker, A. J.; Patel, I.; Williams, A. *J. Org. Process Res. Dev.* **2008**, *12*, 496.

53. Hansen, K. B.; Hsiao, Y.; Xu, F.; Rivera, N.; Clausen, A.; Kubryk, M.; Krska, S.; Rosner, T.; Simmons, B.; Balsells, J.; Ikemoto, N.; Sun, Y.; Spindler, F.; Malan, C.; Grabowski, E. J. J.; Armstrong III, J. D. *J. Am. Chem. Soc.* **2009**, *131*, 8798.

54. (a) Nicola, T.; Brenner, M.; Donsbach, K.; Kreye, P. *Org. Process Res. Dev.* **2005**, *9*, 513. (b) Farina, V.; Shu, C.; Zeng, X.; Wei, X.; Han, Z.; Yee, N. K.; Senanayake, C. H. *Org. Process Res. Dev.* **2009**, *13*, 250. (c) Arumugasamy, J.; Arunachalam, K.; Bauer, D.; Becker, A.; Caillet, C. A.; Glynn, R.; Latham, M.; Liu, J.; Mayes, B. A.; Moussa, A.; Rosinovsky, E.; Salanson, A. E.; Soret, A. F.; Stewart, A.; Wang, J.; Wu, X. *Org. Process Res. Dev.* **2013**, *17*, 811.

55. Wheeler, P.; Phillips, J. H.; Pederson R. L. *Org. Process Res. Dev.* **2016**, *20*, 1182.

56. Activated charcoal was utilized in removal of residual Ru from asymmetric synthesis of chiral acid: Maligres, P. E.; Humphrey, G. R.; Marcoux, J.-F.; Hillier, M. C.; Zhao, D.; Krska, S.; Grabowski, E. J. J. *Org. Process Res. Dev.* **2009**, *13*, 525.

57. Naud, F.; Spindler, F.; Rueggeberg, C.; Schmidt, A. T.; Blaser, H.-U. *Org. Process Res. Dev.* **2007**, *11*, 519.

58. Gallou, F.; Saim, S.; Koenig, K. J.; Bochniak, D.; Horhota, S. T.; Yee, N. K.; Senanayake, C. H. *Org. Process Res. Dev.* **2006**, *10*, 937.

59. Pérez-Balado, C.; Willemsens, A.; Ormerod, D.; Aelterman, W.; Mertens, N. *Org. Process Res. Dev.* **2007**, *11*, 237.

60. Damon, D. B.; Dugger, R. W.; Hubbs, S. E.; Scott, J. M.; Scott, R. W. *Org. Process Res. Dev.* **2006**, *10*, 472.

61. Hawkins, J. M.; Dubé, P.; Maloney, M. T.; Wei, L.; Ewing, M.; Chesnut, S. M.; Denette, J. R.; Lillie, B. M.; Vaidyanathan, R. *Org. Process Res. Dev.* **2012**, *16*, 1393.

62. Davies, I. W.; Marcoux, J.-F.; Corley, E. G.; Journet, M.; Cai, D.-W.; Palucki, M.; Wu, J.; Larsen, R. D.; Rossen, K.; Pye, P. J.; DiMichele, L.; Dormer, P.; Reider, P. J. *J. Org. Chem.* **2000**, *65*, 8415.
63. Hayes, S. T.; Assaf, G.; Checksfield, G.; Cheung, C.; Critcher, D.; Harris, L.; Howard, R.; Mathew, S.; Regius, C.; Scotney, G.; Scott, A. *Org. Process Res. Dev.* **2011**, *15*, 1305.
64. Okauchi, T.; Itonaga, M.; Minami, T.; Owa, T.; Kitoh, K.; Yoshino, H. *Org. Lett.* **2000**, *2*, 1485.
65. Akao, A.; Nonoyama, N.; Mase, T.; Yasuda, N. *Org. Process Res. Dev.* **2006**, *10*, 1178.
66. Tian, Q.; Cheng, Z.; Yajima, H. M.; Savage, S. J.; Green, K. L.; Humphries, T.; Reynolds, M. E.; Babu, S.; Gosselin, F.; Askin, D.; Kurimoto, I.; Hirata, N.; Iwasaki, M.; Shimasaki, Y.; Miki, T. *Org. Process Res. Dev.* **2013**, *17*, 97.

20 Methods for Impurity Removal

Removal of impurities and reducing them to process-acceptable levels in active pharmaceutical ingredients (APIs) are critical, because the presence of these impurities, especially the genotoxic impurities, even in small amounts in the drug substances, may affect drug safety and efficacy. Impurities can be originated from raw materials/solvents or be generated during the process. Depending on the nature of the impurities and the process, impurities in raw materials/solvents are possibly carried through the process to become the impurities in the final drug product. The additives in solvents, for example, butylated hydroxytoluene (BHT) in THF, may potentially contaminate the product. Nevertheless, these impurity issues can be readily resolved by establishing appropriate criteria for the incoming starting materials and grades of solvents that are used in the process. Impurities that are generated during processing are the major concerns. Therefore, various approaches are developed in attempts to control these impurities during workup.

20.1 REMOVAL OF FLUORIDE

A nucleophilic aromatic substitution reaction (S_NAr) is ubiquitous in organic chemistry in which the nucleophile displaces a leaving group such as fluoride to furnish the substitution product along with fluoride as a by-product. At large-scale production, however, the presence of fluoride poses a serious concern as it may generate HF to etch glass-lined reactors, especially under acidic conditions. To address this issue, such fluoride-involved reactions have to be performed using stainless steel reactor or under basic and mild conditions. Methods have been developed to remove fluoride prior to acidification of the reaction mixture or sequester fluoride with scavengers.

20.1.1 Use of Aqueous Wash

The S_NAr reaction of 1-(3,5-difluorobenzyl)indole **1** with alkoxide produced 2 equiv of potassium fluoride as a by-product (Equation 20.1).[1]

$$(20.1)$$

On a pilot-plant scale, the fluoride content needed to be reduced below 20 ppm level before the post-reaction mixture could be acidified. Accordingly, brine wash was implemented during the workup, which brought the fluoride in the organic layer to a level below 0.5 ppm.

Procedure

Stage (a)
- Charge 80.5 kg (717.4 mol, 6.0 equiv) of KOtBu, 130.5 kg of toluene, and 63.5 kg (834.5 mol, 7.0 equiv) of 2-methoxyethanol.
- Stir the mixture at 80 °C for 30 min.

DOI: 10.1201/9781003288411-20   747

- Charge a solution of **1** (50.0 kg, 119.2 mol) in 106 kg of 1,3-dimethyl-3,4,5,6-tetra-hydro-2(1H)-pyrimidinone (DMPU) and 22 kg of toluene.
- Heat at 110 °C for 3 h.

Stage (b)
- Cool to 60 °C.
- Add toluene to bring the batch contents to ca. 10 vol.
- Wash at 60 °C with 5% brine (3×250 kg).
- Wash with 6 N HCl (165 kg, 3.3 vol).
- Charge 170 kg of heptane to the organic layer.
- Cool to 0 °C and filter.
- Wash with 50 kg of isopropanol.
- Dry under vacuum.
- Yield: 52.8 kg (83%, 99.14 area%).

20.1.2   USE OF CACL$_2$

Calcium chloride (CaCl$_2$) is an extremely hygroscopic solid that liberates large amounts of heat during water absorption and dissolution. Calcium fluoride (CaF$_2$), however, has poor solubility in water (0.016 g/L at 20 °C). Due to the enormous solubility difference between CaCl$_2$ and CaF$_2$, CaCl$_2$ can act as a fluoride scavenger by converting the free fluoride in the reaction mixture to an insoluble CaF$_2$.

The synthesis of 3-fluoro-2-(1H-imidazol-2-yl)-pyridine **8** was achieved in a three-step sequence, starting with 3-fluoropyridine-2-carbonitrile **3** (Scheme 20.1).[2] The first step reaction of **3** with sodium methoxide gave the desired imidate **4** along with side products 3-methoxypyridine-2-carbo-nitrile **5** and sodium fluoride (NaF). The subsequent reaction of **4** with aminoacetaldehyde dimeth-ylacetal **6** generated **7**. Imidazole ring closure was realized by acidification of **7** with 5 N HCl followed by heating.

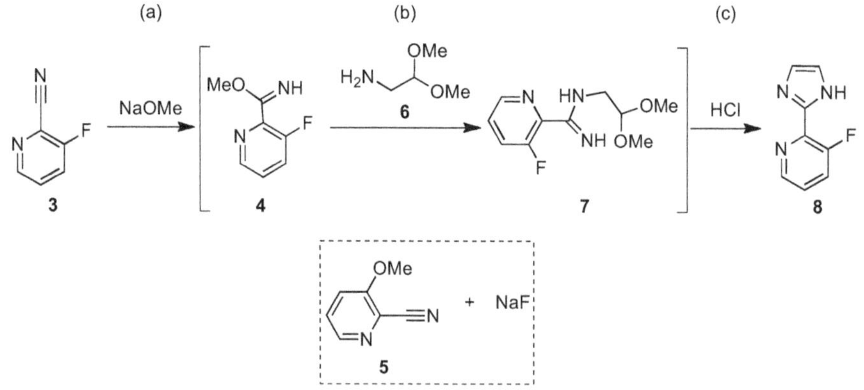

SCHEME 20.1

In Stage (c), the side product NaF formed in Stage (a) will react with HCl to generate HF. The corrosive HF thus formed can severely etch the glass reaction vessel. In order to suppress the HF formation, CaCl$_2$ was utilized to sequester the free fluoride before acidification with HCl.[3]

Procedure

Stage (a)
- Charge 1 kg (8.19 mol) of **3** into a reactor followed by adding MeOH (5.0 L).
- Cool to –20 °C.

- Charge a solution of NaOMe (25%, 1.772 kg, 8.20 mol) in MeOH within 30 min at −20 °C.
- Stir the resulting mixture at −20 °C for 4 h.

Stage (b)
- Charge 775 g (7.376 mol) of **6** at −20 °C within 5 min.
- Charge 978 g (16.3 mol) of AcOH within 5 min.
- Heat the resulting mixture at 65 °C for 30 min.
- Cool to room temperature and stir overnight.

Stage (c)
- Charge 90 g (0.818 mol) of CaCl$_2$ followed by 5N HCl (6 L).
- Heat the resulting suspension at 65 °C for 4 h.
- Distill under reduced pressure to remove 4 L of MeOH.
- Cool to room temperature.
- Add water (4 L) followed by 30% NaOH to pH 12 upon cooling with ice/water bath.
- Add *n*-BuOH (5 L) and separate layers.
- Extract the aqueous layer with *n*-BuOH (2 L).
- Distill the combined organic layers under reduced pressure to dryness.
- Add MeCN (2.5 L) to the residue.
- Heat the resulting suspension at 80 °C.
- Filter while hot.
- Cool the filtrate to 0 °C.
- Filter and wash with MTBE (2×1 L).
- Dry under vacuum.
- Yield: 696 g (52%).

20.1.3 USE OF CACO$_3$

Alternatively, calcium carbonate (CaCO$_3$) was employed to remove fluoride from the S$_N$Ar reaction mixture. As demonstrated in Reaction (20.2),[4] using CaCO$_3$ could reduce the fluoride level in the S$_N$Ar reaction between 2-fluorobenzonitrile **9** and piperidine **10** to <20 ppm. CaCO$_3$ had a dual role as both the base and the fluoride scavenger.

$$(20.2)$$

Procedure

- Charge 165 g (1.65 mol) of CaCO$_3$ to a mixture of 200 g (1.65 mol) of **9** and 387 g (2.47 mol, 1.5 equiv) of **10** in DMSO (800 mL).
- Stir the batch at 120 °C for 7 h.
- Cool to room temperature.
- Filter and wash with EtOAc (2.0 L).
- Wash the filtrate with water (2.0 L).
- Distill the organic layer under reduced pressure.
- Crystallize the residue in a mixture of MeOH (1200 mL) and water (360 mL).

- Filter and wash with water (720 mL).
- Dry under vacuum.
- Yield: 335 g (83%).

REMARK

The use of Ca(OH)$_2$ as the base caused hydrolysis of the ester group in the product **11**.

20.2   REMOVAL OF IODIDE

Ammonium chloride salts are ubiquitous in organic synthesis with many applications such as in phase transfer-catalyzed reactions (PTC) (usually using quaternary ammonium salts), in the purification of reaction products, and in improving water solubility/bioavailability of APIs. In contrast to ammonium chlorides, ammonium iodide salts are less common probably due to their instability of the iodide counterion towards air oxidation. For instance, the quaternary ammonium salt **13**, obtained from a reaction of amine **12** with methyl iodide followed by Boc deprotection (Equation 20.3),[5] was brown in color, which led to severe purification problems in the subsequent steps.

$$(20.3)$$

Solutions

To circumvent these purification issues, an ion-exchange strategy was developed by treating the crude product **13** with Bu$_3$BnNCl, producing **14** as a white crystalline solid (Equation 20.4).

$$(20.4)$$

Procedure

Stage (a) of Reaction 20.3
- Charge 1.18 kg (8.30 mol) of MeI in one portion to a solution of **12** (1.58 kg, 6.47 mol) in 7.9 L of EtOH at room temperature.
- Warm the reaction mixture to 35–40 °C and stir at the temperature for 22 h.
- Distill until 4.7 L distillate is collected.

Stage (b) of Reaction 20.3
- Charge 12.63 kg (41.517 mol) of 12% ethanolic HCl to the above distillation residue.
- Stir the resulting mixture at room temperature overnight.
- Cool to 0 °C and stir for 2 h.
- Filter and wash with EtOH (2.49 L).

Stage (c) of Reaction 20.4
- Suspend **13** as a wet solid obtained in Stage (b) of Reaction 20.3 with 2.27 kg (7.265 mol) of BnBu$_3$NCl in EtOH (6.92 L).
- Stir the slurry at room temperature overnight.
- Cool to 0 °C and stir for 2 h.

- Filter and wash cold EtOH (2×2.5 L).
- Dry at 25 °C under vacuum.
- Yield of **14**: 1.49 kg (93%) as white crystals.

20.3 REMOVAL OF HIGH-BOILING DIPOLAR APROTIC SOLVENTS

20.3.1 WASH WITH AQUEOUS SOLUTION

Dipolar aprotic solvent (DAS) becomes the solvent of choice when conducting substitution reactions of electrophiles with ionic nucleophiles, because DAS can improve the nucleophilicity of anions through solvation of the countercations. Furthermore, the use of DAS allows an S_N1 reaction to be performed at mild conditions because the polarized transition state in the S_N1 reaction can be stabilized by DAS. Despite their exceptional effects on reactivity, DASs are characterized by high boiling points (except acetonitrile) and thus difficult to remove by distillation, which limits their application toward the process development.

Among DASs, DMSO, DMF, N-methyl-2-pyrrolidone (NMP), DMAc, and MeCN are commonly used solvents in chemical transformations. Table 20.1 lists some typical physical data of these solvents.

Based on the solubility difference between products and the impurities, the aqueous (or organic) extractive workup can be effective in the separation of impurities from the product stream. Washing the organic mixture with water or the aqueous solution in the extractive workup can remove water-soluble impurities such as water-soluble amines and acids, and inorganic salts. Besides, the aqueous wash is frequently used to remove water-miscible DASs including DMF, DMSO, and NMP. For instance, during the extractive workup (MTBE/aqueous Na_2CO_3), DMF from the Heck cross-coupling reaction mixture (Equation 20.5)[6] could be efficiently removed from the postreaction mixture by aqueous sodium carbonate wash.

$$(20.5)$$

Procedure

- Add 566 mL (4.36 mol) of **16** to a mixture of **15** (485.0 g, 1.25 mol), Pd(OAc)₂ (2.8 g, 10 mmol), and DPEPhos (13.5 g, 20 mmol) in anhydrous DMF (2 L) at room temperature.
- Add 652 mL (3.75 mol) of DIPEA.

TABLE 20.1
Physical Properties of Dipolar Aprotic Solvents

	Mw	Bp, (°C)	Mp (°C)	Density (g/mL)	Flash point (°C)
MeCN	41.05	81	-46	0.786	2
DMSO	78.13	189	16-19	1.10	87
DMF	73.09	153	-61	0.944	58
DMAc	87.12	166	-20	0.937	70
NMP	99.13	202	-24	1.028	91

- Heat the mixture at 120 °C under N_2 for 2–3 h until complete conversion.
- Cool to room temperature, filter through a pad of Celite with charcoal, and rinse with DMF (2x150 mL).
- Add water (1 L) to the combined filtrate and extract the mixture with MTBE (2 L).
- Wash the organic layer with 10% Na_2CO_3 (1 L) and water (1 L).
- Add a solution of sodium dithionite (15.0 g) in water (70 mL) to the organic layer.
- Stir the mixture at 50 °C for 2 h.
- Cool to room temperature.
- Separate layers and discard the lower aqueous phase.
- Filter the organic phase through a pad of charcoal and rinse with MTBE (2x100 mL).
- Distill at 40 °C/300 mbar).
- Yield: 83.2% yield of **17** (614.4 ppm of Pd).

20.3.2 EXTRACTION WITH HEPTANE

If a product is soluble in a non-polar organic solvent such as heptane, extraction of the desired product with a non-polar organic solvent will leave DAS in the aqueous stream. DMSO is an excellent solvent for nucleophilic displacement reactions, especially for reactions that involve anionic nucleophilic species. A laboratory-scale preparation of benzyl azide **19** was carried out via the reaction of benzyl chloride **18** with sodium azide in DMSO (Equation 20.6).[7] The removal of DMSO was accomplished by extraction of the product **19** with heptane from the postreaction mixture leaving the high boiling DMSO in aqueous waste.

$$(20.6)$$

20.4 REMOVAL OF TRIPHENYLPHOSPHINE OXIDE

Triphenylphosphine (PPh_3) is widely used in organic synthesis. Because of the phosphorus lone pair of electrons, PPh_3 can serve as an effective nucleophile, reductant, and chelating agent. Accordingly, PPh_3 enables participation in various organic transformations such as Wittig reactions, Mitsunobu reactions, and transition metal-catalyzed cross-coupling reactions (as a chelating ligand). The common by-product of these PPh_3-involved reactions is triphenylphosphine oxide (Ph_3PO). Mitsunobu reactions of alcohols with phenols or carboxylic acids in the presence of triphenylphosphine and azodicarboxylate, such as diethyl azodicarboxylate (DEAD) or diisopropyl azodicarboxylate (DIAD) give the corresponding ethers or esters, along with Ph_3PO as the by-product. Removal of the Ph_3PO by-product without resorting to column chromatography is a significant challenge.

20.4.1 WASH WITH ETHYL ACETATE

Washing the aqueous reaction mixture with organic solvent is designed to remove organic-soluble by-products and impurities from the aqueous product phase. The presence of basic or acidic groups in products allows the conversion of the free bases or free acids to their corresponding water-soluble organic salts, which are poised to be separated from organic-soluble by-products and impurities by washing the aqueous product phase with an organic solvent.

The Mitsunobu coupling between phenol **20** and alcohol **21** was carried out using di-*tert*-butyl azodicarboxylate (DTAD) as the activation agent in THF to synthesize compound **22**, a potent SRC kinase inhibitor developed by AstraZeneca (Equation 20.7).[8]

(20.7)

An extractive workup protocol was developed to remove the Ph_3PO by-product from the Mitsunobu coupling product **22** by organic wash. Upon reaction completion, the reaction solvent was swapped to ethyl acetate. After a basic wash to remove by-products associated with DTAD, compound **22** could be extracted into the aqueous phase with aqueous HCl followed by washing the aqueous product phase with ethyl acetate rejecting Ph_3PO into the organic phase.

Procedure

- Charge a solution of triphenylphosphine (3.057 kg, 11.66 mol, 2.47 equiv) in THF (8 L) over 15 min into a 100-L vessel containing a slurry of **20** (2.139 kg at 92 w/w% purity, 4.73 mol) and DTAD (2.771 kg, 12.03 mol, 2.54 equiv) in THF (31 L) at ambient temperature.
- Rinse the line with THF (2 L) and stir for 10 min.
- Cool to 15 °C.
- Charge a solution of **21** (1.050 L, 1.049 kg, 7.27 mol, 1.54 equiv) in THF (3 L) at below 20 °C over 12 min.
- Rinse the line with THF (1 L).
- Stir the resulting mixture at ambient temperature for 50 min.
- Charge a solution of **21** (348.5 mL, 348.2 g, 2.41 mol, 0.51 equiv) in THF (1 L).
- Stir the batch for 1.5 h.
- Distill on a rotary evaporator (40 °C, 100 mbar).
- Add EtOAc (4 L).
- Distill again to give an oil.
- Dissolve the oil in EtOAc (30 L).
- Wash with 0.2 M K_2CO_3 (20 L).
- Extract the organic phase with 0.5 N HCl (20 L) then with 0.5 N HCl (4 L).
- Wash the combined acid layers with EtOAc (10 L).
- Charge 12 L of 37% HCl to the mixture at 19–20 °C over 15 min.
- Stir the resulting mixture at ambient temperature for 2 h.
- Add 3 L of 47% NaOH at <30 °C over 35 min.
- Add 20 L of n-BuOH followed by 3 L of 47% NaOH to pH 13.
- Separate layers.
- Extract the aqueous layer with n-BuOH (10 L).
- Distill the combined n-BuOH layers.
- Dissolve the resulting oil in IPA (4 L).
- Distill to obtain an oil.
- Dissolve the resulting oil in IPA (6 L).
- Filter through Celite and wash with IPA (6 L) (the resulting IPA solution was directly used for the subsequent difumarate preparation).

REMARK

One advantage of using DTAD over commonly used Mitsunobu reagents, such as DEAD and DIAD, is the susceptibility of the *tert*-butyl ester groups to acidic hydrolysis. Utilization of the DTAD, impurity **25** in the crude product **22** was readily removed by treating the acidic aqueous product phase with 37% HCl to convert **25** to a water-soluble **26** (Scheme 20.2):

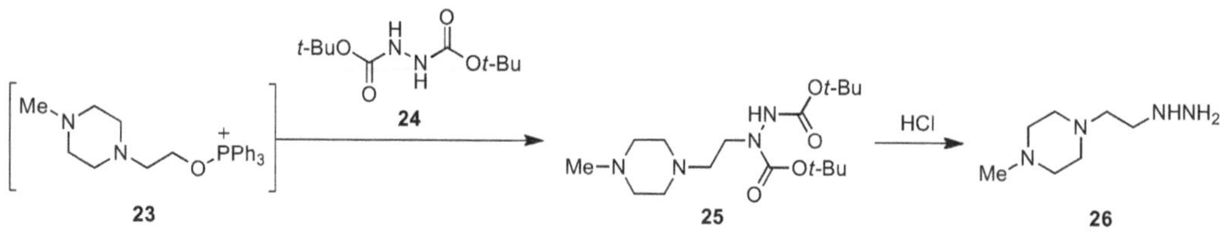

SCHEME 20.2

20.4.2 PRECIPITATION OF PH₃PO WITH MGCL₂

Mitsunobu reaction of chiral alcohols with phenols or acids produces the corresponding alkyl aryl ethers or esters in an inversion of configuration. For example, the Mitsunobu coupling between phenol **27** and *cis*-cyclobutanol **28** furnished ether **29** in *trans*-configuration (Equation 20.8).[9] In order to remove Ph₃PO without resorting to column chromatography, a method was developed by precipitating Ph₃PO with magnesium chloride (MgCl₂).

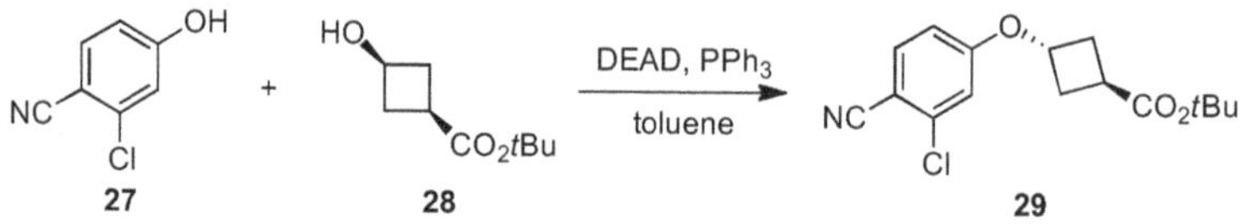

$$(20.8)$$

It was found that Ph₃PO can form a complex with MgCl₂ and the resulting Ph₃PO–MgCl₂ complex has relatively low solubility in non-polar solvents such as toluene. Accordingly, upon the completion of the reaction, MgCl₂ (2 equiv) was added and the mixture was heated at 50–60 °C for 1–2 h prior to cooling to ambient temperature. The Ph₃PO–MgCl₂ complex was then readily removed by filtration.

Procedure

- Cool the mixture of **27** (84.5 g, 0.55 mol), **28** (200 g, 0.58 mol, *cis:trans*=92:8), and Ph₃P (175 g, 0.67 mol) in toluene (700 mL) to 0–5 °C.
- Charge a solution of DEAD (300 g, 0.69 mol, 40% toluene solution) at <25 °C.
- Warm the resulting mixture to 25–30 °C.
- Upon completion of the reaction, charge 130 g (1.34 mol) of MgCl₂ (325 mesh powder).
- Heat the batch to 60 °C.
- Add heptane (700 mL).
- Continue to heat until Ph₃PO in the supernatant is <5%.
- Cool the mixture to ambient temperature.
- Filter and wash with toluene/heptane (700 mL).
- Distill the combined filtrates under reduced pressure.
- Add IPA and cool to <10 °C.
- Filter and wash with IPA/water.
- Dry at 50 °C under vacuum.
- Yield: 145 g (81%).

20.4.3 PRECIPITATION OF PH₃PO WITH ZNCL₂

Removal of Ph₃PO by-product from the reductive cyclization product **31** was achieved by precipitation of Ph₃PO with zinc chloride from alcohol solution (Equation 20.9):[10]

(20.9)

Procedure

- Dissolve the solid product mixture (from 0.60 mole-scale reaction) in 2.0 L of EtOH at 35–40 °C.
- Add a solution of $ZnCl_2$ (202 g, 1.48 mol) in 1.0 L of EtOH.
- Filter to remove the white precipitate ($Zn(Ph_3PO)_2Cl_2$) and rinse with EtOH.
- Concentrate the filtrate and crystallize the residue from methylcyclohexane/toluene.
- Yield: 111 g.

20.4.4 PRECIPITATION OF PH₃PO WITH HEPTANE

Although Wittig olefination is frequently used to convert ketones (or aldehydes) into the corresponding olefins, one of the drawbacks is the generation of a Ph_3PO by-product whose removal often requires column chromatography. If the Wittig product is soluble in non-polar solvents, such as heptane and diethyl ether, removal of Ph_3PO can be achieved by precipitation of Ph_3PO with non-polar solvents followed by filtration. For instance, compound **34**, prepared from the Wittig reaction of methyl (4-iodophenyl)ketone **32** with methyltriphenylphosphonium bromide **33** in the presence of potassium *tert*-butoxide (Equation 20.10),[11] is soluble in a non-polar solvent which was poised for extraction of **34** with heptane to reject Ph_3PO by-product by filtration. Ultimately, using this extraction/precipitation approach the Wittig product **34** was isolated by heptane extraction, rejecting Ph_3PO by-product by precipitation and filtration.

(20.10)

Procedure

- Charge THF (35 L) followed by adding 5.298 kg (21.35 mol) of **32**.
- Charge 8.492 kg (23.77 mol) of **33** followed by adding 14.6 kg (25.85 mol) of 20% KO*t*Bu THF solution at 20–30 °C.
- Cool to 22 °C and stir for 1.5 h.
- Transfer the reaction mixture into a mixture of heptane (30 L) and water (45 L) while maintaining 20–30 °C.
- Stir for 0.5 h and separate layers.
- Wash the organic layer with 1% aqueous NaCl solution (45 L).
- Distill (40–55 °C/250–375 mmHg) to remove 50 L of distillate leading to Ph_3PO precipitation.
- Filter to remove solid Ph_3PO and rinse with heptane (20 L).
- The heptane layer was used directly in the subsequent step.

REMARK

The Ph_3PO by-product from the Wittig reaction could also be precipitated with isohexane and removed by filtration.[12]

20.5   USE OF SODIUM BISULFATE

Extraction is a unit operation that is frequently used to separate (or recover) organic product from the aqueous mixture, rejecting water-soluble impurities in the aqueous phase. Impurity removal via crystallization/precipitation is usually achieved by manipulation of the solubility difference between products and impurities. The use of additives is capable of enlarging the solubility difference; thereby removal of impurities can be conveniently achieved.

20.5.1   Removal of Methacrylic Acid from Acid Product

The alkylation of *o*-cresol **35** with 2-Bromo-2-methylpropanoic acid **36** occurred in the presence of NaOH in methyl ethyl ketone (MEK), providing the desired product **37** in a quantitative yield (Equation 20.11).[13] However, the desired product **37** was contaminated with methacrylic acid **38**, which was difficult to reject by extraction due to the similarity in pKa with **37**.

$$(20.11)$$

The addition of $NaHSO_3$ converted **38** to a more water-soluble adduct **39** which was readily rejected into the aqueous phase during the isolation (Equation 20.12).

$$(20.12)$$

Procedure

- Dissolve 14.5 kg (134 mol) of **35** in 290 L of MEK.
- Charge 26.8 kg (670 mol, 3 equiv) of NaOH pellets.
- Stir the mixture at 50 °C for 2 h.
- Charge the solution of **36** (40.3 kg, 241 mol, 1.8 equiv) in 87 L of MEK over 8 h.
- Stir at 50 °C for 3 h.
- Add 145 L of water.
- Distill under reduced pressure to remove MEK.
- Wash the resulting aqueous mixture with MTBE (145 L).
- Add 59.9 kg of 3 N HCl to pH 6.5.
- Charge 27.9 kg (268 mol) of $NaHSO_3$.
- Stir at room temperature overnight.
- Charge 124 kg of 3 N HCl to pH 1.6.
- Extract with MTBE (2×290 L) (The MTBE product solution was used directly in the subsequent step).
- Yield: quantitative.

REMARK

The reaction of $NaHSO_3$ with methacrylic acid **38** was highly pH sensitive and the optimal pH was between 6 and 7.

20.5.2 REMOVAL OF ALCOHOL FROM ALDEHYDE PRODUCT

$NaHSO_3$ was used for aldehyde product purification via the formation of $NaHSO_3$–aldehydes adduct leaving the side product intact. The reduction of ethyl ester **40** with diisobutyl aluminum hydride (DIBAL-H) in toluene/methylcyclohexane at –90 °C afforded the desired aldehyde **41** along with overreduction of the side product **42** (4.4%) (Scheme 20.3).[14]

Treatment of the reaction mixture (containing both **41** and **42**) with $NaHSO_3$ led to a solid bisulfite adduct **43**, leaving the alcohol **42** in the mother liquor after the centrifugation.

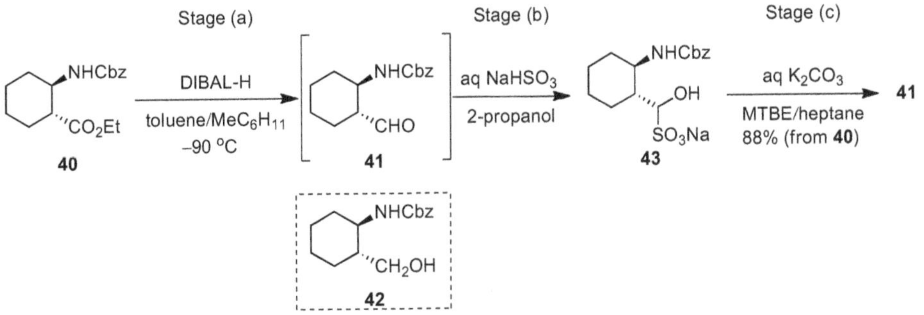

SCHEME 20.3

Procedure

Stage (a)
- Dissolve, in a 600-L reactor, 17.5 kg (57.3 mol) of **40** in a mixture of toluene (167 kg) and methylcyclohexane (27.0 kg).
- Cool to –90 to –96 °C.
- Charge 90.1 kg (127 mol, 2.2 equiv) of 20 wt% solution of DIBAL-H in hexanes at <–90 °C.
- After the addition hold the resulting mixture for an additional 30 min.
- Add a mixture of IPA (13.8 kg) and *n*-heptane (23.0 kg).

Stage (b)
- Transfer the content to a 1000-L vessel containing a solution of citric acid (54.3 kg, 258 mol) and water (112 L) at <30 °C.
- Separate layers.
- Wash the organic layer with 5% aqueous $NaHCO_3$ (147 kg) followed by 10% brine (150 kg).
- Add IPA (27.5 kg) to the organic layer.
- Charge sodium bisulfite (11.9 kg, 62.6 mol) in water (24.5 L).
- Filter to isolate the adduct **43**.

Stage (c)
- Charge the wet cake of **43** back to the 1000-L vessel followed by adding water (140 L), MTBE (90.9 kg), and 50% aqueous K_2CO_3 solution (40.8 kg).
- Stir to dissolve the solids.
- Remove aqueous layer.
- Wash the organic layer with 15% brine (2×180 kg).
- Distill under reduced pressure to switch the solvent to *n*-heptane.
- Cool and filter.

- Dry at 30 °C under vacuum.
- Yield: 12.75 kg (85%).

REMARK

Citric acid was used to remove aluminum.

20.5.3 REMOVAL OF EXCESS FORMALDEHYDE

An Aldol crossed-Cannizzaro reaction of **44** was carried out with 20 equiv of paraformaldehyde in the presence of sodium ethoxide in ethanol to prepare diol **46** (Scheme 20.4).[15]

The workup involved distillation to remove EtOH, extraction with MTBE, and solvent exchange to MeOH, followed by crystallization. The presence of volatile, irritating formaldehyde during the distillation posed a scaling-up challenge. To address this issue, NaHSO₃ (18 equiv) was added to the postreaction mixture converting the excess formaldehyde to the corresponding non-volatile NaHSO₃–formaldehyde adduct, which facilitated the subsequent workup processing.

Procedure

- Charge a slurry of paraformaldehyde (25.0 kg, 832.5 mol) in EtOH (76 L) to a solution of **44** (27.33 kg, 41.8 mol) in EtOH (232 L).
- Heat the mixture to 55 °C.
- Charge 27.11 kg (79.7 mol) of 20 wt% NaOEt in EtOH over 5 min.
- Stir the batch at 55 °C for 4 h (until <5% of **45**).
- Add a solution of NaHSO₃ (78.3 kg, 752.4 mol) in water (272 L) at 55 °C over 20 min.
- Stir the resulting mixture for 30 min before cooling to 35 °C.
- Distill under reduced pressure to remove EtOH.
- Extract the aqueous phase with MTBE (273 L).
- Wash the MTBE layer with water (164 L) and back extract the water wash with MTBE (150 L).
- Distill the combined MTBE phases under reduced pressure to 40 L.
- Add MeOH (109 L).
- Distill under reduced pressure to 40 L.
- Add MeOH (109 L).
- Distill to 91 L.
- Add **46**·MeOH solvate as seeds.
- Granulate at room temperature for 8 h before cooling to 5 °C.
- Stir at 5 °C for 2 h.
- Filter, wash with cold (–15 °C) MeOH (21 L) then with heptane (55 L), and dry at 40 °C under vacuum.
- Yield of **46**: 10.63 kg.

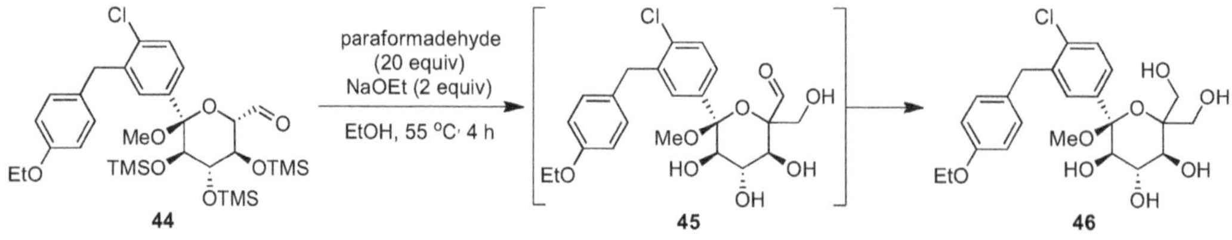

SCHEME 20.4

NOTE

The reactor was vented through an aqueous MeOH scrubber to sequester formaldehyde.

20.5.4 Removal of Ketone Intermediate

The preparation of diol **49** was brought about by the addition of Grignard reagent to methyl ester **47** to give a mixture of the desired diol product **49** and ketone intermediate **48** in a ratio of 7:1 (Scheme 20.5).[16] A complete consumption of ketone intermediate was never achieved, presumably because of competitive enolization.

To remove **48** from **49** without column chromatography, the postreaction mixture was washed with saturated aqueous NaHSO$_3$ solution.

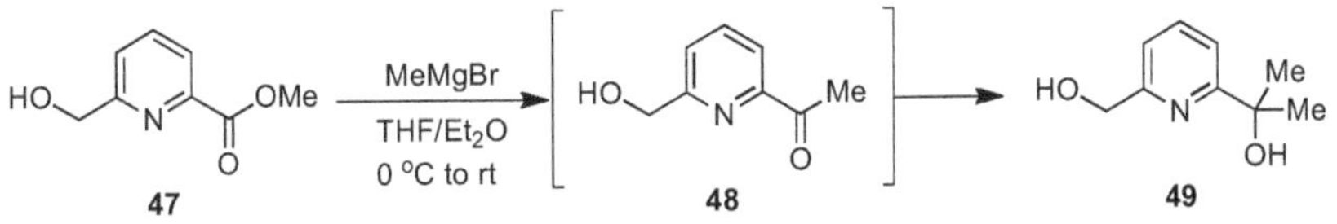

SCHEME 20.5

20.6 REMOVAL OF EXCESS REAGENTS

Two- or multi-component reactions are usually carried out by setting the expensive or hard-to-access material as the limited reagent while using less expensive materials in excess in order to deliver a high throughput process. These excess reagents, however, have to be removed during the workup or they may become impurities to contaminate the intermediates or the final products.

20.6.1 Use of Dimethylamine to Remove Excess Formaldehyde

Besides using NaHSO$_3$, dimethylamine was employed to scavenge the excess formaldehyde in the reductive amination during the one-pot synthesis of tertiary amine **51** (Equation 20.13).[17]

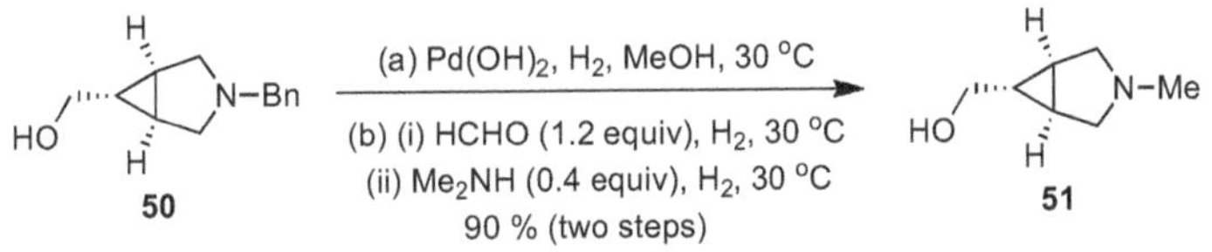

$$(20.13)$$

In this modified process the excess formaldehyde was converted to trimethylamine that is easily removed by distillation.

Procedure

Stage (a)
- Charge, to an inerted 30-L Hastelloy pressure reactor, 275 g (20% Pd content, 50% wet) of Pd(OH)$_2$, 20 L of MeOH, and 2.75 kg (13.5 mol) of **50**.
- Purge and vent three times.
- Warm the batch to 30 °C.
- Pressurize with H$_2$ to 50 psig for 16 h.

Stage (b)-(i)
- Purge the reactor with N$_2$.
- Charge 1.32 kg (37% w/w, 16.2 mol, 1.2 equiv) of aqueous formaldehyde solution.
- Pressurize with H$_2$ to 50 psig and hold at 30 °C for 8 h.

Stage (b)-(ii)

- Purge the reactor with N_2.
- Charge 0.61 kg (40% w/w, 5.4 mol, 0.4 equiv) of aqueous dimethylamine solution.
- Heat the batch at 30 °C under 50 psig of H_2 for 4–8 h.
- Filter and distill to displace MeOH with toluene under reduced pressure.
- Filter and distill the filtrate.
- Yield: 1.55 kg (90%).

20.6.2   USE OF *N*-METHYLPIPERAZINE TO REMOVE BOC ANHYDRIDE

Boc anhydride (Boc_2O) is frequently used to protect functional groups in organic synthesis. The unreacted Boc_2O, however, may become impure in the downstream chemistry if the removal procedure is not effective during the workup. The Boc protection of oxazolidine **54** was accomplished with excess Boc_2O (1.1 equiv) to furnish the corresponding *N*-protected product **55**. An approach for removal of the unreacted Boc_2O was devised by treatment of the postreaction mixture with *N*-methylpiperazine to consume the Boc_2O, resulting in *N*-Boc-*N'*-methylpiperazine that could be removed along with unreacted *N*-methylpiperazine with a dilute aqueous HCl wash (Scheme 20.6).[18]

Procedure

Stage (a)

- Slurry 22.5 kg (1 equiv) of L-serine methyl ester hydrochloride **52** with 16.0 kg (1.1 equiv) of Et_3N and 15.1 kg (1.2 equiv) of **53** in 225 L of 2-methylpentane.
- Heat to reflux under Dean–Stark conditions for 4 h.
- Cool to 25 °C.
- Remove the solid by filtration.
- Distill the filtrate to provide a concentrated solution of **54**.

Stage (b)

- Charge 100 L of THF and 33.8 kg (1.1 equiv) of Boc_2O.
- Stir at 20 °C for 6 h.
- Charge 20 wt% aqueous K_2CO_3 (135 kg) and stir for 14 h.

Stage (c)

- Add 2-methylpentane (113 L) and separate layers.
- Charge 7.2 kg of *N*-methylpiperazine to the organic layer and stir for 1 h.
- Wash the organic layer with 1 N HCl (2×68 kg), followed by 20 wt% aqueous K_2CO_3 (82 kg) and water (67 kg).
- Concentrate the organic layer.
- Dilute the resulting distillation residue of **55** with THF and use directly in the subsequent reaction.

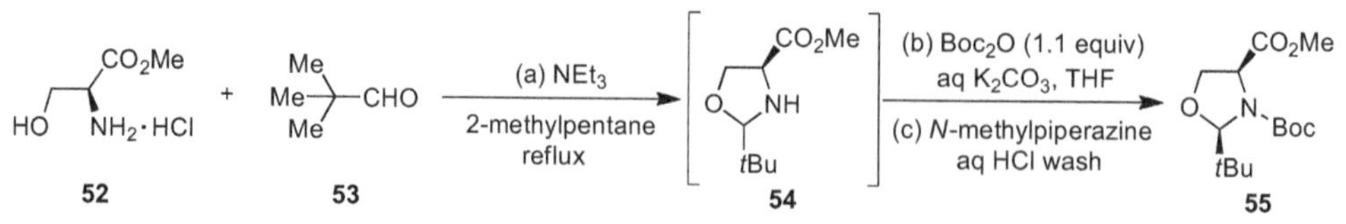

SCHEME 20.6

20.6.3 Use of CO_2 to Remove Excess Piperazine

A two-molecule reaction is frequently conducted by using one of the starting materials in excess in order not only to increase the reaction rate, but also to suppress side reactions. For instance, an alkylation reaction in the preparation of 2-(piperazin-1-yl)acetonitrile **58** was carried out with 2.7 equiv of piperazine **57** in order to limit the formation of dialkylated product (Equation 20.14).[19]

$$(20.14)$$

(I) Problems

Removal of the excess **57** after the reaction proved to be challenging, as the acid wash will not only remove **57** but the product **58** as well.

(II) Solutions

The solution for the isolation of the piperazine product **58** was to treat the reaction mixture with gaseous carbon dioxide to precipitate the excess **57** as its carboxylate salt.

Procedure

- Dissolve 2.0 kg (23.2 mol, 2.7 equiv) of **57** in 10 L of MeCN at 50 °C.
- Charge a solution of **56** (642 g, 8.5 mol) in MeCN (4 L) over 3.5 h.
- After reaction completion, cool the batch to ambient temperature.
- Filter off the solid.
- Bubble CO_2 into the filtrate (the temperature increased from 20 °C to 42 °C) until the temperature starts to drop.
- Filter off the carbonate salt.
- Distill the filtrate at 45 °C/50 Torr.
- Yield of **58**: 1.07 kg (99%) as oil.

20.6.4 Use of Succinic Anhydride to Remove 1-(2-Pyrimidyl)piperazine

With an intention to consume all of the expensive starting material such as chiral nosylate **59**, the S_N2 reaction of **59** was conducted with excess 1-(2-pyrimidyl)piperazine **60** (1.29 equiv), providing the desired product **61** (Equation 20.15).[20] However, the unreacted **60** in the crude product **61** could inhibit the product crystallization.

$$(20.15)$$

Removal of excess **60** was realized by the addition of succinic anhydride and triethylamine to convert **60** into ammonium carboxylate **62** that was rejected into aqueous wastes.

20.6.5　Use of Pivalaldehyde to Remove 4-Chlorobenzylamine

Conversion of ester **63** to amide **65** was effected with 3 equiv of 4-chlorobenzylamine **64** in ethylene glycol at 132 °C (Equation 20.16).[21] However, the excess of amine **64** had to be removed since it readily forms an insoluble carbonate salt.

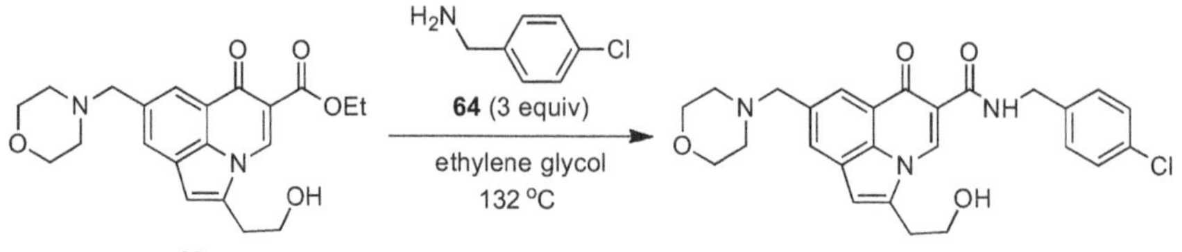

(20.16)

Ultimately, 1 equiv of pivalaldehyde **66** was employed to convert the **64** to aminal **68** which was rejected into the mother liquor (Scheme 20.7).[22]

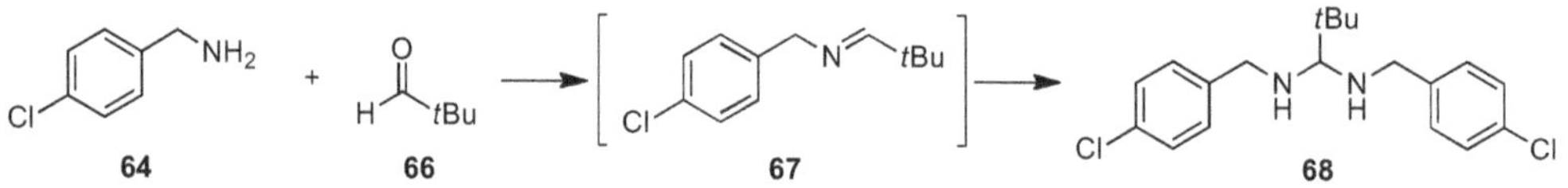

SCHEME 20.7

Procedure

Stage (a)
- Charge 800 g (2.1 mol) of **63** into a 22-L, three-necked flask under nitrogen, followed by adding ethylene glycol (4.8 L) and 755 mL (6.2 mol, 3 equiv) of **64**.
- Heat the resulting suspension at 132 °C for 8 h.
- Cool to 102 °C.
- Add toluene (2.4 L).
- Cool to 75 °C.

Stage (b)
- Add MeCN (1.2 L), followed by adding 215 mL (2.1 mol, 1.0 equiv) of **66**.
- Cool to room temperature and stir the resulting slurry overnight.
- Filter, wash with EtOH (2 L) and MeCN (2×1 L), and dry at 55 °C under vacuum.
- Yield: 1318 g (132%) of crude **65**.

20.6.6　Use of DABCO to Remove Benzyl Bromide

Potential genotoxic impurities (PGI) could damage DNA and lead to mutation. Thus, tight control of the PGI level[23] in the final drug substances a challenging task for process chemists. Factors that have direct impacts on the level of genotoxins include route design,[24] reagent selection,[25] and the optimal form of the drug substance. Torcetrapib **71** was a drug candidate undergoing clinical trials in Pfizer as a cholesteryl ester transfer protein (CETP) CETP inhibitor. The final step for the synthesis of carbamate **71** was N-benzylation using 3,5-bis(trifluoromethyl)benzyl bromide **70** (1.4 equiv) (Equation 20.17).[26] The use of bromide **70** in the final step posed a significant challenge, as **70** was identified as PGI.[27]

(20.17)

Accordingly, attention was given to the removal of **70** from the final product **71**. After completion of the reaction, the excess alkylating agent **70** was converted to quaternary salt **72** by the addition of nucleophilic 1,4-diazabicyclo[2.2.2]octane (DABCO). The quaternary salt was very stable and not readily hydrolyzed under acidic or basic conditions. Thus, aqueous HCl washes removed the excess **70** via the quaternary salt **72**.

Procedure

Stage (a)
- Charge 5.175 kg (13.82 mol) of **69** into a 100-L glass reactor followed by the addition of CH_2Cl_2 (20 L) and 1.551 kg (13.82 mol) of KOtBu.
- Stir for 5 min.
- Charge 3.5 L (19.1 mol, 1.38 equiv) of **70** at 20–25 °C.
- Stir for 2.3 h.
- Charge 46.10 g (0.41 mol) of additional KOtBu.
- Continue to stir for total 4.5 h.

Stage (b)
- Charge 918 g (8.18 mol, 0.6 equiv) of DABCO to the reaction solution.
- Stir for 1 h.
- Add diisopropyl ether (40 L) and 0.5 N HCl (30 L).
- Stir the mixture and separate layers.
- Wash the organic layer with 0.5 N HCl (2×30 L) and brine (15 L).
- Crystallize from diisopropyl ether/ethanol.
- Yield: 73% of **71** (with non-detectable levels of **70**).

20.6.7 USE OF AQUEOUS AMMONIA TO REMOVE DIETHYL SULFATE

The alkylation of 3-methyl-4-nitrophenol **73** with an excess of diethyl sulfate gave the desired ether **74** (Equation 20.18).[28] As $(EtO)_2SO_2$ is a highly toxic alkylation agent, any residual $(EtO)_2SO_2$ left in the reaction intermediate is considered a genotoxic impurity and has to be completely destroyed after the reaction.

(20.18)

Therefore, aqueous ammonia was employed to efficiently decompose the remaining $(EtO)_2SO_2$ in the postreaction mixture.

20.6.8 Use of Hydrogen Peroxide to Oxidize PH₃P

The preparation of diethyl ester **78** was carried out starting with the Wittig reaction of benzyl bromide **75** with 5-bromopentanal, followed by catalytic reduction of the resulting olefin **77**. It was found that residual triphenylphosphine carried over from the Witting reaction can poison the catalyst in the subsequent hydrogenation step (Scheme 20.8).[29]

Solutions

The use of hydrogen peroxide can oxidize the residual triphenylphosphine used in the preparation of phosphonium salt. Thus, an oxidative workup protocol was developed by treating the Wittig reaction mixture with 30% H_2O_2 under reflux to oxidize the residual triphenylphosphine completely to triphenylphosphine oxide.

Procedure

Stage (a)
- Charge 651 g (1.63 mol) of **75** into a reactor, followed by adding 477 g (1.82 mol) of triphenylphosphine and MeCN (5.4 L).
- Heat the batch at reflux for 1 h.
- Cool the content down to 40 °C.

Stage (b)
- Charge 296 g (1.79 mol) of 5-bromopentanal and 8.2 L (95.2 mol) of 1,2-epoxybutane.
- Heat at reflux for 18 h.
- Add 272 mL of 30 wt % H_2O_2 over 10 min.
- Heat the resulting mixture at reflux for 2.5 h while maintaining an N_2 flow.
- Cool to room temperature and isolate the product, then go to Stage (c).

REMARKS

(a) The authors[29] suggested that Cs_2CO_3 should be used to replace 1,2-epoxybutane as an acid scavenger for large-scale preparation.

(b) Oxidation using H_2O_2 inflammable solvent is potentially hazardous. It is important to maintain a constant flow of inert gas to prevent oxygen from building up in the reaction vessel.

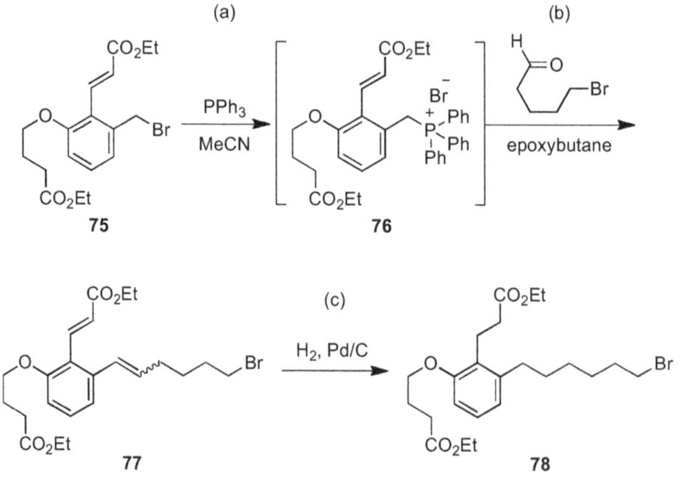

SCHEME 20.8

20.7   CONVERSION OF IMPURITY TO STARTING MATERIAL

Instead of removing the side product from the postreaction mixture during the workup, transformation of the side product into the starting material is a method that is exceptionally useful when the starting material is expensive and not easy to access. In the event, both the product purity and yield would be improved. Methylation of pyrimidine-4,5-diol **79** with methyl iodide under conditions [MeI, $Mg(OMe)_2$, at 60 °C in DMSO] provided a mixture of *N*-Me **80** and *O*-Me **81** in a ratio of 78:22 (Equation 20.19).[30]

(20.19)

It was found that the side product **81** could be demethylated back to the starting material **80** in situ at a higher reaction temperature (65 °C) with a longer reaction time (20 h) resulting in an improved selectivity to 99:1 for the desired *N*-Me product **80**. Ultimately, the methylation was carried out in NMP with trimethylsulfoxonium iodide in the presence of water and magnesium hydroxide. Under the optimized methylation conditions [2 equiv of $Mg(OH)_2$, 2 equiv of $Me_3S(O)I$, NMP, 100 °C, 6 h] the methylation reaction achieved >99% conversion with 92% isolated yield of **80**.

20.8   CONVERSION OF IMPURITY TO PRODUCT

Similarly, in some cases, the side product can be converted to the desired product, an approach that is frequently used by process chemists.

20.8.1   DEOXYCHLORINATION

Deoxychlorination of **82** was carried out using $POCl_3$ (3 equiv) in acetonitrile under reflux. Upon completion of the reaction, the reaction mixture was quenched with ice–water and the precipitated product **83** was collected by filtration (Equation 20.20).[31]

(20.20)

(I) Problems

Upon scaling up, it was observed that the isolated product yield was decreased from 60% down to 45% due to the formation of phosphoramidic dichloride **84** that was observed in the filtrate.

(II) Solutions

In order to address this issue, a hydrolysis approach was developed to convert the side product **84** to the product **83**. Thus, the filtrate was reheated to 55 °C for 5 h and the resulting second crop of product **83** was precipitated and collected by filtration. The combined yield was 67% at a kilogram scale with >98% purity.

Procedure

- Charge 7.54 kg (49.2 mol) of POCl₃ slowly to a mixture of **82** (3.0 kg, 16.4 mol) in anhydrous MeCN (30.0 L) at <50 °C.
- Heat to reflux and stir at reflux for 3–5 h.
- Cool the batch to room temperature.
- Add ice–water (45.0 L) and stir for 45 min.
- Filter and wash with water (15.0 L).
- Dry at 60 °C under vacuum.
- Yield (first crop): 1.80 kg (55%).
- Heat the combined mother liquor at 55 °C for 5 h.
- Cool to room temperature.
- Filter and wash with a mixture of MeCN/water (1:1, 20 L).
- Dry at 60 °C under vacuum.
- Yield (2ⁿᵈ crop): 0.41 kg (12%).

20.8.2 CYCLOADDITION REACTION

An intermolecular cycloaddition between glyoxal **85** and diamine acetate **86** was realized in the presence of hydrochloric acid at 45 °C in *N*-methylpyrrolidine (NMP) (Equation 20.21).³² Under these reaction conditions, hydrolysis of side product **88** was observed in 5%, whose removal proved problematic.

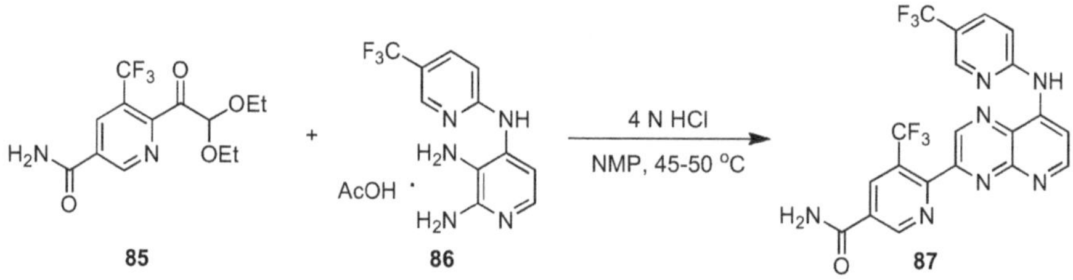

(20.21)

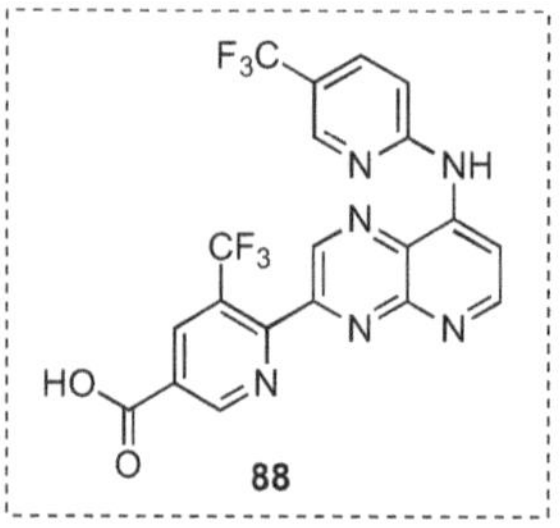

A solution to this problem was to convert **88** into the product **87** by treatment of the crude product **87** containing **88** with 1,1-carbonyldiimidazole (CDI) and ammonia.

Procedure

- Charge 4.23 kg (12.9 mol) of **86** to a solution of **85** (3.16 kg, 93 wt%, 9.19 mol) in NMP (12 L), followed by adding 12 L of 4 N HCl.

- Heat the batch at 45 °C for 40 h and at 50 °C for 8 h.
- Cool to 12 °C overnight.
- Add water (15 L) over 1.5 h and stir for 30 min.
- Filter, wash with THF/water (1:1, 2×11.8 L), and dry at 45–50 °C under vacuum.
- Yield of crude **87**: 3.73 kg.
- Slurry 3.73 kg (6.95 mol) of crude **87** in THF (28 kg).
- Heat to 52 °C.
- Charge 1.13 kg (6.95 mol) of CDI portionwise over 10 min.
- Heat the batch at 55 °C for 1 h and at 58 °C for 3 h.
- Cool the resulting slurry to 15 °C.
- Charge 1.67 kg (29.66 mol) of 16 N aqueous NH_4OH solution over 15 min at <25 °C.
- Heat the batch to 45 °C.
- Add water (44.5 kg) over 1.5 h.
- Filter, wash with water/THF (3:1, 14.8 L), and dry at 50 °C under vacuum.
- Yield of **87**: 3.38 kg (74%).

20.9 REMOVAL OF OTHER IMPURITIES

20.9.1 REMOVAL OF POLYMERIC MATERIAL

The deprotection of 4-methoxyphenethyl group using aqueous HBr (48%) and AcOH was accomplished at 80 °C, giving a relatively clean conversion to amide **90** (Equation 20.22).[33] A problem of using this protocol was that the 4-methoxystyrene by-product polymerized under the reaction conditions.

In order to reject the polymeric material from the product stream, toluene was used as an antisolvent to precipitate the product **90** as acetic acid salt. A subsequent filtration rejected the poly(4-methoxy)styrene into the mother liquor.

Procedure

- Charge 15.42 kg (97.19 wt%, 23.73 mol) of **89** to a reaction vessel, followed by adding 65.2 kg (1085.6 mol, 45.8 equiv) of AcOH and 46.29 kg (131.7 mol, 5.55 equiv) of 48% aqueous HBr.
- Heat the mixture at 80 °C for 7 h.
- Add toluene (64.9 kg) at 80 °C.
- Cool to 25 °C.
- Add a solution of NaOH (11.0 kg, 275.0 mol, 11.59 equiv) in water (75.0 kg) at below 70 °C.
- Heat the mixture to 75 °C.
- Charge a slurry of **90** acetate seed crystals (30.0 g) in toluene (1.5 L).
- Hold the batch at 75 °C for 30 min.
- Cool to 25 °C over 2 h and age for 3 h.
- Filter and rinse with water (61.8 kg) and toluene (26.7 kg).
- Recharge the wet cake into a clean reactor.
- Charge AcOH (35.4 kg).

- Heat the batch to 75 °C to form a solution.
- Add toluene (80.0 kg) at 65–75 °C.
- Charge a slurry of **90** acetate seed crystals (15.0 g) in toluene (1.5 L).
- Hold the batch at 75 °C for 30 min.
- Cool to 25 °C over 2 h and hold at 25 °C for 1 h.
- Filter and wash with toluene (26.7 kg).
- Dry at 79 °C under vacuum.
- Yield of **90**·AcOH: 9.16 kg (67%).

Alternatively, to address the polymerization issue, the addition of an additive approach was applied and anisole was chosen as the 4-methoxyphenethyl carbocation scavenger. Ultimately, the reaction was conducted in 85% aqueous H_3PO_4 in the presence of anisole at 100 °C, giving a clean and complete conversion to **90**. The product **90** was isolated simply by the addition of methyl ethyl ketone (MEK), then water followed by filtration.

Procedure

- Charge 30.0 kg (92.68 wt%, 42.79 mol) of **89** into a reaction vessel, followed by adding 89.4 kg (826.7 mol, 19.3 equiv) of anisole and 151.8 kg (1316.6 mol, 30.8 equiv) of H_3PO_4.
- Heat the batch at 100 °C for 75 min.
- Cool to 80 °C.
- Add MEK (48.3 kg) followed by adding water (150 kg) at 75–83 °C over 1 h.
- Hold the batch at 80 °C for 30 min.
- Cool to 20–25 °C over 1 h and hold for 1 h.
- Filter and wash with water (90 kg), then IPA (70.8 kg).
- Dry at 70 °C with a nitrogen sweep.
- Yield of **90**: 26.17 kg (96%).

20.9.2 USE OF SODIUM PERIODATE TO REMOVE DIOL

Arylboronic acid **92** was prepared via a palladium-mediated borylation with bis(pinacolato)diboron [$B_2(pin)_2$] (Scheme 20.9).[34] Using this protocol, the by-product pinacol can induce equilibrium between **92** and boronate ester **93** during the workup.

To circumvent the pinacol exchange, an oxidative workup was developed to convert pinacol into acetone by treatment of the reaction mixture with sodium periodate.

Procedure

Stage (a)
- Charge 28.0 g (0.11 mol, 1.0 equiv) of $B_2(Pin)_2$, 27.1 g (0.28 mol, 2.8 equiv) of KOAc, and 0.75 g (1 mol%) of $PdCl_2(dppf)$ to a solution of **91** (0.11 mol) in MeTHF.

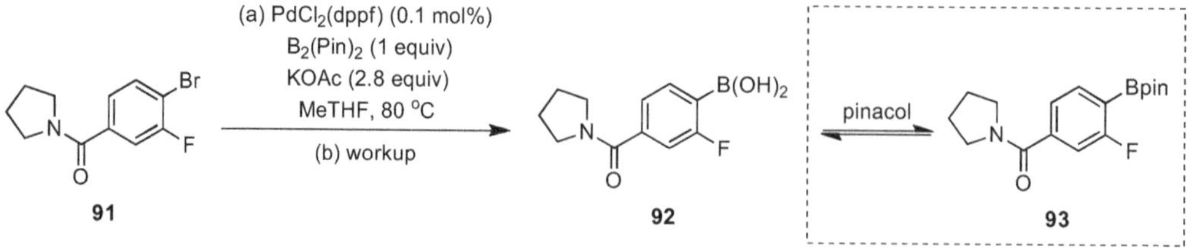

SCHEME 20.9

Stage (a)

- Sparge nitrogen for 5 min.
- Heat to 80 °C and stir for 15 h.

Stage (b)
- Cool the mixture to 20 °C.
- Dilute with water to 10 mL/g.
- Filter through Celite.
- Wash the organic layer with water.
- Filter the organic layer through Darco, wash with 1 N NaOH (10 mL/g) (and discard the organic layer).
- Add MeTHF (10 mL/g) to the combined aqueous layers.
- Cool to 0 °C and adjust pH to 1.5 with 3 N HCl.
- Filter through Celite.
- Add water (10 mL/g) followed by 28.2 g (0.13 mol) of $NaIO_4$.
- Cool to 10 °C and stir for 1.5 h.
- Add 1 N HCl (7 mL/g) and stir for 15 h at 20 °C.
- Wash the organic layer with 20 wt% sodium thiosulfate (5 mL/g), 12 wt% aqueous NaCl (5 mL/g).
- Filter through Celite and concentrate to 4 mL/g.
- Mix the MeTHF solution with heptane (4 mL/g) and stir for 1 h.
- Filter, wash, and dry to give **92** as a white solid.

REMARK

Nucleophilic oxidants such as peroxides are known to oxidatively add to boron, however, electrophilic oxidants, such as $NaIO_4$, are unable to donate an electron pair to boron's empty orbital, thus providing chemoselectivity for the pinacol while leaving the boronic acid functionality intact.

20.9.3 Use of Phenylboronic Acid to Remove Diol

During process development of drug substance **95**, dimeric diol compound **96** was formed from side reactions of methyl aryl ketone intermediate (Equation 20.23)[35]

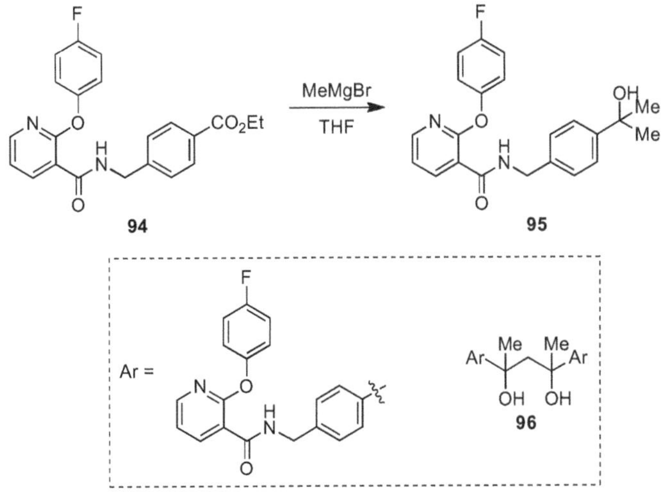

(20.23)

A similar dimeric diol side product **99** was generated during the synthesis of sulfonamide **98**, an advanced intermediate from a drug candidate (Equation 20.24).[35] These diol impurities proved difficult to purge and remained in final drug substance at unacceptably high levels.

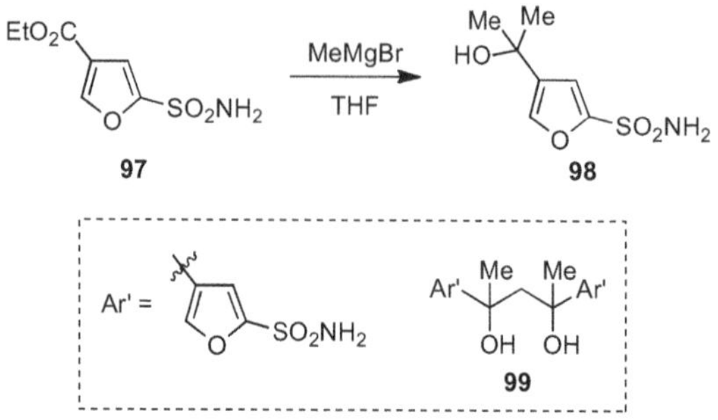

(20.24)

It was found that the 1,3-diol dimeric impurities could be converted into cyclic boronic esters **100** (Equation 20.25) by treating the reaction mixture with phenylboronic acid; the subsequent crystallization allowed for easy purification of the tertiary alcohols leaving the soluble cyclic boronic esters in the mother liquors.

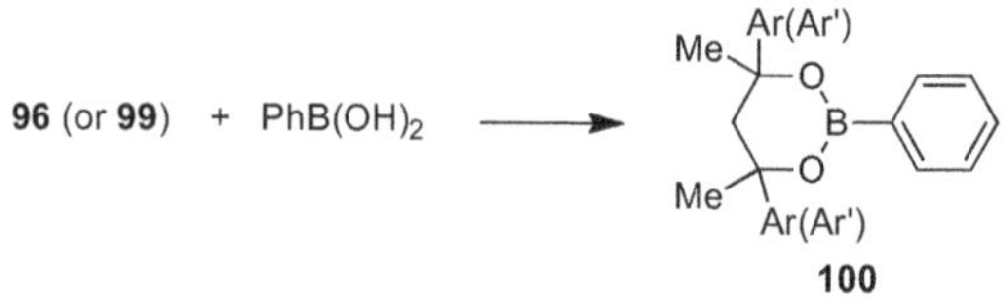

$$\textbf{96 (or 99)} \;+\; PhB(OH)_2 \longrightarrow \quad \textbf{100}$$

(20.25)

REMARK

These diol side products are the result of an alternate reaction pathway which involves the enolization of ketone intermediates, followed by Aldol-type reaction (Scheme 20.10).

Procedure (for purification of **95**)

- Charge 4.97 g (13.07 mmol) of solid **95** (containing 2.6% of **96**) into a 50 mL flask.
- Add 25 mL of toluene, followed by 159 mg (1.3 mmol, 10 mol%) of phenylboronic acid.
- Reflux the content for 20 min.
- Cool to room temperature over 1 h.
- Filter and wash with 5 mL of toluene.
- Yield: 3.13 g of **95** with 98.4% purity (containing 0.03% of **96**).

20.9.4 USE OF SODIUM DITHIONATE TO REDUCE NITRO GROUP

The *N*-amination of pyrrole **101** generates 2–5% of 4-nitrobenzoic acid hydrazide **104** a side product in addition to the desired product **103** (Equation 20.26).[36] Attempts to limit the formation of **104** via a reverse addition of the pyrrole anion of **101** to a solution of **102** gave limited success (reducing the level of **104** to 1–2%).

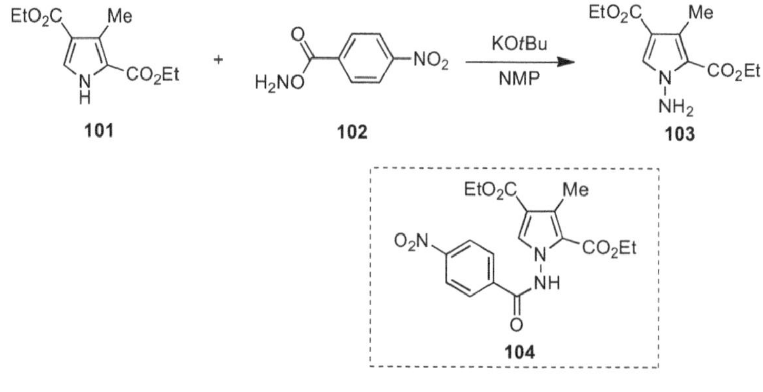

(20.26)

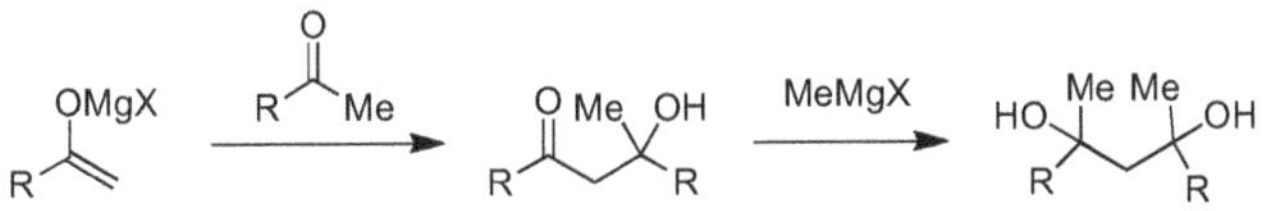

SCHEME 20.10

Instead of suppressing the side reaction, an alternative approach was developed. The addition of sodium dithionate as a reducing agent during the workup converted **104** into a water-soluble aniline **105**, which was rejected into the aqueous phase (Equation 20.27).

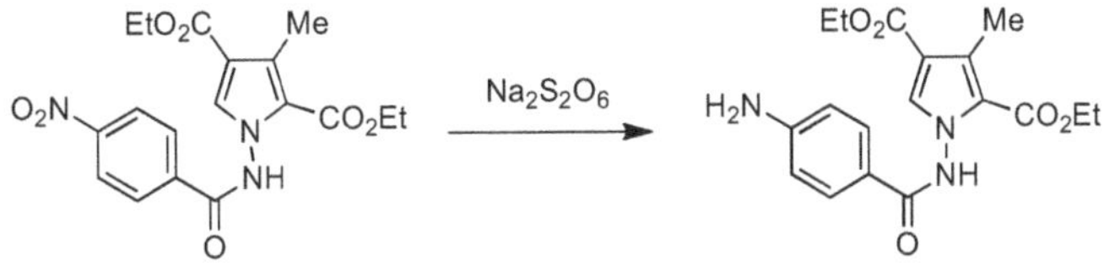

$$(20.27)$$

Procedure

- Charge KO*t*Bu (5.5 kg, 49.0 mol, 1.1 equiv) as solid to a solution of **101** (10 kg, 44.4 mol) in 60 L of NMP.
- Stir at 20–37 °C until all KO*t*Bu dissolved.
- Cool to 29 °C.
- Charge a solution of **102** (8.1 kg, 44.5 mol) in dry THF (30 L, KF<0.1%) at 28–32 °C over 20–30 min.
- Cool the batch to 20–25 °C over 90 min.
- Add a solution of sodium dithionate (6.7 kg, 32.5 mol, 0.7 equiv) in water (93.3 L) at 20–35 °C.
- Stir for 30 min.
- Add toluene (50 L), separate layers.
- Extract the aqueous layer with toluene (2×50 L).
- Wash the combined organic layers with 5% NaHCO$_3$ (60 L) followed by water (60 L).
- Distill at 40–60 °C under reduced pressure to 10 L (and used directly in the subsequent reaction).

DISCUSSION

Solid (KO*t*Bu) addition into a flammable organic solution is not recommended for large-scale operations.

20.9.5 USE OF POLYMERIC RESIN TO REMOVE HYDRAZIDE

Following the coupling reaction of acid chloride **106** with semicarbazide **107**, cyclization of the intermediate **108** furnished a mixture of the desired triazolone **109** and hydrazide **110** in a ratio of 15:1 (Scheme 20.11).[37]

Removal of hydrazide **110** could be realized by the addition of amberlyst 15 sulfonic acid resin followed by simple filtration.

Procedure

Stage (a)
- Charge 1.169 kg (4.247 mol, 1.0 equiv) of **107** into a 22-L flask followed by adding EtOAc (8.75 L).

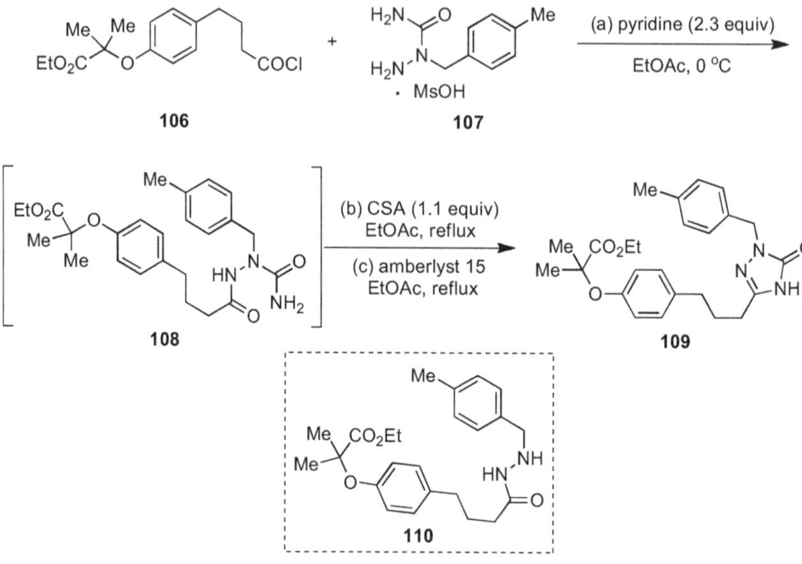

SCHEME 20.11

- Charge 790 mL (9.77 mol, 2.3 equiv) of pyridine.
- Cool the mixture to 0–5 °C.
- Charge a solution of **106** (4.247 mol) in EtOAc (1.0 L) dropwise at <25 °C.
- Stir the reaction mixture at ambient temperature for 1 h.

Stage (b)
- Charge 1.973 kg (8.494 mol, 2.0 equiv) of CSA.
- Stir the batch at reflux for 16.5 h.
- Cool to 20 °C.
- Transfer the batch into 1 N HCl (7.5 L) and stir.
- Separate layers and wash the organic layer with saturated Na_2CO_3 (7.5 L) and water (1.0 L).
- Dry over $MgSO_4$.

Stage (c)
- Transfer the EtOAc solution into a 22-L flask.
- Charge 1.975 kg of amberlyst-15 resin.
- Heat at reflux for 1 h.
- Cool to 20 °C.
- Filter and rinse with EtOAc (2×2.0 L).
- Distill the filtrate under reduced pressure to give a tan solid.
- Dissolve the solid in MTBE (5.0 L) at 45–50 °C.
- Cool to 0–5 °C and hold for 1 h.
- Filter and wash with cold MTBE (1.5 L).
- Dry at 45 °C under vacuum.
- Yield: 1.027 kg (55.2%).

NOTES

1. Bullock, K. M.; Burton, D.; Corona, J.; Corona, J.; Diederich, A.; Glover, B.; Harvey, K.; Mitchell, M. B.; Trone, M. D.; Yule, R.; Zhang, Y.; Toczko, J. F. *Org. Process Res. Dev.* **2009**, *13*, 303.

2. Xu, Y.; Han, B.; Wang, Z.-Q.; Shaw, K. R.; Chenard, B. L.; Maynard, G.; Xie, L. *Org. Process Res. Dev.* **2007**, *11*, 716.

3. Singh, M.; Kanvinde, V. Y. *Adv. Chem. Eng. Nucl. Process Ind.* **1994**, 289.

4. Fujino, K.; Takami, H.; Atsumi, T.; Ogasa, T.; Mohri, S.; Kasai, M. *Org. Process Res. Dev.* **2001**, *5*, 426.

5. Singh, J.; Kim, O. K.; Kissick, T. P.; Natalie, K. J.; Zhang, B.; Crispino, G. A.; Springer, D. M.; Wichtowski, J. A.; Zhang, Y.; Goodrich, J.; Ueda, Y.; Luh, B. Y.; Burke, B. D.; Brown, M.; Dutka, A. P.; Zheng, B.; Hsieh, D.-M.; Humora, M. J.; North, J. T.; Pullockaran, A. J.; Livshits, J.; Swaminathan, S.; Gao, Z.; Schierling, P.; Ermann, P.; Perrone, R. K.; Lai, M. C.; Gougoutas, J. Z.; DiMarco, J. D.; Bronson, J. J.; Heikes, J. E.; Grosso, J. A.; Kronenthal, D. R.; Denzel, T. W.; Mueller, R. H. *Org. Process Res. Dev.* **2000**, *4*, 488.

6. Píša, O.; Rádl, S.; Čerňa, I.; Šembera, F. *Org. Process Res. Dev.* **2022**, *26*, 915.

7. Kopach, M. E.; Murray, M. M.; Braden, T. M.; Kobierski, M. E.; Williams, O. L. *Org. Process Res. Dev.* **2009**, *13*, 152.

8. Ford, J. G.; Pointon, S. M.; Powell, L.; Siedlecki, P. S. *Org. Process Res. Dev.* **2010**, *14*, 1078.

9. Lukin, K.; Kishore, V.; Gordon, T. *Org. Process Res. Dev.* **2013**, *17*, 666.

10. Batesky, D. C.; Goldfogel, M. J.; Weix, D. J. *J. Org. Chem.* **2017**, *82*, 9931.

11. Magnus, N. A.; Aikins, J. A.; Cronin, J. S.; Diseroad, W. D.; Hargis, A. D.; LeTourneau, M. E.; Parker, B. E.; Reutzel-Edens, S. M.; Schafer, J. P.; Staszak, M. A.; Stephenson, G. A. Tameze, S. L.; Zollars, L. M. H. *Org. Process Res. Dev.* **2005**, *9*, 621.

12. Ainge, D.; Ennis, D.; Gidlund, M.; Stefinovic, M.; Vaz, L.-M. *Org. Process Res. Dev.* **2003**, *7*, 198.

13. Oh, L. M.; Wang, H.; Shilcrat, S. C.; Herrmann, R. E.; Patience, D. B.; Spoors, P. G.; Sisko, J. *Org. Process Res. Dev.* **2007**, *11*, 1032.

14. Yue, T.-Y.; Mcleod, D. D.; Albertson, K. B.; Beck, S. R.; Deerberg, J.; Fortunak, J. M.; Nugent, W. A.; Radesca, L. A.; Tang, L.; Xiang, C. D. *Org. Process Res. Dev.* **2006**, *10*, 262.

15. Bernhardson, D.; Brandt, T. A.; Hulford, C. A.; Lehner, R. S.; Preston, B. R.; Price, K.; Sagal, J. F.; St. Pierre, M. J.; Thompson, P. H.; Thuma, B. *Org. Process Res. Dev.* **2014**, *18*, 57.

16. Sharif, E. U.; Miles, D. H.; Rosen, B. R.; Jeffrey, J. L.; Debien, L. P. P.; Powers, J. P.; Leleti, M. R. *Org. Process Res. Dev.* **2020**, *24*, 1254.

17. Berliner, M. A.; Dubant, S. P. A.; Makowski, T.; Ng, Karl, Sitter, B.; Wager, C.; Zhang, Y. *Org. Process Res. Dev.* **2011**, *15*, 1052.

18. Anson, M. S.; Clark, H. F.; Evans, P.; Fox, M. E.; Graham, J. P.; Griffiths, N. N.; Meek, G.; Ramsden, J. A.; Roberts, A. J.; Simmonds, S.; Walker, M. D.; Willets, M. *Org. Process Res. Dev.* **2011**, *15*, 389.

19. Watson, T. J.; Ayers, T. A.; Shah, N.; Wenstrup, D.; Webster, M.; Freund, D.; Horgan, S.; Carey, J. P. *Org. Process Res. Dev.* **2003**, *7*, 521.

20. Lafrance, D.; Caron, S. *Org. Process Res. Dev.* **2012**, *16*, 409.

21. Dorow, R. L.; Herrinton, P. M.; Hohler, R. A.; Maloney, M. T.; Mauragis, M. A.; McGhee, W. E.; Moeslein, J. A.; Strohbach, J. W.; Veley, M. F. *Org. Process Res. Dev.* **2006**, *10*, 493.

22. The imine was proposed in the original paper as the end product during trapping the amine with pivalaldehyde.

23. (a) For control of PGI in APIs, see review: Robinson, D. I. *Org. Process Res. Dev.* **2010**, *14*, 946. (b) Snodin, D. J. *Org. Process Res. Dev.* **2011**, *15*, 1234.

24. Butters, M.; Catterick, D.; Craig, A.; Curzons, A.; Dale, D.; Gillmore, A.; Green, S. P.; Marziano, I.; Sherlock, J.-P.; White, W. *Chem. Rev.* **2006**, *106*, 3002.

25. Challenger, S.; Dessi, Y.; Fox, D. E.; Hesmondhalgh, L. C.; Pascal, P.; Pettman, A. J.; Smith, J. D. *Org. Process Res. Dev.* **2008**, *12*, 575.

26. Damon, D. B.; Dugger, R. W.; Hubbs, S. E.; Scott, J. M.; Scott, R. W. *Org. Process Res. Dev.* **2006**, *10*, 472.

27. Snodin, D. J. *Org. Process Res. Dev.* **2010**, *14*, 960.

28. Peters, R.; Waldmeier, P.; Joncour, A. *Org. Process Res. Dev.* **2005**, *9*, 508.

29. Shu, L.; Wang, P.; Radinov, R.; Dominique, R.; Wright, J.; Alabanza, L. M.; Dong, Y. *Org. Process Res. Dev.* **2013**, *17*, 114.

30. Humphrey, G. R.; Pye, P. J.; Zhong, Y.-L.; Angelaud, R.; Askin, D.; Belyk, K. M.; Maligres, P. E.; Mancheno, D. E.; Miller, R. A.; Reamer, R. A.; Weissman, S. A. *Org. Process Res. Dev.* **2011**, *15*, 73.

31. Zhao, Y.; Zhu, L.; Provencal, D. P.; Miller, T. A.; O'Bryan, C.; Langston, M.; Shen, M.; Bailey, D.; Sha, D.; Palmer, T.; Ho, T.; Li, M. *Org. Process Res. Dev.* **2012**, *16*, 1652.

32. Cleator, E.; Scott, J. P.; Avalle, P.; Bio, M. M.; Brewer, S. E.; Davies, A. J.; Gibb, A. D.; Sheen, F. J.; Stewart, G. W.; Wallace, D. J.; Wilson, R. D. *Org. Process Res. Dev.* **2013**, *17*, 1561.

33. Reeves, J. T.; Fandrick, D. R.; Tan, Z.; Song, J. J.; Rodriguez, S.; Qu, B.; Kim, S.; Niemeier, O.; Li, Z.; Byrne, D.; Campbell, S.; Chitroda, A.; DeCroos, P.; Fachinger, T.; Fuchs, V.; Gonnella, N. C.; Grinberg,

N.; Haddad, N.; Jäger, B.; Lee, H.; Lorenz, J. C.; Ma, S.; Narayanan, B. A.; Nummy, L. J.; Premasiri, A.; Roschangar, F.; Sarvestani, M.; Shen, S.; Spinelli, E.; Sun, X.; Varsolona, R. J.; Yee, N.; Brenner, M.; Senanayake, C. H. *J. Org. Chem.* **2013**, *78*, 3616.

34. Zhu, J.; Razler, T. M.; Xu, Z.; Conlon, D. A.; Sortore, E. W.; Fritz, A. W.; Demerzhan, R.; Sweeney, J. T. *Org. Process Res. Dev.* **2011**, *15*, 438.

35. Tucker, J. L.; Couturier, M.; Leeman, K. R.; Hinderaker, M. P.; Andresen, B. M. *Org. Process Res. Dev.* **2003**, *7*, 929.

36. Shi, Z.; Kiau, S.; Lobben, P.; Hynes, J., Jr.; Wu, H.; Parlanti, L.; Discordia, R.; Doubleday, W. W.; Leftheris, K.; Dyckman, A. J.; Wrobleski, S. T.; Dambalas, K.; Tummala, S.; Leung, S.; Lo, E. *Org. Process Res. Dev.* **2012**, *16*, 1618.

37. Braden, T. M.; Coffey, D. S.; Doecke, C. W.; LeTourneau, M. E.; Martinelli, M. J.; Meyer, C. L.; Miller, R. D.; Pawlak, J. M.; Pedersen, S. W.; Schmid, C. R.; Shaw, B. W.; Staszak, M. A.; Vicenzi, J. T. *Org. Process Res. Dev.* **2007**, *11*, 431.

21 Pharmaceutical Salts

Investigation of pharmaceutical salts has been recognized as an essential task, as the selection of an appropriate salt provides scientists with the opportunity to modify the characteristics of the potential drug substance, such as the aqueous solubility, stability, bioavailability, crystal form, and mechanical properties. Approximately, half of the drug molecules used in medicine are administered as their corresponding salts.

Most pharmaceutical salts are developed to improve the aqueous solubility of drug substances. In general, the aqueous solubility of the drug substances or their salts in the range of 0.1–1.0 mg/mL[1] can meet the dissolution requirements for standard solid and/or oral dosage forms of drugs with moderate to good potency. Occasionally, salt formation is also used to decrease the solubility of basic drug molecules in order to control the drug release rate. Pamoic acid, insoluble in many common solvents, including water, ethanol, and ether, can be utilized to diminish the dissolution of active drugs.[2]

In the development of pharmaceutical salts, the first information, generated for each candidate, is the pKa value of each ionizable group in the molecule. Knowledge of the pKa value would facilitate the selection of potential salt-forming agents (counter-ions). Generally, in order to form a stable salt, it is widely accepted that there should be a minimum three-unit difference in the pKa values between the functional group and its counter-ion, especially when the drug substance is a weak acid or base.[1] Nitrogen atoms play a critical role in salt formation, and the majority (>90%) of small molecule drug substances (<550 MW) contain at least one nitrogen atom and an aromatic ring.[3]

Upon the completion of data collection, the selection of salts is typically based on thermal and light stability, moisture sensitivity, melting point, crystallinity, solubility, ratio (of free base to acid or free acid to base), etc. In addition to good thermal and light stability, the desired salt should have a relatively high melting point and favorable crystallinity. Apparently, the high ratio of free base (or free acid) to acid (or base) is preferred in order to achieve a relatively high potency.

REMARKS

Apart from the transformation of API into its salt in order to overcome the solubility problems during the formulation, another approach is to convert the API into a more water-soluble prodrug by attaching hydrophilic functional group(s) such as the phosphate group. This strategy was applied when the formulation of etoposide **1**, an important drug in the treatment of leukemia, testicular cancer, and small-cell lung cancer. Thus, Etoposide **1** was converted to its prodrug **2** (Figure 21.1)[4] by attaching a phosphate group to the phenolic hydroxyl group. The resulting prodrug, Etoposide 4'-phosphate **2**, has excellent water solubility and activity comparable to **1** in vivo.

21.1 SALTS OF BASIC DRUG SUBSTANCES

Most active pharmaceutical ingredients (APIs) contain basic nitrogen atoms which are poised for the formation of their corresponding salts with appropriate acids. Acids including inorganic acids and organic acids that are frequently used for pharmaceutical salt formation are listed in Table 21.1.

21.1.1 Hydrochloride Salts

Among pharmaceutical salts, hydrochloride salts are the most commonly used salts for basic pharmaceutical compounds. Hydrochloride salts are typically formed through the addition of the requisite amount of hydrochloric acid (HCl) to the solution of the API-free base. These can be carried

DOI: 10.1201/9781003288411-21

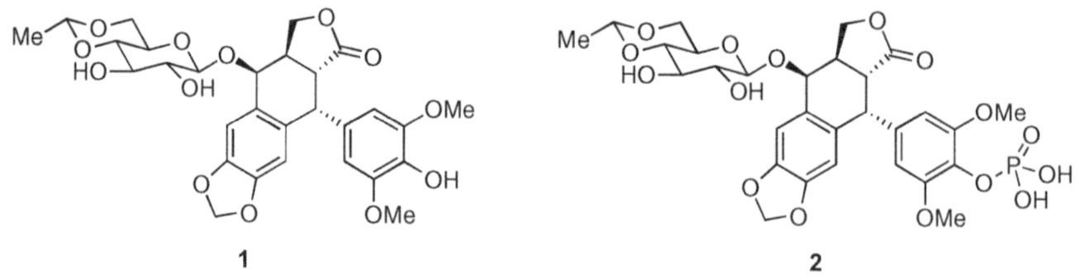

FIGURE 21.1 Structures of Etoposide **1** and Etoposide 4'-phosphate **2**.

out under a range of conditions to suit the maximization of yield, purity, and operational simplicity. Besides HCl, other amine hydrochlorides, such as ammonium chloride, can be used as HCl surrogates.

Fasudil hydrochloride **5**, as an effective inhibitor of Rho-kinase, was used to inhibit the contraction of blood vessels to improve cerebral microcirculation. **5** was prepared by conversion of the crude free base **4** to hydrochloride salt with concentrated hydrochloric acid in ethanol at 0–5 °C followed by recrystallization from methanol/isopropyl ether (Scheme 21.1).[5]

Procedure

- Dissolve crude **4** (as a light-yellow oil, obtained from a reaction of **3** in 3.2 mol scale) in EtOH (2.4 L).
- Add conc. HCl to pH 5.5–6.0 at 0–5 °C.
- Concentrate the resulting mixture under reduced pressure.
- Recrystallize the resultant crude **5** from MeOH/isopropyl ether (2:1).
- Yield: 0.82 kg (78.1% yield over 3 steps and 99.94% purity).

Occasionally, a trace of excess HCl remained in the isolated salts (as an impurity) creating an unexpected isolation problem. Therefore, the removal of the trace amount of HCl from HCl salts is necessary. Reaction (21.1)[6] describes the preparation of HCl salts of 3-amino-quinazoline-2,4-diones, **8** and **9**, potent gyrase/topoisomerase inhibitors as anti-bacterial agents. It was found that the removal of the excess HCl was critical in the isolation of the final HCl salt. The presence of excess HCl made the final product extremely hygroscopic. Thus, the excess HCl was removed by treatment of HCl salts with propylene oxide.

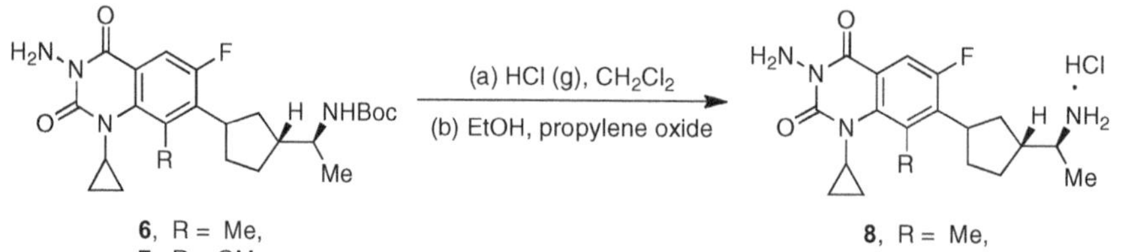

6, R = Me,
7, R = OMe

(a) HCl (g), CH$_2$Cl$_2$
(b) EtOH, propylene oxide

8, R = Me,
9, R = OMe

(21.1)

Procedure

- Charge HCl (g) (by bubbling) into a solution of crude **6** (294 g, 0.89 mol) in CH$_2$Cl$_2$ (2.4 L) at 5–15 °C over 3.5 h.
- Distill azeotropically the resulting mixture under reduced pressure (with 600 mL of toluene) to dryness.
- Dissolve the resulting solid in EtOH (2.6 L).
- Cool to 15 °C.

TABLE 21.1
Commonly Used Acids for Salt Formation

Acid	Formula	Appearance	pK_a	Density (g/cm³)	Melting point, °C	Boiling point, °C	Solubility (in H₂O, g/100 mL)
AcOH Acetic acid	$C_2H_4O_2$	Colorless liquid	4.76	1.049	16–17	118–119	Miscible
Benzene-sulfonic acid	$C_6H_6O_3S$	Colorless crystalline solid	−2.8	1.32	44 (hydrate) 51 (anhydrous)	190	soluble
Camphor-sulfonic acid	$C_{10}H_{16}O_4S$	Solid	1.2	–	195 (Dec)	–	–
Citric acid	$C_6H_8O_7$	Solid	3.13 4.16 6.39	1.665	156	310	147.76 (20 °C)
HCO₂H Formic acid	CH_2O_2	Colorless fuming liquid	3.77	1.220	8.4	100.8	Miscible
Fumaric acid	$C_4H_4O_4$	White solid	3.03 4.44	1.635	287	–	0.63
HBr Hydrogen bromide	HBr	Colorless gas	−9.0	3.6452 g/L	−86.9	−66.8	221 (0 °C) 204 (15 °C) 193 (20 °C) 130 (100 °C)
HCl Hydrogen chloride	HCl	Colorless gas	−7.0	1.490 g/L	−114.22	−85.05	82.3 (0 °C) 72.0 (20 °C) 56.1 (60 °C)
Maleic acid	$C_4H_4O_4$	White solid	1.9 6.07	1.59	135 (Dec)	–	78.8
CH₃SO₃H Methane-sulfonic acid	CH_4O_3S	Liquid	−1.9	1.48	17–19	167/10 mmHg	soluble
(CO₂H)₂ Oxalic acid	$C_2H_2O_4$	White crystals	1.25 4.14	1.90 (anhydrous) 1.653 (dihydrate)	102–103	–	14.3
H₃PO₄ Phosphoric acid	H_3PO_4	White solid or colorless viscous liquid	2.48 7.198 12.319	1.885 (liquid) 1.685 (85% solution) 2.030 (crystal)	42.35 (anhydrous) 29.32 (hemihydrate)	158	446 (15 °C) Miscible (42.3 °C)
Succinic acid	$C_4H_6O_4$	White solid	4.2 5.6	1.56	184	235	5.8
H₂SO₄ Sulfuric acid	H_2SO_4	Colorless and odorless liquid	−3 1.99	1.84	10	337	Miscible
L-tartaric acid	$C_4H_6O_6$	White crystalline solid	2.89 4.40	1.79	171–174	–	133
Toluene-sulfonic acid	$C_7H_8O_3S$	White solid	−2.8 (water) 8.5 (acetonitrile)	1.24	103–106	140/20 mmHg	67

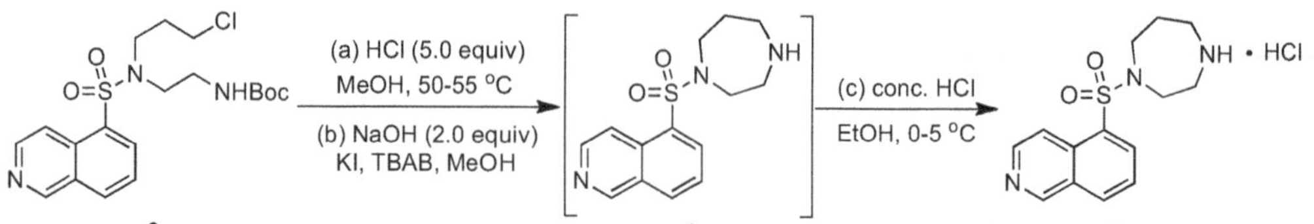

SCHEME 21.1

- Charge 165 mL (2.4 mol, 3.4 equiv) of propylene oxide portionwise at 15 °C to room temperature over 30 min.
- Stir the resulting mixture at room temperature for 4 h.
- Add MTBE (1.3 L) and stir for 1 h.
- Filter, rinse with MTBE (700 mL), and dry at room temperature in the air.
- Dissolve the resulting mustard-colored powder (211 g) in hot water (2.5 L) and filter.
- Add IPA (6.5 L) to the filtrate.
- Filter to give Crop 1.
- Distill the mother liquor to 1.5 L and add IPA (1 L).
- Filter to give Crop 2.
- Combine Crops 1 and 2 and slurry in IPA (1 L) for 30 min.
- Filter and dry at 75 °C/15 mmHg for 3.5 days.
- Yield of **8**: 164 g.

Some pharmaceutical HCl salts present stability issues which may lead to the decomposition of the APIs under certain conditions. This instability of hydrochloride salts should be addressed during the manufacturing process and storage, otherwise, it may lead to the degradation of APIs resulting in impurities. Nitrosamines exhibit mutagenic and carcinogenic activity; therefore, strict regulations and guidelines have been applied to control their presence in pharmaceutical products. Reports[7] discuss the reactivity of N-nitroso-N,N-dimethylamine (NDMA), possible pathways for its formation, and various strategies to mitigate its formation. In September 2019, NDMA was found within ranitidine products in low levels from a number of manufacturers. Investigations by GlaxoSmithKline (GSK) into the cause of NDMA showed that NDMA did not present in freshly prepared batches of drug substances and demonstrated that NDMA was formed through an intermolecular degradation reaction of ranitidine molecules that occurred primarily in the solid state.[8] Scheme 21.2 presents a possible reaction pathway for NDMA formation in the solid ranitidine drug substance.

21.1.2   HEMISULFATE SALT

Lagociclovir valactate **15**, a prodrug of **16**, is a drug candidate with high oral bioavailability in humans and potent activity against hepatitis B virus (HBV) (Figure 21.2).[9]

The final stage of the synthesis consisted of catalytic deprotection of Cbz group in **17** in a mixture of ethyl acetate and acetic acid (Scheme 21.3).[9] To obtain the desired hemisulfate salt, a salt exchange approach was used by treating the hydrogenolysis reaction mixture containing **15** as its acetate salt with sulfuric acid in 2-propanol to give the hemisulfate salt of **15** in 73% yield over 2 steps.

Procedure

Stage (a)
- Charge 4.53 kg (7.9 mol) of **17** to a mixture of EtOAc (130 L) and AcOH (45 L).
- Warm the mixture to 26–30 °C.

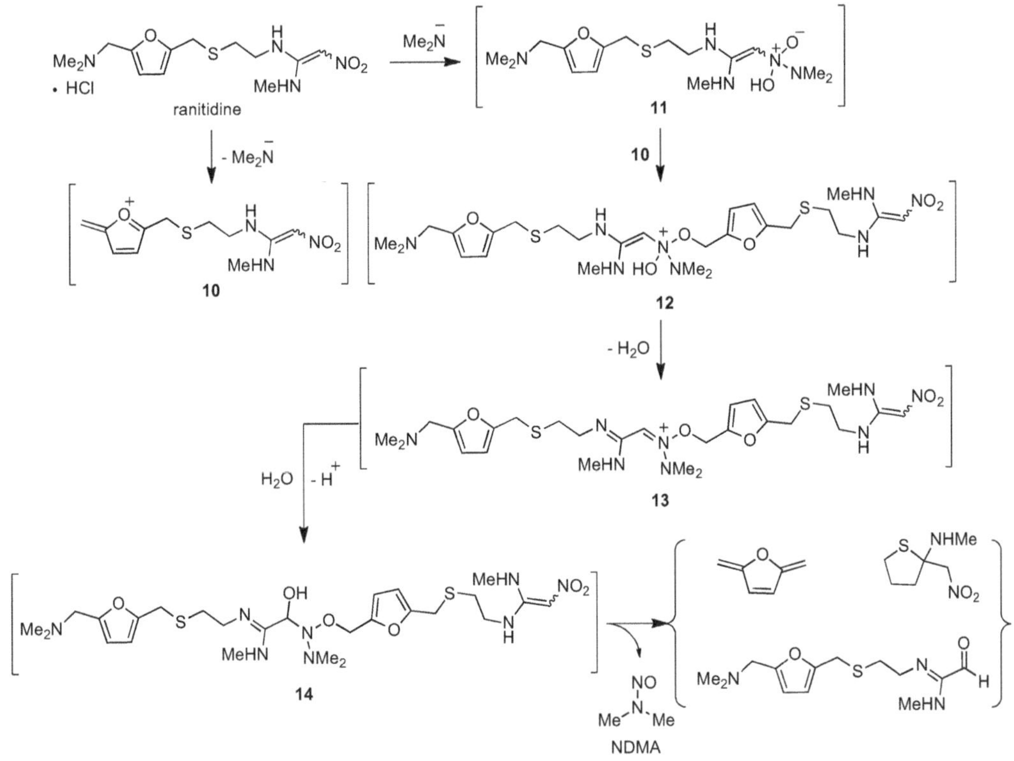

SCHEME 21.2

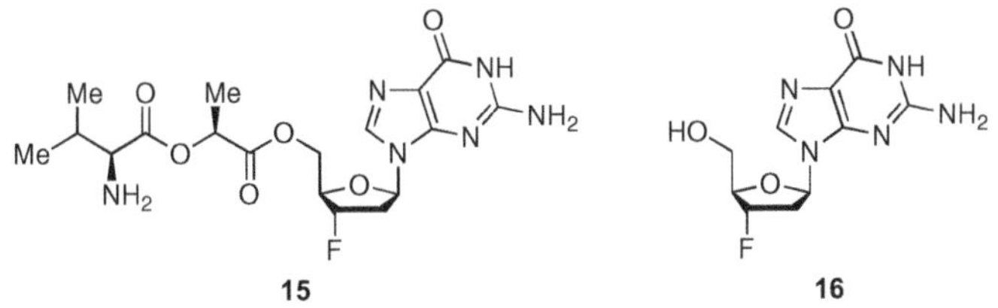

15   16

FIGURE 21.2   The structures of Lagociclovir Valactate **15** and **16**.

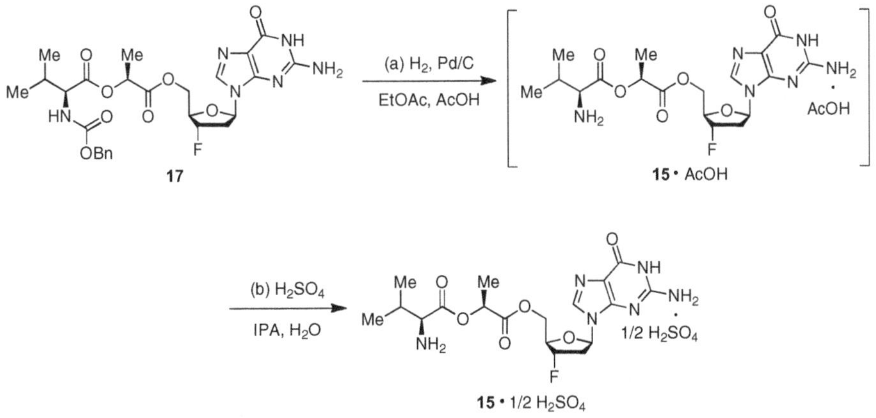

SCHEME 21.3

- Charge 3.9 kg (3.6 mol, 46 mol%) of 10% Pd/C.
- Stir at 30 °C under 3 bar of H_2 for 3 h.
- Filter and rinse with a mixture of EtOAc/AcOH (40 L, 1:3).
- Distill at 55–60 °C/20 mbar until 4–5 L remained.
- Add water (4.4 L) and stir at 22–25 °C for 15 min.
- Add 0.5 kg (50%) aqueous sodium sulfide.
- Stir for 5 min and hold for 15 min without stirring.
- Add a suspension of charcoal (40 g) in water (100 mL).
- Stir for 20 min.
- Filter and transfer the filtrate into a glass-lined reactor.

Stage (b)
- Add IPA (5.4 kg) followed by 2.47 kg (5.04 mol) of 20% aqueous sulfuric acid to pH 2.2–2.3.
- Add IPA (14 L) over 15 min.
- Add aqueous sulfuric acid (150 g) to adjust to pH 2.2–2.3 and stir for 2.5 h.
- Filter and wash with a mixture of IPA/water (3.4 L, 3:1), then with IPA (9 L).
- Dry at 50 °C/8 mbar for 8 h.
- Yield: 3.1 kg (73%) as a white powder.

Note

Aqueous sodium sulfide was used to reduce the level of the residual palladium.

21.1.3 Citric Acid Salt

Compound **18** is a potent 5-HT_4 partial agonist for the treatment of Alzheimer's disease. The physical properties of the free base **18** were less than optimal for pharmaceutical purposes. With a melting point of only 92.6 °C, dry milling was found to be impossible due to the plastic-like nature of the solid, making it difficult to control the particle size for tablet manufacture. Conversion of the free base **18** to the hemi-citrate salt **19** improved the physical properties allowing for easy control of the particle size through milling (Equation 21.2).[10]

$$(21.2)$$

The formation of the hemi-citrate salt **19** also improved the API's stability with a melting point of 165.9 °C.

Due to the low solubility of the gefapixant free base, the preparation of its citrate salt **20** was initially carried out via a slurry-to-slurry process (Route I, Scheme 21.4).[11] This process suffered poor quality control, wherein impurities were not well rejected, and unreacted free base often persisted in the citrate produced.

In order to address the solubility issue, a salt metathesis approach was developed (Route II, Scheme 21.4). The gefapixant free base was transiently converted to a highly soluble glycolate salt

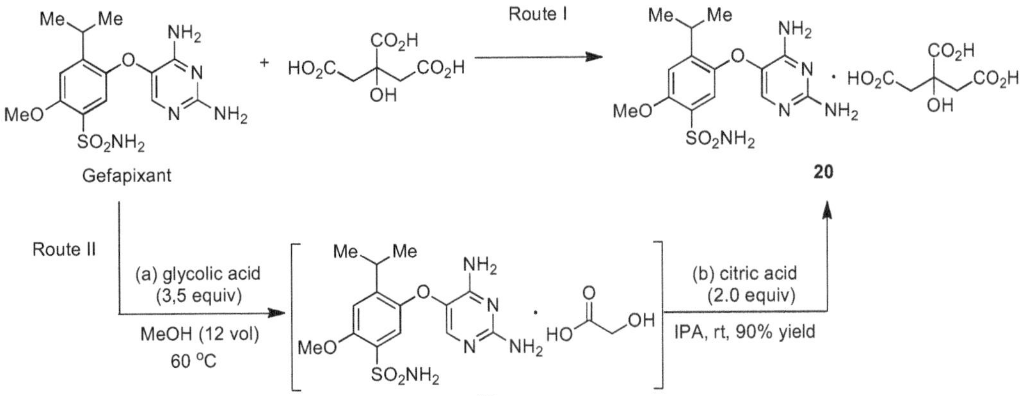

SCHEME 21.4

21, which enabled a polish filtration of the resulting glycolate salt solution prior to salt metathesis. Subsequently, a direct crystallization of the citrate salt **20** was realized by adding citric acid and the desired citrate salt **20** was isolated in a high yield.

REMARKS

The use of IPA significantly reduced the rate of methyl citrate formation in the MeOH/IPA reaction mixture. IPA served also as an anti-solvent to reduce the loss of product into the mother liquor.

21.1.4 VARIOUS PHARMACEUTICAL SALTS

Pharmaceutical salts play a critical role in the biological activity of medicines since they determine the solubility of the drug molecules in biologically relevant fluids. Pharmaceutical salts also contribute to the ability to isolate and purify the APIs and lend to the stability profile of the parent molecules. Furthermore, the choice of salts can lead to a simplification of the drug development process through the minimization of polymorphic forms. Table 21.2 collects various pharmaceutical salts of basic drug substances.

21.2 SALTS OF ACIDIC DRUG SUBSTANCES

For acidic drug substances, salts can be obtained from reactions with either inorganic or organic bases.

21.2.1 USE OF INORGANIC BASES

21.2.1.1 Sodium Salts

Avibactam is a β-lactamase inhibitor and is used for the treatment of various infections by combining with Ceftazidime. The final step in the synthesis of avibactam was salt exchange by swapping the ammonium salt to the sodium salt via a reaction of tetrabutylammonium sulfate **22** with sodium 2-ethylhexanoate in an aqueous ethanol mixture (Equation 21.3).[12]

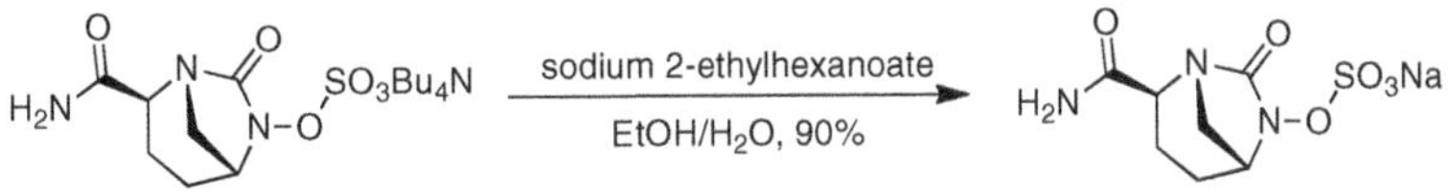

(21.3)

TABLE 21.2

List of API salts.

Entry	Structure	Appearance	Comments
1	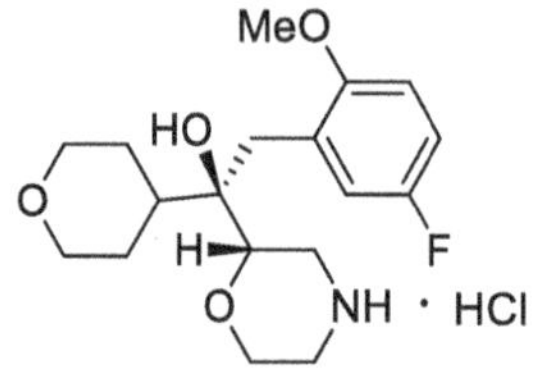	Solid[a]	A selective norepinephrine uptake inhibitor is under development at Eli Lilly for the treatment of depression.
2	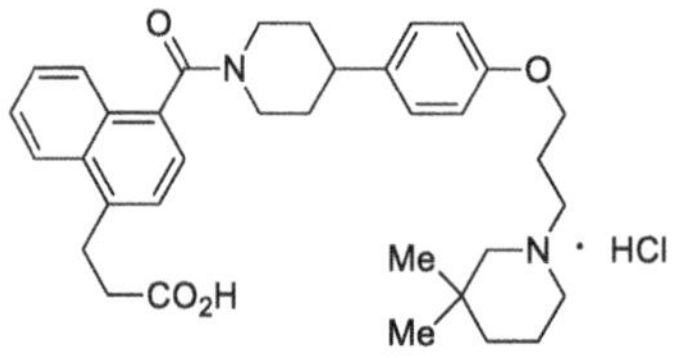	White solid[b]	An H1–H3 antagonist that was developed by GlaxoSmithKline as an oral medicine for the treatment of allergic rhinitis.
3	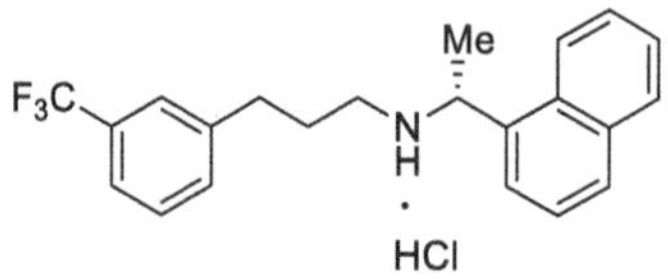	White solid, mp 181–183 °C[c]	Cinacalcet hydrochloride is a calcimimetic agent and calcium-sensing receptor antagonist and is useful for the treatment of secondary hyperparathyroidism in patients with chronic kidney disease and hypocalcaemia in patients with parathyroid carcinoma.
4	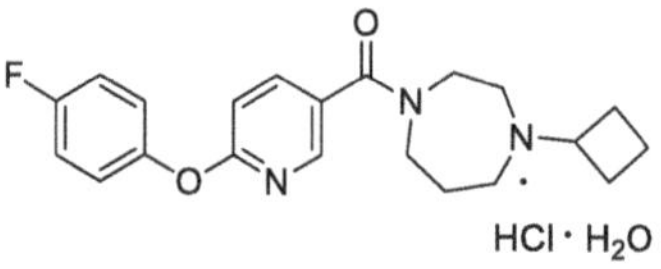	White solid, mp 143.4 °C[d]	The two H_3 receptor antagonists in similar structures could have therapeutic applications for various disease-states, including narcolepsy, ADHD, Alzheimer's disease, and obesity.
5	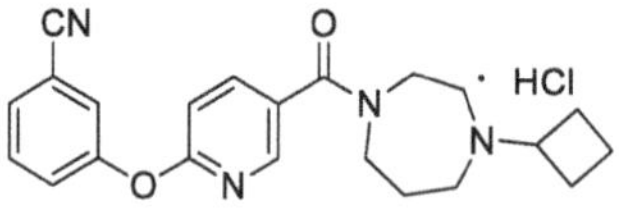	Off-white crystalline solid[15]	
6	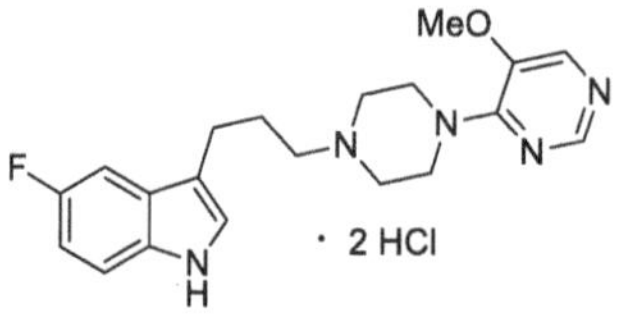	Solid, mp 225 °C[e]	A robust and rapid-onset agent for the treatment of clinical depression.
7	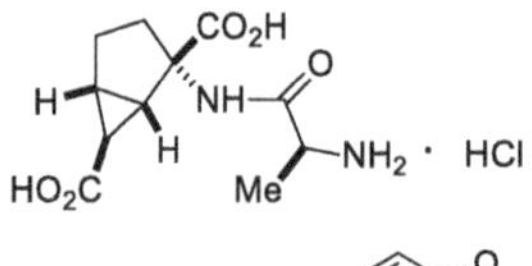	White crystalline solid[f]	A potent, orally active mGluR agonist, developed for the treatment of central nervous system disorders including generalized anxiety disorders.
8	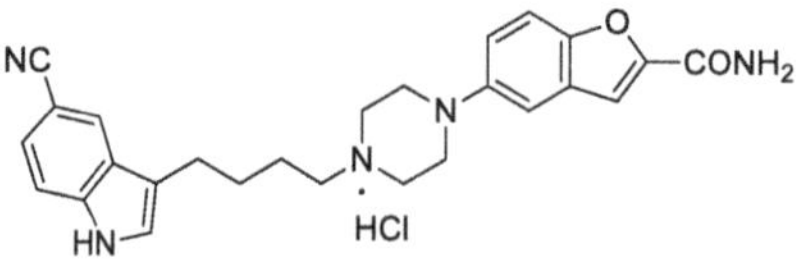	White crystalline solid, mp 234–236 °C[g]	A dual selective serotonin reuptake inhibitor (SSRI) and serotonin 5-HT1A receptor partial agonist for the treatment of major depressive disorder.
9	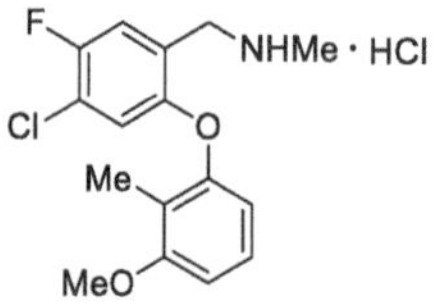	White powder, mp 186.5–188.8 °C[h]	A SRI/5-HT_{2A} antagonist.

(Continued)

TABLE 21.2 (CONTINUED)
List of API salts.

Entry	Structure	Appearance	Comments
10		White solid, mp 275–276 °C[i]	A drug candidate for the treatment of obesity.
11		White solid, mp 299 °C[j]	A potent non-peptidic glycoprotein IIb/IIIa antagonist and consequently inhibits platelet aggregation.
12		Solid[k]	A drug candidate for the treatment of psychiatric disorders such as obsessive-compulsive disorder and attention deficit disorder.
13		Crystalline solid[l]	A drug candidate for the treatment of COPD and asthma.
14		White solid[m]	A clinical candidate for the treatment of cardiovascular disorders.
15		Solid[n]	A potent and competitive M_3 selective receptor antagonist (M3SRA) for oral treatment of urinary incontinence.
16		Solid[o]	A potent p38 MAP kinase inhibitor and for the treatment of autoimmune and inflammatory diseases.
17		Pale brown solid[p]	Approved by FDA in 2002, Eletriptan hydrobromide (trade name is Relpax) is a selective agonist of 5-HT1-receptors and particularly of 5-HT1B/1D receptors and is widely used as an anti-migraine agent.

(*Continued*)

TABLE 21.2 (CONTINUED)
List of API salts.

Entry	Structure	Appearance	Comments
18		Pale yellow solid, mp 111.4 °C[q]	A potent PPARα/γ dual agonistic agent that is used for the treatment of type-2 diabetes.
19		White powder[r]	A prodrug with high oral bioavailability in humans and potent activity against the hepatitis B virus.
20		White to off-white powder[s]	Marketed by Bristol-Myers Squibb and Sanofi-Aventis under the trade name Plavix, clopidogrel bisulfate is a potent oral anti-platelet agent used in the treatment of coronary artery disease and peripheral vascular and cerebrovascular diseases.
21		White crystalline solid[t]	A clinical development candidate for the treatment of chronic pain.
22		Solid[u]	An antagonist of $\alpha_v\beta_3$ for the treatment of cancer.
23		Solid[v]	A PARP inhibitor showed promising results in recent clinical trials for the treatment of advanced solid tumors. The phosphate salt was used for intravenous formulation.
24		Solid[w]	Sitagliptin, a selective, potent DPP-4 inhibitor developed by Merck, is the active ingredient in JANUVIA and JANUMET (a fixed-dose combination with the anti-diabetic agent metformin), both received approval for the treatment of type-2 diabetes by the FDA in 2006.

(Continued)

TABLE 21.2 (CONTINUED)
List of API salts.

Entry	Structure	Appearance	Comments
25	· 2.33 H_3PO_4	White crystalline solid, mp 159.7 °C[x]	A potent, selective, and reversible inhibitor of cathepsin S for the treatment of autoimmune diseases.
26	H_3PO_4 YM758 monophosphate	Solid[y]	YM758 monophosphate is a drug candidate which is useful as a preventive and/or treating agent for diseases of circulatory system such as ischemic heart diseases (e.g., angina pectoris and myocardial infarction), congestive heart failure, arrhythmia, etc.
27		White crystalline solid, mp 303 °C[33]	The camphorsulfonate salt was selected for tablet formulation.
28	· AcOH	White crystal, mp 182–183 °C[z]	A candidate drug that was intended to treat pain and other CNS disorders.
29	· AcOH	Solid[aa]	As a potent and selective antagonist of the platelet glycoprotein IIb/IIIa receptor for the prevention and treatment of a wide variety of thrombotic diseases.
30	· TFA R = H, Et	White solid, mp 152.0–152.1 °C (R = Et)[ab]	Glycoprotein IIb/IIIa receptor antagonists as potent anti-platelet agents; ethyl ester as a prodrug.
31	· $MeSO_3H$	White solid[ac]	A heat shock protein 90 (Hsp90) inhibitor used as a therapeutic target for cancer treatment.
32	· 2 $MeSO_3H$	Off-white solid, mp 288.6 °C[ad]	A PI3K inhibitor, was discovered by Genentech, as an anti-cancer agent.

(Continued)

TABLE 21.2 (CONTINUED)
List of API salts.

Entry	Structure	Appearance	Comments
33	· 2 MeSO₃H	Solid[ae]	A selective inhibitor of p38 mitogen-activated protein kinase for the treatment of human malignancies.
34	· MeSO₃H	White crystalline solid[af]	It is an immuno-stimulant and may serve as a potential adjuvant to standard chemotherapy.
35	· MeSO₃H	Solid[ag]	The mesylate salt was approved by the FDA in 2010 to treat people with metastatic breast cancer.
36	· PhSO₃H	White solid, mp 256.2–259.6 °C[ah]	It was identified by Pfizer Discovery chemists as a promising small molecule CB₁ antagonist for the treatment of obesity.
37	· p-TsOH	Solid[ai]	It was discovered and developed within Pfizer as a potent H₃ antagonist for multiple indications related to neurological disorders including excessive daytime sleepiness and Alzheimer's disease.
38	· p-TsOH	White solid, mp 241–245 °C[aj]	Denagliptin tosylate1 was a dipeptidyl peptidase IV (DPPIV) inhibitor in development for the treatment of type II diabetes.
39	· 2 p-TsOH	Solid[22]	A drug candidate for the treatment of psychiatric disorders such as obsessive-compulsive disorder and attention deficit disorder.

(Continued)

TABLE 21.2 (CONTINUED)
List of API salts.

Entry	Structure	Appearance	Comments
40		Crystals, mp 107–109 °C[ak]	A Ca^{2+} antagonist with potent cardioprotective activity.
41		White crystals, mp 184–186 °C[al]	A growth hormone secretagogue.
42		White solid, mp 1111–114 °C[am]	A potent and selective sphingosine-1-phosphate receptor subtype 1 (S1P1) agonist and potentially for the treatment of diseases or conditions associated with inappropriate immune responses, including transplant rejection and autoimmune diseases such as multiple sclerosis and psoriasis.
43		Solid, mp 103.5 °C[an]	A muscarinic antagonist for the treatment of overactive bladder (is commercialized under the name Toviaz).
44		White solid[ao]	A drug candidate as a hypotensive agent.
45		Pale beige solid, mp 154–155 °C[ap]	A drug candidate for the treatment of various conditions, including depression, anxiety, Fragile X syndrome, addiction, and L-DOPA-induced dyskinesia.
46		Off-white solid, mp 207–213 °C[aq]	A renin inhibitor under clinical studies by Novartis.
47		White to pale yellow solid[ar]	A drug candidate for the treatment of Alzheimer's disease.

TABLE 21.2 (CONTINUED)
List of API salts.

Entry	Structure	Appearance	Comments
48		White crystalline solid, mp 184–196 °C[as]	For the treatment of migraine, anxiety, and age-associated memory impairment.
49		Solid[22]	A drug candidate for the treatment of psychiatric disorders such as obsessive-compulsive disorder and attention deficit disorder.
50		White solid[at]	A drug candidate as a L/T calcium channel blocker potentially for the treatment of hypertension and angina pectoris.
51		White solid, mp 166–167 °C[au]	Studies indicate that the L-tartrate salt is a potent and highly selective H3R antagonist with potent recognitive activity in several animal models suggestive of clinical utility for treatment of attention-deficit hyperactivity disorder (ADHD) or other cognitive disorders.
52		White powder[av]	A drug candidate for the treatment of Alzheimer's disease. The hemigalactarate salt was selected to avoid using hygroscopic hydrochloride salt.
53		Solid[aw]	Sildenafil is the first agent for the treatment of male erectile dysfunction. This new drug was approved for prescription use within the United States and the European Union during 1998.
54		White solid, mp 166.5–168.6 °C[ax]	A drug candidate for potential treatment of the acute and chronic conditions of inflammatory and autoimmune diseases.

(Continued)

TABLE 21.2 (CONTINUED)
List of API salts.

Entry	Structure	Appearance	Comments
55		White crystalline solid[a,y]	A nicotinic partial agonist for nicotine addiction. Varenicline (trade name Chantix in the United States) marketed by Pfizer, usually in the form of tartrate salt.

[a] Slattery, C. N.; Deasy, R. E.; Maguire, A. R.; Kopach, M. E.; Singh, U. K.; Argentine, M. D.; Trankle, W. G.; Scherer, R. B.; Moynihan, H. *J. Org. Chem.* **2013**, *78*, 5955.

[b] Bret, G.; Harling, S. J.; Herbal, K.; Langlade, N.; Loft, M.; Negus, A.; Sanganee, M.; Shanahan, S.; Strachan, J. B.; Turner, P. G.; Whiting, M. P. *Org. Process Res. Dev.* **2011**, *15*, 112.

[c] Tewari, N.; Maheshwari, N.; Medhane, R.; Nizar, H.; Prasad, M. *Org. Process Res. Dev.* **2012**, *16*, 1566.

[d] Pippel, D. J.; Mills, J. E.; Pandit, C. R.; Young, L. K.; Zhong, H. M.; Villani, F. J.; Mani, N. S. *Org. Process Res. Dev.* **2011**, *15*, 638.

[e] Anderson, N. G.; Ary, T. D.; Berg, J. L.; Bernot, P. J.; Chan, Y. Y.; Chen, C.-K.; Davies, M. L.; DiMarco, J. D.; Dennis, R. D.; Deshpande, R. P.; Do, H. D.; Droghini, R.; Early, W. A.; Gougoutas, J. Z.; Grosso, J. A.; Harris, J. C.; Haas, O. W.; Jass, P. A.; Kim, D. H.; Kodersha, G. A.; Kotnis, A. S.; LaJeunesse, J.; Lust, D. A.; Madding, G. D.; Modi, S. P.; Moniot, J. L.; Nguyen, A.; Palaniswamy, V.; Phillipson, D. W.; Simpson, J. H.; Thoraval, D.; Thurston, D. A.; Tse, K.; Polomski, R. E.; Wedding, D. L.; Winter, W. *J. Org. Process Res. Dev.* **1997**, *1*, 300.

[f] Coffey, D. S.; Hawk, M. K. N.; Pedersen, S. W.; Ghera, S. J.; Marler, P. G.; Dodson, P. N.; Lytle, M. L. *Org. Process Res. Dev.* **2004**, *8*, 945.

[g] Hu, B.; Song, Q.; Xu, Y. *Org. Process Res. Dev.* **2012**, *16*, 1552.

[h] Tao, Y.; Widlicka, D. W.; Hill, P. D.; Couturier, M.; Young, G. R. *Org. Process Res. Dev.* **2012**, *16*, 1805.

[i] Ragan, J. A.; Bourassa, D. E.; Blunt, J.; Breen, D.; Busch, F. R.; Cordi, E. M.; Damon, D. B.; Do, N.; Engtrakul, A.; Lynch, D.; McDermott, R. E.; Monggillo, J. A.; O'Sullivan, M. M.; Rose, P. R. *Org. Process Res. Dev.* **2009**, *13*, 186.

[j] Atkins, R. J.; Banks, A.; Bellingham, R. K.; Breen, G. F.; Carey, J. S.; Etridge, S. K.; Hayes, J. F.; Hussain, N.; Morgan, D. O.; Oxley, P.; Passey, S. C.; Walsgrove, T. C.; Wells, A. S. *Org. Process Res. Dev.* **2003**, *7*, 663.

[k] Watson, T. J.; Ayers, T. A.; Shah, N.; Wenstrup, D.; Webster, M.; Freund, D.; Horgan, S.; Carey, J. P. *Org. Process Res. Dev.* **2003**, *7*, 521.

[l] Caine, D. M.; Paternoster, I. L.; Sedehizadeh, S.; Shapland, P. D. P. *Org. Process Res. Dev.* **2012**, *16*, 518.

[m] Han, Z. S.; Xu, Y.; Fandrick, D. R.; Rodriguez, S.; Li, Z.; Qu, B.; Gonnella, N. C.; Sanyal, S.; Reeves, J. T.; Ma, S.; Grinberg, N.; Haddad, N.; Krishnamurthy, D.; Song, J. J.; Yee, N. K.; Pfrengle, W.; Ostermeier, M.; Schnaubelt, J.; Leuter, Z.; Steigmiller, S.; Däubler, J.; Stehle, E.; Neumann, L.; Trieselmann, T.; Tielmann, P.; Buba, A.; Hamm, R.; Koch, G.; Renner, S.; Dehli, J. R.; Schmelcher, F.; Stange, C.; Mack, J.; Soyka, R.; Senanayake, C. H. *Org. Lett.* **2014**, *16*, 4142.

[n] Pramanik, C.; Bapat, K.; Chaudhari, A.; Tripathy, N. K.; Gurjar, M. K. *Org. Process Res. Dev.* **2012**, *16*, 1591.

[o] Yoshida, S.; Hayashi, Y.; Obitsu, K.; Nakamura, A.; Kikuchi, T.; Sawada, T.; Kimura, T.; Takahashi, T.; Mukuta, T. *Org. Process Res. Dev.* **2012**, *16*, 1818.

[p] Kumar, U. S.; Sankar, V. R.; Rao, M. M.; Jaganathan, T. S.; Reddy, R. B. *Org. Process Res. Dev.* **2012**, *16*, 1917.

[q] Lee, H. W.; Ahn, J. B.; Kang, S. K.; Ahn, S. K.; Ha, D.-C. *Org. Process Res. Dev.* **2007**, *11*, 190.

[r] Brodszki, M.; Bäckström, B.; Horvath, K.; Larsson, T.; Malmgren, H.; Pelcman, M.; Wähling, H.; Wallberg, H.; Wennerberg, J. *Org. Process Res. Dev.* **2011**, *15*, 1027.

[s] Aalla, S.; Gilla, G.; Anumula, R. R.; Charagondla, K.; Vummenthala, P. R.; Padi, P. R. *Org. Process Res. Dev.* **2012**, *16*, 1523.

[t] Thiel, O. R.; Bernard, C.; King, T.; Dilmeghani-Seran, M.; Bostick, T.; Larsen, R. D.; Faul, M. M. *J. Org. Chem.* **2008**, *73*, 3508.

[u] Weisenburger, G. A.; Anderson, D. K.; Clark, J. D.; Edney, A. D.; Karbin, P. S.; Gallagher, D. J.; Knable, C. M.; Pietz, M. A. *Org. Process Res. Dev.* **2009**, *13*, 60.

[v] Gillmore, A. T.; Badland, M.; Crook, C. L.; Castro, N. M.; Critcher, D. J.; Fussell, S. J.; Jones, K. J.; Jones, M. C.; Kougoulos, E.; Mathew, J. S.; McMillan, L.; Pearce, J. E.; Rawlinson, F. L.; Sherlock, A. E.; Walton, R. *Org. Process Res. Dev.* **2012**, *16*, 1897.

[w] Hansen, K. B.; Hsiao, Y.; Xu, F.; Rivera, N.; Clausen, A.; Kubryk, M.; Krska, S.; Rosner, T.; Simmons, B.; Balsells, J.; Ikemoto, N.; Sun, Y.; Spindler, F.; Malan, C.; Grabowski, E. J. J.; Armstrong III, J. D. *J. Am. Chem. Soc.* **2009**, *131*, 8798.

(Continued)

TABLE 21.2 (CONTINUED)
List of API salts.

[x] Lorenz, J. C.; Busacca, C. A.; Feng, X.; Grinberg, N.; Haddad, N.; Johnson, J.; Kapadia, S.; Lee, H.; Saha, A.; Sarvestani, M.; Spinelli, E. M.; Varsolona, R.; Wei, X.; Zeng, X.; Senanayake, C. H. *J. Org. Chem.* **2010**, *75*, 1155.

[y] Yoshida, S.; Marumo, K.; Takeguchi, K.; Takahashi, T.; Mase, T. *Org. Process Res. Dev.* **2014**, *18*, 1721.

[z] Malmgren, H.; Cotton, H.; Frøstrup, B.; Jones, D. S.; Loke, M.-L.; Peters, D.; Schultz, S.; Sölver, E.; Thomsen, T.; Wennerberg, J. *Org. Process Res. Dev.* **2011**, *15*, 408.

[aa] Pesti, J. A.; Yin, J.; Zhang, L.; Anzalone, L.; Waltermire, R. E.; Ma, P.; Gorko, E.; Confalone, P. N.; Fortunak, J.; Silverman, C.; Blackwell, J.; Chung, J. C.; Hrytsak, M. D.; Cooke, M.; Powell, L.; Ray, C. *Org. Process Res. Dev.* **2004**, *8*, 22.

[ab] Wityak, J.; Sielecki, T. M.; Pinto, D. J.; Emmett, G.; Sze, J. Y.; Liu, J.; Tobin, A. E.; Wang, S.; Jiang, B.; Ma, P.; Mousa, S. A.; Wexler, R. R.; Olson, R. E. *J. Med. Chem.* **1997**, *40*, 50.

[ac] Duan, S.; Venkatraman, S.; Hong, X.; Huang, K.; Ulysse, L.; Mobele, B. I.; Smith, A.; Lawless, L.; Locke, A.; Garigipati, R. *Org. Process Res. Dev.* **2012**, *16*, 1787.

[ad] Tian, Q.; Cheng, Z.; Yajima, H. M.; Savage, S. J.; Green, K. L.; Humphries, T.; Reynolds, M. E.; Babu, S.; Gosselin, F.; Askin, D.; Kurimoto, I.; Hirata, N.; Iwasaki, M.; Shimasaki, Y.; Miki, T. *Org. Process Res. Dev.* **2013**, *17*, 97.

[ae] Deasy, R. E.; Slattery, C. N.; Maguire, A. R.; Kjell, D. P.; Hawk, M. K. N.; Joo, J. M.; Gu, L. *J. Org. Chem.* **2014**, *79*, 3688.

[af] Gibson, F. S.; Singh, A. K.; Soumeillant, M. C.; Manchand, P. S.; Humora, M.; Kronenthal, D. R. *Org. Process Res. Dev.* **2002**, *6*, 814.

[ag] Kim, S. T.; Park, Y.; Kim, N.; Gu, J.; Son, W.; Hur, J.; Lee, K.; Baek, A.; Song, J. Y.; Kim, U. B.; Lee, K.-Y.; Oh, C.-Y.; Park, S. Shin, H. *Org. Process Res. Dev.* **2022**, *26*, 123.

[ah] Brandt, T. A.; Caron, S.; Damon, D. B.; DiBrino, J.; Ghosh, A.; Griffith, D. A.; Kedia, S.; Ragan, J. A.; Rose, P. R.; Vanderplas, B. C.; Wei, L. *Tetrahedron* **2009**, *65*, 3292.

[ai] Hawkins, J. M.; Dubé, P.; Maloney, M. T.; Wei, L.; Ewing, M.; Chesnut, S. M.; Denette, J. R.; Lillie, B. M.; Vaidyanathan, R. *Org. Process Res. Dev.* **2012**, *16*, 1393.

[aj] Patterson, D. E.; Powers, J. D.; LeBlanc, M.; Sharkey, T.; Boehler, E.; Irdam, E.; Osterhout, M. H. *Org. Process Res. Dev.* **2009**, *13*, 900.

[ak] Kato, T.; Ozaki, T.; Tsuzuki, K.; Ohi, N. *Org. Process Res. Dev.* **2001**, *5*, 122.

[al] Andersen, P.; Ankersen, M.; Jessen, C. U.; Lehman, S. V. *Org. Process Res. Dev.* **2002**, *6*, 367.

[am] Anson, M. S.; Graham, J. P.; Roberts, A. J. *Org. Process Res. Dev.* **2011**, *15*, 649.

[an] Dirat, O.; Bibb, A. J.; Burns, C. M.; Checksfield, G. D.; Dillon, B. R.; Field, S. E.; Fussell, S. J.; Green, S. P.; Mason, C.; Mathew, J.; Mathew, S.; Moses, I. B.; Nikiforov, P. I.; Pettman, A. J.; Susanne, F. *Org. Process Res. Dev.* **2011**, *15*, 1010.

[ao] Michia, M.; Takayanagi, Y.; Imai, M.; Furuya, Y.; Kimura, K.; Kitawaki, T.; Tomori, H.; Kajino, H. *Org. Process Res. Dev.* **2013**, *17*, 1430.

[ap] Gontcharov, A.; Dunetz, J. R. *Org. Process Res. Dev.* **2014**, *18*, 1145.

[aq] Shieh, W.-C.; Du, Z.; Kim, H.; Liu, Y.; Prashad, M. *Org. Process Res. Dev.* **2014**, *18*, 1339.

[ar] Kuroda, K.; Tsuyumine, S.; Kodama, T. *Org. Process Res. Dev.* **2016**, *20*, 1053.

[as] Burks, J. E., Jr.; Espinosa, L.; Labell, E. S.; McGill, J. M.; Ritter, A. R.; Speakman, J. L.; Williams, M. A.; Bradley, D. A.; Haehl, M. G.; Schmid, C. R. *Org. Process Res. Dev.* **1997**, *1*, 198.

[at] Funel, J.-A.; Brodbeck, S.; Guggisberg, Y.; Litjens, R.; Seidel, T.; Struijk, M.; Abele, S. *Org. Process Res. Dev.* **2014**, *18*, 1674.

[au] Pu, Y.-M.; Grieme, T.; Gupta, A.; Plata, D.; Bhatia, A. V.; Cowart, M.; Ku, Y.-Y. *Org. Process Res. Dev.* **2005**, *9*, 45.

[av] Hardouin, C.; Poixblanc, A.; Barière, F.; Tamion, R.; Dubuffet, T.; Hervouet, Y.; Mouchet, P. *Org. Process Res. Dev.* **2018**, *22*, 1419.

[aw] (a) Dunn, P. J. *Org. Process Res. Dev.* **2005**, *9*, 88. (b) Dale, D. J.; Dunn, P. J.; Golightly, C.; Hughes, M. L.; Levett, P. C.; Pearce, A. K.; Searle, P. M.; Ward, G.; Wood, A. S. *Org. Process Res. Dev.* **2000**, *4*, 17. (c) Dunn, P. J.; Wood, A. S. European Patent, EP 0 812845.

[ax] (a) Teleha, C. A.; Branum, S.; Zhang, Y.; Reuman, M. E.; Van Der Steen, L.; Verbeek, M.; Fawzy, N.; Leo, G. C.; Kang, F.-A.; Cai, C.; Kolpak, M.; Beauchamp, D. A.; Wall, M. J.; Russell, R. K.; Sui, Z.; Vanbaelen, H. *Org. Process Res. Dev.* **2014**, *18*, 1622. (b) Teleha, C. A.; Branum, S.; Zhang, Y.; Reuman, M. E.; Van Der Steen, L.; Verbeek, M.; Fawzy, N.; Leo, G. C.; Winters, M. P.; Kang, F.-A.; Kolpak, M.; Beauchamp, D. A.; Lanter, J. C.; Russell, R. K.; Sui, Z.; Vanbaelen, H. *Org. Process Res. Dev.* **2014**, *18*, 1630.

[ay] Breining, S. R.; Genus, J. F.; Mitchener, J. P.; Cuthbertson, T. J.; Heemstra, R.; Melvin, M. S.; Dull, G. M.; Yohannes, D. *Org. Process Res. Dev.* **2013**, *17*, 413.

Procedure

- Add 2 kg of seeds into a solution of **22** (100 kg) in a mixture of ethanol (736 L) and water (12.5 L) at 30 °C.
- Add a solution of sodium 2-ethylhexanoate (65.6 kg, 2 equiv) in ethanol (591 L) at 30 °C over 4 h.
- The batch was held for 2 h.
- Filter, wash with ethanol (2×253 L), and dry.
- Yield: 51.0 kg (90%) as a white crystalline solid.

Dolutegravir sodium **25** is the sodium salt of dolutegravir, an orally bioavailable integrase strand-transfer inhibitor, with activity against human immunodeficiency virus type 1 (HIV-1) infection. The final stage of synthesis of **25** involved the removal of the benzyl-protecting group of **23** and the following sodium salt formation with sodium hydroxide (Scheme 21.5).[13]

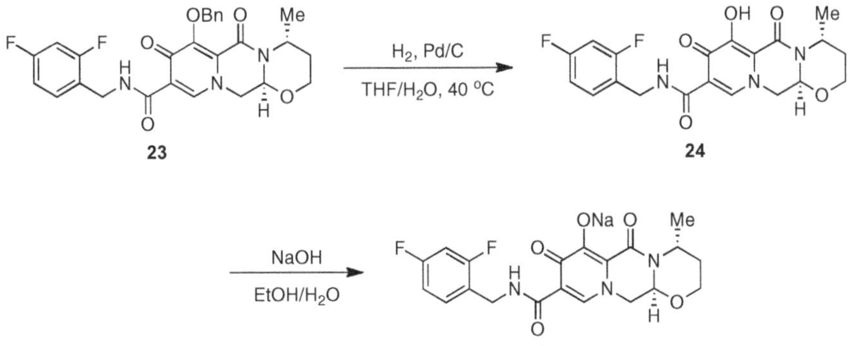

SCHEME 21.5

Procedure

- Add 2.96 kg of 5% Pd/C in water (7.8 kg) to a solution of **23** (3.70 kg, 7.26 mol) in THF (55.9 kg) and water (1.9 kg) under nitrogen.
- Replace the atmosphere with H_2.
- Stir the batch at 40 °C for 2 h.
- Filter, add activated carbon (0.37 kg) to the filtrate, and stir at 40 °C for 2 h.
- Filter and concentrate the filtrate under reduced pressure.
- Add water (74 kg) and stir for 1 h.
- Filter, wash with EtOH, and dry under reduced pressure to give **24** (2.22 kg).
- Dissolve **24** in EtOH (94.6 kg) and water (13.3 kg) at 70 °C.
- Add a solution of NaOH (212 g) in water (2.6 kg) at 60 °C.
- Stir the mixture at 25 °C for 3 h.
- Filter, wash with EtOH (8.8 kg), and dry under reduced pressure.
- Yield of **25**: 2.22 kg (69%, 99.9 HPLC peak area %).

21.2.1.2 Potassium Salts

Acidic organic compounds can form salts via deprotonation with inorganic bases, especially with strong bases. For instance, potassium *tert*-butoxide (KO*t*Bu) was employed to prepare the potassium salt **27** by removing the N–H proton in the neutral drug molecule **26** (Equation 21.4).[14]

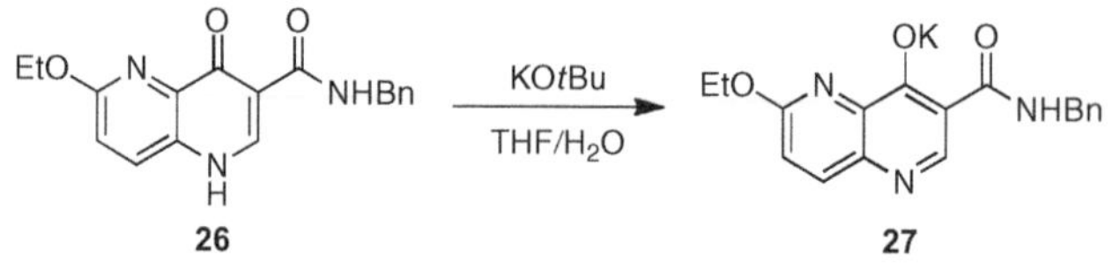

(21.4)

Compound **26** was a subtype-selective GABA-A receptor inverse agonist for the treatment of cognition disorders such as Alzheimer's disease. However, the drug candidate **26** had poor solubility in most solvents which was inappropriate for the toxicology studies. Compared with compound **26**, the potassium salt **27**, generated in THF using KO*t*Bu in the presence of H_2O (9 equiv), was more water soluble and selected for the toxicology studies.

Procedure

- Charge 38 kg (117.5 mol) of **26** into a 500-gallon, glass-lined reactor followed by adding THF (302.8 L), 13.5 kg (120.3 mol, 1.02 equiv) of KO*t*Bu, and 19 L of H_2O.
- Heat to reflux for 1 h.
- Filter the batch.
- Concentrate the filtrate to 208 L.
- Cool and filter.
- Yield: 44.8 kg (as a wet cake).

Macrocycle **28** was identified by Merck as a potent HCV protease inhibitor with a potentially superior therapeutic profile. However, the amorphous nature of **28** posed challenges in both purification and stability during its development.

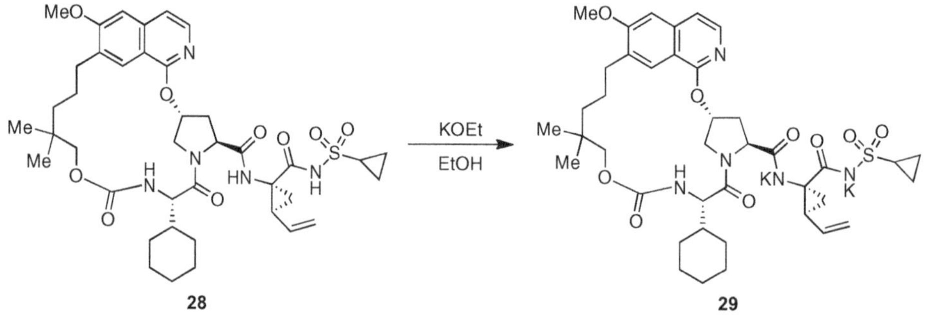

(21.5)

In order to overcome these issues, potassium salt **29** was selected for safety and clinical studies (Equation 21.5).[15] Ultimately, the potassium salt **29** was crystallized from ethanol in 94% yield.

21.2.1.3 Calcium Salts

A salt-exchange approach to generate a crystalline calcium salt **31** was investigated. The sodium salt of **30** was dissolved in a mixture of methanol and water and to this solution was charged with calcium chloride to induce crystallization of the calcium salt (Equation 21.6).[16]

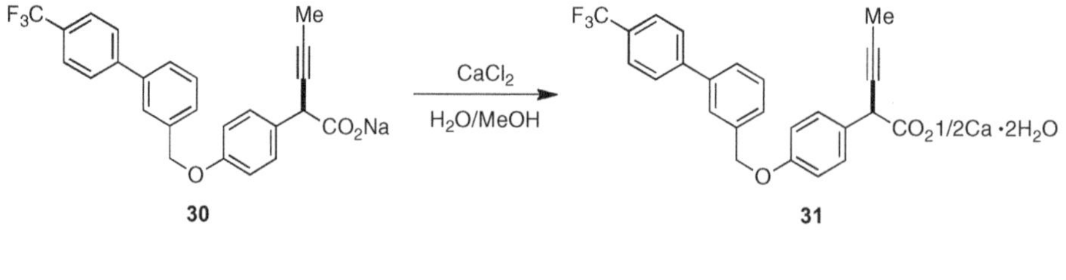

(21.6)

Procedure

- Charge 120.5 g (0.235 mol) of **30** into a mixture of MeOH (578 mL) and H_2O (289 mL) and stir to form a clear solution.
- Charge a solution of $CaCl_2 \cdot 2H_2O$ (17.3 g, 0.118 mol, 0.5 equiv) in H_2O (35 mL) dropwise into the solution of **30** over 50 min at 20 °C.
- Age the resulting slurry for 16 h.
- Filter and wash with MeOH/H_2O (2:1, 2×150 mL).
- Dry at 40 °C under vacuum for 36 h.
- Yield: 121.0 g.

Compound **33**, developed by Bristol Myers Squibb Company, was a member of the drug class known as statins, used for lowering blood cholesterol. This calcium salt formation was affected by adopting the salt-exchange approach as illustrated in Reaction (21.7).[17]

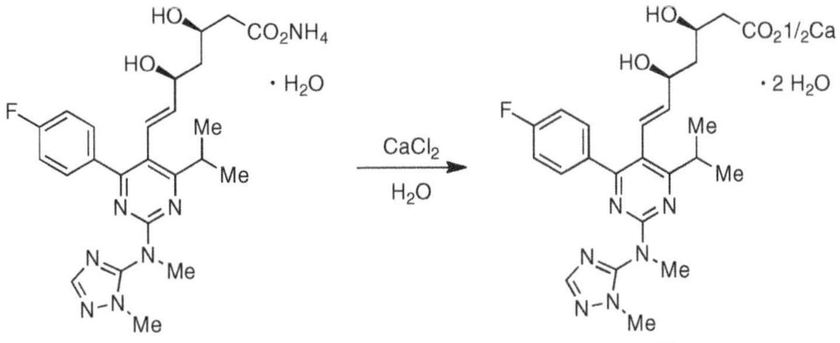

(21.7)

Procedure

- Charge 28.4 kg of **32** to a 1900-L Hastelloy reactor.
- Add 284 kg of water.
- Stir at 20 °C to form a slurry.
- Charge a 13.6 wt% of solution of $CaCl_2$ (44.6 kg, 11 equiv) in 284 kg of water.
- Hold the resulting slurry at 20 °C for ca. 1 h.
- Heat to 50 °C over a 2 h period.
- Add water (170.4 kg).
- Hold the batch at 50 °C for 14 h.
- Cool to 20 °C.
- Filter and wash with heptane.
- Dry at 50 °C under full vacuum.
- Yield: 23.88 kg (79.9%).

21.2.1.4   Various Inorganic Salts

An assortment of pharmaceutical inorganic salts is shown in Table 21.3.

21.2.2   USE OF ORGANIC BASES

The acidic drug substances can also form the corresponding salts with appropriate organic bases. Table 21.4 presents a number of salts of the acidic organic drug substances with organic bases.

TABLE 21.3
List of Inorganic Salts of Some Acidic Drug Substances

Sodium	Potassium	Magnesium	Calcium
Appearance: White solid[a]	Appearance: Solid (mp 208 °C)[65]	Appearance: Crystalline solid[b]	Appearance: Crystalline solid[c]
Appearance: White solid[d]	Appearance: Solid[e]		Appearance: Crystalline solid[67]
Appearance: Light brown solid[f]			Appearance: Crystalline solid[g]
Appearance: Solid[h]			Appearance (Lipitor, atorvastatin calcium): White crystalline solid

(Continued)

TABLE 21.3 (CONTINUED)
List of Inorganic Salts of Some Acidic Drug Substances

Sodium	Potassium	Magnesium	Calcium

Appearance: White powder[i]

Appearance: Off-white powder[68]

Appearance: White solid, mp 170 °C[j]

[a] Ormerod, D.; Willemsens, B.; Mermans, R.; Langens, J.; Winderickx, G.; Kalindjian, S. B.; Buck, I. M.; McDonald, I. M. Org. Process Res. Dev. **2005**, 9, 499.

[b] Engelhardt, F. C.; Shi, Y.-J.; Cowden, C. J.; Conlon, D. A.; Pipik, B.; Zhou, G.; McNamara, J. M.; Dolling, U.-H. J. Org. Chem. **2006**, 71, 480.

[c] Ashcroft, C.; Hellier, P.; Pettman, A.; Watkinson, S. Org. Process Res. Dev. **2011**, 15, 98.

[d] Mueller, R. H.; Wang, S.; Pansegrau, P. D.; Jannotti, J. Q.; Poss, M. A.; Thottathil, J. K.; Singh, J.; Humora, M. J.; Kissick, T. P.; Boyhan, B. Org. Process Res. Dev. **1997**, 1, 14.

[e] Jacks, T. E.; Belmont, D. T.; Briggs, C. A.; Horne, N. M.; Kanter, G. D.; Karrick, G. L.; Krikke, J. J.; McCabe, R. J.; Mustakis, J. G.; Nanninga, T. N.; Risedorph, G. S.; Seamans, R. E.; Skean, R.; Winkle, D. D.; Zennie, T. M. Org. Process Res. Dev. **2004**, 8, 201.

[f] De Koning, P. D.; Gladwell, I. R.; Moses, I. B.; Panesar, M. S.; Pettman, A. J.; Thomson, N. M. Org. Process Res. Dev. **2011**, 15, 1247.

[g] Cohen, J. H.; Bos, M. E.; Cesco-Cancian, S.; Harris, B. D.; Hortenstine, J. T.; Justus, M.; Maryanoff, C. A.; Mills, J.; Muller, S.; Roessler, A.; Scott, L.; Sorgi, K. L.; Villani, J., F. J.; Webster, R. R. H.; Weh, C. Org. Process Res. Dev. **2003**, 7, 866.

[h] Jansen, R.; Knopp, M.; Amberg, W.; Bernard, H.; Koser, S.; Müller, S.; Münster, I.; Pfeiffer, T.; Riechers, H. Org. Process Res. Dev. **2001**, 5, 16.

[i] Lattuada, L.; Argese, M.; Bio, V.; Galimberti, L.; Gazzetto, S. Org. Process Res. Dev. **2014**, 18, 1175.

[j] Hao, Q.; Pan, J.; Li, Y.; Cai, Z.; Zhou, W. Org. Process Res. Dev. **2013**, 17, 921.

TABLE 21.4

Examples of Salts between Acidic Drug Substances and Organic Bases

Entry	Chemical Structure	Comments
1	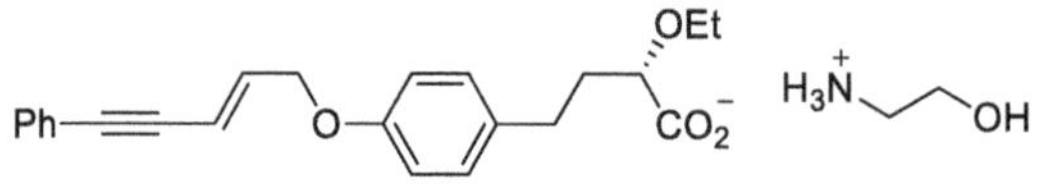	Light yellow crystals.[a]
2	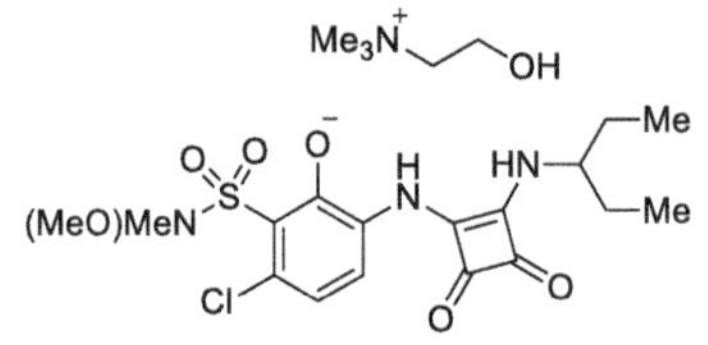	Light yellow solid.[b]
3	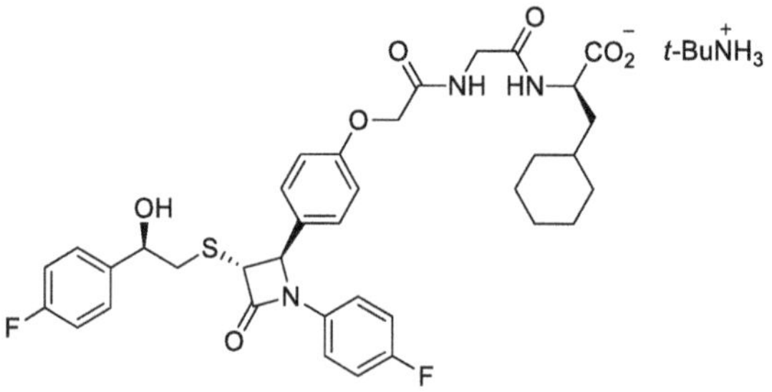	Isolated as a solid.[c]
4	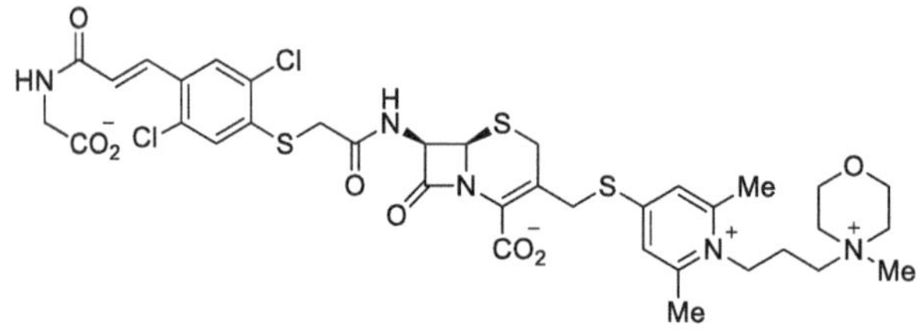	These double zwitterions were solid with mp 180–183 °C.[d]
5	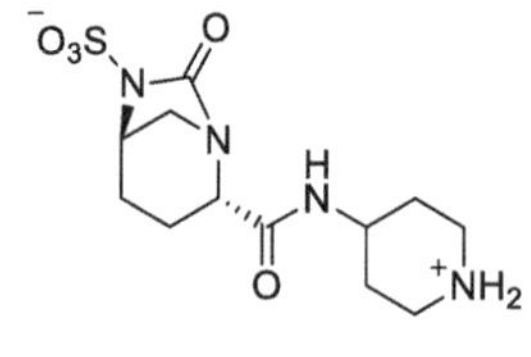	This compound[e] was a crystalline solid and in clinical trials with β-lactam antibiotics for the treatment of serious and antibiotic-resistant bacterial infections.
6	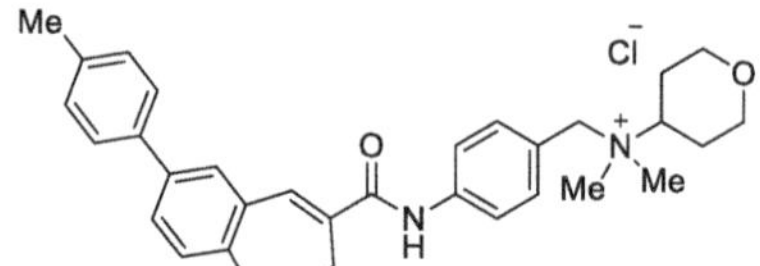	This compound inhibits the replication of macrophage (M)-tropic HIV-1 in both MAGI-CCR5 cells and peripheral blood mononuclear cells.[f]
7	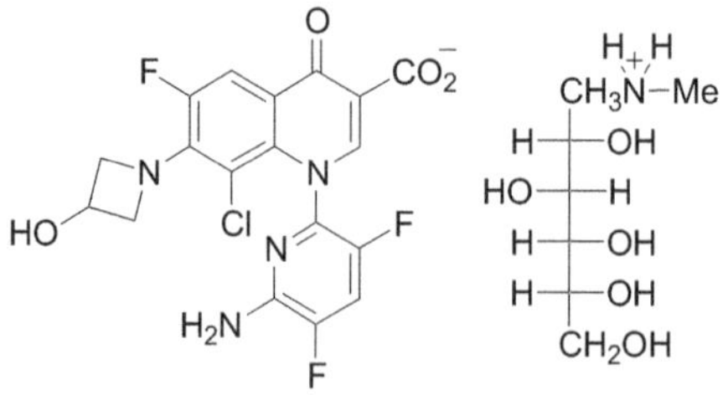	Delafloxacin[g] was an exploratory drug with excellent anti-bacterial activity against Gram-positive organisms, including both methicillin-susceptible *S. aureus* and MRSA.

(Continued)

TABLE 21.4 (CONTINUED)
Examples of Salts between Acidic Drug Substances and Organic Bases

Entry	Chemical Structure	Comments
8		The tris(hydroxymethyl) aminomethane (TRIS) salt of phosphate prodrug was developed by BMS for the treatment of HIV infection.[h]
9		The drug candidate was used for the treatment of type-2 diabetes.[i]

[a] Deussen, H.-J.; Jeppesen, L.; Schärer, N.; Junager, F.; Bentzen, B.; Weber, B.; Weil, V.; Mozer, S. J.; Sauerberg, P. *Org. Process Res. Dev.* **2004**, *8*, 363.

[b] Martin, B.; Lai, X.; Bettig, U.; Neumann, E.; Kuhnle, T.; Porter, D.; Robinson, R.; Hatto, J.; D'Souza, A.-M.; Steward, O.; Watson, S.; Press, N. J. *Org. Process Res. Dev.* **2015**, *19*, 1038.

[c] Karlsson, S.; Sörensen, J. H. *Org. Process Res. Dev.* **2012**, *16*, 586.

[d] Singh, J.; Kim, O. K.; Kissick, T. P.; Natalie, K. J.; Zhang, B.; Crispino, G. A.; Springer, D. M.; Wichtowski, J. A.; Zhang, Y.; Goodrich, J.; Ueda, Y.; Luh, B. Y.; Burke, B. D.; Brown, M.; Dutka, A. P.; Zheng, B.; Hsieh, D.-M.; Humora, M. J.; North, J. T.; Pullockaran, A. J.; Livshits, J.; Swaminathan, S.; Gao, Z.; Schierling, P.; Ermann, P.; Perrone, R. K.; Lai, M. C.; Gougoutas, J. Z.; DiMarco, J. D.; Bronson, J. J.; Heikes, J. E.; Grosso, J. A.; Kronenthal, D. R.; Denzel, T. W.; Mueller, R. H. *Org. Process Res. Dev.* **2000**, *4*, 488.

[e] (a) Miller, S. P.; Zhong, Y.-L.; Liu, Z.; Simeone, M.; Yasuda, N.; Limanto, J.; Chen, Z.; Lynch, J.; Capodanno, V. *Org. Lett.* **2014**, *16*, 174. (b) Kim, J.; Xu, F.; Dance, Z. E. X.; Waldman, J. H.; Wallace, D. J.; Wu, F.; Kats-Kagan, R.; Ekkati, A. R.; Brunskill, A. P. J.; Peng, F.; Fier, P. S.; Oblifacion, J. V.; Sherry, B. D.; Liu, Z.; Emerson, K. M.; Fine, A. J.; Jenks, A. V.; Armenante, M. E. *Org. Process Res. Dev.* **2021**, *25*, 2249.

[f] Hashimoto, H.; Ikemoto, T.; Itoh, T.; Maruyama, H.; Hanaoka, T.; Wakimasu, M.; Mitsudera, H.; Tomimatsu, K. *Org. Process Res. Dev.* **2002**, *6*, 70.

[g] Hanselmann, R.; Johnson, G.; Reeve, M. M.; Huang, S.-T. *Org. Process Res. Dev.* **2009**, *13*, 54.

[h] (a) Chen, K.; Risatti, C.; Bultman, M.; Soumeillant, M.; Simpson, J.; Zheng, B.; Fanfair, D.; Mahoney, M.; Mudryk, B.; Fox, R. J.; Hsiao, Y.; Murugesan, S.; Conlon, D. A.; Buono, F. G.; Eastgate, M. D.; *J. Org. Chem.* **2014**, *79*, 8757. (b) Fox, R. J.; Tripp, J. C.; Schultz, M. J.; Payack, J. F.; Fanfair, D. D.; Mudryk, B. M.; Murugesan, S.; Chen, C.-P. H.; La Cruz, T. E.; Ivy, S. E.; Broxer, S.; Cullen, R.; Erdemir, D.; Geng, P.; Xu, Z.; Fritz, A.; Doubleday, W. W.; Conlon, D. A. *Org. Process Res. Dev.* **2017**, *21*, 1095.

[i] Hyde, A. M.; Liu, Z.; Kosjek, B.; Tan, L.; Klapars, A.; Ashley, E. R.; Zhong, Y.-L.; Alvizo, O.; Agard, N. J.; Liu, G.; Gu, X.; Yasuda, N.; Limanto, J.; Huffman, M. A.; Tschaen, D. M. *Org. Lett.* **2016**, *18*, 5888.

NOTES

1. Bastin, R. J.; Bowker, M. J.; Slater, B. J. *Org. Process Res. Dev.* **2000**, *4*, 427.
2. Haynes, D. A.; Jones, W.; Motherwell, W. D. S. *CrystEngComm.* **2005**, *7*, 538.
3. Carey, J. S.; Laffan, D.; Thomson, C.; Williams, M. T. *Org. Biomol. Chem.* **2006**, *4*, 2337.
4. Silverberg, L. J.; Dillon, J. L.; Vemishetti, P.; Sleezer, P. D.; Discordia, R. P.; Hartung, K. B.; Gao, Q. *Org. Process Res. Dev.* **2000**, *4*, 34.

5. Zhao, J.; Wang, D.; Yang, W.-L.; Niu, J.; Wu, W. *Org. Process Res. Dev.* **2022**, *26*, 91.

6. Beylin, V.; Boyles, D. C.; Curran, T. T.; Macikenas, D.; Parlett, R. V. IV; Vrieze, D. *Org. Process Res. Dev.* **2007**, *11*, 441.

7. (a) King, F. J.; Searle, A. D.; Urquhart, M. W. *Org. Process Res. Dev.* **2020**, *24*, 2915. (b) Borths, C. J.; Burns, M.; Curran, T.; Ide, N. D. *Org. Process Res. Dev.* **2021**, *25*, 1788.

8. López-Rodríguez, R.; McManus, J. A.; Murphy, N. S.; Ott, M. A.; Burns, M. J. *Org. Process Res. Dev.* **2020**, *24*, 1558.

9. Brodszki, M.; Bäckström, B.; Horvath, K.; Larsson, T.; Malmgren, H.; Pelcman, M.; Wähling, H.; Wallberg, H.; Wennerberg, J. *Org. Process Res. Dev.* **2011**, *15*, 1027.

10. Widlicka, D. W.; Murray, J. C.; Coffman, K. J.; Xiao, C.; Brodney, M. A.; Rainville, J. P.; Samas, B. *Org. Process Res. Dev.* **2016**, *20*, 233.

11. Maloney, K. M.; Zhang, S.-W.; Mohan, A. E.; Lee, A. Y.; Larpent, P.; Ren, H.; Humphrey, G. R.; Desmond, R.; DiBenedetto, M.; Liu, W.; Lee, I. H.; Sirota, E.; Di Maso, M. J.; Alwedi, E.; Song, S.; Yao, H.; Chang, D. *Org. Process Res. Dev.* **2020**, *24*, 2498.

12. Ball, M.; Boyd, A.; Ensor, G. J.; Evans, M.; Golden, M.; Linke, S. R.; Milne, D.; Murphy, R.; Telford, A.; Kalyan, Y.; Lawton, G. R.; Racha, S.; Ronsheim, M.; Zhou, S. H., *Org. Process Res. Dev.* **2016**, *20*, 1799.

13. Aoyama, Y.; Hakogi, T.; Fukui, Y.; Yamada, D.; Ooyama, T.; Nishino, Y.; Shinomoto, S.; Nagai, M.; Miyake, N.; Taoda, Y.; Yishida, H.; Yasukata, T. *Org. Process Res. Dev.* **2019**, *23*, 558.

14. Beaudin, J.; Bourassa, D. E.; Bowles, P.; Castaldi, M. J.; Clay, R.; Couturier, M. A.; Karrick, G.; Makowski, T. W.; McDermott, R. E.; Meltz, C. N.; Meltz, M.; Phillips, J. E.; Ragan, J. A.; Ripin, D. H. B.; Singer, R. A.; Tucker, J. L.; Wei. L. *Org. Process Res. Dev.* **2003**, *7*, 873.

15. Song, Z. J.; Tellers, D. M.; Dormer, P. G.; Zewge, D.; Janey, J. M.; Nolting, A.; Steinhuebel, D.; Oliver, S.; Devine, P. N.; Tschaen, D. M. *Org. Process Res. Dev.* **2014**, *18*, 423.

16. (a) Morrison, H.; Jona, J.; Walker, S. D.; Woo, J. C. S.; Li, L.; Fang, J. *Org. Process Res. Dev.* **2011**, *15*, 104. (b) Walker, S. D.; Borths, C. J.; DiVirgilio, E.; Huang, L.; Liu, P.; Morrison, H.; Sugi, K.; Tanaka, M.; Woo, J. C. S.; Faul, M. M. *Org. Process Res. Dev.* **2011**, *15*, 570.

17. Hobson, L. A.; Akiti, O.; Deshmukh, S. S.; Harper, S.; Katipally, K.; Lai, C. J.; Livingston, R. C.; Lo, E.; Miller, M. M.; Ramakrishnan, S.; Shen, L.; Spink, J.; Tummala, S.; Wei, C.; Yamamoto, K.; Young, J.; Parsons, R. L.; Jr. *Org. Process Res. Dev.* **2010**, *14*, 441.

22 Solid Form

A solid organic compound generally exists as either an amorphous or a crystalline state. A crystalline solid is formed by arranging the components in a regular repeating three-dimensional array (a crystal lattice), whereas an amorphous solid is formed by arranging them more or less randomly. Crystalline solids have sharp melting points and well-defined edges, faces, and X-ray powder diffraction (XRPD) patterns. Polymorphism is defined as the property of a substance to exist in more than one crystal structure. The various polymorphic forms (or polymorphs) of a substance have different arrangements (packing polymorphism) and/or conformations (conformational polymorphism) in the crystal lattice[1] and, thereby, different properties, such as density, hardness, chemical stability, solubility, rate of dissolution in different solvents, melting point, and hygroscopicity.[2] Polymorphic, pseudopolymorphic, as well as amorphous transformations, can occur in a reversible or irreversible way.[3]

22.1 POLYMORPHISM

Polymorphism is a widespread phenomenon and, statistically, 85% of active pharmaceutical ingredients (APIs) exhibit (pseudo)polymorphism, and 50% of APIs have multiple forms.[4] Accordingly, polymorphism investigation of new APIs becomes a regulatory and important practice in the pharmaceutical industry. As polymorphic transformations can occur in a reversible or irreversible way which will contribute to significant variability in APIs, it is crucial during the drug development that the most thermodynamically stable form of the API should be selected. Ideally, the polymorph selected in the drug development should have good solubility and dissolution rate profile, be non-hygroscopic, with a high melting temperature, have compact morphology (no needles or plates), and not form hydrates even at high relative humidity levels.

Various testing tools[5] are available such as differential scanning calorimetry (DSC), thermo gravimetric analysis (TGA), XRPD, and dynamic vapor sorption (DVS) to elucidate the melting points, thermal properties, crystallinities, and hygroscopicities, respectively. In addition, several online methods, including Raman spectroscopy,[6] focused beam reflectance measurement (FBRM), and Fourier transform infrared spectroscopy (FTIR), can be used for in situ monitoring the transformation of polymorphs. Raman spectroscopy is a light-scattering technique and has the capability to monitor both the liquid phase and the solid phase at the same time. FBRM is a probe-based high-solid concentration particle characterization tool and can obtain particle dimension measurements over a large size range.

As polymorphism may be route-specific, polymorphs of interest shall be manufactured through carefully controlled conditions, such as temperature, supersaturation, rate of crystallization, seeding, or solvent. There are two proposed mechanisms for polymorphic transformations, that is, solid-state transformation[7] and solution-mediated transformation. During the primary nucleation in an unseeded environment, it is the unstable polymorph or pseudopolymorph in the form of hydrate or solvate that may crystallize out first. However, after the nucleation of a more stable crystal form, a previously prepared meta-stable form could no longer be obtained. This polymorph transformation from meta-stable form to thermodynamically stable form may be catalyzed by the presence of a seed of the latter. A report[8] describes numerous vanishing polymorph examples including epimerization at the anomeric carbon such as mannose and melibiose (Table 22.1).

DOI: 10.1201/9781003288411-22

TABLE 22.1

List of Vanishing Polymorphs

Compounds	Mp	Comments
(Benzylidene-*dl*-piperitone)	69–70 °C	It was found that both α-form (mp 59–60 °C) and β-form (mp 63–64 °C) could be transformed readily into γ-form (mp 69–70 °C) by seeding.
((*E*)-3-Phenyl-1-*p*-tolylprop-2-en-1-one)	76 °C	Once the form (mp 76 °C) was obtained, the lower melting point form (mp 54.6 °C) could not be obtained from recrystallization.
(Benzocaine picrate)	162–163 °C	A metastable form (mp 132 °C) could not be obtained once the more stable form (mp 162–163 °C) was obtained.
(Xylitol)	94–97 °C (commercially available)	Although xylitol was first prepared as syrup in 1891, there was no report of crystallization until fifty years later when the metastable hygroscopic form (mp 61 °C) was prepared. Once a stable orthorhombic form (mp 94 °C) was obtained, it was difficult to prepare the metastable monoclinic form.
(1,2,3,5-Tetra-*O*-acetyl-β-D-ribofuranose)	81–83 °C (available in Sigma–Aldrich)	Initially, the compound was prepared with mp 56–58 °C. However, this polymorph could not be obtained after the appearance of the more stable form (mp 81–83 °C).
α form β form Mannose	–	Mannose exists mainly as two forms of pyranose, the α-form (67%) and the β-form (33%). The β anomer could be obtained from the mixture by extraction at 0 °C with 80% alcohol.
α form β form Melibiose (α form)	182 °C (α form) (dec)	Only the α form is commercially available.

Source: Reprinted with permission from Dunitz, J.D. and Bernstein, J., *Acc. Chem. Res.* 28, 193. Copyright 1995 American Chemical Society.

22.1.1 CONTROL OF POLYMORPH BY SEEDING

22.1.1.1 Use of Direct Addition

Purification of a mixture of polymorphs sometimes can be realized via the transformation of one form to another under certain conditions.[9] The final step synthesis of a cancer drug candidate diol **2**, developed by Takeda California Inc., was initially carried out with sulfuric acid (9 N) in ethanol at an elevated temperature (Equation 22.1).[10] The precipitated product **2** was isolated from the reaction mixture by simple filtration. However, DSC and XRPD analyses indicated that the isolated product was a mixture of amorphous solid and undesired Form H.

(22.1)

Subsequent studies showed that this polymorph mixture could be readily converted to the desired Form A in alcoholic solvents by direct addition of Form A seeds at 55 °C. In the event, a viable process was developed involving deprotection, in situ form transformation, and crystallization. In this process, the deprotection of the penultimate intermediate **1** was accomplished in NMP/MeOH at 55 °C with concentrated HCl. The metastable zone was achieved by diluting with methanol as the anti-solvent, followed by adding seeds. Further aging at 55 °C for 5 h was capable of completing the polymorph transformation and crystal growth.

Procedure

- Heat a mixture of **1** (13.6 kg, 25.0 mol), NMP (44.0 L), MeOH (3.45 L), and conc. HCl (4.08 kg) at 55 °C for 2 h, followed by polish filtration.
- Add MeOH (27 L) at 55 °C, followed by charging seeds (131 g) of Form A.
- Add MeOH (334 L) over 3 h and hold at 55 °C for 2 h.
- Cool to 0 °C over 4 h and hold at 0 °C for 9 h.
- Filter and wash with MeOH (3×60 L).
- Yield of **2**: 10.6 kg (84%) as a white solid.

During the preparation of an API-free base **3** (Figure 22.1),[11] crystallization of the desired anhydrous Form I resulted in a very poor impurity rejection. On the other hand, the dihydrate form provided the desired impurity purge. Ultimately, the API **3** was isolated as dihydrate from THF/water media.

Conversion of the dihydrate to the desired anhydrous Form I by drying in a vacuum oven failed, resulting in an undesired Form III. Eventually, this polymorph transformation from meta-stable Form III to thermodynamically stable Form I was achieved with the direct addition method by adding the seeds of the latter during the recrystallization.

Procedure

- Dissolve 1.64 kg of **3** (in Form III) in iPrOAc (8 L) at 75 °C.
- Cool the resulting solution to 67 °C.
- Charge the Form I seeds.
- Cool the slurry to 10 °C over 3 h.

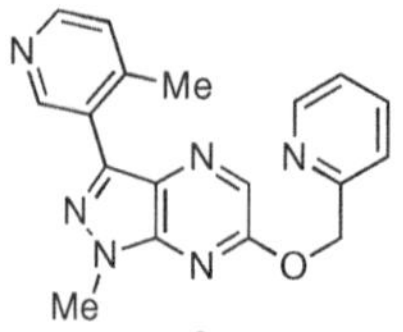

FIGURE 22.1   The chemical structure of API free base **3**.

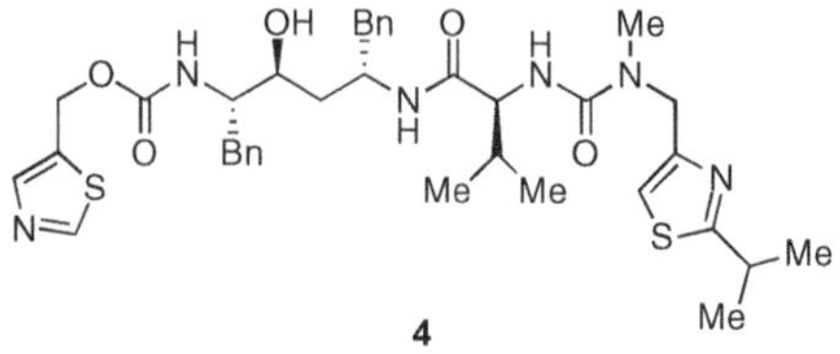

FIGURE 22.2   The structure of ritonavir **4**.

- Granulate at 10 °C for 10 h.
- Filter, wash with THF/H$_2$O (1:2, v/v) (8 L), and dry at 50 °C under vacuum with N$_2$ sweep.
- Yield: 1.50 kg of **3** (in Form I) as a white solid.

22.1.1.2   Use of Reverse Addition

Ritonavir **4** (known as norvir in the market) (Figure 22.2)[12] is a protease inhibitor for the treatment of HIV/AIDS. **4** was discovered by Abbott Laboratories in late 1992 and commercialized in 1996 as a semisolid capsule formulation.

In 1998, an unexpected new polymorph (Form II) emerged, which resulted in the precipitation of a large portion of the drug substance out of the formulated product. Compared with Form I, Form II is thermodynamically more stable but less soluble. In order to selectively generate the less thermodynamically stable Form I, a super seeding approach was used. However, the high seed loading (50%) rendered the throughput of this approach rather poor. Ultimately, a reverse addition technique was developed by adding the solution of **4** to the seed (Form I) slurry.

Procedure

- Charge 1 kg of **4** to Reactor A, followed by adding EtOAc (4 L).
- Reflux to dissolve the solid.
- Charge 5 g of seed crystals (of Form I) to Reactor B, followed by adding heptane (4 L).
- Filter the hot solution of **4** through a 0.2 μm filter cartridge in Reactor A into Reactor B over ≤2 h.
- Cool the resulting slurry in Reactor B to ambient temperature.
- Stir for 3 h.
- Filter, wash with heptane, and dry.

22.1.2   Control of Polymorph by Temperature

Three polymorphic forms of Acitretin **6** are reported,[13] Forms I, II, and III. Form II is thermodynamically more stable (as a cube shape), and the transformation of Form I to Form II could be realized on heating above 200 °C. It was observed that hydrolysis of butyl ester **5** at different temperature could result in either Form III (as a needle shape) (<55 °C) or Form II (>55 °C) (Equation 22.2).[14]

(22.2)

22.1.3 Control of Polymorph by Slurrying

Compound **9** is a drug candidate for the treatment of diabetes, obesity, and obesity-related disorders. The synthesis of **9** was straightforward by coupling carbamate **7** with amine **8** (Equation 22.3).[15]

(22.3)

After the coupling reaction, **9** was obtained as Form D by crystallization from a DMF/H_2O system. However, Form D was not a good candidate for the clinical study due to the absorption/desorption of water reversibly depending on the relative humidity. Therefore, Form D was converted to Form B, a more favorable form, by slurrying Form D in MeCN.

Procedure

- Stir a mixture of **8** (1.68 kg, 7.5 mol), DIPEA (990 mL, 5.67 mol) in DMF (13.9 L) for 0.5 h.
- Charge 1.98 kg (6.8 mol) of **7**.
- Heat the resulting mixture at 40–45 °C for 2 h.
- Cool to ambient temperature.
- Add 119 mL (2.08 mol) of AcOH, followed by adding water (4.2 L).
- Stir the batch for 1 h to initiate the crystallization.
- Add water (5.1 L) over 2–4 h and age the batch for 2 h.
- Filter and wash with DMF/H_2O (1:1) and water.
- Yield: 2.76 kg (97%) as monohydrate of Form D.
- Convert Form D to the anhydrous Form B by suspending Form D in MeCN (7.9 L) overnight to give 2.5 kg of **9**.

22.1.4 Control of Polymorph by Aging

Isolation of the aldehyde product **12** proved challenging because of the slow filtration of the crude mixture (Scheme 22.1).[16]

It was found that the filtration rate was significantly improved through a ripening process. After aging for 1–2 h at 50 °C, the filtration was about 10-time faster. This was the result of a crystal form change confirmed by DSC and XPRD. It was also noticed that the ripening process afforded larger crystals as indicated by the microscopy data. Under the optimal conditions, the desired product was reproducibly produced in 94–98% yield.

Procedure

- Cool a solution of **10** (9.45 kg, 35.0 mol) in anhydrous THF (94.5 L) to <–8 °C.
- Add a 20 wt% solution of *i*PrMgCl in THF (8.64 kg, 16.8 mol) followed by adding a 25 wt% solution of *n*-BuLi in heptane (9.84 kg, 38.4 mol) at <–8 °C.

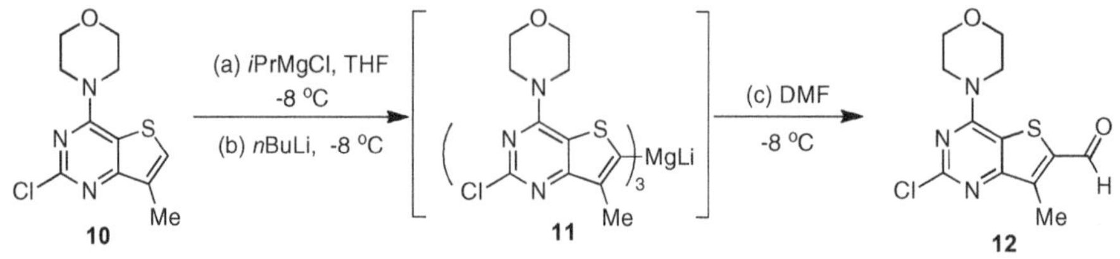

SCHEME 22.1

- Stir the resulting mixture at <−8 °C for 2 h.
- Add anhydrous DMF (5.11 kg, 69.9 mol) at <−8 °C.
- Stir the batch for 1 h.
- Transfer the batch to a mixture of AcOH (23.6 kg), 37% aqueous HCl (9.10 kg), IPA (94.4 L), and water (91.7 L).
- Heat the resulting slurry to 50–55 °C and stir at that temperature for 1 h.
- Concentrate the suspension under reduced pressure to remove THF.
- Cool the suspension to room temperature.
- Filter, wash with water (2x20 L), and dry at 40–60 °C under reduced pressure.
- Yield of **12**: 10.0 kg (96%) as a yellow solid (mp 198.0 °C).

REMARK

Sonication can also cause polymorph transformation.[17]

22.1.5 CONTROL OF POLYMORPH BY ADDING POLYMER

A new crystal form was discovered during the development of the crystallization method for the purification of Sonogashira product **15** by adding a non-adsorbing polymer (copovidone) additive, highlighting the effect of polymer additive on the polymorphic transformation and product purification (Scheme 22.2).[18]

Crystallization is one of the most effective and cost-efficient operations used in pharmaceutical manufacturing. However, conventional crystallization could be ineffective in rejecting impurity **16**

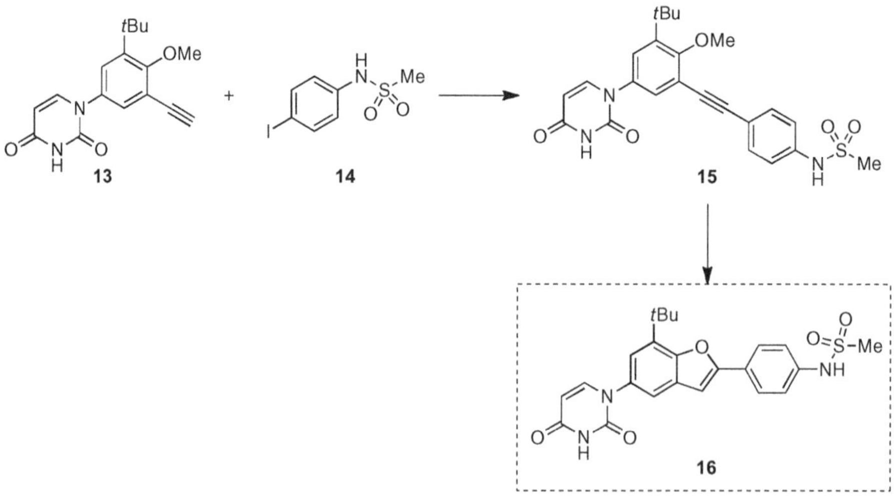

SCHEME 22.2

due to its relatively low solubility. The impurity **16** was generated during the Sonogashira cross-coupling reaction in a pilot plant batch at a level of 3%. In order to remove the impurity **16**, a crystallization strategy was designed to slow down the crystallization kinetics of **16** by adding copovidone as an additive. It was found that the addition of copovidone resulted in not only the delayed crystallization of **16**, but also a new polymorphic form, Form III, of the product **15** crystals. Implementing the optimized crystallization protocol improved the yield of isolated product **15** (as Form III) from ~80% to ~90% with less than 0.1% impurity **16**.

Note

During the cGMP campaign, the content of copovidone in the isolated product **15** had to be controlled. To reduce the residual polymer content, a reslurry process was introduced in which the isolated product was suspended in isopropanol (30 mL/g) at 60 °C for 2 h. This suspension was then cooled to 20 °C and filtered. The resultant product **15** from the reslurry process contained less than 0.1% w/w copovidone determined by ^{1}H NMR.

22.1.6 POLYMORPH TRANSFORMATION

Formoterol (foradil) **19** is a long-acting β_2-agonist used as a bronchodilator in the therapy of asthma and chronic bronchitis. The last two-step synthesis involved deprotection and L-tartaric acid salt formation (Scheme 22.3).[19] The isolated crude product **19** was contaminated with four impurities: **20–23**. Because of structural similarity, impurity **21** could not be removed to the level below 0.2% by crystallization.

Morphology studies on compound **19** identified the presence of two distinct crystal forms, Forms A and B, and a hydrated crystalline intermediate, Form C. Among the three crystal forms, Form A is the most stable form with a melting point of 192 °C. These forms were identified at three stages: (i) the crude product **19**/Form B generated after the addition of L-tartaric acid to a solution of the crude free-base **18**, (ii) the crystalline intermediate **19**/Form C formed after warming the slurry of **19**/Form B to effect the impurity removal, and (iii) recrystallization of **19**/Form C from 25% aqueous isopropyl alcohol to give **19**/Form A (Scheme 22.4).

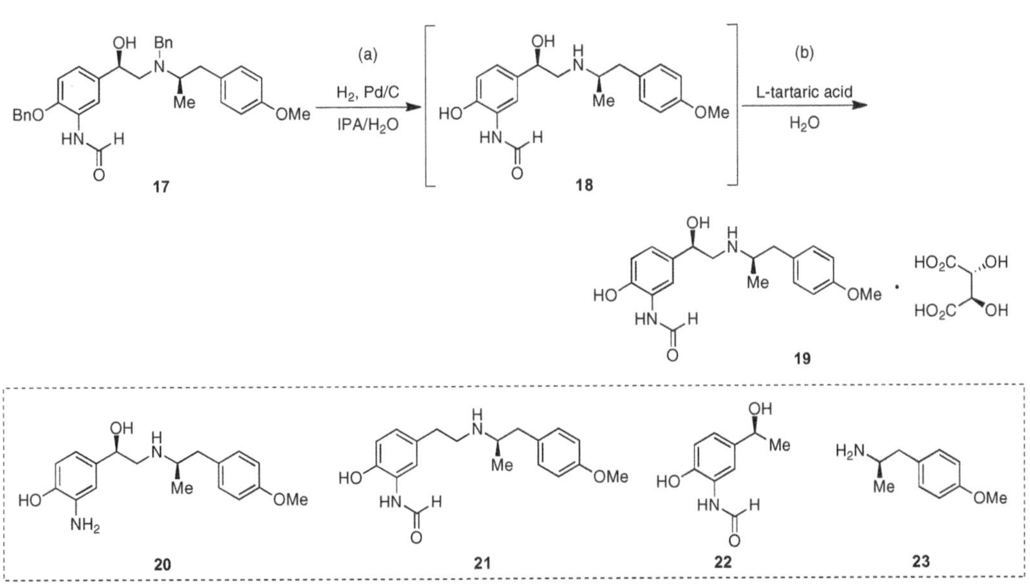

SCHEME 22.3

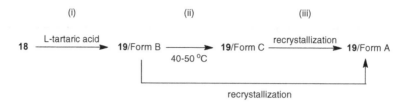

SCHEME 22.4

It was found that the recrystallization of **19**/Form B from the reaction mixture containing 13 wt% or more of toluene at 40–50 °C could achieve the impurity **21** removal and crystalline form transformation concomitantly. Accordingly, a polymorph-controlled purification and crystallization approach was developed based on the fact that impurity removal and crystal form conversion occurred concomitantly.

Procedure

Stage (i)
- Add a solution of L-tartaric acid (40.8 g) in water (237 g) to 460 g of a solution of crude **18** (ca. 164 g of **12** L) in a mixture of IPA/toluene (3.63:1, w/w).
- Stir the resulting solution for 2 h, during which a slurry is formed.

Stage (ii)
- Heat the resulting slurry at 45–50 °C until the level of **21** was <0.15 A% (2–3 h) (concomitant thickening of the slurry occurred due to the conversion of **19**/Form B to **19**/Form C).
- Cool the mixture to 22 °C.
- Filter and dry to give 109 g of **19**/Form C.

Stage (iii)
- Mix 109 g of **19**/Form C with IPA (214 g) and water (272 g).
- Heat the resultant slurry at 50–55 °C until dissolution occurred.
- Add 1.1 g of **19**/Form A followed by 545 g of IPA.
- Cool immediately to 40–45 °C and stir at the temperature for 30 min.
- Cool to 0 °C and stir for 2 h.
- Filter to give 93 g of **19**/Form A as a white solid.

22.2 COCRYSTALS

The formation of pharmaceutical cocrystals has gained increased interest as a means of optimizing the physicochemical property of drug molecules, such as solubility, dissolution rate, and thereby bioavailability and physical and chemical stability. Cocrystals are also useful in the purification of amorphous solids. A report[20] describes applications of cocrystals to systematically modulate the aqueous solubility and melting behavior of an anti-cancer drug. Recent advances in crystal engineering[21] have enabled the design of cocrystals.

Without forming/breaking covalent bonds, the formation of cocrystals can be achieved through evaporation[22] or cooling,[23] co-grinding the components,[24] sublimation,[25] etc.[26] Solution crystallization is the preferred method for cocrystal formation and is amenable to large-scale manufacturing. However, a drawback of this solution-based approach is the formation of undesired solvates/hydrates.[27] Recently, a solid-state grinding approach[28] has been developed to prepare cocrystals by the neat grinding of two (or more) components together with a mortar and pestle or in a mixer mill. In addition to its "green" nature in which it avoids excessive use of organic solvents, solid-state grinding also provides a means of obtaining nearly quantitative product yields.[29] A study shows that theophylline (TP) and nicotinamide (NCT) can form a cocrystal in a 1:1 molar ratio by means of solid-state grinding or slow evaporation from ethanol. Compared with anhydrous theophylline, the TP–NCT cocrystal is more soluble and has higher hygroscopicity.[30] A number of reports[31] demonstrated the preparation and characterization of pharmaceutical cocrystals.

22.2.1 COCRYSTAL WITH L-PHENYLALANINE

Cocrystallization has been applied toward the purification of pharmaceutical intermediates and APIs. A cocrystallization approach was developed by Bristol Myers Squibb Company to purify an amorphous API **24**, a potent and selective hSGLT2 inhibitor potentially for the treatment of diabetes (Equation 22.4).[32] The amorphous solid **24** was purified by mixing **24** with L-phenylalanine, leading to the crystalline product **25** as a monohydrate. After filtration and drying, the cocrystal product **25** was obtained an 80–88% yield (58 kg) with 99.9 A% purity. This material was used for clinical investigations.

(22.4)

24

L-phenylalanine · H$_2$O

25

Procedure

- Heat a slurry of **24** (10.0 g) in EtOH (14.0 mL) to 60 °C to form a clear solution.
- Charge this solution to a suspension of L-phenylalanine (4.6 g) in water (180 mL).
- Rinse with EtOH (3.6 mL).
- Heat the batch to 83 °C and then cool slowly to 52–54 °C over 15 min.
- Add seed (ca. 1%).
- Cool to 40–42 °C and stir for 4 h.
- Cool to 22–25 °C over 2–4 h and stir for 2 h.
- Filter, wash with ice-cold water (25 mL) and MTBE (2×5 mL), and dry at 40 °C under vacuum.
- Yield of **25**: 12.0 g (82%).

22.2.2 COCRYSTAL WITH L-PYROGLUTAMIC ACID

Ertugliflozin **26**, a drug candidate developed by Pfizer for the potential treatment of type 2 diabetes mellitus, exists as a hygroscopic amorphous solid with a low glass transition temperature. Accordingly, to improve the physical properties of API manufacturing, the hygroscopic amorphous solid **26** was converted to the corresponding cocrystal with L-pyroglutamic acid (L-PGA) (Equation 22.5).[33] The complex **27** exhibits high analytical purity with an approximate stoichiometric ratio of 1:1 (**26**:L-PGA).

$$(22.5)$$

Procedure

- Concentrate a solution of **26** (4.80 g) in MeOH (17.3 mL) at 40 °C under 30 torr to a minimal volume.
- Add IPA (18.7 mL) and stir to form a solution.
- Filter through a speck-free filter.
- Heat the filtrate to 60 °C.
- Add water (18.7 mL).
- Charge a solution of L-PGA (3.91 g) in water (56.2 mL).
- Heat the resulting mixture to 80 °C.
- Cool to 40 °C at 3 °C/min.
- Add seeds.
- Granulate the batch at 40 °C for 10 h.
- Cool to 20 °C at 0.1 °C/min.
- Filter, wash with toluene (2×9.8 mL), and dry at 55 °C under vacuum for 4 h.
- Yield: 5.26 g (85.3%) as a white crystalline solid (mp 142.5 °C).

22.2.3 COCRYSTAL WITH PHOSPHORIC ACID

Compound **28** was a drug candidate, developed by Boehringer Ingelheim, as a selective glucocorticoid receptor agonist for the treatment of rheumatoid arthritis and other inflammatory conditions. Compound **28** is a weak free base with low aqueous solubility (4.3 µg/mL). Compared to the free form, the phosphoric acid cocrystal **29** improved the solubility and bioavailability with desirable physicochemical properties. A robust procedure was developed to produce over 100 kg of the cocrystal **29** in multiple batches (Equation 22.6).[34]

$$(22.6)$$

Procedure

- Dissolve 20 kg of **28** (as acetic acid solvate or anisole solvate) in methyl ethyl ketone (MEK) (160 L) at 60 °C.
- Filter the resulting solution to remove any particulates.

- Charge at 50 °C 85% aqueous phosphoric acid (1.05 equiv) followed by adding heptane (26.7 L).
- Add seeds.
- Age for 30 min following the initiation of crystallization.
- Add additional heptane (53.4 L).
- Cool the batch to 20 °C over 2 h and hold for ≥2 h.
- Filter and wash with MEK/heptane (1:2 v/v).
- Dry at <60 °C under vacuum.
- Yield: 94–96% (>99.5%).

22.2.4   COCRYSTAL WITH L-PROLINE

Compound LX2761 was a drug candidate for the treatment of diabetes mellitus. The final step synthesis of LX2761 was base-catalyzed hydrolysis of penultimate intermediate **30** in ethanol. It was challenging to reproducibly obtain crystalline LX2761 directly from the process stream while providing acceptable impurity purging. Additionally, LX2761 free base was poorly soluble in pure water. Salt-screening experiments mostly gave tacky solids that were prone to deliquescence. To address these challenging issues, a process of converting LX2761 free base to cocrystal was developed (Equation 22.7).[35]

LX2761 · L-Proline cocrystal

$$(22.7)$$

Ultimately, a well-defined LX2761·L-proline cocrystal (1:1) was obtained that exhibited acceptable physicochemical properties with a melting point of 150 °C.

Procedure

- Add a solution of NaOMe (25 wt%, 19.4 mL, 85 mmol, 0.05 equiv) in MeOH to a suspension of triacetate **30** (1.30 kg, 1.21 kg active, 1.66 mol, 1.0 equiv) in EtOH (6.0 L).
- Stir the batch at 45 °C until reaction completion (3–5 h).
- Prepare a solution of L-proline (220 g, 1.91 mol, 1.15 equiv) in EtOH (485 mL)/H$_2$O (102 mL).
- Add, to the batch, ca. 50% (ca. 380 mL) of the L-proline solution followed by adding seeds of LX2761· L-proline (11.9 g).
- Stir at 40 °C for 1 h.
- Add slowly the remaining L-proline solution (355 mL) over 1 h.
- Age the resulting thick suspension at 40 °C for 1 h.
- Add MTBE (10 L) at 30 °C over 1 h and age for 1 h.

- Cool to 20 °C and age for 2–5 h.
- Filter, wash with MTBE/EtOH (3.6 L, 2:1) and MTBE (7 L), and dry at 30–45 °C under reduced pressure.
- Yield: 91.0% with 99.7% purity (HPLC area%).

22.2.5 Cocrystal with Adipic Acid

Compound BIIB068 was discovered as a promising drug candidate for the treatment of autoimmune diseases. In order to improve the bioavailability of the free base BIIB068, HCl salt 31 was initially identified as a solid salt form for development. However, chloro impurity 32 was generated when BIIB068 was treated with HCl (Scheme 22.5).[36] Compound 32 was a potential mutagenic impurity (PMI), and efforts were therefore shifted to the development of the hemiadipic acid cocrystal 33.

Ultimately, 33 was produced using 0.9–1.1 equiv of adipic acid at 70–75 °C in the presence of seeds and isolated after cooling to 5 °C in 90% yield with >99% HPLC purity.

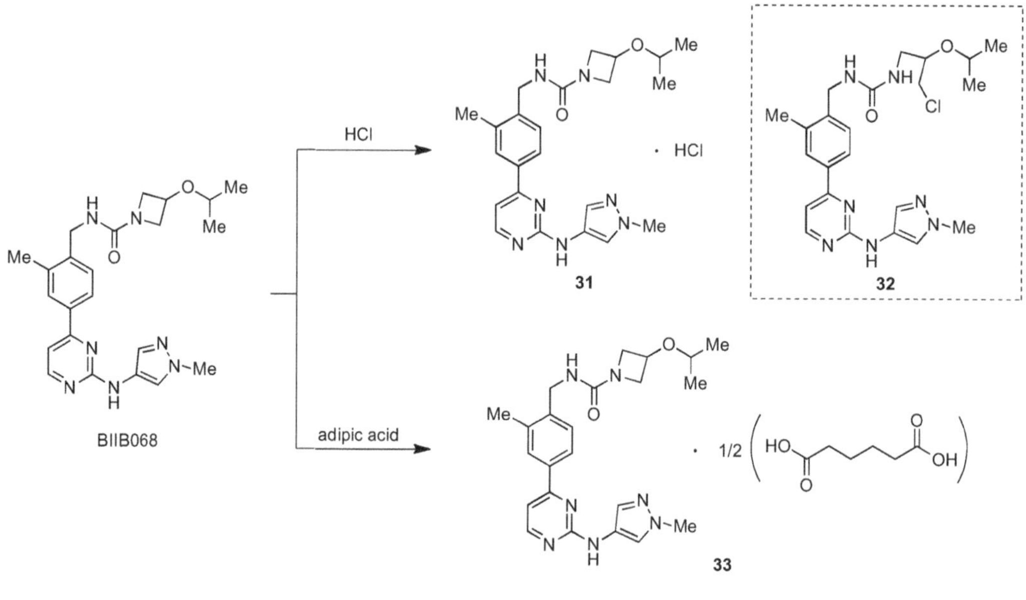

SCHEME 22.5

NOTE

Slow cooling was essential, as rapid cooling caused the monoadipic acid cocrystal to precipitate.

22.3 API PARTICLE SIZE

Big particle size, in many cases, is favorable since it offers a number of benefits in downstream operations, such as fast filtration and good recovery. For pharmaceutical products with low aqueous solubilities, however, small particles with narrow particle size distribution are desirable for enhanced oral bioavailability. In general, these particle sizes can be achieved by different milling devices, for example, pin, air-attrition jet, high shear rotor-stator, ultrasound,[37] or hydrodynamic cavitation. In contrast to the conventional particle breakage mechanism, that is, milling after the crystallization, a semibatch crystallization process, coupled with a high-speed rotor-stator device, was developed for the generation and control of fine particles.[38]

In order to achieve the desired particle size, a relatively slow stirring rate should be implemented during a crystallization process in order to avoid trituration by the agitation blades. To reduce fine particles during crystallization, heat cycle is an effective method to improve crystal size distribution (CSD). An example was presented nicely by a process group from Amgen for the manufacture of a drug substance AMG 925 hydrochloride salt **34** (Equation 22.8).[39]

AMG 925   **34**

(22.8)

(a) HCl (1.05 equiv), AcOH (6 vol)/H_2O (3 vol), 50 °C
100 rpm;
(b) seed, 45 °C, crystalliztion, and two heat cycles
(20 °C/50 °C)
25 rpm;
(c) IPA, 35 rpm;
(d) aging at 20 °C,12 h
25 rpm

In the protocol, three agitation rates were applied: 100 rpm for the HCl salt formation (Stage a); 25 rpm for the cooling and crystallization (Stage b), 35 rpm during the addition of IPA anti-solvent, and 25 rpm for aging. In addition, two heat cycles between 20 and 50 °C were performed over a course of 12 h to reduce the number of fine (<10 μm) particles.

22.4 AMORPHOUS SOLIDS

A large number of new pharmaceutical small molecules under development today are found to have poor water solubility; this in turn may lead to poor bioavailability. The development of compounds with poor bioavailability poses a greater challenge in preclinical toxicology studies which often aim to maximize exposure in order to assess the detrimental effects of the development compounds. In order to improve water solubility, several approaches such as changing polymorphic form, particle size reduction, and salt formation have been attempted. In addition, an amorphous solid is of great interest since it offers the potential for higher solubility and better bioavailability. Amorphous materials are commonly prepared via spray drying, solvent-induced method, or hot-melt extrusion.

22.4.1 Use of Spray Drying

The spray drying method is an efficacious approach for converting poorly soluble crystalline polymorphic drug substances into highly soluble amorphous forms to enhance bioavailability. Because of its relatively mild operation conditions, this method is suitable for thermo-sensitive and sparingly soluble drug substances.

Homoharringtonine **35** (Figure 22.3)[40] is a plant alkaloid isolated from the genus *Cephalotaxus* with anti-leukemic activity.

To access the amorphous solid, compound **35** was dissolved in a mixture of methylene chloride and methanol (98/2, v/v) followed by spray drying to give amorphous **35** with fine particle size. The amorphous solid of **35**, characterized by XRPD and scanning electron microscope (SEM), was more soluble in water than in corresponding crystalline form.

Etravirine **36** (Figure 22.4)[41] is a drug used for the treatment of human immunodeficiency virus (HIV). **36** is slightly soluble in water (0.07 mg/ml, at 25 °C) and practically insoluble in pH

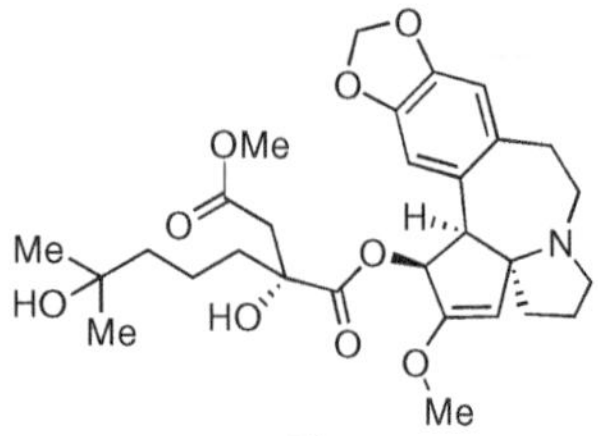

FIGURE 22.3   The Structure of homoharringtonine **35**.

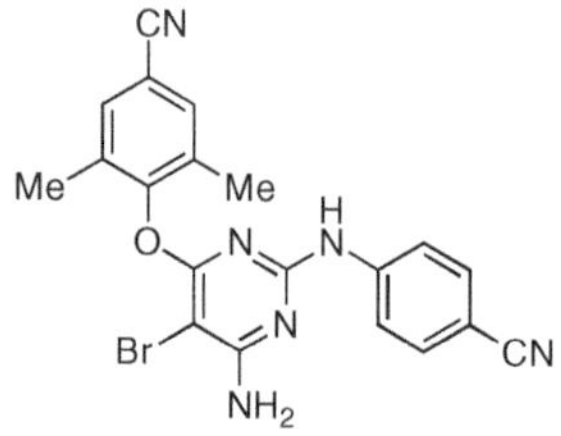

FIGURE 22.4   The structure of etravirine **36**.

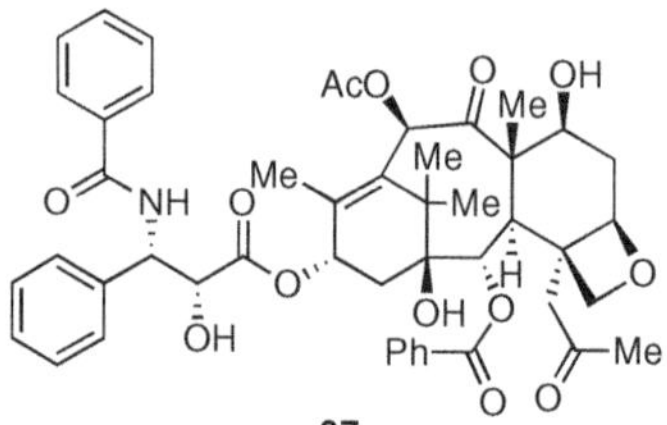

FIGURE 22.5   The structure of paclitaxel **37**.

6.8 phosphate buffer (0.01 mg/ml, at 25 °C), which resulted in poor bioavailability during oral administration.

The spray drying method was chosen for converting the poorly soluble crystalline form of **36** into a highly soluble amorphous form using Soluplus as a carrier (drug to polymer in 1:2 ratio) to increase its aqueous solubility and rate of drug release. As a result, the solubility of **36** increased by almost 4.0 folds in purified water and more than 15 folds in pH 6.8 phosphate buffer as compared to that of the crystalline form of **36**.

22.4.2   Use of Solvent-Induced Method

Paclitaxel **37** (Figure 22.5)[42] is a diterpene alkaloid derived from the epidermis of the yew tree and is an anti-cancer substance used for the treatment of various cancers.

As a class IV drug, identified by the biopharmaceutics classification system (BCS), paclitaxel **37** exhibits low solubility and intestinal permeability. Therefore, a more compatible and efficient delivery system was required. The morphology of APIs is an important factor that greatly affects the solubility and permeability of drugs during formulation. To overcome the poor solubility issue, a solvent-induced method was employed to prepare an amorphous solid of **37** by dissolving **37** in toluene (or methyl tertiary butyl ether) followed by removing the solvent and drying with a rotary evaporator.

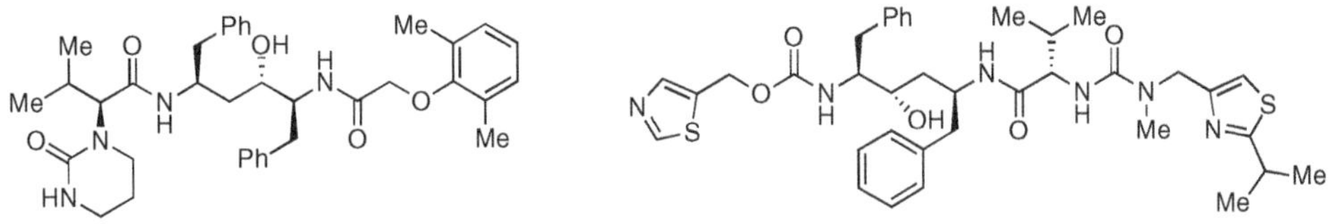

FIGURE 22.6   The structures of lopinavir **38** and ritovavir **39**.

22.4.3   Use of Hot-Melt Extrusion

Kaletra (lopinavir/ritonavir) is a prescription medicine that is used with other anti-retroviral medicines to treat HIV-1 infection in adults and children. Both lopinavir **38** and ritonavir **39** (Figure 22.6)[43] drug substances have very low water solubility, and their crystalline forms have minimal bioavailability. This led to an initial marketed formulation (U.S. approval in September 2000) of a soft gelatin capsule containing 133.3 mg of **38** and 33.3 mg of **39** as a solution in a mixture of oleic acid, propylene glycol, and water, with a storage condition of 2–8 °C.

Hot-melt extrusion is a thermal fusion process used to form amorphous solid dispersions. To improve the formulation and avoid cold storage, a tablet formulation was subsequently developed using hot-melt extrusion technology to render amorphous forms of **38** and **39** in the formulation. Film-coated tablets containing 200 mg of **38** and 50 mg of **39** were approved in the U.S. in October 2005 with room-temperature storage conditions.

NOTES

(a) Because the API is processed at a high temperature and shear under hot-melt extrusion conditions, the possibility of thermally induced degradation may occur, especially when processing less stable APIs.

(b) Because the polymorphic, pseudopolymorphic, as well as amorphous transformations can occur in a reversible pathway, using spray drying or solvent-induced method may compromise the physical stability of the amorphous form due to the possibility of crystallization.

(c) Besides the superior solubility, an amorphous solid[44] may exhibit different in reactivity. During a resolution of the two enantiomers of **40** by means of enzyme-catalyzed hydrolysis of the ester functionality, it was found that the conversion of this hydrolysis did not exceed 40% though the reaction was highly selective for the hydrolysis of the (R)-enantiomer (Equation 22.9).[45] Interestingly, only the amorphous state of **40** participated in the hydrolysis while no reaction occurred when using crystalline **40**.

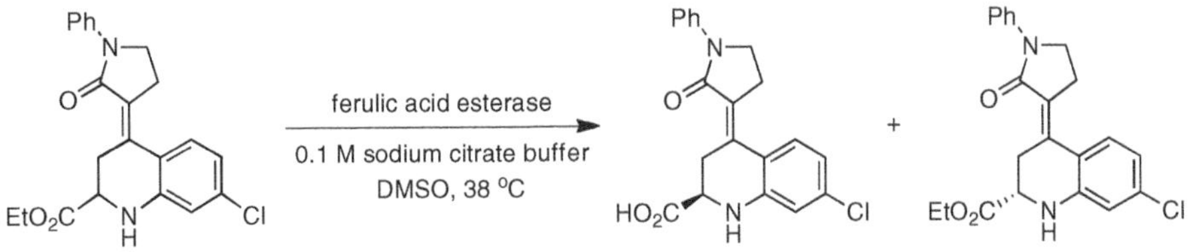

(22.9)

NOTES

1. (a) Bernstein, J. *Polymorphism in Molecular Crystals*; Oxford University Press, New York, 2002. (b) Datta, S.; Grant, D. J. W. *Nat. Rev. Drug Discov.* **2004**, *3*, 42.
2. Doherty, C.; York, P. *Int. J. Pharm.* **1988**, *47*, 141.

3. Andresen, B. M.; Couturier, M.; Cronin, B.; D'Occhio, M.; Ewing, M. D.; Guinn, M.; Hawkins, J. M.; Jasys, V. J.; LaGreca, S. D.; Lyssikatos, J. P.; Moraski, G.; Ng, K.; Raggon, J. W.; Stewart, A. M.; Tickner, D. L.; Tucker, J. L.; Urban, F. J.; Vazquez, E.; Wei, L. *Org. Process Res. Dev.* **2004**, *8*, 643.

4. (a) Sirota, N. N. *Cryst. Res. Technol.* **1982**, *17*, 661. (b) Karpinski, P. H. *Chem. Eng. Technol.* **2006**, *29*, 233.

5. Lewis, N. *Org. Process Res. Dev.* **2000**, *4*, 407.

6. Wang, F.; Wachter, J. A.; Antosz, F. J.; Berglund, K. A. *Org. Process Res. Dev.* **2000**, *4*, 391.

7. A polymorph transformation in solid-stage was observed during drying under vacuum, Takeuchi, H.; Kamata, K.; Tsuji, E.; Mukaiyama, H.; Kai, Y.; Yokoyama, K.; Jo, K.; Terada, K. *Org. Process Res. Dev.* **2012**, *16*, 647.

8. Dunitz, J. D.; Bernstein, J. *Acc. Chem. Res.* **1995**, *28*, 193.

9. The desired, thermodynamically stable polymorph was obtaines from a mixture polymorphs via slurrying in 2-propanol, see: Ragan, J. A.; Bourassa, D. E.; Blunt, J.; Breen, D.; Busch, F. R.; Cordi, E. M.; Damon, D. B.; Do, N.; Engtrakul, A.; Lynch, D.; McDermott, R. E.; Monggillo, J. A.; O'Sullivan, M. M.; Rose, P. R. *Org. Process Res. Dev.* **2009**, *13*, 186.

10. Zhao, Y.; Zhu, L.; Provencal, D. P.; Miller, T. A.; O'Bryan, C.; Langston, M.; Shen, M.; Bailey, D.; Sha, D.; Palmer, T.; Ho, T.; Li, M. *Org. Process Res. Dev.* **2012**, *16*, 1652.

11. Gontcharov, A.; Dunetz, J. R. *Org. Process Res. Dev.* **2014**, *18*, 1145.

12. Chamburkar, S. R.; Bauer, J.; Deming, K.; Spiwek, H.; Patel, K.; Morris, J.; Henry, R.; Spanton, S.; Dziki, W.; Porter, W.; Quick, J.; Bauer, P.; Donaubauer, J.; Narayanan, B. A.; Soldani, M.; Riley, D.; McFarland, K. *Org. Process Res. Dev.* **2000**, *4*, 413.

13. Luciana, M.; Grato, A. M.; Norberto, M.; Angelo, S. *J. Pharm. Sci.* **2005**, *94*, 1067.

14. Sathe, D.; Sawant, K.; Mondkar, H.; Naik, T.; Deshpande, M. *Org. Process Res. Dev.* **2010**, *14*, 1373.

15. Itoh, T.; Kato, S.; Nonoyama, N.; Wada, T.; Maeda, K.; Mase, T.; Zhao, M. M.; Song, J. Z.; Tschaen, D. M.; McNamara, J. M. *Org. Process Res. Dev.* **2006**, *10*, 822.

16. Tian, Q.; Hoffmann, U.; Humphries, T.; Cheng, Z.; Hidber, P.; Yajima, H.; Guillemot-Plass, M.; Li, J.; Bromberger, U.; Babu, S.; Askin, D.; Gosselin, F. *Org. Process Res. Dev.* **2015**, *19*, 416.

17. Morrison, H.; Quan, B. P.; Walker, S. D.; Hansen, K. B.; Nagapudi, K. Cui, S. *Org. Process Res. Dev.* **2015**, *19*, 1842.

18. Czyzewski, A. M.; Chen, S.; Bhamidi, V.; Yu, S.; Marsden, I.; Ding, C.; Becker, C.; Napier, J. *Org. Process Res. Dev.* **2017**, *21*, 1493.

19. Tanoury, G. J.; Hett, R.; Kessler, D. W.; Wald, S. A.; Senanayake, C. H. *Org. Process Res. Dev.* **2002**, *6*, 855.

20. Aakeröy, C. B.; Forbes, S.; Desper, J. *J. Am. Chem. Soc.* **2009**, *131*, 17048.

21. (a) Goldberg, I. *CrystEngComm* **2008**, *10*, 637. (b) Aakeröy, C. B.; Salmon, D. J. *CrystEngComm* **2005**, *7*, 439. (c) Banerjee, R.; Saha, B. K.; Desiraju, G. *CrystEngComm* **2006**, *8*, 680. (d) Saha, B. K.; Nangia, A.; Jaskólski, M. *CrystEngComm* **2005**, *7*, 355. (e) Aakeröy, C. B. *Acta Crystallogr.* **1997**, *B53*, 569. (f) Braga, D.; Brammer, L.; Champness, N. R. *CrystEngComm* **2005**, *7*, 1. (g) Aakeröy, C. B.; Desper, J.; Fasulo, M.; Hussain, I.; Levin, B.; Schultheiss, N. *CrystEngComm* **2008**, *10*, 1817.

22. Shattock, T. R.; Arora, K. K.; Vishweshwar, P.; Zaworotko, M. J. *Cryst. Growth Des.* **2008**, *8*, 4533.

23. Hickey, M. B.; Peterson, M. L.; Scoppettuolo, L. A.; Morrisette, S. L.; Vetter, A.; Guzmán, H.; Remenar, J. F.; Zhang, Z.; Tawa, M. D.; Haley, S.; Zaworotko, M. J.; Almarsson, Ö. *Eur. J. Pharm. Biopharm.* **2007**, *67*, 112.

24. Chadwick, K.; Davey, R.; Cross, W. *Cryst. Eng. Commun.* **2007**, *9*, 732.

25. Palmer, D. S.; Llinás, A.; Morao, I.; Day, G. M.; Goodman, J. M.; Glen, R. C.; Mitchell, J. B. O. *Mol. Pharm.* **2008**, *5*, 266.

26. (a) Seefeldt, K.; Miller, J.; Alvarez-Núñez, F.; Rodríguez-Hornedo, N. *J. Pharm. Sci.* **2007**, *96*, 1147. (b) Zhang, G. G. Z.; Henry, R. F.; Borchardt, T. B.; Lou, X. C. *J. Pharm. Sci.* **2007**, *96*, 990. (c) Takata, N.; Shiraki, K.; Takano, R.; Hayashi, Y.; Terada, K. *Cryst. Growth Des.* **2008**, *8*, 3032.

27. Reddy, L. S.; Bhatt, P. M.; Banerjee, R.; Nangia, A.; Kruger, G. J. *Chem. Asian J.* **2007**, *2*, 505.

28. (a) Shan, N.; Toda, F.; Jones, W. *Chem. Commun.* **2002**, *20*, 2372. (b) Friščić, T.; Trask, A. V.; Motherwell, W. D. S.; Jones, W. *Cryst. Growth Des.* **2008**, *8*, 1605.

29. Trask, A. V.; Motherwell, W. D. S.; Jones, W. *Chem. Commun.* **2004**, *7*, 890.

30. Lu, J.; Rohani, S. *Org. Process Res. Dev.* **2009**, *13*, 1269.

31. (a) Li, Z.; Yang, B.-S.; Jiang, M.; Eriksson, M.; Spinelli, E.; Yee, N.; Senanayake, C. *Org. Process Res. Dev.* **2009**, *13*, 1307. (b) Bhatt, P. M.; Ravindra, N. V.; Banerjee, R.; Desiraju, G. R. *Chem. Commun.* **2005**, *8*, 1073. (c) Trask, A. V.; Motherwell, W. D. S.; Jones, W. *Cryst. Growth Des.* **2005**, *5*, 1013.

32. Deshpande, P. P.; Singh, J.; Pullockaran, A.; Kissick, T.; Ellsworth, B. A.; Gougoutas, J. Z.; Dimarco, J.; Fakes, M.; Reyes, M.; Lai, C.; Lobinger, H.; Denzel, T.; Ermann, P.; Crispino, G.; Randazzo, M.; Gao, Z.; Randazzo, R.; Lindrud, M.; Rosso, V.; Buono, F.; Doubleday, W. W.; Leung, S.; Richberg, P.; Hughes, D.; Washburn, W. N.; Meng, W.; Volk, K. J.; Mueller, R. H. *Org. Process Res. Dev.* **2012**, *16*, 577.

33. Bowles, P.; Brenek, S. J.; Caron, S.; Do, N. M.; Drexler, M. T.; Duan, S.; Dubé, P.; Hansen, E. C.; Jones, B. P.; Jones, K. N.; Ljubicic, T. A.; Makowski, T. W.; Mustakis, J.; Nelson, J. D.; Olivier, M.; Peng, Z.; Perfect, H. H.; Place, D. W.; Ragan, J. A.; Salisbury, J. J.; Stanchina, C. L.; Vanderplas, B. C.; Webster, M. E.; Weekly, R. M. *Org. Process Res. Dev.* **2014**, *18*, 66.

34. Kim, S.; Li, Z.; Tseng, Y.-C.; Nar, H.; Spinelli, E.; Varsolona, R.; Reeves, J. T.; Lee, H.; Song, J. J.; Smoliga, J.; Yee, N.; Senanayake, C. *Org. Process Res. Dev.* **2013**, *17*, 540.

35. Sirois, L. E.; Zhao, M. M.; Lim, N.-K.; Bednarz, M. S.; Harrison, B. A.; Wu, W. *Org. Process Res. Dev.* **2019**, *23*, 45.

36. Li, C.; Franklin, L.; Chen, R.; Mack, T.; Humora, M.; Ma, B.; Hopkins, B. T.; Guzowski, J.; Zheng, F.; MacPhee, M.; Lin, Y.; Ferguson, S.; Patience, D.; Moniz, G. A.; Kiesman, W. F.; O'Brien, E. M. *Org. Process Res. Dev.* **2020**, *24*, 1199.

37. Pasti, J.; Chen, C.-K.; Spangler, L.; DelMonte, A. J.; Benoit, S.; Berglund, D.; Bien, J.; Brodfuehrer, P.; Chan, Y.; Corbett, E.; Costello, C.; DeMena, P.; Discordia, R. P.; Doubleday, W.; Gao, Z.; Gingras, S.; Grosso, J.; Haas, O.; Kacsur, D.; Lai, C.; Leung, S.; Miller, M.; Muslehiddinoglu, J.; Nguyen, N.; Qiu, J.; Olzog, M.; Reiff, E.; Thoraval, D.; Totleben, M.; Vanyo, D.; Vemishetti, P.; Wasylak, J.; Wei, C.; *Org. Process Res. Dev.* **2009**, *13*, 717.

38. Kamahara, T.; Takasuga, M.; Tung, H. H.; Hanaki, K.; Fukunaka, T.; Izzo, B.; Nakada, J.; Yabuki, Y.; Kato, Y. *Org. Process Res. Dev.* **2007**, *11*, 699.

39. Affouard, C.; Crockett, R. D.; Diker, D.; Farrell, R. P.; Gorins, G.; Huckins, J. R.; Caille, S. *Org. Process Res. Dev.* **2015**, *19*, 476.

40. Kim, B.-S.; Kim, J.-H. *Korean J. Chem. Eng.* **2009**, *26*, 1090.

41. Ramesh, K.; Chandrashekar, B.; Khadgapathi, P. *Int. J. Pharm. Pharm. Sci.* **2015**, *7*, 98.

42. Kang, H.-J.; Kim, J.-H. *Biotechnol. Bioprocess Eng.* **2020**, *25*, 86.

43. De Savi, C.; Hughes, D. L.; Kvaerno, L. *Org. Process Res. Dev.* **2020**, *24*, 940.

44. (a) Hancock, B. C.; Parks, M. *Pharm. Res.* **2000**, *17*, 397. (b) Hancock, B. C.; Zograf, G. *J. Pharm. Sci.* **1997**, *86*, 1.

45. Banks, A.; Breen, G. F.; Caine, D.; Carey, J. S.; Drake, C.; Forth, M. A.; Gladwin, A.; Guelfi, S.; Hayes, J. F.; Maragni, P.; Morgan, D. O.; Oxley, P.; Perboni, A.; Popkin, M. E.; Rawlinson, F.; Roux, G. *Org. Process Res. Dev.* **2009**, *13*, 1130.

Index

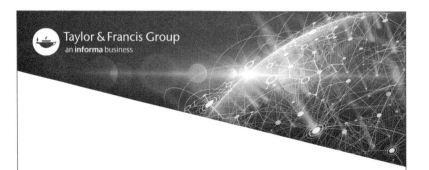